U0213621

An English-Chinese Dictionary of Civil Aviation
Abbreviations and Acronyms

# 英汉民航缩略语词典

祝世兴　主编

中国民航出版社

**图书在版编目（CIP）数据**

英汉民航缩略语词典/祝世兴主编 . —北京 : 中国民航出版社,2012. 4
ISBN 978-7-5128-0064-9

Ⅰ . ①英… Ⅱ . ①祝… Ⅲ . ①民用航空-缩略语-词典-英、汉 Ⅳ . ①V2-61

中国版本图书馆 CIP 数据核字（2012）第 028044 号

责任编辑：刘庆胜 杨玉芹

**英汉民航缩略语词典**

祝世兴 主编

| | |
|---|---|
| **出版** | 中国民航出版社 |
| **地址** | 北京市朝阳区光熙门北里甲 31 号楼 （100028） |
| **排版** | 中国民航出版社照排室 |
| **印刷** | 北京京师印务有限公司 |
| **发行** | 中国民航出版社 （010）64297307　64290477 |
| **开本** | 880×1230　1/32 |
| **印张** | 32. 125 |
| **字数** | 1735 千字 |
| **印数** | 2000 册 |
| **版本** | 2013 年 1 月第 1 版　2013 年 1 月第 1 次印刷 |

**书号** ISBN 978-7-5128-0064-9
**定价** 95. 00 元

## 《英汉民航缩略语词典》编写组

主　编　祝世兴

编　委　（按姓氏笔划排序）

　　　　许　洪　刘桂华　汪　虹

　　　　李运明　杨桢梅　殷恒光

　　　　崔　澍　潘　超

# 序　言

英语缩略语因其言简意赅、表达的信息量大，在文献中使用的频率呈不断上升趋势，并且随着科学技术的发展和进步，新的缩略语不断涌现，已经渗透到科技领域的各个方面，一部分缩略语已成为人们工作交流用语。正确翻译、准确理解英语缩略语的中文表达，不仅关系到对文献内容的理解和使用，更关系到交流的质量与效果。

编撰英汉专业缩略语词典是一项非常烦琐、细致和艰苦的工作。主要表现在词汇的收集、选择和翻译上，词汇选择涉及的知识面广、专业性强，不仅要考虑到词汇本身的含义，更要考虑词汇专业含义以及行业翻译差异，要做到释义简练、明确，需要作者付出大量的时间和具备丰富的专业知识。

多年来，作者及其团队一直在民航从事与民航相关方面的科学研究和教学工作，积累了大量的素材，承担了多项国家自然科学基金、天津市科技基金和民航局科技基金项目以及航空规章方面的科技项目，取得了包括民航科学技术一等奖在内的 5 个奖项。

本书吸纳了作者多年在民航及航空器设计、制造方面积累的文献和研究成果，编撰了这部《英汉民航缩略语词典》，该词典精选了与民航及航空器设计、制造相关的英语缩略语词汇 54000 余条以及七个附录，内容丰富，具有一定权威性，是国内首部公开出版的《英汉民航缩略语词典》，能够满足行业技术人员的需求，具有广泛的应用前景，一定不会辜负人们对它的期待！

中国工程院院士　管德

2012.10.29

# 前　言

航空，是人类在 20 世纪取得的最为重大的科技成就之一，它在很多方面改变了人们的生产和生活方式，提高了人们的工作效率和生活质量。特别是在民用航空领域，它给人们带来的变化和效益是空前的。在航空、海运、公路、铁路和管道五大运输方式中，航空运输方式以其独特的优势越来越被人们所重视。首先是快捷，为人们公务和私人旅行节省了大量的时间，缩短了空间距离，其价值是巨大的；其二是机动，高山、大川、沙漠、海洋等已不再是障碍，可以到达世界上的任何地方，特别是在地震等灾害的救援中发挥着越来越重要的作用；其三是安全，据国际民航组织统计，近年来，世界民航定期航班失事率大大降低，平均每亿公里的死亡人数为0.04 人，是其他运输方式事故死亡人数的几十分之一到几百分之一。因此，民用航空运输已被越来越多的人所了解、接纳，随着民用航空运输业的普及以及我国民用大飞机计划的实施，与国外交流的人数、从事民航运营和民用飞机制造的人数越来越多，通过英文学习来了解民用航空运输领域知识的人也在逐年增多。另外，随着航空科学和技术的发展，在各类科学和技术文献中涌现出的民用航空方面的缩略语词汇也越来越多，新的缩略语与日俱增。为满足广大科技工作者和民用航空从业者的需要，项目组经过多年的积累和收集，编撰了这部《英汉民航缩略语词典》。

本词典收录了民用航空方面的缩略语、简语词汇共计 54000 余条。在词汇选择与收词范围上，我们力求前沿、专业、准确、适用，基本能够覆盖民用航空领域的各个方面。本词典主要以民用航空的飞行、空中交通管制、航空维修工程、适航、机场运营、航空运输管理、通讯导航与信息技术、航空气象、航空安全、民用航空组织机构等为主体，并收录了部分航空器设计与制造、复合材料、计算机技术、航空旅游和有关其他航空服务保障等方面的英语缩略语。另外，本词典还收录了世界机场三字代码 1100 余个。本词典是专为民用航空从业者编撰的工具书，收词一是来源于作者在多年的教学和科研中所编写、阅读、评审过的各种书籍、报告、论文等；二是来源于相关方面的各种技术词典、航空规章及专业网站等。

本词典附录一收录了中国民用航空运输机场（大陆地区）182家，内容包括机场名称、所在地区及省份（直辖市、自治区等）、三字和四字代码、机场等级数、机场英文简称；附录二收录了中国香港、澳门和台湾地区机场12家；附录三收录国内运输类航空公司（大陆地区）36家，香港5家，澳门1家，台湾6家，内容包括航空公司名称、航徽、两字代码、网址；附录四收录了美国50个州的简称及其各州、州首府的中文、英文名称；附录五收录了几种单位制之间的换算表；附录六收录了世界主要城市与北京时差；附录七收录了民用飞机注册号前码表。

本词典在编撰过程中得到了中国民用航空局人事科教司、航空器适航审定司、飞行标准司，以及中国民航大学有关领导和专家的大力支持，他们对本书的编写给予了许多帮助并提出了许多建设性意见，在此表示衷心的感谢！另外，主编的同事孙淑光、闫芳、田静、张宏伟、杨永刚以及主编的研究生王海军、王朔、王鹏等在词典的编撰、资料收集和校对等过程中也做了许多工作，在此一并表示感谢！

本缩略语词典承蒙中国民航大学张艳玲教授、杨新涅教授，中国民航局适航审定司前处长毛祖兴、中国民航局航空事故调查中心正处级高级调查员霍志勤同志审阅，提出了许多宝贵意见和建议，对此表示诚挚的谢意！

由于民用航空科学与技术涉及面广，发展迅速，学科交叉融合度大，航空与民航领域在词汇的翻译上有一定的差别，使得中文翻译和中文词意的选择上具有一定的复杂性和难度，加之编撰者受专业领域和专业水平限制，经验不足，书中的不妥、遗漏甚至错误之处在所难免，恳请广大读者在使用本词典过程中提出宝贵的建议、批评和指正，以便进一步修改和完善。如有修改建议，可直接与编者联系，E-mail 地址是：zhusx_ buffalo@ yahoo. com。

编者

2011 年 6 月于天津

# 编写说明

1. 本词典除收录英文缩略语外，还收录了部分简写词和国际机场三字代码。所有收录的以字母开头的词汇全部按照英文字母的先后顺序排列，但加引号的字母排在最前面，其顺序按加引号字母的顺序排列，如"A"Check，中文释义为 A 检，排在数字开头的前面。以数字开头的词汇排在英文字母的前面，上下角标不参加排序。以希腊字母开头的词汇集中排在字母 Z 的最后面，其顺序按希腊字母的顺序排列。

2. 本词典共收录国际机场三字代码 1100 余条，每个词条的英文释义后均用括号注有（Airport Code）字样，以示区别。美国机场的三字代码，在英文释义的机场名称后、国家名称前均有两字简写字母代表美国的州名。例如 ABI，Abilene，TX，USA（Airport Code），其中的 TX 代表得克萨斯州，上述机场的中文释义是美国得克萨斯州阿比林机场，其中的简称 TX 的中文释义可以在附录四中查到。

3. 在英文原文中，每个单词的首字母大写，如 CAAC，此词条的英文全文及中文译文为 Civil Aviation Advisory Committee 民航咨询委员会。冠词（如 the）、连词（如 and）、介词（如 of）等虚词一般不大写。如 CAGE，其英文全文及中文译文为 Commercial and Government Entity 商业和政府机构。但如该类词是缩写词的首字，则首字母要大写。如 TBC，英文全文及中文译文为 To Be Confirmed 待证实；TIACA，英文全文及中文译文为 The International Air Cargo Association 国际航空货运协会。

4. 对于同一个缩略语，有两个以上释义的，一般用逗号分开。如 CASFO，其英文全文及中文译文为 Civil Aviation Security Field Office 民用航空保安现场办公室，民用航空安全地区办事处。

5. 有些英文缩略语或简语，并不是严格按照首字母缩写成为缩略语的，也不像单词那样有严格的拼写规则，而且不同的领域、专业有各自约定俗成的术语、简语。如 PROP，其英文全文及中文译文为 Propeller Aircraft 螺旋桨航空器；WXR，其英文全文及中文译文为

Weather Radar 气象雷达。

6. 不同的缩略语，虽具有相同的英文原文和释义，但按两条词条收录。如 XMT，Transmitter 发射机；XMTR，Transmitter 发射机。

7. 在有些汉语释义中，为了清楚起见，编者在这些缩略语词前、词后加了一些注释性说明，并将这些说明用括号（）括起来，供读者参考。如 PSN，其英文全文及中文译文为 Packet Switched Network（数据）包交换网络；Y2K，其英文全文及中文译文为 Year 2000 计算机两千年问题（"千年虫"）。

8. 有些缩略语，不是代字与符号，也与原文没有什么缩略语关系，但是业界约定俗成，也就成了缩略语。如 MAP，其英文全文及中文译文为 Aeronautical Charts 航图；SI，其英文全文及中文译文为 International System of Units 国际单位制。

9. 在英文缩略语中，按照读音，常用 X 代替一些发音带 "X" 的单词。如 LHX，其英文全文及中文译文为 Light Helicopter Experimental 轻型试验直升机；或表示侧面、交叉（cross）之意，或传输（trans-）之意，如 XWC，Cross Wind Component 侧风分量；X BLEED，Cross Bleed 交输引气；XMITTER，Transmitter 发射机。

10. 本词典还收录了中国航空、中国民航企事业等单位名称缩写。如 CNAH，其英文全文及中文译文为 China National Aviation Holding Company 中国航空集团公司；CAUC，其英文全文及中文译文为 Civil Aviation University of China 中国民航大学。

# 目　录

# 目 录

# A

"A"    A 检

"C"    C 检

"D"    D 检

¢P    Differential Pressure 压差

°T    True（degrees）真向（度）

1bf/in$^2$    Pounds force per square inch 磅力/平方英寸

3D    Three Dimensional 三维的，立体的

3G    Third Generation 第三代移动通信技术

3P    Three Poles 三极

4Cs    Customer, Cost, Convenience, Communication 客户、成本、便利、沟通

4D    Four Dimensional（Lat, Long, Alt, Time）四维（横、纵、垂向和时间）

4Ps    Product, Price, Place, Promotion 产品、价格、渠道、促销

4S    Stealth, Supersonic cruise, Supermaneuver, Short takeoff 隐身、超音速巡航、超级机动、短距起飞

5S    Smile, Skill, Simple, Speed, Satisfaction 微笑、专业、便捷、迅速和满意服务

6S    Seiri, Seiton, Seiso, Seiketsu, Shitsuke, Security 整理、整顿、清扫、清洁、素养、安全

A    Abeam 正切

    Above 在…之上

    Acceleration 加速（度）

    Acceptance 接受，承认

    Action code 操作码

    Action 行动，活动，作用

    Active 有效的，有作用的

    Actual 实际的，有效的

    Address 地址

    Administrator 署长

    Advisory 报告气象，咨询的

    Aero 航空的，飞行的，空中的

Aeroplane 飞机

Air 空气，大气，航空的

Airbus 空客

Aircraft 航空器，飞机

Airmen 飞行人员

Alarm 警报

Alerting Area 告警区，警戒区

Alternate 备用的，备降场

Altimeter 高度表

Altitude 高度

Amber 琥珀（色）

Ambulance 救护航空器，救护车，救护船

Ampere（s）安培

Angstrom 埃（波长）

Annual 年度

Anode 阳极

Antenna 天线

Application Layer 应用层

Approach Light 进近灯

Area(s) 地区，区域

Arrive/Arrival 到达

Asbestos 石棉

Assemble or Assembly 组件，组装，汇编

At or Above an Altitude 在或高于一个高度

Atmosphere 大气

Automatic 自动的

Axis 轴（线）

Azimuth 方位（角）

Cross-Sectional 横截面积

Nozzle Throat Area 喷口颈区

Speed of Sound 音速

**A&AEE**    Aeronautical and Aircraft Experimental Establishment 航空与飞机实验研究院（英国）

**A&CO**    Assembly and Checkout 装配与检测

**A&E** Airframe and Engine 机体与发动机

Analysis & Evaluation 分析与评价

Architecture and Engineering 结构和工程

**A&EM** Alarm & Event Management 告警与事件管理

**A&I** Accident and Indemnity 事故和损失赔偿

Assembly and Installation 装配与安装

**A&ISD** Aeronautical & Information System Division 航空与信息系统分部

**A&L** Approach and Landing 进近和着陆

Assignment & Location 指配与定位

**A&M** Department of Administration and Management 行政和管理事务部

**A&P** Aircraft & Powerplant 飞机与动力装置

Airframe and Powerplant 机体与动力装置

**A&R** Automation and Robotics 自动化与机器人

**A&T** Assembly and Test 组装与试验

**A&VE** Audio/Video Editor 音频/视频编辑器

**A. C.** Aerodynamic Center 气动中心

**A. ICE** Anti-ice, Anti-icing 防冰

**a. m. p. g** Air Miles per Gallon 每加仑飞行英里

**A/A** Air Abort 中断飞行

Air to Air 空对空

Airdrome to Airdrome 机场间的

Angle of Attack 迎角,攻角

Any Acceptable 任何可接受的

**A/B** Air Bill 空运提单

Autobrake 自动刹车

**A/BCN** Airport Beacon 机场信标

**A/C** Account 账目,账户

Air Conditioning 空调,空气调节

Aircraft 飞机,航空器

Alternating Current 交流

Approach Control 进近管制

**A/COLL** Anti-Collision 防撞

**A/CS** Aircraft Call Signal 飞机呼叫信号

**A/D** Aerodrome 机场

After Date 某日之后

Alarm and Display 警报与显示

Analogue to Digital 模/数(转换)

**A/DC** Analog to Digital Converter 模数转换器

**A/F** Air Freight 空运货物,航空货运

Auto Flight 自动飞行

**A/FD** Airport Facility Directory 机场设施指南

**A/G** Air Guard 空中保安

Air to Ground 空/地

**A/H** Airhostess 乘务员

Alter Heading 改航,改变航向

Altitude/Height 高度

**A/I** Accident Investigation 事故调查

Aircraft Inspector 飞机检查员

Anti-Ice 防冰

Approval of Import 进口许可

**A/L** Airlift 空运

Airline 航线,航空公司

Autoland 自动着陆

**A/M** Auto/Manual 自动/人工

**A/N** Account Number 账号

Alphanumeric 字母数字的

**A/O** Account of 重视,入账,原因

Armed or Operate 准备或工作

At Once 立刻

**A/P** Account Paid 现款支付

Advance Payment 预付款

Air Purser 乘务长

Airplane 飞机

Autopilot 自动驾驶仪

**A/R** Audio Reproducer 音频再生器

**A/RUTG** Actual Routing 实际航路

**A/S** Airspeed 空速

Alongside 伴随

As Stated 如上所述

Auto Stabilization 自动稳定

Autobrake System 自动刹车系统

**A/STAB** Auto-Stabilizer 自动安定面

**A/T** Adjustment/Test 调节/测试

Aircraft Technician 飞机技术员

Auto Throttle 自动油门

**A/THR** Automatic Thrust 自动推力

**A/V** Audio/Video 音频/视频

**A/VP** Audio/Video Panel 音频/视频面板

**A/W** Access Way（模拟机）入口门

Actual Weight 实际重量

Airway 航路

All-Weather 全天候

**A³TC** Advanced Automated Air Traffic Control 先进的自动化空中交通管制

**AA** Absolute Address 绝对地址

Absolute Altitude 真高度，绝对高度

Access Agent 访问代理

Access Authorization 进入许可（证）

Acquisition Aiding 帮助截获

Actual Arrival（Time）实际到达（时间）

Administration Assistant 行政助理

Administrative Agreement 管理协定

Aeronautics Act 航空法

Air Almanac 航空历

Airborne Alert 空中警戒

Aircraft Airworthiness 航空器适航性

Airlines Algeria 阿尔及利亚航空公司

Airlines Argentines 阿根廷航空公司

Airport Approach 机场进近

Airship Association 飞艇协会（英国）

Airworthiness Approval 适航批准书

Airworthiness Authority 适航当局

All After 一切附后

Amended 修订，修正

American Airlines Inc. 美国航空公司

Anti-Aircraft 防空的，高射的，高射兵器

Approach Angle 进近角

Approving Authority 批准机关

Arithmetic/Arithmetical Average 算术平均

Arrival Angle 入射角

Artificial Antenna 仿真天线

Assistance Administrator 助理署长

Auto Flight Status Annunciator 自动飞行方式告示器

Auto-Alarm 自动报警

Autoanalyzer 自动分析仪

Autoflight Annunciator 自动飞行方式信号器

Automated Approach 自动进近，自动进场

Automatic Alarm Call 自动报警呼叫

Automatic Answer 自动应答

Automatic Approach 自动进近

Autopilot Annunciator 自动驾驶信号牌

Aviation Annex 航空附件

Azimuth Angle 方位角

**AA/G** Application Agent/Gateway 应用代理/网关

**AAA** Adaptive Aircraft Assignment 适应性飞机分配

Affordable Acquisition Approach 能探测到的进近

Airport Advisory Area 机场咨询区

Alaska Airmen's Association 阿拉斯加飞行员协会

All American Aviation 全美航空联盟

Allocations, Assessments and Analysis 配置（定位），鉴定与分析

American Accounting Association 美国会计协会

American Airship Association 美国飞艇协会

American Arbitration Association 美国仲裁协会

American Automobile Association 美国汽车学会

Ansett Airlines of Australia 澳大利亚安塞特航空公司

Antique Airplane Association 古老飞机协会（美国）

Association of African Airways 非洲航空公司协会

Australian Automobile Association 澳大利亚汽车协会

Authentication, Authorization and Accounting 验证,授权和账户

Awaiting Aircraft Availability 等待飞机的有

效性

**AAAA** Antique Aeroplane Association of Australia 澳大利亚古老飞机协会

Army Aviation Association of America 美国陆军航空协会

Australian Aerial Agricultural Association 澳大利亚农业航空协会

**AAAE** American Association of Airport Executives 美国机场管理者协会,美国航空站主管人员协会

**AAAF** Airport's and Airline's Forum 机场与航空公司的论坛

Association Aeronautique et Astronautique de France 法国航空航天协会

**AAAH** Airbus Approved Abbreviations Handbook 空客批准的缩略语手册

**AAAI** American Association for Artificial Intelligence 美国人工智能协会

Association of Australian Aerospace Industries 澳大利亚航空航天工业协会

**AAAM** American Association of Aircraft Manufacturers 美国飞机制造商协会

**AAAR** Abbreviated Aviation Accident Report 简要的航空事故报告

Association for the Advancement of Aeronautical Research 航空研究促进会

**AAAS** Alternate/Alternative Audio Alert Selector 备用音频告警选择器,备用音频警报选钮

American Academy of Arts and Sciences 美国艺术和科学研究院

American Association for the Advancement of Science 美国科学促进会

**AAATC** Advanced Automated Air Traffic Control 先进的自动化空中交通管制

**AAB** Adaptive Angle Bias 自适应角偏差

Advanced Airplane Branch 先进飞机部

Aircraft Accident Board 飞机事故委员会,飞机失事管理局

Applications Assistance Branch 应用服务处

Automatic Alternative Billing 自动更换记账

Automatic Answer Back 自动应答

Auxiliary Air Base 备用航空基地

Aviation Advisory Board 航空咨询委员会

**AABB** Axis-Aligned Bounding Box 沿坐标轴的轴向包围盒

**AABSHL** Aircraft Anti-collision Beacon System High-intensity Light 飞机防撞信标系统高强度灯

**AABY** As Amended by 按照…修订

**AAC** Acoustical Absorption Coefficient 吸音系数

Acquisition Advice Code 采集通知码,捕捉消息码

Active Address Code 有效地址码

Adaptive Antenna Control 自适应天线控制

Advanced Adaptive Control 先进自适应控制

Advanced Airspace Concept 先进空域概念

Aerial Ambulance Company 航空救护公司

Aeronatical Approach Chart 航空进近图

Aeronautical Administrative Communication 航空行政管理通信

Aeronautical Advisory Council 航空咨询委员会

Aeronautics Advisory Committee 航空咨询委员会

Air Approach Control 空中进近管制

Airborne Automatic Checkout 空中自动检查

Air-Carbon-Arc-Cutting 空气碳极弧切割

Aircraft Accessories Corporation 飞机附件公司

Aircraft Airworthiness Center 航空器适航中心

Aircraft Airworthiness Certificate 航空器适航证

Airline Administrative Communications 航空公司行政通讯

Airworthiness Advisory Circular 适航咨询通告

All Aluminum Conductor 全铝导线

American Aviation Corporation 美国航空公司

Amplitude Absorption Coefficient 振幅吸收系数

Anti-Aircraft Command 防空司令部

Anti-Aircraft Common 防空公共区

Army Aviation Center 陆军航空中心

Australian Air Academy 澳大利亚空军学院

Australian Aircraft Consortium 澳大利亚飞机联合公司

Automatic Alarm Call 自动报警呼叫

Automatic Amplitude Control 自动幅度控制

Automatic Aperture Control 自动孔径控制

Automatic Approach Control 自动进近管制

Auxiliary Air Control 辅助空气控制

Aviation Advisory Commission 航空咨询委员会

**AACA** Alaska Air Carriers' Association Inc. 阿拉斯加航空公司协会

**AACB** Aeronautics and Astronautics Coordinating Board 航空航天协调委员会(美国)

Aerospace Activity Coordinating Board 航空航天活动协调委员会

**AACC** Air Approach Control Center 空中进近管制中心

Airport·Associations Coordinating Council 机场协会协调理事会,机场协会协调委员会

All-Attitude Control Capability 全方位控制能力

Alternate Avionics Computer Control 备用航空电子计算机控制

American Association for Contamination Control 美国污染控制协会

American Automatic Control Council 美国自动控制委员会

Area Approach Control Center 地区进场控制中心,区域进近管制中心

Automatic Approach Control Coupler 自动进近管制耦合器

**AACD** Aging Airplane Corrosion Prevention and Control Document 老龄飞机防腐和控制文件

**AACE** Aircraft Alerting Communication Electromagnetics 飞机告警通信电磁学

Aircraft Alerting Communication EMP 飞机告警通信电磁脉冲

**AACF** Auburn Automated Control File 奥本自动控制文件

**AACI** Airports Association Council International 国际机场联合委员会

American Association for Conservation Information 美国信息交流协会

**AACO** Arab Air Carriers Organization 阿拉伯航空承运人组织

**AACP** Advanced Airborne Command Post 先进的空中指挥所

Aging Airplane Corrosion Prevention and Control Program 老龄飞机防腐和控制程序

American Association for Correctional Psychology 美国矫正心理学协会

**AACS** Active Attitude Control System 主动姿态控制系统

Aileron Active Control System 副翼主动控制系统

Airborne Astrographic Camera System 机载天文定位照相系统

Airways and Air Communication(s) Service 航路及航空通讯服务

Antiskid/Autobrake Control System 防滑/自动刹车系统

Asynchronous Address Communication System 异步寻址通信系统

Attitude & Articulation Control System 飞行姿态与铰接控制系统

Attitude and Antenna Control System 姿态和天线控制系统

Attitude and Articulation Control Subsystem 姿态和转动关节控制分系统

Automatic Access Control System 自动存取控制系统

Automatic Area Control System 自动地区控制系统

**AACSM** Airways and Air Communication Service Manual 航路和空中通信服务手册

**AACSR** Airways and Air Communication Service Regulation 航路和空中通信服务规章

**AACSS** Automated Access Control Security System 自动进入控制保险系统

**AACTP** American Association for the Certification of Training Program 美国培训认证协会

**AACU** Antiskid/Autobrake Control Unit 防滞/自动刹车控制组件

Area Approach Control Unit 区域进近管制部门

**AACV** Auxiliary Air Control Valve 辅助空气控制活门(阀)

**AAD** Active Acoustic Device 有源声装置

Adaptive Arithmetic Decoder 自适应算术码译码器

Advanced Avionics Development 先进的航空电子系统开发

Aero Acoustic Detection 航空声测

Aircraft Airworthiness Department 适航司

Aircraft Arresting Device 飞机系留装置,轮挡

Alphanumeric Alarm Display 文字警报显示

Assigned Altitude Deviation 指定高度偏差

Average Absolute Deviation 平均绝对偏差

**AADA** Anti-Aircraft Defence Area 防空区域

**AADAA** Autonomous Customs Warehouse Administration 自主保税仓库管理局

**AADC** Advanced Avionics Digital Computer 高级航空电子数字计算机

All Applications Digital Computer 通用数字计算机

American Airlines Data Conversion 美国航空公司数据转换

Analog Air Data Computer 模拟大气数据计算机

Approach and Departure Control 进近与离场控制

**AADHS** Advanced Avionics Data Handling System 先进的航空电子数据处理系统

**AADRP** Aircraft Accident Data Reporting Panel 航空器事故数据报告专家组

**AADS** Advanced Air Defense System 先进的空防系统

Airspeed and Direction Sensor 空速和航向传感器

Area Air Defense System 区域空防系统

Automatic Aircraft Diagnostic System 飞机自动诊断系统

**AAE** Above Airport Elevation 相对机场高度

Adaptive Arithmetic Encoder 自适应算术码编码器

Aerospace Auxiliary Equipment 航空航天辅助设备,宇航辅助设备

American Association of Engineers 美国工程师协会

**AAEC** Attitude Axis Emergency Control 姿态轴应急控制

**AAEE** Aeronautical and Aircraft Experimental Establishment 航空与飞机实验研究院(英国)

**AAERR** Aerophysics and Aerospace Engineering Research Report 航空物理学和航空航天工程研究报告

**AAES** Advanced Aircraft Electrical System 先进的飞机电气系统

**AAEW** Advanced Airborne Early Warning 先进的空中预警

**AAEWR** Advanced Aircraft Early Warning Radar 先进的飞机预警雷达

**AAF** Access Adaption Function 存取自适应功能

African Aviation Federation 非洲航空联合会

Airway Facilities Service 航路设施服务

American Astronautical Federation 美国航天联合会

Analog Antialiasing Filter 模拟防混淆滤波器

Association Astronautique Francaise 法国宇航协会

Atlantic Amphibious Force 大西洋两栖部队

**AAFCS** Advanced Automatic Flight Control System 先进的自动飞行控制系统

**AAFDS** Aircraft Assembly Flow Design System 飞机装配流程设计系统

**AAFE** Advanced Applications Flight Experiment 先进的应用飞行试验

Aero Assist Flight Experiment 空气制动方案飞行实验(美国)

**AAFEA** Australian Airline Flight Engineers Association 澳大利亚航空公司工程师协会

**AAFIF** Automated Air Facility Information File 航空器材信息库(美国)

**AAFIS** Advanced Avionic Fault Isolation System 先进的航空电子故障隔离系统

**AAFM** American Airlines Flight Maintenance 美国航空公司飞行维护

**AAFRA** Association of African Airlines 非洲航空公司协会

**AAG** Aeromedical Airlift Group 航空医疗空运大队

Association of American Geographers 美国地理学家协会

**AAGR** Average Annual Growth Rate 年平均增长率

**AAHA** Awaiting Action Higher Authority 等待上级决定

**AAHK** Airport Authority Hong Kong 香港机场管理局

**AAHS** American Aviation Historical Society 美国航空史学会

**AAI** Air to Air Identification 空—空识别

Aircraft Accident Investigation 飞机事故调查

Airline Avionics Institute 航空公司航空电子研究所

Airport-Acceptance Interval 机场接收间隔

All Attitude Indicator 全姿态显示器

Angle of Approach Indicator 进近角度指示器

Angle of Attack Indicator 迎角指示器

Arrival Aircraft Interval 进港飞机间隔

Austrian Aeronautics Industries 奥地利航空工业

Average Arrival Interval 平均到达间隔

Azimuth Angle Increment 方位角增量

**AAIB** Air Accident Investigation Board 航空事故调查委员会

Air Accident Investigation Branch 航空事故调查部门

Aircraft Accident Investigation Board 飞机事故调查委员会

Aircraft Accident Investigation Bureau 飞机事故调查局

**AAIC** Accounting Authority Identification Code 结算机构识别码

Aircraft Accident Investigation Commission 航空事故调查委员会

**AAIFF** Air-to-Air Identification Friend or Foe 空对空敌我识别

**AAIG** Australian Aviation Insurance Group 澳大利亚航空保险集团

**AAIM** Aircraft Autonomous Integrity Monitoring 飞机自主完好性监视

**AAIP** Advanced Airborne Instrumentation Platform 先进的空中仪表平台

Advanced Avionics Integration Program 先进的航空电子大纲

Analog Autoland Improvement Program 模拟式自动着陆改进程序

Approved Airworthiness Inspection Program 批准的适航检查大纲

Autopilot Autoland Improvement Program 自动驾驶自动着陆增益程序

**AAIR** Annual Airworthiness Information Report 适航信息年度报告

**AAIS** Advanced Airborne Instrumentation System 先进的航空仪表系统

Anti-Aircraft Information Service 防空情报处(美国)

Automated Aerodrome Information Service 自动机场情报服务

**AAIU** Air Accident Investigation Unit 航空事

故调查局

Auxiliary Arrangement Interconnect Unit 辅助装备互连组件

**AAJC**　Automatic Antijam Circuit 自动抗干扰电路

**AAL**　Above Airdrome Level 高出机场场面
Actual Available Load 实际可用载重
Aircraft Approach Light 飞机进近灯光
Aircraft Approach Limitation 飞机进近限制
Aircraft Assignment Letter 飞机分配证
Angle of Approach Light(s) 进近下滑角指示灯,进近灯的角度
Anti-Aircraft Light 防空探照灯,对空探照灯
Arctic Aerospace Laboratory 北极航空航天实验室
Australian Air League 澳大利亚航空协会

**AALC**　Amplified Automatic Level Control 放大自动电平控制

**AALPA**　American Air Line Pilot Association 美国航空公司驾驶员协会

**AALS**　Advanced Approach & Landing System 高级的进近和着陆系统

**AAM**　Ambient Air Monitor 环境空气监测
Area Airworthiness Manager 区域适航经理
Automatic Approach Mode 自动驾驶进场着陆方式
Autopilot Actuator Monitor 自动驾驶作动器监视器

**AAMG**　Airbus Application Management Group 空客应用软件管理小组

**AAMP**　Advanced Architecture Microprocessor 先进的结构微处理器

**AAMRL**　Armstrong Aerospace Medical Research Laboratory 阿姆斯特朗航空航天医学研究实验室(美国)

**AAMS**　Advanced Automatic Monitoring System 先进的自动监测系统
Airborne Auxiliary Memory System 机载辅助存储系统
Area Airspace Management System 地区空域管理系统

Association of Air Medical Services 美国航空医疗服务协会
Automated Aircrew Management System 自动化的机组管理系统

**AAN**　Advance Alteration Notice 更改预告
Airworthiness Approval Note 适航批准通知单

**AAO**　Analog Attitude Output 模拟式姿态输出
Anti-Air Output 防空数据输出
Authorized Acquisition Objective 核准捕获目标

**AAP**　Acceptable Alternative Product 合格代用产品
Additional Attendant Panel 附加乘务员面板
Aerodynamics Advisory Panel 空气动力学咨询小组
Affirmative Action Plan 确认操作计划
Aging Aircraft Program 老龄飞机计划
Aircraft Assembly Plant 飞机装配厂
Aircraft Available Productivity 飞机可用生产率
Ambient Absolute Pressure 外界绝对压力,环境绝对压力
American Aviation Publication 美国航空出版物
Angle of Approach 进近角
Antiair Processing (Program) 防空数据处理程序
Applied Administrative Publications 应用管理出版物
Appollo Applications Program 阿波罗应用计划
Associative Array Processor 相关数组处理程序
Audible Alarm Panel 音响警报信号板
Audible Alert Panel 音响报警控制板

**AAPA**　Association of Asia Pacific Airlines 亚太航空公司协会

**AAPD**　Auto Astro Position Device 自动天文定位器

**AAPP** Airborne Auxiliary Power Plant 机载辅助动力装置

**AAPS** Australian Airspace Policy Statement 澳大利亚空域政策声明
Automated Astronomic Positioning System 自动天文定位系统

**AAPU** Airborne Auxiliary Power Unit 机载辅助动力装置

**AAQS** Ambient Air Quality Standard 环境空气质量标准

**AAR** Aarhus, Denmark (Airport Code) 丹麦奥尔胡斯机场
Advanced Arrangements Required 需做预先安排
After Action Report 事后处理报告
After All Risks 全保险
Aircraft Accident Record 飞机事故记录
Aircraft Accident Report 飞机事故报告
Airport Acceptance Rate 机场容许进场率
Airport Arrival Rate 机场到场率
Air-to-Air Refueling 空中加油
Array Area Ratio 阵列区域比
Automatic Alternative Routing 自动迂回路由
Automatic Altitude Report 自动高度报告

**AARC** Atlantic Air Rescue Center 大西洋航空救援中心

**AARCM** Asia Association of Risk and Crisis Management 亚洲风险与危机管理协会

**AARD** Autonomous Airborne Refueling Demonstration 自主空中加油演示

**AARE** Application Association Response 应用联合响应

**AARG** Average Annual Rate of Growth 年平均增长率

**AARL** Aeronautical and Astronautical Research Laboratory 航空航天研究实验室

**AARPLS** Advanced Airborne Radio Position Location System 先进的机载无线电定位系统

**AARQ** Application Association Request 应用联合请求

**AARS** Attitude and Azimuth Reference System 姿态方位基准系统
Attitude/Altitude Retention System 姿态高度保持系统
Automatic Address Recognition Subsystem 自动寻址识别子系统
Automatic Altitude Report/Reporting System 高度自动报告系统

**AARTS** Automated Audio Remote Test System 自动音频远程测试系统
Automatic Audio Remote Test Set 自动音频远程测试装置

**AAS** Abort Advisory System 飞行失败咨询系统
Adjusted Air Speed 已调空速
Advanced Administration System 先进的管理系统
Advanced Aerial Scout 先进的空中巡逻
Advanced Air Station 高级航空站(英国)
Advanced Antenna System 先进的天线系统
Advanced Automated System 先进的自动系统
Advanced Automatic System 先进的自动系统
Aeronautical Advisory Station 航空咨询电台
Aft Axle Steering 后轴操纵
Airborne Antenna System 机载天线系统
Aircraft Accident Summary 航空器事故摘要
Aircraft Airworthiness Section 航空器适航处
Airport Advisory Service 机场咨询服务
Altitude Alert/Alerting System 高度警报系统
American Astronautical Society 美国航天协会
Analog Alarm Section 模拟警报组
Angular Acceleration Susceptibility 角加速度敏感性
Antiaircraft Searchlight 防空探照灯
Antiaircraft Service 防空勤务
Anti-Icing Advisory System 防冰咨询系统
Apollo Abort System 阿波罗应急停飞系统

Arnold Air Society 阿诺德航空学会

Atlantic Aviation Services 大西洋航空服务

Australian Aircraft Sales 澳大利亚飞机销售公司

Auto-Alignment Angle Sensor 自动校准角度传感器

Automatic Addressing System 自动寻址系统

Automatic Announcement Subsystem 自动通知子系统

Automatic Audio Switching System 自动音频转换系统

Avionics and Aircraft System 航空电子设备和飞机系统

**AASA** Advanced Airborne Surveillance Antenna 先进的机载监视天线

Advanced Avionics System Analysis 先进的航空电子系统分析

Aging Aircraft Safety Act 老龄飞机安全法案

**AASC** Aerospace Application Studies Committee 航空航天(技术)应用研究委员会

**AASF** Advanced Air Striking Force 前线空中打击力量

**AASIR** Advanced Atmospheric Sounder and Imaging Radiometer 先进的大气探测器与图像无线电辐射仪

**AASL** Anti-Aircraft Searchlight 防空(高射)探照灯

Associated Aero Science Laboratories 联合航空科学实验室

**AASP** Advanced Acoustic Signal Processor 先进的声音信号处理器

**AASQ** Aeromedical Airlift Squadron 航空医疗空运中队

**AASR** Advanced Airborne Surveillance Radar 先进的机载监视雷达

Airport and Airways Surveillance Radar 机场与航路监视雷达

**AASS** Advanced Acoustic Search Sensor 先进的声测传感器

Advanced Airborne Surveillance Sensor 先进的机载监视传感器

Advanced Airborne Surveillance System 先进的机载监视系统

**AASSG** Aircraft Avionics Support Subsystem Group 飞机航空电子支援子系统集团

**AAT** Accelerated Aging Test 加速老化试验

Administrative Appeals Tribunal 行政上诉审裁处

Advanced Avionics Technology 先进的航空电子技术

Airports Authority of Thailand 泰国机场当局

Airworthiness Approval Tag 适航批准标签

Asia Airfreight Terminal Limited 亚洲货运中心有限公司

Assigned Arrival Time 指定到达时间

Australian Air Transport 奥地利航空运输公司

Automatic Answer Trunk 自动应答中继

Availability Analysis Tool 可利用的分析工具

Aviation Applied Technology 航空应用技术

**AATC** Aeronautical Apprentice Training Center 航空技工培训中心

American Air-Traffic Controllers' Council 美国空中交通管制员联合会

Anti-Aircraft Training Center 防空培训中心

Australian Air Transport Co. 奥地利航空运输公司

Automatic Air Traffic Control 自动空中交通管制

**AATCC** American Air Traffic Management Control Council 美国空中交通管制员理事会

**AATCFO** Assistant Air Traffic Control Facility Officer 助理空中交通设备管理员

**A-ATCK** Angle of Attack 迎角,攻角

**AATD** Aviation Applied Technology Directorate 航空应用技术理事会

**AATE** Advanced Affordable Turbine Engine 先进的经济性的涡轮发动机

**AATF** Air Assault Task Force 空袭特遣队

Airport and Airway Trust Fund 航空港与航

路信托基金

Airworthiness Assurance Task Force 适航保证任务组

**AATG** Average Annual Traffic Growth 交通平均年增长

**AATH** Automatic Approach to Hovering 自动进入悬停

**AATM** Assistant Air Traffic Manager 助理空中交通管制员

**AATMS** Advanced Air Traffic Management Systems 先进的空中交通管理系统(美国)

**AATP** Anti-Airsickness Training Program 抗空晕病训练计划

**AATPO** Association of African Trade Promotion Organizations 非洲贸易促进组织协会

**AATS** Advanced Automatic Test System 先进的自动测试系统

Advanced Automation Training System 先进的自动化培训系统

Alerting Automatic Telling Status 自动报警状态

Alternate/Alternative Aircraft Takeoff System 飞机备用起飞系统

Automated Air Traffic System 自动化空中交通系统

Automatic Altitude Trim System 自动高度配平系统

Committee on the Application of Aerospace Technology to Society 航空航天技术学会应用委员会

**AATSR** Advanced Along Track Scanning Radiometer 先进的沿航向扫描辐射计

**AATT** Absolute Arrival Time Technique 绝对到达时间技术

Advanced Air Traffic Technologies 先进的空中交通技术

Advanced Air Transportation Technologies 先进的航空运输技术

Advanced Aviation Transportation Technology 先进的航空运输技术

**AATU** Association of Air Transport Unions 空中运输协会联合会

**AAU** Absolute Alignment Update 绝对校准更新

Access Authorization Unit 存取审定组件

Audio Access Unit 音频存取组件

Audio Accessory Unit 音频附件盒

Automatic Answering Unit 自动应答单元

**AAUT** Alternate Attitude Update Techniques 备用姿态更新技术

**AAV** Advanced Aerospace Vehicle 先进的航空航天飞行器

Autonomous Air Vehicle 自主式空袭飞行器

Auxiliary Air Valve 辅助空气活门

**AAVCS** Airborne Automatic Voice Communications System 机载自动音频通信系统

Automatic Aircraft Vectoring Control System 航空器自动引导控制系统

**AAVD** Automatic Alternate Voice/Data 自动语音/数据交替使用

**AAVS** Aerospace Audio-Visual Service 航空航天声像业务处(美国)

Automatic Aircraft Vectoring System 航空器自动引导系统

**AAWC** Anti-Air Warfare Coordinator 防空作战协调官

**AAWF** Auxiliary Aviation Weather Facility 辅助航空气象处

**AAWG** Airworthiness Assessment Working Group 适航评估工作小组

Airworthiness Assurance Working Group 适航保证工作小组

**AAWO** Association of American Weather Observers 美国气象观察员协会

**AAWOS** Aviation Automatic Weather Observing System 航空自动天气观测系统

**AAWS** Airport Aviation Weather Service 机场航空气象服务

Anti-Air Warfare System 防空作战系统

Automatic Aviation Weather Service 自动航空气象服务

**AB** Abbreviated dialing 缩位拨号

Abort 中断飞行,紧急停车

Absolute Pressure 绝对压力

Adapter Booster 连接增压泵

Address Bus 地址总线

Address on Bag 箱包上地址

Advanced Boarding 预先登机

Advisory Board 咨询委员会

Aerodrome Beacon 机场信标,机场灯标

Aeronautical Board 航空委员会

After Body 后机身

Afterburner Control 加力燃烧室控制

Afterburner 加力燃烧室,喷射式燃烧室

Air Bag 空气囊

Air Blast 加力燃烧室

Air Board 航空局,航空委员会

Air Brake 减速板

Airborne 空中运行的,机载的

Air-Braker 空气断路器

Aircraft Bulletin 飞机公报

Airman Basic 飞行员基本训练

Aligned Bundle 定位光纤束

All Before 一切在前

Anchor Bolt 支持杆,支撑杆

Application Block 应用区

Assault Breaker 冲击断路器

Audio Bandwidth 音频带宽

Autobrake 自动刹车

**AB/LD** Airbrake/Lift Dumpers 减速板与减升板

**ABA** American Bar Association 美国律师协会

American,British,Australian 美国、英国、澳大利亚

**ABAA** Australian Business Aircraft Association 澳大利亚公务机协会

**ABAC** APEC Business Advisory Council APEC 工商咨询理事会

Association of British Aero Clubs and Centres 英国航空俱乐部/中心协会

Association of British Aviation Consultants 英国航空顾问协会

**ABACE** Asia Business Aviation Conference Exhibition 亚洲商务航空展览会

**ABAS** Aircraft Based Augmentation System 机载增强系统,航基增强系统

Asia Brand Assessment System 亚洲品牌测评体系

**ABB** Abbreviation 省略,缩略语,缩写

**ABBR** Abbreviated or Abbreviation 缩减或缩略语

**ABC** Abbreviations and Codes 缩写和代码

Address Bus Control 地址总线控制

Advance Booking Charter 预订包机,预订座位的包机航班

Advanced Blade Concept 改进叶片方案,前引桨时升力利用原则

Advanced Broadband Communications 高级宽带通信

Advanced Business Communications 先进的商业通讯公司(美国)

Advancing Blade Concept 前行桨叶概念

Aerial Board of Control 空中管理委员会(英国)

Air-Blast Cooled 气冷却的

Airborne Control 空中控制,机上控制

American,British,Canadian 美国,英国,加拿大

Analog Baseband Combiner 模拟基带合并器

Auto Billing Calling 自动记账呼叫

Automatic Background Control 自动本底控制

Automatic Bandwidth Control 自动带宽控制

Automatic Bass Compensation(Control)自动低音补偿(控制)

Automatic Beam Control 自动波束控制

Automatic Bias Compensation 自动偏压补偿

Automatic Bias Control 自动偏置控制

Automatic Binary Computer 自动二进制计算机

Automatic Block Controller 自动数据块控制器,自动区域控制器

Automatic Boost Control 自动助推控制

Automatic Brake Control 自动刹车控制

Automatic Brightness Control 自动亮度调整

**ABCA** American, British, Canadian, Australian 美国,英国,加拿大,澳大利亚

**A-BCAS** Active Beacon Collision Avoidance System 有源信标避撞系统

**ABCC** Automatic Brightness Contrast Control 自动亮度对比控制

**ABCCC** Airborne Battlefield Command and Control Center 空中战场指挥与控制中心

Airborne Command and Control Center 空中指挥和控制中心

Airborne Command, Control & Communications 空中指挥、控制与通信

**ABCD** Airbus Boeing, Convair, Douglas 空客、波音、康维尔、道格拉斯

**ABCM** Alternate Brake Control Manifold 备用刹车控制总管

**ABCP** Asset-Backed Commercial Paper 资产支持型商业票据

Automatic Bias Compensation 自动偏置补偿

**ABCS** Active Brake Control System 主动刹车控制系统

Advanced Business Communication Satellite 先进的商业通信卫星

Automated Budget Control System 自动预算控制系统

Automatic Broadcast Control System 自动广播控制系统

**ABCU** Alternate Braking Control Unit 备用刹车控制组件

Auto Brake Control Unit 自动刹车控制组件

**ABD** Abadan, Iran (Airport Code) 伊朗阿巴丹机场

Abbreviated Dialing 缩位拨号

Aboard 在机上,登机,机上

Advanced Base Depot 先进的基地检修厂

Airbus Directive 空中客车指令

Answer-Back Device 应答设备

**ABDAS** Airborne Bus Data Acquisition System 机载总线数据采集系统

**ABDC** After Bottom Dead Center 在下死点后

Air Base Defense Center 航空基地防御中心

**ABDG** Advisory Body Dangerous Goods 危险货物咨询委员会

**ABDL** Automatic Binary Data Link 自动二进制数据链路

**ABE** Allentown, PA, USA (Airport Code) 美国宾夕法尼亚州阿伦敦机场

**ABEND** Abnormal End of Task 任务非正常终止

Abnormal End of Test 测试非正常终止

Abnormal End 非正常终止

**ABEP** Advanced Burst Error Processor 高级突发错误处理器

**ABER** Average Bit Error Rate 平均错误比特率

**ABES** Air-Breathing Engine System 空气喷气发动机系统

**ABET** Accreditation Board for Engineering and Technology 工程与技术鉴定委员会

**ABETS** Airborne Beacon Electronic Test Set 机载信标电子测试仪

**ABF** Air Blown Fiber 充气光纤

Assets By Financing 以资产融资

Auto Beam Forming (无源声纳)自动形成波束

Aviation Boatswain's Mate (Fuels) 负责燃油的航空水手长

**ABI** Abilene, TX, USA (Airport Code) 美国得克萨斯州阿比林机场

Advance Boundary Information 高级边界信息

Airbus Industries (France) 空中客车工业公司(法国)

Application Binary Interface 应用二进制接口

Avionics Bus Interface 航空电子设备总线接口

**ABIA** Associacao Brasileira das Industrias Aeronauticas 巴西航空工业协会

**ABILA** Airborne Instrument Landing Approach 机载仪表着陆进近

**ABIM** Advanced Boundary Information Message 高级边界信息报文

**ABIRD** Aircraft-Based Infrared Detector 机载红外探测器

**ABIT** Advisory Body Information Technology 信息技术咨询委员会
Airborne Imagery Transmission 空中成像传送

**ABJ** Abidjan, Cote D'ivoire (Airport Code) 科特迪瓦阿比让机场
Air-Breathing Jet 航空喷气发动机,吸气式发动机

**ABL** Ability 能力
Above Baseline 高于基线
Automatic Brightness Limiter 自动亮度限制器

**ABLD** Automatic Blind Letdown 自动盲目下降

**ABLDG** Abnormal Landing 非正常着陆

**ABLM** Advisory Body Legal Matters 法律事务咨询委员会

**ABM** Abeam 正切,横梁
Activity Based Management 作业制成本管理
APU Build-up Manual 辅助动力装置装配手册
Asynchronous Balanced Mode 异步平衡模式
Bamaga, Queensland, Australia (Airport Code) 澳大利亚昆士兰州巴马加机场

**ABME** Asynchronous Balanced Mode Extended 异步平衡模式扩充

**ABMS** Auburn Business Management System 奥本商务管理系统

**ABMTM** Associated British Machine Tool Makers 英国机床制造业协会

**ABN** Abnormal 非正常
Aerodrome Beacon 机场标志灯,机场灯标
Airborne 机载的,机上的,空中的

**ABNORM** Abnormal 异常,非正常

**ABO** Abolish 废除
Actual Burnoff 实际耗油

Astable Blocking Oscillator 非稳定间歇振荡器
Aviator's Breathing Oxygen 飞行员用氧气

**ABOBA** Asynchronous Bidirectional Optical Branching Amplifier 不对称双向光支路放大器

**ABP** Advance Boarding Pass 提前登机牌
Airborne Beacon Processor 机载信标处理机
Arterial Blood Pressure 动脉血压

**ABPR** Advisory Body Public Relations 公共关系咨询委员会

**ABPS** Air Breathing Propulsion System 喷气推进系统
Airborne Beacon Processing System 机载信标处理系统

**ABPSK** Aviation Binary Phase Shift Keying 航空二进制移相键控

**ABQ** Albuquerque, NM, USA (Airport Code) 美国新墨西哥阿尔伯克基机场

**ABR** Address Buffer Register 地址缓冲寄存器
Agile Beam Radar 捷变装束雷达
Area Border Router 区域边界路由器
Automatic Bit Rate selection 自动比特率选择
Available Bit Rate 可用比特率
Average Bit Rate 平均比特率

**ABRN** Airborne 在空中,离陆

**ABRS** Automated Baggage Reconciliation System 行李自动确认系统

**ABRT** Abortive 中途停航,中止飞行

**ABRT/RAD** Abortive due to Radio out 因无线电故障中止飞行

**ABS** Absence 缺席
Absolute Viscosity 绝对粘度
Absolute 绝对
Abstract 摘要
Abu Simbel, Egypt (Airport Code) 埃及阿布辛贝机场
Access Barred Signal 接入受阻信号
Adaptive Braking System 适配制动系统

Air Braking System 通气系统

Air-Brake Switch 空气断路开关

Alternate Billing Service 可选计费业务

American Bureau of Standards 美国标准局

Antenna Base Spring 天线基座弹簧

Antenna Bridge Structure 天线桥式结构

Anti-Lock Brake System 防抱死刹车系统，防抱死制动系统

Antipodal Baseband Signaling 对应基带信号

Australian Bureau of Statistics 澳大利亚统计局

Automatic Braking System 自动刹车系统

**ABS VAL** Absolute Value 绝对值

**ABSCU** Auto Brake System Control Unit 自动刹车系统控制组件

**ABSD** Advanced Base Sectional Dock 先进的基地分段检修厂

Advanced Base Supply Depot 先进的基地供应车

**ABSELV** Alternate Brake Selector Valve 备用刹车选择活门

**ABSL** Ambient Background Sound Level 环境背景声级

**ABSLT** Absolute 绝对的，完全的

**ABSORB** Absorber 减震器

**ABSR** Abstract Reasoning Test 抽象推理测试

**ABSTR** Abstract 摘要

**ABSV** Air Bypass Solenoid Valve 空气旁通电磁活门(阀)

**ABTA** Association of British Travel Agents 英国旅行社协会

**ABV** Abuja, Nigeria (Airport Code) 尼日利亚阿布贾机场

Air Bypass Valve 空气旁通活门

**ABVT** Advisory Body Vocational Training 职业培训咨询委员会

**ABW** Advise By Wire 电话通知

**ABX** Albury, New South Wales, Australia (Airport Code) 澳大利亚新南威尔士奥尔伯里机场

**ABY** Albany, GA, USA (Airport Code) 美国乔治亚州奥尔巴尼机场

**ABZ** Aberdeen, Scotland, United Kingdom (Airport Code) 英国苏格兰阿伯丁机场

**AC** Above Clouds 在云层上，云上

Absolute Ceiling 绝对升限，绝对云幂高

Access Code 存取码，接入码

Access Concentrator 接入集中器

Access Control 访问控制

Access Coupler 通路耦合器

Access Cycle 存取周期

Accounting Function 计费功能

Accumulator 蓄压瓶，累加器

Adaptive Control 自适应控制

Advice of Charge 计费通知

Advisory Circular 咨询通告

Aerodrome Control 机场指挥，机场调度

Aerodynamic Center 空气动力中心，焦点

Air Canada 加拿大航空公司

Air Cargo 航空货运，空运货物

Air Carrier 航空承运人

Air Charter 包机

Air Cleaner 空气滤清器

Air Compressor 空气压缩机，压气机

Air Condenser 空气冷凝器

Air Conditioning 空气调节

Air Controlman 空中交通管制员

Aircraft Commander 机长

Aircraft Control 飞机操纵

Aircraft 飞机，航空器

Airspace Control 空域管制

Airworthiness Certificate 适航证

All Cargo 全货舱

Altering Course 改变航向，改变航线

Alternating Current (电)交流电

Altitude Coarse 粗高度

Altocumulus 高积云

Analog Computer 模拟计算机

Application Channel 应用信道

Application Context 应用关系

Approach Control 进近管制

Area Code 地区号

Asynchronous Computer 异步计算机

Authentication Center 认证中心

Authentication Code 认证码

Automated Customized 自动客户化的

Automatic Control 自动控制,自动调节

Auxiliary Pilot 助理飞行员

Average Cost 平均成本

Azimuth Compass 方位罗盘

**AC&S** Attitude Control and Stabilization 姿态控制和稳定

**AC/DC** Alternating Current/Direct Current 交直流两用

**ACA** Acapulco, Mexico (Airport Code) 墨西哥阿尔普尔科机场

Accounting Analysis 结算分析

Adaptive Channel Allocation 自适应信道分配

Adjacent Channel Attenuation 相邻信道衰减

Advanced Cargo Aircraft 先进的货机

Advanced Combat Aircraft 先进的作战飞机

Air Crew Association 空勤人员协会(英国)

Airspace Control Authority 空域管制当局

Airspace Coordination Area 空域协调区

Approach Control Area 进近管制区

Arctic Control Area 极地管制区

Attitude Control Assembly 姿态控制装置

Australian Commonwealth Authority 澳洲传播主管当局

Australian Council for Aeronautics 澳大利亚航空委员会

Automated Cable Analysis 电缆自动分析

Automatic Collision Avoidance 自动避撞

**ACAA** Academic Center for Aging Aircraft 老龄飞机学术中心

Air Carrier Association of America 美国航空运载协会

**ACAAI** Air Cargo Agents Association of India 印度航空货运代理人协会

**ACAAR** Aircraft Communications Addressing and Reporting 飞机通信寻址与报告

**ACABQ** Advisory Committee on Administrative and Budgetary Questions 行政和预算问题咨询委员会

**ACAC** Air Control Area Commander 空中交通管制区指挥员

Air-Cooled Air Cooler 气冷空气冷却器

Arab Civil Aviation Commission 阿拉伯民用航空委员会

Automatic Connector Assembly Cell 自动联接器组装室

AVIC-Commercial Aircraft Co. 中航商用飞机有限公司

**ACACS** Air Cycle Air Conditioning System 空气循环空调系统

**ACAD** Academy 学院

Advanced Computer Aided Design 先进的计算机辅助设计

**ACADS** Alarm Communication and Display System 报警通信与显示系统

**ACAH** Attitude Command/Attitude Hold 姿态指令/姿态保持

**ACAKD** Acknowledged 知道的

**ACAMPS** Automatic Communications and Message Processing System 自动通信与信息处理系统

**ACAMS** Aircraft Communications and Management System 飞机通信与管理系统

Aircraft Condition Analysis and Management System 飞机状态分析与管理系统

**ACAP** Advanced Composite Airframe Program 先进的复合材料机体计划

Airplane Characteristics and Airport Planning 飞机特性与机场计划

Application Configuration Access Protocol 应用配置访问协议

Automatic Circuit Analysis Program 自动电路分析计划

Aviation Consumer Action Project 航空消费者行动项目

**ACARE** Advisory Council for Aeronautics Research in Europe 欧洲航空研究咨询委员会

**ACARS** Air Concepts and Requirements Study 航空概念与需求研究

Airborne Communications Addressing and Reporting System 机载通信寻址和报告系统

Airborne Communications and Reporting System 机载通信和报告系统

Aircraft Communications Addressing and Reporting System 飞机通信寻址和报告系统

Aircraft Communications and Reporting System 飞机通信和报告系统

Automatic Communication and Recording System 自动通信与记录系统

**ACAS** Accident Consequence Assessment System 事故后果评价系统

Airborne Collision Avoidance System 机载防撞系统

Aircraft Collision Avoidance System 飞机防撞系统

Automatic Central Alarm System 中央自动警报系统

Automatic Collision Avoidance System 自动避撞系统

**ACASTD** Advisory Committee on the Application of Science and Technology to Development 科学和技术促进发展咨询委员会

**ACATC** Asia Civil Aviation Training Conference 亚洲民用航空培训会议

**ACAU** Air Conditioning Accessory Unit 空调辅助组件

Automatic Calling and Answering Unit 自动呼叫与应答装置

**ACAVS** Advanced Cab and Visual System 高级驾驶舱及目视系统

**ACAW** Aircraft Control and Warning 飞机控制与警告

**ACB** Air Circuit Breaker 空气断路器

Automatic Call Back 自动回叫

Auxiliary Controller Bus 辅助控制器总线

**ACBO** Actual Cumulative Burn-off 实际累积耗油

**ACC** Acceleration 加速度

Accelerator 累加器

Acceptance, Accepted 接受,承兑

Accessory 附件

Accidental 偶然的,意外的

According to 根据,按照

Account Card Calling 计账卡呼叫

Accra, Ghana(Airport Code)加纳阿克拉机场

Accumulator 储压器, 蓄电池,蓄压瓶

Active Clearance Control 主动间隙控制

Administrative Committee on Co-ordination 行政协调委员会

Aft Cargo Carrier 后货盘托架

Air Combat Command 空中联合司令部

Air Conditioning Controller 空调控制中心

Air Coordinating Committee（美国）航空协调委员会

Airborne Control Computer 机载控制计算机

Airspace Coordination Center 空域协调中心

Airworthiness Control Committee 适航管制委员会

Area Control Centre 区域管制中心

Asynchronous Communication Control 异步通信控制

Audio Communications Controller 音频通信控制器

Automatic Carrier Control 载波自动控制

Automatic Congestion Control 自动拥挤控制

**ACCA** Accounting Center of China Aviation 中国航空结算中心

Advanced Composite Cargo Aircraft 先进的复合材料货运飞机

Aircrew Contamination Control Area 机组污染控制区

Association of Chartered Certified Accountants 特许公认会计师公会(英国)

**ACCAD** Advisory Committee on Climate Applications and Data 气候应用和数据咨询委员会

**ACCB** Aircraft Configuration Control Board 飞机构型控制委员会

Aircraft/Airframe Change Control Board 飞机更改控制委员会

**ACCC** Alternate Command and Control Center 备用指挥与控制中心

Area Control Computer Complex 区域管制计算机网

Australian Consumer Competition Commission 澳大利亚消费者竞赛委员会

**ACCD** Aircraft Compatibility Control Drawing 飞机互换性控制图

**ACCDG** According 根据,按照

**ACCDLY** Accordingly 相应地

**ACCEL** Accelerometer 加速度计

Accelerator 加速器

**ACCELN** Acceleration 加速度

**ACCEPTN** Acceptance 接受,承兑

**ACCESS** Accessory 附件,辅助,附加

Aircraft Communication Control & Electronic Signaling System 飞机通信控制和电子信号系统

Aircraft Computerized Equipment Support System 飞机计算机设备支持系统

Automatic Computer Controlled Electronic Scanning System 自动计算机控制电子扫描系统

**ACCH** Associated Control Channel 随路控制信道

**ACCID** Accident 事故

Notification of an Aircraft Accident 航空器失事通知

**ACCLRM** Accelerometer 加速度表

**ACCOM** Accommodation 设备,膳宿,住宿

**ACCPTBL** Acceptable 可接受的,可承兑的

**ACCR** Advanced Customer Care and Retention 下一代客户关系管理

**ACCRD** Accrued 累计的,应记收的

**ACCRY** Accuracy 精确度

**ACCS** Access 入口,进入,通道

Active Contamination Control System 主动污染控制系统

Adapt Communications and Control Subsystem 自适应通信和控制子系统

Advanced Checkout and Control System 先进的检测与控制系统

Advanced Communications Control System 先进的通信控制系统

Aerospace Command and Control System 航空航天指挥与控制系统

Air Cargo Clearance System 空运货物清关系统

Air Command and Control System 空中指挥和控制系统

Association of Copyright for Computer Software 计算机软件著作权协会

Attitude Coordinate Converter System 姿态坐标数据换算系统

Automatic Calling Card Service 自动呼叫卡业务

Automatic Checkout and Control System 自动检测与控制系统

**ACCSRY** Accessory 附件,附加,辅助

**ACCT** Account of 因为,由于

Account 账户,账目,算账

Accountability (有)责任

Accountability Plan 责任计划

Accounting 算账

Automatic Control Centre of Transform 自动转换控制中心

**ACCTBLTY** Accountability 负有责任,可说明

**ACCTS** Automatic Continuous Close Loop Testing System 自动连续闭合环路测试系统

Aviation Coordinating Committee for Telecommunication Services 航空电信服务协调委员会

**ACCU** Accumulator 储压器,蓄能器

Audio Center Control Unit 中央音响控制装置

**ACCUM** Accumulate 累加

Accumulated 累积的

**ACCUR** Accurate 精确的

**ACCY** Accessory 附件,附加,辅助

Accordingly 相应地
Accuracy 精确度,准确度
**ACD** Accord 一致
According 根据
Active Control Devices 主动控制装置
Advanced Control Devices 先进的控制装置
Aircraft Certification Directorate 航空器审定管理局
Aircraft Certification Division 航空器合格审定室
Airframe Certification Document 飞机审定文件
Airways Clearance Delivery 发布放行许可
All Can Do 均可办到
Architectural Control Document 结构控制文件
Attitude Control Document 姿态控制文件
Authorized Certification Department 授权审查部门
Automatic Call Distribution 自动呼叫分配
Automatic Chart Display 自动航图显示器
**ACDB** Aircraft Component Data Base 飞机元件数据库
Aircraft Database 飞机数据库
Airports Characteristics Data Bank 机场特征数据库
**ACDGLY** Accordingly 依照,于是
**ACDGY** Accordingly 相应地
**ACDM** Association for Configuration and Data Management 构型与数据管理协会
**ACDMA** Advanced CDMA 高级码分多址
**ACDO** Air Carrier District Office 航空公司地区办事处
**ACDS** Advanced Command Data System 高级指令数据系统
Aircraft Control Display System 飞机控制显示系统
Automatic Comprehensive Display System 自动综合显示系统
**ACDT** Accident 失事,事故
**ACDTR** Airborne Central Data Tape Recorder

机载中央数据磁带记录器
**ACE** AC Exciter 交流励磁机
Acceptance Checkout Equipment 验收检测设备
Access Connection Element 接入连接单元
Actuator Control Electronics 作动器控制电子装置
Aeroplex of Central Europe 欧洲中部航空网
Air Cargo Equipment 航空货运设备,空中货运设备
Air Conditioning Equipment 空调设备
Airbus Concurrent Engineering 空客合作工程
Aircraft Condition Evaluation 飞机状况评估
Altimeter Control Equipment 高度控制设备
Association of Cost Engineers 成本工程师协会(英国)
Attitude Control Electronics 姿态控制电子设备
Automatic Calling Equipment 自动呼叫设备
Automatic Check-out Equipment 自动测试设备
Auxiliary Control Element 辅助控制单元
Lanzarote, Canary Islands, Spain (Airport Code) 西班牙加那利群岛兰萨罗特机场
**ACEA** Action Committee for European Aerospace 欧洲航空航天行动委员会
**ACEBP** Air-Condition Engine Bleed Pipe 空调发动机引气管
**ACEE** Aircraft Energy Efficiency 飞机能量效率
**ACEM** Aircraft Equipment Modification 飞机设备改装
**ACER** Advisory Committee on Environmental Resources 环境资源咨询委员会
**ACES** Adaptation Controlled Environment System 适应环境控制系统
Advanced Countermeasures Electronic System 先进的对抗电子系统
Aeronautical Communications Evaluation System 航空通信评价系统

Airline Cost Estimation System 航空公司成本估算系统

Airspace Concept Evaluation System 空域概念评价系统

Automatic Control Electronic Switching 自动控制电子开关

Automatically Controlled Electrical System 自动控制电气系统

**ACESA** American Clean Energy and Security Act 美国清洁能源安全法案

**ACESS** Advanced Cabin Entertainment and Service System 先进的客舱娱乐和服务系统

**ACF** Access Control Field 访问控制字段

Advanced Communications Function 高级通信功能

Aircraft Checkout Facility 飞机检查设施

Airline Consumer Forum 航空公司用户论坛

Alternate Communication Facility 备用通信设施

Area Control Facility 区域管制设施,地区管制设施

Arrangement Calibration File 排列校准文件

Authentication Control Function 鉴权控制功能

Automatic Colour Filter 自动彩色过滤器

**ACFA** Asian Clean Fuels Association 亚洲清洁燃料协会

**ACFC** Air Cooled Fuel Cooler 气冷燃油冷却器

Angeles City Flying Club 天使城飞行俱乐部

**ACFD** Advanced Civil Flight Deck 先进的民用飞机驾驶舱

**ACFE** Association of Certified Fraud Examiners 认证欺诈审核师协会

**ACFI** Advisory Committee on Flight Information 飞行情报咨询委员会

**ACFO** Aircraft Certification Field Office 航空器合格审定地方办公室

Association of Commercial Flying Organization 商用飞行组织协会

**ACFOD** Asian Cultural Forum on Development 亚洲发展文化论坛

**ACFS** Aircraft Carrier Flagship 航空母舰旗舰

Axial-Centrifugal Fan Shaft 轴向离心式风扇轴(发动机)

**ACFT** Aircraft Flying Training 飞机飞行训练

Aircraft 飞机,航空器

**ACFTINSP** Aircraft Inspector 飞机检查员

**ACFTSTR** Aircraft Strength 飞机强度

**ACFU** Aircraft Check Follow Up 飞机跟踪检查

**ACG** According to 根据

Address Coding Guide 地址编码指南

**ACGC** Automatic Color Gain Control 自动彩色增益控制

**ACGCF** AC Generators Constant Frequency 恒频交流发电机

**ACGV** Automatically Control and Guided Vehicle 自动控制与制导飞行器

**ACH** Answer Charge 应答计费

**ACHG** Advanced Charge 预付费用

**ACHI** Application Channel Interface 应用信道接口

**ACI** Access Control Information 接入控制信息

Acoustic Comfort Index 容许噪声指数

Acoustic Comfort Indictor 容许噪声指示器

Adjacent Channel Interference 相邻信道干扰

Air Cargo Incorporated 航空货运公司

Airman Certification Inspector 航空人员资格审定检查员

Allocated Configuration Identification 定位构型识别

American Certification Institute 美国认证协会

Asynchronous Communication Interface 异步通信接口

Attitude Controls Indicator 姿态操纵指示器

Automation Center International 国际自动化中心

**ACIA** Asynchronous Communications Interface Adapter 异步通讯接口器

**ACIC** Aeronautical Chart and Information Center 航图与情报中心

**ACID** Advanced Computing Information Database 先进的计算机信息数据库

Aircraft Identification 飞机识别标志

Automatic Classification and Interpretation of Data 数据的自动分类与说明

**ACIP** Aerodynamic Coefficient Identification Package 气动系数识别包

AFI Comprehensive Implementation Programme 非印全面实施方案

Airport Capital Improvement Program 机场资本改进方案

Aviation Career Incentive Pay 航空专业鼓励津贴

**ACIPS** Airfoil and Cowl Ice Protection System 机翼和整流罩防冰保护系统

**ACIR** Aviation Crash Injury Research 航空失事伤情研究

**ACIRU** Attitude Control Inertial Reference Unit 姿态控制惯性基准组件

**ACIS** Advanced Cabin Interphone System 先进的客舱内话系统

Aircraft Crew Interphone System 飞机机组内部通话系统

Avionics Central Information System 航空电子中央信息系统

**ACJ** Advisory Circular Joint 联合咨询通告

**ACK** Acknowledge 承认,收悉

Nantucket, MA, USA ( Airport Code ) 美国马萨诸塞州楠塔基特岛机场

**ACL** Access Control List 访问控制表

Actual 实际

Advanced Communication Link 高级通信链路

Aeronautical Computer Laboratory 航空计算机实验室

Aircraft Circular Letter 飞机间传送的通报

Aircraft Control Link 飞机操纵链

Allowable Cargo Load 允许载货量

Allowance Cabin Load 容许机舱负荷

Allowed Commercial Load 允许商载

Altimeter Check Location 高度表校准位置,高度表校准点

Anti-Collision Light 避撞灯,防撞灯

Application Control Language 应用控制语言

Assembly Connection List 组件连接清单

Authorized Component List 授权成员表

Automatic Crosshair Laying 自动瞄准线瞄准

Aviation Circular Letter 航空通报

**ACLICS** Airborne Communications Location Identification and Collection System 空中通信位置识别和收集系统

**ACLS** Air Cushion Landing System 气垫着陆系统

Aircraft Carrier Landing System 航空母舰着陆系统

American Council of Learned Societies 美国学术团体委员会

Automated Control and Landing System 自动控制和着陆系统

Automatic Carrier Landing System 自动运输机着陆系统,自动舰上降落系统

**ACLT** Actual Calculated Landing Time 实际计算着陆时间

**ACM** Access Control Machine 访问控制机

Access Control Method 访问控制法

Access Control Module 访问控制模块

Additional Crew Member 额外空勤组员

Address Complete Message 完全地址信息

Advanced Composite Materials 高级复合材料

Air Commercial Manual 航空商业手册

Air Cycle Machine 空气循环机

Air-Conditioning Module 空调组件

Aircraft Configuration Matrix 飞机构型矩阵

Aircraft Conversion Manual 飞机改装手册,飞机换算手册

Alarm Control Module 报警控制模块

APU Condition Monitoring 辅助动力装置状

态监控

Association for Computing Machinery 计算机协会(美国)

Associative Communication Multiplexer 关联通信多路复用器

Attendant Control Module 乘务员控制组件

Attitude Control Motor 姿态控制电机

Automatic Clutter Mapping 自动杂波测绘

Automatic Communication Monitor 自动通信监视器

Auxiliary Control Module 辅助控制模块

**ACMA** Air Carrier Mechanic Association 航空公司机械员协会

**ACMB** Aircraft Configuration Management Board 飞机配置管理委员会

**ACMEL** Associated Continental Middle East Lines 欧洲大陆中东航运公会

**ACMF** Airplane Condition Monitoring Function 飞机状态监控功能

**ACMI** Aircraft, Crew, Maintenance and Insurance 飞机,机组,维修和保险

**ACMM** Abbreviated Component Maintenance Manual 简化部件维修手册

**ACMP** Acquisition Career Management Program 采购业务管理计划

Airframe Condition Monitoring Procedure 飞机状态监控程序

Alternating Current Motor Pump 交流马达驱动泵

Annual Cost Monitoring Plan 年度成本监测计划

**ACMR** Aircraft Configuration Management Rules 飞机构型管理规则

**ACMS** Advanced Configuration Management System 高级配置管理系统

Advanced Cost Management Systems 先进的成本管理系统

Aircraft Computerized Maintenance System 飞机计算机化维修系统

Aircraft Condition Monitoring System 飞机状态监控系统

Application Control and Management System 应用控制及管理系统

Automated Communications Management System 自动通信管理系统

Automated Computer Management System 自动计算机管理系统

Aviation Computerized Maintenance System 航空计算机化维修系统

**ACMT** Aircraft Certification Management Team 飞机认证管理小组

Aircraft Component Management Team 飞机部件管理组

Automated Configuration Management Tool 自动配置管理工具

**ACN** Advance Change Notice 提前更改通知

Aircraft Classification Number 航空器等级号,航空器分类号

Aircraft Number 机号

Australian Company Number 澳洲公司编号

**ACO** Administrative Contracting Office 执行合同办公室

Administrative Contracting Officer 执行合同官员

Aerosat Coordination Office 航空卫星协调办公室

Aircraft Certification Office 航空器审定办公室

Alarm Cut-off 报警切断

Assembly & Checkout 组装与检查

**ACOB** Air Carrier Operation Bulletin 航空公司飞行通告

Automatic Call out Box 自动报告盒

**ACOC** Air Cooled Oil Cooler 气冷式滑油冷却器

Area Communication Operations Center 区域通讯操作中心

**ACOE** Automatic Checkout Equipment 自动检测装置

**ACOG** Aircraft on Ground 停在地面的飞机,停飞待用飞机

**ACOM** Automatic Centralized Operations &

Maintenance 集中自动运行与维护

**ACOS** Aircraft Certification Office Subsystem 飞机合格审定办公室分系统

**ACP** AC Pump 交流泵

Achieved Communication Performance 实际通信性能

Agent Creation Point 代理生成点

Aircraft Performance 飞机性能

Airlift Command Post 空运指挥所

Altimeter Check Point 高度表检查点

Area Call Panel 区域呼叫面扳

Audio Control Panel 音频控制面板

Automatic Command Processing 自动指令处理

Auxiliary Control Panel 辅助的控制仪表板,辅助控制盘

Aviation Cooperation Program 航空合作项目

Azimuth Control Pulse 方位控制脉冲

**ACPC** AC Power Center 交流电源中心

Aircraft Camera Parameter Control 飞机照相参数控制

**ACPET** Australian Council for Private Education and Training 澳大利亚私立教育和培训理事会

**ACPL** Aircraft Certification Policy Letter 航空器合格审定政策函

**ACPM** ATM Central Process Module ATM 中央处理模块

**ACPS** AC Power Supplies 交流电源

Attitude Control Propulsion System 姿态控制推进系统

Automated Contract Preparation System 自动合同草拟系统

**ACPT** Accept 接受

**ACPTBL** Acceptable 可接受的

**ACPTD** Accepted 已接受的

**ACPTNC** Acceptance 接受

**ACPY** Accompany 陪同,伴随,附带

**ACQ** Acquire 获取,获得

**ACQN** Acquisition 采集

**ACR** Access Control Register 访问控制寄存器

Accounting Registration 结算登记

Acrobatics 特技飞行

Across 交叉,穿过

Ad-hoc Clearance Request 特殊状况许可要求

Advance Copy Releases 先期复印发放

Advanced Capabilities Radar 性能先进雷达

Aerodrome Control Radar 机场管制雷达

Air Control Regulator 空气控制调节器

Air Control Room 航空管制室

Aircraft Control Room 航空器管制室,飞机管制室

Airfield Control Radar 机场管制雷达,进场着陆控制雷达

Airline Conversion Rate 航空兑换率

Airplane Cost Reporting 飞机成本报告

Allowed Cell Rate 允许的信元率

Approach Control Radar 进近管制雷达

Attenuation-to-Crosstalk Ratio 衰减串扰比

Australian Capital Region 澳大利亚首都区

Automatic Call Recording 自动呼叫记录

Automatic Character Recognition 自动字符识别

Available Cell Rate 可用信元率

Average Cell Rate 平均信元率

Avionics Communication Router 航空电子通讯路由器

**ACRDC** Australian Capital Region Development Council 澳大利亚首都区发展贸易局

**ACRDRTS** Aircraft Remote Diagnosing and Real-Time Tracking System 飞机远程诊断与实时跟踪系统

**ACRE** Airline Composite Repair Evaluation 航空公司复合材料修理评估

Automatic Call Recording Equipment 自动呼叫记录设备

**ACRES** Airborne Communication Relay Station 空中通信中继站,机载通信中继站

**ACRP** Airport Cooperative Research Program 机场合作研究项目

ACRS Accelerated Cost Recovery System 加速成本回收制

Asset Condition Reporting System (Canada) 资产状况报告系统(加拿大)

ACRT Additional Cross Reference Table 附加交叉参考表

Aviation Chief Radio Technician 航空无线电主任技师

ACRU Avionics Cooling Refrigeration Unit 航空电子冷却制冷组件

ACRWE Airborne Cosmic Radiation Warning Equipment 机载宇宙辐射预警装置

ACS Access Control Server 接入控制服务器

Access Control System 接入控制系统

Adaptive Control System 自适应控制系统

Admission Control Service 许可控制服务

Advanced Crew Systems 先进的机组系统

Advanced Cryptographic System 高级密码系统

Aerial Common Sensor 空中通用传感器

Airborne Common System 机载通用系统

Air-Combat Simulator 空战模拟机

Air-Conditioning System 空调系统

Aircraft Certification Service 航空器审定服务

Aircraft Communication System 飞机通信系统

Aircraft Control System 飞机控制系统

Airport Communication System 机场通信系统

Airspace Control System 空域管制系统

Airworthiness Control System 适航性控制系统

Alarm Call Service 报警呼叫业务

All Channel Signaling 全信道信令

Alternating Current Supply 交流供电

Area Control Station 区域管制站

Association of Caribbean States 加勒比国家联盟

Asynchronous Communication Server 异步通信服务器

Attitude Control System 姿态控制系统

Audio Communication System 音频通信系统

Audio Control System 音频控制系统

Auto Chart System 自动制图系统

Automated Classification System 自动分类系统

Automatic Call Sender 自动呼叫发送器

Automatic Coding System 自动编码系统

Automatic Control System 自动控制系统,自动调节系统

Auxiliary Cooling System 辅助冷却系统

ACSC Air Conditioning System Controller 空调系统控制器

Avionic Compartment System Cooling 电子舱系统冷却

ACSE Access Control and Signaling Equipment 接续控制与信号设备

Association Control Service Element 联合控制业务单元

ACSEP Aircraft Certification System Evaluation Program 航空器审定系统评审大纲

ACSI Advanced Computer System Interface 先进的计算机系统接口

Aircraft Certification Staff Instruction 航空器合格审定人员工作指南

Airport Certification Safety Inspector 机场认证安全监督员

American Customer Satisfaction Index 美国客户满意度指数

ACSN Advance Change Study Notice 提前变更研究通知

Auto Cover Serial Number 自动侦控序号

ACSS Aviation Communication and Surveillance Systems 航空通信与监视系统

ACSU Adjustable Cam Switch Unit 可调凸轮电门组件

ACT Acting 代理的,行动

Action 动作,行动,作用

Activator 作动筒

Active Control Technology 主动控制技术

Active Flutter Depression 主动颤振抑制

Active 现用的,主动的

Activity 行动,活动

Actual Temperature 实际温度

Actual 实际的

Adaptive Computing Technology 自适应计算机技术

Additional Center Tank 附加中央油箱

Advanced Composites Technology 先进的复合材料技术

Advisory Council on Technology 技术咨询委员会

Aircraft Configuration Tracking 飞机构型跟踪

Aircrew Classification Test 空勤人员分类测验

Analogical Circuit Technique 模拟电路技术

Associated Container Transportation Ltd. (英国)联合集装箱运输公司

Australian Capital Territory 澳大利亚首都领地

Automatic Checkout Technology 自动检测技术

Automatic Computer Testing 自动计算机测试

Auxiliary Central Tank 辅助中央油箱

Waco, TX, USA(Airport Code)美国得克萨斯州韦科机场

**ACTD** Actuated 作动的

Advanced Concept Technology Demonstration 先进概念技术验证

**ACTG** Acting 代理的,行动

Actuating 作动

**ACTL** Actual 实际

**ACTM** Actual Cumulative Time 实际累积时间

**ACTN** Action 动作,行动,作用

**ACTR** Actuator 作动器,作动筒,激励器

**ACTS** Advanced Communication Technologies and Services 先进的通信技术与业务

Advanced Communications and Technology Satellite 先进的通讯技术卫星(美国)

Air Cargo Terminal System 航空货运终端系统

Aircrew Training System 空勤人员训练系统

**ACTTAB** Action Table 动作表

**ACTVT** Activate 启动,引发

**ACU** Acceleration Control Unit 加速控制组件

Access Control Unit 接入控制单元

Air Conditioning Unit 空调组件

Air Control Unit 空气控制组件

Aircraft Control Unit 飞机控制组件

Airframe Converter Unit 机体变换器组件

Airport Control Unit 机场控制单元

Airshow Control Unit 空中博览控制组件

Annunciator Control Unit 告示器控制组件

Antenna Control Unit 天线控制装置

Antenna Coupler Unit 天线耦合器组件

Apron Control Unit 挡板控制组件

Area Control Unit 区域管制部门

Asian Cleaning Union 亚洲清算联盟

Audio Channel Unit 音频信道单元

Automatic Calling Unit 自动呼叫装置

Autopilot Control Unit 自动驾驶仪控制组件

Avionics Control Unit 航空电子控制组件

**ACUC** Avionic Control Unit Computer 航空电子控制装置计算机

**ACUD** Avionics Conditional Usage Drawing 航空电子设备状态使用图

**ACV** Air Control Valve 空气调节活门(阀)

Air-Cushion Vehicle 气垫飞行器

Arcata, CA, USA (Airport Code)美国加利福尼亚州阿克塔机场

**ACVR** Alternating Current Voltage Ratio 交流电压比

**ACW** Aircraft Control & Warning 飞机控制和警报

**ACWP** Actual Cost of Work Performed 完成工作实际成本

**ACWS** Aircraft Control & Warning Squadron 飞机操纵和警告飞行中队

Aircraft Control and Warning System 飞机操

纵和警告系统

**ACY** Atlantic City , NJ , USA ( Airport Code )
美国新泽西州大西洋城机场

**ACYC** Anticyclone 反气旋

**ACYR** Anti Cycling Relay 反循环继电器

**AD** Access Door 通道门,入口门

Accidental Damage 事故损坏,意外损伤

Actual Departure Time 实际离场时间

Add 增加

Addenda 附录,补遗

Address Bus 地址总线

Address Decoder 地址译码器

Administrative Domain 管理域

Advertisement 广告

Advisory Route 咨询航路

Aerodrome 机场

Aerodynamic Damping 气动阻尼

Aerodynamic Disturbance 气动力扰动

After Date 某日之后

Air Data 大气数据

Air Defense 防空

Air Distance 空中飞行距离,无风距离

Aircraft Direction 飞机引导

Airplane Datum 飞机数据,飞机基准面

Airworthiness Directive 适航指令

Airworthiness Division ( CAA ) 适航处 ( 英国民航局 )

Anno Domoni 公元

Autopilot Disconnect 自动驾驶仪断开

Aviation Daily 每日航空,航空日报

Air Distance 空中距离,空中海里

**ADA** Advisory Area 咨询区

Aeronautical Development Agency 航空工艺发展机构

Air Defence Area 防空区

Airborne Data Automation 机上数据自动化

Architecture Design Analysis 结构设计分析

Automatic Data Acquisition 自动数据采集

Aviation Data and Analysis ( system ) 航空数据分析

**ADAC** Acoustic Data Analysis Center 声学数据分析中心

Analog-Digital-Analog Converter 模拟—数字—模拟转换器

Automatic Data Acquisition Center 自动数据采集中心

Automatic Direct Analog Computer 自动直接模拟计算机

**ADACC** Automatic Data Acquisition and Computer Complex 自动数据采集与计算机综合体

**ADADS** Advanced Distribution & Drawing System 先进的分配与绘图系统

Automated Drawing Accountability & Distribution System 自动绘图责任与分配系统

**ADAF** Air Defense Alert Facility 防空预警设施

Aircraft De-icing and Anti-icing Fluid 飞机防冰与防冰液

Airport Development Aid Fund 机场发展援助基金

**ADAHRS** Air Data Attitude Heading Reference System 大气数据姿态航向基准系统

**ADAM** Advanced Data Access Method 高级数据存取方法

Advanced Design and Manufacturing 先进的设计与制造

Advanced Digital Avionics Map 先进的数字式航空电子设备布局图

Area Damage Assessment Model 区域损坏评估模式

**ADAMS** Airborne Data Acquisition & Monitoring System 机载数据采集与监测系统

Airborne Data Analysis & Monitoring System 机载数据分析监控系统

Aircraft Dispatch and Maintenance System 航空器签派与维修系统

**ADAO** Air Defense Aera Operation 防空区作战

**ADAP** Aerodynamic Data Analysis Program 气动数据分析计划

Airport Development Aid Program 机场发展

援助计划

Automatic Data Acquisition/Processor 自动数据采集处理器

**ADAPS** Automated Design and Packaging Service 自动化设计与组装业务

Automatic Data Acquisition and Processing System 数据自动采集和处理系统

Automatic Display and Plotting System 自动显示和绘图系统

**ADAPT** Adaptive Data Analysis and Processing Technology 自适应数据分析与处理技术

ATS Data Acquisition Processing and Transmission 空管服务数据录取处理传递

Automated Data Analysis and Presentation Techniques 自动化数据分析与显示技术

**ADAR** Advanced Data Acquisition Routine 高级数据采集程序

Advanced Defense Array Radar 改进设计的相控阵雷达

Advanced Design Array Radar 先进的矩阵雷达设计

Advise Arrival ( Reservations Code ) 告知到达班机

**ADARS** Airplane Data and Reporting System 飞机数据和报告系统

Automatic Download and Reporting System 自动下载和报告系统

**ADAS** Advanced Digital Avionics System 先进的数字航空电子系统

Airborne Data Annotation System 空中数据注解系统

Airborne Data-Acquisition System 机载数据采集系统

Airborne Dynamic Alignment System 机载动态校准系统

Automatic Data Acquisition System 自动数据采集系统

Auxiliary Data Acquisition System 辅助数据采集系统

AWOS Data Acquisition System 自动气象观察系统数据获取系统

**ADAT** Abu Dhabi Aircraft Technologies 阿布扎比飞机技术公司

**ADATS** Airborne Digital Avionics Test System 机载数字航空电子测试系统

Automated Data and Telecommunication Service 自动数据和电信服务

**ADAU** Air-Data Acquisition Unit 大气数据采集组件

Auxiliary Data Acquisition Unit 辅助数据采集组件

**ADB** Advise if Duplicated Booking 请告是否重复订座

Aeronautical Data Base 航空数据库

Africa Development Bank 非洲发展银行

Air Defense Board 防空委员会

Airports Database 机场数据库

Area Distribution Box 区域分配盒

Asian Development Bank 亚洲开发银行

Aviation Development Board 航空发展局

Avionics Data Base 航空电子设备数据库

Izmir, Turkey ( Airport Code ) 土耳其伊兹密尔机场

**ADBI** Asian Development Bank Institute 亚洲开发银行研究所

**ADC** Acquisition, Development and Construction 收购、开发与建设

Adaptive Data Compression 自适应数据压缩

Air Data Computer 大气数据计算机

Air Data Computing 大气数据计算

Air Defence Clearance 防空许可

Air Development Center 航空发展中心

Airborne Digital Computer 机载数字计算机，机载数字式计算机

Airbus Delivery Centre 空客交付中心

Aircraft Directives Configuration 飞机指令结构

All Day Calls 全日呼叫

Analog to Digital Converter 模数转换器

Analog-to-Digital Conversion 模数转换

Automatic Digital Computer 自动数字式计算

机

Automatic Digital Control 自动数字式控制

Aviation Development Council 航空发展理事会

**ADCA** Australian Department of Civil Aviation 澳大利亚民航部

**ADCAD** Airways Data Collection and Distribution 航路数据收集和分发

**ADCAP** Advanced Capability 先进性能

**ADCC** Air Defence Control Center 防空管制中心

Aircraft Data Control Center 飞机数据控制中心

Asynchronous Data Communication Channel 异步数据通信信道

Aviation Data Communication Corporation 民航数据通讯公司

**ADCCP** Advanced Data Communications Control Procedure 先进的数据通讯控制程序，高级数据通信控制规程

**ADCL** Airworthiness Directives Compliance List 适航指令符合性清单

**ADCN** Access Delivery Communication Network 接入传输通信网

Advance Design Change Notice 图纸更改的先期通知

Advance Drawing Change Notice 图纸修改通知

Air Data Communications Network 大气数据通信网

Avionics Data Communication Network 航空电子数据通讯网络

**ADCOM** Advanced Communications 先进的通信

Advanced Compound 先进的复合材料

**ADCON** Address Constant 地址常数

Administrative Control 行政管理

Analog to Digital Converter 模拟—数字转换

**ADCP** Advanced Display Core Processor 高级显示核心处理器

Advanced Data Communication Protocol 高级

数据通信协议

**ADCS** Aerodynamics Data and Control Systems 航空力学数据与控制系统

Air Data Computer System 大气数据计算机系统

Airborne Data Collection System 机载数据收集系统

Attitude Determination and Control Subsystem 姿态确定控制子系统

Automated Data Collection Systems 自动数据采集系统

**ADCSIP** Acceptance and Delivery Customer Satisfaction Improvement Programme 验收和交付客户满意度提升计划

**ADCU** Autodeploy Control Unit 自动放出控制组件

**ADD** Addis Ababa, Ethiopia-Bole ( Airport Code)埃塞俄比亚亚的斯亚贝巴博莱机场

Aircraft Description Database 飞机说明数据库

Airstream Direction Detector 气流方向探测器

Airworthiness Documentation Division 适航资料室

**ADDA** Air Defense Defended Area 防空区

**ADDAL** Additional 附加的,额外的

**ADDAR** Automatic Digital Data Acquisition and Recording 数字数据自动收集记录装置

**ADDAS** Automatic Digital Data Assembly System 自动数字数据汇编系统

**ADDC** Air Defence Direction Center 防空指挥中心

**ADDCY** Aerodrome Directory 机场指南

**ADDE** Addressee 收件人

**ADDEL** Advise Delivery 发布许可时通知

**ADDGM** Aerodrome Diagrams 机场图纸

**ADDI** Automatic Digital Data Interchange 自动化数字数据信息交换

**ADDN** Addition 附加,附加物

Additional 附加的,额外的

Automated Defense Data Network 自动防御

数据网络

**ADDNL**　Additional 附加的，额外的

**ADDR**　Address 地址

**ADDS**　Advanced Data Display System 先进的数据显示系统

**ADDSE**　Addressee 收件人

**ADE**　Aden, Yemen ( Airport Code ) 也门亚丁机场

Aeronautical Development Establishment 航空开发院

Automatic Drafting Equipment 自动制图设备

**ADEL**　Application Data Enhanced Loading 应用程序数据增强装载

**ADELT**　Automatically Deployable Emergency Locator Transmitter 自动紧急定位发射器

**ADEM**　Advanced Diagnosis Engine Management 先进的诊断发动机管理

Advanced Diesel Engine Management 先进的柴油发动机管理

**ADEMS**　Airborne Display Electrical Management System 机载显示电管理系统

**ADEOS**　Advanced Earth Observing Satellite 先进的地球观测卫星

**ADEP**　Airport/Aerodrome of Departure 起飞机场

**ADEPT**　Aircraft Data Engine Performance Trending 飞机数据发动机性能发展趋势

**ADES**　Automatic Digital Encoding System 自动数字编码系统

**ADESS**　Automatic Data Editing and Switching System 自动数据编辑与转接系统

**ADEU**　Automatic Data Entry Unit 自动数据输入组件

**ADF**　Adjusted Delay Factor 可调延迟因子

Aerodynamic Data File 气动数据文件

African Development Fund 非洲开发基金

Air Defense Fighter 防空战斗机

Air Direction Finder 空中测向器

Airport Duty Free 机场免税

Anti-icing/De-icing Fluid 防冰除冰液

Authentication Data Function 鉴权数据功能

Automatic Direction Finder 自动定向仪，无线电罗盘

Automation Direction Finding 自动定向

Auxiliary Detector Flow 辅助检测器流程图

**ADFA**　Automatic Direction Finding Approach 自动定向进近

**ADFC**　Air Defense Filter Center 航空情报鉴定中心

**ADFE**　Automatic Direction Finding Equipment 自动测向装置

**ADFHQ**　Air Defense Force Headquarters 防空航空队指挥部

**ADFLIR**　Advanced Forward Looking Infrared 先进的前视红外设备

**ADFR**　Automatic Direction Finder, Remote-Controlled 自动遥控测向仪

**ADFRMI**　ADF Radio Magnetic Indicator 自动定向仪无线电磁指示器

**ADFS**　Active Divergence/Flutter Suppression 主动发散—颤震抑制

Automatic Direction Finder 自动测向系统

**ADFSC**　Automatic Data Field System Command 自动数据场系统指令

**ADG**　Accessory Drive Gear 附件驱动齿轮，附件传动机构

Aircraft Delivery Group 空中交付组

Air Driven Generator 空气驱动发电机

Australian Dangerous Goods 澳大利亚危险货物

**ADGPC**　ADG Power Center 空气驱动发电机电源中心

**ADH**　Asynchronous Digital Hierarchy 异步数字系列

**ADHP**　Air Driven Hydraulic Pump 气动液压泵

**ADHS**　Analog Data Handling System 模拟数据处理系统

Automatic Data Handling System 自动数据处理系统

**ADI**　Aggregate Demand Indicators 集合指令

指示器

Air Data Instruments 大气数据仪表

Alternating Direction Interaction 交替方向叠代

Alternating-Direction-Implicit 交替方向描述

Attitude Director Indicator 姿态方向指示器

Anti-Detonation Injection 注入防爆剂

Altitude Direction Indicator 高度方位指示器

Automatic Director Indicator 自动方位指示器

Aviation Machinist's Mate 航空机械技师

**ADIC** Advanced Digital Information Corporation 先进的数字信息公司

**ADIF** Audio Data Interchange Format 音频数据交换格式

**ADIOS** Automatic Digital Input/Output System 自动数字输入/输出系统

**ADIRS** Air Data Inertial Reference System 大气数据惯性参考系统

**ADIRU** Air Data Inertial Reference Unit 大气数据惯性基准组件

**ADIS** Airborne Digital Instrument System 机载数字仪表系统

Automatic Data Interchange System 自动数据交换系统

**ADISP** Automated Data Interchange Systems Panel 自动数据交换系统专家小组

**ADIT** Analog-Digital Integrating Translator 模数综合转换装置

**ADIX** Advanced Digital Information Exchange 先进的数字信息交换

**ADIZ** Air Defence Identification Zone 防空识别区

Air Defence Intercept Zone 防空拦截区

**ADJ** Adjacent 附近的,邻接的

Adjust, Adjustment 调节,调整

Adjustable 可调节的

Adjustment 调整

Adjutant 助手

Amman, Jordan-Marka(Airport Code) 约旦安曼马尔卡机场

**ADJG** Adjusting 调节,调整

**ADJMT** Adjustment 调节

**ADL** Acceptable Defect Level 容许缺陷水平

Acoustic Delay Line 声延迟线

Addition-Delete List 旅客增减名单报

Adelaide, South Australia, Australia (Airport Code)澳大利亚,南澳大利亚阿德莱德机场

Administration Description Language 管理描述语言

Advanced Development Laboratory 先进的开发实验室

Advisory Light 咨询灯,指示灯

Aeronautical Data Link 航空数据链

Air Data Loader 大气数据加载,大气数据输入器

Airborne Data Loader 机载数据加载器

Aircraft Data Link 飞机数据链

Aircraft Data Loader 飞机数据装载机

Allowable Deficiency List 允许缺陷单

Antenna Dummy Load 天线等效负载

Application Development Language 应用开发语言

Atmospheric Devices Laboratory 大气装置实验室

Authorized Data List 规定的数据清单

Authorized Distribution List 批准的分配清单

Automatic Data Link 自动数据链

Automatic Data Logger 自动数据记录器,自动数据记录仪

Avionics Development Laboratory 航空电子设备开发实验室

**ADLC** Advanced Data Link Control 高级数据链路控制

Analogue-Digital Line Converter 模拟—数字线路变换器

**ADLP** Airborne Data Link Protocol 机载数据链协议

Aircraft Data Link Processor 飞机数据链处理器

**ADLR** Australia Dollar 澳元

**ADLS** Aeronautical Data Link System 航空数

据链系统

**ADM**    Adaptive Delta Modulation 自适应增量调制

Add/Drop Multiplexer 分插复用器

Administration 管理,管理局

Administrator 局长

Advanced Development Model 高级发展模型

Aeronautical Data Maintenance 航空数据维护

Aeronautical Decision Making 航空决策

Aid in Decision Making 辅助决策

Air Data Module 大气数据模块

Automated Data Management 自动数据管理

**ADMA**    Automatic Drafting Machine 自动绘图器,自动绘图机

Aviation Distributors and Manufacturers Association 航空分销商及制造商协会

**ADMAC**    Air Defense Monitor and Control 防空监视与控制

**ADMAR**    Aircraft Meteorological Data Relay 航空器气象资料下传

**ADMC**    Actuator Drive and Monitor Computer 制动器驱动与监测计算机

**ADMIS**    Automated Data Management Information System 自动数据管理信息系统

**ADMS**    Advanced Data Management System 先进的数据管理系统

Aircraft Diagnostic and Maintenance System 飞机诊断与维护系统

Airline Data Management System 航空公司数据管理系统

**ADMSC**    Automatic Digital Message Switching Center 自动数字信息交换中心

**ADMU**    Air Distance Measuring Unit 无风航程测算装置

**ADN**    Administration Data Network 管理数据网络

**ADNC**    Advise if Not Correct 如不正确请告之

**ADNET**    Administration Data Network 管理数据中心

**ADNS**    Automated Digital Network Systems 自动数字网络系统

**ADNSC**    Automatic Digital Network Switching Center 自动数字网交换中心

**ADO**    Advance Development Objective 远景发展目标

Airbus Documentation Office 空客文档办公室

Attribute Driven Option 特征驱动方案

Audio Decade Oscillator 音频十进制振荡器

Automotive Diesel Oil 内燃机柴油

Aviation Diesel Oil 航空柴油

**ADOAP**    Alternative Design Organisation Approval Procedure 设计组织批准的可选程序

**ADOCS**    Advanced Digital Optical Control System 先进的数字光学控制系统

**ADOP**    Attitude Dilution of Precision 姿态精度扩散因子

**ADP**    Acceptance Data Package 验收数据包

Acoustic Data Processing 声数据处理

Acoustic Data Processor 声音数据处理器

Adaptor 转换器,适配器,接合器

Administrative Data Processing 管理数据处理

Advanced Data Processing 先进的数据处理,高级数据处理

Advanced Development Phase 先进的开发阶段

Advanced Development Program 先进的开发程序

Advanced Ducted Propeller 先进的涵道螺旋桨(发动机),先进的冲压式螺旋桨

Advanced Ducted Propulsion 先进的涵道喷气(发动机)

Air Driven Pump 空气驱动泵,气动泵

Airport Development Program 机场开发计划

Associative Data Processing 关联数据处理

Automated Dispatch Panel System 自动调度板系统

Automatic Data Processing 自动数据处理

Automatic Data Processor 自动数据处理器

**ADPC** Automatic Data Processing Center 数据自动处理中心

**ADPCM** Adaptive Delta Pulse Code Modulation 自适应三角脉冲码调制

Adaptive Differential Pulse Code Modulation 自适应差分脉冲码调制

Adaptive Digital Pulse Coded Modulation 自适应数字脉冲编码调制

**ADPCU** Advanced Digital Passenger Control Unit 先进的数字式旅客控制组件

**ADPE** Automatic Data Processing Engineering 自动数据处理工程

Automatic Data Processing Equipment 自动数据处理设备

Auxiliary Data Processing Equipment 辅助数据处理设备

**ADPG** Atmospheric Dynamic Payload Group 大气动力有效载荷组

**ADPI** Air Data Pressure Instrument 大气数据压力仪表

**ADPLL** All Digital Phase-Locked Loop 全数字式锁相环

**ADPM** Aircraft Deactivation Procedures Manual 飞机偏差程序手册

Automatic Data Processing Machine 自动数据处理机

**ADPO** Aircraft Depot 飞机库

**ADPS** Automatic Data Processing System 自动数据处理系统

**ADPT** Adapt 适应,改写

Adopt 采用

**ADPTR** Adapter 转接器,适配器

**ADQ** Audits of Data Quality 数据质量审核

Kodiak, AK, USA (Airport Code) 美国克迪亚科机场

**ADR** Accident Data Recorder 事故数据记录器

Accident Data Recording 事故数据记录

Achievable Data Rate 可达到的数据速率

Add Register 加寄存器

Address Digit Receiver 地址数字接收机

Advance Deviation Report (发生) 偏差的先期报告

Advisory Route 咨询航路

Air Data Recorder 航空数据记录仪

Air Data Reference 大气数据基准

Air Defense Radar 防空雷达

Air Defense Region 防空区

Air Defense Requirement 防空要求

Airborne Digital Recorder 机载数字记录器

Aircraft Direction Room 飞机指挥室

Aircraft Discrepancy Report 飞机缺陷报告

Airport Departure Rate 机场离站率

Analog-Digital Recorder 模拟—数字记录器

Angle Data Record 角度数据记录器

Austin Data Recorder 奥斯汀数据记录器

Automated Digital Relay 自动数字继电器

Auxiliary Data Record 辅助数据记录器

Average Daily Rate 日平均价格

Aviation Design Research 航空设计研究

**ADR&DE** Air Defense Research and Development Establishment 防空研究与发展中心 (英国)

**ADRAC** Automatic Digital Recording and Control 自动数字记录与控制

**ADRAS** Aircraft Data Recovery and Analysis System 飞机数据恢复与分析系统

Airplane Data Recovery and Analysis Software 航空器数据恢复与分析软件

**ADREP** Accident/Incident Data Reporting System 事故/事故征候数据报告系统

**ADREPP** Aircraft Accident Data Reporting Panel 航空器失事数据报告专家小组

**ADRES** Aircraft Data Retrieved System 飞机数据检索系统

**ADRI** Aircraft Design and Research Institute 飞机设计研究所

**ADRIS** Automatic Dead Reckoning Instrument Systems 自动推测领航仪表系统

**ADRN** Advance Document Revision Note 资料更改预先通知

**ADRP** Air Data Reference Panel 大气数据基

准面板

**ADRS** Analog-Digital Recording System 模拟数字记录系统

**ADRT** Advanced Data Recording Technique 先进的数据记录技术

Analog Data Recording Transcriber 模拟数据记录转换器

**ADS** Abbreviated Dialing Service 缩位拨号业务

Accessory Drive System 附件驱动系统

Active Double Star 有源双星

Advance Direction Sign 预告方向标志

Advanced Debugging System 先进的程序调试系统

Advanced Display System 高级显示系统

Aerodynamic Deceleration System 空气动力减速系统

Aeronautical Design Standard 航空设计标准

Air Data Sensor 大气数据传感器

Air Data System 大气数据系统

Airworthiness Directives 适航指令

All Digital Simulation 全数字仿真

American Depositary Shares 美洲航材储存共享

Angular Displacement Sensor 角位移传感器

Application Data Structure 应用数据结构

Application Development System 应用开发系统

Asynchronous Data Service 异步数据业务

Audio Distribution System 音频分配系统

Automatic Data System 自动数据系统

Automatic Dependent Surveillance 自动相关监视

Automatic Digital Switch 自动数字交换机

Autopilot Disengage Switch 自动驾驶断开开关

**ADS-A** Automatic Dependent Surveillance-Addressed 自动相关监视—地址模式

**ADS—ATC** Air Traffic Control based on Automatic Dependent Surveillance 自动相关监视空中交通管制

**ADS-B** Automatic Dependent Surveillance-Broadcast 广播式自动相关监视,自动相关监视—广播

**ADS-C** Automatic Dependent Surveillance-Contract 契约式自动相关监视,合同式自动相关监视

**ADSE** Addresser 发货人

**ADSF** Automatic Dependent Surveillance Function 自动相关监视功能

**ADSG** Airport Design Study Group 机场设计研究组

**ADSHC** Air Data Sensor Heater Controllers 大气数据传感器加温控制器

**ADSI** Analog Display Service Interface 模拟显示服务接口

**ADSIM** Advanced Simulation Missions 先进的模拟任务

**ADSL** Asymmetric Digital Subscriber Line 非对称数字用户线,非对称数字环路

Asymmetric Digital Subscriber Loop 非对称数字用户环路

**ADSP** Automatic Dependent Surveillance Panel 自动相关监视专家组

**ADSS** Aerospace Data Systems Standards 航空宇航数据系统标准

Aircraft Damage Sensing System 飞机损伤传感系统

Automated Dependent Surveillance System 自动相关监视系统

Avionics Development Simulation Systems 航空电子开发模拟系统

**ADST** Advanced Distributed Simulation Technology 先进的分布仿真技术

**ADSU** Automatic Dependent Surveillance Unit 自动相关监视装置

**ADT** Advanced Data Technologies 先进的数据技术

Air Data Tester 大气数据测试仪

Air Data Transducer 大气数据传感器

Airborne Data Terminal 机载数据终端

Airborne Digital Timer 机载数字计时器

Approved Departure Time 批准离场时间

Assigned Departure Time 指定离场时间

Assured Data Transfer 有保证的数据传送

Automated Data Transmission 自动数据传输

Automatic Detection and Tracking 自动探测和跟踪

Average Daily Trips 平均日交通量

Average Download Time 平均下载时间

**ADTA** Air Data Transducer Assembly 大气数据转换器组件

Aircraft Development Test Activity 飞机发展试验活动处

Augmented Docking Target Adapter 增强型对接目标适配器

Aviation Development Test Activity 航空发展试验活动处

**ADTL** Additional 附加的,额外的

**ADTN** Addition 附加,附加物

Administrative Data Telecommunications Network 管理数据通信网络

Administrative Data Transmission Network 管理数据传输网络

**ADTNCHG** Additional Charge 附加费用

**ADTNL** Additional 附加的,额外的

**ADTOD** Advise Time of Delivery 请告交付时间

**ADTOR** Advise Time of Receipt 请告收到时间

**ADTOT** Advise Time of Transmission 请告转交时间

**ADTS** Advanced Detection Technology Sensor 先进的探测技术传感器

Advanced Detection Technology System 先进的探测技术系统

Advanced Digital Terminal System 先进的数字终端系统

Advanced Digital Test Station 先进的数字测试站

Air Data Test System 大气数据测试系统

Automated Digital Terminal System 自动数字终端系统

Automatic Data Test System 自动数据测试系统

Avionics Depot Test System 航空电子基地测试系统

**ADU** Actuator Drive Unit 作动筒驱动组件

Adapter Unit 接头,适配器

Air Data Unit 大气数据组件

Air Drive Unit 空气驱动组件

Align Display Unit 校准显示组件

Alignment Display Unit 校准显示组件

Antenna Drive Unit 天线驱动组件

Automatic Dialing Unit 自动拨号装置

Auxiliary Display Unit 辅助显示器

Avionics Display Unit 航空电子显示组件

**ADV** Adaptive Digital Vocoder 自适应数字音码器

Advance 提前,预先,推进

Advanced Development Vehicle 高级发展装置

Advanced 先进,预先

Adverse 相反

Advisory 咨询

**ADVBL** Advisable 适当的

**ADVC** Analog to Digital Video Converter 模拟到数字视频转换器

**ADVEN** Adventure 冒险

**ADVENT** Adaptive Versatile Engine Technology 自适应通用发动机技术

**ADVFRT** Advance Freight 预付运费

**ADVN** Advance 提前,预先

Advise Name(s)(Reservations Code)告知旅客姓名

**ADVPMT** Advance Payment 预付

**ADVS** Adverse 相反

Advise 通知,告之

Adviser 顾问

Advisory Service 咨询服务

**ADVSBL** Advisable 合理的

**ADVTG** Advantage 利益,优势

Advantageous 有利

**ADVY** Advisory 咨询

**ADW**　Air Defence Warning 防空警报

**ADWCP**　Automated Digital Weather Communication Program 自动气象数据通信计划

**ADWG**　Alcohol and Drugs Working Group 酒精及药物工作组

**ADWKP**　Air Defence Warning Key Point 防空警报中枢

**ADWS**　Automatic Digital Weather Switch 自动数字天气交换

**ADX**　Automatic Data Exchange 自动数据交换

**ADXS**　Automatic Data Exchange System 自动数据交换系统

**ADZ**　Air Defense Zone 防空区

　　Airport Development Zone 机场开发区

　　San Andres Island, Colombia（Airport Code）哥伦比亚圣安德烈岛机场

**AE**　Aero-Electronics 航空电子学

　　Animation Editor 动画编辑器

　　Application Entity 应用实体

　　Assistant Engineer 助理工程师

　　Audio Equipment 音频设备

　　Aviation Electrician 航空电气师

　　Aviation Electronicsman 航空电子技术人员

**AEA**　Aeronautical Engineers Association 航空工程师协会（英国）

　　Air Entraining Agent 引气剂

　　Aircraft Electronics Association 航空器电子协会

　　Aircraft Engineer Association 飞机工程师协会

　　Aircraft Engineering Artificer 飞机工程技工

　　All Electrical Aircraft 全电飞机

　　Association of European Airlines 欧洲航空公司协会

**AEAP**　As Early As Possible 尽可能早

**AEB**　Aft Equipment Bay 飞机尾部设备舱

　　Airline Engineering Bulletin 航空公司工程通告

　　Airworthiness Engineering Branch 适航工程处

　　Avionics Equipment Bay 航空电子设备舱

**AEC**　Aeromedical Evaluation Crew 航空医学评估组

　　After Electronic Center 后电子舱

　　Area Emergency Coordinator 区域紧急协调员

　　Automatic Error Correction 自动误差校正

　　Avionics Engineering Center 航空电子工程中心

**AECC**　Aeromedical Evacuation Control Center 航空医疗转运调度中心

**AECI**　Agencia Espanola de Cooperacion Internacional（Spanish International Cooperation Agency）西班牙国际合作机构

**AECMA**　European Association of Aerospace Industries 欧洲宇航工业协会

**AECS**　Advanced Environment Control System 先进的环境控制系统

**AECSP**　Aeronautical Emergency Communications System Plan 航空应急通讯系统项目

**AECU**　Audio Electronic Control Unit 音频电子控制组件

**AED**　Active Electronic Decoy 有源电子诱惑

　　Alphanumeric Entry Device 字母数字输入设备

　　Automated Engineering Design 自动化工程设计

**AEDB**　Aero-Engine Data Base 航空发动机数据库

**AEDE**　Aircraft Economic Design Evaluation 飞机经济设计评定

**AEDPS**　Automated Engineering Document Preparation System 自动工程文献编制系统

**AEDS**　Advanced Electric Distribution System 先进的配电系统

　　Airbus Enterprise Directory Service 空客企业目录服务

　　Airport Engineering Data Sheet 机场工程数据表

　　Association of Electronic Data Systems 电子数据系统协会

Atmospheric Electric Detection System 大气层电探测系统

**AEE** Aeronautic Electronic Equipment 航空电子设备

**AEEC** Airlines Electronic Engineering Committee 航空电子工程委员会

**AEEL** Aeronautical Electronic and Electrical Laboratory 航空电子与电气实验室

**AEEM** Airborne Electronic Equipment Modification 航空电子设备改装

**AEF** Aerospace Education Foundation 航空航天教育基金会

Airborne Equipment Failure 机载设备故障

Aircraft Engineering Foundation 飞机工程基金,飞机工程基础

**AEG** Active Element Group 有源元件组

Aeromedical Evaluation Group 航空医学评估组

Aircraft Evaluation Group 航空器评估组

Airworthiness Engineering Group 适航工程组

**AEHP** Atmospheric Electricity Hazards Protection 天电防护

**AEI** Aerial Exposure Index 航空曝光指数

Air Express International 国际航空快递

Automatic Equipment Identification 自动设备识别

Average Earnings Index 平均收益指数

Average Efficiency Index 平均效率指数

**AEIS** Aeronautical Enroute Information Service 航空航路情报服务

Association of Electronic Industries in Singapore 新加坡电气行业联合会

Association of Electronic Manufacturers 电子生产商联合会

**AEL** Adaptive Engineering Lab 自适应工程实验室

Aerospace Education Laboratory 航空教育实验室

Aircraft Engine Laboratory 飞机发动机试验室

Aircraft Equipment List 飞机设备清单

Airplane & Engine License 飞机和发动机维修执照

Allowance Equipage List 许可设备清单

**AELS** Airborne Electronic Library System 机载电子程序库系统

Airborne Electronics Laser System 机载电子激光系统

**AEM** Advanced Engine Management 先进的发动机管理

Aerospace Engineering and Maintenance 航天工程和维修

Aerospace Engineering and Mechanics 航天工程与力学

Aircraft Engine Mechanic 飞机发动机机械工

Airplane Energy Management 飞机能源管理

Assembly Evaluation Method 装配评价方法

**AEMA** Aero-Medical Association 航空医学协会

**AEO** Acousto-Electric Oscillator 电声振荡器

Advance Engineering Order 先期工程指令

Air Engineer Officer 航空工程师军官

Air Engineering Officer 航空工程官员

Airborne Electronics Operator 机上电子设备操作员

Airport Environment Officer 机场环境官员

Airworthiness Engineering Organization 适航工程组织

All Engines Operating 全部发动机运行

Armed Escort Operation 武装护送作战

Assistant Experimental Officer 实验军官助理

**AEP** Airport Emergency Plan 机场应急计划

Annual Execution Plan 年度执行计划

Audio Entertainment Player 音频娱乐播放器

Buenos Aires, Argentina-Jorge Newberry (Airport Code) 阿根廷布宜诺斯艾利斯霍尔赫纽贝里机场

**AEPC** APU/External Power Contactor APU/外部电源接头

**AEPS** Aircraft Electrical Power System 飞机

电力系统

Aircrew Escape Propulsion System 机组人员应急离机弹射系统

Airlines Employment Placement Service 航空公司雇员安置服务

Automated Environmental Prediction System 自动化环境预测系统

**AEPT** Air Engineer Procedure Trainer 空中机械程序练习器

**AER** Aeronautical Engineering Report 航空工程报告

Air Expansion Ratio 进气膨胀比

Approach End of Runway 跑道的进近端

**AERA** Air Emissions Risk Analysis 航空排放风险分析

Automated En-Route Air Traffic Control 自动航路空中交通管制

**AEREP** Analysis and Evaluation Report 分析和鉴定报告

**AERIS** Aeronautical En-Route Information Service 航空航路情报服务

Airborne Electronic Ranging Instrumentation System 机载电子测距仪表系统

Airways Environmental RADAR Information System 航路环境雷达信息系统

**AERO** Aerodrome 机场

Aeronautical Events Reports Organizer 航空活动报告组织者

Aeronautics 航空学

Aerospace Education and Research Organization 航空航天教育和研究机构

Air Education and Recreation Organization 航空教育与娱乐组织

Aircraft Engine Repair and Overhaul 飞机发动机修理和大修

Aviation Routine Weather Report 航空定期气象报告

**AEROCOM** Aeronautical Communications 航空通信

**AEROFAC** Aeronautic Facility 航空设施

**AEROMET** Aeronautical Meteorological 航空气象

**AEROSAT** Aeronautical Satellite 航空卫星

**AERP** Advanced Equipment Repair Program 先进的设备维修计划

Airworthiness Panel 适航性专家小组

Alternative Emission Reduction Plan 可供选择的减排计划

Average Effective Radiated Power 平均有效辐射功率

**AES** Active Electromagnetic System 有源电磁系统

Advanced Encryption Standard 高级加密算法标准

Advanced Extravehicular Suit 先进的舱外活动航天服

Aerodrome Emergency Service 机场应急服务

Aerospace Electrical Society 航空航天电气学会(美国)

Aerospace Electronics System 航空航天电子系统

Air Equipment and Support 航空设备与支援

Aircraft Earth Station 飞机地球站,飞机地面站

Aircraft Electrical Society 飞机电气学会

Aircraft Engineering Squadron 飞机工程中队

Aircraft Equipment Simulator 飞机设备模拟机

Airplane Evacuation Simulator 飞机撤离模拟机

Airways Engineering Society 航线(航空公司)工程学会

American Electromechanical Society 美国发动机学会

Array Element Study 阵列元研究

Artificial Earth Satellite 人造地球卫星

Automatic Encoder System 自动编码器系统

Automatic Erection System 自动直立系统

Auxiliary Encoder System 辅助编码系统

Auxiliary Engine Seat 辅助发动机座

**AESA** Active Electronically Scanned Array 有源相控阵

**AESC** American Engineering Standards Committee 美国工程标准委员会

**AESD** Aerospace Electronic Systems Department 宇航电子系统部

**AESO** Aircraft Environmental Support Office 飞机环境保障办公室

**AESS** Aerospace and Electronics Systems Society 航空航天电子系统学会

Aircraft Environmental Surveillance System 飞机环境监视系统

**AEST** Automated Enhanced Security Tool 自动强化安全工具

**AESU** Aircraft Environment Surveillance Unit 飞机环境监视装置

**AET** Aerospace Engineering Technician 航天工程技术人员

Allakaket, AK, USA ( Airport Code ) 美国阿拉斯加州阿拉卡基特机场

Alliance for Environmental Technology 环境技术联盟

Average Execution Time 平均执行时间

**AETMS** Airborne Electronic Terrain Map System 机载电子地形地图系统

**AETS** Air-traffic-controller English Test System of CAAC 中国民航管制员英语等级测试系统

Automatic Engine Trim System 发动机自动配平系统

**AEU** Audio Electronic Unit 音频电子装置

**AEV** Automatic Expansion Valve 自动膨胀活门（阀）

**AEVC** American Electric Vehicle Company 美国电子车辆公司

Avionics Equipment Ventilation Computer 航空电子设备通风计算机

Avionics Equipment Ventilation Controller 航空电子设备通风控制器

**AEW** Air Electronic Warfare 航空电子战

Airborne Early Warning 空中预警

**AEWTS** Advanced Electronic Warfare Trainer System 先进的电子战训练系统

**AF** Access Facilities 接入设施

Accuracy Factor ( Figure ) 精确度因子,精确度指数

Activity Factor ( Propeller Blade ) 有效性因子( 螺旋桨叶)

Adaption Facility 适配设施

Adaption Function 适配功能

Address Field 地址域

Advance Freight 预付运费

Affiliate 分会

Air Field 机场

Air Filter 空气过滤器

Air France 法航

Airway Facilities 航路设施

All Freighter 全货机

Assured Forwarding 可确定的转发

Audio Frequency 音频频率

**AFA** Air Freight Asia 亚洲航空物流技术展览会

Air Freight Association 航空货运协会

Aircraft Finance Association 飞机信贷协会( 美国)

Airline Fleet Assignment 航班机型分配

Association of Flight Attendants 空中乘务员协会

Audio Frequency Amplifier 音频放大器

Automatic Fire Alarm 自动火警

**AFAC** Airborne Forward Air Controller 前方机上空中管制员

**AFAP** Australian Federation of Air Pilots 澳大利亚航空驾驶员协会

**AFB** Air Force Base 空军基地

Air Freight Bill 航空货运单

Airframe Bulletin 飞机体通告

Antifriction Bearing 耐磨轴承

Axial Flow Blower 轴流风机

**AFC** Aerodynamic Flight Control 空气动力飞行控制

After Fuel Crisis 石油危机后

Air Flow Control 空气流量控制

Air Freight Charge 航空运费

Airborne Freight Corp. (USA) 空中运输公司(美国)

Airbus Finance Corporation 空中客车财务公司

Aircraft Flyaway Cost 飞机出厂价

Aircraft Frame Change 飞机机身更改

Airfield Construction 机场建筑

Airframe Change 机体变化

Alkaline Fuel Cell 碱性燃料电池

Area Forecast Center 区域预报中心

Asking for Correction 请求更正

Australian Film Commission 澳大利亚电影委员会

Automatic Feedback Control 自动反馈控制

Automatic Flight Control 自动飞行控制

Automatic Flow Control 自动流量控制

Automatic Following Control 自动跟踪控制

Automatic Frequency Control 自动频率控制

Average Fixed Cost 平均固定成本

Aviation Fire Control 航空火力控制

**AFCAC** African Civil Aviation Commission 非洲民用航空委员会

**AFCAS** Advanced Flight Control Actuation System 先进的飞行操纵作动系统

Automatic Flight Control Augmentation System 自动飞行控制增益系统

**AFCE** Automatic Flight Control Equipment 自动飞行控制设备,自动驾驶仪

**AFCP** Advanced Flight Control Programmer 高级飞行控制程序设计员

Advanced Flow-Control Procedure 先进的流量控制程序

**AFCR** Air Flow Control Regulation 空气流量控制调节

**AFCS** Adaptive Flight Control System 自适应飞行控制系统

Advanced Flight Control System 先进的飞行操纵系统

Aircraft Flight Control System 飞机飞行控制系统

Automatic Flight Control System 自动飞行控制系统

**AFCT** Affect 影响

**AFD** Advanced Flight Deck 先进的驾驶舱

Airport Facility Directory 机场设施指南

Assistant Flight Director 辅助飞行指引仪

Automatic Flight Director 自动飞行指引仪

**AFDAS** Aircraft Fatigue Data Analysis System 飞机疲劳数据分析系统

**AFDC** Autopilot Flight Director Computers 自动飞行指引计算机

**AFDD** Audit Findings and Differences Database 审计结果和差异数据库

**AFDS** Advanced Flight Deck Simulator 先进的驾驶舱模拟器

Aircraft Flight Director System 航空器飞行指引仪系统

Automatic Flight Data System 自动飞行数据系统

Autopilot Flight Director System 自动驾驶飞行指引仪系统

**AFDX** Avionics Fast Switched Ethernet 航空电子快速开关以太网

Avionics Full Duplex Ethernet 航空电子全双工通讯以太网

**AFE** Above Field Elevation 高于机场标高

**AFECU** Automatic Fire Extinguishing Control Unit 自动灭火控制组件

**AFESD** Arab Fund for Economic and Social Development 阿拉伯经济和社会发展基金会

**AFF** Affirmative 肯定,证实

Aircraft Firefighting 救火飞机

Autonomous Formation Flight 自主编队飞行

Aviation Flight Facility 航空飞行设施

**AFFC** Airport Freight Forwarding Centre 机场空运中心

**AFFF** Aqueous Film-Forming Foam 水成膜泡沫灭火剂

**AFFTC** Air Force Field Technical Center 空军机场技术中心

Air Force Flight Test Center 空军飞行试验

中心

**AFG** Analog Function Generator 模拟函数发生器

Antenna Field Gain 天线场强增益

Arbitrary Function Generator 任意函数发生器

**AFGS** Air Force Guide Specification 空军指导规范

Automatic Flight Guidance System 自动飞行引导系统

**AFI** Africa-Indian Ocean 非洲—印度洋地区,非洲—印度洋

African Region 非洲地区

Air Filter Institute 空气过滤学会

Aircraft Friction Index 飞机摩擦指数

Airfreight Institute 空运研究所

Assistant Flying Instructor 飞行助教

Authority and Format Identifier 权限和格式标识符

Automatic Fault Indication 自动错误显示

Automatic Fault Isolation 自动故障隔离

**AFIA** Aerial Firefighting Industry Association 空中消防工业协会

**AFIDB** Auxiliary Flight Information Data Base 辅助飞行信息数据库

**AFIL** Flight Plan Filed in the Air 空中提交的飞行计划

**AFINS** Airways Flight Inspector 航线飞行检查员

**AFIPS** American Federation of Information Processing Society 美国信息处理协会联合会

**AFIRAN** Africa-India Ocean Region Air Navigation 非洲—印度洋地区导航

**AFIS** Advanced Flight Instrument System 先进的飞行仪表系统

Airborne Flight Information Service 机载飞行情报服务

Airbus In-Flight Information Services 空客空中情报服务

Aircraft Fault Identification System 飞机故障识别系统

Airport/Aerodrome Flight Information Service 机场飞行情报服务

Analogue Flight Instrument System 模拟飞行仪表系统

Automatic Flight Information Service 自动飞行情报服务

**AFISO** Aerodrome Flight Information Service Office 机场飞行情报服务室

**AFIT** Accelerated Flight & Instrument Training 加速飞行与仪表训练

Asian Forum for Information Technology 信息技术亚洲论坛

Automatic Fault Isolation Test 自动故障隔离试验

**AFIZ** Aerodrome Flight Information Zone 机场飞行情报区

**AFJ** Axial-Flow Jet 轴流式喷气发动机

**AFL** Above Field Level 高于机场场面

Active Flight List 显示航班表

Actual Flight Level 实际飞行高度

Aileron Force Limiter 副翼操纵力限制器

Air Flow 气流

Air Freight List 空运货单

Aircraft Furnishing Limited 飞机装备有限公司

Avionics Flying Laboratory 航空电子学飞行实验室

**AFLC** Adaptive Fuzzy Logic Controller 自适应模糊控制器

**AFLD** Airfield 机场

**AFM** Affirmative 肯定,证实

Air Flow Measurement 气流测量

Air Freight Manifest 空运货单,空运舱单

Aircraft/Airplane Flight Manual 飞机飞行手册

Airport Firemen 机场消防员

Atomic Force Microscope 原子力显微镜

Automatic Flight Management 自动飞行管理

Aviation Fleet Maintenance 航空机队维护

Aviation Forecast Model 航空预报模型

**AFMA** Anti-Fuel-Misting Additive 燃油防雾化添加剂

**AFMC** Advanced Flight Management Computer 先进的飞行管理计算机

Auxiliary Fuel Management Computer 辅助燃油管理计算机

Avionics Flight Management Computer 航空电子飞行管理计算机

**AFMS** Advanced Flight Management System 先进的飞行管理系统

Aircraft Flight Manual Supplement 航空器补充飞行手册

Automatic Flight Management System 自动飞行管理系统

Auxiliary Fuel Management System 辅助燃油管理系统

Auxiliary Fuel Monitoring System 辅助燃油监控系统

**AFN** ATS Facilities Notification 空中交通服务设施通知

Average Failure Number 平均故障数

**AFO** Airport Fire Officer 机场消防官员

**AFOLT** Automatic Fire/Overheat Logic Test System 自动火警/过热逻辑测试系统

**AFOS** Automation of Field Operations and Services 野外操作与服务自动化

**AFP** Advanced Function Printing 先进的功能打印

Alternate Flight Plan 备份飞行计划

Area Flight Plan 地区飞行计划

Association for Finishing Processes 抛光工程协会,抛光工艺协会

ATC Flight Plan 空中交通管制飞行计划

Australian Federal Police 澳大利亚联邦警察

**AFPD** Air Force Policy Directive 空军政策指示

**AFPL** Air France Partners Leasing 法航合伙航空租赁公司

**AFPR** Adjustable Fuel Pressure Regulator 可调的燃油压力调节器

**AFPU** African Postal Union 非洲邮政联盟

**AFR** Acceptable Failure Rate 容许故障率

Actual Fuel Remaining 实际燃油余量

Africa 非洲

African France 非洲法郎

Air/Fuel Ratio 空气燃油比

Airframe 机身

Automatic Flexible Routing 自动可变路由

Average Failure Rate 平均故障率

**AFRA** Average Freight Rate Assessment 平均运费率的确定

**AFRAA** African Airlines Association 非洲航空公司协会

**AFRC** Air Fuel Ratio Control 空气燃料比控制

Armed Forces Reserve Center 武装部队预备役中心

Automatic Frequency Ratio Control 自动频率比控制

**AFRCU** Air/Fuel Ratio Control Unit 空气燃油比控制装置

**AFRM** Advanced Flight Research Model 高级飞行研究模型

Airframe 机体

**AFRS** Advanced Fighter Radar System 高级战斗机雷达系统

Aerodrome/Airport Fire and Rescue Service 机场消防和救护服务

Aircraft Fault Reporting System 航空器故障报告系统

American Forces Radio Stations 美国军用无线电台

Armed Forces Radio Services 武装部队无线电服务

Auxiliary Flight Reference System 辅助飞行参考系统

**AFRU** Aerodrome Frequency Response Unit 机场频率应答组件

**AFS** Active Flutter Suppression 颤振主动抑制

Active Formboard Sheet List 有效模板清单

Active Fusing System 主动式引信系统

Advanced Flying School 高级飞行学校,高
级航校

Aerial Film Speed 航空胶片感光速度

Aeronautical Fixed Service 航空固定服务

Aeronautical Flight System 航空飞行系统

Aging Fleet Survey 老龄机群调查

Aircraft Flight Standards 航空器飞行标准

Airways Facilities Sector 航路设施部门

Area Forecast System 区域天气预报系统

Arm and Fuse System 打开保险和引信装置

Arming & Fusing System 预位与发射系统

Autoflight System 自动飞行系统

Automated Forecasting System 自动预报系统

Automatic Firing Sequencer 自动发射程序装
置

Automatic Flight System 自动飞行系统

Automatic Frequency Selection 自动频率选
择

Automatic Frequency Stabilization 自动频率
稳定

Aviation Fuel Station 飞机加油站

Avionics Flight Software 航空电子设备飞行
软件

**AFSA** Airbus France Standard Architecture 空
客法国公司标准体系

**AFSC** Auxiliary Fuel System Controller 辅助
燃油系统控制

Aviation Fuel Supply Company（香港）航油
供应公司

**AFSCA** Asian Freight & Supply Chain Award
亚洲货运与供应链奖项

**AFSCP** Automatic Flight System Control Panel 自动飞行系统控制板

**AFSG** Audio Frequency Signal Generator 音
频信号发生器

**AFSK** Audio Frequency Shift Keying 音频移
频键控

**AFSM** Airline File Server Module 航空公司
文件服务器模块

**AFSS** Active Flutter Suppression System 主动
颤振抑制系统

Automated Flight Service Station 自动化飞行
服务站

Automated Forecast Shop Simulator 自动预报
车间模拟器

Automated/Automatic Fuel Service Station 自
动燃油服务站

Autonomous Flight Safety System 自主飞行
安全系统

**AFT** Adapter Fault Tolerance 适配器容错

Advanced Flying Training 高级飞行训练

Afterward 向后

Air Freight Terminal 航空货运集散站

Aircraft 飞机,航空器

Asynchronous Frame Technology 异步帧技术

Automatic Fine Tuning 自动微调

**AFTA** ASEAN Free Trade Area 东盟自由贸
易区

Avionics Fault Tree Analysis 航空电子设备
故障树分析

**AFTAX** Aeronautical Fixed Telecommunication Automatic Exchange 航空固定电信自动
交换

**AFTCOMPT** After Compartment 后舱

**AFTEC** Air Force Test & Evaluation Center
空军测试和评估中心

**AFTI** Advanced Fighter Technology Integration 先进的战斗机技术综合

**AFTK** Available Freight Tonne-Kilometers 可
用货邮吨公里

**AFTN** Aeronautical Feed Telecommunication
Network 航空供给电信网

Aeronautical Fixed Telecommunication Network 航空固定电信网

Afternoon 下午

Automated Fixed Telecommunication Network
自动固定电信网

**AFTO** Air Force Technical Order 空军技术
法规

**AFTP** Acceptance Functional Test Procedure
验收功能试验程序

Additional Flying Training Periods 附加飞行

训练期

Aircraft Flight Test Plan 飞机试飞大纲

Aircrew Flight Training Period 机组人员飞行训练期

Anonymous File Transfer Protocol 匿名文件传输协议

**AFTS** Adaptive Flight Training System 自适应飞行训练系统

Advanced Flying Training School 高级飞行训练学校

Air Fuel Test Switch 空气燃料测试开关

Airborne Flight Test System 机载飞行试验系统

**AFU** Actual Fuel Used 实际使用燃油

Air Filtration Unit 空气过滤装置

Artificial Feel Unit 模拟感觉组件,人工感力装置

Auxiliary Functioning Unit 辅助功能部件

**AFWAL** Air Force Wright Aeronautical Laboratory 空军莱特航空实验室

**AFZ** Aircraft Flight Zone 飞机飞行区

**AG** Agency General 总代理

Air to Ground 空对地

Attitude Gyro 姿态陀螺

Application Gateway 应用网关

Availability Guarantee 可用性保证书

Aviation Gasoline 航空汽油

**AGA** Aerodrome and Ground Aids 机场和地面助航设备

Agadir, Morocco (Airport Code) 摩洛哥阿加迪尔机场

Automatic Gas Analgzer 自动气体分析仪

**AGACS** Automatic Ground-to-Air Communication System 自动地空通讯系统

**AGARD** Advisory Group for Aerospace Research & Development 宇航研发顾问小组

**AGAS** Aviation Gasoline 航空汽油

**AGATE** Advanced General Aviation Transport Experiments 先进的通用航空运输试验

Air-to-Ground Acquisition & Tracking Equipment 空对地数据采集与跟踪设备

**AGB** Accessory Gearbox 附件齿轮箱

Alliance for Global Business 全球商业联盟

Angle Gearbox 角齿轮箱

Automatic Gear Box 自动变速箱

**AGC** Aerojet-General Corporation 通用航空喷气发动机公司

Affinity-Group Charter 合伙包机

African Groundnut Council 非洲花生理事会

Air Ground Chart 航空地形图

Air-Ground Check 空地校验

Air-to-Ground Code 空地电码,空—地电码

Air-to-Ground Communication 空—地通信

Apollo Guidance Computer 阿波罗引导计算机

Associated General Contractor 总关联合同方

Automatic Gage Control 自动轨距控制,自动计量控制

Automatic Gain Control 自动增益控制

Automatic Gauge Control 自动测量调整装置,自动厚度控制

Automatic Gauge Controller 自动测量控制器

Automatic Generation Control 自动生成控制

Automatic Generator Control 发动机自动控制

**AGCA** Automatic Ground-Controlled Approach 地面控制自动进近

**AGCC** Air-Ground Communications Channel 空—地通信波道

**AGCH** Access Grant Channel 接入允许信道

**AGCR** Auxiliary Generator Control Relay 辅助发电机控制继电器

**AGCS** Advanced Guidance and Control Systems 先进的引导和控制系统

Air Ground Communication Service 空地通讯服务

Air Ground Communication System 空地通讯系统

**AGCU** Auxiliary Generator Control Unit 辅助发电机控制组件

**AGCY** Agency 代理机构

**AGD** Agree or Agreed 同意

Air Generator Drive 风力发电机驱动

Axial Gear Drive 轴向齿轮驱动

Axial-Gear Differential 轴向差动齿轮

**AGDLS** Air-to-Ground Data Link System 空地数据链系统

**AGDS** American Gauge Design Standard 美国量规设计标准

**AGE** Aerospace Ground Equipment 航空地面设备

Aircraft Ground Equipment 飞机地面设备,航空器地面设备

**AGERD** Aerospace Ground Equipment Recommendation Data 航空和航天地面设备的推荐数据

**AGES** Air Ground Engagement System 空气地面啮合系统

Air/Ground Engagement Simulation 空/地交战模拟

Air/Ground Equipment Sales 空中/地面设备销售

**AGG** Airborne Gravity Gradiometer 航空重力梯度仪

Airborne Ground Generators 航空地面发电机

Airbus General Guide 空客通用指南

**AGGG** Advisory Group on Greenhouse Gases 温室气体咨询组

**AGGR** Aggregate 总计

**AGH** Helsingborg, Sweden ( Airport Code ) 瑞典赫尔辛堡机场

**AGHE** Aircraft Ground Handling Equipment 飞机地面拖移设备

**AGHME** Aircraft Geometric Height Measurement Element 飞机几何高度测量单元

**AGI** Advanced Ground Instructor 高级地面教员

**AGIL** Airborne General Illumination Light 机载通用照明灯

**AGL** Above the Ground Level 离地高度

Aerodrome Ground Lighting 机场地面灯光

Aircraft Ground Light 飞机地面灯光

Approach Guidance Lights 进近引导灯

**AGM** Air/Ground Message 空/地电报

Aircraft Ground Mishap 飞机地面事故

Annual General Meeting 年会,年度大会,年度股东大会

**AGMA** American Gear Manufacturer's Association 美国齿轮制造商协会

**AGMC** Aerospace Guidance and Metrology Center 航空航天制导与气象中心

**AGMCS** Airport Ground Movement Control System 机场地面活动控制系统

**AGMT** Agreement 协议

Augment 放大,助力

**AGN** Again 又,再

**AGNCY** Agency 代理机构

**AGNIS** Azimuth Guidance Nose-In Stand 机头向内方向引导

**AGNT** Agent 代理人,代理机构

**AGP** Advanced Graphics Processor 先进的图形处理器

Agra, India( Airport Code ) 印度阿格拉机场

Agreement on Government Procurement 政府采购协议

Aircraft Gateway Processor 飞机网关处理器

Aircraft Grounded for lack of Parts 因缺乏零件停飞的飞机

Malaga, Spain ( Airport Code ) 西班牙马拉加机场

**AGR** Agra, India( Airport Code ) 印度阿格拉机场

Agree or Agreed 同意

Arresting Gear 停止索装置

**AGRD** Agreed 已同意

**AGREE** Advanced Ground Receiving Equipment Experiment 先进的地面接受设备实验

Advisory Group on Reliablility of Electronic Equipment 电子设备可靠性咨询组

**AGRMT** Agreement 协议

**AGRT** Adjusted Ground Receipt Time 调节地面接收时间

Automatic Guard Receiver Terminal 自动地

面接收机终端

AGS　Abort Guidance System 应急制导系统

Advanced Graphic System 高级图形系统

Advanced Guidance System 高级制导系统

Air Generation System 空气发生系统

Air/Ground System 空/地系统

Aircraft General Stores 飞机通用(器材)仓库

Aircraft Generation Squadron 飞机设计集团

Airport Ground Service 机场地面服务

Application Generator System 应用生成器系统

Augusta, GA, USA (Airport Code) 美国佐治亚州奥古斯塔机场

Auto Ground Spoiler 自动地面扰流板

Automatic Gain Stabilization 自动增益稳定

AGSD　Advanced Ground Segment Design 先进的地面段设计

AGST　Against 反对

AGSTR　Aircraft Ground Service Tie Relay 飞机地面服务连接继电器

AGT　Advanced Graphic Technologies 先进的图形技术

Advanced Ground Transport 先进的地面运输

Agent 代理人,代理机构

Agreement 协议

Aviation Gas Turbine 航空燃气涡轮

Travel Agent 旅行社代理商或其人员

AGTA　Airline Ground Transportation Association (US) 航空公司地面运输协会(美国)

AGTC　Airport Ground Traffic Control 机场地面交通管制

AGU　Aguascalientes, Mexico (Airport Code) 墨西哥阿瓜斯卡连特斯机场

Air Generation Unit 空气发生装置

American Geophysical Union 美国地球物理学联合会

AGW　Actual Gross Weight 实际总重

Allowable Gross Weight 允许起飞总重

AH　Alert Height 警戒高度,警戒高

Antihunt 阻尼稳定器

Artificial Horizon 航空地平仪,人工地平,地平线

Authentication Header 认证头

AHARS　Airborne Heading-Attitude Reference System 机载姿态与航向参考系统

AHAS　Avian Hazard Advisory System 鸟击危害咨询系统

AHC　Attitude and Heading Computer 姿态和航向计算机

AHCS　Automatic Hovering Control System 悬停自动控制系统

AHD　Ahead 在前面,提前

AHEAD　Aircraft Health Evaluation and Diagnosis 飞机健康评估与故障诊断

AHEM　Association of Hydraulic Equipment Manufacturers 液压设备制造商协会

AHIP　Army Helicopter Improvement Program 陆军直升机改进计划

AHLF　Additional High Level Function 补充高层功能

AHM　Aircraft Handling Manual 飞机操纵手册

Airplane Health Management 飞机健康管理

Airport Handling Manual 机场作业手册

Ampere-Hour Meter 安培小时表

Attitude Hold Mode 姿态保持方式

AHN　Athens, GA, USA (Airport Code) 美国佐治亚州阿森斯机场

AHO　Alghero, Sardinia, Italy (Airport Code) 意大利撒丁岛阿尔盖罗机场

AHP　Analytic Hierarchy Process 层次分析法

Army Helicopter 军用直升机

Army Heliport 军用直升机场

Auxiliary Hydraulic Power 辅助液压动力

AHR　Airport Height Restrictions 机场高度限制

Another 另外

AHRS　Altitude Heading Reference System 高度航向基准系统

Attitude and Heading Reference System 姿态

与航向基准系统

Attitude Heading Reference System 姿态航向基准系统

Automatic Heading Reference System 自动航向基准系统

**AHRU** Attitude and Heading Reference Unit 姿态和航向基准组件

**AHS** American Helicopter Society 美国直升机学会,美国直升机协会

Attitude Heading System 姿态航向系统

Automated Handwritten System 自动手写系统

**AHSA** The Aviation Historical Society of Australia 澳大利亚航空史学会

**AHSR** Air Height Surveillance Radar 空中高度监视雷达

**AHSU** Attitude Heading Sensing Unit 姿态航向传感组件

**AHSW** Aural High-Speed Warning 声响高速警报

**AHT** Average Handling Time 平均处理时间

Average Holding Time 平均保持时间

**AHU** Air Handling Unit 空气处理单元

Air Heating Unit 空气加热单元

**AHW** Atomic Hydrogen Welding 原子氢焊

**AI** Accident Indemnity 事故保险

Accident Injuries 失事损伤,偶然损伤

Action Item 行动项目,措施项目,操作项

Airborne Interceptor 空中拦截

Air-data/Inertial 大气数据/惯性

Airlines Industry 航空运输业

Airport Industry 机场业

Airspeed Indicator 空速指示器

Alarm Indication 报警指示

Anti-Ice/Anti-Icing 防冰

Approval of Import 进口许可

Artificial Intelligence 人工智能

Attitude Indicator 姿态指示仪

Automatic Inspection 自动检验

**AIA** Aerospace Industries Association 宇航工业协会

Airbus Industries Asia 空客工业公司亚洲部

Aircraft Industries Association 飞机工业协会

American Institute of Aeronautics 美国航空协会

American Inventors Association 美国发明家协会

Association of International Accountants 国际会计协会

**AIAA** Aerospace Industries Association of America 美国航空航天工业协会

Aerospace Industry Analysis Association 航空航天工业分析协会(美国)

Aircraft Industries Association of America 美国航空工业协会

American Institute of Aeronautics and Astronautics 美国航空航天研究所

Area of Intense Aerial/Air Activity 稠密空气层活动区

**AIAC** Aerospace Industries Association of Canada 加拿大宇航工业协会

Air Industries Association of Canada 加拿大航空工业协会

**AIAEA** Air-India Aircraft Engineers Association 印度航空飞机工程师协会

**AIAI** Air Intake Anti-Icing 进气道防冰除冰

**AIANZ** The Aviation Industry Association of New Zealand Inc. 新西兰航空工业协会

**AIB** Accident Investigation Branch 事故调查分局

Aeronautical Information Bureau 航空资料局

**AIC** Access Illustration Cards 接近图解卡,存取图解卡

Aerodynamic Influence Coefficient 气动效应系数

Aeronautical Information Circular 航空信息通报

Aircraft Identification Control 飞机识别控制

Aircraft In Commission 现役飞机

Aircraft Industry Conference 飞机工业会议

Air-Inlet Controller 进气调节器

Analog Interface Card 模拟接口卡

Artificial Intelligence Center 人工智能中心

Attack Information Center 攻击情报中心

Avionics Integration Center 航空电子集成中心

**AICC** Air Intercept Control Common 通用空中截击控制

Aviation Industry CBT Committee 航空工业 CBT 委员会

**AICM** Aeronautical Information Conceptual Model 航空情报概念模型

**AICMA** International Association of Aircraft Manufactures 国际航空器制造商协会

**AICO** Action Information Control Officer 行动情报控制官

**AICS** Advanced Interior Communication System 先进的内部通讯系统

Aero-Inertial Control System 机载惯性控制系统 ·

Air Induction Control System 进气控制系统

Air Intake Control System 进气口控制系统

Air Intercept Control School 空中截击控制学校

Aircraft Identification and Control System 飞机识别和控制系统

Automated Industrial Control System 工业自动控制系统

Automatic Inlet Control System 入口自动控制系统

Automatic Intersection Control System 交叉点自动控制系统

**AICT** Automatic Integrated Circuit Tester 自动集成电路测试器

**AICU** Advanced Interface Control Unit 先进接口控制组件

Anti Ice Control Unit 防冰控制装置

**AID** Accident/Incident/Deficiency 事故/事故征候/差错

Aeronautical Inspection Directorate 航空监察署(英国)

Agency for International Development 国际开发署(美国)

Aircraft Inspection Department 飞机检查部

Aircraft Installation Delay 飞机安装延误

Aircraft Integrated Data 飞机综合数据

Aircraft Interface Discrepancy 飞机接口误差(故障)

Airport Information Desk 机场问讯处

Altered Item Drawing 变更项目图表

Automatic Information Distribution 自动信息分配

**AIDA** Automated Inspection of Data 数据自动检查

**AIDAPS** Automatic Inspection Diagnostic and Prognostic System 自动检测诊断与预报系统

**AIDAS** Advanced Instrumentation and Data Analysis System 高级仪表测量与数据分析系统

**AIDC** Aero Industry Development Center 航空工业发展中心

Aerospace Industrial Development Corporation 航空工业发展公司

ATC Interfacility Data Communications 空中交通管制设施间的数据通信

ATS Inter-Facilty Data Communications 空中交通服务设施间的数据通信

Automatic Identification and Data Capture/Collection 自动识别和数据采集

Automotive Industry Development Centre 汽车工业发展中心

**AIDE** Aircraft Installation Diagnostic Equipment 飞机设备故障判断装置

**AIDES** Automated Image Data Extraction System 自动图像数据提取系统

**AIDRS** Aircraft Integrated Data Recording System 飞机综合数据记录系统

**AIDS** Accident/Incident Data System 事故/事件数据系统

Acoustic Intelligence Data System 声学情报数据系统

Acquired Immune Deficiency Syndrome 要求的无缺陷特征群

Advanced Integrated Data System 高级综合数据系统

Advanced Integrated Display System 先进的综合显示系统

Advanced Interactive Display System 高级交互式显示系统

Aerospace Intelligence Data System 航空航天情报数据系统

Airborne Indication Detection System 机载指示探测系统

Airborne Inertial Data System 机载惯性数据系统

Airborne Instrumentation Data System 机载仪表数据系统

Airborne Integrated Display System 机载综合显示系统

Aircraft Integrated Data System 飞机综合数据系统，飞机集成数据系统

Apron Information Display System 停机坪信息显示系统

Automatic Identification Data System 自动识别数据系统

Automatic Incident Detection System 交通事故自动侦察系统

Automatic Integrated Data System 自动综合数据系统

Automatic Integrating Debugging System 自动综合测试系统，自动综合排故系统

**AIDU** Aeronautical Information Document Unit 航行信息文件单位

**AIDUS** Automated Input and Document Update System 自动输入与资料更新系统

**AIDX** Aviation Information Data Exchange 航空信息数据交换

**AIEE** American Institute of Electrical Engineers 美国电气工程师协会

**AIEM** Airlines International Electronics Meeting 国际航空公司电子大会

**AIFC** Airbus Industries Finance Services 空中客车工业公司金融服务

**AIFF** Audio Interchange File Format 音频交换文件格式

**AIFS** Anti Ice Flow Sensor 防冰流量传感器

**AIFTDS** Airborne Integrated Flight Test Data System 机载综合飞行试验数据系统

**AIG** Accident Investigation and Guidance 事故调查与预防

Accident Investigation Group 事故调查组

Address Indicating/Indicator Group 地址指示组

Airbus Industries Group 空中客车工业集团

All-Inertial Guidance 全惯性制导

American International Group 美国国际集团

Augmented Inertial Guidance 增加的惯性制导

**AIGS** Acoustic Intelligence Gathering System 声响情报搜集系统

All-Inertial Guidance System 全惯性制导系统

**AIHS** Aircrew Integrated Helmet System 机组综合头盔系统

**AIIM** Association for Information and Image Management 信息与图像管理协会

**AIIS** Airport Information Integration System 机场信息集成系统

**AIL** Aeronautical Instrument Laboratory 航空仪表实验室

Aileron 副翼

Airborne Instruments Laboratory 机载仪表实验室

Artificial Intelligence Language 人工智能语言

Audio Interface Library 音频接口库

Avionics Integration Laboratory 航空电子综合实验室

**AILAS** Airborne Instrument Landing Approach System 机载仪表着陆进近系统

Automatic Instrument Landing Approach System 自动仪表着陆进近系统

Automatic Instrument Low Approach System 自动仪表低空进近系统

**AILC** Anti-Ice and Leak Detection Controller

防冰与渗漏检测控制器

**AILS** Advanced Instrument Landing System 先进的仪表着陆系统

Automatic Instrument Landing System 自动仪表着陆系统

**AILSA** Aerospace Industrial Life Sciences Association 航空航天工业生命科学协会

**AIM** Address Indexing Method 地址检索方法

Advanced Image Management 高级图像管理

Advanced Information Manager 高级信息经理

Advanced Interface Module 高级接口模块

Aerial Intercept Mobile 空中拦截机动

Aeronautical Information Manual 航空情报手册

Aerospace Information Management 航行情报管理

Airbus Integrated Management 空客综合管理

Aircraft Integrtaed Maintenance 飞机综合维护

Airman's Information Manual 飞行员情报手册

Asynchronous Interface Module 异步接口模块

Automatic Inflation Modulation 自动充气调节

Automatic Information Management 自动信息管理

Automatic Investment Management 自动投资管理

Aviation Information Manual 航空信息手册

Avionics Interface Module 航空电子接口组件

**AIMD** Aircraft Intermediate Maintenance Department 飞机中级维修部门

**AIME** American Institute of Mechanical Engineers 美国机械工程师协会

**AIMI** Airborne Infrared Measurement Instrument 机载红外线测量仪器

Avionics Information Management Interface 电子信息管理界面

**AIMS** Advanced Information Management System 先进的信息管理系统

Aeronautical Information Management System 航行情报管理系统

Airbus Improvement Management System 空客改进管理系统

Airbus Inventory Management System 空客资产管理系统

Aircraft Idenfication Monitoring System 飞机识别监控系统

Aircraft Integrated Management System 飞机综合管理系统

Aircraft Integrated Monitoring System 飞机综合监控系统

Airplane Information Management System 飞机信息管理系统

Airport Information Management System 机场信息管理系统

Airways Integrating and Monitoring System 航路综合监控系统

Attitude Indicator Measurement System 姿态指示器度量系统

Automated Instrumentation and Monitoring System 自动插桩和监测系统

Automated Integrated Manufacturing System 自动综合制造系统

**AIN** Advanced Intelligent Network 高级智能网

Airworthiness Inspection Notice 适航性检验通告

Asia Internet Network 亚洲因特网网络

Aviation International News 国际航空新闻

**AINS** Advanced Inertial Navigation System 先进惯性导航系统

Airborne Inertial Navigation System 机载惯性导航系统

Aircraft Information Network System 飞机信息网络系统

Area-Inertial Navigation System 区域惯性导

航系统

**AIO** Analog Input and Output 模拟输入输出

**AIP** Aeronautical Information Publication 航空资料出版物,航行资料汇编

Air Inlet Pressure 空气进口压力

Airborne Instrumentation Platform 机载仪表平台

Aircraft Interface Pannel 航空设备接口板

Aircraft Interference Program 飞机干扰规划

Airplane Integration Plan 飞机综合规划

Airport Improvement Program /Plan 机场改进项目,机场改进计划

Anti-Ice Panel 防冰控制器

Attendant Indication Panel 乘务员指示面板

Automated Imagery Processing 自动成像处理

Avionics Integration Plan 航空电子设备集成规则

Avionics Integration Program 航空电子设备综合化计划

**AIPA** Accident Initiation and Progression Analysis 事故引发与进展分析

Australian and International Pilots Association 澳大利亚国际飞行员协会

**AIPC** Aircraft Illustrated Parts Catalog 飞机部件图解目录

**AIPS** Astronomical Information Processing System 天文信息处理系统

**AIPU** Arab Inter-Parliamentary Union 阿拉伯议会联盟

Accident Investigation Report 事故调查报告

Additional Information Request 附加信息请求

Aerospace Information Report(s) 航空航天信息报告

Air Interrupt Rate 空中中断率

Airborne 机载

Aircraft Identification Radar 飞机识别雷达

Aircraft Incident Report 飞机事故报告

Aircraft Inflight Report 飞机空中报告

Aircraft Inspection Report 飞机检验报告

Airplane 飞机

Airworthiness Inspection Representative 适航检查代表

Association Internationale des Registres 国际登记协会

Aviation Information Resources 航空信息资源

**AIRAC** Aeronautical Information Regulation and Control 航行资料定期颁发制

**AIRACS** Aircraft Acquisition & Support 飞机搜索和支援,飞机获得和支援

**AIRC** Airlines Industrial Relations Conference 航空公司工业联系会议(美国)

**AIRCGO** Air Cargo 航空货物

**AIRCLNR** Air Cleaner 空气净化器

**AIRCN** Air Communication Network 空中通讯网

**AIRCO** Air Conditioning 空调

**AIRCOM** Airways Communications System 航空公司通信系统

**AIRCOMM** Air Communication 航空通信

**AIRCOND** Air Condition, Air Conditioning 空调

**AIRDES** Automated Interline Revenue Data Exchange Service 联运收入自动数据交换业务

**AIREP** Air Report 航空报告

**AIREPDN** Aircraft Repair Division 航空器修理部,飞机修理部

**AIRF** Aircraft Instrument Repair Facility 飞机仪表修理设施

**AIRFD** Airfield 机场

**AIRFRT** Airfreight 航空货运

**AIRMAN** Aircraft Maintenance Analysis 飞机维修分析软件

**AIRMET** Airman's Meteorological Information 飞行员气象信息

Weather Advisory Service 气象咨询服务

**AIRNAVAID** Air Navigational Aid (空中)导航辅助设备

**AIRNZ** Air New Zealand 新西兰航空公司

**AIRO** Analog Input with Reference Output 具有基准输出的模拟输入

**AIRP** Air Power 航空动力

Aircraft Instrumentation Research Program 飞机仪表研究方案

Airworthiness Panel 适航性专家小组

**AIRPCL** Air Parcel 航空包裹

**AIRPLT** Airplane Pilot 飞机驾驶员

**AIRRB** Accident Investigation Report Review Board 事故调查报告审查委员会

**AIRS** Accident Information Retrieval System 失事信息检索系统

Advanced Inertial Reference Sphere 先进的惯性基准地球仪

Advanced Inertial Reference System 先进的惯性参考系统

Aircraft Systems 飞机系统

Airline Inventory Redistribution System 航空公司库存量再分配系统

Airport Information Retrieval System 机场信息检索系统

ARRL Interference Reporting System 美国无线电中继语言干扰报告系统

Atmospheric Infrared Sounder 大气红外探测器

Automated Image Retrieval System 自动图像检索系统

Automatic Information Retrieval System 自动信息检索系统

Autonomous Infrared Sensor 自主式红外传感器

**AIS** Academic Instructor School 航校教员学校

Advanced Integration Study 高级综合研究

Advanced Integration System 高级综合系统

Aeronautical Information Service 航空情报服务

Airborne Instrumentation Subsystem 机载仪表分系统

Airborne Instrumentation System 机载仪表系统

Aircraft Instrument Subsystem 飞机仪表子系统

Alarm Indication Signal 报警指示信号

Anti-Icing System 防冰系统

Application Interface System 应用接口系统

Artificial Immune System 人工免疫系统

Audio Integrated System 音频综合系统

Automatic Identification System 自动识别系统

Automatic Information System 自动信息系统

Automatic Intercept System 自动截取系统

Avionics Integrated Shop 航空电子设备综合车间

Avionics Intermediate Shop 航空电子设备中间站

**AIS—AIMSG** Aeronautical Information Services-Aeronautical Information Management Study Group 航行情报服务—航行情报管理研究小组

**AISD** Automated Information Systems Division 自动化信息系统部

**AISF** Avionics Integration/Integrated Support Facilities 航空电子综合保障设备

**AISI** American Iron and Steel Institute 美国钢铁协会

**AISS** Airborne Infrared Surveillance Set 机载红外监视装置

**AIST** Advanced Industrial Science and Technology 先进工业科学与技术

Agency of Industrial Science and Technology 工业科学与技术研究院

**AIT** Acoustic Impact Technique 声冲击技术

Advanced Individual Training 高级单人飞行训练

Air Intake Temperature 进气温度

Assembly Integration and Test 部件组装和测试

Assured Intermediate Task 保障的中间任务

Auto-Ignition Temperature 自动点火温度

Automatic Information Test 自动信息测试

**AITA** Air Industries & Transport Association

航空工业与运输协会

**AITAL** Association Internationale de Transport Aéro Latino-Américain 拉丁美洲航空公司协会

**AITF** Avionics Integrated Test Function 航空电子设备综合测试功能

**AITI** Aero Industries Technical Institute 航空工业技术学会,航空工业技术学院

Area Inspection Task Item 区域检查工作项目

Automated Interchange of Technical Information 技术信息自动交换

**AIU** Advanced Instrumentation Unit 高级仪表设备

Aircraft Interface Unit 飞机接口组件

Alarm Interface Unit 告警接口装置

American International Underwriter 美国国际保险公司

Armament Interface Unit 装备接口组件

ATM Interface Unit ATM 接口单元

Audio Interface Unit 音频接口组件

**AIV** Accumulator Isolation Valve 储压器隔离活门

Air Inlet Valve 进气活门

Annulus Inverting Valve 环形转换活门

Anti Ice Valve 防冰活门

**AIX** Advanced Interactive Executive 高级交互执行程序

**AIXM** Aeronautical Information Exchange Model 航空情报交换模型

**AIY** Atlantic City, NJ, USA ( Airport Code ) 美国新泽西州大西洋城机场

**AIZ** Aerodrome Information Zone 机场信息区

Association of Inventors 发明者协会( 前苏联)

**AJ** Anti-Jam 抗干扰

**AJA** Ajaccio, Corsica, France ( Airport Code ) 法国科西嘉岛阿雅克肖机场

**AJAB** Advisory Joint Appeals Board 咨询联合申诉委员会

**AJACS** Advanced Joint Air Combat System 先进的联合航空作战系统

**AJSTMT** Ajustment 调节,调整

**AJT** Advanced Jet Trainer 高级喷气训练机

Amerijet International 亚美利国际喷气机

**AK** Accepting Confirmation 证实接受

Aviation Storekeeper 航空仓库管理员

**AKAD** Acknowledge and Advice 确认并告之

**AKL** Auckland, New Zealand ( Airport Code )新西兰奥克兰机场

**AL** Accuracy Landing 着陆准确性

Address Line 地址线

Aerodynamics Laboratory 航空动力学实验室

Aeronautical Laboratory 航空实验室

Air Launch 空中发射

Air Letter 航空信件

Air Liaison 空中联络

Air Lift 空运,空运量

Air Lock 气锁

Airline( s ) 航空公司,航线

Airlock 密封过渡舱

Alarm Light 报警灯,提醒灯

Albania 阿尔巴尼亚

Alcohol 酒精

Alert Limit 告警门限

Algorithmic Language 算法语言

Alternate 备份,备降

Altimeter 高度表

Altitude 高度,标高

Altocumulus 高积云

Aluminium 铝

Amber Light 琥珀灯

Amended Level 修正后高度层

Amendment List 修正表

Antenna Laboratory 天线实验室

Application Layer 应用层

Approach and Landing 进近和着陆

Approach Light 进场灯

Approach Lighting 进近灯光,进场照明

Arrival Locator 到达定位器

Assemble/Load 装配与载荷

Audio Language 音频语言

Audio Library 音频库
Automatic Landing 自动着陆
Automatic Location 自动定位
Average Life 平均寿命
Aviation Laboratory 航空实验室
Avionics Laboratory 航空电子学实验室

**ALA** Aircraft Landing Area 航空器着陆区
Alighting Area 着陆区域,降落区
Almaty, Kazakhstan ( Airport Code ) 哈萨克斯坦阿拉木图机场
Approach and Landing Accidents 进近和着陆事故
Approved Landing Area 批准的着陆区域
Authorised Landing Area 允许着陆区

**ALAE** Association of Licensed Aircraft Engineers 持证飞机工程师协会(英,澳)

**ALAEA** Australian Licensed Aircraft Engineers Association 澳洲持证飞机工程师协会

**ALALY** Aluminum Alloy 铝合金

**ALAN** Advanced Local Area Network 高级局域网

**ALAR** Approach and Landing Accident Reduction 减少进近和着陆事故

**ALARP** As Low As Reasonably Practical 尽可能降低风险

**ALAs** Approach-and-Landing Accidents 进近与着陆事故

**ALB** Albany, NY, USA ( Airport Code ) 美国纽约州奥尔巴尼机场
Approach Light Beacon 进近灯标
Automatic Loop Back 自动回环

**ALB( N )** Albania 阿尔巴尼亚

**ALC** Adaptive Logic Circuit 适配逻辑电路
Air Lane Compass 航道罗盘
Air Launched Balloon System 空中发射准则
Air Lines Circuit 空中航线网
Airlift Control 空运控制
Airline Link Control 航空公司链路控制
Alcohol 酒精
Alicante, Spain ( Airport Code ) 西班牙阿利坎特机场

Ambient Lighting Condition 室内照明条件
Analog Line Card 模拟用户线插件
Approach and Landing Charts 进场和着陆图
Assigned Labor Code 规定派工码
Automatic Landing Control 自动着陆控制
Automatic Level Control 自动电平控制,自动水平控制
Automatic Light Control 自动光亮控制
Automatic Load Control 自动负载控制,自动载重控制
Automatic Loading Circuit 自动加载控制线路
Automatic Locking Circuit 自动锁定电路

**ALCAC** Air Lines Communications Administrative Council 航空公司通信管理委员会

**ALCC** Airlift Control Center 空运控制中心

**ALCE** Airlift Control Element 空运管制分队

**ALCH** Approach Light Contact Height 进近指示灯光能见高度

**ALCO** Airlift Coordinating Office 航空运输协调处

**ALCS** Active Lift Control System 主动升力控制系统
Airborne Launch Control System 机载发射控制系统

**ALD** Actual Landing Distance 实际着陆距离
Airworthiness Liaison Div. 适航联络处
Arbitrary Landing Distance 任意着陆距离
At a Later Date 在晚些日期
Available Landing Distance 可用的着陆滑跑距离

**ALDB** Airline Loadable Database 航空公司可装载数据库

**ALDCS** Active Lift Distribution Control Subsystem 升力分布主动控制子系统
Active Lift Distribution Control System 升力分布主动控制系统

**ALDY** Already 已经

**ALE** Airport Lighting Equipment 机场照明设备
Annualized Loss Expectancy 年度损失期望

Automatic Laser Encoder 自动激光编码器

Automatic Link Establishment 自动连接功能

**ALEA** Airborne Law Enforcement Association 航空执法人协会,航空法实施协会(美国)

**ALERFA** Alert Phase 警戒阶段

Authorised Landing Area 允许着陆区

**ALERT** Automatic Logging Electric Reporting Telemetry 自动记录电子信息报告与遥测系统

**ALERTS** Airborne Laser Equipment Real Time Surveillance 机载激光设备实时监视

**ALF** Aft Looking Forward 由后向前看

Auxiliary Landing Field 辅助降落机场

**ALFC** Automatic Load-Frequency Control 自动载荷频率控制

**ALFS** Airborne Low Frequency Sonar 机载低频声纳

**ALG** Aeroleasing Group 航空租赁集团

Aircraft Landing Gear 飞机起落架

Algiers, Algeria ( Airport Code ) 阿尔及利亚阿尔及尔机场

Along 沿着

Autonomous Landing Guidance 独立着陆引导

**ALGS** Approach & Landing Guidance Systems 进近和着陆引导系统

**ALH** Advanced Light Helicopter 高级轻型直升机

**ALHP** Airframe Life-History Program 机身寿命—经历提纲

**ALI** Advance Line Information 预先航线信息

Airworthiness Limitation Item 适航限制项目

Asynchronous Line Interface 异步线路接口

**ALIGN** Alignment 对正,校准

**ALIM** ATM Line Interface Module ATM 线路接口模块

**ALIND** Alarm Indicator 报警指示器

**ALIS** Airlines Interactive Service 航空公司交互式服务

Airport Language Identification System 机场行李识别系统

**ALL** Aircraft Landing Lamp 飞机着陆灯

**ALLOC** Allocation, Allocate 分配

**ALLW** Allow 允许

Allowance 允许量,限度

**ALLY** Alloy 合金

**ALM** Aircraft Landing Minima 飞机着陆最低天气标准

Alarm 警报

**ALMS** Aircraft Landing Measurement System 飞机着陆测量系统

Air-Lift Management System 空运管理系统

**ALMT** Allotment 分配

**ALN** Alignment 校准

**ALNOT** Alert Notice 警戒通知

**ALNZ** The Air League of New Zealand Inc. 新西兰公司航空联合会

**ALO** Air Liaison Officer 空中联络官

Automatic Lock-On 自动锁定(目标)

Waterloo, IA, USA ( Airport Code ) 美国爱荷华州滑铁卢机场

**ALOS** Advanced Land Observing Satellite 先进的陆地观测卫星

Airborne Light Observation System 机载灯光观测系统

**ALP** Aleppo, Syria-Nejrab ( Airport Code ) 叙利亚阿勒颇机场

Air Liaison Party 对空联络组

Air Liaison Post 对空联络哨

Alternative Launch Point 交替发射点

Ambulance Loading Post 救护飞机装载机场

Automated Learning Process 自动学习过程

**ALPA** Airline Pilots Association 航线运输驾驶员协会,航空公司驾驶员协会

**ALPHA** Alphabet 字母表

**ALR** Aircraft Load Rating 飞机额定载荷

Alert 警报

Alerting Message 告警报

**ALRDY** Already 已经

**ALREP** Airline Pilot Report 航线驾驶员报告,航空公司驾驶员报告

**ALRN** Aileron 副翼

**ALRT**　Advanced Light Rapid Transit 高级轻型快速运输

Alert 警报

**ALS**　After Limit Switch 后极限电门

Aircraft Landing System 飞机着陆系统

Aircrew Life Support 机组生命支持

Alamosa, CO, USA（Airport Code）美国科罗拉多州阿拉莫萨机场

Alternate Landing Site 备用着陆场

Ambient Light Sensor 周围环境照明传感器

Approach Light System 进近灯光系统

Automated Location System 自动定位系统

Automatic Landing System 自动着陆系统

Azimuth Laying Set 方位瞄准装置

**ALSCU**　Auxiliary Level Sensing Control Unit 辅助水平感觉控制组件

**ALSF**　Approach Light System with Sequenced Flashing Light 带有顺序闪光灯的进近灯光系统

Approach Lighting Sequenced Flashing 进近灯光顺序闪亮

**ALSIP**　Airborne Laser Systems Integration Project 机载激光系统集成项目

Approach Light System Improvement Program 进近灯光系统改进项目

Approach Lighting System Improvement Plan 进近灯光系统改进规划

**ALSS**　Advanced Life Support System 高级生命保障系统

Advanced Location Strike System 高级定位攻击系统

Airline System Simulator 航线系统模拟器

**ALSTG**　Altimeter Setting 高度表拨正

**ALT**　Accelerate Life Test 加速寿命测试

Airborne Laser Tracker 机载激光跟踪器

Alert 警报

Algorithmic Learning Theory 算法学习理论

Alternate Aerodrome 备用机场,备降机场

Alternate 备份,备降场

Altimeter 高度表

Altitude Hold 高度保持

Altitude 海拔,高度

Automatic Line Test 自动线路测试

**ALTF**　Alternate Field 备降机场

**ALTHO**　Although 尽管

**ALTHOLD**　Altitude Hold 高度保持

**ALTM**　Altimeter 高度表

**ALTN**　Alternate 备降机场,备用,备份

Alternate 交换(替)

Alternative 选择的

**ALTNT**　Alternate 交替,备份,备降机场

**ALTP**　Air Line Transport Pilot 航班运输机飞行员

Alter 改变

Alternate 备用的,交变的

**ALTRV**　Altitude Reservation 高度预定

**ALTS**　Altimeter Setting 高度表调定数

Altitude Select 高度选择

**ALTSEL**　Altitude Select 高度选择

**ALTTA**　Application of Hybrid Laminar Flow Technology on Transport Aircraft 混合层流技术在运输机上的应用

**ALTU**　Annunciator Light Test Unit 信号灯测试组件

**ALU**　Arithmetic and Logic Unit 算术和逻辑组件

**ALUM**　Aluminium 铝

**ALW**　Allow 允许

Always 总是

Walla Walla, WA, USA（Airport Code）美国华盛顿瓦拉瓦拉机场

**ALWCE**　Allowance 允许,容差

**ALWD**　Allowed 允许

**ALY**　Alexandria, Egypt（Airport Code）埃及亚历山大机场

Alloy 合金

**AM**　Above Mentioned 上述

Access Module 接入模块

Accounting Management 计费管理

Address Mark 地址标记

Aeronautical Meteorology 航空气象

Air Mail 航空邮件

Air Mass 大气质量

Airport Manager 机场经理

Airspace Management 空域管理

Airway Manual 航线手册

Amber 琥珀色

Amplitude Modulation 调幅

Ante-Meridian 上午

Aviation Medicine 航空医学

**AMA** Academy of Model Aeronautics 航空模型研究院

Action Media Adapter 自动媒体适配器

Advanced Measurement Approaches 高级计量法

Aero-Medical Association 航空医学协会

Aircraft Maintenance Alert 飞机维修警告

Airworthiness Manual Advisory 适航手册咨询

Amarillo, TX, USA（Airport Code）美国得克萨斯州阿马里洛机场

Area Minimum Altitude 区域最低高度

Association for Model Aviation 航空模型协会

Audio Monitor Adapter 音响监控接头

Automatic Message Accounting 自动信息计费

**AMAC** Aircraft Monitoring and Control 飞机监视与控制

**AMAD** Airframe Mounted Accessory Drive 机身安装的辅助驱动器

**AMAL** Amalgamated 混合的,合并的

Aviation Medical Acceleration Lab 航空医学加速度实验室

**AMARS** Automatic Message Address Routing System 自动报文地址路由选择系统

Automatic Multiple Address Relay System 自动多址中继系统

**AMAS** Analysis of Market Research System 市场研究分析系统

Automatic Manoeuvring Attack System 自动机动攻击系统

**AMASS** Airport Movement Area Safety System 机场地面活动区安全系统

**AMAT** Aircraft Multi Purpose Access Terminal 飞机多用途进近终端

Average Memory Access Time 存储器平均访问时间

**AMB** Aircraft Maintenance Base 飞机维修基地

Aircraft Maintenance Bulletin 飞机维护通告

Aircraft Mishap Board 飞机事故委员会

Airways Modernization Board 航路现代化委员会

Ambassador 大使

Amber 琥珀色

Ambient 周围的,相邻的,环绕的

Avionics Management Board 航空电子系统管理局（委员会）

**AMBA** Aviation Master of Business Administration 航空工商管理硕士

**AMBR** Aircraft Maintenance Bulletin Request 飞机维护通告需求

**AMBT** Ambient Temperature 外界温度,环境温度

**AMC** Acceptable Means of Compliance 符合性验证方法

Aerodynamic Mean Chord 气动平均翼弦

Aero-Medical Centre 航空体检中心

Airbus Military Company 空客军用飞机公司

Aircraft Maintenance Control 飞机维修控制

Aircraft Maintenance Costs 飞机维修费用

Aircraft Management Computer 飞机管理计算机

Aircraft Manufacturer's Council 飞机制造商理事会

Aircraft Manufacturing Center 飞机制造中心

Airspace Management Cell 空域管理单元

Alternate Media Center 备用媒体中心

ATM inter-Module Connector ATM 模块间连接器

Automatic Mixture Control 自动混合控制

Aviation Maintenance Certification 航空维修审定

Aviation Museum of China 中国航空博物馆

Avionics Maintenance Conference 航空电子设备维修会议

**AMCM** Aircraft Maintenance Control Manual 航空器维修控制手册

**AMCO** Airline Management Consultants & Operations 航空公司管理咨询与运营

**AMCP** Advanced Multimedia Communication Protocol 高级多媒体通信协议

Aeronautical Mobile Communication Panel 航空移动通信专家小组

**AMCR** Adaptive Multimedia Communication Routing 自适应多媒体通信路由算法

**AMD** Advisory Map Display 咨询图显示

Aerospace Medical Division 航空与航天医学处

Ahmedabad, India ( Airport Code)印度艾哈迈达巴德机场

Air Management Division 航空管理处

Air Movement Data 空运数据

Aircraft Maintenance Department 飞机维修部

Airworthiness Management Document 适航管理文件

Amend 修订

Amendment 修订件

Aviation Maintenance Department 航空维修部

Aviation Medical Division 航空医学处

**AMDA** Airlines Medical Directors Association 航空公司医务经理协会

**AMDAR** Aircraft Meteorological Data Relay 飞机气象资料中继

**AMDB** Aerodrome Mapping Data Base 机场地图数据库

Airport Management Data Base 机场管理数据库

Automated Maintenance Data Base 自动维护数据库

**AMDC** Aircraft Materials Distribution Center 飞机器材分配中心

**AMDP** Aircraft Maintenance Delay for Parts 飞机因缺零件而推迟维修

**AMDT** Amendment 修订件

**AME** Aero-Medical Evaluation 航空医学审定

Aeronautical Maintenance Engineer 航空维修工程师

Aircraft Maintenance Engineer 飞机维修工程师

Amplitude Modulation Equipment 调幅设备

Amplitude Modulation Equivalent 等值调幅

Angle Measuring Equipment 测角仪

Assised Medical Examiner 指定的体检医生

Associated Memory Equipment 相关存储装置

Attitude Measurement Electronics 姿态测量电子设备

Automatic Monitoring Equipment 自动监控设备

Average Magnitude of Error 误差平均值

Aviation Maintenance Evaluation 航空维修评估

Aviation Medical Examiner 航空体检医生

Aviation Mission Equipment 航空任务设备

**AMECO** ( Beijing ) Aircraft Maintenance & Engineering Corporation (北京)飞机维修工程公司

**AMEL** Active Matrix Electro-Luminescent 有源矩阵电致发光

Aeromedical Equipment Laboratory 航空医学设备实验室

Aircraft Maintenance Engineer Licence 飞机维护工程师执照

Airplane, Multi-Engine Land 多发陆地飞机

**AMENDMT** Amendment 修订件

**AMER** American 美国的,美国人

**AMES** Aircraft Maintenance and Engineering System 航空器维护与工程系统

Aircraft Maintenance Event Scheduler 飞机维修活动调度员

ATM Mobility Extension Service ATM 移动性

扩展服务

**AMETS** Aircraft Maintenance Engineering Time Standard 航空器维修工程计时标准

**AMEX** American Express 美国快递

**AMF** Arab Monetary Fund 阿拉伯货币基金组织

Authentication Management Function 认证管理功能

Avionics Maintenance Facility 航空电子维修设施

**AMG** Analysis Model Generator 分析模型发电机

Association Management Group 协会管理组

**AMH** Automated Message Handling 自动信息处理

**AMHS** Aeronautical Message Handling System 航空信息处理系统

Automated Message Handling Service 自动信息处理服务

Automatic Message Handling System 自动电文处理系统

Aviation Message Handling Standards 航空信息处理标准

**AMI** Aircraft Maintenance Item 飞机维护项目

Airline Modifiable Information 航空公司可修改信息

Airline Mutual Insurance Ltd. 航空公司共同保险公司

Airplane Modifiable Information 飞机可变更信息

Airspeed-Mach Indicator 空速马赫数表

Alternative Mark Inversion 信号交替取反

Assistant Maintenance Inspector 助理维修检查员

Mataram, Indonesia（Airport Code）印度尼西亚马塔兰机场

**AMIC** Automated Management Information Center 自动管理信息中心

**AMIDBS** Air Meteorology Information Data Base System 航空气象情报数据库系统

**AMIDD** Aircraft Maintenance Integrated Diagnostics Demonstration 飞机维修综合诊断显示

**AMIDS** Aviation Meteorological Information Dissemination System 航空气象资料发放系统

**AMII** Airline Maintenance Inspection Intervals 航空公司维护检查间隔

**AMIS** Acquisition Management Information System 采购管理信息系统

Advanced Management Information System 高级管理信息系统

Aircraft Management Information System 飞机管理信息系统

Airport Management Information System 机场管理信息系统

Audio Message Interactive Specification 声讯交互规范

Automated Maintenance Information System 自动维修信息系统

Automatic Management Information System 自动管理信息系统

**AMIT** Airline Modifiable Information Tool 航空公司可修改信息工具

**AMJ** Advisory Material-Joint 联合咨询资料

**AML** Aircraft Maintenance Licence 飞机维修执照

Alternate Materials List 替代材料清单

Alternate Mode Light 备用模式灯

Aviation Materials Laboratory 航空材料实验室

**AMLCD** Active Matrix Liquid Crystal Display 主动矩阵式液晶显示器，有源矩阵液晶显示器

**AMLIP** Airfield Marking and Lighting Improvement Program 机场标志和灯光改进程序

**AMM** Airbus Maintenance Manual 空客维修手册

Airplane/Aircraft Maintenance Manual 飞机维修手册

Amman, Jordan-Queen Alia (Airport Code) 约旦安曼机场

Ammeter 安培表

Auxiliary Management Module 辅助维修模块

**AMMC** Aviation Material Management Center 航空器材管理中心

**AMME** Automatic Multi-Media Exchange 自动多媒体交换机

**AMMR** Aircraft Modification Management Rules 飞机改装管理规则

**AMMRL** Aircraft Maintenance Material Readiness List 飞行器维修材料准备清单

**AMND** Amend 修改,修订

**AMNT** Amount 合计,总数

**AMO** Aerodrome Meteorological Office 机场气象局

Air Mail Only 只限航邮

Aircraft Maintenance Officer 飞机维修官

Aircraft Maintenance Office 飞机维修办公室

Airport Meteorological Office 机场气象办公室

Approved Maintenance Organization 批准的维修组织

Asset Management Organization 资源管理组织

Aviation Maintenance Officer 航空维修官

**AMOC** Alternate Method of Compliance 替代符合性方法

Atlantic Meridional Overturning Circulation 北大西洋经圈翻转环流

**AMOFSG** Aerodrome Meteorological Observation and Forecast Study Group 机场气象观测和预报研究组

**AMOS** Aircraft Maintenance Operations Simulation 航空器维护运行模拟

Automated Meteorological Observing System 自动气象观测系统

Automatic Meteorological Observing Station 自动气象观测站

**AMP** Advanced Manned Penetrate 高级有人驾驶突防飞机

Advanced Microelectronics Program 先进的微电子技术计划

Aeronautical Meteorology Programme 航空气象学计划

Aircraft Maintenance Plan 飞机维修计划

Aircraft Maintenance Program 航空器维修计划

Aircraft Modernization Plan 飞机现代化计划

Amperage 电流强度,安培数

Ampere 安培

Amperes 安培

Amplifier 放大器

Amplitude 振幅,幅度

Analysis Methods & Procedures 分析方法与程序

Audio Management Panel 音频管理板

Automated Manufacturing Plan 自动化生产计划

Avionics Modernization Program 航空电子设备现代化计划

**AMPA** Australian Parts Manufacturing Approvals 澳大利亚零部件制造批准书

**AMPC** Automatic Message Processing Center 自动信息处理中心

**AMPE** Application Message Processing Environment 应用信息处理环境

Application Multiple Processing Environment 应用多路处理环境

Automated Message Processing Exchange 自动信息处理交换机

**AMPG** Air-Mile per Gallon 每加仑燃油飞行英里数

**AMPH** Amphibian 水陆两用飞机

**AMPL** Amplifier 放大器

**AMPOS** Aircraft Maintenance Procedure Optimization System 航空器维修程序优化系统

**AMPR** Aircraft Manufacturer's Planning Report 飞机制造商的计划报告

**AMPS** Advanced Mobile Phone System 高级移动电话系统

Assisted Mission Planning System 辅助飞行计划系统

Automated Message Processing System 自动化信息处理系统

**AMR** Adaptive Multiple Rate 自适应多速率

Advanced Material Request 高级材料申请（书）

Automatic Message Recording 自动抄表

**AMRD** Australian Maintenance Requirements Document 澳大利亚维修需求文件

**AMRS** Air Maintenance and Repair System 航空维护和修理系统

**AMS** Acquisition Management System 采购管理系统

Advanced Meteorological Satellite 高级气象卫星

Aeronautical Material Specification 航空材料规范

Aeronautical Mobile Service 航空移动通信服务

Aerospace Material Specifications 宇航材料规范,航天器材规范

Aircraft Maintenance Schedule 飞机维修计划

Aircraft Management Simulator 飞机管理模拟机

Aircraft Management System 飞机管理系统

Aircraft Modification Status 飞机改装状态

American Material Specification 美国器材规范

Amsterdam, Netherlands-Schiphol 荷兰阿姆斯特丹斯希波尔机场

Approved Maintenance Schedule 批准的维修大纲

Automated Maintenance System 自动维护系统

**AMSA** Advanced Manned Stategic Aircraft 先进的有人驾驶战略飞机

**AMSAC** Air Medical Safety Advisory Council 空中医疗安全顾问委员会

Aviation Maintenance Safety Auditing and Consulting 航空维修安全审计及咨询

**AMSC** Automatic Message Switching Center 自动文电交换中心

**AMSCS** Aeronautical Mobile Satellite Communication System 航空移动卫星通信系统

**AMSDL** Acquisition Management Source Data List 采购管理资源数据清单

Acquisition Management System & Data Requirements Control List 采购管理系统和数据需求控制清单

Acquisition Management Systems & Data Requirements List 采购管理系统和数据需求清单

Acquisition Management Systems Data List 采购管理系统数据清单

Authorized Management System and Data Requirement List 审定的管理系统和数据需求清单

**AMSE** Aircraft Maintenance Support Equipment 飞机维修支援设备

**AMSIA** AV Multimedia Service Implementation Agreement 音像多媒体业务执行协议

**AMSIS** Aircraft Multi-Sensor Integration Study 飞机多传感器综合研究

**AMSL** Above Mean Sea Level 高出平均海平面高度

Acquisition Management Systems List 采集管理系统表

Approved Materials Substitution List 批准的代用材料单

**AMSS** Aeronautical Mobile Satellite Service 航空移动卫星服务

Aeronautical Mobile Satellite System 航空移动卫星系统

Airbus Modular Spares Services 空中客车模块化零备件服务

Aircraft Maintenance Support System 飞机维修支持系统

Automated Master Schedule System 自动主控程序系统

Automatic Message Switching System 自动转

报系统,航空电报自动转送系统

Aviation & Maintenance Support Systems 航空与维修支持系统

Aviation Movement Satellite System 航空移动卫星系统

**AMSSP** Aeronautical Mobile Satellite Service Panel 航空移动卫星服务专家组

**AMST** Advanced Medium STOL Transport 先进的中型短距起降运输机

**AMSU** Advanced Microwave Sounding Unit 先进的微波探测装置

Aeronautical Material Screening Unit 航空材料扫描装置

Aircraft Motion Sensing Unit 飞机移动传感组件

Aircraft Motion Sensor Unit 飞机运动传感器部件

Air-Motor Servo Unit 航空发电机伺服机构

**AMT** Accelerated Mission Test 加速的任务试车

Active Maintenance Time 实际维修时间,有效维修时间

Advanced Manufacturing Technology 先进的制造技术

Advanced Medium Transport 先进的军用运输机

Air Mail Transfer 航空信汇

Aircraft Maintenance Technician 飞机维修技术人员

Amount 数额,总数

Area Multi-channel Terminal 区域多路终端

**AMTE** Aviation Maintenance Technical Engineer 航空维修技术工程师

**AMTEC** Automatic Time Element Compensator 自动延时补偿器

**AMTI** Airborne Moving Target Indicator 机载活动目标显示器

Area Moving Target Indicator 区域活动目标显示器

Automatic Moving Target Indicator 活动目标自动显示器

**AMTOSS** Aircraft Maintenance Task Oriented Support System 面向飞机维修任务的保障系统

**AMTS** Aeronautical Message Transfer Service 航空报文移交业务

Aeronautical Mobile Telecommunication Service 航空移动电信服务

**AMU** Air Mileage Unit 飞机里程表

Aircraft Maintenance Unit 飞机维修组件

Audio Management Unit 音频管理组件

**AMUS** Avionics Multiplex System 航空电子多路传输系统

**AMUX** Avionics Multiple Bus 航空电子多路传输总线

**AN** Above-Named 上述的,前述的

Access Network 接入网

Access Node 接入节点

Active Network 有源网络

Active Node 有源节点

Aids to Navigation 导航设备,助航设备

Air Navigation 空中导航,领航

Airworthiness Notice 适航管理通告

Answer 答复,回答

Approach Navigation 进近导航

Arrival Notice 到达通知

**ANA** Air Navigation Act 空中航行法规

All Nippon Airways 全日本航空公司

Association of Naval Aviation 海军航空协会

Automatic Network Analyzer 自动网络分析器

**ANAL** Analysis 分析

**ANAO** Australian National Aerospace Organization 澳大利亚国家航空航天组织

**ANAP** Airport Noise Abatement Plan 机场噪声抑制计划

**ANAS** ATM Network Access Subsystem ATM 网接入子系统

**ANB** Air Navigation Bureau 空中航行局

**ANC** Active Noise Control 主动噪声控制

Adaptive Noise Concellation 自适应噪声对消

Air Navigation Charge 空中导航费用

Air Navigation Commission 空中航行委员会

Air Navigation Conference 空中导航会议

Airborne Navigation Computer 机载导航计算机

All-Number Calling 全号呼叫，全数字呼叫

Anchorage, AK, USA（Airport Code）美国阿拉斯加州安克雷奇特德·史蒂文斯机场

Ancient 古代

Area Navigation Capability 区域导航能力

**ANCAT** Abatement of Noise Caused by Air Transport 降低航空运输噪声

**ANCC** Access Network Control Center 接入网控制中心

**ANCE** Announce 通知，声明

**ANCG** Announcing 宣布，通告，通报

**ANCPTR** Anticipater 预测器

**ANCS** Airborne Night Classification System 机上夜间分类系统

**ANCU** Anti-Noise Control Unit 防噪控制组件

**AND** Advanced Network Design 先进网络设计

Aircraft Data Network 航空器数据网络

Aircraft/Airplane Nose Down 飞机机头向下，飞机俯冲

Automatic Network Dialing 自动网络拨号

**ANDAS** Automatic Navigation and Data Acquisition System 自动导航和数据采集系统

**ANE** Aeronautical and Navigation Electronics 航空与导航电子设备

**ANEF** Australian Noise Exposure Forecast 澳大利亚噪声暴露预报

**ANEP** Airport Noise Evaluation Process 机场噪声评估方法

**ANF** Air Navigation Facility 航空导航设施

Antofagasta, Chile（Airport Code）智利安托法加斯塔机场

**ANFE** Aircraft Not Fully Equipped 装备不全飞机

**ANFTMS** Aircraft Noise and Flight Tracking

Monitoring System 飞机噪音及航迹监视系统

**ANG** Air National Guard 国家空防警卫队

Angle 角度

**ANGB** Air National Guard Base 航空国民警卫队基地

**ANI** Analog Input 模拟输入

Aniak, AK, USA（Airport Code）美国阿拉斯加州阿尼亚克机场

Automatic Number Identification 自动号码识别

**ANIP** Air Navigation Integrated Programme 空中航行综合方案

**ANK** Ankara, Turkey（Airport Code）土耳其安卡拉机场

**ANL** Airplane Nose Left 机头向左

Animal 动物

Automatic Noise Limiter 噪声自动限制器

**ANLG** Analog 模拟

Analogic 模拟逻辑

**ANLYS** Analysis 分析

**ANM** Announcement Machine 录音通知机

Automatic Number Machine 自动编号机

**ANMI** Air Navigation Multiple Indicator 航行多路显示器

Animal 动物

**ANMP** Account Network Management Program 账目网络管理程序

**ANMS** Aircraft Navigation and Management System 飞机导航和管理系统

Automated Network Monitoring System 自动化网络监控系统

**ANN** Announcement 通知，通报

Annual 年度

Annul 注销

Annunciator 信号牌（信号器）

Artificial Neural Network 人工神经网络

**ANN LT** Annunciator Light 信号灯

**ANNCE** Announce 通知，声明，通报

**ANNCMT** Announcement 通知，声明

**ANNFLYQUIRE** Annual Flying Requirements 年度飞行要求

**ANNIV** Anniversary 周年纪念

**ANNMENT** Announcement 通知,通报

Annulment 注销

**ANNUN** Annunciator 信号牌,告示牌

**ANO** Air Navigation Office 领航处

Air Navigation Order 空中领航命令,航行指令,航行规程

Analog Output 模拟输出

**ANOC** Authorized Notice of Change 批准的更改通知

**ANOPP** Aircraft Noise Prediction Program 飞机噪声预测程序

**ANORE** Aircraft Not Operationally Ready for Equipment 飞机因装备不全不能飞行

**ANP** Actual Navigation Performance 实际导航性能

Air Navigation Plan 空中航行计划

**ANPRM** Advanced Notice of Proposed Rule Making 建议立法的预先通告

**ANPT** Aeronautical National Taper Pipe Thread 航空国家锥形管螺纹

Air Navigation Procedures Trainer 空中导航程序训练器

**ANR** Air Navigation Regulation 空中导航规则

Airplane Nose Right 机头向右

Antwerp, Belgium-Deurne ( Airport Code ) 比利时安特卫普德尔讷机场

**ANREP** Annual Report 年度报告

**ANRPC** Association of Natural Rubber Producing Countries 天然橡胶生产国协会

**ANRS** Automatic Noise Reduction System 噪声自动减少系统

**ANS** Access Network System 接入网系统

Advanced Networks and Services 高级网络与服务

Air Navigation Services 空中导航服务

Air Navigation System 飞机导航系统

Airborne Navigation System 机载导航系统

Aircraft Navigation Simulator 飞机导航模拟器,飞机导航模拟机

Alternate Navigation System 备用导航系统

Ambient Noise Sensor 环境噪声传感器

American National Standard 美国国家标准

Ansett Airlines of Australia 澳大利亚安塞特航空公司

Area Navigation System 区域导航系统

Automatic Navigation System 自动导航系统

Automatic Noise Suppressor 自动噪声抑制器

Autonomous Navigation System 独立导航系统

**ANSA** Advanced Network System Architecture 高级网络系统体系结构

**ANSI** American National Standards Institute 美国国家标准协会

**ANSP** Adaptive Network Sensor Processor 自适应网络传感器处理装置

Air Navigation Service Provider 空中航行服务提供商

Air Navigation System Plan 空中导航系统计划

**ANSR** Air Navigation System Requirements 空中导航系统要求

**ANSS** Access Network Support System 接入网支持系统

**ANSU** Aircraft Network Server Unit 飞机网络服务器组件

**ANT** Abstract of New Technology 新技术摘要

Acoustical Noise Test 声学噪声测试

Antenna 天线

**ANTA** Anti-Aircraft 防空的

**ANTC** Advanced Network Test Center 高级网络测试中心

Air Navigation & Traffic Control 空中导航和交通管制

Air Navigation Technical Committee 航行技术委员会

**ANTCP** Anticipated 预计的,预期的

**ANTCPD** Anticipated 预期的

**ANTCPT** Anticipate 预期

**ANTHR** Another 另一个

**ANTLE** Affordable Near Term Low Emissions 经济的近期低排验证机

**ANTICOLL** Anti-Collision 防撞

**ANU** Aircraft Nose Up 飞机机头向上

Australian National University 澳大利亚国立大学

Saint Johns/Antigua, Antigua and Barbuda (Airport Code) 安提瓜和巴布达圣约翰/安提瓜机场

**ANUTY** Annuity 年金

**ANVG** Aviator's Night Vision Goggles 驾驶员夜视镜

**ANVIS** Aviation Night Vision Imaging System 航空夜视成像系统

Aviator's Night Vision Imaging System 飞行员夜视成像系统

**ANWT** Actual Net Weight 实际净重

**ANX** Annex 附录,附件

Anxiety, Anxious 忧虑

Automotive Network Exchange 自动网络交换

**ANXLY** Anxiously 忧虑地,切望

**ANYTHG** Anything 任何事情

**ANZ** Air New Zealand Ltd. 新西兰航空公司

Airport Noise Zone 机场噪音区

**AO** Access Opening 接近开口

Air Observer 空中观察员

Air/Oil 空气/滑油

Airworthiness Office 适航办公室

Announcement of Opportunity 商机公示

Assembly Order 装配指令

Assembly Outline 装配大纲

Automated Office 自动化办公室

**AOA** Aerodrome Owners Association 机场场主协会(英国)

Airborne Optical Adjunct 机载光学附属装置

Aircraft Operating Agency 航空器经营代理

Airport Owners Association 机场主协会

American Overseas Airlines 美国海外航空公司

Amphibious Objective Area 两栖目标地域

Angle of Arrival 到达角

Angle of Attack 迎角,攻角

**AOA/I** Angle of Attack Indicator 攻角指示器

**AOA/S** Angle of Attack System 按攻角调节的自动油门

**AOA/T** Angle of Attack Transmitter 攻角传感器

**AOAI** Indicated Angle of Attack 指示攻角

**AOAO** Advanced Orbiting Astronomical Observatory 先进的轨道天文观测台

**AOAS** Angle of Attack Sensor 迎角传感器

**AOAT** Allowed Off-Aircraft Time 允许离机时间

Angle of Attack Transducer 迎角传感器

**AOAWS** Angle of Attack Warning System 攻角报警系统

**AOB** Angle of Bank 坡度,倾斜角,滚转角

Any Other Business 其他任何事物

At or Below 在或低于

**AOC** Adaptive Optical Camouflage 自适应光学伪装

Advice of Charge 计费通知

Aerodrome Obstacle Chart 机场障碍物图

Aeronautical Operational Control 航空运行管理

Air Operating Certificate 航空使用证书

Air Operator's Certificate 航空营运人许可证

Air/Oil Cooler 空气/滑油冷却器

Aircraft Operating Center 飞机运行中心

Aircraft Operation Control 飞机运行控制

Aircraft out of Commission 飞机停飞,退出现役的飞机

Airlift Operations Center 空运经营中心

Airline Operation Center 航空公司运营中心

Airline Operational Communications 航空公司运营通信

Airline Operational Control 航空公司运营管

理,航空公司运营控制

Airline Operators Committee 航空业者联席会议

Airlines Operators' Committee 航空公司营运人协会

Airport Operational Communications 机场运营通信

Airport Operations Center 机场运行中心

Airport Operators Council 机场营运人协会

All-Optical Communication 全光通信

Approach on Course 沿航线进近

Attitude and Orbit Control 姿态和轨道控制

Automatic Operation Control 自动操作控制

Automatic Output Control 自动输出控制

Automatic Overload Control 自动超载控制

Autonomous Operating Capability 自主操作能力

Autopilot Omni-Coupler 自动驾驶仪总耦合器

**AOCC** Airline Operation Control Center 航空公司运行控制中心

**AOCI** Aircraft Owners Council International 航空器拥有人国际协会

Airport Operators Council International 机场经营人国际协会

**AOCM** Advanced Optical Countermeasures 高级光学对抗

Aircraft out of Commission for Maintenance 因维修而停飞的飞机

**AOCP** Airborne Operation Computer Program 航空运营计算机程序

Aircraft out of Commission for Parts 飞机缺件停飞

**AOCR** Advanced Optical Character Reader 先进的光符阅读器

Aircraft Operating Cost Report 飞机营运费用报告,飞机营运成本报告

**AOCs** Air Operator Certificates 航空营运人许可证

**AOCS** Airline Operational Communication System 航空公司运营通信系统

Airline Operational Control Society 航空公司运营管理学会(美国)

Attitude and Orbit Control System 姿态和轨道控制系统

Automatic Overload Control System 自动超载控制系统

**AOCV** Air/Oil Cooler Valve 空气/滑油冷却器活门

**AOD** Acousto-Optic Deflector 声—光反射器

Active Optical Device 有源光器件

Angle of Descent 下滑角

**AODB** Air Operation Database 航空运行数据库

Airport Operation Database 机场运行数据库

**AOE** Airport/Aerodrome of Entry 入境航站/机场

Appropriate Operational Equipment 适用的运行设备

**AOEC** Airways Operation Evaluation Center 航线经营鉴定中心

**AOF** Active Optical Fiber 有源光纤

**AOFS** Avionics Operational Flight Software 航空电子系统飞行操作软件

**AOG** Aircraft on Ground 飞机停飞待件

Airplane on Ground 飞机停飞待件

Arrival of Goods 货物到达

**AOGCM** Atmosphere-Ocean General Circulation Model 大气海洋环流模式

**AOH** Aircraft Operating Overhaul 进行大修的飞机

Average Operating Hour 平均运行时数

Aviator's Oxygen Helmet 飞行员供氧气罩

Awaiting Overhaul 等待大修

**AOHE** Air/Oil Heat Exchanger 空气/滑油热交换器

**AOI** Aircraft Operating Instruction 航空器操作说明

Airways Operation Instruction 航路使用说明

**AOIL** Aviation Oil 航空滑油

**AOIV** Automatically Operated Inlet Valve 自动操作的进气活门

**AOL** Aircraft Operating Limitations 飞机操作限制

All Operator Letter 致各用户信函

**AOLS** Airbus On-Line Services 空客在线服务

**AOM** Academy of Management 管理学会

Acousto-Optic Modulator 声光调制器

Aerodrome Operating Minimum 机场最低飞行条件

Aircraft Operating Manual 航空器使用手册

Airfield Operation Manual 机场运行手册

Airline Operation Manual 航空公司的运行手册

Annual Operational Maintenance 年使用维护

**AOMT** Airbus Onboard Maintenance Tool 空客机载维护工具

**AON** Active Optical Network 有源光网

All Optical Network 全光网络

Automated Optical Navigation 自动光学导航

**AOO** Altoona/Martinsburg, PA, USA (Airport Code) 美国宾夕法尼亚州阿尔托纳机场

**AOP** Airborne Optical Platform 机载光学平台

Airbus Operational Plan 空客运行计划

Aircraft Operational Program 飞机操作计划

All Other Persons 其他人员

Analog Output 模拟输出

Annual Operating Plan 年度运行计划

Automatic Operations Panel 自动操作仪表板,自动运营专家组

Avionics Operational Program 航空电子设备操作计划

**AOP/SG** Aerodrome Operational Planning Sub-Group 机场运营规划组

**AOPA** Aircraft Owner & Pilot Association 航空器拥有者及飞行员协会

**AOPAA** Aircraft Owners & Pilots Association of Australia 澳大利亚航空器业主与飞行员协会

**AOPG** Aerodrome Operations Group 机场运营组

**AOPR** Auburn Overhead Performance Reporting 奥本顶板性能报告

**AOQ** Average Outgoing Quality 平均输出质量,平均检出质量

**AOQL** Average Outgoing Quality Limit 平均抽检质量极限

**AOR** Area of Responsibility 责任区,责任区域

Atlantic Oceanic Region 大西洋海域

**AORE** Atlantic Ocean Region East 东大西洋区域

**AORW** Atlantic Ocean Region West 西大西洋区域

**AOS** Acquisition of Signal 信号捕获

Advanced Operating System 高级操作系统

Aircrew Oxygen System 机组氧气系统

Airport Operation System 机场运行系统

**AOSC** Administrative and Operational Services Costs 行政和业务服务费用

**AOSE** Aircraft Operational Support Equipment 飞行运行保障设备

**AOSS** Airborne Oil Surveillance System 机载滑油监视系统

**AOT** Acquisition on Target 目标截获

Active on Target 对目标的作用

Airports of Thailand 泰国机场

Angle on Target 目标角度

**AOV** Air Operated Valve 气动活门(阀)

Automatically Operated Valve 自动操纵活门(阀)

**AOWS** Aircraft Overhaul Work Stoppage 飞机翻修工作停止

**AOY** Angle of Yaw 偏航角

**AP** Accelerometer Package 加速计组件

Access Panel 舱盖

Access Point 存取(进入)点

Access Port 接近口,检查窗(口)

Access Protocol 访问协议

Account Paid 付现款,现款已付

Accounts Payable 应付账

Acid Proof 耐酸的

Acquisition Point 发现点,捕获点

Action Point 作用点

Adapter Panel 转接器控制站

Additional Premium 附加保险费

Administrative Procedures 管理程序

Aerial Photography 空中照相术

Aerial Port 航空站,飞机场,航空港

Aeronautical Planning 航行计划

After Peak 高峰(期)后

After Perpendicular 后垂线

Aiming Point 瞄准点

Air Passage 风道,气路

Air Pilot 飞行员,飞行驾驶员

Air Pipe 空气管(道)

Air Police 空中宪兵

Air Policeman 空中警察

Air Pollution 空气污染

Air Position 空中位置,无风位置,无风点

Air Pressure 空气压力

Air Publication 航空刊物,航空出版物

Airborne Printer 机载打印机

Airplane Performance 飞机性能

Airplane 飞机

Airport Park 停机坪

Airport 机场

Alarm Panel 告警信号板

All Passenger 全客运

Analog Processing 模拟处理

Analysis Point 分析点

Appendix 附件,附属物

Applications Processor 应用处理机

Approach Lights 进近灯光

Approach 进近,进场

Area Planning 地区计划

Argument Programming 论证程序

Array Processor 矩阵处理器

Ascent Phase 上升阶段

Associative Processor 联合(信息)处理器

Assumed Position 假定位置,推算位置

Atlantic and Pacific 大西洋和太平洋

Atmospheric Pressure 大气压

Attack Plane 攻击机,强击机

Attack Plotter 攻击标图板

Automatic Parallel 自动平行

Automatic Pilot 自动驾驶

Autopilot 自动驾驶仪

Auxiliary Power 辅助供电

Avionics Processor 航空电子设备处理器

Awaiting Parts 等待零件

Axial Pitch 轴向齿距,轴节

**APA** Air Pathway Analysis 航路分析

Aircraft Performance Analysis 飞机性能分析

Airline Pilot's Association 航空公司飞行员协会

Allied Pilots Association 飞行员联合会,飞行员同盟会

Antenna Pattern Analyzer 天线方向图分析仪

**APAA** Asia-Pacific Airlines Association 亚太航空公司协会

**APAC** Aircraft Production Advisory Council 飞机生产咨询委员会

Asia and Pacific 亚洲和太平洋

Asia Pacific Airport Consultants 亚太机场顾问集团

**APACS** Airplane Position & Attitude Camera System 飞机位置与姿态摄影机系统

**APANPIRG** Asia Pacific Air Navigation Planning and Implementation Regional Group 亚洲/太平洋空中航行规划和实施地区小组

**APAPI** Abbreviated Precision Approach Path Indicator 简易精密进近航道指示器

**APAR** Authorized Program Analysis Report 授权的程序分析报告

Automatic Precision Approach Radar 自动进近精密雷达

Automatic Programming & Recording 自动编程与记录

**APARMO** Asia/Pacific Approvals Registry and Monitoring Organization 亚太地区批准注册和监控系统

**APAS** Aerodynamic Preliminary Analysis System 空气动力学初步分析系统

Automated Pilot Advisory System 自动的飞行员咨询系统

**APASHE** Aircraft Publication Automated Shipping Expedite 飞机出版物自动发货

**APATS** Asia Pacific Airline Training Symposium 亚太区航空公司培训研讨会

**APATSI** Airport/Air Traffic System Interface 机场/空中交通系统界面

**APB** Auxiliary Power Breaker 辅助电源断路器,辅助电源跳开关

**APC** Accelerated Provision/Provisioning Concepts 加速物资供应概念

Advanced Process Control 先进的过程控制

Advanced Professional Certificate 高级资格认证

Aeronautical Passenger Communication 航空旅客通信

Aeronautical Planning Chart 航行计划图

Aeronautical Public Correspondence 航空公用信件

Air Pollution Control 机场污染控制

Aircraft Pilot Control 飞机驾驶员控制

Aircraft Pilot Coupling 飞机飞行员耦合

Aircraft Position Chart 飞机位置图

Airport Club 机场俱乐部

Airport Code 机场代码

Approach Control 进近管制

Approach Power Compensator 进近时功率补偿器

Approach Power Control 进近功率控制

Area Positive Control 绝对管制区,区域绝对管制

Automatic Power Control 自动功率控制

Automatic Phase Control 自动相位控制

Automatic Pressure Controller 压力自动调节器

Auxiliary Power Contactor 辅助电源接触器

Auxiliary Power Control 辅助电源控制

**APCH** Approach 进近

**APCHG** Approaching 进近

**APCM** Adaptive Pulse Code Modulation 自适应脉码调制

**APCNDT** Asian Pacific Committee on Non-Destructive Testing 亚太地区无损检测委员会

**APCOA** Airport Parking Company of America 美国机场停机坪公司

**APCR** Auxiliary Power Control Relay 辅助电源控制继电器

**APCS** Air Photographic and Charting Service 航空摄影和制图服务

Approach Path Control System 进近航路控制系统

**APCU** Auxiliary Power Control Unit 辅助动力控制组件

**APD** Advanced Pneumatic Detector 先进的气动检测器

Air Passenger Duty 航空客运税

Aircraft Performance Data 飞机性能数据

Airlines Prorata Directory 航空公司分摊指南

Airport Planning Document 机场规划文件

Approach Progress Display 进近程序显示,进近过程显示器

Approved 已批准,已通过

Automated Problem Detection 自动问题探测

**APDAS** Airborne Preflight Data Analysis System 机载飞行前数据分析系统

Airborne Programmable Data Aquisition System 机载可编程数据采集系统

**APDF** Aircraft Program Data File 飞机程序数据文件(档案)

**APDL** Advanced Programs Development Laboratory 高级程序开发实验室

**APDML** Airplane Product Data Markup Language 飞机产品数据标记语言

**APDS** Aircraft Power Distribution System 飞机电源分布系统

Auxiliary Power Distribution System 辅助电源分布系统

**APDU** Application Protocol Data Unit 应用协议数据单元

**APE** Adaptive Predictive Encoding 自适应预测编码

Advanced Process Equipment 先进的工艺设备

Advanced Production Engineering 先进的生产工程

Animation Production Environment 动画制造环境

Automated Production Equipment 自动生产设备

Automatic Photomapping Equipment 自动光电绘图设备

Autopilot Engaged 自动驾驶仪已接通

**APEC** Asia-Pacific Economic Cooperation 亚太经济合作组织,亚太经贸合作组织

**APEP** Airplane Performance Evaluation Program 飞行性能评估程序

**APERC** Asia Pacific Energy Research Center 亚太能源研究中心

**APERT** Aperture 孔径

**APES** Airline Performance Evaluation Systems 航线性能评估系统

**APESS** Advanced Passenger Entertainment Service System 先进的旅客娱乐服务系统

**APEX** Advanced Passenger Excursion Fare 优惠的旅客旅游票价,预订订购旅行票价

Analysis of Performance and Expenditure 性能与费用分析

**APF** Accurate Position Finding 精确定位

Airline Policy File 航空公司政策文档

Approach Control Function 进近管制功能

Australian Parachute Federation 澳大利亚跳伞联合会

Naples, FL, USA ( Airport Code) 美国佛罗里达州那不勒斯机场

**APFA** Association of Professional Flight Attendants 职业空乘人员协会

**APFD** Autopilot Flight Director 自动驾驶飞行指引仪

**APFDS** Autopilot Flight Director System 自动驾驶飞行指引仪系统

**APG** Air Pressure Gauge 气压计

Airplane General 飞机概述

**APGC** Auxiliary Power Generator Control 辅助动力发电机控制

**APGCU** Auxiliary Power Generator Control Unit 辅助动力发电机控制组件

**APGS** Auxiliary Power Generation Subsystem 辅助电源产生分系统

**APH** Access Permit Holder 进入许可证持有者

Accounting Procedures Handbook 会计程序手册

**APHMWG** Air Passenger Health Multi-Disciplinary Working Group 航空旅客健康多学科工作组

**API** Accurate Position Indicator 精确位置指示器

Advanced Passenger Information 先进的旅客信息(系统)

Air Pollution Index 空气污染指数

Air Position Indicator 空中位置指示器

Aircraft Parameter Identification 飞机参数识别

Aircraft Proximity Index 航空器进近指数

American Petroleum Institute 美国石油学会

Analog Parameter Input 模拟参数输入

Angle( Angular) Position Indicator 角度位置指示器

Application Implementation 应用软件执行

Application Program( ming) Interface 应用程序接口

Application Programmers Instruction 应用程序说明

Atmospheric Pressure Ionization 大气压电离

Automatic Position Indicator 自动位置指示器

Aviation Partners Inc. 航空伙伴公司

**APIC** Advanced Programmable Interrupt Controller 高级可编程中断信号控制器

Automatic Power Input Controller 自动功率输入控制器

Auxiliary Power International Commission 辅助动力国际协会

**APICS** American Production & Inventory Control Society 美国生产与库存管理协会

**APII** Asia Pacific Information Infrastructure 亚太信息基础设施

**APIL** Avionics Prototype Integration Laboratory 航空电子设备原型集成实验室

**APIP** Airport Passenger Information Platform 机场旅客信息平台

**APIRG** AFI Planning and Implementation Regional Group 非洲—印度洋地区规划和实施地区组

**APIS** Advance Passenger Information System 旅客预检信息系统

Advanced Personnel Information System 高级人才信息系统

Approved Production Inspection System 批准的生产检验系统

Array Processing Instruction Set 矩阵处理指令组

**APISAT** Asia-Pacific International Symposium on Aerospace Technology 亚太航空航天技术国际研讨会

**APIWP** Approach Intercept Way Point 进近切入点

**APK** Air Park, Aircraft Park 停机坪

Amplitude Phase-Shift Keying 幅度相移键控

**APL** Abbreviated Flight Plan 简式飞行计划

Airplane 飞机

Airport Lights 机场灯光

Airworthiness Policy Letter 适航政策函

Allowance Parts List 许可的零部件清单

Applied Physics Laboratory 应用物理实验室

Approach Light 进近灯

Approval 批准

Approved Parts List 已批准的零件目录

April 四月

Assembly Parts List 组装零件清单

Automated Parts List 自动部件清单

Automatic Phase Lock 自动相位锁定

Automatic Program Location 自动程序定位

**APLC** Aircraft Power Line Conditioner 飞机供电线路调节器

Automated Parking Lot Control 自动停车场控制

**APLCBL** Applicable 适用的

**APLCTN** Application 申请,应用

**APLD** Applied 应用,申请

**APLSS** Aircrew Protective Life Support System 机组人员生命保障系统

**APLY** Apply 申请,应用

**APLYG** Applying 应用,申请

**APM** Aircraft Performance Monitoring 飞机性能监控

Aircraft Project Memorandum 飞机设计备忘录

Airport Manager 机场经理

Airport Planning Manual 机场计划手册

All Pilots Meeting 全体飞行员会议

Automated People Movers 自动旅客输送

**APMA** Area of Project Management Application 项目管理的应用领域

**APMCU** Autopilot Monitor & Control Unit 自动驾驶仪监控装置

**APMP** Aircraft Performance Monitoring Program 飞机性能监控方案

**APMS** Advanced Power Management System 高级动力管理系统

Airborne Parametric Measurement System 机载参数测量系统

Airborne Particle Monitoring System 机载质点监控系统

Automated Performance Measurement System 自动性能测量系统

Automated Publications Maintenance System 自动出版物维护系统

Automatic Performance Management System 自动性能管理系统

Automatic Power Management System 自动能

源管理系统

Aviation Performance Measuring System 航空绩效测量系统

**APMT** Apartment 公寓

Appointment 任命

Asia-Pacific Satellite Mobile Telecommunication 亚太卫星移动通信

**APN** Aircraft Pulse Navigation 飞机脉冲导航

Alpena, MI, USA ( Airport Code) 美国密歇根州阿尔皮纳机场

Apron 停机坪

**APNT** Appoint 任命,约定

**APO** Airport Office 机场办公室

Annual Program Objectives 年度计划目标

**APOD** Aerial Port of Debarkation 下机航空港,卸货机场

**APOE** Aerial Port of Embarkation 上机航空港,装货机场

Air Port of Embarkation 装货机场

**APP** Advanced Planetary Probe 先进的行星探测器

Air Parcel Post 航空包裹邮寄

Airport Passenger Processing 机场旅客处理

Angle Pointing Processor 角度瞄准处理机

Apparatus 仪器,器械

Appearance 出现,外貌

Appendix 附录,附属物

Approach Control Office 进近管制室,进近管制处

Approach Control Service 进近管制服务

Approach Control 进近管制

Approach Pattern 进近航线

Approach Service 进近调度勤务

Approach 进近

Automatic Plate Processor 自动板加工器,自动板处理器

Autopilot Panel 自动驾驶仪操纵板

**APPA** Advise Present Position and Altitude 通知现在的位置与高度

American Powered Parachute Association 美国动力跳伞协会

**APPAM** Association for Public Policy Analysis and Management 公共政策分析与管理协会

**APPAR** Apparatus 装置

**APPC** Automatic Power Protecting Circuit 电源自动保护电路

**APPD** Applied, Application 应用,申请

Approved 批准的

Aviation Personnel Planning Data 航空人员计划数据

**APPL** Application, Apply 申请,应用

Approval 批准

**APPMT** Appointment 任命

**APPN** Advanced Peer-to-Peer Networking 高级对等联网

Appropriation 拨款,预算

**APPO** Air Parcel Post 航空邮包

**APPR** Appear 出现

Appoximate 大约

Appraisal 评价,鉴定

Approach 进近

Approver 批准者

Automatic Pre-Programmed Routes 自动预先规划航线

Avionics Project Planning Review 航空电子项目规划评审

**APPRCT** Appreciate 感激,欣赏

**APPROX** Approximately 近似的,大致(约,概)

**APPRX** Approximate 大约

**APPS** Airport Passengers Processing System 机场旅客处理系统

Auxiliary Payload Power System 辅助业载电源系统,辅助有效载荷动力系统

**APPT** Appointment, Appoint, Appointed 任命,指出

**APPTD** Appointed 任命的

**APPTMT** Appointment 任命

**APPU** Asian-Pacific Postal Union 亚洲—太平洋邮政联盟

Asymmetry Position Pick off Unit 不对称位

置传感器组件

**APPV** Approve, Approval 通过,批准,任命

**APPVAL** Approval 通过,批准,任命

**APPX** Appendix 附录,附件

**APPXN** Approximation 大约,近似值

**APR** Acquisition Plan Review 采购计划评审

Airman Performance Report 驾驶员作业报告

Annual Percentage Rate 年百分利率

April 四月

APU Power Relay 辅助动力装置电源继电器

Automated Parts Release 自动化零件发放

Automated Power Reserve 自动功率储备

Automatic Performance Reserve 自动性能储备

Auxiliary Power Relay 辅助电源继电器

**APRA** Air Public Relations Association 航空公共关系协会(英国)

**APREQ** Appreciation Request 请求谅解

**APRI** Air Priority 航空优先权,空中优先权

**APRICOT** Asia Pacific Regional Internet Conference on Operational Technologies 亚太区域互联网操作技术大会

**APRNTC** Apprentice 见习员

**APROP** Appropriate 适当

**APRS** Autopilot Rate Sensor 自动驾驶速率传感器

**APRT** Airport 机场

**APRTLY** Apparently 显然地

**APRV** Approve 批准,通过

**APRVD** Approved 批准的,通过的

**APRVL** Approval 批准,通过

**APRX** Approximately, Approximate 大约

**APS** Absolute Pressure Sensor 绝对压力传感器

Access Points 存取点,进入点

Accessory Power Supply 辅助电源供给

Accessory Power System 辅助电源系统

Air Photography & Survey 空中摄影与观察(测量)

Air Pressure Switch 空气压力开关

Airborne Power Supply 机上电源

Aircraft Performance Sheet 飞机性能表

Aircraft Prepared for Service 待命飞机,准备执行任务的飞机

Airport Planning Standard 机场设计标准

Airway Planning Standard 航路规划标准

Antenna Pointing System 天线定向系统

APU Power Switch APU 电源电门

Automatic Protection Switching 自动保护切换

Auto-Pilot System 自动驾驶系统

Auxiliary Power Supply 辅助动力/电力供给

Auxiliary Power System 辅助动力系统

Auxiliary Propulsion System 辅助推进系统

Aviation Planning Service Ltd. 航空规划服务有限公司

Avionic Printing Specification 航空电子印刷规范

**APSD** Airport Standards Division 机场标准部

**APSG** After Passing 通过之后,飞经之后

**APSK** Amplitude Phase Shift Keying 振幅移相键控

**APSL** Alternate Parts Signature List 备用件签名清单

**APSR** Airport Surveillance Radar 机场监视雷达

Auxiliary Power Slave Relay 辅助电源随动继电器

**APT** Air Passenger Tariff 航空客运票价准则

Airline Passenger Tariff 客运手册,航空旅客票价

Airport 航空港,机场

Alarm Pointer 告警指针

Analog Program Tape 模拟程序带

Apartment 公寓

Appointment 任命

Asia Pacific Telecommunity 亚太电信组织

Automated Program Tool 自动程序设计工具

Automated Programming of Tools 机床自动程序编制

Automatic Picture Transmission 图像自动传

送系统

Automatically Programmed Tool 自动编程工具

**APTA** American Public Transportation Association 美国公共运输协会

**APTC** Airlines Pilot Training Centre 航空飞行员培训中心

Airport Traffic Controller 机场交通管制员

**APTMT** Apartment 公寓

Appointment 任命

**APTS** Aerial Profiling of Terrain System 航空地形剖面系统

Asia-Pacific Training & Simulation 亚太训练和模拟中心

Automatic Picture Transmission System 自动图像传送系统

**APTT** Aircrew Part-Task Trainer 空勤组部分任务练习器

**APTU** Auxiliary Power and Thrust Unit 辅助动力和推力装置

**APU** Accessory Power Unit 附属动力装置

Audio Playback Unit 音频回放组件

Audio Processing Unit 音频处理单元

Auxiliary Power Unit 辅助动力装置

Auxiliary Propulsion Unit 辅助推进组件

**APUC** Auxiliary Power Unit Controller 辅助动力装置控制器

Average Procurement Unit Cost 平均采购单位成本

**APUP** Auxiliary Power Unit Panel 辅助动力装置控制板

**APUT** Auxiliary Power Unit Test 辅助动力装置测试

**APV** Approach Procedures with Vertical Guidance 带垂直引导的进近程序

Approval, Approve 批准, 通过

Approved 批准的

Auto-Piloted Vehicle 自动驾驶运载器

**APVD** Approved 批准的, 通过的

**APW** Apia, Samoa-Faleolo ( Airport Code ) 西萨摩亚阿皮亚法雷奥机场

Area Penetration Warning 区域进入警告

Automatic Pitch Warning 自动俯仰警告

**APWS** Aircraft Proximity Warning System 飞机近地报警系统

Automation Placement Wiring Subsystem 自动化布局布线子系统

**APX** Appendix 附录, 附件

Approximately 接近

**AQ** Aircraft Quality 飞机质量

**AQAB** Air Quality Advisory Board 航空质量咨询委员会

**AQAD** Aviation Quality Assurance Directorate 航空质量保证管理局

**AQAFO** Aeronautical Quality Assurance Field Office 航空质量保证地方办公室

**AQC** Administrative Quality Control 管理质量控制

Airman's Qualification Card 飞行员合格卡

Aviator Qualification Certificate 飞行合格证

Aviator Qualification Course 飞行员训练课程

**AQCR** Air Quality Control Region 空气质量控制区

**AQD** Aeronautical/Aviation Quality Directorate 航空质量管理局

**AQE** Advanced Quality Engineer 先期质量工程师

Air Quality Engineering 空气质量工程

Airman Qualification Examination 飞行员资格考试

**AQI** Air Quality Index 空气质量指数

**AQL** Acceptable Quality Level 可接受质量标准, 可接受质量水平

Agreed Quality Level 协议质量水平

Average Quality Limit 平均质量限

**AQP** Advanced Qualification Programme 高级资格审定项目

Airworthiness Qualification Plan 适航资格审定计划

Approval Quality Procedure 品质承认程序

Arequipa, Peru ( Airport Code ) 秘鲁阿雷基

帕机场

**AQPSK** Aviation Quadrature Phase Shift Keying 航空90°相移键控

**A-QPSK** Aviation Quadrate Phase Shift Keying 航空四相相移键控

**AQS** Advanced Quality System 先进的质量体系

Airworthiness Qualification Specification 适航合格规范

Ambient Quality Standard 环境质量标准

**AQT** Airman Qualifying Test 飞行员合格测试

Aviation Qualification Test 航空审定试验

**AR** Acceptance Requirements 验收要求

Acceptance Review 验收评审

Accidental Report 事故报告

Address Register 地址寄存器

Aeronautical Radionavigation 航空无线电导航

Aeronautical Requirement 航空要求

Air Refueling 空中加油

Air Rescue 空中救援

Air Route 空中航路

Airplane Repair 飞机修理

Alternate Route 迂回路由

Altitude Rate 高度变化率

Annunciator Relay 信号转接器

Area 区域

Arrival Route 进场航路

Arrival 到达

Artificial Reality 人工现实

As Required 按需

Aspect Ratio 展弦比

Audio Reproducer 放音机

**AR&O** Airplane Requirements and Objectives 飞机要求与目标

**ARA** Aerospace Research Association 航空航天研究协会(美国)

Airborne Radar Approach 机载雷达进近着陆

Airborne Radar Attachment 机载雷达附件

Aircraft Replaceable Assembly 飞行器更换装配

Attitude Reference Assembly 姿态参考组件

Office of Research and Acquisition 研究与数据采集办公室

**ARAA** Airport Radar Approach Aids 机场雷达进近设备

**ARAC** Acid Rain Advisory Committee 酸雨咨询委员会

Atmospheric Release Advisory Capability 大气排放咨询能力

Aviation Rulemaking Advisory Committee 航空规章制定咨询委员会

**ARAIC** Aircraft and Railway Accidents Investigation Commission 航空与铁路事故调查委员会

**ARB** Air Register Board 航空登记处

Air Registration Board 航空注册局(英国)

Air Research Bureau 航空研究局

Air Reserve Base 航空备用基地

Air Resources Board 大气资源委员会(美国)

Aircraft Recovery Bulletin 飞行器回收公报

Airport Rotating Beacons 机场旋转灯标

Analysis Review Board 分析评审委员会

Ann Arbor, Michigan (Airport Code)(美国)密歇根州安阿伯机场

Application Review Board 应用评审委员会

Arbiter 仲裁人

Arbitration 仲裁

**ARBS** Automatic Radar Beacon Sequencer 自动雷达信标程序装置

**ARC** Acquisition Review Council 采购评审委员会

Administrative Radio Conference 无线电管理会议

Advanced Research Center 预研中心(美国)

Aeronautical Research Council 航空研究委员会(英国)

Air Refueling Computer 空中加油计算机

Air Rescue Centre 航空援救中心

Airborne Radio Communication 航空无线电通信

Airline Reporting Corporation 航空报告公司

Airworthiness Review Certificate 适航复审合格证

Amateur Radio Club 业余无线电俱乐部

American Radio Co. 美国无线电公司

Area Reprogramming Capability 分区再编程能力

Area Reprogramming Concept 分区再编程原理

Authorized Release Certificate 批准放行许可证书

Automated Routing Control 自动程序控制

Automatic Range Control 自动距离控制

Automatic Remote Control 自动遥控控制

Automatic Rollout Control 自动滑跑控制

Autopilot Rate Control 自动驾驶速率控制

Auxiliary Rollout Control 辅助滑跑控制

Aviation Research Centre 航空研究中心

Aviation Rulemaking Committee 航空立法委员会

**ARCAL** Aircraft Radio Control of Aerodrome Lighting 航空无线电控制式机场灯光

**ARCN** Aeronautical Radio of Canada 加拿大航空无线电台

**ARCO** Airspace Reservation Coordination Office 空域保留协调办公室

**ARCON** Automatic Rudder Control 方向舵自动控制

**ARCP** Aerodrome Reference Code Panel 机场参考码专家组

Air Refueling Check Point 空中加油检查站

Air Refueling Control Point 空中加油控制点

**ARCS** Airbus Representative Communications System 空客典型通信系统

Automated Routing Connection Service 自动航程衔接服务

**ARCT** African Regional Centre for Technology 非洲地区技术中心

Air Refueling Control Team 空中加油控制小组

Air Refueling Control Time 空中加油控制时间

**ARCTS** Automated Radar-Controlled Terminal System 自动化雷达控制终端系统

**ARD** Acceptance Requirement Document 验收要求文件

Advanced Research and Development 预先研究与开发

Airplane Recovery Document 飞机恢复文件

Applied Research Department 应用研究部

Approach Reference Datum 进近基准点

**ARDS** Advanced Remote Display System 先进的遥控显示系统

Airborne Radar Demonstration System 机载雷达演示系统

Aviation Research and Development Service 航空研究与发展局

**ARE** Aircraft Reactor Equipment 航空反应堆设备

Aircraft Reactor Experiment 航空反应堆实验

**AREP** Atmospheric Research and Environmental Programme 大气研究和环境计划

**ARES** Airplane Responsive Engine Selection 飞机发动机适应性选择

**ARF** Access Link Relay Function 接入链路转接功能

Airline Risk Factor 航空公司风险因素

Airport Reservation Function 机场预定使用任务

Andean Reserve Fund 安第斯储备基金会

ASEAN Regional Forum 东盟地区论坛

Aviation Research Foundation 航空研究基金

**ARFCN** Absolute Radio Frequency Channel Number 绝对无线频率信道数

**ARFF** Aircraft Rescue and Fire Fighting 飞机救援和消防

Aircraft Rescue and Firefighting Foam 飞机救援与消防泡沫

Aircraft Rescue Firefighting Facility 飞机救

援消防设施

Airport Rescue and Fire Fighting 机场救援与消防

Aviation Rescue and Fire Fighting 航空救援和消防

**ARFL** Air Refuel 空中加油

**ARFOR** Area Forecast 区域(天气)预报

**ARG** Accident Review Group 事故审查组

Aerial Refueling Group 空中加油组

Air Rescue Group 航空救援组

Amphibious Ready Group 两栖备用小组

Argent 银色

Argentina 阿根廷

Aviation Readiness Group 航空备用小组

**ARGF** Ancillary Revenue Generation Fund 辅助创收基金

**ARH** Active Radar Homing 主动雷达自导引

Arkhangelsk, Russia ( Airport Code ) 俄罗斯阿尔汉格尔斯克机场

Armed Reconnaissance Helicopter 武装侦察直升机

**ARI** Aileron/Rudder Interconnect 副翼/方向舵互连

Airborne Radio Installation 机载无线电装置

Arica, Chile ( Airport Code ) 智利阿里卡机场

Automatic Rudder Interconnect 自动方向舵交连

**ARIA** Advanced Range Instrumentation Aircraft 高级测距仪表飞机

**ARIAL** Automatic Radar Identification Analysis & Alarm 自动雷达识别分析和报警

**ARINC** Aeronautical Radio Incorporated 航空电信公司

Aeronautical Research Incorporated 航空研究公司

Aircraft Radio Incorporated 航空器无线电公司

**ARINF** Arrival Information 到达信息

**ARIP** Air Refueling Initial Point 空中加油起始点

**ARJ** Advanced Regional Jet 先进的支线喷气飞机

**ARJ21** Advanced Regional Jet for the 21st Century 面向 21 世纪的先进的涡扇支线飞机

**ARJS** Airborne Radar Jamming System 机载雷达干扰系统

**ARK** Air Rescue Kit 航空急救包

Arusha, Tanzania ( Airport Code ) 坦桑尼亚阿鲁沙机场

**ARL** Acceptable Reliability Level 可接受可靠性标准

Aerospace Research Labs 空间研究实验室

Aircraft Readiness Log 飞机备用状态记录

Applied Research Laboratory 应用研究实验室

Aviation Research Laboratory 航空研究实验室

**ARLB** Automatic Radio Location Beacon 自动无线电定位信标

**ARM** Active Risk Manager 活跃的风险经理

Aircraft Recovery Manual 航空器恢复手册

Aircrew Resource Management 机组资源管理

Airworthiness Review Meeting 适航评审会议

Arming 准备,预位

Asynchronous Response Mode 异步响应模式

Automated Release Module 自动释放模块

Availability, Reliability and Maintainability 可用性,可靠性和可维护性

**ARMAC** Airport Resource Manager Air Canada 加拿大航空公司机场资源经理

**ARMC** American Risk Management Company 美国的风险管理公司

Area Regional Maintenance Center 区域支线维修中心

**ARMD** Aeronautics Research Mission Directorate 航空研究任务理事会

Armed 预位的

**ARMG** Arming 预位,待命

**ARML** Airmail 空邮

**ARMLD**　Airmailed 空邮的

**ARMMS**　Automatic Reliability and Maintenance Management System 自动可靠性和维修管理系统

**ARMS**　Accounting Record Management System 会计记录管理系统

Aircraft Reporting and Monitoring System 飞机的报告和监控系统

Automated Requirements Management System 自动化要求管理系统

Automatic Records Management System 自动记录管理系统

**ARMTS**　Advanced Radar Maintenance Training Set 先进的雷达维护训练设备

**ARN**　Aeronautical Radio Navigation 航空无线电导航

Air Request Net 空中救护联系网

Air Traffic Services Route Network 空中交通服务航路网络

Airborne Radio Navigation 机载无线电导航

Aircraft Registration Number 航空器注册号

Area Route Network 地区航路网

Arlanda, Stockholm, Sweden ( Airport Code ) 瑞典斯德哥尔摩阿兰达机场

**ARNAV**　Area Navigation 区域导航

**ARND**　Around 周围, 在附近

**ARNG**　Arrange, Arrangement 安排

**ARNK**　Arrival Unknown 抵达班机不详

**ARNS**　Aeronautical Radio Navigation Service 航空无线电导航服务

**ARO**　Aerodrome Reporting Officer 机场报告员

After Receipt Order 收到指令之后

Airport Reporting Office 机场报告办公室

Airport Reservation Office 机场订票处

Annualized Rate of Occurrence 年度发生率

ATS Reporting Office 空中交通服务报告室

Automatic Range Only 自动测距仪

**AROC**　Air Rescue Operation Center 航空救援中心

**AROD**　Aerodrome Runway and Obstruction Data 机场跑道和障碍数据

**AROG**　Automatic Roll-out Guidance 自动着陆滑跑制导

**AROL**　Aviation Red Obstruction Light 红色航空障碍灯

**ARP**　Address Resolution Protocol 地址解析协议

Aerodrome Reference Point 机场基准点

Aeronautical Recommended Practice 航空推荐施工法

Air Report Point 空中报告点

Airplane Related Problem 飞机相关问题

Airport Reference Point 机场基准点

Alert Reliability Program 警戒可靠性程序

Analytical Rework Program 分析性返工程序

Argentine Peso 阿根廷比索

Avionics Reference Plate 航空电子基准板

Azimuth Reference Pulse 方位基准脉冲

**ARPA**　Advanced Research Projects Agency 高级研究规划局

**ARPANET**　Advanced Research Projects Agency Network 高级研究规划局网络

**ARPS**　Advanced Radar Processing System 先进的雷达处理系统

Advanced Real-Time Processing System 高级实时处理系统

Advanced Regional Prediction System 先进的区域预报系统

**ARPT**　Airport 机场

**ARQ**　Annual Review Questionnaire 年度评审调查表

Automatic Error Correction 自动误差修正

Automatic Repeat Request 自动重复请求

**ARR**　Absolute Risk Reduction 绝对风险降低

Accounting Rate of Return 投资报酬率

Airborne Radio Relay 航空无线电中继通信

Airborne Reference Radar 机载基准雷达

Aircraft Radio Regulations 飞机无线电规章

Aircraft Radio Relay 飞机无线电中继通信

Annunciator Reset Relay 信号复位转接器

Arrangement 安排

Arrival Message 到达电报

Arrival 进港,到达,进场

Arriving 进港,到达

**ARRB** Audit Results Review Board 审计结果审查委员会

**ARREC** Accident Report Recommendations 事故报告介绍

**ARRGT** Arrangement 安排

**ARRNG** Arrange 安排

**ARROW** Aircraft Routing Right of Way(System) 飞机航线修正(系统)

Airworthiness Certificate, Radio Station Licence, Registration Certificate, Operating Limitations,Weight and Balance 适航证、电台执照、登记证、操作限制、重量与平衡

**ARRT** Aviation Regulation Review Taskforce 航空规章评审工作组

**ARS** Air Report Special 特殊空中报告

Airborne Ranging System 机载测距系统

Aircraft Recovery System 飞机搬运系统

Aircraft Relay Subsystem 飞机中继分系统

Airline Reservation System 航空电脑订位系统

Airworthiness Review Sheet 适航性审查单

Attitude Reference System 姿态参考系统

Auto Rate Selection 自动速率选择

Auto Response System 自动响应系统

Automatic Recovery System 自动复原系统

Automatic Reporting System 自动报告系统

Special Air-Report 特殊空中报告

**ARSA** Aeronautical Repair Station Association 航空修理站协会

Airport Radar Service Area 机场雷达服务区

**ARSAG** Aerial Refueling Systems Advisory Group 空中加油系统咨询组

**ARSD** Advanced Reentry System Deployment 高级再进入系统配置

Aviation Repair Supply Depot 航空修理补给站

**ARSR** Air Rescue Service Regulation 航空救援勤务条令

Air Route Surveillance Radar 空中航路监视雷达

**ARST** Arresting 系留,止步,拦截

**ARSTA** Air Station 航站

**ART** Abstract Reasoning Test 抽象推理测试

Active Repair Time 实际修理时间

Actuator Remote Terminal 作动器远程终端

Advanced Rotorcraft Technology 先进的旋翼技术

Airborne Radiation Thermometer 机载辐射温度计

Article 条款,项目,物品

Artificial 人工的,人造的

Automated Reasoning Tool 自动推论工具

Automatic Reverse Thrust 自动反推

Average Restoration Time 平均恢复时间

Aviation Radio Technician 航空无线电技师

Watertown, NY, USA (Airport Code)美国纽约沃特敦机场

**ARTAS** ATC Radar Tracker and Server 空中交通管制雷达跟踪和服务器

**ARTC** Aerospace Research & Testing Committee 宇航研究与试验委员会(美国)

Air Route Traffic Control 空中航路交通管制

Aircraft Research & Testing Committee 飞机研究与试验委员会(美国)

Automatic Radar Target Classification 雷达目标自动分类

**ARTCC** Air Route Traffic Control Center 空中航路交通管制中心

**ARTCL** Articles 物品

**ARTCLD** Articulated (用关节)连接的

**ARTCS** Advanced Radar and Traffic Control System 先进的雷达和空中交通管制系统

**ARTE** Above Runway Threshold Elevation 超出跑道入口高度

**ARTEMIS** Automated Reporting, Tracking and Evaluation Management Information System 自动报告,跟踪和评估管理信息系统

**ARTF** Artificial 人工的,人造的

**ARTI** Advanced Rotorcraft Technology Integration 先进的旋翼机技术一体化

**ARTIF** Artificial 人工的,人造的

**ARTS** Advanced Real-Time Simulation (System)先进的实时模拟(系统)

Air Route Traffic System 航路交通系统

Approach Radar Tracking System 进场雷达跟踪系统

Automated Radar Tracking System 自动雷达跟踪系统

Automated Resource Tracking System 自动资源跟踪系统

Automatic Radar Terminal System 自动雷达终端系统

**ARTU** Automatic Range Tracking Unit 自动距离跟踪装置

**ARU** Address Recognition Unit 地址识别装置

Aircraft Reception Unit 飞机验收单位

Aircraft Router Unit 飞机路由器装置

Altitude Rate Unit 高度变化率组件

Attitude Retention Unit 姿态保持装置

Automatic Range Unit 自动距离组件

**ARV** Acquisition Review Committee 采集审查委员会

Air Release Valve 放气阀

Alternate Refill Valve 备用加油活门

Arrive 到达

**ARVD** Arrived 已到达,抵达

**ARVL** Arrival 抵达

**ARW** Air Refueling Wing 空中加油联队

Arad, Romania( Airport Code) 罗马尼亚阿拉德机场

**ARWG** Aviation Regulatory Watch Group 航空法规监视组

**ARWR** Advanced Radar Warning Receiver 高级雷达警报接收机

Aviation Routine Weather Report 日常航空天气报告

**ARWS** Advanced Radar Warning System 先进的雷达报警系统

**ARWY** Airways 航线,航路,空路

**AS** Absent-Subscriber 缺席用户

Activity Scanning 活动扫描

Adapter Section 过渡段

Aerospace Standards 航宇标准

Air Service 航空服务

Air Start 空气启动

Air Station 航空站

Aircraft Standard 飞机标准

Airport Secretary 机场秘书

Airport Service 机场服务

Airscoop 集气口

Airspeed 空速

Altostratus 高积云

American Standard 美国标准

Antiskid 防滞

Asia 亚洲

Assured Service 确保服务

Authentication Server 认证服务器

Autonomous System 自治系统

Aviation Standards 航空标准

**ASA** Abort Sensor Assembly 失事传感装置

Advanced Surveillance Aircraft 先进的侦察飞机

Advanced System Avionics 先进的航空电子系统

Air Service Agreement 航空服务协定,航空服务协议

Airborne Separation Assurance 空中间隔保证

Aircraft Separation Assurance 飞机分离保障

Aircraft Surveillance Applications 航空器监视应用

Airline Suppliers Association 航空供应商协会

All Speed Aileron 全速副翼

Alternative Standard Approach 另外一种标准法

American Standard Association 美国标准协会

Anti-Static Additive 抗静电添加剂

As Soon As 立刻

Audio and Sign Adapter 音频与符号适配器

Austrian Space Agency 奥地利航天局

Autoflight Status Annunciator 自动飞行状态指示器

Autoland Status Annunciation 自动着陆状态通告

Autoland Status Annunciator 自动着陆状态信号牌

Automatic Spectrum Analyzer 自动光谱分析仪

Automatic Steerable Antenna 自动可操纵天线

Automatic System Analysis 自动系统分析

Automatic Ticketing Arrangement 自动购票装置

Aviation Safety Advisory 航空安全咨询

Aviation Security Audit(ASA) Section 航空保安审计处

Aviation Studies Atlantic 大西洋航空研究公司

Azimuth Servo Assembly 方位(角)伺服装置

**ASAAC** Allied Standard Avionics Architecture Council 标准航空电子结构联合协会

**ASAC** Airborne Surveillance and Area Control 空中监视和区域控制

Antisubmarine Air Controller 反潜空中控制员

Australian Statistics Advisory Council 澳大利亚统计咨询委员会

Automatic Supervisory Adaptive Control 自动监管适应控制系统

Aviation Security Advisory Committee 航空安全咨询委员会

Aviation System Analysis Capability 航空系统分析能力

**ASAE** American Society for Aerospace Education 美国航空航天教育学会

**ASAIB** Aerospace Safety Accident Investigation Board 航空航天安全事故调查委员会

**ASAP** Aerospace Safety Advisory Panel 航空航天安全咨询委员会

Aircraft Synthesis Analysis Programme 飞机综合分析计划

Application Service Access Point 应用业务接入点

As Soon As Possible 尽快

Automated Shipboard Aerological Programme 自动船载高空计划

Aviation Safety Action Partnership 航空安全行动合作系统

Aviation Safety Action Program 航空安全行动计划

Aviation Safety Analysis Program 航空安全分析程序

**ASAR** Advanced Synthetic Aperture Radar 先进的合成孔径雷达

**ASARS** Advanced Synthetic Aperture Radar System 先进的合成孔径雷达系统

**ASAS** Academic Standards and Assessment System 学术标准和评估体系

Aerodynamic Stability Augmentation System 气动稳定性增强系统

Airborne Separation Assurance System 机载间隔保障系统

All Source Analysis System 全源分析系统

Aviation Safety Analysis Subsystem 航空安全分析子系统

Aviation Safety Analysis System 航空安全分析系统

**ASAT** Anti-Satellite 反卫星

**ASB** Advisory Service Bulletin 咨询服务通告

Air Safety Board 航空安全局

Aircraft Safety Beacon 飞机安全信标

Aircraft Service Base 飞机维修基地

Alert Service Bulletin 警示服务通告

Area Service Box 区域服务电子盒

Asymmetric Switched Broadband 非对称宽带交换

Auditing Standards Board 审计准则委员会

Turkmenistan-Ashkhabad(Airport Cord) 土库曼斯坦阿什哈巴德机场

**ASBAA** Asian Business Aviation Association

亚洲商务航空协会

**ASBL** Assemble 集合,聚集,装配

**ASBLW** As Below 如下

**ASBR** Autonomous System Border Router 自治域边界路由器

**ASBU** Arab States Broadcasting Union 阿拉伯国家广播联盟

**ASC** Advanced Scientific Computer 先进的科学计算机

Advice of Schedule Change 航班计划更改通知

Air Supply Control 供气控制

Aircraft System Controller 飞机系统管制员

Aircraft Systems Computer 飞机系统计算机

Airspace Control 空域管制

American Standard Committee 美国标准委员会

Anti Skid Control 防滑控制

Audio Synthesizer Card 音频综合器卡

Aural Synthesizer Card 音频合成卡

Automatic Selectivity Control 选择性自动调节

Automatic Sensitivity Control 灵敏度自动调节

Automatic Switching Center 自动交换中心

Automatic Systems Control 自动系统控制

Aviation Safety Council 航空安全委员会

Aviation Statistic Center(加拿大)航空统计中心

**ASCB** Avionics Standard Communication Bus 航空电子标准通信总线

Avionics Synchronized Control Bus 航空电子同步控制总线

**ASCC** Aeronautical Satellite Communications Center 航空卫星通信中心

Aeronautical Services Communications Centre 航空局通信中心

Air Standards Co-ordinating Committee 航空标准协调委员会

Automatic Sequence Controlled Calculator 自动程序控制计算机

**ASCE** American Society of Civil Engineers 美国土木工程师学会

**ASCII** American Standard Code for Information Interchange 美国信息交换标准码

**ASCO** Airbus Service Company Inc. 空客服务有限公司

**ASCP** Air System Control Panel 大气系统控制板

Automatic System Checkout Program 自动系统检查程序

**ASCPC** Air Supply Cabin Pressure Controller 供气客舱压力控制器

**ASCS** Air Supply Control System 供气控制系统

Automatic Stabilization and Control System 自动稳定和控制系统

Aviation Satellite Communications System 航空卫星通信系统

**ASCTS** Air Supply Control & Test System 供气控制与试验系统

**ASCTU** Air Supply Control and Test Unit 供气控制和测试组件

**ASCU** Advanced System Control Unit 先进的系统控制组件

Air Support Control Unit 空中支援指挥组

Anti Skid Control Unit 防滑控制组件

Automatic Scanning Control Unit 自动扫描控制组件

**ASCV** Anti-Skid Control Valve 防滑控制活门

**ASD** Accelerate Stop Distance 加速停止距离

Aeronautical Systems Division 航空系统部

Airborne Situation Display 机载空情显示

Aircraft Situation Display 飞机状态显示

Aircraft Statistical Data 飞机统计数据

Assembly Start Data 装配开始日期

Automatic Synchronized Discriminator 自动同步鉴别器

**ASDA** Accelerate Stop Distance Available 可用加速停止距离

**ASDAR** Aircraft to Satellite Data Relay 飞

机—卫星数据中继

**ASDC** Airline Service Data Collection 航空公司服务数据采集

**ASDE** Air Situation Data Exchange 大气状况数据交换

Airborne Support Data Extension 机载支持数据的延伸

Airport Surface Detection Equipment 机场地面探测设备

Assistant State Design Engineer 设计工程师助理状态

**ASDL** Aeronautical Satellite Data Link 航空卫星数据链

Aerospace Systems Design Laboratory 航空系统设计实验室

**ASDR** Accelerate Stop Distance Required 所需停止加速距离

Aeronautical Systems Division Regulation 航空系统部条例

Airport Surface Detection Radar 机场地面探测雷达

**ASDS** Advanced Sensor Distribution System 先进的传感器分布系统

Air Situation Display System 空中状况显示系统

Aircraft-Sound Description System 飞机噪声描述系统

**ASDSS** Aviation Safety Data Sharing System 航空安全数据共享系统

**ASDU** Application-layer Service Data Unit 应用层服务数据单元

**ASE** Advanced Sensor Exploitation 高级传感器开发

Advanced System Engineering 高级系统工程

Airborne Search Equipment 机载搜索设备

Airborne Support Equipment 机载支援设备，机载辅助设备

Airbus Supplied Equipment 空中客车公司提供的设备

Aircraft Stabilization Equipment 飞机稳定设备

Aircraft Survivability Equipment 飞机救生装备

Allowable Steering Error 允许操纵误差

Altimetry System Error 测高系统误差

American Science & Engineering 美国科学与工程公司

Amplified Spontaneous Emission 放大自发辐射

Application Service Element 应用业务单元

Application System Entity 应用系统实体

Aspen, CO, USA ( Airport Code ) 美国科罗拉多州阿斯彭机场

Auto Slat Extension 自动缝翼伸出

Automatic Stabilization Equipment 自动稳定设备

**AS-E** Arrival Service-Extended 过站维护—扩展

**ASEAN** Association of Southeast Asian Nations 东南亚国家联盟(东盟)

**ASEB** Aeronautic and Space Engineering Board 航空航天工程局(美国)

Audio Seat Electronics Box 音频座椅电子盒

**ASECNA** Agency for Air Navigation Safety in Africa and Madagascar 非洲和马达加斯加空中航行安全机构

Agency for Air Navigation Safety in Africa 非洲航行安全局

**ASEL** Airplane, Single-Engine Land 单发陆地飞机

Applied Science and Engineering Laboratories 应用科学与工程实验室

**ASEP** Accident Sequence Evaluation Program 事故序列评价程序

Advanced Science Education Program 先进的科学教育计划

Advanced Skills Education Program 高级技能教育计划

Aircrew Survivability Enhancement Program 空勤机组生存力强化计划

Automatic Sequence Execution and Processor 自动顺序执行与处理机

**ASEQ** Aircraft & Special Equipment Qualification 飞机及专用设备质量检查,飞机及专用设备合格证明

**ASES** Airplane, Single-Engine Sea 单发水上飞机

Aviation Safety Exchange System 航空安全信息交换系统

**ASESS** Aviation Safety Exchange Support System 航空安全信息交换支持系统

**ASET** Advanced Sensor Evaluation and Test 先进的传感器评估与试验

Advanced System and Evaluation Technique 先进的系统与评估技术

Aeronautical Satellite Earth Terminal 航空卫星地面终端

Aeronautical Services Earth Terminal 航空勤务地面终端

Aircrew Standardization and Evaluation 机组标准化及评估

**ASF** Accelerometer Scale Factor 加速度计标度因数

Additional Selection Factor 附加选择因素

Advanced Simulation Facility 高级模拟设备

Aero Shell Fluid 壳牌滑油

Air Safety Foundation 航空安全基金(会)

Aircraft Services Facility 飞机维修设施

Aircraft Servicing Flight 飞机维修飞行(英国用法)

Airport Security Force 机场保安部队

Air-Supported Fiber 空气间隙光缆

Amperes per Square Foot 每平方英尺安培数

Application Support Facilities 应用保障设施

Auto Start Flag 自动启动旗标

Aviation Safety Forum 航空安全论坛

Aviation Support Facility 航空保障设施

Avionics Simulation Facility 航空电子模拟设施

**ASFC** Aero Sports Federation of China 中国航空运动协会

**ASFG** Aircraft Sales & Finance Group (USA) 飞机销售与财务集团(美国)

Atmospheric Sound Focusing Gain 大气声聚焦增益

**ASG** Advisory Services Guide 咨询服务指导

Aeronautical Standards Group 航空标准组

Airbus Security Group 空客安全小组

Application Software Group 应用软件组

Assign, Assignment 指定,分派

Automation Systems Group 自动化系统组

Auxiliary Starter Generator 辅助启动发电机

Aviation Security Group 航空保安组

Avionics Support Group 航空电子支援组

**ASGD** Assigned 指派

**ASGN** Assign 指定,分派

**ASH** Advanced Scout Helicopter 高级侦察直升机

**ASHI** After-Service Health Insurance 离职后健康保险

American Safety and Health Institute 美国安全与健康研究所

**ASI** Actuator Sensor Interface 执行器传感器接口

Advance Service Information 提前的服务信息

Air Speed Indicator 空速表,空速指示器

Airworthiness Staff Instruction 适航人员工作指南

Alternate Space Inversion 隔位空号翻转

Application Specific Input 应用指定输入

Automatic Source Interpolation 自动源插入

Aviation Safety Inspector 航空安全监察员

Aviation Safety Institute 航空安全研究所

Aviation Service, Inc. 航空服务公司

Avionics System Integration 航空电子系统集成

Azimuth-Speed Indicator 方位—速度指示器

**ASIA** Airlines Staff International Association 国际航空公司职工协会

Aviation Security Improvement Act 航空安全改进法案

**ASIAC** Aerospace Structures Information & Aeronautical Center 航空与航天结构信息分

析中心

**ASIAS** Aviation Safety Information Analysis and Sharing 航空安全信息分析及共享

**ASIC** Application Specific Integrated Circuits 特定用途集成电路

Area Security Information Center 区域安全信息中心

Aviation Security Identification Card 航空安全身份识别卡

Avionics Subsystems Interface Contractor 航空电子分系统接口合同方

**ASID** American Society of Industrial Designers 美国工业设计师协会

**ASII** American Science Information Institute 美国科学情报研究所

**ASIL** Avionics Systems Integration Laboratory 航空电子系统集成实验室

**ASIM** Automated Security Incident Measurement 自动安全事件评估

Automated Systems Information Management 自动系统信息管理

Automatic Security Incident Management 自动安全事件管理

**ASIP** Aircraft Structural Improvement Program 飞机结构改进程序

Aircraft Structural Integrity Plan 飞机结构完整性计划

Aircraft Structural Integrity Program 飞机结构完整性大纲

**ASIR** Aerodrome Special Information and Regulations 机场特殊资料和规定

Airspeed Indicator Reading 空速指示读数

**ASIS** Abort Sensing and Implementation System 故障传感与处理系统

American Society for Industrial Security 美国工业安全协会

American Society for Information Science 美国情报科/信息学学会

**ASIST** Advanced Scientific Instruments Symbolic Translator 高级科学仪器符号译码器

**ASIT** Adaptable Surface Interface Terminal 适配面接口终端

Advanced Security and Identification Technology 高级安全与识别技术

Airplane and Services Integration Team 飞机与服务整合组

**ASK** Amplitude Shift Keying 移幅键控

Available Seat Request 空位请求

Available Seat-Kilometre 可用座公里

Awareness, Skill and Knowledge 意识,技能和知识

**ASKD** Antiskid 防滑

**ASKG** Asking 询问

**ASL** Above Sea Level 海拔,在海平面以上,高海平面高度

Aeronautical Structures Laboratory 航空结构实验室

Atmospheric Science Laboratory 大气科学实验室

Average Service Life 平均使用寿命

Average Stage Length 平均航段长度

Aviation Systems Laboratory 航空系统实验室

Azimuth Steering Line 方位操纵线

**ASLC** Analog Subscriber Line Circuit 模拟用户线电路

**ASM** Advanced Server Management 先进的服务器管理

Advanced Systems Management 先进的系统管理

Air Separation Module 空气分离模块

Aircraft Schematics Manual 航空器简图手册

Aircraft Serviceability Message 飞机可用信息

Aircraft Structural Maintenance 飞机结构维修

Aircraft Systems & Manufacturing Inc. 飞机系统与制造公司

Airspace Management 空域管理

American Society for Metals 美国金属协会

Application Systems Maintenance 应用系统的维护

Area Sales Manager 地区销售经理

Asmara, Eritrea ( Airport Code ) 埃塞俄比亚阿斯马拉机场

Assembly 装配

Association of Systems Management 系统管理协会

Autothrottle Servo Motor 自动油门伺服马达

Available Seat Miles 可用座英里

Aviation Security Manual 航空安全保卫手册

**ASMA** Aerospace Medical Association 航空航天医学协会

Aircraft Systems Maintenance Aids 飞机系统维护援助

**ASME** Airport Surface Movement Equipment 机场地面活动目标显示设备

American Society of Mechanical Engineers 美国机械工程师协会

Aviation Support Material and Equipment 航空支援器材与设备

**ASMES** Air Space Management and Evaluation System 空域管理与评估系统

**ASMGCS** Airport Surface Movement Guidance & Control System 机场地面活动引导和控制系统

Advanced Surface Movement Guidance & Control System 先进的地面活动引导和控制系统

**ASMI** Airfield Surface Movement Indication 机场地面活动情况显示

Airfield Surface Movement Indicator 机场地面运动目标指示器

**ASMIS** Airport Security Management Information System 机场安全管理信息系统

Aviation Safety Management Information System 航空安全管理信息系统

**ASMO** Arab Standardization and Metrology Organization 阿拉伯标准化和计量组织

Environmental Assessment and Monitoring Committee 环境评估和监视委员会

**ASMS** Advanced Synchronous Meteological Satellite 先进的气象同步卫星

**ASMU** Automatically Stabilized Maneuvering Unit 自动稳定机动装置

**ASN** Allotment Serial Number 分配序列号

Assigned Subject Number 指定工程项目号

Aviation Safety Network 航空安全的网络

**AS-N** Arrival Service-Normal 过站维护—正常

**ASNT** American Society for Nondestructive Testing 美国无损检测协会

**ASO** Aeronautical Station Operator 航空站经营人

Aeronautics Supply Office 航空供应处

And So On 等等

Aviation Supply Office 航空供应处

**ASOC** Air Sovereignty Operations Centers 空中统治运行中心

Air Support Operations Center 空中支援运行中心

**ASOP** Automatic Scheduling and Operating Program 自动编排与操作程序

Aviation Safety Oversight Program 航空安全监督大纲

**ASOR** Aircraft Safety Occurrence Report 飞机安全事故报告

Allocation of Safety Objectives and Requirements 安全目标与需求分配

**ASOS** Automated Surface Observing System 自动场面观测系统

Automatic Storm Observation Service 风暴自动观测服务

Automatic Storm Observation Station 风暴自动观测站

**ASOSAI** Asian Organization of Supreme Audit Institution 亚洲最高审计组织

**ASOT** Airline Selectable Option Tool 航空公司可选择项工具

**ASP** Active Server Pages 动态服务器主页

Advanced Single Point 先进的单支点

Advanced Speech Processing 高级语音处理

Aircraft Self Protection 飞机自保护

Airfield Smart Power 机场智能电源

Analog Signal Processor 模拟信号处理器

Application Service Provider 应用服务提供商

Arrival Sequencing Program 顺序进场计划

Attendant Switch Panel 乘务员开关面板

Audio Selector Panel 音频选择面板

**ASPA** Advanced Strategic Penetrating Aircraft 先进的战略突防飞机

Aircraft Service Period Adjustment 飞机服务时间调整

Aircraft Structive Periodic Adjustment 飞机结构周期性调整

Annual Safety Performance Award 年度安全绩效奖励

Association of South Pacific Airlines 南太平洋航空公司协会

**ASPC** Analysis of Spare Parts Change 备件更换分析

**ASPEC** Aircraft Static Pressure Error Corrector 航空器静压误差校正器

**ASPH** Asphalt 沥青

**ASPJ** Advanced Self Protection Jammer 高级自我保护干扰发射机

Advanced Self-Protection Jamming 先进的自卫干扰

Airborne Self-Protection Jammer 机载自卫干扰机

**ASPM** Average Seats per Miles 平均每英里座位数

**ASPSU** Autonomous Standby Power Supply Unit 自主备用电源组件

**ASQ** Airport Service Quality 机场服务质量

**ASQP** Airline Service Quality Performance 航空公司服务质量性能

**ASQS** Application Specific Quality of Service 专用质量服务

**ASR** Access Service Request 访问服务请求

Advance Seat Reservation 机上座位预订

Aerodrome Surveillance Radar 机场监视雷达

Air Search Radar 空中探索雷达

Air Surveillance Radar 对空监视雷达

Airborne Surveillance Radar 机载监视雷达

Airport Surveillance Radar 机场监视雷达

Altimeter Setting Region 高度表拨正区

Annual Safety Report 年度安全报告

Approach Surveillance Radar 进场监视雷达

Area Surveillance Radar 区域监视雷达

Automated Speech Recognition 自动语音识别

Automatic Send-Receiver 自动发送接收机

Automatic Speech Retrival 自动语音检索

Aviation Safety Regulation 航空安全条例

**ASRD** Aircraft Shipment Readiness Date 飞机装载准备日期

**ASRS** Aerospace Safety Research System 航天安全研究系统

Air Surveillance and Range System 空中监视和区域系统

Air-Surveillance Radar System 航空监控雷达系统

Automated Software Reporting System 自动软件报告系统

Automated Support Requirements System 自动支援需求系统

Aviation Safety Reporting System 航空安全报告系统

**ASR-WSP** Airport Surveillance Radar Weather System Processor 机场监视雷达气象系统处理机

Airport Surveillance Radar Wind Shear Processor 机场监视雷达风切变处理机

**ASS** Air Signallers' School 空中报务学校

Air Suction System 空气吸入系统

Airborne Surveillance System 机载监视系统

Aircraft Security System 飞机警戒系统,飞机安全系统

Assembly 装配

Assess 估价

Assist, Assistance 帮助,协助

Assistant 助手

Attitude Sensing System 姿态传感系统

Audio Selector System 音频选择系统

Audio Subsystem 音频子系统

Automatic Synchronized System 自动准同期装置

Average Signal Strength 平均信号强度

**ASSA** Airport Surface Situational Awareness 机场场面态势感知

**ASSCE** Assurance 保险

**ASSD** Assessed 估价的

Assigned 指派的

Assorted 分类的

Assured 担保的

**ASSE** Air Servicing and Standard Equipment 航空服务与标准设备

Aircraft State Sensing/Estimating 飞机状态传感/估算

American Society of Safety Engineers 美国安全工程师协会

**ASSET** Advanced System & Software Engineering Technologies 先进的系统与软件工程技术

Aerothermodynamics Structural System Environmental Test 空气热力学结构系统环境测试

Airplane and System Integration Team 飞机及系统整合组

**ASSIMT** Assignment 指派,转让

**ASSIST** Assistance 帮助

Assistant 助手,副职

**ASSMT** Assessment 估价

Assignment 任务

Assortment 分类

**ASSN** Association 协会

**ASSOC** Association 协会,联合会

**ASSP** Aerospace System Susceptibility Program 航空与航天系统敏感性计划

Application Specific Standard Product 专用标准产品

**ASSR** Airport Surface Surveillance Radar 机场地面监视雷达

**ASSS** Airport Surface Surveillance System 机场地面监视系统

Avionics System Segment Specification 航空电子系统分区规范

**ASST** Advanced Supersonic Transport 先进的超音速运输机

Assent 同意

Assistant 助手,副职

Avionics Subsystem 航空电子分系统

Avionics System 航空电子系统

**ASSTC** Aerospace Simulation and Systems Test Center 宇航模拟与系统测试中心

**ASSTD** Assented 同意的

Associated 联合的

Assorted 各式各样的

**ASSV** Alternate Source Selection Valve 备用源选择活门

**ASSW** Associated with 有关于

**ASSY** Assembly 组装,组件

**AST** Advanced Simulation Technology 先进的模拟技术

Advanced Subsonic Technology 先进的亚音速工艺

Advanced Supersonic Transport 先进的超音速运输机

Airbus Supper Transporter 空客超级运输机

Aircraft System Trainer 飞机系统训练器

Asynchronous Shared Terminal 异步共享终端

At Same Time 同时

Atlantic Standard Time 大西洋标准时间

Authoring Software Tools 多媒体编辑软件工具

**ASTA** Advanced Software Technology and Applications 先进的软件技术及应用

Advanced Strategic Transport Aircraft 先进的战略运输机

Aerial Surveillance and Target Acquisition 航空监测和目标锁定

Aircrew Synthetic Training Aids 机组综合训练保障设备

Airport Surface Traffic Automation 机场地面交通自动化

Automatic System Trouble Analysis 自动系统故障分析

**ASTAR** Airborne Search Target Attack Radar 机载目标搜索袭击雷达

**ASTB** Advanced Survivability Test Bed 先进的抗毁试验台

American Standards Testing Bureau 美国标准测试局

Australian Safety Transport Bureau 澳大利亚安全运输局

**ASTC** Airport Surface Traffic Control 机场地面交通管制

Astronaut Selection and Training Centre 航天员选拔训练中心

**ASTCs** Aviation Security Training Centres 航空保安培训中心

**ASTD** Advanced Systems and Technology Department 高级系统与技术部

American Society for Training and Development 美国培训与发展学会

Assented 同意

**ASTE** Aerospace Systems Test Environment 航空航天系统试验环境

American Society of Test Engineers 美国试验工程师协会

American Society of Transportation Engineers 美国运输工程师协会

**ASTI** Airport Surface Traffic Indicator 机场地面交通指示器

**ASTM** American Society for Testing and Materials 美国材料与试验协会

American Standard of Testing Materials 美国材料试验标准

American Standard Test Manual 美国标准试验手册

American Standard Test Method 美国标准试验方法

**ASTO** Arab Satellites Telecommunications Organization 阿拉伯卫星电信组织

**ASTOR** Airborne Standoff Radar 机载防区外雷达

**ASTOVL** Advanced Short Take-Off & Vertical Landing Aircraft 先进的短距起飞和垂直着陆飞机

**ASTP** Aviation Security Training Package 航空保安成套培训材料

**ASTR** Astronomy 天文学

**ASTRAP** Panel on Application of Space Techniques Relating to Aviation 空间技术应用于航空问题研究小组

**ASTRO** Astrodynamics 宇宙飞行力学

**ASTRSK** Asterisk 星号

**ASTS** Airport Surface Traffic Simulator 机场地面交通模拟器

Automatic Stabilizer Trim System 自动安定面配平系统

**ASTT** Airborne Surveillance and Tracking Technology 机载监视与跟踪技术

**ASTU** Automatic Stabilizer Trim Unit 自动安定面配平组件

**ASTZCR** Average Short Time Zero Crossing Rate 短时平均过零率

**ASU** Aircraft Starting Unit 飞机启动装置

Aircraft Storage Units 航空器存储装置

Asuncion, Paraguay-Silvio Pettirossi (Airport Code) 巴拉圭亚松森锡尔维奥佩蒂罗西机场

Attendant Service Unit 乘务员服务组件

**ASUMD** Assumed 假设

**ASV** Acceleration Switching Valve 加速度转换门

Advanced Safety Vehicle 先进安全的飞行器

Aerothermodynamic Structural Vehicles 空气热动力结构(试验)飞行器(美国)

Air Solenoid Valve 空气电磁活门(阀)

Air Switching Valve 空气交换活门(阀)

Amboseli, Kenya (Airport Code) 肯尼亚安博塞利机场

Angle Stop Valve 角式截止活门

ATM-based Scalable Video 基于 ATM 的可控视频

Automatic Shuttle Valve 自动关闭活门(阀)

**ASVD**  Analog Synchronous Voice Data 模拟同步语音数据

**ASVR**  Aircraft System Validation Rig 飞机系统验证设施

**ASW**  Acoustic Surface Wave 声表面波

Aswan, Egypt ( Airport Code ) 埃及阿斯旺机场

**ASWACS**  Aerospace Surveillance Warning and Control System 航空航天预警和控制系统

Airborne Surveillance and Warning Control System 机载监视与警告控制系统

**ASWC**  Advanced Stall Warning Computer 先进的失速警告计算机

**ASWG**  American Standard Wire Gage 美国标准线规

Audit Software Working Group 审计软件工作小组

**ASYI**  Asynchronous Interface 异步接口

**ASYM**  Asymmetric, Asymmetry 不对称

Asymmetrical 非对称的

**AT**  Acceptance Test 验收试验

Acceptance Trial 接收试验

Action Time 操作时间, 作用时间

Advanced Trainer 高级教练机

Air Tariff 航空运价

Air Traffic 空中交通

Air Transport 空运

Air Turbine 空气涡轮发动机

Ambient Temperature 环境温度, 周围温度

Analog Trunk 模拟中继

Assembly and Test 装配测试

Atlantic 大西洋

Autothrottle 自动油门

Autothrust 自动推力

**AT&T**  American Telephone and Telegraph 美国电话电报公司

**ATA**  Actual Time of Arrival 实际到港时间

Advanced Tactical Aircraft 先进的战术飞机

Advanced Technology Aircraft 先进技术航空器

African Technical Association 非洲技术协会（技协）

Air Training Advisor 航空训练顾问

Air Transport Association 航空运输协会

Airport to Airport 机场到机场（服务）

Air-to-Air 空对空

American Transport Association 美国运输协会

Analog Terminal and Access 模拟终端和访问

Antenna Train Angle 天线方向角

Asynchronous Terminal Adapter 异步终端适配器

Automatic Ticketing Arrangement 自动购票装置

Automatic Trouble Analysis 自动故障分析

Aviation Training Association 航空训练协会

Avionics Test Article 航空电子试验项目

**ATAA**  Air Transport Association of America 美国航空运输协会

**ATAC**  Air Transport Advisory Council 航空运输咨询委员

Air Transport Association of Canada 加拿大航空运输协会

**ATACC**  Advanced Tactical Aircraft Control Capability 先进的战术飞机控制能力

**ATAD**  Advanced Technologies ASW Display 高技术美国线规显示器

Air Transport Avionics Division 航空运输航空电子部

Air Transportation & Delivery 空运与空投

Automatic Target Acquisition Detection 自动目标搜索探测

**ATAG**  Air Training Advisory Group 航空训练顾问团

Air Transport Action Group 航空运输执行组

**ATAM**  Airbus Takeoff Analysis Module 空客起飞分析组件

**ATAP**  ATA Procurement Data System 航空运输协会采购数据系统

**ATAR**  Acquisition Tracking and Recognition

采集跟踪与识别

Air Traffic Approval Regulations 空中交通批准条例

Automated Travel Agents Reservation 旅行社自动订票

**ATARF** ATA Reference Form 航空运输协会标准表格

**ATARS** Advanced Tactical Air Reconnaissance System 先进的战术空中侦察系统

Air Traffic Activity Reporting System 空中交通活动报告系统

Aircrew Training and Rehearsal Support 机组训练和演习支持

Aircrew Training and Rehearsal System 机组训练和演习系统

Air-to-Air Recognition System 空空识别系统

Anti-Terrain Avoidance Radar System 反地形回避雷达系统

Automated Traffic Advisory and Resolution Service 交通自动咨询和鉴别服务

**ATAS** Advanced Tactical Attack System 先进的战术攻击系统

Advanced Target Acquisition Sensor 高级目标搜索传感器

Advanced Technology Assessment System 先进的技术评估系统

Aerodynamic Threat Analysis System 气动威胁分析系统

Air Traffic Advisory Service 空中交通咨询服务

Air Traffic Advisory Supervisor 空中交通咨询监察员

Air Transport Auxiliary Service 航空运输辅助业务

Airborne Target Acquisition System 机载目标采集系统

Airborne Target Assessment System 机载目标评估系统

Automatic Test Analysis System 自动测试分析系统

Automatic Tracking Antenna System 自动跟踪天线系统

**ATASCC** Aviation Technical and Safety Co-operation Committee 航空技术安全协作委员会

**ATASS** Air Traffic Audio Simulation System 空中交通音频模拟系统

**ATB** Air Transport Bureau 航空运输局

Air Transportation Board 航空运输委员会

Air Turn Back 空中返航

Amphibious Training Base 水陆训练基地

Automated Ticket and Boarding Pass 自动打印客票和登机牌

Aviation Technical Bulletin 航空技术通报

**ATBM** Average Time Between Maintenance 平均维修间隔时间

**ATC** Accelerometer Test Complete 加速计测量完成

Active Transfer Command 有效转移指令

Actual Time of Completed 实际完成时间

Adaptive Transfer Code 自适应变换编码

Adaptive Transform Coding 自适应变换编码

Additional Technical Conditions 附加技术条件

Advanced Technology Center 先进技术中心

Aerodynamic Torque Converter 气动力扭矩变换器

Air Traffic Control 空中交通管制

Air Traffic Controller 空中交通管制员

Air Training Center 航空训练中心

Air Transport Committee 航空运输委员会

Air Transport Conference 空中运输会议

Air Travel Card 航空旅行卡

Airborne Test Conductor 机载试验导航

Airbus Training Center 空客培训中心

Aircraft Technical Committee 飞机技术委员会

Airport Traffic Control 机场交通管制

Airway Traffic Control 航线交通管制

Alert Transmit Console 警报发送控制台

Alloy-Tin Couple 铝合金电偶

Ambient Temperature Condition 环境温度条

件

Amended Type Certification 型号合格证更改

Annual Traffic Census 交通统计周年报告

Annular Turbojet Combustor 环行涡轮喷气发动机燃烧室

Approved Type Certificate 型号合格证,机型批准书

Atlantic Test Center 大西洋试验中心

Australian Transport Council 澳大利亚运输委员会

Automated Technical Control 自动技术控制

Automatic Target Counting 自动目标计数

Automatic Temperature Compensator 自动温度补偿器

Automatic Temperature Control 自动温度控制

Automatic Temperature Controller 自动温度控制器

Automatic Threshold Control 自动阀控制,自动临界值控制

Automatic Throttle Control 自动节流控制

Automatic Timing Control 自动计时控制

Automatic Timing Corrector 自动计时校正器

Automatic Tone Correction 自动音调调整

Automatic Tracking Control 自动跟踪控制

Automatic Tuning Control 自动调谐控制

Autothrottle Control 自动油门控制

Average Total Cost 平均总成本

Avionics Team Communication 航空电子组通信

**ATCA** Advanced Tanker Cargo Aircraft 先进的可空中加油货运飞机

Agreement on Trade-in Civil Aircraft 民用飞机贸易协定

Air Traffic Control Assistant 空中交通管制助理

Air Traffic Control Association 空中交通管制协会

Air Traffic Controllers Association 空中交通管制员协会

Air Training Corps of America 美国航空训练团

Attitude and Translation Control Assembly 姿态与平移控制装置

**ATCAA** Air Traffic Control Assigned Airspace 空中交通管制指定空域

**ATCAC** Air Traffic Control Advisory Committee 空管咨询委员会

**ATCALS** Air Traffic Control and Landing Systems 空中交通管制与着陆系统

**ATCAP** Air Traffic Control Automation Panel 空中交通管制自动化专家小组

**ATCAS** Air Traffic Control Automated System 空中交通管制自动化系统

**ATCB** Air Traffic Control Board 空中交通管制委员会

**ATCBI** Air Traffic Control Beacon Interrogator 空中交通管制信标询问机

**ATCC** Air Traffic Control Center 空中交通管制中心

Air Traffic Control Communication 空中交通管制通信

Airline Technical Control Center 航空公司技术控制中心

Airway Traffic Control Centers 航路交通管制中心

**ATCCC** Air Traffic Control Command Center 空中交通管制指挥中心

Air Traffic Control Coordination Center 空中交通管制协调中心

**ATCCMS** Air Traffic Control Command Monitoring System 空中交通管制指挥监测系统

**ATCCS** Air Traffic Command and Control System 空中交通指令与控制系统

**ATCD** Advanced Technology Concept Demonstration 高技术概念演示

Automatic Telephone Call Distribution 自动电话呼叫分配

**ATCDB** Aircraft Technical Characteristics Data Base 飞机技术特性数据库

**ATCE** Attitude and Translation Control Electronics 姿态与转换控制电子装置

Automatic Test & Checkout Equipment 自动测试设备

Automatic Test and Control Equipment 自动测试与控制设备

**ATCEU**   Air Traffic Control Evaluation Unit 空中交通管制评估单位

**ATCF**   Air Traffic Control Facility 空中交通管制设备

Air Traffic Control Flight 空中交通管制飞行

**ATCFAS**   ATC Flight Advisory Service 空中交通管制飞行咨询服务

**ATCFTSO**   Air Traffic Control Facility Training and Standardization Officer 空中交通管制设备培训和标准化官员

**ATCH**   Attach 附上,附后

Attachment 附件

**ATCHD**   Attached 附上,附后

**ATCHMT**   Attachment 附件

**ATCI**   Air Traffic Control and Information 空中交通管制和情报

**ATCL**   Air Traffic Control Line 空中交通管制线

Air Traffic Controller License 空中交通管制员执照

**ATCO**   Air Taxi and Commercial Operator 出租飞机与商业运输公司

Air Traffic Control Officer 空中交通管制员

Air Traffic Coordinating Office 空中交通协调办公室

Air Traffic Coordination Officer 空中交通协调员

Aviation Transportation Coordination Office 航空运输协调办公室

**ATCOMS**   Air Traffic Control Operations Management System 空中交通管制运行管理系统

ATC Operations Management System 空中交通管制运行管理系统

**ATCP**   Air Traffic Control Procedure 空中交通管制程序

**ATCPNL**   Air Traffic Control Panel 空中交通管制专家组

**ATCR**   Air Traffic Control Radar 空中交通管制雷达,航空雷达

**ATCRB**   Air Traffic Control Radar Beacon 空中交通管制雷达信标

**ATCRBS**   Air Traffic Control Radar Beacon System 空中交通管制雷达信标系统

**ATCS**   Advanced Train Control Systems 先进的训练控制系统

Air Traffic Communication Service 空中交通通信服务

Air Traffic Communication Station 空中交通通信站

Air Traffic Communication System 空中交通通信系统

Air Traffic Control Service 空中交通管制服务

Air Traffic Control Specialist 空中交通管制专家

Airway Traffic Control Stations 航路交通管制台(站)

**ATCSCC**   Air Traffic Control System Command Center 空中交通管制系统指挥中心

**ATCSS**   Air Traffic Control Signaling System 空中交通管制信号系统

**ATCT**   Air Traffic Control Tower 空中交通管制塔台

Airport Traffic Control Tower 机场交通管制塔台

ATC Transponder 空中交通管制应答机

**ATCTS**   ATC Transponder System 空中交通管制应答机系统

**ATCU**   Airway Traffic Control Unit 航路交通管制单位

ATC Unit 空中交通管制单位

**ATD**   Actual Time of Departure 实际离港时间,实际起飞时间

Advanced Technology Demonstration 先进技术验证

Advanced Technology Development 先进技术

发展

Advanced Training Device 先进的训练设备

Air Transportable Dock 可空运附件检修架

Air Turbine Drive 航空涡轮驱动

Aircraft Technical Definition 飞机技术定义

Aircraft Time & Depart 飞机时间与离港

Aircrew Training Devices 机组训练器件

Airport to Door 机场到用户服务

Along-Track Distance 沿航线距离

**ATD/C** Aided Target Detection/Classification 辅助目标探测与分类

Automatic Target Detection/Cueing 自动目标探测与识别

**ATDM** Asynchronous Time-Division Multiplexing 异步时分复用

**ATDMA** Advanced Time-Division Multiple Access 先进的时分多址

**ATDP** Air Traffic Data Processor 空中交通数据处理器

**ATDR** Aeronautical Technical Directive Requirement 航空技术指示性要求

**ATDS** Airborne Tactical Data System 机载战术数据系统

Aircraft Time & Departure System 飞机时间与离场系统

Aviation Technical Data System 航空技术数据系统

**ATDT** Attendant 乘务员

**ATE** Advanced Technology Engine 先进技术发动机

Air Terminal Equipment 航空终端设备

Airborne Test Equipment 机载测试设备

Annual Training Evaluation 年度培训评估

Automatic Test Equipment 自动测试设备，自动测试装置

**ATEC** Air Transport Electronic Council 航空运输电子会议

Automatic Test Equipment Complex 自动测试成套设备

Automatic Testing Equipment Computer 自动检测设备计算机

Aviation Technician Education Council 航空技术教育委员会

**ATEGG** Advanced Turbine Engine Gas Generator 先进的涡轮发动机气体发生器

**ATEM** Aircraft Test Equipment Modification 飞机试验设备的改进

**ATEMS** Automatic Tracking Equipment Management System 自动跟踪设备管理系统

**ATERM** Air Terminal 候机楼

**ATF** Actual Time of Fall 实际下降时间

Advanced Tactical Fighter 先进的战术战斗机

Advanced Technology Fighter 先进的技术战斗机

Aerodrome Traffic Frequency 机场交通频率

Aircraft Torque Factor 飞机扭矩系数

Auto Transformer 自耦合变压器

Automatic Terrain Following 自动地形匹配

Automatic Test Facility 自动测试设施

Aviation Trust Fund 航空信托资金

Aviation Turbine Fuel 航空涡轮燃油

**ATFCM** Air Traffic Flow and Capacity Management 空中交通流量与容量管理

**ATFI** Advanced Technology Fan Integrator 先进技术的风扇集成器

**ATFM** Air Traffic Flow Management 空中交通流量管理

Analytical Trial Function Method 分析试验函数法

**ATFMU** Air Traffic Flow Management Unit 空中交通流量管理单位

**ATG** Air Training Group 航空训练组

Air Transport Group 空运组

Air Turbine Generator 空气涡轮发动机

Antenna Test Group 天线测试组

Approval Test Guide 鉴定试验指南

Automatic Test Generator 自动测试信号发生器

**ATH** Advanced Training Helicopter 先进的训练直升机

Athens, Greece ( Airport Code ) 希腊雅典海

利尼肯机场

**ATHEANA** A Technique for Human Error Analysis 人误分析技术

**ATHR** Autothrust System 自动推力系统

**ATHS** Automatic Target Handoff System 自动目标交换系统

**ATI** Aero Transporti Italiani 意大利航空货运公司

Aero Turn Indicator 飞行器转弯指示器

Air Technical Index 航空技术(资料)索引

Air Transport Indicator 空运指示器

Air Transport Industry 航空运输业

Air Transport Instrument 航空运输机仪表

Air Transport Intelligence 航空运输情报所

Airline Training Institute ( USA ) 航空公司培训协会,航空公司培训学院(美国)

Alarm Transmission Interface 报警传输接口

Average Total Inspection 平均总检查

**ATIC** Advanced Technology Innovation Center 先进技术创新中心

Advanced Technology Integration Center 先进技术集成中心

Air Technical Intelligence Center 航空技术情报中心

Air Traffic Incident Commission 空中交通事故委员会

**ATIDS** Airport Surface Target Identification System 机场移动目标识别系统

**ATIGS** Advanced Tactical Inertial Guidance System 先进技术的惯性制导系统

**ATIIS** Air Telegram and Information Interchange System 航空电报和情报交换系统

**ATIMS** Air Traffic and Information Management System 空中交通和情报管理系统

**ATIR** Air Traffic Incident Report 空中交通事故报告

**ATIS** Advanced Thermal Imaging Scanner 先进的热成像扫描器

Air Technical Intelligence Service 航空技术情报服务

Air Technical Intelligence Study 航空技术情报研究

Air Traffic Information Service 空中交通情报服务

Air Traffic Integration System 空中交通综合系统

Airborne Test Instrumentation System 机载测试仪表系统

Airbus Technical Information System 空客技术信息系统

Airline Traffic Information System 航空公司交通信息系统

Airport Terminal Information Systems 机场终端信息系统

Airport Traffic Information Service 机场交通信息服务

Automatic Terminal Information Service 自动终端情报服务

Automatic Terminal Information System 终端区自动信息系统

**ATK** Available Tonne-Kilometre 可用吨公里,有效吨公里

**ATL** Active Template Library 活动样板库

Actual Time of Landing 实际着陆时间

Actual Total Loss 实际总损耗(亏损)

Atlanta, GA, USA ( Airport Code)美国佐治亚州亚特兰大机场

Atlantic 大西洋

**ATLA** Air Transport Licensing Authority 航空运输执照管理局,空运牌照局

**ATLAS** Abbreviated Test Language for All System 所有系统的简化测试语言

Abbreviated Test Language for Avionics System 航空电子系统的简化测试语言

Air Traffic Land & Airborne System 空中交通地面和机载系统

Automated Testing and Load Analysis System 自动化测试与负载分析系统

Automated Tool Location & Accountability System 自动工具定位与责任系统

Automated Tool Location and Storage 自动工具定位与存储

**ATLB** Air Transport Licensing Board 航空运输执照管理局

**ATLIS** Airborne Tracking Laser Identification System 机载激光跟踪识别系统
Airborne Tracking Laser Illumination System 机载激光跟踪照明系统

**ATLN** Alternate 备份,备降场

**ATLSS** Advanced Target Location Suppression Systems 先进的目标位置抑制系统

**ATM** Acceptance Test Manual 验收测试手册
Aerial Turning Motor 天线转动电动机
Air Traffic Management 空中交通管理
Air Transport Management 航空运输管理
Air Transport Movement 空中运输动态
Air Turbine Motor 航空涡轮发动机
Aircraft Transport Movement 航空运输动态
Aircraft Transportability Manual 飞机运输性手册
Aircrew Training Manual 飞行人员训练手册
Airspace and Traffic Management 空域和空中交通管理
Air-Turbine Motor 空气涡轮马达
Apollo Telescope Mount 阿波罗望远镜架
Assumed Temperature Method 假设温度方法
Asynchronous Transfer Mode 异步传输模式
Asynchronous Transmission Mode 异步传输模式
Asynchronous Trunk Module 异步中继模块
Atmosphere 大气
Automated Teller Machine 自动取款机
Automatic Test Methods 自动测试方法
Automatic Thrust Management 自动推力管理
Automatic Transmitting Measuring 自动传输测量值
Automatic Trim Monitor 自动配平监控
Available Ton-Mile 可用吨—英里
Axial Turbo Machinery 轴流式涡轮机械(装置)

**ATMAC** Air Traffic Management Advisory Committee 空中交通管理咨询委员会
Air Traffic Management Automated Center 空中交通管理自动化中心

**ATMB** Air Traffic Management Bureau 空中交通管理局

**ATMC** Air Traffic Movement Control Center 空中交通管制中心

**ATMCP** Air Traffic Management Concept Panel 空中交通管理概念专家组
Air Traffic Management Operational Concept Panel (ICAO)空中交通管理运行概念专家组

**ATMD** Air Traffic Management Division 空中交通管理部

**ATMDC** Air Traffic Management Development Centre 空中交通管理开发中心

**ATME** Automatic Transmission Measuring Equipment 自动传输测量设备

**ATMG** Airbus Technical Management Group 空客技术管理小组
Airspace and Traffic Management Group 空域和交通管理小组

**ATMIN** Air Traffic Management Information Notices 空中交通管理信息通告

**ATML** Automatic Test Markup Language 自动测试的标记语言

**ATM-NIC** ATM-Network Interface Card ATM 网络接口卡

**ATMOS** Air Traffic Management and Operations Simulator 空中交通管理运行模拟器
ATM Optical Switching ATM 光交换

**ATM-PON** ATM-Passive Optical Network ATM 无源光网络

**ATMPT** Attempt 试图

**ATMS** Advanced Test Management System 先进的测试管理系统
Advanced Text Management System 先进的文档管理系统
Advanced Traffic Management System 先进的交通管理系统
Assembly Tracking Management System 装配跟踪管理系统
Automated Thrust Management System 自动

推力管理系统

Automatic Teletype Message Switching 自动电传打字转报

**ATMSO** Air Traffic Management Standards Office 空中交通管理标准办公室

**ATN** Addition Transition Network 增强转移网络

Aeronautical Telecommunications Network 航空电信网

Augmented Transition Network 扩充转移网络

**ATND** Attend 出席,参加

**ATNP** Aeronautical Telecommunication Network Panel 航空电信网专家组

**ATNS** Air Traffic Navigation Services 空中交通导航服务

**ATO** Aborted Take-off 中断起飞

Action Technical Order 技术操作规程

Actual Time Over 实际经过时间,实际飞临目标上空时间

African Timber Organization 非洲木材组织

After Take-off 起飞后

Air Traffic Organization 空中交通组织

Air Transport/Transportation Office 航空运输办公室

Aircraft Technical Order 飞机技术规程

Along Track Offset 沿航迹偏置

Assisted Take off 辅助起飞

**ATOA** Air Taxi Operators Association 出租飞机经营人协会

**ATOCA** Advanced Tanker Outside Cargo Aircraft 先进的外油箱货机

**ATOG** Allowable Take-Off Gross (Weight) 容许起飞总重量

**ATOL** Air Travel Organiser's License 航空旅行组织者执照

Assisted Takeoff & Landing 辅助起飞和着陆

**ATOMS** Air Traffic Operational Management System 空中交通运行管理系统

**ATON** Aids to Navigation 导航设备

**ATOPS** Advanced Transport Operating System 先进的运输操作系统

**ATOS** Air Traffic Operations Service 空中交通运行服务

Air Transportation Oversight System 航空运输监督系统

Assisted Takeoff System 起飞助推系统

Automated Technical Order System 自动技术指令系统

**ATOT** Actual Take off Time 实际起飞时间

Actual Time Over Target 在目标上空的实际时间

Angle Track on Target 目标角跟踪

**ATOVS** Advanced TIROS Operational Vertical Sounder 高级 TIROS 业务垂直探空仪

**ATP** Acceptance Test Plan 验收测试计划

Acceptance Test Procedure 验收测试程序

Accepted Test Procedure 可接受试验程序

Admissions Testing Program 进入测试大纲,进入接纳测试大纲

Advance Test Procedure 预测试程序

Advanced Technology Program 先进技术计划

Advanced Turbo-Prop 先进的涡桨飞机

Air Test Port 大气测试口

Air Test Pressure 空气测试压力

Airline Transport Pilot 航线运输驾驶员

Assembly Test Program 汇编测试程序

At Time or Place 在…时间或地点

At Time, At Point 在…时间,在…地点

Authority/Authorized to Proceed 允许进行设计

Available to Promise 可承诺量

**ATP&C** Air Traffic Passenger & Cargo 航空旅客与货物运输

**ATPAC** Air Traffic Procedures Advisory Committee 空中交通飞行程序咨询委员会

**ATPASS** Automated Tool & Production Assembly Sequence System 自动工具与产品组装程序系统

**ATPCO** Airline Tariff Publishing Company 航空运价出版公司

**ATPE**  Airline Transport Pilot Examiner 航线运输机驾驶员考核员

**ATPG**  Automatic Test Pattern Generator 自动测试码生成程序

**ATPGS**  Automated Test Program Generation System 自动测试程序产生系统

**ATPL**  Air Transport Pilot's Licence 航空运输驾驶员执照

Airline Transport Pilot's Licence 航线运输驾驶员执照

**ATR**  Acceptance Test Report 验收测试报告

Actual Time of Return 实际返航时间

Actual Time Remaining 实际剩余时间

Advanced Test Reactor 高级试验反应堆

Aided Target Recognition 辅助目标识别

Air Traffic Regulation 空中交通规则

Air Transport Racking 运输机无线电设备机尺寸标准

Air Transport Radio 航空运输无线电台

Air Transport Report 航空运输报告

Air Transportability Racking 航空可运输性考核

Aircraft Trouble Report 航空器故障报告

Airline Transport Rating 航空公司空运额度

Ambient Temperature Recorder 环境温度记录仪

Analog Tape Recorder 模拟式磁带记录仪

Annual Technology Report 年度技术报告

Antitransmit-Receive 反收发

Attenuated Total Reflectance 衰减全反射

Automatic Target Recognition 目标自动识别

Automatic Target Recognizer 目标自动识别器

Automatic Throttle Retarder 自动收油门装置

Aviation Transmitter and Receiver 航空收发机

**ATRA**  Advanced Technology Reconnaissance Aircraft 技术先进的侦察飞机

Advanced Technology Regional Airliner 技术先进的支线飞机

**ATRAC**  Adaptive Transform Acoustic Coding 自适应变换域声音编码

Automotive Tracking Reporting Analysis and Control 自动跟踪报告分析与控制

**ATRAN**  Automatic Terrain Recognition & Navigation 地形自动识别与导航

**ATRC**  Air Traffic Regulation Center（美国）航空运输条例中心

**ATRCC**  Automatic Turbine Rotor Clearance Control 自动涡轮转子间隙控制

**ATRCCS**  Automatic Turbine Rotor Clearance Control System 自动涡轮转子间隙控制系统

**ATRCTV**  Attractive 吸引力

**ATRH**  Advanced Tandem Rotor Helicopter 先进的纵列式直升机

**ATRK**  Along-Track Error 沿航线误差

**ATRS**  Assembly Test-Recording System 装配试验记录系统

Automated Travel Record System 自动航行记录系统

Automatic Target Recognition System 自动目标识别系统

Aviation Training Research Simulator 航空训练研究模拟器

**ATRU**  Auto-Transformer Rectifier Unit 自动变压整流器

**ATS**  Acceptance Test Specification 验收试验规程

Advanced Technological Satellite 技术先进的卫星

Aeronautical Telecommunication System 航空电信系统

Aeronautical Information System 航空情报系统

Air to Surface 空对地面,空对水面

Air Traffic Services 空中交通服务

Air Traffic Systems 空中交通系统

Air Turbine Starter 空气涡轮启动机

Airbus Technical Specification 空客技术规范

Aircrew Training System 机组训练系统

Attitude Retention System 姿态保持系统

Audible Traffic Signal 行人过路发声装置

Austrian Schilling 奥地利先令

Automated Trade System 自动交易系统

Automatic Test Station 自动试验台

Automatic Test/Testing System 自动测试系统

Automatic Throttle System 自动节流系统, 自动油门系统

Automatic Thrust System 自动推力系统

Automatic Transfer Service 自动转账服务

Automatic Translation System 自动翻译系统

Automatic Tuning System 自动调谐系统

Autothrottle System 自动油门系统

Autothrust System 自动推力系统

Office of Air Traffic Services 空中交通服务办公室

**ATSA** Advanced Tactical Support Aircraft 先进的战术支援飞机

Air Traffic System Analysis 空中交通系统分析

Aircraft Technical Support Association 飞机技术保障协会

**ATSAC** Air Traffic Safety Assessment Committee 空中交通安全评估委员会

**ATSB** Air Transportation Stabilization Board 航空运输稳定委员会

Australian Transport Safety Bureau 澳大利亚运输安全局

**ATSC** Air Traffic Security Coordinator 空中交通安全协调员

Air Traffic Services Communication 空中交通服务通信

Aircraft Tank Service Company 飞机油箱服务公司

**ATSCCP** Air Traffic Service Contingency Command Post 空中交通服务应急指挥所

**ATSCV** Air Turbine Starter Control Valve 空气涡轮启动机控制活门

**ATSD** Air Traffic Services Department 空中交通服务部

Air Traffic Situation Display 空中交通情况显示器

**ATSG** Acoustic Test Signal Generator 声学测试信号发生器

Advanced Technology System Group 先进的技术系统集团

Air Transportation Support Group 空中运输保障组

**ATSMP** Air Traffic Service Message Processor 空中交通服务信息处理器

**ATSORA** Air Traffic Service Outside Regulated Airspace 规定空域以外的空中交通服务

**ATSP** Air Traffic Service Provider 空中交通服务提供者

**ATSPH** Atmosphere 大气

**ATSS** Advanced Tactical Support System 高级战术支援系统

Advanced Tactical Surveillance System 高级战术监视系统

Air Traffic Service System 空中交通服务系统

Air Transportation Support System 空中运输的支持系统

Airway Transportation System Specialist 空运运输系统专家

Automatic Test Support System 自动测试支持系统

**ATSU** Air Traffic Service Unit 空中交通服务组件

Air Travel Security Unit 航空旅行安全装置

**ATT** Address Translation Table 地址转换表

Advanced Tactical Transport 先进的战术运输机

Advanced Training Technologies 先进的培训技术

Advanced Transport Technologies 先进的运输技术

Air Traffic Transponder 空中交通应答机

Attached 附上, 附后

Attachment 附件

Attenuator 衰减器
Attitude Reference 姿态基准
Attitude 姿态
Automatic Target Tracking 自动目标跟踪
Automatic Test Terminal 自动测试终端
Automatic Turbine Testing 自动涡轮测试
**ATTAS** Advanced Technologies Testing Aircraft System 先进的技术试验飞机系统
**ATTC** Aviation Technical Training Center 航空技术训练中心
**ATTCS** Automatic Take-off Thrust Control System 起飞推力自动控制系统
**ATTD** Advanced Technology Test Demonstrator 高级技术测试示范器
Advanced Technology Transfer Demonstration 高级技术传送示范
Advanced Technology Transition Demonstration 先进技术的转移演示
Attached 附上,附后
Attendant 服务员,维护人员
**ATTEN** Attenuation 衰减,减少
**ATTEND** Attendant 乘务员
**ATTFM** Air Traffic Tactical Flow Management 空中交通战术流量管理
**ATTITB** Air Transport and Travel Industry Training Board 航空运输和旅游业训练委员会
**ATTMA** Advanced Transport Technology Mission Analysis 先进运输机技术任务分析
**ATTN** Attendant 乘务员
Attention 注意
**ATTND** Attend to, Attendance 出席,参加
Attendant 乘务员
**ATTNDC** Attendance 出席,参加
**ATTR** Attractive 有吸引力
**ATTRIB** Attributed 属性的,归属的
**ATTS** Assembly Target Time System 装配目标时间系统
Automated Terminal Tracking System 自动终端跟踪系统
**ATU** Acceptance Test Unit 验收测试组件

African Telecommunications Union 非洲无线电通信联盟
Antenna Tuning Unit 天线调谐组件
Auto Transformer Unit 自动变压器
Automatic Test Unit 自动测试组件
Automatic Tracker Unit 自动跟踪组件
Auxiliary Tracker Unit 辅助跟踪组件
**ATUC** Air Transport Users Committee 航空运输用户委员会
**ATV** Aerodynamic Test Vehicle 空气动力试验飞行器
Amateur Television 业余电视
Automatic Threshold Variation 自动门限调整
Automatic Ticket Vendor 自动售票机
**ATW** Advanced Technology Workshop 先进技术工作组
Advanced Training Wing 高级训练机翼
Air Transport World 世界航空运输
Aircraft Tail Warning 飞机机尾报警
Appleton, WI, USA（Airport Code）美国威斯康星州阿普尔顿机场
**ATWIT** Air Traffic Workload Input Technique 空中交通工作负荷输入技术
**ATWS** Automatic Track While Scan 扫描时自动跟踪
**ATX** Air Taxi 出租飞机
**ATY** Watertown, SD, USA（Airport Code）美国南达科他州瓦特镇机场
**ATYP** Aircraft Type 机型
**ATZ** Aerodrome Traffic Zone 机场交通地带,机场交通区
Air Traffic Zone 空中交通区
Airport Traffic Zone 机场交通区
**ATZD** Authorized 认可的
**AU** Access Unit 访问(接入)单元
Adapter Unit 适配单元
Administrative Unit 管理单元
African Union 非洲联盟
Aircraft Utilization 飞机利用率
Arithmatical Unit 运算单位,运算组件

Astronomical Unit 天文单位

Atomic Units 原子单位

Australia 澳大利亚

Authentication of User 用户认证

**AUA** Airport Urban Area 机场都市区

Aruba, Aruba (Airport Code) 阿鲁巴阿鲁巴机场

**AUAP** AU Access Point 接入单元的接入点

**AUC** Airline Users Committee 航空公司用户委员会

Authentication Center 认证中心

**AUCS** Advanced UHF Communication System 先进的特高频通信系统

**AUD** Audio 音频

Australian Dollar 澳元(货币单位)

Available for Use Data 可供使用的数据

**AUF** Autonomous Functional Unit 自给功能装置

**AUG** Augment, Augmentation 扩大,增加

Augmentation 加力

August 八月

**AUH** Abu Dhabi, United Arab Emirates (Airport Code) 阿拉伯联合酋长国阿布扎比机场

**AUI** Attachment Unit Interface 连接单元接口

**AUK** Alakanuk, Alaska (Airport Code) (美国)阿拉斯加州阿拉克纳克机场

**AUM** Asynchronous User Module 异步用户模块

**AUP** Acceptable Use Policy 可接受使用策略

Airspace Use Plan 空域使用计划

Average Unit Price 平均单位价格

**AUPTR** Administration Unit Pointer 管理单元指针

**AUR** Aircraft Utilization Report 飞机利用情况报告

**AUS** Austin, TX, USA (Airport Code) 美国得克萨斯州奥斯汀机场

**AUSSAT** Australian Communication Satellite 澳大利亚通信卫星

**AUST** Austria 奥地利

**AUSTL** Australia 澳大利亚

**AUT** Automatic 自动

Automobile 汽车

**AUTH** Authentic 正式的

Authority 当局

Authorization 授权,委托

Authorize 授权

Authorized 授权的,认可的

**AUTHD** Authorized 授权的,认可的

**AUTHTY** Authority 当局

**AUTM** Autumn 秋季

**AUTO** Automatic, Automatically 自动的,自动地

**AUTOCAL** Autocalibration 自动校验

**AUTODIN** Automated Digital Information Network 自动数字信息通讯网

Automatic/Automated Digital Network 自动化数字网络

**AUTOL** Autoland 自动着陆

**AUTOLand** Automatic Landing 自动着陆

**AUTOM** Automated 自动化

**AUTOMBL** Automobile 汽车

**AUTOPSY** Automatic Operating System 自动化操作系统

**AUTOVON** Automatic Voice Network 自动化话音网络

**AUTZC** Authorization Code 鉴别码

**AUW** All Up Weight 全重,总重,起飞重量,起飞全重

**AUX** Auxiliary 辅助的

**AUXFSS** Auxiliary Flight Service Station 辅助飞行服务站

**AUXY** Auxiliary 辅助

**AV** Actual Velocity 实际速度

AD Valorem 按价计算运费

Aerospace Vehicle 航空航天飞行器

Air Vehicle 航空器

Air Vent 通气道,出气孔

Analog Video 模拟视频

Analog Voice 模拟话音

Analysis of Variance 方差分析

Annealed in Vacuum 真空退火

Audio-Visual 视听

Authorized Version 标准版本

Average Variability 平均变化率

Average 平均

Aviation 航空

Aviator 航空员,飞行员

Avionics 航空电子设备,航空电子学

**AV/CM** Ampere-Volt/Centimeter 安伏/厘米（磁场强度单位）

**AV/D** Alternate Voice/Data 语音—数据交换

**AVA** Audio/ Visual Annunciator 音频与视频信号机

Audio-Frequency Voltage Amplifier 音频电压放大器

Automatic Voice Advice 自动语言通告

Automatic Voice Alarm 自动声音报警

Aviation 航空

Azimuth/Versus Amplitude 方位—振幅,方位与振幅的关系

**AVAA** Active Van Attar Array 有源范阿塔天线阵

**AVAD** Automatic Voice Altering Device 自动话音告警装置

**AVAIL** Availability 适用

**AVAPS** Airborne Vertical Atmospheric Profiling System 空中垂直大气廓线系统

**AVASI** Abbreviated Visual Approach Slope Indicator 简易目视进近坡度指示器

**AVASIS** Abbreviated Visual Approach Slope Indicator System 简易目视进近下滑指示系统

**AVBL** Available 适用的,可用的

**AVBLTY** Availability 适用,有效

**AVC** Advanced Video Coding 先进视频编码

Audio for Video-Conferencing 会议电视用的音频

Audio Visual Connection 音像连接

Aural and Visual Code 听觉与视觉信号编码

Automatic Volume Control 自动音量控制

Average Variable Cost 平均可变成本

**AVCAL** Aviation Consolidated Allowance List 航空部件统一定额表

**AVCC** Aviation Crash Crew 飞机失事抢救人员,机场抢救组

**AVCD** Audio Video Compression Disk 音频视频压缩光盘

**AVCL** Air Vehicle Center Line 飞行器中心线

**AVCO** Average Cost 平均成本

**AVD** Alternate Voice and Data 交变语音和数据

Audio Video Driver 音视频驱动器

**ARDC** Aviation Resources & Development Corporation (USA) 航空资源与开发公司（美国）

**AVDS** Automatic Voice Data Switching 自动语音数据交换

**AVDSP** Audio Video Data Signal Processor 音视频数据信号处理器

**AVE** Aerospace Vehicle Equipment 空间飞行器设备

Air Vehicle Equipment 飞行器设备

Airborne Vehicle Equipment 机载设备

Audio Video Engine 音视频引擎

Average Velocity Error 平均速度误差

Aviation Enterprises (USA) 航空事业（美国）

**AVER** Average 平均

**AVFUEL** Aviation Fuel 航空燃油

**AVG** Asset Value Guarantee 资产价值保证

Availability Status Message 航班座位情况电报,旅馆可用状态讯息

Average 平均

**AVGAS** Aviation Gasoline 航空汽油

**AVHRR** Advanced Very High-Resolution Radiometer 先进的甚高分辨率辐射计（美国）

**AVI** Air Vehicle Instrumentation 航空器仪表

Audio Video Interactive 音视频交互作用

Audio Video Interlaced 音视频数据交叉(交织)

Audio Video Interleave 音视频交替

**AVIC** Aviation Industries Corporation of China 中国航空工业集团公司

**AVIC Ⅰ** China Aviation Industry Corporation Ⅰ 中国航空工业第一集团公司

**AVIC Ⅱ** China Aviation Industry Corporation Ⅱ 中国航空工业第二集团公司

**AVICOM** Aviation Communication 航空通信

**AVID** Advanced Visual Information Display 高级视频信息显示

**AVIP** Avionics Integrity Program 航空电子设备综合计划

**AVIS** Audio-Visual Information System 视听信息系统

Audio-Visual Interaction Service 视听交互业务

**AVIT** Audio-Visual and Information Technology 声像及信息技术

**AVK** Audio Video Kernel 音频视频核心

**AVL** Air Vehicle Limit 航空器极限

Approved Vendor List 认可的供货商清单

Asheville , NC, USA (Airport Code) 美国北卡莱罗纳州阿什维尔机场

Audio Video Library 音频视频库

Automatic Vehicle Location 自动车辆定位

Available 适用的,可用的

**AVLAB** Aviation Material Laboratory 航空材料实验室

**AVLAN** Avionics Local Area Network 航空电子局域网

**AVLC** Aviation VHF Link Control 航空甚高频链路控制

**AVLF** Airborne Very Low Frequency 机载甚低频

**AVLN** Automatic Vehicle Location and Navigation 车辆自主定位和导航

**AVM** Airborne Vibration Monitor 机载振动监视器

Airborne Vibration Monitoring 机载振动监视

Aircraft Vibration Meter 飞机振动仪

Airplane Vibration Monitoring 飞机振动监视

Application Value Management 应用价值管理

Audio-Visual Modulator 音像调制器

**AVMAP** Air Vehicle Maintenance Audio Panel 航空器音频维护板

**AVMSC** Anti-Vibration Monitor System Control 防振动监控系统控制

**AVMUX** Audio Video Multiplexer 音频视频多路器

**AVN** Aviation System Standards office 航空系统标准办公室

Aviation 航空

Avignon , France (Airport Code) 法国阿维尼翁机场

**AVNC** Avionics 航空电子

**AVOD** Audio Video on Demand 音频视频点播

**AVOIL** Aviation Oil 航空油料

**AVOL** Aerodrome Visibility Operational Level 机场能见度运行等级

**AVP** Advanced Video Processor 高级视频处理器

Advanced Video Products 先进的视频产品

Analog Video Processor 模拟视频处理器

Wilkes Barre/Scranton , PA, USA 美国威尔克斯巴里机场

**AVPAC** Aviation Packet Communication 航空数据包通信

**AVPLD** Average Payload 平均业载

**AVR** Audio/Video Recording System 音频/视频记录系统

Automatic Voice Recognition 自动话音识别

Automatic Voice Relay 自动话音中继

Automatic Voltage Regulation 自动电压调整

Automatic Voltage Regulator 自动电压调节器,自动稳压器

**AVRI** Altitude/Vertical Rate Indicator 高度/垂直速率指示器/升降速度指示器

**AVRS** Airborne Video Recording System 机载

视频记录系统

Audio/Video Recording System 音频/视频记录系统

Automatic Vehicle Recognition System 车牌自动辨认系统

**AVS** Aided Video System 辅助视频系统

Air Vehicle Simulation 航空器模拟

Application Visualization System 应用可视化系统

Audio Visual System 视听系统

Audio/Video Server 音像服务器

Aviation Standards 航空标准

Avionics Ventilation System 电子通风系统

**AVSAT** Aviation Satellite 航空卫星

**AVSEC** Aviation Security Study Group 航空保安

Aviation Security 航空保安

**AVSECP** Aviation Security Panel 航空保安专家组

**AVSS** Audio Video SubSystem 音频视频子系统

Audio Video Support System 音频视频支撑系统

**AVSTATS** Aviation Statistics 航空数据

**AVT** Audio Visual Trainer 视听训练器

**AVTR** Airborne Video Tape Recorder 机载录像机

**AVTS** Automatic Valve Transfer System 活门（阀门）自动转换系统

**AVTUR** Aviation Turbine Fuel 航空涡轮燃油

Aviation Turbine 航空涡轮机

**AVUM** Aviation Unit Maintenance 航空组件维修

**AVVI** Altitude Vertical Velocity Indicator 高度升降速度表

**AVW** Aviation West 西部航空公司（美国）

**AW** Actual Weight 实际重量

Airway(s) 航空公司,航路

All-Weather 全天候

Area of Wing 机翼面积

**AW&ST** Aviation Week & Space Technology 航空周刊与空间技术

**AWA** Advanced Work Authorization 预先工作授权

ATM Wireless Access ATM 无线接入

Aviation Writers Association（USA）航空作家协会

**AWAC** Airborne Warning and Control 机载预警与控制

**AWACS** Airborne Warning and Control System 机载预警与控制系统

ATM Wireless Access Communication System ATM 无线接入通信系统

Space Warning and Control System 空间警告与控制系统

**AWADS** Adverse Weather Aerial Delivery System 恶劣天气空投系统

**AWAM** Association for Women in Aviation Maintenance 航空维修妇女联合会

**AWANS** Aviation Weathers and Notice to Airmen System 航空天气和航行通告系统

**AWARAU** All-Weather Automatic Radio Aids Unit 全天候无线电导航试务

**AWARE** Advanced Warning Equipment 先进的报警装置

Advanced Weather and Reporting Enhancement 先进的天气和报告改进

Automatic Warning and Recording Equipment 自动报警与记录设备

**AWARS** Airborne Weather and Reconnaissance System 机载气象和侦察系统,机载气象和侦察系统

All Weather Airborne Reconnaissance System 全天候机载侦察系统

**AWAS** Air Warfare Analysis Section 空战分析组（英国）

Airborne Wind-Shear Alert Sensor 机载风切变告警传感器

All-Weather Approach System 全天候进近系统

Ansett Worldwide Aviation Services 安赛特

环球航空服务公司

Automated Weather Advisory Station 自动化气象咨询站

**AWB** Air Waybill 航空货运单

Airworthiness Bulletin 适航通告

**AWC** Airworthiness Certification 适航证

Aviation Weather Center 航空气象中心

**AWCLS** All Weather Carrier Landing System 全天候航空母舰着陆系统

**AWD** Award 授予

**AWEL** Aircraft Wiring Equipment List 飞机布线设备清单

**AWF** Aviation Weather Facility 航空气象设施

**AWG** American Wire Gauge 美国线规,美制线材规格

Audible Warning Generator 音频警告发生器

**AWGN** Additive White Gaussian Noise 附加的高斯白噪声

**AWGP** Assumed Worse Glide Path 假定的较差下滑道

**AWI** Aircraft Weight Indicator 飞机重量指示器

Airframe Wiring Interface 飞机连线接口

**AWIB** Aerodrome Weather Information Broadcast 机场气象情报广播

**AWIPS** Advanced Weather Interactive Processing System 先进的交互式气象处理系统

**AWIS** Advanced Wafer Imaging System 先进的晶片成像系统

Aerodrome Weather Information Service 机场气象情报服务

Aircraft With Initial Storage 出厂缺件的飞机

Airport Weather Information System 机场气象信息系统

All-Weather Identification Sensor 全天候识别传感器

Automated Weather Information Systems 自动气象信息系统

**AWL** Aircraft Wiring List 飞机布线清单

Airworthiness Limitation 适航性限制

Automated Wire List 自动线路表

**AWLAR** All Weather Low Altitude Route 全天候低空航路

**AWLS** All-Weather Landing System 全天候着陆系统

**AWM** Aircraft Wiring Manual 航空器线路图手册

Airworthiness Manual 适航手册

Awaiting Maintenance (飞机)等待维修

**AWO** Air Weather Office 航空气象室

All Weather Operations 全天候飞行

**AWOP** All Weather Operation(s) Panel 全天候运行专家组,全天候飞行专家组

**AWOS** Automated Weather Observation System 自动气象观测系统

Automated Weather Observing Station 自动气象观测站

**AWP** Aviation Weather Processor 航空气象处理器

Awaiting Parts 维修部件

**AWR** Airborne Weather Radar 机载气象雷达

**AWRP** Aviation Weather Research Program 航空气象研究计划

**AWRS** Airborne Weather Radar System 机载气象雷达系统

Airborne Weather Reconnaissance System 机载气象探测系统

All-Weather Reconnaissance System 全天候侦察系统

Automatic Weather Reporting System 自动气象报告系统

Aviation Weather Reporting Station 航空气象报告台站

**AWS** Advanced Warning System 高级预警系统

Air Weather Service 航空气象服务

Air Weather Station 航空气象站

All-Wheel Steering 全轮转向

America Welding Society 美国焊接协会

Automatic Weather Station 自动气象站

Aviation Weather Service 航空气象服务

Aviation Weather Station 航空气象站

**AWSM** Air Weather Service Manual 航空气

象服务手册

**AWSS** Aviation Weather Sensor System 航空气象探测系统

**AWSTR** Air Weather Service Technical Report 航空气象服务技术报告

**AWT** Actual Work Time 实际工作时间

Average Wait Time 平均等待时间

Average Work Time 平均工作时间

**AWTC** Airport World Trade Center 机场世贸中心

Amman World Trade Center 安曼世界贸易中心

Awaiting 等待

**AWY** Airway 航路

**AX** Access Authorized 授权存取，委托授权，授权接近

Axis 轴

Longitudinal Acceleration 纵向加速度

**AXE** Autothrottle Servo Enable 自动油门伺服使能

**AXPS** Air Express 航空快递

**AY** Lateral Acceleration 横向加速度

**AYGAS** Aviation Gasoline 航空汽油

**AYI** Angle of Yaw Indicator 偏航角指示器

**AYMED** Aviation Medicine 航空医学

**AYMS** Airlines Yield Management System 航空公司效益管理系统

**AYRQ** As Your Request 如所要求，根据贵方要求

**AYT** Antalya, Turkey ( Airport Code ) 土耳其安塔利亚机场

**AZ** Azimuth Station 方位台

Azimuth Transmitter 方位发射机

Azimuth 方位，方位角

Vertical Acceleration 垂向加速度

**AZFW** Actual Zero Fuel Weight 实际无燃油重量

**AZM** Azimuth 方位，方位角

**AZO** Kalamazoo, MI, USA ( Airport Code ) 美国密执安州卡拉马索机场

**AZRAN** Azimuth & Range 方位与距离

**AZS** Automatic Zero Set 自动调零

Samana International Airport, Dominican Republic ( Airport Code ) 多米尼加共和国萨马纳机场

# B

**B**   At or Below an Altitude 在或低于一个高度

B Class Accommodation 二等舱位

Baggage 行李

Bar 巴(压强单位)

Biplane 双翼机

Blue 蓝色的

Bomber 轰炸机

Byte 字节

**B/A**   Bank Angle 坡度角

Beta Angle 贝塔角

Block Altitude (空中)封锁高度

Breaking Action 刹车动作

Bundle Assemble 导线束组件

Bureau of Aeronautics 航空局(美国)

**B/B**   Back-Beam 反向波束,后梁

**B/C**   Back Course 背台航道

Base Check 基地检查

Battery Charger 电瓶充电器

Bills for Collection 托收票据

Braking Coefficient 刹车系数

Bulk Cargo 散货舱

Business Class 公务舱

**B/CAST**   Broadcast 广播

**B/CRS**   Back Course 背,反航道,回程

**B/D**   Backdrive 反驱动

Bank Draft 银行汇票

Bottom of Descent 下降底点

Brought Down 转下页

**B/DFT**   Bank Draft 银行汇票

**B/E**   Bill of Entry 报关单

Bill of Exchange 汇票

**B/F**   Brought Forward 承前页

**B/H**   Bill of Health 健康证书

**B/L**   Bill of Lading/Loading 提货单

Black Label 黑标签

Bleed 引气

Blue Line 蓝线

**B/M**   Back Marker 背台指点标

Bill of Material 材料单

**B/N**   Bank Note 银行承兑票据

**B/O**   Burn off 烧掉,耗油

**B/P**   Bill of Parcels 包裹单

Bill of Payment 支付票据

**B/R**   Beam Rate 波束变化率

Bill Receivable 应收票据

**B/S**   Behind Schedule 班期时刻表后

Bill of Sale 销售单

**B/SEC**   Bits per Second 比特/秒

**B/T**   Block Time 轮挡时间

**B/W**   Black and White 黑白

**B2B**   Business to Business 企业到企业(电子商务)

**B2C**   Business to Consumer 企业到消费者(电子商务)

**BA**   Bandwidth Allocation 带宽分配

Barometric Altimeter 气压高度表

Basic Access 基本接入

Basic Airplane 基本型飞机

Beam Approach 波束引导进近着陆

Bell Aerosystems Company 贝尔航空系统公司

Blind Approach 仪表进近着陆

Boarding Advisory 登机通知

Braking Action 刹车作用,刹车效应

British Airways 英国航空公司

Business Audio 商用音频

**BAA**   Bilateral Airworthiness Agreement 双边适航协议

British Airport Authority 英国机场管理公司

British American Airways (USA) 英美航空公司(美国)

**BAABI** Basic Approved ATA Breakdown Index 批准的基本 ATA 分解索引

**BAAE** Bachelor of Aeronautical and Astronautical Engineering 航空与航天工程学士

**BAAI** Business Aviation Association for India 印度公务机协会

**BAAR** Board for Aviation Accident Research 航空事故调查委员会

**BAASA** Business Aviation Association of Southern Africa 南非公务航空协会

**BAB** Budget Advisory Board 预算咨询委员会

**BABS** Beam Approach Beacon System (飞机)进场波束信标系统

Blind Approach Beacon System 仪表进近信标系统

**BAC** Bangkok Aviation Centre 曼谷航空中心

Bell Aerospace Corporation 贝尔航空航天公司

Blood Alcohol Concentration 血液中酒精浓度

Boeing Aerospace Company 波音宇航公司

Boeing Airplane Company 波音飞机公司

Boeing Assigned Color 波音指定的颜色

Booster Assembly & Checkout 增压器组装与检查

Booster Assembly & Contractor 增压器组装与合同方

British Aircraft Corporation 英国飞机公司

Bus Adapter Control 总线适配器控制

By-Pass Air Control 旁通气动控制

**BACA** Baltic Air Charter Association 波罗的海包机公司协会

British Advisory Committee for Aeronautics 英国航空咨询委员会

**BACAN** British Association for the Control of Aircraft Noise 英国飞机噪声控制协会

**BACD** Boeing Airplane Company Design 波音飞机公司设计

**BACE** Basic Automatic Check-out Equipment 基本自动检验设备

**BACEA** British Airport Construction & Equipment Association 英国机场建设和设备协会

**BACFD LAB** Boeing Aerospace Computational Fluid Dynamics Laboratory 波音宇航流体动力学计算实验室

**BACO** Boeing Airplane Company 波音飞机公司

**BACP** Broadband Assign Control Protocol 带宽分配控制协议

Broadcasting Authorization Control Protocol 广播授权控制协议

**BACS** Bleed-Air Control System (发动机)引气控制系统

Boeing Airport Control System 波音机场管理系统

Boeing ATLAS Compiler System 波音通用系统简略测试语言程序编制系统

**BADG** British Aerospace Dynamics Group 英国宇航动力集团

**BAe** British Aerospace Public Limited Co. 英宇航公众有限公司

**BAE** Bachelor of Aeronautical Engineering 航空工程学士

Baroom Aviation Establishment (Saudi Arabia) 沙特巴若姆航空研究院(沙特阿拉伯)

Boeing Aerospace & Electronics 波音宇航与电子公司

Boeing Electronics Incorporated 波音电子公司

Bundle Effectivity Index (导)线束有效性索引

Bundle Equipment Index (导)线束设备索引

Bundle Equipment Input (导)线束设备输入

**BAF** Baffle 隔板

British Air Ferries, Ltd. 英国空中渡运有限公司

Bunker Adjustment Factor 燃油附加费

**BAFO** Best and Final Offer 最优和最后报价

**BAG** Baggage 行李

Bandwith Allocation Gap 带宽分配间隔

Baguio, Philippines (Airport Code) 菲律宾碧瑶机场

Burst Address Generator 脉冲群地址发生器

**BAGCK** Baggage Check 行李票

**BAGG** Baggage 行李

**BAGTG** Baggage Tag 行李牌

**BAH** Bahrain, Bahrain (Airport Code) 巴林国巴林机场

British Airways Helicopters 英国直升机公司

**BAHO** British Association of Helicopter Operators 英国直升机驾驶员协会

**BAI** Barometric Altitude Indicator 气压高度指示器

Battlefield Air Interdiction 战场空中封锁

Boeing Aerosystems International Incorporated 波音航空系统国际公司

Boeing Airborne Instrumentation 波音机载仪表装置

Buttler Aviation International (USA) 布特勒国际航空公司(美国)

**BAIC** British Aviation Insurance Company 英国航空保险公司

**BAICD** Blind Anchored Interference-Cancelling Detector 盲区固定消干扰探测器

**BAIG** British Aviation Insurance Group 英国航空保险集团

**BAIR** British Airports Information Retrieval 英国机场信息检索

**BAIS** Blind Approach Instrument System 仪表进近系统

**BAK** Baku, Azerbaijan-Baku (Airport Code) 阿塞拜疆巴库机场

Basic Aeronautical Knowledge 基本航空知识

**BAL** Balance 平衡

Ballast Hold Loaded 压舱物

Basic Assembly Language 基本汇编语言

Batman, Turkey (Airport Code) 土耳其白特曼机场

**BALPA** British Air Line Pilots Association 英国航空公司驾驶员协会

**BALS** Blind Approach Landing System 仪表进近着陆系统

**BAM** Bird Avoidance Model 避鸟模型

**BAMCL** British Aircraft Manufacturing Company Ltd. 英国飞机制造有限公司

**BAMP** Boeing Applied Meteorology Program 波音气象应用程序

**BAP** Bank Angle Protection 坡度角保护

Basic Audio Processor 基本音频处理器

Beacon Aircraft Position 飞机信标定位

Boeing Associated Products 波音相关产品

Broadband Access Point 宽带接入点

**BAPA** Bangladesh Airline Pilots' Association 孟加拉国航空公司飞行员协会

**BAPC** British Aircraft Preservation Council 英国飞机维护协会

**BAPDMS** Boeing Airplane Performance Data Management System 波音飞机性能数据管理系统

**BAPE** Base Plate 底板

**BAPMN** Background Air Pollution Monitoring Network 本底空气污染监测网

**BAPTA** Bearing and Power Transfer Assembly 轴承和功率输出组件

**BAQ** Barranquilla, Colombia (Airport Code) 哥伦比亚巴兰基亚机场

Basic Allowance for Quarters 方位基本修正量

**BAR** Barometer 气压表

Base Address Register 基地址寄存器

Bleed Air Regulator 引气调节器

Board of Airline Representatives Hong Kong 香港航空公司代表协会

**BARC** Barometric Altitude Rate Computer 气压高度速率计算机

British Aeronautical Research Committee 英国航空研究委员会

**BARCIS** British Airport Rapid Control and Indication System 英国机场快速控制和指

示系统

**BARD** Back Azimuth Reference Datum 反方位进近基准点

**BARGN** Bargain 交易,讨价还价

**BARL** Barrel 桶

**BARO** Barometer 气压表
Barometric 气压的

**BARO ALT** Barometer Altitude 气压高度

**BARS** Backup Attitude Reference System 备用姿态基准系统
Basic Aviation Risk Standard 基本航空风险标准
Boeing Access Request System 波音接入请求计划

**BAS** Basic Airspeed 基本表速
Bit-rate Allocation Signal 比特率分配信号
Blind Approach System 仪表进近系统
Boeing Advanced Systems 波音先进系统
Boeing Airplane Services 波音飞机服务
Broadband Access Server 宽带接入服务器
Building Automation System 楼宇设备自动化系统
Bureau of Aviation Safety 航空安全局

**BASA** Bilateral Aviation Safety Agreement 双边航空安全协定

**BASC** Beijing Aviation Simulator Center 北京航空模拟中心
British Aerial Standards Council 英国航空标准委员会

**BASE** Cloud Base 云底

**BASH** Bird Aircraft Strike Hazard 鸟机相撞危害
Bird/wildlife Aircraft Strike Hazard 鸟/野生动物撞航空器危害

**BASI** Bureau of Air Safety Investigation 航空安全调查局

**BASIC** Basic Aviation Sub-System Integration Concept 基本航空子系统综合方案

**BASIS** British Airways Safety Information System 英国航空公司安全信息系统

**BASOPS** Base Operations 基地航行调度

**BASS** Backup Avionics Subsystem Software 备份航空电子分系统软件

BCAG Automated Scheduling System 波音商用飞机集团自动计划制定系统

Bleed Air Supply System 引气系统

Boeing Avionics System Simulator 波音航空电子系统模拟机

**BAT** Bandwidth Allocation Threshold 带宽分配阈值
Basic Air Temperature 基础气温
Battery 电瓶,电池,蓄电池
Battery(Electrical) 电瓶(电源)
Beam-Approach Training 波束进近着陆训练
Blind Approach Training 仪表进近训练
British Airtours,Ltd.(Great Britain) 航空旅游有限公司(英国)

**BATC** Battery Contactor 电瓶接触器
Boeing Atlantic Test Center 波音大西洋试飞中心
Business & Advanced Technology Centre 商务与先进技术中心

**BATCC** Backup Air Traffic Control Centre 备用空中交通管制中心

**BATCX** Backup Air Traffic Control Complex 备用空中交通管制大楼

**BATE** Baseband Adaptive Transversal Equalizer 基带自适应横向均衡器

**BATH** Best Available True Heading 可用最佳航向

**BATM** Bureau of Air Traffic Management 空中交通管理局

**BATOA** British Air Taxi Operators' Association 英国出租飞机经营人协会

**BATS** Basic Attributes Testing System 基本属性测试系统
Business Air Transport Service 商业航空运输服务

**BATT** Battery 电瓶

**BAU** Broadband Access Unit 宽带接入设备

**BAUA** Business Aircraft Users' Association 公务飞机用户协会

**BAV**  Basic Air Vehicle 基本航空器
Bleed Air Valve 引气活门,放气活门

**BAZ**  Back Azimuth Station 反方位台
Back Azimuth 后方位(背航道)
Barcelos,Brazil(Airport Code)巴西巴塞卢斯机场

**BB**  Back Beam 背台波束
Ball Bearing 滚珠轴承
Broad Band 宽频带,宽波带
Building Block 预制件
Buy Back 回购

**BBA**  BroadBand Access 宽带接入

**BBC**  British Broadcasting Corporation 英国广播公司

**BBCC**  Broad Band Communication Channel 宽带通信信道

**B-BCC**  Broadband Bearer Connection Control 宽带承载连接控制

**BBDC**  Before Bottom Dead Center 下死点前

**BBE**  Background Block Error 背景误块
Better than Best-Effort Service 优于尽力型业务

**BBER**  Background Block Error Ratio 背景误块比

**BBGA**  British Business and General Aviation Association 英国公务和通用航空协会

**BBJ**  Boeing Business Jets 波音公务机

**BBL**  Barrel 桶
Beacon and Blind Landing 信标和仪表进近着陆
Body Buttock Line 机身纵剖线

**BBM**  BaseBand Modem 基带调制解调器

**BBML**  Baby Food 婴儿餐食

**BBN**  BroadBand Network 宽带网

**BBO**  Buy-Build-Operate (机场)购买—建设—经营

**BBR**  Bankers' Buying Rate 银行买入价,银行货币买进费率

**BBRG**  Ball Bearing 滚珠轴承

**BBS**  Behavior-Based Safety 基于行为的安全
Bulletin Board Service 公告板服务
Bulletin Board System 公告板系统

**BBU**  Bandwidth Based Unit 基于带宽的单元
Baneasa, Bucharest, Romania ( Airport Code)罗马尼亚布加勒斯特巴纽莎机场

**BBX**  Base Band Exchange 基带交换

**BBYCY**  Bow Buoyancy 机头浮力

**BC**  Back Course 背台航道,后航道
Baggage Container Train 集装箱拖车
Basic Charge 基价
Bearer Capability 承载能力
Bearer Channel 承载信道
Bearer Circuit 承载电路
Bearer Control 承载控制
Between Centers 中心距
Billing Center 计账中心
Boundary Control 边界控制
Broadcast 广播
Bus Controller 总线控制器

**BCA**  Best Cruising Altitude 最佳巡航高度
Bit Count Appendage 位计数附属部分
Boarding Control Assistant (旅客)登机控制助理系统
Boeing Commerical Airplanes 波音商用飞机
British Continental Airways 英国大陆航空公司

**BCAA**  British Civil Aviation Authority 英国民航局

**BCAC**  Boeing Commercial Airplane Company/Corp 波音商用飞机公司

**BCACSBS**  BCAG Computing Services Budget System 波音商用飞机集团计算服务预算系统

**BCAG**  Boeing Commercial Airplane Group 波音商用飞机集团

**BCAQIPS**  BCAG Quality Improvement Project System 波音商用飞机集团质量改进计划系统

**BCAR**  British Civil Airworthiness Requirements 英国民用适航规范,英国民航适航要求
British Civil Aviation Regulation 英国民航规

章

British Civil Aviation Requirements 英国民航要求

**BCARs** British Civil Airworthiness Requirements 英国民用航空适航要求

**BCAS** Beacon Collision Avoidance System 信标避撞系统,信标防撞系统

British Compressed Air Society 英国压缩空气学会

Bureau of Civil Aviation Security 民用航空安全局

Business and Commuter Aviation Systems 商务与转运航空系统

**BCASC** British Civil Aviation Standing Conference 英国民航常务会议

**BCAT** Boeing Computing Architecture Team 波音计算设计组

**BCATS** Boeing Computer-Assisted Time Standard 波音计算机辅助的时间标准

**BCBP** Bar Coded Boarding Pass 条形码登机牌

**BCC** Beacon Control Console 信标控制台

Bearer Channel Connection 承载信道连接

Bearer Connection Control 承载连接控制

Beijing Climate Centre 北京气象中心

Binary Coded Checklist 二进制编码检查单

Bleed Control Computer 引气控制计算机

Block Character Check 块特性校验

Boarder Crossing Card 出入境卡

Boeing Capital Corporation 波音金融公司

Border Crossing Card 过境卡

British Copyright Council 英国版权委员会

Broadband Communication Channel 宽带通信信道

Business Centre Club 商务中心俱乐部

**BCCB** BAC Configuration Control Board 波音飞机公司配置控制局

**BCCH** Broadcast Control Channel 广播控制信道

**BCCM** Baggage, Cargo, Comat, Mail 行李、货物、公司物资、邮件

**BCCP** Bearer Channel Control Protocol 承载信道控制协议

**BCD** Bacolod, Philippines (Airport Code) 菲律宾巴科洛德机场

Bad Conduct Discharge 不良接触放电

Behind Completion Date 延期完工

Binary Coded Decimal 二—十进制代码

Buyer Committed Date 购方指定日期

**BCDS** BITE Centralized Data System 自检装置中央数据系统

Broadband Connectionless Data Service 宽带无连接数据服务

**BCF** Bearer Control Function 载体控制功能

Brake Cooling Fan 刹车冷却风扇

Bulk Cargo Fan 散货舱风扇

**BCFG** Fog Patches 雾团

**BCFMS** Bar-Coding File Management System 电脑条码档案管理系统

**BCFRC** Backup Centralised Fault Report Centre 备用综合仪器故障报告中心

**BCGD** Background 背景

**BCH** Beach 海滩,海滨

Broadcast Channel 广播信道

**BCHG** Battery Charger 蓄电池充电器

**BCIA** Beijing Capital International Airport 北京首都国际机场

**BCIU** Bus Control Interface Unit 总线控制接口装置

**BCK** Back 反面,背向,向后

**BCL** Bank Confirmation Letter 银行确认信

Battery Charge Limiter 电瓶充电限制器

**BCLM** Baggage Cargo Loading Manual 行李货物装载手册

**BCM** Back Course Marker 后航道指点标

Backup Control Module 备用控制模块

Baggage Control Message 行李管制文电

Basic Call Management 基本呼叫管理

Beyond Capability of Maintenance 超出维修能力

Business Continuity Management 业务持续管理

**BCN** Barcelona, Spain (Airport Code) 西班牙巴塞罗纳机场

Beacon 信标,灯标

Boeing Control Number 波音控制号码,波音控制数量

Broadband Communication Network 宽带通信网

**BCOB** Broken Clouds or Better 多云或少云,碎云或疏云

**BCP** Basic Call Processing 基本呼叫处理

Boundary Control Point 边境管制站

Boundary Crossing Point 过境站

Break Cloud Procedure 穿云程序

Business Continuity Plan 业务持续计划

**BCPA** British Copyright Protection Association 英国版权保护协会

**BCPI** Bar Code Part Identification 条形码零件识别

**BCPN** Business Customer Premises Network 商业用户驻地网

**BCR** Baggage Check-in Record 行李受理记录

Baggage Claim Request 行李认领要求

Bar Code Reader 条形码阅读器

Blade Cooling Report 叶片冷却报告

Borocarbon Resistor 硼碳膜电阻

Branch and Count Register 转移与记数寄存器

Byte Count Register 字节计数寄存器

**BCRC** Bulk Crew Rest Compartment 散装货舱机组休息舱

**BCREF** Boeing Company Reference 波音公司基准

**BCRU** Battery Charge and Rectifier Unit 电瓶充电与整流组件

**BCS** Backup Control System 备用控制系统

Bar Code Scanner 条形码扫描器

Bar Code Sorters 条形码选卡机(分捡员)

Baseline Comparison System 基线比较系统

Basic Commitment Summary 基本协议要点

Beam Control System 波束控制系统

Block Check Decimal 十进制块检查

Block Check Sequence (数据)块检查顺序

Boeing Computer Services 波音计算机服务

Bore Cooling System 内腔冷却系统

Brake Control System 刹车控制系统

British Computer Society 英国计算机协会

Burner Control System 燃烧器控制系统

Business Computer System 商用计算机(网络)系统

**BCSC** Boeing Computer Services Company 波音计算机服务公司

**BCSM** Basic Call State Model 基本呼叫状态模型

**BCSOS** Boeing Computer Services Office Solutions 波音计算机服务处解法

**BCST** Broadcast 广播

**BCT** Bandwidth Coding Technique 带宽编码技术

Bandwidth Compression Technique 带宽压缩技术

Basic Color Term 基本色彩项

BITE and Configuration 自检和布局终端

**BCTR** Battery Charger Transfer Relay 蓄电池充电器转换继电器

**BCTS** Boeing Computer Terminal Services 波音计算机终端服务

Boeing Conversational Terminal System 波音会话式终端系统

**BCU** Basic Communication Unit 基本通信单元

Brake Control Unit 刹车控制组件

Buffer Control Unit 缓冲器控制组件

Bus Controller Unit 汇流条控制器组件,总线控制器装置

**BCUSM** Basic Call Unrelated State Model 呼叫不相关的状态模型

**BCV** Ball Check Valve 球形单向活门

Belly Cargo Volume 机腹货舱容量

Bleed Control Valve 引气控制活门

Bore Cooling Valve 内腔冷却活门

**BCWP** Budgeted Cost of Work Performed 工

作完成的费用预算

**BCWS**　Budgeted Cost of Work Scheduled 计划工作的费用预算

**BD**　Band 波段

Bangladesh 孟加拉国

Block Diagram 方框图

Blowing Dust 扬尘

Board 板,局,委员会

Boarding 登机

Bottom of Descent 下降底点

Bound 限制,边界

Boundary 边界

**BDA**　Backdrive Actuator 反驱动作动器

Battle Damage Assessment 战斗损伤评估

Business Decision Analysis 商务决策分析

Hamilton, Bermuda ( Airport Code ) 西印度群岛哈密尔顿机场

**BDAR**　Battlefield Damage Assessment & Repair 战场抢修

**BDB**　Bidirectional Data Bus 双向数据总线

Boeing Design/Build 波音设计/建造

**BDC**　Boeing Development Center 波音开发中心

**BDCAST**　Broadcast 广播

**B-DCN**　Backbone Data Communication Network 骨干数据通信网

**BDCO**　Boeing Designated Compliance Organization 波音指定的符合性组织

**BDCS**　Backup Departure Control System 离港备份控制系统

**BDD**　Binary Decision Diagram 二元决策图

Bus Description Document 总线说明文件

**BDE**　Beam Deflection Error 波束反射误差

**BDFC**　Bi-Directional Fiber-Coax 双向光纤—同轴线

**BDG**　Bridge 桥

**BDGP**　Broadcasting Data packet Grouping Protocol 广播数据包分组协议

**BDGT**　Budget 预算

**BDH**　Basic Decision Height 基本决断高度

**BDHI**　Bearing Distance Heading Indicator 方位距离航向指示器

**BDI**　Bearing Deviation Indicator 方位偏差指示器

Bearing Distance Indicator 方位距离指示器

Business Data Interchange 商务数据交换

**BDL**　Bundle 包,束

Hartford, CT, USA ( Airport Code ) 美国康涅狄格州哈特福德机场

**BDLD**　Bundled 包好的

**BDLI**　German Aerospace Industries Association 德国宇航工业协会

**BDM**　Boeing Design Manual 波音设计手册

**BDO**　Bandung, Indonesia ( Airport Code ) 印度尼西亚万隆机场

**BDP**　Bhadrapur, Nepal ( Airport Code ) 尼泊尔巴德拉普尔机场

**BDPSK**　Binary Differential Phase Shift Keying 二相差分相移键控

**BDR**　Border 国境

**BDRY**　Boundary 边界

**BDS**　Boeing Data Standard 波音数据标准

Boeing Design Standard 波音设计标准

Boeing Drafting Standard 波音图纸标准

Business & Decision System 经营与决策系统

**BDSL**　Broadband DSL 宽带数字用户线

Broadcast DSL 广播数字用户线

**BDU**　Bright Display Unit 高亮度显示器

**BDY**　Boundary 边界

**BE**　Belgium 比利时

**BEA**　British Europe Airways Corp. 英国欧洲航空公司

**BEAC**　Bank of the Central African States 中非国家银行

Boeing Engineering Analog Computer 波音工程模拟计算机

**BEAF**　Deaf Passenger 丧失听觉的旅客

**BEAM**　Boeing Electronic Assembly Modularization 波音电子组件模块化

Boeing Engagement Analysis Model 波音衔接分析模型

**BEAR**　Bearing 轴承,方位

**BEC**　Backward Error Correction 后向纠错
Base Extension Course 基地延长航线
Boeing Electronics Company 波音电子公司
British Electrotechnical Committee 英国电气技术委员会

**BECC**　Boeing Engineering & Construction Company 波音工程与建造公司

**BECN**　Backward Explicit Congestion Notification 后向显示拥塞控制

**BECO**　Booster Engine Cut-Off 助推发动机停车

**BECTL**　Boeing Engineering Computer Tools Library 波音工程计算机工具程序库

**BED**　Bedford, MA, USA（Airport Code）美国马萨诸塞州贝德福德机场
Stretch Installed 安装在客舱的担架

**BEDS**　Boeing Electronic Distribution of Software 波音软件电子分发
Boeing Electronics Delivery System 波音电子传输系统

**BEF**　Belgian Franc 比利时法郎

**BEFOR**　Before 之前

**BEG**　Belgrade, Yugoslavia（Airport Code）南斯拉夫贝尔格莱德机场

**BEGNG**　Beginning 开始

**BEH**　Benton Harbor, MI, USA（Airport Code）美国密执安州本顿港机场

**BEHC**　Boeing Equipment Holding Company 波音设备储备公司

**BEL**　Basic Equipment List 基本设备清单
Belem, Brazil（Airport Code）巴西贝伦机场
Belgrade 贝尔格莱德
Belong 属于
Below 低于

**BELF**　Breakeven Load Factor 盈亏平衡点载客率,盈亏平衡负载因子

**BELG**　Belgian 比利时人
Belgium 比利时

**BELW**　Brake Energy Limit Weight 刹车能量极限重量

**BEM**　Boundary Element Method 边界元方法

**BEMS**　Boeing Electronic Mail Service 波音电子邮件服务

**BEN**　Benghazi, Libya-Benina（Airport Code）利比亚班加西贝尼纳机场
Benin 贝宁

**BENEF**　Beneficiary, Benefit 利益

**BENG**　Beginning 开始

**BEP**　Bit Error Probability 误码概率

**BEPS**　Backup Electrical Power System 备用电源系统

**BER**　Basic Encoding Rules 基本编码规则
Berlin, Germany（Airport Code）德国柏林机场
Beyond Economical Repair 非经济修理
Bit Error Rate 误码率

**BERT**　Bit Error Rate Testing 误比特率测试

**BERTH**　Berthing 停机位

**BES**　Booster Exhaust Stream 助推器排气流
Brest, France（Airport Code）法国布雷斯特机场
Business Expansion Scheme 企业扩展计划

**BESA**　British Engineering Standard Association 英国工业标准协会

**BESD**　Boeing Electronics Support Division 波音电子支援部门

**BESG**　Boeing Electronic Systems Group 波音电子系统集团

**BET**　Best Estimate Trajectory 最佳估计轨道
Bethel, AK, USA（Airport Code）美国阿拉斯加州贝塞尔机场
Boeing Equivalent Thrust 波音等效推力
Bundle Effectivity Tab 电缆有效性卡片

**B-ET**　Broadband Exchange Termination 宽带交换终端

**BETA**　Battlefield Exploitation and Target Acquisition 战场侦察和目标捕获
Boeing Engineering Thermal Analyzer 波音工程热试验器（分析员）

**BETW**　Between 在…中间,在…之间

**BEV**　Bevel 斜角的

**BEW**   Basic Empty Weight 基本净重,基本空重

Beira, Mozambique ( Airport Code ) 莫桑比克贝拉机场

**BEX**   Broadband Exchange 宽带交换

**BEY**   Beirut, Lebanon ( Airport Code ) 黎巴嫩贝鲁特机场

Beyond 超出,不属于

**BF**   Back Feed 反馈

Base Frame 基帧

Beat Frequency 拍频

Block Fuel 轮挡耗油

Business Flight 公务飞行,商业飞行

**BFAANN**   British Federation Against Aircraft Nuisance 英国反飞机噪声损害联盟

**BFC**   Braking Force Coefficient 制动摩擦系数

**BFCFC**   Broadcast Feedback Channel Flow Control 广播反馈信道流控制

**BFCS**   Brief Case 文件包

**BFD**   Bradford, PA, USA ( Airport Code ) 美国宾西法尼亚州布拉德福德机场

**BFDAS**   Basic Flight Data Acquisition System 飞行基本数据采集系统

**BFE**   Basic Flight Equipment 基本飞行设备

Blind Flying Experiment 仪表飞行试验

Boeing Furnished Equipment 波音提供的设备

Buyer Finished Equipment 买方选装设备

Buyer Furnished Equipment 买方准备设备

**BFF**   Body Freedom Flutter 机身自由度颤振

**BFI**   Basic Flight Instrument 基本飞行仪器

Boeing Flight International 波音国际试飞站

**BFL**   Bakersfield, CA, USA ( Airport Code ) 美国加利福尼亚州贝克斯菲尔德机场

Balanced Field Length 平衡场长

**BFM**   Basic Flight Module 主飞行舱

**BFN**   Bloemfontein, South Africa-Jbm Hertzog ( Airport Code ) 南非布隆方丹机场

**BFO**   Beat Frequency Oscillator 拍频振荡器,差频振荡器

**BFP**   British Flying Permit 英国飞行许可证

**BFR**   Before 之前

Belgium Franc 比利时法郎

Buffer 阻尼,缓冲器

Biennial Flight Review 两年一次的飞行检查

**BFRL**   Basic Facilities Requirements List 基本设备需求单

**BFS**   Basic Flying School 基础飞行学校,初级飞行学校

Bureau of Flight Standards 飞行标准局

**BFSK**   Binary Frequency Shift Keying 二进制频移键控

**BFSR**   Boeing Field Service Representative 波音驻场服务代表

**BFT**   Binary File Transfer 二进制文件传送

Boundary Functional Test 边界功能测试

Buffet 快餐部,冷餐部

**BFTP**   Broadcasting File Transfer Protocol 广播文件传送协议

**BFTU**   Boeing Field Test Unit 波音外场测试仪

**BG**   Background 背景

Bag 包

Baggage 行李

Base Group 基群

Bearing 轴承,方位

Big 大

Body Gear 机身起落架

Border Gateway 边界网关

Build Group ( Assembly Group ) 装配组( 组装组)

**BGA**   British Gliding Association 英国滑翔协会

**BGC**   Beginning Climb 开始爬升

Build Group Component 装配组部件

**BGCOMPT**   Baggage Compartment 行李舱

**BGD**   Background 背景

Beginning Descent 开始下降

Budget 预算

**BGF**   Bangui, Central African Republic ( Airport Code）中非共和国班吉机场

Basic Global Function 基本全局功能

**BGI**   Basic Ground Instructor 基础地面课程教员

Bridgetown, Barbados Grantley Adams（Airport Code）巴巴多斯布里奇敦机场

**BGL**   Bulgarian Lev 保加利亚列弗

**BGM**   Background Music 背景音乐

Binghamton, NY, USA（Airport Code）美国纽约宾汗姆顿机场

Boarding Music 登机音乐

**BGN**   Begin, Beginning 开始

**BGO**   Bergen, Norway（Airport Code）挪威卑尔根机场

**BGP**   Boarding Gate Printer 登机证印表机

Border Gateway Protocol 边界网关协议

**BGP4**   Border Gateway Protocol version 4 边界网关协议第 4 版

**BGPT**   Beginning Procedure Turn 开始程序转弯

**BGR**   Bangor, ME, USA（Airport Code）美国缅因州班戈尔机场

Boarding Gate Reader 登机门登机证读取机, 磁条登机牌阅读器

**BGRV**   Boost-Guide Reentry Vehicle 助推滑翔重入式飞行器

**BGS**   Beacon Ground S-Band 地面 S 波段信标

Build Group Stack-up 装配组件排列

**BGSIA**   Beginning Standard Instrument Approach 开始标准仪表进近

**BGT**   Budget 预算

**BGTC**   Budgetary Control 预算控制

**BGTF**   Budgetary Forecast 预算预测

**BGV**   Back Ground Video 背景视像

**BGW**   Baghdad, Iraq（Airport Code）伊拉克巴格达机场

**BGY**   Milan, Italy-Orio Al Serio（Airport Code）意大利米兰贝加莫机场

**BH**   Block Hours 轮挡小时

Boeing Helicopters 波音直升机

**BHAB**   British Helicopter Advisory Board 英国直升机咨询委员会

**BHC**   Backhaul Check 回拽检查

Bell Helicopter Company 贝尔直升机公司

**BHCA**   Busy Hour Call Attempt 忙时试呼

Busy Hour Calling Amount 忙时呼叫次数

**BHD**   Belfast City, United Kingdom（Airport Code）英国贝尔法斯特城市机场

Bulkhead 加强隔框

**BHG**   Bristow Helicopters Group, Inc.（Great Britain）伯瑞斯托直升机集团公司（英国）

**BHM**   Basic Hypermedia Model 基本超媒体模型

Birmingham, Al, USA（Airport Code）美国阿拉巴马州伯明翰机场

**BHMD**   Binocular Helmet Mounted Display 双目头盔显示器

**BHN**   Brinell Hardness Number 布氏硬度值

**BHND**   Behind 之后

**BHP**   Brake Horsepower 制动马力

British Horse Power 英制马力

**BHRA**   British Hydraulics Research Association 英国液压研究协会

British Hydromechanics Research Association 英国流体力学研究协会

**BHS**   Baggage Handling System 行李处理系统, 行李分检系统

**BHT**   Baggage Handling Terms 行李处理文电

**BHTC**   Boeing Holloman Test Center 波音霍乐门测试中心

**BHWT**   Boeing Hypersonic Wind Tunnel 波音高超音速风洞

**BHX**   Birmingham, England, United Kingdom（Airport Code）英国伯明翰机场

**BI**   Brand Information 商标信息

Bus Interface 总线接口

Business Intelligence 商业情报

Royal Brunei Airlines 文莱皇家航空公司

**BIA**   Bahrain International Airport 巴林国际机场

Business Impact Assessment 业务影响分析

**BIAM** Beijing Institute of Aeronautical Materials 北京航空材料研究院

**BIAS** Beijing International Aviation Summit 北京国际航空峰会

**BIATA** British Independent Air Transport Association 英国独立航运输协会

**BIB** Backward Indicator Bit 反向指示比特

**BIC** Baggage Identification Chart 行李识别表

Bearer Identification Code 承载信道识别码

Boeing International Corporation 波音国际公司

Built-In Check 自校验

**BIDS** Baggage Information Display System 行李信息显示系统

Broadband Integrated Distribution Star Network 宽带综合分配星形网

**BIDT** Billing Information Data Tape 订座开账信息数据带

Booking Information Data Tape（机票）订购数据（记录）带

**BIEPP** Boeing Installed Engine Performance Program 波音安装的发动机性能程序

**BIFA** British International Freight Association 英国国际货运协会

**BIG** BCS Interactive Graphics System 波音计算机服务交互式图形系统

**BIGFON** Broadband Integrated Fiber Optic Network 宽带综合光纤通信网

**BIL** Billings, MT, USA（Airport Code）美国蒙大拿州比灵斯机场

Billion 十亿（美国），万亿（英国）

Bus Interface Logic 总线接口逻辑

**BIM** Blade Indicating Method 桨叶裂纹报警系统

Blade Inspection Method 桨叶检验法

British Institute of Management 英国管理学会

**BIN** Bamiyan, Afghanistan（Airport Code）阿富汗巴米扬机场

Binary 二进制

Business Information Network 商业信息网络

**BINGO** Bearing Indicator and Navigator to Ground Operator 地面管制员方位指示器和导航仪

**BIO** Bilbao, Spain（Airport Code）西班牙毕尔巴鄂机场

**BIP** Baggage Improvement Programme 行李改善计划

Biparting Doors 双拉门，双向滑动门

Bi-Phase 双相

Bit Interleaved Parity 比特间插奇偶校验

**BIP-ISDN** Broadband, Intelligent and Personalized ISDN 宽带化、智能化和个人化的ISDN

**BIPL** Bi-Phase Level 双向电平

**BIPM** Backplane Interface/Power Monitor 后板接口电源监控器

International Bureau of Weight and Measures 国际计量局

**BIPS** Billion Instructions Per Second 十亿条指令/秒

**BIR** Baggage Irregularity Report 行李运输事故记录

**BIRE** British Institute of Radio Engineers 英国无线电工程师学会

**BIS** Bank for International Settlements 国际结算银行

Bismarck, ND, USA（Airport Code）美国北达科他州俾斯麦机场

Board of Inspection and Survey 检查和鉴定委员会

Business Information System 商业信息系统

**B-ISDN** Broadband Integrated Services Digital Network 宽带综合业务数字网

**BIST** Built-in Self Test 机内自测

Built-in Self Tester 内装自检器

**BIT** Built-in-Test 机内自检测

**BIT/FI** Built-in Test and Fault Isolation 自检测和故障隔离

**BITA** Bilateral Interline Traffic Agreement 双边联运协议

**BITE** Built-in Test Equipment 机内自检设备

**BITEM** Built-in Test Equipment Manual 自检手册

**BITG** Boeing Industrial Technology Group 波音工业技术集团

**BITS** Building Integrated Timing System 大楼综合定时系统

**BITUM** Bituminous 沥青的

**BIU** Basic Information Unit 基本信息单元
BITE Interface Unit 机内测试设备接口组件

**BIX** Boeing Information Exchange 波音信息交换

**BN** Benin 贝宁

**BJI** Bemidji, MN, USA(Airport Code) 美国明尼苏达州比米齐机场

**BJM** Bujumbura, Burundi (Airport Code) 布隆迪布琼布拉机场

**BJX** Leon, Mexico (Airport Code) 墨西哥莱昂机场

**BK** Back 反面,向后,背后
Bank 银行
Black 黑色的
Book 书
Brake 制动,刹车,刹住(车)
Break 中断,打断

**BKBLE** Bookable 可订购的

**BKCY** Bankruptcy 破产

**BKD** Background 背景
Booked 预定的

**BKDN** Breakdown 中断,故障,分解,分离

**BKG** Booking 订座

**BKGD** Background 背景

**BKGOFC** Booking Office 售票处

**BKGRD** Background 背景

**BKHD** Bulkhead 隔舱,隔板

**BKI** Kota Kinabalu, Sabah, Malaysia (Airport Code) 马来西亚沙巴州哥打基纳巴卢机场

**BKK** Bangkok, Thailand (Airport Code) 泰国曼谷机场

**BKLOG** Backlog 积压货物

**BKLT** Booklet 小册子

**BKN** Broken 破损,裂开,裂云

**BKO** Bamako, Mali-Senou (Airport Code) 马里巴马科塞努机场

**BKPT** Bankrupt 破产

**BKR** Breaker 断电器

**BKT** Basket 托架

**BKUP** Backup 备份,备用

**BKW** Beckley, WV, USA (Airport Code) 美国西弗吉尼亚州贝克里机场

**BKWD** Backward 倒退

**BKX** Brookings, SD, USA(Airport Code) 美国南达科他州布鲁金斯机场

**BL** Barrel 桶
Baseline, Base Line 基线,基准线
Between Layer 层间
Bill of Lading 提货单
Bill 账单,票据
Bleed 放气,引气
Blue 蓝色
Buttock Line 纵剖线

**BLA** Barcelona, Venezuela(Airport Code) 委内瑞拉巴塞罗纳机场

**BLAC** British Light Aviation Center 英国轻型飞机航空中心

**BLC** Basic Lines Catalog 基本航线目录
Boundary Layer Control 附面层控制

**BLCS** Boundary Layer Control System 附面层控制系统

**BLCU** Boundary Layer Control Unit 附面层控制器

**BLD** Bleed 引气,放气

**BLDG** Building 建筑物

**BLER** BLock Error Rate 误块率

**BLES** Brake Life Extension System 刹车寿命延长系统

**BLEU** Blind Landing Experimental Unit 盲降实验装置,仪表着陆实验装置

**BLF** Bluefield, WV, USA (Airport Code) 美国西弗吉尼亚州布尔菲尔德机场

**BLG** Body Landing Gear 机身起落架

**BLI** Bellingham, WA, USA (Airport Code)

美国华盛顿州贝灵汉机场

**BLID**　Blind Passenger 盲人旅客

**BLK**　Black 黑色的

　　Blank 空白

　　Block or blocking 封锁,断路

　　Block Time 轮挡时间

　　Block 块,阻塞,地段

　　Bulk 散货

**BLKD**　Blocked seats ( load control terms ) 保留机位

　　Blocked 限制出售,受阻,封锁

　　Bulkhead 隔板,隔框

**BLKT**　Blanket 地毯

**BLL**　Below Lower Limit 低于下限

**BL-L**　Butt Line Left 左纵剖线

**BLN**　Balloon 气球

**BLNC**　Balance 平衡

**BLND**　Blind 盲人

**BLNK**　Blank 空白

**BLNKT**　Blanket 地毯

**BLNT**　Broadband Local Network Technology 宽带局部网技术

**BLO**　Below Clouds 云下

　　Buttock Line Zero 零纵剖线

**BLOFF**　Off Block Time 取轮挡时间

**BLOM**　Booster Lift off Mass 助推器起飞质量

**BLON**　On Block Time 上轮挡时间

**BLOSA**　Body Line-of-Sight Angle 机体视线角

**BLOW**　Blower 鼓风机

　　Booster Lift off Weight 助推器起飞重量

**BLP**　Bypass Label Processing 旁路标号处理

**BLQ**　Bologna,Italy( Airport Code ) 意大利博洛尼亚机场

**BLR**　Bangalore,India( Airport Code ) 印度班加罗尔机场

**BL-R**　Butt Line Right 右纵剖线

**BLR**　BLocking Request 阻塞请求

**BLS**　Beacon Landing System 信标着陆系统

　　Bezel Light Sensor 仪表(遮光)板光传感器

　　Broadband Local Switch 宽带市话(本地)交换

**BLSN**　Blown Snow 吹雪

　　Boundary Layer Self Noise 附面层自身噪声

**BLSS**　Base Level Support Sufficiency 基地级满足供应量

**BLST**　Ballast 配重,镇流器

**BLT**　Bolt 螺栓

**BLTN**　Bulletin 通告,公告,通报

**BLU**　Basic Link Unit 基本链路单元

　　Blue 蓝色的

**BLV**　Bleed Valve 引气活门,放气活门

**BLZ**　Blantyre,Malawi( Airport Code ) 马拉维布兰太尔机场

**BM**　Back Marker 背台指点标,后航道指点标

　　Bandwidth Manager 带宽管理器

　　Beam 波束,梁

　　Bio-Manufacturing 生物制造

　　Boundary Marker 跑道边界标,(机场)界标

　　Breakdown Maintenance 破损维修,分解维修

　　Breech Mechanism 尾部机构

　　Buffer Memory 缓冲存储器

　　Buffer Multiplexer 缓冲器多路选择器

　　Buffet Margin 抖动裕度,振动系数

**BMA**　Block Matching Algorithm 块匹配算法

　　Boeing Management Association 波音管理协会

　　Boeing Military Airplanes 波音军用飞机

　　British Manufacturers Association 英国制造商协会

　　Stockholm,Sweden-Bromma Airport ( Airport Code ) 瑞典斯德哥尔摩布罗马机场

**BMAA**　British Microlight Aircraft Association 英国微型飞机协会

　　British Minimum Aircraft Association 英国小型飞机协会

**BMAC**　Boeing Military Airplane Company 波音军用飞机公司

**BMAD**　Boeing Military Airplane Development

波音军用飞机研制

**BMAG** Boeing Military Airplane Group 波音军用飞机集团

**BMAP** Boeing Mission Analysis Program 波音飞行(任务)分析程序

**BMAPDB** Boeing Mission Analysis Program Database 波音飞行(任务)分析程序数据库

**BMC** Bleed Management Computer 引气管理计算机

Bleed Monitoring Computer 引气监控计算机

Budget and Management Committee 预算管理委员会

Bus Master Controller 总线主控制器

**BMDH** Basic Minimum Descent Height 基本最低下降高度

**BMEP** Brake Mean Effective Pressure 制动平均有效压力

**FMEP** Friction Mean Effective Pressure 摩擦平均有效压力

**BMI** Banded Matrix Iteration 分段矩阵迭代

Bloomington, IL, USA(Airport Code) 美国伊利诺州布鲁明顿机场

**BML** Boeing Meteorology Laboratory 波音气象学实验室

**BMM** Block Maxima Method 区块最大分组法

Boarding Music Machine 登机音乐播放机

**BMP** Batch Message Processing 成批信息处理

Best Management Practices 最佳管理措施

Billing Management Point 计费管理点

Brake Mean Pressure 平均刹车压力

**BMR** Body-Mounted Radiator 机体安装的服务器

**BMS** Basic Meteorological Services 基本气象服务

Billing and Management System 计费和管理系统

Boeing Management System 波音管理系统

Boeing Marine Systems 波音海洋系统

Boeing Material Specification 波音材料规范

Boeing Material Standards 波音器材标准

Bulletin Meteo Special 特殊气象通告

Burner Management System 燃烧器管理系统

Business Management Software 商务管理软件

Business Management System 商务管理系统

**BMT** Boeing Materials Technology 波音材料技术

**BMTC** Boeing Military Training Center 波音军用训练中心

Boeing Mojave Test Center 波音莫杰夫测试中心

**BMTS** Bandwidth Manager and Traffic Scheduler 带宽管理和业务调度程序

**BMU** Basic Measurement Unit 基本测量单位

**BMV** Brake Metering Value 刹车配油活门,刹车调节活门

**BN** Backbone Network 骨干网

Balancing Network 平衡网络

Bank 银行

Blind Navigation 仪表导航

Blowing Sand 扬沙

Brunei 文莱

**BNA** Boeing Network Architecture 波音网络结构

Nashville, TN, USA (Airport Code) 美国田纳西州纳什维尔机场

**BNACC** Boeing Network Access 波音网络存取

**BNCH** Bench 工作台,操作台

**BNCHBD** Benchboard 工作台,操作台

**BNCSR** British National Committee on Space Research 英国国家航天研究委员会

**BND** Band 波段

Bend 弯曲

Bonded 贴接的

Bound 飞往

**BNDDRY** Boundary 边界

**BNDG** Bonding 贴接

**BNDRY** Boundary 边界,边缘

**BNDY** Boundary 边界

**BNE** Brisbane, Australia ( Airport Code) 澳大利亚布里斯班机场

**BNFT** Benefit 利益

**BNJ** Bonn, Germany ( Airport Code) 德国波恩机场

**BNK** Bank 银行

**BNN** Boundary Network Node 边界网络结点

**BNR** Binary Numerical Reference 二进制数字基准

Binary Words 二进制数据制

Binary 二进制

Burner 燃烧室

**BNS** Binary Number System 二进计数制

Broadband Network Services 带宽网络服务

**B-NT** Broadband Network Termination 宽带网络终端

**BO** Back out 关断, 收回

Blocking Oscillator 间歇振荡器

Body Odour 身体气味

Boeing Objective 波音目标

Boeing 波音

Boundary Light 边界灯

Burn off 烧掉, 耗油

By Operational Requirement 按运行要求

**BOA** Basic Object Adapter 基本对象适配器

Basic Ordering Agreement 基本订购协议

Break-off Altitude 出云高度, 航向突变高度, 中断高度

Breakout Altitude 转入目视飞行高度

**BOAC** British Overseas Airways Corporation 英国海外航空公司

**BOAN** Business-oriented Optical Access Network 面向企业的光接入网

**BOB** Bureau of Budgets 预算局

**BOC** Bank of China 中国银行

Base Operations Center 基地( 航行) 调度中心

Blowout Coil 烷弧线圈

Body on Chassis 底盘上机身

Boeing of Canada 波音加拿大公司

Boeing Operations Center 波音运营中心

Bottom of Climb 爬升起始点

**BOCAPS** Boeing Computerized Automated Publishing System 波音计算机化自动出版系统

**BOD** Bandwidth on Demand 按需提供带宽

Basic Operational Data 基本运行数据

Beneficial Occupancy Date 受益占用日期

Biological Oxygen Demand 生物需氧量

Board of Directors 董事会, 理事会

Bordeaux, France ( Airport Code) 法国波尔多机场

Bottom of Descent 下降底部

**BOE** Basis of Estimate 评价基础

Boeing 波音公司

**BOE BFE** Boeing Owned Buyer Furnished Equipment 波音拥有用户提供设备

**BOECOM** Boeing Communications Computing System 波音通信计算系统

**BO-EOC** Bit-Oriented Embedded Operations Channel 面向比特的嵌入式工作信道

**BOF** Beacon off 信标台关闭

Boeing Fixture 波音夹具

Burn off 烧掉, 耗油

**BOG** Bogota, Colombia-Eldorado ( Airport Code) 哥伦比亚波哥大埃尔多拉多机场

**BOGART** Boeing Graphic Art 波音图形工艺

**BOH** Break-off Height 中断高度, 航向突变高度, 出云高度

**BOI** Basis of Issue 发行根据

Boise, ID, USA ( Airport Code) 美国爱达荷州博伊西机场

Break of Inspection 检查中断, 检查间隔, 检验间隔

**BOL** Beginning of Lifetime 寿命初期

Bill of Lading 提货单

Bolivia 玻利维亚

Bus-Only Lane 巴士专用线

**BOLD** Boeing Online Data 波音在线数据

**BOLDS** Burroughs Optical Lens Docking System 保鲁夫视镜泊机系统

**BOM** Beginning of Message 报文开始, 底部,

基部
Beginning-of-Month 月初
Bill of Materials 器材账单,器材清单,物料清单
Bombay,India(Airport Code)印度孟买机场
**BOMARC** Boeing-Michigan Aeronautical Research Center 波音密执安航空研究中心
**BON** Beacon on 信标台开放
Bound 飞往
Building-out Network 附加网络
**BONI** Basic Optical Network Interface 基本光网接口
Broadband Optical Network Terminal 宽带光网络终端
**BOO** Bodo,Norway(Airport Code)挪威博德机场
**BOOT** Build-Own-Operate-Transfer（机场）建设—拥有—运营—转让
**BOOTP** BOOTstrapping Protocol 引导协议
**BOP** Balance of Power 能源平衡
Beginning of Packet 分组信息开始
Bill of Processing 工艺清单
Bit Oriented Protocol 面向位的协议
Boeing of Portland 波音波特兰公司
**BOPACE** Boeing Plastic Analysis Capability for Engines 发动机波音塑性分析能力
**BOS** Basic Operating System 基本操作系统
Boston,MA,USA-Logan（Airport Code）美国马萨诸塞州波士顿洛根机场
Brief of Service 简历
Bureau of Safety 飞行安全局
**BOSSCO** Boeing Optical Shaped Scan Correlator 波音光学成形扫描相关器
**BOSTID** Board on Science and Technology for International Development（USA）国际开发科学技术委员会
**BOT** Beginning of Tape 磁带始端
Bottom 底,底部
Bought 购买
Build-Operate-Transfer（机场）建设—经营—转让

**BOTT** Bottle 瓶
**BOV** Bar out of View 指令杆消失,指令杆偏出视野
**BOVC** Below Overcast 密云下飞行
**BOW** Basic Operating Weight 基本使用重量
Brake-off Weight 开始滑行时的重量,松刹车重量
**BP** Back Propagation 反向传播
Bag Phone 行李电话
Bandpass 带通
Barometric Pressure 大气压力,气压,气压计压力
Bid Price（机票）最低可售价
BITE Panel 自检面板,机内测试设备专家组
Blueprint 蓝图
Board President 董事长
Boilerplate 模拟舱
Booster Pressure 增压泵
Bottom Plug 下电嘴
Budget Program 预算计划
**BPA** Back Propagation Algorithm 反向传播算法
Basic Pressure Altitude 修正压力高度,标准修正表高
Blanket Purchase Agreement 一揽子采购协议
Blower Plate Assembly 风扇叶片组件
Business Process Analysis 业务流程分析
Business Process Assessment 业务流程评估
**BPB** Bank Post Bills 银行邮寄汇票
**BPC** Back Pressure Control 背压控制
Barometric Pressure Control 气压控制
**BPCU** Bus Power Control Unit 汇流条电源控制组件,汇流条功率控制装置
**BPD** Boost Pressure Difference 增升压差
**BPDU** Bridge Protocol Data Units 桥式协议数据单元
**BPE** Boom-Person Exposure 音爆对人体影响指数
**BPF** Band Pass Filter 带通滤波器

**BPFC** Backup Power Frequency Converter 备用电源频率转换器

**BPG** Break Pulse Generator 断路脉冲发生器

**BPI** Bits per Inch 位/英寸

Boeing Process Instruction 波音流程指引

Brake Pressure Indicator 刹车压力指示器

Business Process Instruction 业务流程指引

**BPL** Basic Parts List 基本零件清单

Birthplace 出生地

**BPM** Baggage Processed Message 行李已处理信息

Business Process Management 业务流程管理

Business Process Modeling 业务流程建模

**BPMI** Business Process Management Initiative 业务流程管理联盟

**BPML** Business Process Modeling Language 业务流程建模语言

**BPMS** Boeing Project Management System 波音项目管理系统

**BPN** Boeing Partner Network 波音合作伙伴网络

Business Partner Network 商业合作伙伴网络

**BPOC** Before Proceeding on Course 进入航线之前

**BPON** Broadband Passive Optical Network 宽带无源光网络

**BPP** Boarding Pass Printer 条码登机牌打印机

Bus Protection Panel 汇流条保护板

**BPQL** Business Process Query Language 业务流程查询语言

**BPR** Business Process Redesign 业务过程再设计

Business Process Reengineering 业务流程重组,业务流程再造

Bypass Ratio 涵道比,流量比

**bps** Bits per Second 比特/秒,每秒传送位数,每秒比特数,位/秒

**BPS** Balanced Pressure System 均衡压力系统

Beacon Processing System 信标处理系统

Boeing Parts Standards 波音零件标准

Boeing Petroleum Service 波音加油服务

Boeing Process Specification 波音工艺规范

**BPSK** Biphase Shift Keying 两相相移键控

**BPSS** Basic Packet Switched System 基本的分组交换系统

**BPSU** Brake and Position Sensor Unit 刹车和位置传感器组件

**BPT** Beaumont, TX, USA (Airport Code) 美国得克萨斯州博蒙特机场

Begin Procedure Turn 开始程序转弯

British Petroleum (Great Britain) 英国汽油公司

**BPTU** Brake Pedal Transmitter Unit 刹车脚蹬传感组件

**BPU** Basic Processing Unit 主处理装置

Battery Protection Unit 电池保护单元

**BPV** Back Pressure Valve 背压活门

**BPX** Broadband Packet Exchange 宽带分组交换

**BQ** A Reply to a Request 回答所求问题

**BQA** British Quality Association 英国质量协会

**BQC** Basic Quality Control 基本质量控制

**BQK** Brunswick, GA, USA (Airport Code) 美国佐治亚州布伦瑞克机场

**BQS** Basic Quality System 基本质量体系

Blagoveschensk, Russia (Airport Code) 俄罗斯布拉格维申斯科机场

**BR** Backward Reporting 后向报告

Bandwidth Renegotiation 带宽协调

Battery Relay 蓄电池继电器

Battlefield Repair 战场抢修

Bearing 轴承,方位

Bedroom 卧室

Boeing Requirement 波音要求

Border Router 边界路由器

Branch 分支

Brand 商标

British 英国人

Brown 棕色的
Bus Request 总线请求
Mist 轻雾
Radiocommunication Bureau 无线电通信局

**BRA** Basic Rate Access 基本速率接入
Bench Replacement Assembly 台架上可替代装配
Bergen Aviation (Norway) 伯极航空公司（挪威）
Brazil 巴西
Business Requirement Analysis 商务需求分析

**BRAMS** Beacon Range Altitude Monitor System 信标距离高度监控系统

**BRAN** Braking Action None 无制动作用
Broadband Radio Access Network 宽带无线接入网

**BRAS** Bench Replaceable Assemblies 台架上可替代装配

**BRAVIA** British-Russian Aviation Co. 英俄航空公司

**BRC** Brace 支柱,张线

**BRCH** Branch 分支

**BRD** Board 板,局,委员会
Boarding 登机
Braid 编网
Brainerd, MN, USA (Airport Code) 美国明尼苏达州布雷纳德机场
Broad 宽

**BRDCST** Broadcast 广播

**BRDD** Boarded 已登机

**BRDG** Boarding 登机
Bridge 桥,电桥

**BRDMUSIC** Boarding Music 登机音乐

**BRE** Bremen, Germany (Airport Code) 德国不莱梅机场

**BREL** Boeing Radiation Effects Laboratory 波音辐射效应实验室

**BRF** Bandwidth Request Field 带宽要求字段

**BRFP** Baseline Reference Flight Plan 基准参考飞行平面

**BRG** Bearing 方位,轴承
Bridge 桥,搭桥
Bring 带来

**BRI** Bari, Italy (Airport Code) 意大利巴里机场
Basic Rate Interface 基本速率接口（界面）

**BRIL** Brilliance 亮度

**BRIT** Britain 不列颠
British 英国人

**BRITE** Bright Radar Indicator Tower Equipment 塔台高亮度雷达显示设备
Bright 明亮

**BRK** Brake 刹车,刹车毂
Broken 破碎的

**BRKDWN** Breakdown 破坏,故障

**BRKFST** Breakfast 早餐

**BRKG** Braking 刹车

**BRKGE** Brokerage 回扣,佣金

**BRKNG** Braking 刹车

**BRKR** Breaker 跳开关,断路器

**BRKS** Brakes 刹车

**BRKT** Bracket 托架

**BRL** Barrel 桶,筒
Bearing Range Line 方位距离线
Building Restriction Line 建筑物限制线
Burlington, IA, USA (Airport Code) 美国依阿华州布灵顿机场

**BRM** Barometer 气压表
Barquisimeto, Venezuela (Airport Code) 委内瑞拉巴基西梅托机场
Basic Reference Model 基本参考模型
Business Research Method 商业研究方法
Business Resource Management 商业资源管理
Business Risk Management 商业风险管理

**BRN** Berne, Switzerland (Airport Code) 瑞士伯尔尼机场
Brown 棕色
Brunei 文莱

**BRNAV** Basic Area Navigation 基本区域导航

**BRNG** Bearing 方位

**BRO** Brownsville, TX, USA（Airport Code）美国得克萨斯州布朗斯维尔机场

**BROC** Best Rate of Climb 最佳爬升率

**BROK** Broken 破损的
Brokerage 回扣,佣金

**BROT** Brought 带来,转接

**BRP** Basic Research Plan 基础研究计划
Basic Reservation Protocol 基本预留协议
Beacon Ranging Pulse 信标测距脉冲
Body Reference Plane 机身参考平面

**BRQ** Brno, Czech Republic（Airport Code）捷克布尔诺机场

**BRS** Baggage Reconciliation System 行李再确认系统
Best Range Speed 最佳速度
Brass 黄铜
Bristol, England, United Kingdom（Airport Code）英国布里斯托尔机场

**BRST** Broadcast 广播

**BRSU** Broadband Remote Switch Unit 宽带远端交换单元

**BRT** Bright 明亮,发亮的
Brightness 亮度

**BRTH** Berth 卧铺,铺位

**BRU** Brussels, Belgium-National（Airport Code）比利时布鲁塞尔国家机场

**BRW** Barrow, AK, USA（Airport Code）美国阿拉斯加州巴罗机场
Boeing Retention Warehouse 波音保管仓库
Brakes Release Weight（飞机）开始起飞重量

**BRWT** Boeing Research Wind Tunnel 波音研究风洞

**BRZ** Binary Return to Zero 二进制回零
Breeze 微风
Bronze 青铜色

**BS** Bags 包
Base Station 基站
Battery Switch 蓄电池开关
Beam Splitter 分光器

Blowing Snow 吹雪
Body Station 机身站位
Boeing Specification 波音规范
Boresight 瞄准线
Both Sides 双边
British Standard 英国标准
Broadcast Satellite 广播卫星
Broadcast/Broadcasting Station（Commercial）广播电台（商用）
Bus System 总线系统
Bypass Switch 旁通转换

**BSAE** Bachelor of Science in Aeronautical Engineering 航空工程理学学士
Bachelor of Science in Aerospace Engineering 航空和航天工程理学学士

**BSAP** Broadband Service Access Point 宽带业务接入点

**BSAS** Beta Stability Augmentation System β稳定性增益系统

**BSB** Boeing Service Bulletin 波音服务通告
Brasilia, Brazil（Airport Code）巴西巴西利亚机场

**BSC** Balanced Score Card 平衡记分卡
Base Station Controller 基站控制器
Basic 基本,基础
Beam Steering Computer 波束控制计算机
British Safety Council 英国安全委员会

**BSCCY** Basic Currency 基础货币

**BSCE** Bird Strike Committee Europe 欧洲鸟撞委员会

**BSCF** Basic Fare 基础价格

**BSCI** Bird Strike Committee Italy 意大利鸟撞事件委员会

**BSCR** Basic Rate 基本运费,基本运价

**BSCS** Baggage Security Check Systen 行李安全检查系统
Braking and Steering Control System 刹车和转弯控制系统

**BSCU** Brake System Control Unit 刹车系统控制组件
Braking and Steering Control Unit 刹车和转

弯控制组件

**BSD** Base Station Demodulator 基站解调器

Beside 旁边

Boeing Services Division 波音服务部

Both Sideband 双边带

**BSEMT** Basement 地下室

**BSFC** Brake Specific Fuel Consumption 制动比油耗

**BSG** Boeing Support Group 波音支援组

**BSHG** Bushing 垫套

**BSI** Boeing Services International, Inc. 波音国际服务公司

Borescope Inspection 孔探仪检查

British Standards Institution 英国标准学会

**BSIAP** Beginning Straight in Approach 开始直线进近

**BSIC** Base Station Identification Code 基站标识码

**BSIP** Bird Strike Identification Program 鸟击鉴定计划

**BSL** Basel, Switzerland ( Airport Code ) 瑞士巴塞尔机场

Body Station Line 机身站位线

Buried Service Lightguide cable 埋式光缆

**BSLS** Boeing Space Launch Services 波音空间发射服务

**BSM** Backward Setup Message 反向建立消息

Baggage Source Message 行李来源信息

Base Station Modulator 基站调制器

Basic Security Module 基本安全模块

**BSMS** Boeing Safety Management System 波音安全管理系统

Broadcasts Short Message Service 广播短信息服务

**BSN** Backward Sequence Number 反向序号

Broadband Service Node 宽带业务节点

Broadcast and Select Network 广播及选择网络

**BSNS** Business 交易,商业,业务,事物

**BSP** Bank Settlement Plan 银行清账计划

Billing and Settlement Plan 开账与结算计划

Board Support Package 登机支援箱

Boeing Standard Performance 波音标准性能

Boeing Standard Program 波音标准程序

Bulk-Synchronous Parallel 大容量同步并行

**BSR** Banker's Selling Rate 银行货币卖出费率

Basra, Iraq ( Airport Code ) 伊拉克巴士拉机场

Bus Sensing Relay 汇流条敏感继电器

**BSRC** Boeing Scientific Research Center 波音科学研究中心

**BSRL** Boeing Scientific Research Laboratory 波音科学研究实验室

**BSRM** Boeing Small Research Module 波音小型研究模块

**BSS** Base Station Subsystem 基站子系统

Base Station System 基站系统

Best Signal Selection 信号优选

Boeing Support Services 波音支援服务

Boeing Support Standards 波音支援标准

British Standard Specification 英国标准规范

Broadband Switching System 宽带交换系统

Broadcast Satellite Service 广播卫星业务

Bulk Storage System 大容量存储系统

**BST** Beijing Summer Time 北京夏季时间

Boost 冲压,升压,增压

Bost, Afghanistan ( Airport Code ) 阿富汗博斯特机场

British Summer Time 英国夏季时间

**BSTA** Body Station 机身站位

**BSTD** Bastard 非标准的

**BSTLP** Boeing Standard Thrust Limit Program 波音标准推力极限程序

**BSTS** Boost Surveillance and Tracking System 增强监视与跟踪系统

**BSU** Base Station Unit 基站单元

Beam Steering Unit 波束操纵组件

Bearer Switchover Unit 承载切换单元

Bypass Switch Unit 旁通转换组件,旁路转换单元

**BSV** Burner Selection Valve 燃烧选择活门
Burner Staging Valve 分级燃烧活门

**BSVC** Broadcast Signaling Virtual Channel 广播信令虚通道

**BSWS** Boeing Software Standard 波音软件标准

**BSWT** Boeing Supersonic Wind Tunnel 波音超音速风洞

**BSY** Busy 占线

**BT** Baggage Transfer 转港行李
Bought 买
Bus Tie, Bustie 汇流条连接

**BTA** Beijing Telecommunications Administration 北京电信管理局
Beijing Terminal Area 北京终端区
Berth Tenancy Agreement 泊位租用合约
Best Technical Approach 最佳技术途径
Bilateral Trade Agreement 双边贸易协定
Business Technology Association 商业技术协会

**B-TA** Broadband Terminal Adaptor 宽带终端适配器

**BTAC** Bus Tie Alternating Contactor 汇流条连接转换接触器

**BTAO** Bureau of Technical Assistance Operations 技术援助业务局(技援局)

**BTB** Bus Tie Breaker 汇流条连接断路器
Business to Business 企业对企业的电子商务

**BTBR** Bus Tie Breaker Relay 汇流连接接触器继电器

**BTC** Basic Training Center 基础(飞行)训练中心
Brake Temperature Compensator 刹车温度补偿器
Breath Testing Centre 呼吸测试中心
Bus Tie Contactor 汇流条连接接触器
Business to Customer 企业对个人的电子商务

**BTCC** Bus Tie Continuous Contactor 汇流条持续连接接触器

**BTCM** Brake Temperature Compensation Module 刹车温度补偿组件

**BTCP** Broadcasting Transfer Control Protocol 广播传送控制协议

**BTDC** Before Top Dead Center 在上死点前

**B-TE** Broadband Terminal Equipment 宽带终端设备

**BTEC** Business and Technology Education Council 工商及技术教育委员会

**BTED** Bundle Termination Equipment Date 电缆终止设备日期

**BTH** Bathroom 盥洗室

**BTHE** Brake Thermal Efficiency 制动热效率,刹车热效率

**BTI** Balanced Technology Initiative 平衡技术倡议
Boeing Technology International, Inc. 波音国际技术公司

**BTL** Between Layers 云层间飞行,云层之间
Bottle 氧气瓶,灭火瓶

**BTM** Baggage Transfer Message 行李转运报
Bottom 底部
Brake Temperature Monitor 刹车温度监视器
Butte, MT, USA (Airport Code) 美国蒙大拿州比尤特机场

**BTMA** Busy Tone Multiple Access 忙音多址访问

**BTMS** Brake Temperature Monitoring System 刹车温度监视系统

**BTMU** Brake Temperature Monitoring Unit 刹车温度监控组件

**BTN** Between 之间
Boeing Telecommunications Network 波音电信网
Boeing Telephone Network 波音电话网
Button 按钮,按键

**BTNCC** Boeing Telecommunications Network Control Center 波音电信网控制中心

**BTO** Build-Transfer-Operate (机场)建设—转让—经营
Time Between Overhaul 大修间隔期限

**BtoB**  Business to Business 企业对企业
**BtoC**  Business to Consumer 企业对用户
**BTP**  Baggage Tag Printer 行李牌打印机
Bulk Transfer Protocol 成批转移协议
**BTR**  Baton Rouge, LA, USA(Airport Code) 美国路易斯安那州巴吞鲁日机场
Bus Tie Relay 汇流条连接继电器
Butter 黄油
**BTRY**  Battery 电瓶
**BTS**  Base Transceiver Station 基地收发信机站
Bleed Temperature Sensor 引气温度传感器
Blue Tool Steel 蓝色工具钢
Boeing Technology Services 波音技术服务
Boeing Test Support 波音测试支援
Brake Temperature Sensor 刹车温度传感器
Bratislava, Slovakia-Ivanka (Airport Code) 斯洛伐克布拉迪斯发机场
**BTSI**  Boeing Technology Services International Inc. 波音国际技术服务公司
**BTTPU**  Brake Temp/Tire Pressure Unit 刹车温度/轮胎压力组件
**BTU**  Basic Transmission Unit 基本传输单元
British Thermal Unit 英制热量单位
**BTV**  Blast Test Vehicle 爆破试验飞行器
Burlington, VT, USA(Airport Code) 美国佛蒙特州布灵顿机场
**BTWT**  Boeing Transonic Wind Tunnel 波音跨音速风洞
**BTY**  Battery 电瓶
**BTZ**  Bursa, Turkey (Airport Code) 土耳其布尔萨机场
**BU**  Back Up, Backup 备用
Battery Unit 电瓶组件
Branching Unit 支路单元
Buffer Unit 缓冲组件
Build Up 建造, 装配, 组合, 增长
Bureau 局, 处
Burma 缅甸
Bushel 蒲式耳
**BUA**  British United Airways 英国联合航空公司

**BUAA**  Beijing University of Aeronautics and Astronautics 北京航空航天大学
**BUCCS**  Boeing Uniform Classification & Coding System 波音统一分类与编码系统
**BUCS**  Back Up Component System 备用组件系统
**BUD**  Budapest, Hungary (Airport Code) 匈牙利布达佩斯机场
Budget 预算
**BUE**  Buenos Aires, Argentina(Airport Code) 阿根廷布宜诺斯艾利斯机场
**BUEC**  Backup Emergency Communications 备用紧急通信
**BUECE**  Back-up Emergency Communications Equipment 备用紧急通信设备
**BUF**  Buffalo, NY, USA(Airport Code) 美国纽约州布法罗机场
Buffer 缓冲
**BUG**  Backup Generator 备用发电机
**BUH**  Bucharest, Romania (Airport Code) 罗马尼亚布加勒斯特机场
**BUKD**  Bulkhead 隔板
**BUL**  Bulletin 公报, 通告
**BULK**  Bulky Baggage 大型行李
**BUM**  Baggage Unload Message (baggage handling terms) 行李卸下电文
**B-UNI**  Broadband User Network Interface 宽带用户网接口
**BUR**  Burbank, CA, USA (Airport Code) 美国加利福尼亚州伯班克机场
Bureau 局, 处
**BUS**  Broadcast and Unknown Server 广播与未知地址服务器
Busbar 汇流条导电条
Bushel 蒲式耳
Business 业务, 事物, 商业, 交易
**BUT**  Broadband Unbalanced Transformer 宽波段不平衡变量器
Butter 黄油
Button 按钮

**BUTE** Bent-Up Trailing Edge 上弯后缘

**BUTICS** Building Total Information and Control System 楼宇全信息及控制系统

**BUTT** Button 按钮

**BV** Ball Valve 球阀
Bleed Valve 引气活门,放气活门
Boundary Value 边界值
Breakdown Voltage 击穿电压
Butterfly Valve 蝶形活门
Bus Voltage 汇流条电压
Bypass Valve 旁通活门

**BVA** Bleed Valve Actuator 引气(放气)活门作动器

**BVC** Boeing Vertol Company 波音垂直起降飞机公司

**BVCU** Bleed Valve Control Unit 引气活门控制组件

**BVD** Beacon Video Digitalizer 信标视频数字器

**BVG** Beverage 饮料

**BVI** Blade Vortex Interaction 桨—涡干扰

**BVID** Barely Visible Impact Damage 不可见撞击损伤

**BVMS** Boeing Voice Mail Service 波音语音邮件服务

**BVWT** Boeing Vertol Wind Tunnel 波音垂直起降机风洞

**BVR** Beyond Visual Range 超出视程范围

**BVRG** Beverage 饮料

**BVRR** Beyond Visual Range Radar 超视距雷达

**BW** Bandwidth 宽带
Basic Weight 基本重量(指有效载重时的飞机重量)
Beam Width 波束宽度
Biweekly 隔周飞行
Black and White 书面文字为凭
Bonded Warehouse 存关行李仓库
Both Way 双向
Botswana 博茨瓦纳

**BWA** Bhairawa, Nepal (Airport Code) 尼泊尔派勒瓦机场
Botswana 博茨瓦纳
Broadband Wireless Access 宽带无线接入

**BWB** Blended Wing Body 翼身融合体

**BWC** Boeing Wichita Company 波音威切达公司

**BWD** Backward 倒退,向后

**BWDM** Bandpass Wavelength Division Multiplexer 带通波分复用器

**BWG** Baseline Working Group 基线工作组

**BWI** Baltimore/Washington International Airport (Airport Code) 巴尔的摩—华盛顿国际机场

**BWL** Belt Work Line 传送带流水线
Body Waterline 机身水线

**BWLL** Broadband Wireless Local Loop 宽带无线本地环路

**BWN** Bandar Seri Begawan, Brunei Darussalam (Airport Code) 文莱斯里巴加湾机场

**BWPA** Backward-Wave Parametric Amplifier 回波震荡器参数放大器
British Women Pilots' Association 英国女飞行员协会

**BWR** Bandwidth Radio 带宽比

**BWS** Base Weather Station 基地气象站
Beginning of Word Synchronization 同步数据字开始,字词同步开始
Body Wheel Steering 机身轮子转弯
Boeing Wichita Standard 波音威切达标准

**BWSC** Broadband Wireless Service Center 宽带无线业务中心

**BWT** Boeing Wind Tunnel 波音风洞
BothWay Trunk 双向中继

**BWU** BandWidth Utilization 带宽利用

**BX** Box 盒,箱

**BXD** Boxed 装箱的

**BYD** Beyond 不属于,超出

**BYDU** Backup Yaw Damper Unit 备用偏航阻尼组件

**BYND** Beyond 超出

**BYP** Bypass 旁通

Bypaths 侧管,旁路

**BYR** Buyer 买方

**BZE** Belize City, Honduras (Airport Code) 洪都拉斯伯利兹城机场

**BZN** Bozeman, MT, USA (Airport Code) 美国蒙大拿州波兹曼机场

**BZV** Brazzaville, Congo (Airport Code) 刚果布拉扎维机场

# C

C　Business Class 公务舱
　　C Class C 舱
　　Carat 克拉
　　Cargo 货物
　　Caution 注意
　　Celsius, Centigrade 摄氏
　　Cent 分
　　Center 中心,中央
　　Charter 包机
　　Chord of Wing 翼弦
　　Circle Trip 环程
　　Civil 民用
　　Clear 消除
　　Climb 爬升
　　Close 关闭,闭合
　　Cold 冷
　　Commercial Pilot 商用飞机飞行员
　　Comparator 比较器
　　Continuous 连续
　　Convertible 可转换的
　　Customs 海关
　　Cyan 深蓝色
　　Cycle 循环
℃　Degrees Celsius, Degrees Centigrade 摄氏度
C of A　Certificate of Airworthiness 适航证
　　Condition of Aircraft 飞机状态
C of G　Center of Gravity 重心
C of M　Certificate of Maintenance 维护证
C of R　Certificate of Release 移交许可,放行许可
C&C　Command & Control 指令和控制
　　Commitment & Control 委托和控制
　　Communications and Control 通信与控制
　　Computer & Communication 计算机与通信

C&D　Collected and Delivered (Shipping) 运费收讫和货物交毕
C&DC　Control and Display Console 控制和显示台
C&DH　Communication and Data Handling 通讯和数据处理
C&DM　Communication and Data Management 通讯和数据管理
C&I　Cost and Insurance 货价加保险
C&M　Care and Maintenance 维护保养
C&N　Communication & Navigation 通信与导航
C&SM　Communications and System Management 通信和系统管理
C&T　Communications and Tracking 通信和跟踪
C&W　Caution and Warning 提醒和警告
　　Control and Warning 控制和告警
C.G.　Center of Gravity 重心
C.P.　Center of Pressure 压力中心
C/A　Cabin Attendant 乘务员
　　Cable 电极,电缆
　　California 加利福尼亚(美国)
　　Coarse Acquisition Code 粗捕获码
　　Course/Acquisition 航向截获
　　Current Account 现金账户
C/B　Circuit Breaker (电)断路器,断路开关,电路跳闸开关
C/C　Capability & Capacity 能力与容量
　　Combustion Chamber 燃烧室
　　Computer Complex 计算机设备
　　Control Code (Functional Test) 控制编码(功能测试)
　　Coordinate 坐标系,坐标的,协调
C/D　Countdown 倒计时,倒数

Customs Declaration 报关单

**C/E** Course Error 航道误差

**C/F** Carriage Forward 转下页

Cost and Freight 成本加运费

**C/H** Certificate of Health 健康证

**C/I** Carrier-to-Interference Ratio 信号干扰比

Certificate of Inspection 检验证书

Certificate of Insurance 保险证书

**C/L** Center Line 中心线

Check List 检查单

Coefficient of Lift 升力系数

Cumulative Line 附加线

**C/M** Cold Meal 冷食

Command/Monitor 指令/监控

Coordinate Memo 协调摘要

Counter Measure 对策,措施,干扰

Crew Member 机组成员

**C/N** Carrier-to-Noise Ratio 信噪比

Change Notice 改变通知

Circular Note 旅行支票

Code Number 代号

Construction Number (飞机)制造厂编号

**C/O** Care of 转交

Carry out 实行

Cash Order 现金汇单

Change of Order 更换订单

Change Over 转换

Check Order 检查指令

Company 公司,陪同

Consist of 由…组成

**C/P** Cabin Purser (Chief Purser)客舱长

Cleaning/Painting 清除/喷涂

**C/R** Certificates of Registration 注册证书

**C/S** Call Sign 呼号

**C/SSR** Cost/Schedule Status Report 成本/计划状态报告

**C/T** Cable Transfer 电汇

Contract 合同

**C/W** Center Wing 中央翼

Continuous Wave 等幅波

Control Wheel 驾驶盘

**C2** Command and Control 命令与控制

**C2B** Customer to Business 消费者到企业(电子商务)

**C3** Command, Control, and Communications 指挥、控制与通信

**C3(3C)** Computer, Communication, Control 计算机、通信、控制

**C3CM** Command, Control, and Communications Countermeasures 指挥、控制与通信干扰

**C3I** Command, Control, Communication and Intelligence 指挥、控制、通信和情报

**C4I** Command, Control, Communication, Computer and Intelligence 指挥、控制、通信、计算机和情报

**C5** Computer, Communication, Content, Customer and Control 计算机、通信、容量、顾客和控制

**CA** Cabin Altitude 座舱高度,客舱高度

Cabin Attendant 机舱乘务员

Cable 电极,电缆

Calibrated Altitude 校正高度

California 加利福尼亚

Call Address 调用地址

Call Agent 呼叫代理

Canada 加拿大

Cargo Attendant Discount 货运人员折扣

Caution Annunciator 警戒信号牌

Center for Astrophysics 天体物理学中心

Certificate Authority 认证中心

Certificate of Airworthiness 适航认证

Certification Authority 核证机关,颁照当局

Charging Analysis 计费分析

Circuit Analog 线路模拟

Civil Aviation 民用航空,民航

Collision Avoidance 防止碰撞

Commercial Aviation 商业航空,商用航空

Communication Automation 通信自动化

Conflict Alert 冲突告警

Congestion Avoidance 防止拥塞

Continued Airworthiness 持续适航

Contract Award 签订合同,合同裁决
Controlled Airspace 管制区域
Controller, Aircraft 管制员
Controlling Authority 管制当局
Conversion Angle 转换角,外形变换角
Cost Analysis 成本分析
Course Acquisition Code 航向截获码
Course Alignment 航线校准
Course Angle 航线角
Crab Angle 偏流修正角
Criticality Analysis 临界状态分析
Cruise Altitude 巡航高度
Current Assets 流动资产

**CAA** Cabin Attendant 客舱服务员
Central Aviation Archive 航空中心档案文件
Change Accountability Audit 更改责任审查
Change Action Authorization 更改措施批准
Civil Aeronautics Administration (USA) 民用航空管理局(美国)
Civil Aeronautics Association 民用航空管理协会(美国)
Civil Aviation Administration 民用航空局
Civil Aviation Agency 民用航空办事处
Civil Aviation Authority (UK) 民用航空局(英国)
Clean Air Act 净化空气法案
Commercial Aviation Association 商业航空协会
Common Aviation Areas 共同航空区域,公共航空区域
Computer Assisted Animation 计算机辅助动画
Computer-Aided Analysis 计算机辅助分析
Computer-Aided Assemble 计算机辅助装配
Conformal-Array Aerial 保形阵列天线
Current Account 现金账户

**CAAA** China Academy of Aerospace Aerodynamics 航天空气动力技术研究院
Commuter Airline Association of America 美国短途航空公司协会

**CAAC** Civil Aviation Administration of China 中国民用航空局
Civil Aviation Advisory Committee 民航咨询委员会
Commission African de Aviation Civil 非洲民航委员会

**CAACSRI** CAAC Second Research Institute 中国民航局第二研究所

**CAAD** Computer Aided Architecture Design 计算机辅助结构设计
Computer-Aided Aircraft Design 计算机辅助飞机设计

**CAADRP** Civil Aircraft Airworthiness Data Recording Program 民用飞机适航数据记录程序
Civil Aviation Airworthiness Data Requirement Programme 民用航空适航资料要求纲要(英国)

**CAAFI** Commercial Aviation Alternative Fuels Initiative 商用航空替换燃料倡议

**CAAIP** Civil Aircraft Airworthiness Inspection Procedures 民用航空器适航检查程序

**CAAM** Civil Aeronautics Administration Manual 民用航空管理手册
Civil Aviation Authority of Macao 澳门民航局
Continuing Airworthiness Assessment Methodologies 持续适航评估方法

**CAARC** Commonwealth Advisory Aeronautical Research Council 英联邦航空咨询研究委员会

**CAAS** Civil Aviation Authority of Singapore 新加坡民用航空局,新加坡民航局
Computer Assisted Approach Sequencing 计算机辅助进场程序
Computer-Aided Alerting Subsystem 计算机辅助报警分系统
Computer-Aided Approach Spacing 计算机控制的进近间隔时间
Computer-Assisted Animation System 计算机辅助动画系统

**CAASA** Commercial Aviation Association of

Southern Africa 南非商业航空协会

**CAASD** Center for Advanced Aviation System Development 先进的航空系统发展中心

**CAASG** Commercial Airplane Avionics Support Group 商用飞机航空电子设备支援集团

**CAAT** Computer-Assisted Audit Technique 计算机辅助审计技术

**CAATC** Civil Aviation Administration Type Certificate 民航局飞机型号证书

**CAATS** Canadian Automated Air Traffic System 加拿大自动空中交通系统

China Aerospace and Aviation Technology Show 中国航空航天技术展

Computer Assisted Aircraft Trouble Shooting 计算机辅助航空器故障诊断

**CAB** Cabin 客舱

Cable 电极,电缆

Civil Aeronautics Bureau 民用航空局

Civil Aviation Board 民用航空委员会

Civil Aviation Bureau of Ministry of Transport 运输部民用航空局

Civil Aviation Bureau 民用航空局

Collision Avoidance Beacon 防撞灯标

Customer Advisory Board 客户咨询委员会

**CABADD** Cable Address 电报挂号

**CABD** Common Antenna Broadcast Distribution system 公共天线广播分配系统

**CABEI** Central American Bank of Economic Integration 中美洲经济一体化银行

**CABINAIR** Cabin Air Distribution Model 客舱空气分配模式

**CABS** Cockpit Air Bag Systems 驾驶舱气袋系统

**CAC** Cabin Air Compressor 客舱空气压缩机

Call Admission Control 呼叫允许控制

Cargo Accounting Advice 货运财务通知

Carrier Aircraft Contractor 运载机合同商

Caution Advisory Computer 警戒咨询计算机

Center Accessory Compartment 中央附属舱

Centralised Approach Control 集中进近管制

Civil Aviation College of Ministry of Transport 运输部民用航空学院

Climate Analysis Center 气候分析中心

Commonwealth Aircraft Corporation 联邦飞机公司

Computer Aided Creating 计算机辅助创意

Computer Assisted Composition 计算机辅助创作(作曲)

Computer-Assisted Counseling 计算机辅助咨询

Connection Admission Control 连接允许控制

Contact Approach Control 能见进近管制

Customer Administration Center 客户管理中心

**CACAC** Civil Aircraft Control Advisory Committee 民用飞机操纵咨询委员会

**CACAS** Civil Aviation Council of Arab States 阿拉伯国家民航委员会,阿拉伯国家民用航空理事会

Civil Aviation Caretaker Authority of Somalia 索马里民用航空代管机构

**CACC** Cargo Air Conditioning Card 货舱空调插件卡

Cargo Air Control Center 航空货运管制中心

Civil Aviation Communication Center 民航通信中心

**CA-CFAR** Cell Averaging Constant False Alarm Rate 单元平均恒虚警率

**CACL** Computer Assisted Computer Language 计算机辅助计算机语言

**CACM** Central American Common Market 中美洲共同市场

**CACON** Cargo Container 货运集装箱

**CACP** Cabin Area Control panel 座舱区域控制面板

**CACRC** Commercial Aircraft Composite Repair Committee 商用飞机综合修理委员会

**CACRS** Canadian Advisory Committee on Remote Sensing 加拿大遥感咨询委员会

**CACS** Cargo Air Conditioning System 货舱空调系统

Chinese American Chemical Society 中美化学与化工学会

Comprehensive Airport Communications System 综合性机场通信系统

Compressor Air Control System 压气机气流控制系统

Computer Assisted Communication System 计算机辅助通信系统

**CACT** Civil Air Carrier Turbojet 民用航空涡轮喷气机

**CACTCS** Cabin Air Conditioning and Temperature Control System 客舱空调与温度控制系统

**CAD** Calendar 日历,日程表

Canadian Dollar 加元(货币单位)

Cash Against Documents 凭单付款

Central Air Data 中央大气数据

Central Aircraft Dispatch 飞机调度中心,飞机签派中心

Chief, Airworthiness Standards 适航标准总监

Civil Aviation Department 民用航空部

Civil Aviation Directorate 民用航空董事会

Civil Aviation Directive 民航指令

Clean Airspeed Departure 清洁速度起飞

Commercial Airplane Division 商用飞机分部

Computer-Aided Design 计算机辅助设计

Computer-Aided Detection 计算机辅助探测

Computer-Aided Diagnosis 计算机辅助诊断

Computer-Aided Draughting 计算机辅助制图

**CADA** Computer Assisted Design and Analysis 计算机辅助设计与分析

**CADAC** Computer-Aided Design and Construction 计算机辅助设计与结构

**CADAM** Computer-Aided Design and Manufacturing 计算机辅助设计与制造

**CADAR** Computer-Aided Design, Analysis and Reliability 计算机辅助设计,分析与可靠性

**CADC** Central Air Data Computer 中央大气数据计算机

China Civil Aviation Development Corporation 中国民航开发服务公司

**CADD** Computer-Aided Design Development 计算机辅助设计开发

Computer-Aided Drafting and Design 计算机辅助草图与设计

**CADDIS** Cranfield Airborne Digital Data Instrumentation System 克伦菲尔德机载数字数据测量系统

**CADE** Client/Server Application Development Environment 客户伺服器应用开发环境

Computer-Aided Design and Engineering 计算机辅助设计与工程

**CADES** Computer-Aided Design and Evaluation System 计算机辅助设计与评估系统

Computer-Aided Development and Evaluation System 计算机辅助开发与评估系统

**CADF** Centralized Airspace Data Function 集中化空域数据功能

China Aviation Development Foundation 中国航空发展基金会

Common Avionics Data Format 通用航空电子设备数据格式

Commutated-Aerial Directing Finder 变向天线测向器

**CADI** ChengDu Aircraft Design & Research Institute 成都飞机设计研究所

**CADIZ** Canadian Air Defence Identification Zone 加拿大防空识别区

**CADORS** Civil Aviation Daily Occurrence Report System 民航日常事件报告系统

**CADP** Central Annunciator Display Pannel 中央信号显示板

Computer-Aided Design Package 计算机辅助设计程序包

**CADRE** Center for Aerospace Doctrine, Research and Education 航空航天概则研究教育中心(美国)

**CADRP** Civil Airworthiness Data Recording Programme 民用飞机适航性数据记录程序

**CADS**　Central Air Data System 中央大气数据系统

Central Attitude Determine System 中央姿态测定系统

Computer Aided Design System 计算机辅助设计系统

Computer Analysis and Design System 计算机分析和设计系统

**CADU**　Channel Access Data Unit 信道访问数据单元

Control and Display Unit 控制和显示装置

**CAE**　Carrier Aircraft Equipment 载机设备，母机设备

Columbia, SC, USA（Airport Code）美国南卡来罗纳州哥伦比亚机场

Computer-Aided Education 计算机辅助教育

Computer-Aided Engineering 计算机辅助工程

Control Area Extension 管制区延伸，管制区域扩展

Customer Application Engineering 用户应用工程

**CAECES**　Carrier Aircraft Equipment Computer Environmental Simulation 载机设备计算机环境模拟

**CAED**　Computer-Aided Engineering Design 计算机辅助工程设计

**CAEDM**　Community/Airport Economic Development Model 社团/机场经济发展模型

**CAEE**　Committee on Aircraft Engine Emissions（ICAO）航空器发动机排放委员会

Computer-Aided Engineering Education 计算机辅助工程教育

**CAEM**　Cargo Airline Evaluation Model 货运航空公司评估模型

Commission for Aeronautical Meteorology（WMO）航空气象委员会（世界气象组织）

**CAEMIS**　China Air Express Management Information System 中国民航快递管理信息系统

Civil Aviation Express Management Informa-tion System 民航快递综合管理信息系统

**CAEP**　Committee on Aviation Environmental Protection 航空环境保护委员会

**CAESAR**　Core and Engine Structural Assess-ment Research 发动机核心和结构评估研究

**CAET**　Civil Aviation Economics&Technology 民航经济与技术

**CAEU**　Council of Arab Economic Unity 阿拉伯经济统一体理事会

**CAF**　Calibration Flight 校正飞行

Cancel and File 取消并存档

Cost and Freight 成本加运费

**CAFAC**　Commission Africaine de l'Aviation Civile 非洲民航委员会

**CAFD**　Contact Analog Flight Display 目视模拟飞行显示器

**CAFH**　Cumulative Aircraft Flight Hours 飞机累计飞行小时

Cumulative Airframe Flight Hours 飞机机体累计飞行小时

**CAFI**　Computer-Aided Fault Isolation 计算机故障隔离

**CAFIT**　Computer Assisted Fault Isolation Test 计算机辅助故障隔离试验

Computer-Aided Fault Isolation Testability 计算机辅助故障隔离试验

**CAFM**　Commercial Air Freight Movement 商业航空货运

**CAFOS**　Comprehensive Automated Flight Op-erations System 全自动飞行运行系统

**CAFU**　Civil Aviation Flying Unit 民航飞行小组

**CAFUC**　Civil Aviation Flight University of China 中国民航飞行大学

**CAG**　Cagliari, Sardinia, Italy（Airport Code）意大利撒丁岛卡利亚里机场

Carrier Air Group 航母飞行大队

Civil Air Guard 民用防空队

Collective Address Group 集址组

Commercial Airplane Group 商用飞机集团

Constant Altitude Glide 等高滑翔

Customer Access Gateway 客户接入网关

**CAGD** Computer Aided Geometric Design 计算机辅助几何设计

**CAGE** Commercial and Government Entity 商业和政府机构

Computerized Aerospace Ground Equipment 计算机化航空航天地面设备

**CAGEC** Commercial and Government Entity Code 商业和政府机构代码

**CAgM** Commission for Agricultural Meteorology 农业气象学委员会

**CAGR** Compound Annual Growth Rate 复合年增长率

Cumulative Average Growth Rate 累积平均增长率

**CAGS** Central Attention Getting System 中央注意获得系统

**CAGWA** Commercial Airplane Group Work Authorization 商用飞机集团工作审定

**CAH** Cabin Attendant Handset 客舱服务员手机

Capital Airports Holding Company 首都机场集团公司

**CAHC** Cabin Attendant Handset Cradle 客舱乘务员手机挂钩

**CAHS** Canadian Aviation Historical Society 加拿大航空历史协会

**CAI** Cabin Altitude Indicator 客舱高度指示器

Cairo, Egypt (Airport Code) 埃及开罗机场

Canadian Aeronautical Institute 加拿大航空研究所

Canadian Airlines International 加拿大国际航空公司

Civil Aeromedical Institute 民航医学研究所

Civil Aeronautics Institute 民航航空研究所

Combustion Area Inspection 燃烧区域检查

Common Air Interface 公共空中接口

Computer-Aided Instruction 计算机辅助教学

Computer-Assisted Interrogation 计算机辅助咨询

Confederation of Aerial Industries 航空工业联盟

Constant Altitude Indicator 等高显示器

Customs Affairs Institute 海关事物研究所

**CAIG** Cost Analysis Improvement Group 成本分析改进组

**CAIMS** Civil Aviation Information Monitoring System 民航信息监控系统

**CAINS** Carrier Aircraft Inertial Navigation System 航母舰载机惯性导航系统

**CAIP** Continuous Airworthiness Inspection Plan 持续适航检查计划

**CAIRS** Computer-Assisted Information Retrieval System 计算机辅助情报检索系统

Computerized Accident/Incident Reporting System 计算机故障/事故报告系统

Confidential Aviation Incident Reporting System 航空事件秘密报告系统

**CAIS** Computer Assisted Introduction System 计算机辅助介绍系统

**CAIV** Cost As Independent Variable 将费用作为独立变量

Cowl Anti-Ice Valves 整流罩防冰活门

**CAK** Akron, OH, USA (Airport Code) 美国俄亥俄州阿克伦机场

**CAL** Calendar 日历

Calibrate, Calibration 校准

Calibration, Calibrated 校验、校验的

Calorie 卡, 卡路里

Canadian Astronautics Ltd. 加拿大航天公司

China Airlines, Taiwan 中华航空公司(台湾)

Computer Assisted Learning 计算机辅助学习

**CALC** Calculate 计算

Cargo Acceptance and Load Control 货物验收和装载控制

**CALD** Calculated 计算的

**CALEL** Call Select 选择呼叫

**CALIB** Calibration 校准

**CALIC** China Aviation Life-Support Industry

Corporation 中国航空救生工业公司

**CALI-DAL** Centralized Alarm Interface-Dependent Alarm Equipment 集中告警接口,有关告警设备

**CALN** Calculation 计算

**CALPA** Canadian Air Line Pilots Association 加拿大航空飞行员协会

**CALS** Commerce At Light Speed 光速商务

Computer-Aided Acquisition and Logistics Support 计算机辅助采购和物流保障

Computer-Aided Logistics Support 计算机辅助后勤保障

Continuous Acquisition and Life-Cycle Support 持续采购与寿命周期保障

**CALSEL** Air-To-Ground Selective Calling System 空地选择呼叫系统

Call Select 呼叫选择

**CAM** Cabin Assignment Module 客舱分配模块

Charter Advisory Message 包机咨询电报

Civil Aeronautics Manual 民用航空手册,民用航空条例

Civil Air Movement 民航运输

Commercial Air Movement 商业航空运输

Commission for Aeronautical Meteorology 航空气象委员会

Committee on Aviation Medicine 航空医学委员会

Computer Aided Manufacturing 计算机辅助制造

Computer Address Matrix 计算机地址矩阵

Computer-Aided Management 计算机辅助管理

Computer-Aided Manufacturing 计算机辅助制造

Computer-Aided Measurement 计算机辅助测量

Computer-Aided Module 计算机辅助组件

Consolidated Aircraft Maintenance 联合飞机维护

Content Addressable Memory 内容寻址存储器

Control Account Manager 控制账目经理

Core Access Module 核心接入模块

Crew Alertness Monitor 机组警告监视器

Customer Acceptance Manual 客户验收手册

**CAMA** Centralized Automatic Message Accounting 集中式自动信息记数

Chinese Aeronautical Meterological Association(中国台湾)"中华航空气象协会"

Civil Aviation Medical Association 民航医学协会

Computer-Assisted Mass Appraisal System 计算机辅助质量评价体系

**CAMAC** Civil Aviation Maintenance Association of China 中国民用航空维修协会

Computer-Aided Measurement and Control 计算机辅助测量和控制

**CAMC** Customer Access Maintenance Center 用户接入维护中心

**CAMEL** Customized Application for Mobile Enhanced Logic 移动增强逻辑的特定用户应用

**CAMI** Civil Aeromedical Medical Institute 民用航空医学研究所

Computer-Aided Manufacturing and Inspection 计算机辅助制造和检验

**CAMIC** Civil Aviation Management Institute of China 中国民航管理学院

**CAMIS** Commercial Activities Management Information System 商业活动管理信息系统

**CAML** Cargo Aircraft Mine Laying 货机资源布置

**CAMNET** Computer-Aided Manufacturing Network 计算机辅助制造网络

**CAMO** Continuous Airworthiness Management Organization 持续适航管理机构

**CAMP** Civil Aviation Master Plan 民用航空主计划

Computer-Assisted Maintenance Program 计算机辅助维修计划

Computer-Assisted Message Processor 计算机

辅助信息处理器

Computerized Aircraft Maintenance Program 计算机化飞机维修计划

Continuous Air Monitoring Program 连续空中监控计划

Continuous Airworthiness Maintenance Program 持续适航维修大纲

Cost Accumulation for Multiple Purposes 多用途成本积累

**CAMPER** Computer Aided Movie perspectives 计算机辅助电影画面制作

**CAMPP** Computer-Aided Material Planning & Purchasing 计算机辅助材料计划与采购

**CAMR** Camera 照相机,摄像机

**CAMRAD** Comprehensive Analytical Model of Rotorcraft Aerodynamics and Dynamics 旋翼机气动力和动力学综合分析模型

**CAMS** Communication Area Master Station 通讯区主控站

Computer Aided Measurement Systems 计算机辅助测量系统

Computer-Aided Maintenance System 计算机辅助维修系统

Control of Aircraft Maintenance and Serving 飞机维护和使用控制

**CAMTN** Central American Meteorological Telecommunications Network 美国中部气象电信网

**CAMU** Central Airspace Management Unit 中央空域管理装置

Central Avionics Management Unit 中央航电管理组件

Communication and Audio Management Unit 通讯和音频管理组件

**CAN** Advance Change Notice 提前更改通知

Airborne Communications Node 机载通信节点

Aircraft Classification Number 飞机等级号,航空器分类号

Aircraft Number 机号

Cable Area Network 缆区网络

Campus Area Network 校园区域网络

Canada 加拿大

Cancel 取消,注销

Canister 装运箱,罐

Canopy 座舱罩

City Area Network 市区网络

Climate Action Network 气候行动网

Committee on Aircraft Noise 航空器噪音委员会

Compact Access Node 密集的接入结点

Controller Action Notice 控制器操作通知

Controller Area Network 控制器局域网

Correlation Air Navigation 相关空中导航

Customer Access Network 用户接入网

**CANAC** Computer-Assisted National Air Traffic Control Center 计算机辅助的国家空中交通管制中心

**CANC** Cancel 取消,注销

**CANCLG** Cancelling 取消,注销

**CAND** Candidate 候选人,候补

Canned 罐装的

**CandF** Cost and Freight 成本加运费,货价加运费

**CANES** Centre for Advanced Numerical Engineering Simulations 先进的数字工程模拟中心

**CANGO** Committee for Air Navigation and Ground Organization 空中导航和地面组织委员会

**CANM** Civil Aircraft National Marking 民用飞机国籍(识别)标志

**CANP** Civil Air Notification Procedure 民航通知程序

Civil Aircraft Notification Procedure 民用飞机通知程序

**CANPA** Constant Angle Non-Precision Approach 固定下滑角的非精密进近

**CANS** Civil Air Navigation School 民用空中领航学校

**CANSO** Civil Air Navigation Services Organisation 民用空中航行服务组织

**CANT** Can Not 不能

**CANTIL** Cantilever 悬臂

**CAO** Cargo Aircraft Only 仅限货运飞机(仅限货机)

Civil Aviation Order 民航法典

Computer-Aided Optimization 计算机辅助优化

Computer-Assisted Ordering 计算机辅助订货

Contract Administration Office 合同管理处

**CAOA** Canadian Academy of Aeronautics 加拿大航空研究院

Corrected Angle of Attack 修正迎角

**CAOC** Combined Air Operations Center 联合空中运作中心

**CAOHC** China Aviation Oil Holding Company 中国航空油料集团公司

**CAOM** Cabin Attendant Operating Manual 客舱乘务员使用手册

**CAOS** Completely Automatic Operating System 全自动操作系统

**CAOSC** China Aviation Oil Supply Corporation 中国航空油料公司

**CAP** Cable Access Point 纤缆接入点

Capacitor 电容器

Capacity 容量,能力

Captain 机长

Capture 截获

Carpet 毯

Carrierless Amplitude Phase 无载波幅相(调制)

Channel Assignment Problem 信道分配问题

Chief Aviation Pilot 机长,首席驾驶员,主任驾驶员

Civil Air Patrol 民航巡逻队

Civil Aviation Publications 民用航空出版物

Common Alerting Protocol 公共警报协议

Competitive Access Provider 相互竞争的接入提供商

Computer-Aided Production 计算机辅助生产

Computer-Aided Programming 计算机辅助程序设计

Computer-Aided Purchasing 计算机辅助采购

Continuing Airworthiness Panel 持续适航专家小组

Continuing Airworthiness Problems 持续适航问题

Control Account Plan 控制账目计划

Control Anticipation Parameter 操纵期望参数

Controlled Acceleration Propulsion 受控加速推进

Cost Account Plan 成本账目计划

Crew Activity Plan 乘员活动计划

Crew Alerting Panel 机组警告面板

**CAPA** Central Airborne Performance Analyzer 中央机载性能分析器

Centre for Asia Pacific Aviation 亚太航空中心

Computer Aided Problem Analysis 计算机辅助问题分析

Corrective and Preventive Actions 纠正和预防措施

**CAPC** Civil Aviation Planning Committee 民航计划委员会

Computer-Aided Production Control 计算机辅助生产控制

**CAPD** Computer-Aided Process Design 计算机辅助工艺设计

**CAPDI** China Aeronautical Project & Design Institute 中国航空工业规划设计研究院

**CAPLTY** Capability 能力

**CAPM** Computer-Aided Production Management 计算机辅助产品管理

**CAPMS** Computer Aided Production Management System 计算机辅助生产管理系统

**CAPP** Computer Aided Process Planning 计算机辅助工艺规程设计

Computer-Aided Production Planning 计算机辅助生产计划

**CAPPBL** Capable 有能力的

**CAPPI** Constant Altitude Plan Position Indicator 等高平面位置显示器

**CAPPS** Computer Aided Project Planning System 计算机辅助计划规划系统

**CAPS** Civil Aviation Purchasing Service 民用航空采购服务

Computer Antenna Pointing System 计算机控制天线指向系统

Computer Assisted Passenger Screening 计算机辅助乘客扫描

Computerized Aircraft Performance System 计算机化飞机性能系统

**CAPSCA** Cooperative Arrangement for the Prevention of Spread of Communicable Disease through Air Travel 防止传染性疾病通过航空旅行传播的合作安排

**CAPSIN** Civil Aviation Packet-Switched Integrated Network 民航成套交换综合网

**CAPSS** CA Passenger Service System 国航个性化旅客服务系统

**CAPT** Captain 机长

Caption 标题

Computer-Aided Procedures Trainer 计算机辅助飞行动作练习器

**CAPTAIN** Character and Pattern Telephone Access Information Network System 图文电话信息网络系统

**CAPTC** Chief Airport Traffic Controller 机场交通管制主任

**CAPTL** Capital 首都,资本

**CAPY** Capacity 容量,能力

**CAR** Canadian Aviation Regulations 加拿大航空条例(规章)

Carat 克拉

Cargo 货物

Caribbean 加勒比

Caribean Region 加勒比地区

Carriage 运费

C-band Angle Receiver C 波段角度数据接收机

Certification Approval Report 核准报告

Changed Airspace Restriction 已变化的空域限制

Channel Address Register 通道地址寄存器

Channel Assignment Register 信道分配寄存器

Civil Airworthiness Regulations 民用适航条例

Civil Aviation Regulations 民用航空规章

Civil Aviation Requirements 民用航空要求

Cloud Altitude Radiometer 云高辐射计

Committed Access Rate 承诺的接入速率

Computer-Assisted Retrieval 计算机辅助检索

Configuration Audit Report 布局审计报告

Corrective Action Record 排故记录

Corrective Action Report 排故报告

Corrective Action Request 排故请求

**CARA** Cargo and Rescue Aircraft 货运和救援飞机

Combined-Altitude Radar Altimeter 组合高度雷达高度表

Computer-Aided Requirements Analysis 计算机辅助需求分析

**CARAC** Canadian Aviation Regulation Advisory Council 加拿大航空规章咨询委员会

Civil Aviation Radio Advisory Committee 民航无线电咨询委员会

**CARAD** Computer Aided Reliability and Design 计算机辅助可靠性与设计

**CARAM** Content Addressable Random Access Memory 按内容寻址随机存取存储器

**CARB** Carburetor 汽化器

**CARBD** Cardboard 硬纸板

**CARC** Central Air Rescue Center 中央航空救护中心

Commercial Aircraft Requirements Committee 商用飞机技术条件委员会

**CARD** Channel Allocation and Routing Data 信道分配与路由选择数据

Civil Aeronautics Research and Development

Policy Study 民用航空研究与发展政策研究[美国]

Civil Aviation Research & Development 民航研究与发展

Computer-Aided Remote Driving 计算机辅助遥控驱动

**CARDIS** Cargo Data Interchange System 货物数据交换系统

**CARDS** Calman's Automated Routing and Design System 卡尔曼自动路由选择和设计系统

Computer-Assisted Radar Display System 计算机辅助雷达显示系统

**CARE** Center for Aviation Research and Education 航空研究与教育中心

Computer Aided Reliability Evaluations 计算机辅助可靠性评估

Continuing Airframe Health Review and Evaluation 机身健康连续检查和评估

Continuous Aircraft Reliability Evaluation 飞机可靠性连续鉴定

Critical Action Request Expedite 关键行动申请超期,措施申请超期

Customer Acceptance & Review Evaluation 用户接受度和审查评估

**CAREC** China National Aero-Engine Corporation 中国航空发动机总公司

**CARERI** Chinese Aeronautical Radio Electronics Research Institute 中国航空无线电电子研究所

**CARF** Careful 小心

Center Airman Record File 飞行员记录档案中心

Central Altitude Reservation Facility 中央高度保留设施

Central Altitude Reservation Function 中央飞行高度保留功能

Control Altitude Reservation Facility 专用空域控制设备

**CARFM** Conference on Airport & Route Facilities Management 机场和航路设施管理委员会(会议)

**CARI** Civil Aeromedical Research Institute 民用航空医学研究所

Computer-Assisted Radar Identification 计算机辅助雷达识别

**CARMS** Civil Aviation Radio Measuring Station 民航无线电测试站

**CARNF** Charges for Airport and Route Navigation Facilities 机场及航路导航设施费用

**CARNOC** Civil Aviation Resource Net of China 民航资源中国网(简称民航资源网)

**CARO** Cargo Aircraft Only 仅限货机

Civil Aviation Regulatory Office 民用航空管理办公室

**CARP** Carpet 地毯

Carriage 运费

**CARS** Climate Applications Referral System 气候应用查询系统

Community Aerodrome Radio Station 地区性公共机场无线电台

Computer-Aided Routing System 计算机辅助路由选择系统

**CARSAMA** Caribbean/South America Monitoring Agency 加勒比/南美洲地区监测机构

**CARSRA** Computer-Aided Redundant System Reliability Analysis 计算机辅助冗余系统可靠性分析

**CARTR** Cartridge 磁带盒

**CARTS** Cargo Tracking System 货物追踪系统

**CAS** Calibrated Air Speed 校准空速(用符号 $V_C$ 表示)

Calibrated Airspeed, Calibrated Air Speed 校正空速、修正表速

Canadian Applications Satellite 加拿大应用卫星

Canadian Astronatical Society 加拿大航天学会

Casablanca, Morocco-Anfa (Airport Code) 摩洛哥卡萨布兰卡安法机场

Casing 套管,装箱

Channel Associated Signaling 随路信令

China Air Service Ltd. 中国航空服务公司

Civil Aid Service 民众安全服务队

Close Air Support 近距空中支援

Cockpit Avionics System 座舱航空电子系统

Collision Avoidance System 飞机防撞系统

Command Augmentation System 指令增强系统

Commercial Airplane Support 商用飞机支援

Commercial Avionics System 商用航空电子系统

Commission for Atmospheric Sciences 大气科学委员会

Computed Air Speed 计算空速

Computer-Aided Software 计算机辅助软件

Computer-Aided System 计算机辅助系统

Continuous Analysis and Surveillance 持续分析和监督

Control Actuation Section 操纵面偏转部分

Control-Augmentation System 控制增稳系统

Cooperative Application Satellite 合作应用卫星

Cost Accounting System 成本会计系统

Crew Alerting System 机组警告系统

**CASA** Citizen's Alliance for Saving the Atmosphere and the Earth 公民拯救大气层和地球联盟

Civil Aviation Safety Authority Australia 澳大利亚民航安全局

Civil Aviation Safety Authority 民航安全局

Computer Aided Situation Awareness 计算机辅助态势感知

Computer Aided Slot Allocation 计算机辅助时刻分配

**CASAP** China Aviation Safety Audit Program 中国民航安全审计程序

**CASB** Canadian Transportation Safety Board 加拿大运输安全委员会

**CASC** China Aviation Suppliers Corporation 中国航空器材公司

China Aviation Supplies Imp. & Exp Corporation 中国航空器材进出口公司

Civil Aviation Safety Center 民航安全中心,民用航空安全中心

Civil Aviation Safety Centre (Beirut) 民用航空安全中心(贝鲁特)

Combined Acceleration and Speed Control 加速度和速度综合控制

**CASD** Computer-Aided System Design 计算机辅助系统设计

**CASE** Common Application Service Element 公共应用服务单元

Computer Applications and Software Engineering 计算机应用与软件工程

Computer Automated Support Equipment 计算机自动化支援设备

Computer-Aided Software Engineering 计算机辅助软件工程

Computer-Aided System Engineering 计算机辅助系统工程

Computer-Aided System Evaluation 计算机辅助系统评估

Controller & Signal Evaluation 控制器与信号评估

Coordinated Aerospace Supplier Evaluation 航空航天供应商合作评估

Coordinating Agencies for Suppliers Evaluation 供应商评估合作机构

Crew Accommodations and Support Equipment 乘员居住和保障设备

**CASF** Composite Air Strike Force 组合空中打击力量

**CASFO** Civil Aviation Security Field Office 民用航空保安现场办公室,民用航空安全地区办事处

**CASGC** China Aviation Supplies Imp. & Exp. Group Corporation 中国航空器材进出口集团公司

**CASH** Cashier 出纳

Comprehensive Allocation Space Handling 综合机位配合作业设定

Paid by Cash 现金支付

**CASI** Commission Aeronautique Sportive Internationale 国际航空体育委员会(法)

**CASID** Committee for Aviation and Space Industry Development 航空和宇航工业发展委员会

**CASMA** Computerised Airline Sales and Marketing Association 计算机化航空销售与市场协会

**CASO** Cataloging and Standardization Office 编目与标准化处

Centralized Aviation Scheduling Office 航空综合计划处

Civil Aviation Security Office 民用航空保安办公室

**CASP** Computer-Assisted Search Planning 计算机辅助搜寻计划

Cooperative Aviation Security Programme 合作航空保安计划

**CASP-AP** Cooperation Aviation Security Plan-Asia/Pacific 合作航空保安计划—亚洲/太平洋地区

Cooperative Aviation Security Programme for the Asia and Pacific Region 亚太地区互助航空保安计划

**CASR** Civil Aviation Safety Regulations 民用航空安全规章

**CASRP** Canada Aviation Safety Reporting Program 加拿大航空安全报告项目

**CASS** Canadian Airport Security System 加拿大机场保安系统

Cargo Accounts Settlement System 货物账目结算系统

Centralized Aircraft Support Systems 集中式飞机支援系统

Chinese Academy of Social Sciences 中国社会科学院

Commercial Air Service Standards 商业航空服务标准

Committee on Aeronautical and Space Sciences 航空和航天科学委员会

Computer-Assisted Surveillance System 计算机辅助监视系统

Consolidated Automated Support System 强化自动支援系统

Continuous Analysis and Surveillance System 持续分析与监督系统

**CASSOA** Civil Aviation Safety and Security Oversight Agency 民用航空安全和保安监督机构

**CASST** Civil Aviation Safety Strategy Team 民用航空安全战略小组

Commercial Aviation Safety Strategy Team 商业航空安全战略小组

**CAST** Cast Aluminum Structures Technology 铸铝结构技术

Center for Aviation & Space Technology 航空宇航技术中心(中国台湾)

Center of Aviation Safety Technology 航空安全技术中心

China Air Service & Travel (Japan) Ltd. (日本)中国航空旅行服务有限公司

China Association for Science and Technology 中国科学技术协会

Chinese Academy of Space Technology 中国空间技术研究院

Civil Aircraft Study Team 民用飞机研究组

Commercial Aviation Safety Team 商业航空安全小组

**CASTE** Collision Avoidance System Technical Evaluation 防撞系统技术鉴定

**CASTS** Composites for Advanced Space Transportation System 先进的空间运输系统组合件

**CASTWG** Converging Approach Standards Technical Working Group 汇聚进近标准技术工作组

**CAT** Carbureter Air Temperature 汽化器气温

Cargo attendant on Cargo Aircraft 货机押运员

Categories for Landing Minimums 机场运行最低标准

Category or Catalog or Catalogue 类目,目录

Civil Air Transport 民航运输,民用空运

Clear Air Turbulence 晴空湍流,晴空颠簸

Cockpit Automation Technology 座舱自动化技术

Combat Aircraft Technology 作战飞机技术

Comite de l'Assistance Technique 技术援助委员会

Common Administrator Tool 通用管理员工具

Compressed-Air Tunnel 压缩空气风洞

Computer Aided Test/Testing 计算机辅助测试

Computer Aided Trainer 计算机辅助训练器

Computer Aided Translation 计算机辅助翻译

Computer Aided Typesetting 计算机辅助排版

Configuration Administration Tool 构型管理工具

**CAT Ⅱ**　Category Ⅱ Testing Ⅱ类测试

Category Ⅱ. Ⅱ类

Category Two ILS 二类盲降

Operational Performance Category Ⅱ Ⅱ类仪表着陆系统运行性能

**CAT Ⅲ**　Category Three ILS 三类盲降

**CAT Ⅲa**　Category Ⅲa 三类 a 级仪表着陆

**CAT Ⅲ A**　Operational Performance Category ⅢA ⅢA类仪表着陆系统运行性能

**CAT Ⅲb**　Category Ⅲb 三类 b 级仪表着陆

**CAT Ⅲ B**　Operational Performance Category ⅢB ⅢB类仪表着陆系统运行性能

**CAT Ⅲc**　Category Ⅲc 三类 c 级仪表着陆

**CAT Ⅲ C**　Operational Performance Category ⅢC ⅢC类仪表着陆系统运行性能

**CAT A**　Category A Aerodrome A 类机场

**CAT B**　Category B Aerodrome B 类机场

**CAT C**　Category C Aerodrome C 类机场

**CAT Ⅰ**　Category One ILS 一类盲降

**CAT1/2/3**　Bad Weather Landing Category 不良天气的着陆等级

**CAT Ⅰ**　Category Ⅰ approach Ⅰ类进近

**CAT Ⅱ**　Category Ⅱ approach Ⅱ类进近

**CAT Ⅲ**　Category Ⅲ approach Ⅲ类进近

**CATA**　Canadian Air Transportation Administration 加拿大航空运输管理局

China Air Transport Association 中国航空运输协会

Civil Air Transport Agreement 民用航空运输协定

Computer Assisted Traditional Animation 计算机辅助传统动画显示

**CATAL**　Catalogue 目录

**CATB**　Canadian Air Transport Board 加拿大空运局(委员会)

Cooperative Avionics Test Bed 航空电子设备协同测试平台

**CATC**　Civil Aviation Training Center 民用航空培训中心

College of Air Traffic Control 空中交通管制学院

Commonwealth Air Transport Council 英联邦空运委员会

**CATE**　Centre for Appropriate Technology Exchange 适用技术交流中心

Computer-Controlled Automatic Test Equipment 计算机控制自动测试设备

Conference on Coordination of Air Transport in Europe 欧洲航空运输协调会议

**CATG**　Capital Assets Task Group 资本资产任务组

China Aviation Tourism Guide 中国航空旅游指南

Computer Aided Test Generator 计算机辅助测试发电机

**CATIA**　Computer-Aided Three-Dimensional Interactive Analysis 计算机辅助三维互动分析

Computer-Aided Three-Dimensional Interactive Application 计算机辅助三维交互应用

**CATIC**　China National Areo-Technology Import and Export Corporation 中国航空技术进

出口公司

**CATM** Civil Air Traffic Management 民用空中交通管理

**CATMAC** Co-operation Air Traffic Management Concept 空中交通管理合作方案

**CATO** Civil Air Traffic Operations 民航空中交通管理

**CATS** Centralized Automatic Testing System 集中自动测试系统

Common Automatic Test System 通用自动测试系统

Communication and Tracking System 通信和跟踪系统

Computer Aided Training Systems 计算机辅助训练系统

Computer-Aided Telemetry System 计算机辅助遥测系统

Computer-Aided Trouble Shooting 计算机辅助故障查找

Corporate Air Travel Survey 商务航空出行调查

Crew Activity Tracking System 机组活动跟踪系统

**CATSA** Canadian Air Transport Security Agency 加拿大航空运输安全局

**CATTC** China Aero-Technology Translation Corporation 中国航空科技翻译公司

**CATV** Cabin Air Temperature Valve 座舱气温调节活门

CAble TeleVision 有线电视

Community Antenna TeleVision 共用天线电视系统

**CAU** Cell Antenna Unit 小区天线单元

Cold Air Unit 空气冷却装置

Crypto Auxiliary Unit 加密辅助设备

**CAUC** Civil Aviation University of China 中国民航大学

**CAUT** Caution 警戒，注意

**CAUT LT** Caution Light 警戒灯

**CAV** Common Aero Vehicle 通用航空飞行器

Common Aerospace Vehicle 通用航空飞行器

Computer Aided Verification 计算机辅助验证

Constant Air Volume 定风量

Constant Angular Velocity 恒角速度

**CAVE** Cellular Authentication Voice Encryption 蜂窝鉴权和语音加密

**CAVG** Computer Assisted Video Generation 计算机辅助影像生成

**CAVOK** Ceiling and Visibility OK 天气晴朗，好天

Cloud and Visibility OK 云高和能见度好

Visibility, Cloud and Present Weather Better than Prescribe 能见度、云和目前天气比预报的好

**CAVU** Ceiling and Visibility Unlimited 晴朗和能见度极好

**CAWA** Civil Aviation Wireless Association 民航无线电协会

**CAWP** Central Annunciator Warning Panel 中央显示器警告面板

**CAWS** Caution and Warning System 提醒和警告系统

Caution, Advisory and Warning System 警诫、警示和警告系统

Central Aural Warning System 中央声响警告系统

Common Aviation Weather Subsystem 通用航空气象子系统

**CB** Cargo Bay 货舱

Circuit Board 电路板

Circuit Breaker 电路断路器，电路跳开关

Citizens' Band Radio 民用波段无线电

Citizen's Band 民用电台频带

Communication Bus 通信总线

Compass Bearing 罗盘方位

Crew Back-up 后备机组

Cubic 立方的

Cumulonimbus 积雨云

**CB(S)** Circuit Breaker(s) 跳开关

**CBA** Capability Based Assessment 基于能力

的评价

Cargo Boarding Advisory 装货通知

Cargo Booking Advice 订舱单

Command Broadbeam Antenna 指令性宽波束天线

Cost Benefit Analysis 成本效益分析

Cross Border Area 交叉边界区（跨国界区域）

**CBAA**　Canadian Business Aircraft Association 加拿大商务飞机协会

**CBAL**　Counterbalance 平衡,移轴补偿

**C-Band**　Approx. 5000MHz C 波段

**CBAS**　Combined British Astronautical Societies 英国联合宇航学会

Commercial Bank Application System 商业银行应用系统

**CBBG**　Cabin Baggage 舱内行李（指占用额外座位的行李）

**CBCF**　Circuit Breaker Configuration Function 电路跳开关构型功能

**CBD**　Cash Before Delivery 交货前付款

Central Business District 商业中心区

Commerce Business Daily 贸易商务日报

Component Based Development 基于组件的开发

**CBDS**　Chemical and Biological Detection System 化学与生物检测系统

Connectionless Broadband Data Service 无连接宽带数据业务

**CBE**　Computer Based Education 计算机辅助教育

Cosmic Background Explorer 宇宙背景探测器

**CBERS**　China-Brazil Earth Resources Satellite 中巴地球资源卫星

**CBF**　Central Board of Finance 中央财政委员会

Cubic Feet 立方英尺

**CBFT**　Cross Boundary Ferry Terminal 跨境客运渡轮码头

**CBI**　Cabin Service Items 客舱服务项目

Component Burn-in 元件预先强化

Computer Based Instruction 计算机基本指令,计算机化教学

**CBIC**　Circuit Breaker Interface and Control 电路跳开关界面与控制

**CBIT**　Center for Business Information Technologies 商务信息技术中心

Computerized Built-in Test 计算机机内测试

Continuous Built in Test 连续机内测试

**CBK**　Call Back 回叫

Circuit Breaker 跳开关

**CBL**　Cable 电缆,钢索

Cargo Boarding List 装货单

Commercial Bill of Lading 商业提货单

**CBM**　Certified Business Manager 注册管理师

Circuit Breaker Map 电路跳开关连接图

Circuit Breaker Monitoring 电路跳开关监控

Client Business Manager 客户业务经理

Condition Based Maintenance 基于状态的维修

Cubic Meter 立方米

**CBMA**　Computer-Based Music Analysis 基于计算机的音乐分析

**CBMF**　Circuit Breaker Monitoring Function 电路跳开关监控功能

**CBMIS**　Computer-Based Management Information System 计算机化管理信息系统

Circuit Breaker Monitoring System 电路跳开关监视系统

**CBMU**　Circuit Breaker Monitoring Unit 电路跳开关监控组件

Current BIT Monitor Unit 电流二进位数字监控器

**CBN**　Cabin 客舱

Code Beacon 信号灯标

Container Booking Note 集装箱订舱单

Cubic Boron Nitride 立方氮化硼

**CBNLD**　Cabin Load 客舱装载

**CBNSVC**　Cabin Service 客舱服务

**CBO**　Channel Bus out 通道总线输出

Congressional Budget Office 国会预算办公室(美国)

Cost Based Overbooking 依据成本超售(机票)法

Cumulative Burn-off 累积耗油

Cycles Between Overhaul 大修间隔循环

**CBORE** Counterbore 平头钻

**CBP** Circuit Breaker Panel 跳开关面板

Compressor Bleed Pressure 压缩机引气压力

Conceptual Build Plan 概念性创建计划

US Customs and Border Protection 美国海关及边境保护局

**CBPM** Computer Based Performance Monitoring 基于计算机的效能监控

Continuous Business Process Management 持续业务流程管理

**CBPS** Combined Braking/Correction Propulsion System 制动/修正联合推进系统

**CBQ** Class-Based Queuing 基于级别的排队

**CBR** California Bearing Ratio 加州轴承比

Canberra, Australian Capital Territory, Australia (Airport Code)澳大利亚堪培拉机场

Case-Based Reasoning 基于案例的推理

Chemical, Biological and Radiological 化学/生物学/放射源

Constants Bit Rate 固定码率

Constraint Based Routing 基于约束的选路

Content Based Retrieval 基于内容的检索

Continuous Bit Rate 连续比特率

Controlled Burn Rate 可控燃烧速率

**CBS** Call Box Station 电话亭

Columbia Broadcasting Systems Inc. 哥伦比亚广播公司(美国)

Commission for Basic Systems 基本系统委员会

Common Business Systems 公用商务系统

Computer Business Systems 计算机商务系统

Continuous-Beam Steering 连续波束控制

Controlled Blip Scan 受控尖峰脉冲扫描

**CBSD** Component-Based Software Development 基于构件的软件开发

**CBSRP** Capacity-Based Session Reservation Protocol 基于容量的对话保留协议

**CBSV** Cycles Between Scheduled Visits 计划进厂间隔循环

**CBT** Cabinet 箱,小室

Computer Based Training 计算机辅助教学训练

Computer-Based Trainer 计算机辅助教学训练器

**CBUI** Circuit Breaker User Interface 电路跳开关用户界面

**CBV** Cross Bleed Valve 交叉引气活门

**CBW** Cross Bleed Wing Valve 交叉引气机翼(防冰)活门

**CBX** Cam Box 凸轮盒

Computerized Branch Exchange 计算机化支线交换机

**CC** Call Center 电话服务中心

Call Collision 呼叫冲突

Call Control 呼叫控制

Cam Chair 轮椅

Camp Chair 轻便折椅

Carbon Copy 复写纸副本

Cargo Carrier 货运公司,货机

Center Conductor 中央导体

Central Computer 中央计算机

Central Control 中央控制

Channel Check 信道检查,通道检查

Channel Controller 通道控制器

Character Code 字符码

Charges Collect 运费到付

Chinese Character 中文字符

Circuit Control 电路控制

Cirrocumulus 卷积云

Clock Chronometer 时钟计时器

Close Coupled 紧密耦合的

Closing Coil 闭合线圈

Cloud Ceiling 云幕高,云高

Cluster Controller 群集控制器

Coaxial Cable 同轴电缆

Code Check 代码检验

Code Converter 代码转换器
Coded Command 编码命令
Color Code 彩色编码
Color Compensation 彩色补偿
Color Correction 彩色校正
Color Sensitive 色敏感的
Combustion Chamber 燃烧室
Command Center 指挥中心
Common Collector 共集电极
Communication Center 通信中心
Communication Controller 通信控制器
Communications 通讯设备
Compass Course 罗盘航向
Composite Command 复合指令（美国）
Compression and Coding 压缩与编码
Computation Center 计算中心
Computer Center 计算机中心
Condition Code 状态码
Configuration Control 构型控制,组合控制
Connection Center 连接中心
Connection Confirm 连接确认
Control Center 控制中心,管制中心
Control Console 控制台,操纵台
Control Coolant 控制冷却剂
Control Counter 控制计数器
Convolutional Coding 卷积编码
Co-ordination Center 协调中心
Cost Center 成本中心
Cotton-Covered 纱包的
Country Code 国码
Coupled Control 耦合控制
Credit Card 信用卡
Crew Cost 机组人员成本
Critical Condition 临界条件,临界状态
Cross Channel 交叉通道
Cross Couple 交叉耦合
Cubic Centimeter 立方厘米
Current Comparator 电流比较器
Cursor Control 游标控制
Cyclic Check 循环检查
Cyclic Code 循环码

**CC：COPE** Climate Convention：Co-operation Programme 气候公约合作方案
**CCA** Call Control Agent 呼叫控制代理
Capital Consumption Allowance 资本消耗扣除
Charges Correction Advice 货物运费更改通知
Carrier-Controlled Approach 载波控制方法
Central Control Actuator 中央控制作动器
Charges Collect Advice 运费到付通知
China Civil Aviation 中国民用航空
China Commercial Aircraft CO. LTD 中国商用飞机有限责任公司
Circuit Card Assembly 电路卡组件
Cluster Compression Algorithm 集聚压缩算法
Cold Cranking Amperes 冷启动安培
Collective Corrective Action 集中校正措施,集中排故
Collins Component Avionics 柯林斯航空电子部件
Common Cause Analysis 共有原因分析
Common Communication Adapter 公用通信适配器
Common Cryptographic Architecture 公用密码结构
Continental Control Area 大陆管制区
**CCADSC** China Civil Aviation Development Service Corporation 中国民航开发服务公司
**CCAF** Call Control Access Function 呼叫控制接入功能
**CCAM** Commission of Climatology and Applications of Meterology 气候学和应用气象学委员会
Computer Communications Access Method 计算机通信存取方法
**CCAPS** Circuit Card Assembly & Processing System 电路板组合件与处理系统
**CCAQ** Consultative Committee on Administrative Questions（UN）行政问题协商委员会（联合国）

**CCAR** China Civil Aviation Regulations 中国民用航空规章

**CCARCS** Canadian Civil Aircraft Register Computer System 加拿大民用航空器登记计算机系统

**CCAS** Central Crew Alerting System 中央机组警告系统

**CCATS** Committee on Communication and Air Traffic Service 通信和空中交通勤务委员会

Computer Controlled Automatic Test Set 计算机控制自动测试组件

**CCB** Call Control Block 呼叫控制块

Change Control Board 变更管理委员会

Checkpoint Control Block 检查点控制(数据)区

Configuration Control Board 构型控制委员会

Connection Control Block 连接控制块

Control Center Building 控制中心大楼

Converter Circuit Breaker 变频器电路断电器

Converter Control Breaker 变频器控制断路器

Customer Care and Billing 客户服务和计费

Customer Cost Benefit 客户成本效益

**CCBD** Configuration Control Board Directive 构型控制局指令

**CCBS** Completion of Call to Busy Subscriber 遇忙呼叫完成

**CCC** Camera Controller Combiner 照相机控制组合器

Cellular Communication—Navigation—Surveillance Concept 蜂窝式通信—导航—监视概念

Central Communications Controller 中央通讯控制器

Central Computer Complex 中央计算机综合设备

Change Control Committee 更改控制委员会

Class & Code Catalog 分类编码目录

Color Cell Compression 色彩单元压缩

Color Code Chart 彩色编码图

Command and Control Center 指令与控制中心

Command Control Console 指令控制台

Command, Control, Communication 指挥、管制、通信

Common Channel Control 公用信道控制

Communication Control Center 通信控制中心

Communications Command Center 通信指令中心

Communications/Command & Control 通信/指挥与控制

Component Change Card 部件更换卡

Component Change Control 部件更换控制

Computer Communication Converter 计算机通信转换器

Configuration Control Committee 构型控制委员会

Container Control Computer 集装箱控制计算机

Converter Circuit Breaker 转换器跳开关

Core Compartment Cooling 核心舱冷却

Council for Cultural Cooperation 文化合作理事会

Crash Crew Chart 飞机坠毁机组图

Credit Card Calling 信用卡呼叫

Customer Coordination Center 用户协调中心

Customs Cooperation Council 海关合作理事会,海关合作委员会

**CCCB** Component Change Control Board 组件变更控制委员会

Configuration Control Coordination Board 配置管理协调委员会

**CCCCISR** Command, Control, Communication, Computer, Intelligence, Surveillance and Reconnaissance 指挥、控制、通信、计算机、情报、监视与侦察

**CCCG** Climate Change Co-ordinating Group 气候变化协调组

**CCCH** Common Control Channel 公共控制信道

**CCCI** Communication, Command, Control and Intelligence 通信、指挥、控制与情报
Computer Control Company, Inc. 计算机控制有限公司

**CCCO** Joint Committee on Climatic Changes and the Ocean 气候变化与海洋联合委员会

**CCCP** Compressor Cavity Control Pressure 压气机内腔控制压力

**CCCS** Central Command and Control System 中央指挥与控制系统
Central Control Computer System 中央控制计算机系统
Command Control Communication System 指挥控制通信系统
Communications Computation Control System 通信计算控制系统
Compatibility of Command and Control System 指挥与控制系统兼容性
Computer Control Cooling System 计算机控制制冷系统

**CCD** Charge-Coupled Device 电荷耦合器件
Computer Controlled Display 计算机控制显示
Consolidated Cab Display 综合机舱显示器
Cursor Control Device 光标控制装置

**CCDL** Common Channel Data Link 公共信道数据链路
Cross-Channel Data Link 跨信道数据链路

**CCDN** Corporate Consolidation Data Network 共同统一数据网络

**CCDP** Continuously Computing Delivery Point 连续计算投放点

**CCDR** Contractor Cost Data Reporting 承包商成本数据报告

**CCDU** CES Channels Dispatch Unit 电路仿真业务信道分配单元

**CCE** Charge Composition Explorer 电荷成分探测器
Communication Control Equipment 通信控制设备
Cooperative Computing Environment 协作计算环境

**CCF** Call Control Function 呼叫控制功能
Central Control Function 中央控制功能
Charter Class Fare 租赁价,租赁费
Communication Control Field 通信控制字段
Connection Control Function 接续控制功能
Credit Conversion Factor 信用转换系数

**CCFA** Common Cause Failure Analysis 共因失效分析

**CCFG** Compact Constant-Frequency Generator 小型恒频发电机

**CCFP** Center for Civil Force Protection 民事武力保护中心
Collaborative Convective Forecast Product 协同对流预报产品
Critical Care Flight Paramedic 重症监护飞行医务人员

**CCG** Cabin Configuration Guide 客舱布局指南
Cargo Center of Gravity 货物重心

**CCGE** Cold Cathode Gauge Experiment 冷阴极真空规实验(美国)

**CCH** Control Channel 控制信道

**CCI** Co-Channel Interference 同信道干扰
Commission for Climatology 气候学委员会
Communication Control Information 通讯管制情报
Conference Call Indicator 会议电话指示器

**CCIA** Computer and Communication Industry Association 计算机与通信工业协会

**CCIL** Continuously Computed Impact Line 连续计算命中线

**CCIP** Command Center Improvement Program 指挥中心改进计划
Command Control Information Processing 指令控制信息处理
Common Configuration Implementation Program 通用布局实施计划
Contamination Control Implementation Plan 污染控制实施计划
Continuously-Computed Impact Point 连续计

算命中点

**CCIR** Consultative Committee International Radio 国际无线电咨询委员会

**CCIRID** Charge Coupled Infrared Imaging Device 电荷耦合红外成像设备

**CCIS** Coaxial Cable Information System 同轴电缆信息系统

Common Channel Interoffice Signalling 共路局间信令

**CCITT** Consultative Committee on International Telephone and Telegraph 国际电报电话咨询委员会

**CCKW** Counter-Clockwise 逆时针方向的

**CCL** Compliance Checklist 符合性检查单

Customs Clearance 通关

**CCLN** Council for Computerized Library Networks 计算机化图书馆网络委员会

**CCM** Cargo Control Module 货运控制组件

Configuration Control Module 配置控制模块

Counter-Countermeasure 反电子对抗

Crew/Cargo Module 乘员/货物舱

**CCMD** Continuous Current Monitoring Device 连续电流监控器件

**CCMIS** Certification and Compliance Management Information System 许可证及符合性管理信息系统

**CCMRs** Candidate Certification Maintenance Requirements 候选审定维修要求

**CCMS** Checkout, Control, and Monitoring Sub-system 检测、控制和监视分系统

**CCMU** Centralized Crypto Management Unit 集中密码管理组件

**CCN** Cabin Computer Network 客舱计算机网络

Cloud Condensation Nuclei 云凝结核

Contract Change Notice 合同更改通知

**CCO** Contract Change Order 合同更改规程

Crystal-Controlled Oscillator 晶体控制振荡器

**CCOC** Combustion-Chamber Outer Casing 燃烧室外壳

**CCOIC** China Chamber of International Commerce 中国国际商会

**CCOM** Cabin Crew Operating Manual 客舱乘务组操作手册

**CCOS** China Climate Observing System 中国气候观测系统

Computer Controlled Optical Surfacing 计算机控制光学表面成型

Computer-Controlled Operating System 计算机控制操作系统

**CCP** Cabin Control Panel 客舱控制面板

Call Confirmation Procedure 呼叫确认过程

Call Control Procedure 呼叫控制过程

Call Control Processor 呼叫控制处理器

Central Control Panel 中央控制面板

Closed Cherry-Picker 封闭式升降舱(美国)

Common Control Physical Channel 公共控制物理信道

Communication Control Package 通信控制程序包

Communication Control Processor 通信控制处理器

Compass Control Panel 罗盘控制面板

Consolidated Cryptologic Program 统一密码程序

Contingency Command Post 应急指挥站

Contract Change Proposal 合同更改建议书

Critical Control Point 关键控制点

Crop Classification Performance 作物分类特性

Currency of Country Payment 付款国货币

**CCPC** Civil Communications Planning Committee 民用通信设备计划会议

**CCPCH** Cross-Connection Point 交叉连接点

**CCPD** Charge-Coupled Photodiode Device 电荷耦合光电二极管设备

**CCPIT** China Council for the Promotion of the International Trade 中国国际贸易促进委员会

**CCPP** Combined Cycle Power Plant 蒸汽联合循环发电站

**CCPR** Cruise Compressor Pressure Ratio 巡航压缩机压力比

**CCQ** Cataloging & Classification Quarterly 编目与分类季刊

Creative Climate Questionnaire 创新气氛问卷

Cross Crew Qualification 交叉机组驾驶资格

**CCR** Call Congestion Ratio 呼损率

Central Control Room 中央控制室

Circulation-Controlled Rotor 环流控制旋翼

Commodity Classification Rates 等级货物运价

Common Computing Resource 公共计算资源

Concord, CA, USA (Airport Code) 美国加利福尼亚州康科德机场

Credit Card Reader 信用卡读卡机

Customer Change Register 客户变更登记

**CCRA** Risk Analysis and Management 风险分析与管理

**CCRAC** Combined Center Radar Approach Control 雷达进近管制联合中心

**CCRC** Cabin Crew Rest Compartment 客舱机组休息舱

China Commercial Relations Committee 对华贸易关系委员会

Communications Control and Restoral Center 通信控制和修复中心

**CCRM** Cabin Crew Rest Module 客舱机组休息间

**CCRP** Continuously Computed Release Point 连续计算投放点

**CCRS** Canada Center for Remote Sensing 加拿大遥感中心

**CCRZ** Climb to... and Cruise 爬升到···巡航

**CCS** Cabin Communications System 客舱通信系统

Call Connected Signal 呼叫接通信号

Caracas, Venezuela-Simon Bolivar (Airport Code) 委内瑞拉加拉加斯西蒙玻利瓦尔机场

Carbon Capture and Sequestration 碳(二氧化碳)捕获和封存

Cargo Community System 货物信息自动传递系统,货运公共系统

Centi-Call Seconds 百秒呼

Central Computer System 中央计算机系统

Central Control Systems 中央控制系统

Command Control Subsystem 指令控制分系统

Commercial Communications Satellite 商业通用卫星

Common Channel Signaling 共路信令

Common Computing System 公共计算系统

Common Core Services 公共核心服务

Communications Control System 通信设备控制系统

Computer Cargo System 计算机货舱预定系统

Computer Command Subsystem 计算机指令分系统

Control Coordination System 控制协调系统

Coordinated Control System 协调控制系统

Cruise Control System 巡航控制系统

**CCS7** Common Channel Signaling No. 7 七号共路信令

**CCSA** Common-Control Switching Arrangement 公用控制交换方案

**CCSD** Configuration Change Support Data 组合(配置)更改支持数据

**CCSDS** Consultative Committee for Space Data System 空间数据系统协商委员会

**CCSE** Common Channel Signaling Equipment 公共信道信令设备

**CCSI** China Commercial Service International 中国华联国际服务联合公司

**CCSM** Common Channel Signaling Module 公共信道信令模块

**CCSN** Common Channel Signaling Network 公共信道信令网

**CCSP** Carbon Cycle Science Plan 碳循环科学计划

Center for Communications and Signal Processing 通信与信号处理中心

Certified Cargo Screening Program 验证货物扫描计划

Climate Change Science Program 气候变化科学计划

Communications, Control & Signal Processing 通信、控制与信号处理

Credit Card Settlement Plan 信用卡结账计划

**CCSS** Canada Center for Space Science 加拿大空间科学中心

Common Channel Signaling System 公共信道信令系统

Control Center of the Symphonie Satellite System 交响乐卫星系统控制中心

**CCST** Center for Computer Science and Technology 计算机科技中心

Consultative Committee on Satellite Telecommunications 卫星通信协商委员会

Satellite Telecommunications Coordinating Committee 卫星通信合作委员会

**CCT** China Coast Time 中国沿海时间

Circuit 电路

Computer Compatible Tape 计算机兼容磁带

Correct 修正,修改

**CCTA** Central Computer and Telecommunication Agency 中央计算机与电信局

**CCTM** Cabin Configuration Test Module 客舱配置测试模件

**CCTR** Concentration 集中,集结

Concentrator 集中器

**CCTS** Cabin Cordless Telephone System 客舱无绳电话系统

Change Control Tracking System 更改控制跟踪系统

**CCTV** Closed Circuit Television 闭路电视

**CCU** Calcutta, India (Airport Code)印度加尔各答机场

Cargo Control Unit 货舱控制组件

Central Control Unit 中央控制组件

Cockpit Control Unit 座舱控制组件

Communication Control Unit 通信管制单元

**CCV** Control-Configured Vehicle 随控布局飞行器

**CCW** Circulation-Controlled Wing 环量控制机翼

Counter Clock Wise, Counter Clockwise 逆时针方向

**CCWS** Common Controller Workstation 通用管制员工作站

**CD** Call Deflection 呼叫改向

Call Distribution 呼叫分配

Candela 烛光,坎德拉

Canned 罐装的

Card 卡片

Carried 运送的

Carrier Detect 载波检测

Case Drain 壳体回油

Cell Delay 信元延时

Certification Demonstration 适航演示

Coefficient of Drag 阻力系数

Cold Drawn 冷拉伸

Collision Detection 冲突检测

Common Digitizer 通用数字化仪设备

Compact Disk 压缩磁盘,激光唱盘,光盘

Compass Deviation 罗盘偏差

Compressed Data 压缩数据

Consular Declaration 领事声明

Contract Definition 合同规定标准

Control Desk 控制面板,操纵(控制)台

Control Display 控制显示

Controlled Difusion 可控扩散度

Convergent/Divergent 收缩扩张(喷管)

Cumulative Distance 累积距离

Cycle-Dependent 飞行次数相关的

Drag Coefficient 阻力系数

**CD-2** Common Digitizer 2 通用数字读出器-2

**CDA** Central Design Authority 中央设计管理局(欧)

Collision Detection and Avoidance 防撞系统

Command and Data Acquisition 指令和数据

采集

Concept Demonstrator Aircraft 概念展示飞机

Continuous Descent Approaches 持续下降进近

Current Data Array 当前数据阵列

**CDAF** Central Data Acquisition Facility 中央数据采集设备

Content Delivery Agent for Flight Bag 飞行数据包传输代理

**CDAM** Centralized Data Acquisition Module 集成数据采集模块

**CDAS** Central Data Acquisition System 中央数据采集系统

Command and Data Acquisition Station 指令发送与数据收集站

Control & Data Acquisition System 控制与数据采集系统

**CDAU** Centralized Data Acquisition Unit 集成数据采集组件

**CDB** Central Data Bank 中央数据库

Common Data Base 公用数据库

Common Data Bus 公共数据总线

Configuration Data Base 构型数据库, 布局数据库

**CDBP** Command Data Buffer Program 指令数据缓冲程序

**CDBT** Component Design and Build Team 部件设计与制造组

**CDC** Cargo Door Controller 货舱门控制器

Center for Document Control 文件控制中心

Central Digital Computer 中央数字计算机

CITS Dedicated Computer 中央综合测试系统专用计算机

Classified Document Control 保密文件控制

Commercial Delivery Center 商用交付中心

Communication Dispatch Center 通信发送中心

Computer Data Corporation 计算机数据公司

Computer Display Channel 计算机显示波道, 计算机显示通道

Configuration Design Computing 组合设计计算

Construction Design Criteria 结构设计准则

Control Data Corporation 控制数据公司

Control Display Cooling 控制显示冷却

Course and Distance Computer 航向距离计算机

**CDCCL** Critical Design Configuration Control Limitations 关键设计构型控制限制

**CDCE** Central Data Conversion Equipment 中央数据转换装备

**CDCS** Central Data Collection System 中央数据收集系统

Centralized Digital Control System 集中数字控制系统

Computerized Data Collection System 计算机化数据收集系统

**CDD** Central Data Display 中心数据显示

**CDT** Central Daylight Time 中部夏令时间

**CDU** Central Display Unit 中央显示装置

**CEC** Central Economic Committee 中央经济委员会

**CDDA** Compact Disc Digital Audio 数字音频光盘

**CDDI** Cable Distributed Data Interface 电缆分布式数据接口

Coaxial Distributed Data Interface 同轴分布式数据接口

Common Distributed Data Interface 通用分布式数据接口

Copper Distributed Data Interface 铜缆分布式数据接口

**CDDN** Cabin Data Distribution Network 客舱数据分配网络

**CDDP** Cataloging Distribution Data Processing 编目分发数据处理

Console Digital Display Programmer 控制台数字显示程序装置

**CDDS** Competent Documentation Data System 部件文件数据系统, 部件归档数据系统

**CDDT** Countdown Demonstration Test 倒计时演示试验

**CDDV** Compact Disc Digital Video 数字视频光盘

**CDE** Code 代码

Collaboration Development Environment 协同开发环境

Compact Disc Erasable 可擦光盘

Concept Design and Evaluation 概念设计和评价

**CDEK** Computer Data Entry Keyboard 计算机数据输入键盘

**CDES** Chemical Defense Experimental Station 化学防御实验站

Computer Data Entry System 计算机数据输入系统

**CDF** Channel Definition Format 频道定义格式

Coded Digital FAX 编码数字传真

Common Dode Failure 共模故障

Communication Data Field 通信数据字段

**CDFA** Continuous Descent Final Approach 连续下降的最后进近

**CDFF** Common Document File Format 公用资料文件格式

**CDFM** Compact Disc File Manager 光盘文件管理系统

**CDFS** Cascade Dual Frequency Smoothing 级联双频平滑滤波

Compact Disc File System 光盘文件系统

**CDG** CDMA Development Group CDMA 开发组

Charles De Gaulle, Paris, France (Airport Code) 法国巴黎戴高乐机场

Configuration Database Generator 布局数据库发生器

**CD-G** Compact Disc-Graphics 图形光盘

**CDH** Camden, AR, USA (Airport Code) 美国阿肯色州卡德恩机场

Clearance Delivery Head 放行主任,放行许可发布席主任

Command and Data Handling 指令与数据处理

**CDHS** Communications and Data Handling System 通信和数据处理系统

Comprehensive Data Handling System 综合数据处理系统

**CDI** Canadian Airlines International Ltd. 加拿大国际航空有限公司

Capacitor Discharge Ignition 电容式点火

Collector Diffusion Isolation 集电极扩散隔离

Compact Disc Interactive 交互式光盘

Compass Director Indicator 航道罗盘指示器

Control Direction Indicator 方向控制指示器

Conventional Defense Initiative 常规防御倡议

Course Deviation Indicator 航道偏离指示器

**CDIM** Component Data Instruction Manual 部件数据说明手册

**CDIO** Conceive, Design, Implement, Operate 构思、设计、实施、运行,构想、设计、实施、操作

**CDIS** Common Data Interface System 公共数据接口系统

Customization Definition Information System 客户化定义信息系统

**CDL** Capability Demonstration Laboratory 能力验证实验室(美国)

Configuration Data Loader 状态数据输入器

Configuration Deviation List 构型偏离清单,构型偏差单

**CDLA** Condition Lever Angle 状态控制杆角度

**CDLCU** Cockpit Door Locking Control Unit 驾驶舱门锁控制组件

**CDLI** Called Line Identity 被叫线路鉴别

**CDLR** Canadian Dollar 加元

**CDLS** Cockpit Door Locking System 驾驶舱门锁系统

**CDM** Cash Deposit Machine 自动存款机

Clean Development Mechanism 清洁发展机制

Code Division Multiplex 码分复用

Collaborative Decision Making 协同决策

Continuous Data Modulation 连续增量调制

Corona Diagnostic Mission 日冕探测任务

**CDMA** Code Division Multiple Access 码分多址

**CDM-EB** Executive Board of Clean Development Mechanism 清洁发展机制执行理事会

**CDMF** Clean Development Mechanism Fund 清洁发展机制基金

**CDMLS** Commutated-Doppler Microwave Landing System 转换式多普勒微波着陆系统

**CD-MO** Compact Disk Magnet Optical 磁光盘

**CDMS** Central Data Management System 中央数据管理系统

Color Display Management System 色彩显示管理系统

Command and Data Management Subsystem 指令与数据管理分系统

Commercial Data Management System 商用数据管理系统

Communication and Data Management System 通信和数据管理系统

Contract Data Management System 合同数据管理系统

**CDN** Cabin Data Network 驾驶舱数据网络

Common Data Network 客舱数据网络

Condition 情况,条件

Content Delivery Network 内容发布网络

**CDNC** Chinese Domain Name Consortium 中文域名协调联合会

Computer-Aided Design Numerical Control 计算机辅助设计数字控制

**CDNM** Cross-Domain Network Manager 跨域网络管理程序

**CDO** Continuous Descent Operations 持续下降运行

**CDOL** Customer Data and Operations Language 用户数据和操作语言

**CDOS** Customer Data and Operations System 用户数据和操作系统

**CDP** Cartographic Digitizing Plotter 制图数字化绘图机

Central Data Processor 中央数据处理器

Centralized Data Processing 集中数据处理

Compact Disc Player 光盘机

Compressor Delivery Pressure 压气机交输压力

Compressor Discharge Pressure 压气机排气压力

Contract-Definition Phase 合同条款拟订阶段

Counter-measures Dispenser Pod 电子对抗投放吊舱

Critical Decision Point 临界判定点

Customer Data Processing 用户数据处理

Parasite Drag Coefficient 废阻力系数

**CDPC** Central Data Processing Center 中央数据处理中心

Central Data Processing Computer 中央数据处理计算机

**CDPD** Cellular Digital Packet Data 蜂窝数字分组数据

Customer Driven, Previously Delivered 客户发起,以前交付的

**CDPI** Crash Data Position Indicator 失事位置数据指示器

**CDPIE** Command Data Processor Interface Equipment 指令数据处理器接口设备

**CDPS** Central Data Processing System 中央数据处理系统

**CDPSB** CDP Switch Box 光盘机转换盒

**CDR** Call Data Recording 呼叫数据记录

Call Detail Record 呼叫详细记录

Call Detail Recorder 呼叫细节记录器

Capacity to Demand Ratio 容量需求比

CD Recordable 可记录光盘

Central Distribution Room 中央配电室

Chadron, NE, USA (Airport Code) 美国内布拉斯加州查德隆机场

Changed Data Report 更改的数据报告

Charging Data Recording 计费数据记录

Command Destruct Receiver 指令自毁接收器

Commander 指挥长

Conditional Route 条件航路

Conflict Detection and Resolution 冲突探测与解脱

Crash Data Recorder 坠毁数据记录器

Critical Data Request 关键数据请求

Critical Design Review 关键设计评审

Cumulative Default Rate 违约率

Current Directional Relay 电流定向继电器

**CDRAM** Cache Dynamic Randon Access Memory DRAM 高速缓存

**CDRB** Cross-Divosonal Review Board 跨部门评审委员会

**CDRD** Computation and Data Reduction Division 计算和数据处理部

Contract Data Requirements Document 合同数据要求文件

**CDRL** Contract data Requirements List 合同数据要求清单

**CDRM** Critical Design Review Meeting 关键性设计审查会议

**CD-ROM** Compact Disc-Read Only Memory 致密盘只读存储器

**CDRS** Command and Data Retrieval System 指令与数据检索系统

Container Data Retrieval System 集装箱数据检索系统

Control and Data Retrieval System 控制与数据检索系统

**CDRTOS** Compact Disc Real Time Operating System 光盘实时操作系统

**CDRW** CD-Rewritable 可重复刻录光盘

**CDS** Central Data Subsystem 中央数据分系统

Central Display System 中央显示系统

Central Dynamic Store 中央动态存贮器

Cockpit Dynamic Simulator 驾驶舱动态模拟机

Cold Drawn Steel 冷拉伸钢

Command and Data System 指令和数据系统

Commitment Development Schedule 承担改进计划

Common Display System 公共显示系统

Compatible Duplex System 兼容双工系统

Component Documentation Status 部件文档状态

Comprehensive Display System 综合显示系统

Compressed Data Storage 压缩数据存储器

Computerized Documentation Service 计算机化的文献服务

Container Delivery System 集装箱运送系统

Control and Display System 控制和显示系统

Controlled Delivery System 可控空投系统

Controls and Dislays Subsystem 控制和显示分系统

**CDSE** Computer-Driven Simulation Environment 计算机驱动的模拟环境

**CDSF** Commercially Developed Space Facility 商业应用空间设施(美国)

Customer Data Services Facility 用户数据服务设施

**CDSMS** China Development Strategic Modeling System 中国发展战略模型系统

**CDSS** Cockpit Door Surveillance System 驾驶舱门监视系统

Compressed Data Storage System 压缩数据存储系统

**CDSU** Cockpit Door Surveillance Unit 驾驶舱门监测组件

**CDT** Cabin Door Trainer 客舱门培训员

Calendar Date 日历日期

Central Data Terminal 中央数据终端

Central Daylight Time 中部夏令时,中部白昼时间

Command Destruct Transmitter 指令自毁发射机

Compressor Discharge Temperature 压缩机出口温度

Computer Data Terminal 计算机数据终端

Controlled Departure Time 管制离场时间

**CDTI** Cockpit Display of Traffic Information 驾驶舱交通信息显示

Compartment Duct Temperature Indicator 机舱通风管温度指示器

**CDTO** Carrier Detect Timeout 载波检测暂停

**CDTS** Compressor Discharge Temperature Sensor 压缩机出口温度传感器

**CDU** Center Drive Unit 中央驱动组件

Central Display Unit 中央显示器

Cockpit Display Unit 座舱显示装置

Control Display Unit 控制显示组件

**CD-UDF** Compact Disc Unified Disk Format 统一格式的光盘

**CDV** Cell Delay Variation 信元迟延变化

Compressed Digital Video 压缩数字视频

Cordova, AK, USA ( Airport Code ) 美国阿拉斯加州科尔多瓦机场

**CD-V** Compact Disc-Video 视频光盘

**CDVT** Cell Delay Variation Tolerance 信元迟延变化容差

**CDW** Common Data Warehouse 公共数据仓库

**CDWD** Component Data Working Document 部件数据工作文件

**CE** Call Establishment 呼叫建立过程

Certificate of Experience 飞机驾龄证书

Chief Engineer 总工程师

Chip Enable 航程测验板使能

Circuit Emulation 电路仿真

Circular Error 循环误差

Common Emitter 共射极

Communications Electronics 通信电子设备

Commutator End 整流子端

Compass Error 罗差

Computing Element 计算单元

Computing Environment 计算环境

Concurrent Engineering 并行工程

Conducted Emissions 传导辐射

Connecting Element 连接单元

Convolutional Encoder 卷积编码器

Customer Engineer 客户工程师

Customer Engineering 客户工程

**CE&A** Configuration Engineering and Analysis 构型工程与分析

**CAC** Cessna Aircraft Company 赛斯纳飞机公司

**CEA** Circular Error Average 平均循环偏差

Code of European Airworthiness 欧洲适航代码

Combined Electronics Assembly 组合电子设备

Cost-Effectiveness Analysis 成本效益分析

Council of Economic Advisors 经济顾问委员会

**CEAC** Committee for European Airspace Coordination 欧洲空域协调委员会

Cost & Economic Analysis Center 成本和经济分析中心

Cost Evaluation and Analysis Center 成本评估和分析中心

**CEACR** Committee of Experts on the Application of Conventions and Recommendations 公约和建议书实施专家委员会

**CEAH** China Eastern Air Holding Company 中国东方航空集团公司

China Eastern Aviation Holding Company 中国东方航空集团公司

**CEANS** Conference on the Economics of Airports and Air Navigation Services 机场和空中航行经济学会议

**CEAO** West African Economic Community 西非经济共同体

**CEAT** French National Aeronautical Test Center 法国国家航空试验中心

**CEATS** Central European Air Traffic Services 中欧空中交通服务组织

China and European Air Traffic Services 中欧空中交通服务组织

**CEB** Cebu, Philippines ( Airport Code ) 菲律宾宿雾机场

Chief Executives Board 首席执行委员会

Controller Evaluation Board 管制员评价委员会

Cooperation Board of Executives 行政首长协调理事会

Curve of Equal Bearings 等方位曲线

Customer Engineering Bulletin 用户工程通告

**CEC** Cabin Electronics Compartment 客舱电子舱

Cell Error Control 信元差错控制

Central Equipment Center 中央设备中心

Change Event Code 更改事件码

Combined Error Cross-section 组合误差抽样

Common Equipment Card 普通设备卡

Communications Equipment Cabinet 通信设备柜

Computer Evaluation Center 计算机评估中心

Consolidated Electrodynamics Cooperation 电动力学联合公司

Crescent City, CA, USA (Airport Code) 美国加利福尼亚州克雷申特城机场

**CECAM** Centralized Cabin Monitoring 集中客舱监控

**CECS** Conference on European Communication Satellite 欧洲通信卫星会议

**CED** Competitive Engineering Definition 竞争性工程设计

Continued Engineering Development 连续工程研制

Cooling Effect Detector 冷却效果探测器

**CEDDA** Center for Experimental Design and Data Analysis 实验设计与数据分析中心

**CEDS** Civil Engineering Data System 民用工程数据系统

Component Electronic Data Sheet 电子元件数据单

Comprehensive Engine Diagnostic System 发动机综合诊断系统

**CEET** Cabin Emergency Evacuation Trainer 客舱紧急撤离训练器

**CEF** Cost Effective Flight 低成本高效飞行

**CEFH** Cumulative Engine Flight Hours 累计发动机飞行小时

**CEI** Contract End Item 合同终止项

Critical Engine Inoperative 关键发动机不工作

**CEIL** Ceiling 云高，云底高

**CEIS** Center for Environmental Information and Statistics 环境与信息统计中心

Computer Economic Incorporated Sizing 计算机经济型公司化配置

Cost and Economic Information System 成本与经济情报系统

**CEL** Celsius 摄氏

Chemical Engineering Laboratory 化学工程实验室

Common Equipment List 通用设备清单

Communications Engineering Laboratory 通讯工程试验室

Component Evolution List 部件发展清单

**CELSS** Closed Ecologic Life Support System 密闭生态生命保障系统

Closed Environmental Life Support System 密闭环境生命保障系统

Controlled Ecological Life Support System 受控生态生命保障系统

**CEM** Channel Electron Multipliers 通道式电子倍增器

Computing Electromagnetics 计算电磁学

Cost & Effectiveness Method 成本和效益法

Cost Estimation Model 成本估算模型

**CEMA** Consumer Electronics Manufacturing Association 消费者电子制造业协会

Conveyor Equipment Manufacturers Association 输送机设备制造协会

**CEMS** Central Electronic Management System 中央电子管理系统

Chemical Engineering & Materials Science 化学工程与材料科学

Civil Engineer Management System 土木工程师管理系统

Communication Electronics Management System 通信电子管理系统

Comprehensive Engine Management System 发动机综合管理系统

Constituent Electronic Mail System 电子邮件子系统

Continuous Emission Monitoring System 烟气排放连续监测系统

**CEN** Canadian Environment Network 加拿大环境网络

Cell Error Number 信元差错数

Ciudad Obregon, Mexico（Airport Code）墨西哥欧布雷贡城机场

Communication, Electronics, and Navigation 通信,电子与导航

**CENELEC** European Committee for Electrotechnical Standardization 欧洲电工标准化委员会

**CENEUR** Central Europe 中欧

**CENPAC** Central Pacific 太平洋中部

**CENT** Center 中心

Centigrade 百分度

Century 世纪

**CEO** Chief Executive Officer 首席执行官,总经理

**CEOS** Committee on Earth Observation Satellites 国际卫星对地观测委员会

**CEP** Call set-up Error Probability 呼叫建立差错概率

Circular Error Probability 圆概率误差

Circular Error Problem 循环误差问题

Communications Equipment Program 通信设备计划

**CEPGL** Economic Community of the Great Lakes Countries 大湖国家经济共同体

**CEPS** Chinese Electronic Periodicals Service 中文电子期刊服务

**CEPT** Commission of European Post and Telecommunications 欧洲邮电管理委员会

Conference of European Postal & Telecommunications 欧洲邮电管理会议

Council of European Posts and Telegraphs 欧洲邮政和电信理事会

**CER** Cell Error Rate 信元错误率

Ceramic 陶瓷

Certified Emission Reduction 经核证的减排量

Climb Enroute 航路爬升

Complete Engine Repair 发动机大修

Cost Estimating Relationship 成本预算关系

**CERAC** Combined Center Radar Approach Control 雷达进近管制联合中心

**CERAP** Combined Center/Radar Approach Control 联合中心/雷达进近管制

**CERES** Coalition for Environmentally Responsible Economies 环境负责经济体联盟

**CERM** Center for Environmental Resource Management 环境资源管理中心

Certified Enterprise Risk Manager 企业风险管理师证书

**CEROR** Certificate of Origin 始发地证明书

**CERT** Certificate or certified 证书

Computer Emergency Response Team 计算机应急事件响应小组

**CES** Circuit Emulation Service 电路仿真业务

Coast Earth Station 海岸地球站

Cognitive Environment Simulation 认知环境模型

Cognitive Environment Simulator 认知环境模拟器

Communication Engineering Standards 通信工程标准

Computer Engineering Systems 计算机工程系统

**CESC** Conference on European Security and Cooperation 欧洲安全合作会议（简称"欧安会"）

**CESE** Center for Environmental Sciences and Engineering 环境科学与工程中心

**CESI** Centrifugal Engine Speed Indicator 离心式发动机转速表

Chinese Electronic Standardization Institute 中国电子技术标准化研究所

**CESM** Commercial Engine Service Memorandum 商用发动机服务备忘录

**CESPN** Chinese Environmental Scholars and Professionals Network 华人环境学者工程师协会

**CEST** Coast Earth Station Telex 海岸地球站用户电报

**CET** Clearance Expiry Time 放行限期时间
College English Test 大学英语等级考试

**CETS** Contract Engineering Technical Services 合同工程技术服务
Contract Engineering Technical Support 合同工程技术支援

**CEU** Cargo Electronic Unit 货舱电子组件

**CEV** Clutch Electro Valve 离合器电活门
Crew Exploration Vehicle 机组探测飞行器

**CEZ** Cortez, CO, USA (Airport Code) 美国科罗拉多州科特兹机场

**CF** Call Forwarding 呼叫前转
Captive Flight 载飞试验
Carriage Free 免付运费
Centrifugal Force 离心力
Change Frequency to 改频到
Charter Flight 包机航班
Check Flight 检验飞行,技术检查飞行
Commodities Finance 商品融资
Commuter Flight 通勤飞行
Computer File 机读文档
Concept Feasibility 概念可行性
Concept Formulation 概念公式化
Confer 对照
Confirm 确认,参见,比较
Constant Frequency 恒频
Constructed Fare 组合票价
Conversion Facility 转换设施
Core Function 核心功能
Course to a Fix 沿航路到定位点
Cumulative Fuel 累积燃油

**CFA** Capacity and Flow Assignment 容量和流量分配
Combination Fabrication and Assembly 组合件制造和装配

**CFAE** Contractor Furnished Aerospace Equipment 合同方提供的宇航设备
Contractor-Furnished Aeronautical Equipment 合同方提供的航空设备

**CFAR** Constant False Alarm Rate 恒虚警率

**CFB** Call Forwarding Busy 呼叫前转忙

**CFC** Carbon Fibre Composites 碳纤维复合材料
Central Fire Control 中央点火(发射)控制
Central Flow Control 中央流量管制
Chlorofluorocarbon 氯氟碳化合物
Cost of Facilities Capital 基本设施成本

**CFCC** Central Flow Control Computer 中央流量管制计算机

**CFCF** Central Flow Control Facility (Function) 中央流量管制设施(功能)

**CFD** Computational Fluid Dynamics 计算流体动力学

**CFDIU** Centralized Fault Display Interface Unit 集中故障显示接口组件

**CFDL** Confidential 机密的

**CFDPS** Compact Flight Data Processing System 小型飞行数据处理系统

**CFDS** Centralized Fault Display System 集中故障显示系统

**CFE** Certified Fraud Examiner 认证欺诈审核师
Contractor Furnished Equipment 合同方提供设备
Customer Furnished Equipment 客户提供设备

**CFG** Configuration (Aircraft) 机舱座位配置
Configuration 结构,布局
Constant Frequency Generator 恒定频率发生器
Context-Free Grammar 无上下文的语法

**CFH** Conductive Fibre Heater 导电纤维加热器

Cubic Feet per Hour 立方英尺/每小时

**CFI** Central Fuel Injection 中央燃料喷射

Certificated Flight Instructor 合格的飞行教员

Certified Flight Instructor 持照飞行教员

Chief Flight or Flying Instructor 飞行主任教官

Cost, Freight and Insurance 货价包括运费及保险费

**CFIT** Commission for Integrated Transport (UK) 综合运输委员会

Controlled Flight Into Terrain 可控撞地飞行

**CFIUS** Committee on Foreign Investment in the U. S. 在美国的外国投资委员会

**CFS** Call Failure Signal 呼叫失败信号

**CFL** Check-in Flight List 值机航班显示

Cleared Flight Level 许可飞行等级,许可飞行高度层

Critical Field Length 临界机场长度

**CFLP** China Federation of Logistics & Purchasing 中国物流与采购联合会

**CFM** Cable Fabrication Manual 钢索制造手册

CFM International CFM 国际发动机公司

Confirm 证实,确认

Cubic Feet per Minute 立方英尺/分钟

**CFMD** Confirmed 确认的

**CFME** Continuous Friction Measuring Equipment 连续摩擦测量设备

**CFMI** CFM International CFM 国际发动机公司

**CFMU** Central Flow Management Unit 中央流量管理装置

**CFN** Carrier Frequency Net 载频网

Confine 范围,限制

**CFNR** Call Forwarding No Reply 呼叫无应答前转

**CFO** Chief Finance/Financial Officer 首席财务官

**CFOT** Clean Fuel Oil Tank 清洁燃油箱

**CFP** Cargo Fire Panel 货舱火警板

Change Flight Plan 改变飞行计划

Cold Frontal Passage 冷锋过境

Company Flight Plan 公司飞行计划

Computerized Flight Plan 电子化飞行计划

Concept Formulation Phase 概念制定阶段

Contractor Furnished Property 合同方提供参数(器材)

Customer Furnished Parts 用户提供的零件

Customer Furnished Property 用户提供参数(器材)

**CFPS** Card File Power Supply 插件卡电源

Cargo Front-end Processing System 前期货运处理系统

**CFR** Caen, France(Airport Code)法国卡昂机场

Cell Failure Ratio 信元失效比

Center for Future Research 未来研究中心[(美)加利福尼亚大学]

Channel Failure Ratio 信道失效比

Code of Federal Regulations 联邦规章法典

Confer 比较,参见

Constant Failure Rate 常数故障率

Contact Flight Rules 目视飞行规则

Cost and Freight 成本加运费

Crash, Fire and Rescue 失事,消防和救援

**CFRP** Carbon Fiber Reinforced Plastic 碳纤维增强塑料

**CFS** Cabin File Server 客舱文件服务器

Call Failure Signal 呼叫故障信号

Centrifuge Flight Simulation 离心机飞行模拟

Chronic Fatigue Syndrome 慢性疲劳综合症

Cold Finished Steel 冷加工成品钢

Container Freight Station 集装箱货运站

Contract Field Services 合同外场服务

Cubic Feet per Second 立方英尺/秒

**CFSA** China Flight Simulation Association 中国航协模拟飞行委员会

**CFSR** Contract Funds Status Report 合同基金情况报表

**CFT** Captive Flight Test 载飞测试

Craft 飞机,飞艇

Cubic Foot 立方英尺

**CFTP** Certification Flight Test Program 合格审定飞行试验项目

**CFTS** Computerized Flight Test System 计算机化飞行试验系统

**CFU** Call Forwarding Unconditional 无条件呼叫转送

Kerkyra, Greece(Airport Code) 希腊克基拉机场

**CFVCU** Cargo Fire Verification Control Unit 货舱火警测试控制组件

**CFVS** Cargo Fire Verification System 货舱火警测试系统

**CFWP** Central Flow Weather Processor 中央流量气象处理机

**CFWSU** Central Flow Weather Service Unit 中央流量气象服务单元(组件)

**CFY** Clarify, Clarification 澄清

**CG** Cargo 货物

Center of Gravity 重心

Charging Gateway 计费网关

Coast Guard 海岸防卫队

Computer Graphics 计算机图形学

Congo 刚果

**CGAA** China General Aviation Association 中国通用航空协会

**CGAC** China General Aviation Corporation 中国通用航空公司

Customs General Administration of China 中国海关总署

**CGAE** Common Ground Aircraft Equipment 通用地面飞机设备

**CGAS** Coast Guard Air Station 海岸警卫队航空站

**CGB** Central Gearbox 中央减速器,中央齿轮箱

**CGC** Charge Generation Control 计费生成控制

Circuit-Group-Congestion 电路群拥塞

Customer Ground Check 客户地面检查

**CGCS** Center of Gravity Control System 重心控制系统

**CGE** Carriage 运费

Cockpit Geometry Evaluation 飞机座舱几何尺寸评价

**CGH** Computer-Generated Holograms 计算机生成的全息图

Sao Paulo, Brazil(Airport Code)巴西圣保罗机场

**CGI** Cape Girardeau, MO, USA 美国开普吉拉多机场

Certified(or Chief)) Ground Instructor 主任地面教员

Common Gateway Interface 通用网关接口

Computer Graphics Interface 计算机图形接口

Computer-Generated Imagery 计算机产生的图标

Cruise Guide Indicator 巡航制导指示器

**CGIVS** Computer Generated Image Visual System 计算机生成图像目视系统

**CGK** Jakarta, Indonesia (Airport Code) 印尼雅加达机场

**CGL** Center of Gravity Locator 重心定位器

Circling Guidance Light(s) 盘旋引导灯

**CGM** Computer Graphics Metafile 计算机绘图文件

**CGMS** Coordination Group for Meteorological Satellites 气象卫星协调小组

**CGN** Cologne/Bonn, Germany (Airport Code) 德国科隆/波恩机场

**CGO** Cargo 货物

**CGP** Chittagong, Bangladesh(Airport Code) 孟加拉国吉大港机场

Computervision Graphics Processor 计算机视觉图像处理器

**CGR** Campo Grande, Mato Grosso Do Sul, Brazil (Airport Code) 巴西南马托格罗索州大坎普机场

**CGRS** Central Gyro Reference System 中心陀螺基准系统

**CGS** Centimeter, Gram, Second System of U-nit 厘米克秒制

Centimeter-Gram-Second 厘米—克—秒单位制

Combined Guidance System 联合制导系统

Command Guidance System 指令制导系统

**CGT** Consolidated Ground Terminal 统一的地面终端

**CGWIC** China Great Wall Industry Cop. 中国长城工业公司

**CH** Call Handler 呼叫处理器

Center Hatrack 中行李架

Channel 频道,波道,通道

Chapter 章节

Character 字母,特性,人物

Charge 充电,费用

Chief 主任,首长,首席

Chromel 镍铬合金

Compass Heading 罗盘航向

Critical Height 临界高度,临界高

Customhouse 海关

**CHA** Character 品质,角色

Chattanooga, TN, USA(Airport Code)美国田纳西州查塔努加机场

**CHAIR** Chairman 主席

**CHAM** Chamfer 弯度,倒角,斜角

**CHAMP** Champion 冠军,优胜者,战士

**CHAN** Channel 频道,通道,电路

Charge Analysis 计费分析

**CHAP** Chapter 章

**CHAPI** Compact Helicopter Approach Path Indicator 严密的直升机进近航路指示器

**CHAR** Character 字母,人物,特性

Charter 包机

**CHAS** Channel Associated Signaling 随路信令

Chassis 底盘

**CHAT** Chart 航图

**CHC** Channel Controller 通道控制器

Christchurch, New Zealand (Airport Code)新西兰基督城机场

Container Handling Committee 货柜设施发展委员会

**CHCK** Check 检查

**CHCL** Change of Class 等级改变

**CHCS** Cargo Handling Control System 货舱装载控制系统

**CHD** Call Handling 呼叫处理

**CHDO** Certificate-Holding District Office 许可证持有区域办公室

**CHE** Cargo Handling Equipment 货物搬运设备

**CHEM** Chemistry, Chemical 化学,化学的

**CHEML** Chemical 化学的

**CHESA** Cargo Handling Equipment, Systems and Appliances 货物装卸设备、系统、装置

**CHF** Chief 主任,首领,最高的,主要的

Swiss Franc 瑞士法郎

**CHG** Change 更改,改变

Charge 充电,费用

Critical Hydraulic Gradient 临界水力梯度

**CHGD** Changed 改变的

Charged 费用已付

**CHGR** Charger 充电器,加载装置

**CHGVAL** Charge According to Value 按价值计费

**CHGWT** Charge According to Weight 按重量计费

**CHI** Channel Interface 通道接口

Cloud Height Indicator 云高指示器

Computer-Human Interaction 人机交互

Computer-Human Interface 人机接口

**CHINML** Chinese Meal 中餐

**CHIRP** Confidential Hazardous Incident Reporting Programme 危险事故秘密报告项目

Confidential Human-Factors Incident Reporting Program 人为因素事故秘密报告项目

**CHIS** Center Hydraulic Isolation System 中央液压隔离系统

**CHK** Check 检查,校对,支票

**CHKD** Checked 检查过的

**CHKG** Checking 检查

**CHKIN** Check in 输入检查字

**CHKL** Checklist 检查单

**CHKPNT** Check Point 检查点

**CHKRD** Check Read 读出检查字

**CHM** Chairman 主席

Chamber 室,容器

Chemical 化学的

**CHM** Chime 谐音

**CHMBR** Chamber 室

**CHML** Child Meal 儿童餐食

**CHN** Chain 链子

Chairman 主席

China 中国

**CHNGS** Changes 变化,改变

**CHNL** Channel 波道,通道

**CHNT** Change Name To 改名为

**CHO** Charlottesville, VA, USA（Airport Code）美国弗吉尼亚州夏洛茨维尔机场

**CHOC** Chocolate 巧克力

**CHOCL** Change of Class 改变等级

**CHP** Charges Prepaid 运费预付

**CHPR** Cheaper 较便宜

**CHQ** Chania, Greece（Airport Code）希腊干尼亚机场

Cheque 支票

**CHR** Checker 检查员

Chromel 镍铬合金

Chronograph 计时器,记时器

Chronometer 精密计时器,航行表

Commission on Human Rights（联合国）人权委员会

**CHRFLAG** Charging FLAG 计费标记

**CHRG** Charge 费用,充电

**CHRGR** Charger 充电器,加载装置

**CHRM** Chairman 主席,主持

**CHRMT** Charging Method 计费方法

**CHRO** Chronometer 天文钟,计时器

**CHS** Call Hold Service 呼叫保持业务

Cargo Handling System 货舱处理系统

Charleston, SC, USA（Airport Code）美国南卡罗来纳州查尔斯顿机场

Check Sum 检查和

**CHSUM** Check Sum 检查和

**CHT** Call Holding Time 呼叫保持时间

Chart 航图

Chest 箱子

Cylinder Head Temperature 气缸头温度

**CHTR** Channel Translator 信道变换器

Charter 包机

**CHY** Commission of Hydrology 水文学委员会

**CI** Cabin Interphone 客舱内话

Cargo Insurance 货物保险

Carry In 输入,装入

Cast Iron 铸铁

Cell Identity 信元识别

Change Item 更改项目

Cirrus 卷云

Cluster Interface 群接口

Combined Index 综合索引

Command Identifier 指令标识符

Community Interface 社区接口

Computer Indicator 计算机显示器

Computer Interconnect 计算机互联

Configuration Index 结构标记（或结构索引）

Configuration Item 构型项目,结构项目

Congestion Indication 拥塞指示

Conservation Internation 养护国际

Consular Invoice（Shipping）领事签证

Continued Item 保留项目

Continuous Improvement 持续改进

Contract Item 合同项目

Contractor Inquiry 合同方查询

Control Indicator 控制指示器

Control Interface 控制接口

Controlled Item 控制项目

Conversion Instruction 转换说明

Corporate Identity 企业形象,公司形象

Cost Index 成本指数

Counter Intelligence 反情报

Course Indicator 航道指示器,航向指示器

Course to an Intercept 到切入点的航道

Critical Item 关键项

Crossbar Interconnection 纵横制互联

Cruise 巡航

Customer Installation 用户装置

Customs and Immigration 海关和移民局

**CIA** Capital International Airways, Inc. (USA) 首都国际航空公司(美国)

Check-In Assistant 值机助理系统

Ciampino, Rome, Italy (Airport Code) 意大利罗马钱皮诺机场

Communication Interface Adaptor 通信接口适配器

Contractor Interface Agreement 承包商接口协议

**CIAS** Changi International Airport Services 樟宜国际机场服务公司

**CIASOR** Covered with Ice and Snow on Runway 冰雪覆盖跑道

**CIAT** Communications Installation Advisory Team 通讯设备安装咨询队

**CIB** Climb 爬升

**CIC** Cabin Interphone Controller 客舱内话控制器

Carrier Identification Code 载波识别码

Chico, CA, USA (Airport Code) 美国加利福尼亚州奇克机场

Circuit Identification Code 电路识别码

Communications Intelligence Channel 通信智能信道

**CICERO** Centre for International Climate and Energy Research-Oslo (Norway) 奥斯陆国际气候和能源研究中心(挪威)

**CICS** Customer Information Control System 客户信息控制系统

**CID** Call Instance Data 呼叫实例数据

Caller IDentification 主叫识别

Cedar Rapids, IA, USA (Airport Code) 美国依阿华州锡特拉皮兹机场

Channel Identifier 信道标识

Component Identification 部件标识

**CIDA** Canadian International Development A-

gency 加拿大国际开发署

**CIDFP** Call Instance Data File Point 呼叫实例数据文件段

**CIDIN** Common ICAO Data Interchange Network 国际民航组织公用数据交换网

**CIDR** Class Inter-Domain Routing 分级域间路由选择

Classless Inter Domain Routing 无级域间路由选择

**CIDS** Cabin Intercommunication Data System 客舱内部数据通讯系统

Cabin Interphone Distribution System 客舱内话分配系统

Contracting Information Database System 合同信息数据库系统

**CIE** Center for Independent Education 独立教育中心

Company 公司,陪同

**CIEC** China International Exhibition Center 中国国际展览中心

Conference on International Economic Cooperation 国际经济合作会议

**CIF** Common Intermedia Format 通用中间格式

Cost, Insurance & Freight 成本、保险费和运费(离岸价格)

**CIFC** Cost, Insurance, Freight and Charges 成本、保险费、运费和捐税

Cost, Insurance, Freight and Commission 成本、保险费、运费及手续费

**CIFFA** Canadian International Freight Forwarders Association 加拿大国际货运代理协会

China International Freight Forwarders Association 中国国际货运代理协会

**CIFMS** Computer Integrated Flexible Manufacturing System 计算机集成柔性制造系统

**CIFP** Cancel IFR Flight Plan 取消仪表飞行规则的飞行计划

**CIFR** Cancel IFR (Instrument Flight Rules) 取消仪表飞行规则

**CIFRR** Common IFR (Instrument Flight Rules) Room 通用仪表飞行规则室

**CIFS** Common Internet File System 通用因特网文件系统

**CIG** Ceiling 云底高

Cell Interconnection Gateway 信元互联网关

Cigarette 香烟

Computer Image Generator 计算机图像发生器

Computerized Interactive Graphics 计算机化交互图形学

**CIGAC** China International General Aviation Convention 中国国际通用航空大会

**CIH** Cabin Interphone Headset 客舱内话耳机

**CII** Configuration Identification Index 组合(配置)标识符索引

**CIIF** China International Industry Fair 中国国际工业博览会

**CIIMS** Central Integrated Information Management System 中央信息集成管理系统

**CIIP** Critical Information Infrastructure Protection 关键信息基础设施保护

**CIL** Call Identification Line 呼叫识别线路

**CILOP** Conversion In Lieu of Procurement 改装取代采购

**CILS** Computer Integrated Logistics System 计算机综合物流系统

**CILT** Chartered Institute of Logistics and Transport 英国皇家物流与运输学会,特许物流运输协会

**CIM** Cockpit Information Management 驾驶舱信息管理

Cockpit Integrated Management 驾驶舱综合管理

Common Information Model 公共信息模型

Computer Input Media 计算机输入媒体

Computer Input Microfilm 计算机输入微缩胶片

Computer Integrated Manufacturing 计算机集成制造,计算机综合制造

Continuous Image Microfilm 连续成像微缩胶片

Control Interface Module 控制接口模块

**CIMO** Commission for Instruments and Methods of Observation 仪器和观测方法委员会

**CIMS** Cockpit Information Management System 座舱信息管理系统

Common Information Management System 公共信息管理系统

Computer Integrated Manufacturing System 计算机集成制造系统

Customer Information Management System 客户信息管理系统

**CIMSS** Cooperative Institute for Meteorological Satellite Studies 气象卫星合作研究所

**CIN** Change Identification Number 更改识别号

**CINS** Cabin Information Network System 客舱信息网络系统

Cryogenic Inertial Navigating System 低温惯性导航系统

**CIO** Cash in Order 订货时付款

Chief Information Officer 首席信息官

**CIOS** Communication Institute for Online Scholarship 在线学者交流协会

**CIP** Call Information Processing 呼叫信息处理

Cast Iron Pipe 铸铁管

Change Implementation Plan 更改执行计划

Chief Information Officer 首席信息官

Combined Instrument Panel 综合仪表板

Commercial Important Person 商业要人

Common Integrated Processor 通用综合处理器

Component Improvement Program 部件改进程序

Compressor Inlet Pressure 压缩机进口压力

Continuous Improvement Program 持续改进计划

Control Interface Processor 控制接口处理器

Control Inlet Panel 控制输入的仪表板

Cost Improvement Program 成本改进程序

**CIPEC** Center for the Study of Institutions, Population, and Environmental Change 制度，人口和环境变化研究中心

Intergovernmental Council of Copper Exporting Countries 铜出口政府间委员会

**CIPFS** Configuration Item Product Function Specification 配置项目产品功能规范

Critical Item Product Function Specification 关键项目产品功能规范

**CIPR** Continuous In-flight Performance Recorder 连续飞行性能记录器

**CIPS** Centralized Industry Prorata Service 集中的工业分派服务

Chartered Institute of Purchasing and Supply 英国皇家采购与供应学会

Cockpit Instrument Panel Space 驾驶舱仪表板空间

Computer Integrated Process Systems 计算机集成过程系统

**CIQ** China Entry-Exit Inspection and Quarantine 中国出入境检验检疫

Chiquimula, Guatemala (Airport Code) 危地马拉奇基穆拉机场

Customer Information Quality 客户信息的质量

Customs, Immigration and Quarantine 海关，移民和检疫

**CIR** Calling-line-Identity-Request 主叫线路识别请求

Change Impact Report 更改效果报告

Change Incorporation Record 变更公司记录

Change Incorporation Request 更改合并要求

Circle 圆，环，循环，周期

Circuit 线路，电路

Circular 通报，圆形的

Circulate 传播，流通

Circumference 周围

Committed Information Rate 议定（提交）信息率

Cost Information Reports 成本信息报告

Customer Inspection Record 用户检验报告

**CIR-1** Connector Installation Report 连接器安装报告

**CIR-2** Connector Identification & Installation Report 连接器标识与安装报告

**CIRA** Committee International Reference Atmosphere 国际基准大气委员会

Cooperative Institute for Research in the Atmosphere 大气合作研究所

Critical Issue Review and Analysis 重要问题审查与分析

**CIRC** Circular 通报，圆形的

Circulate 循环

Circulator 循环器

**CIRCS** Circumstance 环境，状况

**CIS** Cabin Interphone System 客舱内话系统

Chinese Industrial Standards 中国工业标准

Cockpit Interface Simulation 驾驶舱接口模拟

Command Instrument System 指令仪表系统

Common Interface Standard 通用接口标准

Computer Information Systems 计算机信息系统

Cooperative Independent Surveillance 合作式独立监视

Customer Integrated System 客户一体化系统

Customer Introduction/Record System 客户介绍/录音系统

**CISA** Certified Information System Auditor 注册信息系统审计师

**CISCE** International Committee for European Security and Co-operation 欧洲安全与合作国际委员会

**CISCO** Commercial & Industrial Security Corporation 工商保安机构

**CISD** Critical Incident Stress Debriefing 危急事件应激报告

**CISM** Critical Incident Stress Management 危急事件应激管理

**CISS** Crew Information System Services 机组信息系统服务

Communication and Information Systems Security 通信信息系统安全

**CIT** Captain's Instrument Terminal 机长仪表终端

Compressor Inlet Temperature 压气机进口温度

Computer Information Technology 计算机信息技术

Critical Incident Technology 关键事件技术

Customer Integration Team 客户整合小组

Near or Over City 接近或飞越城市上空

**CITA** Canadian Industrial Transportation Association 加拿大工业交通协会

Communications & Information Technology Association ( UK ) 通信与信息技术协会(英国)

**CITE** Cargo Integration Test Equipment 货舱综合测试设备

Computer Integrated Test Equipment 计算机综合测试设备

**CITIC** China International Trust and Investment Corporation 中国国际信托投资公司

**CITL** Community Independent Transaction Log 欧共体独立交易日志

**CITS** China International Travel Service 中国国际旅行社

**CIU** Camera Interface Unit 摄像机接口组件

Cell Input Unit 信元输入单元

Computer Interface Unit 计算机接口组件

Sault Ste Marie, MI, USA ( Airport Code ) 美国密执安州苏圣玛丽机场

**CIV** Centre Interconnect Valve 中心互连活门

Certificate Issue Voucher 证书发放单

Critical Ionization Velocity 临界电离速度

Crossbleed Isolation Valve 交叉排气隔离活门

**CIW** Customer Information Warehouse 用户信息仓库

**CIWS** Central Instrument Warning System 中央仪表警告系统

**CIX** Commercial Internet Exchange 商用因特网交换

**CJC** Cold Junction Compensation 冷接补偿

**CJI** Controller Jurisdiction Indicator 管制员权限指示

**CJU** Cheju, South Korea ( Airport Code ) 韩国济州岛机场

**CK** Check 检查,核对

Cheque 支票

**CKB** Clarksburg, WV, USA ( Airport Code ) 美国西弗吉尼亚州克拉克斯堡机场

**CKD** Checked 检查过的

**CKE** Chinese Keypad Entry 中文键盘输入

**CKI** Check-in 办理登机手续

**CKO** Chief Knowledge Officer 知识工程主管

**CKPT** Cockpit 驾驶舱

**CKTL** Cocktail 鸡尾酒

**CKTS** Circuits 线路

**CKW** Clockwise 顺时针方向

**CKY** Conakry, Guinea-Gbessia ( Airport Code ) 几内亚科纳克里格贝西亚机场

**CL** Call 呼叫

Center Left 中左

Center Line 中心线

Center of Lift 升力中心

Centerline Lights 跑道中线灯

Check List 检查单

Class 等级

Classification 分类,类别

Clause 条款,规定

Climb 爬升

Clip 卡子

Clock 时钟

Close 接近,接通,关闭,精密的,近距的

Closed Loop 封闭环路

Clutch 离合器

Coefficient of Lift 升力系数

Component Location 部件位置

Container Load 集装箱装载(量)

Course Line 航线

Runway Center Line Lights 跑道中心线灯

**CLA**　Center Line Average 平均中心线

Clearance Array 间隔矩阵

Condition Lever Angle 状态改平角

Coupled Loads Analysis 耦合负载分析

**CLAC**　Closed Loop Approach Control 闭合回路进近控制

**CLAIM**　Claiming 索赔

Claim 认领,要求

Cleared Level Adherence Monitoring 放行高度保持监视

**CLAMP**　Closed Loop Aeronautical Management Programme 闭路航空管理程序

Closed Loop Aeronautical Materials Program 闭路航空材料计划

**CLAMTI**　Clutter Locked Airborne Moving Target Indicator 杂波锁定机载动目标显示器

**CLASS**　Classification 分类,类别

Coherent Radar Airborne Shear Sensor 相参激光雷达机载风切变传感器

Custom Local Area Signaling System 用户局域信令系统

Customized Local Access Signaling Service 定制的本地接入信令服务

**CLB**　Cabin Log Book 客舱记录本

Center Line of Bend 弯曲中心线

Climb 爬升

Crash Locater Beacon(飞机)坠毁自动定位信标,应急定位信标

**CLBM**　CLassical Broadcasting Model 传统的广播模型

**CLBR**　Caliber 口径,测径仪,量规

Calibrate, Calibration 校准,校正

**CLBT**　Closed Loop Bench Test 闭环测试台

**CLC**　Course Line Computer 航线计算机

**CLCD**　Color Liquid Crystal Displays 彩色液晶显示器

**CLCK**　Clock 时钟

**CLCP**　Central Lateral Control Package 中央横向控制组件

**CLCRS**　Cross-Lane Communication Registers 交叉通道通信寄存器

**CLD**　Called 称为,叫做

Cancelled 取消的,注销的

Carlsbad, CA, USA(Airport Code)美国加利福尼亚州卡尔斯巴德机场

Cell Loss Detection 信元损失检测

Cleared 已允许的,已报关的

Closed 关闭的,闭合的

Cloud 云

Component Location Data 部件配置数据

Cooled 冷却的

**CLDB**　Component Location Data Bank 部件位置数据库

**CLDS**　Cockpit Laser Designation System 座舱激光指示系统

Cost & Labor Distribution System 成本与劳力分配系统

**CLDY**　Cloudy 多云

**CLE**　Central Location Element 中心配置元件

Cleveland, OH, USA (Airport Code) 美国俄亥俄州克利夫兰机场

Closed End 终端关闭

**CLEAR**　Clearance 间隔,间歇,许可,清算

**CLEC**　Competitive Local Exchange Carrier 竞争性的本地交换运营商

**CLET**　Collet 套筒

**CLFI**　Close File 结案归档

**CLFS**　Closed Loop Frequency Set-on 闭环频率设定

**CLFY**　Clarify 澄清

**CLG**　Calling 访问,要求,呼叫

Ceiling 顶板,天花板,升限,云底高

Centerline Landing Gear 中心线起落架(中主起落架)

Closing 关闭

Cooling 冷却

**CLI**　Calling Line Identification 主叫线路识别

Command Line Interface 指令线路接口

**CLIAS**　Climbing Indicated Airspeed 爬升表

速

**CLICOM** Climate Computing 气候计算
Comprehensive Climate Data Management and
Application 综合气候数据管理应用

**CLIF** Called Line Identification Facility 被叫
线路识别设备

**CLIM** Climate 气候

**CLIN** Contract Line Item Number 合同行项
目号
Contract Line Item 合同行项目

**CLIP** Calling Line Identification Presentation
主叫用户线识别提示
Classical over IP IP 承载传统业务
Connected Line Identification Presentation 被
叫连接线识别提示

**CLIPS** Climate Information and Prediction
Services 气候信息和预测服务

**CLIR** Calling Line Identification Restriction
主呼线路识别限制
Connected Line Identification Restriction 被
叫连接线识别限制

**CLIRI** Calling Line Identification Request Indication 主叫线路识别请求指示

**CLIS** Called Line Identification Signal 被叫
线路识别信号

**CLIVAR** Climate Variability and Predictabili-
ty 气候变异和可预测性
Climate Variability Research Program 气候变
异研究计划

**CLJ** Cluj, Romania ( Airport Code ) 罗马尼
亚克鲁日机场

**CLK** Clerk 职员
Clock 时钟

**CLL** Center Line Lights 中心线灯
College Station, TX, USA ( Airport Code ) 美
国得克萨斯州克利奇站机场

**CLLM** Consolidated Link Layer Management
message 综合链路层管理信息

**CLLR** Councillor 顾问,议员

**CLM** Cargo Load Message 货载电报
Cargo Loading Manual 货物装载手册

Climb 爬升

Column 栏

Component Location Manual 部件位置手册

**CLMG** Claiming 索赔,宣称

**CLN** Clinometer 倾斜仪

**CLNAP** Connectionless Network Access Protocol 无连接网络接入协议

**CLNC** Clearance 放行许可

**CLNG** Ceiling 天花板,顶板,升限
Cleaning 清除,清扫
Cooling 冷却

**CLNP** Connectionless Network Protocol 无连
接网络规程(协议)

**CLNS** Connectionless Network Service 无连
接网络服务

**CLO** Cali, Colombia ( Airport Code ) 哥伦比
亚阿拉贡机场

**CLODA** Closing Date 截止日期

**CLOE** Common Logistics Operating Environment 通用保障操作环境

**CLOFI** Close File 结案

**CLOG** Clogging 堵塞

**CLP** Call Processor 呼叫处理机
Cell Loss Priority 信元丢失优先权
Cell Loss Probability 信元丢失概率
Clamp 卡环
Container Load Plan 装箱单

**CL-PDU** Connectionless Protocol Data Unit
无连接协议数据单元

**CLPI** Cell Loss Priority Indication 信元丢失
优先级指示

**CLPR** Clapper 阻尼凸台,铃锤,拍板

**CLQ** Colima, Mexico ( Airport Code ) 墨西哥
科利马机场

**CLR** Cell Loss Rate 信元丢失率
Cell Loss Ratio 信元丢失比
Clear 变干净,变清楚,清除,放行
Clearance 许可,间隙,放行许可,容差
Colour 颜色
Computer Language Recorder 计算机语言记
录装置

Flight Clearance Message 航班放行电报

**CLR ALT** Clearance Altitude 许可的高度，净空高度

**CLRD** Cleared 已清除，放行

**CLS** Cabin Lighting System 客舱照明系统

Cargo Loading System 货物装载系统

Channel Load Sensing 信道负载检测

Class 等级

Clause 条款

Close 关闭

Continuous Level Sensor 连续平面传感器

Contractor Logistics Support 合同方后勤支援

Control Loading System 装载管制系统

Controlled Load Service 受控负载业务

Cost Leadership Strategy 低成本战略

Customer Link Service 用户链路业务

**CLSD** Closed 关闭的，封闭的，闭合的，接近的

**CLSF** Connectionless Service Function 无连接服务功能

**CLSG** Closing 关闭

**CLSM** Cargo Loading System Manual 货舱装载系统手册

Controlled Low Strength Material 控制性低强度材料

**CLSP** Channel Load Sensing Protocol 信道负载检测协议

**CLSS** Communication Link SubSystem 通信链路子系统

**CLSU** Chime/Light Sensor Unit 谐音/灯光传感器组件

**CLT** Charlotte, NC, USA ( Airport Code ) 美国北卡莱罗纳州夏洛特机场

Classical Lamination Theory 古典层板理论

Client 客户

Closed Loop Test 闭环测试

Closed-Loop Tester 闭环测试器

Collect 收集，收账，托收

Command Legality Test 指令合法性测试

Communication Line Terminal 通信线路终端

**CLTM** Component Location Training Manual 部件位置培训手册

**CLV** Clevis 挂钩，吊环

**CLWY** Clearway 净空道

**CLZ** Close 关闭

**CM** Cable Modem 纤缆调制解调器

Calibrated Magnification 校准放大

Call Maintenance 要求维护

Call Manager 呼叫管理器

Cap Missing 盖帽脱落

Cell Merger 信元归并

Center Matched 中心匹配的

Center of Mass 质量中心

Centimeter 厘米

Certificate Management 证件管理

Certification Manager 审定主任

Change Management 更改管理

Change to Metal 改用金属

Chemical Milling 化学蚀刻

Circling Minima 反向最低标准

Circuit Master 电路主控器

Coherence Multiplexing 相干复用

Collective Modification 集体改装

Command Module 指令模块

Command 指令，司令部

Common Memory 公共存储器

Communications Module 通信舱

Company Material 公司物品

Condition Monitoring 条件监控

Configuration Management 配置管理，构型管理

Confusion Matrix 混淆矩阵

Connection Manager 连接管理器

Connection Matrix 连接矩阵

Continuation of Message 持续报文

Continuous Media 连续性媒体

Continuous Monitoring 条件监控

Control Main 控制总线

Conversion Manual 换算手册

Converter-Multiplexer 转换器，多路选择器

Coordinate Memo 协调摘要

Core Memory 磁芯存储器

Corrective Maintenance 校正维修,修复性维修

Countermeasure 干扰,对抗

Crew Member 机组成员

Current Message 目前讯息

**CMA** China Meteorological Administration 中国气象局

Coherence Multiple Access 相干多址访问

Common Mode Analysis 共有模式分析

Context Management Application 范围管理应用

Continuous Monitoring Approach 连续监测做法

**CMAC** Control Mobile Attenuation Code 控制移动衰减码

**CMAS** Christmas 圣诞节

Computer-Based Maintenance Aids System 基于计算机的维修辅助系统

Configuration Management Application Service 配置管理应用服务

Control Monitoring Automated System 控制监控自动系统

Council for Military Aircraft Standards 军用飞机标准委员会

Crisis Management Action System 危机管理行动系统

**CMB** Climb 爬升

Colombo, Sri Lanka (Airport Code) 斯里兰卡科伦坡机场

Combustor 燃烧室

Concord Management Board 协和式飞机管理局

Crisis Management Briefings 危急管理讲解

**CMBD** Combined 组合的

**CMBSTR** Combustor 燃烧室

**CMC** Central Maintenance Computer 中央维护计算机

Coherent Multi-Channel 相干多信道

Concurrent Media Conversion 并行媒体转换

Conference Management Center 会议管理中心

Constant Mach Cruise 恒定马赫数巡航

Copyright Management Center 版权管理中心

Crew Member Certificates 空勤人员证书

Crisis Management Center 危机管理中心

CUG Management Center CUG 管理中心

Customer Management Center 客户管理中心

**CMCB** CoMmunication Control Block 通信控制功能块

**CMCC** Central Management Communication Control 中央管理通信控制

Certification Maintenance Coordination Committee 审定维修协调委员会

China Mobile Communications Corporation 中国移动通讯公司

Classified Matter Control Center 密件控制中心

Combined Movements Control Center 联合行动控制中心

**CMCF** Central Maintenance Computing Function 中央维护计算功能

**CMCS** Central Maintenance Computing System 中央维护计算系统

Communications Monitoring and Control Subsystems 通信监听与控制分系统

Communications Monitoring and Control System 通信监听与控制系统

COMSEC Material Control System 通信保密器材控制系统

**CMD** Command 指令,命令

Configuration Management Division 组合(配置,构型)管理分部

Contract Management District 合同管理区

Contract Management Division 合同管理分部

**CMDB** Computerized Meteorological Data Base 计算机气象数据库

**CMDO** Consolidated Material Distribution Objectives 综合物资分发目标

**CMDR** Coherent Monopulse Doppler Radar 同调单脉冲多普勒雷达

**CMDS** Centralized Message Data System 集中

式信息数据系统

Cockpit Management Display System 驾驶舱管理显示系统

Command Manpower Data System 指挥部人力数据系统

Countermeasures Dispenser System 干扰分配器系统

**CMDTY** Commodity 商品

**CME** Ciudad Del Carmen, Campeche, Mexico（Airport Code）墨西哥卡门城机场

Communication Management Entity 通信管理实体

Connection Management Entity 连接管理实体

**CMEA** Council for Mutual Economic Assistance 经济互助委员会

**CMEC** Combined Material Exploitation Center 联合物资利用中心

**CMEE** Carbon Management Energy Efficiency 碳管理能源效率

Chief Mechanical and Electrical Engineer 机械及电气总工程师

**CMEU** Cabin Management Expansion Unit 客舱管理扩展组件

**CMF** Central Maintenance Function 中央维护功能

Coherent Memory Filter 相关存储滤波器

Communications Management Function 通信管理功能

Compressor Mid Frame 压缩器中框架

**CMFLR** Cam Follower 凸轮随动器

**CMG** Coming 来到

Control Moment Gyroscope 控制力矩陀螺仪

Course Made Good 正确航向,规定航线

**CMH** Columbus, OH, USA（Airport Code）美国俄亥俄州哥伦布机场

**CMI** Call Management Information 呼叫管理信息

Champaign, IL, USA（Airport Code）美国伊利诺州尚佩恩机场

Coding Method Identifier 编码方式标识符

Computer Managed Instruction 计算机管理说明书

Continuous Maximum Icing 连续最大结冰

Customer Modifiable Information 客户修订信息

**CMID** Commodity Manager Input Data 商品管理员输入数据

Crew Member Identification 机组成员识别

**CMIP** Common Management Information Protocol 公共管理信息协议

Cost Management Improvements Plan 成本管理改进计划

**CMIS** Common Management Information Service 公共管理信息服务

Computer Management Information System 计算机管理信息系统

Costing and Management Information System 成本及管理资料系统

**CMIS/P** Common Management Information Service/Protocol 公共管理信息服务/协议

**CMISE** Common Management Information Service Element 公共管理信息业务单元

**CMIT** Component Management Integration Team 部件管理综合组

**CMIV** Cabin Management and Interactive Video 客舱管理及交互式视频

**CMK** Current Market Outlook 当前市场展望

**CML** Cargo Master Loader 货物主要装货人

Commercial 商务的

Consumable Material List 消耗材料清单

**CMM** Calibration Memory Module 校准存储模块

Capability Maturity Model 功能成熟度模型

Cell Management Module 信元管理模块

Commission of Marine Meteorology 海洋气象学委员会

Common Mode Monitor 普通方式监视器

Component Maintenance Manual 部件维修手册

Computerized Modular Monitoring 计算机化模组监控

Condition Monitored Maintenance 维修状况监控

Cross-Modality Matching 跨通道匹配法

Cubic Meter per Minute 立方米每分钟

**CMMD** Command 命令,指令

**CMMM** Component Maintenance Manual Manufacturer 生产厂家的部件维修手册

**CMMV** Component Maintenance Manual Vendor 供应商的部件维修手册

**CMN** Casablanca, Morocco-Mohamed ( Airport Code ) 摩洛哥卡萨布兰卡穆罕默德机场

Cell Misinsertion Number 信元错插数

Column 栏,列

Commission 手续费

Communication 通讯

Control Motion Noise 操纵动作噪声

**CMNC** Commence 开始

**CMNPS** Canadian Minimum Navigation Performance Specification 加拿大最低导航性能规范

**CMNT** Comment 评论

**CMO** Certificate Management Office 合格证管理办公室

Chief Marketing Officer 高级营销经理职位

Configuration Management Office 配置管理办公室

Contract Management Office 合同管理办公室

**CMOD** Commodity 商品

**CMOS** Complementary Metal Oxide Semiconductor 互补性金属氧化物半导体

Complementary Metal Oxide Silicon 互补性金属氧化硅

**CMP** Certificate Management Protocol 证书管理协议

Certificate of Maintenance Review ( 英国民航局 ) 维修审查证

Compound Helicopter 复合直升机

Compromise 折中,妥协

Condition Monitoring Program 状态监控程序

Configuration Deviation List 构型缺损清单

Configuration Maintenance Procedure 结构维修程序

Configuration Management Plan 构型管理计划

Content Management and Protection 内容管理和保护

Customer Maintenance Program 客户维修方案

Customer Management Point 客户管理点

**CMPETR** Competitor 竞争者

**CMPL** Critical Material Parts List 关键性器材零件清单

**CMPLMT** Complement,Complementary 补充

**CMPLNT** Complaint 投诉

**CMPLRY** Complementary 补充的

**CMPNSN** Compensation 补偿

**CMPNT** Component 元件,组件

**CMPP** Controlled Materials Plan Priority 控制物资管理优先计划

**CMPR** Compare 比较

**CMPS** Centimeters per Second 厘米/秒

Centre for Management and Policy Studies 管理、政策研究中心

Compass 罗盘

Composites Technology 合成技术

**CMPSN** Composition 组成,构成

**CMPT** Composite Technology 复合技术

**CMPTG** Computing 计算

**CMPTR** Computer 计算机

**CMPTV** Competitive 竞争的

**CMPX** Complex 复合的,成套装置

**CMR** Call Modification Request message 呼叫改变请求消息

Cell Misinsertion Rate 信元误插率

Camera 照相机,摄像机

Certificate Maintenance Review 维修审查证

Certification Maintenance Requirement 审定维修要求

Continuous Maximum Rating 最大连续额定值

**CMRB** Condition Monitoring Review Board 状态监控审查委员会

**CMRLR** Cam Roller 凸轮滚子

**CMRM** Call Modification Reject Message 呼叫改变拒绝信息

**CMRS** Countermeasures Receiving System 对抗接收系统

**CMRTS** Cellular Mobile Radio Telephone System 蜂窝移动无线电话系统

**CMS** Cabin Management System 客舱管理系统，机舱管理系统

Call Management System 呼叫管理系统

Central Maintenance System 中央维修系统

Central Monitoring System 中央监控系统

Cluster Management System 群集管理系统

Code Management System 编码管理系统

Combat Mission Simulator 战斗任务模拟机

Component Maintenance Sheet 部件维修单

Configuration Management System 构型(配置)管理系统

Conversational Monitor System 对话监测系统

Cost Management Systems 成本管理系统

Crew Management and Scheduling 机组管理与排班

**CMSC** Common Motor Start Controller 通用马达启动控制器

**CMSD** Commissioned 投产，已工作，已委任

**CMSR** Charter Movement Service Request 包机动态服务要求

**CMT** Cabin Management Terminal 客舱管理终端

Cellular Message Telecommunications 蜂窝消息电信业务

Comment 评论

Commissioning and Maintenance Terminal 入网和维修终端

Computer Managed Training 计算机管理训练

Corrected Mean Temperature 修正的平均温度

**CMTE** Committee 委员会

**CMTP** Calibration and Maintenance Test Procedures 标准与维护测试程序

Common Medium Term Plan 通用中期计划

Crisis Management Training Program 危机管理培训计划

**CMTS** Cable Modem Termination System 线缆调制解调器端接系统

Centralized Maintenance Test System 集中式维护测试系统

**CMTT** Committee for Television Transmission 电视广播委员会

**CMU** Cabin Management Unit 客舱管理组件

Central Management Unit 中央管理组件

Communications Management Unit 通信管理组件

**CMV** Concentrator and Multiplexer for Video 视频集线器和多路转换器

**CMW** Common Middle Ware 公共中间件

**CMX** Hancock, MI, USA (Airport Code) 美国密执安州汉考克机场

**CN** Celestial Navigation 天文导航

Change Number 改变号

China 中国

Consignment Note 发货通知单

Contract Number 合同号

Core Network 核心网

Cover Note 认保单

Credit Note 贷款通知单

**CNA** Common Nozzle Assembly 通用喷嘴组件

Communication Network Architecture 通信网络体系结构

Cooperative Networking Architecture 协作式联网体系结构

**CNAC** China National Aviation Corp. 中国航空公司

**CNAEC** China National Association of Engineering Consultants 中国工程咨询协会

**CNAF** China National Aviation Fuel Group Corporation 中国航空油料集团公司

**CNAH** China National Aviation Holding Company 中国航空集团公司

**CNAL** China National Accreditation Board for Laboratories 中国实验室国家认可委员会

**CNAP** Communications Network Analysis Program 通信网络分析程序
Computerized Network Analysis Program 计算机化网络分析程序
Connectionless Network Access Protocol 无连接网络接入协议

**CNAS** Civil Navigation Aids System 民用导航辅助系统
Core Network Application Services 核心网络应用服务

**CNC** Computer Numerical Control 计算机数字控制
Congestion Notification Cell 拥塞通知信元

**CNCC** China National Convention Center 中国国家会议中心
Customer Network Control Center 用户网控制中心

**CNCE** Communication Nodal Control Element 通信节点控制器
Communications Network Control Element 通信网络控制分部

**CNCL** Cancel 取消
Council 委员会

**CNCLG** Cancelling 取消, 注销

**CNCLR** Councillor 参赞, 理事, 地方议员

**CNCLSN** Conclusion 结论, 结果

**CNCR** Converse Network Chain Reaction Model 逆向网络链式反应模型

**CNCS** Central Navigation and Control School 中央导航和控制学校

**CNCT** Connect, Connection 连接, 衔接

**CNCTD** Connected 连接的

**CNCTN** Connection 连接

**CNCTR** Connector 连接器

**CNCTRC** Concentric 同心的

**CNCV** Concave 凹形的

**CND** Canadian Dollar 加拿大元

Condition 条件

Conduit 导管

Constanta, Romania(Airport Code) 罗马尼亚康斯坦萨机场

Could Not Duplicate 不能复现

**CNDB** Customized Navigation Database 用户化导航数据库

**CNDI** Commercial Non-Developmental Items 商业非发展项目

**CNDP** Communication Network Design Program 通信网络设计程序

**CNDS** Committee of National Defence and Security 国防和安全委员会

**CNEL** Community Noise Equivalent Level 社区等效噪声级

**CNF** Computer Navigation Fix 计算机导航定位点
Confine 限制
Confirm 证实, 确认

**CNFD** Confidential 机密的, 亲信的

**CNG** Compressed Natural Gas 压缩天然气

**CNGATM** Chinese New-Generation Air Traffic Management System 中国新一代空中交通管理系统

**CNI** Certified Network Instructor 认证网络专家
Common Network Interface 通信网络接口
Communication, Navigation and Identification 通讯、导航和识别
Composite Noise Index 综合噪声指数

**CNIA** Controls Not In Agreement 控制不一致

**CNII** China National Information Infrastructure 中国国家信息基础设施

**CNIR** Communication Navigation Identification and Reconnaissance 通信、导航、识别和侦察

**CNK** Cause Not Know 原因不明

**CNL** Cancel or Canceled 取消
Constructive Negligence Liability 推定过错责任制

**CNLP** Connectionless Network Layer Protocol

无连接模式网络层协议

**CNM** Centralized Network Management 集中式网络管理

Customer Network Management 用户网管理

**CNMI** Communications Network Management Interface 通信网络管理接口

**CNMS** Common Network Management System 公共网管理系统

Computer Network Management System 计算机网络管理系统

Customer Network Management System 客户网络管理系统

**CNN** Cable News Network 美国有线电视台

Cellular Neural Network 蜂窝神经网络

**CNNS** Connectionless Node Network Service 无连接节点网络服务

**CNOM** Committee of Network Operation and Management 网络营运与管理专业委员会

**CNP** Communications Network Processor 通信网络处理器

Construction Noise Permit 建筑噪音许可证

**CNR** Carrier Noise Ratio 载噪比

Carrier Signal Noise Ratio 载波信噪比

Composite Noise Rating 综合噪声系数

Corner 角落

**CNRT** Concrete 混凝土,具体的

**CNS** Cairns, Queensland, Australia ( Airport Code) 澳大利亚凯恩斯机场

Cargo Network Services Corporation 航空货运公司

China National Standards 中国国家标准

Common Nozzle Assembly 通用喷嘴组件

Communication, Navigation and Surveillance 通信、导航和监视

Communications Networking System 通信网络系统

Computer Network System 计算机网络系统

Consolidated NOTAM System 综合航行通告系统

Continuous 连续的

**CNS/ATM** Communications, Navigation, Surveillance and Air Traffic Management 通信、导航、监视和空中交通管理

**CNSA** China National Space Administration 中国国家航天局

**CNSDR** Consider 考虑

**CNSL** Console 控制台

**CNSP** Consolidated NOTAM System Processor 综合航行通告系统处理器

**CNST** Constant 常量,恒定的

**CNSU** Cabin Network Server Unit 客舱网络服务组件

**CNTL** Control 控制,管制

**CNTN** Contain 包括

**CNTNR** Container 集装箱

**CNTOR** Contactor 电流接触器

**CNTR** Container 货柜

Counter 记数器

**CNTRCT** Contract 合同

**CNTRL** Central 中心

Control 控制,管制

**CNTRTR** Concentrator 聚能器,集中器

**CNTRY** Country 国家

**CNTY** Continuity, Continuous 连续

**CNVT** Convert 转换,兑换

**CNX** Cancelled 取消,废除,中止

Chiang Mai, Thailand ( Airport Code) 泰国清迈机场

**CNY** Chinese New Year 春节,农历新年

Chinese Renminbi Yuan 人民币元

**CO** Carry out 执行,实现

Carry Over 归入,转入

Central Office 中心局,中心办公室

Change Order 改装指令,更改订单,更改规程

Check out 检查,校正,检验程序,结账

Clean out 清除

Color 颜色

Connection Oriented 面向连接

Contracting Office 合同部

County 县,郡

Customer Objective 用户目标

Cut-out 终止,切断,关闭,结束工作

$CO_2$　Carbon Dioxide 二氧化碳

**COA**　Certificate of Airworthiness 适航证

Change Over Acknowledge 倒换证实信号

Collect on Arrival 到达付费

Constant Output Amplifier 恒定输出放大器

Contract of Affreightment 运输合同

Cost of Ownership Analysis 业主成本分析

**COACT**　Company Account 公司账户

**COAT**　Coherent Optical Adaptive Technique 相干光自适应技术

Corrected Outside Air Temperature 修正外界大气温度

**COAX**　Coaxial 同轴的

**COB**　Close of Business 停止营业

**COBIT**　Control Objectives for Information and Related Technology 信息和相关技术控制目标

**COBOL**　Common Business Oriented Language 面向商业的通用语言

**COBRA**　Consultative, Objective and Bifunctional Risk Analysis 咨询、目标双重风险分析

**COBUS**　Company Business 公司事务

**COC**　Cabin Operation Center 客舱操作中心

Central Office Connection 中心局连接

Change of Contract 合同更改

Computer Operations Center 计算机操作中心

Country of Commencement of Travel 旅行始发地国家

Customer Originated Change 客户发起更改

**COCESNA**　Central American Safety Services Corporation 中美洲安全服务公司

**COCF**　Connection-Oriented Convergence Function 面向连接的会聚功能

**COCOM**　Contextual Control Model 情景控制模型

Coordinating Committee for Export to Communist Countries 巴黎统筹委员会

**COCT**　Cabin Operations Consultation Tool 客舱操作咨询工具

**COD**　Carrier Onboard Delivery 集装箱运输

Cash on Delivery 货到收款

Certificate of Deposit 存款单

Chemical Oxygen Demand 化学需氧量

Coding 编码

Collect on Delivery 货到付费

Correction of Deficiency 故障排除

Coupon Origin and Destination 起迄联程票

**CODAS**　Consolidated Operations and Delay Analysis System 综合运行和延迟分析系统

Customer-Oriented Data Retrieval and Display System 基于客户的数据搜索和显示系统

**CODATA**　Committee on Data for Science and Technology 科学和技术数据委员会

**CODEC**　Coder/Decoder 编码器/解码器

**CODEVMET**　Cooperative Development of Aviation Meteorology 航空气象服务合作开发

**COEF**　Coefficient of Lift 升力系数

**COEFF**　Coefficient 系数

**COF**　Centrifugal Oil Filter 离心滑油滤

**COFW**　Certification of Flight Worthiness 飞行适航性审定

**COG**　Center of Gravity 重心

Congo 刚果

Course over Ground 对地运动方向

Crab-Oriented Gyro 侧航专用陀螺仪

**COH**　Connection OverHead 连接开销

**COHO**　Coherent Oscillator 相干振荡器

**COI**　Central Office Interface 中心局接口

**COIS**　Computed Operated Instrument System 计算机控制仪表系统

**COIU**　Central Office Interface Unit 中心局接口单元

**COK**　Cochin, India (Airport Code) 印度科钦机场

**COL**　Collate, Collation 核对

Color or Colored 颜色,有色的

Column 列, 栏

Computer-Oriented Language 面向计算机的语言

**COLI** Cost of Living Index 生活费指数

**COLL** Collect, Collection 收集,搜集
College 学院

**COM** Cam out Monitor 凸轮脱开监视器
Centralized Operation and Maintenance 集中的操作和维护
Cockpit Operating Manual 驾驶舱操作手册
Command 指令,指挥,控制
Commerce 商业
Commercial 商业的,商务的
Committee of Management 管理委员会
Committee 委员会
Common Object Model 公用对象模型
Common 通用的,共同的
Communication 通信,交流
Company Mail 公司邮件
Company Operation Manual 公司运行手册
Company Organization Manual 公司组织手册
Compressor 压缩机
Computer Output Microfilm 计算机输出微缩胶带
Continuation of Message 报文继续

**COM/NAV** Communication/Navigation 通信/导航

**COMAC** Commercial Aircraft Corporation of China 中国商用飞机有限责任公司
Communication Advisory Committee 通信咨询委员会
Continuous Multiple-Access Collator 连续多址校验机

**COMAT** Company Material 公司材料
Computer Assisted Training 计算机辅助训练

**COMB** Combination or Combined 组合,组合的
Combined 联合的,客货两用的
Combustion 燃烧

**COMBI** Combination Passenger Cargo Airplane 客货混装飞机

**COMBL** Combustible Liquid 易燃液体

**COMC** Centralized Operations and Mainte-nance Center 集中的操作和维护中心

**COMCL** Commercial 商务的,商业的

**COMD** Command 指挥,指令

**COMEDS** Continental Meteorological Data System 大陆气象数据系统

**COMESA** Committee on Meteorological Effect of Stratospherical Aircraft 高空飞机气象效应委员会
Common Market for Eastern and Southern Africa 东非和南部非洲共同市场

**COMETS** Communication and Broadcasting Engineering Test Satellite 通信和广播工程试验卫星
Computerized Operational Maintenance and Evaluation Tracking System 计算机化运行维护与评估跟踪系统
Computer-Operated Multifunction Electronic Test Station 计算机操作的多功能电子测试站

**COMFY** Comfortable 舒适的

**COMI** Commercial Invoice 发票

**COMINT** Communications Intelligence 通信智能化,通信情报

**COML** Commercial 商务的,商业的

**COMLO** Compass Locator 罗盘定位器,罗盘示位信标,罗盘示位台

**COMM** Communication 通讯,通信

**COMM/NAV** Communication and Navigation 通信与导航

**COMMD SPD** Command Speed 指令速度

**COMMN** Commission 委员会

**COMMS** Communications 通信
Customer Oriented Manufacturing Management System 面向客户的制造管理系统

**COMN** Common 共同的

**COMP** Companion 伙伴
Company 公司,陪同
Compare 比较,对照
Comparative 相当的
Compartment 舱,室
Compass 罗盘

Compensate 补偿

Compensator 补偿器

Complete 完成

Comply 同意

Component 元件

Compressor 压缩机,压气机

Compromise 折中,妥协

Computer 计算机

Wind Component 风分量

**COMPASS** BeiDou Navigation Satellite System 北斗导航卫星系统

Condition Monitoring and Performance Analysis Software System（发动机）状况监控和性能分析软件系统

**COMPD** Compound 混合的

**COMP-FULL** Compact-Full 全部压缩

**COMPLN** Completion 完工

**COMPLT** Complete 完成

**COMP-PART** Compact-Partial 部分压缩

**COMPR** Compare 比较

Compressor 压缩机

Compromise 妥协

**COMPSN** Compensation 补偿

**COMPT** Compartment 舱,室,隔间

**COMPT TEMP** Compartment Temperature 舱温

**COMPTR** Comparator 比较器,比较仪

Competitor 竞争者

Computer 计算机

**COMPTV** Comparative 相当的

Competitive 竞争的

**COMSAT** Communication Satellite 通讯卫星

Communications Satellite Corp. 通信卫星公司

Congressional Committee on Science and Astronautics 国会科学及航天委员会

**COMSEC** Communication Security 通信保密措施

**COMTN** Commonwealth Secretariat 联邦秘书处

Commutation 交换,换算,转换,替换

**CON** Consol Beacon 库索尔归航台

Contingency Fuel 应急燃油

Continuous 连续的,不断的

Maximum Continuous Thrust 最大连续推力

**CONC** Concentrate 集中,浓缩

Concrete Surfaced Runway 水泥面跑道

Concrete 混凝土,钢筋水泥,具体的

**CONCD** Concerned 有关的,涉及到的

**CONCR** Concrete 钢筋混凝土,具体的

**COND** Condenser 电容器,冷凝器

Condition 条件,状况,状态

Conditional Reservation 有条件订座

Conditional 有条件的

Conditioned,Conditioning 制约的,制约

Conditioning 调节

**CONDTR** Conditioner 调节器

**CONEX** In Connection With 关于

**CONF** Confer 协商,给予,比较

Confidence,Confident 信任

Confidential Conference 机密会议

Confine 限制

Confirm 证实,确认

**CONF（IG）** Configuration 构型,形态,布局,结构

**CONFD** Confirmed 确认的

**CONFDNT** Confident 坚信的

**CONFED** Confederation 同盟,联盟

**CONFG** Configuration 结构,表面配置

**CONFIG** Configuration Warning Module 构型警告模块

Configuration 构形,构型,组合,布局

**CONFTY** Conformity 符合

**CONG** Congress 会议

**CONJ** Conjunction 联结

**CONN** Connect 连接,结合

Connected,Connection 联结,连接

Connector 连接器

**CONP** Connection Oriented Network layer Protocol 面向连接的网络层协议

**CONQ** Consequently 因而,所以

**CONS** Connection-Oriented Network Service

面向连接的网络服务
Consider 考虑,认为
Consign 委托,托运
Consumption 消耗,消耗量
Continuous 连续的

**CONSGT** Consignment 委托,寄交的货品
**CONSID** Consideration 考虑
**CONSL** Consul 领事
Consulate 领事馆
Consult 咨询
**CONSR** Consumer 用户,消费者
**CONST** Consent 同意
Constant 恒定的,稳定的,不变的,常量
Constraint 限制,约束
Construct, Construction 建筑,建设,结构
**CONSTN** Construction 结构,建设
**CONT** Contact 接触,联系
Contactor 接触器
Containing 包括
Content 内容,容量
Continue, Continuous, Continued 继续,连续,连续的
Contour 轮廓
Contract 合同
Control 控制
Controller 控制器,管制员
**CONT PNL** Control Panel 控制板
**CONTBD** Contraband 走私
**CONTD** Contacted 接触的
Contained 包括,包含
Continued 连续的,延续的
Controlled 受控的,受管制的
**CONTG** Containing 含有
**CONTL** Continental 大陆的
Control 控制,管制
**CONTR** Contract 合同
Contrast 对比
**CONTS** Contents 内容,目录
**CONUS** Continental United States 美国本土
**CONV** Convenient, Convenience 方便,方便的

Conventional 常规的,惯例的
Convergence 收敛,汇聚
Conversion 转换,转变
Converter 转换器
Convertible 可变的,可转换的
Convert 转换,变换
**CONVERG** Convergence 聚焦,收敛
**CONVL** Conventional 惯例的
**CONVN** Conversion 转换,兑换
**CONVT** Convert 转换,兑换
**COO** Chief Operating Officer 首席运营官
Cost of Ownership 拥有成本
Cotonou, Benin (Airport Code) 贝宁科托努机场
**COOL** Coolant 冷却剂
Cooling, Cooler 冷却,冷却器
**COOP** Cooperation, Cooperative 合作,合作的
Customer Order On-Line Process 用户订货联机处理
**CO-OP** Cooperation 合作
**COOR** Coordinate, Coordination 协调,配合
**COORDS** Coordinates 坐标
**COP** Change Over Point 转换点
Character Oriented Protocol 面向字符协议
Coherent Optical Processor 相干光处理器
Condition of Pilot 飞行员状况
Continuation of Packet 分组的连续性
Co-pilot 副驾驶
Copper 铜
Coppered 包铜的
Country of Payment 付款国家
**COPAC** Committee on Pollution Abatement and Control 污染减轻与控制委员会
**COPE** Climate System Observational and Prediction Experiment 气候系统观测与预测试验计划
**COPS** Centre of Policy Studies 政策研究中心
Common Open Policy Service 通用开放策略服务

**COPT** Complete Procedure Turn 完全程序转弯

**COPTER** Helicopter 直升机

**COQ** Channel Optimized Quantizer 信道最佳化量化器

**COR** Carrier Operated Relay 载波控制继电器

Certificate of Registration 注册执照

Contactor 接触器,触点

Cordoba, Cordoba, Argentina (Airport Code) 阿根廷科尔多瓦市科尔多瓦机场

Corner 角

Corrector 修正器,校正器

Corrosive 腐蚀剂,腐蚀性的

Corse Air International (France) 柯尔斯国际航空公司(法国)

**CORA** Conflict Resolution Assistant 冲突解决助手

Cost of Risk Analysis 风险成本分析

Customer Order Administration 用户订货管理

**CORBA** Common Object Request Broker Architecture 公共对象请求代理结构

**CORCTD** Corrected 修正的

**CORD** Coordinate, Coordination 协调,配合

**CORP** Corporation, Corporate 公司,团体,法人

**CORR** Correct, Correction 更正,纠正

Corridor 走廊

**CORS** Continuously Operating Reference Station 连续运行基准站

**CORTE** Company Route 公司航路

**COS** Cash on Shipment 装货付款

Cell Output Switch 信元输出交换

Class of Service 机舱等级

Colorado Springs, CO, USA (Airport Code) 美国科罗拉多州科罗拉多斯普林机场

Communications-Oriented Software 面向通信的软件

Conical Surface 锥面,锥形面

Cooperation for Open Systems 开放系统协作

Cosine 余弦

**COSCAP** Cooperative Development of Operational Safety and Continuing Airworthiness Project 运行安全及持续适航合作发展方案

**COSINE** Cooperation for Open System Interconnection Networking in Europe 欧洲开放系统互联网络协会

**COSL** Component Operating and Storage Limits 附件使用储存极限

**COSNA** Composite Observing System for the North Atlantic 北大西洋综合观测系统

**COSP** Continued Operational Safety Program 持续运行安全项目

**COSS** Common Object Services Specifications 公共对象服务规范

**COT** At the Coast 海岸边

Central Office Terminal 中心局终端设备

Class of Traffic 业务种类

Class of Trunk 中继类别

Climb on Prescribed Track 沿规定航迹爬升

Cotter 开口销

**COTS** Commercial Off-the-Shelf 商业货架产品

Commercial Orbital Transportation Services 商业轨道运输服务

Connection Oriented Transfer Service 面向连接的传送业务

**COU** Columbia, MO, USA (Airport Code) 美国密苏里州哥伦比亚机场

**COUN** Counseller 参赞,顾问

**COUNT** Counter 记数器

**COUR** Commercial Courier 商务信使

**COV** Cover 罩,盖,覆盖,包括

**COVD** Covered 有盖的,覆盖的,涂敷的

**COVQ** Channel-Optimized Vector Quantization 信道最佳化矢量量化

**COWL** Cowling 整流罩

**CP** Carriage Paid 运价已付

Center of Pressure 压力中心

Central Processor 中央处理器

Chief Pilot 正驾驶

Clock Pulse 时钟脉冲

Collision Presence 出现冲突

Command Post 指挥所

Common Part 公共部分

Compare 比较

Connection Point 连接点

Consolidation Point 集合点

Constant Potential 恒定电压

Content Provider 内容供应商

Control Panel 控制面板

Co-Pilot 副驾驶员

Critical Point 临界点

Customer Premise 用户所在地

**CPA** Canadian Pacific Airlines 加拿大太平洋航空公司

Cathay Pacific Airways 国泰航空公司

Certified Public Accountant 注册会计师

China Postal Airlines 中国邮政航空

Circular Protected Airspace 圆形保护空域

Closed Packed Array 密集矩阵

Closest Point of Approach 最近进近点

Cocoa Producer's Alliance 可可生产者联盟

Computer Performance Analysis 计算机性能分析

Continuous Patrol Aircraft 持续巡逻飞机

Cost and Price Analysis 成本与价格分析

Critical Path Analysis 关键路径分析

Customer Price Agreement 客户价格协议

**CPAC** Cabin Pressure Automatic Controller 客舱压力自动控制器

**CPAF** Cost Plus Award Fee 成本加奖励费

**CPAM** Cabin Pressure Acquisition Module 座舱压力采集组件

Committee of Purchasers of Aviation Materials 航材购买人委员会

**CPAR** Contract Performance Assessment Report 合同履行评估报告

Cooperative Pollution Abatement Research (Canadian) 消除污染研究协会(加拿大)

Cooperative Program for Educational Administration 教育管理合作计划

**CPARS** Compact Programmed Airline Reservation System 简单程序航空公司订票系统

Contract Performance Assessment Reporting System 合同履行评估报告系统

**CPAS** Composite Primary Aircraft Structure 飞机复合材料基本结构

**CPB** Channel Program Block 信道程序块

**CPBL** Capable 有能力的

**CPBLY** Capability 能力,容量

**CPBX** Centralized Private Branch Exchange 集中式电话小交换机

**CPC** Cabin Pressure Control 客舱压力控制

Cabin Pressure Controller 客舱压力控制器

Call Processing Control 呼叫处理控制

Collaborative Product Commerce 产品协同商务

Computer Program Components 计算机程序元件

**CPCEI** Computer Program Contract End-Item 计算机合同终结项

**CPCH PROP** Controllable Pitch Propeller 可控螺距螺旋桨

**CPCI** Computer Program Configuration Items 计算机程序组合项

**CPCP** Cabin Pressure Control Panel 座舱压力控制面板

Corrosion Prevention & Control Program 腐蚀预防与控制大纲

**CPCS** Cabin Pressure Control System 客舱压力控制系统

Common Part Convergence Sublayer 公共部分会聚子层

**CPCU** Cabin Pressure Control Unit 客舱压力控制组件

Cabin Pressure Controller Unit 客舱压力控制器组件

**CPD** Cocurrent Product Definition 共电流生成定义

Collaborative Product Development 协同产品开发

Computer Preliminary Design 计算机初步设

计

Contact Potential Difference 触点电位差

Continuing Product Development 持续产品开发

Continuing Professional Development 持续专业发展

Cut-off & Pierce Die 切断与冲孔模

**CPDLC** Controller-Pilot Data Link Communication(s) 管制员—驾驶员数据链通信

**CPDP** Computer Program Development Plan 计算机程序开发计划

**CPE** Central Processing Element 中央处理单元

Chief Project Engineer 项目主任工程师

Circular Probable Error 圆概率误差

Company Purchased Equipment 公司购买的设备

Contractor Performance Evaluation 合同方性能评估

Customer Premises Equipment 用户端设备

**CPE/DPG** Computer Programmer Evaluator/Distributor Processor Group 计算机程序编译器鉴别器/分配器处理器组

**CPESS** Connectorless Passenger Entertainment & Service System 无连接器旅客娱乐与服务系统

**CPFF** Cost Plus Fixed Fee (Contract) 成本加固定费用(合同)

**CPFSK** Continuous Phase Frequency Shift Keying 连续相位频移键控

**CPG** Call Progress 呼叫进展

Central Processor Group 中央处理器组

Cents Per Gallon 分/加仑

Clock Pulse Generator 时钟脉冲发生器

Command Processing Group 指令处理组

Gunner 射手

Customer Protection Group 客户保护组

**CPGS** Cassette Preparation Ground Stations 任务资料磁带地面准备站

**CPH** Calling Party Handling 呼叫方处理

Copenhagen, Denmark-Kastrup (Airport Code) 丹麦哥本哈根凯斯楚普机场

**CPHCH** Common Physical Channel 公共物理信道

**CPI** Characters Per Inch 字符数/英寸

Common Part Indicator 公共部分指示

Conference Papers Index 会议论文索引

Consumer Price Index 消费者物价指数

Cost Performance Index 成本绩效指数

Crash Position Indicator 失事位置指示器

**CPICH** Common Pilot Channel 公共导频信道

**CPIF** Cost Plus Incentive Fee 成本加奖励费用

**CPIFC** Cost Plus Incentive Fee Contract 成本加奖励费用合同

**CPILS** Correlation Protected Instrument Landing System 抗干扰仪表着陆系统

**CPIN** Computer Program Identification Number 计算机程序识别号

**CPIOM** Central Processor Input/Output Module 中央处理器输入/输出模块

**CPIT** Cabin Procedural Investigation Tool 客舱程序性调查工具

**CPL** Commercial Pilot Licence 商用飞机驾驶员执照

Computer Program Library 计算机程序库

Computer Programming Language 计算机程序语言

Couple 连接，耦合

Current Flight Plan Message 现行飞行计划电报(国际民航组织用语)

Current Flight Plan 现行飞行计划

**CPL/H** Commercial Pilot's Licence/Helicopter 商用直升机驾驶员执照

**CPL/IR** Commercial Pilot's Licence/Instrument Rating 带仪表飞行等级的商用直升机驾驶员执照

**CPL/SEL** Commercial Pilot Licence/Single Engine Limitation 单发动机商用飞机驾驶员执照

**CPLD** Complex Programmable Logic Device

可编程逻辑器件
Coupled 连接的，耦合的
**CPLG** Coupling 连接，耦合
**CPLMT** Complement 补充
**CPLR** Coupler 耦合器，连接器
**CPLT** Complete 完成，填写
Completion 完成
Copilot 副驾驶
**CPLX** Complex 复合，成套
**CPM** Call Processing Model 呼叫处理模式
Call Processor Module 呼叫处理机模块
Call Progress Message 呼叫进行消息
Call Protocol Message 呼叫协议信息
Call Protocol Module 呼叫协议模块
Cards per Minute 卡/分
Central Processing Module 中央处理模块
Cognitive Perceptual Motor 认知感知运动
Computer Performance Management/Measurement 计算机性能管理/测量
Computer Performance Monitor 计算机性能监控器
Computer Program Module 计算机程序模块
Continuous Phase Modulation 连续相位调制
Contract Performance Management 合同绩效管理
Control Protocol Message 控制协议消息
Core Packet Module 核心分组模块
Core Processor Module 核心处理器模块
Corrosion Prevention Manual 腐蚀防护手册
Critical Path Method 关键路径方法，临界途径法
Cross Phase Modulation 交叉相位调制
Customer Profile Management 客户类型管理
Cycles Per Minute 转/分，每分钟循环次数
Maximum Practical Load 最大实际重量
**CPM/BASIC** Core Processor Module/Basic 核心处理器组件/基本
**CPM/COMM** Core Processor Module /Communications 核心处理器组件/通信
**CPM/GG** Core Processor Module /Graphics Generator 核心处理器组件/图形发生器

**CPMS** Cabin and Passenger Management System 客舱和旅客管理系统
**CPMU** Cabin Passenger Management Unit 客舱旅客管理组件
**CPN** Closed Private Network 专用闭环网络
Collins Part Number 柯林斯件号
Corporation 公司，法人
Coupon 票根，联票的一张
Critical Path Network 关键路径网络图
Current Plan 目前计划
Customer Premises Network 用户驻地网
**CPOST** Computer Power-on Self-Test 计算机电源接通后自检
**CPP** Call Processing Program 呼叫处理程序
Calling Party Pays 主叫付费
Certification Program Plan 合格审定项目计划
Cost Per Passenger 每一旅客之费用
**CPPC** Cost Plus Percentage of Cost Contract 成本加成本百分比合同
**CPPS** Permanent Commission for the South Pacific 南太平洋国家常设委员会
**CPR** Casper, WY, USA ( Airport Code) 美国怀俄明州卡斯珀机场
Compressor Pressure Ratio 压气机增压比
Copper 铜
Cost Performance Ratio 成本绩效比率
Cost Performance Report 成本绩效报告
Critical Problem Report 关键问题报告
**CPRMA** Centralized PRMA 集中式分组预留多址
**CPRS** Carbon Pollution Reduction Scheme 减少碳污染计划
Common Problem Reporting System 常见问题报告系统
**CPRSR** Compressor 压缩机
**CPRV** Cabin Pressure Relief Valve 客舱压力安全活门(阀)，客舱释压活门(阀)
**CPS** Cabinet Pressurization System 客舱增压系统
Call Privacy Service 呼叫保密业务

Call Processing System 呼叫处理系统

Central Processing System 中央处理系统

Certification Practice Statement 认证操作规定

Characters Per Second 每秒字符

Common Part Sublayer 公共部分子层

Cycles Per Second 周/秒,每秒循环周期

**CPSE** Crew and Passenger Support Equipment 机组和旅客保障设备

**CPS-PDU** Common Part Sublayer Protocol Data Unit 公共部分子层协议数据单元

**CPS-PH** Common Part Sublayer Packet Header 公共部分子层分组头

**CPS-PP** Common Part Sublayer Packet Payload 公共部分子层净荷

**CPT** Cape Town, South Africa (Airport Code) 南非开普敦机场

Captain (Pilot-in-Command) 机长

Capture 截获

Cellular Paging Telecommunications 蜂窝寻呼电信业务

Check Point 检查点

Civil Pilot Training 民航驾驶员训练

Clearance (Pre Taxi Procedure) 放行许可 (滑行前程序)

Cockpit Procedures Trainer 驾驶舱程序训练器

Cockpit Procedures Training 驾驶舱程序训练

Compartment 舱,室

Compatibility Test 兼容性测试

Control Packet Transmission 控制报文分组传输

Cost Per Thousand 每千人次访问收费

Crew Procedures Training 机组程序训练

Critical Path Technique 临界路径技术

**CPTP** Civil Pilot Training Program 民航飞行员训练计划

**CPTY** Capacity 容量,能力

**CPU** Central Processing Unit 中央处理器

**CPV** Campina Grande, Pernambuco, Brazil

(Airport Code) 巴西帕拉伊巴州大坎皮纳机场

**CPWDM** Chirped-Pulse Wavelength Division Multiplexing 脉冲波分复用

**CPY** Copy 复制,抄件,复写

**CQA** Customer Quality Assurance 客户品质保证

**CR** Calling Rate 呼叫率

Cargo 货舱

Carriage Return (Teletype) 字盘返回

Center Right 中右

Change Request 变更要求,更改请求

Change Requisite 改装要求

Channel Reservation 信道预订

Close Range 近距离

Cold Rolled 冷轧

Common Rail 高压共轨

Connect Request/Connection Request 联接申请

Contra-Rotating 反向旋转

Contrast Ratio 对比率

Control Relay 控制继电器

Conversion Rates 兑换率,换算率

Credit Rating 信用等级

Credit 信用,赊销

Crew 机组

Cruise 巡航

**CR/M** Crew Member 机组成员

**CRA** Comparative Risk Analysis 比较风险分析

Comparative Risk Assessments 比较风险评估

Composite Research Aircraft 复合材料研究飞机

Conflict Resolution Advisory 冲突解决咨询

Crew Rest Area 机组休息区

Customerized Record Announcement 客户规定的记录通知

**CRAC** Channel Reservation for Ahead Cell 前信元信道预留

Cosmic Radiation and Aviation Committee 宇

宙辐射及航空委员会

**CRACU** Critical Avionics Control Unit 关键航空电子设备控制组件

**CRAD** Contract Research & Development 合同研究与开发

**CRAF** Civil Reserve Air Fleet 民航储备机队

**CRAM** Computerized Reliability Analysis Method 计算可靠性分析方法

Conditional Route Availability Message 条件航路可用性信息

**CRAS** Computer Reliability Analysis System 计算机可靠性分析系统

**CRB** Caribbean Air Cargo, Ltd. 加勒比航空货运有限公司

Change Review Board 更改审查委员会

Crash Rescue Boat 坠机救生艇

**CRBA** Common Request Broker Architecture 公共请求代理体系结构

**CRC** Consumer Research Center 消费者研究中心

Camera Ready Copy 摄像即用拷贝

Centralized Radio Control 集成无线电控制

Centralized Resource Control 集中资源控制

Common Routing Channel 公共路由选择信道

Communication Research Center 通信研究中心（加拿大）

Continuous Repetitive Chime 连续重复谐音

Control and Reporting Center 管制与报告中心

Crew Chief 乘务长

Crew Rest Compartment 机组休息舱

Cyclic Redundancy Check 循环冗余校验

Cyclic Redundancy Code 循环冗余代码

**CRCA** Cyclic Redundancy Code Accumulator 循环冗余码累加器

**CRCI** Cabin Rate of Climb Indicator 客舱爬升率指示器

**CRCO** Central Route Charges Office 中央航路费结算室

Customer Requested Change Order 客户要求更改订货

**CRCT** Correct 改正, 修正

**CRCTN** Correction 改正, 修正

**CRD** Call Rerouting Distribution 重选呼叫路由分布

Card 卡片

Clock Recovery Device 时钟恢复设备

Collision Resolution Device 碰撞检测设备

Control Rod Drive 操纵杆传动

**CRDA** Converging Runway Display Aid 交汇跑道显示设备

**CRDC** Customer Requested Design Change 客户要求更改设计

**CRDM** Control Rod Drive Mechanism 操纵杆传动机构

**CRDT** Credit 信用, 赊销

**CRE** Certified Reliability Engineer 合格的可靠性工程师

Class Rating Examiner 级别等级考试员

Component Reliability Engineering 部件可靠工程

Corrosion-Resistant 不锈的, 耐蚀的

**CREAM** Cognitive Reliability and Error Analysis Method 认知可靠性与差错分析方法

**CRED** Center for Research on Economic Development 经济发展研究中心

Credit Card Calling 信用卡呼叫

Credit 信用, 赊销

**CREP** Compulsory Reporting Point 强制报告点

**CRES** Corrosion-Resistant Steel 不锈钢

**CRF** Channel Repetition Frequency 信道重复频率

Composite Rear Fuselage 复合材料后机身

Compressor Rear Frame 压气机后框架

Cosmic Ray Flux 宇宙射线通量

**CRG** Cargo 货物

Carriage 托架

Central Resource Group 中央资源小组

**CRI** Call Request with Identification 识别呼叫请求

Certification Review Item 审定复查项目

Class Rating Instructor 级别等级教员

Collective Routing Indicator 集群路由选择标志

Compass Repeater Indicator 航向重复指示器

Critical Review Item 关键评审项目

**CRIAQ** Consortium for Research and Innovation in Aerospace in Quebec 魁北克航空航天创新研究联合体

**CRIB** Computer Resources Information Bank 计算机资源信息库

**CRIC** Commercial Radio International Committee 国际商业无线电委员会

**CRIs** Certification Review Items 合格审定条款

**CRIS** Current Research Information System 最新研究情报系统

**CRISD** Computer Resources Integrated Support Data 计算机资源综合支援数据

**CRISP** Computer Resources Integrated Support Plan 计算机资源综合支援计划

**CRIT** Critical 临界的,关键的

**CRJ** Canadian Regional Jets 加拿大支线喷气机

**CRK** Clark Field, Philippines (Airport Code) 菲律宾克拉克机场

Crank 风转,曲柄

**CRL** Certificate Revocation List 证书撤销清单

Collar 套管

Crew Luggage 机组行李

**CRLCMP** Computer Resource Life Cycle Management Plan 计算机资源生命周期管理计划

**CRM** Call Recording Monitor 呼叫记录监视器

Cockpit Resource Management 驾驶舱资源管理

Collision Risk Modeling 碰撞风险模型

Computer Resources Management 计算机资源管理

Contact Relationship Management 合同关系管理

Continuous Risk Management 持续风险管理

Credit and Risk Management 信用和风险管理

Credit Risk Mitigation 信用风险缓释技术

Crew Resource Management 机组资源管理

Customer Relations Management 客户关系管理

**CRM C** C Reference Model C 参考模式

**CRMP** Computer Resources Management Plan 计算机资源管理计划

Computer Resources Management Program 计算机资源管理方案

**CRMS** Control Room Management System 控制室管理系统

**CRN** Call Return 呼叫返回

**CRNGRPH** Chronograph 计时器

**CRO** Central Radio Office 无线电通讯枢纽

**CROSS** Center for Reliable Operational System Security 可靠的运行系统安全中心

**CRP** Call Request Packet 呼叫请求分组

Capacity Requirements Planning 能力需求计划

Compulsory Reporting Point 强制性位置报告点

Continuous Replacement Program 连续补充库存计划

Control and Reporting Post 控制与报知站

Controllable Reversible Pitch 可变桨距,可控可逆桨距

Corpus Christi, TX, USA (Airport Code) 美国得克萨斯州科珀斯科里斯蒂机场

Currently Recommended Alternate Path 当前推荐的迂回路由

**CRPA** Controlled Reception Pattern Antenna 手控图形接收天线

**CRR** Component Reliability Report 部件可靠性报告

**CRS** Calibration Requirements Summary 校正

要求摘要

Call Redirection Server 呼叫再定向服务器

Call Redirection Supervisor 呼叫再定向监视器

Call Routing System 呼叫路由选择系统

Cell Relay Service 信元中继业务

Certificate of Release to Service（飞机）批准放行使用证

Coarse 粗糙的

Cold Rolled Steel 冷轧钢

Compliance Record Sheets 符合性记录单

Component Repair Squadron 部件修理中队

Computer Reservations System 计算机订座系统

Course 方向,航道,航迹,课程

**CRS ERROR** Course Error 航道误差

**CRSN** Corrosion 腐蚀

**CRSS** Call Related Supplementary Services 呼叫相关辅助业务

Crisis 危机

**CRT** Call Request Time 呼叫请求时间

Cathode Ray Tube 阴极射线管

Computer Remote Terminal 计算机远程终端

Contour 外形

Correct, Correction 更正,改正

**CRTP** Compressed Real-Time Protocol 实时压缩协议

**CRU** Card Reader Unit 读卡器,读卡组件

**CRV** Crew Return Vehicle 机组返回交通工具

Crotone, Italy（Airport Code）意大利克罗托内机场

**CRW** Charleston, WV, USA（Airport Code）美国西弗吉尼亚州查尔斯敦机场

**CRW** Crew 机组

**CRWG** Computer Resources Working Group 计算机资源工作组

Customer Relations Working Group 客户关系工作组

**CRWS** Change Request Work Statement 更改要求工作说明

**CRY** Crystal 晶体

**CRYPTO** Cryptographic 密码

Cryptography 密码单

**CRYST** Crystal 水晶,透明

**CRZ** Cruise, Cruising 巡航

**CS** Call Sign 呼号

Call Switch 呼叫开关

Capability Set 能力集

Cargo System 货运系统

Case 案件,事件,箱

Cell Station 小区站

Center Section 中段

Central Station 中心站

Certification Specification 审定规范

Certification Standard 审定标准

Channel Selector 信道选择器

Channel Switching 信道交换

Character Strings 字符串

Circuit Switch 电路交换

Cirrostratus 卷层云

Client-Server 客户机—服务器

Close Support 闭环支援

Coil Sketch 线圈简图

Commercial Standard 商业标准

Common Stock 公共仓库

Communication Satellite 通讯卫星

Communication Station 通讯站台

Compass Error 罗盘误差

Compass System 罗盘系统

Compression System 压缩系统

Computer Simulation 计算机仿真

Conducted Susceptibility 传导敏感率

Connection Server 连接服务器

Constant Speed 等速,恒速

Convergence Sublayer 会聚子层

Course Shift 航向偏离

Customer Satisfaction 用户满意,客户满意

Czechoslovakia 捷克斯洛伐克

Segment Closed 航段关闭

**CS/T** Combined Station/Tower 机场管制联合塔台

**CS-1** Capability Set-1 能力组 1

**CSA** Canadian Space Agency 加拿大航天局

Cargo System Allocation 货舱系统配置

Carrier Service Area 载波服务区

Channel of Standard Accuracy 标准信道准确度

Chinese Society of Astronautics 中国宇航学会

Client Server Architecture 客户机服务器体系结构

Committee on Science and Astronautics 科学和航天委员会

Communication Satellite Act 通信卫星法案

Communications Systems Agency 通信系统局（机构）

Communications Systems Analyzer 通信系统分析员

Configuration Status Accounting 结构状态统计,技术状态状况登记

Contractor Support Area 供货商支援区

Standard Accurate Channel 标准精度通道

**CSAA** Chinese Society of Aeronautics and Astronautics 中国航空学会

**CSAH** China Southern Air Holding Company 中国南方航空集团公司

**CSAMC** Southwest Airlines Maintenance Company 西南航空维修公司

**CSAR** Circular Synthetic Aperture Radar 圆迹合成孔径雷达

Configuration Status Accounting Report 结构状态统计报告

Critical Safety Analysis Report 临界安全分析报告

**CSAS** Command and Stability Augmentation System 指令与稳定性增强系统

Conditioned Service Air System 空气系统调节勤务

Configuration Status Accounting System 配置状态会计系统,结构状态统计系统

Cooled Service Air System 空气系统冷却勤务

Criticality Safety Analysis Sequence 临界安全分析序列

**CSATA** Centre for Scientific and Technological Applications 科学和技术应用中心

**CSB** Chinese Science Bulletin 中国科学通报

**CSBPC** Control Stick Boost and Pitch Compensator 操纵杆助推与俯仰控制器

**CSC** Cargo System Controller 货舱系统控制器

Circuit Supervision Control 电路监控

Circuit Switching Center 电路交换中心

Common Signaling Channel 公共信令信道

Common-channel Signaling Controller 公共信道信令控制器

Communication Satellite Corporation 通信卫星公司（美国）

Computer Sciences Corp. , Computer Sciences Corporation 计算机科学公司

Control Signaling Code 控制信令码

Course and Speed Computer 航程与速度计算机

Customer Service Center 客户服务中心

**CSCD** Computers Supported Cooperative Design 计算机支持下的协同设计

**CSCE** Centralized Supervisory and Control Equipment 集中监控设备

Commission on Security and Cooperation in Europe 欧洲安全暨合作委员会

**CS-CELP** Conjugate-Structure Coded-Excited Linear Predication 共轭结构码激励线性预测

**CSCF** Constant Speed Constant Frequency 恒速恒频

**CSCFPS** Constant Speed Constant Frequency Power System 恒速恒频电源系统

**CSCI** Computer Software Component Interface 计算机软件元件接口

Computer Software Configuration Item 计算机软件配置项目

**CSCM** Centre for Supply Chain Management 供应链管理中心

Coherent Subcarrier Multiplexing 相干副载波复用

**CSCP** Cabin Service Control Panel 客舱勤务控制面板

Cabin System Control Panel 客舱系统控制面板

**CSCS** Common Signaling Channel Synchronizer 公共信令信道同步器

**CSCT** Circuit-Switched Connection Type 电路交换连接类型

**CSCU** Cargo Smoke Control Unit 货舱烟雾控制组件

**CSCW** Computer Supported Cooperative Work 计算机支持的协同工作

**CSD** Call Set-up Delay 呼叫建立延迟

Circuit Switched Data 电路交换数据

Commission on Sustainable Development 可持续发展委员会

Constant Speed Design 恒速设计

Constant Speed Drive 恒速驱动装置

Contract Start Date 合同开始日期

Control Surface Display 操纵面显示

Customer Service Department 客户服务部

**CSDB** Centralized Scheduling Data Base 集中调度数据库

Collins Serial Data Bus 柯林斯串行数据总线

Commercial Standard Digital Bus 商用(飞机)标准数字总线

Common Source Data Base 通用资源数据库

**CSDD** Chinese Science Document Database 中国学科文献数据库

Common Source Data Dictionary 通用资源数据词典

**CSDM** Concession Support Decision Manual 特许支援决策手册

**CSDN** Circuit Switched Data Network 电路交换数据网

Circuit Switched Digital Network 电路交换数字网络

**CSDS** Cargo Smoke Detection System 货舱烟雾探测系统

Command Ship Data System 指挥舰数据系统

Communication Signal Distribution System 通信信号分配系统

Constant Speed Drive Starter 恒速驱动启动机

**CSE** Common Support Equipment 公共支援设备

Communication Satellite Experiment 试验型通信卫星

Computer Science and Engineering 计算机科学与工程

Control System Engineeing 控制系统工程

Course Setting Error 航道设定误差

Critical Safety Evaluation 临界安全评价

**CSEI** China Special Equipment Inspection and Research Institute 中国特种设备检测研究院

Common Channel Signaling Equipment Interface 公共信道信令设备接口

**CSEL** Consolidated Support Equipment List 联合支援设备清单

**CSEU** Control System Electronic Unit 控制系统电子组件

**CSEUPS** Control System Electronic Unit Power Supply 控制系统电子组件供电

**CSF** Cell Site Function 信元位置功能

Channel Selection Filter 信道选择滤波器

Critical Success Factors 关键的成功因素

**CSF/L** Continuous Safe Flight/Landing 连续安全飞行/着陆

**CSFDR** Crash Survivable Flight Data Recorder 抗坠毁的飞行数据记录器

**CSG** Capital Systems Group 资本系统组

Casing 装箱

Columbus, GA, USA (Airport Code) 美国佐治亚州哥伦布机场

**CSH** Called Subscriber Hold 被叫用户保持

Combat Support Helicopter 战斗支援直升机

**CSI** Called Subscriber Identification 被叫用

户识别

Carrier Scale Internetworking 载波级互通

Channel State Information 信道状态信息

Container Security Initiative 货柜安全倡议

Convergence Sublayer Indication 会聚子层指示

Critical Safety Item 关键的安全项目

Customer Satisfaction Index 顾客满意指数

**CSIC** Customer Specific Integrated Circuit 用户专用集成电路

**CSIM** Crew Station Information Management 机组站信息管理

**CSIP** Collaborative Signal and Information Processing 协同信号与信息处理

Customer Satisfaction Improvement Programme 客户满意度改进计划

**CSIRT** Computer Security Incident Response Team 计算机安全事件响应小组

**CSK** Code Shift Keying 码移键控

**CSL** Computer Structure Language 计算机结构语言

Console 控制台,操作台

**CSLIP** Compressed Serial Line Internet Protocol 压缩的串行线路因特网协议

**CSM** Cabin System Module 客舱系统组件

Call Segment Model 呼叫段模型

Call Set-up Message 呼叫建立消息

Call Supervision Message 呼叫监控信息

Central Subscriber Multiplex 中心用户复用

Climate System Monitoring 气候系统监测

Clock Supply Module 时钟供给模块

Commission for Synoptic Meteorology 天气(图)学委员会

Computer Software Manual 计算机软件手册

Computer Software Module 计算机软件模块

Current Service Messages 当前服务信息

Customer Service Management 用户服务管理

**CSM/G** Constant Speed Motor/Generator 恒速马达/发电机

**CSMA** Carrier Sense Multiple Access 载波侦听多路访问

**CSMA/CD** Carrier Sense Multiple Access with Collision Detection 带冲突检测的载波多路监听

**CSMC** Common Start Motor Controller 通用启动马达控制器

**CSMS** Control Surface Measurement System 操纵面测量系统

**CSMU** Cabin System Management Unit 客舱系统管理组件

Crash Survivable Memory Unit 冲撞存活记忆单元,抗撞存储器组件

**CSN** Catalog Sequence Number 目录顺序号

Circuit Switched Network 电路交换网

Common Services Network 公共服务网

**CSNAS** Computer Supported Network Analysis System 计算机支持的网络分析系统

**CSNC** China Satellite Navigation Conference 中国卫星导航学术年会

**CSNP** Complete Sequence Numbers Protocol data unit 完整序号协议数据单元

**CSO** City Sales Office 城区营业办事处

Composite Second Order 复合二次失真

Cycles Since Overhaul 翻修后周期,大修后循环次数

**CSOC** Consolidated Space Operations Center 联合空间操作中心

**CSP** Call Signal Processing 呼叫信号处理

Commerce Service Provider 商业性服务供应商

Common Signal Processor 通用信号处理机

Constant Speed Propeller 恒速螺旋桨

Control Signal Processor 控制信号处理机

**CSPA** Closely Spaced Parallel Approach 近间隔平行进近

**CSPDN** Circuit Switched Public Data Network 电路交换公共数据网

**CSPM** Call & Signaling Process Module 呼叫和信令处理模块

Customer Support Procedure Manual 客户支持程序手册

**CSPR** Closely Spaced Parallel Runways 近间

隔平行跑道

Committee on Scientific Planning and Review 科学计划和评估委员会

**CSR** Cell Start Recognizer 信元起始识别程序

Cell Switch Router 信元交换路由器

Centre Station Rearrangement 中心站调整

Communications Satellite Relay 卫星通信中继

Construction Service Request 结构维护请求

Continuous Speech Recognition 连续语音识别

Control Shift Register 控制移位寄存器

Coordination Sheet Release 协调页颁发

Covert Strike Radar 隐蔽攻击雷达

Critical Study Review 关键研究审查

Crossair-Switzerland 瑞士横越航空公司

Cursor 游标,光标

Customer Service Representative 客户服务代表

Customer Signature Required 要求用户签字

**CSS** Cabin Services System 客舱服务系统

Cell Site Switch 信元位置转换

Channel Signaling System 信道信令系统

Cleveland Soaring Society 克利夫兰滑翔社团

Cockpit System Simulator 驾驶舱系统模拟器

Collins Support Services 柯林斯支援服务

Control Stick Steering 操纵杆操纵

**CSSCI** Chinese Social Science Citation Index 中文社会科学引文索引

**CST** Cabin Service Trainer 客舱服务训练器

Central Standard Time 中部标准时间(美国)

Coast 海岸

Computer Science and Technology 计算机科学与技术

Critical Solution Temperature 临界溶解温度

Cross Software Test 交叉软件测试

**CSTA** Center for Scientific and Technical As-

sessment 科学和技术评估中心

Chief Scientific and Technical Advisors 首席科学和技术顾问

China Science and Technology Association 中国科学和技术学会

Computer Systems Technology Association 计算机系统技术协会

Computer-Supported Telecommunications Applications 计算机支持的电信应用

**CSTG** Casting 铸造

**CSTM** Customs 海关

**CSTMR** Customer 用户,客户,顾客

**CSTOL** Cargo Short Take off & Landing 货机短距起飞与着陆

Combined Short Take-off and Landing 联合短距起飞与着陆

Controlled Short Take off & Landing 受控短距起飞与着陆

**CSTP** Committee for Scientific and Technological Policy 科学与技术政策委员会

**CSTR** Canister 罐,装运箱

Constraint 限制,抑制,束缚

Customs Regulations 海关规定

**CSU** Central Switching Unit 中央转换组件

Channel Service Unit 信道服务单元

Circuit Switching Unit 电路转换组件,电路交换单元

Command Sensor Unit 指令感应组件

Common Service Unit 公共业务单元

Computer Support Unit 计算机支援组件

Computing Service Unit 计算机服务组件

Configuration Strapping Unit 构型捷联组件

Crew Station Unit 机组位组件

Critical Switching Unit 临界转换组件

Customer Service Unit 用户业务单元

**CSUBANS** Called Subscriber Answer 被叫用户应答

**CSUD** Call Set-Up Delay 呼叫建立延迟

**CSVF** Constant Speed Variable Frequency 恒速变频

**CSW** Channel Status Word 信道状态字

CT　Call Transfer 呼叫转移
　　Carat 克拉
　　Center Tape 中带,中间带
　　Central Terminal 中央终端
　　Central Time 中部时间(美国)
　　Circle Trip 环程
　　Circuit 电路,线路
　　Coat 涂层,镀层
　　Colour/Type 箱包颜色/样式
　　Communications Technology 通讯技术
　　Compass Track 罗盘航迹
　　Containerized Traffic 集装箱运输
　　Control Transformer 控制变压器
　　Cumulative Time 累积时间
　　Current Transformer 电流互感器
CT2　Cordless Telephone 2 第二代无绳电话
CT3　Cordless Telephone 3 第三代无绳电话
CTA　Air Transport Certificate 航空运输合格证
　　Air Transport Certification (法)航空运输合格证
　　Calculated Time of Arrival 计算到达时间
　　Catania, Sicily, Italy (Airport Code) 意大利西西里岛卡塔尼亚机场
　　Chicago Transit Authority 芝加哥飞越管理局
　　Chief Technical Adviser 首席技术顾问
　　Consolidated Tool Accountability & Certification System 强化工具责任与取证系统
　　Control Terminal Area 终端控制区,终端管制区
　　Controlled-Time of Arrival 到达时间控制
　　Cordless Terminal Adaptor 无绳终端适配器
　　Cowl Thermal Anti-ice 整流罩加热防冰
　　Current Transformer Assembly 电流互感器组件
CTAF　Common Traffic Advisory Frequency 通用交通咨询频率
CTAI　Cowl Thermal Anti-Icing 整流罩热防冰
CTAM　Climb to and Maintain 爬升后保持高度

Continental Tropical Air Mass 大陆热带气团
CTAS　Central TRACON Automation System 中央终端雷达进近管制自动化系统
　　Continued Tracon Automation System 连续自动终端雷达控制系统
CTB　Composite Triple Beat 复合三次拍频
CTB/L　Combined Transport Bill of Lading 联合运输提单
CTC　Cabin Temperature Controller 客舱温度控制器
　　Calibration & Testing Center 计量检测中心
　　Cam Timing Contact 凸轮定时触点
　　Cell Type Checker 信元类型检测器
　　Centralized Traffic Control 集中交通管制
　　Channel Traffic Control 信道业务量控制
　　Command Transmitter Controller 指令发射机控制器(管理员)
　　Communication Transistor Corporation 通信晶体管公司
　　Contact or Contacted or Contacting 联系
　　Contract Target Cost 合同目标成本
　　Cross Talk Cancellation 串扰消除
　　Customer Training Centre 客户培训中心
CTC(A)　Canadian Transport Commission (Air Transport Committee) 加拿大运输委员会(空运委员会)
CTCA　Channel To-Channel Adaptor 信道间适配器
　　Contact Address 联系地址
CTCCC　Close Type Control Circuit Contact 闭合型控制电路触点
CTCF　Channel and Traffic Control Facility 信道与通信量控制设备
CTCP　Client-To-Client Protocol 客户对客户协议
　　Commission for Trade and Consumer Protection 贸易和消费者保护委员会
　　Contact Phone 联系电话
CTD　Cell Transfer Delay 信元传送迟延
　　Checklist Technical Documentation 技术文件

检查单
Coated 涂层的,镀层的
Column Timer Decoder 柱状定时器译码器
Common Technical Document 通用技术文件
Continued 连续的
Cross Track Deviation 交叉航迹偏离
Cross Track Distance 偏航迹

**CTDC** Civil Transport Development Corporation 民用航空运输开发公司

**CTDG** Command Track and Distance to Go 给定航迹和待飞距离

**CTDM** Cell Time Division Multiplexing 信元时分复用

**CTDP** Cabin Trainer Data Packages 客舱训练器数据包

**CTDS** Code-Translation Data System 码转换数据系统

**CTE** Cable Termination Equipment 电缆终端设备
Channel Translating Equipment 信道转换设备
Critical Technology Element 关键技术元素
Customer Terminal Equipment 用户终端设备

**CTEC** Certified Technical Education Center 认证技术培训中心
Chemical Transportation Emergency Center 化学运输急救中心
Computer Training and Education Center 计算机培训和教育中心
Customer Technical Education Center 客户技术培训中心

**CTEE** Committee 委员会

**CTF** Central Test Facilities 中央测试设施
Certificate 证明书
Certify 证明
Correction To Follow 下列更正

**CTFQ** Central Test Facility Questions 中央测试设施询问

**CTFR** Central Test Facility Response 中央测试设施反应

**CTG** Carriage 车费,运费

**CTH** China Travel Sky Holding Company 中国民航信息集团公司

**CTI** Computer Telephony Integration 计算机电话集成

**CTK** Capacity Tonne Kilometre 吨千米容量
Cargo Tonne Kilometers 货运吨公里

**CTL** Central 中心的,中央的
Certified Tool List 合格工具列表

**CTP** Certified Trager Practitioner 注册律师

**CTL** Charleville, Queensland, Australia (Airport Code) 澳大利亚昆士兰州查尔维尔机场
Control Panel 控制板
Control 控制,管制

**CTLG** Catalog, Catalogue 目录

**CTLR** Controller 控制器,管制员
Cross Tie Lockout Relay 交叉连接闭锁继电器

**CTLRS** Cross Tie Lockout Relay Slave 交叉连接闭锁从继电器

**CTLZ** Control Zone 管制地带

**CTM** Cabin Telephone Maintenance 客舱电话维护
Circle Trip Minimum 环城最低限额检查
Circuit Transfer Mode 电路转移模式
Cumulative Time 累积时间
Cycle Time Monitor 循环时间监视器
Cargo Transfer Manifest 货物转港舱单

**CTME** Cumulative Time 累积时间

**CTMIS** Canadian Technical Management Information System 加拿大技术管理信息系统

**CTMO** Centralized Air Traffic Flow Management Organization 中央空中交通流量管理组织
Centralized Traffic Management Organization 中央交通管理组织

**CTN** Carton 纸箱,纸板
Caution 注意

**CTNR** Container 容器,集装箱

**CTO** Chief Technology Officer 首席技术官,

首席技术主管

City Ticket Office 城区票务办事处

Conventional Take-off 常规起飞

**CTOF** Calculated Time Over Fix 飞越航线控制站的计算时间

**CTOL** Cataloging Tools on Line 在线编目工具

Conventional Takeoff & Landing 常规起飞与着陆

**CTOT** Calculated Take-off Time 计算的起飞时间

Constant Thrust on Takeoff 恒推力起飞

**CTP** Communications Timing Procedure 通信时序程序

Connection Terminal Point 连接终端点

Contract Target Price 合同目标价格

Cyclic Time Processor 循环时间处理器

**CTR** Center 中心,中央

Contour 轮廓

Control Zone 管制区域,管制地带

Controllable Twist Rotor 可控扭转旋翼

Cost-Time Resource Sheet 成本时间资源表

Counter 柜台

(critical temperature resistor) 临界温度电阻器

Cross Tie Relay 交叉汇流条继电器

Cumulative Time Remaining 累积剩余时间

**CTRD** Centered 居中的

**CTRL** Control 管制,控制

**CTS** Cabin Temperature Selector 客舱温度选择器

Cabin Temperature System 客舱温度系统

Cents 分

Certificates 证书

Clear to Send 允许发送

Clear-to-Send Protocol 清除发送协议

Clear-to-Send 清除发送

Communications Technology Satellite 通信技术卫星

Compass Tilt Signal 罗盘倾斜信号

Computer Telegram System 计算机电报系统

Computer Training System 计算机训练系统

Contractor Technical Services 合同方技术服务

Control Tracking Station 控制跟踪站

Conversational Terminal Service 终端通话服务

Credit Telephone Service 信用卡电话业务

Sapporo, Japan (Airport Code) 日本札幌机场

**CTSB** Canada Transportation Safety Board 加拿大运输安全委员会

Critical Task Selection Board 重要任务选举委员会

**CTSK** Countersunk 埋头的

**CTSO** Chinese Technical Standard Order 中国技术标准规定(项目)

**CTSOA** Chinese Technical Standard Order Authorization 中国技术标准规定项目批准书

**CTTD** Cross Tie Time Delay 交叉汇流条时间延迟

**CTTE** Common TDMA Terminal Equipment 通用 TDMA 终端设备

**CTTEE** Committee 委员会

**CTTL** Course Training Tasks List 课程培训任务清单

Critical Training Task List 关键培训任务清单

**CTU** Cabin Telecommunications Unit 机舱通信装置

Central Telemetry Unit 中央遥测单元

Central Terminal Unit 中央终端装置

Central Timing Unit 中央定时组件

**CTV** Cable TeleVision 有线电视

Colour Television 彩电

**CTVS** Cockpit Television Sensor 驾驶舱电视传感器

**CTWT** Counterweight 配重

**CTX** Customer Telephone Exchange 用户电话交换机

**CTY** City 城市

Contingency Fuel 应急燃油

**CTZ** Control Zone 管制地带

**CU** Cabin Unit 客舱组件

Can You 能否

Computer Unit 计算机组件

Control Unit 控制组件,管制单位,控制单元

Cubic 立方

Cumulus 积云

Current 电流,当前

Custom Union 关税同盟

**CUA** China United Airlines 中国联合航空公司

Circuit Unit Assembly 电路单元组件

Control Unit Assembly 控制单元组件

**CUB** Cubic 立方

**CUBF** Cubic feet 立方英尺

**CUD** Could 能够

**CUDU** Current Unbalance Detection Unit 电流不平衡探测组件

**CUF** Channel Utilization Factor 信道利用因素

Cumuliform 积状云

**CUFT** Cubic Feet 立方英尺

**CUG** Closed User Group 闭合用户群

**CUGOA** Closed User Group with Outgoing Access 带向外访问口的闭合用户群

**CUI** Character User Interface 字符用户接口

**CUID** Called User Identification number 被叫用户识别号

**CUL** Culiacan, Sinaloa, Mexico ( Airport Code) 墨西哥库利亚坎锡那罗亚机场

**CUM** Cumulative 积累的

**CUMD** Cumulative Distance 累积距离

**CUMT** Cubic Meter 立方米

**CUN** Cancun, Mexico ( Airport Code) 墨西哥坎昆机场

Charter of the United Nations 联合国宪章

Common User Network 公共用户网络

**CUR** Current 电流,当前

**CURR** Currency 货币

**CURT** Current 电流,当前

**CUS** Customs Available 有海关

Customs 海关

**CUSS** Common Use Self Service 通用自助值机服务,通用旅客自服务

**CUST** Customs 海关

Custumer 用户,顾客,买主

**CUSTD** Custody 保管,监管

**CUT** Company Utility Channel 公司使用波道

**CUTE** Common Use Terminal Equipment 公用终端设备

Common-User Terminal Equipment 公共用户终端设备

**CUTS** Consumer Unity and Trust Society (印度)消费者团结与信任协会

**CUU** Chihuahua, Mexico ( Airport Code) 墨西哥奇瓦瓦机场

**CV** Check Valve 单向活门

Computer Virus 计算机病毒

Constant Velocity 恒速

Constant Voltage 恒定电压

Continuously Variable 连续变化

Control Valve 控制活门

**CVA** Characteristic Vector Analysis 特征矢量分析

Charted Visual Approach 图示目视进近

Customer Value Analysis 客户价值分析

**CVCF** Constant Voltage and Constant Frequency 恒定电压与恒定频率

**CVD** Chemical Vapor Deposition 化学气相沉积

**CVDC** Voluntary Civil Defence Committee 志愿民防委员会

**CVDP** Conventional Vehicle Development Program 常规运载工具开发程序

**CVDR** Cockpit Video Digital Recorder 驾驶舱视频数字记录器

**CVE** Certification Verification Engineer 合格审定确认工程师

Compliance Verification Engineer 符合性验证工程师

**CVF** Controlled Visual Flight 受控目视飞行,

管制目视飞行
Convair Freighter 康维尔货机

**CVFDR** Cockpit Voice and Flight Data Recorder 驾驶舱话音和飞行数据记录仪

**CVFP** Charted Visual Flight Procedure 图示目视飞行程序

**CVFR** Controlled Visual Flight Rules 可控的目视飞行规则

**CVG** Cincinnati, OH, USA (Airport Code) 美国俄亥俄州辛辛那提机场

**CVI** Chemical Vapour Infiltration 化学气象浸渗

**CVIP** Computer Vision and Image Processing 计算机视觉与图像处理

**CVIS** Cabin Video Information System 客舱视频信息系统

**CVL** Configuration Variation List 构型偏差清单
Control Value List 控制活门清单

**CVM** Cabin Video Monitoring 客舱视频监控
Card Verification Method 信用卡检验方法
Comparative Vacuum Monitoring 比较真空监控
Control Valve Module 控制活门模块
Customer Value Management 客户价值管理

**CVMS** Cabin Video Monitoring System 客舱视频监控系统

**CVN** Clovis, NM, USA (Airport Code) 美国新墨西哥州克洛维斯机场

**CVOL** Control Volume 控制体积

**CVOR** Conventional VHF Omnidirectional Range 常规甚高频全向信标

**CVP** Container Vehicle Park 货柜车停车场

**CVPTV** Encrypt-Vision Pay TeleVision 加密收费电视

**CVR** Change Verification Record 变更核实记录
Cockpit Voice Recorder 驾驶舱话音记录器
Computer Voice Response 计算机语音响应
Controlled Visual Rules 管制目视飞行规则
Convair Aircraft 康维尔全系列飞机

Cost Variance Report 成本差异报告
Cost, Value and Risk 成本、价值和风险
Cover 盖子,盖住

**CVRCP** Cockpit Voice Recorder Control Panel 驾驶舱话音记录器控制板

**CVRS** Cockpit Voice Recorder System 驾驶舱话音记录器系统

**CVRSN** Conversion 转换,兑换

**CVS** Cargo Video Surveillance 货舱视频监测
Creating Virtual Studios 虚拟制作室

**CVSD** Continuously Variable Slope Deltamodulation 连续可变斜率增量调制

**CVSM** Conventional Vertical Separation Minimum 常规垂直间隔最低标准

**CVT** Center Vent Tube 中央通气管
Circuit Validity Testing 电路有效性测试
Constant-Voltage Transformer 恒压变压器
Continuously Variable Transmission 无级变速传动
Controlled Variable Time 可控变量定时
Convertible 可变性,敞篷车

**CVTBL** Convertible 可变换的,可兑换的

**CVTD** Converted (Aircraft) 转向,兑换

**CVU** Crypto Voice Unit 秘密话音组件

**CVX** Convex 凸状面,钢卷尺

**CW** Call Waiting 呼叫等待
Carrier Wave 载波
Caution and Warning 注意和警告
Clockwise 顺时针
Clockwise 顺时针方向的
Continuous Wave 连续波

**CWA** Wausau, WI, USA (Airport Code) 美国威斯康星州沃索机场

**CWAR** Continuous Wave Acquisition Radar 连续波探测雷达

**CWAS** Clean Wing Advisory System 净形机翼咨询系统

**CWB** Central Weather Bureau 中央气象局
Curitiba, Parana, Brazil (Airport Code) 巴西巴拉那河库里提巴机场

**CWC** Cross Wind Component 侧风分量

Customer Web Center 用户网络中心

**CWD** Change Working Directory 变更工作目录

**CWF** Cross Wind Force 侧风力

**CWG** Charges Working Group（货运）计费工作组

Chinese Wire Gauge 中国线规

Cockpit Working Group 驾驶舱工作组

**CWGT** Change Weight to 重量改为

**CWI** Call Waiting Indication 呼叫等待指令

Call Waiting Indicator 呼叫等待指示器

Centralized Warning Indicator 中央警告指示器

Continuous Wave Illuminator 连续波照明器

Continuous Wave Interference 连续波干扰

**CWL** Cardiff, United Kingdom（Airport Code）英国加的夫机场

Cross-Wind Landing（飞机）侧风着陆

**CWLU** Cabin Wireless LAN Unit 客舱无线局域网组件

**CWO** Cash With Order 现款订货

**CWP** Central Weather Processor 中央气象处理器，中央气象处理机

Central West Pacific Ocean Region 中西太平洋地区

Computer Word Processing 计算机文字处理

Crew Workload Profile 机组工作量分布图

**CWQC** Company Wide Quality Control 全公司的质量控制

**CWR** Civil Work Requirements 民用工作要求

Colored Weather Radar 彩色气象雷达

Continuous Wave Radar 连续波雷达

**CWS** Caution and Warning System 提醒和警告系统

Control Wheel Steering 用驾驶盘操控

**CWSIU** Combined Wheel Speed Interface Unit 组合轮速接口组件

**CWSU** Center Weather Service Unit 中心气象服务装置（中央气象服务单元）

**CWT** Center Wing Tank 中央翼燃油箱

Cowra, New South Wales, Australia（Airport Code）澳大利亚新南威尔士科拉机场

**CWY** Clearway 净空道

**CXI** Christmas Island, Kiribati（Airport Code）基里巴斯圣诞岛机场

**CXN** Correction 改正，修正

**CXT** Circuit 电路，线路

**CY** Calendar Year 日历年

City 城市

Container Yard 集装箱货场

Copy 复制，复印

Currency 货币

Cycle 循环

**CYA** See You Again 再见

**CYC** Cycle 循环，周期，周

Cyclic 周期的

Cyclone 气旋

**CYL** Cylinder 汽缸，圆桶

See You Later 再见

**CYS** Cheyenne, WY, USA（Airport Code）美国怀俄明州夏延机场

**CZ** Caution Zone 警戒区域

**CZM** Cozumel, Quintana Roo, Mexico（Airport Code）墨西哥科苏梅尔机场

# D

D　Danger Area 危险区
Day 天,白天,昼间,日
Delete 删去
Departure 起飞,离港
Depot 仓库,机库,航空站
Derated 减功率
Diameter 直径
Digit 数字
Direct Aircraft 引导机,指挥机
Direct Route 直飞航路
Discounted Fare 折扣票价
Drag Force 阻力
Electric Flux Density 电通量密度

**D&D**　Distress and Diversion 应变和偏转

**D&O**　Description and Operation 描述与使用

**D. O**　Dayton, Ohio ( Boeing Company ) 波音戴登公司

**D/A**　Danger Area 危险区
Designated Apron 指定停机坪
Digital-to-Analog 数/模转换
Drift Angle 偏流角

**D/B**　Data Base 数据库
Date of Birth 出生日期
Day Book 流水账

**D/C**　Discount 折扣
Down Converter 下变频器

**D/D**　Date of Delivery 交货日期
Deferred Delivery 延期交货
Departure Delay 起飞延误
Engine out Drift Down Point 发动机失效漂降点

**D/DT**　Differential of Decimal 十进制阶差

**D/FD**　Direction Finder 定向机

**D/I**　De-Icing Fluid 除冰液
De-Icing 除冰

**D/O**　Delivery Order ( Insurance ) 交货单,出货单,栈单
Description and Operation 说明与工作

**D/P**　Delivery Against Payment 付款交货

**D/RDR**　Doppler Radar 多普勒雷达

**D/S**　Delivered Sound 交货完好

**D/V**　Demonstration and Validation 演示和批准
Digital-to-Video 数字视频

**DA**　Danger Area 危险区
Data Access 数据存取
Data Acquisition 数据获取,数据采集
Data Administrator 数据管理员
Decision Altitude 决断高度
Delayed Action 延期作用,延发(引信)
Delayed Arming 延期打开保险
Delta Airlines 达美航空公司,三角航空公司
Delta Amplitude 幅度增量
Demand Assignment 按需分配
Demand-Altitude 指令分配
Density Altitude 密度高度
Destination Address 目的地址
Detonation Altitude 起爆高度
Deviation Altitude 偏流高度
Deviation Angle 落后角,偏角(流过叶片气流的)
Dip Angle ( 磁 )倾角
Direct Access 直接存取,直接采集
Direct Action 瞬发引信
Direct Ascent 直接上升
Director Aircraft 指挥机,引导机
Disassemble 拆卸
Distributed Amplifier 公布参数放大器
Document Attached 附件

Double-Acting 双动的,双重作用,复动式
Drift Angle 偏航角,漂移角,下滑角
Duct Airflow 管道气流
Dummy Antenna 假天线,仿真天线

**D-A** Digital-Analog 数模

**DA（H）** Decision Altitude（Height）决断高度（高）

**DA/H** Decision Altitude/Height 决断高度/决断高

**DA/TS** Data Accumulation/Transmittal Sheet 数据累积/传送单

**DAA** Data Access Arrangement 数据存取装置

Data Acquisition Assembly 数据采集组件

Digital Analog Adapter 数字模拟转接器

Divisional Assistance Authorization 师支援核准

**DAA&AM** Defense Aid Aircraft and Aeronautical Material 防御支援飞机及航空器材

**DAAS** Demonstration Advanced Avionics System 先进的航空电子系统论证

**DAASM** Doppler Arrival Angle Spectra Measurements 多普勒着陆角光谱测量

**DAAT** Digital Angle-of-Attack Transmitter 数字式攻角发送器

**DAB** Daytona Beach, FL, USA（Airport Code）美国佛罗里达州代托纳比奇机场

Digital Audio Broadcast 数字音频广播

**DABNS** Discrete Address Beacon and Navigation System 离散地址信标和导航系统

**DABS** Decision Addressing Beacon System 决断寻址信标系统

Direct Access Beacon System 直接存取信标系统

Direct Address Beacon System 直接地址信标系统

Discontinued Address Beacon System 不连续地址信标系统

Discrete Address Beacon System 离散寻址信标系统

**DAC** Damage Assessment Center 损坏鉴定中心

Data Acquisition Camera 数据采集相机

Data Acquisition Center 数据采集中心

Data Acquisition Computer 数据采集计算机

Data Acquisition Controller 数据采集控制器

Data Analysis and Control 数据分析和控制

Defensive Aids Computer 防御辅助计算机

Derived Air Concentration 导出空气浓度

Dhaka, Bangladesh-Zia（Airport Code）孟加拉国达卡奇亚国际机场

Digital Audio Control 数字式音频控制

Digital Autopilot Controller 数字式自动驾驶仪控制器

Digital to Analog Converter 数—模转换器

District Advisory Council 地区顾问理事会

Double Annular Combustor 双环型燃烧室

Douglas Aircraft Corporation 道格拉斯飞机公司

Drawing Aperture Card 图纸穿孔卡

**DACAN** Douglas Aircraft Company of Canada 加拿大道格拉斯飞机公司

**DACC** Dangerous Air Cargoes Committee 危险品空运委员会

Direct Access Communication Channel 直接接通通信信道

Display and Computer Console 显示与计算机装置

Drug and Alcohol Control Committee 禁毒,戒酒委员会

**DACG** Departure Airfield Control Group 出发机场管制组

**DACI** Direct Adjacent Channel Interference 相邻信道的直接干扰

**DACL** Discretionary Access Control List 无条件访问控制列表

Dynamic Analysis and Control Laboratory 动力分析与控制实验室

**DACN** Desk Area Computer Network 桌域计算机网

**DACON** Data Controller 数据控制器

**DACS** Data & Analysis Center for Software 软

件数据与分析中心

Data Acquisition and Control System 数据采集和控制系统

Data Acquisition and Correction System 数据采集与校正系统

Data Acquisition, Control and Simulation 数据采集、控制与模拟

Digital Access & Cross-connect System 数字接入交叉连接系统

Digital Acquisition and Control System 数字化采集与控制系统

Digital Attitude Control System 数据姿态控制系统

**DACU** Data Acquisition and Control Unit 数据获取与控制装置

**DAD** Da Nang, Vietnam ( Airport Code ) 越南岘港机场

Data Acquisition Display 数据采集显示

Delivery Authorization Document 核定交货单据

Design Approval Data 设计批准数据

Design Approval Delegate 设计审批委任

Digital Audio Disk 数字音频磁盘

**DADC** Digital Air Data Computer 数字式大气数据计算机

**DADR** Digital Audio Data Recorder 数字音频数据记录器

**DADS** Data Acquisition and Display System 数据获取和显示系统

Deployable Air Data Sensor 可部署空气资料感测器

Digital Air Data System Display System 数据获取和显示系统

Digital Air Data System 数字式大气数据系统

**DADU** Data Accumulation & Distribution Unit 数据积累与分配装置

**DAE** Data Acquisition Equipment 数据采集设备

Distributed Agent Environment 分布式代理环境

Dubai Aerospace Enterprise 迪拜航空航天集团

**DAEMO** Data Adaptive Evaluator and Monitor 自适应数据鉴定器和监视器

**DAES** Digital Audio Entertainment System 数字式音频娱乐系统

**DAF** Damping Augmentation Function 阻尼放大功能

Data Acquisition Facility 数据收集设备

Delivered at Frontier 边境交货

Departure Airfield 起飞机场，飞离机场

Destination Address Field 目的地址字段

**DAFC** Departure Airfield Control 出发机场控制，离场控制

Digital Automatic Flight Control 数字自动飞行控制

Digital Automatic Frequency Control 数字自动频率控制

Digital Flight Control System 数字式飞行控制系统

**DAFCS** Digital Advance Flight Control System 数字式先进飞行控制系统

Digital Automatic Flight Control System 数字式自动飞行控制系统

**DAFO** District Aviation Forecast Office 地区航空预报室

Division Accounting and Finance Office 财务与金融办公室分部

**DAG** Data Analysis Group 数据分析组

Defensive Aids Group 防御辅助组

**DAGC** Delayed Automatic Gain Control 延迟自动增益控制

Digital Automatic Gain Control 数字自动增益控制

**DAGE** Depot Aerospace Ground Equipment 航天基地地面设备

**DAGMAR** Database of Aeronautical Agreements and Arrangements 航空协议和安排数据库

Drift and Ground Speed Measuring Airborne Radar 机载偏航角和地速测定雷达

**DAGS** Dry Air Generation System 干空气发生系统

**DAI** Data Adapter Interface 数据适配器接口

Distributed Artificial Intelligence 分布式人工智能

Drift Angle Indicator 漂移角指示器

**DAIM** Data Analysis Information Memorandum 数据分析信息备忘录

**DAIR** Direct Altitude and Identify Readout 高度与识别直接读出

**DAIS** Digital Airborne Information System 机载数字信息系统

Digital Avionics Information System 数字式航空电子信息系统

**DAISY** Data Acquisition and Interpretation System 数据的采集和解释系统

Database & Airport Information System 数据库与机场信息系统

Decision Aided Information System 辅助决策信息系统

Digital Accessible Information System 数字接入信息系统

**DAIU** Digital to Analog Interface Unit 数字—模拟接口装置

**DAIW** Danger Area Interference Warning 危险区闯入告警

**DAJS** Distributed Area Jamming System 分布区域干扰系统

**DAL** Dallas, TX, USA-Love Field (Airport Code) 美国得克萨斯州达拉斯拉夫菲尔德机场

Data Access Line 数据存取线路

Data Acquisition List 数据采集清单

Design Assurance Level 设计保障级别

Development Assurance Level 开发保证等级

**DALGT** Daylight 日光,白天

**DALR** Digital Audio Legal Recorder (FAA) 数字音频法定记录仪(FAA)

Dry Adiabatic Lapse Rate 干燥绝热下降率

**DALS** Data Acquisition Logging System 数据采集记录系统

**Distress** Alerting and Locating System 失事警报定位系统

Downed Aircrew Locator System 飞行机组向下定位系统

Downed Aviator Locator System 飞行员向下定位系统

**DAM** Damage 损坏

Damascus, Syria (Airport Code) 叙利亚大马士革机场

Data Acquisition Method 数据采集方法

Data Acquisition & Monitoring 数据采集和监测

Data Addressed Memory 数据定址存储器

DECT Authentication Module DECT 认证模块

Digital Asset Management 数字资产管理

Dollars per Aircraft-Mile 美元/飞机英里

**DAMA** Demand Assigned Multiple Access 按需分配多址访问

**DAMAS** Deconvolution Approach for the Mapping of Acoustic Sources 声源成像图反馈积法

**DAME** Damage Assessment Methods 损坏评估方法

Data Acquisition and Monitoring Equipment 数据的采集和监视设备

Designated Aviation Medical Examiner 法定航医检验员,指定的航空医学考官

Digital Automatic Measuring Equipment 数字式自动测量设备

Distance Azimuth Measuring Equipment 距离方位测量设备

Distributed Aircraft Maintenance Environment 分布式航空器维护环境

**DAMI** Designated Aircraft Maintenance Inspector 委任的航空器维修检查员

**DAMPS** Digital Adanced Mobile Phone System 数字高级移动电话系统

**DAMR** Director of Aircraft Maintenance and Repair 飞机维修主任

**DAMS** Data Acquisition and Management System 数据采集与管理系统

Direct Access Management System 直接存取
管理系统

**DAN** Danish 丹麦的

Load(DecaNewton) 负载(十牛顿)

**DANAC** Decca Area Navigation Airborne
Computer 台卡区域导航机载计算机

**DANGCGO** Dangerous Cargo 危险货物

**DAO** Design Approval Organization 设计审批
机构

District Aviation Office 地区航空局

**DAP** Data Access Point 数据接入点

Data Access Protocol 数据存取协议

Data Acquisition and Processing 数据获取与
处理

Data Acquisition Processor 数据采集处理器

Departure and Approach Procedures 离场和
进场程序

Depot Acceptance Procedures 仓库验收手续

Digital Audio Playback 数字式音响播放

Digital Avionics Processor 数字式航空电子
处理器

Directory Access Protocol 目录访问协议

Documents Against Payment 付款交单跟单
付款

**DAPCA** Development and Production Costs
for Aircraft 飞机研制和生产成本

**DAPF** Data Analysis and Processing Facility
数据分析与处理设施

**DAPM** Design Approval Procedures Manual
设计审批程序手册

**DAPP** Data Acquisition and Processing Program 数据获取与处理计划

**DAPS** Data Access Protocol System 数据访问
协议系统

Direct Access Programming System 直接存取
编程系统

**DAPU** Data Acquisition and Processing Unit
数据采集与处理单元

**DAQCP** De-Icing/Anti-Icing Quality Control
Pool 除冰/防冰质量控制联盟

**DAR** Dar Es Salaam, Tanzania (Airport
Code)坦桑尼亚达累斯萨拉姆机场

Data Access Register 数据存取寄存器

Data Acquisition Request 数据采集要求

Data Address Register 数据地址寄存器

Data Analysis Report 数据分析报告

Data Article Requirements 数据条款的要求

Data Automation Requirements 数据自动化
要求

Defense Acquisition Radar 防御目标截获雷
达

Defense Acquisition Regulation 防御获取规
则

Deficiency Action Report 故障过程报告,事
故报告

Design Approval Representative 设计审批代
表

Designated Airworthiness Representative 委任
适航代表

Digital Aids Recorder 数字辅助记录器

Digital Angle Readout 数字式角度读数

Digital Audio Recorder 数字音频记录器

Directional Antenna for Receiver 接收机方向
性天线

Doppler Acoustic Radar 多普勒声雷达

Drawing Authentication Record 图纸鉴定记
录

Drive Annunciator Relay 驱动报警继电器

Dynamic Alternate Routing 动态迂回路由

**DARA** Defence Aviation Repair Agency (英
国)国防航空修理机构

**DARAS** Direction and Range Acquisition System 方向和距离探测系统

**DARB** Distressed Airman Recovery Beacon 遇
难飞行员救援信标

**DARC** Data Radio Channel 数据无线信道

Direct Access Radar Channel 直接存取雷达
信道

**DARE** Document Abstract Retrieval Equipment 文件摘要检索设备

Documentation Automatic Retrieval Equipment 文件整理自动化检索设备

Doppler and Range Evaluation 多普勒导航和航程估计

Doppler Automatic Reduction Equipment 多普勒自动处理装置

**DARP** Dynamic Air Route Planning 动态航线计划，动态航路计划

Dynamic Airborne Reroute Procedure 动态空中改航流程

**DARPA** Defense Advanced Research Projects Agency 美国国防部高级研究规划局

**DARPS** Dynamic Aircraft（Air）Route Planning Study 动态飞机航线计划研究

**DARS** Data Acquisition and Reduction System 数据采集与简化系统

Data Adaptive Recording System 数据自适应记录系统

Data-Acquisition and Recording System 数据采集与记录系统

Digital Altitude Reference System 数字高度基准信号

Digital Audio Radio System 数字音频无线电系统

Digital Audio Reference Signal 数字音频参考信号

Drogue Air Refueling System 锥套式空中加油系统

**DART** Data Analysis Risk Technology 数据分析风险技术

Direct Action Response Team 直接行动响应小组

Direct Airline Reservation Ticketing 直接向航空公司预订机票

Director and Response Tester 导向仪和响应测试仪

Dynamic Automatic Radar Tester 动态自动雷达测试器

**DARTS** Derivative Airplane Requirements & Tabulation System 改型飞机要求和制表系统

**DAS** Data Acquisition System 数据收集系统

Data Analysis Software 数据分析软件

Data Analysis System 数据分析系统

Data Automation System 数据自动处理系统

Design Assurance Software 设计保证软件

Design Assurance System 设计保证系统

Designated Alteration Station 委任改装站，指定的备用电台

Digital Address System 数码广播系统

Digital Analysis System 数字分析系统

Directional Autopilot Servo 自动驾驶方向伺服

Directorate of Aerodrome Standards 飞机场标准管理局

**DASA** Defense Atomic Support Agency 国防原子能支援局

**DASC** Defense Administrative Support Center 国防行政支持中心

Direct Air Support Center 直接空中支援中心

District Air Support Center 地区空中支援中心

**DASD** Direct Access Storage Device 直接存取存储装置

**DASF** Direct Air Support Flight 直接空中支援飞行

Drone Anti-Submarine Helicopter 无人驾驶反潜直升机

**DASI** Digital Altimeter Setting Indicator 数字高度设置指示器

**DASP** Discrete Analog Signal Processing 离散类比讯号处理

**DASR** Daily Aircraft Status Report 每日飞机状况报告

Defense Analysis Special Report 防御分析特别报告

Digital Airport Surveillance Radar 数字式机场监视雷达

Direct Air-to-Satellite Relay 直接空中卫星中继站

**DASS** Defence Aids Support System 防御辅助保障系统

Digital Air Traffic Service System 数字化空管系统

Digital-Analog Servo System 数字—模拟伺服系统

Direct Air Support System 直接空中支援系统

**DAST** Digital Aircraft System Trainer 数字式航空器系统训练器

**DAT** Data Analysis Team 数据分析组

Data Assessment Tool 数据评估工具

Delta Air Transport（Belgium）达美航空运输公司（比利时）

Digital Audio Tape 数据音频磁带

**DATAC** Development Areas Treasury Advisory Committee 发展区财政咨询委员会

Digital Automatic Target Analysis and Classification 数字式自动目标分析与分类

Digital Automatic Test and Classifier 数字式自动测试和分类机

Digital Autonomous Terminal Access Communication 数字式自主终端存取通讯

**DATACOM** Data Communications 数据通讯

**DATAR** Detection and Tactical Alert of Radar 探测与战术雷达警报

Digital Automatic Tracking and Ranging 数字自动跟踪与测距

Digital Automatic Tracking and Remoter 数字自动跟踪与遥控

**DATC** Development and Training Center 发展与训练中心

Director of Air Traffic Control 空中交通管制主任

**DATDC** Data Analysis and Technique Development Center 数据分析和技术开发中心

**DATIS** Digital Airport Terminal Information System 数字式机场终端信息系统

**D-ATIS** Data Link Automatic Terminal Information System 数据链自动终端情报系统

Digital Air Traffic Information Service 数字空中交通情报服务

Digital Automated Terminal Information Service 数字自动终端情报服务

**DATIS** Direct Access Technical Information System 直接存取技术信息系统

**DATR** Design Approval Test Report 设计批准测试报告

Digital Audio Tape Reproducer 数字式音频磁带放音机

**DATS** Data Automated Tower Simulator 数据自动塔台模拟器

Despun Antenna and Test Satellite 反旋转天线实验卫星

Dynamic Accuracy Test System 动态准确测试系统

**DAU** Data Access & Update 数据存取与更新

Data Acquisition Unit 数据采集单元

Data Analysis Unit 数据分析组件

**DAV** Drift Angle Valid 有效偏流角

**DAVC** Delayed Automatic-Volume Control 延时自动音量控制

Digital Audio/Video International Council 国际数字音频/视频理事会

**DAVID** Digital Audio/Video Interactive Decoder 数字音频/视频交互式解码器

**DAWS** Design of Aircraft Wing Structures 机翼结构设计

Digital Advanced Wireless Service 数字高级无线服务

**DAY** Dayton, OH, USA（Airport Code）美国俄亥俄州代顿机场

**dB** Decibel 分贝

**DB** Data Base, DataBase 数据库

Data Bus 数据总线

Denied Boarding 拒绝登机

Diffusion Bonding 扩散焊接

Double Bonding 双键

Double 两倍

Dynamic Braking 动力制动

**dB（A）** A-Weighted Decibel A 分贝（噪音）

**DBA** Database Administrator 数据库管理人员，数据库管理程序

Decibels Adjusted 调整分贝

Design Basis Accident 设计基准事故

Dynamic Bandwidth Allocation 动态带宽分配

**DBBS** Data Base Bulletin Service 数据库通告服务

**DBC** Data Base Check 数据库检查

Data Bus Check 数据总线检查

Data Bus Communication 数据总线通讯

Data Bus Coupler 数据总线耦合器

Decibel Below Carrier 低于载波分贝数

Decimal to Binary Converter 十进制到二进制的转换器

Denied Boarding Compensation 让座补偿金

Diameter Bolt Circle 圆堵塞直径

Dynamic Bandwidth Controller 动态带宽控制器

**DBCS** Data Base Control System 数据库控制系统

**DBD** Data Base Description 数据库描述

Data Basis for Design 设计数据基础

Database Design 数据库设计

Decision Basis Documents 决定基础文件

**DBDD** Data Base Design Document 数据库设计文件

**DBF** Database 数据库

Digital Beam-Forming 数字波束形成

Doppler Best Frequency 多普勒最佳频率

**DBHP** Designed Brake Horse Power 设计制动马力

**DBHS** Data Base Handling System 数据库处理系统

**DBI** Database Index 数据库索引

**D-BITE** Digital BITE 数字式机内测试设备

**DBK** Debark 卸货

Drawback 回收,退税

**DBL** Data Base Loader 数据装载器,数据库输入器

Double 两倍,双

**DBLB** Double Room with Bath 有洗澡间的双人房间

**DBLE** Double 两倍,双

**DBLK** Deblocked 解除控制,清除堵塞

**DBLR** Doubler 倍增器,倍频器

**DBLS** Double Room with Shower 有淋浴设备的双人房间

**DBM** Database Management 数据库管理

Decibels Reference to Milliwatt 毫瓦分贝

**DBML** Data Base Management Language 数据库管理语言

Database Markup Language 数据库标记语言

Diabetic Meal 糖尿病患者餐食

**DBMS** Database Management Software 数据库管理软件

Database Management System 数据库管理系统

**DBMT** Database Build and Modify Tool 数据库建立和调试工具

**DBN** Database Network 数据库网络

**DBO** Dubbo, Australia (Airport Code) 澳大利亚达博机场

**DBPSK** Differentially Coherent Binary PSK 差分相干二进制相移键控

**DBQ** Dubuque, IA, USA (Airport Code) 美国依阿华州迪比克机场

**DBR** Dead Bus Relay 死区汇流条继电器

Deterministic Bit Rate 确定性比特率

**DBRAT** Data Base Related to Air Transport 航空运输相关数据库

**DBRITE** Digital Bright Radar Indicator Tower Equipment 数字照明雷达指示塔台装置

**DBS** Data Base Storage 数据库存储

Depth Below Slots 定子在槽下的深度

Direct Broadcast Satellite 直接广播卫星

Direct Broadcasting Satellite 直接广播卫星

Domestic Base Station 国内基站

Doppler Beam Sharpening 多普勒锐化波束

Double Break Switch 双断开关

Double Side-band 双边带

**DBSR** Dead Bus Slave Relay 死区汇流条从继电器

**DBT** Debit 借方

Debt 债务

Design Build Team 设计建造组

Diffusion-Bonded Titanium 扩散结合钛

Doubt 怀疑

**DBTF** Duct-Burning Turbofan 外函加力燃烧涡轮风扇发动机

**DBTR** Debtor 债务人

**DBV** Dubrovnik, Croatia（Airport Code）克罗地亚杜布罗夫尼克机场

**DC** Data Compression 数据压缩

Departure Clearance Delivery 离场许可传送

Departure Clearance 离场放行许可

Descent 下降

Digital Computer 数字计算机

Digital Control 数字控制

Digital Convergence 数字汇聚

Direct Cost 直接成本

Direct Current 直流电

Discard 放弃

Discount 折扣

Dispersion Compensation 色散补偿

Domain Coordinator 区域协调员

Down Compatibility 向下兼容

Drag Coefficient 阻力系数

Driving Circuit 驱动电路

Drop Cable 引入缆

Dust Collector 吸尘器

**DCA** Data Correction Amplifier 数据校正放大器

Decimal Classification Astronautics 航天文献十进制分类法

Defense Communications Agency 国防通信局（美国）

Deflection Coil Amplifier 偏转线圈放大器

Department of Civil Aviation Malaysia 马来西亚民航局

Department of Civil Aviation 民航局，民航部

Design Change Authorization 设计更改核准权限

Diagnostic Communication Ability 诊断通讯能力

Digital Computer Assembly 数字式计算机组件

Digital Computers Association 数字计算机协会

Digital Control Attenuator 数字式控制衰减器

Director of Civil Aviation 民航管理局局长

Directorate of Civil Aviation（Denmark）丹麦民用航空局

Directorate of Civil Aviation 民航管理局

Discrepancy Control Area 故障控制区，缺陷控制区

Distance Counter Assembly 距离计算器组件

Distance of Closest Approach 最接近距离

Distributed Communication Architecture 分布式通信结构

Document Change Analysis 文件修改分析

Document Content Architecture 文件内容结构

Drift Correction Angle 偏航修正角

Dual-Capable Aircraft 双功能飞机，双重任务飞机

Dynamic Channel Allocation 气动态通道装置

Dynamics Corporation of America 美国动力公司

Washington, DC, USA-Ronald Reagan Washington National Airport（Airport Code）美国罗纳德·华盛顿国立机场

**DCAA** Defense Contract Audit Agency 国防合同审计局（美国）

**DCAC** Define and Control Airplane Configuration 飞机构型定义与控制

**DCAS** Data Collection and Analysis System 数据收集与分析系统

Defense Contract Administration Service 国防合同管理局

Defense Contract Administration Services（Now DCMC）（美国国防部）国防合同管理处

Digital Control Audio System 数字式控制音

频系统

Digitally Controlled Audio System 数字式控制音频系统

**DCATS** Dynamic Carrier Avionics Test Set 运输机航空电子设备动态测试仪

**DCB** Design Change Board 更改设计委员会

Display Colored Bar 显示彩条

**DCC** Data Communication Channel 数据通信信道

Data Country Code 数据国家码

Digital Control Channel 数字控制信道

Digital Cross Connect 数字交叉连接

Display Channel Complex 显示器通道综合

Drone Control Center 无人机控制中心

**DCCH** Dedicated Control Channel 专用控制信道

**DCCN** Distributed Computer Communication Network 分布式计算机通信网络

**DCCS** Data Communications Control System 数据通信控制系统

DC Current Sensor 直流电流传感器

Defense Communications Control System 国防通信控制系统

Digital Camera Control System 数字相机控制系统

Digital Command Communications System 数字指令通信系统

**DCCU** Data Communication Control Unit 数据通信控制装置

Data Correlation Control Unit 数据相关控制单元

Digital Communications and Control Unit 数字通信与控制装置

Direct Current Control Unit 直流控制组件

**DCD** Data Carrier Detect 数据载波检测

Data Carrier Detected 检测到数据载波

Data Communication Device 数据通信装置

Data Control and Display 数据控制和显示

Decode 译码

Directorate of Communications Development (英国)通讯发展管理局

Double Channel Duplex 双通道解调器

**DCDI** Digital Course Deviation Indicator 数位偏航指示器

**DCDL** Digital Control Design Language 数字控制设计语言

**DCDR** Decoder 译码器

**DCDU** Data Control and Display Unit 数据控制和显示组件

Data-Link Control Display Unit 数据链控制显示组件

**DCE** Data Circuit Equipment 数据电路设备

Data Circuit-Terminating Equipment 数据电路终端设备

Data Communications Equipment 数据通讯设备

Data Connection Equipment 数据连接设备

Data Control Equipment 数据控制设备

Deputy Chief Engineer 副总工程师

Differential Compound Engine 差动复合发动机

Digital Circuit-terminating Equipment 数据电路端接设备

Display Central Equipment 中央显示器

Distributed Computing Environment 分布式计算环境

Domestic Credit Expansion 国内信用膨胀

Drive Command Electronics 驱动指令电子设备

**DCF** Data Communication Function 数据通信功能

Discounted Cash Flow 现金流量贴现

Dispersion Compensation Fiber 色散补偿光纤

**DCGF** Data Conversion Gateway Function 数据转换网关功能

**DCGS** Distributed Common Ground System 分布式通用地面系统

Distributed Common Ground Station 分布式通用地面站

**DCI** Data Communications Interface 数据通信接口

Dataflow Control Interface 数据流控制接口

Design Change Instruction 设计变更说明

**DCIR** DC Isolation Relay 直流隔离继电器

**DCIU** Data Control Interface Unit 数据控制接口单元

**DCKG** Docking 停靠

**DCL** Data Link Departure Clearance Service 数据链式离场许可服务

Declaration 申报

Departure Clearance Delivery 起飞许可传送

Diagram Check List 线路图检查清单

Digital Cabin Logbook 数字化客舱日志

Digital Channel Link 数字信道链路

Digital Control Loading 数字式控制加载

Doctor of Civil Law 民法博士

Door Closer 门关锁器

**DCLE** Data Conversion Language Engine 数据转换语言引擎

**DCLN** Declaration 申报

Decline 拒绝

**DCLU** Digital Carrier Line Unit 数字载波线路单元

**DCM** Data Communication Module 数据通信模块

Data Communication Multiplexer 数据通信复用器

Dispersion Compensator Module 色散补偿模块

Dynamic Capacity Management 动态容量管理

Dynamic Connection Management 动态连接管理

**DCMA** Data Communication Mesh Architecture 数据通信网状结构

**DCMC** Data Communication Message Controller 数据通信报文控制器

Defense Contract Management Command 国防合同管理部

Display Control Management Computer 显示控制管理计算机

**DCME** Digital Circuit Multiplication Equip-ment 数字线路多路复用器

**DCMED** Decommissioned 停止工作,停止使用

**DCMF** Data Communication Management Function 数据通讯管理功能

**DCMS** Data Communication Management System 数据通讯管理系统

Defense Communications Monitoring System 国防通信监控系统

Display Control Management System 显示控制管理系统

Doors Control and Monitoring System 舱门控制和监控系统

**DCMT** Decrement 缩减量

Document 文件

**DCMU** Doors Control and Monitoring Unit 舱门控制和监控组件

**DCN** Data Communication Network 数据通信网

Design Change Notice 设计更改通知

Digital Communication Network 数字通信网

Distributed Computer Network 分布式计算机网络

Drawing Change Notice 图纸更改通知

**DCNA** Data Communication Network Architecture 数据通信网络体系结构

**DCNI** Digital Course Navigation Indicator 数字航向偏离指示器

**DCNSD** Decommissioned 停止工作,停止使用

**DCOM** Distributed Component Object Model 分布式构件对象模型

**DCP** Data Central Processor 中央数据处理器

Data Collection Platform 数据收集平台

Data Communication Processor 数据通信处理器

Data Coordinating Point 数据协调点

Decision Coordination Paper 决定协调书

Design Change Proposal 设计更改建议书

Designated Check Point 指定检查点

Development Concept Paper 开发方案文件

Digital Communication Protocol 数字通信协议

Digital Computer Programming 数字计算机程序设计

Display Control Panel 显示控制板

Distributed Communication Processor 分布式通信处理器

**DCPA** Distributed Call Processing Architecture 分布式呼叫处理结构

**DCPC** Direct Controller Pilot Communication 管制员驾驶员直接通信

**DCPCS** Digital Cabin Pressure Control System 数字式座舱压力调节系统

**DCPN** Domestic Customer Premises Network 国内用户驻地网

**DCPS** Data Collection Platform System 数据收集平台系统

Data Compression Processing System 数据压缩处理系统

Data Control Panel Submodule 数据控制面板组件

Direct Current Power System 直流电源系统

Dynamic Crew Procedure Simulator 机组人员工作程序动态模拟器

**DCPSK** Differentially Coherent Phase-Shift Keying 差分相干相移键控

**DCR** Data Conversion Receivers 数据转换接收器

Data Coordinator and Retriever 数据配合器和检索器

Decrease 减少,减退

Delivery Control Room 交付控制室

Design Certification Review 设计取证审查

Design Change Request 设计变更请求

Destination Call Routing 按目标选择路由

Digital Coded Radar 数字编码雷达

Digital Conversion Receiver 数字式转换接收器

Direct Current Resistance 直流电阻

Direct Current Restorer 直流恢复器

Drawing Change Request 图纸更改请求

Dynamically Controlled Routing 动态控制路由

**DCRB** Design Change Review Board 设计修改审查委员会

**DCRC** Dock on Crew Rest Compartment 机组休息舱

**DCRS** Data Collection and Reduction System 数据收集与整理系统

Digital Communication Recording System 数据通讯记录系统

**DCS** Data Collection System 数据收集系统

Defense Communication System 国防通信系统

Departure Control System 离港控制系统

Design Control Specification 设计检验规范

Designated Certification Specialist 指定颁证专家

Desktop Conferencing System 桌式会议系统

Digital Cellular System 数字蜂窝系统

Digital Code Squelch 数字编码静噪

Digital Communication System 数字通信系统

Digital Computer Subsystem 数字计算机分系统

Digital Computing System 数字计算系统

Digital Cross-Connect System 数字交叉连接系统

Direct Current Supply 直流供电

Distributed Control System 分布式控制系统

Double Channel Simplex 双信道单向通信

Drawing Control System 图纸控制系统

Dynamic Channel Selection 动态信道选择

Dynamic Channel Stealing 动态信道挪用

**DCT** Data Collection Tool 数据收集工具

Data Communication Terminal 数据通信终端

Direct 直飞,直达,直线飞行

Discrete Cosine Transformation 离散余弦变换

Display Compression Technology 显示压缩技术

Domestic Connecting Time 国内线转机时间

**DCTCU** Direct Current Tie Control Unit 直流

汇流条控制组件

**DCTE** Data Circuit Terminating Equipment 数据电路终接设备

**DCTU** Directly Connected Test Unit 直接连接测试单元

**DCU** Data Channel Unit 数据信道单元

Data Concentration Unit 数据集中组件

Data Control Unit 数据控制组件

Data Converter Unit 数据转换器组件

Decimal Counting Unit 十进制计数组件

Digital Computer Unit 数字计算机组件

Digital Connection Unit 数字连接单元

Digital Counting Unit 数字计算组件

Direction Control Unit 方向控制组件

Distribution and Control Unit 分配和控制单元

Dual Carrier Unit 双载波单元

**DCV** Design Change Verification 设计更改检验

Direct Current Voltage 直流电压

Directional Control Valve 方向控制活门

**DCVR** Digital Cockpit Voice Recorder 驾驶舱数字式话音记录器

Direct Current Voltage Ratio 直流电压比

**DD** Date of Departure 起飞日期

Dated Departure 注明日期

Department of Defense 国防部

Deputy Director 副处长,副厂长

Detail Design 详细设计

Digital Display 数字显示

Drift Down 漂降

**D-D** Digital-Digital 数数

**DDA** Data Decoder Assembly 数据译码器装置

Depth-Duration-Area 深度—持续时间—面积

Differential Digital Analysis 微分数字分析

Digital Differential Accumulation 数字差累加

Digital Differential Analyzer 数字微分分析器

Digitally Directly Analog 数字直接模拟

Display and Decision Area 显示与判定区

Downdrought Drift Angle 低于下滑道偏离角

Drawing Departure Authorization 图纸发行审批

**DDAPS** Digital Data Acquisition and Processing System 数字数据采集与处理系统

**DDAS** Digital Data Acquisition System 数字数据采集系统

**DDB** Digital Data Bus 数字数据总线

Distributed Database 分布式数据库

**DDBMS** Distributed Database Management System 分布式数据库管理系统

**DDC** Defense Documentation Center 国防部文献中心

Digital Data Communication 数字数据通信

Digital Data Converter 数字数据转换器

Digital Display Console 数字显示操纵台

Direct Digital Control 直接数字控制(系统)

**DDCE** Digital Data Conversion Equipment 数字化数据转换设备

**DDCMP** Digital Data Communication Message Protocol 数字式数据通讯信息协议

**DDCS** Digital Data Calibration System 数字数据校准系统

**DDCU** Dedicated Display and Control Unit 专用显示与控制组件

Display Digital Computer Unit 显示数字计算机组件

Display Digital Control Unit 显示数字控制组件

**DDD** Data Download and Display 数据下载与显示

Deadline Delivery Date 交货截止日期

Demand Driven Dispatch 需求驱动签派

Digital Data Demodulator 数字数据解调器

Direct Distance Dialing 长途直接拨号,长途自动拨号

**DDE** Dynamic Data Evaluation 动态数据评价

Dynamic Data Exchange 动态数据交换

**DDF**　Demand Data File 需求数据文档

　　Digital Distribution Frame 数字配线架

　　Dispersion-Decreasing Fiber 低色散光纤

**DDFS**　Digital Direct-Frequency Synthesizer 直接式数字频率合成器

**DDG**　Digital Display Generator 数字显示发生器

　　Dispatch Deviations Guide 放行偏差指南

**DDHB**　Distributed Dynamic Hypermedia Browser 分布式动态超媒体浏览器

**DDI**　Data Display Indicator 数据显示指示器

　　Depth Deviation Indicator 深度偏差指示器

　　Design Drawing Instruction 设计绘图说明

　　Digital Data Interface 数字式数据交联装置

　　Digital Display Indicator 数字式显示器

　　Direct-Dialing-in 直接拨入

　　Documentary Data Insert 公文数据引入

**DDL**　Database Definition Language 数据库定义语言

　　Digital Data Line 数字数据线

　　Digital Data Link 数字数据链路

　　Dispatch Deviation List 缺件放行单

　　Down Data Link 下行数据链

**DDM**　Data Demand Module 数据要求模块

　　Data Display Module 数字显示模块

　　Data Display Monitor 数据显示监视器

　　Design Data Management 设计数据管理

　　Design Decision Memo 设计决定备忘录

　　Design Decision Memorandum 设计决策备忘录

　　Difference in Depth of Modulation 调制深度差

　　Digital Data Module 数字式数据组件

　　Direct Digital Manufacturing 直接数字制造

　　Distributed Data Management 分布数据管理

**DDMS**　Digital Data Measuring System 数字数据测量系统

　　Digital Drawing Management System 数字制图管理系统

**DDN**　Digital Data Navigation 数字化数据导航

Digital Data Network 数字化数据网络

　　Digital Distribution Network 数据分配网络

　　Distributed Data Network 分布式数据网

　　Documentation Development Notification 文件改进通知

　　Drawing Discrepancy Notice 图纸缺陷通知

**DDNS**　Dynamic Domain Name System 动态域名系统

**DDOS**　Distributed Deny of Service 分布式拒绝服务

**DDOV**　Digital Data Over Voice 话音传送数字数据

**DDP**　Datagram Delivery Protocol 数据报投递协议

　　Declaration of Design and Performance 设计和性能声明

　　Design Development Plan 设计开发计划

　　Digital Data Processor 数字数据处理器

　　Distributed Data Processing 分布数据处理

　　Distributed Data Processor 分布式数据处理器

　　Diversion Decision Point 改航决断点

　　Dynamic Data Presentation 动态数据显示

**DDPE**　Digital Data Processing Equipment 数字数据处理装置

**DDPG**　Dispatch Deviation Procedure Guide 放飞偏离程序指南,派遣偏差程序指南

**DDPS**　Digital Data Processing Subsystem 数字数据处理子系统

　　Digital Data Processing System 数字数据处理系统

　　Distribution Data Processing System 分布数据处理系统

**DDR**　Data Descriptive Record 数据描述记录

　　Decision -Directed Receiver 直接判定的接收器

　　Decoy Discrimination Radar 假目标识别雷达

　　Design Development Record 设计发展记录

　　Detailed Design Review 详细设计评审

　　Detection Difficulty Ranking 检测难度等级

Determination of Direction and Range（超声波）定向和测距装置

Digital Data Receiver 数字式数据接收器

Digital Display and Recorder 数字显示与记录

Direct Drive 直接驱动,直接传动

Doppler Deadreckon 多普勒推算航行法

Double-Drift-Region 双漂移区

Drawing Data Record 图纸数据记录

Drawing Data Required 所需的图纸数据

**DDR&E** Defense Development Research & Engineering 国防研究和工程部

Deputy Director for Research & Engineering 研究和工程部副主任

**DDRMI** Digital Distance and Radio Magnetic Indicator 数字式距离和无线电磁指示器

**DDRS** Defense Data Reporting System 国防数据报告系统

Digital Data Recording System 数字数据记录系统

**DDS** Demand Data Sheet 需求数据表

Design Data Sheet 设计数据表

Digital Data Service 数字数据业务

Digital Data System 数字数据系统

Digital Display Scope 数字显示器

Direct Digital Satellite 直播数字卫星

Direct Digital Service 直接数字服务

Direct Digital Synthesis 直接数字合成

Direct Digital Synthesizer 直接数字合成器

Display and Debrief Subsystem 显示与分析子系统

Doppler Detection System 多普勒探测系统

Drawing Data Sheet 图纸数据表

**DDT** Department of Development and Technology 开发与技术部（欧洲航天局）

Detailed Design Phase 详细设计阶段

**DDT&E** Design, Development, Test & Evaluation 设计、研发、测试与评估

**DDTS** Digital Data Technology Standards 数字数据技术标准

Digital Data Transmission System 数字数据传输系统

Direct Dialing Telephone System 直拨电话系统

Distributed Data and Telecommunications System 分布数据与电信系统

**DDU** Deceleration Detector Unit (RR engine) 减速探测器组件(罗·罗发动机)

Disc Drive Unit 磁盘驱动器

**DDV** Direct Drive Valve 直接驱动活门

**DE** Date of Entry 进港日期,入界日期

Defer 延期

Designated Engineer 指定工程师

Development Estimate 研制预计

**DEA** Data Exchange Aggreement 数据交换协议

Defense Exchange Aggreement 防务交换协议

Display Electronics Assemblies 显示电子设备

Drive Electric Assembly 电子驱动组件

**DEAF** Deaf Passenger 耳聋旅客

**DEB** Debrecen, Hungary(Airport Code) 匈牙利德布勒森机场

**DEBK** Debark 卸载

**DEC** Decatur, GA, USA (Airport Code) 美国佐治亚州迪凯特机场

Decca 台卡导航系统

Deceleration, Decelerate 减速

December 十二月

Decimal 十进制的

Declaration 申报

Declination 偏差,倾斜,偏角,偏斜

Decompose 分解

Decrease/Decrement 减少

Decrement 减少率

Descent 下降

Digital Equipment Corporation 数字设备公司(美国)

District Emergency Coordinator 区紧急协调程序(协调员)

Document Effected Code 文件有效代码

**DECA** Display/Electronic Control Assembly 显示器电子控制组件

**DECAN** Distance Measuring Equipment Command and Navigation 指挥与领航用测距设备

**DECCA** Decca 台卡(导航设备)

**DECEA** Department of Air Space Control 空域管制部

**DECEL** Decelerate(d) 减速,降低速度
Deceleration 减速度,负加速度,减速

**DECL** Declaration 申报
Decline 拒绝

**DECM** Deceptive Electronic Countermeasures 防御电子对抗
Decommissioned 停止工作,解除的

**DECN** Decision 决定

**DECR** Decrease 降低,减少

**DECT** Digital Enhanced Cordless Telecommunications 数字增强型无绳电信

**DECU** Digital Engine Control Unit 发动机数字控制装置

**DECUS** Digital Equipment Computer Users Society 数字化设备计算机用户协会
Digital Equipment Corporation Users Society 数字化设备公司用户协会

**DED** Data Entry Display 数据输入显示器
Dedendum 齿高
Dedicated 专用的
Deduct 扣除,减去
Deduction 减少
Descent 降下,下坡
Designate 指定,标志
Designation 指明,规定
Designator 指定者,标志符
Destroy 摧毁,破坏
Development Engineering Division 研制工程处,发展工程处
Dust Extraction System 除尘系统
Dynamically Established Data Link 动态建立数据链路

**DED GEN** Dedicated Generator 专用发动机

**DEDE** Density-Depth 密度—深度

**DEDF** Distributed Erbium-Doped Fiber 分布式掺铒光纤

**DEDFA** Distributed Erbium-Doped Fiber Amplifier 分布式掺铒光纤放大器

**DEDP** Data Entry and Display Panel 数据输入和显示面板

**DEDS** Data Entry and Display Sub-system 数据输入与显示分系统
Data Entry and Display System 数据输入与显示系统
Data Exchange and Distribution System 数据交换与分配系统
Digital Enhancement Database System 数字增强数据库系统

**DEEC** Digital Electronic Engine Control 发动机数字电子控制

**DEES** Dynamic Electromagnetic Environment Simulator 动态电磁环境模拟器

**DEET** Deflection 偏转,偏差

**DEF** Defense 防护,保卫
Defer 延期
Deficit 赤字
Definate 肯定的,明确的
Define, Definition 规定,定义

**DEFDARS** Digital Expandable Flight Data Acquisition & Recording System 数字式可扩展飞行数据获得和记录系统

**DEFI** Digital Electronic Fuel Injection 数字电子燃油喷射

**DEFL** Deflect, Deflection 偏转

**DEFNT** Definite 下定义,详细说明

**DEFT** Deficit 赤字
Definite 肯定的,明确的
Display Evaluation Flight Test 显示鉴定飞行试验

**DEFUEL** Defueling 抽油

**DEG** Degree(s) 度数,程度,等级,度

**DEG C** Degrees Centigrade 摄氏度

**DEG F** Degrees Fahrenheit 华氏度

**DEH** Digital Electrohydraulic 数字电液

**DEI** Design Engine Inspection 设计发动机检查

**DEIS** Draft Environmental Impact Statement 气流环境冲击报告

**DEK** Data Encryption Key 数据加密密钥

**DEL** Delaware 特拉华州(美国)

Delay Message 延误电报

Delay 延误

Delegate 代表

Delegation 代表团

Delete by 被…取消,被…删除

Delete or Deleted or Deleting 删除

Delhi, India ( Airport Code ) 印度新德里机场

Delivery 送交,递交

**DELD** Deliver 递交,交送

Delivered 交付的,交送的

**DELE** Delete 删除,取消

**DELEG** Delegate 代表

Delegation 代表团

**DELFO** Delivery by Telephone 电话交付,电话传达

**DELMIA** Digital Enterprise Lean Manufacturing Interactive Application 数字企业精益制造的交互式应用

**DELT** Default 错误,缺点,空缺

Delete 删去,取消

Difficult 困难的

**DELTAP** Differential Pressure 压差

**DELTN** Deflection 偏差,偏转

**DELV** Deliver 送交,递交

Delivery 送交,递交

**D-ELV** Destination Elevation 终点标高,降落机场标高

**DELVD** Delivered 送交,递交,交货

**DELY** Delay 延误

Delivery 交付,传送,递交

**DEM** Demand 要求,需求

Digital Electronic Module 数字电子组件

Digital Elevation Model 数字标高模型

**DEMD** Digital Engine Monitor Display 发动机数字监控显示器

**DEMO** Democratic 民主的

Demonstration 示范,表现,展示

**DEMOD** Demodulator 解调器,检波器

**DEMUX** Demultiplexer 信号分离器

**DEMVAL** Demonstration and Validation 演示与验证

**DEN** Data Enable 数据启动

Data Exchange Network 数据交换网络

Denmark 丹麦

Denote 指示,表示,意味着

Density 密度

Denver, CO, USA ( Airport Code ) 美国科罗拉多州丹佛机场

Directory-Enabled Networking 基于目录的联网

**DENEB** Fog Dispersal Operations 驱雾作业

**DENG** Diesel Engine 内燃机

**DENL** Denial 否认

**DENOM** Demonstration 表演,展示

**DENS** Density 密度

**DENT** Dentist 牙科医生

**DEO** Development Engineering Operations 研制工程运作

**DEOS** Digital Engine Operating System 数字引擎操作系统

**DEP** Data Entry Panel 数据输入面板,(登记)数据记录板

Data Entry Plug 数据输入插头

Data Exchange Program 数据交换程序

Deflection Error Probable 可能的偏转误差,偏转概率误差

Depart or Departed or Departure 离场

Depart 离港,起飞

Department of Environmental Protection 环境保护部

Department 部,处

Departure Airfield 出发机场

Departure Airport 出境地机场

Departure Control 离场管制

Departure End of Runway 离场跑道末端

Departure Message 出发电报
Departure Point 出发点,起航点
Departure 离场,起飞
Dependency 从属,从属物
Dependent 从属的,依赖的
Deployment 开展,开伞
Depot 仓库,停车场
Depressurize 减压,释压
Depth of Penetration 穿透,深度
Depth 深度
Deputy 副的,代理人
Design Eye Position 设计(飞行员)眼睛位置
Domestic Emergency Plan 紧急民防计划

**DEP/ARR**　Departure/Arrival 离港/进港

**DEPA**　Deportee（Accompanied by an Escort）遣返旅客(有押送人员)

**DEPARTURE**　Departure Procedure 离场程序

**DEPCON**　Departure Control 离场管制

**DEPLD**　Deployed 放出,伸出
Dependent 从属,依赖

**DEPO**　Deportee 被驱逐出境者,遣返旅客

**DEPR**　Departure 离场
Depressurization 减压
Depressurize 减压,释压

**DEPRESS**　Depressurization 释压

**DEPT**　Department 部,处
Departure 离开,出发,起飞

**DEPU**　Deportee（Unaccompanied）遣返旅客（无押送人员）

**DEPUTN**　Deputation 代表团

**DEPY**　Deputy 副的,代理人

**DER**　Departure End of Runway 跑道起飞端,跑道起飞末端
Departure End of the Runway 跑到起飞末端
Departure Message 起飞电报
Designated Engineering Representative 委任工程代表
Distinguish Encoding Rule 区分编码规则

**DERA**　Defense Research Agency 防御研究机构

**DERD**　Display of Expand Radar Data 扩展雷达数据显示

**DES**　Data Encryption Standard 数据加密标准
Data Entry System 数据登录系统
Data Exchange Standards 数据交换标准
Descent 下降
Desert 沙漠
Design and Evaluation System 设计与评估系统
Design Environment Simulator 设计环境模拟器
Design 设计
Desire 愿望
Destination 目的地,降落站
Digital Equipment Simulator 数字设备模拟器
Digital Expansion System 数字扩充系统
Document Exchange System 文件交换系统
Douglas Engineering Standard 道格拉斯工程标准

**DESC**　Defense Electronics Supply Center 国防电子供应中心
Describe, Description 说明,描述,绘制

**DESGNTR**　Designator 符号,种类,指定人

**DESIG**　Designate, Designation 指定,称呼,名称

**DESIGN**　Designate 指定,称呼,名称

**DESP**　Dispatch 派遣,发送,调度

**DESPCH**　Dispatch 派遣,发送,调度
Dispatcher 调度员,签派员

**DESS**　Digital Electronic Switching System 数字电子交换系统
Discrete Event Stochastic System 离散事件随机系统

**DEST**　Destination Airport 目的地机场
Destination 目的地,降落站

**DESTRC**　Destruction 毁坏,破坏

**DESV**　Deserve 应受到

**DET**　Design Evaluation Test 设计评估测试

Detached 拆下的

Detachment 分离，拆下，取下

Detail 详情

Detect 探测

Detection 发现，探测

Detector 探测器

Determine 决定

**DETD**    Determined 决定，决心

**DETING**    Determining 决定，确定

**DETM**    Determine 决定

**DETN**    Determination 决定

**DETNT**    Detent（机械上的）止动装置，棘爪，制动器

**DETR**    Detract 损坏，降低

**DETRESFA**    Distress Phase 遇险阶段，遇险状态

**DEU**    Data Encryption Unit 数据加密单元

Data Entry Unit 数据登录装置

Data Exchange Unit 数据交换装置

Decoder Encoder Unit 编码译码组件

Detector Electronics Unit 探测器电子设备单元

Digital Evaluation Unit 数字评价单元

Display Electronics Unit 电子显示组件

**DEV**    Developed Height 展开高度

Developed 发展的，展开的

Development 发展，开发

Deviation or Deviating 偏差，偏向，偏离，偏移

Device 装置，器件，设备

**DEV EQL**    Deviation Equalizer 偏离均衡器

**DEVAL**    Devaluation 贬值

**DEVCO**    Committee for Standardization in the Developing Countries 发展中国家标准化委员会

Committee on Developing Country Matters 发展中国家计划

**DEVN**    Deviation 偏差，偏离，偏移

**DEVP**    Develop 发展，开发

**DEVPT**    Development 发展，开发

**DEVTN**    Deviation 偏航，绕航

**DEW**    Delivery Empty Weight 交机空重，交付空机重量

Distant Early Warning 远程预警

**DEWIZ**    Distance Early Warning Identification Zone 远程早期警报识别区

**DEX**    Access Demanded 要求接近

**DEXAN**    Digital Experimental Airborne Navigator 实验机载数字导航仪

**DF**    Dead Freight 空舱费

Direct Flight 直接飞行

Direct to a Fix 直飞到定位点

Direction Finder 定向机，无线电罗盘

Direction Finding 测向，定向

Draft 汇票，草稿，草案

Duty Free 免税

**DFA**    Data Flow Analysis 数据流分析

Data Flow Analyzer 数据流程分析器

Delayed Flaps Approach 延迟襟翼进近

Design For Assembly 面向装配的设计

Designated Field Activity 指定范围活动

Direction Finding Antenna 定向天线

**DFAD**    Digital Feature Analysis Data 数字特性分析数据

Digital Feature Attribute Data 数字特征属性数据

**DFAS**    Distributed Frame Alignment Signal 分布式帧排列信号

**DFB**    Decided-Feed-Back 判决反馈

Distributed Feed Back 分布反馈

**DFBL**    Digital Flight By Light 数字光传操纵系统

**DFBW**    Digital Fly-By-Wire 数字电传操纵系统

**DFC**    Data Flow Control 数据流控制

Design For Cost 面向成本设计

Digital Flight Controller 数字飞行控制器

Direct Force Control 直接力控制

Dynamic Feedback Control 动力反馈控制

**DFCC**    Digital Flight Control Computer 数字飞行控制计算机

**DFCF**    Dispersion Flat Compensation Fiber 色

散平坦补偿光纤

**DFCLS** Digital Flight Control and Landing System 数字飞行控制与着陆系统

**DFD** Data Flow Diagram 数据流程图
Digital Flight Display 数字飞行显示器

**DFDAC** Digital Flight Data Acquisition Card 数字式飞行数据获得卡

**DFDAF** Digital Flight Data Acquisition Function 数字式飞行数据获得功能

**DFDAMU** Digital Flight Data Acquisition and Management Unit 数字飞行数据采集和管理组件

**DFDAU** Digital Flight Data Acquisition Unit 数字式飞行数据采集装置

**DFDMU** Digital Flight Data Management Unit 数字式飞行数据管理组件

**DFDR** Digital Flight Data Recorder 数字式飞行数据记录器

**DFDRS** Digital Flight Data Recording System 数字式飞行数据记录系统

**DFDU** Digital Flight Data Unit 数字式飞行数据组件

**DFE** Decision Feedback Equalizer 判决反馈均衡器
Design For the Environment 面向环境的设计
Duty Free 免税

**DFEE** Designated Flight Engineer Examiner 委任飞行工程师考核员

**DFF** Dispersion Flattened Fiber 色散平坦光纤

**DFG** Differential-Frequency-Generation 差分频率产生

**DFGC** Digital Flight Guidance Computer 数字飞行制导计算机

**DFGS** Digital Flight Guidance System 数字飞行制导系统

**DFI** Development Flight Instrumentation 研制飞行仪表设备
Development Flight Instruments 研制试飞仪器

Digital Facility Interface 数字设备接口

Digital Fax Interface 数字传真接口

**DFIB** Data Function Information Book 数据函数资料册

**DFIDU** Dual Function Interactive Display Unit 双功能交互式显示组件

**DFIS** Data-Link Flight Information Service 数据链飞行情报服务
Designated Flight Instrument System 指定的飞行仪表系统
Digital Flight Inspection System 数字飞行检查系统

**DFIU** Digital Flight Instrument Unit 数字式飞行仪表组件

**DFL** Data Flow Controller 数据流控制器
Debt Financing Lease 贷款融资租赁
Duplex 双工的,双重的

**DFLD** Definitely Loaded 肯定已装上飞机

**DFLT** Default 错误,缺陷,空缺

**DFM** Deputy for Financial Management 财政管理代理
Design for Maintenance 面向维修的设计
Design for Manufacturability 可制造性设计
Design for Manufacturing 面向制造的设计

**DFMA** Design for Manufacturing and Assembly 面向制造和装配的设计

**DFMC** Design for Mass Customization 面向大规模定制生产的产品设计

**DFMS** Digital Fuel Management System 数字式燃油管理系统

**DFNE** Designated Flight Navigator Examiner 委任飞行领航员考核员

**DFOS** Distributed Fiber Optic Sensing 分布式光纤传感

**DFP** Desired Flight Path 所需飞行航迹
Deviate Flight Plan 偏离飞行计划
Distributed Function Plane 分布式功能平面

**DFPC** Decoupled Flight Path Control 去耦飞行航迹控制

**DFR** Decreasing Failure Rate 故障率递减型
Defer 延期

Departure Flow Regulation 离港(飞机)流量控制

Detail Fatigue Rating 细节疲劳额定值

Digital Flight Recorder 数字式飞行记录器

**DFRC** Dryden Flight Research Center 德赖登飞行研究中心

**DFRNCE** Deference 依从

Difference 差别,余额

**DFRS** Digital Flight Recorder System 数字式飞行记录器系统

**DFRT** Demonstration Flight Rating Tests 示范飞行人员等级测试

**DFS** Data Formatting Statement 数据格式语句

Decision Feedback System 判决反馈系统

Dedicated File Server 专用文件服务器

Digital Formatting System 数据格式化系统

Digital Frequency /Function Selection 数字频率/功能选择

Digital Frequency Synthesizer 数字频率合成器

Direct Flooding System 直接注水系统

Director of Fire Service 消防负责人

Director of Flight Safety 飞行安全处处长(主任)

Directorate of Flight Safety 飞行安全局

Discrete Fourier Series 离散傅立叶级数

Display Format Software 显示格式软件

Distance Finding Station 测距站,测距台

Distributed Fiber Sensor 分布式光纤传感器

Dual Frequency Smoothing 双频平滑

Dynamic Flight Simulator 飞行动态模拟器

**DFSK** Differential Frequency Shift Keying 差分频移键控

Double Frequency Shift Keying 双频移键控

**DFSM** Dispersion Flattened Single Mode 色散平坦单模

**DFSR** Directorate of Flight Safety Research (美国)飞行安全研究局

**DFT** Deaerating Feed Tank 通风供油箱

Delayed-First-Transmission 延迟优先传输

Department for Transport (英国)运输部

Direct Flight Test 直接飞行试验

Discrete Fourier Transform 离散傅立叶变换

Draft 汇票,草案,草稿

Drift Angle 偏流角

Drift 偏流,偏航,漂移

Droppable Fuel Tank 副油箱

Dynamic Fault Tree 动态故障树

**DFTA** Dynamic Fault Tree Analysis 动态故障树分析

**DFTI** Distance from Touchdown Indicator 距接地面指示器的距离

**DFTP** Detailed Flight Test Plan 详细试飞计划

**DFTR** Deflector 导流片,转向器,偏转板

**DFU** Data Facilities Unit 数字设备单元

**DFUC** Direct Fare Undercut Check 优先使用直达运价检查

**DFW** Dallas/Fort Worth, TX, USA (Airport Code) 美国得克萨斯州达拉斯/沃思堡机场

**DFWD** Discrete Flight Warning Display 离散式飞行告警显示器

**DG** Dangerous Goods 危险品

Defense Guidance 防御指导方针

Degree 度数

Differential Generator 差动式传感器,差动振荡机

Diplomatic or Government 外交或政府人员

Directional Gyro 陀螺半罗盘,航向陀螺仪,陀螺方向仪

Director General 局长,董事长

**DGAC** Directorate General of Air Communication 航空通信局长

French Directorate General of Civil Aviation 法国民航总局

**DGCA** Director General of Civil Aviation 民航局长

Directorate General of Civil Aviation (印度)民航总局

**DGD** Shipper's Declaration for Dangerous

Goods 托运人危险品申报单

**DGI**  Digital Input 数字输入

**DGM**  Dangerous Goods Management 危险品管理

Dangerous Goods Manual 危险品手册

**DGMDB**  Digital Ground Map Database 数字地面地图数据库

**DGMM**  Dangerous Goods Transportation Management Manual 危险物品运输管理手册

**DGN**  Diagnosis 诊断

**DGNSS**  Differential Global Navigation Satellite System 差分全球导航卫星系统

**DGO**  Digital Output 数字输出

**DGOM**  Dangerous Goods Transportation Operation Manual 危险物品运输操作手册

**DGON**  German Institute of Navigation 德国导航协会

**DGP**  Dangerous Goods Panel 危险货物运输专家组,危险货物标志牌

Digitizing Graphics Package 数字化图形包

Directional Gyroscope Position 航向陀螺仪位置

**DGPS**  Differential Global Positioning System 差分全球定位系统

**DGR**  Danger 危险

Dangerous Goods Regulations 危险物品规则

Designated Government Representative 指定政府代表

**DGRADE**  Downgrade 降级

**DGS**  Docking Guidance System (飞机门与登机门)对接制导系统

**DGSI**  Drift and Ground Speed Indicator 航流和地速指示器

**DGT**  Dangerous Goods Training 危险品训练

**DGTL**  Digital 数字的

**DGU**  Directional Gyro Unit 航向陀螺装置

**DGVS**  Doppler Ground Velocity System 多普勒地速系统

**DH**  Decision Height 决断高度

Desired Heading 要求航向,希望航向,规定航向

Desired Height 要求高度,目标高度

**DHA**  Dhahran, Saudi Arabia (Airport Code) 沙特阿拉伯达兰国际机场

**DHCP**  Dynamic Host Configuration Protocol 动态主机配置协议

Dynamic Host Control Protocol 动态主机控制协议

**DHDD**  Digital High Definition Display 数字化高分辨率显示

**DHF**  Data Handling Function 数据处理功能

**DHFS**  Defense Helicopter Flying School 防卫直升机飞行学校

**DHI**  Daewoo Heavy Industries 大宇重工业公司

Directional Horizon Indicator 水平方向指示器

**DHN**  Digital Home Network 数字家庭网

**DHOO**  Daily Hours of Operation 每日运营小时数

**DHS**  Department of Homeland Security 国土安全部

**DHV**  Deutscher Hangegleiter Verband 德国滑翔飞行协会

**DHW**  Down High Way 下行公共信道

Data Input 数据输入

Data Item 数据项目

Defense Industry 国防工业

De-Icing 除冰

Digital Interface 数字接口

Direction Indicator 航向指示器,方向指示器

Dispersion-Increasing 色散增加

Document Identifier 文件识别器

**DIA**  Diameter 直径

**DIAC**  Department of Immigration and Citizenship 移民和公民部

**DIAD**  Delivery Information Acquisition Database 交货信息采集装置

**DIAG**  Diagonal 对角线

Diagram 图表

**DIAL**  Direct Information Access Line 直接信

息存取线

**DIAM** Diameter 直径

Diamond 钻石

**DIANE** Distance Indicating Automatic Navigation Equipment 指示距离自动导航设备

**DIAPH** Diaphragm 隔膜,振荡片

**DIAS** Differential GNSS Instrument Approach System 差分 GNSS 仪表进近系统

Digital Integrated Avionics System 数字集成航空电子系统

**DIB** Data Input Bus 数据输入总线

Department Information Bulletin 部门情报通报

Device Independent Bitmap 与设备无关位图

Directory Information Base 目录信息库

**DIBU** Door Illumination Ballast Unit 舱门照明镇流器组件

**DICH** Dedicated Information Channel 专用信息信道

**DICS** Digital Image Correction System 数字图像校正系统

**DICT** Dictionary 辞典

**DICU** Display Interface Control Unit 显示接口控制组件

**DID** Data Input Display 数据输入显示

Data Insertion Device 数据插入装置

Data Item Description 数据项目描述

Design Implementation Document 设计实施文件

Device Identification 数字标识符

Device Identifier 设备标识符

Digital Image Data 数字图像数据

Digital Information Display 数字信息显示,数字信息显示器

Direct in Dialing 直接拨号

Direct Inward Dial/Direct Inward Dialing 直接内部拨号

Drum Information Display 磁鼓信息显示

**DIDC** Data Input Display Console 数据输入显示控制台

**DIDS** Decision Information Distribution System 决策信息分发系统

Defense Integrated Data System 国防综合数据系统

Digital Information Display System 数字信息显示系统

Digital Information Distribution System 数字信息分发系统

Domestic Information Dispatching System 国内情报发送系统

**DIE** Document of Industrial Engineering 生产工程文件

**DIF** Difference 不同,差别

Different 不同的,有差的

Difficult 困难的

Difficulty 困难

Diffuse 扩散,传播

**DIFAR** Directional Finding and Ranging 测向和测距

**DIFF** Difference 不同,差额,差异

Differential 差异的,不同的

Difficult 困难的

**DIG** Digit 数字

Digital 数字的

**DIGACC** Digital Guidance and Control Computer 数字制导和控制计算机

**DIGS** Delta Inertial Guidance System 三角惯性制导系统

Digital Inertial Guidance System 数字惯性制导系统

**DII** Dynamic Invocation Interface 动态调用接口

**DIJ** Dijon, France (Airport Code) 法国第戎机场

**DIL** Derived Intervention Level 导出干预水平

Digital Integrated Logic 数字综合逻辑

Dili, Indonesia (Airport Code) 印度尼西亚帝力机场

**DIM** Dim 减光

Dimension 尺寸

Dimmer 减光器

Dimming 变暗

Direction Input Module 方向输入组件

**DIMA** Direct Imaging Mass Analysis 直接图像质谱分析

Distributed Integrated Modular Avionics 分布式综合模块化航电体系

**DIME** Direct Memory Execute 直接内存执行

**DIMM** Dual-Part Integrated Memory Monitor 双组件综合存储监控器

**DIMS** Dimensions 容积,尺寸

**DIN** Data Input 数据输入

Dinar 第纳尔(货币单位)

**DINA** Digital Network Analyzer 数字网络分析程序

Direct Noise Amplifer 直接噪声放大器

**DINAC** Directorate of Civil Aeronautics 民用航空局

**DINAMO** Dynamic Inventory and Maintenance Optimize 动态库存和维修最佳化

**DINE** Digital Inertial Navigation Equipment 数字惯性导航设备

**DINOS** Distributed Network Operating System 分布式网络操作系统

**DINS** Digital Inertial Navigation System 数字惯性导航系统

**DIP** Data Input Panel 数据输入板

Data Interrupt Program 数据中断程序

Decisions in Progress 进程中决断

Digital Image Processing 数字图像处理

Diplexer 双工器,天线分离滤波器

Diplomacy 外交

Diplomatic Baggage 外交行李

Diplomatic Mail 外交函件

Diplomatic 外交的

Display Information Processor 显示信息处理机

Double In-Line Package 双列直插式组件

Dual In-Line Package 双列直插式组件

**DIPEC** Defense Industrial Plant Equipment Center (美国)国防工业工厂设备中心

**DIPL** Digital Information Processing Lab 数字信息处理实验室

Diplomatic 外交的

Diplomatic Courier (reservations and ticket code) 外交身份信差

**DIPP** Defence Industry Productivity Program (Canada) (加拿大)国防工业生产力计划

**DIR** Department of Industrial Research 工业研究部

Design Information Release 设计信息请求

Direct 直飞,直达

Direction 方向

Directional 有方向性的

Director 指引仪,主任,局长

Distance Remaining 剩余距离,待飞距离

**DIRCM** Directional Infrared Countermeasure 方向性红外线计算测量

**DIRCOUP** Directional Coupler 定向耦合器

**DIRFLT** Directional Filter 定向滤波器

**DIRMA** Digital Impulse Radio Multiple Address 数字脉冲无线多址

**DIRRUT** Direct Route 直飞航路

**DIRS** Damage Information Reporting System 损伤信息报告系统

**DIRSN** Direction 方向

**DIRTLY** Directly 直接地

**DIRTO** Direct to 指向

**DIS** Digital Identification Signal 数字识别信号

Discharge 卸,释放,放电

Discount 折扣

Discrete 分立的,离散的

Dispatch 调度,签派

Distance 距离

Doppler Inertial System 多普勒惯性系统

**DISAP** Disappointed 失望的

Disapprove 不允许,不批准

**DISASSY** Disassembly 分解

**DISBS** Disbursement 交付

**DISC** Data Information Service Center 数据信息服务中心

Defect Information and Serviceability Control

缺陷信息和可使用性控制

Digital International Switching Center 国际数字交换中心

Disconnect 使分离,分开,断开

Disconnected 断开的,脱开的

Discontinued 中断的,不连接的

Discount 折扣

Discover 发现

Discrete 离散

Discussion 讨论

**DISC SOL** Disconnect Solenoid 断开电磁线圈

**DISCH** Discharge 放电,卸货,释放,排出(液体,气体等)

Discharged 排出的

**DISCHG** Discharge 放电,卸货

**DISCN** Discussion 讨论

**DISCO** Data Interchange for Shipping Companies 航运公司数据交换计划

Disconnect 不连接

**DISCONT** Discontinued 不连续,停止,中断

**DISCR** Discrepancy 偏差,差异

Discrete 离散

Discretion 斟酌决定

**DISCT** Discount 折扣

**DISCTN** Disconnection 断开,脱开

**DISENG** Disengage 解开,解除,使脱离

Disengaged 脱开

**DISHD** Dishonoured 拒付的

**DISHR** Dishonour 拒付

**DIS-IN** Discrete Input 离散输入

**DISN** Defense Information Systems Network 国防系统信息网

**DISP** Defense Industrial Security Program 国防工业安全计划(程序)

Defense Industry Studies Program 国防工业研究计划

Directory Information Shadowing Protocol 目录信息隐匿协议

Dispatch 遣派,调度

Dispense 分配

Displace 位移

Dispose 处理

Document Information Service Program 文件信息服务程序

**DISPL** Displace 移动,取代

Displaced 调换

Displacement 替换,位移

Display 显示

**DISPO** Disposal 处理,安排,支配

Disposition 处理,安排,布置

**DISRGD** Disregard 不理睬

**DISRMD** Disarmed 解除预位

**DIST** Discount 折扣

Distance 距离,路程

Distant 远的

Distribute 分配

Distribution 分配,分派

Distributor 分配器,分电器,分配者

District 区域,行政区

**DISTN** Distortion 失真

**DISTR** Distribute, Distribution, Distributor 分发,分配,分配器

**DIT** Direct Image Technical 直接成像技术

Directory Information Tree 目录信息树

Dynamic Integrated Test 动力学综合试验

**DITCH** Ditching 水上迫降

**DITS** Digital Information Test System 数字信息测试系统

Digital Information Transfer Set 数字信息传输装置

Digital Information Transfer System 数字信息传输系统

**DIU** Data Interface Unit 数据接口组件

Digital Interface Unit 数字接口组件

Display Interface Unit 显示接口组件

**DIV** Diversion Message 飞机改降文电

Diversion 转换,转向,变更

Divert or Diverting or Diversion 改航

Diverter 分流,转向器

Divide 划分,除,分开

Division Air Defense 防空部

Division 界分,分界处,部门

Flight Diversion 航班转换,改航

**DIVD** Divide 分开,分为

**DIVOT** Digital-to-Voice Translator 数字—语音转换器

**DIVT** Divert 改变,改航

**DJ** Drill Jig 钻模,夹具

**DJTA** Development Joint Test Assembly 研制联合测试组件

**DJTU** Development Joint Test Unit 研制联合测试组件

**DK** Dark 黑暗

**DKK** Danish Krone 丹麦克朗(货币单位)

**DKPG** Depth Keeping 保持深度

**DKR** Dakar, Senegal-Yoff ( Airport Code ) 塞内加尔达喀尔约夫机场

**DL** Damage Limitation 损坏极限

Data Language 数据语言

Data Link 数据链,数据传输线路

Data List 数据清单,数据表

Data Loader 数据装载器

Daylight 日光,白昼

Dead Light 断电灯

Dead Load 自重,静载

Delay Line 延迟线

Delay 延期,延误

Deleted 删除的

Deliver on 交付

Deliver 交付,传送

Direct Line Interphone 直通线内话机

Direct Line 直通线

Distribution List 分配清单

Double Acetate 双醋酸

Down Link 下行链路

Draw Language 制图语言

Drawing List 制图清单

Driver's License 驾驶员执照

Dynamic Link 动态链接

U. S. Dollar 美元

**DLA** Data Link Address 数据链地址

Defense Logistics Agency 美国国防后勤局

Delay Message 延误电报

Delay 延误

Douala, Cameroon ( Airport Code ) 喀麦隆杜阿拉机场

**DLAC** Data Link Applications Coding 数据链应用编码

**DLC** Data Link Control 数据链路控制

Data-Link-Connection 数据链路连接

Digital Loop Carrier 数字环路载波

Direct Lift Control 直接升力控制

Direct Load Control 直接负载控制

Dynamic Load Control 动态负载控制

**DLCI** Data Link Connection Identifier 数据链路连接标识符

Digital Loop Carrier Interface 数字环路载波接口

**DLCM** Data Loading Configuration Manager 数据装载配置管理员

**DLCP** Data-Link Control Protocol 数据链路控制协议

**DLCS** Data Loading and Configuration System 数据装载和配置系统

Direct Lift Control System 升力控制系统

**DLCU** Data Link Control Unit 数据链路控制单元

**DLD** Deadline Date 截止日期

**DLE** Data Link Entity 数据链路实体

Digital Local Exchange 本地数字交换

Distributed LAN Emulation 分布式局域网仿真

**D-Level** Depot Level 航空站等级

**DLGF** Data Load Gateway Function 数据装载网关功能

**DLGN** Delegation 代表团

**DLGT** Delegate 代表

**DLH** Direct Labor Hours 直接工时

Duluth, MN, USA ( Airport Code )美国明尼苏达州德卢斯机场

**DLI** Data Link Interface 数据链路接口

**DLIC** Data Link Initiation Capability 数据链启用功能

**DLIR** Downward-Looking Infra-Red 下视红外设备

**DLIS** Digital Link Interface Software 数字链路接口软件

**DLIVY** Delivery 交送,运送

**DLL** Data Link Layer 数据链路层
Delay Lock Loop 延迟锁定环路
Design Limit Load 设计极限载荷
Digital Leased Line 数字租用线路
Digital Local Line 数字本地线路
Dynamic Linked Library 动态链接程序库

**DLLF** Design Limit Load Factor 设计极限载荷因数

**DlM** Dimming 变暗

**DLM** Data Loader Module 数据装载模块
Dead Load Moment 恒载力矩
Depot Level Maintenance 维修基地级维修,定检

**DLMF** Depot Level Maintenance Facility 工厂级维修设施,大修级维修设施

**DLMS** Data Link Management System 数据链管理系统

**DLMU** Data Link Management Unit 数据链管理组件

**DLN** Double Loop Network 双环网络

**DLOAD** Data Loading 数据装载

**DLODS** Duct Leak and Overheat Detection System 管道渗漏和过热探测系统

**DLP** Data Link Processor 数据链接处理器
Data Link Protocol 数据链路协议
Data Load Package 数据装入程序包
Deck Landing Practice 甲板降落训练
Digital Light Projector 数字式光投影仪
Digital Line Printer 数字式串行打印机,数字式打印机

**DLPAR** Dynamic Logical Partitioning 动态逻辑分区

**DLR** Data Loader Receptable 数据装载接收器
Dealer 商人,经纪人
Delay Line Register 延迟线寄存器

Depot Labor Rate 维修基地工时价格
Disconnect Location Report 脱开位报告
District Materials Center 区域材料中心
Division Level Review 划分等级评审
Dollar 美元
Doppler Laser Radar 多普勒激光雷达

**DLRB** Data Loader Routing Box 数据装载器发送盒

**DLS** Data Link Service 数据链路服务
Data Link Subsystem 数据链子系统
Data Link System 数据链系统
Data Loader System 数据装载系统
Data Loading Selector 数据装载器选择器
DME Landing System 测距仪着陆系统
DME-Based Landing System 基于测距仪的着陆系统

**DLSAP** Data Link Service Access Point 数据链路服务接入点

**DLSDU** Data Link Service Data Unit 数据链路服务数据单元

**DLT** Data Link Terminal 数据链路终端
Delete 删除
Digital Link Tester 数字链路测试器

**DLTU** Digital Line and Trunk Unit 数字线路和中继线单元

**DLU** Data Loader Unit 数据输入器组件

**DLV** Delivery 交送,递交

**DLVR** Deliver 送交,递交

**DLVY** Delivery 交付,传送,递交

**DLW** Design Landing Weight 设计着陆重量

**DLX** Deluxe 豪华的

**DLY** Daily 每天
Delay, Delayed 延迟,延误

**DLYD** Delayed 延期的,延误的

**DLZ** Drop Landing Zone 空投区

**DM** Data Management 数据管理
Data Mining 数据挖掘
Data Module 数据模块
Data Multiplexer 数据多路复用器
Delay Modulation 延迟调制
Delta Modulation 增量调制

Deputy Manager 副经理

Design Manual 设计手册

Deutsche Mark 马克(货币单位)

Disconnected Mode 断开方式

Dispersion Management 色散管理

**DMA** Data Management Analysis 数据处理分析

Defense Mapping Agency 国防制图局

Direct Memory Access 直接存储器存储

**DMAC** Distributed Multi-Agent Coordination 分布式多代理协作

**DMAP** Data Management and Analysis Plan 数据管理与分析计划

**DMAWG** Depot Maintenance Activation Working Group 机库维修启动工作组

**DMB** Dead Main Bus 主汇流条失效

Digital Media Broadcasting 数字媒体广播

**DMBS** Digital Measuring Borescope System 数字计量式管道镜系统

**DMC** Data Module Code 数据模块编号

Data Multiplex Channel 数据复用信道

Design Modification Committee 设计改装委员会

Digital Media Center 数字媒体中心

Direct Maintenance Cost 直接维修费用,直接维修成本

Direct Minimum Check 方向性限额检查

Display Management Computer 显示管理计算机

**DMCI** Digital Media Control Interface 数字媒体控制接口

**DMCWS** Distributed Multimedia Cooperative Writing System 分布式多媒体协同编著系统

**DMD** Diagram Manual Drawing 线路手册图纸

Diamond 钻石

Differential Mode Delay 差分模式时延

Digital Message Device 数字信息设备

Digital Micromirror Device 数字式微型镜器件

**DMDLTR** Demodulator 解调器,检波器

**DMDOR** Designated Modification Design Organization Representative 改装设计委任代表

**DMDT** Diagnostic Model Development Tool 诊断模型开发工具

**DME** Depot Maintenance Equipment 机库维修设备

Design Margin Evaluation 设计安全裕度评价

Designated Machanic Examiner 委任机械师考核员

Designated Medical Examiner 指定体检医生

Digital Multiplex Equipment 数字复用设备

Direct Measurement Explorer (空间)直接测量探测器

Distance Measuring Equipment 测距仪,测距设备

Distance Monitoring Equipment 远距离监控装备

Distributed Management Environment 分布式管理环境

Moscow, Russia-Domodedovo (Airport Code) 俄罗斯莫斯科多莫杰多沃机场

**DME/N** Distance Measuring Equipment/Narrow Pulse 狭脉冲测距仪

Distance Measuring Equipment/Normal 标准测距设备

**DME/P** Distance Measuring Equipment/Precision 精密测距设备

**DME/T** Distance Measuring Equipment/Terminal 机场测距装置

**DMEA** Damage Mode and Effects Analysis 损坏模式和影响分析

Defence Micro-Electronics Activity 国防微电子活动

Distance Measuring Equipment Antenna 远距测量设备天线

**DMEI** DME Interrogator 测距仪应答器

**DMEP** Data Management Entry Panel 数据管理引进板

**DME-P** DME Precision 精确测距装置

**DMET** Defense Management Educating and Training 国防管理教育与训练
Distance Measuring Equipment Touch-Down 测距装置接地
Distance Monitoring Equipment TACAN 塔康距离监控器

**DMF** Data Models and Formats 数据模型和格式
Dispersion-Managed Fiber 色散管理光纤
Dual-Mass Fly Wheel 双质量飞轮

**DMG** Damage 损坏

**DMGC** Digital Map Generation Computer 数字地图产生计算机

**DMGD** Damaged 被损坏的

**DMGS** Digital Map Generation System 数字地图产生系统

**DMH** Decimeter Height Finder 分米波测高器

**DMHS** Data and Message Handling System 数据和信息处理系统

**DMI** Deferred Maintenance Item 推迟维修项目,保留维修项目

**DMIR** Data Management and Information Retrieval 数据管理与信息检索
Data Mining and Information Retrieval 数据挖掘与信息检索
Designated Manufacturing Inspection Representative 委任的制造检验代表

**DMIS** Distributed Multimedia Information System 分布式多媒体信息系统

**DMK** Distributed Multimedia Kiosk 分布式多媒体触摸屏服务

**DML** Data Manipulation Language 数据操作语言
Demolish 拆坏

**DMLS** Doppler Microwave Landing Guidance 多普勒微波着陆导航,多普勒微波降落引导
Doppler Microwave Landing System 多普勒微波着陆系统

**DMM** Data Memory Module 数据存储组件

Damman, Saudi Arabia-King Fahad (Airport Code) 沙特阿拉伯达曼法赫德国王机场
Digital Multimeter 数字万用表

**DMMH** Direct Maintenance Man Hour 直接维修工时

**DMMM** Direct Maintenance-Man-Minutes 直接维修人工分钟

**DMN** Data Multi-address Network 数据多址网络
Dimension 量纲,维,因次,尺度,尺寸
Dimensional 量纲的,维的,因次的
Distributed Memory Network 分配的存储器网络

**DMO** Dependent Meteorological Office 附属气象台
Diode Microwave Oscillator 二极管微波振荡器
Domodedovo, Moscow, Russia (Airport Code) 俄罗斯莫斯科多莫杰多沃机场

**DMOS** Diffusive Mixing of Organic Solutions 有机溶液扩散混合实验

**DMP** Display Management Processor 显示管理处理器
Dump 倾倒,自卸

**DMPDD** Depot Maintenance Planning Data Document 机库维修计划数据文件

**DMPG** Dumping 倾倒,倾销

**DMPP** Display and Multi-Purpose Processor 显示和多功能处理器

**DMPR** Damper 阻尼器

**DMR** Data Management Routines 数据常规管理
Defective Material Report 材料缺陷报告
Dimmer 减光器
Dual-Mode Ramjet 双模发动机
Dual-Mode Receiver 双模接收器

**DMRS** Data Management & Retrieval System 数据管理与检索系统
Direct Mode Rate Sensor 直接模式速率传感器
Dual Mode Radar System 双模雷达系统

**DMS**　Daily Management System 日常管理系统

Data Maintenance Systems 数据维修系统

Data Management System 数据管理系统

Data Memory Structure 数据存储器结构

Data Memory Subsystem 数据存储分系统

Data Monitoring System 数据监视系统

Data Multiplexing System 数据复用系统

Debris Monitoring Sensor 碎片监测传感器

Debris Monitoring System 碎片监控系统

Defense Management Subsystem 防御管理分系统

Demand Multimedia System 多媒体点播系统

Design Maneuvering Speed 设计机动速度

Deviation Mean Standard 平均标准偏差

Digital Mapping System 数字绘图系统

Digital Measuring System 数字式测量系统

Disc Monitor System 硬盘监控系统

Display Management System 显示管理系统

Distributed Multimedia Service 分布式多媒体服务

Document Management System 文件管理系统

Doppler Measurement System 多普勒测量系统

Dynamic Mission Simulator 动态任务模拟机

Dynamo Management System 电机管理系统

**DMSD**　Digital Multi Standard Decoder 数字多标准译码器

Digital Multi Standard Decoding 数字多制式解码

**DMSH**　Diminish 缩小

**DMSP**　Defense Meteorological Satellite Program 国防气象卫星计划

**DMSS**　Data Multiplexing Subsystem 数据复用子系统

**DMST**　Demonstration 展示,演示

**DMT**　Data Monitor Tool 数据监控工具

Design Maintenance Test 设计维护测试

Design Management Team 设计管理组

Design Maturity Test 设计成熟度测试

Discrete Multitone 离散多音复用

**DMTI**　Digital Moving Target Indicator 数字移动目标显示

**DMU**　Data Management Unit 数据管理装置

Decision Making Unit 决策装置(单元)

Detector Monitoring Unit 传感器监测装置

Digital Measurement Unit 数字式测量装置

Digital Memory Unit 数字式记忆装置

**DMUX**　Demultiplexer 信号分离器

**DMZ**　Demilitarized Zone 非军事区

**DN**　Destination Network 目标网络

Directory Number 电话簿号码

Distributed Network 分布式网络

Distribution Network 分配网

Domain Name 域名

Down 下面,向下,(口语)停飞,禁止飞行

Downstream Node 下行节点

Dozen 一打

**DNA**　Air Navigation Directorate 空中航行管理局

Data Network Address 数据网络地址

Defense Nuclear Agency 国防核工业局

Destination Node Address 目标节点地址

Digital Network Architecture 数字网络结构

Directorate of Naval Aviation 海军航空局

Distributed Network Architecture 分布式网络结构

Does Not Apply 不适用,不能应用

Domain Name Address 域名地址

**DNBR**　Dialing Number 拨号号码

**DNC**　Direct Numerical Control 直接数字控制

Dynamic Network Collection 动态网络收集

**DNCC**　Data Network Control Center 数据网络控制中心

Develop New Cabin and Cargo 开发新的客舱和货舱

**DNCS**　Distributed Network Control System 分布式网络控制系统

**DN-CTL**　Down Control 向下控制

**DNDFT**　Down Draft 下沉气流

**DNE**　Danger 危险

Dangerous 危险的

Data Network Element 数据网络单元

Data Network Equipment 数据网络设备

Dedicated Network Environment 专用网络环境

Digital Network Engineer 数字网络工程师

Digital News Gathering 数字新闻采访

Digital Noise Filter 数字噪音过滤器

Do Not Enter 请勿进入

**DNGR** Danger 危险

Dangerous 危险的

Downgrade 降级

**DNH** Dunhuang 敦煌

**DNHR** Dynamic Non-Hierarchical Routing 动态无级路由选择

**DNI** Data Network Interface 数据网接口

Desktop Network Interface 桌面网络接口

Digital Non-Interpolated 数字非内插

Distributed Network Interface 分布式网络接口

Drawing Numerical Index 绘图数字索引

**DNIC** Data Network Identification Code 数据网标识码

**DNK** Denmark 丹麦

Dniepropetrovsk, Ukraine (Airport Code) 乌克兰第聂伯罗彼得罗夫斯克机场

**DNL** Daily News Live 每日直播新闻

Day-Night Average Sound Level 日夜平均噪声级

Differential Non-Linearity 微分非线性

Dynamic Noise Limiter 动态噪声限制器

**DNLK** Down Lock 放下锁

**DNM** Distributed Network Management 分布式网络管理

**DNMA** D590 New Model Aircraft D590 新型号飞机

**DNMEP** Data Network Modified Emulator Program 数据网络改进型仿真程序

**DNP** Digital Net Work Panel 数字化中枢控制板

**DNR** Data Not Received 未收到数据

Digital Noise Rejection 数字噪声抑制

**DNS** Data Network Service 数据网络业务

Distributed Network System 分布式网络系统

Domain Name Server 域名服务器

Domain Name Service 域名服务

Domain Name System 域名系统

Doppler Navigation System 多普勒导航系统

**DNSS** Defense Navigation Satellite System 防御导航卫星系统

Doppler Navigation Satellite System 多普勒导航卫星系统

**DNTK** Downtrack 下航迹

**DNTKFX** Downtrack Fix 下航迹定位点

**DNTS** Data Network Test System 数据网络测试系统

**DNZ** Denizli, Turkey (Airport Code) 土耳其代尼兹利机场

**DO** Data Output 数据输出

Delivery Order 交货单

Design Office 设计办公室

District Office 地区办事处

Ditto 同上

Document 记录(文件)

**DOA** Data out from the A Register Instruction A 指令寄存器数据输出指令

Date of Appointment 聘任日期

Date of Arrival 到达日期

Date of Availability 有效期,可用性

Degree of Angle 角的度数

Delegation of Authority 授权

Delegation Option Authorization 委任代表权

Description of Aircraft 飞机说明书

Design Organization Approval 设计机构批准书

Dictionary of Acronyms 缩写词词典

Dominant Obstacle Allowance 飞越主要障碍物高度

**DOAL** Directorate of Airlift 空运局,空运处

**DOB** Date of Birth 出生日期

**DOC** Data Optimizing Computer 数据优化计算器

Department of Commerce 商务部

Department of Communications 通信部

Designated Operational Coverage 指定操作范围

Direct Operating Cost 直接使用费用,直接营运成本

U. S. Department of Commerce 美国商业部

**DOCCHG** Document Charge 文件费

**DOCT** Doctor 博士,医生

**DOCU** Document 文件

**DOD** Data on Demand 按需提供数据

Department of Defense (美国)国防部

Direct Outward Dialing 国外直拨电话

Direct Outward Dial 国外直拨电话

Domestic Object Damage 国内目标损坏

**DOE** Department of Energy 能源部

**DOF** Degree-of-Freedom 自由度

Direction of Flight 飞行方向

**DOG** Days of Grace 宽限日期

**DOH** Doha, Qatar (Airport Code) 卡塔尔多哈机场

**DOI** Department Operating Instruction 部门运行指令

Descent/Docking Orbit Insertion 降落(或对接)入轨

Descent Orbit Insertion 进入下降轨道

Digital Operation Interpreter 数字操作翻译机

Docking Orbit Insertion 进入对接轨道

**DOJ** Department of Justice 司法部

**DOL** Deployed Operating Location 展开的运作位置

Direct-on-Line 直接接通到同路上的,直接联机

Dispersed Operating Location 分散的操作站

Dollars 美元

Dynamic Octal Load 动态八进制负载

U. S. Department of Labor 美国劳工部

**DOM** Design Organisation Manual 设计机构手册

Director of Maintenance 维修主管

Document Object Model 文档对象模型

Domestic 国内的,本地

**DOMF** Domestic Fare 国内票价

**DOMSAT** Domestic Satellite 国内卫星

**DOMSVC** Domestic Service 国内航线(服务),国内航班(服务)

**DON** Doppler Optical Navigation 多普勒光导航

**DONA** Decentralized Open Network Architecture 分散开放式网络体系结构

**DONS** Doppler Optical Navigation System 多普勒光学导航系统

**DOORS** Dynamic Object Oriented Requirements System 动态目标导向需求系统

**DOP** Developing-out Paper 想像纸

Dilution of Precision 精度扩散因子

Division Operation Plan 部门工作计划

Doppler 多普勒(导航设备)

**DOPS** Department of Public Safety 公共安全部

**DOR** Date of Request 申请日期

Digital Output Relay 数字式输出中继

**DORA** Directorate of Research and Analysis 研究与分析管理局

**DORAN** Doppler Range And Navigation 多普勒测距与导航

**DOS** Daily Operating Schedule 每日工作计划

Date of Sale 出售日期

Date of Shipment 装船日期

Date of Supply 供应日期

Denial of Service 拒绝服务

Disk Operating System 磁盘操作系统

Display Operator Console 显示器操作台

Distributed Operating System 分配操作系统

**DOT** Date of Termination 终止日期

Department of Transport 运输部

Designated Optical Tracker 指定光学跟踪装置

Differential Oil Temperature 滑油温差

Document 文件

**DOTS** Dynamic Ocean Tracking System 动态海洋跟踪系统

**DOTWCC** Declaration of the World Climate Conference 世界气候会议宣言

**DOUB** Double 两倍

**DOV** Data Over Voice 话音传数据

**DOW** Dry Operating Weight 干使用重量

**DOZ** Dozen 打(12 个)

**DP** Data Processing 数据处理

Decision Point 决断点

Deep 深

Departure Procedure 离场程序

Departure 起飞,离场

Detection Point 检测点

Dew Point Temperature 露点温度

Dew Point 露点

Differential Pressure 压差

Differential Protection 差动保护

Display Panel 显示面板

Distribution Point 分配点

Duty Paid 付过税的

**DPA** Data Processing Activity 数据处理活动

Data Processing Area 数据处理范围

Dependent Parallel Approach 相关平行进近

Design Proposal Approval 设计建议的批准

Destructive Parts Analysis 破坏性零件分析

Destructive Physical Analysis 破坏性物理分析

Digital Pre-Assembly 数字式预汇编,数字式预组装

Direct Planning Action 直接下达的任务

Drawing Proposal Authorization 制图建议批准

Dual Processing Authorization 双型处理审核

**DPC** Data Processing/Process Center 数据处理中心

**DPCA** Dynamic Principal Component Analysis 动态主元分析

**DPCC** Data Processing Control Center 数据处理控制中心

Data Processing Control Console 数据处理控制台

**DPCCH** Dedicated Physical Control Channel 专用物理控制信道

**DPCM** Differential Pulse Code Modulation 差分脉码调制

**DPCN** Digital Private Circuit Network 数字专用电路网

**DPCT** Differential Protection Control Transformer 差动保护控制变压器

Differential Protection Current Transformer 差动保护电流变压器

**DPCU** Digital Passenger Control Unit 数字式旅客控制组件

Digital Processing and Control Unit 数字处理与控制装置

**DPCV** Differential Pressure Control Valve 压差控制阀(活门)

Dual Plate Check Valve 双板止回阀

**DPD** Data Processing Digest 数据处理文摘

Data Product Definition 数字产品定义

Department of Planning and Development 规划与发展司

**DPDCH** Dedicated Physical Data Channel 专用物理数据信道

**DPDS** Data Processing and Display System 数据处理与显示系统

**DPDT** Double-Pole, Double-Throw 双极,双掷开关

**DPE** Designated Pilot Examiner 委派的飞行员考试官,指定的飞行员考试官

Distributed Processing Environment 分布式处理环境

**DPF** Data Processing Function 数据处理功能

**DPFS** Data Processing and Forecasting System 资料加工和预报系统

**DPG** Digital Pair Gain 数字线对增益

**DPGS** Data Processing Ground Station 地面数据处理站

**DPHCH** Dedicated Physical Channel 专用物理信道

**DPI** Data Processing Installation 数据处理设

备

Differential Pressure Indicator 压差表,差压指示器

Dot per Inch 每英寸点数,每英寸可打印的点数

**DPKO** Department of Peacekeeping Operations 维持和平行动部

**DPL** Data Protection Layer 数据保护层
Detailed Parts List 详细零部件清单

**DPLR** Doppler 多普勒(导航设备)

**DPLY** Deeply 深深地

**DPM** Design Progress Meeting 设计进程会议

**DPMA** Data Processing Management Association 数据处理管理协会

**DPMO** Defects Per Million Opportunities 每百万次缺陷次数

**DPN** Dual-Private Network 双向专用网络
Dynamical Path Network 动态路径网

**DPON** Domestic Passive Optical Network 国内无源光网络

**DPPM** Differential Pulse Position Modulation 微分脉冲位置调制

**DPR** Differential Pressure Recorder 差压记录器
Differential Protection Relay 差动保护继电器

**DPRK** Democratic People's Republic of Korea 朝鲜民主主义人民共和国

**DPRM** Delivery Planning Review Meeting 交付计划评审会议

**DPRT** Depart 离场,离开

**DPS** Data Processing Service 数据处理服务
Data Processing System 数据处理系统
Denpasar Bali, Indonesia(Airport Code) 印度尼西亚巴里岛机场(机场代码)
Descent Propulsion System 下降推进系统
Distributed Packet Switching 分布式分组交换
Douglas Process Standard 道格拉斯处理标准
Dynamic Packet State 动态分组状态

**DPSK** Differential Phase Shift Keying 差分相移键控
Digital Phase Shift Keying 数字移相键控

**DPST** Deposit 存储,寄存,付(保证金)
Double Pole Single Throw 双极单置

**DPT** Depart 离场,起飞
Department 部门,部
Deposit 存款
Depth 深度
Dynamic Packet Transport 动态分组传送

**DPTH** Depth 深

**DPTR** Departure 离场,起飞

**DPTV** Digital Particle Tracking Velocimetry 数字化粒子迹线测速技术

**DPTY** Deputy 副的,代理人

**DPU** Data Processing Unit 数据处理装置
Detector Powering Unit 传感器动力装置
Digital Processor Unit 数字处理器单元
Distributed Processing Unit 分布式处理单元

**DPV** Directional Pilot Valve 导向活门

**DPX** Duplex 双工的,双重的

**DQA** Data Quality Assessment 数据质量评估
Data Quality Assurance 数据质量保证
Design Quality Assurance 设计质量保证

**DQCM** Data Quality Control Monitor 数据质量控制监视器

**DQDB** Distributed Queue Dual Bus 分布式队列双总线

**DQE** Detective Quantum Efficiency 探测器量子效率

**DQM** Data Quality Management 数据质量管理
Data Quality Monitoring 数据质量监测

**DQN** Design Query Note 设计查询注释

**DQPSK** Differential Quadrature Phase Shift Keying 差分四相相移键控

**DQV** Design Quality Verification 设计质量验证

**DR** Dead Reckoning 推测领航
Deficiency Report 缺陷报告,故障报告
Deflection Routing 改向路由选择

Deviation Range 偏移范围

Direct Route 直达路由

Director 处长,理事

Discrepancy Report 偏差范围,偏离报告

Dispatch Release 签派放行

Dispatch Reliability 放行可靠性

Distance Remaining 剩余距离

Doctor 医生,博士

Door 门

Drive 驱动,推进,开车

Dynamic Routing 动态选路

**DRA** Dead Reckoning Analyzer 推算分析器,领航分析器

Delayed Departure due to … 由于…原因延误起飞

**DRAF** Data Reduction and Analysis Facility 数据简化与分析设备

**DRAI** Dead Reckoning Analyzer Indicator 推测分析器指示器

**DRAM** Dynamic Random Access Memory 动态随机存取存储器

**DRAO** Design Requirement and Objectives 设计要求与目标

**DRAS** Door and Ramp Actuation System 舱门与登机梯作动系统

**DRAT** Operational Readiness and Airport Transfer 运行准备和机场移交

**DRCT** Direct 直接的,直飞的

**DRCTY** Directly 直接地

**DRDB** Dual Redundant Data Bus 双余度数据总线

**DRDP** Detection Radar Data Processing 探测雷达数据处理

**DRDPS** Digital Radar Data-Processing System 数字雷达数据处理系统

**DRDT** Digital Radar Data Transimission 数字雷达数据传输

**DRDTO** Detection Radar Data Takeoff 按探测雷达数据起飞

**DRE** Dead Reckoning Equipment 推测领航设备

Delayed Departure due to Crew 由于机组原因延误起飞

**DREM** Distance Remaining 剩余距离,待飞距离

**DRF** Directional Radio Frequency 定向射频

**DRFM** Digital Radio-Frequency Memory 数字射频存储装置

**DRFT** Draft 草稿,草案

Drift 漂移,偏航

**DRG** Drag Force 阻力

Drawing 绘画,提款

**DRI** Dead Reckoning Indicator 推测领航指示器

Disaster Recovery Infrastructure 灾案恢复基础设施

Dynamic Response Index 动态响应指数

**DRID** Destination Routing Identifier 目的路由选择标识符

**DRIP** Dead Reckoning Information Processor 推测领航信息处理器

Design Response Intended for Playback 征求反馈设计提案

Dispatch Reliability Improvement Program 签派可靠性改进大纲

**DRIPS** Dynamic Real-Time Information Processing Systems 动态实时信息处理系统

**DRK** Dark 黑暗

**DRL** Data Requirements List 数据需求表

Data Retrieval Language 数据检索语言

Daytime Running Lamps 日间运转灯,白天行驶灯

Defense Research Laboratories 国防研究实验室

Direct Ranging LORAN 直接测距罗兰系统

Directional Reference Locator 方向基准定位器

Documentation Requirement List 文件要求清单

Drill 钻头,钻孔

**DRLMS** Digital Radar Landmass Simulation 数字雷达大陆模拟

**DRM** Delayed Departure due to Technical Reason 因技术原因而延误起飞

Digital Rights Management 数字信息版权管理

Disk Resoure Management 磁盘资源管理

Drawing Room Manual 绘图室手册

Duct Repair Manual 管道修理手册

**DRMI** Distance Radio Magnetic Indicator 无线电磁式距离指示器

**DRMS** Data Record Management System 数据记录管理系统

Data Resource Management System 数据资源管理系统

Defense Resource Management System 国防资源管理系统

Distance Root Mean Square 距离均方根值

**DRN** Directorate of Radio Navigation 无线电导航处

Document Release Notice 文件发放通告

Document Revision Notice 文件修改通知

**DRNG** During 在…期间

**DRNS** Doppler Radar Navigation Set 多普勒雷达导航装置

**DRO** Data Readout 数据读出

Delayed Departure due to ATC 因空中交通管制原因而延误起飞

Durango, CO, USA（Airport Code）美国科罗拉多州杜兰戈机场（机场代码）

**DRP** Delayed Departure due to Waiting for Passenger 由于等待旅客而延误起飞

Disaster Recovery Plan 灾难恢复计划

Distribution Resource Planning 分销资源计划,配送资源计划

Dynamic Routing Protocol 动态路由协议

**DRR** Delayed Departure due to Lack of Stand-by Aircraft 由于无备用飞机而延误起飞

Data Reduction System 数据简化系统

Data Relay Satellite 数据中继卫星

Digital Radio System 数字无线电系统

Digital Recording System 数字记录系统

Document Retrieval System 文件检索系统

Doors 舱门

Doppler Radar System 多普勒雷达系统

Dresden, Germany（Airport Code）德国德累斯顿机场

**DRSM** Dynamic RPV System Model 动态遥控飞行器系统模型

**DRSN** Drifting Snow 飘雪

**DRSR** Drive Running Slave Relay 驱动运行从继电器

**DRSS** Data Relay Satellite System 数据中继卫星系统

**DRT** Dead Reckoning Tracer 推测（跟踪）描图仪

Del Rio, TX, USA（Airport Code）美国得克萨斯州德尔里奥机场

Delayed Departure due to Weather 由于天气原因而延误起飞

Direct 直接,直飞

**DRU** Data Recovery Unit 数据回收装置,数据恢复装置

Data Retrieval Unit 数据复原单元

Desert Rescue Unit 沙漠援救队

Digital Register Unit 数字寄存装置

Digital Remote Unit 数字遥控组件

Direct Reporting Unit 直接报告单元

Drive Unit 传动装置,驱动装置,激励装置

Drum Unit 磁鼓

Dynamic Reference Unit 动态基准组件

**DRV** Developmental Reentry Vehicle 试验性再入飞行器

Double Regulating Valve 双重调节阀门

**DRVG** Driving 驱动

**DRVN** Driven 驱动的

**DRVR** Driver 驱动器

**DRW** Darwin, Northern Territory, Australia（Airport Code）澳大利亚北领地达尔文机场

Drawing 绘图,提款

**DRZ** Delayed Departure due to Pantry Catering 由于食品供应原因而延误起飞

**DRZL** Drizzle 毛毛雨

**DS**　Damaged Seal 损坏的密封

　　Data of Service 维修(服务)数据

　　Data Set 数据组

　　Data Sheet 数据表

　　Data Stream 数据流

　　Data Switching 数据交换

　　Data System 数据系统

　　Decade Scaler 十进制换算电路

　　Decanning Scuttle 非密封舱口

　　Decision Science 决策科学

　　Departure Station 起飞站,出发站

　　Design Specification 设计规范

　　Design Standards 设计标准

　　Destination Select 目的地选择

　　Destination Station 终点站,目的地机场

　　Dial Service 自动电话业务

　　Dial Setting 拨号装置

　　Differentiated Service 区分业务

　　Digital Sequence 数字序列

　　Diode Switch 二极管开关

　　Direct Sequence 直接序列

　　Direct Shipment 直接装运

　　Direct Support 直接支援,直接支承

　　Directionally Solidified 定向固化,定向结晶

　　Directory Services 目录服务

　　Discard 报废

　　Discarding Sabot 报废的衬套

　　Disconnect Switch 断开电门

　　Disconnecting Switch 切断开关

　　Discrete Signal 离散信号

　　Dispersion Shift 色散位移

　　Documentation Storage 文件存储

　　Dust Storm 尘暴

**DSA**　Data Set Assembly 数据组汇编

　　Data System Architecture 数据系统体系结构

　　Dataroute Serving Area 数据供应服务区

　　Decision Science Applications 决策科学应用

　　Decision Support Assistant 辅助决策

　　Deep-Space Antenna 深空天线

　　Defense Supply Agency 国防供应局

　　Defense System Analysis 防御系统分析

　　Defense System Avionics 防御系统航空电子设备

　　Design Schedule Analysis 设计进度(表)分析

　　Development Signature Approval 研制工作签字批准

　　Digital Signature Algorithm 数字签名算法

　　Directorate of Aviation Safety 飞行安全局(英国)

　　Directory System Agent 目录系统代理

　　Distributed System Architecture 分布式系统结构

　　Doppler Spectrum Analyzer 多普勒频谱分析仪

　　Dynamic Signal Analyzer 动态信号分析仪

　　Dynamic Storage Area 动态存储器面积(区域)

**DSAA**　DECT Standard Authentication Algorithm DECT 标准认证算法

**DSABL**　Disable 无能力的,故障的

**DSAL**　Decelerated Steep Approach and Landing 减速急剧下降进近着陆

**DSAMA**　Dynamic Slot Allocation Multiple Access 动态时隙分配多址接入

**DSAP**　Data Link Service Access Point 数据链路业务接入点

**DSAPP**　Disappearance 消失

**DSAR**　Data Storage and Retrieval 数据存储与检索

　　Document Storage and Retrieval 文件存储与检索

**DSARC**　Defense Systems Acquisition Review Council 国防系统采办审查委员会

**DSASBL**　Disassemble 拆散

**DSB**　Defense Science Board 防御科学局

　　Direct Satellite Broadcast 直接卫星广播

　　Display Site Equipment 位置显示设备

　　Double Side Band 双边带

**DSB – AM**　Double Sideband Amplitude 双边带调幅

**DSBG**　Disbursing 支付

**DSBL** Disable 无能力的,故障的,不适用的

**DSBS** Disburse 支付的

**DSBSC** Double Side-Band Suppressed Carrier 双边带抑制载波

**DSC** Decision Support Center 决策支持中心
Direct Satellite Communications 直接卫星通信

**DSCA** DECT Standard Cipher Algorithm DECT 标准密码算法

**DSCD** Discard 报废

**DS-CDMA** Direct Sequence CDMA 直接序列 CDMA

**DSCF** Dispersion Slope Compensating Fiber 色散斜率补偿光纤

**DSCNCT** Disconnect 断开

**DSCR** Debt Service Coverage Ratio 偿债率

**DSCRM** Discriminator 鉴别器,鉴(相,频)器

**DSCRP** Description 描写,说明书

**DSCRT** Discrete 离散的,不连续的

**DSCRT IO** Discrete I/O Error 离散输入/输出偏差

**DSCS** Data Sharing Control System 数据共享控制系统
Defense Satellite Communications System 防御卫星通信系统
Doors and Slides Control System 舱门和撤离滑梯控制系统
Doors Slide Control System 侧滑门控制系统

**DSCTN** Discontinue 停止,中断

**DSCV** Discover 发现

**DSDD** Double Sided Double Density 双边双密度

**DSDL** Dedicated Serial Data Link 专用系列数据链,专用串行数据链
Digital Systems Design Laboratory 数字系统设计实验室

**DSDU** Data Signal Display Unit 数据信号显示单元
Digital Signal Display Unit 数字信号显示装置

**DSE** Data Storage Equipment 数据存储设备
Data Switching Equipment 数据交换设备
Data Switching Exchange 数据交换机
Digital Switching Equipment 数字式交换机

**DSEB** Digital Seat Electronic Box 数字式座椅电子盒

**DSEC** Data Systems Evaluation Center 数据系统评估中心

**DSENGD** Disengaged 脱开

**DSF** Design Safety Factor 设计安全系数
Dispersion Shifted Fiber 色散位移光纤
Display System Function 显示系统功能

**DSG** Design Service Goal 设计服务目标

**DSGN** Design 计划,设计

**DSGNT** Designate 指定

**DSI** Digital Speech Interpolation 数字话音内插
Discrete Input 离散输入
Divertless Supersonic Inlet 无边界层隔道超音速进气道

**DSIP** Development Support and Integration Program 供应发展综合规划
Domestic Science Information Program 国内科学情报计划
Door/Slide Indication Panel 舱门/滑梯指示面板

**DSL** Data Set Label 数据集符号
Deep Scattering Layer 深放射层
Depot Storage Level 仓库储存水平
Design Safety Limit 设计安全极限
Diesel 内燃机,柴油机
Digital Service On-Line 联机数字业务,联机码业务
Digital Simulation Language 数字模拟信号,数字模拟语言
Digital Subscriber Line 数字用户线路
Direct Static Logic 直接静态逻辑

**DSLAM** Digital Subscriber Line Access Multiplexer 数字用户线接入复用器

**DSM** Demand Side Management 需求方管理
Des Moines, IA, USA (Airport Code) 美国

依阿华州得梅因机场

Digital Switch Module 数字交换模块

Direct-Sequence Modulation 直接序列调制

Distributed Storage Management 分布式存储管理

Dynamic Single Mode 动态单模

**DSMA** Digital Sense Multiple Access 数字侦听多重访问

**DSMC** Direct Seat-Mile Cost 座英里直接成本

**DSMCU** Doors and Slides Management Control Unit 舱门与滑梯管理控制组件

**DSMD** Dismissed 解雇, 解散

**DSMS** Doors and Slides Management System 舱门与滑梯管理系统

**DSMTD** Dismounted 拆卸的

**DSN** Defense Switching Network ( Telephone ) 国防转换网络(电话)

Design 设计

Digital Switching Network 数字交换网络

Distributed Switching Node 分布式交换节点

**DSO** Discrete Output 离散输出

**DSOT** Daily System Operational Test 系统的操作准备状况每日检验

**DSP** Data Link Service Provider 数据链服务提供商

Departure Sequencing Program 起飞排序计划, 离港排序计划

Design Services Procedures 设计服务程序

Digital Signal Processing 数字信号处理

Digital Signal Processor 数字信号处理器

Directory System Protocol 目录系统协议

Dispatch 调度, 遣派

Dispatcher 调度员, 签派员

Display Select Panel 显示选择面板

Domain Specific Part 域专用区

**DSPL** Display 显示, 展示

Disposal 处理, 配置

**DSPN** Disposition 处理, 配置

**DSPS** Digital Singal Process System 数字信号处理系统

**DSPT** Descent Point 下降点

**DSPY** Display 显示

**DSR** Desire 愿望, 要求

Desired 所需的, 打算的, 要求的

Digital Stepping Recorder 数字步进记录器

Display System Replacement 显示系统替代

Dynamic Source Routing 动态源路由

**DSRC** Dedicated Short Range Communication 短距离微波通信

**DSRR** Disregard 不管, 不予理睬

**DSRS** Data Signal Rate Selection 数据信号速率选择

**DSRTK** Desired Track Angle 所需航迹角

**DSS** Data Select Switch 数据选择电门

Data Storage System 数据存储系统

Decision Support System 决策支持系统

Deep Space Station 深空站, 航天站, 太空站

Digital Signature Standard 数字签名标准

Digital Subscriber Service 数字用户业务

Direct Satellite System 直播卫星系统

Directory Service System 目录服务系统

**DSS1** Digital Subscriber Signaling System No. 1 1 号数字用户信令系统

**DSS2** Digital Subscriber Signaling System No. 2 2 号数字用户信令系统

**DSSG** Decision Support System Generator 决策支持系统生成器

**DSSMAN** Data Service Specific MAN 数据业务专用城域网

**DS-SMF** Dispersion Shifted Single Mode Fiber 色散位移单模光纤

**DSSP** Deep Submergence System Program 深潜系统程序

**DSSS** Direct Sequence Spread Spectrum 直接顺序发散频谱

**DSSSMA** DSSS Multiple Access 直接序列扩频多址接入

**DST** Daylight Saving Time 日光节约时间, 夏令时间(将时针拨快一小时)

Desktop Simulation Trainer 台式(计算机)模拟训练器

Destroy 毁坏

Dispersion Supported Transmission 色散支持传输

Display Storage Tube 显示存储管

Division for Science and Technology (UNCTAD) 科学和技术司

Dynamic Soliton Transmission 动态孤子传输

**DSTM** Dynamic Synchronous Transfer Mode 动态同步转移模式

**DSTN** Destination 目的地,降落站

**DSTO** Defense Science and Technology Organization 防御科学和技术组织

**DSTR** Distance Remaining 待飞距离,剩余距离

Dual Stage Target Recognizer 双级目标识别器

Dynamic System Test Rigs 动态系统试验设备

**DSTRK** Desired Track 所需航迹

**DSTS** Data Storage and Transfer Set 数据贮存与转换装置

Digital-Signal Transfer Unit 数字信号转换装置

**DSU** Data Service Unit 数据服务单元

Data Storage Unit 数据存储装置

Data Switch Unit 数据交换单元

Data Synchronization Unit 数据同步装置

Deployment Sensor Unit 部署传感器单元

Device-Switching Unit 装置转换器件,设备开关装置

Digital Service Unit 数字业务单元

Digital Switching Unit 数字交换单元

Direct Support Unit 直接保障单位

Disk Storage Unit 磁盘存储装置

Dynamic Sensor Unit 动态传感装置

**DSUG** Drawing Set User Guide 绘图仪使用指南

**DSV** Data Steal into Voice 数据插入语音

**DSVD** Digital Simultaneous Voice & Data 语音和数据同时数字传输

**DSVMA** Data Steal into Voice Multiple Access 数据插入语音多址接入

**DSW** Door Switch 门开关

**DSX** Digital Singals Cross-connect 数字信号交叉连接

**DT** Data Tape 数据磁带

Data Terminal 数据终端

Data Tip 数据上端

Data Transfer 数据传送

Data Transmission 数据传输

Data Transmittal & Information Form 数据发送和信息格式

Data Transmittal & Information 数据发送和信息

Data 数据

Date 日期

Daylight Time 夏令时间

Decay Time 衰变时间

Dedicated Tanker 专用加油机

Deep Tank 深油箱

Delay Time 延迟时间

Delivery Time 交货时间

Design Tool 设计工具

Destination Report 目的地机场报告

Development Test 研制试验

Development Testing 研制试验

Digital Terminal 数字终端

**DT&E** Design, Test & Engineering 设计、测试和工程

Detailed Test & Evaluation 详细测试和评估

Development Test and Evaluation 研制测试和评估

Developmental Test & Engineering 开发性试验和工程

**DT&OT** Developmental Test/Operational Test 开发试验/运行试验

**DTA** Damage Tolerance Analysis 损伤容限分析

Data Transfer Agreement 数据传输协议

Desired Track Angle 所需航迹角

Development Test Article 开发测试条款

Differential Thermal Analysis 热差分析

Door to Airport 用户到机场(服务)

**DTACCS** Defense Telecommunications and Command and Control Systems 防御电信、指挥与控制系统

**DTAM** Descend to and Maintain 下降至并保持

Digital Telephone Answering Machine 数字电话答录机

Document Transfer Access Method 文件传输存取方法

**DTB** Dead Tie Bus 连接汇流条断电

**DTC** Data Transfer Controller 数据传输控制器

Defense Trade Controls 国防贸易管制

Design to Cost 按成本设计，按费用设计

Domain Technology Committee 域名技术委员会

**DTCS** Data Transmission and Control System 数据传输与控制系统

Digital Telephone Circuit Service 数字电话电路业务

Digital Test Command System 数字试验指挥系统

Drone Tracking and Control System 无人机跟踪和控制系统

**DTCTR** Detector 探测器

**DTD** Damage Tolerance Design 损伤容限设计

Data Terminal Display 数据终端显示

Date Transfer Device 数据传送设备

Dated 注明日期的

Digital Terrain Data 数字地形数据

Directorate of Technical Development（英国）技术开发局

Document Type Definition 文件类型定义

Door to Door 门到门(服务)

**DTE** Damage Tolerance Evaluation 损伤容限评估

Data Terminal Equipment 数据终端设备

Data Transmission Equipment 数据传输设备

Date 日期

**DTED** Dated 注明日期的

Department of Trade and Economic Development 贸易与经济发展部

Digital Terrain Elevation Database 数字地形高程数据库

**DTEO** Defence Test & Evaluation Organization 防御测试及评估机构

**DTF** Data Test Facility 数据检测设备

Delayed Transfer 延迟转移

Digital Tape Format 数码录音格式

Digital Test Data Format 数字测试数据格式

Digital Transmission Facility 数字传输设备

Dispersion-Tailored Fiber 色散预定光纤

Dispersion-Tapered Fiber 色散锥形光纤

**DTFF** Differential 差,差值

**DTFT** Discrete Time Fourier Transform 离散时间傅立叶变换

**DTG** Date-Time Group 时序分组,日期时间分组

Distance to Go 待飞距离

**DTH** Direct to Home 直接到户

**DTI** Data Transmission Interface 数据传输接口

Department of Trade and Industry 贸易与工业部(英国)

Digital Terminal Interface 数字终端接口

Digital Transmission Interface 数字传输接口

Digital Trunk Interface 数字中继接口

**DTIC** Defense Technical Information Center 防御技术信息中心

**DTIM** Digital Transmission Interface Module 数字传输接口模块

**DTK STS** Desired Track Status 期望航迹状态

**DTL** Data Transmission Line 数据传输线路

Detail, Detailed 细节, 详细的

**DTLCC** Design to Life Cycle Cost 设计至寿命周期成本

**DTLM** Digital Trunk Line Module 数字中继线路模块

**DTM** Data Transfer Medium 数据传递装置

Data Transfer Module 数据传输模件

Diagnostic Test Mode 诊断测试模式

Digital Terrain Matrices 数字地形模型

Digital Terrain Model 数字地形模型

Digital Trunk Module 数字中继模块

Dortmund, Germany ( Airport Code ) 德国多特蒙德机场

Douglas Tooling Material 道格拉斯工装材料

Duration Time Modulation 持续时间调制

Dynamic Synchronous Transfer Mode 动态同步转移模式

**DTMB**　Defence Traffic Management Board 国防交通管理委员会

**DTMC**　Direct Ton-Mile Cost 吨英里直接成本

**DTMF**　Dual Tone Multiple Frequency 双音多频

**DTMO**　Development, Test & Mission Operations 研制、测试和飞行任务运行

**DTMS**　Damage Tolerance Monitoring System 损伤容限监控系统

Defense Traffic Management Service 国防交通管理局

Digital Test Monitor System 数字式测试监控系统

**DTMU**　Damage Tolerance Monitoring Unit 损伤容限监控组件

**DTN**　Data Transport Network 数据传输网络

**DTO**　Data Transfer Operation 数据转移操作

Derated Take off 减功率起飞

Digitally Tuned Oscillator 数字调谐振荡器

Direct Turn Over 直接翻转,直接移交

**DTOA**　Differential Time of Arrival 抵达时间差

Due Time of Arrival 正点到达时间

**DTP**　Data Transfer Phase 数据转移阶段

Data Transfer Protocol 数据转移协议

Detailed Test Plan 详细的测试计划

**DT-PDU**　DaTa Protocol Data Unit 数据协议数据单元

**DTR**　Damage Tolerance Rating 损伤容限率

Data Terminal Ready 数据终端就绪

Digital Trunked Radio 数字无线中继

**DTRIOR**　Deteriorate 变坏,品质下降

**DTRM**　Determine 决定,确定,决心

Dual-Thrust Rocket Motor 双推力火箭发动机

**DTRT**　Deteriorate 恶化,变化

Deteriorating 变坏

**DTS**　Data Test Station 数据测试站

Data Transfer Set 数据传输装置

Data Transfer System 数据传输系统

Data Transmission Subsystem 数据传输辅助系统,数据传输子系统

Data Transmission System 数据发送系统,数据传输系统

Defense Technologies Study 防御技术研究

Defense Transportation System 防御运输系统

Deferred Tooling System 延迟工装系统

Detail Test Specification 详细试验规范

Detail Type Specification 细则形式的标准规范

Development Test Satellite 研制试验卫星

Dialog Terminal System 对话终端系统

Digital Telemeter/Telemetering System 数字式遥测系统

Digital Termination Service 数字终端勤务

Documents Transmission System 文件传送系统

Dynamic Test System 动力试验系统

**DTS(W)**　Double Throw Switch 双掷开关

**DTSA**　Defense Technology Security Administration 国防技术保密局

Dynamic Time Slot Allocation 动态时隙分配

**DTSC**　Digital Transmissions Standards Committee 数字传输标准委员会

**DTSWCH**　Digital Trunk Switch 数字中继交换

**DTT**　Data Timing Transfer 数据定时传送

Data Transmission Technique 数据传输技术

Desk Top Trainer 桌面训练器

Domestic Technology Transfer 国内技术转让

Dynamic Test Target 动态试验目标

**DTTN** Digital Trade and Transportation Network 数字贸易与运输网络系统

Distributed Tactical Test Network 分布式战术试验网络

**DTU** Data Table Update 数据表更新

Data Terminal Unit 数据终端单元

Data Transfer Unit 数据传输组件

Data Transformation Unit 数据变换器

Digital Telemetry Unit 数字遥测装置

Digital Test Unit 数字式测试组件

**DTUC** Data Transfer Unit Cartridge 数据传输装置磁带

Design to Unit Cost 单位成本设计

**DTUPC** Design to Unit Production Cost 按单件生产的成本设计

**DTV** Desktop Video 桌面视频

Digital Television 数字电视

Driver's Thermal Viewer 驾驶员热视仪

Dynamic Test Vehicle 动力试验器

**DTW** Detroit, MI, USA (Airport Code) 美国密执安州底特律机场

Dry Tank Weight 无油重量

Dual Tandem Wheel Landing Gear 四轮小车式起落架

Dynamic Time Warping 动态时间规整算法

**DT-WDMA** Dynamic Time Wavelength Division Multiple Access 动态时间波分多址接入

**DTY** Duty 工作,税

**DU** Dispersion-Unshifted 非色散位移光纤

Display Unit 显示组件

Dust 尘土,灰尘

**DUA** Data Use Agreement 数据使用协议

Directory User Agent 目录用户代理

Don't Use Acronyms 不要使用缩略词

**DUAT** Direct User Access Terminal 直接用户访问终端,直接用户存取终端

**DUATS** Direct User Access Terminal Service 直接用户存取终端服务

Direct User Access Terminal System 直接用户存取终端系统

**DUB** Dublin, Ireland (Airport Code) 爱尔兰都柏林机场

**DUC** Data Utilization Center 数据利用中心

Demand Used in Computation 计算机使用要求

Dense Upper Cloud 高空浓云

**DUD** Dunedin, New Zealand (Airport Code) 新西兰达尼丁机场

**DUDB** Display Unit Data Base 显示组件数据库

**DUI** Data Unit Interface 数据单元接口

Data Use Identifier 数据使用识别器

**DUJ** Du Bois, PA, USA (Airport Code) 美国杜波依斯机场

**DUL** Design Ultimate Load 设计载荷极限

**DUM** Dummy 缓冲器

**DUP** Data User Part 数据用户部分

Destination User Prompter 目的用户提示器

Duplicate 重复,复印,复制,副本

**DUPE** Duplicate Message 复制电报

Duplicate 复制

**DUR** Duration 持续时间

Durban, South Africa (Airport Code) 南非德班机场

**DURG** During 在…期间

**DURN** Duration 持续时间

**DUS** Design Unit Specification 设计装置规格

Diagnostic Utility System 诊断实用系统

Dusseldorf, Germany (Airport Code) 德国杜塞尔多夫机场

**DUT** Dutch Harbor, AK, USA (Airport Code) 美国阿拉斯加州荷兰港机场

**DV** Design Verification 设计验证

Digital Video 数字视频

**DV(C)** Device 装置,器件

**DVB** Digital Video Broadcast 数字视频广播

**DVC** Desktop Video Conference 桌面型视频会议系统

Digital Video Compression 数字视频压缩

Digital Video Controller 数字视频控制器

**DVD** Delta Velocity Display 速度增量显示（器）

Differential Velocity Display 差动速度显示（器）

Digital Versatile Disc 数字通用光盘

Digital Video Disc 数字式影碟，数字视频光盘

Divide 分开，除以

**DVDAF** Digital Flight Data Acquisition Function 数字飞行数据采集功能

**DVDR** Digital Voice Data Recorders 数字音频数据记录器

Divide 分压器，分频器

**DVDS** Digital Video Display System 数字视频显示系统

**DVE** Digital Video Effect 数字视频效果

Distributed Virtual Environment 分布可视环境

Driver's Vision Enhancer 驾驶员视野增效器

**DVFR** Daylight Visual Flight Rule 日间目视飞行规则

Defense Visual Flight Rule 防空目视飞行规则

**DVHT** Digital Video Home Terminal 数字电视家庭终端

**DVI** Digital Video Interactive 交互式数字视频

Digital Visual Interface 数字式视觉界面

Direct Voice Input 直接语音输入

**DVIP** Designated Very Important Person 指定的非常重要的人物

Digital Video Integrating Processor 数字视频积分处理器

Distinguished Very Important Person 杰出的非常重要的人物

**DVLP** Develop 发展，扩大

**DVM** Data Voice Multiplexer 数据语音复用器

Digital Voltmeter 数字式电压表

**DV-MCI** Digital Video-Media Control Inter-

face 数字视频媒体控制接口

**DVMP** Data Voice MultiPlex 数据话音多路复用

**DVMRP** Distance Vector Multicast Routing Protocol 距离向量多目路径协议

**DVN** Digital Video Network 数字视频网

Division 部门，处，划分

**DVO** Davao, Philippines(Airport Code) 菲律宾达沃机场

Direct View Optics 直视光学仪器

Direct Voice Output 直接话音输出

**DVOR** Doppler VHF (Very-High-Frequency) Omnidirectional Range 多普勒甚高频全向信标

Doppler VHF Omnidirectional Radio Range 多普勒甚高频全向无线电信标

**DVP** Delta Velocity Planet 有速度增量的行星

Deterministic Virtual Path 确定性虚通路

Digital Video Processor 数字式视频处理机

**DVP&R** Design Verification Plan and Report 设计验证计划和报告

**DVR** Design Validation Report 设计确认报告

Driver 司机

**DVRS** Digital Voice Recorder System 数字语言录音系统

**DVS** Desktop Video Studio 桌面视频演播室

Digital Video System 数字视频系统

Doppler Velocity Sensor 多普勒速度传感器

Dynamic Vertical Sensor 动态垂直传感器

**DVST** Direct View Storage Tube 直观存储管

**DVU** Design Verification Unit 设计批准单位

**DVVI** Data Voice Video Integration 数据话音视频集成

**DVXS** Digital Visual Exchange Service 数字可视交换业务

**DW** Data Warehouse 数据仓库

Data Word 数据字

Dual Wheel Undercarriage 双向起落架

Dual Wheel 双轮

**DWC** Dubai World Central 迪拜世界中心

**DWCT** Dead Weight Cargo Tonnage 净载重吨

**DWDM** Data Warehousing and Data Mining 数据仓库和数据挖掘

Dense Wavelength Division Multiplexing 密集波分复用

**DWG** Daily Writing Group 日常写作组

Data Working Group 数据工作组

Dependability Working Group 可靠性工作小组

Designated Work Group 指定的工作小组

Drawing 图纸,制图,牵引,绘图

**DWMS** Data Warehouse Management System 数据仓库管理系统

**DWMT** Discrete Wavelet Multi-Tone 离散小波多音

**DWN** Down 向下

**DWR** Doppler Weather Radar 多普勒气象雷达

Drawing Change Request 草图变更需求

**DWRR** Dynamic Weighted Round-Robin 动态加权循环法

**DWS** Department of Water Supply 供水部

Dialable Wideband Service 可拨号的宽带业务

Drinking-Water Standards 饮用水标准

**DWT** Discrete Wavelet Transform 离散小波变换

**DX** Access Demanded 接近要求

Design Exchange 设计交换

Distance 距离

Duplex 双的,二重的

**DXB** Dubai, United Arab Emirates ( Airport Code)阿拉伯联合酋长国迪拜机场

**DXC** Digital Cross Connect System 数字交叉连接系统

Digital Cross Connection 数字交叉连接

**DXI** Data Exchange Interface 数据交换接口

**DY** Day 日

Delivery 交货

Duty 税

**DYB** Dynamic Braking 动力刹车,动力制动

**DYN** Dynamic 动力

Dynamics 动力学

**DYNMTC** Dynamometric 测力的

**DYRT** Duststorm 尘暴

**DYS** Days 日(复数)

**DYU** Dushanbe, Tajikistan( Airport Code )塔吉克斯坦杜尚别机场

**DZ** Algeria 阿尔及利亚

Dead Zone 盲区

Dozen 一打

Drizzle 毛毛雨

Dropping Zone 空投地区

Dust Haze 尘霾

**DZA** Drop Zone Area 空投区面积

# E

E Early 早的
East 东,东方
Eastern Longitude 东经
Eastern 东方的
Electric Field 电场
Emergency 紧急
Empty 空
End 末端,完结
Endurance 续航时间
Equipment 器材设备
Error 错误
Estimate 预计,估计
Excursion 游览
Export 出口
Program Manager 项目经理

**E&EQT** Environmental & Engineering Qualification Test 环境与工程鉴定试验

**E&I** Equip & Install 装备与安装

**E/A** Error in Address 地址有误

**E/B** East-Bound 向东

**E/C** Engine Change 换发

**E/D** End of Descent 下降终点

**E/E** Electrcial/Electronic 电子/电子的
Electronic Equipment 电子设备

**E/E COMP** Electric/Electronic Compartment 电气/电子设备舱

**E/EEP** Electrical/Electronic Equipment Packaging 电气/电子设备装箱

**E/G** Engine 发动机

**E/O** Electro-Optical Conversion 电/光转换
Engine Out 发动机停车

**E/R** Emitter/Receptor 发送器/接收器
En Route 在航路上
End Route 终端航线

**E/S** Engine Speed 发动机转速

**E/T** Electrical Trim 电配平

**E/WD** Engine/Warning Display 发动机警告显示

**E/Z** Electrical Zero 零电位

**E3** Electromagnetic Environmental Effects 电磁环境效应,电磁环境影响

**EA** Each 每个
Earth 土地,地球
Electronic Asseembly 电子装置
Engine Alliance 发动机联盟
Engineering Authority 工程部门
Engineering Authorization 工程授权
Evolutionary Algorithms 进化算法
Expedited Data Acknowledgement 加速数据确认

**EAA** East Asia Airlines Limited 亚太航空有限公司
Eastern Australia Airlines, Limited 澳大利亚东方航空有限公司
Education Amendment Act 教育修正案
Encyclopedia of American Associations 美国社团百科全书
Engineer in Aeronautics and Astronautics 航空航天工程师,航空和宇宙航行工程师
Environmental Assessment Association 环境评估协会
Europe, Africa & Asia 欧洲、非洲和亚洲
Experimental Aircraft Association 试验飞机协会,试验航空器协会
Experimental Aviation Association 试验航空协会
Export Airworthiness Approval 出口适航批准书

**EAAC** East African Airways Corporation 东非

航空公司

European Aviation Air Charter 欧洲航空公司包机公司

**EAAI** European Association of Aerospace Industries 欧洲航空航天工业协会

**EAAP** European Association for Aviation Psychology 欧洲航空心理学协会

**EAAPS** European Association of Airline Pilots' Schools 航空公司飞行员学校欧洲协会

**EAAWG** European Aging Aircraft Working Group 欧洲老龄飞机工作组

**EAC** East African Community 东非共同体

Emergency Action Committee 应急行动委员会

Estimate at Completion 完工成本计算

Estimated Approach Control 估计进近控制

Executive Alarm Control 超限警报监控

Expected Approach Clearance Time 预计许可进近时间

Expected Approach Clearance 预计进近许可

Export Airworthiness Certificate 出口适航证

**EACS** Electronic Automatic Chart System 电子自动测绘系统

**EACT** Emergency Action and Coordination Team 应急行动与合作组

**EAD** Electronic Airworthiness Directives 电子设备适航指令

Electronic Attitude Director 电子姿态指引仪

Emergency Airworthiness Directive 紧急适航指令

Engine's Alert Display 发动机告警显示器

Estimated Availability Date 估计可用性日期,估计有效日期

European Aeronautical Database 欧洲航空数据库

Exposure At Default 违约风险暴露

**EADB** Enhanced Area Distribution Box 增强型区域分配盒

**EADF** Electronic Automatic Direction Finder 电子自动定向仪

**EADI** Electronic Airborne Data Indicator 机载电子数据指示器

Electronic Attitude Director Indicator 电子姿态指引仪

Electronic Attitude Display Indicator 电子姿态显示指示器

**EADS** Electronic Airworthiness Directives 电子设备适航指令

Engineering Administrative Data System 工程管理数据系统

Engineering Analysis & Data System 工程分析与数据系统

Environmental Assessment Data System 环境评估数据系统

European Aeronautic Defence and Space Company 欧洲航空防务航天公司

**EAE** European Air Express 欧洲航空快递

**EAEC** European Airlines Electronic Commission 欧洲航空电子委员会

**EAEM** Energy and Environmental Management 能量与环境管理

European Airlines Electronics Meeting 欧洲航空公司电子学会议

**EAF** Emery Air Freight Corp. (USA) 金刚砂航空货运公司

Engineering Analysis Facility 工程分析设施

**EAFR** Enhanced Airborne Flight Recorder 增强型机载飞行数据记录器

**EAGE** Electronic Aerospace Ground Equipment 航空航天地面电子设备

**EAGLE** Elevation Angle Guidance Landing Equipment 仰角引导着陆装置

**EAH** Emergency Artificial Horizon 应急地平仪

**EAI** Engine Air Intake 发动机进气系统

Engine Anti-Ice 发动机防冰

Enterprise Application Integration 企业应用集成

**EAIC** Engine Air Inlet Controls 发动机进气道调节机构

Engine Inoperative Altitude Capability 发动

机失效高度能力

**EAIM** E1 ATM Interface Module E1 ATM 接口模块

**EAL** Engine Assignment Letter 发动机调拨书

Evaluation Assurance Level 评估安全级别

**EALA** European Air Law Association 欧洲航空法协会

**EALAF** East Asia-Latin America Forum 东亚—拉美论坛

**EAM** Electric/Electronic Accounting Machine(s) 电子计算机

Emergency Action Message 紧急措施信息

Evaluation, Audit and Management 评估、审计和管理

**EAMDS** European Airline Medical Directors Society 欧洲航空公司医务经理协会

**EAMU** Emergency Audio Management Unit 紧急音频管理组件

**EAN** Electronic Aids to Navigation 电子助航设备

External Access Network 外部接入网络

**EANA** European Article Numbering Association 欧洲条码协会

**EANF** Establishment of Air Navigation Facilities 空中导航设备中心

**EANPG** European Air Navigation Planning Group 欧洲空中航行规划组，欧洲航行规划小组

**EAO** Environmental Assessment Office 环境评估办公室

**EAP** Effective Air Path 有效航线，有效航空线

Emergency and Accident Procedure 紧急和事故程序

Employee Assistance Program 员工帮助计划

Engine Alert Processor 发动机警戒处理器

Executive Airplane Program 飞机执行程序

Executive Airplane 专机

Experimental Aircraft Program 试验飞机项目

Experimental Aircraft Project 试验飞机计划

**EAPAS** Enhanced Airworthiness Program for Airplane Systems 改进的飞机系统适航计划

**EAPT** Unified Air Passenger Tariff 统一航空旅客票价

**EAR** Engineering Action Request 工程措施要求

Engineering Analysis Report 工程分析报告

Electronic Array Radar 电子阵列雷达

Electronic Audio Recognition 电子声音识别

Electronically Agile Radar 电子捷变雷达

Experimental Array Radr 试验阵列雷达

**EARB** Engineering Associates Registration Board 工程协会登记处

European Airlines Research Bureau 欧洲航空公司研究局

**EAROM** Electrically Alterable Read Only Memory 电控可变只读存储器

**EARS** Emergency Airborne Reaction System 机载应急反应系统

**EARTS** Enhanced Aircraft Radar Test Station 增强的飞机雷达测试站

Enroute Automated Radar Tracking System 航路自动雷达跟踪系统

**EAS** Eastern Aircraft Services, Ltd. (USA) (美国)东方飞机服务有限公司

Electronique Aero Spatiable (法)宇航电子设备公司

Engine Air Supply 发动机供气

Engineering Administration Services 工程管理服务

Equivalent Air Speed 等效空速

Essential Air Service 必要的航空服务

Estimated Air Speed 估计空速

Express Air Services 快速航空服务

San Sebastian, Spain(Airport Code) 西班牙圣塞瓦斯蒂安机场

**EASA** European Aviation Safety Agency 欧洲航空安全机构

European Aviation Safety Authority 欧洲航空安全局

European Aviation Security Association 欧洲

航空安全协会

**EASE** Electronic Analog and Simulation E-quipment 电子模拟与仿真设备

Emergency Air Supply Equipment 应急航空保障设备

**EASCON** Electronics and Aerospace System Conference 电子设备和宇航系统会议

**EASE** Escape and Survival Equipment 安全救生设备

**EASP** Essential Air Service Program 普遍航空服务计划

**EASS** Electronic Anti-Skip System 电子防滑系统

Engine Automatic Start System 发动机自动启动系统

**EASTCON** Electronics and Aerospace System Technical Convention 电子设备和航天系统技术会议

**EASY** Efficient Assembly System 有效装配系统

Engine Analysis System 发动机分析器系统

Evasive Aircraft System 飞机规避飞行系统

Exception Analysis System 异常分析系统

**EAT** Earliest Arrival Time 最早到达时间

Electronic Angle Tracking 电子角度跟踪

Entering Air Temperature 进气温度

Estimated Approach Time 预计进近时间

Expected Approach Time 期望进近时间

Expected Arrival Time 期望到达时间

Wenatchee, WA, USA ( Airport Code ) 美国华盛顿州文纳奇机场

**EATC** Electronic Automatic Temperature Controller 电子自动温度调节器

**EATCHIP** European Air Traffic Control Harmonisation and Integration Programme 欧洲空中交通管制协调和综合程序

European ATC Harmonization Implementation Program 欧洲空中交通管制协调实施计划

**EATCHP** European ATC Harmonization Implementation Program 欧洲空中交通管制协调实施计划

**EATMP** European Air Traffic Management Programme 欧洲空中交通管理项目

**EATMS** Europe ATM System 欧洲航空运输管理系统

European Air Traffic Management System 欧洲空中交通管理系统

**EATS** Electronic Altimeter Test Set 电子高度表测试仪

European Air Transport Service 欧洲航空运输处

European Air Transport Station 欧洲航空运输站

Extended Area Test System 扩展区域测试系统

**E-ATS** Electronic Audible Traffic Signal 电子发声交通信号

**EAU** Eau Claire, WI, USA ( Airport Code ) 美国威斯康星州欧克莱尔机场

Engine Accessory Unit 发动机附件组件

European Accounting Unit 欧洲货币结算单位

**EAV** Effective Angular Velocity 有效角速度

Engine Assembly Vehicle 发动机装配车

**EAWG** Electrical/Avionics Working Group 航空电气/电子工作组

**EAX** Electronic Automatic Exchange 电子自动交换机

**EB** Early Brake 早期制动

Electron Beam, Electronic Beam 电子束

Emergency Brake 紧急制动器

Engine Bleed 发动机引气

Engine Bulletin 发动机通报

Engineering Bulletin 工程通告

Equipment Bay 设备舱

Errored Block 误块

Essential Bus 主汇流条

**EBA** Emergency Breathing Apparatus 紧急排气装置

Engine Bleed Air 发动机放出的空气

**EBA/H** Engine Bleed Air and Hydrazine 发动机放出的空气和肼

**EBAA** European Business Aviation Association 欧洲商业航空协会

**EBAC** Electronic Braking Actuator Controller 电刹车作动器控制器

**EBADS** Engine Bleed Air Distributor System 发动机引气分配系统

**EBAS** Engine Bleed Air System 发动机引气系统

**EBB** Entebbe, Uganda (Airport Code) 乌干达恩德培机场
Extra Best 特优

**EBC** Electronic Business Communications 电子商业通信

**EBCDIC** Extended Binary Coded Decimal Interchange Code 广义二进制编码的十进制交换码

**EBCU** Electronic Brake Control Unit 电刹车控制组件
Emergency Brake Control Unit 紧急刹车控制组件

**EBE** Electron Beam Evaporator 电子束蒸发器

**EBER** Excessive Bit Error Rate 过高误比特率

**EBF** Externally Blown Flap 外吹襟翼

**EBG** Electron Beam Gun 电子束枪

**EBGO** Embargo 禁运

**EBGP** External Border Gateway Protocol 外部边界网关协议

**EBH** Equivalent Baseline Hours 等效基线小时

**EBHA** Electrical Backup Hydraulic Actuator 电动备用液压作动器

**EBHP** Equivalent Brake Horse Power 等效制动马力

**EBI** Electronic Business Infrastructure 电子商务基础设施

**EBIC** Electron-Bombardment-Induced Conductivity 电子轰击感应电导率

**EBIT** Earnings before Interest and Taxes 利息和税前利润

**EBJF** Externally Blow Jet Flap 外吹喷气襟翼

**EBM** E-Business Model 电子商务模型
Electron Beam Machining 电子束加工
Electron Beam Melted 被电子束熔化的
Electron Beam Method 电子束法
Electron Beam Microanalysis 电子束微量分析
Electronic Battle Management 电子作战管理
Engineering Business Management 工程商务管理

**EBMA** Engine Booster Maintenance Area 发动机助推器保养场

**EBMD** Electron Beam Mode Discharge 电子束放电

**EBMF** Electron Beam Micro Fabricator 电子束微加工机

**EBMR** Electron Beam Metal Removal 电子束

**EBN** East by North 东偏北

**EBND** Eastbound 向东航行的

**EBO** Emerging Business Opportunities 新兴商业机会
Evaluation by Objective 客观评价

**EBOP** Emergency Bearing Oil Pump 应急备用润滑油泵

**EBP** Electric Bilge Pump 舱底电力泵
Electron Beam Perforated 电子束穿孔的
Electron Beam Pumping 电子束抽运
Exhaust Back Pressure 排气反压力

**EBPA** Electron Beam Parametric Amplifier 电子束参量放大器

**EBPM** Electronic Brake Power Module 电刹车控制组件

**EBPSU** Electronic Brake Power Source Unit 电刹车电源组件

**EBPVD** Electron Beam Physical Vapor Deposition 电子束物理气相沉积

**EBR** Electron Beam Recorder 电子束记录器
Electron Beam Recording 电子束录像
Electron Beam Regulator 电子束调节器
Engineering Business Report 工程工作报告

**EBRM** Electronic Bearing & Range Marker 电子方位和距离刻度标志

**EBRS** Electron Beam Recorder System 电子束记录系统

**EBS** East by South 东偏南

Educational Broadcasting Satellite 教育广播卫星

Electron Beam Semiconductor 电子束半导体

Electron Beam System 电子束系统

Engine Balancing System 发动机平衡系统

Engine Bleed System 发动机引气系统

Engine Breather Separator 发动机通气孔隔板

Engine Build Specification 发动机组装规范

Errored Block Second 误块秒

European Broadcast Satellite 欧洲广播卫星

Extruded Bar Solder 挤压焊条

**EBSC** European Bird Strike Committee 欧洲防鸟(撞)击委员会

**EBSR** Errored Block Second Rate 误块秒率

**EBSS** Electron Beam Scanning System 电子束扫描系统

**EBSV** Engine Bleed Shutoff Valve 发动机泄漏断开阀

**EBT** Effective Brightness Temperature 有效亮度温度

Electron Beam Technology 电子束工艺,电子束技术

Excess Baggage Ticket 超额付李付费票联

**EBTS** Electron Beam Test System 电子束检测系统

**EBU** Engine Build-up Unit 发动机装配组件

Engine Build-up 发动机装配/安装

European Broadcasting Union 欧洲广播联盟

**EBW** Effective Band Width 有效带宽

Electron Beam Welding 电子束焊

Exploding Bridge Wire 电桥式传爆线

**EBYN** East By North 东偏北

**EBYS** East By South 东偏南

**EC** Earth Current 接地电流

Echo Cancellation 回声消除

Economic Commission 经济委员会

Eddy Current 涡流

Edge Connector 边缘连接器

Effective Concentration 有效浓度

Elasticity Coefficient 弹性系数

Electrical Conductivity 导电性,电导率

Electrical Current 电流

Electro Chemical 电化学

Electrolytic Cell 电解电池

Electrolytic Corrosion 电解腐蚀

Electron Coupling 电子耦合

Electronic Calibration 电子校准

Electronic Coding 电子编码

Electronic Combat 电子战

Electronic Commerce 电子商务

Electronic Conductivity 电子导电率

Electronic Control 电子控制

Electronic Counter 电子计数器

Electronically Controlled 电子控制的

Electronics and Control 电子设备和控制

Electronics Card 电子设备卡片,电子设备插件

Electrostatic Collector 静电收集器

Elevation Console 仰角操纵台

Elevation Correction 仰角修正,高度修正

Eliminating-Charging 消除充填

Emergency Capability 应急能力

Enamel-Covered Wire 漆包线

Enamel-Covered 上瓷漆的,上珐琅的

Engine Change 换发

Engine Contractor 发动机合同商

Engine Control 发动机控制

Engineering Change 工程更改

Environmental Control 环境控制

Equipment Component 设备部件

Equipment Controller 设备控制器

Error Control 错误控制,误差控制

Error Correcting 误差校正,误差修正

European Commission 欧洲委员会

European Community 欧洲共同体

Example Cause (For Example) 举例说明,

例如

Exhaust Closes 排气(阀)关闭

Experimental Certificate 试验许可证

Extra Control 额外控制,附加控制

**ECA** Earth Central Angle 地心角

Economic Commission for Africa 非洲经济委员会

Electrical Control Assembly 电子控制组件

Electronic Classified Advertising 电子分类广告

Electronics Control Amplifier 电子控制放大器

Electronics Corporation of America 美国电子公司

Engineering Change Action 工程更改措施

Engine-out Control Augmentation 发动机停车增稳控制

Enter Control Area 进入控制区

Equipment Condition Analysis 设备状态分析

European Civil Aviation 欧洲民航组织

European Combat Aircraft 欧洲战斗机

Export Credit Agency 出口信贷机构

**ECAA** Egyptian Civil Aviation Authority 埃及民用航空局

European Commom Aviation Area 欧洲共同航空区

**ECAC** Electromagnetic Compatibility Analysis Center 电磁兼容性分析中心

Electromagnetic Computer Analysis Center 电磁计算机分析中心

European Civil Aviation Commission 欧洲民用航空委员会

European Civil Aviation Conference 欧洲民用航空会议

**ECAD** Electronic Computer-Aided Design 电子计算机辅助设计

**ECAE** European Committee for Aviation Electronics 欧洲航空电子委员会

**ECAFE** Economic Commission for Asia and the Far East 亚洲及远东经济委员会

**ECAI** External Credit Assessment Institution 外部信用评估机构

**ECAM** Electronic Central Aircraft Monitoring System 飞机电子中央监控系统

Electronic Centralized Aircraft Monitor 飞机电子集中监控器

Electronic Centralized Aircraft Monitoring 飞机电子集中监控

Electronic Crew Alerting & Monitoring 电子机组警告与监视

Electronically Computer Aided Manufacturing 电子计算机辅助制造

Energy Conservation and Management 能量守恒和管理

Enterprise Communications Analysis Module 企业通讯分析模块

**ECAN** Entertainment Center Area Network 娱乐中心局域网络

**ECAP** Electronic Circuit Analysis Program 电子电路分析程序

Electronic Control Assembly Pitch 俯仰电子控制组件

Environmental Compatibility Assurance Program 环境适应性保证计划

**ECAPS** Emergency Capability System 应急能力系统

**ECAR** Electronic Control Assembly-Roll 横滚电子控制装置

European Civil Aviation Requirement 欧洲民用航空需求

**ECARS** Electronic Coordinator Graph and Readout System 电子坐标仪读出系统

**ECASS** Electronically Controlled Automic Switching System 电子控制的自动转换系统

**ECAT** Equipment Category 设备种类

**ECAY** Electronic Control Assembly Yaw 偏航电子控制组件

**ECB** Economic Cruising Boost 经济巡航助推

Eddy Current Brake 涡流制动器

EDIFACT Coordination Board 管理、商业和运输电子数据交换协调委员会

Electronic Circuit Breaker 电路跳开关

Electronic Code Book 电子码书

Electronic Components Board 电子元件部

Electronic Control Box 电子控制盒

Electronically Controlled Birefringence Mode 电控双折射模式

Etched Circuit Board 刻蚀电路板

European Central Bank 欧洲中央银行

Event Control Block 事件控制程序块

**ECC** Eccentric 偏心的

Embedded Communication Channel 嵌入式通信信道

Embedded Control Channel 嵌入式控制信道

Engine Control Centre 发动机控制中心

Engineering Change Classification 工程更改分类

Environmental Campaign Committee（香港）环境运动委员会

Environmental Compatibility Certificate 环境兼容性合格审定

Equipment Cooling Card 设备冷却卡

Equipment Cooling Controller 设备冷却控制器

Error Checking and Correction 误差检验与校正

Error Control Code 差错控制码

Error Correction Circuit 误差校正电路

Error Correction Code 纠错码

European Community Commission 欧洲共同体委员会

**ECCAIRS** European Coordination Centre for Accident and Incident Reporting Systems 欧洲事故和事故征候报告系统协调中心

European Coordination Centre for Aviation Incident Reporting System 欧洲航空事故征候报告系统协调中心

**ECCANE** East Coast Conference on Aerospace and Navigational Electronics 航空航天及导航电子设备东海岸会议

**ECCAS** Economic Community of Central African States 中非国家经济共同体

**ECCB** Remote Control Circuit Breaker 远程控制断路器

**ECCCS** Emergency Command Control Communications System 应急指挥、控制、通信系统

**ECCM** Electronic Counter-Countermeasures 电子反干扰，电子反对抗

**ECD** Electron Capture Detector 电子捕获检测器

Electron-Chemical Drilling 电化学钻孔

Estimated Completion Date 预计完成日期

Excusable Contract Delay 允许的合同延误

**ECDB** Equipment Central Data Bank 设备中央数据库

**ECDES** Electronic Combat Digital Evaluation System 电子战斗数字评定系统

**ECDG** Electrochemical Discharge Grinding 电化放电研磨

**ECDL** Engineering Change Differences Lists 工程更改偏差清单

**ECDM** Electrochemical Discharge Machining 电化与放电加工

**ECE** Economic Commission for Europe 欧洲经济委员会

Electron Cyclotron Emission 电子回旋加速器发射

Environmental Control Equipment 环境控制设备

External Combustion Engine 外燃发动机

**ECEF** Earth Centered Earth Fixed 地心地固坐标，地心固连

**ECEMP** Electron Caused Electromagnetic Pulse 电子设备引起的电磁脉冲

**ECET** Electronics and Computer Engineering Technology 电子与计算机工程技术

European Centre for Education and Training 欧洲教育与训练中心

**ECETCH** Electrolytic Corrosion Etch 电刻蚀

**ECF** Effective Cutoff Frequency 有效切断频率

Electro-Chemical Forming 电化学成形

Enhanced Connective Facility 加强结合装置

Envelope Check Fixture 包络检查夹具

Environment and Conservation Fund（香港）环境保护基金

Externally Caused Failure 外因造成的故障

**ECFA** Economic Cooperation Framework Agreement 两岸经济合作框架协议

**ECG** Effective Center of Gravity 有效重心

Electrocardiogram 心电图

Electronic Component Group 电子元件组

Emergency Coordination Group 应急合作组

**ECGD** Exports Credit Guarantee Department 出口信贷保证局

**ECH** Earth Coverage Horn 地球覆盖喇叭天线

Echo Cancelled Hybrid 回波抵消混合

Engine Compartment Heater 发动机舱加热器

**ECI** Echo Control Indicator 回波控制指示器

Electronic Communication Instrument 电子通讯仪表

Engine Component Improvement 发动机部件改进

Environmental Condition Indicator 环境状况指示器

**ECIC** Earth-Centered Inertial Coordinates 地心惯性坐标

**ECIS** Earth-Centered Inertial System 地心惯性系统

Error Correction and Information System 纠错与信息系统

**ECI-turbo** Electronic Control Injection Turbo 电子控制燃油喷射涡轮增压发动机

**ECL** Eddy Current Loss 涡流损耗

Electronic Check List 电子检查单

Electronic Components Laboratory 电子元件实验室

Emitter Coupled Logic （发）射极耦合逻辑

Engineering Change Letter 工程更改函

Equipment Check List 设备检查单

Equipment Component List 设备元件明细表

Exchange Control Logic 交换控制逻辑

Exercise Caution at Landing 着陆时注意

External Cavity Laser 外腔激光器

External CLock 外部时钟

**ECLA** Economic Commission for Latin America 拉美经济委员会

Europe, Canada, Latin America 欧洲、加拿大和拉丁美洲

**ECLO** Emitter Coupled Logic Operator （发）射极耦合逻辑算子

**ECLS** Electrical Cargo Loading System 电货物装载系统

**ECLSS** Environmental Control and Life-Support System 环境控制与生命保障系统

**ECM** Electric Coding Machine 电动编码机

Electro-Chemical Machining 电化学加工，电子加工

Electrochemical Machining 电解加工，电化学加工

Electron Cyclotron Maser 电子回旋微波激射器

Electronic Card Modules 电子插件

Electronic Configuration Management 电子构型管理

Electronic Control Module 电子控制模块

Electronic Counter Measures 电子干扰，电子对抗

Electronic Countermeasures Mission 电子对抗任务

Emergency Changeover Message 紧急转换消息

Emergency Conservation Measures 应急保存措施

Engine Condition Monitoring 发动机状态监控

Engineering Change Management 工程更改管理

Engineering Change Memorandum 工程更改备忘录

Engineering Coordination Memorandum 工程协调备忘录

Enterprise Content Management 企业内容管

理

Equipment Change Management 设备变更管理

Error Correction Mode 纠错模式

European Common Market 欧洲共同市场

**ECMA** European Computer Manufacturing Association 欧洲计算机制造协会

**ECME** Electronic Checkout Maintenance Equipment 电子检测维修设备

Electronic Countermeasures Equipment 电子对抗设备

**ECMEA** European Conference of Meteorological Expert for Aeronautics 欧洲航空气象专家会议

**ECMJ** Electronic Counter Measures Jammer 电子对抗干扰机

**ECMP** Electronic Countermeasure Program 电子对抗计划

**ECMR** Engineering Configuration Management Requirements 工程构型管理要求

**ECMRITS** Electronic Counter Measures Resistant Information Transmission System 抗电子对抗信息传输系统

**ECMS** Electrical Contractor Management System 电气接触器管理系统

Electronic Component Management System 电子部件管理系统

Electronic Countermeasure Subsystem 电子干扰子系统

Engine Condition Monitoring System 发动机状态监控系统

Engine Configuration Management System 发动机配置管理系统

**ECMT** European Conference of Ministers of Transport 欧洲运输部长会议,欧洲交通部长会议

**ECMU** Electrical Contractor Management Unit 电接触器管理组件

**ECMWF** European Center for Medium-range Weather Forecast 欧洲中期天气预报中心

**ECN** Electronic Communication Network 电子通信网

Emergency Communications Network 应急通信网络

Engineering Card Notice 工程工卡通知

Engineering Change Note 工程更改说明

Engineering Change Notice 工程更改通知

Engineering Change Notification 工程变更通知

Engineering Change Number 工程更改目录号

Explicit Congestion Notification 显性拥塞通知

**ECNL** Equivalent-Continuous Noised Level 当量的持续噪声级

**ECO** Economic 经济的

Economy 经济

Electron Coupled Oscillator 电子耦合振荡器

Electronic Checkout 电子检验

Emergency Changeover Order 紧急转换命令

Engine Certification Office 发动机审定办公室

Engine Cut-off, Engine Cutoff 发动机停车

Engineering Change Order 工程变更指令

**ECOMS** Jeppesen Explanation of Common Minimum Specifications 杰普逊共用最低标准规范的说明

**ECON** Economic 经济的

Economics 经济学

Economize 节约

Economizer 节油器,节热器

Economy(Cruise) 经济(巡航)

**ECOSEC** European Cooperation Space Environment Committee 欧洲联合空间环境委员会

**ECOWAS** Economic Community of West African States 西非国家经济共同体

**ECP** ECAM Control Panel 飞机电子中央监控控制面板

EICAS Control Panel 发动机指示和机组提警系统控制板

Electrical Contact Plate 电接触板

Electronic Circuit Protector 电子电路保护装置

Engine Control Program 发动机控制大纲

Engineering Change Procedure 工程更改程序

Engineering Change Proposal 工程变更建议

Engineering Cost Proposal 工程成本建议

Equipment Change Proposal 设备更改建议

Evaluation &Control Processor 评估和控制处理器

Experimental Communication Payload 实验通讯载荷

**ECPAI** Evaluation Control Processor for Avionics Integration 航空电子设备集成评估控制处理器

**ECPD** Engineers Council for Professional Development 工程师专业发展理事会

**ECPE** External Combustion Piston Engine 外燃活塞发动机

**ECPI** Electronic Computer Programming Institute 电子计算机程序设计研究所

**ECPLC** Engineering Computer Program Library Control 工程计算机程序库控制

**ECPMU** Enhanced Cabin Passenger Management Unit 增强型客舱旅客管理组件

**ECPNL** Effective Continuous Perceived Noise Level 有效连续探测噪声级

Equivalent Continuous Perceived Noise Level 等值连续探测噪声级

**ECPP** Effective Critical Parts Plan 有效关键部件计划

**ECPT** Engineering Change Proposal Types 工程更改建议类型

**ECR** Efficient Consumer Response 有效客户反应

Electrical Contact Resistance 电子接触阻力

Electrochemical Reduction 电解还原

Electron Cyclotron Resonance 电子回旋共振（推进技术）

Electronic Cash Register 电子收款机

Embedded Computer Resources 灌封的计算机设备

Engine Condition Review 发动机状态评审

Engineering Change Request 工程更改申请

Equipment Create-Delete Report 设备建立—删除报告

Error Car Rejection 错误卡排除

Explict Cell Rate 显性信元率

**ECRA** Engineering Change Request/Authorization 工程更改申请/批准

**ECRI** Engineering Change Revision Instruction 工程更改修定指令

**ECROS** Electrically Controllable Read-Only Storage 电可控只读存储器

**ECRS** Engineering Change Record System 工程更改记录系统

**ECS** Electronic Cable Specialists 电子设备电缆专家

Electronic Chronometer System 电子计时系统

Electronic Configuration System 电子组合系统，电子构型系统

Electronic Cooling System 电子冷却系统

Embedded Computer System 嵌入式计算机系统

Emergency Coolant System 紧急冷却剂系统

Encryption Control Signal 加密控制信号

Engine Control System 发动机操纵系统，发动机控制系统

Engineering and Computer Science 工程与计算机科学

Engineering Compiler System 工程编译程序系统

Engineering Computer Systems 工程计算机系统

Engineering Construction Specification 工程结构规范

Engineering Control Systems 工程控制系统

Engineering Coordination Sheet 工程协调单

Enterprise Communication System 企业通信系统

Environmental Control System 环境控制系统

European Communication Satellite 欧洲通信卫星

Experimental Control Software 试验性控制软件

**ECSA** Enviromental Control Systems Analysis 环境控制系统分析

**ECSB** Embedded Computer System Bus 嵌入式计算机系统总线

**ECSC** European Communication Satellite Committee 欧洲通信卫星委员会

**ECSD** Enhanced Circuit-Switched Data 增强型电路交换数据

**ECSEC** European Center for Scientific and Engineering Computing 欧洲科学与工程计算中心

**ECSMC** Environmental Control System Miscellaneous Card 环境控制系统多功能卡

**ECSO** European Communication Satellite Organization 欧洲通信卫星组织

**ECSS** Energy Conversion and Storage System 能量转换与储存系统

Equipment Communication Subsystem 设备通信分系统

European Communication Satellite System 欧洲通信卫星系统

Executive Control and Subordinate System 行政管理与附属系统

**ECT** Echo Cancellation Technique 回波消除技术

Electronic Cash Technology 电子货币技术

Engine Cutoff Timer 发动机停车计时器

Exercise Caution Takeoff 起飞时注意

**ECTL** Emitter Coupled Transistor Logic 发射极耦合晶体管逻辑

**ECTM** Engine Condition Trend Monitoring 发动机状态趋势监控

Engine Continuous Trend Monitoring 发动机持续状态趋势监控

**ECTMS** Environmental Control Thermal Management System 环境控制热管理系统

**ECTS** Echo Canceller Testing System 回声消除器测试系统

Electric Circuit Test Set 电路测试装置

Environmental Control Test Set 环境控制测试仪

**ECU** Electrical Control Unit 电气控制组件

Electronic Conversion Unit 电子转换装置

Elevator Control Unit 升降舵控制组件

Engine Change Unit 发动机更换装置

Engine Control Unit 发动机控制组件

Environmental Control Unit 环境控制设备,环境控制装置

European Currency Unit 欧洲货币单位

Exercise Control Unit 演习指挥组,演练控制分队

External Compensation Unit 外部补偿装置

**ECUCV** Electronic Control Unit Cooling Valve 电子控制组件冷却活门

**ECUD** Equipment Conditional Usage Drawing 设备状态使用图表

**ECV** Economy Cooling Valve 节能冷却活门

Environmental Control Valve 环境控制活门

**ECW** Engineering Control Word 工程控制字

**ECWA** Economic and Social Commission for Western Asia 西亚经济社会委员会

**ECWAS** Economic Community of West African States 西非国家经济共同体

**ECYC** Engine Cycles 发动机循环

**ED** Edge Device 边缘设备

Edge Distance 边缘距离

Edge 边,端

Edit, Editor 编辑

Edition 版本

Education, Educational 教育,教育的

EICAS Display 发动机指示和机组提醒系统显示器

Electrostatic Discharger 静电放电器

Emergency Distance 紧急距离

End of Descent 下降底点,下降终点,下降结束

End-Data Code 数据结束代码

Ending Delimiter 结束定界符

Engine Drive 发动机驱动

Engineering Dataset 工程数据集

Engineering Design 工程设计

Engineering Development 工程开发

Engineering Division 工程部门

Engineering Document 工程文件

Envelop Drawing 包络图

Environmental Damage 环境破坏

Environmental Deterioration 环境损伤

Error Detection 错误检测

Estimated Date 预订日期

Expanded Display 扩充显示器

Expedited Data 加速数据

Extended Density 扩展密度磁场

**ED SK** Engineering Department Sketch 工程部门简图

**EDA** Early Departure Authorized 批准提前起飞

Effective Disc Area（直升机旋翼的）有效桨盘面积

Electronic Design Aids 电子设计辅助设备

Electronic Design Automation 电子设计自动化

Electronic Digital Analyzer 电子数字分析器

Embedded Document Architecture 嵌入式文档结构

Emergency Distance Available 可用应急距离

**EDAC** Electronic Dive Angle Control 电子俯冲角控制

Engineering Design Advisory Committee 工程设计咨询委员会

Equipment Distribution and Condition 设备分配与条件

Error Detection and Correction 误差检测与修正

**E-DARC** Enhanced Direct Access Radar Channel 增(加)强的直接存取雷达信道

**EDAU** Engine Data Acquisition Unit 发动机数据采集装置

**EDAX** Energy Dispersive Absorption of X-rays X 射线能量扩散吸收

Energy Dispersive Analysis of X-rays X 射线能量扩散分析

**EDB** ECAM Data Book 飞机电子中央监控数据手册

Economic Development Board 经济发展委员会

Engineering Data Bank 工程数据库

Engineering Development Board 工程开发委员会

Environment Data Bank 环境数据库

Environmental Database 环境数据库

Evolvable Data Base 可展开的数据库

Extensional Data Base 扩充型数据库

**EDBMS** Engineering Data Base Management System 工程数据库管理系统

**EDBS** Engineering Data Bank System 工程数据库系统

**EDC** Electronic Development Corporation 电子开发公司

Electronic Digital Computer 数字电子计算机

Engineering Data Control 工程数据控制

Engineering Design Control 工程设计更改

Engineering Document Control 工程文件控制

Engineering Drawing Control 工程图纸控制

Error Detection and Correction 差错检测与纠正

Error Detection Circuit 差错检测电路

Error Detection Code 差错检测码

**EDCARS** Electro-optional Detection Classification & Ranging System 光电检测分类和测距系统

**EDCC** Environmental Detection Control Center 环境检测控制中心

Environmental Dispute Coordination Commission 环境争论协调委员会

**EDCP** Engineering Design Change Proposal 工程设计变更建议

**EDCS** Electrical Drive and Control System 电气驱动和控制系统

Evolutionary Design of Complex Software 复

杂软件环境设计

**EDCT** Electronic Data Collection Tool 电子数据集合工具

Estimated/Expected Departure Clearance Time 预计离港起飞放行时间，预计批准离港时间

**EDD** Electronic Data Display 电子数据显示

Engineering Data Depository 工程数据存储

Engineering Delivery Documents 工程交付文件

Envelop Delay Distortion 包线延迟失真

**EDDF** Error Detection & Decision Feedback 错误检测和决策反馈

**EDDS** Electronic Document Distribution System 电子文档分配系统

**EDEN** Electronic Documents and Enquiry Network 电子文件和查询网络

**E-DET** Error DETector 误码检测器

**EDF** EICAS Display Function 发动机指示和机组提醒系统显示功能

Equipment Data Form 设备数据格式

Estimated Date of Flight 估计飞行日期，预定飞行日期

**EDFA** Electronic Differential Analyzer 电子微分分析器

**EDFPS** Expended Delta Fast Packet Switching 扩展增量快速分组交换

**EDG** Electrical-Discharge Grinding 放电研磨，放电抛光

Electronically Drive Gyro 电动陀螺仪

Engineering Design Graphics 工程设计图纸

**EDGE** Electronic Data Gathering Equipment 电子数据采集设备

Enhanced Data GSM Environment 增强型数据 GSM 环境

**EDHP** Engine Driven Hydraulic Pump 发动机驱动液压泵

**EDI** Edinburgh, Scotland, United Kingdom (Airport Code) 英国苏格兰爱丁堡机场

Electronic Data Interchange 电子数据交换

Electronic Document Interchange 电子文件交换

Engine Data Interface 发动机数据接口

Enterprise Data Integration 企业数据集成

**EDIAU** Electronic Data Interchange Access Unit 电子数据交换存取单元

**EDIF** Electronic Data Interchange Format 电子数据交换格式

Electronic Document Interchange Format 电子文档交换格式

Electronics Design Interchange Format 电子设计交换格式

Engine Data Interface Function 发动机数据接口功能

**EDIFACT** Electronic Data Interchange For Administration, Commerce and Transport 用于行政、商业和运输业电子数据交换

**EDIG** European Defence Industries Group 欧洲国防工业集团

**EDIM** Electronic Data Interchange Management 电子数据交换管理

Electronic Data Interchange Message 电子数据交换信息

**EDIME** Electronic Data Interchange Messaging Environment 电子数据交换信息环境

**EDIMS** Electronic Data Interchange Message Storage 电子数据交换信息存储

Electronic Data Interchange Messaging System 电子数据交换信息处理系统

Environmental Data and Information Management Systems 环境数据和信息管理系统

**EDIN** Electronic Data Interchange Notification 电子数据交换通知

**EDIP** European Defence Improvement Programme 欧洲防务改进计划

**EDIU** Engine Data Interface Unit 发动机数据接口组件

**EDIUA** Electronic Data Interchange User Agent EDI 用户代理

**EDL** Electrical Discharge Laser 放电激光器

Engineering Development Laboratory 工程开发试验室

Equipment Data List 设备数据清单

**EDLC** Ethernet Data Link Control 以太网络数据链接控制

**EDM** Electrical Discharge Machining 电火花加工

Electro-Discharge Machining 电火花加工

Electromagnetic Distance Measuring 电磁测距

Electronic Data Management 电子数据管理

Electronic Distance Measurement 电子测距

Engine Data Multiplexer 发动机数据多路复用器

Engine Data Multiplexing 发动机数据多路传输

Engine Divergence Monitoring 发动机参数偏差监控

Engineeering Department Manual 工程部门手册

Engineering Development Mission 工程研制任务

**EDMC** Engineering Drawing Microfilm Card 工程绘图微缩卡

**EDME** Electronic Distance Measuring Equipment 电子测距设备

**EDMI** Electronic Distance Measuring Instruments 电子测距仪

**EDMS** Electronic Data Management System 电子数据管理系统

Electronic Document Management System 电子文件管理系统

Engineering Data Management System 工程数据管理系统

Engineering Data Microreproduction System 工程设计微型复制系统

Engineering Document Management System 工程文件管理系统

Enterprise Document Management System 企业文件管理系统

Environmental Data Management System 环境数据管理系统

Experimental Data Management System 试验数据管理系统

**EDN** Electronic Distribution Network 电子分销网络

Emergency Data Network 应急数据网络

Engineering Department Notice 工程部门通知

Engineering Document Number 工程文件编号

Enterprise Data Network 企业数据网络

Event Detection and Notification 事件检测和通知

Experimental Data Network 实验数据网络

**EDNA** Enhanced Diagnostic Navigational Aid 增强型诊断导航设备

**EDO** Engineering Design Outsourcing 工程设计外包

Engineering Deviation Order 工程偏离指令

Environmental Development Office 环境开发办公室

**EDOC** Effective Date of Change 更改有效日期

Effective Date of Contract 合同有效日期

**EDP** Electric Diffusing Process 电渗处理

Electronic Data Processing 电子数据处理

Electronic Data Processor 电子数据处理器

Embedded Data Processor 嵌入式数据处理机

Engine Driven Pump 发动机驱动泵

Engineering Data Package 工程数据包

Engineering Design Plan 工程设计计划

Engineering Development Phase 工程开发阶段

Event Detection Point 事件检测点

Experimental Data-Processor 实验数据处理机

Experimental Dynamic Processor 实验动态处理机

Extended Delivery Points 扩展分运点

**EDPA** Er-Doped Planar Amplifier 掺铒平面放大器

**EDPC** Electronic Data Processing Center 电子

数据处理中心

**EDPE** Electronic Data Processing Equipment 电子数据处理设备

**EDPF** Experimental Distributed Processing Facility 试验分配处理设施

**EDPIS** Electronic Data Processing and Information Systems 电子数据处理和信息系统

**EDPM** Electronic Data Processing Machine 电子数据处理机

Electronic Data Processing Manager 电子数据处理经理

**EDP-N** Event Detection Point-Notification 事件检测点通知

**EDPR** Engineering Design Process Requirements 工程设计工艺要求

Engineering Development Part Release 工程开发零件发放

**EDP-R** Event Detection Point-Request 事件检测点请求

**EDPS** Electronic Data Processing System 电子数据处理系统

Engine Driven Pump System 发动机驱动泵系统

**EDPT** Electronic Data Processing Test 电子数据处理试验

**EDQA** Electronic Devices Quality Assurance 电子设备质量保证

**EDR** Engine Data Retrieval 发动机数据回收

Engineering Data Record 工程数据记录

Engineering Data Release 工程数据发放

Engineering Data Requirement 工程数据要求

Engineering Design Review 工程设计审查

Engineering Design Rules 工程设计规则

Engineering Drawing Release 工程图发放

Environmental Damage Rating 环境损伤等级

Environmental Data Record 环境数据记录

Equivalent Direct Radiation 等效的直接辐射

Estimated Date of Return 估计返回日期

Expected Departure Release 预期允许分离，预期放行

Experiment Data Record 实验数据记录

**EDRB** Engineering Design Review Board 工程设计审查委员会

**EDRL** Engineering Data Release List 工程数据发布清单

Engineering Data Requirement List 工程数据要求清单

**EDRMS** Electronic Document and Record Management System 电子文件和记录管理系统

**EDRS** Electronic Data Reporting System 电子数据报告系统

Engineering Data Reproduction System 工程数据复制系统

Engineering Data Retrieval System 工程数据检索系统

European Data Relay Satellite 欧洲数据中继卫星

**EDS** Electrical Distribution Subsystem 配电分系统

Emergency Detection System 应急故障探测系统

Engine Diagnostic System 发动机故障判断系统

Engineering Data System 工程数据系统

Engineering Department Specification 工程设计部门规范

Engineering Drawing System 工程图纸系统

Environmental Data Service 环境数据资料服务处

European Distribution System 欧洲分配系统

Explosive Detection System 爆炸物探测系统

**EDSC** Engineering Data Service Center 工程数据服务中心

Engineering Data Support Center 工程数据支援中心

**EDSD** Electrostatic Discharge Sensitive Device 静电释放传感装置

**EDSE** Electronic Data Standard Exchange 电子数据标准交换

**EDSG** Electro-optical & Data Systems Group

光电和数据系统组

**EDSP** EICAS Display Select Panel 发动机指示和机组提警系统显示选择板

**EDSV** Enamel Double Silk Varnish 双丝漆包线

**EDT** Eastern Daylight Time 东部夏令时间（美国），东部白昼时间

Electrical Discharge Tube 放电管

Enforcement Decision Tool 执法决策工具

Engineering Design Test 工程设计试验

Engineering Development Test 工程开发试验

Expected Departure Time 预计离场时间

**EDTCC** Electronic Data Traffic Control Center 电子数据通信控制中心

Electronic Data Transmission Communication Center 电子数据传输通信中心

Electronic Data Transmission Control Center 电子数据传输控制中心

**EDTO** Extended Diversion Time Operations 延程改航时间运行

**EDTV** Enhanced Digital Television 增强的数字电视

Extended Definition Television 扩展清晰度电视

**EDU** EICAS Display Unit 发动机指示和机组提警系统显示器

Electronic Display Unit 电子显示组件

Engine Diagnostic Unit 发动机诊断组件

Engine Display Unit 发动机显示器

Engineering Development Unit 工程开发组件

**EDUC** Eductor 喷射器，排放（气）管

**EDUCP** Electronic Display Unit Control Panel 电子显示组件控制板

**EDUR** Engineering Drawing Usage Record 工程图使用记录

**EDV** Electric Depressurization Valve 电气减压活门

**EDW** Education Data Warehouse 教育数据仓库

Enterprise Data Warehouse 企业数据仓库

**EDWA** Erbium Doped Waveguide Amplifier 掺铒波导放大器

**EE** Earth Entry 进入地球大气层

Eastern Europe 东欧（东欧航图代号）

Electric Eye 电眼

Electrical and Electronic 电气和电子

Electrical Equipment 电气设备

Electrode Effect 电极效应

Electrodynamic Explorer 电动力学探测卫星

Electronic Engineer 电子工程师

Electronic Equipment 电子设备

Environmental Engineering 环境工程

Errors Excepted 允许误差

Expected Errors 预期误差

External Environment 外部环境

**E-E** End-to-End 端到端

**EE&C** Engineering Estimating & Cost 工程预算和成本

**EEA** Environmental Effect Analysis 环境影响分析

European Economic Area 欧洲经济区

European Environment Agency 欧洲环境署

**EEAM** Element Event Analysis Method 基元事件分析方法

**EEBR** End to End Brassboard Radar 端对端实验雷达

**EEC** Electric Echo Canceller 电回波消除器

Electronic Engine Control 电子发动机控制

Electronic Engine Controller 发电机电子控制器

Electronic Equipment Committee 电子设备委员会

Electronic Equipment Compartment 电子设备舱

Engine Electronic Control 发动机电子控制

Eurocontrol Experimental Centre 欧控试验中心

European Economic Community 欧洲经济共同体

Extendible Exit Cone 扩展式尾锥管

External Equipment Connector 外部设备连

接器

**EECA**　Engineering Economic Cost Analysis 工程经济费用分析

　　Engineering Evaluation Cost Analysis 工程评估费用分析

**EECM**　Electronic Engine Control Monitor 发动机电子控制监控器

**EECOOL**　Electronic Equipment Cooling Card 电子设备冷却插卡

　　Electronic Equipment Cooling 电子设备冷却

**EECP**　Emergency Engineering Change Proposal 紧急工程更改建议书

**EECS**　Electrical Engineering and Computer Science 电气工程与计算机科学

　　Electronic Engine Control System 发动机电子控制系统

**EECU**　Engine Electronic Control Unit 发动机电子控制装置

**EED**　Engine Exceedance Data 发动机超限数据

**EEDR**　Engineering Electronic Data Retention 工程电子数据保存

**EEDS**　Electrical and Electronic Distribution Systems 电气与电子分布系统

　　Electronic Engine Display System 发动机电子显示系统

**EEE**　Earth and Environmental Engineering 地球与环境工程

　　Electrical & Electronics Engineering 电气和电子工程

　　Electrical, Electronic & Electromechanical 电气、电子和机电的

　　Electromagnetic Environment Experiment 电磁环境试验

　　Electromagnetic Environmental Effects 电磁环境的影响

　　Energy Efficient Engine 节能发动机

**EEFAE**　Efficient and Environmentally Friendly Aero Engine 高效和环境友好的航空发动机

**EEHS**　External Event Handling Service 外部事件处理服务

**EEHWG**　Electromagnetic Effects Harmonization Working Group 电磁影响协调工作组

**EEL**　Emergency Egress Lighting 紧急出口照明

**EE-LED**　Edge-Emitting LED 边发射发光二极管

**EELQMS**　European Engine Lubricants Quality Management System 欧洲发动机润滑油质量管理系统

**EEM**　Enterprise Engineering Management 企业工程管理

　　Extended Memory Management 扩展内存管理

　　External Error Mode 外在差错模式

**EEMB**　Electrical Equipment Mounting Base 电气设备安装基座

**EEO**　Experimental Engineering Order 试验性工程指令

**EEPDC**　Emergency Electrical Power Distribution Center 应急电源分配中心

**EEPGS**　Enhanced Electrical Power Generation System 增强型发电系统

**EEPROM**　Electrically Erasable Programmable Read-Only Memory 电可擦可编程只读存储器

　　Electronically Erasable Programmable Read Only Memory 电子可擦除可编程只读存储器

**EER**　Energy Efficiency Ratio 能效比

　　Equipment Evaluation Report 设备评估报告

　　Explosive Echo Ranging 爆炸回波测距

**EEROM**　Electronically Erasable Read-Only Memory 电可擦只读存储器

**EERPS**　Electronic Equipment Reliability Prediction System 电子设备可靠性预测系统

**EES**　Electrical Engineering Society 电气工程学会

　　Electrical Engineering Squadron 电力工程队

　　Electronic Environment Simulator 电子环境模拟器

Emergency Evacuation System 应急撤离系统

Environmental Engineering Society 环境工程学会

**EESA** End to End Security Assessment 端到端的安全评估

**EET** Energy Efficient Transport 能量效率运输,能量效率运输机

Estimated Elapsed Time 预计航程(经过)时间,预计实耗的时间

Estimated Enroute Time 预计航路时间

**EETC** Energy and Environment Technology Center 能源与环境技术中心

**EETDN** End-to-End Transit Delay Negotiation 端对端传输延迟协商

**EETP** Engineering Evaluation Test Program 工程评估测试计划

**EEU** Easten Europe 东欧

Electronic Equipment Unit 电子设备装置

Engineering Evaluation Unit 工程评价单元

European Economic Union 欧洲经济同盟

Experiment Environmental Unit 试验环境的单位

**EE-VPC** End-to-End Virtual Path Connection 端对端虚拟通路连接

**EEZ** Exclusive Economic Zone 专属经济区

**EF** Elementary Function 基本功能

Expedited Forwarding 加速转发

Experimental Flight 试验飞行

External Flap 外部襟翼

Extended Function 扩展功能

**EFA** Engineering Failure Analysis 工程失效分析

Engineering Flight Activity 工程飞行活动

European Fighter Aircraft 欧洲战斗机

Event Frequency Analysis 事件频率分析

Experiment Flight Applications 试验飞行申请

**EFAG** Emergency Field Arresting Gear 应急机场制动装置

**EFAL** Electronic Flash Approach Lighting 电子闪光进场照明

**EFAS** Electronic Flash Approach System 电子闪光进近系统

Engine Failure Assist System 发动机故障(时使用的)助推系统

Enhanced Fault Alarm System 增强的故障警告系统

Enroute Flight Advisory Service 航路飞行咨询服务

Extended Final Approach Segment 扩展最后进近段

**EFATO** Engine Failure after Take-off (飞机)起飞后发动机停车

**EFB** Electronic Flight Bag 电子飞行包

**EFBGL** Erbium Fiber Bragg Grating Laser 掺铒光纤布拉格光栅激光器

**EFC** Electric Frequency Control 电子频率控制

Electrical Flight Control 电动飞行操纵

Electronic Fuel Control 电子燃油控制

Elevator Feel Computer 升降舵传感计算机

Expected Further Clearance 预期进一步放行许可

**EFCC** Electronic Flight Control Computer 电子飞行控制计算机

**EFCI** Explicit Forward Congestion Indication 显性前向拥塞指示

**EFCL** Everett Flight Control Lab 埃伏瑞特飞行操纵实验室

**EFCP** EFIS Control Panel 电子飞行仪表系统控制板

**EFCS** Electrical Flight Control System 电子飞行控制系统

Emergency Flight Control System 应急飞行控制系统

**EFCT** Effect 影响,作用

Expected Further Clearance Time 预计下次放行时间

**EFCTV** Effective 有效的,生效的

**EFCTV VIS** Effective Visibility 有效能见度

**EFCU** EFC Unit 升降舵传感组件

Electronic Flight Control Unit 电子飞行控制

组件

**EFD** Early Failure Detection 早期故障探测
Electronic Flight Display 电子飞行显示器
Event Forwarding Discriminator 事件前向识
别器
Houston, TX, USA（Airport Code）美国得
克萨斯州休斯敦机场

**EFDARS** Electronic Flight Data and Recording System 电子飞行数据和记录系统
Expandable Flight Data Acquisition and Recording System 可扩展的飞行数据采集和记
录系统

**EFDC** Early Failure Detection Center 早期故
障检测中心

**EFDM** Engine Flight Data Monitoring 发动机
飞行数据监控

**EFDR** Expanded Flight Data Recorder 扩展
的飞行数据记录器

**EFDS** Engine Fault Display System 发动机故
障显示系统

**EFDT** Early Failure Detection Test 早期故障
检测试验

**EFDU** Engine Fault Display Unit 发动机故障
显示装置

**EFE** Electronic Flight Engineer 电子飞行工
程师

**EFEO** European Flight Engineers Organization 欧洲飞行工程师组织

**EFF** Effectivity 有效性
Electronic Flight Folder 电子飞行资料夹
Engine Fuel Flow 发动机燃油流量

**EFFE** European Federation of Flight Engineers 欧洲飞行工程师联合会

**EFFY** Efficiency 效率

**EFH** Engine Flight Hour 发动机飞行小时

**EFI** Electronic Flight Instruments 电子飞行
仪表

**EFIC** Electronic Flight Instrument Controller
电子飞行仪表控制器

**EFID** Electronic Flight Instrument Display 电
子飞行仪表显示器

**EFIP** Electronic Flight Instrument Processor
电子飞行仪表处理器

**EFIS** Electronic Flight Information Systems 电
子飞行信息系统
Electronic Flight Instrument System 电子飞
行仪表系统

**EFISCP** Electronic Flight Instrument System
Control Panel 电子飞行仪表系统控制面扳

**EFISG** Electronic Flight Instrument Symbol
Generator 电子飞行仪表符号产生器

**EFISLS** Electronic Flight Instrument System
Light System 电子飞行仪表灯光系统

**EFISS** Electronic Flight Instrument Symbol
System 电子飞行仪表符号系统

**EFISSG** Electronic Flight Instrument System
Signal Generator Unit 电子飞行仪表系统信
号产生器组件
Electronic Flight Instrument System Signal
Generator 电子飞行仪表系统信号发生器

**EFM** Electromagnetic Frequency Management
电磁频率管理
Enterprise Feedback Management 企业反馈
管理
Exchange of Foreign Money 外币兑换

**EFMS** Electronic Fuel Management System 电
子燃油管理系统
Experimental Flight Management System 试验
飞行管理系统

**EFNDT** European Federation for Non-Destructive Testing 欧盟无损检测联合会

**EFOA** Estimated Fuel of Arrival 预计到达时
的油量
European Fuels Oxygenates Association 欧洲
燃料氧化剂协会

**EFOB** Estimated Fuel on Board 预计机上剩
余燃油

**EFOD** Electronic Filing of Differences 差异电
子归档

**EFP** Enroute Fuel Planning 航路燃油计划
Equivalent Fare Paid 实付等值票价

**EFPAC** Engine Fleet Planning and Costing 发

动机机队计划与成本分析

**EFPHB** Expedited Forwarding PHB 加速转发 PHB

**EFPMS** Engine Fuel Pump and Metering System 发动机燃油泵与计量系统

**EFR** Enhanced Full Rate 增强型全速率
Equipment Failure Rate 设备故障率
Equivalent Fare Paid 等值费用支付

**EFRR** Executive Flight Readiness Review 执行飞行准备就绪检查

**EFS** Effective Flight Strength 现役飞行力量
Efficient Filtration System 高效过滤系统
Electric Field Strength 电场强度
Elevator Feel Shift 升降舵感觉变换
End of Frame Sequence 帧序列结尾

**EFSM** Elevator Feel Shift Module 升降舵感觉变换组件

**EFSSS** Engine Failure Sensing and Shutdown System 发动机失灵感知与停车系统

**EFT** Electronic Funds Transfer 电子资金转账,无纸财会系统
Engine Fuel Tank 发动机燃油箱
Estimated Flight Time 预计飞行时间
Experimental Flight Test 试验飞行测试
External Function Translator 外部函数翻译器

**EFTA** European Free Trade Association 欧洲自由贸易协会,欧洲自由贸易联盟

**EFTC** Engine Fuels Technical Committee 发动机燃料技术委员会

**EFTI** Engineering Flight Test Inspector 技术试飞监察员

**EFTP** Ethernet File Transfer Protocol 以太网文件传送协议

**EFTS** Electronics Fund Transfer System 电子资金转移系统
Elementary Flying Training School 初级飞行训练学校

**EFTWA** Engineering Flight Test Work Authorization 工程性飞行试验工作核准

**EFVS** Electronic Fighting Vehicle System 车载电子作战系统
Enhanced Flight Vision System 增强飞行视觉系统

**EG** Engineering Guide 工程指南
Experimental Group 试验组

**EGA** Enhanced Graphics Adapter 增强型图形适配器

**EGAL** Elevation Guidance for Approach and Landing 进近着陆仰角制导

**EGAO** Egyptian General Aero Organization 埃及航空总署

**EGB** Engine Generator Breaker 发动机发电机断路器

**EGC** Electronic Gyro Compass 电子陀螺罗盘
Emergency Generator Contactor 应急发电机接触器
Equal Gain Combining 等增益组合

**EGD** Equal Gain Diversity 等增益分集

**EGE** Vail/Eagle, CO, USA (Airport Code) 美国科罗拉多州韦尔机场

**EGEMS** Engine Gas-path Electrostatic Monitoring System 发动机气路静电监测系统

**EGH** Engine Ground Handling 发动机地面操纵

**EGIS** Engineering Geographic Information System 工程地理信息系统
Environmental Graphical Information System 环境地理信息系统

**EGIU** Electrical Generation Interface Unit 发电接口组件

**EGM** Enhanced Graphics Module 增强的图形模块

**EGMS** Electrical Galley Management System 厨房电气管理系统

**EGNOS** European Geostationary Navigation Overlay Service 欧洲同步卫星导航覆盖服务
European Geostationary Navigation Overlay System 欧洲同步卫星导航覆盖系统
European Global Navigation Overlay System 欧洲全球导航覆盖系统

**EGP** English for General Purposes 一般用途英语

Exterior Gateway Protocol 外部网关协议

**EGPRS** Enhanced General Packet Radio Service 增强型通用分组无线服务

**EGPWM** Enhanced Ground Proximity Warning Module 增强型近地警告组件

**EGPWS** Enhanced Ground Proximity Warning System 增强型近地警告系统

**EGR** Equitable Geographical Representation 公平地域代表性

Exhaust Gas Recirculation 废气再循环

**EGSE** Electrical Ground Support Equipment 电气地面保障设备

**EGT** Elapsed Ground Time 地面已停留时间

Engine Gas Temperature 引擎烟雾温度

Estimated Ground Time 估计地面停留时间

Exhaust Gas Temperature 排气温度

**EGTT** Expert Group on Technology Transfer 技术转让专家组

**EGV** Exit Guide Vane 出口导流叶片

**EGY** Egypt 埃及

**EH** Eastern Hemisphere 东半球

Engine Hoods 发动机罩

**EHA** Electro Hydrostatic Actuator 电动静液作动器,电动液压传动装置

European Helicopter Association 欧洲直升机协会

Evergreen Helicopters of Alaska, Inc. (USA) 阿拉斯加长青直升机公司(美国)

**EHAC** Electro-Hydrostatic Actuator Contactor 电—液作动筒接触器

**EHAS** Environmental Health and Safety 环境健康与安全

**EHCU** Electromechanical Handforce Control Unit 电动机械臂力控制系统

**EHD** Engine Handling Document 发动机处理文件

**EHDI** Electronic Horizontal Director Indicator 电子指引地平仪指示器

**EHF** Extremely High Frequency (30000 to 300000 MHz) 极高频

**EHL** Electronic Home Library 电子家庭图书馆

**EHM** Engine Health Management 发动机健康管理

Engine Health Monitoring 发动机健康监控

Engine Heavy Maintenance 发动机大修

**EHMP** Engine Health Monitoring Program 发动机状态监控方案

**EHMS** Engine Health Monitoring System 发动机健康状态监控系统

**EHOC** European Helicopter Operators' Committee 欧洲直升机承运人委员会

**EHP** Equivalent Horse Power 当量马力

**EHPS** Electronic Hydrostatic Power Steering 电子液压助力转向

**EHR** Equipment History Record 设备经历记录

**EHRS** Engine on Hours 发动机工作时间

Environmental Health and Radiation Safety 环境卫生和辐射安全

**EHS** Electrical Heater System 电加热器系统

Extra High Shear 极高剪力

Extremely Hazardous Substance 特别危险物质

**EHSD** Electronic Horizontal Situation Display 电子水平姿态显示器

**EHSI** Electronic Horizontal Situation Indicator 电子水平姿态指示器

Elevator Horizontal Situation Indicator 升降舵水平姿态指示器

**EHSID** Electronic Horizontal Situation Indicator Display 电子水平姿态指示器显示

**EHSV** Electro-Hydraulic Servo Valve 电—液伺服活门(阀)

**EHV** Electro-Hydraulic Valve 电液活门(阀)

Extra High Voltage 超级高压,极高压,超高压

**EHW** Extreme High Water 最高水位

**EI** Electrical Integration 电子集成

Electronic Interface 电子接口

Emissions Index 排烟指数

End Item 最后项目,结束项目

Engine Instruments 发动机仪表

Engineer Inspection 工程师检查

Engineering Index 工程引文检索,美国工程技术文献索引

Engineering Information 工程情报

Engineering Installation 工程安装

Engineering Instruction 工程说明

Erroneous Indication 错误指示

Exchange Identification 交换机标识

**EIA** Early Impact Analysis 早期影响分析

Electronic Industries Association 电子工业协会

Electronics Institute of America 美国电子学会,电子工业联合会

Engineering Index Annual 美国工程技术文献索引年刊

Environmental Impact Analysis 环境影响分析

Environmental Impact Assessment 环境影响评估

Evergreen International Airlines 美国常青国际航空公司

**EIAP** European Integrated Aeronautics Plan 欧洲综合航空计划

**EIB** European Investment Bank 欧洲投资银行

**EIC** Electrically Insulated Coating 电绝缘涂层

Engineer-in-Chief 总工程师

Engineering & Installation Center 工程与安装中心

Engineering Item Code 工程条形码

Equipment Identification Code 设备识别编码

Equipment in Compartment 货舱设备

Equipment Installation and Checkout 设备安装与测试

**EICAS** Engine Indicating and Condition Advisory System 发动机指示和状态咨询系统

Engine Indication and Crew Advisory System 发动机指示和机组咨询系统

Engine Indication and Crew Alerting System 发动机指示和机组警告系统

Engine Instrument and Crew Alerting System 发动机仪表和机组警告系统

**EICASC** Engine Indication and Crew Alerting System Controls 发动机指示和机组报警系统控制元件

**EICD** Electrical Interface Control Document 电气界面控制文件

**EICMS** Engine In-Flight Condition Monitoring System 发动机空中状况监控系统

**EID** Electrically Inspectorate Directorate 电气检查局

Electron Impact Desorption 电子碰撞脱附

Emergency Isolation Device 应急隔离装置

Engineering Installation Drawing 工程安装图纸

**EIDI** Electro-Impulse De-Icing 电脉冲除冰

**EIDS** Electronic Information Delivery System 电子信息分发系统

**EII** Enterprise Information Integration 企业信息集成

Environmental Impact Index 环境影响指标

**EIIP** Employment Intensive Investment Programme 就业密集型投资计划

**EIL** Electrical Integration Laboratory 电气集总实验室

**EILC** Emergency Interlock Line Contactor 紧急互锁线路接触器

**EIM** Engineering Information Management 工程信息管理

Exchange Interface Module 交换机接口模块

**EIMS** End Item Maintenance Sheet 最终项目维修分析单

**EIN** Eindhoven, Netherlands ( Airport Code ) 荷兰艾恩德霍芬机场

Environmental Information Network 环境信息网

Excitation Voltage 励磁电压

**EINS** Enterprise Information Network System

企业信息网络系统

**EIO** Error in Operation 操作差错

**EIP** Enterprise Information Portal 企业信息门户

**EIPC** Engine Illustrated Parts Catalogue 发动机图解零件目录

**EIPM** Engine Interface Power Management 发动机接口电源管理

**EIPMU** Engine Interface Power Management Unit 发动机接口电源管理组件

**EIR** Equipment Identification Register 设备身份登记器

Excessive Information Rate 过大的信息速率

Exhibit Investigation Request 展览调查申请

**EIRD** Equipment Installation Requirement Document 设备安装要求文件

**EIRP** Equivalent Isotropic Radiate Power 等效全向辐射功率

**EIS** Electronic Instrument System 电子仪表系统

End Item Specification 最终项目技术规格

Engine Indicating System 发动机指示系统

Engine Instrument System 发动机仪表系统

Engineering Information System 工程信息系统

Entry into Service 投入使用，投入运营

Environmental Impact Statement 环境影响报告

Environmental Information Systems 环境信息系统

**EISA** Enhanced Industry Standard Architecture 增强型工业标准架构

Extended Industry Standard Architecture 扩展的工业标准架构

**EIT** Electrical Information Test 电气信息测试

Electromagnetic Interference Testing 电磁干扰试验

Electronic Information Technology 电子信息技术

Engineer in Training 受训工程师

European Institute of Technology 欧洲技术学会

External Interface Testing 外部接口测试

**EIU** Electronic Interface Unit 电子接口组件

Engine Interface Unit 发动机接口组件

Experimental Interaction Unit 实验互动装置

**EIVMU** Engine Interface and Vibration Monitoring Unit 发动机接口和振动监控组件

**EJS** Engineering Job Sheet 工程工作单

**EKF** Extended Kalman Filter 扩展卡尔曼滤波器

**EKO** Elko, NV, USA (Airport Code) 美国内华达州埃尔科机场

**EL** Early 早的

Echo Loss 回声损耗

Element Layer 单元层

Elevation 标高

Emergency Lighting 应急灯

Empty Load 空载

Engineering Laboratory 工程实验室

Equipment List 设备清单

Error Level 误差等级

Expected Loss 预期损失

Export Licence 出口许可证

Reliability 可靠性

**ELA** Electrical Load Analysis 电负载分析

Enroute Low Altitude 航线最低安全高度

Equipment Leasing Association 英国设备租赁协会

**ELAC** Electroacoustic 电声的

Elevator and Aileron Computer (Airbus A320) 升降舵与副翼计算机

**ELAN** Ethernet Local Area Network 以太局域网

**ELAR** Electrical Load Analysis Report 电气负载分析报告

**ELB** Electronic Log Book 电子飞行日记

Emergency Location Beacon 紧急定位信标

**ELBA** Emergency Location Beacon-Aircraft 航空器紧急定位信标

**ELC** Electronic Library Computer 电子文件

库计算机

Emergency Line Contactor 应急线路接触器

**ELCA**  Earth-Landing Control Area 地面降落控制区

**ELCC**  Electrical Load Control Contactor 电负荷控制接触器

**ELCF**  Electrical Load Control Function 电负荷控制功能

**ELCM**  Electrical Load Control Module 电负荷控制模块

**ELCT**  Electrical 电的,电气的
Electronic 电子的

**ELCU**  Electrical Load Control Unit 电气负载控制组件

**ELD**  El Dorado, AR, USA (Airport Code)美国阿肯色州埃尔多拉多机场
Elder 年长的
Eldest 最年长的
Electronic Load Detector 电子负载检测器
Extra Load Devices 额外装载设备

**ELE**  Elevation, Elevator 标高,升降舵
Emergency Lighting Equipment 应急照明设备

**ELEC**  Electric, Electrical, Electricity 电的,电气的,电学
Electrical 电的,电气的 与电有关的,电气科学的

**ELEK**  Electronic 电子的

**ELEKS**  Electronics 电子学

**ELEM**  Element 元件

**ELEV**  Elevation 标高,高度,海拔
Elevator 升降舵,升降机

**ELEV STA**  Elevator Station 升降舵站位

**ELEVTR**  Elevator 升降舵

**ELEX**  Electronic 电子
Electronics 电子学

**ELF**  El Fasher, Sudan (Airport Code) 苏丹法舍尔机场
Electronic Library Function 电子图书馆功能
Elevator Load Feel 升降舵负载感觉
Extremely Low Frequency (3~30 Hz) 极低

频

**ELFA**  Elevator Load Feel Actuator 升降舵载荷探测执行机构

**ELFC**  Engine Lease Finance Corporation 引擎租赁融资公司

**ELG**  Emergency Landing Ground 应急降落场

**ELGE**  Emergency Landing Gear Extension 起落架应急放下

**ELI**  Equipment List Identification 设备识别清单

**ELIM**  Eliminate 消除

**ELINT**  Electronic Intelligence 电子智能,电子情报

**ELM**  Electrical Load Management 电负载管理
Element 元件
Elmira, NY, USA (Airport Code)美国纽约州艾尔迈拉机场
Extended Length Messages 扩展长度的讯息码

**ELMC**  Electrical Load Management Centers 电气负载管理中心

**ELMF**  Electrical Load Management Function 电气负载管理功能

**ELMS**  Electrical Load Management System 电气负载管理系统

**ELMT**  Element 元件

**ELMU**  Electrical Load Management Unit 电气负载管理组件

**ELP**  El Paso, TX, USA(Airport Code) 美国得克萨斯州埃尔帕索机场

**ELR**  Engineering Liaison Request 工程联系要求
Extra Long Range 超远程

**ELS**  Earth Landing System 地面着陆系统
Electric Loads Simulator 电子负载模拟器
Electronic Library System 电子资料库系统
Emergency Landing Strip 应急着陆跑道
Emergency Lighting System 应急灯光系统
Emitter Location System 发射机定位系统

Equipment List Summary 设备清单摘要,设备清单总结

**ELSI** Extremely Large Scale Integrated Circuit 超大规模集成电路

**ELSP** Ethernet LAN Simulation Package 以太网局域网模拟分组信息

**ELT** Electronic Technician 电子设备技师

Emergency Locator Transmitter 应急定位器发射机

Emergency Locator 应急定位器

**ELV** Elevation 标高

Elevator 升降舵,升降机

**ELW** Extreme Low Water 极低水位

**EM** Echo Monitor 回波监视器

Economy and Management 经济与环境

Electro Magnetic 电磁

Electronic Mail 电子邮政

Electronic Mailbox 电子信箱

Electronic Manual 电子手册

Emergency Maintenance 紧急维护

Emergency Management 应急管理

Emigrant 移民

Emigration 移民局

Emission 发射

Engine Manual 发动机手册

Engineering Manual 工程手册

Engineering Model 工程模型

Environmental Management 环境管理

Evaluation Model 评估模型

Expectation Maximization 期望最大化

Explanatory Memorandum 工程备忘录

**EMA** Electro-Mechanical Actuator 机电作动器

Electronics Manufacturers Association 电子制造商协会

Engineering Method Analysis 工程方法分析

Enterprise Management Association 企业管理协会

Environmental Management Association 管理环境协会

Ethernet Media Adapter 以太网卡适配器

Excess Mileage Allowance 超里程优惠

**EMAD** Engine Maintenance, Assembly and Disassembly 发动机维修、装配和拆卸

**E-mail** Electronic Mail 电子邮件

**EMAS** Energy Management Accounting System 能源管理会计系统

Engineered Materials Arresting System 工程材料阻拦系统

Equipment Marking Accounting System 装备标记计算系统

**EMB** Embargo 禁运

Embark 登机

Embassy 大使馆

**EMBA** European Masters of Business Administration 欧洲工商管理硕士

Executive Master of Business Administration 高级工商管理硕士

**EMC** Electro Magnetic Compatibility 电磁兼容性

Electro-Mechanical Controller 机电控制器

Engine Maintenance Center 发动机维修中心

Engine Management Center 发动机管理中心

**EMCAB** Electromagnetic Compatibility Advisory Board 电磁兼容性咨询委员会

**EMCD** Electromagnetic Compatibility Directive 电磁兼容指令

Equipment Maintenance Cost Dispatch 设备维修成本分派

**EMCEQ** Emergency Call Equipment 紧急呼叫设备

**EMCF** European Monetary Cooperation Fund 欧洲货币合作基金组织

**EMCM** Engineering Material Control Manual 工程物料控制手册

**EMCON** Electromagnetic Emission Control 电磁辐射控制

Emergency Control 紧急控制

Emission Control 辐射控制

**EMCU** Electronic Motor Control Unit 电动马达控制组件

Electro-Mechanical Control Unit 电动机械控

制组件

Emergency Manual Control Unit 应急手动控制组件

**EMD** Electric Motor Driven 电动机驱动的

Electronic Map Display 电子地图显示

Electronic Miscellaneous Document 电子杂费凭证,电子收费凭证

Emerald, Queensland, Australia ( Airport Code)澳大利亚昆士兰州埃默拉尔德机场

Emergency Management Division 紧急情况管理部

Engine Management Display 发动机管理显示器

Engineering Manufacturing Design 工程制造设计

Engineering Manufacturing Development 工程制造及开发

Environmental Management Division 环境管理部

Equilibrium Mode Distribution 稳态模分布

**EMDB** Electrical Main Data Bank 电子主数据库

Electro-Magnetic Data Base 电磁数据库

Equipment Maintenance Data Base 设备维修数据库

**EMDP** Electric Motor Driven Pump 电马达驱动泵

**EMDT** Executive Management Development Training 高级管理人员发展培训

**EME** Electrical & Mechanical Engineers 电气与机械工程师

Electromagnetic Environment 电磁环境

**EMED** Engine Manufacturing Engineering Definition 发动机制造工程定义

**EMER** Emergency 应急,紧急

Emergent 应急的,紧急的

**EMERG** Emergency 紧急情况,紧急,应急

**EMF** Electromagnetic Field 电磁场

Electromagnetic Force 电磁力

Electromagnetic Frequency 电磁频率

Electro-Motive Force 电动势

European Monetary Fund 欧洲货币基金组织

**EMFI** Electro-Mechanical Flight Instruments 电动—机械飞行仪表

**EMG** Eastern Management Group 东方管理组

Electro-Magnetic Generator 电磁发电机

Emergency 紧急

Environment Management Group 环境管理组

Environmental Management Guidelines 环境管理准则

**EMGE** Electronic Maintenance Ground Equipment 电子维修地面设备

**EMH** Expedited Message Handler 加速报文处理器

**EMHB** Expedited Message Handler Buffer 加速报文处理器缓冲区

**EMHP** Electronic Motor Hydraulic Pump 电马达液力泵

**EMI** Electro-Magnetic Interference 电磁干扰

**EMIS** Education Management Information System 教育管理信息系统

Electronic Management Information System 电子管理信息系统

Emergency Management Information System 应急管理信息系统

Environmental Management Information System 环境管理信息系统

Equipment Management Information System 设备管理信息系统

**EMISM** Electro Magnetic Interference Safety Margin 电磁干扰安全裕度

**EMISN** Emission 发射,放射,污染

**EMIT** Engineering Management Information Technique 工程管理信息技术

Engineering, Manufacturing, Information Technology 工程、制造与信息技术

**EMITS** Electromagnetic Interference Test System 电磁干扰测试系统

**EML** Element Management Layer 单元管理层

Environmental Measurements Laboratory

（USA）环境测量实验室

Equipment Modification List 设备更改清单

Estimated Maximum Loss 估计最大损失

**EMLS**　Emergency Lighting System 应急灯系统

**EMM**　Engine Maintenance Manual 发动机维修手册

**EMMIS**　Electronics Maintenance Management Information System 电子设备维护管理信息系统

**EMO**　Electromechanical Optical 电机械光学装置

Emergency off 紧急脱离,应急切断

Equipment Management Office 设备管理办公室

Equipment Move Order 设备迁移指示

**EMOS**　Environmental Management Overview Strategy 环境管理综览策略

European Meteorological Satellite 欧洲气象卫星

**EMP**　Electromagnetic Pulse 电磁脉冲

Electronic Motor Pump 电动马达泵

Empennage 尾翼

Evaluated Maintenance Programming 经过评价的维修计划

**EMPC**　Electro-Mechanical Power Controller 机电混合式功率控制器

**EMPL**　Employee 雇员

**EMPLMT**　Employment 雇用

**EMPPTYDT**　Empty Seat 空座位

**EMPS**　Emergency Power Supply 应急电源

**EMR**　Electromagnetic Radiation 电磁辐射

Expenditure Management Report 支出管理报告

**EMS**　Elastic Mode Suppression 弹性模态抑制

Electromagnetic Spectrum 电磁频谱

Electromagnetic Surveillance 电磁监视

Electronic Mail Service 电子邮政服务

Electronic Mail System 电子邮政系统

Element Management System 网元管理系统

Emergency Medical Services 紧急医疗服务

Energy Management System 能量管理系统

Engine Management Support 发动机管理支援

Engine Management System 发动机管理系统

Engine Master Switch 发动机主电门

Engine Monitoring Section 发动机监控科

Engineering Mainline Systems 干线工程系统

Enterprise Message System 企业信息系统

Environment Maintenance Squadron 环境保护中队

Environmental Management Systems 环境管理系统

Equipment Management System 设备维修系统

Equipment Modification Sheet 设备改装单

European Mobile Satellite System 欧洲移动卫星系统

European Mobile Satellite 欧洲移动通讯卫星

Excess Mileage Surcharge 超里程附加费

Express-Mail Services 快递服务,邮政特快专递

**EMSC**　Emergency Monitoring and Support Centre 紧急事故监察及支援中心

Engine Monitoring System Computer 发动机监控系统计算机

**EMSD**　Electrical and Mechanical Services Department（Hong Kong）（香港）机电工程署

**EMSG**　European Maintenance System Guide 欧洲维修系统指南

**EMSR**　Expected Marginal Seat Revenue 期望边际座位收入

**EMSTF**　Electrical and Mechanical Services Trading Fund 机电工程营运基金

**EMT**　Elapsed Maintenance Time 维修历经时间

Elapsed Mission Time 任务历经时间

Engine Materials Technology 发动机材料技术

**EMTBF**　Estimated Mean Time between Failures 预估故障平均间隔时间

**EMTR** Emitter 发射极,发射器

**EMU** Engine Maintenance Unit 发动机维护组件

Engine Monitoring Unit 发动机监测装置

European Monetary Union 欧洲货币联盟

European Monetary Unit 欧洲货币单位

**EUMS** Engine Usage Monitoring System 发动机使用监控系统

**EMV** Electromagnetic Volume 电磁电容

**EMWS** Electro Mechanical Window Shade 电气机械遮光板

**EN** Engineering Notice 工程通知

English 英国的,英语,英国人

Enterprise Network 企业网

Equipment Number 设备号

Extended Nomenclature 扩展名称

**ENA** Exhaust Nozzle Area 排气喷口面积

Kenai, AK, USA(Airport Code)美国阿拉斯加州基奈机场

**ENBL** Enable 使能够

**ENC(L)** Enclosure 信内附件

**ENCD** Encode 编码

**ENCDR** Encoder 编码器

**ENCL** Enclose 附寄

Enclosure 外壳,随函附入,包围,内附的

**ENCOMS** Engine Control Monitoring System 发动机监控系统

**END** Endorsement 签转,担保

Endurance 续航时间,持久性

**ENDCE** Endurance 续航时间

**ENDG** Ending 最后,末尾

**ENDRS** Endorsement 签转,担保

**ENDVR** Endeavour 努力,尽力

**ENE** East North East 东北东

**ENG** Engaging 接合,连接

Engine 发动机

Engineer 工程师

Engineering 工程

England 英国

English 英国的,英语,英国人

**ENG LIM** Engine Limit 发动机极限

**ENG OUT** Engine out 发动机失效

**ENG S/N** Engine Series Number 发动机序号

**ENG STA** Engine Station 发动机站位

**ENGA** Engage 啮合,连接

**ENGD** Engaged 衔接

**ENGL** England 英国

English 英国的,英语,英国人

**ENGMT** Engagement 约定,义务

**ENGS** Engines 发动机

**ENGV** Engine V-belt 发动机 V 形带

**ENISA** European Network and Information Security Agency 欧洲网络和信息安全局

**ENM** Electrical Network Management 电网管理

**ENMF** Electrical Network Management Function 电网管理功能

**ENMS** Electrical Network Management System 电网管理系统

**ENOC** Engineering Network Operations Center 工程网络运行中心

**ENP** Estimated Normal Payload 预计正常载荷

**ENQ** Enquiry 询问,查询

**ENR** En Route 在航路上,在途中

**ENRGZ** Energize 通电,供电

**ENRI** Electronic Navigation Research Institute 电子导航研究所

**ENRL** Enroll 注册,登记

**ENRPC** Environmental and Natural Resources Protection Committee 环境和自然资源保护委员会

**ENRT** En Route 在途中,在航路上

**ENS** E-mail Notifying Server (电子)邮件通知服务器

Ensure 担保,保证

Enterprise Network Services 企业网络服务

Enterprise Network Strategy 企业网络战略

**ENT** Engineering Note 工程要点

Enter 输入

Entertainment 娱乐

Entrance 入口,门口

Entry 进入,入口,输入,开始

Exhaust Nozzle Temperature 排气喷管温度

**ENTC** Engine Negative Torque Control 发动机负扭矩控制

**ENTMT** Entertainment 娱乐

**ENTRY** Entry 入境

**ENV** Envelope 信封,包线,范围

Environment 环境

**ENVLP** Envelope 信封,包线,范围

**EO** End Office 端局

Engine Oil 发动机滑油

Engine out 发动机停车

Engineering Order 工程指令

Executive Order 执行次序

**EOA** End of Address 地址末尾

**EOAC** Engine out Altitude Capability 发动机失效高度能力

**EOAR** European Office of Aerospace Research 欧洲宇航研究处

**EOB** End of Block 数据块结束符

**EOBT** Estimated off Block Time 预计撤掉轮挡时间

**EOC** Embedded Operation Channel 嵌入式操作通道

Emergency Operations Center 应急运行中心

End of Contract 合同终止

End of Course 航线终了,课程终了

End of Cruise 巡航结束

Engineering Order Change 工程指令更改

Engine-out Capability 发动机停车性能

Equivalent Operational Capability 等效作战能力

Error of Closure 闭合误差

**EOCM** Electro-Optical Countermeasures 电子光学对抗

Environment-Oriented Cost Management 面向环境的成本管理

**EOCMS** Electro-Optical Countermeasures System 电光对抗系统

**EOCP** Engine out of Commission for Parts 发动机因缺少零件而停止使用

**EOCR** Engineering Order Control Record 工程指令控制记录

**EOCTS** Electro-Optical Contract Test Set 电光接触器试验装置

**EOD** Education on Demand 教育点播

End of Data 数据末端

End of Date 截止日期

Entrance on Duty 执行任务,进入值勤岗位

Every Other Day 每隔一天

**EODD** Electro-Optical Digital Deflector 电光数字式偏转器

**EOE** Earth Orbit Ejection 地球轨道入轨

**EOEF** Engineering Order Evaluation Form 工程指令评估单

**EOF** Emergency Operations Facility 紧急运行设施

End of File 卷尾

End of Flight 飞行终止

End of Frame 帧结束

**EOFC** Electro-Optical Fire Control 电子光学火控

**EOFP** End of Flight Plan 飞行计划终止

**EOGO** Eccentric Orbiting Geophysical Observatory 偏心轨道地球物理观测台

**EOI** Electro-Optical Instrumentation 电子光学仪表

End of Image 图像的结束

End of Inquiry 询问结束

Engine Operating Instructions 发动机使用手册

Engineering Operating Instruction 工程工作说明

**EOIATS** Electro-Optical Identification and Tracking System 电子光学识别与跟踪系统

**EOIEC** Effects of Initial Entry Conditions 进入大气层的最初条件影响

**EOIFS** Engineering Order Implementation Feedback Sheet 工程指令执行反馈单

**EOIS** Electro-Optical Imaging System 光电成像系统

**EOL** Effective Operational Length 有效工作长度(时间)

End of Life 寿命末期

End of Line 线的尽头,线端

End of List 清单终止

Engine-off Landing 发动机关机着陆

Expected Operating Life 预计使用寿命

Expression-Oriented Language 措辞专用语言,表述专用语言

**EOLB** End of Line Block 行数据块终止

**EOLM** Electro-Optical Light Modulation 电—光光调制

Electro-Optical Light Modulator 电—光光调制器

**EOM** Earth Observation Mission 地球观测任务

Earth Orbital Mission 地球轨道飞行任务

End of Message 电文完,信息结束,电极末尾

End of Mini-up 小型操作结束

End of Mission 任务结束

End of Month 月底

Equation of Motion 运动方程式

Equations of Motion Program 运动方程程序

Every Other Month 每隔一个月

**EOMA** Emergency Oxygen Mask Assembly 应急氧气面罩组件

**EOMF** End of Machine Functions 机器作用停止,机器停止

**EOP** Emergency Oxygen Pack 应急氧气包

End of Packet 分组结尾

End of Part 部分终止,零件端头

Engine Oil Pressure 发动机滑油压力

Engine Operating Point 发动机工作点

Engineering Operating Procedures 工程工作程序

Equipment Operating Procedure 设备操作程序

Experiment of Opportunity Payload 搭载有效载荷实验

Experimental Operating Procedure 实验操作程序

**EOPF** End of Powered Flight 动力飞行结束

**EOPP** Earth Observation Preparatory Program 地球观测筹备计划

**EOQ** Economic Ordering/Order Quantity 经济订货量

**EOQC** European Organization for Quality Control 欧洲质量控制组织

**EOR** Earth-Orbit Rendezvous 地球轨道会合

End of Record 记录结束

Engine Order 发动机订货单

Engineering Order Report 工程指令报告

Exclusive or Register 异寄存端,异—或寄存器

Extend off Retract 放下

**EOS** Earth Observation Satellite 地球观测卫星

Earth Observation System 地球观测系统

Electrical Over Stress 电超载

Electro Optical Scanner 电光学扫描仪

Electro Optical System 光电系统,电子光学系统

Electronic Order System 电子订货系统

Electrophoresis Operations in Space 空间电泳作业

Embedded Operating System 嵌入式操作系统

Emergency Operations System 应急操作系统

Emergency Oxygen System 应急氧气系统

End of Signaling 信令结束

**EOSAT** Earth Observation Satellite (Company) 地球观测卫星(公司)

**EOSID** Engine out Standard Instrument Departure 发动机失效标准仪表离港

**EOSP** Electro-Optical Signal Processor 电子光学信号处理器

**EOSPC** Electro-Optical Signal Processing Computer 电子光学信号处理器计算机

**EOSS** Earth Orbiting Space Station 地球轨道空间站

Electro-Optical Sensor System 光电传感器系

统

Electro-Optical Simulation System 光电仿真系统

Electro-Optical Surveillance System 光电监视系统

**EOST** Emergency Operations Simulation Technique 应急操作模拟技术

**EOT** End of Tape 录音结束

End of Text 文本结束

End of Transmission 传输结束,发射完毕

Engine Oil Temperature 发动机润滑油温度

**EOTAD** Electro-Optical Target Acquisition and Designation 电光目标截获与识别

**EOTADS** Electro-Optical Target Acquisition and Display System 电光目标截获与显示系统

**EOTDS** Electro-Optical Target Detection System 电光目标探测系统

**EOTS** Electro-Optical Threat Sensor 电光威胁传感器

Electron Optic Tracking System 电光跟踪系统

**EOTV** Electric Orbit Transfer Vehicle 电推进轨道转移飞行器

**EOU** Electro Optical Unit 电光装置

**EOV** Electrically Operated Valve 电操纵活门,电控阀

**EOVL** Engine out Vertical Landing (发动机)停车垂直降落

**EOW** Electro Optical Warfare 电子光学战

End of Week 周末

Engine-over-the-Wing 翼上发动机

**EOY** End of Year 年终

**EP** Earth Plate 接地板

Eastward Position 向东位置

Electric Panel 仪表配电板

Electric Polarization 电极化

Electric Primer 电火帽

Electrical Power 电力

Electronic Purse 电子钱包

Electron-Photon 电子光子的

Electroplate 电镀,电镀物

Electro-Pneumatic 电动—气动的

Electrostatic Powder 静电粉末

Element Processor 单元处理机

Elongated Punch 延长穿孔

Engine Performance 发动机性能

English Patent 英国专利

Entry Panel 输入面板

Entry Point 进入点

Estimated Position 预计位置,估计位置

Etched Plate 蚀刻板

European Parliament 欧洲议会

Executive Program 执行程序

Expanded Pitch 延伸俯仰刻度盘

Explosion Proof 防爆的

External Power 外电源

External Pressure 外部压力,外压

Extreme Pressure 最大压力,极限压力

**EP AVAIL** External Power Available 外电源可用

**EP&D** Electrical Power and Distribution 电力与分配

**EPA** Eastern Provincial Airways (Canada) 东部省航空公司(加拿大)

Economic Planning Agency 经济企划厅(日),经济计划局(新加坡)

Education and Public Affairs Inc. 教育和公共事务公司

Engineering Parts Approval 工程零部件批准

Environmental Prediction Assessment 环境预测评估

Environmental Protection Act 环境保护法案

Environmental Protection Agency 环境保护局,环境保护署

European Part Approval 欧洲零部件批准

Extended Power Aging 扩大功率衰减

**EPAD** Effective Project Approval Document 有效项目批准文件

Electrically Powered Actuation Design 电动作动设计

**EPAS** Emergency Power Assist System 应急

动力辅助系统

**EPB** Engine Power Board 发动机动力委员会

**EPC** Electrical Power Center 电源中心

Electronic Power Conditioner 电子功率调节器

Engine-Performance Computer 发动机性能计算机

Error Producing Condition 差错诱发条件

Experiment Package Console 实验仪器组件操纵台

External Power Connection 外部电源连接

External Power Contactor 外部电源接触器

**EPCC** Engine Performance Control Committee 发动机性能控制委员会

**EPCG** Engine Performance Control Group 发动机性能控制组

**EPCM** Engineering Process Control Manual 工程工艺控制手册

**EPCO** Emergency Power Cutoff 紧急断电,紧急切断动力

**EPCR** External Power Control Relay 外电源控制继电器

**EPCS** Electric Power Control System 电源控制系统

Electronic Propulsion Control System 电子推进控制系统

Engine Performance Control System 发动机性能控制系统

Engine Propulsion Control System 发动机推力控制系统

**EPCU** Electric Power Control Unit 电动力控制组件

Electric Propulsion Control Unit 电推力控制组件

Electrical Power Control Unit 电源控制装置

**EPD** Earliest Possible Date 尽早的日期

Early Packet Discard 早期包(分组)丢弃

Electric Power and Distribution 电源与分配

Electrical Power Distribution 电力分配

Electrophoretic Display 电泳显示

Emergency Phone Dialer 紧急电话拨号机

Energetic Particle Detector 高能粒子探测器

Engine Parameter Display 发动机参数显示器

Environmental Protection Department 环境保护署

Environmental Protection Devices 环境保护装置

Extended Principle Diagram 展开的原理图

**EPDC** Electrical Power Distribution and Control 电源分配和控制

Electrical Power Distribution Center 电源分配中心

**EPDD** Engineering Product Definition Data 工程产品定义数据

**EPDP** Engine Preliminary Design Program 发动机初步设计大纲

**EPDS** Electronic Processing and Dissemination System 电子处理与分发系统

**EPE** Emergency Passenger Exit 旅客应急出口

Energetic Particle Explorer 高能粒子探测器

Enhanced Performance Engine 增强性能发动机

Estimated Position Error 估计位置误差

**EPEC** Electric Program Evaluator Controller 电程序装置,鉴定器,控制器

**EPED** Everett Production Engineering Document 埃弗瑞特生产工程文件

**EPEFE** European Programme on Emissions, Fuels and Engine Technologies 欧洲排放,燃料及发动机技术计划

**EPERA** Extractor Parachute Emergency Release Assembly 降落伞应急释放分高装置

**EPES** Ejector-Powered Engine Simulator 喷射驱动的发动机模拟器

**EPESC** Enhanced Passenger Entertainment System Controller 增强型旅客娱乐系统控制器

**EPF** Electric and Percussion Fuse 电发和击发两用引信

**EPFCS** Electrical Primary Flight Control System 初级飞行电子控制系统

**EPFP** External Power Fault Protection 外部电源故障保护

**EPG** Electronic Program Guide 电子节目指南

Electronic Proving Ground 电子验证场

European Participating Government 欧洲参加国政府

**EPGDS** Electrical Power Generation and Distribution System 电源产生和分配系统

**EPGS** Electrical Power Generation System 发电系统

External Power Generation System 外电源系统

**EPI** Electronic Position Indicator 电子位置指示器

Elevation Position Indicator 仰角位置指示器

Elevator(Surface) Position Indicator 升降舵位置指示器

Engine Performance Indicator 发动机性能指示器

Engineering Problem Item 工程问题项目

Engineering Program Integration 工程项目集成

Evaluator Program IC 鉴定程序编制器集成电路

Expanded Position Indicator 扩展的位置指示器

**EPIA** European Photovoltaic Industry Association 欧洲光电设备工业协会

**EPIC** Earth Pointing Instrument Carrier 指向地球的仪器运载卫星

Electronic Preassembly Integration on CATIA 计算机图形辅助三维交互应用工程预组装

Electronic Properties Information Center 电子特性信息中心

Emergency Photovoltaic Industry Association Center 应急措施信息中心

Engineering and Production Information Control 工程和生产信息管理

Engineering Preassembly in CATIA 计算机图

形辅助三维交互应用工程预组装

Epitaxial Passivity Integrated Circuit 外延被动化集成电路

Extended Performance and Increased Capability 扩大的性能及增加的能力

**EPICS** Engine Production and Information Control System 发动机生产和信息控制系统

Enlisted Personnel Individualized Career Systems 单独征召人员专业系统

**EPIL** Experimental Preflight Inspection Letter 实验性飞行前检查书

**EPIRB** Electronic Position Indicating Radio Beacon 电子定位指示无线电信标

Emergency Position Indicating Radio Beacons 紧急定位指示无线电信标

**EPL** Engine Power Lever 油门杆

Extreme Pressure Lubricant 极高压润滑剂

**EPLB** Engine Performance Log Book 发动机性能记录本

**EPLD** Electronically Programmable Logic Device 电可控程序逻辑装置

Erasable Programmable Logic Device 可擦除可编程逻辑装置

Estimated Payload 预计业载

**EPLRS** Electronic Position Location and Reporting System 电子定位报告系统

Enhanced Position Locating Reporting System 增强型定位报告系统

**EPM** Electric Pulse Motor 电脉冲电动机

Electronic Postmark 电子邮戳

Engineering Procedures Manual 工程程序手册

Engineering Program Manager 工程项目经理

Enterprise Performance Management 企业绩效管理

Enterprise Project Management 企业项目管理

Eternal Power Monitor 外部电源监视器

**EPMA** Electronic Parts Manufactures Association 电子部件制造商协会

**EPMS** Electrical Power Management System

电源管理系统

Electronic Power Management System 电子电源管理系统

Emergency Power Management System 应急电源管理系统

Engine Performance Monitoring System 发动机性能监控系统

Engineering Project Management System 工程项目管理系统

Environment Protection Management Systems 环境保护管理系统

**EPMU** Engine Power Management Unit 发动机电源管理组件

**EPN** Effective Perceived Noise 有效感觉噪声

**EPNDB** Effective Perceived Noise Decibels Level 有效感觉噪声分贝级

**EPNdB** Effective Perceived Noise Decibels 有效感觉噪声分贝

**EPNL** Effective Perceived Noise Level 有效感觉噪声级

**EPO** Earth Parking Orbit 地球停泊轨道

Emergency Power off 紧急切断电源

Engine Propeller Order 发动机螺桨指令

**EPP** Emergency Power Package 应急动力包,应急电源包

Error Performance Parameter 差错性能参数

**EPPI** Electronic Plan Position Indicator 电子平面位置指示器

**EPPT** Electrical Power Production Technician 电力生产技术员

**EPR** Electronic Parts Reliability 电子部件可靠性

Engine Power Rating 发动机功率额定值

Engine Pressure Ratio 发动机增压比,发动机压力比

Engine Pressure Regulator 发动机压力调节器

Equivalent Parallel Resistance 等效并联电阻

Error Pattern Register 错误模式寄存器

Essential Performance Requirements 基本性能要求

Eternal Power Relay 外部电源继电器

Exhaust Pressure Ratio 排气压力比

External Power Receptacle 外接电源插座

External Power Relay 外部电源继电器

**EPRACT** Engine Pressure Ratio Actual 实际发动机压力比

**EPRCMD** Engine Pressure Ratio Command 指令发动机压力比

**EPRF** Environmental Prediction Research Facility 环境预报研究中心

**EPRI** Electric Power Research Institute 电源研究所

**EPRIDL** EPR Command at an Idle Condition 慢车指令状态的发动机压力比

**EPRL** Electric Power Research Laboratory 电力研究实验室

Engine Pressure Ratio Latch 发动机压力比锁存器

Engine Pressure Ratio Limit 发动机压力比限度,推力比范围

**EPRMA** Extended Packet Reservation Multiple Access 扩充分组预留多址访问

**EPRMAX** Value of EPR Command at Maximum TRA 最大油门杆解算器角度下指令发动机压力比值

**EPRMCT** Engine Pressure Ratio of Maximum Continuous 最大连续状态的发动机压力比

**EPRMTO** Engine Pressure Ratio of Maximum Take off 最大起飞状态的发动机压力比

**EPROM** Electrically Programmable Read-Only Memory 电可编程序只读存储器

Erasable Programmable Read-Only Memory 可清除程序可控只读存储器

**EPRTRA** EPR Commanded by TRA Input 推力手柄解算器角度输入指令发动机压力比

**EPRTRIM** EPR Trim 发动机压力比配平

**EPS** Earnings per Share 每股收益

Electric Power Storage 蓄电池

Electrical Power Supply 电源供应

Electrical Power System 电源系统

Electronic Payment System 电子付款系统

Electronics Parts Store 电子备件库存

E-Mail Paging System 电子邮件播叫系统（全球呼—呼全球）

Emergency Power Subsystem 应急电源子系统

Emergency Power Supply 应急电源

Emergency Power System 应急电源系统

Emergency Pressurization System 应急加压系统

Energetic Particle Sensor 高能粒子传感器

Engine Preparation Schedule 发动机准备程序

Engineered Performance Standards 技术性能标准

Enterprise Project Structure 企业项目结构

**EPSCS** Enhanced Private Switched Communications Service 增强型专用交换通信业务

Enhanced Private Switched Communications System 增强型专用交换通信系统

**EPSL** Engineering Purchased Standards Listing 工程订购标准清册

**EPSM** Engineering Purchasing Specification Manual 工程采购工程手册

**EPSOC** Earth Physics Satellite Observation Campaign 卫星地球物理学观测运动

**EPSP** Electrical Power Services Panel 电源勤务面板

**EPSR** Equipment Purchase and Spares Report 设备采购和备件报告

**EPSRC** Engineering and Physical Sciences Research Council 英国工程和自然科学研究委员会

**EPSS** Equipment Power Supply System 设备电源系统

**EPS-TC** Electric Power Systems Technical Committee 电力系统技术委员会

**EPSTF** Electrical Power System Test Facility 电力系统试验设备

**EPSU** Emergency Power Supply Unit 应急电源供电组件

European Public Service Union 欧洲公共业务联盟

**EPT** Electrical Power Transmission 电力传输

Environmental Proof Test 耐环境试验

External Pipe Thread 外管螺纹

**EPTA** Expanded Program of Technical Assistance 技术援助发展计划

**EPTC** Essential Power Transfer Contactor 重要电源转换接触器

**EPTR** Emergency Power Transfer Relay 应急电源转换继电器

Engineering Parts Test Release 工程零件测试发放

**EPU** Electrical Power Unit 电源组件,电功率单位

Electronic Power Unit 电子电源组件

Emergency Power Unit 应急电源装置

Environmental Protection Unit 环境保护装置

Essential Power Unit 基本电源;基本动力装置

Estimated Position Uncertainty 估计位置的不确定性

Expandable Processor Unit 可扩展的信息处理器单元

**EPUT** Events per Unit Time 单位时间内发生的事件

**EPV** External Pressure Vessel 外部受压容器

**EPVS** Emergency Propellant Venting System 推进剂应急排气系统

**EPW** Electric Pressure Wave 电压波

Elliptically Polarized Wave 椭圆偏振波

Equivalent Planar Waveguide 等效平面波导

**EPWR** Emergency Power 应急电源

**EPXC** Electrical Path Cross Connect System 电通路交叉连接系统

**EPZ** Electron Polar Zone 电极区

Emergency Planning Zone 应急计划区

Export Processing Zone 出口加工区

**EQ** Emotional Quotient 情商

Enquiry 询问,查询

Equal or Equivalent 同等或等于

Equal 相等
Equalize 使相等
Equation 等式,方程式
Equipment Identification 设备标识符

**EQA** Environmental Quality Act 环境质量法
Environmental Quality Assessment 环境质量评价
Equipment Quality Analysis 设备质量分析
European Quality Alliance 欧洲质量联盟

**EQBM** Engineering/Qualification Backup Model 工程评定备用样机

**EQC** Ecological Quality Control 生态质量控制
Environmental Quality Control 环境质量控制
Environmental Quality Council 环境质量理事会
External Quality Control 外部质量控制

**EQDR** Engineering Quality Deficiency Report 工程质量缺陷报告

**EQI** Environmental Quality Index 环境质量指数

**EQIR** Engineering Query Investigation Report 工程查询调查报告

**EQIVNT** Equivalent 等于

**EQL** Equal 相等,相当
Equalizer 均衡器,计偿器
Expected Quality Level 期望质量水平,期望质量级

**EQL SP** Equally Spaced 等距间隔

**EQLAMP** Equalizing Amplifier 均衡放大器,补偿放大器

**EQLTY** Equality 同等

**EQLZ** Equalization 相等,平衡,补偿
Equalize 平衡,补偿

**EQM** Excellent Quality Management 精益质量管理

**EQN** Extended Queuing Network 扩展的排队网络

**EQP** Environmental Quality Program 环境质量计划

**EQPMT** Equipment 设备,装置,仪器

**EQPT** Equipment 设备

**EQS** Environmental Quality Standard 环境质量标准
Environmental Quality Systems 环境质量体系

**EQSP** Equal Space or Equally Spaced 等间隔(的)

**EQT** Engine Qualification Test 发动机合格测试
Engineer Qualification Table 工程师资格表
Engineering Qualification Test 工程合格试验
Environment Qualification Test 环境合格试验,环境鉴定试验
Equation of Time 时差

**EQUAL** Equalize 使均等,平衡,补偿

**EQUIL** Equilibrium 平衡,平衡状态,相称性

**EQUIP** Equipment 设备

**EQUIV** Equivalency 等效性,等值性
Equivalent 相当于,等量的,当量

**EQV** Equivalent 相等的,等效的,当量的

**EQVT** Equivalent 相当于,等量的

**EQX** Equator Crossing 赤道交叉线,跨越赤道

**ER** Early Release 过早投放
Echo Ranging 回波测距
Economic Risk 经济风险
Edge Router 边缘路由器
Electrical Resistance 电阻
Electronic Reconnaissance 电子侦察
Electronic Release 电子释放
Electron-Recording Tube 电子记录管
Elevator Rib 升降舵翼肋
Emergency Request 应急需要
Emergency Rescue 紧急救护
Emergency Room 应急室
Emergency Routing 应急路由选择
End Routing 结束路由
Engineering Release 工程放行
Engineering Review 工程审查
En-Route 航线飞行中
Entity Relationship 实体关系

Equipment Regulation 设备规则

Equipment Requirement 设备要求

Erroneous 错误的

Error Report 误差报告

Error 误差

Established Reliability 确凿可靠性

Etch Rate 侵蚀率,刻蚀率

Event Record 事件记录

Exchange Ratio 兑换率,更换率

Explicit Route 显式路由

Extended Range 延伸航程

Extension Register 扩充寄存器

External Register 外部寄存器

External Resistance 外阻力,外电阻

**ER&C** Engineering Resources and Control 工程资源和控制

**ERA** Electronic Reading Automation 电子阅读自动化

Electronic Retailing Association 电子零售协会

Environmental Risk Assessment 环境风险评估

European Recovery Act 欧洲复兴法案

**ERAA** European Regions Airlines Association 欧洲支线航空公司协会

**ERAB** Energy Research Advisory Board 能源研究咨询委员会

**ERADCOM** Electronics Research and Development Command 电子设备研究和开发司令部

**ERASE** Electromagnetic Radiation Source Elimination 电磁辐射源清除

Electronic Reconnaissance Access Set 电子侦察存取装置

**ERAU** Embry-Riddle Aeronautical University 安柏瑞德航空大学

**ERB** Event Request Broker 事件请求代理

**ERBS** Earth Radiation Budget Satellite 地球辐射收支卫星

**ERC** Economic Research Council 经济研究委员会

Electric Repair Center 电气修理中心

Electronic Rack Cooling 电子设备架冷却

Electronics Research Center 电子学研究中心

Engineering Research Center 工程研究中心

Enroute Chart 航路图

Environmental Research Council 环境研究委员会

Equal-Ratio Channel 等比率信道

**ERC-H** Enroute Chart-Upper Airspace 航路图/高空

**ERC-L** Enroute Chart-Lower Airspace 航路图/低空

**ERCN** End Routing Change Notice 结束程序更改通知

**ERCR** Engineering Release Change Request 工程公布更改申请

Engineering Release Completion Record 工程公布完成记录

**ERD** Earth Resources Data 地球资源数据

Electronic Research Directorate 电子研究局

Emergency Return Device 紧急返回装置

End Routing Domain 端点路由选择范围

Energy Recovery Device 能量回收装置

Equipment Requirements Document/Documentation 设备要求文件

Error-Rate Detector 误码率检测器

Estimated Receipt Date 预计收到日期

Experimental Research Design 实验研究设计

**ERDL** Electronic Research and Development Laboratories 电子研究与开发实验室

Engineering Research and Development Laboratory 工程研究开发实验室

Extended-Range Data Link 增程数据链路

**ERDY** External Ready 外部准备

**ERE** Echo Range Equipment 回波测距设备

**EREC** European Renewable Energy Council 欧洲可再生能源理事会

**ERECT** Erection 安装,装配,竖立

**EREP** Environment Record Edit and Print 环

境记录编辑和打印

**ERF** Eliminating Rectifier 分离整流器
Error Function 误差函数
Event Report Function 事件报告功能
Exponential Reliability Function 指数可靠性函数

**ERFC** Error Function Complementary 补余误差函数

**ERFE** External Radio Frequency Environment 无线电频率外部环境

**ERG** Electrical Resistance Gauge 电阻仪
Engine Reliability Group 发动机可靠性小组

**ERGA** Engine Rotor Governor Assessment 发动机旋翼调速器评估

**ERGS** Electronic Route Guidance System 电子航路导航系统
Experimental Route Guidance System 试验航路引导系统

**ERGY** Energy 能量,电源

**ERH** Electronic Reference Handbook 电子参考手册

**ERI** Electric Revolution Indicator 电(动)转数指示器
Erie, PA, USA (Airport Code) 美国宾夕法尼亚州伊利机场
Extended Repeat Interval 延长重复间隔

**ERICA** Explicit Rate Indication for Congestion Avoidance 避免拥塞的显式速率指示

**ERIE** Engineering Requirements Information Exchange 工程需求信息交换

**ERIR** End Routing Investigation Request 终端航路调查请求

**ERIS** Earth Resources Information System 地球资源信息系统
Extro-Atmospheric Reentry (Vehicle) Intercept System 大气层外再入(飞行器)截击系统

**ERISA** Employee Retirement Income Security Act 雇员退休收入保障法案

**ERJ** Embraer Regional Jet 巴西支线喷气飞机

External Ramjet 外部冲压发动机

**ERL** Economic Repair Life 经济修理寿命
Environmental Research Laboratories 环境研究试验室
Erosion Resistant Lacquer 防腐漆
Event Record Log 事件记录本

**ERLY** Early 早的

**ERM** Ejection Resistant Mechanism 喷射抑制机构
Emergency Response Manual 应急响应手册
Emitter-Receiver Module 辐射—接收模块
En-Route Metering 飞行途中计量,航路计量管制,航路中测量
Enterprise Relationship Management 企业关系管理
Enterprise Resource Management 企业资源管理
Enterprise Risk Management 企业风险管理
Environmental Resources Management 环境资源管理
European Mediterranean 欧洲,地中海地区
Explicit Rate Marking 显式速率标记

**ERMAC** Electromagnetic Radiation Management Advisory Council 电磁辐射管理咨询理事会

**ERMF** Event Report Management Function 事件报告管理功能

**ERMS** Educational Resource Management System 教育资源管理系统
Electronic Records Management System 电子记录管理系统
Equipment Resource Management System 设备资源管理系统

**ERMU** Experimental Remote Maneuvering Unit 实验用遥控机动单元

**ERN** Earth Referenced Navigation 大地基准导航,大地参考导航
Engineering Release Notice 工程发放通知

**ERO** Enterprise Resource Optimization 企业资源的优化
External Relations Office (ICAO) 对外关系

处(国际民航组织)

**EROM** Electron Readout Measurement 电子读出测量

Electronic Readout Machine 电子阅读机

Erasable Read Only Memory 可擦只读存储器

**EROPS** Extended Range Operations 延长航路飞行

Extended Range Twin Engine Operations 续航双发操作

**EROS** Earth Resource Observation Satellite 地球资源观测卫星

Earth Resources Observation System 地球资源观测系统

Engine Repair and Overhaul Squadron 发动机检修和大修中队

Environment and Radar Operations Simulator 环境与雷达运行模拟器

**ERP** Effective Radiated Power 有效辐射功率

Electronic Requirement Plan 电子需求计划

Emergency Recovery Plan 应急恢复计划

Emergency Response Plan 应急响应计划

Emergency Response Procedure 应急响应程序

Engine Requirement Program 发动机要求大纲

Enterprise Resource Planning 企业资源计划

Environmental Resource Planning 环境资源计划

Error Repair Process 错误修改过程

Event Related Potential 事件相关电位

External Ramjet Program 外置冲压发动机计划

Eye Reference Point 目视基准点

**ERPA** Emission Reduction Purchase Agreement 排放减少购买协议

Error-Rate Performance Analysis 误码特性分析

**ERPD** Eye Reference Position Datum 目视基准位置标志

**ER-PDU** Error Protocol Data Unit 差错协议数据单元

Error Report Protocol Data Unit 差错报告协议数据单元

**ERPL** Equipment Repair Parts List 设备修理零件清单

**ERPM** Emergency Response Procedures Manual 应急处置程序手册

**ERPR** Effectiveness Report-Performance Report 有效性报告—性能报告

**ERPS** Enterprise Resource Planning Systems 企业资源规划系统

**ERQ** Endorsement Request 签转要求

**ERR** Engine Removed Report 发动机拆卸报告

Engineering Reliability Review 工程可靠性检查

Erroneous 错误的

Error 误差

**ERRC** Error Correction 误差修正

Expendability, Recoverability, Repairability Code 消耗性、可恢复性、可修理性代码

**ERRDF** Earth Resources Research Data Facility 地球资源研究数据设施

**ERRSAC** Eastern Regional Remote Sensing Applications Center 东部地区遥感应用中心

**ERRSTOP** Error Stop 错误停机

**ERS** Earth Resource Satellite 地球资源卫星

Earth Resources Survey 地球资源勘测

Electronic Remote Switching 电子遥控转换

Electronic Retailing System 电子零售系统

Emergency Recovery Section 紧急抢修工段

Engine Repair Section 发动机修理工段

Engine Repair Shop 发动机修理车间

Environmental Research Satellite 环境研究卫星

Erase 消除,清除

Erased 消除的,清除的

European Remote Sensing Satellite 欧洲遥感卫星

Experimental Radar System 试验雷达系统

External Regulation System 外部调节系统

**ERS-1** European Remote Sensing Satellite1 欧空局遥感卫星 1 号

**ERS-2** European Remote Sensing Satellite2 欧空局遥感卫星 2 号

**ERSA** Extend Range Strike Aircraft 增大航程攻击机

**ERSATS** Earth Resources Survey Satellites 地球资源勘测卫星

**ERSCP** End Refueling and Start Climb Point 加油结束和开始爬升点

**ERSOS** Earth Resources Survey Operational Satellite 工作型地球资源勘测卫星

**ERSP** Earth Resources Survey Program 地球资源勘测计划

Electronic Reservations Service Providers 电子预订服务提供商

**ERSPRC** Earth Resources Survey Program Review Committee 地球资源勘测计划评审委员会

**ERSR** Engine Running Signal Relay 发动机行车信号继电器

**ERSS** Extended Range Surveillance System 扩大范围监视系统

**ERST** Error State 异常状态,错误状态,误差状态

**ERSTA** Erase State 擦除状态

**ERT** Electrical Resistance Temperature 电阻温度

Electrical Resistance Test 电阻试验

Engine Rotor Tester 发动机转子试验器

**ERTS** Earth Resource Technical Satellite 地球资源技术卫星

Earth Resources Technology Satellite 地球资源技术卫星

**ERU** Earth-Rate Unit 地球速率单位

Ejector Release Unit 弹射器释放装置

Emission Reduction Unit 排放减少单位

Engine Relay Unit 发动机继电器组件

Equipment Replaceable Unit 设备可替换单元

**ERV** Earth Return Vehicle 地球返回飞行器

Extended Range Vehicle 延长航路飞行器

**ERVC** Efficient Reservation Virtual Circuit 有效预留虚电路

**ERWE** Enhanced Radar Warning Equipment 改进型雷达警戒设备

**ERWTS** Enhanced Return Wave Tracker System 增强回波跟踪器系统

**ES** Earth Station 地球站

Earth Switch 接地开关

Echo Sounding 回声探测

Echo Suppressor 回波抑制器

Effective Span (代号) 有效翼展

Electrical Simulation 电模拟

Electromagnetic Stirring 电磁扰动

Electromagnetic Storage 电磁存储器

Electronic Synchrotron 电子同步加速器

Electronic Standard 电子标准

Electronic Switch 电子开关,电子转换器

Electrostatic Spraying 静电喷涂

Electrostatic Storage 静电存储器

Element Synchronism 码(网)元同步

Element Synchronization 码(网)元同步化

Enamel Single-Silk-Covered 单丝漆包的(导线)

End System 终端系统

Engineering Schedule 工程计划

Engineering Specifications 工程规范

Engineering Standards 工程标准

Engineering Suppliers 工程供应商

Engineering System 机务工程系统

Engine-Sized 机器上胶的

Environment Subsystems 环境子系统

Equal Section 相等截面,相等段

Errored Second 差错秒

Estimated 预计的,估计的

Expert System 专家系统

**ESA** Electronic Security Association 电子安全协会

Emergency Safe Altitude 紧急安全高度

Engineering Support Activity 工程支援效率

Entertainment Software Association 娱乐软件

协会

Environmental Services Association 环境服务协会

European Space Agency 欧洲空间局

**ESAD** Experiment Safety Assessment Document 试验安全评估文件

Equivalent Still-Air Distance 等效静风距离

**ESAF** Eastern and Southern African 东南非

**ESAR** Extended Subsequent Applications Review 扩展后续应用评审

**ESAS** Enhanced Situational Awareness System 增强的情况探测系统

**ESB** Emergency Service Bureau 应急服务机构

Emergency Switch Board 应急配电盘

Engine Storage Building 发动机仓库

Enterprise Service Bus 企业业务总线

Ankara, Turkey-Esenboga（Airport Code）土耳其安卡拉埃森博阿机场

Evaluation Service Bulletin 服务通告评估

**ESC** Engine Supervisory Control 发动机监控控制

Entertainment Service Controller 娱乐服务控制器

Environmental System Controller 环境系统控制器

Escanaba, MI, USA（Airport Code）美国密执安州埃斯卡纳巴机场

Escape 撤离, 救生, 逃出

**ESCAT** Emergency Security Control of Air Traffic 空中交通紧急安全管制

**ESCC** Electrical Supply and Control Center 供电控制中心

**ESCI** Enlarged Storage Contemporary Interior 加大型新式客舱顶行李架箱

**ESCM** Electronic Supply Chain Management 电子供应链管理

**ESCV** Eleventh Stage Control Valve 第十一级控制活门

**ESD** Electro Sensitive Device 电子感应装置

Electro Static Discharge 静电放电

Electro Static Discharger 静电放电器

Electronic Sensing Device 电子感测器件

Electronic Support Division 电气支援部分

Electronic System Display 电子系统显示

Electronic System Division（IAEA）电子系统司

Electrostatic Discharge 静电泄放, 静电放电

Electrostatic Sensitive Device 静电敏感器件

Energy Spectral Density 能谱密度

Estimated Shipping Date 预计装运日期

Event Sequence Diagram 事件序列图

**ESDAC** European Space Data Center 欧洲航天数据中心

**ESDB** Element Standard Data Bank 零件标准数据库

**ESDD** Electronic Software and Data Distribution 电子软件与数据分配

**ESDE** Electronic Systems Design Engineering 电子系统设计工程

Electrostatic Discharge Effects 静电放电效应

**ESDL** Expert System Definition Language 专家系统定义语言

**ESDN** Extended System/Structure Description Note 扩展系统/结构说明记录

**ESDS** Electrostatic Discharge Sensitive 静电放电灵敏度

**ESE** East-South-East 东南东

Electrical Support Equipment 电气支援设备

**ESF** Alexandria, LA, USA（Airport Code）美国路易斯安那州亚历山德里亚机场

Electronic System Facility 电子系统设施

Electrostatic Focusing 静电聚焦

Epitaxy Stacking Fault 外延堆垛层差错

Estimated Fabrication 概算的制造

European Science Foundation 欧洲科学基金会

Expert System Framework 专家系统结构

**ESFDD** Enhanced Security Flight Deck Door 增强型驾驶舱门

**ESG** Evaluated Service Goal 评估服务目标

**ESGN** Electrically Suspended Gyro Navigator

电子支撑式陀螺导航仪

**ESH** Environmental, Safety and Health 环境、安全与健康

**ESHP** Effective Summed Horse Power 总有效马力

Engine Shaft Horse Power 发动机轴马力

Equivalent Shaft Horse Power 当量轴马力，等效轴马力

Estimated Shaft Horse Power 估计轴马力

**ESH-PDU** End System Hello Protocol Data Unit 端点系统呼叫协议数据单元

**ESI** Engine Start Inhibit 发动机启动抑制

Ethernet Serial Interface 以太网络串行接口

Extremely Sensitive Information 极敏感情报

**ESIC** Earth Science Information Center 地球科学信息中心

Electrical System Indication and Control 电气系统指示与控制

Environmental Science Information Center 环境科学情报中心

External Structural Influence Coefficient 外部结构影响系数

**ESID** Engine and System Indication Display 发动机与系统指示显示器

**ESIMS** Engineering and Services Information Management System 工程和服务信息管理系统

**ESIP** Exhaust System Interaction Program 排气系统相互作用程序

**ESIS** Engine and System Indication System 发动机与系统指示系统

**ES-IS** End System-to-Intermediate System Routing 端系统对中间系统的路由选择

**ESK** Engineering Sketch 工程草图，工程简图

**ESL** Electronic Systems Laboratory 电子系统实验室

Emergency Service Level 应急服务水平

English as a Second Language 英语为第二语言

**ESLB** ECAM System Logic Book 飞机电子中央监控系统逻辑手册

**ESM** Electronic Support Measures 电子支持措施

Electronic Surveillance Measures 电子监视措施

Electronic System Management 电子系统管理

Electronics Supply and Manufacturing 电子供应与制造

Engine Shop Manual 发动机车间手册

Engineering Systems Management 工程系统管理

Ethernet Switching Module 以太交换模块

**ESMD** Exploration Systems Mission Directorate 探测系统任务执行委员会

**E-SMR** Enhanced Specialized Mobile Radio 增强型专用移动无线电

**ESMTP** Extended Simple Mail Transfer Protocol 增强型简单邮件传送协议

**ESN** Electronic Security Number 电子保密号码

Electronic Serial Number 电子序列号

Electronic Switched Network 电子交换网络

Emergency Service Number 应急服务号码

Engine Serial Number 发动机序列号

Engineering Services Network 工程服务网络

Enterprise Storage Network 企业存储网络

European Scientific Notes 欧洲科学通报

**ESNTL** Essential 重要的，主要的

**ESO** Electronic Supply Office 电子设备供应处

**ESOP** Employee Stock Ownership Plan 雇员配股计划

**ESOS** Education Services for Overseas Students 海外学生教育服务

**ESP** Electrical Standard Practices 标准电气施工

Encapsulating Security Protocol 封装安全协议

Energy System Planning 能源系统计划

Engineering Support Program 工程支持项目

English for Specific Purpose 特殊用途英语

Enroute Spacing Program 航路间隔计划

Enterprise Strategic Plan 企业战略计划

Especially 格外的

Extended Service Plan 扩展服务计划

Spain 西班牙

**ESPH** Engine Services per Hour 按小时付费的发动机服务

**ESPI** Electronic Speckle Pattern Interferometry 电子散斑干涉测量

**ESPM** Electrical Standard Practices Manual 标准电气施工手册

**ESPO** Engine Services per Operation 按运行收费的发动机服务

**ESPSI** Electronic Speckle Pattern Shearing Interferometry 电子散斑剪切干涉测量

**ESQ** Environmental, Safety and Quality Assurance 环保、安全、质量保证

**ESR** Effect Severity Ranking 影响严酷度等级

Electronic Systems Review 电子系统评审

Enroute Surveillance Radar 航路监视雷达

Environmental Surveillance Report 环境监视报告

Error Second Ratio 差错秒比率

**ESRI** Environmental Sciences Research Institute 环境科学研究所

**ESRP** European Supersonic Research Program 欧洲超音速航空器研究计划

**ESS** Electronic Self-Protection System 电子自卫系统

Electronic Switching System 电子交换系统

Electronic Systems Simulator 电子系统模拟器

Emergency Survival System 紧急救生系统

Engine Start Signal 发动机启动信号

Engine Start System 发动机启动系统

Engineering Standards Specification 工程标准规范

Entry Survival System 进入(大气层)救生系统

Environmental Sensor System 环境监测系统

Environmental Stress Screening 环境应力筛选,环境应力屏蔽

Equivalent Sensor (Sub)System 等效传感器(子)系统

Erection Subsystem 垂直安装子系统

**ESSA** Electronic Scanning and Stabilizing Antenna 电子扫描与稳定天线

Electronic Systems Security Assessment 电子系统安全评估

Environmental Science Service Administration 环境科学服务局(美国)

Environmental Science Students Association 环境科学学生协会

**ESSBR** Electronically Scanned Stacked Beam Radar 电子扫描多波束雷达

**ESSC** Emergency Support Schedule Changes 紧急支援计划更改

**ESSDE** European Space Software Development Environment 欧洲航天软件开发环境

**ESSI** European Strategic Safety Initiative 欧洲战略安全举措

**ESSL** Emergency Speed Select Lever 应急速度选定杆

**ESSP** Earth System Science Pathfinder 地球系统科学勘探者

Environmental Stress Screening Procedures 环境应力筛选程序

European Space Station Program 欧洲空间站计划

**ESSS** External Stores Support System 外挂物支撑系统

**EST** Eastern Standard Time 东部标准时间(美国)

Eastern Summer Time 东部夏季时间(美国)

Electrical Safety Testing 电气安全测试

Elevation, Slope, Temperature 海拔、坡度、温度

Engine System Test 发动机系统试验

Environmental Science and Technology 环境科学与技术

Environmental System Test 环境系统测试
Equipment Status Telemetry 设备状态遥测
Establish 建立
Established 既定的,建立的
Estimate Message 预计飞越报

**ESTA** Earth Science Technologies Association 地球科学技术协会
Electronically Scanned Tacan Antenna 电扫描塔康天线

**ESTAB** Establish 建立

**ESTAD** Estimated Arrival Date 预计到达日期

**ESTAMT** Establishment 建立,企业

**ESTAR** Electronically Scanned Thin Array Radiometer 电扫描薄阵列辐射计

**ESTB** Establish 既定的,建立的

**ESTBD** Established 设立

**ESTC** European Space Technology Center 欧洲航天技术中心

**ESTD** Estimated 预计的,估计的
Exchange Signaling Transfer Delay 交换机信令传送时延
Extend 延伸

**ESTE** Electrical Special Test Equipment 电气专用测试设备

**ESTEC** European Space Research and Technology Centre 欧洲空间研究和技术中心

**ESTF** Electronic System Test Facility 电子系统测试设备

**ESTG** Estimating 预计,估计

**ESTING** Extinguisher 灭火器
Extinguishing 灭火

**ESTL** Electronic System Test Laboratory 电子系统测试实验室(美国航空航天局)

**ESTMT** Establishment 建立,企业

**ESTN** Estimation 预计,估计

**ESTP** Electronic Satellite Tracking Program 电子卫星跟踪计划(国际大气测量协会)

**ESTRACK** European Space Tracking 欧洲太空跟踪

**ESTS** Electronic System Test Set 电子系统测试装置

**ESTU** Electronic System Test Unit 电子系统试验单元

**ESTWT** Estimated Weight 估计重量

**ESU** Electronic Service Unit 电子服务单元
Electrostatic Unit 静电组件
Emergency Supply Unit 应急电源设备
Engine Safety Unit 发动机安全组件
Ethernet Switch Unit 以太网转换组件
Exchange Signaling Unit 交换机信令单元

**ESV** Earth Satellite Vehicle 地球卫星运载器
Electrostatic Voltmeter 静电伏特计(电压表)
Emergency Shutoff Valve 应急关断活门(阀)
Experimental Safety Vehicle 实验性安全飞行器

**ESVS** Enhanced Synthetic Vision Systems 增强型合成视景系统

**ESWL** Equivalent Single-Wheel Load 等效单轮负载

**ESWR** Engineering Schedule Work Report 工程计划工作报告

**ET** Eastern Time 东部时间(美国)
Echoplex Technique 回送技术
Effective Temperature 有效温度
Elapsed Time 消失时间,所飞时间
Electro Tube 电子管
Electronic Technican 电子技术员
Electronic Technology 电子技术
Electronic Test(s) 电子测试
Electronic Ticket 电子客票
Electro-Thermal 电热的
Emerging Technology 新生技术
Emission Trading 排放交易
End Terminal 末端的
Endurance Test 耐久性试验
Energy Transfer 能量转移
Engineering Technology 工程技术
Engineering Test 工程测试
Entry Table 项目表

Environmental Technology 环境技术

Environmental Test 环境测试

Equation of Time 时差

Estimated Time 预计时间,估计时间

Evaluation Test 鉴定试验,评估试验

Exchange Termination 交换终端

External Tank 副油箱

**ET&E** Engineering Test and Evaluation 工程试验与评价

European Test and Evaluation 欧洲实验和鉴定

**ETA** Effective Turn Angle 有效转折角

Ejector Thrust Augmenter 引射器加力装置

Electronic Transfer Account 电子转账

Engine Thermodynamic Analyzer 发动机热力学分析仪

Engineering Technology Analysis Inc 工程技术分析公司

Estimated Time of Acquisition 预计截获时间

Estimated Time of Arrival 预计到港时间

Event Tree Analysis 事件树分析

Exception Time Accounting 例外时间计时

Expected Time of Arrival 预计到达时间

**ETAADS** Engine Technical and Administrative Data System 发动机技术和管理数据系统

**ETAC** Environmental Technical Applications Center 环境技术应用中心

**ETACCS** European Theater Air Command and Control Study 欧洲战区空中指挥与控制研究

**ETACS** Extended Total Access Communication System 扩展式全向通信系统

External and Taxiing Aid Camera System 外部和滑行辅助摄像系统

**ETADS** Engine Technical and Administrative Data System 发动机技术和管理数据系统

Enhanced Transportation Automated Data System 扩大的运输自动化数据系统

**ETAI** Engine Thermal Anti-Ice 发动机热防冰

**ETAS** Effective True Airspeed 有效真空速,假定真空速

Elevated Target Acquisition Sensor 高目标截获传感器

Elevated Target Acquisition System 高目标截获系统

**ETB** Engineering Test Base 工程试验基地

Engineering Test Basis 工程测试基础

Estimated Time of Boundary 预计边界时间

**ETBA** Energy Trace and Barrier Analysis 能量踪迹障碍分析

**ETC** Earth Terrain Camera 地球地形相机

Electronic Telephone Circuit 电子电话电路

Electronic Temperature Control 电子温度控制

Electro-Thermal Circuit 热电电路

Electro Turbine Control 电子涡轮控制

Emergency Training Center 应急训练中心

Engine Technical Committee 发动机技术委员会

Engine Test Certification 发动机试车合格证

Environment Test Center 环境实验中心(美国)

Environmental Tectonics 环境构造学

Environmental Thermal Control 环境热控制

Estimated Time of Commencement 预开装/卸期

Estimated Time of Completion 预计完成时间

Exchange Terminal Circuit 交换机终端电路

**ETCL** Estimated Time of Commence/Complete Loading 预开/完装货期

**ETCO** External Tank Corporation 外贮箱公司(美国)

**ETD** Effective Transfer Date 有效转移日期

Electronic Travel Distribution 电子旅游分销

Engine Trend Data 发动机趋势资料

Engineering Test Directive 工程试验指南

Estimated Time of Delivery 预计交运时间

Estimated Time of Departure 预计离港时间

Expected Time of Departure 预定起飞时间,预定离港时间

**ETDM**   Electrical TDM 电时分复用

**ETDN**   Electronic Ticket Delivery Network 电子客票传输网络

**ETDP**   Emergency Traffic Disposition Plan 应急交通安排计划

**ETDRS**   Experimental Track and Data Relay Satellite 实验型跟踪和数据中继卫星(日)

**ETDS**   Engineering Technical Data Service 工程技术数据服务

**ETDZ**   Economic and Technological Development Zone 经济技术开发区

**ETE**   End to End 端到端

Energy, Technology and Environment 能源、技术与环境

Engineering Test Equipment 工程测试设备

Environmental Test and Evaluation 环境实验和评定

Environmental Test Engineering 环境测试工程

Estimated Time En-Route 预计航线飞行时间

Estimated Time Entry 预计进场时间

**ETEB**   Engineering Test and Evaluation Board 工程测试与评估委员会

**ETEC**   End-to-End-Circuit 端到端电路

Expendable Turbine Engine Concept 扩展的涡轮发动机概念

**ETES**   Early Training Estimation System 早期培训评价系统

**ETF**   Eglin Test Facility 埃格林基地试验设备

Electro Thermal Furnace 电热炉

Electronic Time Fuse 电子定时引信

Electronic Tuning Fork 电子音叉

Electronically-Tunable Filter 电子可调滤波器

Engine Test Facility 发动机试车台

Error Threshold Firing 点火误差极限

Estimated Time of Flight 预计飞行时间

Estimated to Fly 预计飞行

**ETG**   Electrically Heated Generator 电气加热器

Electronic Turbine Governor 电子涡轮调节器

Enhanced Target Generator 增强的显示目标产生器

European Tripartite Group 欧洲三方小组

**ETH**   Elat, Israel ( Airport Code ) 以色列埃拉特机场

Electronic Trading Hours 电子交易时间

Engine Torque History 发动机扭矩历史

Extra Terrestrial Hypothesis 宇宙假说, 外地假说

**ETI**   Elapsed Time Indicator 累积时间指示器, 经历时间指示器

Electric Test Installation 电气试验装置

Engine Test Information 发动机试验情报

Engine Thrust Indicator 发动机推力指示器

Estimated Information 估计情报

Estimated Time of Interception 估计拦截时间

Extra Territorial Income 境外租赁收入

**ETIC**   Engineering Technology Industry Council 工程技术行业理事会

Enterprise Turnaround Initiative Corp 企业再生支援机构

Estimated Time in Commission 预计执行任务时间

**E-Ticket**   Electronic Ticket 电子客票

**E-TIME**   Execution Time 执行时间

**ETKT**   Electronic Ticket 电子客票

**ETL**   Electro Technical Laboratory 电子技术实验室

Emergency Tolerance Limit 应急容许极限

Endurance Test Level 耐久试验等级

Engineering Team Leader 工程队负责人

Engineering Test Laboratory 工程测试实验室

Environmental Technology Laboratory 环境技术实验室

Environmental Test Laboratory 环境测试实验室

Etching by Transmitted Light 透射光刻蚀

**ETLT** Equal to or Less Than 等于或小于

**ETM** Elapsed Time Measurement 经历时间测量

Elapsed Time Meter 使用时间指示器

Electronic Tactical Map 电子战术地图

Electronic Test and Maintenance 电子仪器试验和技术维护

Embedded Transmission Module 嵌入式传输模块

Energy Transfer Module 能量转移模量

Engine Trend Monitoring 发动机趋势监控

Engineering Test Module 工程试验组件

Enhanced Thematic Mapper 改进型主题测绘仪

**ETMA** Engineering Tooling Manufacturing Aid 工程工具加工制造辅助设备

**ETMS** Earth Terminal Measurement System 地面终端测量系统

Education Training Management System 教育训练管理系统

Electronic Territory Management System 电子化区域管理系统

Enhanced Traffic Management System 增强型交通管理系统

**ETMWG** Electronic Trajectory Measurements Working Group 电子运动轨道测量工作组

**ETN** Estimated True North 估计正北

**ETO** Electronic Temperature Offset 电子温度补偿

Electronics Technology Office 电子技术办公室

Estimated Time of Operation 估计运转时间

Estimated Time off 估计飞离时间

Estimated Time Over Significant Point 预计飞越重要点的时间

Estimated Time Over 预计飞越时间

Express Transport Organization 急运机构

**ETOC** Estimated Time of Correction 估计校正时间

**e-TOD** Electronic Terrain Obstacle Database 电子地形和障碍物数据库

**ETOM** Electro Trapping Optical Memory 可重写光盘(电刻光存贮器)

**ETOP** Engineering Technical Operating Procedure 工程技术操作程序

Extended Overwater Operation 延程跨水飞行

**ETOPS** Extended Twin-Engine Overwater Operations 双发飞机延程跨水飞行

Extended-Range Twin-Engine Operational Performance Standards 双发飞机延程运行性能标准

**ETOT** Estimated Time over Target 估计飞越目标时间

**ETOW** Estimated Takeoff Weight 预计起飞重量

**ETP** Electron Temperature Probe 电子式温度传感器

Engine Test Panel 发动机试车台,发动机试验操纵台

Engineering Technical Procedure 工艺规程

Engineering Test Pilot 技术试飞员

Engineering Test Plan/Program 工程测试计划

Environmental Test Plan 环境试验大纲

Environmental Test Procedures 环境试验程序

Equal Time Point 等时点

Estimated Time of Penetrating 估计突防时间,估计穿越时间

Experimental Test Procedure 实验性试验程序

External Tracking Processor 外部跟踪处理器

**ETPR** Engineering Test Parts Release 工程测试零件发放

**ETPS** Empire Test Pilots School 帝国试飞员学校

**ETR** Electron Tube Rectifier 电子管整流器

Electronic Trouble Report 电子故障报告

Electronically Tuned Radio 电子调谐无线电台

Emergency Test Request 紧急试验申请单

Engineering Test Request 工程试验申请（书）

Enter 进入，记入

Entertainment Tape Reproducer 娱乐放像机

Estimated Time for Refueling 预计加油时间

Estimated Time of Recovery 预计回收时间，预计恢复时间

Estimated Time of Repair 预计修理时间

Estimated Time of Return 预计返航时间

Estimated Time Release 预计放行时间，预计发出时间

Estimated Time Remaining 预计剩余时间

External Technical Report 外部技术报告

**ETRA** Engineering Target Revision Authorization 工程目标修改批准

**ETRAC** Electronic Thrust Reverser Actuation Controller 电子反推作动控制器

**ETRAS** Electrical Thrust Reverser Actuation System 电子反推作动系统

**ETRC** Expected Taxi Ramp Clearances 预计的滑行机坪间隔

**ETRG** Engineering Test Requirements Group 工程测试要求小组

**ETRO** Estimated Time of Return to Operation 估计恢复操作（运转）时间

**ETS** Edwards Test Station 爱德华兹试验站

Electronic Test Set 电子测试设备

Electronic Test Station/Stand 电子测试站

Electronic Timekeeping System 电子计时系统

Electronic Timing Set 电子定时器

Electronic Trading System 电子交易系统

Electronic Translator System 电子译码器系统，电子转换器系统

Emergency Trip System 汽轮机紧急跳闸系统

Emission Trading System 排放交易体系

Emissions Trading Scheme 排放交易计划（机制）

Endless Tangent Screw 正切无终点蜗杆

Engagement Tracking Station 衔接跟踪站

Engine Test Stand 发动机试车台

Engineering and Technical Services 工程和技术服务

Equipment Technical Specification 设备技术规范

European Telecommunication Standard 欧洲电信标准

European Telephone System 欧洲电话系统

Evaluator Trainer System 评价教练仪系统

Engineering Test Satellite 工程试验卫星

**ETSAL** Electronic Terms for Space Age Language 航天时代语言的电子术语

**ETSG** Elevated Temperature Strain Gauge 高温应变仪

**ETSI** European Telecommunication Standards Institute 欧洲电信标准学会

**ETSO** European Technical Standard Orders 欧洲技术标准规定

**ETSQ** Electrical Time Superquick 超快电动计时

**ETST** Engineering Test-Service Test 工程试验—使用试验

**ETT** Electro Thermal Thrusters 热电推力器

Electron Tube, Triode 三极电子管

Electronically Tuned Tuner 电子调谐的调谐器

Equipment Task Time 设备工作时间

Estimated Time for Takeoff 预计起飞时间

Estimated Time of Tracking 估计跟踪时间

Expected Test Time 预计试验时间

Extreme Temperature Tests 极限温度试验

Extrusion Trim Template 挤压型材修正模板

**ETU** Engineering Test Unit 工程测试组件，工程测试单位

**ETV** Educational Television 教育电视

Election Test Vehicle 弹射试验飞行器

Elevating Transfer Vehicle 升降装卸车

Engine Test Vehicle 发动机试验车，发动机试验飞行器

**ETVC** Environmental Test Vacuum Center 真空环境试验中心

**ETVM** Electrostatic Transistorized Voltmeter 静电型晶体管电压表

**ETVS** Enhanced Terminal Voice Switch 增强型的终端话音交换

**ETW** European Transonic Windtunnel 欧洲跨音速风洞

**ETX** End of Text 电文结束
End of Transmission 发送结束

**EU** Electronic Unit 电子部件
End User 最终用户
European Union 欧洲联盟

**EUC** Engineering Unit Conversion 工程单位转换

**EUCARE** European Confidential Aviation Safety Reporting Network 欧洲航空安全秘密报告网络

**EUCT** Electronic Unit Cooling Trail 电子部件冷却底板

**EU-ETS** European Union Emission Trading System 欧盟排放交易体系

**EUFALDA** European Federation of Airline Dispatchers Association 欧洲航线签派同盟组织

**EUFT** Engine Underground Fuel Tank 发动机地下油箱

**EUG** Eugene, OR, USA ( Airport Code ) 美国俄勒冈州尤金机场

**EUIR** European Upper Flight Information Region 欧洲高空飞行情报区

**EULMS** Electrical Utilities Land Management System 通用电气设备着陆管理系统
Engine Usage Life Monitoring System 发动机使用寿命监控系统

**EUM** European Mediterranean Region 欧洲地中海地区

**EUMET** European MET Broadcast 欧洲气象广播

**EUMETNET** Network of European Meteorological Services 欧洲气象服务网络

**EUO** Emergency Use Only 仅供紧急情况使用

**EUPB** Electronic Unit Protection Box 电子组件保护盒

**EUPS** External Uninterruptible Power Supply 外部不间断供电源

**EUR** Effective Upon Receipt 收到时生效
Emergency Unsatisfactory Report 不满意紧急报告, 应急措施不利报告
Engineering Unsatisfactory Report 工程缺陷情况报告
Euro ( European Monetary Unit ) 欧元( 欧洲货币单位)
Europe, European 欧洲

**EURACA** European Air Carrier Assembly 欧洲航空公司大会

**EURATOM** European Atomic Energy Community 西欧原子能共同体

**EURECA** European Retrievable Carrier 欧洲可回收重复使用的运载工具

**EURFCB** European Frequency Co-Ordinating Body 欧洲频率协调体

**EURICAS** European Research Institute for Civil Aviation Safety 欧洲民航安全研究所

**EURO** European Currency Unit 欧洲货币单位( 欧元)

**EUROCAE** European Organisation for Civil Aviation Equipment 欧洲民用航空设备组织
European Organization for Civil Aviation Electronics 欧洲民用航空电子学组织
Minimum Operational Performance Specifications for Crash Protected Airborne Flight Recorder 防毁机载飞行记录器最低操作性能规范

**EUROCON** European Conference on Electro Technology 欧洲电工技术会议

**EUROCONTROL** European Organisation for the Safety of Air Navigation 欧洲空中航行安全组织( 欧安局)

**EUROSAT** European Application Satellite System 欧洲应用卫星系统
European Communications Satellite Corporation 欧洲通信卫星公司

**EUROSPACE** European Industrial Space Study Group 欧洲航天研究工业集团

**EUTAS** European Utility Tactical Transport Aircraft System 欧洲通用战术运输机系统

**EUVE** Extreme Ultraviolet Explorer 极远紫外(线)探测器

**EV** Earned Value 利润值,所得值

East Velocity 东向速度

Electron Valve 电子活门

Electron Volt 电子伏特

Entry Visa 入境签证

Equivalent Velocity 当量速度

Evaluated 求值的,评价的,被鉴定的

Event 事件

Every 每一个

External Vision 外部视线

**EVA** Early Valve Actuation 阀门提前动作

Economic Value Added 经济附加值,经济价值增值

Electronic Velocity Analyzer 电子速度分析器

Elevation Versus Amplitude 仰角与振幅比较关系

Enhanced Visual Approach 增强目视进近

**EVAC** Evacuate 撤离

Evacuated 抽空的

Evacuation 撤离,抽真空

**EVAL** Earth Viewing Applications Laboratory 地球观测应用实验室

Evaluate 鉴定

Evaluation 评价,鉴定

Evaporation 蒸发(作用)

**EVALN** Evaluation 评价,鉴定

**EVAP** Evaporate 蒸发

Evaporation 蒸发,蒸发器

**EVAPS** Evaporation Process 蒸发过程

**EVAR** Experiment Vehicle for Avionics Research 用于航空电子学研究的实验飞行器

**EVAS** Extravehicular Activity System 舱外活动系统

**EVBC** Engine Vane and Bleed Control 发动机叶片与引气控制

**EVC** Engine Vane Control 发动机导向叶片控制

Error Vector Computer 误差向量计算机

**EVCO** Electronic Vibration Cutoff 电子振荡终止

**EVCS** Enhanced Video Connector Standard 强化视频连接器标准

Extravehicular Communications System 舱外通信系统

**EVCTD** Evacuated 抽空的

**EVD** Elementary Vortex Distribution 涡流基本分布

**EVDE** External Visual Display Equipment 外部视觉显示设备

**EVDP** Engine Vibration Diagnostic Program 发动机振动诊断程序

**EVEA** Extra Vehicular Engineering Activities 舱外工程活动

**EVF** Electromagnetic Vibrating Feeder 电磁振动进料器

**EVG** Electric Vacuum Gyro 电动真空陀螺仪

Electrostatically Supported Vacuum Gyro 静电支承真空陀螺

**EVI** Engine Vibration Indicator 发动机振动指示器

External Video Interface 外部视频接口

**EVID** Evidence 证据,凭据

**EVLN** Evolution 开方(数),发展,演变,进化

**EVLTN** Evaluation 估价,评价

**EVM** Earned Value Management 实现价值管理

Earth Viewing Module 对地观测舱

Economic Value Management 经济价值管理

Electronic Voltmeter 电子伏特表

Electrostatic Voltmeter 静电伏特表

Engine Vibration Monitor 发动机振动监视器

Engineering Verification Model 工程检验模型

Enterprise Value Management 企业价值管理

Evacuation Module 评估模块

**EVMS** Earned Value Management Standard 挣值管理标准

Earned Value Management System 挣值管理系统

**EVMU** Engine Vibration Monitoring Unit 发动机振动监控组件

**EVN** Yerevan, Armenia (Airport Code) 亚美尼亚埃里温机场

**EVP** Ejector Vacuum Pump 喷射真空泵

Experimental Version Prototype 实验型样机

**EVR** Electronic Video Recorder 电子视频记录装置

Electronic Video Recording 电子视频记录

**EVRC** Enhanced Variable Rate Coder 增强型可变速率编码器

**EVRMT** Environment 环境

**EVRS** Electro-Optical Viewing and Ranging Set 电光观察与测距装置

**EVS** Effectivity Verification Sheet 生效性(范围)验证单

Electro-Optical Viewing System 光电观察系统

Engine Vertical Scale 发动机垂直标度

Enhanced Vision System 增强型目视系统

Extravehicular Suit 舱外活动服

**EVSC** Engine Vibration Signal Conditioner 发动机振动信号调节器

Enhanced Video System Control Unit 增强型录像系统控制组件

Extravehicular Suit Communications 舱外活动服中的通话设备

**EVSD** Electronic Vertical Situation Display 电子式垂直位置显示器,电子式垂直情况显示器

**EVSS** Extravehicular Space Suit 舱外活动航天服

**EVSTC** Extravehicular Suit Telecommunications 舱外活动服中的电信设备

**EVSTCS** Extravehicular Suit Telemetry and

Communications System 舱外活动服中的遥测和通信系统

**EVT** Earned Value Technique 实现价值技术

Equi-Viscous Temperature 等粘度温度

Event 事件

Extravehicular Transfer 舱外转运

Extreme Value Theory 极值理论

**EVV** Evansville, IN, USA (Airport Code) 美国印第安那州埃文斯维尔机场

**EVVA** Extravehicular Visor Assembly 舱外活动盔装置

**EW** Early Warning 预警

East-West 东西(向)的

Electrical Welding 电焊

Electronic Warfare 电子战争

Empty Weight 空重

**EW/GW** Empty Weight to Gross Weight Ratio 空重和总重比,空载与总载重比

**EWA** East-West Airlines, Ltd. (Australia) 东西航空公司(澳大利亚)

Engineering Work Assignment 工程任务分配

Engineering Work Authorization 工程权

**EWAMS** Early Warning and Monitoring System 预警监视系统

**EWAS** Electronic Warfare Analysis System 电子战分析系统

Electronic Wide-Angle Camera System 电子广角照相系统

En-Route Weather Advisory Service 航路气象咨询服务

**EWAU** Electronic Warfare Avionics Unit 电子战航空电子设备(美国)

**EWB** Electronic Welding Beam 电子束焊

**EWC** Early-Warning Coverage 预警有效区

Either-Way Communication 半双工通信

Electric Water Cooler 电动水冷却器

**EWCL** Electromagnetic Warfare and Communication Laboratory 电磁战与通讯实验所

**EWCM** Electronic Warfare Control Module 电子战控制模块

**EWCS** Electronic Warfare Combat System 电

子战作战系统

**EWD** Electronic Window Dimmer 电子窗口调光器

Engine/Warning Display 发动机/警告指示

**EWDTC** Electronic Warfare Design to Cost 电子战成本设计

**EWEDS** Electronic Warfare Evaluation Display System 电子战判断显示系统

**EWES** Electronic Warfare Environmental Simulator 电子战环境模拟器

Electronic Warfare Evaluation Simulator 电子战评价模拟器

**EWF** Early Warning Flight 预警战斗机

**EWITS** Early Warning Identification and Transmission System 预警识别与传输系统

**EWJT** Electronic Warfare Joint Test 电子战联合试验

**EWN** New Bern, NC, USA ( Airport Code ) 美国北卡莱罗纳州新伯尔尼机场

**EWO** Emergency War Operations 紧急作战计划

Emergency War Order 紧急作战命令

**EWOS** Electronic Warfare Operational System 电子作战系统

**EWPA** European Women Pilots' Association 欧洲女飞行员协会

**EWR** Early Warning Radar 预警雷达

Early Warning Report 预警报告

Eastern/Western Region 东西部地区

Newark, NJ, USA ( Airport Code ) 美国新泽西州纽瓦克机场

**EWS** Early-Warning Satellite 预警卫星, 远程警戒卫星

Early-Warning Station 预警 ( 雷达 ) 站

Early-Warning System 早期警报系统

Electronic Warfare Support 电子战支援

Electronic Warning System 电子预警系统

Emergency Warning System 紧急报警系统

Emergency Water Supply 紧急情况供水

End of Word Sync 示码同步终端

Engineering Working Station 工程师工作站

Estimate Work Sheet 估计工作单

**EWSM** Early-Warning Support Measures 预警保障措施

Electronic Warfare Support Measures 电子战支援措施

**EWT** Eastern Winter Time 东部冬季时间 ( 美国 )

Expandable Wing Tank 可扩展翼下油箱

Expected Waiting Time 预计等待时间

**EWTAD** Early Warning Threat Analysis Display 预警威胁分析显示器

**EWTES** Electronic Warfare Threat Environmental Simulator 电子战战区环境

**EWTS** Expandable Wing Tank Structure 模拟器可扩展翼下油箱结构

**EWWS** Electronic Warfare Warning System 电子战报警系统

**EX** Example 例子

Examination 检查, 考试

Example 例如, 例子

Exchange 汇票, 交换, 兑换

Exciter 激励器, 励磁机

Excluding 扣除, 除外

Execute 执行

Executed 已执行

Executive Committee 执行委员会

Exhaust 排气

Exit from 从…出来

Exit 出口

Expanding 膨胀

Exponential Smoothing 指数校平

Export 出口

Extra 附加的, 额外的

Extract 抽出, 摘录

**EXA** Executive Aircraft, Ltd. ( USA ) 公务机有限公司 ( 美国 )

**EXAGT** Executive Agent 执行代理人

**EXAM** Examination 考试

Experimental Aerospace Multiprocessor 实验航空航天多处理机

**EXAMD** Examined 检查过的

**EXAMN**　Examination 考试,检查

**EXBAG**　Excess Baggage 超重行李

**EXBAGCHG**　Excess Baggage Charge 超重行李费

**EXC**　Excellency 阁下

Except 除…外

Excessive 过量的

Excitation 激励,励磁

Excite 产生励磁

Exclude 排除,除去

Excuse 原谅

Exhaust Closed 排气门关闭

**EXCAP**　Exercise Capability 锻炼能力

Expanded Capability 扩展能力型

**EXCD**　Exceedance 超过数额

Exceeding 超过

Excited 加励磁

**EXCELS**　Expended Communication Electronic System 扩展的通讯电子系统

**EXCESS**　Excessive 超过的,过大的

**EXCG**　Exciting 励磁的

**EXCH**　Exchange 交换,兑换

**EXCHR**　Exchange Rate 比价

**EXCIT**　Excitation 激励,励磁

**EXCL**　Exclude 除外,不包括

Excluding 除外的,排除的,除外

**EXCL or EX**　Exclusive or Exclude or Excluded or Excluding 排除

**EXCLD**　Exclude 除外

**EXCLDG**　Excluding 除…外

**EXCN**　Excitation 励磁

**EXCODOP**　Externally Coherent Doppler 外部相干多谱勒(雷达)

**EXCON**　Executive Component 执行部件

**EXCP**　Except 除…外

**EXCPN**　Exception 例外

**EXCT**　Excitation 激励,励磁

**EXCTR**　Exciter 励磁机

**EXD**　Examined 检查过的

**EXDAMS**　Extendible Debugging and Monitoring System 扩大排除故障与监控系统

**EXEC**　Execute 执行,完成

Executive Control System 执行控制系统

Executive 执行的,主管人员

**EXER**　Exercise 练习,演习

**EXF**　Extra Fare 额外票价

**EXGA**　External Gage 外径规

**EXH**　Exhaust 排出,排气

Exhibit 展览,附件

Exhibition 展览会,展示

**EXH TEMP**　Exhaust Gas Temperature 排气温度

**EXHIB**　Exhibition 展览会,展示

**EXHV**　Exhaust Vent 排气孔

**EXIM**　Export and Import 出口与进口

Export and Import Bank of India 印度进出口银行

**EXIST**　Existing 出口

**EXL**　Execute Local 本机执行

**EXLI**　Export License 出口许可证

**EXM**　Executive Monitoring 执行监测

**EXMIL**　Excess Mileage 超过的里程

**EXO**　European X-Ray Observatory 欧洲 X 射线观测卫星

Exhaust Open 排气门打开

Extra-Atmospheric 大气层外的

**EXP**　Expand 扩展,延伸

Expanded 扩展的,延长的

Expander 扩展器

Expansion 扩展,扩充

Expect 期待,预期

Expected 预期的

Expend 消耗,消费

Expenses 费用,支出

Expire 期满

Export 出口

Express 表示;快递,特快

**EXP JT**　Expansion Joint 伸缩接缝,伸缩接头,涨缩接合

**EXP/IMP**　Export-Import 进出口

**EXPD**　Expedite 加速,促进

**EXPDT**　Expedite 快速

**EXPER** Experience 经验
Experiment 实验

**EXPIR** Expiration 截止期,有效期

**EXPL** Explosive 易爆的

**EXPLN** Explain 解释,说明费
Explanation 解释,说明

**EXPN** Expansion 扩展,延长
Expenses 费用,支出

**EXPND** Expenditure 经费

**EXPNT** Exponent 解说者;样品,试样,典型

**EXPO** Experimental Order 实验性指令

**EXPR** Experiment 实验,试验
Experimental 实验的,试验的,经验的
Experimentation 实验,实验过程
Express 快递,特快

**EXPRN** Expiration 截止日期,有效期

**EXPRS** Express 快递,特快

**EXPS** Expenses 费用,支出

**EXPSR** Exposure 暴露,曝光

**EXPT** Expect 预期,期待
Experiment 实验
Expert 专家
Export 出口

**EXRE** Exercise 演习

**EXS** Expandable Switching System 可扩展的
交换系统

**EXSEC** Extra Section 加班,附加部分

**EXST** Existing 现有的,存在的
Extra Seat 额外座位

**EXT** Exeter, United Kingdom ( Airport Code )
英国埃克塞特机场
Exhaust 排气
Existing 存在的,现有的
Extend 延长,加长
Extend, Extension 延伸,伸出
Extension 延长,加长,电话分机
Exterior, External 外部,外部的
External Register 外部寄存器
External 外部的
Extinguish 灭火
Extinguisher 灭火器

Extra 加的,额外的
Extract 抽出,摘录
Extreme 极端的

**EXTC** Extract 提取,排出

**EXTCD** Extracted 排出的

**EXTCG** Extracting 排出

**EXTD** Extend 伸出,延伸
Extender 延长器,伸展器,补充剂

**EXTDIA** External Diameter 外径

**EXTERA** Extra-Terrestrial Research Agency
地球外空间研究处

**EXTERN** External 外部的

**EXTF** Extension Filter 增设滤波器

**EXTFIX** External Fix 外部固定,外部定位

**EXTG** Extending 延伸,伸出

**EXTIG** Extinguishing 熄灭

**EXTIN** Extinction 熄灭
Extinguish, Extinguished 灭火,熄灭的

**EXTIN（G）** Extinguish, Extinguishing 灭火,
熄灭,扑灭

**EXTIR** Extinguisher 灭火器

**EXTN** Extension 延长,加长,电话分机

**EXTR** Extractor 提取器
Extraordinary 非常的,特别的
Extrusion 挤出,挤压

**EXTRACT** Extraction, Extractor 提取,提取器

**EXTSN** Extension 伸长,扩大,延长,延期

**EY** Emergency 紧急,应急

**EYCL** Economy Class 经济舱

**EYLT** Eyelet 小孔,小眼

**EYPC** Eyepiece 目镜

**EYW** Key West, FL, USA ( Airport Code ) 美
国佛罗里达州基韦斯德机场

**EZ** Electrical Zero 电零位
Extraction Zone 投伞地带

**EZE** Buenos Aires, Argentina Ezeiza ( Airport
Code ) 阿根廷布宜诺斯艾利斯埃塞萨机场

**EZFW** Estimated Zero Fuel Weight 预计无油
重量

**EZS** Elazig, Turkey ( Airport Code ) 土耳其埃
拉齐格机场

# F

F Aerodrome Weather Forecast 机场天气预报
Field 机场
Fighter 战斗机,歼击机
Filter 滤子
First Class Baggage 头等舱行李
First Class 头等舱
Fixed 固定的
Flashing Light 闪光灯
Flight 飞行,航班
Floor 地板,楼层
Florin 弗罗林,一种货币
Fog 雾
Force 力,部队
Forward 前方,向前,传递
Franc 法郎(货币单位)
Frequency 频率
Friday 星期五
Fuel 燃油
Fueler 加油车
Full 满
Function 功能
Fuse 引信,保险丝
Minimum Flap Retract Speed 最小收襟翼速度
Thrust ("F"refers to thrust"Force") 推力
°F Degrees Fahrenheit 华氏度
F&D Freight and Demurrage 运费加滞期费
F&E Facilities and Equipment 设施和设备
F&F Fire and Forget 发射后免控,发射后不管
F&R Functional and Reliability Testing 功能与可靠性试验
F,E&D Facilities, Engineering and Development 设施、工程和发展
F.A.S.T. Fast Address System Trunking 快

速寻址集群
F/A Field Activities 机场勤务,战地活动
First Aid 急救
Final Assembly 总装
Flight Attendant 空中乘务员
F/A RATIO Fuel-to-Air Ratio 燃油—空气比
F/C First Class 头等舱
Flight Compartment 机组舱
Flight Crew 飞行人员,空勤组
F/CTL Flight Control 飞行管制,飞行操纵
F/D Flight Deck 驾驶舱
Flight Director 飞行指引仪,飞行引向器
F/E Flight Engineer 随机工程师,飞行工程师
Fuel Flow per Engine 每台发动机燃油流量
F/F First Flight 首航
Flip Flop 触发器
Fuel Flow 燃油流量
F/H Fueling Handbook 加油手册
F/LDG Forced Landing 强迫着陆,迫降
F/N Flight Navigation 空中领航(航行)
F/O First Observer 第一观察员
First Officer 副驾驶
F/P Flight Purser 随机乘务长
Fuel Policy 燃油政策
F/PLN Flight Plan 飞行计划
F/Q Fuel Quantity 燃油量
F/R Final Run 进入第五边
Flight Recorder/Recording 飞行记录器/记录
Fuel Remaining 剩余燃油
F/S Fast Set 快调
Fast Slew 快转
Fast/Slow 快/慢

Fine Stabilized 微调的

First Stage 第一阶段

Flight Steward 空中服务员

Stick Force 操纵杆受力

**F/T** Flight Time 飞行时间

From/To 自…至…

Functional Test 功能性测试

**F/U** Follow-up 随动,跟踪报告

**F/W** Failure Warning 失效警告

Fire Wall 防火墙

**F3** Form-Fit Function 壳式功能

**FA** Area Forecast 区域预报

Aviation Forecast 航空预报

Failure Analysis 故障分析

Feeder Automation 馈线自动化

Fiber Adaption 光纤适配

Final Approach 最终进近

Fire Alarm 火警

Flight Attendant 飞机客舱服务员

Flying Accident 飞行事故

Frame Aligner 帧定位器

Frame Alignment 帧同步

Free-Air 开口式风洞

**FAA** Federal Aviation Administration 联邦航空管理局

Federal Aviation Agency 联邦航空管理署

Federal Aviation Authority 联邦航空当局(美国)

Final Approach Area 最后进近区

Flight Advisory Area 飞行咨询区

**FAAD** Forward Area Air-Defense 前方地区防空

**FAAH** Federal Aviation Administration Handbook 联邦航空管理手册

**FAAO** Finance and Accounts Office 财务与会计处

**FAAP** Federal-Aid Airport Program 联邦支援机场计划

**FAAR** Forward Area Acquisition Radar 前方地区搜索雷达

**FAAST** FAA's Aviation Safety Team 联邦航空局航空安全小组

Forward Area Alerting Radar 前方地区警戒雷达

**FAAT** Flighter Attack Avionic Technology 战斗机攻击航空电子技术

**FAATC** FAA Technical Center 联邦航空局技术中心

**FAB** Fiber Array Block 光纤阵列块

First Aid Box 急救箱

Fixed Acoustic Buoy 固定音响浮标

Functional Airspace Block 功能性空域区块

Functional Auxiliary Block 功能辅助舱

**FABM** Fiber Amplifier Booster Module 光纤放大器增强模块

**FABR** Fabricate 制作,构成;捏造,伪造,虚构

**FABU** Fuel Additive Blender Unit 燃料添加剂搅拌器

**FABX** Fire Alarm Box 火警箱

**FAC** Facility 设备

Facsimile 传真,复制

Factor 系数,因素

Fast as Can 尽快

Federal Airport Corporation 联邦机场公司

Federal Aviation Commission 联邦航空委员会(美国)

Final Approach Course 最后进近航迹

Flight Augmentation Computer 飞行增稳计算机

Forward Acting Code 前向作用码

Forward Air Control 前方空中指挥

Forward Air Controller 前方空中管制员

**FACC** Ford Aerospace and Communication Corporation 福特航空空间与通信公司

**FACCH** Fast Associated Control Channel 快速相关控制信道

**FACE** Facilities and Communication Evaluation 设施与通讯鉴定

**FACET** Future ATM Concepts Evaluation Tool 未来空中交通管理概念评价工具

**FACF** Final Approach Capture Fix 最后进近

截获定位点

Final Approach Course Fix 最后进近航道定位点

**FACH** Forward Access Channel 前向接入信道

**FACI** First Article Configuration Inspection 首件产品构型检验

**FACO** Final Assembly and Check-out 总装与检验

**FACRI** Flight Automatic Control Research Institute 飞行自动控制研究所(中国)

**FACS** Fine Attitude Control System 精确姿态控制系统

Flight Augmentation Computer System 飞行增稳计算机系统

Flight Augmentation Control System 飞行增稳控制系统

**FACT** Facility of Automation, Control and Test 自动化、控制和测试设备

Factor 因素,因子,系数

Factual Compiler 真实编译程序

Fast Asymptotic Coherent Transmission 快速渐近相干传输

Fault Analysis Central Maintenance Computer Tool 中央维护计算机故障分析工具

Flexible Automatic Circuit Tester 通用自动电路测试仪

Flight Acceptance Composite Test 飞行验收综合试验

Forecast and Control Technique 预报和控制技术

Fully Automatic Compiler Translator 全自动编译翻译程序

**FACTY** Factory 工厂

**FAD** Fast-Action Device 速动装置

Flight Aerodynamics Development 飞行空气动力学研究

Free Air Delivery 免费空运

**FADA** Federation Argentina De Aeroclubes 阿根廷航空俱乐部联合会

**FADD** Fatigue and Damage Data 疲劳和损伤

数据

Fatigue And Defect Data 疲劳和缺陷数据

**FADE** FAA Airline Data Exchange FAA 的航空公司数据交换

**FADEC** Full Authority Digital Electronic Control 数字式全权电子控制

Full Authority Digital Engine Control 数字式全权限发动机控制,数字式全权限引擎控制

Full Authority Digital Engine Controller 数字式发动机全权电子控制器

**FADS** Flexible Air-Data System 接口可变型大气数据系统

**FAE** Final Approach Equipment 最后进近设备

**FAEEC** Full Authority Electronic Engine Control 全权电子发动机控制

**FAEI** Federation of Aerospace Enterprises in Ireland 爱尔兰宇航企业联盟

**FAEJD** Full Automatic Electronic Judging Device 全自动电子判定器

**FAER** First Article Engineering Review 首件工程评审

**FAF** Final Approach Fix 最后进近定位点,最终进近坐标

**FAFC** Forward Air Freight Center 空运货物中转中心

Full Authority Fuel Control 全权燃油控制

**FAGC** Fast Automatic Gain Control 快速自动增益控制

Forward Area Ground Control 前方地区地面控制

**FAH** Degrees Fahrenheit 华氏度

**FAI** Fairbanks, AK, USA (Airport Code) 美国阿拉斯加州费尔班克斯机场

Federation Aeronautique Internationale 国际航空联合会

First Article Inspection 首件检验

International Aeronautical Federation 国际航空联合会

**FAIL** Failed 失效的

Failure 失效

**FAIP** First Article Inspection Plans 首件检验大纲

First-Assignment Instructor Pilot 第一(指令)飞行教官

**FAIR** Facility Annual Inspection Report 设备年度检验报告

Failure Analysis Information Retrieval 故障分析信息检索

Fairing 整流

Fast Access Information Retrieval 快速存取信息检索

First Article Inspection Report 首批制品检验报告

**FAIT** First Article Inspection Test 首批制品检验测试

**FAITH** Fiber Almost Into the Home 准光纤到家

**FAK** Fly Away Kit 随机维修包,随机成套工具

**FAL** Final Assembly Line 总装线

Frame Alignment Loss 帧定位丢失

Frequency Allocation List 频率分配表

**FAM** FAA Assessment Model FAA 评估模型

Familiarization 熟悉,掌握技能

Family of Frequency 频率族(种类)

Family 家族

Financial Audit Manual 财务审计手册

Fixed Asset Management 固定资产管理

Flight Advisory Message 飞行咨询电报

Flight Attendant Manual 乘务员手册

Frame Alignment Module 帧定位模块

**FAMAS** Flutter and Matrix Analysis System 颤振与矩阵分析系统

**FAME** Final Approach Monitoring Equipment 最后进近监控设备

Forecasting, Analysis and Modeling Environment 预测、分析与建模环境

**FAMIS** Full Aircraft Management/Inertial System 全飞机管理/惯性系统

**FAMOS** Fleet Air Meteorological Observation Satellite 机队航空气象观测卫星

**FAMS** Family Service 家庭服务

Fixed Assets Management System 固定资产管理系统

Fine Attitude Measurement System 最佳飞行姿态测量系统

Fuels Automated Management System 燃油自动化管理系统

**FAMV** Fan Air Modulating Valve 风扇空气调节活门

**FAN** Fiber in the Access Network 接入网光纤

**FANS** Future Air Navigation System 新航行系统,未来空中导航系统

Future Airspace Navigation System 未来空间导航系统

**FANSTIC** Future ATC New System and Technologies Impact on the Cockpit 未来空中交通管制新系统和新技术对驾驶舱的影响

**FAO** Faro, Portugal (Airport Code) 葡萄牙法鲁机场

Food and Agricultural Organization 联合国粮食及农业组织

**FAOM** Flight Attendant Operating Manual 乘务员使用手册

**FAP** Failure Analysis Program 故障分析程序

Final Approach Point 最后进近点

First Aid Post 急救站

Fleet Average Performance 机群平均性能

Flight Attendant Panel 乘务员面板

**FAPA** First Air Pilots Association 第一个航空飞行员协会

Future Airline Pilots of America 美国未来航空公司飞行员

Future Aviation Professionals of America 美国未来航空专业人员

**FAPAA** Federation of Asia-Pacific Aircargo Associations 亚太航空货运协会联盟

**FAQ** Fair Average Quality 良好平均品质,中等货

Foundation Aerovision Quebec 魁北克航空

基金会

Frequently Answer Question 常见问题解答

Frequently Asked Questions 常遇到的问题

**FAR** Failure Analysis Report 故障分析报告

False Alarm Rate 虚警率

False Alarm Ratio 伪报警比率

Fargo, ND, USA ( Airport Code ) 美国北达科他州法戈机场

Fatigue Analysis Report 疲劳分析报告

Federal Acquisition Regulation 联邦征购条例(美国)

Federal Airworthiness Requirements 联邦适航要求(美国)

Federal Aviation Regulations 联邦航空规章（美国）

First Article Report 首件品报告

Flight Acceptance Review 飞行验收评审

Forced Action Rapid 快速反应部队(法)

Fuel-to-Air Ratio 燃料与空气比

**FARA** Formula Air Racing Association 方程式飞机竞赛协会

**FAROA** Final Approach and Runway Occupancy Awareness 最终进近和跑道占用情况感知

**FAROS** Final Approach Runway Occupancy Signal 最终进近跑道占用信号

**FARP** Fully Automatic Radar Plotter 全自动雷达绘图仪

**FARS** Failure Analysis Report Summary 故障分析报告摘要

Fatal Accident Reporting System 重大事故报告系统

First Airplane Reporting System 首机报告系统

Flight Assurance Review System 飞行保险检查系统

**FART** Frame Alignment Recovery Time 帧同步恢复时间

**FAS** Federal Air Surgeon 联邦航空医生

Federal Aviation Services 联邦航空服务

Fiber Access System 光纤接入系统

Final Approach Segment 最后进近段

Final Assembly Scheduling 总装计划

Financial Accounting System 财务会计系统，财务结算系统

Fire Alarm System 火灾报警系统

Flexible Access System 灵活接入系统

Flexible Assembly System 柔性装配系统

Flight Accessories Services 飞行附件维修公司

Flight Augmentation System 飞行增稳系统

Flight-Attendant Station 飞行乘务员舱位

Foreign Airlines Service Corporation 外航服务公司

Forward Acquisition Sensor 前视探测传感器

Forward Acquisition System 前向探测系统

Frame Alignment Signal 帧定位信号

Freight Automated System 货运自动系统

Frequency Agile Subsystem 频率捷变分系统

Fueling at Sea 海上加油

**FASA** Final Approach Spacing Assignment 最后进场间隔分配

Final Approach Spacing of Aircraft 飞机最后进场间隔

**FASB** Financial Accounting Standard Board of the United States 美国财务会计准则委员会

**FASCO** Foreign Airlines Service Corporation 外航服务公司

**FASE** Fast Auroral Snapshot Explorer 极光速摄探测器(美国)

**FASER** Frequency Amplification by Simulated Emission of Radiation 通过受激辐射发射的信频

**FASID** Facilities and Services Implementation Document(s) 设施和维修实施文件，设施和服务实施文件

**FASM** Final Approach Separation Management 最后进近间隔管理

**FASO** Field Aviation Supply Office 机场航空供应处

Fleet Aviation Support Office 机队航空支援办公室

**FASR**　Frequency-Agile Solar Radiotelescope 频率可调的太阳射电望远镜

**FASS**　Forward-and-Aft Scanner System 前后扫描系统

**FAST**　FAA Acquisition System Toolset 联邦航空局采购系统工具库

Facility for Automated Simulation Test 自动模拟试验设施

Facility for Automatic Sorting and Testing 自动分类与探测设备

Fan and Supersonic Turbine 风扇和超音速涡轮

Fast Acquisition Search and Track 快速截获搜索和跟踪

Fast Automatic Shuttle Transfer 航天飞机快速自动输送

Fence Against Satellite Threats 防卫星雷达情报网

Fiber at Subscriber Terminal 用户端光纤

Field Data Applications, Systems and Techniques 数据组数据应用、系统和技术

Final Approach Spacing Tool 最后进近间隔工具

Flight Advisory Service Test 飞行咨询服务测试

Forward Airborne Surveillance and Tracking 前线机载监视与跟踪

Forward Area Support Team 前方地域保障队

Free and Secure Trade 自由与安全贸易

Fuel and Sensor Tank 燃油与传感器箱

Fuel Assembly Stability Test 燃油与稳定性测试

Future Aircraft Supersonic Transport 未来超音速运输飞机

Future Aviation Safety Team 未来航空安全团队

**FASTNR**　Fastener 紧固件

**FAT**　Factory Acceptance Test 工厂验收试验

Far Eastern Air Transport 远东航空运输

File Allocation Table 文件分配表

Film and Television 电影与电视

Final Approach Track 最后进近航迹

Final Assembly Test 总装试验

First Article Test/Testing 首件产品试验

Flexible Access Termination 灵活接入终端

Flight Acceptance Test 飞行验收测试

Flight Attendant Training 空中乘务员训练

Flow Allocation Table 流量分配表

Free Air Temperature 大气温度

Fresno, CA, USA（Airport Code）美国加利福尼亚州弗雷斯诺机场

**FATDL**　Frequency and Time-Division Data Link 频分与时分数据链(路)

**FATE**　Factory Acceptance Test Equipment 工厂验收试验设备

Future Aircraft Technology Enhancements 未来飞机技术提高(计划)

**FATG**　Fixed Air-to-Ground 空对地固定目标

**FATMI**　Finnish Air Traffic Management Integration 芬兰空中交通管理一体化

**FATO**　Final Approach and Take-off Area 最后进近和起飞区

**FATSL**　Flight Acceptance Test Support Laboratory 飞行验收测试支持实验室

**FAUSST**　Franco-Anglo-United States SST Committee 法英美超音速运输机委员会

**FAV**　Fan Air Valve 风扇空气活门

Favor 帮助

**FAW**　Frame Alignment Word 帧定位字

**FAWP**　Final Approach Waypoint 最后进近航路点

**FAWS**　Flight Advisory Weather Service 飞行天气咨询服务

**FAX**　Aeronautical Fixed Station 航空用地面固定无线电台

Facsimile System 传真, 传真系统

Facsimile/Transmission 传真

**FAXIWF**　FAX InterWorking Function 传真互通功能

**FAY**　Fayetteville, NC, USA（Airport Code）美国北卡莱罗纳州费耶特维尔机场

**FB**　Feed Back, Feedback 反馈
　　Fiber Booster 光纤增强器
　　Fiber Bundle 光纤束
　　Freight Bill 货运单
　　Function Button 功能按钮
**FBAG**　Free Baggage 免费行李
**FBB**　Fiber Backbone 光纤干线
**FBCN**　Fuzzy Backward Congestion Notification 模糊反向拥塞通知
**FBE**　Foreign Bill of Exchange 外国汇票
**FBG**　Fiber Bragg Grating 光纤布拉格光栅
**FBI**　Federal Bureau of Investigation 联邦调查局(美国)
**FBL**　Fly by Light 光传操纵
**FBM**　Foot Board Measure 英尺板测量
　　Lubumbashi, Zaire (Airport Code) 扎伊尔卢本巴希机场
**FBO**　Fan Blade off 风扇叶片高速安全性
　　Federal Budget Outlays 联邦预算支出一览表
　　Fixed-Base Operation 固定基地运营, 专供公务机使用的候机楼
　　Fixed-Based Operator 机场服务商, 通用航空服务站, 固定基地经营者
　　Flights Between Overhauls (两次) 翻修间飞行次数
**FBOE**　Frequency Band of Emission 发射频段
**FBOOST**　Fuel Booster Pump 燃油增压泵
**FBR**　Feedback Resistance 反馈电阻
　　Fiber Bragg Reflector 光纤布拉格反射器
　　Fiber 纤维
　　Fixed Bit Rate 固定比特率
**FBRBD**　Fiberboard 纤维板
**FBRS**　Fibrous 纤维的
**FBS**　Fixed Base Simulator 固定基模拟机
**FBT**　Feedback Technology 反馈技术
　　Form Block Template 成型块模块
　　Forward Ballast Tank 前压舱油箱
**FBTS**　Fixed Base Training Simulator 固定基训练仿真器
**FBU**　Oslo, Norway (Airport Code) 挪威奥斯陆机场
**FBV**　Force Balance Valve 力平衡活门
　　Fuel Bleed Valve 放油活门
**FBW**　Fly by Wire 电传操纵, 飞机电传操纵
**FBWS**　Fly-by-Wire System 电传操纵系统
**FC**　Fan Cowl 风扇整流罩
　　Fiber Channel 光纤通道
　　Fire Control 防火控制
　　First Class 头等舱
　　Flight Change 航班变更, 飞行更改
　　Flight Charts 飞行图
　　Flight Clearance 飞行许可, 放行
　　Flight Condition 飞行状况, 飞行条件
　　Flight Configuration Change 飞行形态改变
　　Flight Configuration 飞行配置, 飞行构型
　　Flight Control 飞行管制, 飞行操纵
　　Flight Coupons 承运票联
　　Flight Crew 飞行人员, 空勤组
　　Flight Cycle 飞行起落
　　Forward Compatibility 前向兼容性
　　Franc 法郎(货币单位)
　　Fuel Cell 燃油箱, 燃油室, 燃油电池
　　Fully Closed 全关
　　Functional Check 功能检查
　　Funnel Cloud 漏斗云
**FCA**　Fault Containment Area 故障包容区
　　Fixed Channel Allocation 固定信道分配
　　Flight Control Actuator 飞行控制作动器
　　Flight Control Assembly 飞行控制组件
　　Flow Control Assembly 流量控制组件
　　Free Carrier 货交承运人
　　Fuel Control Assembly 燃油控制组件
　　Function Configuration Audit 功能配置审计
　　Functional Configuration Audit 功能构型审核
　　Future Cycle Accumulation (发动机零件的) 未来周期储存
　　Kalispell, MT, USA (Airport Code) 美国蒙大拿州卡利斯佩尔机场
**FCAA**　Finnish Civil Aviation Administration 芬兰民航局

Flight Control Actuator Assembly 飞行控制
作动器组件

Foreign Civil Aviation Authority 外国民用航
空当局

**FCAC** Forward Cargo Air Conditioning 前货
舱空调(系统)

Frequency Control and Analysis Center 频率
控制与分析中心

Fuzzy Channel Allocation Controller 模糊通
道分配控制器

**FCAL** Fiber Channel Arbitrated Loop 光纤信
道仲裁环路

**F-CAPICH** Forward-Common Auxiliary Pilot
Channel 前向公共辅助导频信道

**FCAS** Fleet Corrective Action Status 机队纠
正措施状况

Flight Control Actuator Subassembly 飞行操
纵作动器分组件

Flight Control Augmentation System 飞行控
制增稳系统

Foreign Contract Administration Services 外
国合同管理处

**FCAST** Forecast 预报

**FCB** File Control Block 文件控制块

Frequency Coordinating Body 频率协调机构
(国际民航组织)

**FCC** Facsimile Control Channel 传真控制信
道

Federal Communication Commission 联邦通
信委员会

Fire Control Computer 火控计算机

Flat Conductor Cable 扁导体带状电缆

Flight Communication Center 飞行通信中心

Flight Control Center 飞行控制中心

Flight Control Computer 飞行控制计算机

Flight Coordination Center 飞行协调中心

Frequency Channel Code 频道编码

**FCCC** Flight Coordination Control Center 飞
行协调控制中心

**FCCCH** Forward Common Control Channel 前
向公共控制信道

**FCCS** Federal Cost-Control Survey 联邦成本
控制调查

Flight Control Computer System 飞行控制计
算机系统

**FCCSET** Federal Coordinating Council for
Science, Engineering, and Technology 联邦科
学、工程、技术协调委员会(美国)

**FCCU** Flight Control Computer Unit 飞行操
纵计算机组件

**FCD** Fan Cowl Door 风扇整流罩舱门

Flight Controls Development 飞行操纵面研
制

Full Color Display 全色显示

**FCDC** Flight Computer Data Controller 飞行
计算机数据控制器

Flight Control Data Concentrator Computers
飞行控制数据集中器计算机

Flight Control Data Concentrator 飞行控制数
据集中器

Flight Control DC System 飞行直流电系统

Flight Control Digital Computer 飞行控制数
字计算机

Flight Control Direct Current 飞行操纵直流
电

Flight Critical DC 重要飞行直流电

**FCDL** Flight Control Development Laboratory
飞行控制开发实验室

**FCDR** Failure and Consumption Data Report
故障和消耗数据报告

Failure Cause Data Report 故障原因数据报
告

**FCE** Flight Control Electronics 飞行控制电
子设备

**FCEI** Facility Contract End Item 设备合同最
后条款

**FCEP** Flight Control Electronics Package 飞
行操纵电子装置

**FCES** Flight Control Electronic System 飞行
操纵电子系统

**FCEU** Flight Control Electronic Unit 飞行控
制电子组件

**FCF**   Functional Check Flight 功能检查试飞

**FCFP**   Functional Check Flight Procedures Manual 功能检验飞行程序手册

**FCFS**   First Come, First Served 先来先服务

**FCG**   Facing 朝向

Fatigue Crack Growth 疲劳裂纹扩大（增大）

Fuel Contents Gauge 油量表

**FCGES**   Flight Control Group Electronic System 飞行控制组电子系统

**FCGU**   Flight Control and Guidance Unit 飞行控制与指引装置

**FCH**   Flight Controller Handbook 飞行管制员手册

**FCI**   False Color Image 假彩色图像

Flight Command Indicator 飞行指令指示器

Fluid Control Institute 流体控制学会

Fuel Consumed Indicator 燃油消耗指示器

Functional Configuration Identification 功能结构识别

**FCIF**   Flight Crew Information File 飞行人员信息档案

**FCIM**   Flexible Computer Integrated Manufacturing 计算机集成柔性制造

**FCL**   Flight Control Laboratory 飞行控制实验室

Flight-Crew Licensing 飞行人员执照

Full Container Load 整箱货物

**FCLP**   Field Carrier-Landing Practice 陆上着舰练习

**FCLT**   Freeze Calculated Landing Time 固定计算着陆时间

**FCLTP**   Flight Crew Licensing and Training Panel 飞行机组执照和培训专家组

**FCLTY**   Facility 设备，设施

**FCM**   Fault Containment Module 故障包容组件

Fiber Composite Matter 纤维合成材料

Financial Controls Management 财务控制管理

Fluid Condition Monitor 流体工况监测仪

Formal Confidential Memorandum 正式保密备忘录

Frequency Control Module 频率控制组件

Frequency Controlled Motor 频率控制马达

Fuzzy Cognitive Map 模糊认知图

Fuzzy Control Model 模糊控制模型

**FCMC**   Flight Control Mechanical Characteristics 飞行控制机械特性

Fuel Control and Monitoring Computer 燃油控制和监控计算机

**FCMS**   Fuel Control and Monitoring System 燃油控制和监控系统

**FCN**   Frequency-Converting Network 变频网络

Full Connected Network 全连接网络

**FCO**   Fire Control Operator 防火控制操作员

Flight Communications Operator 飞行通信员，空中报务员

Formal Change Order （合同的）正式更改通知书

Rome, Italy-Leonardo Da Vinci/Fiumicino (Airport Code) 意大利罗马菲乌米奇诺机场

**FCOC**   Fuel Cooled Oil Cooler 燃油冷却的滑油散热器

**FCOM**   Flight Crew Operations Manual 飞行机组使用手册，飞行机组操作手册

**FCOS**   Flight Computer Operating System 飞行计算机操作系统

Flight Control Operating System 飞行控制操作系统

**FCP**   Fare Construction Points 票价组合点，运价构成点

Fatigue Crack Propagation 疲劳断裂蔓延

Final Control Point 最后控制点

Flight Control Panel 飞行控制面板

Fluid and Chemical Processing 流体和化学处理

Frequency Control Program 频率控制程序

Fuel Cell Powerplant 燃油电池动力装置

Fuel Consumption Projections 燃料消耗量预测

Fuel Control Panel 燃油控制面版

Function Control Package 操作控制包

**FCPC** Flight Control Primary Computer 主飞行控制计算机

Flight Crew Plane Captain 飞行机组组长

**FCPF** Foreign Commodity Production Forecasting 国外商品产量预报

**F-CPHCH** Forward Common Physical Channel 前向公共物理信道

**FCPI** Flight Control Position Indicator 飞行操纵位置指示器

**FCPSA** Flight Control Power Supply Assembly 飞行操纵电源组件

**FCR** Facility Change Request 设备更改申请

Fault Containment Region 故障抑制区

Fire Control Radar 火控雷达

Flight Control Room 飞行控制室

Flight Crew Report 飞行机组报告

Flight Crew Rest 机组休息室

**FCRC** Flight Crew Rest Compartment 飞行机组休息舱

**FCRL** Flight Crew Reading Light 飞行机组阅读灯

**FCS** Embassy of the USA Foreign Commercial Service 美国驻华使馆商务处

Facsimile Communications System 传真通信系统

Fast Circuit Switching 快速电路交换

Fiber Channel Standard 光纤信道标准

Field Bus Control System 现场总线控制系统

Fire Control System 火控系统

Flight Control System 飞行操纵系统,飞行管制系统

Flight Crew Systems 飞行机组系统

Frame Check Sequence 帧校验序列

Fuel Control System 燃油调节系统

Future Communication Study 未来通讯研究

**FCSC** Flight Control Secondary Computer 辅助飞行操纵计算机

**FCSOV** Flow Control and Shutoff Valve 流量控制和关断活门

**FCSS** Flight Control Stabilization System 飞行控制稳定系统

Flight Control System Simulator 飞行控制系统模拟器

Flight Control Systems Section 飞行控制系统组

**FCST** Federal Council for Science and Technology（美国）联邦科学技术委员会

Forecast 预报

Frontier of Computer Science and Technology 计算机科学与技术的前沿

**FCT** Factor 系数,因素

Fatigue Cracking Test 疲劳断裂试验

First Configuration Test 初级构型试验

Fixed Cellular Terminal 固定蜂窝终端

Fixed Cruise Thrust 恒定巡航推力

Flight Certification Test 飞行合格审定测试

Flight Control Team 飞行控制小组

Foreign Comparative Testing 外国产品比较测试

Function Test 功能测试

**FCTL** Flight Controls 飞行操纵

**FCTM** Flight Crew Training Manual 飞行机组训练手册

**FCTN** Function 功能

**FCTR** Fan/Core Thrust Radio 风扇与核心发动机推力比

Flight Controls Test Rig 飞行操纵测试台

**FCTS** Flying Crew Trainer Simulator 飞行人员训练模拟器

**FCTY** Factory 工厂

**FCU** Fare Calculation Unit 费用计算单元

Flap Control Unit 襟翼操纵组件

Flight Control Unit 飞行控制组件

Flush Control Unit 冲水控制组件

Fuel Control Unit 燃油控制组件（装置）

**FCV** Flow Control Valve 流量调节活门

Fuel Cell Vehicle 燃料电池汽车

Fuel Control Valve 燃料控制活门

**FCWBS** Final Contract Work Breakdown Structure 最终合同工作分解结构

**FD** Failure Detection 故障检测

Fatigue Damage 疲劳损伤

Fault Detection 故障检测

Fiber Duct 光纤管道

Final Development 最后改进

Fixed Displacement 固定位移

Flap Down 襟翼放下

Flight/Data 航班号/日期

Flight Deck 驾驶舱

Flight Despatcher 飞行调度员,飞行签派员

Flight Despatching 航行调度,航行签派

Flight Director 飞行指引仪

Forbush Decrease 福布什下降

Forced Discharge 强行卸货

Forced Draughts 强制通风,压力通风

Found 发现,找到

Frame Disassembler 帧分解器

Free Delivery 免费交运

Frequency Domain 频率范围

Frequency Duplex 频道双工制

Fuel Drain 放油

Fund 资金

Fuselage Datum 机身基准面

Winds Aloft Forecast 高空风预测

**FDA** Failure Detection and Annunciation 故障探测和显示

Flight Data Analysis 飞行数据分析,航班数据分析

Food and Drug Administration 食品和药品管理局

Functional Data Analysis 功能数据分析

**FDAC** Flight Data Acquisition Card 飞行数据采集卡

**FDAF** Flight Data Acquisition Function 飞行数据采集功能

**FDAI** Flight Direction and Attitude Indicator 飞行方向和姿态指示仪

Flight Director Attitude Indicator 飞行指引仪姿态指示器

**FDAMS** Flight Data Acquisition and Management System 飞行数据采集与管理系统

Flight Data Acquisition and Monitoring System 飞行数据采集与监测系统

**FDANA** Frequency-Domain Automatic Network Analyzer 频域自动网络分析仪

**FDAP** Flight Deck Audio Panel 驾驶舱音频板

**F-DAPICH** Forward Dedicated Auxiliary Pilot Channel 前向专用辅助导频信道

**FDARS** Flight Data Acquisition Recording System 飞行数据采集记录系统

**FDAS** Flight Data Acquisition System 飞行数据采集系统

Flight Data Analysis System 飞行数据译码分析系统

**FDAU** Flight Data Acquisition Unit 飞行数据采集组件

**FDB** Floor Disconnect Box 地板脱开电子盒

Full Data Blocks 全数据控制块

**FDBK** Feedback 反馈

**FDC** Facility Design Criteria 设施设计准则,设施设计标准

Failure Detection Circuit 故障探测电路

Fault Development Control 故障发展控制

Fire Department Connection 防火部门联系手段

Fire Detection Center 火警检测中心

Flexible Disk Controller 软磁盘控制器

Flight Data Cards 飞行数据卡

Flight Data Center 飞行数据中心

Flight Data Company 飞行数据公司

Flight Data Control 航班数据控制

Flight Director Computer 飞行指引仪计算机

Flight Dispatch Center 飞机签派中心

Formation Drone Control 编队的无人驾驶飞机控制

Forward Direction Center 前方导引中心

Frequency-Digital Converter 频率—数字转换器

Frequency-to-Digital Converter 频率—数字转换器

Fuel Data Concentrator 燃油数据集中器

**FDCA** Found Cargo 已找到的货物

**F-DCCH** Forward Dedicated Control Channel 前向专用控制信道

**FDCCP** Flight Director Computer Control Panel 飞行指引仪计算机控制板

**FDCD** Facility Design Criteria Document 设备设计标准文件

**FDCE** Flight Deck and Cabin Effects 驾驶舱与客舱效应

**FDCF** Fire Detection Card File 火警检测卡存储柜

Flight Deck Communication Function 飞行驾驶舱通讯功能

**FDCM** Flight Director Computer/Monitor 飞行指引计算机/监控器

**FDCT** Forward Discrete Cosine Transform 前向离散余弦变换

**FDCU** Fire Detection Control Unit 火警检测控制组件

**FDD** Flight Description Document 飞行说明文件

Flight Dynamics Division 飞行动力学处

Floppy Disk Drive 磁盘驱动器

Frequency Division Duplex 频分双工

**FDDI** Fiber Digital Data Interface 数字式光纤数据接口

Fiber Distributed Data Interface 分布式光纤数据接口

**FDDL** Frequency Division Data Link 分频数据链

**FDDRC** Flight Deck Digital Radio Control 驾驶舱数字式无线电控制

**FDDU** Floppy Disk Drive Unit 软盘驱动组件

**FDDVA** Flow Divider and Drain Valve Assembly 流量分配器泄放活门组件

**FDE** Fault Detection and Exclusion 故障检测与排除

Field Decelerator 机场减速器

Fire Detection and Extinguishing 火警探测与灭火

Flight Data Entry 飞行数据输入

Flight Deck Effect 驾驶舱效应

Flight Deck Equipment 驾驶舱设备

Flight Dynamics Engineer 飞行动力学工程师

**FDEP** Flight Data Entry and Printout 飞行数据输入和打印输出

Flight Data Entry Panel 飞行数据输入面板

Flight Data Entry Position 飞行数据输入位置

**FDERAS** Flight Data Event Risk Assessment System 飞行数据事件风险评估系统

**FDES** Fault Diagnostic Expert Systems 故障诊断专家系统

Flight Data Entry System 飞行数据输入系统

**FDEVSS** Flight Deck Entry Video Surveillance System 驾驶舱入口视频监视系统

**FDF** Flight Data File 飞行数据文件

Flight Dynamics Facility 飞行动力学设施

Forced Draft Fan 强制通风机,压力送风机

Fort De France, Martinique ( Airport Code ) 马提尼克岛法兰西堡机场

Full-Duplex Data Flow 全双工数据流

**FDG** Flight Director Group 飞行指挥组

Founding 建立

Funding 资助

**FDGW** Flight Design Gross Weight 飞行设计总重量

**FDH** Failure Detection and Handling 故障检测和处理

Flight Deck Handset 驾驶舱手提式话筒

Friedrichshafen, Germany ( Airport Code ) 德国腓特烈斯港机场

**FDI** Failure Detection and Isolation 故障检测和隔离

Feeder Distribution Interface 馈线分配接口

Flight Deviation Indicator 航迹偏差指示器

Flight Direction Instrument 飞行指引仪表

Flight Director Indicator 飞行指引指示器

Foreign Direct Investment 跨境外来直接投资

**FDIAF** Flight Data Interface Acquisition

Function 飞行数据接口采集功能

**FDIMU** Flight Data Interface and Management Unit 飞行数据接口及管理组件

**FDIO** Flight Data Information Operation (FAA) 飞行数据信息操作

Flight Data Input/Output 飞行数据输入/输出

**FDIR** Failure Detection, Isolation, and Reconfiguration 故障检测、隔离和重新配置

Failure Detection, Isolation, and Recovery 故障检测、隔离和恢复

**FDIS** Flight Displays and Interface System 飞行显示与接口系统

**FDIU** Flight Data Interface Unit 飞行数据接口组件

**FDL** Fiber Delay Line 光纤延迟线

Flight Datum Line 飞行基准线

Flight Dynamics Laboratory 飞行动力学实验室

Frequency Double Laser 倍频激光器

Full-Dracon Line (在显示器上)实线标志

**FDLS** Failure Detection Location System 故障检测定位系统

**FDM** Feedback Data Management 反馈数据管理

Finite-Difference Method 有限差分方法

Flight Data Monitoring 飞行数据监控

Flight Dispatch Manual 飞行签派手册

Flight Dynamics Model 飞行动力学模型

Frequency Division Modulation 频率划分调制,分频调制

Frequency Division Multiplex 频分复用

Frequency Division Multiplexing 频分复用

**FDMA** Factory Design and Manufacturing Analysis 工厂设计与制造分析

Frequency Division Multiple Access 频分多址,频分多路访问

**FDME** Frequency Division Multiplex Equipment 频分多路传输设备

**FDMS** Flight Data Management System 飞行数据管理系统

Frequency Division Multiplexing System 频分多路传输系统

**FDO** Flight Duty Officer 飞行值日官

**FDP** Fiber Distribution Point 光纤分布点

Flight Data Processing 飞行数据处理

Flight Data Processor 飞行数据处理器

Flight Data Programme 飞行数据程序

Flight Demonstration Program 飞行演示大纲

Flying Duty Period 飞行值勤期间

Funded Delivery Period 长交付期

**FDPS** Flight Data Processing System 飞行数据处理系统

**FDR** Fatigue Damage Range 疲劳损伤范围

Fault Detection Rate 故障检测率

Feeder 馈线

Final Design Review 最后设计审查

Final Development Review 最终研制审查

Final Drawing Release 最后图纸放行

Fire Door 防火门

Fixed Depression Reticle 固定压低角光环

Flight Data Recorder 飞行数据记录器,黑匣子

Flight Data Review 飞行数据评审

Formal Design Review 正式设计评审

Forward Deflection Routing 前向改向路由选择

Frequency Diversity Radar 频率分集雷达

**FDRPL** Flight Data Recording Parameter Library 飞行数据记录参数数据库

**FDRS** Finders 寻线机,探测器

Flight Data Recording System 飞行数据记录系统

**FDS** Fastener Damaged Seal 损坏的紧固件封严(件)

Fault Detection Specification 故障探测说明

Final Data System 最终数据系统

Final Discrete Station 最终离散站

Financial Decision Support 财务决策支持

Flight Data Subsystem 飞行数据分系统

Flight Data System 飞行数据系统

Flight Deck Simulator 驾驶舱模拟机

Flight Design Systems 飞行设计系统

Flight Director System 飞行指引仪系统,飞行定向器系统

Flight Dynamics System 飞行动力学系统

Function Distributed System 功能分布系统

**FD-SS** Frequency-Diversity Spread Spectrum 频率分集扩频

**FDSU** Flight Data Storage Unit 飞行数据储存组件

**FDT** Failure Diagnostic Team 故障诊断组

Fault Detection Time 故障检测时间

Field Definition Table 外场定义表

Flight Data Transmitter 飞行数据发射器

Flight Dispatcher Training 飞行签派员训练

Flight Dynamics Team 飞行动力学组

**FDTD** Finite Difference Time Domain 有限时间差定义域

**FDTN** Foundation 基础

**FDTS** Fault Diagnosis Test Subsystem 故障诊断测试分系统

Field Data Tracking System 外场数据跟踪系统

Firing Device Test Set 点火装置测试设备

**FDTT** Full Duplex Teletype 全双工电传打字机

**FDU** Bandundu, Zaire（Airport Code）扎伊尔班顿杜机场

Fire Detection Unit 火警探测组件

Flight Data Unit 飞行数据装置

Flight Development Unit 飞行发展部

Floppy Drive Unit 软盘驱动装置

Flux Detector Unit（磁）通量探测组件

Frequency Divider Unit 分频器

**FDV** Fault Detect Verification 故障（损伤）检测证明

Fuel Distribution Valve 燃油分配活门

**FDW** Feed Water 供水

**FDWD** Flight Deck Window Design 驾驶舱窗户设计

**FDWL** Fiberboard Double Wall 双层纤维板墙

**FDX** Full Duplex 全双工

**FE** Far East 远东

Flight Engineer 随机工程师,飞行工程师

Flight Envelope 飞行包线

Flight Examiner 飞行考试员

For Example 例如

Function Element 功能单元

Function Entity 功能实体

**FEA** Failure Effect Analysis 故障影响分析

Federal Energy Administration（美国）联邦能源管理局

Finite Element Analysis 有限元分析

Function Entity Action 功能实体作用

**FEAM** Functional Entity Access Management 功能实体接入管理

**FEAP** Far East and Pacific 远东及太平洋地区

**FEAR** Failure-Effect Analysis Report 故障后果分析报告

**FEAS** Feasibility 可行性

Feasible 可行的

Foreign Enterprise Air Service Corp. 外企航空服务公司

**FEATM** Future European Air Traffic Management 未来欧洲空中交通管理

**FEATS** Future European Air Traffic System 未来欧洲空中交通系统

**FEB** February 二月

Forward Equipment Bay 前设备舱

Function Electronic Block 电子功能块

**FEC** Failure Effect Category 故障影响类型

Fan Exhaust Case 风扇排气匣

Flight Envelope Computer 飞行包线计算机

Foreign Exchange Certificate 外汇兑换券

Forward Equipment Center 前设备中心

Forward Error Conrol 前向差错控制

Forward Error Correction 前进误差修正,前向纠错

Forwarding Equivalence Class 转发等价类型

**FECC-F** Forward Error Correction Count-Fast Data 前向纠错快速计数数据

**FECN** Forward Explicit Congestion Notification 前向显式拥塞通知

**FECP** Facility Engineering Change Proposal 设备工程更改建议

**FECU** Flap Electronic Control Unit 襟翼电子控制组件

**FED** Federal Express 联邦快递
Federal or Federation 联邦
Forward Error Detection 前向检错

**FEDC** Federal Equipment Data Center 联邦设备数据中心

**F-EDFA** Forward Pumped EDFA 前向泵激励掺铒光纤放大器

**FEDN** Federation 联合会

**FEDR** Feeder 馈线

**FEDS** Flight Environment Data System 飞行环境数据系统

**FEDSPEC** Federal Specification 联邦规格

**FEDSTD** Federal Standard 联邦标准

**FEF** Flight Evaluation Folder 飞行评定卷宗

**FEFET** Ferro-Electric Field Effect Transistor 铁电场效管

**FEFI** Flight Engineer Fault Isolation 空勤工程师故障隔离
Flight Environment Fault Indicator 飞行环境故障指示器

**FEFM** Flight Engineer Flight Manual 随机工程师飞行手册

**FEFO** First Ended First Out 先结束先送

**FEGV** Fan Exit Guide Vane 风扇出口导向叶片

**FEIA** Flight Engineers International Association 国际飞行工程师协会

**FEICRO** Federation of European Industrial Cooperative Research Organizations 欧洲工业合作研究组织联合会

**FEID** Flight Equipment Interface Device 飞机设备接口装置

**FEL** Flight Engineering Licence 飞行工程师执照

**FELC** Field Effect Liquid Crystal 场效液晶

**FEM** Female 女性的
Finite Element Method 有限元方法
Finite Element Model 有限元模型

**FEMA** Fast Ethernet Media Adapter 快速以太网卡
Federal Emergency Management Agency （美国）联邦紧急事物管理局

**FEMIS** Federal Emergency Management Information System 联邦应急管理信息系统
Flight Efficiency Management and Information System 飞行效率管理及信息系统

**FEMS** Federation of European Materials Societies 欧洲材料学会联盟

**FENG** Flight Engineer 飞行工程师

**FEO** Federal Energy Office （美国）联邦能源局

**FEP** Freeport, IL, USA （Airport Code）美国伊利诺州弗里波特机场
Front End Processor 前端处理器，前端处理机

**FEPC** Flight Equipment Processing Contract 飞行设备处理合同

**FEPS** Facility and Equipment Planning System 设施、设备计划系统
Federal Electronic Processing Standard 联邦电子处理标准

**FER** Forward Engine Room 前发动机舱
Frame Erasure Rate 帧删除率
Frame Error Rate 误帧率

**FERA** Federal Emergency Relief Administration 联邦紧急救济局

**FERD** Facility and Equipment Requirements Document 设施和设备要求文件

**FERF** Far End Receive Failure 远端接收失效

**FERPA** Family Education Rights and Privacy Act 家庭教育权利和隐私权法

**FES** Fire Extinguishing System 灭火系统
Fixed Earth Station 固定地球站
Flight Environment Sensor 飞行环境传感器
Fluids Experiments System 流体试验系统

**FESC** Forward Electrical/Electronic Server Center 前方电气/电子服务中心

**FEST** Festival 节日

**FET** Far East Time 远东时间
Field Effect Transistor 场效应晶体管
Flight Environment Test 飞行环境试验

**FETF** Flight Engine Test Facility 发动机飞行试验设备

**FETT** First Engine to Test 首台测试发动机
First Engine Type-Test 第一次发动机型号试验

**FEW** Fleet Empty Weight 机队空重

**FEWG** Flight Experiment Working Group 飞行试验工作组

**FEWP** Federation of European Women Pilots 欧洲女飞行员联合会

**FEXT** Far-End Cross Talk 远端串音,远端串扰
Fire Extinguisher 灭火器

**FEZ** Fez, Morocco（Airport Code）摩洛哥非斯机场

**FF** Fan Frame 风扇框架
Feeder Fault 馈电线故障
First Flight 首次飞行,处女飞行
Flight Follower 签派员
Flight Forecast 航班预报
Flip Flop 双稳态多谐振荡器
Folding Fin 折叠式尾翼
Free Flight 自由飞行
Friction Force 摩擦力
Fuel Flow 燃油流量

**FF/E** Fuel Flow per Engine/Hour 每台发动机燃油流量/小时

**FFA** Frequent Flyer Awards 常旅客奖励计划

**FFAR** Feel Force/Stick Angle Relation 感力与驾驶杆偏角的关系
First Flight Article Review 首次飞行项目检查

**FFBD** Functional Flow Block Diagram 功能流程方块图

**FFBW** Full Flying By Wire 全电传操纵

**FFC** Fan-Failure Clutch 风扇故障离合器
Final Flight Certification 最终飞行鉴定
Flight Facilities Checking 飞行设施检查
Flight Following Service 飞行跟踪服务

**FFCC** Forward Facing Crew Cockpit Concept 前视机组驾驶舱概念
Forward Facing Crew Cockpit Configuration 前视机组驾驶舱布局
Forward-Facing Crew Cockpit 前向机组驾驶舱

**FFCCV** Fan Frame Compressor Case Vertical 风扇框架压气机机匣垂面
Fan Frame Compressor Case Vertical（Sensor）风扇框架压气机机匣垂面（传感器）
Forward Flange Compressor Casing Vibration 压气机机匣前凸边振动,前向法兰压缩机匣振动

**F-FCH** Forward Fundamental Channel 前向基本信道

**FFCM** Free Fall Control Module 重力放下控制模块

**FFD** Forward Flight Deck 前驾驶舱
Functional Flow Diagram 功能流程图

**FFDC** First Failure Data Capture 首次故障数据捕获

**FFES** Fixed Fire Extinguishing System 固定灭火系统

**FFF** Flight Flutter Facility 飞行振动地面设施

**FFG** Fuel Flow Governor 燃油流量调节器

**FFH** Fast Frequency-Hopping 快速跳频
Fixed Flight Hours 固定飞行小时
Fleet Flight Hours 机队飞行小时

**FFI** Fuel Flow Indicator 燃油流量指示器

**FFL** Field Failure 机场故障
Flip Flop Latch 触发锁

**FFLY** Faithfully 诚挚地

**FFM** Future Flight Management 未来飞行管理

**FFNN** Feed Forward Neural Network 前馈神经网络,前向神经网络

**FFOS**   Forward Flying Observation System 先进的飞行观测系统

**FFP**   Fiber Fabry-Perot 光纤法布里—珀罗

Fixed Firm Price 固定公司价格

Frequent Flyer Program 常旅客计划

**FFPF**   Fiber Fabry-Perot Filter 光纤法布里—珀罗滤波器

**FFPI**   Fiber Fabry-Perot Interferometer 光纤法布里—珀罗干涉仪

**FFP-TF**   Fiber Fabry-Perot Tunable Filter 光纤法布里—珀罗可调滤波器

**FFR**   France Franc 法国法郎

Fuel Feasibility Review 燃料可行性评审

Fuel Flow Regulator 燃油流量调节器

Full Flight Regime 全飞行阶段

Full Frequency Resolution 全频分辨率

**FFRAT**   Full Flight Regime Autothrottle 全飞行状态自动油门

**FFRAT/SC**   Full Flight Regime Auto-Throttle Speed Control 全航程自动油门/速度控制

**FFRATS**   Full Flight Regime Auto-Throttle System 全航程自动油门系统

**FFRN**   Four-Fiber Ring Node 四纤环节点

**FFS**   For Further Study 供进一步探讨

Formatted File System 格式化文档系统

Fuel Flow Summation 燃油流量累加

Fuel Flow Summator 燃油流量累加器

Fuel Flow System 燃油流量系统

Full Flight Simulator 全动飞行模拟机

**FFSP**   Full Function Signal Processor 全功能信号处理器

**FFT**   Fast Fourier Transform 快速傅立叶变换

Full Functional Test 全功能测试

**FFTA**   Fast Fourier Transform Analysis 快速傅里叶变换分析

Fuzzy Fault Tree Analysis 模糊故障树分析

**FFW**   Failure-Free Warranty 无故障保险期

Feeder Fault Warning 馈电线故障告警

**FG**   Flight Guidance 飞行制导

Fog 雾

**FG/FDS**   Flight Guidance and Flight Deck System 飞行引导与驾驶舱系统

**FGC**   Flight Guidance and Control 飞行制导与控制

Flight Guidance Computer 飞行引导计算机

Functional Group Code 功能组代码

**FGCC**   Federal Geodetic Control Committee 联邦大地测量管理委员会

**FGES**   Flight Guidance and Envelope System 飞行指引与包络系统

**FGGE**   First GARP Global Experiment 第一次全球大气研究项目全球试验

**FGM**   Functionally Graded Material 功能梯度材料

**FGN**   Foreign 外来的,外国的

**FGPFL**   Fixed Group Flashing Light 固定闪光灯组

**FGS**   Flight Guidance System 飞行制导系统,飞行指引系统

**FGT**   Flight 飞行,航班

Freight 货物,运费

**FH**   Flight Hours 飞行小时

Frame Handler 帧处理器

Frame Header 帧头

Frequency Hopping 频率跳变

Fuel Heater 燃油加热器

**FHA**   Fault Hazard Analysis 故障危险性分析

Functional Hazard Analysis 功能危害分析

Functional Hazard Assessment 功能危害性评估

**FHC**   Fire Hose Cabinet 防火水管柜

**FHI**   Fuji Heavy Industries Ltd 富士重工业公司

**FHR**   Fixed Hierarchical Routing 固定等级选路

**FHS**   Factory Headcount System 工厂主统计系统

Flight Hour Service 飞行小时勤务

Fluid Handling System 流体处理系统

**FHSP**   Frame Handler Subport 帧处理程序子端口

**FHSS**   Frequency Hopping Spread Spectrum

跳频扩频

**FI** Fault Isolation 故障隔离

Finland 芬兰

Flight Information 航班信息发布

Flight Inspection 飞行检查,飞行校验

Flight Instructor 飞行教员

Flight Instruments 飞行仪表

Format Identifier 格式标识符

**FIA** Farnborough International Airshow 范堡罗国际航空航天展览会

Flight Information Area 飞行情报区

**FIAC** Flight Information Advisory Committee 飞行信息咨询委员会

**FIAS** Flight-Inspection Aircraft System 空中检查飞机系统

**FIATA** International Federation of Freight Forwarders Associations 国际货运代理协会联合会

**FIAU** Flight Instrument Accessory Unit 飞行仪表附件盒

**FIB** Freight Investigation Bureau（India）运费调查局

**FIC** Fiber Interface Card 光纤接口卡

Flight Information Centre 飞航情报中心

Flow Indicating Controller 流量指示控制器

Frequency Interference Control 频率干扰控制

**FICLS** First Class 头等舱

**FICNO** Field Closed to Night Operations 机场夜航关闭

**FICO** Flight Information and Control of Operations 飞行信息和飞行控制

**FICS** Facsimile Intelligent Communication System 传真智能通信系统

**FID** Fault Isolation Detection 故障隔离探测

Fault Isolation Diagnostics 故障隔离诊断

Flight Inspection Data 飞行检验数据

Free Induction Decay 自由诱导衰减

**FIDAS** Flight Inspection Data Archiving System 飞行检查数据档案系统

**FIDO** Flight Inspection District Office 飞行检

查区办公室

Fog Investigation and Dispersal Operation 消雾作业

**FIDS** Fault Isolation and Detection System 故障隔离和探测系统

Flight Information Display System 飞行信息显示系统,航班信息显示系统

**FIE** Flight Instructor Examiner 飞行教员考试员

Fuel Injection Equipment 燃油喷入设备

**FIFO** First Input,First Output 先入先出

Flight Information Field Office 飞行情报地面站

Flight Inspection Field Office 飞行检查区域办公室

**FIFOR** Flight Forecast 飞行预报

**FIFS** First In First Served 先进先服务

**FIG** Figure 图像,数字

Flight Inspection Group 飞行检查组

**FIGV** Fan Inlet Guide Vane 风扇进口导向叶片

**FIH** Flight Information Handbook 飞行情报手册

Kinshasa,Zaire（Airport Code）扎伊尔金沙萨机场

**FII** Federal Item Identification 联邦物品识别,联邦物品标志

Flight International,Inc.（USA）国际飞行公司（美国）

**FIIG** Federal Item Identification Guide 联邦物品识别指南

**FIIN** Federal Item Identification Number 联邦物品识别号

**FIL** Filament 灯丝,阴极

Fillet 整流片

Filter 滤波器,过滤器

Fuel Injection Line 喷油线

Undeveloped Film/Unexposed Film 未冲洗胶卷

**FILG** Filing 归档

Filling 填料

**FILH** Fillister Head 带槽螺钉头

**FILO** First in Last out 先进后出

**FILS** Flarescan Instrument Landing System 闪耀扫描仪表着陆系统

**FILT** Filter 滤波器,过滤器

**FIM** Fault Identification Manual ( ATA 100 Specification) 故障鉴定手册

Fault Isolation Manual 故障隔离手册

Fault Isolation Monitoring 故障隔离监控

Fiber Interface Module 光纤接口模块

Field-Ion Microscope 场离子显微镜

Flight Information Manual 飞行信息手册

Flight Interruption Manifest 班机异常旅客转机名册,中断飞行舱单

**FIM/CM** Feature Interactive Management and Calling Management 特征交互管理与呼叫管理

**FIMA** Future International Military Airlifter 未来国际军事运输机

**FIMIS** Flight Inspection Management Information System 飞行检查管理情报系统

**FIMS** Fault Isolation and Monitoring System 故障隔离与监控系统

Feature Interaction in Multimedia System 多媒体系统特征交互

Flight Information Management System 飞行情报管理系统

**FIN** Final 最后的

Finance 财政

Financial 财政的

Finish 完成

Finland 芬兰

Full Interconnection Network 全互联网络

Function Item Number 功能项目号

**FINCL** Financial 财政的

**FINL** Final 最后的

**FIO** Flight Information Office 飞行情报室

For Information Only 仅供参考

**FIOG** Foreign IATA Operators Group 国际航空运输协会外部会员组

**FIP** Fault Isolation Procedure 故障隔离程序

Fleet Indoctrination Program 机队入门计划

Fuel Injection Pressure 注油压力

Fuel Injection-Pump 燃油喷射泵

**FIPS** Federal Information Processing Standards 联邦信息处理标准

Flight Information Processing System 飞行信息处理系统

Flight Inspection Position System 飞行检查定位系统

**FIR** Facility Installation Review 设备安装检查

Facility Interference Review 设备干扰检查

Far Infra-Red 远红外(线)

Fault Isolation Rate 故障隔离率

Finite Impulse Response 有限冲击响应

Fired 着火的,发射的

Flight Incident Recorder 飞行事故记录器

Flight Information Region 飞行情报区

Flight Information Report 飞行信息报告

Flight Irregularity Report 班机异常报告

Fuel Indicator Reading 油量表读数

Full Indicator Reading 全指示器读数

**FIRB** Flight Information Boundary 飞行情报区边界

Foreign Investment Review Board 外商投资审核委员会

**FIRE** Flight Investigation Reentry Environment 再入大气层环境飞行研究

**FIRR** Financial Internal Rate of Return 财务内部收益率

**FIRS** Flight Information Regions 飞行情报区

Framing Infra-Red Sensor 成帧红外传感器

**FIS** Fixed-Interval System 定期订货方式

Flight Information Service ( s ) 航班信息服务,飞行情报服务

Flight Instrument System 飞行仪表系统

Flight Interphone System 飞行内话系统

**FISA** Automatic/Automated Flight Information Service 自动飞行情报服务

**FIS-B** Flight Information Services-Broadcast 飞行信息服务—广播

**FISC** Fast Instruction Set Computer 快速指令集计算机

Flight Instrument Signal Converter 飞行仪表信号转换器

Fuels Information Service Center 燃油信息服务中心

**FISDL** Flight Information Services Data Link 航班信息服务数据链

**FISS** Flight Information Service System 飞行情报服务系统

**FISSC** Flight Instrument System Signal Converter 飞行仪表系统信号转换器

**FISSP** Flight Information Service Strategic Plan 飞行情报服务战略计划

**FIST** Fault Isolation by Semiautomatic Techniques 半自动故障隔离技术

**FIT** Fabrication, Integration Test 制作、集成与试验

Failure Information Team 故障信息小组

Fault Isolation Test 故障隔离测试

Fault Isolation Time 故障隔离时间

Fault Isolation Tree 故障隔离树

Filter 滤波器,滤波

Frequency, Intensity and Time 频率、强度和时间

Frequent Individual Traveler 个人旅客（散客）

**FITD** Far Infrared Target Detector 远红外目标探测器

**FITI** Far Infrared Target Indicator 远红外目标指示器

**FITL** Fiber in the Loop 环路光纤,光纤环路

**FITMP** Flight Instrument Test Modes Panel 飞行仪表测试方式组（操纵台）

**FITR** Filter 过滤器

**FITVC** Fluid-Injection Thrust Vector Control 喷流推力向量控制

**FIU** Facilities Interface Unit 设备接口单元

Fingerprint Identification Unit 指纹识别器

Frequency Identification Unit 频率识别装置

**FIVS** Italian Federation of Simulated Flight 意大利模拟飞行联合会

**FIX** Fault Isolation Extract 故障隔离提取

Fault Isolation Manual 故障隔离手册

Fixed 固定的

Fixture 固定设备

**FIZ** Flight Information Zone 飞行情报地带

**FJCC** Fall Joint Computer Conference 秋季联合计算机会议（美国）

Fuel Jettison Control Card 燃油紧急放油控制卡

**FKL** Franklin, PA, USA (Airport Code) 美国宾夕法尼亚州富兰克林机场

**FKT** Flight Kit 飞行备件箱

**FL** Fatigue Limit 疲劳极限

Feel 感觉

Finance Lease 融资租赁

Flashing Light 闪光灯

Flashing 闪光

Flight Leg 航段

Flight Level (Altitude) 飞行高度层（高度）

Flight Line 起飞线,起机线

Flood 泛光,泛滥

Floor 地板,楼层

Flow 流动,流量

Fluid Level 液位

Frequency Flyer ID 常旅客编号

Fluid 液体

Fuel 燃油

Funnel Cloud 漏斗云

**FL/CH** Flight Level Change 飞行高度层改变

**FLA** Fault Location Algorithm 故障定位算法

Flammable 易燃的

Future Large Aircraft 未来大型航空器

Future Large Airlifter 未来大型空运工具

**Fla. G** Flammable Compressed Gas 易燃压缩气体

**Fla. L** Flammable Liquid 易燃液体

**Fla. S** Flammable Solid 易燃固体

**FLAC** Free Lossless Audio Code 自由无损音频解编码

**FLACS** Flight Line Automatic Checkout System 航线自动检查系统

**FLAG** Fiber Link around the Globe 环球光纤链路

Floor Level above Ground 飞机地板离地高度

**FLAIR** Fleet Location and Information Reporting 机群定位和信息报告

**FLAM** Fault Location and Monitoring 故障定位与监控

Flammable 易燃的

**FLAR** Fault Location and Repair 故障定位与修理

Forward Looking Airborne Radar 前视机载雷达

**FLAT** Flight Plan Aid Tracking 飞行计划辅助跟踪

**FLB** Flight Log Book 飞行记录本

**FLC** Feel and Limitation Computer 感觉和限制计算机

Flight Level Controller 飞行高度控制器

**FLCH** Flight Level Change 飞行高度层变更

**FLCS** Flight Control System 飞行控制系统

Force Level Control System 兵力水平控制系统

**FLD** Fairlead 导片,导火锁,引火管

Fault Logic Diagram 故障逻辑图

Field 场地,机场,磁场

Flight Dynamics (USA) 飞行动力公司(美国)

Fluid Dynamics 流体动力学

Fluid 液体

Fond du Lac, Wisconsin, US (Airport Code) 美国威斯康星州丰迪拉克机场

Full Load 满载

**FLDK** Flight Deck 驾驶舱

**FLDR** Flight Loads Data Recorder 飞行负载数据记录器

**FLETC** Federal Law Enforcement Training Center 联邦执法培训中心

**FLEX** File Exchange System 文件交换系统

Flexible 灵活的,变动的

**FLEXAR** Flexible Adaptive Radar 机动自适应雷达

**FLEXTO** Flexible Take off 灵活推力起飞,减推力起飞

**FLFT** Forklift 叉式起重机

**FLG** Flagstaff, AZ, USA (Airport Code) 美国亚利桑那州费拉格斯塔夫机场

Flange 法兰盘,凸边

Flashing 闪光的,闪光

Following 以下的,下述的,随后的

Forward Landing Ground 前方降落机场

Fueling Landing Ground 加油降落机场

**FLH** Final Limit Hoist 最终上升极限

Flat Head 平头

**FLICON** Flight Controls 飞行操纵面

**FLID** Flight Identification 飞行(航班)识别

**FLIDEN** Flight Data Entry 飞行数据输入

**FLIDRAS** Flight Data Replay and Analysis System 飞行数据解读与分析系统

**FLIGA** Flight Incident or Ground Accident 飞行事故与地面事故征候

**FLIP** Flight Information Planning 飞行情报计划

Flight Information Publication 飞行资料汇编

**FLIR** Forward-Looking Infrared Detection 前视红外线探测

Forward-Looking Infrared Radiation 前视红外线辐射

Forward-Looking Infrared Radiometer 前视红外辐射计

Forward-Looking Infrared 前视红外线

**FLITE** Flying Laboratory for Integrated Testing and Evaluation 综合测试和评估飞行实验室

**FLL** Fiber in the Local Loop 局域环路光纤

Flap Load Limiter 襟翼载荷限制器

Flight Line Level 航线高度层

Fort Lauderdale, FL, USA (Airport Code) 美国佛罗里达州劳德尔堡机场

Frequency Locked Loop 频率锁定回路

**FLLT**　Flashing Light 闪光信号灯,手电筒

**FLLW**　Field Length Limit Weight 场长极限重量

**FLLWS**　Follows 下述

**FLM**　Fiber Loop Mirror 光环路镜像
Film 胶卷,影片
Flight Line Maintenance 外场维修
Flight Line Mechanic 外场机械员

**FLMPRF**　Flameproof 耐火的

**FLMS**　Flight Log Monitoring System 飞行记录监控系统

**FLMTS**　Flight Line Maintenance Test Set 航线维修测试装置

**FLO**　Apron Flood Light 机坪照明灯
Flood Light 泛光灯
Florence,SC,USA(Airport Code) 美国南卡来罗纳州弗洛伦斯机场

**FLOLS**　Fresnel Lens Optical Landing System 菲涅耳透镜光学着陆系统

**FLOPS**　Flight Operations 飞行运行,飞行操作

**FLP**　Flap 襟翼
Flight Line Planning 飞行路线计划

**FLPA**　Flight Level Pressure Altitude 飞行压力高度

**FLPRN**　Flaperon 襟副翼

**FLPS**　Flight Loads Preparation System 飞行载荷准备系统

**FLR**　Filler 填料
Flap Load Relief 襟翼负荷释放
Flare 拉平;火焰
Flight Line Recorder 飞行线记录器
Flight Line Reference 航线基准
Florence,Italy(Airport Code) 意大利佛罗伦萨机场
Forward-Looking Radar 前视雷达

**FLS**　Field Level Service 外场级维修
Flight Structures(USA) 飞行结构公司(美国)
Future Launch System 未来(航天)发射系统

**FLSA**　Fuel Level Sensing Amplifier 燃油平面感应放大器

**FLSCU**　Fuel Level Sensing Control Unit 燃油液位感应控制组件

**FLT**　Fault Locating Test 故障定位测试
Fault Location Technology 故障定位技术
Fault 故障
Filter 滤波器,滤器
Fleet 机队
Flight Line Tester 飞行航线测试器
Flight 飞行,航班
Float 浮动
Free Flight 自由飞行
Full Load Takeoff 满载起飞
Full Load Torque 满载转矩

**FLTCK**　Flight Check 飞行检查,飞行校验

**FLTCPN**　Flight Coupon 乘机联

**FLTDIR**　Flight Director 飞行指引仪

**FLTDK**　Flight Deck 驾驶舱

**FLTG**　Floating 浮动的

**FLTL**　Flight Line 外场,航线

**FLTNR**　Flight Number 航班号

**FLTP**　Flight Path 飞行航道

**FLTPAR**　Flight Parameters 飞行参数

**FLTPLN**　Flight Plan 飞行计划

**FLTR**　Filter 滤波器,滤器
From Left to Right 从左到右

**FLTS**　Fault-Locating Test 故障定位测试
Flight Line Test Set 航线测试装置

**FLTSATCOM**　Fleet Satellite Communication 机群卫星通信

**FLU**　Fault Locating Unit 故障定位设备

**FLUC**　Fluctuating 波动,摆动

**FLUOR**　Fluorescent 荧光的,发亮的

**FLW**　Follow 跟随,接下去
Following 跟随
Follows or Following 跟随

**FLWG**　Following 下述的,随后的

**FLWS**　Follows 下述

**FLX**　Flexible 灵活的,流动的

**FLXTO**　Flexible Takeoff 灵活起飞

**FLY** Fly, Flying 飞行, 飞航
**FM** Facilities Management 设备管理
　Fan Marker 扇形指点标
　Fault Exclusion 故障排除
　Fault Management 故障管理
　Fault Monitoring 故障监测
　Field Manual 外场手册, 机场手册
　Flexible Multiplexer 灵活复用器
　Flight Management 飞行管理
　Flight Manual 飞行手册
　Flight Mechanic 随机机械师
　Flight Models 飞行模型
　Flight Monitor 飞行监测
　Form 形式, 格式, 表格
　Forward Monitoring 前向监控
　Fracture Mechanics 断裂力学
　Frequency Modulation 调频
　From 来自…, 自, 从
**FMA** Failure Mode Analysis 故障模式分析
　Financial Management Association 财务管理协会
　Flight Mode Annunciations 飞行方式通告
　Flight Mode Annunciator 飞行状态指示器
　Forward Maintenance Area 前维护区
　Frequency Modulation Association 调频协会
　Frequency Modulator Altimeter 调频高度表
　Fuselage Mounted Antenna 机身装天线
**FMAN** Fireman 消防队员
　Foreman 工头, 监工
**FMAS** Flutter and Matrix and Analysis System 颤振和矩阵分析系统
**FMAV** Flapping-Wing Micro Air Vehicle 微型扑翼飞行器
**FMBD** Formboard 模板
**FMBS** Frame Mode Bearer Service 帧模式承载业务
**FMC** Failure Mode Center (Reliability Laboratory) 故障模式中心
　FDAU Master Controller 飞行数据获取(采集)组件主控制器
　Federal Magistrate Court 联邦裁判司法庭

Federal Manufacturers Code 联邦制造商代号
Fixed-Mobile Covergence 固定移动融合
Fleet Management Center 机队管理中心
Flight Management Center 飞行管理中心
Flight Management Computer 飞行管理计算机
Flight Mode Controller 飞行方式控制器
Flutter Mode Control 颤振模态控制
Fuel Management Computer 燃油管理计算机
Full Mission Capable 全程可使用
**FMCDU** Flight Management Control Display Unit 飞行管理控制显示器
**FMCF** Flight Management Computer Function 飞行管理计算机功能
**FMCS** Fatigue Monitoring Computer System 疲劳监测计算机系统
　Flight Management Computer System 飞行管理计算机系统
　Flight Management Computing System 飞行管理计算系统
　Freight Movement Control System 货物移动控制系统
**FMCU** Flight Management Computer Unit 飞行管理计算机组件
**FMCW** Frequency Modulated Carrier Ware 调频载波
　Frequency Modulation Continuous Wave 调频连续波
**FMD** Failure Message Description 故障信息描述
　Financial Management Division 财政管理处
　Flight Management Display 飞行管理显示器
　Flow Management Division 流量管理处
　Frame Mode Data 帧模式数据
　Function Management Data 功能管理数据
**FMDI** Function Management Data Interpreter 功能管理数据解释程序
**FMDS** Flight Management Data System 飞行管理数据系统

**FME** Floor Mounted Equipment 装在地板上的设备
Flow Metering Equipment 流量计量设备

**FMEA** Failure Mode and Effects Analysis 失效模式及效应分析
Frequent Malfunction Emergency Analysis 常见故障应急分析

**FMECA** Failure Mode, Effect and Criticality Analysis 故障模式、影响和危险度分析

**FMES** Failure Modes and Effects Summary 故障模式与效应概要
Full Mission Engineering Simulator 全任务工程模拟器

**FMET** Failure Modes and Effects Testing 失效模式及效果测试
Functional Management Engineering Team 职能管理工程团队

**FMF** File Micro Program Flags Register 文件微程序标记寄存器
Flight Management Function 飞行管理功能

**FMG** Frequency Modulation Generator 调频信号发生器, 频率调制发生器

**FMFB** Frequency Modulation Feedback 频率调制反馈, 具有反馈的频率

**FMGC** Flight Management and Guidance Computer 飞行管理和指引计算机

**FMGEC** Flight Management Guidance and Envelope Computer 飞行管理指引和包线计算机

**FMGES** Flight Management Guidance and Envelope System 飞行管理指引和包线系统

**FMGS** Flight Management and Guidance System 飞行管理和指引系统

**FMGST** Flight Management Guidance System Trainers 飞行管理引导系统训练器

**FMH** Fault Message History 故障信息记录

**FMI** Flexible Modular Interface 可变模块化接口
Fonds Monetary International 国际货币基金组织
Functional Management Inspection 功能管理检查
Future Margin Income 未来利差收入

**FMIC** Flight Manual Interim Changes 飞行手册临时更改

**FMICW** Frequency Modulated Intermittent Continuous Wave 调频间断连续波

**FMIS** Facilities Management Information System 地面设施管理信息系统

**FMK** Markka 芬兰马克

**FML** Flight Management Library 飞行管理程序库
Formal 正式的

**FMM** Flight Manual Manager 飞行手册管理员

**FMMEA** Failure Mode, Mechanism and Effect Analysis 故障模式、机理及影响分析

**FMMS** Fixed Media Mass Storage 固定媒体大容量存储器
Fuel Measurement and Management System 燃油测量与管理系统

**FMN** Farmington, NM, USA (Airport Code) 美国新墨西哥州法明顿机场

**FMO** Facilities Management Office 设施管理办公室
Flight Management Office 飞行管理办公室
Frequency Management Office 频率管理办公室
Muenster, Germany (Airport Code) 德国曼斯特机场

**FMOF** First Manned Orbital Flight 首次有人驾驶轨道飞行

**FMOP** Frequency Modulation on Pulse 脉冲调频

**FMP** Field Maintenance Party 外场维修组
Fleet Maintenance Program 机队维修项目
Fleet Management Program 机队管理项目
Fleet Modernization Program 机队现代化项目
Flight Manual Program 飞行手册项目
Flight Mode Panel 飞行方式面板
Flow Management Position 流量管理席位

Frequency Management Plan 频率管理计划

Fuel Maintenance Panel 燃料补给操纵台

Fuel Manifold Pressure 燃油总管压力

**FMPS** Flier Mission Payload System 航空器任务配载系统

**FMQGS** Fuel Measurement and Quantity Gauging System 燃油管理和油量计量系统

**FMR** Field Maintenance Reliability 机场维护可靠性

Field Material-Handling Robot 外场(机场)器材自动操作机

Former 以前的,前面的

Frequency Modulated Radar 调频雷达

**FMRB** Flight Manual Review Board 飞行手册审查委员会

**FMS** Famous 著名的

Field Maintenance Squadron 外场维修中队

File Management Subsystem 文件管理子系统

File Management System 文件管理系统

Fleet Management Services 机队管理服务

Flexible Manufacturing System 柔性制造系统,弹性制造系统

Flight Management System 飞行管理系统

Flight Motion Simulation 飞行动作模拟

Fluid Management System 流体管理系统

Frequency Management System 频率管理系统

Frequency Monitoring System 频率监控系统

Fuel Management System 燃油管理系统

Fuel Manifold Staging 燃油总管支架

Full Mission Simulation 全任务模拟

Fuse Maintenance Spare 燃油维护备件

**FMSG** Frequency Management Study Group 频率管理研究组

**FMSP** Flight Management System Procedure 飞行管理系统程序

Flight Mode Selector Panel 飞行模式选择器面板

**FMSR** FP-Mode Suppression Ratio 法布里—珀罗特模式抑制比

**FMSS** Facilities Maintenance Support System 设施维修支持系统

**FMST** Flight Management System Trainer 飞行管理系统训练器

**FMSTS** Flight Management System Test System 飞行管理系统测试系统

**FMT** Format 格式,格式化

Full Mission Trainer 全任务训练器

**FMTS** Field Management Test Set 外场管理测试仪

**FMU** Flight Management Unit 飞行管理组件

Flow Management Unit 流量管理单元

Fuel Metering Unit 燃油计量组件

**FMUX** Flexible Multiplexer 灵活复用器

**FMV** Fuel Metering Valve 燃油计量活门

Full-Motion Video 全运动视频

**FMVRES** Fuel Metering Valve Resolver 燃油计量活门解算器

**FMVSS** Federal Motor Vehicle Safety Standards 美国联邦机动车安全标准

**FMY** Ft. Myers, FL, USA (Airport Code) 美国佛罗里达州迈尔斯堡机场

**FN** Fiber Node 光纤节点

Flight Navigation 航空导航,航行

Function 功能

Functional Network 功能网络

Net Thrust 净推力

**FNA** Federation National Aeronautique 全国航空联合会(法)

Final Approach 最后进近,最终进近

Flexible Networking Architecture 灵活的网络结构

Florida National Airlines (USA) 佛罗里达国家航空公司(美国)

Free Network Address 空闲网络地址

Freetown, Sierra Leone Lungi (Airport Code) 塞拉利昂弗里敦隆吉机场

**FNAE** Free Network Address Element 空闲网络地址元素

**FNAS** Frame Relay Network Access Subsystem 帧中继网络接入子系统

**FNBA** Federation National Belgium Aviation 比利时全国航空联合会

**FNC** Favoured-Nation Clause 最惠国条款
Fiber Nickel Cadmium 镍镉纤维
Funchal, Madeira Islands, Portugal（Airport Code）葡萄牙马德拉群岛沙尔机场

**FNCP** Flight Navigation Control Panel 飞行导航控制面板

**FND** Found 发现，找到，建立

**FNDN** Foundation 基础

**FNF** Net Thrust of the Fan 净推力

**FNJ** Pyongyang, North Korea（Airport Code）朝鲜平壤机场

**FNL** Final 最后

**FNN** Fuzzy Neural Network 模糊神经网络

**FNP** Federal Navigation Plan 联邦导航计划
Frontend Network Processor 前端网络处理机

**FNPT** Flight and Navigation Procedures Trainer 飞行和导航程序训练器
Fusion Point 熔点

**FNRL** Final Release 最后放行

**FNSG** Flight Navigation Symbol Generator 飞行导航符号产生器

**FNSH** Finish 完成

**FNSS** Flight Normal Statistics System 航班正常性统计系统

**FNT** Fleet Number Table and Report 机群号码表与报告
Flint, MI, USA（Airport Code）美国密执安州弗林特机场

**FO** Fiber Optics 光纤
Field Operations 外来工作，野外作业
First Officer 副驾驶
Flight Operator 随机报务员
Flight Order 飞行指令
For Orders 等待指示
Foreign Object 外来物
Foreign Officer 外交人员
Fuel Oil 燃油
Fully Open 全开

**FOA** Fiber Optic Amplifier 光纤放大器
First Operational Aircraft 首架可运行飞机
Forced Oil and Air 压力油与空气
Fuel Oil A A 级燃油

**FOAN** Fiber Optic Access Network 光纤接入网络

**FOB** Forward Operating Base 前方飞行基地
Free on Board 离岸价格
Front of Board 飞机前部，船舶前部
Fuel on Board 机载燃油，机上存油

**FOBS** First Observer 第一观察员

**FOC** Fiber Optic Cable 光缆
Fiber Optic Communication 光纤通信
Final Operation Configuration 最终操作构型
Final Operational Capability 最终运行能力，最终飞行能力
Flight Operating Costs 飞行运营成本
Flight Operations Center 飞行调度中心
Free of Charge 免费
Fuel/Oil Cooler 燃油/滑油冷却器
Full Operation Capability 全运行能力
Full Operational Capability 完全操作能力

**FOCA** Federal Office of Civil Aviation 民航联邦办公室

**FOCAC** Forum for China Africa Cooperation 中非合作论坛

**FOCAS** Fiber Optics Communications for Aerospace System 航空航天光纤通信系统

**FOCC** Flight Operation Control Center 飞行运行管制中心
Forward Control Channel 前向控制信道

**FOCE** Fibre Optic Controlled Engine 纤维光学控制发动机

**FOCN** Fiber Optic Communication Network 光纤通信网

**FOCP** Flight Operational Computer Programs 飞行运行计算机程序

**FOCT** Flight Operations Consultation Tool 飞行操作咨询工具

**FOCUS** Fiber Optic Connection Universal System 光纤连接通用系统

**FOD** Flight Operation Domain 飞行操作范围
Flight Operations Data 飞行运行数据
Foreign Object Damage (飞机,发动机)外来物损坏,外物损害
Foreign Object Debris 机场道面异物
Fort Dodge, IA, USA (Airport Code) 美国依阿华州道奇堡机场
Fuel Over Destination 到达目的地上空时的燃油量

**FODC** Fire Overheat Detection Card 火警过热探测卡

**FODCS** Fiber Optics Digital Control System 光纤数字控制系统

**FODT** Fiber Optics Data Transmission 光纤数据传输

**FOE** Fiber Optic Extender 光纤延长器

**FOEB** Flight Operations Evaluation Board 飞行运行评审委员会

**FOEC** Flight Operation Estimation Committee 飞行运行评估委员会

**FOF** First Orbital Flight 首次轨道飞行
Fluorescent Optical Fiber 发光光纤

**FOG** Fiber Optic Gyroscope 光纤陀螺仪
Flight Operations Group 飞行操作小组

**FOHE** Fuel Oil Heat Exchanger 燃油滑油热交换器

**FOI** Fiber Optic Isolator 光纤隔离器
Flight Operation Inspector 飞行操作检查员
Forced Oil Injection 强迫注油

**FOID** Fiber Opitc Interface Device 光纤接口设备

**FOIP** Fax over IP IP 传真

**FOIRL** Fiber Optic Inter-Repeater Link 中继器间光纤链路

**FOL** Fiber Optic Laser 光纤激光器
First Order Level 第一指令高度
Forward Operating Location 前操作位置

**FOLAN** Fiber Optic LAN 光纤局域网

**FOM** Fiber Optic Modem 光纤调制解调器
Figure of Merit 性能指数,品质因数
First of a Model 原型

Flight Operations Manual 飞行运行手册
Flight Operations Monitoring 飞行运行监控

**FOMAU** Fiber Optic Medium Attachment Unit 光纤媒介附属单元

**FOMF** Fuel Oil Margin Factor 燃油安全系数

**FOMMS** Flight Operations Maintenance Management System 飞行运行维修管理系统

**FONE** Telephone 电话

**FOO** Flight Operation Office 航行室,航务办公室
Flight Operations Officer 飞行指挥官员,航务管理员

**FOOT** Follow-on Operational Training 后继操作训练,后继运行训练

**FOP** Failure of Protocol 协议失效
Flight Operations Planning 航班运营计划
Forward Observation Post 前进观测站
Forward Operating Pad 直升机前方起降场
Free on Plane 飞机上交货价格

**FOPB** Flight Operations Policy Board 飞行运行政策委员会

**FOPF** Fixed Operating Point Flight 固定作业点飞行

**FOPMA** Fiber Optic Physical Medium Attachment 光纤物理媒体装置

**FOQA** Flight Operational Quality Assurance 飞行运行质量保证
Flight Operations Quality Assurance Programs 飞行运行质量保证程序

**FOR** Fail-Operative Redundant 故障后保持工作的余度
Field of Regard 视场,视野
Flight Operations Review 飞行运行评审
Foreign 外国的,外来的
Fortaleza, Brazil (Airport Code) 巴西福塔雷萨机场
Fuel Oil Remaining 剩余燃油
Fuel Oil Return 燃油回路

**FORAS** Flight Operation Risk Assessment System 航班运行风险评估系统

**FOREM** Force Exchange Model 力交换模型

**FORF** Forfeit, Forfeiting, Forfeiture 罚款, 没收, 丧失, 罚金

**FORG** Forging 锻造

**FORGE** Format Generator 格式产生器

**FORGN** Foreign 外国的, 外部的

**FORMAC** Formula Manipulation Compiler 公式处理编译程序

**FORMDIG** Formboard Digitization 数字化成型板

**FORSC** Freight Operation and Reservation System Computer 货运操作系统和预订系统计算机

**FORTRAN** Formula Translation Language 公式翻译语言
Formula Translator System 公式转换系统

**FOS** Fiber Optic Sensor 光纤传感器
File Organization System 文件组织系统
Flight Operations Support 飞行运行保障
Fuel Operator Station 加油站
Functional Operational Specification 功能操作规范

**FOSD** Field Operations Support Division 外场运行保障分部

**FOSGEN** Fog Oil Smoke Generator 雾油烟产生器

**FOSS** Fiber Optic Sensor System 光纤传感器系统
Follow-on System Support 随动系统支援

**FOST** Flight Operations Support Team 飞行运行支援组

**FOT** Flight Operation Telex 飞行运营电传
Flight Operations Team 航班运营团队
Forced Oscillation Technique 强迫振动技术
Free of Tax 免税
Frequency of Optimum Operation 最佳工作频率
Frequency of Optimum Traffic 最佳通讯量的频率
Frequency of Optimum Transmission（雷达）最佳发射频率
Fuel Oil Tank 燃油箱

Fuel Oil Transfer 燃油传输

**FOT&E** Follow-on Operational Test and Evaluation 后续飞行试验和评定

**FOTC** Fiber Optic Trunk Cable 干线光缆

**FOTIC** Fiber Optic Transmitter Integrated Circuit 光纤发射机集成电路

**FOTM** Flight Operations Training Manual 飞行运行训练手册

**FOTN** Fiber Optic Transmission Network 光纤传输网络

**FOTR** Fiber Optic Transmitter/Receiver 光纤发送器/接收器

**FOTS** Fiber Optic Temperature Sensor 光纤温度传感器

**FOTSE** Follow-on Test and Evaluation 随动试验与评估

**FOU** Field Operation Unit 现场运行单位

**FOUO** For Official Use Only 仅供正式使用, 仅供官方使用

**FOV** Field of View 视场, 视野

**FOVE** Flight Operations Versatile Environment 飞行运行通用环境

**FOVH or FOVHT** Fire Overheat 过热着火

**FOWD** Forward 向前, 前方, 传递

**FOX** Fiber Optic Extender 光纤扩展器

**FP** Field Protective 保护性机场
Financing Proposal 融资方案
First Pilot 机长
Fixed Price 固定价格
Flight Path 飞行航迹
Flight Pilot 飞行员
Flight Plan 飞行计划
Flight Progress 飞行进展情况
Flight Phase 飞行阶段
Freezing Point 凝固点, 冰点
Fuel Petroleum 汽油燃料
Fuel Pressure 燃油压力
Fuel Pump 燃油泵
Fully Paid 全部付讫
Function Processor 功能处理机

**FPA** Failure Probability Analysis 故障概率

分析

Federal Preparedness Agency 联邦准备局

Final Power Amplifier 末级功率放大器

Fire Protection Additional Special Apparatus 特殊防火装置

Fire Protection Association（英国）消防协会

Fixed Pre-Assignment 固定预分配

Flight Path Accelerometer 飞行轨迹加速度

Flight Path Angle 飞行航迹角

Flight Plan Analysis 飞行计划分析

Flight Plan Approval 飞行计划批准

Flight Power Assembly 飞行动力组件

Floating Point Arithmetic 浮点运算

Flying Physicians' Association（美国）航空医生协会

Focal Plane Array 聚焦平面天线阵

Focal Plane Assembly 焦平面组件

Forward Pitch Amplifier 前俯仰放大器

Functional Process Assessment 功能过程评估

**FPAC** Flight Path Acceleration 加速飞行航迹，飞行路径加速度

**FPAD** Facsimile Packet Assembly/Disassembly 传真分组组合/拆卸

**FPAR** Flight Path Angle Rate 飞行航迹角变化率

**FPAS** Flight Performance Advisory System 飞行性能咨询系统

Flight-Profile Advisory System 飞行剖面咨询系统

**FPAT** Flight Plan Aided Tracking 飞行计划辅助跟踪

**FPB** Flight Progress Board 航班动态牌

Forward Pressure Bulkhead 前压力隔框

Fuel Preburner 燃料预燃室

**FPBS** Fiber Polarization Beam Spliter 光纤偏振分束器

**FPC** File Parity Checks Register 文件奇偶校验寄存器

File Parity Checks 文件奇偶校验

Fixed Point Calculation 定点计算

Flight Plan Code 飞行计划码

Flight Program Computer 飞行程序编制计算机

Front Passenger Cabin 前客舱

Fuel Pump Controller 燃油泵控制器

**FPCB** Flexible Printed Circuit Board 柔性印刷电路板

Flight/Propulsion Command Blending 飞行/推力指令混合

**FPCC** Flight Propulsion Control Coupling 飞行动力控制连接器

**F-PCH** Forward-Paging Channel 前向寻呼信道

**FPD** Flight Path Deviation 飞行航道偏离

Flight Plan Data 飞行计划数据

Flight Plan Designator 飞行计划设计

Flight Planning Document 制定飞行计划的文件

Fully Paid 全部付讫

**FPDB** Flight Plan Data Base 飞行计划数据库

**FPDI** Flight Path Deviation Indicator 飞行轨迹偏差指示器

**FPDP** Flight Plan Data Processing 飞行计划数据处理

**FPEEPMS** Floor Proximity Emergency Escape Path Marking System 地板附近紧急撤离通道标记系统

**FPFL** Flight Plan Fuel Load 飞行计划规定燃油载荷

**FPG** Fuel Pressure Gauge 燃油压力表

**FPGA** Field Programmable Gate Array 现场可编程门阵列

**FPGL** Flight Plan Gas Load 飞行计划汽油载量

Flight Plan Gasoline Load 飞行计划汽油载量

**FPH** Failures per Hour 故障/小时

Fast Patrol Hovercraft 高速巡逻气垫艇

Feet per Hour 英尺/小时

First Pilot Hour 正驾驶员（飞行）小时

Floor Panel Heating 地板加热

**FPI** Fixed Price Incentive Contract 固定价格奖励合同

Flap Position Indicator 襟翼位置指示器

Flight Parameter Indicator 飞行参数指示器

Fluorescent Particle Inspection 荧光粉检查

Fluorescent Penetrant Inspection 荧光渗透（剂）检查

Fuel Pressure Indicator 燃油压力指示器

**F-PICH** Forward-Pilot Channel 前向导频信道

**FPIF** Fixed Price Incentive Fee 固定价格奖励费

**FPIS** Forward Propagation by Ionospheric Scatter 电离层散射前向传播

**FPL** Fire Plug 防火插头

Flight Plan 飞行计划

Forest Products Laboratory 林产品实验室

Frequency Phase Lock 频率相位锁定

Full Performance Level 全性能水平,全性能高度

Full Power Level 全功率电平,全功率高度

**FPLMTS** Future Public Land Mobile Telecommunication Systems 未来公用陆地移动通信系统

**FPM** Facility Power Monitor 设备功率监察器

Federal Personnel Manual 联邦人员手册

Feet per Minute 英尺/分钟

Flight Path Management 飞行航迹管理

Flight Path Marker 航迹指示器

Flight Path Miles 飞行轨迹的英里数

Flight Position Module 飞行位置模块

Four Photon Mixing 四光子混合

Fuel Pump Monitor 燃油泵监视器

**FPMC** Flight Path Management Control 飞行航迹管理控制

**FPMH** Failures per Million Hours 每百万小时故障数

**FPMU** Flight Path Management Utilities 飞行航迹管理效用

Fuel Properties Measurement Unit 燃油特性测量组件

Fuel Pump Monitoring Unit 加油监控器

**FPN** Fixed Pattern Noise 固定模式噪声

Fixed Pulse Radar Navigation Aid 固定脉冲辅助雷达导航

**FPNC** Family Plan Noise Test Certification System（飞行）噪音测试审定系统系列计划

**FPP** Fixed Pitch Propeller 定距螺旋桨

Flight Procedure Programme 飞行程序项目

**FPPC** Flight Plan Processing Center 飞行计划处理中心

**FPPM** Flight Performance and Plan Manual 飞行性能与计划手册

**FPPS** Flight Plan Processing System 飞行计划处理系统

**FPPU** Feedback Position Pickoff Unit 反馈位置传感组件

**FPR** Fan Pressure Ratio 风扇压比

Federal Procurement Regulation 联邦采购条例

Flight Plan Request 飞行计划申请

Flight Plan Routing 飞行计划航路

Flight Planning Requirements 飞行计划要求

Floating Point Register（计算机）浮点寄存器

Fuel Pressure Regulator 燃油压力调节器

**FPRA** Forecasting, Planning and Resource Allocation 预测、规划与资源分配

**FPRF** Fireproof 防火的

**FPRM** Flight Phase Related Mode 飞行阶段有关方式

Fuel Pipe Repair Manual 燃油管修理手册

**FPS** Fast Packet Switching 快速分组交换

Feet per Second 英尺/秒

Fine-Pitch Stop（螺旋桨）小距限动钉

Fire Protection System 防火系统

Fix Program Send 固定节目发送

Fleet Planning System 机群计划系统

Flight Path Stabilization 飞行轨迹稳定性

Flight Planning Software 飞行计划软件

Flight Progress Strip 飞行进程单

Fluid Power Society 流体动力学会(美国)

Focal Plane Structure 焦(点)平面结构

Focus Projection and Scanning 焦点投影与扫描

Foot-Pound-Second 英尺—磅—秒

Foreign Procurement System 国外订购系统

**FPSLA** Fabry-Perot Semiconductor Laser Amplifier 法布里—珀罗半导体激光放大器

**FPSM** Foreign Procurement System-Material 器材国外订购系统

**FPSO** Floating Production Storage and Off-Loading 浮式油轮生产储油卸油

**FPSOV** Flow Regulating Shutoff Valve 流量调节关断活门

**FPSR** Foot-Pound-Second-Rankine 英尺—磅—秒—绝对华氏度(传统的英国工程单位制)

**FPSS** Finance Performance Summary System 财务效能总结系统

**FPSWBS** Final Project Summary WBS 最终工程项目简要工作分解结构

**FPT** Fan Pressure Turbine 风扇压缩涡轮机

Fast Packet Transfer 快速分组传送

Flap Position Transmitter 襟翼位置传感器

Flight Path Target 飞行航径目标

Flight Proof Test 飞行验证测试

Functional Prototype Testing 原型机功能测试

**FPTRP** Facility Portable Tool Request Program 便携式工具设备申请程序

**FPTRS** Facility Portable Tool Request System 便携式工具设备申请系统

**FPTS** Forward Propagation Tropospheric Scatter 对流层前向散射传播

**FPV** Flight Path Vector 飞行航迹矢量,飞行航径引导

Floating Point Value 浮点值

**FPWBS** Final Project WBS 最终工程项目工作分解结构

**FPX** Fuel Petroleum 汽油燃料

**FQ** Flight Qualities 飞行品质

Fuel Quantity 燃油量

**FQ&P** Flight Qualities and Performance 飞行质量与性能

**FQC** Final Quality Control 最终质量检验

**FQCY** Frequency 频率

**FQDB** Flight Query Database 航班查询数据库

**FQDC** Fuel Quantity Data Concentrator 燃油油量数据集中器

**FQDU** Fuel Quantity Display Unit 燃油量显示组件

**FQGS** Fuel Quantity Gauging System 燃油量测量系统

**FQI** Fuel Quantity Indicating/Indication 燃油油量指示

Fuel Quantity Indication 燃油量指示

Fuel Quantity Indicator 油量表

**FQIC** Fuel Quantity Indication Computer 燃油量指示计算机

**FQIS** Fuel Quantity Indicator Switch 燃油量指示开关

Fuel Quantity Indicating System 燃油量指示系统

**FQM** Flight Performance Model 飞行性能模型

**FQMS** Fluid Quantity Measurement System 流量测量系统

Fuel Quality Monitoring System 燃油质量监控系统

Fuel Quantity and Management System 燃油油量与管理系统

Fuel Quantity Measurement System 燃油量测量系统

Future Quality Management System 未来质量管理系统

**FQP** Fuel Quantity Processor 燃油量处理器

**FQPU** Fuel Quantity Processing Unit 燃油油量处理组件

**FQR** Formal Qualification Review 正式合格

证评审

**FQS** Fixed-Quantity System 定量订货方式
Fuel Quantity System 燃油量系统

**FQT** Final Qualification Test 最后合格测试
Formal Qualification Testing 正式批准测试
Formal Qualification Training 正式合格培训
Fuel Quantity 燃油量
Functional Qualification Test 功能鉴定测试

**FQTI** Fuel-Quantity Totalizer Indicator 总燃油量指示器

**FR** Failure Rate 故障率
Failure Report 故障报告
Fan Reverser 风扇反推器
Financial Regulations 财务条例
Flight Radius 飞行半径
Flight Recorder 飞行记录器
Flight Refueling 空中加油
Frame Relay 帧中继
Frame 结构,隔框
Franc 法郎(货币单位)
Freight Reservation 货运预定吨位
French 法语,法国人
Friday 星期五
Front 前方,锋面
Fuel Remaining 剩余燃油
Full Rate 全速率

**FR/VT** Fixed Receiving/Variable Transmission 收端固定/发端可变
Fixed Receiving/Variable Transmission/A 收端固定/发端可变/A 型
Fixed Receiving/Variable Transmission/B 收端固定/发端可变/B 型

**FRA** Failure Rate Average 平均故障率
File Relative Address 文件相关地址
Fixed Radio Access 固定无线接入
Forecast Requirement and Actuals 预计要求和实况
France Aviation 法兰西航空
Frankfurt, Germany (Airport Code) 德国法兰克福机场
Frequency Response Analysis 频率响应分析

Friction Reducing Agent 减摩剂

**FRACA** Failure Reporting, Analysis and Corrective Action 故障报告、分析和排除措施

**FRACAS** Failure Reporting, Analysis and Corrective Action System 故障报告、分析和纠正措施系统
Failure Reporting and Corrective Action System 故障报告及纠正行为系统

**FRAD** Frame Relay Access Device 帧中继接入设备

**FRAG** Fragile 易碎的

**FRAM** Failure Rate Assessment Machine 故障率评估机

**F-RAMA** Fair Resource Assignment Multiple Address 合理资源分配多址访问

**FRAPs** Fraud Risk Assessment Programmes 欺诈风险评估程序

**FRAV** First Available (Reservations Code) 首先可利用的
First Available 最早能起飞的(航班)

**FRB** Fibre Rotor Blade 纤维旋翼桨叶
Fibreglass Rotor Blade 玻璃钢旋翼桨叶

**FRC** Facility Review Committee 设备评审委员会
Federal Radio Commission 联邦无线电委员会(美国)
Federal Records Center 联邦资料中心
Federal Resource Center 联邦资源中心
Flight Research Center 飞行研究中心
Request Full Route Clearance 请求全航路放行

**FRCMC** Fiber Reinforced Ceramic Matrix Composite 纤维增强陶瓷基复合材料

**FRCP** Flight Recorder Control Panel 飞行记录器控制板

**FRCU** Fractocumulus 碎积云

**FRD** Friend 朋友
Functional Requirement Document 功能要求文件

**FRDA** Failure Rate Data Analysis 故障率数据分析

**FRDI**   Flight Research and Development Instrumentation 飞行研究与发展仪表设备

**FRDLY**   Friendly 友好的

**FRDR**   Failure Rate Data Record 故障率数据记录

**FRDSHP**   Friendship 友谊

**FRDTS**   Frame Relay Data Transmission Services 帧中继数据传输业务

**FRED**   Flight Recorder Electronic Documentation 飞行记录器电子文档

**FREDI**   Flight Range and Endurance Data Indicator 航程和续航数据指示器

**FREM**   Fuel Remaining 剩余燃油

**FREMEC**   Frequent Traveler's Medical Card 残障旅客适航证书

**FREQ**   Frequency 频率
Frequent 频繁的

**FRET**   Freight 货物,货运费

**FRETCH**   Freight Charges 货用费

**FRF**   Flight-Readiness Firing（发动机）飞行前检查启动
French Franc 法国法郎

**FRFI**   Fuel-Related Fare Increase 燃油涨价引起飞机票涨价

**FRG**   Federal Republic of Germany 德意志联邦共和国

**FRGL**   Fragile 易碎的

**FRGT**   Freight 货物,货运费

**FRI**   Frame Relaying Information 帧中继信息
Friday 星期五

**FRIG**   Floated Rate-Integrated Gyro 浮式速率积分陀螺仪

**FRIT**   Freight 货物,货运费

**FRL**   Flight Reference Line 飞行基准线
Fuselage Reference Line 机身基准线

**FRM**   Fairmont, MN, USA（Airport Code）美国明尼苏达州费尔蒙特机场
Fault Reporting Manual 故障报告手册,错误报告手册
Fiber-Reinforced Metal 纤维增强金属
Financial Resource Management 金融资源管理

Financial Risk Manager 金融风险管理人
Form 形式,格式,表格
Frame 帧,框
Frequency Meter 频率计
From 自,从

**FRMG**   Forming 形成
Fuel Remaining 剩余燃油

**FRMP**   Fatigue Risk Management Plan 疲劳风险管理计划

**FRMR**   Frame Reject 帧拒绝

**FRMS**   Fatigue Risk Management System 疲劳风险管理系统

**FRNG**   Firing 射击,点火

**FRNT**   Front 前方,锋面

**FROF**   Freight Office 货运处

**FRP**   Fast Reservation Protocol 快速保留协议
Fast Resolution Protocol 快速分辨协议
Federal Radio-Navigation Plan 联邦无线电导航计划（美国）
Fiberglass Reinforced Plastic 玻璃纤维增强塑料,玻璃钢
Fibre Reinforced Plastic 纤维增强塑料
Flap Reference Plane 襟翼基准平面
Flight Refuelling Probe 空中受油探管
Fuselage Reference Plane 机身水平基准面

**FRPA**   Fixed Reception Pattern Antenna 固定型接收天线,固定接受区天线

**FRPH**   Frame Relay Packet Handler 帧中继分组处理程序

**FRQ**   Frequency 频率
Frequent 频繁的,时常

**FRR**   Fiber Reinforced Rubber 纤维增强橡胶
Flight Readiness Review 待飞状态评审,飞行准备状态评审

**FRRC**   Flow-Recording Ratio Controller 流量记录比例控制器

**FRS**   Failure Reporting System 故障报告系统
False Rear Spar 虚拟后梁
Fan Rotation Speed 风扇转速
Fault Repair Signal 故障修理信号

Flammability Reduction System 可燃性缩减系统

Fleet Reliability Summary 机队可靠性总结

Frame Relay Service 帧中继业务

Frame Relay Switch 帧中继交换机

Frame Relay Switching 帧中继交换

Functional Requirement Summary 功能要求总览

**FRSE**　Frame-Relay Switching Equipment 帧中继交换设备

**FRSF**　Frame Relay Service Function 帧中继业务功能

**FRSS**　Flight Reference Stabilization System 飞行基准稳定系统

**FRT**　Failure Rate Test 故障率试验

　　Fortnight 两周

　　Frame Relay Terminal 帧中继终端

　　Freight 货物,货运费,货物运输

　　Front 前方,锋面

**FRTE**　Frame-Relay Terminal Equipment 帧中继终端设备

**FRTP**　Fiber-Reinforced Thermo-Plastic 玻璃纤维增强热塑材料

**FRTR**　Freighter 运输机

　　Freighter-Aircraft,Cargo 货机

**FRTT**　Fixed Round-Trip Time 固定往返时间

**FRTV**　Forward Repair and Test Vehicle 前方修理测试车

**FRU**　Bishkek, Kyrgyzstan-Bishkek Airport (Airport Code) 吉尔吉斯斯坦比什凯克机场

　　Field Replaceable Unit 现场可换部件

　　Frequency Reference Unit 频率基准组件

**FRV**　Fuel Return Valve 燃油回油活门

**FRWK**　Framework 骨架

**FRY**　Ferry 调机,运送

**FRZ**　Flight Restricted Zone 飞行限制区

　　Freeze 冷冻,冻结

　　Fuel Reduction Zone 燃料减少区

**FRZN**　Frozen 冻结的

**FS**　Facsimile 传真

Factor of Safety 安全系数

Fail Safety 破损安全,故障自动防护

Fan Stator(Cases) 风扇静子(机匣)

Fast Slew 快转

Fax Server 传真服务器

Fiber Sensor 光纤传感器

Field Service 外场保养

Flight Schedule 航班计划

Flight Service 航班服务

Flight Simulator 飞行模拟机

Flight Status 航班状况,飞行状态

Flight Surgeon 航空医生

Flying Safety 飞行安全

Fracto-Stratus 碎层云

Frame Start Single 帧起始信号

Frame State 帧状态

Frame Station 隔框站位

Frame Switching 帧交换

Frame Synchronizer 帧同步器

Free Sale 自由出售

Front Spar 前梁

Fuel Saving 节油

Functional Statement 功能描述

Fuselage Station 机身站位,机身测量点

**FSA**　Federal Security Agency 联邦安全局

　　Final Squint Angle 最后瞄准偏斜角

　　Flight Safety Analyses 飞行安全性分析

　　Fuel-Saving Advisory 节油咨询

**FSAA**　Flight Simulator for Advanced Aircraft 先进航空器飞行模拟器

**FSAN**　Full Service Access Network 全业务接入网

**FSAR**　Failure Summary and Analysis Report 故障总结和分析报告

　　Fuel System Analysis Report 燃油系统分析报告

**FSAS**　Flight Service Automation System 飞行服务自动化系统

　　Fuel Savings Advisory System 节油咨询系统

**FSAT**　Full Scale Aerial Targets 全尺寸空中靶标

**FSB**　Fan Stream Burning 风扇涵道气流燃烧

Fasten Seat Belts 系好安全带

Federal Specifications Board 联邦技术规范局(美)

Field Service Bulletin 外场服务通告

Flight Standard Board 飞行标准委员会

Flight Standardization Board 飞行标准化委员会

Flight Standards Bulletin 飞行标准通告

**FSC**　Fault Simulation Comparator 故障模拟比较器

Fault Symptom Code 故障征兆码

Federal Stock Class 联邦仓库等级,联邦仓库分类

Federal Stock Code 联合仓库代码

Federal Supplier Code 联合供货代码

File Server Card 文件服务器插卡

Flight Safety Canada 加拿大飞行安全公司

Flight Safety Critical 飞行安全临界值

Flight Service Centre 飞行服务中心

Flight Standard Committee 飞行标准委员会

Foreign Sale Corporation 对外销售公司

Fuel Summary Computer 燃油累加计算机

Fuel Surcharge 燃油附加费

Fuel System Computer 燃油系统计算机

Fuel System Controller 燃油系统控制器

**FSCAP**　Flight Safety Critical Aircraft Parts 飞行安全的重要飞机部件

**FSCC**　Flap/Slat Control Computers 襟翼/缝翼控制计算机

Flight Simulator Computer Complex 飞行模拟机计算机成套设备

**F-SCH**　Forward Supplemental Channel 前向辅助信道

**FSCI**　Frequency Space Characteristic Impedance 频率间隔特性阻抗

**FSCL**　Flight Spoiler Control Lever 飞行扰流板操纵杆

**FSCM**　Federal Supply Class Management 联邦补给品分类管理

Federal Supply Code for Manufacturers 联邦补给品制造厂代号

**FSCP**　Flight Safety Critical Part 飞行安全关键机件

**FSCU**　Fuel System Computer Unit 燃油系统计算机组件

**FSD**　Fabrication Services Division 制造与维护分部,制造与服务分部

Final Submission Date 最终提交日期

Flight Standards Department 飞行标准部

Flight Standards Division 飞行标准室

Flying Spot Digitizer 飞点(扫描点)数字转换器

Forecast Support Date 预报支援数据

Full Scale Deflection 满刻度偏差

Full Scale Development 全尺寸研制

Inflight Shutdown 空中停车

Sioux Falls, SD, USA ( Airport Code ) 美国南达科他州苏福尔斯机场

**FSDA**　Facilities Standards Database Systems 设施标准数据库系统

Fail Safe Design Analysis 故障安全设计分析

**FSDB**　Flight Status Database 飞行状态数据库

**FSDG**　Fan Shaft Driven Generator 风扇轴驱动发电机

Flight Symbology Development Group 飞行符号开发组

**FSDO**　Flight Standard District Officers 飞行标准区域官

Flight Standards District Office 飞行标准区域办公室

**FSDP**　Final Safety Data Package 最终安全数据包

Flight Simulation Development Plan 飞行模拟开发计划

**FSDPS**　Flight Service Data Processing System (联邦航空局)飞行服务数据处理系统

**FSE**　Facilities System Engineering 设施系统工程

Factory Support Equipment 工厂支援设备

Field Service Engineer 外场服务工程师

Field Support Equipment 外场保障设备

Fire Safety Engineering 防火安全工程

Flight Support Equipment 飞行保障设备

Flight Systems Engineering 飞行系统工程

**FSED** Full Scale Engineering Development 全尺寸工程研制

**FSEIC** Fuel System EICAS Interface Card 燃油系统 EICAS 连接卡

**FSEU** Flap Slat Electronic Unit 襟翼缝翼电子装置

Flap System Electronic Unit 襟翼系统电子装置

**FSF** First Systems Flight 第一飞行系统

Flight Safety Foundation 飞行安全基金

Flight Service Facility 飞行服务设备

Fuselage Side Fairing 机身侧面整流罩

**FSFT** Full-Scale Fatigue Test 全尺寸疲劳试验

**FSG** Fluid Sphere Gyro 液浮陀螺

**FSHFSG** Flight Safety and Human Factors Study Group 飞行安全与人为因素研究组

**FSI** Fault Symptom Index 故障征兆指数,故障征兆(码)索引

Field Service Instruction 外场维护细则

Flight Safety International 国际飞行安全

Flight Safety Irrelevant 与飞行安全无关

Frame Sync Indication 帧同步指示

Full Screen Interface 全屏幕接口

Functionally Significant Item 重要功能项目

**FSIB** Flight Standard Information Bulletin 飞行标准信息通告

**FSII** Fuel System Icing Inhibitor 燃油系统防冻剂

**FSIMS** Flight Standards Information Management System 飞行标准信息管理系统

**FSIX** Flight Safety Information Exchange 飞行安全信息交流

**FSK** Frequency Shift Keying 频率转换键控

**FSL** Flexible System Link 灵活系统链路

Flight Simulation Laboratory 飞行模拟实验室

Flight Systems Laboratory 飞行系统试验室

Fluid Science Laboratory 流体科学实验室

Full Service Level 全套服务水平

Full Stop Landing 全停着陆

**FSM** Field Strength Meter 场强表

File Server Module 文件服务器模块

Firmware Support Manual 固件保障指南

Flight Safety Maintenance 飞行安全维护

Flight Schedule Monitor 航班时刻监控器

Flight Simulation Monitor 飞行模拟监视器

Fort Smith Municipal, AR, USA （Airport Code）美国阿肯色州史密斯堡机场

Frequency Shift Modulation 频移调制

Fuel Supply Manifold 燃油供应总管

Full Scale Mockup 全尺寸样件

**FSMC** Fuel System Management Card 燃油系统管理卡

**FSMIS** Flight Standard Management Information System 飞行标准管理信息系统

**FSMR** Fire Safety Management Review 消防安全管理审查

**FSMS** Flight Structural Monitoring System 飞行结构监控系统

Food Safety Management System 食品安全管理系统

**FSMT** Financial Systems Master Tables 财务系统主表

Fuel System Maintenance Trainer 燃油系统维修训练器

**FSMWO** Field Service Modification Work Order 外场维修改装工作单

**FSN** Factory Serial Number 制造厂批号,出厂序列号

Federal Stock Number 联邦库存号

Field Service Notice 野外作业注意事项,外场维护通告

Field Service Nozzle 机场供油嘴,机场喷水枪

Fleet Serial Number 机队序列号

Forward Sequence Number 前向顺序号码

French-Speaking Nation 法语国家

Fuel Spray Nozzle 燃油雾化喷嘴

Full Service Network 全业务网

**FSO** Field Service Operation 机场勤务操作

Flight Safety Officer 飞行安全官

Fuel Shut off 燃油关断

**FSOP** Field Standard Operating Procedure 场地标准运行程序

Flying Standing Operation Procedures 常规飞行运行程序

Full Scale Output 满量程输出

**FSOV** Fire Shutoff Valve 火警关断活门

Fuel Shutoff Valve 燃油关断活门

**FSP** Field Support Personnel 外场支援人员

Financial Security Plan 财务安全计划

Flight Safety Plan 飞行安全计划

Flight Strip Printer 飞行进程单打印机

Frequency Standard Primary 主频率标准

Full Screen Processing 全屏幕处理

**FSPM** Flap and Slat Position Monitor 襟翼缝翼位置监视器

Flap/Stabilizer Position Module 襟翼/安定面位置组件

**FSR** Field Service Representative 外场服务代表

Field Strength Radio 无线电场强

Flight Safety Region 安全飞行区

Flight Safety Research 飞行安全研究

Flight Simulation Report 模拟飞行报告

Flight Since Renew 翻修后的飞行时间

Flight-Safety Reporting 飞行安全报告

Frequency Scan Radar 频率扫描雷达

Frequency Set-on Receiver 给定频率接收机

**FSRB** Flight Safety Review Board 飞行安全检查委员会

**FSS** Arequipa, Peru ( Airport Code) 秘鲁阿雷基帕机场

Federal Specification Standards 联邦规范标准

Fenwal Safety Systems 芬沃安全系统

Fixed Satellite Service 固定卫星业务

Fixed Service Structure 固定服务结构（塔架）

Flap Speed Schedule 襟翼速度计划

Flight Safety Section 飞行安全处

Flight Safety Seminar 飞行安全讨论会

Flight Schedule System 航班计划系统

Flight Service Station 飞行服务站

Flight Standard Service 飞行标准处（美国联邦航空局）

Flight Station Simulator 飞行站模拟机

Flight Support System 飞行支援系统

Fluid Support System 流体供应系统

Flying Spot Scanner 飞点扫描器

Frame Synchronous Scrambling 帧同步扰码

Frequency Selective Surface 频率可选面

Front Spar Station 前梁站位

Fuel Safety System 燃料安全系统

Fuel Systems Simulator 燃料系统模拟器

**FSSG** Force Service Support Group 兵力勤务支援组

**FSSR** Federal Safety Standard Regulation 联邦安全性能标准

Flight System Software Requirement 飞行系统软件要求

**FSSS** Flight Software Support System 飞行软件支持系统

Furnace Safeguard Supervisory System 炉膛安全监控系统

**FST** Fire Safety Technology 点火安全技术

Fire/Smoke/Toxicity 火情/烟雾/毒性

Flight Safety Technologies, Inc 飞行安全技术公司

Flight Simulation Test 飞行模拟试验

Flight Station Trainer 飞行站教员

Flight Systems Technology 飞行系统技术

Flight Systems Test 飞行系统试验

Frequency Shift Transmission 频移发射

**FSTD** Fault Selector Time Delay 故障选择时间继电器

Flight Simulation Training Device 飞行模拟训练装置

**FSTE** Factory Special Test Equipment 工厂特殊测试设备

**FSTM** Fire Service Training Manual 消防服务训练手册

Fire System Testing and Maintenance 消防系统试验与维护

**FSTN** Fastener 紧固件

Federal Security Telephone Network 联邦政府保安电话网络

**FSTNR** Fastener 金属紧固件

**FSTS** Flightline System Test Set 航线系统测试仪

Future Space Transportation System 未来航天运输系统

**FSU** Feel Simulation Unit 感觉模拟组件

File Server Unit 文件服务器装置

Fixed Subscriber Unit 固定用户单元

Flight Sequence Unit 飞行顺序装置

Flight Service Unit 飞行服务单位

Frequency Source Unit 频率源组件

Fuel Summation Unit 燃料求和器

**FSVS** Flight Software Verification System 飞行软件校验系统

**FSW** Flight Software 飞行软件

Forward-Swept Wing 前掠翼

Frame Synchronization Word 帧同步字

Friction Stir Welding 搅拌摩擦焊

**FSWFS** Field Standard Weight and Force System 外场标准称重与测力系统

**FSWT** Factory Standard Warranty Terms 工厂标准担保条件

**FSY** Flight Schedule Year 年航班时刻表

**FSYN** Frame Synchronization Signal 帧同步信号

**F-SYNC** Frame Synchronizer 帧同步器

**F-SYNCH** Forward-Synchronous Channel 前向同步信道

**FT** Fast Track 高速线路

Fault-Tolerant 故障容限的, 容错的

Feet 英尺 (复数)

Fiber Termination 光纤终端

Fiber Transmission Standards 光纤传输标准

Fixed Radio Terminal 固定式无线电终端

Flight Termination 飞行终止

Flight Test 飞行试验

Flight Time 飞行时间

Flying Training 飞行训练

Foot 英尺 (单数)

Forint 福林 (匈牙利的货币单位)

Freight Traffic 货 (物) 运 (输)

Full Throttle 全油门, 最大油门

Function Test 功能测试

Functional Test 功能测试

**FT** Feet, Foot 英尺

**FT/MIN** Feet per Minute 英尺每分钟

**FT/R** Functional Test Retrofit Reports 改型功能测试报告

**FTA** Fault Tree Analysis 故障树分析

Flight Test Aircraft 测飞飞机

Flight Test Analysis 飞行试验分析

Free Trade Agreement 自由贸易协议

Free Trade Area 自由贸易区

Freight Transport Association 货物运输协会

**FTAM** File Transfer Access and Management 文件传送存取和管理

File Transfer and Access Method (计算机) 文件传送与存取方法

**FTAMS** File Transfer Access and Management Services 文件传递访问及管理服务

**FTAN** Frequency-Time Analysis 频率—时间分析

**FTB** Fan Trim Balance 风扇配平

Flight Test Base 试飞基地

Flying Test Bed 飞行试车台

For the Time Being 当时, 暂时

**FTC** Facsimile Transport Channel 传真传送信道

Fast Time Constant 短时间常数

Fast Time Control 快时间控制

Fault Tolerant Computer 容错计算机

Fault Tolerant Computing 容错计算

Federal Trade Commission 联邦贸易委员会

（美国）

Federal Transportation Center 联邦运输中心

Flight Test Center 飞行试验中心

Flight Test Configuration 试飞构型,试区布局

Flight Time Capability 最大航时

Flight Training Center 飞行培训中心

**FTCS** Flight Test Computing System 试飞计算系统

**FTD** Field Training Detachment 机场训练支队

Flight Test Diagnostics 飞行测试诊断

Flight Tracking Data 飞行跟踪数据

Flight Training Device 飞行训练器,飞行训练设备

**FTDC** Flight Test Data Center 试飞数据中心

**FTDS** Flight Test Data System 试飞数据系统

**FTE** Factory Test Equipment 工厂试验设备

Fixed Trailing Edge 固定翼后缘

Flight Technical Error 飞行技术误差

Flight Test Engineer 飞行试验工程师

Flight Test Equipment 试飞设备,飞行试验设备

Full Time Engineer 全职工程师

Functional Test Engineering 功能测试工程

**FTESN** Flap Trailing Edge Separation Noise 襟翼后缘(气流)分离噪声

**FTF** Fiber Termination Frame 光纤终端架

Flared Tube Fitting 喇叭管接头

Flygtekniska Foreningen 航空协会(瑞典)

Functional Test Flight 功能试验飞行

**FTG** False Target Generator 假目标产生器

Fitting 连接件,接头

Floated-Type Gyro 液浮陀螺

**FTH** Full-Throttle Horsepower 发动机最大马力

**FTHR** Further 进一步

**FTI** Fixed Target Indicator 固定目标指示器

Fixed Time Interval 固定时间间隔

Flight Test Installation 飞行测试安装

Flight Test Instrumentation 飞行试验仪表

Flight Test Introduction 飞行试验介绍

Flight Training International 国际飞行训练

Frequency Time Indicator 频率时间指示器

Frequency Time Intensity 频率时间强度

Fuel Temperature Indicator 燃油温度指示器

**FTIC** Flight Test Instrumentation Configuration 飞行试验仪表配置,飞行试验仪表布局

**FTINS** Fault Tolerant Inertial Navigation System 容错惯性导航系统

**FTIR** Flight Test Instrumentation Requirements 飞行试验仪表要求

**FTIS** Flight Test Instrumentation System 飞行试验测量仪表系统

Fuel Tank Inerting System 燃油箱惰性气体系统

Function Test Information Sheet 功能测试信息单

**FTIT** Fan-Turbine Inlet Temperature 风扇涡轮进口温度

**FTK** Freight Tonne Kilometer 货物吨公里

Fuel Tank 燃油箱

**FTL** Faster Than Light 超光速

Federal Telecommunications Laboratory 联邦电信实验室

Flight Time Limit 飞行时间限制

Flight Transportation Laboratory 空运实验室

Full Term Licence 全条件许可证

**FTLA** Fiber-to-the-Last Amplifier 光纤到末级放大器

**FTLB** Flight Time Limitation Board 飞行时间限制委员会(英国)

**FTLO** Fast-Tuned Local Oscillator 快速调谐本机振荡器

**FTLWG** Flight Time Limitation Working Group 飞行工作时间限制工作组

**FTM** FDM Transmitter Module FDM 发送器模块

Fiber Terminal Module 光纤终端模块

Fiber Transfer Module 光纤传送模块

File Transfer Manager 文件传送管理器

Flight Test Manual 试飞手册

Flying Training Manual 飞行训练手册

Frequency/Time Modulation 频率/时间调制

**FTMM** Flight Technical Management Manual 飞行技术管理手册

Flight Training Management Manual 飞行训练管理手册

**FTMS** Flight Test Management System 试飞管理系统

**FTMV** Flight Test Miniature Vehicle 小型航空器试飞

**FTN** Facsimile Transmission Network 传真传输网

Four-Terminal Network 四端网络

**FTNS** Flight Test Notification System 飞行测试通知系统

Flight Track Navigation System 飞行轨迹导航系统

**FTO** Flexible Take-off 减推力起飞

Flight Test Operations 试飞操作

Flight Training Officer 飞行培训专员

Flight Training Organisation 飞行培训机构

**FTOF** Face to Face 面对面

**FTP** File Transfer Protocol 文件传输协议

Flight Test Procedure 飞行试验程序

Flight Test Program 试飞计划

Fly-to-Point 飞至目的地

Fuel Tank Pressure 燃油箱压力

Fuel Test Plan 燃油测试计划

Fuel Test Port 燃油测试口

Fuel Transfer Pump 燃油传输泵

Full Throttle Position 最大油门位置

Functional Test Plan 功能测试计划

Functional Test Procedure 功能测试程序

Functional Test Program 功能试验计划

Functional Test Progress 功能试验进程

**FTPR** Flight Test Problem Report 试飞问题报告

**FTR** Failed to Return 未能返回

Fighter 战斗机

Flight Test Report 飞行试验报告

Flight Test Request 试飞申请

Flight Test Requirement 飞行试验要求

Fuel-to Remain 剩余油量,保留燃油

Full Text Retrieval 全文检索

Functional Test Requirement 功能测试要求

Future Transport Rotorcraft 未来运输旋翼航空器

**FTRD** Functional Test Requirements Document 功能测试要求文件

**FTRG** Flight Test Report Guide 飞行试验报告指南

**FTRNG** Feathering 顺桨

**FTS** Fast Track Selector 快速磁道选择器

Federal Telecommunication System 联邦电信系统

Finish Two Sides 两边抛光

Flexible Track System 灵活航迹系统,可选航迹系统

Flight Telemetry Subsystem 飞行遥测分系统

Flight Termination System 飞行终端系统

Flight Test Sketch 试飞概要

Flight Test Specification 飞行测试规范

Flight Test Support 试飞支援

Flight Test System 试飞系统

Flight Traffic Specialist 空中交通管理专家

Flightline Test Set 航线测试仪

Flying Training School 飞行训练学校

Flying Training Squadron 飞行训练中队

Foundation for Traffic Safety 交通安全基金会

Frame Time Select 帧像时间选择

Frame Transport System 帧传送系统

Freight Transport Study 货运研究

Frequency Timing Subsystem 频率定时分系统

Fuel Tank Safety Rule 燃油箱安全规则

Fuel Temperature Sensor 燃油温度传感器

Fuel Transfer System 燃油传输系统

Functional Test Set 功能测试仪

Functional Test Specification 功能试验规范

Functional Test System 功能测试系统

Future Transportation Systems 未来交通系统

**FTSA** Fiber-to-the-Service Area 光纤到服务区

**FTSC** Federal Telecommunications Standards Committee 联邦电信标准委员会

Flight Test Safety Committee 飞行试验安全委员会

**FTSMSTR** Frame Transport System Master 帧传送系统主程序

**FTSP** Floor Temperature Setting Panel 地板温度设置面板

**FTSRR(D)** Flight Test System Requirements Review (Design) 试飞系统要求审查(设计)

**FTSRR(R)** Flight Test System Requirements Review (Requirements) 试飞系统要求审查(要求)

**FTSW** Flight Test Section Worksheet 试飞部分工作单

**FTSWG** Flight Test Support Working Group 试飞保障工作小组

**FTT** Fixed Target Track 固定目标航迹

Flight Technical Tolerance 飞行技术允许误差

Flight Training Team 飞行训练小组

**FTTA** Fiber to The Amplifier 光纤放大器

**FTTB** Fiber to The Bridge 光纤到桥梁

Fiber to The Building 光纤到楼宇

**FTTC** Fiber to The Curb 光纤到路边

**FTTD** Fiber to The Desk 光纤到桌面

**FTTF** Fiber to The Feeder 光纤到馈送器

Fiber to The Floor 光纤到楼层

**FTTH** Fiber to The Home 光纤到户

**FTTK** Fiber to The Kerb 光纤到路边

**FTTN** Fiber to The Node 光纤到节点

**FTTO** Fiber to The Office 光纤到办公室

Flight Training and Testing Office 飞行训练与测试办公室

**FTTP** Fiber to The Premises 光纤到房屋

**FTTR** Fiber to The Radio 光纤到无线电

**FTTS** Fiber to The Subscriber 光纤到用户

**FTTSA** Fiber to The Service Area 光纤到服务区

**FTTV** Fiber to The Village 光纤到村

**FTTZ** Fiber to The Zone 光纤到小区

**FTU** Flight Test Unit 试飞分队,试飞组件

Force Transducer Unit 力传感组件

**FTV** Flight Test Vehicle 试飞飞行器

Functional Test Vehicle 功能测试飞行器

**FTWS** Flight Test Work Sheet 试飞工作单,飞行测试工作单

Flight Test Work Statement 试飞说明,飞行测试说明

**FTX** Field Training Exercise 现场训练实习

**FTY** Factory 工厂

**FTZ** Foreign Trade Zone 对外贸易区

**FU** Feedback Unit 反馈组件

File Upload 文件卸载

Flap Up 襟翼收上

Flight Unit 飞行组件

Follow Up 跟踪

Fuel Uplifted 已装上燃油,加油量

Fuel Used 已耗燃油

Smoke (Aviation Meteorology) 烟(航空气象)

**FUA** Flexible Use of Airspace 空域的灵活使用

Fuel Use Act 燃油使用法案

**FUBAR** Fouled-up Beyond All Repair 坏得无法修理

Fouled-up Beyond All Recognition 错乱得无法辨认

**FUBX** Fuse Box 保险丝盒

**FUE** Fuerteventura, Spain (Airport Code) 西班牙富埃特文图拉岛机场

**FUFO** Fly under, Fly out 低空穿云飞行

**FUJ** Fukue, Japan (Airport Code) 日本福江机场

**FUJA** Future of JAA JAA 未来计划

**FUK** Fukuoka, Japan (Airport Code) 日本福冈机场

**FUL** Full(y) 全部(地)

**FULF** Fulfil 履行

**FUMC** Fuel Management Computer 燃油管理计算机

**FUN** Funafuti, Tuvalu (Airport Code) 图瓦卢富纳富提机场

**FUNC** Function 功能
Functional 功能的

**FUNI** Frame User Network Interface 帧用户网络接口

**FUR** Failure or Unsatisfactory Report 故障或不满意情况(报告)
Failure Unsatisfactory or Removal (Report) 故障不满意或排除情况报告
Future Utility Rotorcraft 未来实用旋翼飞机

**FUS** Firing Unit Simulator 发射装置模拟器
Fuse 保险丝
Fuselage 机身

**FUSLG** Fuselage 机身

**FUSS** Flaps-Up Safety Speed 襟翼上偏安全速度

**FUT** Functional Unit Test 功能单元测试
Future 将来

**FUV** Far Ultra Violet 超紫外线

**FV** Final Valve 最终活门
Flight Vector 飞行矢量
Flux Valve 溢流活门
Forward Visibility 前方能见度

**FVD** Fluorescent Vacuum Display 荧光真空显示器

**FVR** Favour 有利,赞同
Favourable 有利的,赞同的
Flight Vector 飞行向量基准

**FVRBL** Favorable 赞成的

**FVS** Flighting Vehicle System 飞行器系统
Flight Vehicle Simulator 飞行器模拟器
Flight Vehicle Structure 飞行器结构
Flight Verification System 飞行验证系统

**FVT** Function Validation Test 功能鉴定实验

**FVV** Fan Variable Vane 可变风扇叶片

**FW** Face Width 面宽
Fire Warning 失火警告
Firewall 防火墙
Fiscal Week 会计周,财务周,财周
Fixed Wheel (水上飞机的)固定式机轮
Fixed Wing 固定翼
Flight Worthiness 飞行适航性
Fresh Water 淡水
Full Wave 全波
Full Word 全字

**FW&A** Fraud, Waste and Abuse 劣质品、浪费和滥用

**FWA** First West Airlines (USA) 西部第一航空公司(美国)
Fixed Wing Aircraft 固定翼飞机
Fixed Wireless Access 固定无线接入,固定无线访问
Flight Watch Area 空中警戒区
Fort Wayne, IN, USA (Airport Code) 美国印第安那州韦恩堡机场

**FWAN** Fixed Wireless Access Network 固定无线接入网

**FWAS** Failure Warning and Analysis System 故障警报和分析系统

**FWC** Factory Work Code 工厂工作编码
Fault Warning Computer 故障警告计算机
Filament-Wound Cylinder 纤维缠绕圆柱
Flight Warning Computer 飞行警告计算机
Frequency and Optical Wavelength Converter 频率和光波长变换器

**FWD** Foreword 前言
Forward Box Front of Aircraft 飞机前部接线盒
Forward or Forwarded or Forwarding 送出
Forward 前,向前,前面的

**FWDCOMPT** Forward Compartment 前舱

**FWF** Flight Weather Forecast 飞行气象预报

**FWG** Facilities Working Group 设施工作组
Financial Working Group 财务工作组

**FWH** Feed Water Heater 给水加热器
Flexible Working Hours 弹性工作时间
Frigate with Helicopter 直升机护卫舰

**FWHM** Full Wave, Half Modulation 全波半

调制

Full Width at Half Maximum 半峰值全宽度，半峰值全带宽

**FWI**　Fuel Weight Indicator 燃油总量指示器

**FWK**　Field Weakening 电(磁)场减弱

**FWL**　Firewall 防火墙

Fuel Warning Light 燃油报警灯

Fuselage Water Line 机身水线

**FWM**　Four-Wave Mixing 四波混频

**FWME**　Fixed Wing Multi Engine 固定翼多发飞机

**FWN**　Flap Wake Noise 襟翼尾流噪声

**FWO**　Flight Work Order 飞行工作指令

**FWP**　Feed Water Pump 供水泵

Fresh Water Pump 淡水泵

**FWPCS**　Future Wireless PCS 未来无线个人通信系统

**FWPLN**　Failurewater Planes 费尔沃特型飞机

**FWR**　Final Wiring Release 最后布线放行

Full Wave Rectifier 全波整流

**FWS**　Fair Weather Strip 良好天气地带

Fast-Wavelength-Switched 快速波长交换

Fixed Wireless Station 固定无线电台

Flight Warning System 飞行警告系统

Flight Watching Service 飞行监视服务

**FWSD**　Flight Warning and System Display 飞行警告和系统显示

**FWSE**　Fixed Wing Single Engine 固定翼单发飞机

**FWSOV**　Firewall Shutoff Valve 防火墙关断活门

**FWT**　Flight Worthiness Test 适航性试飞，适航性测试

Folding Wing Tip 折叠翼尖

**FWTCM**　Folding Wing Tip Control Module 可折叠翼尖控制模件

**FWTEC**　Folding Wing Tip Electronic Card 可折叠翼尖电子插卡

**FWTFA**　Folding Wing Tip Fold Actuator 可折叠翼尖折叠作动器

**FWTT**　Fixed Wing Tactical Transport 固定翼战术运输机

**FWV**　Fixed Wing Vehicle 固定翼飞行器

**FWW**　Food, Water and Waste Management 食物、水和废物管理

**FWWS**　Food, Water and Waste Subsystem 食物、水和垃圾子系统

**FWY**　Freeway 高速公路,快车道

**FX**　Fix 固定,点

Fixed 固定的,不变的,定位的

Foreign Exchange 外汇

**FXD**　Fixed 固定的

**FXP**　Fixed Point 定点

**FXR**　Flash X-Ray 瞬时 X 射线

**FXTR**　Fixture 夹具,固定器

**FY**　Fiscal Year 财政年度，会计年度

**FYA**　For Your Attention 请你注意

**FYDP**　Five-Year Defence Plan 国防五年计划

**FYIG**　For Your Information and Guidance 望参照执行

**FYP**　Five-Year Plan 五年计划

**FYR**　For Your Reference 供你参考

**FYV**　Fayetteville, AR, USA（Airport Code）美国阿肯色州费耶特维尔机场

**FZ**　Free Zone 自由区

Freezing 冷冻,冻结

Fuse 保险丝

**FZDZ**　Freezing Drizzle 冻毛毛雨

**FZFG**　Freezing Fog 冻霜

**FZL**　Freezing Level 结冰层

**FZRA**　Freezing Rain 冻雨,雨凇

# G

G Acceleration of Gravity 重力加速度

British Virgin Islands 英属维尔京群岛

Earth Acceleration（Symbol）重力加速度（符号）

Gage Atmosphere 表压,显示大气压

Gain 获得,增益

Galley 厨房

Gallon 加仑

Gate 整流栅,控制极

Gear 齿轮,装置

Geared 齿轮传动的

Girder 横（大）梁

Glass 玻璃

Glide Configuration 滑翔位形

Glider 滑翔机

Gliding 滑翔

Good 好,情况良好

Gram 克

Gravity Force 重力

Gravity 重力

Graze 低掠,掠地飞行,轻擦

Green 绿色(的)

Greenwich 格林威治

Grid 栅极,铅扳(蓄电池),门,栅

Ground Control 地面管制

Ground 地面,接地

Group 大队

Guide 制导,导杆

Gyro 陀螺

Load Factor 载重比

On Ground 在地面

Shear Modulus 剪切模量

Single Glass 单层玻璃

Stabilization Bypass 稳定旁路

Surface Target（代号）地面目标

Unit of Gravitational Acceleration 重力加速度组件

g Gravitational Acceleration 重力加速度

G&C Guidance and Control Unit 制导和控制组件

G&CE Guidance and Control Electronics 制导和控制电子组件

G&CI Guidance and Control Integration 制导和控制集成

G. S. Gain Schedule 增益程序

G/A Go Around 复飞

Ground to Air 地对空

Ground/Air 陆/空

G/A/G Ground to Air and Air to Ground 地/空/地

G/D General Declaration 总申报单

G/E Ground Engineer 地面机械师

G/G Ground Guidance 地面引导

Ground to Ground 地对地

G/P Glide Path 下滑道,下滑轨迹

G/S Glide Slope,Ground Slope 下滑道,下滑坡度

Ground Speed 地速

G/T Ground Test 地面测试

G3 Group 3 三类(传真)

G-7 Group of Seven（西方）七国集团

GA Gabon 加蓬(国家名)

Gage 表

Gas Analysis 气体分析

Gauge 仪表

General Agent 总代理

General Application 一般应用

General Arrangement 总安排

General Assembly 总装

General Aviation 通用航空

General-Arrangement(Drawing) 总图,总装图

Gimbal Angle 万向架转角

Glide Angle 下滑角

Go Ahead 请讲,向前走

Go-Around 复飞,重飞

Ground Antenna 地面天线

Gust Alleviation 阵风缓和,阵风减缓

United Nations General Assembly 联合国大会

**GAA** Gain Adjuster Adaptor 增益调整器适配器

General Airline Assistance (Switzerland) 通用航空公司辅助设备(瑞士)

General Aviation Association 通用航空协会

Glasgow Airport Authority 格拉斯哥机场管理局

**GAACC** General Aviation Airworthiness Consultative Committee 通用航空适航性顾问委员会

**GAAP** General Aviation Airport Procedures 通用航空机场程序

Generally Accepted Accounting Principles 通常公认的会计原则

**GAAS** Generally Accepted Auditing Standards 通常公认的审计标准

Ground Airborne Avionics System 地面机载航空电子系统

**GAATAA** General Aviation and Air Taxi Activity and Avionics 通用航空、航空出租活动和航空电子

**GAB** Gabon 加蓬

Global Address Book 全球的通讯录

**GAC** General Aircraft Corporation 通用飞机公司

General Automatic Control 全自动操纵(飞行操纵系统的)

General Aviation Centre 通用航空中心

Global Area Coverage 全球覆盖

Go-Around Computer 复飞计算机

Grumman Aircraft Corporation 格鲁曼飞机公司

Gust Alleviation Control 阵风衰减控制

**GACA** General Administration of Civil Aviation 民用航空总局

**GACW** Gust Above Constant Wind 大于定常风的阵风

**GAD** General Assembly Drawing 总装配图

**GADL** Ground-to-Air Data Link 地空数据链

**GADO** General Aviation District Office 通用航空区办公室

**GADS** Generator Availability Data System 发电机可用性数据系统

Generic Aircraft Display System 通用航空器显示系统

**GAEO** General Assembly Engineering Order 通用装配工程指令

**GAFG** General Aviation Flight Guide 通用航空飞行条例

**GAFOR** General Aviation Forecast 通用航空预报

**GAFPG** General Aviation Facilities Planning Group 通用航空设施计划组

**GAG** Ground-Air-Ground 地对空对地

**GAGAN** GPS Aided GEO Augmented Navigation GPS 辅助型近地轨道增强导航

GPS and GEO Augmented Navigation GPS 和地球同步轨道增强导航,GPS 和 GEO 增强导航

**GAGDT** Ground to Air to Ground Data Terminal 地—空—地数据终端

**GAGT** General Agent 总代理

**GAHMM** Global Aircraft Health Monitoring and Management 全球飞机健康监控与管理

**GAIL** Glide Angle Indicator Light 下滑角指示灯

**GAIN** Global Analysis and Information Network 全球安全分析与信息网

**GAINS** Global Airborne Integrate Navigation System 全球机载综合导航系统

**GAIR** General Assembly Inspection Report 总装检查报告

**GAIT** Ground-based Augmentation and Integrity Technique 陆基增强和完好性技术

**GAL** Galley 厨房

Gas Analysis Laboratory 气体分析实验室

Generic Array Logic 通用阵列逻辑

US Gallon 美加仑

**GALCIT** Guggenheim Aeronautical Laboratory, California Institute of Technology 加州理工学院古根海姆航空实验室

**GALV** Galvanize 电镀

**GAM** General Aeronautical Material 通用航空材料

General Annual Meeting 一般的年度会议

Graphic Access Method 图形存取方法

**GAMA** General Aviation Manufacturers Association 通用航空制造商协会

**GAMC** Guangdong Airport Management Corporation 广东省机场管理集团公司(中国)

**GAMECO** Guangzhou Aircraft Maintenance Engineering Company Ltd. 广州飞机维修工程有限公司

**GAMP** Global Atmospheric Measurement Program 全球大气测量计划

**GAMS** Ground Asynchronous Messaging Service 地面异步信息服务

**GAMTA** General Aviation Manufacturers and Traders Association 通用航空制造商和销售商协会

**GAN** Global Area Network 全球网

Gyro Automatic Navigation 陀螺自动导航

**GANP** Global Air Navigation Plan 全球空中航行计划

**GAO** General Accounting Office 会计总署(美国)

General Auditing Office 审计总署(美国)

Government Accountability Office 政府问责办公室

Government Accounting Office (联邦)政府会计署

**GAP** General Assembler Program 通用汇编程序

General Aviation Propulsion 通用航空推进

Government Aircraft Plant 国营飞机工厂

Ground Avionics Prototype 地面航空电子设备样品

**GAPA** Ground-to-Air Pilotless Aircraft 地面起飞无人驾驶飞机

**GAPAN** Guild of Air Pilots and Air Navigators 飞行员与领航员行业协会(英国)

**GAPCU** Ground and Auxiliary Power Control Unit 地面和辅助动力控制组件

**GAPE** General Aviation Pilot Education 通用航空飞行员培训

**GAPL** Ground Avionics Prototype Laboratory 地面航空电子设备样机实验室

Group Assembly Parts Lists 组装零件表

**GAR** Garage 车库

Generator Annunciator Relay 发电机报警继电器

Global Atmospheric Research (Program) 全球大气研究(计划)

Growth Analysis and Review 增长分析和评定

**GARBD** Garboard 龙骨翼板

**GARC** General Astronautics Research Corporation 通用宇宙航空研究公司

**GARD** General Address Reading Device 通用地址读出装置

**GAREX** Ground Aviation Radio Exchange 地面航空无线电交换机

**GARMI** General Aviation Radio-Magnetic Indicator 通用航空无线磁指示器

**GARN** Garnishment 装饰,修饰

**GARP** Global Atmospheric Research Programme 全球大气研究项目(方案)

A Global Atmospheric Research Programme 全球大气研究计划

**GARS** General Aviation Restraint System 通用航空保护系统

Gyrocompassing Attitude Reference System 陀螺罗盘姿态基准系统

**GARTEUR** Group for Aeronautical Research

and Technology in Europe 欧洲航空研究技术组

**GAS** Gasoline 汽油
Gateway Access Service 网关接入服务
Global Address Space 全局地址空间
Ground Analysis Station 地面分析站

**GASCO** General Aviation Safety Committee 通用航空安全委员会

**GASP** General Aviation Strategy Program 通用航空战略计划
General Aviation Synthesis Program 通用航空综合计划
Generalized Aircraft Sizing Program 飞机定尺寸综合计划
Geometric Acoustic Solver Program 几何声学解算器程序
Global Atmosphere Sampling Program 全球大气取样计划
Global Aviation Safety Plan 全球航空安全计划
Global Aviation Security Program 全球航空安全计划

**GASR** Global Aviation Safety Roadmap 全球航空安全路线图

**GAT** General Air Traffic 一般空中交通
General Aviation Terminal 通用航空终点站
General Aviation Trainer 通用航空教练机
General Aviation Training 通用航空训练
General Aviation Transponder 通用航空应答机
General Aviation Transportation 通用航空运输
Global Air Traffic 全球空中交通
Greenwich Apparent Time 格林威治视时
Ground to Air Transmitter 地对空发射机

**GATCA** Greek Air Traffic Controllers' Association 希腊空中管制员协会

**GATCO** Guild Air Traffic Control Officers 空中交通管制员协会(英国)

**GATE** General Aviation Turbine Engine 通用航空涡轮发动机

General-Purpose Automatic Test Equipment 通用自动测试设备
Glossary of Airbus Terms and Expressions 空客公司术语和表述法汇编

**GATM** Global Air Traffic Management 全球空中交通管理

**GATO** Gas at Takeoff 起飞时油量
Gasoline at Takeoff 起飞载油量
Global Air Traffic Operations 全球空中交通运行

**GATR** Ground to Air Transmitter Receiver 地对空收发机

**GATS** General Agreement on Trade in Services 服务贸易总协定
General Aviation Technical Service 通用航空技术处
Generation Air Transport System 航空运输系统
Graphics Automated Template System 图形自动模板系统

**GATT** General Agreement on Trades and Tariffs 关税及贸易总协定

**GATU** Galley Automatic Transformer Unit 厨房自动变压器组件

**GATWY** Gateway 通道,出入口

**GAU** GPS Antenna Unit 全球定位系统天线装置

**GAVC** Ground Air Visual Code 地空视频通信编码

**GAVRS** Gyrocompassing Altitude and Velocity Reference System 陀螺罗盘高度和速度参考系统

**GAW** Global Atmosphere Watch 全球大气监测

**GAY** Green and Yellow 绿和黄

**GB** Gas Bearing 气体轴承
Gear Box 齿轮箱
Generator Breaker 发电机断路器
Gigabyte 千兆字节
Glide Bearing 滑翔方向
Good-Bye 再见

Great Britain 英国

Grid Bias 栅偏压

Guidance Buffer 制导缓冲器

**Gb** Gigabit 千兆比特

**GBA** Geostationary Broadcast Area 静止卫星广播区域

Global Business Analysis 全球商业分析

**GBAA** German Business Aviation Association 德国公务航空协会

**GBARC** Great British Aeronautical Research Council 英国航空研究理事会

**GBAS** Ground-Based Augmentation System 陆基增强系统

**GBE** Gaborone, Botswana (Airport Code) 博茨瓦纳哈博罗内机场

**GBH** Graphics-Based Hypermedia 基于图形的超媒体

**GBL** Garble 篡改

Government Bill of Lading 美国政府提货单

**GBLD** Garbled 电文不清楚

**GBMD** Guaranteed Bandwidth Minimum Delay 保证带宽的最小延迟

**GBP** Great Britain Pound 英镑(货币单位)

**GBPS** GigaBits per Second 千兆位/秒

**GBR** General Business Requirements 通用商务协议

Great Britain 英国

Gross Bandwidth Request 总带宽请求

Ground-Based Radar 地基雷达

Guaranteed Bit Rate 保证比特率

**GBS** General Business System 通用商业系统

Global Broadcast Service 全球广播服务

Global Broadcast System 全球广播系统

**GBSAS** Ground-Based Scanning Antenna System 陆基扫描天线系统

**GBST** Ground-Based Software Tool 基于地面的软件工具

**GBSVC** General Broadcast Signaling Virtual Channel 通用广播信令虚信道

**GBTA** Guild of Business Travel Agent (British) 公务旅行代理人协会(英国)

**GBW** Guaranteed Bandwidth 保证带宽

**GBX** Greatbox 齿轮箱

**GC** Gaussian Channel 高斯信道

Generator Control 发电机控制

Gigacycle 吉周

Great Circle 大圆圈,大圆航线

Ground Control 地面控制,地面管制

Group Coupler 群耦合器

Guidance Computer 制导计算机

Guidance Control 制导控制

Gyrocompassing 陀螺定向

**GCA** Global Cybersecurity Agenda 全球网络安全议程

Ground Controlled Aircraft 地面控制的飞机

Ground Controlled Approach 地面控制进场,地面管制进近

**GCAA** General Civil Aviation Authority 民用航空总局

**GCAC** Guangzhou Civil Aviation College 广州民航职业技术学院

**GCAR** Generator Control Annunciator Relay 发电机控制报警继电器

**GCAS** Ground Collision Avoidance System 地面防撞系统

Ground Controlled Approach System 地面控制进近系统

**GCAU** Ground Control Approach Unit 地面控制进近装置

**GCB** Gas Circuit Breaker 气体电路断路器

Generator Circuit Breaker 发电机断路器

Generator Control Breaker 发电机控制断路器

**GCC** Gillette, WY, USA (Airport Code) 美国怀俄明州吉莱特机场

Ground Control Center 地面控制中心

Guidance and Control Computer 制导和控制计算机

Guidance Control Console 制导控制台

Gulf Cooperation Council 海湾合作委员会

**GCCS** Global Command and Control System 全球指挥和管制系统

Ground Control and Communication Station 地面控制和通信站

Ground Crew Call System 地面人员呼叫系统

**GCD** Gain Control Driver 增益控制驱动器

GAS Control Decoder 有效载荷搭载容器控制译码器

Generator Contactor Driver 发电机接触器驱动器

Great Circle Distance 大圆(飞行)距离

**GCE** Ground Checkout Equipment 地面检测设备

Ground Communication Equipment 地面通信设备

Ground Control Equipment 地面控制设备

**GCF** Ground-Conditioning Fan 地面空调风扇

**GCG** Guidance Control Group 制导控制组

**GCI** General Communication Interface 通用通信接口

General Command Interface 通用命令接口

Ground Control Information 地面控制信息

Ground Control Intercept 地面控制拦截

**GCL** Great Circle 大圆圈,大圆航线

Ground Control Landing 地面指挥降落,地面控制着陆

**GCM** Global Call Model 全球呼叫模型

Graphic Color Monitor 图形彩色监视器

Ground Checkout Module 地面检查模件

**GCN** Gateway Connection Network 有网关的连接网络

Generalized Connection Network 通用连接网络

Grand Canyon, AZ, USA (Airport Code) 美国亚利桑那州大峡谷机场

**GCO** Ground Communication Outlet 地面通信分站

**GCOS** Global Climate Observing System 全球气候观测系统(气候观测系统)

**GCP** General Compliance Program 通用符合性程序

General Conditions of Purchase 购买通用条件

General Contour Package 通用仿形设备

Generator Control Panel 发电机控制面板

Glareshield Control Panel 遮光控制面板

Global Carbon Project 全球碳项目

Ground Control Point 地面控制点

Ground Control Processor 地面控制处理器

Group Change Program 群更改程序,组更改程序

**GCR** General Cargo Rate 普通货物运价

Generator Control Regulator 发电机控制调节器

Generator Control Relay 发电机控制继电器

Great Circle Route 大圆航线

Ground Controlled Radar 地面控制雷达

**GCR AUX** Generator Control Relay Auxiliary 发电机控制辅助继电器

**GCR** General Cargo Rates 普通货物运价

**GCRA** Generic Cell Rate Algorithm 类属信元率算法

**GCS** Gate Controlled Switch 门控开关

Global Communication System 全球通信系统

Ground Clutter Suppression 地面杂波抑制

Ground Communication System 地面通信系统

Ground Control Station(s) 地面控制站

Ground Control System 地面管制系统

Guidance Control Section 制导控制舱

**GCSS** Global Communication Satellite System 全球通信卫星系统

**GCT** Generator Coolant Temperature 发电机的冷却液的温度

Global Communication Technology 全球通讯技术

Greenwich Civil Time 格林威治民用时

Guaranteed Completion Time 保证完成时间

**GCTE** Global Change and Terrestrial Ecosystems 全球变化与陆地生态系统

Global Climate and Terrestrial Ecosystems 全球气候与陆地生态系统

Guidance Computer Test Equipment 制导计算机实验装置

**GCTS** Ground Communication Tracking System 地面通信跟踪系统

**GCU** Galley Cooling Unit 厨房冷却组件

Gear Control Unit 起落架控制组件

General Control Unit 通用控制单元

Generator Control Unit 发电机控制组件

Ground Checkout Unit 地面检测装置

Ground Control Unit 地面控制装置,地面控制设备

Gyroscope Coupling Unit 陀螺仪耦合装置

**GCW** General Continuous Wave 普通连续波

Global Chart of the World 世界航空图

Gross Cargo Weight 货物总重量

**GD** Gas Detector 气体检漏器,气体分析仪,毒气侦察器

Gear Down 起落架放下

General Declaration 总申报单

General Dynamics 通用动力公司

Grade 等级

Green Dot Speed 绿点速度

Ground 地面

Guard 警卫

Guide 指引,引导

**GD&T** Geometric Dimensioning and Tolerancing 几何尺寸及精度

**GDAAS** Guangdong Academy of Agricultural Sciences 广东省农科院

**GDAS** General Dynamics Aviation Services 一般动力学航空服务

Global Data Assimilation System 全球数据同化系统

Graphical Data Analysis System 图形数据分析系统

Ground Data Acquisition System 地面数据采集系统

**GDBMS** Generalized Data Base Management System 综合数据库管理系统

**GDBS** Generalized Database System 综合数据库系统

Global Data Base System 全局数据库系统

**GDC** Graphical Display Controller 图形显示控制器

Guidance Display Computer 引导显示计算机

**GDCI** General Data Communication Interface 通用数据通信接口

**GDCS** Ground Distributed Control System 地面分配控制系统

**GDE** Graphic Display Equipment 图像显示设备

Ground Data Equipment 地面数据设备

Guide 指引,引导

**GDF** Gas Dynamic Facility 气体动力设备

Ground Decommutation Facility 地面译码设备,地面解密设备

**GDH** Generalized Data Handler 综合数据处理机

Ground Data Handling 地面数据处理

**GDHS** Generalized Dynamic Hypermedia System 广义动态超媒体系统

Ground Data Handling System 地面数据处理系统

**GDI** Graphical Device Interface 图形设备接口

**GDIL** General Data Interchange Language 通用数据交换语言

**GDL** Gas Dynamics Laboratory 气体动力实验室

Gas-Dynamic Laser 气体动力激光器

Gear Down Latch 起落架下位锁

Global Data Link 全球数据链路

Guadalajara, Mexico (Airport Code) 墨西哥瓜达拉哈拉机场

**GDLP** Ground Data Link Processors 地面数据链处理机

**GDLU** Game Download Unit 游戏下载组件

**GDM** Generalized Development Model (全球定位系统接收机的)通用发展型

**GDMO** Guidelines for Definition of Managed Objects 被管对象定义的准则

**GDMS** Ground Data Management System 地面数据管理系统

**GDN** Gdansk, Poland (Airport Code) 波兰格旦斯克机场

**GDNC** Guidance 指引, 引导

**GDNT** Gradient 梯度, 变化率

**GDO** Gate Dip Oscillator 栅陷振荡器

Ground Door Opening 地面舱门打开

**GDOP** Geodesic Degradation of Performance 短程性能下降

Geometric Dilution of Position 位置几何衰减因子

Geometric Dilution of Precision 几何精度衰减因子

**GDP** Graphics Display Processor 图形显示处理器

Gross Domestic Product 国内生产总值

**GDPFS** Global Data-Processing and Forecasting Systems 全球资料加工和预报系统

**GDPS** Global Data Processing System 全球数据处理系统, 全球资料处理系统

Ground Data Processing System 地面数据处理系统

**GDR** Ground-Based Radar 地基雷达

**GDS** Gas Deployed Skirt 气体展散边缘

Gas Dynamic System 气体动力系统

General Declassification Schedule 通用解密程序

Global Distribution System 全球分销系统

Goods 货物

Graphysical Display System 图解显示系统

Ground Data System 地面数据系统

Group Display System 群显示系统

GSE Data Sheet 地面支援设备数据单

Guards 警卫

**GDSS** Generic Data Switching System 通用数据交换系统

Global Decision Support System 全球决策支持系统

Global Defense Support System 全球防御支持系统

Global Digital Satellite Services 全球数字卫星服务

Group Decision Support System 群体决策支持系统

**GDSU** Global Digital Service Unit 全球数字业务单元

**GDT** Geographic Data Technology 地理数据技术

Ground Differential Transmitters 地面差动发送器

**GDU** Graphic Display Unit 图解显示器

**GE** Gateway Exchange 关口局

General Electric 通用电气(美国)

General 总则, 一般

Generator Electronic 发电

Gigabit Ethernet 千兆位以太网

Graphite-Epoxy 石墨纤维树脂

Ground Equipment 地面设备

**GEAE** General Electric Aircraft Engines 通用电气飞机发动机集团(美国)

**GEAG** General Electric Airborne Guidance 通用电气公司机载导引

**GEANS** Gimbaled Electrostatic Aircraft Navigation System 环架式静电飞机导航系统

**GEC** General Electric Company 通用电气公司

Ground Earth Station (卫星)地面站

**GECAS** General Electric Capital Aviation Service 通用电气资本航空服务公司

General Electric Commercial Aviation Services 通用电器商业航空服务公司

**GECC** Gasoline Engine, Close Coupled 紧耦合汽油发动机

**GECOS** General Electric Comprehensive Operating System 通用电气公司综合操作系统

**GECS** Ground Environmental Control System 地面环境控制系统

**GED** Gasoline Engine Driven 汽油发动机驱动

**GEDL** General Electric Data Link 通用电气公司数据链

**GEEREF** Global Energy Efficiency and Renewable Energy Fund 全球能源效率和可再生能源基金

**GEF** Global Environment Facility 全球环保设施

Global Environment Fund 全球环保基金

Ground Equipment Failure 地面设备故障

**GEG** Spokane, WA, USA（Airport Code）美国华盛顿州斯波坎机场

**GEH** Graphite-Epoxy Honeycomb 石墨纤维树脂复合蜂巢

**GEIA** Government Electronics and Information Technology Association 政府电子与信息技术协会

**GEJ** General Electric of Japan 日本通用电气公司

**GEM** Generic Electronic Module 通用电子设备舱

Ground Equipment Manual 地面设备手册

Ground-Based Engine Monitoring 地基的发动机监控

**GEMCOR** General Electromechanical Corporation 通用机电公司

**GEMM** Generalized Electronics Maintenance Model 通用电子设备维修模型

**GEMS** General Energy Management System 总能量管理系统

General Errors Model System 通用失误模型系统

Generic Error Modeling System 通用差错建模系统

Global Environment（al）Monitoring System 全球环境监测系统

Grouped Engine Monitoring System 组合发动机监测系统

**GEN** General Air Services, Inc.（USA）通用航空服务公司（美国）

General Information 总述，概况

General 总的，通用的，一般的，通用

Generation 一代

Generator 发电机

**GENG** Gasoline Engine 汽油发动机

**GENI** Global Environment for Network Innovations 全球网络创新环境

Global Environment for Networking Investigations 全球网络环境调查

**GENL** General 总的，通用的，一般的

**GEO** Geographic 地理的

Geography 地理学

Georgetown, Guyana（Airport Code）圭亚那乔治敦机场

Geostationary Earth Orbit 静止地球轨道

Geostationary Orbit 同步轨道

Geostationary 静止的

Geosynchronous Earth Orbit 地球同步卫星轨道，同步地球轨道

Group on Earth Observation 地球观测组

**GEOG** Geographic 地理的

Geography 地理学

**GEOL** Geologic 地质的

**GEOREF** World Geographic Reference System 世界地理参考系统，世界地理基准系统

**GEOS** Geodetic Earth Observation Satellite 大地测量地球观测卫星

Geodynamics Experimental Ocean Satellite 地球动力学实验海洋卫星

Geostationary Earth Orbiting Satellite 对地静止轨道卫星

Geosynchronous Earth Orbit Satellite 同步地球轨道卫星

Global Earth Observing System 全球对地观测系统

Graphic Environment Operating System 图形环境操作系统

**GEOSS** Global Earth Observation System of Systems 全球综合地球观测系统

**GEP** Global Entry Program 全球通关计划

Ground Environment Program 地面环境计划

**GEPEJTA** Group of Experts on Policies, Economics and Legal Matters in Air Transport 航空运输政策、经济和法律问题专家工作组

**GER** German 德国人,德语
Germany 德国

**GERT** Graphical Evaluation and Review Technique 图形评审技术

**GERTS** General Remote Terminal System 通用远程终端系统
General Remote Transmission Supervisor 通用远程传输管理程序

**GES** General Santos, Philippines (Airport Code) 菲律宾桑托斯将军城机场
Ground Earth Station 地面地球站,卫星地面站
Ground Engineering System 地面工程系统

**GET** Ground Elapsed Time 地面经历时间(从起飞开始计)

**GETOL** Ground Effect Takeoff and Landing 地面效应起飞和着陆

**GEV** Generalized Extreme Value 广义极值

**GEWEX** Global Energy and Water Cycle Experiment 全球能量与水循环实验

**GF** General Function 全局功能
Generator Field 发电机励磁
Germanium Content Fiber 含锗光纤

**GFA** Gliding Federation of Australia 澳大利亚滑翔联盟
Government Freight Agent 政府货运代理商
Government Furnished Articles 政府供给物品

**GFAC** Ground Forward Air Controller 前方地面空中管制员

**GFAE** Government Furnished Aircraft Equipment 政府装备机载设备
Government Furnished Aeronautical/Aerospace Equipment 政府提供的航空/航天设备

**GFB** Government Furnished Baseline 政府装备基本设备

**GFC** Generic Flow Control 一般流量控制
Global Flow Control 总流量控制

**GFDL** Geophysical Fluid Dynamics Laboratory 美国国家地球物理流体动力实验室

**GFE** Government Furnished Equipment 政府提供的设备

**GFEL** Government Furnished Equipment List 政府提供的设备清单

**GFI** Government Furnished Information 政府提供信息
Ground Fault Interrupter 地面故障中断器
Ground Fault Interruption 地面故障中断

**GFK** Grand Forks, ND, USA (Airport Code) 美国北达科他州大福克斯机场

**GFLD** Generator Field 发电机磁场

**GFM** Government Furnished Material 政府提供器材

**GFMD** Ground Friction Measurement Devices 地面摩擦测量装置

**GFP** General Function Platform 总功能平面
Global Function Plane 全局功能平面
Government Furnished Parts 政府提供零件
Government Furnished Property 政府提供的资产
Ground Fine Pitch 地面小距(螺旋桨变距)

**GFPT** General Flight Proficiency Test 一般飞行能力测试
General Flying Progress Test 基本飞行过程测试,基本飞行进度测试

**GFR** German Federal Republic 德意志联邦共和国

**GFRP** Glass Fiber Reinforced Plastic 玻璃纤维增强塑料

**GFS** Global Forecasting System 全球天气预报系统
Government Flying Service 政府飞行服务队
Government Furnished Software 政府提供软件

**GFSE** Government Furnished Support Equipment 政府提供的支援设备

**GFSK** Gauss Frequency Shift Key 高斯频移键控

**GFT** General Flight Test 一般飞行测试
Generalized Fast Transform 广义快速变换
Ground Fault Test 地面故障测试

Ground Functional Test 地面功能测试

**GFTM** General Familiarization Training Manual 一般熟悉培训手册

**GFY** Glider Flying/Glider Flight 滑翔机飞行

**GG** Gas Generator 燃气发生器,气体发生器

Graphics Generator 图形产生器

Gravity Gradient 重力梯度

Ground Guidance 地面制导

**GGC** Greenhouse Gas Concentration 温室气体的浓度

Ground Guidance Computer 地面制导计算机

**GGE** Gauge 仪表

Group of Governmental Experts 政府专家小组

**GGG** Longview,TX,USA(Airport Code)美国得克萨斯州朗维尤机场

**GGP** Gateway to Gateway Protocol 网关到网关协议

General Graphics Package 通用图像程序包

**GGPCU** Generator and Ground Power Control Unit 发电机及地面电源控制组件

**GGS** Ground Guidance System 地面制导系统

**GGSN** Gateway GPRS Support Node 网关GPRS支持节点

**GGW** Glasgow,MT,USA(Airport Code)美国蒙大拿州格拉斯哥机场

**GH** Grid Heading 网格坐标航向

Ground Handling 地面操作,地勤作业

Gyro Horizon 陀螺地平仪

**GHA** General Hazard Analysis 一般危害分析

Greenwich Hour Angle 格林威治时角

Ground Hazard Area 地面危险区

**GHB** Ground Handling Breaker 地面操纵断电器

Ground Handling Bus 地面操作汇流条

**GHE** Ground Handling Equipment 地面运作设备

**GHG** Greenhouse Gas 温室气体

**GHR** Ground Handling Relay 地面支持继电器

**GHS** Ground Handling Simulator 地面运作模拟器(机)

**GHTR** Ground Handling Transfer Relay 地面运作转换继电器

**GHz** Gigahertz 千兆赫兹

**GI** General Issue 总发行量

Gill 吉耳(液量单位)

Graded Index fiber 渐变折射率光纤

Ground Idle 地面慢车

Guilder 盾(货币单位)

**GIAA** Guam International Airport Authority 关岛国际机场管理局

**GIACC** Group on International Aviation and Climate Change 国际航空和气候变化小组

**GIB** Generic Information Block 通用信息模块

Gibraltar,Gibraltar(Airport Code)直布罗陀机场

GNSS Integrity Broadcast 全球导航卫星系统完好性数据广播

Guy in Back 后座员

**GIC** General Instruction Card 通用说明卡

GNSS Integrity Channel 全球导航卫星系统完好性通道

GPS Integrity Channel GPS完好性通道

**GICB** Ground-Initiated Comm-B 地面启动的B类通信

**GIDEP** Government-Industry Data Exchange Program 政府和工业部门数据交换程序

**GIE** Global Information Environment 全球信息环境

Ground Instrumentation Equipment 地面仪表设备

**GIF** Graphic Interchange Format 图形交换格式,图形内部转换格式

Guy in Front 前座员

**GIFAS** French Aeronautical and Aerospace Industry Association 法国宇航工业协会

**GIFTS** Geosynchronous Imaging Fourier Transform Spectrometer 傅立叶对地静止成

像转换光谱仪

**GIG**   Global Information Grid 全球信息栅格
Rio De Janeiro, Brazil (Airport Code) 巴西里约热内卢机场

**GIGO**   Garbage in, Garbage out 无用输入，无用输出

**GII**   Global Information Infrastructure 全球信息架构，全球信息基础设施

**GIM**   General Introduction Manual 通用简介手册
Generalized Information Management 综合信息管理，广义信息管理

**GIMADS**   Generic Integrated Maintenance Diagnostics 通用综合维修诊断

**GINO**   Graphical Input and Output 图形输入与输出

**GIOP**   General-Purpose Input/Output Processor 通用输入输出处理机

**GIOS**   Graphical Input/Output System 图形输入与输出系统

**GIP**   Graphic Information Processing 图形信息处理
Ground Instructor Pilot 地面飞行教员
Guidance Improvement Program 制导改进程序

**GI-PCF**   Graded-Index Plastic-Cladding Fiber 渐变折射率塑料包层光纤

**GI-POF**   Graded-Index Plastic Optical Fiber 渐变折射率塑料光纤
Graded-Index Polymer Optical Fiber 渐变折射率聚合物光纤

**GIPSY**   General Image Processing System 通用图像处理系统
General Information Processing System 通用信息处理系统
General Integrated Publication System 通用综合出版系统

**GIRAS**   Geographic Information Retrieval and Analysis System 地理信息检索与分析系统

**GIRL**   Graph Information Retrieval Language 图形信息检索语言

**GIRU**   Ground Interrogator Receiver Unit 地面应答机接收单元

**GIS**   Gas Insulated Switchgear 气体绝缘开关设备
General Information Service 通用信息服务
General/Generalized Information System 通用信息系统
Geographical Indexing Systems 地理索引系统
Geographical Information System 地理信息系统
Geoscience Information Society 地球科学情报学会（美国）
Global Information System 全球情报系统
Graphic Input System 图形输入系统
Grid Information Service 网格信息服务

**GIS/VS**   Generalized Information System/Virtual Storage 综合信息系统/虚拟存储器

**GISC**   General Insurance Standards Council 一般保险标准委员会
Geographic Information Science Center 地理信息科学中心
Geographic Information System Center 地理信息系统中心
Global Innovation and Strategy Center 全球创新和策略中心

**GISP**   Global Invasive Species Programme 全球入侵物种方案

**GIST**   Global Information System Test 全球信息系统试验

**GISW**   Graded-Index Slab Waveguide 渐变折射率平面波导

**GIT**   Generic Intelligent Center 通用智能中心
Geographic Information Technology 地理信息技术
Group Inclusive Tourfare 集体综合游览票价

**GITH**   Gigabit Internet to Home 千兆位因特网到户

**GIU**   Graphic Interface Unit 图形界面组件

**GIX**   Global Internet Exchange 全球因特网交

换机

**GJ** Grown Junction 生长结

**GJP** Graphic Job Processor 图形作业处理器

**GJT** Grand Junction, CO, USA (Airport Code) 美国科罗拉多州大章克申机场

**GK** General Knowledge 常识

Greek 希腊的

**GKS** Graphical Kernel System 图形核心系统

**GL** Gallon 加仑(容量单位)

Gill 吉耳(液量单位)

Ground Level 地面

Guiding Layer 引导层

**GLA** Glasgow, Scotland, United Kingdom (Airport Code) 英国格拉斯哥机场

Gust Load Alleviation 阵风载荷减缓

**GLC** Generator Line Contactor 发电机线路接触器,主接触器

**GLCG** Generator Line Contactor-Generator Section 发电机线路接触器—发电机部分

**GLCR** Generator Line Contactor Relay 发电机线路接触器继电器

**GLCT** Generator Line Contactor-Transfer 发电机线路接触器—转换

**GLD** Glad 愉快的

Glidepath Tracking 下滑道跟踪

Glider 滑翔机

Gold 金

Ground Laser Designator 地面激光指示器

Ground Lift Dumping 地面减升

Gulden 盾(货币单位)

**GLH** Greenville, Mississippi, USA (Airport Code) 美国密西西比州格林维尔机场

**GLL** German Levertage Lease 德国杠杆租赁

**GLM** General List of Modifications 改装综合清单

**GLO** Ground Liaison Officer 地面联络官

**GLOBECOM** Global Communication System 全球通信系统

Global Telecommunications Conference 全球电信会议

**GLONASS** Global Navigation Satellite System (Russian) 全球导航卫星系统(俄罗斯)

Global Orbit/Orbiting Navigation Satellite System 全球轨道导航卫星系统

**GLOTRACK** Global Tracking Network 全球跟踪网络

**GLOW** Gross Lift-off Weight 起飞总重

**GLR** Galley Load Relay 厨房负载继电器

Gateway Location Register 网关位置注册

Generator Line Relay 发电机线路继电器

**GLS** Gas-Liquid-Solid Chromatograph 气—液—固体色谱法

Glass 玻璃

Glide Slope 下滑道

GPS Landing System 全球定位着陆系统

**GLT** Ground Leak Test 地面渗漏测试

Guide Light 导航灯

**GLU** Global Land Use 全球陆地使用

GNSS Landing Unit 全球卫星导航系统着陆组件

**GLY** Galley 厨房

**GM** General Manager 总经理

Gram 克

Green Maintenance 绿色维修

Greenwich Meridian 格林威治子午线

Guidance Material 指导材料

Metacentric Height 定额高度

**GMAT** Graduate Management Admission Test 研究生管理科入学考试

Greenwich Mean Astronomical Time 格林威治平均天文时间

**GMC** General Modifications Committee 通用改装委员会

Ground Movement Control 地面活动管制

**GMDA** Ground Miles and Drift Angle Computer 对地速度和偏流角计算机

**GMDSS** Global Maritime Distress and Safety System 全球海事安全系统

**GMECO** Guangzhou Aircraft Maintenance Engineering Corporation 广州飞机维修工程公司

**GMF** Global Market Forecast 全球市场预测

**GMG** General Motors Gearbox 主电机齿轮箱

**GMI** General Messaging Interface 通用信息接口

Generic Management Information 一般管理信息

**GMIS** GEC Marcony in-flight System 美国通用电气公司马可尼飞行中娱乐系统

General Multimedia Information System 通用的多媒体信息系统

**GML** Generalized Markup Language 通用标记语言

**GMM** Gaussian Mixture Model 高斯混合模型

General Maintenance Manual 通用维护手册,综合维修手册

Global Multimedia Mobility 全球多媒体移动性

**GMN** Global Maintenance Network 全球维护网络

Greenwich Mean Noon 格林威治平午

**GMP** Gimpo, South Korea(Airport Code) 韩国金浦机场

**GMPCS** Global Mobile Personal Communications by Satellite 全球卫星移动个人通信

**GMR** Ground Movement Radar 地面移动雷达

**GMS** Geostationary Meteorological Satellite 同步气象卫星

Ground Monitoring Station 地面监控站

**GMSC** Gateway Mobile Switching Center 网关移动交换中心

**GMSK** Gaussian-Filtered Minimum Shift Keying 高斯滤波最小移位键控

**GMSS** Global Mobile Satellite Service 全球移动卫星服务

**GMT** Greenwich Mean Time 格林威治标准时间

Group Multiplexer Terminator 群复用器终端

**GMTBF** Guaranteed Mean Time Between Failures 保证的平均故障间隔时间

**GMT-DAT** Greenwich Mean Time and Data 格林威治时间和数据

**GMTI** Ground Moving Target Indication 地面移动目标指示

Ground Moving Target Indicator 地面运动目标显示器

**GMTT** Global Management Technology Training 全球管理技术培训

Ground Moving Target Track 地面移动目标跟踪

**GMW** Generic Maintenance Workstation 通用维修工作站

**GN** General Note 总说明,总纪要

Green 绿色

Ground Navigation 地面导航

Guidance and Navigation 制导与导航

**GN&C** Guidance, Navigation and Control Subsystem 制导、导航和控制分系统

**GNACU** General Navigation Avionics Control Unit 通用导航电子控制组件

Guidance Navigation Avionics Control Unit 制导导航电子控制组件

**GNAS** General National Airspace System 综合国家空域系统

**GNC** General Navigation Computer 通用导航计算机

Global Navigation and Planning Chart 全球导航与计划图

Global Navigation Chart 全球导航图

Government Non-Conformance Item 政府不一致项目

Graphic Numerical Control 图形数字控制

Guidance and Navigation Control 制导与导航控制

Guidance and Navigation Computer 制导与导航计算机

**GNCS** Guidance, Navigation and Control System 制导、导航和控制系统

**GND** Ground Control 地面管制

Ground 地面

**GNDCK** Ground Check 地面检查,地面校验

**GNE** Gateway Network Element 网关网元

Gross Navigation Error 总导航误差

Guidance and Navigation Electronics 制导和导航电子设备

**GNIS** Geographic Names Information System 地理名称信息系统

**GNL** General 总的,通用的,一般

**GNLU** Global Navigation and Landing Unit 全球导航和着陆组件

GNSS Navigation and Landing Unit 全球导航卫星系统导航和着陆组件

**GNM** General Motors,Inc.(USA) 通用汽车公司(美国)

**GNN** Global Network Navigator 全球网络导航器

**GNP** Gross National Product 国民生产总值

**GNR** General Aircraft Supplies(USA) 通用飞机供应公司(美国)

Global Navigation Receiver 全球导航接收机

GNSS Navigation Radio 全球导航卫星系统无线电导航

**GNS** Global Navigation System 全球导航系统

Guidance and Navigation System 制导与导航系统

**GNSS** Global Navigation Satellite System 全球导航卫星系统

**GNSSP** Global Navigation Satellite Systems Panel 全球导航卫星系统专家组

**GNT** Grant 批准,准许

**GNTR** Generator 发电机

**GNV** Gainesville,FL,USA(Airport Code) 美国佛罗里达州盖恩斯维尔机场

**GOA** Genoa,Italy(Airport Code) 意大利热那亚机场

**GOAL** Ground Operations Aerospace Language 航空航天地面操作语言

**GOALS** General Operations and Logistics Simulation 通用运作和后勤模拟

**GOARND** Go Around 复飞

**GOCO** Government Owned/Contractor-Operated 政府所有的/合同方运营的

**GOD** Game on Demand 交互式游戏(游戏点播)

Guided-Wave Optical Device 光波导器件

**GOE** Government Owned Equipment 政府所有设备

Ground Operating Equipment 地面运行设备

**GOES** Geostationary Operational Environmental Satellite 静地运行环境卫星

**GOF** Glass Optical Fiber 玻璃光纤

**GOFIR** Global Operators Flight Information Resource 全球飞行运营信息网

**GOI** Ground Objectives Identification 地面目标识别

Nizhniy Novgorod,Russia(Airport Code) 俄罗斯高尔基机场

**GOMS** Goals,Operators,Methods and Selection Rules 目标、操作员、方法与选择规则

Goals,Operators,Methods,Selection Model 目标、操作员、方法、选择模型

**GOP** Group of Pictures 图片组

**GOR** General Operational Requirement 一般使用要求,总使用要求

Gradual Onset Rate 慢增长率

Ground Operations Review 地面操作评审

Guy on the Right 右座员

**GORM** Global Ozone Research and Monitoring 全球臭氧研究与监测

**GOS** Gate Operation System 登机门操作系统

General Operating Specification 通用操作规程

Global Observing System 全球观测系统

Government Open Systems Implementation Profile 政府开放系统实施概况

Government Open Systems Intercommunication Profile 政府开放系统互联概况

Grade of Service 服务等级

**GOSIP** Government Open System Interconnection Profile 政府开放系统互联简介

Government Open System Interconnection Protocols 政府公开系统互联协议

Government Open Systems Intercommunications Profile 政府开放系统双向通讯简介

**GOT** Gothenburg, Sweden-Landvetter（Airport Code）瑞典哥德堡兰德维特机场

**GOV** Govern 调节,管理,控制
Governor 调节器,总督

**GOVT** Government 政府

**GP** Gain Program 增益程序
General Purpose System 通用系统
General Purpose 一般用途
Geographical Position 地理位置
Glide Path, Glide-Path 下滑道,下滑轨迹
Group Lights 群光
Group 群,组,团体
Guiding Principle 指导原则

**GPA** Gas Pipeline Analysis 气路分析方法
General Purpose Amplifier 通用放大器

**GPAC** General Purpose Airborne Computer 通用机载计算机
General Purpose Analogue Computer 通用模拟计算机

**GPACK** General Utility Package 通用实用程序包

**GPAP** General Purpose Associative Processor 通用相联处理机

**GPAS** General Purpose Airborne Simulator 通用机载模拟机
Ground-based Regional Augmentation System 地基区域增强系统

**GPATE** General Purpose Automatic Test Equipment 通用自动测试设备

**GPATS** General Purpose Automatic Test System 通用自动试验系统
General Purpose Automatic Test Set 通用自动测试设备

**GPB** General Purpose Buffer 通用缓冲器
Ground Power Breaker 地面电源断路器

**GPBNTF** Global Performance-Based Navigation Task Force 基于性能导航的全球工作组

**GPC/P** General Purpose Controller/Processor

通用控制器与处理机

**GPCS** General Purpose Control System 通用控制系统

**GPCT** Ground Power Current Transformer 地面电源变压器

**GPCU** Ground Power Control Unit 地面电源控制组件

**GPDC** General Purpose Digital Computer 通用数字计算机

**GPE** General Purpose Equipment 通用设备

**GPES** Ground Proximity Extraction System 贴地飞行拖出（空投）系统

**GPETE** General Purpose Electronic Test Equipment 通用电子测试设备

**GPF** General Protection Fault 一般性保护错误

**GPFL（G）** Group Flashing 组闪光

**GPH** Gallons per Hour 加仑/小时

**GPI** General Precision Instruments 通用精密仪表
General Purpose Input 通用输入
General Purpose Interface 通用接口
Glide-Path Indicator 下滑道指示器
Ground Position Indicator 地面位置指示器

**GPIB** General Purpose Information Bus 通用信息总线
General Purpose Interface Bus 通用接口总线

**GPIO** General Purpose Input/Output 通用输入输出

**GPIP** Glide-Path Intercept Point 下滑道截获点,下滑道切入点

**GPIRS** Global Positioning Inertial Reference System 全球定位惯性基准系统（美国）

**GPL** Ground Proximity Light 地面近地灯光

**GPLS** Glide-Path Landing System 下滑道着陆系统

**GPM** Gallons per Minute 加仑/分钟
General Processing Module 通用处理组件

**GPMS** General Purpose Microprogram Simulator 通用微程序模拟器
General Purpose Multiplex System 通用多路

传输系统

Global Price Management System 全球价格管理系统

Green Product Management System 绿色产品管理系统

Gross Performance Measuring System 总性能测量系统

**GPO**  General Post Office 邮政总局

General Purpose Output 通用输出

Government Printing Office 政府印刷处

**GPP**  General Purpose Processor 通用处理机

**GPRA**  Government Performance and Results Act 政府绩效与成果法案

Government Performance and Review Act 政府绩效与评估法案

**GPROX**  Ground Proximity 近地

**GPRS**  General Packet Radio Service 通用分组无线电业务

**GPS**  Gallons per Second 加仑/秒

General Purpose Simulator 通用模拟器

Geophysical Processor System 地球物理处理机系统

Global Positioning Satellite 全球定位卫星

Global Positioning System 全球定位系统

Global Professional Services 全球专业化服务

Graphic Programming Services 图形编程服务

Ground Playback Station 地面重放站

Ground Processing System 地面处理系统

**GPSR**  Generator Phase Sequence Relay 发动机相序继电器

Global Positioning Satellite Receiver 全球定位卫星接收机

Global Positioning System Receiver 全球定位系统接收机

**GPSS**  General Process Simulation System 一般程序模拟系统

General Purpose Simulation Software 通用模拟软件

General Purpose Simulation Studies 通用模拟研究

General Purpose Simulation System 通用模拟系统

General Purpose Systems Simulator 通用系统模拟器

Global Positioning Satellite System 全球定位卫星系统

**GPSSU**  Global Positioning System Sensor Unit 全球定位系统传感器组件

**GPST**  Global Positioning System Tester 全球定位系统检测器

**GPSU**  Global Positioning System Unit 全球定位系统组件

**GPT**  Gallons per Ton 加仑/吨

General Purpose Technology 通用目的性技术

General Purpose Transport 通用运输

Glide Path Tracking 下滑道跟踪

Ground Path Track 地面轨迹跟踪

Guidance Position Tracking 制导位置跟踪

Gulfport, MS, USA ( Airport Code ) 美国密西西比州格尔夫波特机场

**GPTE**  General Purpose Test Equipment 通用测试设备

**GPU**  General Processing Unit 通用处理器

General Purpose User 普通使用者

Ground Power Unit 地面动力装置,地面电源组件

**GPW**  Global Point Warning 全球定点报警

Ground Proximity Warning 近地告警

**GPWC**  Ground Proximity Warning Computer 地面进近告警计算机

**GPWM**  Ground Proximity Warning Module 近地告警组件

**GPWR**  Ground Power 地面电源

**GPWS**  General Purpose Work System 通用工作系统

General Purpose Workstation 通用工作站

Ground Proximity Warning System 近地警告系统

**GQTS**  Global Quality Tracking System 全球

质量跟踪系统

**GR** Gear 齿轮,起落架

Generator Relay 发电机继电器

Geographical Rotation 地理顺序

Goods Received 收到的货物

Grade 等级

Grain 谷物,晶体

Green 绿

Gross 毛(重)

Ground Relay 地面中继

Group 组

Hail or Soft Hail 冰雹或松冰雹

**GRAD** Global Risk Assessment Device 全球风险评估装置

Gradient 倾斜度,坡度,升降率

Graduation 刻度,分度

**GRADU** Gradually 渐渐地

**GRAM** Global Reference Atmospheric Model 全球参考大气模型

Graduated Resolution and Maintenance 刻度分辨率和维护

**GRAS** Ground-Based Regional Augmentation System 陆基区域增强系统

**GRASP** General Rotorcraft Aeromechanical Stability Program 广义旋翼机航空力学安定性计划

**GRASS** Geographic Resources Analysis Support System 地理资源分析支持系统

**GRB** Green Bay, WI, USA (Airport Code) 美国威斯康星州格林贝机场

**GRBL** Garble 干扰,错乱

**GRC** General Research Corporation 综合研究公司

German Cargo 德国货运公司

Glass Reinforced Composite 玻璃纤维增强复合材料

Glass Reinforced Concrete 玻璃纤维增强混凝土

Glenn Research Center 格伦研究中心

Greece 希腊

Ground Radio Communications 地面无线电通信

**GRD** General Requirements Document 一般要求的文件

Grind or Grinding 研磨,磨光

Ground Resolved Distance 地面能分辨的距离

Ground 地面

Guaranteed 担保的,保证的

**GRDL** Griddle 筛子

**GRDN** Guardian 管理人,保管人

**GRDTN** Graduation 刻度,分度,毕业

**GRE** Gamma Ray Explorer 伽马射线探测器

Gas Release Event 气体释放事件

Glass Reinforced Epoxy 玻璃纤维增强树脂

Graduate Record Examination 研究生入学资格考试

Ground Radar Equipment 地面雷达设备

**GREATS** Gimbaled Radiators and Environment Avoidance Tracking System 万向支架辐射器和环境回避跟踪系统

**GREPECAS** Caribbean/South American Planning and Implementation Regional Group 加勒比/南美洲计划和实施区域小组

**GRF** Ground Relay Facility 地面中继设备

Group Repetition Frequency 分组重复频率

**GRG** Ground-Roll Guidance 着陆滑跑引导

**GRI** Government Reports Index 政府报告索引

Grand Island, NE, USA (Airport Code) 美国内布拉斯加州格兰德艾兰机场

Group Repetition Interval 组重复间隔

**GRID** Global Resource Information Database 全球资源信息数据库

Global Risk Information Database 全球风险信息数据库

Grid Navigation 网格导航

**GRIMS** Ground-based Regional Integrity Monitoring System 地基区域完好性监视系统

**GRIP** Global Reach Improvement Program 全球到达改进计划

Graphics Interactive Programming 交互式图

形编程

**GRJ** George, South Africa ( Airport Code ) 南非乔治城机场

**GRL** General Instruments Corporation 通用仪表公司

Gross Reference List 参考资料总目录

**GRM** Government Resources Management 政府资源管理

**GRN** Green 绿色的

Ground Radio Navigation System Tacan "塔康" 地面无线电导航系统

Ground Reference Navigation 地标导航

Group Routing Node 群路由节点

**GRND** Ground 地面

**GRO** Gerona, Spain ( Airport Code ) 西班牙赫罗纳机场

**GRP** Geographic Reference Point 地理坐标基准点

Glass-Fibre Reinforced Plastic 玻璃纤维增强塑料

Glass-Reinforced Plastic 玻璃增强塑料

Gross Regional Product 地区生产总值

Ground Reference Point 地面基准点

Group 组,团体,集团

**GRR** Grand Rapids, MI, USA ( Airport Code ) 美国密执安州大急流机场

Ground Refuelling Relay 地面补给燃料继电器

**GRS** Generator Relay Slave 发电机从继电器

Global Reference System 全球参考系统

Ground Reference Station 地面基准站

Ground Reference System 地面基准系统

Gust Response Suppression 阵风反应抑制

Gyro Reference System 陀螺基准系统

**GRT** Gamma Ray Telescope 伽马射线望远镜

Garett Turbine Engine Company 盖瑞特涡轮发动机公司

Grant 准许,批准

Great 大的,伟大的

Gross Registered Tonnage 注册(容积)总吨

Ground-Received Time 地面接收时间

Guarantee 保证

**GRTD** Guaranteed 有保证的,担保的

**GRTS** General Real-Time System 通用实时系统

**GRU** Galley Refrigeration Unit 厨房制冷组件

Ground Refrigeration Unit 地面制冷组件,地面冷冻车

Gyro Reference Unit 陀螺基准组件

Sao Paulo, Brazil-Guarulhos ( Airport Code ) 巴西圣保罗瓜鲁柳斯机场

**GRV** Gateway Routing Vector 网关路由向量

Gaussian Random Variable 高斯随机变量

Groove 槽

**GRVD** Grooved 打槽的

**GRVL** Gravel 石,砂石

**GRVTY** Gravity 重力

**GRWT** Gross Weight 全重

**GRX** Granada, Spain ( Airport Code ) 西班牙格拉纳达机场

**GRY** Grey 灰色

**GRZ** Graz, Austria ( Airport Code ) 奥地利格拉茨机场

**GS** Glide Slope Receiver 下滑坡度接收机

Glide Slope 下滑坡度,下滑道,下滑台

Ground Services 地勤服务

Ground Speed 地速

Ground Stabilized 地面稳定的

Ground Supply 地面电源

Group Selector 群选择器

Guaranteed Service 担保服务

Guidance System 制导系统

**GSA** General Sales Agent 销售总代理

General Service Administration 综合服务管理局(联邦政府下属)

Ground Service Agreement 地面服务协议

Ground Spoiler Actuator 地面扰流板作动筒

General Sales Agents 总销售代理

**GSAPR** Ground Service Auxiliary Power Relay 地面勤务辅助电源继电器

**GSARS** Global Search and Rescue System 全球搜寻与救援系统

**GSB** Ground Service Bus 地面勤务汇流条

**GSC** Ground Station Control 地面站控制
Ground Switching Center 地面交换中心

**GSD** General System Description 通用系统描述
Ground Station Data 地面台数据

**GSDI** Global Spatial Data Infrastructure 全球空间数据基础设施

**GSE** Ground Service Equipment 地面服务设备
Ground Support Equipment 地面保障设备，地面支持设备

**GSEPR** Ground Service External Power Relay 地面勤务外接电源继电器

**GSERD** Ground Support Equipment Recommendation Data 地面保障设备推荐数据

**GSFC** Goddard Space Flight Center（NASA）高达德宇宙飞行中心（美国国家航空航天局）

**GSFSR** Ground Safety and Flight Safety Requirements 地面安全和飞行安全要求

**GSGP** Glide Slope Gain Program 下滑道增益程序

**GSHLD** Glareshield 遮光板

**GSI** Glide Slope Indicator 下滑道指示器
Glide Slope Intercept 下滑道截获
Government Safety Inspector 政府安全检查员
Government Source Inspection 政府资源调查
Grid Security Infrastructure 网格安全基础设施

**GSIS** Government Service Insurance System 政府服务机构保险系统

**GSIU** Ground Standard Interface Unit 地面标准接口设备

**GSKT** Gasket 垫圈

**GSL** General Support Laboratory 综合保障实验室
Global Service Logic 全局服务逻辑

**GSM** General Structure Manual 通用结构手册
General System Management 通用系统管理
Government Standard Manual 政府标准手册

**GSMP** General Switch Management Protocol 通用交换管理协议

**GSMT** Global System for Mobile Telecommunication 全球移动通信

**GSN** Gateway Service for Network 网关服务
Gigabyte System Network 千兆字节系统网络
Global Multi-Satellite Network 全球多卫星网
GPRS Support Node GPRS 支持节点
Group Selection Network 群选择网络
Group Switching Network 群交换网络

**GSO** Greensboro, WI, USA（Airport Code）美国威斯康星州格林斯伯勒机场
Ground System Operations 地面系统操作

**GSP** Glare Shield Panel 遮光板
Government Support Part 政府支援零件
Graphic System Processor 图形系统处理器
Greenville, SC, USA（Airport Code）美国南卡罗来纳州格林维尔机场
Ground Service Panel 地面勤务面板
Ground Service Personnel 地面服务人员
Ground Service Plug 地面（电源，气源等）接头
Ground Service Provider 地面维修提供者

**GSPU** GPS Sensor Processor Unit 全球定位系统敏感处理部件

**GSQ** Guaranteed Service Queue 保证业务队列

**GSQA** Government Source Quality Assurance 政府提供质量保证（美国）

**GSR** Gigabit Switch Router 千兆位交换路由器
Global Shared Resource 全球共享资源
Ground Service Relay 地面勤务继电器
Ground Surveillance Radar 地面监视雷达

**GSS** Geodetic Stationary Satellite 静止测地卫星
Global Security System 全球保安系统

Global Strategy Summit 全球战略峰会

Global Surveillance System 全球监视系统

Ground Station Simulator 地面站模拟器

Ground Support System 地面保障系统

Ground Surveillance System 地面监视系统

**GSSC** General System Simulation Center 一般系统模拟中心

Ground Support Simulation Computer 地面支援模拟计算机

**GSSR** Ground Service Select Relay 地面勤务选择继电器

**GSSV** Ground Spoiler Selector Valve 地面扰流板选择活门

**GST** Ground Select Test 地面选择测试

Guest 客人

**GSTN** General Switched Telephone Network 通用交换电话网

**GSTR** Ground Service Transfer Relay 地面勤务转换继电器

**GSTS** Ground-Based Surveillance and Tracking System 陆基监视与跟踪系统

Guidance System Test Equipment 制导系统测试设备

**GSU** Group Switch Unit 群交换单元

**GSV** Globe Stop Valve 球心节流活门

Guided Space Vehicle 制导空间航行器

**GSW** Gross Shipping Weight 总载货量

**GSZ** Ground Safety Zone 地面安全区

**GT** Gas Temperature 气体温度

Grid Track 网格航迹

Gross Tonnage 总吨

Ground Time（飞机）在地面时间

Ground Transit 地面过境

Ground Transmission 地面传输

Group Technology 成组技术

**GTA** General Terms Agreement 通用项目协议

**GTAT** Ground Turnaround Time 回程地面准备时间

**GTC** Gain Time Control 增益时间控制

Gas Turbine Compressor 燃气涡轮压气机

Generator Transfer Contactor 发电机转换接触器

Good Till Cancelled 未取消之前有效

Gyro Time Constant 陀螺时间常数

**GTCP** Gas Turbine Compressor Power 燃气涡轮压气机电源

Gas Turbine Compressor and Power Unit 燃气涡轮压气机与动力装置

Gas Turbine Compressor Pneumatic 燃气涡轮压气机空气

**GTE** Gas Turbine Engine 燃气涡轮发动机

Ground Test Equipment 地面测试设备

**GTEE** Guarantee 保证,担保

**GTF** Geared Turbo Fan 齿轮传动涡扇（发动机）

Great Falls, MT, USA（Airport Code）美国蒙大拿州大瀑布城机场

**GTHRG** Gathering 积累

**GTI** Ground Test Instructions 地面测试说明

**GTIN** Global Trade Identification Number 全球贸易识别号码

Global Trade Item Number 全球贸易项目代码

**GTM** Good This Month 当月有效

**GTN** Global Trunk Network 全球干线网

**GTO** Geostationary Transfer Orbit 同步卫星转移轨道

Gorontalo, Indonesia（Airport Code）印度尼西亚哥伦达洛机场

**GTOS** Global Terrestrial Observing System 全球陆地观测系统

Ground Terminal Operations Support 地面终端操作保障

**GTOW** Gross Takeoff Weight 起飞总重,起飞总质量

**GTP** General Test Plan 通用试验计划

Graphic Transform Package 图形变换程序包

Ground Test Plan 地面测试计划

Ground Test Power 地面测试电源

**GTPN** Global Trade Point Network 全球贸易点网络

**GTPU** Gas Turbine Power Unit 燃气涡轮动力装置

**GTR** Columbus, MS, USA（Airport Code）美国密西西比州哥伦布机场
General Technical Requirements 通用技术要求
Giant Transistor 巨型晶体管
Global Tooling Resources 全球工具资源
Greater 较大的
Ground Test Requirements 地面测试要求

**GTRE** Gas Turbine Research Establishment 燃气涡轮发动机研究组织

**GTRFM** Global Tooling Resource Forecasting Module 全球工具资源预测模块

**GTS** Gas-Turbine Starter 燃气涡轮启动机
General Traffic Shaping 通用业务整形
Global Telecommunication System 全球电信系统
Graphics Terminal System 图形终端系统
Ground Terminal System 地面终端系统

**GTT** Ground Test Time 地面试验时间

**GTTH** Gigabit to the Home 千兆位到户

**GTU** Gas Turbine Unit 燃气涡轮组件
General Terminal Unit 通用终端装置
Group Terminal Unit 群终端单元

**GTV** Ground Test Vehicle 地面试验飞行器

**GTW** Gate Way 联络道
Gateway Aircraft Leasing Co.（USA）通路租机公司（美国）
Gross Take-off Weight 起飞总重量, 起飞全重

**GTY** Gettysburg, PA, USA（Airport Code）美国宾夕法尼亚州葛底斯堡机场

**GU** Gear Up 收轮, 起落架收上

**GUA** Guatemala City, Guatemala（Airport Code）危地马拉危地马拉城机场

**GUAR** Guarantee 保证, 担保

**GUC** Group Unit Center 群单元中心
Gunnison, CO, USA（Airport Code）美国科罗拉多州甘尼森机场

**GUI** Graphical User Interface 图形用户界面

**GUID** Globally Unique Identifier 全局唯一标识符
Guidance 制导, 引导

**GUM** Guam, Guam（Airport Code）美国关岛机场

**GUP** Gallup, NM, USA（Airport Code）美国新墨西哥州盖洛普机场

**GUY** Guyana 圭亚那（拉丁美洲）

**GVA** Geneva, Switzerland（Airport Code）瑞士日内瓦机场
Gross Value Added 总增加值

**GVD** Group Velocity Dispersion 群速色散

**GVE** Give 给
Graphics Vector Engine 图形向量发动机

**GVI** General Visual Inspection 一般目视检查

**GVNS** Global Virtual Network Service 全球虚拟网业务

**GVPF** Geared Variable-Pitch Fan 齿轮传动变距风扇

**GVPN** Global Virtual Private Network 全球专用虚拟网

**GVS** Gas Vortex System 气涡系统
Ground Vibration Study 地面振动研究
Ground Vibration Survey 地面振动测量

**GVT** Gated Video Tracker 门控视频跟踪器
Government 政府
Ground Vibration Test 地面振动测试

**GVTA** Ground Vibration Test Article 地面振动测试件

**GVW** Gross Vehicle Weight 飞行器总重, 车辆总重

**GW** Gale Warning 大风警报
Gateway 网关
Gross Weight 总重, 全重量

**GW/CG** Gross Weight and Center of Gravity 总重和重心

**GWC** Global Weather Center 全球气象中心
Global Weather Control 全球气象控制
Great Wall of China 中国长城
Gross Weight Chart 全重表

**GWD** Guide Wave Demultiplexer 波导分路器

**GWDI** Global Weather Dynamics Inc. 全球气象动态公司

**GWDS** Graphic Weather Display System 天气图形显示系统

**GWDU** Galley Waste Disposal Unit 厨房废物处理组件

**GWE** Global Weather Experiment 全球气象试验

**GWIP** Global Weather Intercept Program 全球气候拦截程序
Government Wideband Internet Protocol 政府宽带互联网协议

**GWP** Gateway Processor 网关处理器

Global Warming Potential 全球变暖潜势
Gross World Product 世界生产总值

**GWR** General Working Rules 一般工作规则

**GWS** Graphic Weather Service 图形气象服务

**GWT** Gross Weight 全重

**GWTO** Gross Weight Takeoff 起飞全重

**GWVSS** Ground Wind Vortex Sensing System 地面风涡遥感系统

**GY** Gray, Grey 灰色, 灰色的

**GYE** Guayaquil, Ecuador（Airport Code）厄瓜多尔瓜亚基尔机场

**GYRO** Gyroscope 陀螺仪

**GYY** Gary, IN, USA（Airport Code）美国印第安那州加里机场

# H

**H** Altitude 高度

Hard Surface 硬道面

Hardware 硬件,硬设备

Hatch 舱口,舱门

Heading 航向

Heater 加温器

Height 高度

Helicopter 直升机

High Altitude 高空

High 高

Hold 等待

Homing Radio Beacon 归航无线电信标

Honey Comb 蜂窝结构

Hooded 模拟仪表飞行的

Horizontal 水平的

Hot 热

Hotel 旅馆

**h** Hour 小时

**H/B** Handbook 手册

**H/F** Half 一半

**H/L** High/Low 高/低(空)

**H/O** Hard Overs 过量,急偏

**H/STAB** Horizontal Stabilizer 水平安定面

**H/T** Hub/Tip Ratio 毂端半径比

**H/U** Hook Up 悬挂,试验线

**H/V** Height/Velocity 高度—速度

**H/W** Hardware 硬件

Head Wind 逆风,顶风

Herewith 与此一起

**HA** Height of Apogee 远地点高度

High Altitude 高空

Hour Angle 时角

Hybrid Access 混合接入

Hydraulic Actuator 液压作动器

**HAA** Hazardous Area Classification 危险区域分类

Height above Airport/Aerodrome 高于机场的高

Helicopter Association of America 美国直升机协会

Helicopter Association of Australia 澳大利亚直升机协会

High Angle of Attack 大迎角

History Aircraft Association 飞机发展史研究协会(英国)

Hungarian Aeronautical Association 匈牙利航空组织

**HAARS** Heading and All-Attitude Reference System 航向和全姿态基准系统

**HAAT** Height Above Average Terrain 高于平均地形高度

**HAATC** High Altitude Air Traffic Control 高空空中交通管制

**HAB** Hazards Analysis Board 险情分析委员会

Hot Air Balloon 热气球

House Air Bill 航空托运单

**HABD** Helicopter Aircrew Breathing Device 直升机乘员使用的应急呼吸器

**HAC** Hawaii Aeronautics Commission 夏威夷航空委员会

Heading Alignment Circle 航向校准圆

Heavy Attack Aircraft Commander 重型强击机机长

Heavy Cargo 重货物

Helicoper Aircraft Commander 直升机机长

Heliport and Airways Committee 直升机场与航空公司委员会

High Acceleration Cockpit 高加速度座舱,高过载座舱

High Altitude Calibration 高空校准

High Altitude Chart 高空图

Hover/Approach Coupler 悬停/自动进场着陆耦合器

Hughes Aircraft Company 休斯飞机公司

**HACCP** Hazard Analysis and Critical Control Point 危害分析及关键控制点

**HACI** Honda Aircraft Company, Inc. 本田飞机公司

**HACIS** Hong Kong Air Cargo Industry Services 香港空运服务有限公司

**HACMP** High Availability Cluster Multi-Processing 高可用性群集多处理

**HACR** Heating, Air Conditioning and Refrigeration 加热、空调和制冷

Helicopter Active Control Rotor 直升机主动控制旋翼

High Airplane Climb Rate 飞机高爬升率

**HACT** Helicopter Active Control Technology 直升机主动控制技术

**HACTL** Hong Kong Air Cargo Terminals Limited 香港空运货站有限公司

**HAD** Hardware Data Acquisition 硬件数据采集

Health Assessment Document 健康评估文件

Hybrid Analog/Digital 模/数混合

**HADC** Hybrid Air Data Computer 混合式大气数据计算机

**HADR** Hughes Air Defense Radar 休斯防空雷达

**HADS** Helicopter Air Data System 直升机大气数据系统

High-Accuracy Digital Sensor 高精度数字传感器

Hypersonic Air Data Sensor 高超音速大气数据传感器

**HADTS** High-Accuracy Data Transmission System 高精度数据传输系统

**HAE** High-Altitude Emergency 高空紧急情况

**HAECO** Hong Kong Aircraft Engineering Company Limited 香港飞机工程有限公司

**HAES** High Altitude Effects Simulation 高空效应模拟

**HAESL** Hong Kong Aero Engine Services Limited 香港航空发动机维修服务有限公司

**HAF** Heavy Aircraft Fuel 重质飞机燃料

Helicopter Assault Force 直升机突击部队

**HAFFA** Hongkong Association of Freight Forwarding and Logistics 香港货运物流业协会

**HAGB** Helicoper Association of Great Britain 英国直升机协会

**HAH** Hot and High 高温高原

Moroni, Comoros (Airport Code) 科摩罗莫罗尼机场

**HAHST** High Altitude High Speed Target 高空高速目标

**HAI** Helicopter Association International 国际直升机协会

Hellenic Aerospace Industry 希腊航空航天工业

**HAIL** Hail 雹

**HAINS** High Accuracy Inertial Navigation System 高精度惯性导航系统

**HAIR** High Accuracy Instrumentation Radar 高精度仪表雷达

High Altitude IR 高空红外线

**HAISC** Hughes Aircraft International Service Company 休斯飞机国际服务公司

**HAISS** High Altitude IR Sensor System 高空红外线感测系统

**HAJ** Hannover, Germany (Airport Code) 德国汉诺威机场

**HAL** Hardware Abstraction Layer 硬件抽象层

Height above Landing 着陆上空高度

Holding and Approach-to-Land 等待进近着陆

Horizontal Alert Limit 水平告警限

**HALE** High Altitude, Long Endurance 高空长时间续航

**HALO** High Altitude Low Opening 低空开伞
（高空跳伞）

High-Altitude Long Operation 高空长距离运行

**HALS** Helicopter Autonomous Landing System 直升机自主着陆系统

High Approach Landing System 高高度进近着陆系统

Hydrographic Airborne Laser Sounder 机载激光水文探测器

**HAM** Fuhlsbuttel, Hamburg, Germany（Airport Code）德国汉堡富尔斯布特机场

Hardware Associative Memory 硬件相关存储器

High-Altitude Mission 高空（飞行）任务

High-Speed Automatic Monitor 高速自动监控器

Hypertext Abstract Machine 超文本抽象机

**HAMC** Harbin Aircraft Manufacturing Company 哈尔滨飞机制造公司

**HAMD** Helicopter Ambulance Medical Detachment 直升机救护医疗队

**HAMMS** High Altitude Multi Mission Study 高空多任务研究

**HAMOTS** High Accuracy, Multiple Objects Tracking System 高精度多目标跟踪系统

**HAMS** High Altitude Mapping System 高海拔测绘系统

**HAMT** Human Aided Machine Translation 人力辅助机器翻译

**HAN** Hanoi, Vietnam（Airport Code）越南河内机场

Home Area Network 家用网络

**HANS** High Altitude Navigation System 高空导航系统

**HAO** High Altitude Observatory 高空观测站

**HAP** Hazardous Air Pollutant 危险的空气污染物

High Altitude Platform 高空平台

High Altitude Probe 高空探测（器）

**HAPI** Helicopter Approach Path Indicator 直升机进近航道指示器

**HAPPI** Height and Plan Position Indicator 高度和平面位置指示器

**HAPS** Helmet-Angle Position Sensor 头盔角度位置感测器

**HAR** Hazard Analysis Report 风险分析报告

High Altitude Redesign 高空空域重组

High Aspect Ratio 大展弦比

**HARA** High-Altitude Radar Altimeter 高空雷达测高仪

**HARC** High-Altitude Radar Controller 高空雷达管制员

**HARD** Helicopter and Airplane Radar Detection 直升机和飞机雷达探测

**HARDS** High Accuracy Radar Data System 高精度雷达数据系统

High-Altitude Radiation Detection System 高空辐射探测系统

**HARDTS** High Accuracy Radar Data Transmission System 高精度雷达数据传输系统

**HARN** Harness 线束，导线

High Accuracy Reference Network 高精度参考网络

**HARP** Helicopter Advanced Rotor Program 直升机先进旋翼计划

Helicopter Airworthiness Review Panel 直升机适航性审查组

High Altitude Relay Point 高高原中继点

High Altitude Research Probe 高空研究探测器

High Altitude Research Project 高空研究计划

Hughes Advanced Rotor Program 休斯先进旋翼研究计划

**HARS** Hazardous Area Reporting Service 危险区域报告服务

Heading-Altitude Reference System 航向姿态基准系统

High Altitude Route System 高空航路系统

**HARV** High Altitude Research Vehicle 高空研究飞行器

**HARWAS** Horizontal-Axis Rotating-Wing Aeronautical System 水平轴旋翼航空系统

**HAS** Heading and Attitude System 航向和飞行姿态系统

Height Above Site 站点之上高度

Helicopter Anti-Submarine 反潜直升机(英国用法)

Helicopter Avionics System 直升机航空电子系统

High Altitude Services 高纬度飞行服务

Hong Kong Airport Services Ltd 香港机场地勤服务有限公司

Horizontal Stabilizer Actuator 水平安定面作动器

Houston Airport System 休斯敦机场系统

Hover Augmentation System 悬停增稳系统

Hughes Aviation Services 休斯航空服务公司

Hydraulic Actuator System 液压作动系统,液压助力系统

Hydraulic Analysis System 液压分析系统

**HASA** Helicopter Association of Southern Africa 南非直升机协会

**HASCO** Helicopter and Airplane Service Corporation 直升机与飞机服务公司

**HASP** High Level Automatic Scheduling Program 高级自动调度程序

High-Altitude Sampling Plane 高空采样飞机

High-Altitude Sampling Program 高空取样程序

High-Altitude Space Platform 高空航天平台

High-Altitude Surveillance Platform 高空监视平台

**HASPA** High-Altitude Superpressured Power Aerostat 高空超压动力气球

**HAST** High-Altitude Supersonic Target 高空超音速目标

**HASTE** Helicopter Ambulance Service to Emergencies 直升机紧急救护服务

**HAT** Height above Terrain 真实高度,相对地形高度

Height above Terrain 地形以上高度

Height above Touchdown 着陆点上空高度

High Altitude Target 高空目标

High Altitude Temperature 高空温度

**HATCDS** High Altitude Terrain Contour Data Sensor 高空地形轮廓数据传感器

**HATOL** Horizontal Attitude Takeoff and Landing 水平姿态起飞和着陆

**HAV** Havana,Cuba(Airport Code) 古巴哈瓦那机场

**HAW** Homing All the Way 全航路归航

**HAWB** House Air Waybill 航空托运单,航空分运单

**HAZ** Hazardous 危险的

Heat-Affected Zone 热影响区

**HAZ(D)** Hazard 危险

**HAZOP** Hazard and Operability Studies 危害和可操作性研究

Hazards and Operability Analysis 危害和可操作性分析

**HB** Brinnell Hardness Test Value 布氏硬度测试值

Hand Book 手册

High Blower 增压泵

Home Banking 家庭银行

Homing Beacon 归航信标

Hot Balloon 热气球

**HBA** Hobart, Tasmania, Australia (Airport Code) 澳大利亚塔斯马尼亚岛霍巴特机场

**HBAR** Stress (Hectobar) 压力(百巴)

**HBI** Heat Budget Instrument 热量收支测量仪

**HBN** Hazard Beacon 危险信标

**HBPR** High Bypass Ratio 高涵道比,大涵道比

**HBR** Harbour 港口

Harvard Business Review 哈佛商业评论

High Bit Rate 高比特率

Human Behavioral Representation 人类行为表示

**HBS** Helicopter Blade Slope 直升机桨叶倾

斜角

Home Base Station 民用基站

**HBY** Here by 特此

**HC** Hand Control 手操纵

Health Certificate 健康证书

Held Covered 继续承保

Helicopter Council 直升机委员会

Hexachloroethane 六氯乙烷

High Capacity 高容量

Host Computer 主计算机

**HCA** Helicopter Club of America 美国直升机俱乐部

High-Cycle Aircraft 高使用率飞机

Historic Cost Accounting 过去的成本核算

Hot Comprssed Air 热压缩空气

**HCAA** Hellenic Civil Aviation Authority 希腊民航当局

**HCAP** Handicap 障碍(物)

**HCC** Helicopter Control Center 直升机管制中心

High Pressure Case Cooling 高压机匣冷却

**HCCM** High Performance CCS7 Module 高性能七号公共信道信令模块

**HCCS** High Capacity Communication System 大容量通信系统

**HCD** Hydraulic Component Design 液压元件设计

**HCE** Helicopter Control Element 直升机控制元件

Human Caused Error 人为失误

**HCF** Hardware Confidence Flight 硬件信任飞行

Height Correction Factor 高度修正因数

Hermetically Coated Fiber 密封涂覆光纤

High Cycle Fatigue 高循环疲劳,高周疲劳

**HCGB** Helicopter Club of Great Britain 英国直升机俱乐部

**HCI** Hardness Critical Item 硬度关键项目

Human Computer Interaction 人机交互

Human Computer Interface 人机界面,人机接口

**HCL** High Cloud and Cumulus 高云和积云

Horizontal Center Line 水平中心线

**HCM** Human Capital Management 人力资本管理

**HCMM** Heat Capacity Map Mission 热容量地图测绘飞行

Heat Capacity Mapping Mission 热容量制图任务

Human Capital Management Market 人力资本管理市场

**HCMR** Heat Capacity Mapping Radiometer 热容量测绘辐射计

**HCMTS** High-Capacity Mobile Telecommunications System 大容量移动通信系统

**HCN** Heterogeneous Computer Network 多机种计算机网络

Homogeneous Computer Network 同机种计算机网络

**HCP** Host Command Processor 主机命令处理机

**HCR** High Compression Ratio 高压缩比

Human Cognitive Reliability Model 人的认知可靠性模型

Hybrid Communication Routing 混合通信路由选择

**HCRST** Hardware Clipping, Rotation, Scaling and Translation 硬件制约、旋转、换算和转换

**HCS** Header Check Sequence 信头检查序列

Helicopter Computer System 直升机计算机系统

Hierarchical Control System 分级控制系统

High Capacity System 大容量系统

High-Carbon Steel 高碳钢

Hovering Control System 悬停控制系统

Humidification Control System 增湿控制系统

Hybrid Computation and Simulation 混合计算和模拟

Hydromechanical Control System 液压机械控制系统

**HCSDS** High Capacity Satellite Digital Service 大容量卫星数字业务

**HCSM** Human Capital Strategic Management 人力资本战略管理

**HCSOF** Hard Clad Silica Optical Fiber 硬包层石英光纤

**HCT** Home Communication Terminal 家庭通信终端

**HCTDS** High Capacity Terrestrial Digital Service 大容量地面数字业务

**HCU** Hand-Held Control Unit 手持控制装置
Helicopter Control Unit 直升机控制组件
High-Capacity Computer Unit 大容量计算机装置
Homing Comparator Unit 归航比较器组件
Hydraulic Control Unit 液压控制组件
Hydraulic Cycling Unit 液压循环组件

**HCV** Header Convertor 信头变换器
Hydraulic Control Valve 液压控制活门
Hypersonic Cruise Vehicle 高超音速巡航飞行器

**HD** Hand 手
Hard 硬
Head 端头,题目,头,顶
Height Difference 高度差
Horizontal Distance 水平距离

**HDA** Hard Disk Assembly 高密度组件
Heading and Drift Angle 航向及偏移角

**HDAS** Hybrid Data Acquisition System 混合数据采集系统

**HDATZ** High Density Air Traffic Zone 空中交通高密度区

**HDB** Heterogeneous Database 异构数据库
High Density Bipolar 高密度双极性码
Home Data Base 家庭数据库
Hypermedia Data Base 超媒体数据库

**HDBG** Hand Baggage 手提行李

**HDBK** Handbook 手册

**HDC** Hardware Design Center 硬件设计中心
Helicopter Direction Center 直升机指挥中心
High Density Control 高密度管制
High Speed Data Channel 高速数据通道

**HDCD** High Density Compact Disc 高密度光盘

**HDCU** Hydraulic Duplicated Control Unit 液压复合控制装置

**HDD** Hard Disk Drive 硬盘驱动器
Head-Down Display 俯视显示器,下视显示器
High Density Data 高密度数据

**HDDA** Hard Disk Drive Array 硬盘驱动矩阵

**HDDR** Head-Down Display Radar 下视雷达
High Density Digital Recorder 高密度数据记录器

**HDDS** High Density Data System 高密度数据系统

**HDDT** High Density Data Transmitters 高密度数据发送机

**HDF** High Dispersion Fiber 高色散光纤

**HDFPA** High-Density Focal Plane Array 密集焦面阵列

**HDG** Heading 航向,标题

**HDG HOLD** Heading Hold 航向保持

**HDG SEL** Heading Select 航向选择

**HDG/DA** Heading/Drift Angle 航向/偏流角

**HDG/S** Heading Selected 航向选定

**HDI** Horizontal Display Indicator 水平显示器

**HDIP** Hazardous Duty Incentive Pay 危险性值勤奖金

**HDL** Handle 手柄,处理,操作
Handling 处理,操作
Hardware Description Language 硬件描述语言
High Density Lipoprotein 高密度脂蛋白

**HDLC** High-Level Data Link Communications 高级数据链通信
High-Level Data Link Control 高级数据链路控制

**HDLG** Handling 处理,操作

**HDLMS** Hybrid Data Link Management System 混合数据链管理系统

**HDLS** High Density Landing System 高密度着陆系统

**HDMCD** High Density Multimedia Compact Disc 高密度多媒体光盘

**HDML** Handheld Device Mark-up Language 手持设备标记语言

**HDMR** High-Density Multitrack Recording 高密度多磁道记录

**HDN** Hayden,CO,USA（Airport Code）美国科罗拉多州海登机场

**HDOC** Hourly Direct Operating Cost 小时直接运营成本

**HDOP** Horizontal Dilution of Precision 水平精度扩散因子

**HDP** Hardware Development Processor 硬件开发处理器

Hardware Diagnosis Program 硬件诊断程序

Harpoon Data Processor 鱼叉数据处理器

Horizontal Data Processing 水平数据处理

**HDPC** High Density Positive Control 高密度绝对管制

**HDPCM** Hybrid Differential Pulse Code Modulation 混合差分脉冲编码调制

**HDQ** Headquarters 总部,指挥部

**HDR** Header 标题,顶盖

High Data Rate 高速数据传输率

High Density Rule 高密度条例

**HDRP** Hybrid Dynamic Reservation Protocol 混合动态预留协议

**HDS** Height Definition Systems 高度定义系统

Helicopter Delivery Service 直升机投送勤务,直升机空运勤务

High-Density Satellite 高密度卫星

High-Speed Digital Switch 高速数字开关

Horizontal Display System 水平显示系统

Horizontal Drive Shaft 水平驱动轴

Hybrid Dynamic System 混合动态系统

**HDSHK** Handshake 信号交接,握手

**HDSL** High-bit-rate Digital Subscriber Line 高比特率数字用户传输线

High-level Data Specification Language 高级数据说明语言

High-Speed Digital Subscriber Line 高速数字用户传输线

**HDSP** Hardships 苦难,难处

**HDST** Headset 头戴式耳机

**HDT** Host Digital Terminal 局用数字终端

**HDTA** High-Density Traffic Airport 高密度交通机场,高密度交通航空港

**HDTIC** Host Data Transmission Interface Controller 主数据传输接口控制器

**HDTM** Half-Duplex Transmission Module 半双工传输模块

**HDTP** Handheld Device Transfer Protocol 手持设备传送协议

**HDTV** High Definition Television 高清晰度电视

**HDU** Hard Disk Unit 硬盘装置

Helmet Display Unit 头盔显示器

Hose Drum Unit 软管绞盘装置

Hydraulic Drive Unit 液压驱动装置

**HDUE** High-Dynamic User Equipment 高动态用户设备

**HDVS** High Definition Video System 高清晰度视频系统

**HDVTR** High-Definition Video Tape Recorder 高清晰度磁带录像机

**HDW(A)** Hardware 硬件

**HDWC** Hardware C 金属布,金属织物

**HDWD** Hardwood 硬木

**HDWDM** High-Density Wavelength Division Multiplexing 高密度波分复用

**HDWHL** Handwheel 手轮

**HDX** Half Duplex 半双工

**HE** Handling Equipment 装卸设备,辅助设备

Head-End 前端

Heat Exchange 热交换

Heat Exchanger 热交换器

Heavy Enamel 厚漆(包线)

Helicopter Escort 直升机护航

Here 此处,这里

High Efficiency 高效率

High Energy 高能

Human Engineering 人机工程学

Hydraulic Equipment 液压设备

Hydrostatic Equilibrium 流体静力平衡

**HE BYP** Heat Exchanger Bypass 热交换器旁通

**HEA** Health Effects Assessment 健康状态评估

Heavy 重的

Higher Education Authority 高等教育局

Human Error Analysis 人为误差分析

**HEACO** Hong Kong Engineering Airplane Co. 香港工程飞机公司

**HEAF** Heavy End Aviation Fuel 重质航空燃油

High Efficiency Air Filter 高效空气过滤器

High-Energy Aircraft Fuel 高能飞机燃料

**HEAO** High-Energy Astronomical Observatory 高能观测天文台

**HEAR** Human Error Analysis Record 人为差错分析记录

**HEARS** Helicopter Acoustic Receiver System 直升机音响接收系统

**HEART** Health Evaluation and Risk Tabulation 健康评估与风险表

Human Error Assessment and Reduction Technique 人误评估与减少技术

**HEAT** Hazardous Energy Assessment Team 危险能源评估小组

Heating 加热

Helicopter Electro Actuation Technology 直升机电作动技术

**HEC** Header Error Check 信头差错校验

Header Error Control 信头差错控制

Header Error Correction 信头纠错

Heading Error Control 航向误差控制

Helicopter Element Coordinator 直升机小队协调员

Helicopter Employment Coordinator 直升机雇用协调员

Hybrid Error Correction 混合纠错

**HED** Head-End for Distribution Services 分配业务前端

**HEDA** Helicopter Enroute Descent Areas 直升机航线下降区

**HEDM** High Energy Density Materials/Matter 高能密度材料/物质

**HEDR** High-Energy Dynamic Radiography 高能射线照相术

**HEETE** Highly Efficient Embedded Turbine Engine 高效嵌入式涡轮发动机

**HEF** High-Energy Fuel 高能燃料

High-Expansion Foam 高膨胀泡沫(灭火剂)

**H-EGPWS** Helicopter-Enhanced Ground Proximity Warning Systems 直升机增强型近地告警系统

**HEGS** Helicopter External Gondola System 直升机外部吊舱系统

Hydraulic Electrical Generating System 液压发电系统

**HEH** Hydraulic External Hatch 液压外部舱

**HEI** Heli-Europe Industries 欧洲直升机工业公司

**HEIAS** Human Engineering Information and Analysis Service 人机工程学信息分析服务

**HEIST** Human Error Identification in Systems Technique 系统中的人为差错识别技术

Human Error Identification Software Tool 系统中的人为差错识别工具

**HEIU** High Energy Ignition Unit 高能点火装置

**HEL** Header Extension Length 信头扩展长度

Helicopter 直升机

Heliservices (Hong Kong) Limited 直升机服务(香港)有限公司

Helsinki, Finland (Airport Code) 芬兰赫尔辛基机场

High Energy Laser 高能激光器

Human Engineering Laboratory 人体工程学实验室

**HELCAR** Helicopter Collision Avoidance Radar 直升机防撞雷达

**HELCIS** Helicopter Command Instrument System 直升机控制仪表系统

**HELCM** HEL Countermeasure 高能激光反制

**HEL-H** Heavy Helicopter 重型直升机,远程直升机

**HEL-L** Light Helicopter 轻型直升机

**HEL-M** Medium Helicopter 中型直升机

**HELMS** Helicopter Multifunction System 直升机多功能系统

**HELO** Heavy Lift Operability 大升力可操作性

Heliport 直升机机场

**HELP** Helicopter Electronic Landing Path 直升机电子控制降落线路

Helicopter Emergency Lifesaver Plan 直升机应急救援计划

**HELRAS** Helicopter Long Range Active Sensor 直升机远距主动传感器

**HEMP** High Altitude Electromagnetic Pulse 高空电磁脉冲

High Altitude EMP 高空电磁脉冲

**HEMS** Helicopter Emergency Medical Services 直升机应急医疗服务

**HEMT** High Electron Mobility Transistor 高电子迁移率晶体管

**HEN** Heat Exchanger Network 热交换器网络

**HENCE** Heterogeneous Network Computing Environment 异构型网络计算环境

**HEO** High Earth Orbit 高地球轨道

High Elliptical Orbit 高椭圆率轨道

**HEP** Human Error Probability 人为差错概率

**HEPL** High-Energy Pulse Laser 高能脉冲激光

**HER** Education and Human Resources 教育与人力资源

Environmental Health Review 环境卫生审查

High Energy Rotor 高能旋翼

Higher Education and Research 高等教育与研究

**HERA** Human and Environmental Risk Assessment 人体健康和环境影响的风险

Human Error in ATM 空中交通管理中的人为因素

**HERC** Hazard Evaluation and Risk Control 危害评价和风险控制

Higher Education Research Centre 高等教育研究中心

**HERF** Hazard of Electromagnetic Radiation to Fuel 电磁辐射对燃料的危害

High Energy Radio Frequency 高能无线电频率

High Energy Radiation Field 高能辐射场

**HERJ** High Explosive Ramjet 高冲压式喷气发动机

**HERO** Hazards of Electromagnetic Radiation to Ordnance 电磁辐射对军火的危害

**HERP** Hazard of Electromagnetic Radiation to Personnel 电磁辐射对人员的危害

**HERS** High Energy Rotor System 高能旋翼系统

**HES** Health, Environment and Safety 健康、环境与安全

**HESS** Human Engineering System Simulation 人机工程系统模拟

**HEST** HEAF Emergency Service Tanks 重质航空燃油紧急服务油箱

**HET** Heavy Equipment Transport 重型装备运输

Helicopter Environmental Technique 直升机环境适应技术

Human Engineering Test 人机工程实验

Human Environmental Training 人机环境培训

**HETR** High Efficiency Tilt Rotor 高效倾转旋翼机

**HEU** Helicopter Experimental Unit 直升机试验装置

**HEX** Hexadecimal 十六进制的

Hexagon 六角

Hexagonal 六角形的

Heat Exchanger 热交换器

**HEXP** Heli Express Limited 空中快线有限公司(香港)

**HF** Height Finding 测高

Height to a Fix 到定位点的高度

High Frequency (3~30MHz) 高频

Home Freight 回程货运

Human Factor(s) 人为因素

**HF/DNS** High Frequency Data Network System 高频数据网络系统

**HFA** Haifa, Israel (Airport Code) 以色列海法机场

High Frequency Antenna 高频天线

Hydrogen Fueled Aircraft 氢燃料的飞机

**HFAA** High Frequency Airborne Antenna 高频机载天线

**HFACS** Human Factors Analysis and Classification System 人为因素分析与分类系统

**HFAJ** High-Frequency Anti-Jam 高频反干扰

**HFBC** High Frequency Broadcast Conference 高频广播会议

**HFC** High Frequency Communication 高频通信

High Frequency Computer 高频计算机

High Frequency Controller 高频控制器

High Frequency Coupler 高频耦合器

High Frequency Current 高频电流

Hybrid Fiber Coax 混合光纤同轴

Hydraulic Fuel Controller 液压燃油控制器

**HFCDF** High-Frequency Detecting and Finding 高频探测(器)

**HFCS** Helicopter Fire-Control System 直升机火控系统

Helicopter Flight Control System 直升机飞行控制系统

**HFCUR** High Frequency Current 高频电流

**HFD** Horizon Flight Director 指引地平仪

Human Factors Design 人为因素设计

Hydro Form Die 液压成型模

**HFDF** High Frequency Direction Finder 高频定向机

High Frequency Direction Finding 高频定向

**HFDL** High Frequency Data Link 高频数据链

**HFDR** High Frequency Data Radio 高频数据无线电通讯

**HFDU** High Frequency Data Unit 高频数据单元

**HFE** Heat Flow Experiment 热流实验

Heavy Fuel Engine 重油发动机

Human Factors Engineering 人为因素工程, 人为因素工程学

**HFEC** High Frequency Eddy Current 高频涡流(检查)

**HFES** Human Factors and Ergonomics Society 人为因素工程学会, 人为因素与人机工程协会

Hypersonic Flight Environmental Simulator 高超音速飞行环境模拟器

**HFF** High-Pressure Fuel Filter 高压燃油过滤器

Hypervelocity Flow Field 超高速流场

**HFFF** Hypervelocity Free Flight Facility 超高速自由飞行设备

**HFG** High Frequency Generator 高频发生器

**HFI** Height Finding Instrument 高度仪

Helicopter Foundation International 国际直升机基金会

**HFM** High-Fidelity Magic 高保真变幻

Human Factor in Maintenance 维修中的人为因素

**HFML** High Fiber Meal 高纤维餐食

**HFNPDU** High Frequency Network Protocol Data Unit 高频网络协议数据装置

**HFO** Heavy Fuel Oil 重质燃油

Heterodyne Frequency Oscillator 外差频率振荡器

High Frequency Oscillator 高频振荡器

Home Fuel Oil 民用燃油

**HFOM** Horizontal Figure of Merit 水平特征图, 水平品质因数

**HFP** Held for Part 适用于零件

**HFPA** Horizontal Flight Path Angle 水平航迹倾角

**HFR** Height Finding Radar 测高雷达

**HFRT** High Frequency Resonance Technique 高频谐振技术

**HFS** High Fidelity Simulator 高逼真度模拟器

High Frequency System 高频系统

Higher Frequency Subband 高频波段

Hot Fuel Storage 热油储存

Human Factors Society 人为因素协会

**HFSS** High Frequency Sounder System 高频探测器系统

**HFSSB** High Frequency Single Side Band 高频单边带

**HFT** Hands-Free Telephone 免提电话

Held for Tools 适用于工具

Historical Flight Tag 飞行历史记载标签

Horizontal Flight Testing 水平飞行试验

**HFW** Hybrid Fiber Wireless 光纤无线混合系统

**HFXR** High Frequency Transceiver 高频收发机

**HG** Heading Gyro 航向陀螺仪

Horizontal Gyro 水平陀螺仪

**Hg** Mercury 水银柱

**HGA** High Gain Antenna 高增益天线

High-Speed Graphics Assembly 高速图形组件

**HGAS** High-Gain Antenna System 高增益天线系统

**HGC** Head-up Guidance Computer 平视引导计算机

**HGD** High Gain Device 高增益装置

Hourglass Device 沙漏装置(一种计时装置)

**HGL** Helgoland, Germany (Airport Code) 德国黑尔戈兰机场

High Gain Link 高增益链路

**HGM** Hot Gas Manifold 热气总管

**HGR** Hagerstown, MD, USA (Airport Code) 美国马里兰州黑格斯敦机场

Hangar 机库

**HGS** Head-up Guidance System 平视引导系统

Holographic Guidance System 全息制导系统

**HGSW** Hom Gap Switch 霍姆间隙电门

**HGT** Height 高,高度

**HGU** Horizon Gyro Unit 陀螺地平仪

**HGY** Highway 高速公路

**HH** Half-Hard 半硬的

Hand Hold 扶手

Hand Hole 扶手孔

Header Hub 信头中心

Heavy Hydrogen 重氢

**HHA** Health Hazard Assessment 健康危害评估

**HHC** Hand-Held Computer 手持计算机

Higher Harmonic Control 高阶谐波控制

**HHCU** Hand-Held Control Unit 手持式控制组件

**HHDLU** Hand-Held Down Load Unit 手持式卸载组件,手持式下载组件

**HHE** Health Hazard Evaluation 健康风险评估

**HHEP** Health Hazard Evaluation Program 健康风险评估规划

**HHH** Hilton Head, SC, USA (Airport Code) 美国南卡来罗纳州希尔顿黑德机场

**HHL** Horizontal Hinge Line 水平铰轴线

**HHLD** Heading Hold 航向保持

**HHMU** Hand-Held Maneuvering Unit 手控机动飞行装置

**HHP** Hand-Held Phone 手持电话

High Hydrostatic Pressure 超高压

Hydraulic Horse Power 液压马力

**HHT** Hand-Held Terminal 手持终端

**HHX** Heavy-Lift Helicopter Experimental 实验性重型空运直升机

Hydrogen Heat Exchanger 氢热交换器

**HI** Hawaii 夏威夷(美国)

Health Index 健康指数

High 高
High Impact 高冲力
High Intensity 高强度
Horizontal Interval 水平间隔
Hovering Indicator 悬停指示器
Humidity Index 湿度指数
Hydraulic Institute 液压研究所

**HIAC** High Accuracy 高精确度
High Accuracy Radar 高精度雷达
High Altitude Camera 高空照相机

**HIAD** Handbook of Instructions for Airplane Designers 飞机设计师指导手册
High Altitude Defense 高空防御

**HIAGSED** Handbook of Instruction for Aircraft Ground Support Equipment Designers 飞机地面支援设备设计师手册

**HIAL** High Intensity Approach Lights 高强度进近灯光
Hyderabad International Airport Limited 海得拉巴国际机场有限公司

**HIALS** High Intensity Approach Lighting/ Light System 高强度进近灯光/系统

**HIAS** HK International Airport Services 香港国际机场服务有限公司

**HIB** Hibbing, MN, USA (Airport Code) 美国明尼苏达州西宾机场
High-Speed Integrated Bus 高速综合总线

**HIC** High Integrity Computer 高集成度计算机
Highest Incoming Channel 最高入局信道
Hybrid Integrated Circuit 混合集成电路

**HICAT** High Altitude Clear Air Turbulence 高空晴空湍流

**HIDAM** Hierarchical Indexed Direct Access Method 分层索引直接存取法

**HIDAN** High Density Air Navigation 高密度空中导航

**HIDD** Hardware Interface Definition Document 硬件接口定义文件

**HIDSS** Helmet Integrated Display and Sight System 头盔综合显示瞄准系统

Helmet Interface Display Sighting Sub-System 头盔接口显示瞄准子系统

**HIE** Helicopter Installed Equipment 直升机机载设备

**HIF** Higher Intermediate Fare 中间较高票价

**HIFEN** High Fidelity Environment 高保真度环境

**HIFI** High Fidelity 高保真

**HIFR** Helicopter In-Flight Refueling 直升机空中加油
Hover In-Flight Refueling 空中盘旋加油

**HIG** Hermetic Integrating Gyro 密封积分陀螺仪

**HIGE** Hover in Ground Effect 地面效应悬停

**HII** Lake Havasu City, AZ, USA (Airport Code) 美国亚利桑那州莱克哈瓦苏城机场

**HIJ** Hiroshima, Japan (Airport Code) 日本广岛机场

**HIL** Hardware in the Loop 环路中硬件
High Intensity Light 高强度灯光
High Intensity Lighting 高强度照明
Hold Item List 保持项目清单
Horizontal Integrity Limit 水平综合性极限

**HIM** Host Interface Module 主机接口模件

**HIMAT** Highly Maneuverable Aircraft Technology 高度可机动飞机技术

**HIN** Hybrid Integrated Network 混合综合网络
Sacheon, South Korea (Airport Code) 韩国泗川机场

**HIO** High Input/Output 高输入输出

**HIOC** Hourly Indirect Operating Cost 按小时计的间接运转费用

**HIP** Higher Intermediate Point 中间较高点
Hot Isostatic Pressure 各向热等压
Hot Isostatically Pressing 热等静压, 冲压

**HIPAR** High Power Acquisition Radar 高功率探测雷达

**HIPO** Hierarchy Input Processing Output 分层输入处理输出

**HIPOT** High Potential 高电位, 高电势, 高潜

力

**HIPOTT**  High Potential Test 高电位测试

**HIPPI**  High Performance Parallel Interface 高性能并行接口

**HIPRI**  High Priority 高优先权

**HIPRNS**  High-Performance Navigation System 高性能导航系统

**HIPS**  Helmet Initiated Pointing System 头盔瞄准系统

High Integrity Protection System 高度完整防护系统

Hybrid Image Processing System 混合图形处理系统

Hyperspectral Image Processing System 高光谱影像处理系统

**HIQ**  High-Quality 高品质因数,高质量因数

**HIQNet**  High Intelligence Quotient Network 高智商网络

**HIR**  Handbook of Inspection Regulations 检查规则手册

Helicopter Instrument Rules 直升机仪表飞行规则

Honiara, Solomon Islands ( Airport Code ) 所罗门群岛霍尼亚拉机场

**HIRAN**  High Intensity Radar Aids to Navigation 高密度飞行区导航雷达

High Precision Short Range Navigation 高精度短程导航

**HIRD**  High Information Rate Display for Aircraft Cockpits 航空器驾驶舱高信息率显示

High Information Rate Display 高信息率显示

High-Intensity Radiation Device 高强度辐射装置

**HIREL**  High Reliability 高可靠性

**HIRES**  High Resolution 高分辨率

Hypersonic In-Flight Refueling System 高超音速飞机空中加油系统

**HIRF**  High Intensity Radiated Field 高强度辐射场

High Intensity Radiated Frequency 高强度辐射频率

High Intensity Radio Frequency 高强度无线电频率

**HIRL**  High Intensity Runway Edge Lights 高强度跑道边灯

High Intensity Runway Lighting 高亮度跑道照明

High Intensity Runway Lights 高亮度跑道信号灯

**HIRLAM**  High-Resolution Limited Area Model 高分辨率有限区域模式

**HIRLS**  High-Intensity Runway Light System 高亮度跑道灯光系统

**HIRS**  Helicopter Infra-Red System 直升机红外系统

High-Resolution Infrared Sounder 高分辨率红外探测器

High-Resolution Infrared Radiation Sounder 高分辨率红外辐射探测器

Holographic Information Retrieval System 全息照相情报检索系统

**HIRT**  High Reynolds Number Tunnel 高雷诺数风洞

**HIRU**  Hughes Inertial Reference Unit 休斯惯性参考单元

**HIS**  Hardware Interrupt System 硬件中断系统

Hazard Information System 危害情报系统,危险信息系统

Home Information System 家庭信息系统

Hospital Information System 医院信息系统

**HISAM**  Hierarchical Indexed Sequential Access Method 分层索引顺序存取法

**HISL**  High-Intensity Strobe Light 高亮度频闪光灯

**HIT**  Highway Information Terminal 高速公路信息终端

Home Intelligent Terminal 家用智能终端

Home Interactive Terminal 家庭交互终端

Hybrid/Inertial Technology 混合/惯性技术

**HITS**  High Intensity Transient Signals 高强度

短暂信号

Highway in the Sky 空中高速公路

**HIV** Hydraulic Isolation Valve 液压隔离活门

**HIWA** Highway 高速公路

**HIWAS** Hazardous Inflight Weather Advisory Service 飞行中危险天气咨询服务

**HIWOL** High Intensity White Obstruction Lights 高强度白色障碍灯

**HK** Hong Kong 香港

**HKA** Hong Kong Airways 香港航空公司

**HKAA** Hong Kong Airport Authority 香港机场当局

**HKAC** Hong Kong Aviation Club 香港航空俱乐部

**HKAECO** Hong Kong Aircraft Engineering Company 香港飞机工程公司

**HKAIP** Hong Kong Aeronautical Information Publication 香港航空资料汇编

**HKALPA** Hong Kong Airline Pilot's Association 香港航空飞行员协会

**HKAOA** Hong Kong Aircrew Officers Association 香港空勤人员协会

**HKAR** Hong Kong Airworthiness Requirement 香港适航要求

**HKASP** Hong Kong Aviation Security Programme 香港航空安全计划

**HKATCA** Hong Kong Air Traffic Control Association 香港航空交通管制协会

**HKBAC** Hong Kong Business Aviation Centre 香港商用航运中心

**HKCPL** Hong Kong Commercial Pilot Licence 香港商用飞机驾驶执照

**HKD** Hakodate, Japan (Airport Code) 日本函馆机场

Hong Kong Dollar 港元 (货币单位)

**HKDCA** Hong Kong Department of Civil Aviation 香港民航局

**HKEA** Hong Kong Express Airways Limited 香港快运航空有限公司

**HKG** Hongkong, China 中国香港机场

**HKIA** Hong Kong International Airport 香港国际机场

**HKIEC** Hong Kong Industrial Estates Corporation 香港工业资产公司

**HKP** Height Keeping Performance 高度保持性能

**HKSAR** Hong Kong Special Administrative Region 香港特别行政区

**HKSCO** Hong Kong Schedule Coordination Office 香港计划协调办公室

**HKSQC** Hong Kong Society of Quality Control 香港质量管理学会

**HKT** Phuket, Thailand (Airport Code) 泰国普吉岛机场

**HKY** Hickory, NC, USA (Airport Code) 美国北卡来罗纳州希里机场

**HL** Have Listed 已列入名单

Heavily Loaded 重负荷

Heavy lift 超重

Height Loss 高度损失

High Level 高空, 高度层

Hole 孔

Hypertext Links 超文本链路

**HLA** Heavy Lift Aircraft 重型运输飞机

Heavy Lift Airship 重型运输飞艇

Helicopter Landing Area 直升机降落区

High Level Architecture 高层体系结构

High Level Assembly 高级别组装

**HLAC** Helicopter Landing Area Certificate 直升机着陆区域证书

High Level AC Voltage 高压交流电

High Level Advisory Committee 高级顾问委员会

**HLAN** High-speed Local Area Network 高速局域网络

**HLCL** Helical 螺线

**HLCM** High Level Committee on Management 管理问题高级别委员会, 高级别管理委员会

**HLCP** High Level Committee on Programmes 方案问题高级别委员会

**HLCPS** Helical Compression 螺旋压缩

**HLCS** Heat Limiter Control Switch 限热器控制开关

High Lift Control System 增升操纵系统

**HLD** High Level Design 高水平设计

High Lift Device 高升力装置

High-Level Dialogue 高层对话

Hold 保持

**HLDG** Holding 等待,保持

**HLDLC** High Level Data Link Control 高级数据链路控制

**HLDR** High Level Design Review 高级设计评审

Holder 保持架,夹具

**HLDY** Holiday 假日

**HLEG** High Level Experts Group 高级专家组

**HLF** Half 一半

High Layer Function 高层功能

**HLFC** Hybrid Laminar Flow Control 混合层流流动控制

**HLH** Heavy Lift Helicopter 重型起重直升机,重型运输直升机

**HLIPS** High Level Image Processing System 高级图像处理系统

**HLL** High Level Language 高级语言

**HLLO** High Level Language Overwriting 高级语言改写

**HLLV** Heavy-Lift Launch Vehicle 大载重发射载具

**HLM** Heterogeneous LAN Management 异机种局域网管理

High Level Meeting 高级别会议

**HLN** Helena,MT,USA(Airport Code)美国蒙大拿州赫勒纳机场

High Level Network 高级网络

**HLP** Help 帮助

High Level Protocol 高级协议

Jakarta,Indonesia(Airport Code)印度尼西亚雅加达机场

**HLPI** High Layer Protocol Identifier 高层协议标识符

High Layer Protocol Interworking 高层协议互通

**HLPR** Helper 助手

**HLR** Home Location Register 归属位置登记器

**HLS** Helicopter Landing Site 直升机着陆场地

Helicopter Landing System 直升机着陆系统

**HLSL** Hybrid LSL 混合大规模集成电路

**HLSTO** Hailstone 雹

**HLTY** Healthy 健康的

**HLWE** Helicopter Laser Warning Equipment 直升机激光告警设备

**HLZ** Helicopter Landing Zone 直升机着陆区

**HM** Hazardous Materials 危险品

Health Monitor 健康监控

Holographic Memory 全息照相存储器

Hybrid Mail 混合邮件

Hydromechanical 液压机械的

Hyper Media 超媒体

**HM&E** Hull, Mechanical and Electrical 机身、机械和电气

**HMA** Human-Machine Adaptation 人机适配

**HMC** Heading Marker Correction 航向指示器校准

Health-Monitoring Computer 健康监视计算机

Hydromechanical Control 液压机械控制

**HMCM** Hazardous Material Control Management 危险物质控制管理

**HMD** Helmet-Mounted Display 头盔显示器

Humid 潮湿的

**HMFL** Harmonically Mode-Locked Fiber Laser 谐波锁模光纤激光器

**HMG** Homing 归航,向台

Hydraulic Motor Generator 液压马达发电机

**HMGS** Hydraulic Motor Generator System 液压马达发电机系统

**HMI** Handbook of Maintenance Instructions 维修说明手册

Heavy Maintenance Inspection 大修检验

Human Machine Interface 人机界面,人机接口

**HMIC** Hybrid Microwave Integrated Circuit 混合微波集成电路

**HMM** Hardware Maintenance Manual 硬件维修手册

Hazardous Materials Management 危险品管理

Heavy Maintenance Manual 大修手册

Hidden Markov Model 隐马尔可夫模型

**HMMP** Hazardous Material Management Plan 危险材料管理计划

Hazardous Material Management Process 危险物品管理过程

Hazardous Material Management Program 危险品管理程序

**HMMWV** High-Mobility Multipurpose Wheeled Vehicle 高机动性多用途轮式车

**HMO** Hermosillo, Mexico ( Airport Code ) 墨西哥埃莫莫西约机场

**HMOM** Hyper Media Object Manager 超媒体对象管理器

**HMP** Host Monitoring Protocol 主机监控协议

**HMR** High-Speed Multimedia Ring 高速多媒体环

Homer 归航台

**HMS** Health-Monitoring System 良好状态监视系统

Heat Management System 热管理系统

Helmet-Mounted Sight 头盔瞄准具

High-Modulus Sheet 高模数板

Hyperbase Management System 超级数据库管理系统

**HMSO** Her Majesty's Stationery Office 英国皇家出版局

**HMT** Hand-Held Multimedia Terminal 手持式多媒体终端

Health Management Technology 健康管理技术

Helicopter Marine Training 直升机海上训练

Hyper Media Tool 超媒体工具

**HMU** Health Monitoring Unit 健康监控设备

Height Measuring Unit 高度测量组件

Height Monitoring Unit 高度监控组件

Helicopter Maintenance Unit 直升机维护组件

Hydromechanical Metering Unit 液压机械计量组件

Hydromechanical Unit 液压机械组件

**HMV** Heavy Maintenance Visit 重大维修检查

**HND** Hand 手

Haneda, Tokyo, Japan ( Airport Code ) 日本东京羽田机场

Hundred 一百

**HNDBK** Handbook 手册,笔记本

**HNDL** Handle 手柄

**HNDLG** Handling 处理,操作

**HNG** Hinge 铰链

**HNIL** High Noise-Immunity Logic 高抗噪声逻辑电路

**HNL** Honolulu, HI, USA ( Airport Code ) 美国夏威夷州檀香山(火奴鲁鲁)机场

**HNM** Helicopter Noise Model 直升机噪音模式

**HNMO** Host-Nation Management Office 东道国管理处

**HNN** Hybrid Neural Network 混合神经网络

**HNS** Haines, AK, USA ( Airport Code ) 美国阿拉斯加州海恩斯机场

Helicopter Navigation System 直升机导航系统

Host-Nation Support 东道国支援

Hybrid Navigation System 组合导航系统

Hyperbolic Navigation System 双曲导航系统

**HNTF** Hot Nozzle Test Facility 热涵道测试设备,热喷管测试设备

**HNVS** Hughes Night Vision System 休斯夜视系统

**HO** Hard Overs 急转弯

Head Office 总公司,总办事处,总办公室

**HOA**  Heads of Agreement 协议签订人

**HOB**  Homing on Offset Beacon 偏移信标台归航

**HOC**  Height Overlap Coverage 高度重叠范围

High Oil Consumption 高滑油消耗

Highest Outgoing Channel 最高输出信道

Horizon Camera 水平照相机

**HODC**  Hand Over Dedicated Channel 转移专用信道

**HOE**  Holographic Optical Element 全息光元件

Homing Overlay Experiment 归航重叠试验

**HOGE**  Hovering out of Ground Effect 无地效悬停

**HOGEN**  Hold off Generator 闭锁发电机，延迟发电机

**HOJ**  Homing on Jamming 干扰寻的

**HOL**  High-Order Language 高级语言

Holiday 假日

Holland 荷兰

**HOLD**  Holding 保持，(空中)等待

**HOM**  High Orbit Mission 高轨道任务

Homer, AK, USA（Airport Code）美国阿拉斯加州荷马机场

**HOMP**  Helicopter Operations Monitoring Program 直升机运行监控程序

**HOMS**  Hubble Opto-Mechanical Simulator 哈勃光学机械模拟器

Human-Oriented Manufacturing System 以人为本的生产体系

**HON**  Honorable 阁下

Huron, SD, USA（Airport Code）美国南达科他州休伦机场

**HOND**  Honduras 洪都拉斯

**HOP**  Hazardous Operating Procedure 危险操作程序

Helicopter Operations 直升机操作

Horse Power 马力

**HOPR**  High Overall Pressure Ratio 高总压比

**HOR**  Hold-off Reset 保持关断复位

Horizontal 水平的

**HORES**  Hotel Reservation 饭店订房

**HORIZ**  Horizontal, Horizon 水平的, 水平

**HORZ**  Horizon 地平面

**HOS**  High-Order System 高阶系统

Human Operation Simulator 操作员模拟器

**HOSP**  Hospital 医院

Hospital Aircraft 医疗飞机

Hospitalization 住院治疗

**HOST**  Hardened Optical Sensor Testbed 加固光学传感器试验台

Harmonically Optimized Stabilization Technique 调谐优化稳定技术

Hybrid Open Systems Technology 混合开放系统技术

**HOT**  High Oil Temperature 高滑油温度

Hot Oil Temperature 热燃油温度

Hot Springs, AR, USA（Airport Code）美国阿肯色州温泉城机场

**HOTAC**  Helicopter Optical Tracking and Control 直升机光跟踪与控制

**HOTAS**  Hands on Throttle and Stick 油门与驾驶杆手动操纵

**HOTCC**  Hands on Throttle, Collective and Cyclic（直升机）三杆操纵技术

**HOTOL**  Horizontal Take-off and Landing 水平起飞和着陆

**HOU**  Houston, TX, USA-Houston Hobby（Airport Code）美国得克萨斯州休斯敦霍比机场

**HOVI**  Handbook of Overhaul Instructions 检修指导手册

**HOW**  Hand-Over Word 转换字符

**HOWR**  However 然而

**HP**  Hewlett-Packard 惠普公司

High Performance 高性能

High Pressure 高压

Holding Pattern 等待航线

Holding Procedure 等待程序

Horizontal Plane 水平面

Horizontal Polarization 水平极化

Horse Power 马力

Hot Press 热压

**HPa** Hectopascal ( one hectopascal = one millibar) 百帕(1 百帕 = 1 毫巴)

**HPA** High Performance Aircraft 高性能飞机

High Power Amplifier 高功率放大器

High Pressure Air 高压空气

Human-Powered Aircraft 人力飞机

Hydraulic Power Assembly 液压动力组件

Hydraulic Project Approval 液压项目批准

**HPAA** High Performance Antenna Assembly 高性能天线装置

High-Pressure Air Accumulator 高压空气瓶,高压集气瓶

**HPAC** Hazard Prediction Analysis Code 危险预测分析代码

Hazard Prediction Assessment Capability 风险预测评估能力

High Performance Aircraft 高性能飞机

High-Pressure Air Compressor 高压空气压缩机

**HPB** Hinged Plotting Board 铰链型标图板

**HPBV** High Pressure Bleed Valve 高压引气活门

**HPBW** Half-Power Beam Width 半功率点波束宽度

**HPC** Handheld PC 手持式个人电脑

High Performance Computer 高性能计算机

High Performance Computing 高性能计算

High Pressure Compressor 高压压气机

High Pressure Controller 高压控制器

High-order Path Connection 高阶通道连接

Hydraulic Pump Controller 液压泵控制器

**HPCC** High Performance Computing and Communication 高性能计算和通信

High Performance Computing Center 高性能计算中心

High Performance Control Center 高性能控制中心

High Pressure Combustion Chamber 高压燃烧室

**HPCR** High Pressure Compressor Rotor 高压压气机转子

**HPCS** High Pressure Compressor Stator 高压压气机定子

**HPD** Hard Point Defense 硬点防御

**HPDL** Hydraulic Oil Pump Discharge Line 液压油泵泄放管路

**HPDR** High-Precision Dynamic-Range 高精度动态范围

**HPE** Heat Pipe Experiment 热管试验

**HPF** Hazardous Processing Facility 事故处理设备

High Pass Filter 高通滤波器

Highest Possible Frequency 最高可用频率

Horizontal Position Finder 水平位置探测仪

**HPFAC** High Pressure Fan Air Controller 高压风扇空气控制器

**HPFG** Hydraulic Pressure Function Generator 液压函数发生器

**HPFP** High Pressure Fuel Pump 高压燃油泵

**HPFS** High Performance File System 高性能文件系统

**HPFTP** High-Pressure Fuel Turbo Pump 高压燃油涡轮泵

**HPG** High Power Generator 大功率发电机

High Precision Gear 高精度齿轮

High-Pressure Gas 高压气体

**HPGC** High Pressure Ground Connection 高压地面接头

**HPH** High-Performance Helicopter 高性能直升机

**HPHR** Horse Power-Hour 马力—小时

**HPI** High Power Illuminator 大功率照明装置,大功率照射器

High Probability of Intercept 高截获率

Homing Position Indicator 自导引位置显示器

Hydraulic Pressure Indicator 液压指示器

**HPIP** High Pressure Intensifier Pump 高压增压泵

**HPIR** High-Power Illuminator Radar 高功率

照射雷达

**HPIS** High Performance Insulation System 高性能绝缘系统

**HPJ** High Power Jammer 大功率干扰器
High Pressure Jet 高压射流

**HPL** High Power Laser 大功率激光器
High-Level Programming Language 高级程序设计语言
Horizontal Protection Level 水平保护级
Human Performance Laboratory 人类绩效实验室

**HPLMN** Home Public Land Mobile Network 本地公用陆地移动网络

**HPLR** Hinge Pillar 铰(链)柱

**HPM** Hazard Process Management 风险过程管理
High Performance Motor 高性能发动机
High-Power Microwave 高功率微波
Hydraulic Pressure Module 液压增压组件

**HPN** High Pass Network 高通网络
Horsepower Nominal 标称马力, 额定马力
Westchester County, NY, USA (Airport Code) 美国纽约州韦斯特切斯特机场

**HPO** High Performance Option 高性能任选方案
High Performance Organization 高性能机构
High-Pressure Oxygenation 高压氧化
Hourly Post Flight 每小时邮政飞行

**HPOT** High-Pressure Oxidizer Turbopump 高压氧化剂涡轮泵

**HPOX** High-Pressure Oxygen 高压氧
High Pressure Oxygen Service 高压氧气服务

**HPPASS** High-Performance Packet-Switching System 高性能分组交换系统

**HPPI** High Performance Parallel Interface 高性能并行接口

**HPPL** Hybrid Passive Phonic Loop 混合无源声音环路

**HPR** Hardware Problem Report 硬件问题报告
High Peak Rate 高峰率

High Power Radar 大功率雷达
High Power Relay 高功率继电器
High-Performance Routing 高性能路由选择

**HPRF** High Power Radio Frequency 高功率射频
High Pulse Repetition Frequency 高脉冲重复频率
Hypersonic Propulsion Research Facility 高超音速推进研究设备

**HPRL** Human Performance Research Laboratory 人类行为特性研究试验室

**HPRP** High Price Rotatable Parts 高价周转件
High-Power Reporting Point (雷达) 大功率报知点

**HPRV** High Pressure Regulating Valve 高压调节活门
High Pressure Relief Valve 高压溢流活门

**HPS** High Performance Switch 高性能开关
High Pressure Steam 高压蒸气
Hybrid Power System 混合电源系统
Hybrid Propulsion System 混合推进系统
Hydraulic Power Supply 液压动力源
Hydraulic Power System 液压动力系统
Hydraulic Pressure Switches 液压开关

**HPSA** Hydraulic Package Servovalve Actuator 液压组件伺服阀作动器

**HPSC** Heading per Standard Compass 按标准罗盘的航向(航行)
Hydraulic Package Storage Container 液压组件存储容器

**HPSD** High Power Switching Device 大功率开关器件

**HPSG** High Pressure Starter Generator 高压启动发电机

**HPSN** High-Performance Scalable Networking 高性能可伸缩网络

**HPSOV** High Pressure Shutoff Valve 高压关断活门

**HPSP** Heat Pipe Sandwich Panel 有热管的夹层结构板

**HPSS** Helmet Position Sensing System 头盔位置感觉系统

High Performance Storage System 高性能存储系统

High Pressure Single Spool 高压单转子(发动机)

**HPT** High Point 高点

High Pressure Test 高压试验

High Pressure Turbine 高压涡轮,高压透平

High-Order Path Termination 高阶通道终端

**HPTACC** High Pressure Turbine Active Clearance Control 高压涡轮主动间隙控制

**HPTC** High Pressure Turbine Clearance 高压涡轮间隙

**HPTCC** High Pressure Turbine Case Cooling 高压涡轮机匣冷却

High Pressure Turbine Clearance Control 高压涡轮间隙控制

**HPTCV** High Pressure Turbine Cooling Valve 高压涡轮冷却活门

HP Turbine Clearance Valve 高压涡轮间隙活门

**HPTE** High Performance Turbine Engine(s) 高性能涡轮发动机

**HPTR** High Pressure Turbine Rotor 高压涡轮转子

**HPU** Hydraulic Power Unit 液压动力组件

Hydraulic Pumping Unit 液压泵

**HPV** High Pressure Valve 高压活门

Human Powered Vehicle 人力交通工具

**HPW** High Pressure Water 高压水流(机场除雪用)

**HPWS** High Pressure Water Separator 高压水分离器

**HPZ** Helicopter Protected Zone 直升机防护区

**HQ** Headquarters 总部,指挥部

High Quality 优质,高质量

**HQI** Hydraulic Quantity Indicator 液压油量指示器

**HQMS** Harrington Quality Management Sys-tem 哈林顿质量管理系统

Hydraulic Quantity Monitoring System 液压油量监测系统

**HQR** Handling Qualities Ratings 操纵品质额定值

**HR** Half Rate 半速率

Handling Room 操纵室

Hazard Report 险情报告

Hear 听到

Heavy Route 繁忙航线

Height Radio 无线电高度

Height Range 高度范围

Helicopter Route 直升机航路

High Reverse 高反推

Hook Rail 导轨钩

Hose Rack 软管支架

Hot Rolled 热轧的

Human Resources 人力资源

**HRA** Hazard and Risk Assessment 风险评估

Health Risk Assessment 健康风险评估

Human Reliability Analysis 人为因素可靠性分析

Human Reliability Assessment 人为因素可靠性评估

Human Resources Association 人力资源协会

Hypersonic Research Airplane 高超音速研究飞机

**HRAC** Human Resources Advisory Committee 人力资源咨询委员会

**HRB** Hazardous Review Board 危险性评审委员会

Human Resources Branch 人力资源处

**HRC** Historical Records Container 历史记录箱,档案柜

Human Resources Committee 人力资源委员会

Hybrid Ring Control 混合环路控制

Hypothetical Reference Circuit 假想参考电路

**HRD** Hardware Requirements Document 硬件要求文件

Human Resources Development 人力资源开发

**HRDP** Hypothetical Reference Digital Path 假想参考数字通道

**HRDW** Hardware 硬件

**HRDWG** Human Resources Development Working Group 人力资源开发工作组

**HRE** Harare, Zimbabwe (Airport Code) 津巴布韦哈拉雷机场

Health Risk Evaluation 健康风险评估

Hypersonic Ramjet Engine 高超音速冲压式喷气发动机

**HRF** Height-Range Finder 测高测距仪

**HRG** Hemispherical Resonant Gyro 半球谐振陀螺

High Resolution Graphics 高分辨率图

High Risk Group 高风险组

Hurghada, Egypt (Airport Code) 埃及赫尔格达机场

**HRGM** High Resolution Ground Map 高分辨率地图, 高清晰度地图

**HRI** Hazard Risk Index 灾害风险指数

Height-Range Indicator 高度—距离指示器

High Resolution Instrument 高分辨率的仪器

Human Readable Interpretation 人可读解释

**HRIR** High-Resolution Infrared Radiometer 高分辨率红外探测仪

**HRIS** Human Resource Information System 人力资源信息系统

**HRK** Kharkov, Ukraine (Airport Code) 乌克兰哈尔科夫机场

**HRL** Hardware Requirements List 硬件要求清单

Harlingen, TX, USA (Airport Code) 美国得克萨斯州哈灵根机场

Horizontal Reference Line 水平基准线

**HRM** High Rate Multiplexer 高速多路传输器

Human Resource Management 人力资源管理

**HRMFB** High Reliability, Maintenance-Free Battery 高可靠性免维护电池

**HRMS** Human Resource Management Systems 人力资源管理体系

**HRN** Human Resource Network 人力资源网络

**HRO** Harrison, AR, USA (Airport Code) 美国阿肯色州哈里森机场

**HRP** High Resolution Plate 高分辨率板

Horizontal Reference Plane 水平基准面

**HRPD** High Rate Packet Data 高速率分组数据

High Resolution Pulse Doppler Radar 高分辨率脉冲多普勒雷达

**HRPI** High Resolution Pointable Imager 可控高分辨率成像器

High Resolution Pointing Imager 高分辨率定向成像仪

**HRPT** High Resolution Picture Transmission 高分辨率图像传送

**HRR** High Resolution Radar 高分辨率雷达

**HRS** Hot Rolled Steel 热轧钢

Hours 小时数

**HRSCMR** High-Resolution Surface Composition Mapping Radiometer 高解析度地面组成照相辐射计

**HRT** Heat Rejection and Transport 热损耗和热传输

High Resolution Tracker 高分辨率跟踪器

**HRTEM** High Resolution Transmission Electron Microscopy 高分辨率透射电镜

**HRTZ** High Intensity Radio Transmission Zone 高强度无线电发射区

**HRV** High-Pressure Relief Valve 高压溢流活门

Hypersonic Research Vehicle 高超音速研究飞行器

**HRVBS** High Resolution Video Bus System 高分辨率视频总线系统

**HS** Hard Standing 停机坪

Hardness 硬度

Heat Shield 隔热罩, 隔热屏

High Speed 高速

**HSAC** Health and Safety Advisory Committee 健康和安全咨询委员会

Health Service Advisory Committee 健康服务咨询委员

Helicopter Safety Advisory Conference 直升机安全咨询会议

High-Speed Analog Computer 高速模拟计算机

**HSAS** Hardened Stability Augmentation System 强化增稳系统

**HSB** High-Speed Buffer 高速缓冲器

High-Speed Bus 高速总线

**HSBL** Horizontal Stablilizer Buttock Line 水平安定面纵剖线

**HSC** High Safety Character 高安全特性

High Speed Cruise 大速度巡航

Home Shopping Channel 家庭购物信道

Hydraulic System Controllers 液压系统控制器

**HSCP** Hydraulic System Control Panel 液压系统控制板

**HSCSD** High Speed Circuit-Switched Data 高速电路交换数据

**HSCT** High Speed Card-Teletype 高速卡片电传打字机

High Speed Civil Transport 高速民用运输机

High Speed Commercial Transport 高速商业运输机

High Speed Compound Terminal 高速混合终端,高速复式终端

Hughes Satellite Communications Terminal 休斯卫星通信终端(站)

**HSD** Hard Surface, Dry 有硬道面的干燥跑道

High Speed Data 高速数据

Horizontal Situation Display 水平位置显示器,水平情况显示器

**HSDARS** High Speed Data Acquisition and Reduction System 高速数据采集和简化系统

**HSDB** High Speed Data Buffer 高速数据缓冲器

High Speed Data Bus 高速数据总线

**HSDC** High Speed Data Card 高速数据卡

High Speed Data Communications 高速数据通信

High Speed Digital Chart 高速数字海图

**HSDL** High Speed Data Link 高速数据链路

**HSDP** High Speed Data Processor 高速数据处理机

**HSDS** High-Speed Data System 高速数据系统

**HSDU** High Speed Data Unit 高速数据装置

**HSE** High-Speed Encoder 高速编码器

**HSEEP** Homeland Security Exercise and Evaluation Program 国土安全演习和评估项目

**HSF** High-Speed Flight 高速飞行

**HSFR** Hydraulic System Frequency Response 液压系统频率响应

**HSFS** High Speed Flight Station 高速飞行站

**HSG** High Speed Gearbox 高速齿轮箱

Housing 外罩,壳体

**HSI** Handbook of Service Instructions 勤务规程手册

Hastings, NE, USA (Airport Code) 美国内布拉斯加州赫斯廷斯机场

Horizontal Situation Indicator 水平姿态指示器,水平位置指示器

Hot Section Inspection 热部件检查,热区检查

Human System Integration 人力系统综合

**HSIF** Hardware/Software Integration Facility 硬件和软件合成设施

**HSK** Chinese Proficiency Test 汉语水平考试

Heading Select Knobs 航向选择按钮

**HSL** Hazard Significant Level 风险显著水平

Hazardous Substance List 危险物品清单

Heading Select 航向选择

High Speed Launch 高速发射

**HSLV** High-Speed Low-Voltage 高速低电压（集成电路）

**HSM**    Hardware Security Module 硬件安全模块

Health and Status Monitor 健康和状态监控

Hierarchical Storage Management 分级存储管理

Human Systems Management 人力系统管理

**HSMD**    High Speed Multiplexer/Demultiplexer 高速多路调制器/分离器

**HSMS**    Hazardous Substance Management System 危险物质管理系统

Health and Safety Monitoring Systems 健康与安全监控系统

High-Speed Measurement Station 高速测量站

High-Speed Microwave Switch 高速微波开关,高速微波转换

**HSMU**    Hydraulic Systems Monitoring Unit 液压系统监控装置

**HSN**    Hierarchically Synchronized Network 分级同步网络

High Speed Network 高速网络

**HSOTN**    High-Speed Optical Transfer Network 高速光传送网络

**HSP**    Host Service Provider 主机服务提供商

**HSPCM**    High Speed Pulse Code Modulation 高速脉冲码调制

**HSPD**    High-Speed Packet Data 高速分组数据

**HSPFN**    High-Speed Plastic Fiber Network 高速塑料光纤网

**HSR**    Helicopter Supported Radar 直升机支援雷达

High Sink Rate 高下降速率

High Speed Relay 高速继电器

High Speed Research 高速研究

High Speed Rotor 高速旋翼

High Speed Radar 高速雷达

High Speed Reader 高速阅读器

Horizontal Scan Radius 水平扫描半径

**HSRA**    High Speed Research Aircraft 高速研究飞机

Hollow Shaft Rotary Actuator 空心轴旋转式作动器

**HSS**    High Speed Steel 高速钢

High Speed Switch 高速交换机

Hub-Spoke-System 枢纽辐射式航线系统

Hybrid Simulation System 混合模拟系统

**HSSV**    High-Speed Servo Valve 高速伺服活门

**HST**    Helicopter Support Team 直升机支援分队

High Speed Taxiway 高速滑行道

High Speed Train 高速火车

High Speed Transmitter 高速发送器,高速发射机

Hoist 起重机,吊车

Horizontal Stabilizer Tank 水平安定面油箱（液压）

Hubble Space Telescope 哈勃空间望远镜

Hypersonic Transport 高超音速运输机

**HSTA**    Horizontal Stabilizer Trim Actuator 水平安定面配平作动筒

**HSTCS**    Horizontal Stabilizer Trim Control System 水平安定面配平控制系统

**HSTCU**    Horizontal Stabilizer Trim Control Unit 水平安定面配平控制组件

**HSTL**    Hospital 医院

Hostel 招待所

**HSTP**    High Signaling Transport Point 高级信令转接点

**HSTR**    High Speed Token Ring 高速令牌环网

**HSTS**    Horizontal Stabilizer Trim Setting 水平尾翼配平调整

**HSTV-L**    High Survivability Test Vehicle-Lightweight 轻型高存活率实验车

**HSV**    Huntsville,AL,USA（Airport Code）美国阿拉巴马州汉斯维尔机场

Hydraulic Selector Valve 液压选择器活门（阀）

**HSVD**    Horizontal Situation Video Display 水平位置视频显示器,水平情况视频显示器

**HSWR** Hardware 硬件

**HSWT** High Speed Wind Tunnel 高速风洞

**HT** Hand-Held Terminal 手持终端

Hard Time 定时

Head of Training 训练主管

Heat 热,加热

Heat Treat 热处理

Height 高,高度

High Tension 高压电,强压力

High Turbine 高压涡轮

Home Trade 国内贸易

Horizontal Tail 水平尾翼

Hyper Text 超文本

**HTA** Heavier Than Air 重于空气的(飞行器)

Hierarchical Task Analysis 层次任务分析

**HTAS** High True Air Speed 真实高空速

**HTAWS** Helicopter Terrain Awareness and Warning System 直升机地形提示与警告系统

**HTBP** Hyper Text Broadcasting Protocol 超文本广播协议

**HTBSL** Hypermedia Time Base Structuring Language 超媒体时基结构语言

**HTC** Half Time to Count 时间计半

Hybrid Technology Computer 混合技术计算机

**HTCU** Hover Trim Control Unit 悬停配平控制机构

**HTDE** High Technology Demonstrator Engine 高技术演示验证发动机

**HTDM** High Time Depot Maintenance 适时机库维修

**HTG** Helicopter Training Group 直升机培训集团

**HTH** Heavy Transport Helicopter 重型运输直升机

**HTL** Hotel Accommodation 旅馆住宿,饭店住宿

**HTML** Hyper Text Mark-up Language 超文本标记语言

**HTO** Horizontal Takeoff 水平起飞

**HTOL** Horizontal Takeoff and Landing 水平起飞和降落

**HTOVL** Horizontal Takeoff, Vertical Landing 水平起飞,垂直降落

**HTP** Horizontal Tail Plane 水平尾翼,水平安定面

**HTPC** Home Theater Personal Computer 家庭影院个人电脑

**HTR** Hardware Trouble Report 硬件问题的报告

Heater 加热器

High Technology Rotor 高技术转子

High Temperature Reactor 高温反应堆

**HTS** Height Telling Surveillance 高度报告监视

High Temperature Superconducting 高温超导

High Temperature Superconductor 高温超导体

**HTSG** Helicopter Transportation System Group 直升机运输系统组

**HTTP** Hypertext Transfer Protocol 超级文本传输协议,超文本传送协议

**HTU** Handheld Terminal Unit 手持式终端设备

Heat Transfer Unit 传热装置

Helicopter Training Unit 直升机训练装置

Home Terminal Unit 家庭终端设备

**HTV** Hypersonic Technology Vehicle 高超音速技术飞行器

Hypersonic Test Vehicle 高超音速试验飞行器

**HTX** Heat Exchanger 热交换器

**HTZ** Helicopter Traffic Zone 直升机交通地带

**HUCP** Highest Useful Compression Pressure 最高有效压缩力

**HUCR** Highest Useful Compression Ratio 最高有效压缩比

**HUD** Head Up Display 平视显示

**HUDC** Head Up Display Computer 平视显示

计算机

**HUDLS** Head Up Display Landing System 平视显示着陆系统

**HUET** Helicopter Underwater Egress Training 直升机水下逃生训练

Helicopter Underwater Escape Training 直升机水下逃逸训练

**HUGS** Head-Up Guidance System 平视仪制导系统

**HUM** Health and Usage Monitoring 完好性与使用监测

Humidity 湿度

**HUMINT** Human Intelligence 人力情报

**HUMS** Health and Usage Monitoring System 状态与使用状况监控系统

Health and Usage Management System 健康与使用状况管理系统

**HUN** Hungary 匈牙利

**HUND** Hundred 百

**HURCN** Hurricane 飓风

**HUT** Head-Up Tilt 头高位倾斜

**HUY** Humberside, England, United Kingdom（Airport Code）英国亨伯赛德机场

**HV** Hardware Virtualizer 硬件虚拟器

Heading Verification 航向校对

Heavy 重的

High Vacuum 高真空

High Velocity 高速度

High Voltage 高电压

Horizontal Visibility 水平能见度

**HVA** Horizontal Viewing Arc 平视弧线

**HVAC** Heating, Ventilating and Air Conditioning 加热、通风和空调

High Vacuum 高真空

High Velocity Air Conditioning 高速空调

High Voltage Alternating Current 高压交流电

High Volume Air Conditioning 大容量空调

**HVAR** High Velocity Aircraft Rocket 高速飞机火箭

**HVC** High Velocity Clouds 高速云团

High Voltage Capacitor 高压电容器

High Voltage Control 高压控制

**HVCRE** High-Volatility Commercial Real Estate 高波动性商业房地产

**HVDC** High Voltage Direct Current 高压直流电源

**HVDF** High and Very High Frequency Direction-Finding Station 高频与甚高频定向台

**HVEM** High-Voltage Electron Microscopy/Microscope 高压电子显微镜

**HVG** Having 财产，财物

High Voltage Generator 高压发电机

**HVI** High Velocity Impact 高速碰撞

High Viscosity Index 高粘度指数

High Visibility Image 高可见度图像

**HVIC** High Voltage Integrated Circuit 高压集成电路

**HVM** Hardware Virtual Machine 硬件虚拟机

High Velocity Maintenance 高速维修

**HVN** New Haven, CT, USA（Airport Code）美国康涅狄格州纽黑文机场

**HVOF** High Velocity Oxygen Fuel 超音速火焰喷涂

**HVP** High Vision Projector 高清晰度投影仪

**HVPS** High Voltage Power Source 高压电源

High Voltage Power Supply 高电压供电

**HVR** Havre, MT, USA（Airport Code）美国蒙大拿州阿夫雷机场

Helicopter Visual Rules 直升机目视飞行规则

Hover 悬停，盘旋

**HVRS** Heading and Vertical Reference System 航向和垂直基准系统

**HVS** Human Vision System 人类视觉系统

**HVSS** Horizontal Volute Spring Suspension 水平螺旋弹簧悬挂系统

**HVTS** High-Voltage Test System 高压测试系统

**HVY** Heavily 沉重地

Heavy 重的

**HWBPT** Hardware Breakpoint 硬件断点

**HWCI** Hardware Configuration Item 硬件布局项目

**HWD** Heavy Weather Damage 因天气造成的严重损失

**HWRP** Hydrology and Water Resources Programme 水文和水资源计划

**HWS** Horizontal Wind Shear 水平风切变
Hyper Works System 超级作业系统

**HWT** Hypersonic Wind Tunnel 超高音速风洞

**HWVR** However 然而

**HWY** Highway 高速公路

**HX** Holding to Fix 定位点等待

**HYA** Hyannis, MA, USA（Airport Code）美国马萨诸塞州海恩尼斯机场

**HYBMED** Hybrid Microelectronic Device 混合微电子器件

**HYD** Hyderabad, India（Airport Code）印度海得拉巴机场
Hydraulic 液压的, 液力的

Hydrogen 氢

**HYDAT** Hydrodynamics Analysis Techniques 液体动力学分析技术

**HYDIM** Hydraulic Interface Module 液压接口组件

**HYIP** High Yield Investment Program 高收益投资项目

**HYQUIM** Hydraulic Quantity Indicating Module 液压油量指示组件
Hydraulic Quantity Interface Module 液压接口模块

**HYRAT** Hydraulic Ram Air Turbine 冲压空气涡轮驱动的液压泵

**HYSTU** Hydrofoil Special Trials Unit 水翼专用试验组件

**Hz** Hertz（cycles per second）赫兹（周/秒）

**HZD** Hazard 危险

**HZRN** Horizontal Reaction 水平反应, 水平反冲

**HZS** Horizontal Surface 水平面

# I

I　Immigration 移入
　Index 指数
　Inertial 惯性的
　Instrument 仪器,仪表
　Interest 利息,兴趣
　Intermediate 中间物,中间体
　Invoice Card 收据卡,发票卡
　Island 岛屿,岛
I&C　Identification and Control 识别与控制
　Installation and Calibration 安装和校准
　Installation and Checkout 安装和(工作完成)检查
　Installation and Control 安装和控制
I&CO　Integration and Checkout 集成和(工作完成)检查
I&M　Installation and Maintenance 安装和维修
I&S　Interchangeability and Substitutability 互换性和可替换性
I. E. R. A　Instrument, Electrical, Radio and Avionics 仪表、电气、无线电和电子
I/B　Inboart 内侧,机内
I/C　Inspection/Check 检验/检查
　Intercommunication 内部通讯,双向通信
　Interphone Communication 内话通讯
I/E　Import and Export 进出口
I/F　Image to Frame Ratio 像帧比
　Instructors Facility 教员便利设备
　Instrument Flight 仪表飞行
　Interface 接口
　Interfacilities 设施间的
I/I　Inspection/Interval 检查/间隔
I/L　Import Licence 进口执照
I/LDG　Instrument Landing 仪表着陆
I/N　Inventory Number 清单,号码

I/O　In/out 进/出
　Instead of 代替
I/P　Input 输入
　Intercept Point 切入点,交叉点
　Intercept Profile 切入剖面
I/R　Instrument Restricted 限于仪表飞行
　Interchangeability/Replaceability 互换性/替换性
I/S　In-Sequence 按顺序
I/V　Instrument/Visual 仪表/目视
IA　In Absentia 缺席
　Inactive Aircraft 非现役飞机,退役飞机
　Indicated Altitude 指示高度
　Indirect Address 间接地址
　Inertial Altitude 惯性高度
　Inertial Autopilot 惯性自动驾驶仪
　Information Access 信息存取
　Information Adaption 信息适配
　Initial Appearance 初现,问世
　Initial Approach 开始进近
　Input Axis 输入轴
　Inspection Authorization 检查核准权
　Installation Approval 装机批准书
　Instruction Address 指令地址
　Intelligence Appliance 智能家电
　Intelligent Agent 智能代理
　Interface Adapter 接口适配器
　Interim Agreement 临时协议
　Intermediate Altitude 中间高度
　Internal Authentication 内部认证
　International Airport 国际机场
　Internet Address 因特网地址
　Interstage Assembly 级间组件
　Inventory Accounting 库存统计
IAA　Initial Aircraft Avionics 飞机初始电子

设备

Initial Approach Altitude 起始进近高度

International Academy of Astronautics 国际宇宙航行学会,国际宇航科学院

International Advertising Association 国际广告协会

International Aerospace Abstracts 国际航空航天文摘

International Aviation Affairs 国际航空事务

Irish Aviation Authority 爱尔兰航空管理局

**IAAA** Integrated Advanced Avionics for Aerospace 集成高级航天电子技术设备

Integrated Advanced Avionics for Aircraft 飞机用高级集成航空电子设备

International Airforwarder and Agents Association 国际航空运输行和代理商协会

International Airfreight Agents Association 国际航空货运代理商协会

**IAAAC** International Airline Advanced Aircraft Committee 国际航空公司先进飞机委员会

**IAAC** International Agricultural Aviation Centre 国际农业航空中心

International Association of Aircargo Consolidators 国际航空运输行协会

**IAAE** Institution of Automotive and Aeronautical Engineers 汽车与航空工程师学会(美国)

International Association of Airport Executives 国际机场管理者协会

**IAAEES** International Association for Advancement of Earth and Environmental Sciences 国际地球与环境科学促进会

**IAAEM** International Association of Aircraft Equipment Manufacturers 国际飞机设备制造商协会

**IAAI** Indonesian Aeronautical and Astronautical Institute 印度尼西亚航空航天研究会

International Airport Authority of India 印度国际机场管理局

**IAAP** Identification of Aircraft Aerodynamic

Parameter 飞机气动力参数识别

**IAARC** International Administrative Aeronautical Radio Conference 国际航空无线电管理会议

**IAASM** International Academy of Aviation and Space Medicine 国际航空航天医学院

**IAAWG** International Airworthiness Assurance Working Group 国际适航保证工作小组

**IAB** Internet Activities Board 因特网活动委员会

Internet Advisory Board 因特网咨询委员会

Internet Architecture Board 因特网体系委员会

**IABA** International Aircraft Brokers Association 国际飞机经营人协会

**IABCS** Integrated Aircraft Brake Control System 飞机刹车综合控制系统

**IAC** Image Attenuation Coefficient 图像衰减系数

Immediate Action Command 立即作用指令,快速行动指令

Indian Airlines Corporation 印度航空公司

Initial Approach Course 初始进近航线

Instructor Aircraft Commander 飞机指挥员教官

Instrument Approach Chart 仪表进近图

Integrating Associate Contractor 集成联合合同方

Integration Assembly and Checkout 集成装配和检验

Intelligence Analysis Center 情报分析中心(美国)

Intelligence Assessment Committee 情报评价委员会

International Aerological Commission 国际高空气象学委员会

International Air Commission 国际航空委员会

International Air Convention 国际航空协定,国际航空公约

International Astronautical Congress 国际宇航会议

**IACA** International Air Cargo Association 国际航空货运协会

International Air Carrier Association 国际航空承运人协会

International Air Charter Association 国际航空包机协会

**IACAC** International Air Cargo Association of Chicago 芝加哥国际航空货运协会

International Association of Civil Aviation Chaplains 国际民航牧师协会

**IACC** International Air Cargo Conference 国际航空货运会议

**IACES** International Air Cushion Engineering Society 国际气垫工程学会

**IACG** Inter-Agency Consultative Group 机构间协商组

**IACGEC** Inter-Agency Committee on Global Environmental Change 机构间全球环境变化问题委员会

**IACO** Integrated Assembly and Checkout 综合装配和检测

Inter-African Coffee Organization 中非国家咖啡组织

International Airplane Co. 国际飞机公司

**IACP** Independent Association of Continental Pilots 大陆飞行员独立协会

International Association of Computer Programmers 国际计算机程序编制者协会

**IACS** Inertial Attitude Control System 惯性姿态控制系统

Instrument and Control System 仪表与控制系统

Integrated Avionics Communication System 航空电子综合通信系统

Integrated Avionics Computer System 航空电子综合计算机系统

Integrated Avionics Control System 集成航空电子控制系统

Intermediate Altitude Communication Satellite 中等高度通信卫星

International Annealed Copper Standard 国际韧铜标准

**IACSP** International Aeronautical Communication Service Provider 国际航空通信业务提供商

**IACZ** Inter Airline Club Zurich 航空公司苏黎世联合俱乐部

**IAD** Initiation Area Discriminator 初始区域鉴别器

Integrated Access Device 集成接入设备

Washington, DC, USA-Washington-Dulles (Airport Code) 美国华盛顿州华盛顿杜勒斯机场

**IADB** Inter-American Defense Board 泛美防务委员会

**IADM** Improved Advanced Development Model 改进的先进开发模型

**IADP** INTELSAT Assistance and Development Program 国际通信卫星组织的协助和开发计划

**IADS** Integrated Air-Defense System 联合空防系统

**IADT** Integrated Automatic Detection and Tracking 综合自动检测和跟踪

**IAE** Infrared Astronomy Explorer 红外天文探测卫星

Institute for Advancement in Engineering 工程改进研究所(美国)

Institution of Aeronautical Engineers 航空工程师协会

International Aero Engines 国际航空发动机公司

**IAEA** Indian Air Engineer's Association 印度航空工程师协会

International Atomic Energy Agency 国际原子能机构(联合国)

**IAEC** International Atomic Energy Committee 国际原子能委员会

**IAEDS** Integrated Advanced Electronic Display System 先进的综合电子显示系统

**IAES**  IEEE Aerospace and Electronic System IEEE 航空航天与电子系统

Institute of Aeronautical Sciences 航空科学学会(英国)

**IAF**  Image Analysis Facility 图像分析设备

Initial Approach Fix 初始进场定位,初始进近点(坐标)

Interactive Facility 交互式地面设施

Intermediate Approach Fix 中间进近定位点

International Aeronautical Federation 国际航空联合会

International Astronautical Federation 国际太空联盟

**IAFA**  International Airfreight Forwarders' Association 国际航空货运商协会(日本)

**IAFIS**  Integrated Automated Fingerprint Information System 集成的自动指纹信息系统

**IAG**  International Association of Geodetical 国际测地协会

**IAGC**  Instantaneous Automatic Gain Control 瞬时自动增益控制

**IAGS**  Inter-American Geodetic Survey 泛美大地测量

**IAH**  Houston, TX, USA-George Bush (Airport Code) 美国得克萨斯州休斯敦乔治·布什机场

**IAI**  Information Acquisition and Interpretation 信息获取和判读

Intake Anti-Ice 进气道防冰

Israel Aircraft Industries 以色列飞机工业公司

**IAIN**  International Aerospace Information Network 国际航空信息网

International Association of Institutes of Navigation 国际导航研究所协会

**IAIS**  Industrial Aerodynamics Information Service 工业空气动力学情报资料处

International Association of Insurance Supervisors 国际保险监督官协会

**IAL**  Instrument Approach and Landing 仪表进近和着陆

International Aeradio Ltd. 国际航空无线电有限公司

International Air Leases Inc. (USA) 国际航空租赁公司(美国)

International Airtraffic League 国际空中交通联合会

**IALA**  Inter-Agency Letter of Agreement 机构间协议书

International Association of Lighthouse Authorities 国际灯塔当局协会

**IALC**  Instrument Approach and Landing Chart 仪表进近和着陆图

**IALPA**  Irish Airline Pilots' Association 爱尔兰航空飞行员协会

**IAM**  Immediate Access Memory 直接存取存储器

Index Addressing Mode 索引寻址模式

Information Asset Management 信息资产管理

Initial Address Management 初始地址管理

Initial Address Memory 初始地址存储器

Initial Address Message 初始地址消息

Institute of Aviation Medicine 航空医学研究所(英国)

Instrument Approach Minima 仪表进场着陆最低气象条件

Integrated Avionics Manager 综合航电管理器

International Association of Machinists 国际机械师协会

International Association of Meteorology 国际气象协会

**IAMAP**  International Association of Meteorology and Atmospheric Physics 国际气象和大气物理学协会

**IAMAS**  International Association of Meteorology and Atmospheric Sciences 国际气象学和大气科学协会

**IAMSAR**  International Aeronautical and Maritime Search and Rescue 国际航空及海事搜索及救援

**IAMTI**    International Aviation Management Training Institute 国际航空管理培训学院

**IAN**    Incomplete Assembly Notice 未完成组装通知

Integrated Analog Network 综合模拟网

Irregularly Activated Network 不规则激活网络

**IANA**    Internet Assigned Numbers Authority 互联网号码分配当局

**IANC**    International Airline Navigators Council 国际航空公司领航员理事会

**IANS**    Institute of Air Navigation Services 航行服务学院

**IAO**    In and out of Clouds 间断云中

Instructor Avionics Officer 航空电子教员

Internal Automation Operation 内部自动化操作

**IAOA**    Indicated Angle of Attack 指示迎角

**IAOPA**    International Aircraft Owner and Pilot Associations 国际私用飞机拥有者及驾驶员协会

International Council of Aircraft Owner and Pilot Associations 私用飞机拥有者和驾驶员协会国际委员会

**IAP**    Immediate After Passing 紧随飞越后

Improved Accuracy Program 改进精确度计划

Initial Aiming Point 起始瞄准点

Initial Approach Procedure 起始进近程序

Inlet Absolute Pressure 进口绝对压力

Instrument Approach Procedure 仪表进近程序

Integrated Actuator Package 集成作动器包

Integrated Aeronautical Program 综合航空计划

International Aero Press 国际航空出版社

Internet Access Point 网际访问点

Internet Access Provider 因特网接入服务供应商

**IAPA**    Industrial Accident Prevention Association 工业事故预防协会

International Airline Passengers' Association 国际航空公司旅客协会

International Aviation Photographers' Association 国际航空照相员协会

**IAPC**    International Airport Planning Consortium 国际航空港规划财团(英国)

**IAPCH**    Initial Approach 起始进近

**IAPO**    Interchangeable at Attach Points Only 只在连接点可以互换

International Association of Physical Oceanography 国际物理海洋学协会

**IAPP**    Initial Approach 起始进近

Instrument Approach 仪表进近

**IAPS**    Integrated Actuator Package System 综合作动器组件系统

Integrated Avionics Processing System 综合航空电子数据处理系统

Ion Auxiliary Propulsion System 离子辅助推进系统

**IAPSO**    International Association for the Physical Sciences of the Ocean 国际海洋物理科学协会

**IAPT**    International Association of Public Transport 公共交通国际联合会

**IAQG**    International Aerospace Quality Group 国际航空航天质量组

**IAQS**    Indoor Air Quality Standard 室内空气质量标准

**IAR**    Independent Assessment Report 独立评估报告

Intelligent Automatic Rerouting 智能型自动重选路由

Intersection of Air Routes 航路交叉点

Investment Analysis Report 投资分析报告

**IARU**    International Amateur Radio Union 国际业余无线电联合会

**IAS**    Immediate Access Storage 快速(立即)存取存储器

Impact Attenuation System 冲击缓减系统

Independent Airlift Support 独立空运保障

Indicated Air Speed 指示空速

Institute for Atmospheric Sciences 大气科学研究院

Institute of the Aeronautical Sciences 航空科学研究院所

Institute of the Aerospace Sciences 航空航天科学学会

Instrument Approach System 仪表进近系统

Integrated Access Server 综合接入服务器

Integrated Acoustic Structure 集成声学结构

Interactive Application Server 交互应用服务器

International Accounting Standards 国际会计标准

International Aircraft Standards 国际飞机标准

International Application Satellite 国际应用卫星

Interplanetary Automated Shuttle 行星间自动穿梭飞行器

**IASA** International Air Safety Association 国际航空安全协会

International Aviation Safety Assessment 国际航空安全评估

**IASB** International Aircraft Standards Bureau 国际航空器标准局

**IASC** International Air Services Commission 国际航空服务委员会

**IASEL** International Automatic Space Ecological Laboratory 国际自动空间生态学实验室

**IASI** Infrared Atmospheric Sounding Interferometer 红外大气干涉探测器

**IASL** Integrated Aircraft System Laboratory 综合飞机系统实验室

**IASS** International Air Safety Seminar 国际航空安全研讨会

**IASSC** Instrument Approach System Steering Computer 仪表进近系统操纵计算机

**IASTA** International Air Services Transit Agreement 国际航班过境协定

**IASY** International Active Sun Year 太阳活动期国际观测年

**IAT** Indicated Air Temperature 指示气温

International Association for Testing Materials 国际考试资料协会

**IATA** International Air Transport Association 国际航空运输协会

International Airline Telecommunications Association 国际航空系统电信协会

International Aviation Traffic Association 国际航空业务协会

**IATAN** International Airlines Travel Agent Network(USA) 国际航空公司旅行代理网（美国）

**IATC** International Air Traffic Communication 国际空中交通通信

International Air Transport Conference 国际空运会议

International Air Transport Convention 国际航空运输公约

**IATCS** International Air Traffic Communication Station 国际空中交通通信站

International Air Traffic Communication System 国际空中交通通信系统

**IATE** Intermediate Automated Test Equipment 中级自动测试设备

**IATF** International Airline Training Fund 国际航空公司培训基金

**IATP** Individual Airplane/Aircraft Tracking Program 单机跟踪程序

International Airlines Technical Pool 国际航空公司技术联合会

Internet Access and Training Program 互联网接入服务和培训项目

**IATS** Intermediate Automatic Test System 中间自动测试系统

**IATSC** International Aeronautical Telecommunication Switching Center 国际航空电信转换中心

**IAU** Interface Adapter Unit 接口匹配器

International Accounting Unit 国际核算机构

International Astronomical Union 国际天文联合会

**IAVC** Instantaneous Automatic Volume Control 瞬时自动音量控制

**IAVW** International Airways Volcano Watch 国际航路火山监视

**IAW** In Accordance with 依照,依据

**IAWG** Industrial Avionics Working Group 工业航空电子设备工作组(英国)

**IAWP** Initial Approach Waypoint 起始进近航路点

**IAWPRC** International Association on Water Pollution Research and Control 国际水污染研究与控制协会

**IB** Identification Beacon 识别信标
Inboard 内侧的,机内的
Inbound 入境的,进港的
Industrial Base 工业基地
Information Briefing 信息简报
Internet Booking 网上购票,网上订座
Ion Beam 离子束

**IBA** Igniter Booster Assembly 点火器助推器组件
Inbound Boom Avoidance 避免进港音爆
International Bauxite Association 国际铝矾土协会

**IBAA** International Bussiness Aircraft Association 国际公务飞机协会(欧洲)
Italian Business Aviation Association 意大利公务航空协会

**IBAC** International Business Aviation Council 国际商业航空理事会

**IBC** Industrial Base Capability 工业基础能力
Information Bearer Channel 信息承载信道
Integrated Broadband Communication 综合宽带通信
International Boundary Commission 国际边界委员会
Ion Beam Coating 离子束镀膜

**IBCN** Integrated Broadband Communication Network 综合宽带通信网
International Broadband Communication Network 国际宽带通信网络

**IBD** Ion Beam Deposition 离子束淀积

**IBDCA** Interference Based Dynamic Channel Allocation 基于干扰的信道动态分配
Interference Based Dynamic Channel Assignment 基于干扰的信道动态分配

**IBDN** Integrated Building Distribution Network 楼宇综合布线网络

**IBE** Institute of British Engineers 英国工程师协会
International Bureau of Education 国际教育局
Internet Booking Engine 互联网订座引擎
Ion Beam Epitaxy 离子束外延

**IBEC** International Bank for Economic Cooperation 国际经济合作银行

**IBF** Internally Blown Flap 内吹(式)襟翼

**IBG** Interblock Gap 数据区间隔,磁带区间隔

**IBGP** Internal Border Gateway Protocol 内部边界网关协议

**IBI** Integrated Building Intelligent 综合大楼智能化
Intergovernmental Bureau for Information 政府间信息局

**IBIS** ICAO Birdstrike Information System 国际民航组织鸟击信息系统
Image Based Information System 图像信息系统

**IBIT** Initiated Built in Test 启动的机内自检
International Business Information Technology 国际商务信息技术

**IBJ** Industrial Bank of Japan 日本工业银行

**IBLS** Integrity Beacon Landing System 整体信标着陆系统

**IBM** Integrated Business Management 综合业务管理
International Business Machines Corporation 国际商业机器公司
International Business Management 国际商务管理

**IBMS** Intelligent Building Management System

智能大厦管理系统

**IBN** Identification Beacon 识别信标, 识别灯标

Integrated Broadband Network 综合宽带网

**IBO** Islamic Broadcasting Organization 伊斯兰广播组织

**IBOM** Integrated Bill of Material 器材综合账单

**IBP** Initial Boiling Point 初始沸点

**IBR** Integrally Bladed Rotor 整体叶片转子

Intra-Base Radio 基地内部电台

Investment Balance Review 投资平衡审查

Investment Budget Review 投资预算审查

**IBRD** Inflated Ballute Retarding Device 充气伞减速装置

International Bank for Reconstruction and Development 国际复兴开发银行

**IBRL** Initial Bomb Release Line 起始投弹线

**IBS** Intelligent Building System 智能大厦系统

Intelsat Bussiness Service 国际通信卫星商用业务

Internet Booking Service 网上订票服务

Ionospheric Beacon Satellite 电离层信标卫星

**IBSS** Infrared Background Signature Survey 红外背景特征探测器

**IBT** Internet Browsing Terminal 因特网浏览终端

**IBU** Independent Backup Unit 独立备用组件

Input Buffer Unit 输入缓冲单元

Instrumentation Buffer Unit 测量仪表缓冲器组件

Integrated Ballast Unit 集成式镇流器组件

**IBVSU** Instrument Bus Voltage Sensing Units 仪表总线电压感应组件

**IBW** Impulse Bandwidth 脉冲带宽

Information Bandwidth 信息带宽

**IBWN** Indoor Broadband Wireless Network 室内宽带无线网络

**IBZ** Ibiza, Spain(Airport Code) 西班牙伊比萨机场

**IC** Identification Code 识别码

Image Check 图像检验

Image Compression 图像压缩

Implementing Card 施工卡

Impulse Conductor 脉冲导线

Impulse Counter 脉冲计数器

Incentive Contract 鼓励合同

Incremental Cost 递增成本

Index of Correction 校正系数

Indicate and Control 指示与控制

Indicating Controller 指示控制器

Indication Cycle 指示周期

Indirect Cycle 间接周期

Inductance Capacitance 电感电容

Industrial Collaboration 工业合作

Industrialized Countries 工业化国家

Influence Coefficients 影响系数

Initial Condition 初始状态, 初始条件

Initial Course 起始航向

Inlet Contract 进口接触点

Input Current 输入电流

Instruction Card 指令卡

Instruction Cell 指令单元

Instruction Character 指令字符

Instruction Code 指令码

Instruction Counter 指令计数器

Instruction Cycle 指令周期

Instrument and Controls 仪表和操纵机构

Intake Closes 过气门关闭

Integrated Circuit 集成电路

Intercabinet 机柜间, 座舱间

Interchangeability Code 可互换码

Intercommunications 内部通信, 机内通信

Interconnection 互连

Interface Control 界面控制, 接口控制

Interim Change 临时更改

Interior Communication 内部通信

Interlock Code 互锁码

Intermediate Circuit 中间电路

Internal Combustion Engine 内燃机

Internal Combustion 内部燃烧

Internal Connection 内部连接

Internal Conversion 内部转换

Invoice Control Card 收据控制卡,发票控制卡

Ion Chamber 电离箱

Ion Chromatography 离子色层法

Item Code 项目代码

Item Condemnation 项目报废

**ICA** Independent Cost Analysis 独立成本分析,单独成本分析

Initial Cruise Altitude 起始巡航高度

Inner-City Airport 城市机场

Instructions for Continuing Airworthiness 持续适航性指南

Instrument Compressed Air 仪器压缩的空气

International Caribbean Airways 加勒比国际航空公司

International Commodity Association 国际合作社联盟

International Communication Association 国际通信协会

**ICAA** International Civil Airports Association 国际民用机场协会

International Civil Aviation Authority 国际民用航空局

International Committee of Aerospace Activities 国际航空航天活动委员会

Investment Capability Analysis and Assessment 投资能力分析和评价

Israel Civil Aviation Authority 以色列民用航空局

**ICAAS** Integrated Control and Avionics for Air Superiority 空中优势集成控制与航空电子

**ICAC** Independent Commission Against Corruption 廉政公署

Initial Cruise Altitude Capability 起始巡航高度能力

International Civil Aviation Committee 国际民航委员会

International Cotton Advisory Committee 国际棉花咨询委员会

**ICACGP** International Commission on Atmospheric Chemistry and Global Pollution 国际大气化学和全球污染委员会

**ICACTE** International Conference on Advanced Computer Theory and Engineering 高级计算机理论与工程国际会议

**ICAD** Integrated Control and Display 综合控制和显示

**ICAEA** International Civil Aviation English Association 国际民航英语协会

**ICAF** International Committee on Aeronautical Fatigue 航空疲劳国际委员会

**ICAI** Intelligent Computer-Assisted Instruction 人工智能计算机辅助教学

**ICAM** Incident Cause Analysis Method 事故原因分析方案

Integrated Computer-Aided Manufacturing 综合计算机辅助制造

International Civil Aircraft Marking 国际民用航空器标志

**ICAMD** International Conference on Advanced Materials and Devices 先进的材料和装置国际会议

**ICAN** International Commission for Air Navigation 空中航行国际委员会

**ICAO** International Civil Aviation Organization 国际民航组织,国际民用航空组织

**ICAOPA** International Council of Aircraft Owners and Pilots Association 国际飞机业主和飞行员协会理事会

**ICAOTAM** ICAO Technical Assistance Mission 国际民航组织技术援助任务

**ICAP** Improved Capabilities 改进的能力

Interagency Committee for Aviation Policy 航空政策机构间委员会

Interface Control Action Plan 接口控制行动计划

International Civil Aviation Policy 国际民航政策

Internet Content Adaptation Protocol 互联网内容适配协议

**ICAPP** Inventory Control and Purchase Planning 航材库存控制及订购计划

**ICARD** International Codes and Routes Designators 国际代码和航路代号

**ICARE** Illustration, Constraint, Activity, Rule, Entity 插图、约束、行为、规则、实体

**ICAS** Independent Collision-Avoidance System 自主式避撞系统

Integrated Condition Assessment System 集成状态评估系统

Interdepartmental Committee of Atmospheric Science 大气科学跨部委员会

Interface Control Action Sheet 接口控制操作单

Intermittent Commercial and Amateur Service 断续商业和业余服务

International Congress of the Aeronautical Sciences 国际航空科学代表大会

International Council of the Aeronautical Sciences 国际航空科学理事会

**ICASS** International Confidential Aviation Safety System 国际航空安全保密系统

**ICAT** International Center for Air Transport 航空运输国际中心

International Civil Aviation Treaty 国际民航条约

**ICATS** International Centre of Advanced Tourism Studies 国际旅游研究中心

**ICAU** International Civil Aviation University 国际民航大学

**ICAW** Instruction for Continued Airworthiness 持续适航指令

Integrated Cautions and Warnings 综合提示与警告

**ICB** Incoming Call Barred 来话加锁

Indicated Control Box 指示器控制盒

Integrated Circuit Board 集成电路板

Intercabinet Bus 机柜间总线

Interface and Controller Board 接口与控制器板

International Competitive Bidding 国际竞争性投标

**ICBD** Ionized-Cluster Beam Deposition 离子团束淀积

**ICBE** Ionized-Cluster Beam Epitaxy 离子团束外延

**ICBI** Inter-Channel Inter-Block Interference 信道间信息组间的干扰

**ICBR** Ice Cuber 冰块机

**ICBUS** Internal Control Bus 内部控制总线

**ICC** IAPS Card Cage 航空电子综合处理系统插卡箱

Information Control Center 情报管制中心

Inner Combustion Chamber 内燃烧室

Installation, Checkout and Calibration 安装、检查和校准

Instantaneous Channel Characteristics 信道瞬态特性

Instrument Control Center 仪表控制中心

Integrated Change Control 综合变更控制

Integrated Circuit Component 集成电路元件

Integrated Circuit Computer 集成电路计算机

Integrated Collection Center 综合采集中心

Integrated Communication Control 综合通信控制

Integrated Communication Controller 综合通信控制器

Intelligence Collection Center 情报采集中心

Inter-Computer Coupler 计算机间连接器

Interface Control Contractor 接口控制合同方

Intergovernmental Copyright Committee 政府间版权委员会

Interior Communication Channel 内部通信信道

Interior Communication Control 内部通信管理

International Chamber of Commerce 国际商会

International Commercial Council 国际商业协会

International Computer Center 国际计算机中心

International Computing Center 国际计算中心

International Conference on Communications 国际通信会议

International Control Commission 国际控制委员会

Internet Call Center 因特网呼叫中心

Item Category Code 项目分类代码

**ICCA** Interactive Computer-Controlled Analysis 交互式计算机控制分析

International Conference on Computer Applications 计算机应用国际会议

**ICCAIA** Inter Coordinating Council of Aerospace Industries Association 航空航天工业协会内部协调理事会

International Coordinating Council of Aerospace Industries Associations 宇航工业协会国际协调理事会

**ICCB** Intermediate Configuration Control Board 中间配置控制板

Internet Control and Configuration Board 控制与配置委员会

Internet Control and Configuration Board 互联网管理和配置委员会

Interoperability Configuration Control Board 互用性配置控制板

**ICCC** International Conference on Computer Communication 国际计算机通信会议

**ICC-CAPA** International Chamber of Commerce Commission on Asia and Pacific Affairs 国际商会亚洲及太平洋事务委员会

**ICCCM** Inter-Client Communications Conventions Manual 客户程序间通信惯例手册

**ICCD** Intensified Charge Coupled Device 增强型电荷耦合器件

**ICCL** International Council of Cruise Lines 国际航线委员会

**ICCO** International Cocoa Organization 国际可可组织

**ICCP** Institute for the Certification of Computer Professionals 计算机专业证书学会

**ICCS** IATA Currency Clearance Service 国际航空运输协会货币结算机构

Integrated Communication Control System 综合通信管制系统

**ICCSIT** International Conference on Computer Science and Information Technology 计算机科学与信息技术国际会议

**ICCSS** Integrated Communication Control System Simulator 综合通信管制系统模拟器

**ICCT** International Conference on Communication Technology 通信技术国际会议

International Council on Clean Transportation 清洁交通国际委员会

**ICD** Inlet Control Device 进气控制装置

Installation Control Drawing 安装控制图

Integrated Control/Display 综合控制/显示器

Interactive Design Center 交互式设计中心

Interface Control Document 界面控制文件，接口控制文件

Interface Control Drawing 接口控制图纸

**ICDA** Interactive Computer Data Analysis 交互计算机数据分析

International Conference on Design Automation 国际设计自动化会议

International Cooperative Development Association 国际合作发展协会

**ICDCS** International Conference on Distributed Computing Systems 分布式计算系统国际会议

**ICDN** Integrated Communication Data Network 综合通信数据网络

**ICDO** International Civil Defence Organization 国际民防组织

**ICDR** Initial Critical Design Review 初始临界设计评审

Internal Critical Design Review 内部关键设计审查

**ICDRUM** Intercommunication Drum 内部通

信磁鼓

**ICDS** Integrated Color Display System 综合彩色显示系统

Integrated Control and Display System 综合控制与显示系统

**ICDSC** International Conference on Digital Satellite Communication 国际数字卫星通信会议

**ICDT** Incident 事件,事故征候

**ICDU** Inertial Coupling Display Unit 惯性耦合显示装置

Integrated Control and Display Unit 综合控制和显示装置

**ICE** Carbon Dioxide, Solid (dry ice) 固体二氧化碳(干冰)

Iceland 冰岛

Immigration and Customs Enforcement 移民与海关执法署

In Circuit Emulation 集成电路仿真,在线仿真

In Circuit Emulator 回路仿真器

Independent Cost Estimate 独立成本核算

Input Checking Equipment 输入校验设备

Instrument Checkout Equipment 仪表检验设备

Inter Connect Equipment 互连设备

Interconnection Element 连接舱

Interface Configuration Environment 接口配置环境

Interference Cancellation Equipment 干扰消除设备

Internal-Combustion Engine 内燃机

International Cometary Explorer 国际彗星探测器

International Cosmos Explorer 国际宇宙探测器

**ICED** Interface Control Envelope Drawing 接口控制包线图纸

**ICEM** Integrated Computer-aided Engineering and Manufacturing 综合计算机辅助工程与制造

Intergovernmental Committee for European Migrations 政府间欧洲移民委员会

**ICENT** International Convention for the Evaluation of New Technologies 新技术评估国际大会

**ICES** International Committee for Earth Sciences 国际地球科学委员会

International Council for the Exploration of the Sea 国际海洋考察理事会

Inventory Cost Estimating System 库存成本估算系统

**ICF** Inertial Confinement Fusion 惯性密封熔合

Information Converting Function 信息转换功能

Initial Contact Frequency 开始接触的频率

Interconnection Files 互联文件

International Conference on Fracture 国际断裂学会议

International Cryptography Framework 国际密码框架

**ICFD** Interdisciplinary Computational Fluid Dynamics 多学科计算流体动力学

**ICG** Icing 结冰

Industries Consultative Group 工业咨询集团

Integrated Command Generator 综合指令发生器

International Committee on GNSS 全球导航卫星系统国际委员会

**ICGIC** Icing in Clouds 云中积冰

**ICGICIP** Icing in Clouds and Precipitation 云中或降水中积冰

**ICH** Improved Cargo Helicopter 改进的货运直升机

Incoming Channel 来话信道

**ICHCA** International Cargo Handling Co-ordination Association 国际货运协调协会

**ICHE** Intercooler Heat-Exchanger 中间散热器热交换器

**ICI** Initial Capability Inspection 起始能力检查

Intelligent Communications Interface 智能通信接口

Inter-Carrier Interference 载波间干扰

Inter-Channel Interference 信道间干扰

International Commission for Illumination 国际照明委员会

**ICID** International Commission on Irrigation and Drainage 国际排灌委员会

**ICINS** International Conference on Integrated Navigation Systems 综合导航系统国际会议

**ICISF** International Critical Incident Stress Foundation 国际危机事件应激基金会

**ICJ** International Court of Justice（联合国）国际法院,海牙国际法庭

**ICL** Inspection Characteristics List 性能检查表

**ICLECS** Integrated Closed-Loop Environmental Control System 综合闭环环境调节系统

**ICM** Idealized Cognitive Model 理想化的认知模式

Ignition Control Module 点火控制模块

Image Compression Manager 图像压缩管理器

Incoming Call Management 来话呼叫管理

Information Classification and Management 信息分类与管理

Information Coordination Memo 信息协调备忘录

Integrated Communications Management 综合通信管理

Intelligent Composite Material 智能复合材料

Inter Console Mark 内仪表板标志,内仪表操纵台标志

Intercommunications 内部通信

Interface Conversion Module 接口转换模块

Interface Coordination Memo 界面协调备忘录

Intergovernmental Committee for Migration 政府间移民委员会

International Capital Market 国际资本市场

**ICMA** International Crisis Management Association 国家危机管理协会

**ICMP** Integrated Communications Management Platform 综合通信管理平台

Interchannel Master Pulse 通道间主脉冲

Interface Control Management Plan 接口控制管理计划

Internal Control Message Protocol 内部控制信息协议

Internet Control Message Protocol 因特网控制信息协议

**ICMPS** Induction Compass 电磁感应罗盘

**ICMR** Instrument Calibration and Maintenance Record 仪表校准与维修记录

**ICMS** Indirect Cost Management System 间接成本管理系统

Integrated Control and Monitoring System 综合控制和监控系统

Internal Countermeasures 内部干扰

International Committee on Meteorological Services 气象勤务国际委员会

Inventory Cost Management System 库存成本管理系统

**ICMUA** International Commission on the Meteorology of the Upper Atmosphere 国际高层大气气象学委员会

**ICN** Indicator Coupling Network 指示器耦合网络

Interface Change Notice 接口更改通知

Interrupt Control Network 中断控制网络

Seoul Inchon ,South Korea( Airport Code) 韩国首尔仁川机场

**ICNA** Integrated Communication and Navigation Avionics 综合通信和导航电子设备

**ICNDT** International Committee for Non-Destructive Testing 国际无损检测委员会

**ICNI** Improved Communications, Navigation and Identification 改进的通信、导航和识别

Integrated Communications, Navigation and Identification 综合通信、导航和识别

**ICNIA** Integrated Communication, Navigation and Identification Avionics 综合通信导航与

识别航空电子系统

**ICNIRP** International Commission on Non-Ionizing Radiation Protection 国际非电离辐射防护委员会

**ICNIS** Integrated Communication Navigation and Identification System 综合通信导航与识别系统

**ICNS** Integrated Communications and Navigation System 综合通信和导航系统

**ICO** Ignition-Cut-off 点火断开

Instinctive Cut-out 本能切断

Intermediate Altitude Circle Orbit 中高度圆轨道

International Commission for Optics 国际光学学会

International Communications Operations 国际通信业务

**ICP** Improved Climb Performance 改进爬升性能

Incoming Call Packet 呼入分组信息

Integrated Control Panels 综合控制面板

Integrated Core Processor 综合核心处理机

Interface Change Package 接口更改（软件）包

Internal Connection Protocol 内部连接协议

Internet Cache Protocol 互联网缓存协议

Internet Communications Protocol 因特网通信协议

Internet Content Provider 因特网内容服务供应商

Interphone Control Panel 内话控制面板

Interworking Control Protocol 互通控制协议

Inventory Control Point 编目控制点

**ICPA** Indian Commercial Pilots' Association 印度商业飞机驾驶员协会

International Cooperative Program Activity 国际合作计划活动

**ICPC** International Cable Protection Committee 国际电缆保护委员会

**ICPI** Integrated Cabin Pressure Indicator 客舱压力综合指示器

**ICPMS** Integrated Cabin Passenger Management System 综合客舱旅客管理系统

**ICPO** International Criminal Police Organization 国际刑事警察组织

**ICPOM** International Conference of Production and Operation Management 国际生产与运营管理学术会议

**ICPS** Incremental Control Processor System 增量控制处理器系统

**ICR** In-Commission Rate 可使用率，服役率，先用的比率

Initial Cell Rate 初始信元率

Internal Control Register 内部控制寄存器

International Collision Regulation 国际防撞规章

Iron Core Reactor 铁芯电感线圈

**ICRC** Interiors Configuration Review Committee 内部构型评审委员会

International Committee of the Red Cross 国际红十字会

**ICRD** Installation Control Requirements Drawings 安装控制要求图纸

Interface Control Requirements Drawings 接口控制要求图纸

**ICRH** Ion Cyclotron Resonance Heating 离子回旋共振加热

**ICROA** International Carbon Reduction and Offset Alliance 国际碳减量与抵换联盟

**ICRP** International Commission on Radiological Protection 国际辐射防护委员会，国际放射防护委员会

**ICRU** International Commission on Radiation Units and Measurements 国际辐射单位与测量委员会

**ICS** Idle Corrected Speed 慢车修正速度

Improved Communications System 改进的（或综合的）通信系统

Improved Composite Structure 改进的复合材料结构

Incident Command System 事故指挥系统

Integrated Checkout System 综合检查系统

Inter Cargo Services（France）国内货运服务（法国）

Inter Communication System 内部通讯系统

Intercom Control Set 内部通信控制装置

Intercom System 内部通信系统

Interconnecting Computer System 互选计算机系统,航空公司订座系统

Interface Change Summary 接口更改总结

Interim Contractor Support 过渡性承包商保障

Interim Cover Sheet 临时封面页

Intermediate Capacity System 中等载客量铁路系统

Internal Countermeasures System 内部对抗措施系统

International Commuter System 国际短程飞机系统

International Conference on Software 国际计算机软件会议

Interphone Control Station 内话控制台

Inventory Control System 座位控制系统,航班控制系统

Inverse Conical Scan 逆圆锥形扫描

Item Count Summary 项目计数总结

**ICSA** International Center for Security Analysis 国际安全分析中心

International Computer Security Association 国际计算机安全协会

International Customer Service Association 国际客户服务协会

**ICSC** Interim Communication Satellite Committee 临时通信卫星委员会

International Civil Service Commission 国际公务员制度委员会

International Communications Satellite Corperation 国际通信卫星公司

**ICSEAF** International Commission for the Southeast Atlantic Fisheries 东南大西洋渔业国际委员会

**ICSM** International Committee of Scientific Management 国际（企业）科学管理委员会

International Conference on Software Maintenance 软件维修国际会议

**ICSMA** Integrated Communications System Management Agency 综合通信系统管理机构

**ICSS** Integrated Communications Switching System 综合通信转换系统

Integrated Control and Safety System 综合控制与安全系统

**ICST** Institute for Computer Science and Technology 计算机科学和技术学会

**ICSU** International Council for Scientific Union 国际科学联盟理事会

International Council of Space Unions 国际空间联盟理事会

**ICT** Implementation Coordination Team 实施协调组

In Coming Trunk 来话中继

Information and Communication Technology 信息和通信技术

Interface Control Tooling 接口控制仪

Wichita, KS, USA（Airport Code）美国堪萨斯州威奇托机场

**ICTA** Instrument Current Transformer Assembly 仪表电流互感器组件,仪表变流器组件

**ICTAC** International Confederation for Thermal Analysis and Calorimetry 国际热分析及量热学协会

**ICTSD** International Centre for Trade and Sustainable Development 贸易和可持续发展国际中心

**ICTV** Integrated Circuit Television 集成电路电视

**ICU** Instrument Comparator Unit 仪表比较器组件

Interface Control Unit 接口控制组件

International Chemical Union 国际化学联合会

Interphone Control Unit 内话控制组件

Interrupt Control Unit 中断控制单元

Isolation Control Unit 隔离控制组件

**ICUG** International Closed User Group 国际闭合用户群

**ICV** Isolation Control Valve 隔离控制活门

**ICVM** ICAO Coordinated Validation Missions 国际民航组织的协调验证任务

**ICW** Independent Carrier Wave 独立载波

Interactive Courseware 交互式多媒体课件

Intermittent Continuous Wave 间歇等幅波

Internet Call Waiting 因特网呼叫等待

**ICWAR** Improved Continuous Wave Acquisition Radar 改进型连续波探测雷达

**ICWG** Interface Control Working Group 接口控制工作组

**ICX** Inter-Cartridge Exchange 部件间交换

**ICY** Interchangeability 互换性

**ID** Dinar 第纳尔(货币单位)

Identification 识别

Identifier 标识码

Identify 鉴定,辨认

Inadvertent Disconnect 意外断开(空中加油时)

Inside Dimensions 内部尺寸

Instrument Departure 仪表离场

Interconnecting Device 互接设备,互接装置

Internal/Inside Diameter 内径

**IDA** Idaho Falls, ID, USA (Airport Code) 美国爱达荷州爱达荷福尔斯机场

Information, Decisions, Actions 信息、决策、动作

Input Data Assembler 输入数据组合器

Institute for Defense Analyses 国防分析研究所

Integrated Digital Access 综合数字接入

Integrated Digital Avionics 数字式综合航空电子系统

Interactive Design Applications 交互式设计应用

Interchange of Data between Administrations 机构间的数据交换

International Development Association 国际开发协会

Internet Direct Access 因特网直接接入

Intrusion Detection Agent 入侵检测代理

**IDAC** Interconnecting Digital Analog Converter 互联数模转换器

**IDAP** Integrated Defense Avionic Platform 综合防御性航空电子装备平台

Integrated Defensive Avionic Program 综合防御性航空电子装备计划

**IDAPS** Image Data Processing System 图像数据处理系统

**IDARA** Improved Distributed Adaptive Routing Algorithm 改进的分布式自适应路由算法

**IDARS** Integrated Data Acquisition and Recorder System 综合数据采集和记录系统

Integrated Document Archive and Retrieval System 综合文件归档和检索系统

**IDAS** Instrument Data Acquisition System 仪表数据获取系统

Integrated Data Acquisition System 综合数据采集系统

Integrated Defensive Aids System 综合防御辅助系统

Integrated Design Automation System 综合设计自动化系统

Integrated Digital Avionics System 综合数字航空电子设备系统

**IDAST** Integrated Digital Avionics for a STOL Transport 短距起降运输数字式综合航空电子设备

**IDB** Integrated Database 综合数据库

Inter-American Development Bank 泛美开发银行

Islamic Development Bank 伊斯兰发展银行

**IDC** Imperial Defence College 帝国国防学院(英国)

Instrument Display Catalog 仪表显示目录

Interactive Design Center 交互式设计中心

International Data Company 国际数据公司

Internet Data Center 因特网数据中心

Internet Database Connector 互联网数据库

连接器

**IDCC** Integrated Data Communication Channel 综合数据通信信道

**IDCSP** Initial Defense Commmunication Satellite Program 初级国防通信卫星计划

**IDCT** Inverse Discrete Consine Transform 离散余弦逆变换

**IDD** Interface Design Document 接口设计文件

International Direct Dialing 国际直拨电话，国际直拨

**IDDS** Image Data Digitizer System 图像信息数字化系统

Information Distribution and Display System 信息发送和显示系统

Instrumentation Data Distribution System 仪表测量数据分配系统

Integrated Data Display System 综合数据显示系统

International Digital Data Service 国际数字数据业务

**IDE** Integrated Development Environment 集成开发环境

Integrated Drive Electronics 集成驱动电子组件

**IDEA** Intelligence Diagnosis Expertise Administration 智能诊断专家管理

International Data Encryption Algorithm 国际数据加密算法

International Data Exchange Association 国际数据交流协会

**IDEAS** International Data Exchange for Aviation Safety 有关航空安全的国际数据交换

**IDECS** Image Discrimination Enhancement and Combination System 图像识别、增强与合成系统

**IDENT** Identification 识别，识别代号，识别标志

Identifier 识别器

Identify 识别

**IDF** Installation Data File 安装数据文件

Instrument Data File 仪器数据文件

Intelligent Data Fusion 人工智能数据合成

International Distress Frequency 国际遇险呼叫频率

**IDFT** Inverse Discrete Fourier Transform 反演离散傅立叶变换

**IDG** Individual Drop Glider 单人空投滑翔机

Industrial Development Group 工业发展小组

Integrated Drive Generator 综合驱动发电机

Internal Drive Generator 内驱动发电机

International Data Group 国际数据集团

**IDGS** Integrated Drive Generator System 综合驱动发电机系统

**IDHS** Intelligence Data Handling System 侦察数据处理系统

**IDI** Initial Domain Identifier 初始域标识符

**IDIP** Intelligence Data Input Package 情报数据输入包

**IDL** Indirect Labor 间接人工

Interactive Data Language 交换式数据语言

Interactive Distance Learning 交互式远程学习

Interface Definition Language 接口定义语言

International Data Line 国际数据线路

**IDLC** Integrated Digital Loop Carrier 综合数字环路载波

**IDLH** Immediately Dangerous to Life or Health 立即危及生命或健康

**IDM** Identity Management 身份管理

Impact Delay Module 冲突延迟模块

Industry Discount Message 行业折扣报

Information and Data Management 信息和数据管理

Integrated Data Management 综合数据管理

Integrated Development Model 综合发展模式

Integrative Decision Making 综合决策

Investment Decision Management 投资决策管理

Ion Drift-Meter 离子漂移计

**IDMS** Improved Deep Moored Sweep 改进的深锚定扫描

Information and Data Management System 信息和数据管理系统

Integrated Database Management System 综合数据库管理系统

**IDN** Indonesia 印度尼西亚

Integrated Data Network 综合数据网络

Integrated Digital Network 综合数字网络,综合数字网

Intelligent Data Network 智能数据网络

Interactive Data Network 交互式数据网络

International Directory Network 国际目录网络

Internationalized Domain Name 国际化域名

**IDO** Identification Officer 识别军官

International Development Organization 国际开发组织

**IDP** Individual Development Programme 个别发展计划

Integrated Data Processing 综合数据处理

International Development Programme 国际发展计划

Internet Directory Provider 因特网目录服务供应商

Internetwork Datagram Protocol 网间数据包协议

**IDPA** International Deaf Pilots Association 国际聋人飞行员协会

**IDPC** Integrated Data Processing Center 综合数据处理中心

**IDPL** Identification Plate 识别牌

**IDPM** Institute of Data Processing Management 数据处理管理学会

**IDPR** Inter-Domain Policy Routing 域间策略路由选择

**IDPS** Image Data Processing Station 图像数据处理站

**IDR** Indonesian Rupiah 印度卢比(货币单位)

Instrument Departure Route 仪表离场航线

Interim Design Review 中间设计评审,临时设计评审

Intermediate Data Rate 中等数据速率

**IDRP** Inter Domain Routing Protocol 域间路由选择协议

**IDS** Ice Detection System 结冰探测系统

Image Data-Sets 图像数据集

Image Display System 图像显示系统

Improved Data Set 改进的数据组

Information Display System 信息显示系统

Infrared Detection Set 红外探测仪

Infrared Detection System 红外探测系统

Integrated Data Storage 综合数据存储

Integrated Data System 综合数据系统

Integrated Decision Support 综合决策支持

Integrated Diagnostic System 综合诊断系统

Integrated Display System 综合显示系统

Integrated Dynamic System 综合动力系统

Interactive Display System 相互作用显示系统

Interface Design Specification 接口设计规范

International Documents Service 国际文献局

Intrusion Detection System 入侵检测系统

Isochronous Data Services 等时数据业务

**IDSCS** Initial Defence Satellite Communications System 初级国防卫星通信系统

**IDSE** International Data Switching Exchange 国际数据交换机(局)

Internetworking Data Switching Exchange 互联网数据交换机(局)

**IDSL** Insurance Database Services Ltd. 保险数据库服务公司

**IDSP** Intelligent Dynamic Service Provisioning 智能型动态业务提供

**IDSS** Integrated Diagnostics Support System 综合检查分析支援系统

Intelligent Decision Support System 智能决策支持系统

**IDT** Identification Disposition Tag(ging) 鉴定处理标签

Industrial Detergents Trade 工业清洁剂贸易

Integrated Digital Terminal 综合数字终端

Intelligent Data Terminal 智能数据终端

Interactive Data Terminal 交互数据终端

**IDTC** International Digital Transmission Center 国际数字传输中心

**IDTS** Instrumentation Data Transmission System 仪表测量数据传输系统

**IDU** Input and Display Unit 输入与显示装置
Integrated Display Unit 综合显示组件
Interactive Display Unit 交互式显示组件
Interface Data Unit 接口数据单元

**IDW** Information and Documentation Work 信息资料工作,信息文献工作

**IE** Initial Equipment 初始装备
Institution of Electronics 电子学会
Instrument Error 仪表误差
Internet Explorer 网络浏览器
Ionosphere Explorers 电离层探险者卫星
Ireland 爱尔兰

**IEA** Industrial Environmental Association 工业环境协会
International Energy Agency 国际能源机构
International Environmental Agreements 国际环境协定
International Ergonomics Association 国际人机工程学协会,国际工效学协会

**IEC** Integrated Environmental Controls 综合环境控制
Integrated Ethernet Chip 集成以太网电路芯片
Inter-Exchange Carrier 局间载波
International Electrotechnical Commission 国际电工技术委员会

**IECM** Induced Environmental Contamination Monitor 诱导环境污染检测器

**IECMS** Inflight Engine Condition Monitoring System 飞行中发动机状态监控系统

**IECO** Inboard Engine Cut-off 内侧发动机关闭

**IECQ** International Electrotechnical Commission Quality 国际电子元器件质量评定委员会

**IED** Improvised Explosive Device 简易爆炸装置

Institute for Educational Development 教育发展协会

**IEE** Institution of Electrical Engineers 电气工程师学会

**IEEE** Institute of Electrical and Electronics Engineers 电气及电子工程师协会(美国)

**IEFCS** Integrated Electrical Flight Control System 综合电气飞行控制系统

**IEI** Imbedded Element Identify 嵌入元件识别

**IEIS** Integrated Engine Instrument System 综合发动机仪表系统

**IELTS** International English Language Testing System 国际英语测试系统

**IEM** Industrial Engineering and Management 工业工程与管理
Information Engineering Methodology 信息工程技术
Initial Evaluation Memorandum 初始评估备忘录
Integrated Equipment Manager 集成设备经理
Interim Examination and Maintenance 临时检查和维修
International Express Mail 国际快邮
Interpretative and Explanatory Manual 解释和说明手册

**IEMA** International Export Market Analysis 国际出口市场分析

**IEMCAP** Intrasystem Electromagnetic Compatibility Analysis Program 系统内电磁兼容性分析程序

**IEME** Inspectorate of Electrical and Mechanical Equipment 电气与机械设备检查组

**IEMP** Internal Electromagnetic Pulse 内部电磁脉冲
International Engine Management Program 国际发动机管理计划

**IEN** Internal Engineering Notice 内部工程简报

Internet Experiment Note 因特网实验备忘录

**IEOF** Ignore End of File 忽略文件末尾

**IEP** Internet Equipment Provider 因特网设备供应商

**IEPG** Independent European Program Group 独立欧洲项目组

Internet Engineering and Planning Group 因特网工程和规划组

**IEPPL** Integrated Engineering Planning Parts List 综合工程计划零件清单

**IEPR** Integrated Engine Pressure Ratio 综合发动机压力比

**IER** Independent Evaluation Report 独立评估报告

Independent Evaluation Review 独立评估审查

Information Exchange Request 信息交换需求

Inherent Equipment Reliability 设备固有可靠性

Institute of Engineering Research 工程研究所

Institutes for Environmental Research 环境研究协会

**IERE** Institution of Electronic and Radio Engineers 电子和无线电工程师协会(英国)

**IES** Industrial Engineering Services 工业工程服务

Industrial Evaluation Sheet 产业评价表,工业评价表

Institute of Environmental Societies 环境科学研究所

Interface Equipment Subsystem 接口设备分系统

Ion Engine System 离子发动机系统

ISDN Earth Station 综合业务数字网络地球站

**IESD** Instrumentation and Electronic Systems Division 仪表和电子系统局

**IESG** Internet Engineering Steering Group 因特网工程指导组

**IESI** Integrated Electronic Standby Instrument 集成电子备用仪表

**IESNA** Illumination Engineering Society of North American 北美照明工程协会

**IESS** Integrated Electromagnetic System Simulator 综合电磁系统模拟器

**IET** Independent Evaluation Teams 独立评估小组

Initial Engine Test 发动机初始试车

Initial Entry Training 起始进入训练

Institution of Engineering and Technology 国际工程技术学会

Interline E-Ticket 联航电子机票

**IETA** International Emissions Trading Association 国际排放交易协会

**IETC** International Environment Technology Centre 国际环境技术中心

**IETD** Interactive Electronics Technical Data 交互式电子技术数据

**IETF** Initial Engine Test Facility 初始发动机试验设备

Initial Engine Test Firing 发动机初期点火试验

Internet Engineering Task Force 因特网工程任务组

**IETM** Integrated Electronic Technical Manual 综合电子技术手册

Integrated Engine Technical Manual 综合发动机技术手册

Interactive Electronic Technical Manual 交互式电子技术手册

**IEU** Inertial Electronics Unit 惯性电子组件

Interface Electronics Unit 入口电子元件

Interphone Electronic Unit 内话电子组件

**IEV** Kiev, Ukraine-zhulhany (Airport Code) 乌克兰基辅茹良尼机场

**IEVD** Integrated Electronic Vertical Display 综合电子垂直显示器

**IEW** Integrated Electronic Warfare 综合电子战

Intelligence and Electronic Warfare 智能电子战

**IEWS** International Electronic War-Fare System 国际电子战系统

**IF** Inflation Factor 放大因子
Inflight 在飞行中
Initial Fix 起始定位点
Inside Frosted 内部有霜
Instruction Folder 文件夹
Instrument Flight 仪表飞行
Intermediate Approach Fix 中间进近定位点
Intermediate Fix 中间定位点
Intermediate Frequency 中频
Internetworking Facility 互联网络设施

**IFA** Integrated Fault Analyzer 综合故障分析器
Intermediate Frequency Amplifier 中频放大器
International Federation of Airworthiness 国际适航联盟

**IFAA** International Flight Attendants' Association 国际机上服务人员协会

**IFAC** International Federation of Automatic Control 国际自动控制联合会

**IFAD** International Fund for Agriculture Development 国际农业发展基金会

**IFALPA** International Federation of Air Line Pilots' Associations 航空公司驾驶员协会国际联合会

**IFAN** International Federation of Standards Users 国际标准用户联盟

**IFAPA** International Foundation of Airline Passengers Association 民航旅客协会国际基金会

**IFARS** Individual Flight Activity Reporting System 个人飞行活动报告系统

**IFAS** International Flow and Analysis System 信息流和分析系统
International Fund of Aviation Safety 航空安全国际基金会

**IFAST** Integrated Facility for Avionics System Test 航空电子系统综合测试设施

**IFATCA** International Federation of Air Traffic Controllers' Associations 空中交通管制员协会国际联合会

**IFATE** International Federation of Airworthiness Technology and Engineering 国际适航技术与工程联合会

**IFATSEA** International Federation of Air Traffic Safety Electronics Association 国际空中交通安全电子协会联合会

**IFB** Integrated Forebody 总装的前机身
Intention for Bid 投标意向书
Invitation for Bid 投标邀请函,招标

**IFBP** Inflight Broadcast Procedure 飞行广播程序

**IFC** In-Flight Calibration 飞行校准
In-Flight Collision 空中相撞
In-Flight Computer 飞行用计算机
In-Flight Control 飞行控制
Instantaneous Frequency Correction 瞬时频率修正
Instrument Flight Center 仪表飞行中心
International Financial Center 国际金融中心
International Financial Corporations 国际金融公司

**IFCN** Interfacility Flow Control Network 设施(单位)间流量管制网络

**IFCS** Inflight Checkout System 飞行检查系统
Integrated Flight Control System 综合飞行控制系统

**IFD** Interface Device 接口设备

**IFDAPS** Integrated Flight Data Processing System 飞行数据综合处理系统

**IFE** In-Flight Entertainment 机上娱乐
In-Flight Emergency 空中紧急情况
Instantaneous Fuel Economy 瞬间耗油量
Instrument Flight Examiner 仪表飞行考官
Integrated Fluids Engineering 综合流体工程

**IFEC** In-Flight Entertainment Center 机上娱乐中心

**IFEO** International Flight Engineers Organism 国际随机工程师组织

**IFES** In Flight Entertainment System 机上娱乐系统

**IFF** Identification Friend or Foe 敌我识别
Institute of Freight Forwarders 航空货运公司协会
International Flying Farmers 国际飞行承包商

**IFFAS** International Financial Facility for Aviation Safety 国际航空安全财务机制

**IFFC** Integrated Flight and Fire Control 综合飞行和火力控制

**IFFCS** Integrated Flight and Fire Control System 综合飞行和火力控制系统

**IFFFA** International Federation of Air Freight Forwarders Association 航空货运代理商协会国际联合会

**IFFPC** Integrated Flight, Fire and Propulsion Controls 综合飞行、火力与推进控制

**IFH** Intelligent Frequency Hopping 智能跳频

**IFIM** International Flight Information Manual 国际飞行信息手册

**IFIP** Integrated Flight Instrument Panel 综合飞行仪表板
International Federation for Information Processing 国际信息处理联合会

**IFIS** Independent Flight Inspection System 独立飞行检查系统
Independent Frequency-Isolation System 独立频率隔离系统
Integrated Flight Instrument System 综合飞行仪表系统

**IFITL** Integrated Fiber in the Loop 综合光纤环路

**IFLA** International Federation of Library Associations 国际图书馆协会联合会

**IFM** In-Flight Maintenance 飞行中维修
In-Flight Monitor 机载监控器
Instantaneous Frequency Measurement 瞬时频率测量
Instrument Flight Manual 仪表飞行手册
Integrated Finance Measurement 综合财务计量

Integrated Flow Management 综合流量管理
Interface Memorandum 接口备忘录
Internal Flow Management 内部流量管理

**IFMA** In-Flight Mission Abort 飞行中断任务,中途停止执行任务
International Facility Management Association 国际设施管理协会
International Foodservice Manufacturers Association 国际食品制造商协会

**IFMC** Integrated Flight Management Computer 综合飞行管理计算机

**IFMN** Information 情报,消息,信息,资料

**IFMP** Integrated Flight Management Program 综合飞机管理规划
Integrated Fuel Management Panel 综合燃油管理板

**IFMR** Instantaneous Frequency Measurement Receiver 瞬时测频接收机

**IFMS** Integrated Fault Management System 综合故障管理系统
Integrated Finance Management System 综合财务管理系统
Integrated Fire Management System 综合消防管理系统
Integrated Flight Management System 综合飞行管理系统

**IFNC** Integrated Flight/Navigation Control 综合飞行及导航控制

**IFO** Intermediate Fuel Oil 燃油
International Field Office 国际地区办公室

**IFOG** Interferometric Fiber-Optic Gyro 光纤干涉式陀螺

**IFORS** International Federation of Operational Research Societies 国际运筹学学会联合会

**IFOV** Instantaneous Field of View 瞬时视场
Instrument Field of View 仪表视场

**IFP** Bullhead City, AZ, USA ( Airport Code ) 美国亚利桑那州布尔黑德城机场
Image Formation Processing 图像形成处理
In Flight Performance 飞行中性能

Initial Flight Path 起始飞行路线

Instrument Flight Procedure 仪表飞行程序

**IFPC** Integrated Flight and Propulsion Control 综合飞行和推力控制

**IFPM** In-Flight Performance Monitor 飞行中性能监视器

In-Flight Performance Monitoring 飞行中性能监控

**IFPPS** Initial Flight Plan Processing System 初始飞行计划处理系统

**IFPPU** Initial Flight Plan Processing Unit 初始飞行计划处理单元

**IFPS** Integrated Flight Plan System 综合飞行计划系统

International Fluid Power Society 国际流体动力协会

**IFPTE** International Federation of Professional and Technical Engineers 专业技术工程师国际联合会

**IFPU** Initial Flightplan Processing Unit 初始飞行计划处理单元

**IFPWC** Integrated Flight, Propulsion and Weapon Control 综合飞行、推力和武器控制

**IFQA** Integrated Flight Quality Assurance 综合飞行质量保证

**IFQC** Individual Fuel Quantity Channel 单独燃油量通道

**IFR** Increasing Failure Rate 故障率递增型

In-Flight Refueling 空中加油

In-Flight Repair 空中修理,飞行中修理

Infrared 红外的

Instrument Flight Recovery 恢复仪表飞行

Instrument Flight Rules 仪表飞行规则

International Flight Research Corp. (USA) 国际飞行研究公司(美国)

**IFRB** International Frequency Registration Board 国际频率注册委员会,国际频率登记委员会

International Frequency Regulation Board 国际频率规章委员会

**IFRS** International Financial Reporting Standards 国际财务报告准则

**IFRU** In-Flight Replaceable Unit 飞行中可替换装置

Interference Frequency Rejection Unit 干扰频率滤波器,干扰频率抑制器

**IFS** Instrument Flight Simulator 仪表飞行模拟机

Instrument Flight System 仪表飞行系统

Integrated Facilities System 综合设施系统

Integrated Flight System 综合飞行系统

Interactive Financial Services 交互式金融服务

Interface Specification 接口规范

International Freephone Service 国际免费电话(被叫付费电话)

**IFSA** Inflight Food Service Association 航空食品服务协会

Inflight Service Association 空中勤务协会

**IFSAU** Integrated Flight System Accessory Unit 综合飞行系统附属装置

**IFSD** In-Flight Shutdown (Rate) 空中停车

International Financial Services District 国际金融服务区

**IFSM** In-Flight Status Monitoring 飞行中状态监控

**IFSO** Independent Flight Safety Office 独立的飞行安全办公室

In-Flight Security Officers 机上保安员

**IFSR** In-Flight Status Report 飞行状态报告

Integrated Flight Software Requirement 综合飞行软件需求

**IFSS** Inboard Front Spar Station 内侧前梁站位

Instrumentation and Flight Safety System 仪表测量和飞行安全系统

International Flight Service Station 国际飞行服务站

**IFT** In-Flight Training 飞行训练

Instrument Flight Trainer 仪表飞行训练器

Integral Fuel Tank 整体油箱

Integrated Functional Test 综合功能测试

Interactive Flight Technologies 交互式飞行技术

Intermediate Frequency Transformer 中频变压器

**IFTA** In Flight Thrust Augmentation 飞行中增大推力

**IFTB** In Flight Turn Back 空中返航

**IFTC** Integrated Flight Trajectory Control 综合飞行轨迹控制

International Film and Television Council 国际电影电视理事会

**IFTD** Instrument Flight Training Division 仪表飞行训练处

**IFTM** In Flight Test and Maintenance 空中试车与维护

**IFTO** International Federation of Tour Operators 旅行社国际联合会

**IFTP** In Flight Test Program 空中测试计划

**IFTS** Integrated Functional Test System 综合功能检测系统

**IFTSA** Integrated Flight Test Simulate Aircraft 综合飞行测试模拟飞机

**IFTU** Intensive Flying Trails Unit 加强飞行轨迹组件

**IFU** Information Functional Unit 信息功能单元

Interface Unit 接口单元

Interworking Functional Unit 互通功能单元

**IFUN** If Unable 如不可能

**IFV** In-Flight Visibility 飞行中能见度

Integral Fill Valve 整体加注活门

**IFWC** Integrated Flight and Weapon Control 飞行和武器综合控制

**IFZ** Independent Fault Zone 独立故障区

Industrial Free Zone 工业免税区

**IG** Imperial Gallon 英加仑(合 4. 546 升)

Inaugural Guests 被邀请参加首航典礼的客人

Inertial Guidance 惯性制导

Interactive Graphics 交互式图形

International Gateway 国际网关

**IGA** Integrated Gate Array 集成网关矩阵

Integrated Graphics Array 综合图形阵列

Intergovernmental Agreement 政府间合作协议

International General Aviation 国际通用航空

**IGAC** International Global Atmospheric Chemistry 国际全球大气化学

**IGACS** Integrated Guidance and Control System 综合制导和控制系统

**IGB** Inlet Gear Box 进口齿轮箱

**IGBP** International Geosphere-Biosphere Programme 国际地球生物圈计划

**IGBP-DIS** International Geosphere-Biosphere Program Data and Information System 国际地球生物圈计划数据与信息系统

**IGBT** Insulated Gate Bipolar Transistor 绝缘栅双极型晶体管

Integrated Gate Bipolar Transistor 集成门双极型晶体管

**IGD** Interaction Graphics Display 交互式图形显示

**IGDS** Inert Gas Distribution System 惰性气体分配系统

**IGE** In Ground Effect 受地面效应作用,有地效

Instrumentation Graphics Environment 仪表制图环境

Instrumentation Ground Equipment 地面仪表设备

**IGEC** International General Electric Corporation 国际通用电气公司

**IGES** Initial Graphics Exchange Specification 初始图像交换规范

Initial Graphics Exchange Standard 初始图像交换标准

Institute for Global Environmental Strategies 全球环境战略研究所

International Graphics Exchange Specification 国际图像交换规范

**IGFET** Insulated Gate Field Effect Transistor 绝缘栅场效应晶体管

**IGFOV** Instantaneous Geometric Field of View 瞬时几何视场

**IGGS** Inert Gas Generating Systems 惰性气体发生系统

**IGI** Instrument Ground Instructor 仪表飞行地面教员

**IGIA** Interagency Group on International Aviation 部门间国际航空事务组

**IGIG** Inertial Guidance Integrating Gyro 惯性制导积分陀螺仪

**IGL** Interactive Graphics Library 交互式图形库

**IGM** Implementation Guide Manuals 执行规则手册
Kingman, AZ, USA(Airport Code) 美国亚利桑那州金曼机场

**IGMP** Internet Group Management Protocol 因特网组管理协议

**IGN** Ignition 点火, 发火, 点火开关
Ignore 不理睬

**IGOS** Integrated Global Observing Strategy 一体化全球观测战略伙伴关系

**IGOSS** Integrated Global Ocean Station System 全球联合海洋台站网

**IGP** Ignition Point 燃点
Inertial Guidance Platform 惯性制导平台
Interior Gateway Protocol 内部网关协议

**IGPM** Imperial Gallon per Minute 英加仑/分

**IGRF** International Geomagnetic Reference Field 国际参照地磁场

**IGRP** Interior Gateway Routing Protocol 内部网关路由协议

**IGS** Inertial Guidance System 惯性制导系统
Information Group Separator 信息组分隔符
Instrument Guidance System 仪表引导系统
Integrated Graphic System 综合图像系统

**IGSO** Inclined Geosynchronous Satellite Orbit 倾斜地球同步卫星轨道

**IGTC** Inertial Guidance Test Center 惯性制导试验中心

**IGU** Iguassu Falls, Parana, Brazil ( Airport Code) 巴西巴拉那河伊瓜苏福尔斯机场
International Geographical Union 国际地理学协会

**IGV** Inlet Guide Valve 进气管导向活门
Inlet Guide Vanes 进口导向叶片

**IGVA** Inlet Guide Vane Actuator 进气管导向叶片作动器

**IGW** Increased Gross Weight 增加全重

**IGY** International Geophysical Year 国际地球物理年

**IH** Inhibition Height 禁止高度(近地警告系统)
Initial Heading 初始航向

**IHADSS** Integrated Helmet and Display Sighting System 综合头盔与显示瞄准系统

**IHAS** Integrated Hazard Avoidance System 综合危害规避系统
Integrated Helicopter Avionics System 直升机综合航空电子系统
Integrated Helmet Assembly Sub-System 综合头盔装配子系统

**IHC** Information Handling Committee 信息处理委员会
Integrated Handset Controller 综合手机控制器

**IHCS** Integrated Helicopter Control System 直升机综合操作系统

**IHDL** Input Hardware Descriptive Language 输入硬件描述语言

**IHDP** International Human Dimensions Programme on Global Environmental Change 国际全球环境变化人文因素计划

**IHFF** Inhibit Halt Flip-Flop 禁止暂停触发器

**IHL** Internet Header Length 因特网报头长度

**IHO** International Hydrographic Organization 国际海道测量组织

**IHOC** International Helicopter Operations Committee 国际直升机运营委员会

**IHP** Indicated Horse Power 指示马力

**IHPH** Indicated Horse Power per Hour 指示马力/小时

**IHPTET** Improved High Performance Turbine Engine Technology 改进的高性能涡轮发动机技术

Integrated High Performance Turbine Engine Technology 综合高性能涡轮发动机技术

**IHR** Increased Hazard Rate 增长的危险率

Infrared Heterodyne Radiometer 红外差辐射仪

International Health Regulations 国际卫生条例

**IHSS** International Helicopter Safety Symposium 国际直升机安全会议

**IHST** International Helicopter Safety Team 国际直升机安全团队

**IHV** Independent Hardware Vendor 独立硬件商

**II** Image Information 图像信息

Image Intensifier 图像增强器

**IIA** Interactive Instructional Authoring 交互式教学写作

Internet Image Appliance 网络影像家电

**IIAL** Institute of International Air Law 国际航空法学会

**IIAS** Interactive Instructional Authoring System 交互式教学写作系统

**IIASA** International Institute for Applied Systems Analysis 国际应用系统分析研究所

**IIB** International Investment Bank 国际投资银行

**IIC** Incoming International Center 收入国际中心

International Institute of Communications 国际通信协会

Investigator-in-Charge 主管调查员

Item Identification Code 项目识别码

**IICL** Institute of International Container Lessors 国际集装箱出租商学会

**IID** Image Intensifier Device 图像增强设备

Independent and Identically Distributed 独立同分布

Integrated Information Display 综合信息显示

**IIDS** Integrated Instrument Display System 集成仪表显示系统

**IIF** Image Interchange Facility 图像交换设备

Immediate Interface 直接接口

Inserted in Flight（数据，目标资料等）在飞行中插入

**IIIN** Intelligent Integrated Information Network 智能综合信息网络

**IILS** International Institute for Labour Studies 国际劳工问题研究所

**IIMS** Integrated Information Management System 综合信息管理系统

**IINS** Integrated Inertial Navigation System 综合惯性导航系统

**IIO** Interface/Input/Output 接口/输入/输出

**IIP** Information in Progress 处理中的信息

Instantaneous Impact Prediction 连续弹着点预测，瞬时弹着点预测

Interface Implementation Plan 接口执行计划

Interface Information Processor 接口信息处理器

**IIPACS** Integrated Information Presentation and Control System 综合信息显示与控制系统

**IIPD** Illustrated Initial Provisioning Document 图解初始提供备件文件

**IIR** Imaging Infrared 红外成像

Integrated Instrumentation Radar 综合测量仪表雷达

**IIRA** Integrated Inertial Reference Assembly 集成惯性基准组件

**IIRS** Instrument Inertial Reference Set 仪表惯性基准装置

**IIS** Integrated Information System 综合信息系统

Integrated Instrument System 综合仪表系统

Internet Information Service 因特网信息服务

IR Imaging System 红外线照相系统

**IISA** Integrated Inertial Sensor Assembly 综合惯性敏感器组件

**IISD** International Institute for Sustainable Development 国际可持续发展研究所

**IISL** International Institute of Space Law 国际太空法律协会

**IISS** Integrated Information Support System 综合信息支援系统

**IISSP** Interim Inter Switch Signaling Protocol 临时交换机间的信令协议

**IIT** Individual It Fare 单个游览票价

Information Integration Technology 信息集成技术

Integrated Information Technology 综合信息技术

Intelligent Interface Technology 智能接口技术

Internal Information Transfer 内部信息传递

**IITA** Information Infrastructure Technology and Application 信息基础设施技术及应用

**IITF** Information Infrastructure Task Force 信息基础设施任务组

**IIU** Inertial Instrument Unit 惯性仪表组件

Input Interface Unit 输入接口组件

**IIVM** Integrated Intelligent Vehicle Management 综合智能飞行器管理

**IJA** International Jet Air, Ltd. (Canada) 国际喷气航空有限公司(加拿大)

**IKA** Imam Khomeini, Iran (Airport Code) 伊朗伊玛姆科梅尼机场

**IKBS** Intelligent Knowledge Based System 基于知识的智能系统

**IKE** Internet Key Exchange 因特网密钥交换

**IKF** Iterated Kalman Filter 迭代卡尔曼滤波器

**IKPT** Initial Key Personnel Training 关键人员初级训练

**IKT** Irkutsk, Russia (Airport Code) 俄罗斯伊尔库茨克机场

**IL** Index List 索引清单

Insertion Loss 插入损耗

Instrument Landing 仪表着陆

**ILA** Instrument Landing Approach 仪表着陆进近

Instrument Low Approach 仪表低空进近

International Lauguage for Aviation 国际航空用语

International Law Association 国际法律协会

**ILAC** International Laboratory Accreditation Conference 国际实验室审定委员会

**ILAF** International Lawyer's Association of Finland 芬兰国际律师协会

**ILAS** Instrument Landing Approach System 仪表着陆进近系统

Instrument Low-Approach System 仪表低空进近系统

**ILC** Intelligent Line Card 智能线路卡

International Labour Conference 国际劳工大会

United Nations International Law Commission 联合国国际法委员会

**ILD** Injection Laser Diode 注入式激光二极管

Insertion Loss Deviation 插入损耗偏差

**ILDP** Instrument Let-Down Procedure 仪表穿云下降程序

Integrated Logistics Data Program 综合后勤(资源)数据程序

**ILDS** Integrated Logistics Data System 综合后勤(资源)数据系统

**ILE** Integrated Language Environment 综合语言环境

**ILEC** Incumbent Local Exchange Carrier 在业的本地交换运营公司

**ILEF** Inboard Leading Edge Flaps 内侧前缘襟翼

**ILES** Inboard Leading Edge Spar 内侧前缘梁

Inboard Leading Edge Station 内侧前缘站位

**ILF** Integrated Logistics File 综合后勤(资源)文档

**ILFC** International Lease Finance Corporation 国际租赁金融公司

**ILI** Idle Line Indicating 空闲线路指示

**ILL** Illustrated or Illustration or Illustrator 说明

**ILLUM**   Illuminate, Illuminated 照亮
Illumination 照明, 点亮

**ILLUMD**   Illuminated 点亮的

**ILLUS**   Illustrate, Illustration 图解

**ILM**   Independent Landing Monitor 独立着陆监视器
Information Lifecycle Management 信息生命周期管理
Integrated Logistics Management 综合物流管理
Intermediate Level Maintenance 中级维护
Wilmington, NC, USA (Airport Code) 美国北卡罗来纳州威尔明顿机场

**ILMA**   Independent Lubricant Manufacturers Association 独立润滑剂制造商协会

**ILMI**   Integrated Local Management Interface 综合本地管理接口
Interim Local Management Interface 临时本地管理接口

**ILO**   Iloilo, Philippines (Airport Code) 菲律宾伊洛伊洛机场
International Labour Office 国际劳工局
International Labour Organization 国际劳工组织

**ILP**   Integrated Logistics Provider 综合物流商

**ILS**   Instrument Landing System 仪表着陆系统
Instrumentation Level Simulator 仪表等级模拟器
Integrated Landing System 综合着陆系统
Integrated Learning System 综合学习系统
Integrated Lodging Service 综合住宿服务
Integrated Logistic Support 综合后勤保障

**ILSAC**   International Lubricant Standardisation and Approval Committee 国际润滑剂标准化和批准委员会

**ILSAP**   ILS Approach 仪表着陆系统进近

**ILSCP**   Instrument Landing System Control Panel 仪表着陆系统控制板

**ILSLA**   Injection Locked Semiconductor Laser Amplifier 注入锁定半导体激光放大器

**ILSMP**   Integrated Logistic Support Management Plan 综合逻辑支援管理计划

**ILSMT**   Integrated Logistics Support Management Team 综合后勤保障管理组

**ILSP**   Integrated Logistics Support Plan 综合后勤保障计划

**ILSPPS**   Integrated Logistics Support Plan for Preoperational Support 预运作支持综合后勤支持计划

**ILSS**   Integrated Life Support System 综合生命支持系统

**ILSWG**   Integrated Logistics Support Working Group 综合后勤保障工作组

**ILT**   Imprecisely Located Targets 非精确定位目标

**ILZSG**   International Lead and Zinc Study Group 国际铅锌研究小组

**IM**   Image Mixing 图像混合
Import 进口
Improvement Maintenance 改进性维修
Inner Marker 近台, 内指点标, 内部标记
Instant Messaging 即时传信
Instrument Measurement 仪表测量
Integrated Modem 集成式调制解调器
Interface Module 接口模块
Interline Message 航空同业文电
Inventory Manager 库存经理
Inverse Multiplexing 反向复用
Item Manager 项目经理

**IMA**   Integrated Modular Avionics 综合模块化航空电子
Interactive Multimedia Association 交互式多媒体协会
Intermediate Maintenance Activity 中级维修机构

**IMAA**   International Miniature Aircraft Association 国际小型航空器协会

**IMAD**   Integrated Multisensor Airborne Display 机载多传感器综合显示器

**IMAL**   Integrity Monitor Alarm 完好性监视告警

**IMAP** Interactive Mail Access Protocol 交互邮件访问协议

Internet Message Access Protocol 因特网信息存取协议

**IMAS** Industrial Management Assistance Survey 工业管理支援调查

Influence Modeling and Assessment System 影响建模与评估系统

International Marine and Shipping Conference 国际海洋和航运会议

**IMAWP** Initial Missed Approach Waypoint 起始复飞航路点

**IMBAL** Imbalance 不平衡

**IMC** Image Motion Compensation 图像运动补偿

Indirect Maintenance Cost 间接维修成本

Information Management Committee 信息管理委员会

Instrument Meteorological Conditions 仪表气象条件

Integrated Marketing Communications 整合营销传播

Inter-Module Connector 模块间连接器

International Maintenance Center 国际维护中心

International Meteorological Center 国际气象中心

Investment Management Committee 投资管理委员会

Item Management Code 项目管理编码

**IMCC** Inter-Module Communication Controller 模块间通信控制器

**IMCS** Interactive Manufacturing Control System 交互式制造业控制系统

Interim Monitoring/Control Software 过渡性监视/管制软件

**IMCSS** Interim Military Communication Satellite System 临时军事通信卫星系统

**IMCT** Intermodule Compatibility Testing 模块间兼容性测试

**IMD** Immediate(ly) 立即

India Meteorological Department 印度气象局

Intermodulation Distortion 交叉调制失真,互调失真

Internal Management Document 内部管理文件

**IMDB** Imagery Management Database 图像管理数据库

Integrated Maintenance Database 综合维修数据库

**IMDC** Integrated Mission Design Center 综合任务设计中心

**IMDS** Integrated Maintenance Data System 维护数据集成系统

Integrated Mechanical Diagnostics System 综合机械诊断系统

Interior Materials Data System 内部器材数据系统

International Material Data System 国际资料数据系统

**IMEI** International Mobile Equipment Identification 国际移动设备识别

International Mobile Equipment Identity 国际移动设备标识

**IMEP** Indicated Mean Effective Pressure 指示平均有效压力

**IMERP** Integrated Monitoring, Evaluation and Research Plan 综合监控、评价和研究计划

**IMF** Imphal, India (Airport Code) 印度英帕尔机场

Integrated Maintenance Facility 综合维修设施

Integrated Management Framework 综合管理框架

Intermediate Fiber 中间光纤

International Monetary Fund 国际货币基金组织,国际货币基金会

**IMG** Imperial Gallons 英制加仑

Inertial Measurement Group 惯性测量组

Information Management Group 信息管理组

**IMHO** International Medical Health Organisation 国际医疗卫生机构

**IMI** Imbedded Message Identifier 嵌入的信息标志符

Installation/Maintenance Instructions 安装及维修说明

Instrument Memory Indicator 仪表存储指示器

Interim Manned Interceptor 临时有人驾驶截击机

Intermediate Maintenance Instruction 中间维修说明

Intermediate Manned Interceptor 中型有人驾驶截击机

**IMICS** Integrated Management Information and Control System 综合管理信息与控制系统

**IMIF** International Maritime Industry Forum 国际航运业论坛

**IMIP** Industrial Management Improvement Program 工业管理改进计划

Industrial Modernization Incentive Program 产业现代化激励计划

**IMIS** Integrated Maintenance Information System 综合维修信息系统

Integrated Management Information System 综合管理信息系统

**IMIT** Imitation 模仿,仿造,伪造

Institute of Management and Information Technology 管理和信息技术研究所

**IML** International Microgravity Laboratory 国际微重力实验室

**IMLI** Import Licence 进口执照

**IMM** Immediate 立即的,即刻的

Inflight Maintenance Message 飞行中维护信息

Institute of Marketing Management (South Africa) 营销管理学院(南非)

Integrated Management Model 综合管理模型

Integrated Material Management 综合航材管理

Intelligent Information Management 智能信息管理

Interacting Multiple Model 交互多模型

Intermediate Maintenance Manual 中级维修手册

Intermodule Memory 模块间存储器

International Marketing and Management 国际市场营销和管理

International Monetary Market 国际货币市场

**IMMH** Indirect Maintenance Man Hour 间接维修工时

**IMMIG** Immigrant 移民

**IMMP** Information Management Master Plan 信息管理主计划

Information Mission Management Plan 信息任务管理计划

Integrated Maintenance and Modernization Planning 综合维修与现代化计划

**IMMS** Installation Maintenance Management System 安装维修管理系统

Integrated Maintenance Management System 综合维修管理系统

Integrated Maintenance Monitor System 综合维修监控系统

Interim Maintenance Management System 临时维修管理系统

Interim Manpower Maintenance System 临时人力维修系统

**IMMT** Integrated Maintenance Management Team 综合维修管理组

Intelligent Memory Management Technology 智能内存管理技术

**IMMTE** Integrated Mix Manifold Temperature Error 整体混合总管温度误差

**IMN** Indicated Mach Number 指示马赫数

**IMNI** Internal Multimedia Network Infrastructure 多媒体网络内部基础设施

**IMO** International Maritime Organization 国际海事组织

International Meteorological Organization 国际气象组织

International Money Order 国际汇票

Inventory Management Organization 库存管

理机构

**IMP** Impact 碰撞

Impedance 阻抗

Imperatriz, Brazil（Airport Code）巴西因佩拉特里斯机场

Imperial 英制

Improve 改进,变好

Individual Mission Plan 单独飞行任务计划

Industrial Management Program 工业管理程序

Information Management Plan 信息管理计划

Integrated Maintenance Procedure 综合维护程序

Integrated Mechanism Problem 综合机构问题

Integrated Mechanism Program 综合机构程序

Integrated Memory Processor 综合存储处理机

Integrated Mission Processor 综合任务处理机

Integrated Motor Propulsor 集成电机推进器

Integration Management Plan 综合管理计划

Interface Message Processor 接口信息处理器,接口报文处理器

Interface Module Processor 接口模块处理器

Inter-Message Processor 内部信息处理程序

**IMPA** International Maritime Pilots Association 国际海上飞行员协会

**IMPACT** Improvement Manufacturer Program for Airbus Customer Training 改进的空客客户培训厂家计划

Integrated Management Planning and Control Technique 综合管理计划与控制技术

Inventory Management/ Procurement and Control Technique 库存管理/订购和控制技术

**IMPCE** Importance 重要性

**IMPD** Interactive Multipurpose Display 交互式多功能显示器

**IMPER** Imperative 紧急的,强制的

**IMPO** Impossible 不可能

**IMPP** Integrated Maintenance Planning Program 综合维修计划程序

**IMPR** Improve 改进

**IMPROV** Improvement 改进,变好

**IMPS** Integrated Manufacturing Production System 综合制造生产系统

Integrated Master Programming and Scheduling 综合主计划和时间安排

Integrated Multi-Purpose System 多功能综合系统

**IMPT** Import 进口

Important 重要的

**IMPV** Imperative 紧急的,强制的

**IMPVD** Improved 改进的,变好的

**IMPVMT** Improvement 改进,改善

**IMR** Indicated Mach Reading 指示马赫数

Interlock Monitoring Relay 互锁监控继电器

**IMRL** Increasing Mean Residual Life 增加产品平均剩余寿命

Individual Material Readiness List 个人器材准备单

Integrated Materials Research Laboratory 综合材料研究实验室

Intermediate Maintenance Requirements List 中级维修需求单

**IMS** Inertial Measuring Set 惯性测量装置

Inertial Measuring System 惯性测量系统

Inflight Management System 飞行中的管理系统

Information Management System 信息管理系统

Integrated Management System 综合管理系统

Integrated Multiplex System 综合多路转换系统

Integrity Monitoring System 完好性监视系统

Interactive Multimedia Service 交互式多媒体服务

**IMSI** International Mobile Subscriber Identifier 国际移动用户标识符

**IMSO** International Mobile Satellite Organization 国际移动卫星组织

**IMSS** Inflight Medical Support System 空中医疗保障系统

**IMT** Information Management Team 信息管理团队

Intelligent Multimode Terminal 智能多模式终端

Intermediate Maintenance Trainer 中级维护训练器

International Mobile Telecommunications 国际移动通信

Iron Mountain, MI, USA ( Airport Code ) 美国密执安州艾恩芒廷机场

**IMT2000** International Mobile Telecommunications2000 国际移动通信 2000

**IMTC** International Multimedia Television Committee 国际多媒体电视委员会

**IMTNAP** International Meteorological Telecommunication Network in Asia and Pacific 亚太地区国际气象电信网

**IMTNE** International Meteorological Teleprinter Network in Europe 欧洲国际气象电传电报网

**IMTS** Improved Mobile Telephone Service 改进型移动式电话业务

**IMTV** Interactive Multimedia Television 交互式多媒体电视

**IMU** Inertial Measurement/Measuring Unit 惯性测量部件,惯性测量装置

Internal Measuring Unit 内部测量仪

**IMUX** Input Multiplex 输入复用

**IMV** Instrumented Measurement Vehicle 仪表测量飞行器

**IN** Inch, Inches 英寸

Inertial Navigation 惯性导航

Inertial Navigator 惯性导航仪

Infant 婴儿

Information Need 信息需求

Inspection 检验,检查

Instructor Navigator 领航教官

Integrated Network 综合网络

Intelligent Network 智能网络

Interconnected Network 互联网络

Interest 利息,兴趣

Internal Node 内节点

**INA** Information Network Architecture 信息网体系结构

Initial Approach 起始进近

Integral Network Arrangement 整体网络布局

Integrated Network Architecture 综合网络体系结构

**INABL** Inability 无能力的

**INAC** Inaccuracy 不准确性

**INAD** Inadmissible Passenger 拒绝入境旅客

**INAP** Intelligent Network Application Part 智能网应用部分

Intelligent Network Application Protocol 智能网应用协议

**INAS** Inertial Navigation and Attack System 惯性导航和攻击系统

Integrated Night Attack Sensor System 综合夜间攻击传感系统

**INB** Inboard 机内,内侧

Inbound 进场,向台

**INBD** Inboard 归航的,向台(电)飞行的

**INC** In Cloud 云中

Include, Including 包括

Integrated Network Connection 综合网络连接

Integrated Numerical Control 集成数字控制

Item Name Code 项目名称代码

**INCA** Integrated Navigation and Communication Automatic 综合导航和自动通信

**INCAS** Integrated Navigation and Collision Avoidance System 综合导航与防撞系统

**INCC** International Network Controlling Center 国际网络控制中心

**INCDT** Incident 事件

**INCDU** Integrated Navigation Control/Display Unit 综合导航控制/显示组件

**INCE** Institute of Noise Control Engineering

噪声控制工程研究所

**INCERFA** Incertitude for Aircraft 飞机情况不明

**INCL** Inclose 封入

Inclosure 信封内

Inclusive or Include or Included or Including 包括

**INCM** Intelligent Network Conceptual Model 智能网概念模型

**INCOMG** Incoming 进来的,进场的

**INCON** Interconnect 交连

**INCOR** Incorrect 不正确

**INCOSE** International Council on Systems Engineering 国际系统工程学会

**INCOT** Including Over Time 包括加班时间

**INCP** Incapacitated Passenger 残障或病患旅客

Incapacitated Passenger ( Non-Medical Clearance ) 行动不便的旅客 ( 无需提供医疗放行证明 )

Incomplete 未完成的,不完全的

**INCR** Increase 增加,增大

Increment 增量

**INCU** Inertial Navigation Control Unit 惯性导航控制装置

**INCVENCE** Inconvenience 不方便

**IND** Independent 独立的

Index 指数

India 印度

Indianapolis, IN, USA ( Airport Code ) 美国印第安纳州印第安纳波利斯机场

Indicator 指示器

Inductance 电感,感应 ( 系数 )

Induction 感应器,电感线圈

Industry 工业

**INDB** Intelligent Network Database 智能网数据库

**INDBMS** Intelligent Network Database Management System 智能网数据库管理系统

Interior Noise Database Management System 内部噪声数据库管理系统

**INDCU** Inertial Navigation Display and Control Unit 惯性导航显示和控制装置

**INDEL** Indefinite Delay 不定期延期

**INDEM** Indemnity 赔偿,保障

**INDENT** Indenture 合同,契约,凭单,双联合同

**INDEP** Independent 独立的,单独的

**INDET** Indeterminate 未定的,不定的,模糊的

**INDG** Indicating 指示

**INDIR** Indirect 不直接的,间接的

**INDIV** Individual 个别的

**INDL** Inflight Data Link 飞行中数据链

**INDR** Indoor 室内

**INDREG** Inductance Regulator 电感调节器

**INDSENS** Indicator Sensitivity 指示器灵敏度

**INDT** Indent 压痕,缩进

**INDTR** Indicator-Transmitter 指示器—发送器

**INDV** Individually 单独地,个别地

**INDX** Index 索引,指数,标志

**INE** Inertial Navigation Element 惯性导航元件

Inertial Navigation Equipment 惯性导航设备

Institute for New Energy 新能源研究所

Intelligent Network Element 智能网元素

**INEA** International Electronics Association 国际电子学协会

**INERU** International Noise Exposure Reference Unit 国际噪声暴露基准单位

**INEV** Inevitable 不可避免的

**INEWS** Integrated Electronic Warfare System 综合电子战系统

**INF** Indefinite 无限的,不定的

Inferior 低劣的

Information 情报,资料,信息

Inland Navigation Facility 内陆导航设施

Interface 交连,交接面

**INFD** Informed 接到通知的,消息灵通的

**INFI** Integrated Navigation and Flight Inspection 综合导航和飞行检查

**INFLT** Inflight 在飞行中,在空中

**INFM** Inform 通知,告知

Intelligent Network Functional Model 智能网功能模型

**INFML** Informal 非正式的

**INFO** Information 信息,情报,资料

Integrated Network Using Fiber Optics 采用光纤的综合网

**INFORSAT** Information Transfer Satellite 信息传输卫星

**INFR** Infrared 红外的

**ING** Inertial Navigation and Guidance 惯性导航和制导

**INH** Inhibit 抑制

**INHAB** Inhabitants 居民

**INHIB** Inhibition or Inhibited 抑制(的),禁止(的)

**INI** Intelligence Network Interface 智能网络接口

**INIT** Initialization 初始化

Initiation 入门,开始实施

**INJC** Injection 注射

**INL** International Falls, MN, USA (Airport Code) 美国明尼苏达州国际瀑布城机场

**IN-LBS** Inch-Pounds 英寸—磅

**INLK** Interlock 连锁

**INLS** Integrated Navigation and Landing System 综合导航与着陆系统

**INM** Integrated Network Management 综合网络管理

Integrated Network Monitoring 综合网络监视

Integrated Noise Model 综合噪声模型

**INMARSAT** International Maritime Satellite Organizaion 国际海事卫星组织

**INMC** International Network Management Center 国际网络管理中心

**INMOS** In Service Management and Operation System 智能网业务管理及运行系统

**INMS** Information Network Management System 信息网络管理系统

Integrated Network Management System 综合网络管理系统

Intelligent Network Management System 智能网络管理系统

**INN** Innsbruck, Austria (Airport Code) 奥地利因斯布鲁克机场

Intermediate Network Node 中间网络节点

**INNO** In Network Operator 智能网运营商

**INOP** Inoperation 不工作

Inoperative 不工作的,不工作

**INP** If Not Possible 如不能

Input or Inputting 输入

Integrated Network Processor 综合网络处理器

Intelligence Processor 情报处理器

Intelligent Network Processor 智能网络处理器

International Network Planning 国际网络计划

Introduction of New Products 新产品介绍

**INPH** Interphone 内话机,对讲机

**INPR** In Progress 在进行中

**INQ** Inquire 询问

Inquiry 询问,询价

**INR** Indian Ruppee 印度卢比(货币单位)

Indicator Register 指示寄存器

Inertial Navigation Reliability 惯性导航可靠性

Inner 内部的,内侧的

Item Number Responsibility 项目数可供性,项目数可靠性

**INREQ** Information Request 信息要求

**INRO** International Natural Rubber Organization 国际天然橡胶组织

**INRTFLR** Inert Filler 中性充填剂

**INRTG** Inert Gas 惰性气体

**INRTL** Inertial 惯性的

**INS** Immigration and Naturalization Service 移民和归化局

Inertial Navigation Sensor 惯性导航传感器

Inertial Navigation System 惯性导航系统

Information Network System 信息网络系统

Insert 插入

Instrument 仪表

Integrated Navigation System 综合导航系统

Intelligent Network Service 智能网络服务

International Navigation System 国际导航系统

International Numbering System 国际编号系统

**INSACS** Interstate Airway Communication Stations 洲际航路通信台

**INSARST** International Search and Rescue Satellite 国际搜寻援救卫星

**INSAT** International Satellite 国际卫星

**INSCS** Integrated Navigation Steering and Control System 综合导航操纵与控制系统

**INSDU** Inertial Navigation System Display Unit 惯性导航系统显示组件

**INSES** In Services Emulation System 智能网业务仿真系统

**INSIG** Insignificant 无意义的

**INSL** Insulated 绝缘的

**IN-SL** In Service Logic 智能网业务逻辑

**IN-SM** Intelligent Network Switching Manager 智能网交换管理器

**INSOS** In Service Operation System 智能网业务操作系统

**INSP** Inspect 检验,检查

Intelligent Network Service Provider 智能网服务供应商

**INSPR** Inspector 检察员

**INSS** Intelligent Network Service Subscriber 智能网业务用户

**INSSCC** Interim National Space Surveillance Control Center 临时国家空间监视控制中心

**IN-SSM** Intelligent Network Switching State Manager 智能网交换状态管理器

Intelligent Network Switching Status Model 智能网交换状态模型

**INST** Install 安装

Installation 安装,设备

Instant (Immediate) 立即,紧急的

Instinctive 本能的

Instruction 说明,细则

Instrument 仪器,仪表

**INSTD** Instead 代替

**INSTL** Install, Installation 安装

**INSTLLD** Installed 安装的

**INSTM** Instrumentation 仪表化

**INSTN** Institution 设立,公共设施

**INSUF** Insufficient 不足的

**INSUL** Insulate 隔离

Insulated 绝缘的,保温的

Insulation 绝缘,保温

Insulator 绝缘体

**INT** Intake 入口,进气道

Integral 整体的

Integrating 整体的,综合的

Intensity 强度

Interactive News Television 交互式电视新闻

Interim 暂时的

Interior 内部

Intermittent 间歇的

Internal 内部的,国内的

International 国际的

Interphone 对讲机,内话机

Interrogator 询问机

Interrupt 中断

Intersection 交叉点,相交,交叉,横截

**INT/F** International Fare 国际票价

**INTC** Intercept 截获,切入

Interception 切入,相交

Interconnect 交连

**INTCHBLE** Interchangeable 可交替的,可互换的

**INTCHG** Interchange 互换,交替

**INTCO** International Code of Signals 国际信号码

**INTCON** Interconnection 互联

**INTEG** Integrator 积分器

**INTELSAT** Intelligence Satellite 间谍卫星

International Satellite Service 国际卫星服务

International Telecommunications Satellite Organization 国际通信卫星组织

**INTEN** Intensity 强度,亮度

**INTER** Intermittent 间歇的

Internal 内部的

**INTERCO** International Code of Signals 国际信号码

**INTERCOM** Intercommunication 内部通讯,双向通信

**INTERN** International 国际的

**INTERP** Interpreter 翻译

**INTFC** Interface 接口,交接面

Interference 干扰

**INTIMN** Intimation 通告

**INTIP** Integrated Information Processing 综合信息处理

**INTIPS** Integrated Information Processing System 综合信息处理系统

**INTK** Intake 入口,进气

**INTL** Initial 起始的,开始的

Internal 内部的

International 国际的

**INTLK** Interlock 互锁

**INTMD** Intermediate 中间的

**INTN** Intention 打算,意向

International 国际的

**INTOSAI** International Organization of Supreme Audit Institutions 国际最高审计机关组织

**INTR** Interruption 中断

**INTRF** Interference 干涉,干扰

**INTRG** Interrogate 询问

Interrogator 询问机

**INTRLN/R** Interline Rate 联运运费

**INTRO** Introduction 介绍,引进

**INTRP** Interrupt 中断,干扰

Interruption 中断,干扰

**INTRP(T)** Interrupt 中断,干扰

**INTRPTD** Interrupted 阻抗或间断

**INTRST** Interest 利息,兴趣

**INTS** Integrated National Telecommunications System 国家综合通信系统

International Switch 国际交换

Inter-Network Time Slot 网络内部时隙

**INTSE** Intelligent System Environment 智能系统环境

**INTSF** Intensify 加强,加剧,强化

Intensifying 增强

**INTST** Intensity 强度

Interest 利息,兴趣

**INTSTD** Interested 有关的,感兴趣的

**INTSV** Intensive 密集的

**INTSYS** Inter System 内部系统

**INTVL** Interval 间隔,周期

**INTVW** Interview 会见,访问

**INU** Inertial Navigation Unit 惯性导航组件

Nauru, Nauru(Airport Code) 瑙鲁瑙鲁岛机场

**INURACK** Inertial Navigation Unit Rack 惯性导航组件架

**INV** In Vain 无效

Invalid 无效的

Invent 发明

Inventory 清单

Invert 倒的,相反的,逆的

Inverter 变流机

Invoice 发票

**INVEST** Investigate, Investigation 调查,考察

**INVIT** Invitation 邀请

**INVMT** Investment 投资

**INVO** Invoice 发票

**INVRN** Inversion 逆温

**INVTR** Inverter 变流机,变压器

**INWP** Intermediate Waypoint 中间航路点

**IO** In Order 有顺序

Integrated Optics 集成光学

**IOA** Infrared Optical Assembly 红外光学组合件

Initial Outfitting Allowance 最初装备量

Instrument Operating Assembly 仪器操作组件

**IOAS** Intelligence Office Automatic System 智

能办公室自动化系统

**IOAT** Indicated Outside Air Temperature 指示的外界大气温度

**IOAU** Input/Output Access Unit 输入输出存取装置

**IOB** Input-Output Buffer 输入输出缓冲器

**IOBB** Input Output Broad Band 宽带输入输出

**IOBD** Integrated On-Board Diagnostics 综合的机上诊断

**IOC** Indirect Operation/Operating Cost 间接运营成本

Initial Operation Configuration 初期操作构型

Initial Operational Capability 初始工作能力

Input/Output Channel 输入/输出信道

Input/Output Controller 输入/输出控制器

Integrated Optical Circuit 集成光路

INTELSAT Operations Center 国际通信卫星操作中心

Inter Office Channel 局间信道

Intergovernmental Oceanographic Commission 政府间海洋学委员会

International Olympic Committee 国际奥林匹克委员会

**IOCA** Image Object Content Architecture 图像对象内容体系结构

**IOCC** Input/Output Control Center 输入/输出控制中心

**IOCE** Input/Output Control Element 输入/输出控制元件

**IOCP** Input/Output Control Program 输入/输出控制程序

**IOCS** Input/Output Control System 输入/输出控制系统

**IOD** Information on Demand 信息点播

Inner Object Damage 内部物损伤

**IODC** Input-Output Data Controller 输入—输出数据控制器

International Operator Direct Calling 国际运营商直接呼叫

**IOE** Initial Operating Experience 初期营运经验

**IOEU** Inboard Overhead Electronics Unit 内侧顶部电子组件

**IOG** Integrated Optic Gyroscope 集成光学陀螺

**IOIC** Integrated Operational Intelligence Center 综合作战情报中心

**IOIS** Input-Output Interface System 输入—输出接口系统

Integrated Operational Intelligence System 作战情报综合系统

**IOL** Inboard Open/Loop 内开/环路

Initial Outfitting List 初次装备表

Instantaneous Overload 瞬时超载

**IOLA** Input/Output Link Adapter 输入/输出链路适配器

**IOLC** Input/Output Link Control 输入/输出链路控制

**IOM** Image-Oriented Memory 面向图像的存储器

Input/Output Module 输入/输出模块

Input/Output Multiplexer 输入/输出多路转换器

Integrated-Optic Modulator 集成光学调制器

**IOMS** Input/Output Management System 输入输出管理系统

**ION** Integrated On-demand Network 综合按需服务网络

Institute of Navigation 导航学会

**IONI** ISDN Optical Network Interface ISDN 光网络接口

**IOOC** International Olive Oil Council 国际橄榄油理事会

**IOP** Input/Output Processor 输入/输出处理机

**IOPDS** Integrated-Optic Position/Displacement Sensor 集成光学位置/位移传感器

**IOR** Immediate Operational Requirement 立即操作需求,即刻运营要求

Indian Ocean Region 印度洋区域

**IOS** Input/Output System 输入/输出系统

Instructor Operating System 教员操纵台

Integrated Observation System 综合观测系统

Integrated Office System 集成办公室系统

Intelligent Office System 智能办公室系统

Interactive Operating System 交互式操作系统

International Organization for Standardization 国家标准化组织

Internet Operating System 因特网操作系统

Internetwork Operating System 网间操作系统

**IOSA** IATA Operational Safety Audit 国际航协运行安全审计

Integrated Optical Spectrum Analyzer 综合光谱分析仪

**IOSB** Input/Output Status Block 输入/输出状态块

**IOSC** Input/Output Switching Channel 输入/输出交换通道

**IOSN** Intelligent Optical Shuttle Node 智能光信息往返节点

**IOT** Image Output Terminal 图像输出终端

Inductive Output Tube 感应输出管

Instrument Operations Team 仪表操作组

**IOT&E** Independent Operational Test and E-valuation 独立运营测试与评估

Initial Operational Test and Evaluation 初始运营测试与评估

**IOTB** Input/Output Transfer Block 输入/输出传送块

**IOU** Input/Output Unit 输入/输出装置

**IP** Identification of Position 位置识别

Identification Point 识别点

ILS Panel 仪表着陆系统控制板

Image Processing 图像处理

Impact Point 碰撞点

Import Permit 进口许可

Index of Performance 性能指数

Inertial Platform 惯性平台

Information Processing 信息处理

Input 输入

Input Power 输入功率

Inspection Plan 检验计划

Inspection Program 检查大纲

Installation Package 安装包

Installation Page 安装页

Instruction Pilot 飞行教员

Instruction Plate 指令板

Instruction Pulse 指令脉冲

Instructor's Panel 教员控制板

Instrument Panel 仪表板

Instrumentation Port 仪器使用端口

Intellectual Property 知识产权

Intelligent Peripheral 智能外设

Intercept Point 切入点，截获点

Intercept Profile 切入剖面，截获剖面

Intermediate Pressure 中压

International Priority 国际优先快递服务

Internet Protocol 网际协议，因特网协议

Internetwork Protocol 网络间协议，网际协议

Internetworking Protocol 组网协议

Interrupt Point 中断点

Interrupt Priority 中断优先(权)

Interrupt Processing 中断处理

Interworking Protocol 互通协议

Inventory Pricing 库存计价

Invoice Packing Card 收据包装卡

Invoice Price 发票价格

Iron Pipe 铸铁管

**IPA** Image Processing Algorithm 图像处理算法

Independent Parallel Approach 独立平行进近

Interworking by Port Access 端口接入的互通

Inspection Program Approval 检验程序批准书

International Phonetic Alphabet 国际音标

International Phonetic Association 国际语音协会

International Publishers Association 国际出

版商协会

**IPACG** Informal Pacific Air Traffic Control Coordination Group 太平洋空中交通管制非正式协调小组

**IPACS** Integrated Power Attitude Control System 综合动力姿态控制系统

**IPAD** Integrated Program for Aerospace Vehicle Design 宇宙飞船设计综合计划

**IPADAE** Integrated Passive-Action Detection and Acquisition Equipment 综合被动探测采集设备

**IPARS** Integrated Programmed Airlines Reservation System 国际定期航线订座系统

**IPAS** Integrated Pressure Air System 综合压缩空气系统

**IPB** Illustrated Parts Breakdown 图解零件分解

**IPC** Illustrated Parts Catalog 图解零件目录
Information Processing Code 信息处理代码
Integrated Peripheral Channel 集成外围通道
Intelligent Peripheral Controller 智能外设控制器
Intermediate Pressure Bleed Check Valve 中压引气单向活门
Intermediate Pressure Compressors 中压压气机
Intermittent Positive Control 不连续无反向力操纵
Inter-Personal Communications 人际通信
Inter-Process Communications 进程间通信
Inter-Processor Communications 处理器间通信

**IPCC** Information Processing in Command and Control 指令与控制信息处理
Intergovernmental Panel on Climate Change 政府间气候变化专门委员会, 政府间气候变化问题小组
International Publishers Copyright Council 国际出版商版权委员会

**IPCCS** Information Processing Command and Control System 信息处理指令和控制系统

**IPCDN** IP over Cable Data Network 电缆数据网传送 IP

**IPCE** International Path Core Element 国际通路核心单元

**IPCP** IP Control Protocol IP 控制协议

**IPCS** Integrated Personnel Communication Subsystem 综合人员通信子系统
Integrated Propulsion Control System 综合推进控制系统
Interactive Problem Control System 交互式问题控制系统
Internet Protocol Communications Security 互联网协议通讯安全

**IPCSM** Input Port Controller Submodule 输入端口控制器子模块

**IPCT** Initial Provisioning Coordination Team 初始备件供应协调队, 初始配置协调小组

**IPCU** Ice Protection Control Unit 结冰保护控制组件

**IPCV** Intermediate Pressure Check Valve 中间压力检查活门

**IPD** Illustrated Provisioning Document 图解配置文件, 图解备件供应文本
Industrial Products Division 工业产品部
Inertial Performance Data 惯性性能数据
Initial Performance Data 起始性能数据
Initial Provisioning Data 初始配置数据, 初始备件供应数据
Insertion Phase Delay 插入相位延迟
Insertion Phase Difference 插入相位差
Integrated Product Development 综合产品研制
Invalid Products Display 无效产品展示
Issue Priority Designator 发布优先权指定器

**IPDC** IP Device Control IP 设备控制

**IPDD** Initial Project Design Description 初期项目计划描述
Integrated Product Design and Development 综合产品设计与开发

**IPDDS** Integrated Product Definition Data Sheet 综合产品定义数据单

**IPDP** Integrated Product Development Process 综合产品开发程序

Intervals of Pulsations of Diminishing Period 减弱期的脉动间隔

**IPDR** Inhouse Preliminary Design Review 内部初步设计审查

**IPDS** Integrated Personal Development System 综合个人发展系统

Integrated Product Development System 集成产品开发系统

Integration Processing Demonstration System 综合处理演示系统

**IPDT** Integrated Product Development Team 综合产品研制组

**IPDU** Ice Protection Data Unit 结冰保护数据组件

**IPE** Improved Performance Engine 改进性能的发动机

In-band Parameter Exchange 带内参数交换

Industrial Plant Equipment 工业工厂设备

International Petroleum Exchange 国际石油交换

**IPEC** International Programme on the Elimination of Child Labour 消除童工劳动国际计划

**IPEI** International Portable Equipment Identity 国际便携式设备标识

**IPEM** Integrated Power Electronics Module 集成电力电子模块

**IPEX** Immediate Purchase Excursion 临时购票旅行

**IPF** Image Processing Facility 图像处理设备

Initial Production Facilitization 初始生产促进，初始生产促销

Integrated Processing Facility 综合处理设施

**IPG** Interactive Program Guide 交互式节目指南

Inter-Packet Gap 分组信息间隙

**IPH** Ipoh, Malaysia（Airport Code）马来西亚怡保机场

**IPI** Initial Protocol Identifier 初始协议标识符

Intelligent Peripheral Interface 智能外围接口

International Patent Institute 国际专利学会

**IPIP** IPAD Information Processor 宇宙飞船设计综合计划信息处理器

**IPL** El Centro, CA, USA（Airport Code）美国加利福尼亚州埃尔森特罗机场

Illustrated Parts List 图解部件清单

Initial Program Load 初始程序装入

Initial Program Loader 初始程序装载机

Intense Pulse Light 高强度脉冲灯

IPAD Program Library 宇宙飞船设计综合计划程序库

**IPLB** IP Load Balancing IP 负载平衡

**IPLC** International Public Leased Circuit 国际公用出租线路

**IPLI** Internet Private Line Interface 因特网专用线接口

**IPLTC** International Private Leased Telecommunication Circuit 国际专用租线通信电路

**IPM** Inspection Procedure Manual 检查程序手册

Intelligent Power Module 智能功率模块

International Project Management 国际项目管理

Inter-Personal Messaging 人际传信

**IPMA** International Project Management Association 国际项目管理协会

**IPME** Inter-Personal Messaging Environment 人际传信环境

**IPM-EOS** Inter-Personal Message Element of Service 人际报文业务单元

**IPMF** International Project Management Forum 国际项目管理论坛

**IPMS** Inter-Personal Messaging Service 人际传信业务

Inter-Personal Messaging System 人际传信系统

**IPMS-MS** Inter-Personal Messaging System Message Store 人际传信系统信息存储

**IPMS-UA** Inter-Personal Messaging System User Agent 人际传信系统用户代理

**IPMT** Integrated Product Management Team 综合产品管理组

Integrated Project Management Team 综合项目管理组

**IPM-UA** Inter-Personal Messaging User Agent 人际传信用户代理

**IPN** Instant Private Network 瞬时专用网络

Inter-Personal Notification 人际通知

**IPNG** Internet Protocol Next Generation 下一代因特网协议

**IPNPB** Intermittent Positive Negative Pressure Breathing 间歇性正负压呼吸

**IPNPV** Intermittent Positive Negative Pressure Ventilation 间歇性正负压通气

**IPO** Initial Public Offering 初始公开销售

Integrated Purchase Order 综合购买订货

International Public Offering 国际公开销售

**IPOA** International Peace Operations Association 国际维和活动协会

**IPP** Intellectual Property Protection 知识产权保护

Internet Payment Provider 因特网支付业务提供商

Internet Platform Provider 因特网平台供应商

Investment Protection Program 投资保护计划

**IPPC** Integrated Pollution Prevention Control 综合污染预防控制

**IPPD** Integrated Product and Process Design 综合产品与过程设计

Integrated Product and Process Development 综合产品与过程开发

**IPPR** Image Processing and Pattern Recognition 图像处理和模式识别

**IPPU** Instrumentation Position Pick-off Unit 仪表位置传感组件

Intermediate Position Pick-off Unit 中间位置传感组件

**IPR** In-Progress Review 过程中审查

Intellectual Property Rights 知识产权

**IPRAM** Integrated Prerecorded Announcement Module 预录广播综合组件

**IPRE** Income-Producing Real Estate 创造收入的房地产

**IPRO** International Patent Research Office 国际专利研究处

**IPS** Image Processing Simulation 图像处理模拟

Image Processing System 图像处理系统

Inches per Second 英寸/秒

Incorrect Phase Sequence 逆相序

Inertial Positioning System 惯性定位系统

Information Presentation System 信息表达系统

Information Processing System 信息处理系统

Information Protection System 信息保护系统

Inlet Particle Separator 入口颗粒分离器

Instrument Pointing System 仪表定向系统

Integrated Power System 综合动力系统

Integrated Processing Schedule 综合处理计划

Integrated Processing System 综合处理系统

Integrated Program Summary 综合程序摘要

Intelligent Protection Switching 智能保护交换

International Pipe Standard 国际管件标准

Interruptions per Second 中断次数/秒

Iron Pipe Size 铸铁管尺寸

**IPSAS** International Public Sector Accounting Standards 国际公共部门会计标准

**IPSE** Integrated Programming Support Environment 综合规划支持环境

Integrated Project Support Environment 综合项目支持环境

IP Service Function IP 业务功能

**IPSIM** Instrument Preflight Simulator 飞行前仪表模拟器

**IPSS** Ice Penetrating Sensor System 穿冰传感系统

Information Processing System Simulator 信息

处理系统模拟器

Initial Preplanned Supply Support 预先计划初始供给支援

International Packet Switching Service 国际分组交换业务

**IPST** Integrated Process and System Testing 集成处理和系统测试

**IPT** Inadvertent Power Transfer 偶发电源转换,意外电源转换

Information Processing Technique 信息处理技术

Information Providing Terminal 信息提供终端

Initial Production Test 初始产品检测

Integrated Product Team 综合产品小组,一体化产品小组

Intermediate Pressure Turbine 中压汽轮机

Internal Pipe Thread 内管螺纹

Iron Pipe Thread 铸铁管螺纹

Williamsport, PA, USA（Airport Code）美国宾西法尼亚州威廉波特机场

**IPTA** International Patent and Trademark Association 国际专利与商标协会

**IPTOS** Intermediate Pressure Turbine Overspeed System 中压涡轮超转系统

**IPTS** Initial Point Turn Short 小转弯起点

Instrumentation Payload Test Set 仪表有效负载测试仪

**IPTV** Internet Protocol Television 互联网协议电视

**IPU** Instruction Processor Unit 指令处理组件

Integrated Power Unit 集成组合动力装置

Interface Processor Unit 接口处理器组件

**IPUI** International Portable User Identity 国际便携式用户标识

**IPV** Improve 改进,变好

Indicator Para Visual 帕拉视觉指示器

**IPX** Internet Packet Exchange 因特网数据包交换

Interprocess Packet Exchange 进程间分组交换

**IPY** International Polar Year 国际极地年

**IQ** Information Query 信息查询

Intelligence Quotient 智商

**IQL** Interactive Query Language 交互式查询语言

**IQP** Integrated Quality Plan 综合质量计划

**IQR** International Quarantine Regulation 国际检疫规定

**IQS** Integrated Queuing System 综合排队系统,综合行列系统

**IQTG** International Qualification Test Guide 国际资格测试指南

**IR** Ice on Runway 跑道积冰

ILS Receiver 仪表着陆系统接收机

Implementation Rules 实施规则

Incident Report 事故征候报告

Incoming Route 入路由

Indian Rupee 印度卢比(货币单位)

Inertial Reference 惯性基准

Information Retrieval 信息检索

Infra-Red 红外(线的)

Inspection Report 检查结果报告

Instrument Rating 仪表飞行等级

Intelligent Robot 智能机器人

Internal Router 内部路由器

**IRA** Inertial Reference Assembly 惯性参考组件

Intermediate Range Aircraft 中程飞机

**IRAC** International Radio Advisory Committee 国际无线电咨询委员会

**IRADS** Infra-Red Acquisition and Detection System 红外捕获与探测系统

**IRAM** Individual Records Administration Manual 个人记录管理手册

**IRAN** Inspection and Repair as Necessary 根据需要检查和修理

**IRAR** Infrared Airborne Radar 机载红外雷达

**IRASA** International Radio Air Safety Association 国际无线电航空安全协会

**IRATE** Interim Remote Area Terminal Equipment 中间遥控区终端设备

**IRATS** Infrared Algorithm Test Simulator 红外算法测试模拟器

**IRB** Industrial Review Board 工业审查委员会

Insurance Rating Board 保险定级委员会

Intelligence Review Board 智能审查委员会

Internal Review Board 内部审查委员会

Investment Review Board 投资审查委员会

**IRC** Inbound Radar Control 进场雷达管制

Information Retrieval Center 信息检索中心,情报检索中心

Infrared Remote Control 红外遥控

Initial Rate of Climb 初始爬升率

Instrument Remote Control 仪表遥控

Interdisciplinary Research Centre 交叉学科研究中心

Internal Radial Clearance 内部径向间隙

International Radiation Commission 国际辐射委员会

International Red Cross 国际红十字会

International Rescue Committee 国际救援委员会

International Route Charge 国际航路费用

Internet Relay Chat 因特网中继交谈

**IRCA** Immigration Reform and Control Act 移民改革与管理法案

Instrument Restricted Controlled Airspace 仪表限制控制空域

**IRCC** International Radio Consultative Committee 国际无线电咨询委员会

**IRCM** Infra-Red Countermeasures 红外干扰,红外对抗

**IRCS** Inertial Reference and Control System 惯性参考与控制系统

International Research Communications System 国际研究通信系统

Intrusion-Resistant Communications System 抗干扰通信系统

**IRD** Independent Research and Development 独立研究与开发

Information Resource Dictionary 信息资源词典

Infrared Detector 红外探测器

Installation Requirements Document 安装需求文件

Interface Requirements Document 接口要求文件

Internal Research and Development 国内研究与发展

International Requirements Document 国际需求文件

International Resource Development 国际资源开发

**IRDA** Infrared Data Association 红外线数据协会

Infrared Detection and Acquisition 红外探测与采集

Integrated Reliability Design Assessment 可靠性设计综合评估

**IRDC** Industry Research and Development Committee 工业研究和发展委员会

Interface Remote Data Concentrator 接口遥控数据集中器

**IRDS** Information Resources Dictionary System 信息资源词典系统

Infrared Detection Set 红外探测器

Infrared Detection System 红外探测系统

**IRE** Institution of Radio Engineers 无线电工程师协会

Instrument Rating Examiner 仪表飞行等级检查员

International Radio Engineers 国际无线电工程师

Ireland 爱尔兰

Irregular 不规则的,不定期的

**IREW** IR Electronic Warfare 红外线电子战

**IREX** Instrument Rating Examination 仪表等级考试

International Research and Exchanges Board 国际研究与交流委员会

**IRF** Impulse Response Function 脉冲响应函数

Integrated Radio Frequency 综合无线电频率

Integrated Research Facility 综合研究设备

**IRFI** International Runway Friction Index 国际跑道摩擦指数

**IRFIS** Inertial Reference Flight Inspection System 惯性基准飞行检查系统

**IRFITS** Infra-Red Fault Isolation Test System 红外故障隔离测试系统

**IRFU** Integrated Radio Frequency Unit 综合无线电频率单位

**IRG** Inter-Record Gap 记录间隙

**IRGAV** Change in Aircraft Type 改换机型

**IRGEQ** Change in Equipment or Layout 改变设备或客舱布局

**IRGHO** Change in Schedule Time 改变班期时刻

**IRI** Infra-Red Image 红外图像

International Reference Ionosphere 国际参考电离层

International Research Institute for Climate Prediction 国际气候预测研究所

Iran Rial 伊朗里亚尔(货币单位)

**IRIG** Inertial Reference Integrating Gyroscope 惯性基准积分陀螺仪

Inter-Range Instrumentation Group 靶场间仪表小组

**IRIS** Infrared Information System 红外信息系统

Infrared Interferometer Spectrometer 红外干涉分光计

Integrated Platform for Regional Information System 地区信息系统综合平台

Integrated Resources Information System 综合资源信息系统

Interactive Retail Information System 交互式零售信息系统

**IRJ** IR Jammer 红外线干扰器

**IRL** Infrared Lens 红外透镜

Inter-Repeater Link 中继器间链路

**IRLAP** Infra-Red Link Access Protocol 红外链接存取协议

**IrLAP** IrDA Link Access Protocol IrDA 链路接入协议

**IRLS** Infra-Red Line Scan System 红外线行扫描系统

Infrared Line Scanner 红外线行扫描器

Interrogation Recording and Location System 询问记录与定位系统

**IRM** Information Resources Managerment 信息源管理

Integrated Reference Model 综合参考模型

Intelligent Robotics Manufacturing (人工)智能机器人制造

**IRMC** Infrared Mobile Communication 红外移动通信

**IRMP** Inertial Reference Mode Panel 惯性基准方式选择板

Infrared Measurement Program 红外测量计划

Integrated Risk Management Process 综合风险管理程序

**IRMS** Information Resource Management Software 信息资源管理软件

Information Retrieval and Management System 信息检索及管理系统

Integrated Resource Management System 整合资源管理系统

Integrated Risk Management Strategy 集成风险管理策略

**IRN** Information Resource Network 信息资源网络

Intermediate Routing Node 中间路由选择节点

Iran 伊朗

**IRNSS** Indian Regional Navigational Satellite System 印度区域导航卫星系统

**IRP** Information Resource Planning 信息资源计划

Integrated Refuelling Panel 综合加油控制板

Integrated Resource Plan 综合资源计划

Intermediate Rated Power 中级额定功率

Internal Reference Point 内部参考点

International Routing Plan 国际路由规划

Irregularity Report 异常报告

**IRQ** Information Repeat Request 信息重传请求

Interrupt Request 中断请求

Interworking Service Request 协作服务请求

**IRR** Instrument Rating Renewal 仪表飞行等级证书(执照更新)

Internal Rate of Return 内部收益率

Investment Return Ratio 投资产出比

Iran Rial 伊朗里亚尔(货币单位)

Irrevocable 不可改变的

**IRRE** International Road Roughness Experiment 国际路面不平整度试验

**IRREG** Irregular 不正常,不规则,不定期

**IRRESP** Irrespective of 无关

**IRS** Idle Reset Solenoid 慢车复位电磁线圈

Incident Reporting System 事故报告系统

Inertial Reference System 惯性基准系统

Informational Retrieval System 信息检索系统

Inspection Reporting System 检查报告系统

Interface Requirements Specification 接口要求规格

International Radar Symposium 国际雷达讨论会

**IRSG** International Rubber Study Group 国际橡胶研究小组

Internet Research Steering Group 因特网研究指导组

**IRSM** IR Surveillance Measure 红外线监视

**IRSS** Inertial Reference Stabilization System 惯性基准稳定系统

Information and Resource Support System 信息资源支持系统

Infrared Sensor System 红外传感器系统

Interactive Radar System Simulator 交互式雷达系统仿真模拟器

**IRST** Infra-Red Search and Track 红外搜索与跟踪

**IRSU** ISDN Remote Subscriber Unit ISDN 远端用户单元

**IRT** Instrument Rating Test 仪表额定(性能)测试,仪表等级测试

**IRTF** Internet Research Task Force 因特网研究任务工作组

**IRTLH** International and Regional Transportation and Logistics Hub 国际和地区运输及物流枢纽

**IRTP** Integrated Reliability Test Program 综合可靠性试验计划

**IRTS** Infra-Red Target Seeker 红外目标导引头

Infra-Red Target Sounder 红外目标探测仪

Infra-Red Temperature Sounder 红外温度探测仪

Initial Radar Training Simulator 初级雷达训练模拟器

**IRTU** Intelligent Remote Terminal Unit 远程智能终端装置

**IRU** Inertial Reference Unit 惯性基准组件

Infrared Receiver Unit 红外线接收装置

Integrated Recovery Unit 综合回收装置

International Rescue Union 国际救援联合会

International Road Transport Union 国际公路运输联盟

**IRUE** Inertial Reference Unit Electronics 惯性基准电子元件

**IRVR** Instrumented Runway Visual Range 仪表的跑道能见度

**IS** Imaging System 成像系统

Index Service 索引业务

Information Science 信息科学

Information System 信息系统

Instrumentation System 仪表系统

Integrated Service 综合业务

Intelligence System 智能系统

Interactive Service 交互式业务

Interactive Signal 交互信号

Interface Specification 接口规范

Interim Standard 临时标准

Intermediate System 中间系统

International Satellite 国际卫星

Island(s) 岛,岛屿

**ISA**    If Space Available 如有空(座)位

Industry Standard Architecture 工业标准结构,工业标准体系结构

Inertial Sensor Assembly 惯性传感器装置

Information System Architecture 信息系统结构

Instruction Set Architecture 指令集结构

Instrument Society of America 美国仪表学会

Interim Standard Architecture 临时标准体系

International Standard Atmosphere 国际标准大气

International Standardization Association 国际标准化协会

International Standards Association 国际标准协会

**ISAA**    Institute of Space and Aeronautical 宇航科学院

International Steward(ess) and Airliner Association 国际空中乘员及航空公司从业人员协会

**ISADS**    Integrated Strapdown Air-Data System 综合捷联式大气数据系统

**ISAE**    International Society of Airbreathing Engineers 国际空气喷气式发动机工程师协会

**ISAGO**    IATA Safety Audit for Ground Operations 国际航空运输协会地面运行安全审计

**ISAMI**    Improvement of Structure Analysis through Multi-Disciplinary Integration 通过多次综合培训进行结构提高分析

**ISAN**    Integrated Service Analog Network 综合业务模拟网

**ISAP**    Interactive Speech Application Platform 交互语言应用平台

**ISAPI**    Internet Server Application Programming Interface 因特网服务器应用编程接口

**ISAR**    Information Storage and Retrieval 信息存储和检索

Institute Space Atmosphere Research 空间大气研究所

Inverse Synthetic Aperture Radar 逆向合成孔径雷达

**ISARP**    International Standard Audit Recommended Procedures 国际标准审计建议程序

**ISASI**    International Society of Air Safety Investigators 国际航空安全调查员协会

**ISA-SL**    International Standard Atmosphere at Sea Level 海平面国际标准大气

**ISAW**    International Society of Aviation Writers 国际航空作家协会

**ISB**    Independent Side Band, Independent Sideband 独立边带

Intelligent Signaling Bus 智能信令总线

Inter System Bus 内部系统汇流条

Interface Scheduling Block 接口调度块

Islamabad, Pakistan (Airport Code) 巴基斯坦伊斯兰堡机场

**IS-BAO**    International Standard for Business Aircraft Operations 国际公务机运行标准

**ISBN**    Integrated Satellite Business Network 综合卫星商用网络

International Standard Book Number 国际标准图书编号

**ISBO**    Islamic States Broadcasting Organization 伊斯兰国家广播组织

**ISC**    Industry Steering Committee 工业指导委员会

Instrument System Corporation 仪表系统公司

Integrated Semiconductor Circuit 集成半导体电路

International Switching Centre 国际转接中心

Internet Software Consortium 因特网软件联盟

Interstellar Communications 星际通信

**ISCAN**    International Sanitary Convention for Air Navigation 国际空中航行环境卫生公约

**ISCC**    International Service Coordination Center 国际业务协调中心

**ISCCI**    International Standard Commerical

Code for Indexing 国际标准商用索引代码

**ISCII** International Standard Code for Information Interchange 国际标准信息交换代码

**ISCOS** Intergovernmental Standing Committee on Shipping 政府间航运常设委员会

**ISCP** ISDN Signaling Control Part ISDN 信令控制部分

**ISCS** International Satellite Communications System 国际卫星通信系统

**ISD** Implementation Support and Development 实施支援和发展

Initial Search Depth 初始搜索深度

International Standard Depth 国际标准深度

Ionization Smoke Detector 离子感烟探测器

Issued 发行的,发布的

**ISDAS** In-Service Data Acquisition System 服务中数据采集系统

Internet Scan Data Acquisition System 因特网扫描数据采集系统

**ISDB** Integrated Services Digital Broadcasting 综合服务数字广播

Integrated Statistical Database 综合统计数据库

**ISDCN** Integrated Service Digital Center Network 综合业务数字中心网

**ISDN** Integrated Services Data Network 综合业务数据网络

Integrated Services Digital Network 综合业务数字网络

**ISDOS** Information System Design and Optimization System 信息系统设计与优化系统

Information System Development and Optimization System 信息系统开发与优化系统

**ISDR** International Strategy for Disaster Reduction 国际减灾战略

**ISDS** Information, Space and Defense Systems 信息、空间和防御系统

Integrated Switched Data Service 综合交换数据业务

**ISDT** Information Services Development Team 信息服务开发小组

Information Systems Design Theory 信息系统设计理论

Instructional Systems Development Team 指令系统研究发展小组

Integrated Service Digital Terminal 综合业务数字终端

**ISDU** Inertial Sensor Display Unit 惯性感应器显示组件

Inertial System Display Unit 惯性系统显示组件

**ISDX** Integrated Service Digital Exchange 综合业务数字交换

**ISE** Information Security Evaluation 信息安全评估

Integrated Service Exchange 综合业务交换

Integrated Switch Element 综合交换单元

Intelligent Synthesis Environment 智能综合环境

Interconnected Stabilizer/Elevator 安定面/升降舵铰接

**ISEA** Industrial Safety Equipment Association 工业安全设备协会

Integrated Safety Engineering Analysis 综合安全工程分析

International Safety Equipment Association 国际安全设备协会

**ISEC** Information System Engineering Command 信息系统工程指令

Internet Service and Electronic Commerce 因特网服务和电子商务

**ISFD** Industrial Security Facilities Database 工业安全设施数据库

Integrated Standby Flight Display 综合备用飞行显示

**ISG** Industry Steering Group 工业指导组

Inflatable Survival Gear 充气救生器材

Information Strategies Group 信息战略组

Information Systems Group 信息系统组

Instrument Symbol Generator 仪表符号发生器

Interchangeable and Substitute Group 互换和

替代小组

**ISGU** Instrument Symbol Generator Unit 仪表符号发生器组件

**ISH** Information Super Highway 信息高速公路

**ISHEM** Integrated System Health Engineering and Management 综合系统健康工程与管理

**ISHM** Integrated Systems Health Management 系统健康综合管理

**ISI** Industry Standard Item 工业标准项目

Institute for Scientific Information 科学情报研究所

Integrated Standby Instrument 集成备用仪表

International Sales Indicator 国际销售代号

International Statistical Institute 国际统计学会

**ISIC** International Standard Industrial Classification 国际工业标准分类

**ISIDE** Interactive Satellite Integrated Data Exchange 交互式卫星综合数据交换

**ISIS** IATA Statistical Information System 国际空运协会统计信息系统

Integrated Security and Identification System 综合安全与识别系统

Integrated Spar Inspection System 综合翼梁检查系统

Integrated Standby Instrument System 综合备用仪表系统

International Security Information Service 国际安全信息处

**IS-IS** Intermediate System to Intermediate System 中间系统到中间系统

**ISL** Iceland 冰岛

Inactive-Status List 停用状态单

Information Systems Laboratory 信息系统实验室

Installation Sequence List 安装顺序清单

Intelligent Systems Laboratory 智能系统实验室

Inter-Satellite Link 卫星之间的链路

Island 岛屿

**ISLAN** Integrated Services Local Area Network 综合业务局域网

**ISLN** International Standard Lawyer Number 国际标准律师编号

Isolation 隔离

**ISLS** International Society for Language Studies 国际语言研究学会

Interrogator Side Lobe Suppression 询问器旁瓣抑制

**ISM** Igniter Safety Mechanism 点火器安全机构

Illumination Sensor Module 照明传感器模块

Industrial, Scientific and Medical (Equipment) 工业、科学和医药(设备)

Industrial Security Manual 工业安全手册

Inertial System Mount 惯性系统架

Integrated Safety Management 综合安全管理

Intelligent Synchronous Multiplexer 智能同步复用器

Interactive Storage Media 交互式存储媒体

Interface Subscriber Module 用户接口模块

International Standard Method 国际标准方法

Internet Server Manager 因特网服务器管理器

**ISMA** Idle Signal Multiple Access 空闲信号多址访问

Internet Streaming Media Alliance 互联网流媒体联盟

**ISMAN** Integrated Services Metropolitan Area Network 综合业务城域网

**ISMC** International Switching Maintenance Center 国际交换维护中心

**ISMG** Internet Short Message Gateway 互联网短信网关

**ISMLS** Interim Standard Microwave Landing System 临时标准微波着陆系统

**ISMP** International Safety Management Process 国际安全管理过程

**ISMS** Image Store Management System 图像存储管理系统

Information Security Management System 信

息安全管理体系

**ISN** Information System Network 信息系统网络

Integrated Services Network 综合业务网

Integrated Synchronous Network 综合同步网

International Signaling Network 国际信令网

Internet Shopping Network 因特网购物网络

Internet Support Node 因特网支持节点

Williston, ND, USA（Airport Code）美国北达科他州威利斯顿机场

**ISO** Independent Service Organization 独立服务组织

Industry Standards Organization 工业标准化组织

International Organization for Standardization 国际标准化组织

International Satellite Organization 国际卫星组织

International Science Organization 国际科学组织

International Standards Organization 国际标准组织

Isolation 隔离

**ISOC** Internet Society 因特网学会

**ISODE** ISO Development Environment ISO 开发环境

**ISOL** Isolate, Isolation 隔离

Isolated 独立的, 隔离的

**ISOPA** ISO Protocol Architecture 国际标准化组织协议体系

**ISORM** International Standards Organization Reference Model 国际标准化组织参考模型

**ISP** Inertially Stabilized Platform 惯性稳定平台

Integrated Static Probe 综合静电探测, 综合静压管

Integrated Support Plan 综合保障计划

Interactive Session Protocol 交互式会晤协议

Intermediate Service Part 中间业务部分

International Signaling Point 国际信令点

International Standardized Profile 国际标准化规格

Internet Service Provider 因特网服务供应商

Interoperable Systems Project 可互操作系统计划

Islip, NY, USA（Airport Code）美国纽约州艾斯利普机场

**ISPACG** Informal South Pacific ATS Co-Ordination Group 非正式南太平洋空中交通服务协调小组

**ISPBX** Integrated Services Private Branch Exchange 综合业务专用局端交换

**ISPC** International Signaling Point Code 国际信令点码

**ISPEMI** International Symposium on Precision Engineering Measurements and Instrumentation 精密工程测量与仪器国际学术研讨会

**ISPSS** In-Seat Power Supply System 座椅电源系统

**ISPSU** In-Seat Power Supply Unit 座椅电源组件

**ISR** Initial Services Requirements 初始使用要求

Initial Submission Rate 初始提供速率

Interim System Review 临时系统评审

International Simple Resale 国际简单转卖

Interrupt Service Routine 中断服务程序

**ISRC** Information Security Research Center 信息安全研究中心

Information Systems Resource Center 信息系统资源中心

International Search and Rescue Convention 国际搜寻与救援公约

International Standard Recording Code 国际标准记录码

International Synthetic Rubber Company 国际合成橡胶公司

**ISRO** Indian Space Research Organization 印度宇航研究组织

In-Service Reportable Occurrence 在岗报告的事件

**ISRS** Inertial Space Reference System 惯性空

间基准系统

Integrated Status Reporting System 综合状况报告系统

**ISRU** International Scientific Radio Union 国际无线电科学联合会

**ISS** Inertial Sensor System 惯性传感系统

Information Systems Services 信息系统服务

Infrared Sensor System 红外传感器系统

Instruction Set Simulator 指令组模拟器

Instrument Subsystem 仪表子系统

Integrated Safety System 综合安全系统

Integrated Sensor System 综合传感器系统

Integrated Surveillance System 综合监视系统

Intelligent Support System 智能支持系统

International Space Station 国际空间站

Issue 发行,发布

**ISSCC** International Solid-State Circuits Conference 国际固态电路会议

**ISSG** Industry Safety Strategy Group 行业安全战略小组

Information Systems Support Group 信息系统支援组

**ISSI** Inter-Switching System Interface 相互交换系统间接口

**ISSLL** Integrated Services over Specific Link Layer 专用链路层上的综合业务

**ISSN** International Standard Serial Number 国际标准序号

**ISSP** Information Systems Standardization Program 信息系统标准化计划

Instrument Source Select Panel 仪表源选择面板

**ISSPP** Information System Security Practices and Procedures 信息系统安全实践与程序

Integrated System Safety Program Plan 综合系统安全项目计划

**ISSPU** Integrated Surveillance System Processor Unit 集成监视系统处理器单元

**ISSR** Independent Secondary Surveillance Radar 独立的二次监视雷达

**ISSS** Information Systems Strategy Study 资讯系统策略研究

Initial Sector Suite Subsystem 起始扇区管制席位分系统

Instrument Source Select Switch 仪表电源选择电门

Interactive Subscriber Service Subsystem 交互式服务子系统

**ISSWG** Information Systems Security Working Group 信息系统安全工作组

**IST** Information Sciences and Technology 信息科学与技术

Integrated Sciences Technology 综合科学技术

Interstellar Travel 星际航行

Istanbul, Turkey-Ataturk (Airport Code) 土耳其伊斯坦布尔阿塔图尔克机场

**ISTAR** Image Storage, Translation and Reproduction 图像存储、解释和再现(系统)

**ISTAT** International Society of Transport Aircraft Trading 国际运输机贸易协会

**ISTC** Integrated System Test Complex 复杂的集成系统测试

International Satellite Transmission Center 国际卫星传输中心

International Switching and Testing Center 国际交换和测试中心

**ISTIP** Information Systems Technical Integration Panel 信息系统技术综合委员会

**ISTN** Interfacility Satellite Telecommunication Network 设备连接卫星电信网

**ISTP** Index to Scientific and Technical Proceedings 科技会议录索引

**ISTV** Integrated Service Television 综合业务广播电视

**ISU** Idle Signal Unit 空闲信号单元

Inertial Sensor Unit 惯性感应器组件

Inertial Signal Unit 惯性信号组件

Integrated Sight Unit 综合瞄准装置

Integrated Surveillance Unit 综合监视组件

Interface Surveillance Unit 接口监视装置

International System of Units 国际单位制

Isochronous Slot Utilization 等时隙利用

Issue 发行,填开,发布

**ISUP** Integrated Service User Part 综合业务用户部分

**ISV** Independent Software Vendor 独立软件销售商

Isolation Valve 隔离活门

**ISVR** Inter Smart Video Recorder 灵巧型视频录像机

**ISWAP** International Society of Women Airline Pilots 国际妇女飞行员协会

**ISWL** Isolated Single Wheel Load 隔离单轮负载,单轮当量载荷

**IT** In Transit 在运输中

Information Technology 信息技术

Information Theory 信息论

International Transit 国际转接

Interruption 中断

Italian 意大利人,意大利语,意大利的

Items 项目

**ITA** Indicated True Altitude 指示真高

Information Technology Agreement 信息技术协议

Information Technology Assessment 信息技术评估

Information Technology Association 信息技术协会

International Telegraph Alphabet 国际电报字母表

International Touring Alliance 国际旅游联盟

International Trade Administration 国际贸易局

**ITA2** International Telegraph Alphabet No. 2 国际电报字母表第二版

**ITAA** Information Technology Association of America 美国信息技术协会

**ITAR** International Trade and Arms Regulations 国际贸易与武器规章

**ITAS** Indicated True Air Speed 指示真空速

**ITAV** Information Technology Audio Visual 信息技术视听

**ITB** Industrial Training Board 工业培训委员会

Integrated Test Bed 综合试验台

International Time Bureau 国际时间局

**ITC** Inclusive Tour Charter 包价旅游包机

Information Transfer Channel 信息传递信道

Intelligent Terminal Controller 智能终端控制器

International Tea Committee 国际茶叶委员会

International Telecommunication Center 国际电信中心

International Telephone Center 国际电话中心

International Television Center 国际电视中心

International Tin Council 国际锡理事会

International Transit Center 国际转接中心

International Transmission Center 国际传输中心

Intertoll Communication 长途局间通信

Investment Tax Credit 投资抵税,投资税减免

**ITCAS** International Training Center for Aerial Survey 国际航测培训中心

**ITCC** International Telecommunication Control Center 国际电信控制中心

**ITCL** Incremental Control 增量控制

**ITCM** Initial Technical Coordination Meeting 初始技术协调会议

**ITCS** Integrated Target Command System 综合目标指挥系统

Integrated Traffic Control Systems 综合交通管制系统

**ITCW** International Tropical Cyclone Watch 国际热带旋风监视

**ITCZ** Intertropical Convergence Zone 热带辐合带

**ITDM** Intelligent Time-Division Multiplexer 智能时分多路复用

**ITDP** International Technology Demonstrator Programme 国际技术示范项目

**ITE** Information Technology Equipment 信息技术设备

Inner Thermal Enclosure 内热控隔层

International Telephone Exchange 国际电话交换台

**ITEA** International Technology Education Association 国际技术教育协会

International Test and Evaluation Association 国际试验与鉴定协会

**ITEC** International Turbine Engine Corporation 国际涡轮发动机公司

Involute Throat and Exit Cone 内旋锥

**ITEL** Illustrated Tool Equipment List 图解工具设备清单

**ITEM** Illustrated Tool and Equipment Manual 图解工具和设备手册

**ITEP** Improved Turbine Engine Program 改进涡轮发动机计划

Integrated Testing and Evaluation Plan 综合测试与评估计划

**ITF** Information Technology Facility 信息技术设施

Information Transport Function 信息传送功能

International Transport Forum 国际交通论坛

International Transport Workers Federation 国际运输工人联合会

**ITFA** Integrated Turbulence Forecast Algorithm 颠簸预报综合算法

**ITG** Industry Technology Group 工业技术组

Information Technology Group 信息技术组

Integrated Terminal Guidance 综合终端制导

**ITGS** Integrated Track Guidance System 综合跟踪制导系统

**ITH** Ithaca, NY, USA ( Airport Code ) 美国纽约州伊萨卡机场

**ITI** Inertia Turn Indicator 惯性转弯指示器

Inspection and Test Instruction 检验与试验指示

International Technical Institute of Flight Engineers 国际飞行工程师技术学会

**ITIP** International Technical Integration Panel 国际技术综合委员会

**ITL** Intent to Launch 准备发射

International Transaction Log 国际交易日志

Italian Lira 意大利里拉(货币单位)

**ITM** Information Technology Management 信息技术管理

Installation, Testing and Maintenance 安装、测试与维护

Integrated Test and Maintenance 综合测试与维护

Integrated Transport Management 综合交通管理

Items 项目

Osaka, Japan-Itami ( Airport Code ) 日本大阪伊丹机场

**ITMC** Information Technology Management Council 信息技术管理委员会

International Transmission Maintenance Center 国际传输维护中心

**ITMS** Integrated Track Management System 综合跟踪管理系统

Integrated Training Management System 综合训练管理系统

International Traffic Management System 国际交通管理系统

**ITN** Integrated Teleprocessing Network 综合远程处理网络

Intelligent Telecommunication Node 智能电信节点

**ITNL** Internal 内部的

**ITNRY** Itinerary 航程

**ITO** Hilo, HI, USA ( Airport Code ) 美国夏威夷州希洛机场

Independent Test Organization 独立试验组织(软件的验证与审定)

Instrument Take-off 仪表起飞

International Technology Office 国际技术办公室

International Trade Organization 国际贸易组织

**ITOUTC** Internal Timeout Cycle 内部暂停循环,内部暂停周期

**ITP** Initial Technical Proposal 初始技术建议
Instruction to Proceed 操作规程
Intention to Proceed 操作意图

**ITPA** International Tea Promotion Association 国际茶叶促进协会

**ITPC** International Television Program Center 国际电视节目中心

**ITPS** International Test Pilots School 国际试飞员学校

**ITR** Incremental Tape Recorder 增量磁带记录器
Initial Trouble Report 最初故障报告
Instantaneous Transmission Rate 瞬时传输速率
International Transit Route 国际中转航线
Internet Talk Radio 因特网无线对话

**ITRMC** Information Technology Resource Management Council 信息技术资源管理委员会

**ITRP** Institute of Transportation and Regional Planning 运输区域规划院
Integrated Technology Rotor Program 综合技术旋翼计划

**ITS** Independent Television Service 独立电视服务
Inertial Timing Switch 惯性定时开关
Information Technology Security 信息技术安全
Information Technology Services 信息技术服务
Information Technology System 信息技术系统
Information Transfer Satellite 信息传送卫星
Information Transfer System 信息转换系统
Information Transmission System 信息传输系统
Insertion Test Signal 插入测试信号

Instrumentation and Telemetry System 仪表和遥测系统

Integrated Teleprocessing System 综合远程信息处理系统,综合遥控处理系统

Integrated Test Specification 综合测试规格

Integrated Training System 综合训练系统

Integrated Trajectory System 综合弹道系统

Intelligent Transportation System 智能交通系统

Internal Technical Specification 内部技术说明规范

International Telecommunication Service 国际电信服务,国际电信业务

International Temperature Scale 国际温标

**ITSA** Information Technology Services Agency 信息技术服务机构
Institute for Traffic Safety Analysis 交通安全分析研究所
Intelligent Transportation Society of America 美国智能交通学会
International Transportation Safety Association 国际运输安全协会

**ITSC** International Telephone Service Center 国际电话业务中心

**ITSD** Information Technology Services Department 信息技术服务部门

**ITSE** Information Technology Systems Engineering 信息技术系统工程
Integrated Test and Support Environment 综合测试与支持环境

**ITSEC** Information Technology Security Evaluation Criteria 信息技术安全评估准则

**ITSH** Internal Transport, Storage and Handling 国内运输、储藏和装卸

**ITSO** International Telecommunications Satellite Organization 国际电信卫星组织

**ITSP** Internet Telephony Service Provider IP 电话业务提供商

**ITT** Information Technology Team 信息技术组
Integrated Test Team 综合试验组

Inter Toll Trunk 长途电话中继线

International Telephone and Telegraph Corp.
国际电话电报公司

Interstage Turbine Temperature 涡轮级间温度

Inter-Turbine Temperature 涡轮内燃气温度

**ITTB** Integrated Technology Testbed 综合技术实验平台

**ITTCC** International Telegraph and Telephone Consultative Committee 国际电话电报顾问委员会

**ITTP** Intelligent Terminal Transfer Protocol 智能终端转换协议

**ITTS** Intelligent Target Tracking System 智能目标跟踪系统

**ITU** International Telecommunication Union 国际电信联盟(联合国)

International Telephone Union 国际电话联盟

**ITU-R** ITU-Radio Communications Sector 国际电信联盟无线电通信组

**ITURR** ITU Radio Regulations 国际电信联盟无线电规则

**ITU-T** ITU-Telecommunication Standardization Sector 国际电信联盟电信标准化组

**ITV** Idling Throttle Valve 慢车油门开关,慢车节流活门

Integrated Test Vehicle 综合测试飞行器

Interactive Television 交互式电视

Internal Transfer Vehicle 场内搬运车

**ITVF** Integration, Test and Verification Facility 组装、测试和检验设施

**ITWG** International Technology Working Groups 国际技术工作组

**ITWS** Integrated Target Weather System 综合目标气象系统

Integrated Terminal Weather Service 综合终端气象服务

Integrated Terminal Weather System 综合终端气象系统

International Tsunami Warning System 国际海啸预警系统

**ITX** Inclusive-Tour Excursion 全包旅游团

**IU** Index Unit 指数单位

Input Unit 输入器,输入装置

Interface Unit 接口单元

**IUAI** International Union of Aviation Insurers 国际航空保险商联合会,航空保险人国际联盟

**IUAPPA** International Union of Air Pollution Prevention and Environmental Protection Associations 国际防止空气污染与环境保护协会联合会

International Union of Air Pollution Prevention Associations 国际防止空气污染协会联合会

**IUC** Initial User Capability 初始用户能力

**IUG** Involuntary Up-Grade 意外升等

**IUGG** International Union of Geodesy and Geophysics 国际测地和地球物理协会

**IUI** Intelligent User Interface 智能用户接口

Inter-User Interference 用户间干扰

**IUIN** International Union for Inland Navigation 国际内陆航行联盟

**IUKF** Iterated Unscented Kalman Filter 迭代无偏卡尔曼滤波器

**IUMS** International Union of Marine Science 国际海洋科学联合会

**IUO** Intelligent Underlay Overlay 智能双层网

**IUOTO** International Union of Tourist Organizations 国际旅游组织联合会

**IUPS** Internal Uninterruptable Power Supply 内部不间断电源

**IUR** Internet Usage Record 因特网使用记录

**IURAP** International Users Resource Allocation Panel 国际用户资源分配委员会

**IUS** International Union of Students 国际学生联合会

**IUT** Instructor Under Training 见习教官

**IUTAM** International Union of Theoretical and Applied Mechanics 国际理论与应用力学联合会

**IV** Initial Velocity 初速

Interactive Video 交互式视频

Interface Vector 接口向量

Invioce Value 发票价值金额

Isolation Valve 隔离活门

**IV&V** Independent Verification and Validation 独立检验和认证

**IVA** Initial Video Address 初始视频地址

Input Video Amplifier 输入视频放大器

Intravehicular Activities 太空载具内活动

**IVAD** Integrate Voice and Data 综合话音和数据(通信数据链)

**IVALA** Integrated Visual Approach and Landing Aid System 目视进近着陆综合助航系统

**IVANS** Insurance Value Added Network Services 保险业增值网络服务

International Value-Added Network Service 国际增值网络服务

**IVAO** International Virtual Aviation Organization 国际虚拟航空组织

**IVAP** Internal Videotex Application Provider 内部可视图文应用供应商

**IVAS** Integrated Video Audio System 综合视频音频系统

**IVASEB** Integrated Video Audio Seat Electronics Box 综合视频音频座椅电子盒

**IVBC** International Videoconference Booking Center 国际电视会议登记中心

**IVC** Independent Virtual Channel 独立虚拟信道

International Venture Capital 国际风险投资

International Videoconference Center 国际会议电视中心

**IVCP** Integrated Value Creation Project 综合价值创造计划(方案)

**IVD** Interactive Video Disk 交互式视频盘

Interactive Video Downloader 交互式视频下载器

Interpolated Voice Data 内插语音数据

**IVDS** Interactive Video Database Services 交互式视频数据库业务

**IVDT** Integrated Voice Data Terminal 综合话音数据终端

**IVE** International Videotex Equipment 国际可视图文设备

**IVF** Integrated Validation Facility 综合验证设施

**IVG** Interactive Video Game 交互式视频游戏

**IVHM** Integrated Vehicle Health Management 综合飞行器健康管理

Integrated Vehicle Health Monitoring 综合飞行器健康监控

**IVHMS** Integrated Vehicle Health Management System 综合飞行器健康管理系统

Integrated Vehicle Health Monitoring System 综合飞行器健康监控系统

**IVHS** Intelligent Vehicle and Highway System 智能车辆和公路系统

**IVI** Instant Visibility Index 瞬时能见度指示

Interchangeable Virtual Instruments 可交换虚拟仪表

**IVIS** Interactive Video Information System 交互视频信息系统

**IVLD** Invalid 无效的

**IVMM** Interactive Virtual Maintenance Model 交互式虚拟维修模型

**IVMS** Integrated Voice-Messaging System 综合语音信息系统

**IVN** Interactive Video Network 交互式视频网络

**IVOD** Interactive Video on Demand 交互式视频点播

**IVOT** Inter-Network Televoting 网间电子投票业务

**IVPN** International Virtual Private Network 国际虚拟专用网

**IVR** Instrumented Visual Range 仪表目视距离

Integrated Voice Response 综合语音响应

Interactive Voice Response 交互式语音应答

**IVRS** Interim Voice Response System 过渡性

话音响应系统

**IVS** Inertial Vertical Speed 惯性垂直速度

Intelligent Video Smoother 智能视频平滑器

Intelligent Visual Surveillance 智能视频监控

Interactive Video Service 交互式视频业务

Interactive Videodisc System 交互式视盘系统

Interactive Voice System 交互语音系统

**IVSI** Instantaneous Vertical Speed Indicator 瞬时垂直速度指示器

**IVV** Instantaneous Vertical Velocity 瞬时垂直速度

**IVVC** Instantaneous Vertical Velocity Computer 瞬时垂直速度计算机

**IVVI** Instantaneous Vertical Velocity Indicator 瞬时垂直速度指示器

**IW** Information War 信息战

**IWAN** Integrated Services Wireless-Access Network 综合业务无线接入网络

**IWBS** Integrated Weight and Balance System 载重平衡综合系统

**IWC** Indoor Wireless Channel 室内无线信道

Instrument Weather Conditions 仪表气象条件

Integrated Wideband Communication 综合宽带通信

Interferometric Wavelength Converter 干涉波长变换器

**IWCS** Intigrated Wideband Communication System 综合宽带通信系统

**IWF** Interworking Facility 互通设备

Interworking Function 互通功能

**IWG** Intelligence Working Group 情报工作组

Interagency Working Group 跨部门联合工作组

Interface Working Group 接口工作组

International Working Group 国际工作组

Inter-system Working Group 系统间工作组

**IWK** Issuer Working Key 发行卡的工作密钥

**IWL** Institute Warranty Limits 协会保证权限

**IWM** Institution of Works Managers 工厂经理协会(英国)

**IWP** Interim Working Party 临时工作组

**IWRM** Integrated Water Resources Management 综合水资源管理

**IWS** Intelligent Work Station 智能工作站

Intelligent Workstation Support 智能工作站支持

With Itemized Work Statement 逐条施工说明

**IWSDB** Integrated Weapon System Database 综合武器系统数据库

**IWU** Inter Working Unit 互通单元

**IXC** Inter-Exchange Carrier 交互运营商

**IXP** Internet Exchange Point 因特网交换点

**IXR** Intersection of Runway 跑道交叉点

Ranchi, India (Airport Code) 印度兰契机场

Runway Intersection 跑道联络道

**IYB** International Year Book 国际年鉴

**IYK** Inyokern, CA, USA (Airport Code) 美国加利福尼亚州因约肯机场

**IYMS** IATA Yield Management System 国际航空运输协会收益管理系统

# J

J  Jet Airway 喷气机航路

Jet( Airway Designator) 喷气机(航路指示代码)

Joule 焦耳

Junction 连接,接头

**J/BOX**  Junction Box 接线盒

**J/S**  Jam/Signal Ratio 扰讯比

**JA**  Japan 日本

Java Applications Java 应用程序

Joint Account 共同账户,联合账

**JAA**  Jabara Award for Airmanship 飞行技术杰巴拉奖

Japan Aeronautic Association 日本航空协会

Japan Asia Airways 日本亚细亚航空

Joint Aerospace Applications 联合航空航天应用

Joint Airworthiness Authorities 联合适航局

Joint Aviation Authorities 联合航空当局

**JAAA**  Japan Ag-Aviation Association 日本农业航空协会

**JAAB**  JAA Board 联合航空局委员会

Judicial Appointments Advisory Board 司法任命咨询委员会

**JAAC**  JAA Committee (欧)联合航空局委员会

Joint Aircraft Allocations Committee 飞机联合分配委员会

**JAA TO**  Joint Aviation Authorities Training Organisation JAA 培训机构

**JAC**  Jackson Hole, WI, USA 美国密歇根州杰克逊机场

Japan Aircraft Company 日本飞机公司

Joint Aircraft Committee 联合飞机委员会

Joint Airworthiness Code 联合适航性法规

Joint Analysis Center 联合分析中心

Journal of Atmospheric Chemistry 大气化学杂志

**JACC**  Joint Airborne Communications Center 联合机载通信中心

**JACCP**  Joint Airborne Communications Command Post 联合机载通信指挥所

**JACG**  Joint Aeronautical Commanders Group 联合航空指挥官小组

**JACIS**  Japan Association of Air Cargo Information Systems 日本航空货运信息系统协会

**JACMAS**  Joint Approach Control Meteorological Advisory Service 联合进近管制气象咨询勤务

**JACOLA**  Joint Airports' Committee of Local Authorities (UK) 联合地方当局机场委员会(英国)

**JACS**  Jet Attitude Control System 喷气飞机姿态控制系统

Joint Air Communications System 联合空中通信系统

**JACV**  Jet Air Control Valve 空气喷射控制活门(阀)

**JAD**  Joint Application Design 联合应用程序设计

**JADC**  Japan Aircraft Development Company 日本飞机开发公司

**JAEC**  Japan Atomic Energy Commission 日本原子能委员会

Japanese Aero Engines Corporation 日本航空发动机公司

Joint Atomic Energy Committee 联合原子能委员会

**JAF**  Japan Automobile Federation 日本汽车联合会

**JAFE** Joint Advanced Fighter Engine 联合先进战斗机发动机(美国)

**JAFEP** Joint Advanced Fighter Engine Program 新战斗机发动机联合发展计划

**JAI** Japanese Aircraft Industry 日本飞机工业

Jet Associates International (USA) 国际喷气联合公司(美国)

**JAIC** Japanese Aircraft Industry Council 日本飞机工业理事会

**JAIMA** Japan Analytical Instruments Manufacturers Association 日本分析仪器制造商工业会

**JAIN** Java Advanced Intelligent Network Java 高级智能网

**JAIR** Journal of Artificial Intelligence Research 人工智能研究学报

**JAL** Jammer Allocation Logic 干扰机定位逻辑

Japan Airlines 日本航空公司

Jet Approach and Landing 喷气机进近和着陆

Jet Approach Letdown 喷气机进近下降

**JALC** Jet Approach and Landing Chart 喷气机进近和着陆图

Joint Aeronautical Logistics Commanders 联合航空后勤司令官

**JALMACS** Japan Airlines Maintenance Administration Control System 日本航空公司维护管理控制系统

**JALOS** Japanese Lubricating Oil Society 日本润滑油协会

**JAM** Jammed, Jamming 卡阻的,卡阻,干扰

Job Assignment Memorandum 工作委派备忘录

**JAMAC** Joint Aeronautical Materials Activity 联合航空器材活动

**JAMC** Japan Aircraft Manufacturing Corporation 日本飞机制造公司

**JAMCO** Japan Aircraft Manufacturing Co. 日本飞机制造公司

**JAMG** Jamming 卡阻,干扰

**JAN** Jackson, MS, USA (Airport Code) 美国密西西比州杰克逊机场

January 一月

Jet Aircraft Noise 喷气飞机噪声

**JANS** Jet Aircraft Noise Survey 喷气机噪声测量

**JAP** Japan 日本

**JAPIB** Joint Air Photographic Intelligence Board 航空照相联合情报局

**JAR** Joint Airworthiness Regulations 联合适航规章

Joint Airworthiness Requirements 联合适航要求

Joint Airworthiness Rules 联合适航规则

Joint Aviation Requirements 联合航空要求

Joint Aviation Rules 联合航空规章(欧洲)

**JARC** Jet Aircraft Runway Cleaner 喷气机跑道清洁器

**JAR-E** Joint Airworthinese Requirements, Europe 欧洲共同适航性要求

**JARP** Joint Advanced Research Projects 先进的联合研究项目

**JART** Joint Assessment and Ranking Team 联合评审定级小组

**JAS** Japan Air System 日本航空系统

Japan Amateur Satellite 日本业余卫星

Jet Air System 喷射空气系统

Job Analysis System 作业分析系统

Joint and Survivor 联合和救生

Joint Airmiss Section 空中相撞联合工作组

Journal of Aeronautical Science 航空学会(科学)学报

**JASDF** Japanese Air Self-Defense Force 日本航空自卫队

**JASG** Joint Advanced Study Group 联合高级研究组

**JASP** Jet Air Suction Port 进气口二次空气喷射

**JASS** Jamming Analysis Strategic System 干扰分析战略系统

Joint Anti-Satellite Study 联合反卫星研究

Journal of the Aerospace Science 航空与航天科学杂志(美国)

**JAST** Joint Advanced Strike Technology 联合先进打击技术

Joint Air Support Tactics 联合空中支援战术

**JASU** Jet Aircraft Starting Unit 喷气机启动装置

**JAT** Joint Acceptance Test 联合验收测试

Joint Action Team 联合行动小组

Joint Application Testing 联合应用测试

**JATAT** Joint Avionics Test Administration Team 航空电子联合测试管理小组

**JATCC** Joint Air Traffic Control Center 联合空中交通管制中心

Joint Air Traffic Control Course 联合空中交通管制航线

Joint Aviation Telecommunications Coordination Committee 联合航空电信协调委员会

**JATCRU** Joint Air Traffic Control Radar Unit 联合空中交通管制雷达设备(英国)

**JATE** Joint Air Transport Establishment 联合航空运输公司(英国)

**JATO** Jet-Assisted Takeoff 喷气助推起飞

**JATS** Jamming Analysis and Transmission Selection 干扰分析与传送挑选

Joint Air Transportation Service 联合空中交通服务

**JAVA** A Type of Object-oriented Programming Language 一种面向对象的编程语言

**JAWG** Joint Airmiss Working Group (英国空中交通局)空中相撞联合工作组

**JAWS** Jamming and Warning System 干扰和预警系统

Jet Advance Warning System 喷气飞机预警系统

Joint Airport Weather Studies 联合机场气象研究

Joint All Weather Seeker 共用全天候导引头

**JAX** Jacksonville, FL, USA(Airport Code) 美国佛罗里达州杰克逊维尔机场

**JB** Jack Box 配电箱

Jet Barrier 喷气机阻拦栅

Junction Box 接线盒

**JBAA** Japan Business Aviation Association 日本公务航空协会

**J-BAR** Jet Runway Barrier 喷气机跑道阻拦栅

**JBD** James Brake Decelerometer(Canada) 詹姆斯刹车减速表(加拿大)

Jet Blast Deflector 喷流偏向板,喷气拆流板

**JBI** James Brake Index (Canada) 詹姆斯刹车指数(加拿大)

**JBIG** Joint Bi-Level Image Group 联合双值图像编码标准

**JBOD** Just a Bunch of Disks 磁盘连续捆束阵列

**JBR** Jonesboro, AR, USA(Airport Code) 美国阿肯色州琼斯伯勒机场

**JBS** Japanese Biochemical Society 日本生化学会

**JC** Jack Connection 插孔连接

Jettison Control 放油控制

Jitter Compensation 抖动补偿

Job Control 作业控制(程序)

Joint Compound 黏合剂

**JCA** Joint Cargo Aircraft 联合货运飞机

**JCAA** Joint Council of Aging Aircraft 老旧飞机联合委员会

**JCAB** Japan Civil Aviation Bureau 日本民用航空局

**JCAHO** Joint Commission on Accreditation of Healthcare Organizations 医疗机构评审联合委员会

**JCB** Job Card Booklet 工卡手册

Joint Coordination Board 联合协调委员会

**JCC** Joint Communications Center 联合通讯中心

Joint Computer Center 联合计算机中心

Joint Consultative Committee 联合咨询委员会(英国)

Joint Coordination Center 飞行管制联合协调中心

**JCL**　Job Control Language 职业控制语言

**JCN**　Job Control Number 作业控制编号

　　Joint Control Number 联合控制号

　　Junction 联结

**JCP**　Jitter Compensation Priority 抖动补偿优先

　　Job Control Program 作业控制程序

　　Junction Call Processing 中继呼叫处理

**JCPS**　Jitter Compensation Processor Sharing 抖动补偿处理器共享

**JCR**　Jamming Cancellation Ratio 干扰对消比

**JCS**　Jet Charter Services, Inc. (USA) 喷气包机服务公司(美国)

　　Joint Human-Machine Cognitive System 联合人机认知系统

**JCSAT**　Japan Communication Satellite 日本通信卫星

**JCT**　Junction 接合

**JCTD**　Joint Capability Technology Demonstration 联合能力技术演示

**JD**　Jig Design 夹具设计

　　Joggle Die 压接模

　　Jointed 连接的,联合的

　　Judging Distance 目测距离

**JDA**　Japanese Defense Agency 日本防卫署

**JDBC**　Java Database Connectivity Java 数据库兼容

**JDK**　Java Development Kit Java 开发工具

**JDP**　Joint Definition Phase 联合定义阶段

　　Joint Departmental Publication 联合部门出版物

　　Joint Development Plan 联合开发计划

　　Joint Development Project 联合开发项目

　　Junior Development Programme 初级开发计划

**JDPO**　Joint Development Program Office 联合开发计划办公室

**JDRS**　Joint Deficiency Reporting System 联合短缺报告系统

**JDT**　Jig Drill Template 钻模模板

**JE**　Junction Equipment 连接设备

**JEBM**　Jet Engine Base Maintenance 喷气发动机基地维修

**JED**　Jeddah, Saudi Arabia (Airport Code) 沙特阿拉伯吉达机场

**JEDEC**　Joint Electronic Device Engineering Council 联合电子器件工程协会

**JEH**　Joint European Helicopter Company 欧洲联合直升机公司

**JEM**　Jam Evaluation Model 干扰评估模型

　　Jet Engine Modulation 喷气发动机调制

**JEMTOSS**　Jet Engine Maintenance Task Oriented Support System 面向喷气发动机维护任务的保障系统,喷气发动机维修工作支持系统

**JEOS**　Janus Earth Observation Satellite 贾纳斯地球观测卫星

**JEP**　Jeppesen 杰普逊公司

**JEPI**　Joint Electronic Payment Initiative 联合电子转账

**JEPO**　Joint Engine Project Office 联合发动机设计室

**JER**　Joint Employment Report 综合就业报告

**JERS**　Japanese Earth Resources Satellite 日本地球资源卫星

**JES**　Japan Engineering Standards 日本工程标准

　　Job Entry Subsystem 作业输入辅助系统

　　Job Entry System 作业输入系统

**JESC**　Japan Engineering Standards Committee 日本工程标准委员会

　　Japan Environmental Sanitation Center 日本环境卫生中心

　　Joint Electronics Standardization Committee 电子设备标准化联合委员会

　　Joint Equipment Standardization Committee 设备标准化联合委员会

**JET**　Jet Power, Inc. (USA) 喷气动力公司(美国)

　　Joint Engineering Team 联合工程小组

　　Joint Evaluation Team 联合评估小组

**JATCC**　Joint Air Traffic Coordination Center

联合空中交通协调中心

**JETDS** Joint Electronics Type Designation System 联合电子装备代号系统

Japan External Trade Organization 日本外贸组织

**JETS** Jetfleets' Statistics 喷气机队统计

Joint En-Route Terminal System 联合航路终端系统

**JETT** Jettison 抛油,投弃

**JF** Jet Fuel 喷气发动机燃油

Joint Financing (ICAO) 财务联营(国际民航组织)

Jumbo Fiber 巨型光纤

**JFC** Jet Flow Control 喷气(发动机燃油)流量控制

Jet Fuel Control 喷气发动机燃油控制

**JFET** Junction Field Effect Transistor 结型场效应晶体管

**JFGW** Jettison Final Gross Weight 最终弃物总重

**JFI** Jet Flight Information 喷气机飞行资料

**JFK** Kennedy, NY, USA (Airport Code) 美国纽约州肯尼迪机场

**JFPM** Joint Frequency-Phase Modulation 联合频率相位调制

**JFS** Jet-Fuel Starter 喷气(燃油)启动机

**JFTO** Joint Flight Test Organization 联合飞行试验组织

**JFTOT** Jet Fuel Thermal Oxidation Test 喷气燃油热氧化试验

**JG** Jitter Gain 抖动增益

**JGAB** Joint Government Agencies Board 政府机构联合委员会

**JGAIA** Japan Gas Appliances Inspection Association 日本气体用具检验协会

**JGPP** Joint Group on Pollution Prevention 污染防治联合小组

**JGSS** Jittered Grid Super-Sampling 移动式栅格超级采样

**JHAT** Japan Helicopter Air Transport 日本直升机航空运输

**JHB** Johor Bahru, Malaysia (Airport Code) 马来西亚新山(柔佛巴鲁)机场

**JHL** Joint Heavy Lift 联合重型运输飞行器

**JHM** Kapalua, HI, USA (Airport Code) 美国夏威夷州卡帕鲁阿机场

**JHMCS** Joint Helmet Mounted Cueing System 联合头盔指引系统

**JHSAT** Joint Helicopter Safety Analysis Team 直升机安全分析联合组

**JHSIT** Joint Helicopter Safety Implementation Team 直升机安全执行联合组

**JHSU** Joint Helicopter Support Unit 联合直升机支援部队

**JHW** Jamestown, NY, USA (Airport Code) 美国纽约州詹姆斯敦机场

**JIA** Jet I Aviation (USA) I 号喷气航空(美国)

Jetstream International Airlines 喷流国际航空公司

**JIAWG** Joint Integrated Avionics Working Group 联合综合航电工作小组

Joint Interface Avionics Working Group 联合接口航电工作小组

**JIB** Bujumbura, Burundi (Airport Code) 布隆迪布琼布拉机场

**JIC** Jet-Induced Circulation 喷射引致环流

Job Instruction Card 工作说明卡

Joint Industry Council 工业联合委员会

Joint Intelligence Center 联合情报中心

Joint Intelligence Committee 联合情报委员会

**JID** Jeppesen Inflight Database 杰普逊飞行数据库

**JIM** Japan Institute of Metals 日本金属学会

**JIMDAT** Joint Implementation Measurement Data Analysis Team 联合实施测量数据分析组

**JINR** Joint Institute for Nuclear Research 联合原子能研究所

**JIO** Joint Intelligence Organization 联合情报组织

**JIS**　Japanese Industrial Standard 日本工业标准

Job Information System 作业信息系统

**JISS**　Japan Ionosphere Sounding Satellite 日本电离层探测卫星

**JIT**　Japanese Industrial Technology 日本工业技术

Job Information Test 就业信息测试

Job Instructional Training 教育培训工作

Joint Investigation Team 联合调查组

Just in Time 准时

**JITP**　Just-in-Time-Paper 准时文件,看板作业文件

**JIU**　Joint Inspection Unit 联合检查组

**JJPTP**　Joint Jet-Pilot Training Programme 喷气式飞机驾驶联合训练计划

**JK**　Jack 插孔,千斤顶

**JKL**　Jack Key Lamp Panel 插孔电键(指示)灯盘

**JKSCR**　Jack Screw 起重螺杆

**JKT**　Jacket 外套,上衣

Jakarta, Indonesia（Airport Code）印度尼西亚雅加达机场

**JLA**　Jammer Logic A 干扰机 A 逻辑

Japanese Lease Association 日本租赁协会

**JLB**　Jammer Logic B 干扰机 B 逻辑

**JLL**　Japanese Leveraged Lease 日本杠杆租赁

**JLS**　Jet Lift System 喷气机升力系统

**JLU**　Jammer Logic Unit 干扰机逻辑组件

**JM**　Air Jamaica 牙买加航空

**JMA**　Japan Meteorological Agency 日本气象厅

**JMAPI**　Java Management Application Programming Interface Java 管理应用程序接口

**JMENS**　Joint Mission Elements Need Statement 联合任务单元需求说明

**JMF**　Joint Message Form 联合信息表格

**JMIC**　Joint Maritime Intelligence Centre 联合海上情报中心

**JMM**　Joint Maintenance Management 联合维修管理

**JMPS**　Japanese Mobile Phone System 日本移动电话系统

**JMS**　Jamestown, ND, USA（Airport Code）美国北达科他州詹姆斯顿机场

**JMSAC**　Joint Meteorological Satellite Advisory Committee 气象卫星联合咨询委员会

**JMSNS**　Justification for Major System New Start 主系统重新启动可行

**JNB**　Johannesburg, South Africa（Airport Code）南非约翰内斯堡机场

**JNC**　Jet Navigation Chart 喷气机导航图

Joint Navigating Council 联合导航委员会

Joint Negotiating Committee 联合谈判委员会

**JNS**　Jet Noise Survey 喷气飞机噪声测量

**JNSC**　Joint Navigation Satellite Committee 导航卫星联合委员会

**JNT**　Joint 连接,联合,加入

**JNU**　Juneau, AK, USA（Airport Code）美国阿拉斯加州朱诺机场

**JNWPU**　Joint Numerical Weather Prediction Unit 联合数值天气预报单位(组)

**JO**　Job Order 工作指令

**JOA**　Joint Operation Agreement 联合操作协议

**JOAP**　Job Opportunity Awareness Program 工作机遇意识程序

Joint Oil Analysis Program 联合油液分析程序

**JOD**　Joint Occupancy Date 联合占用日期

**JOE**　Joint Operational Environment 联合运行环境

**JOEB**　Joint Operation Evaluation Board 联合运行评估委员会

**JOG**　Joggle 榫接

**JOM**　Journal of Operations Management 运营管理杂志

**JONA**　Joint Office of Noise Abatement 消减噪声联合办公室

**JOP**　Joint Operating Procedures 联合操作程

序
Jupiter Orbiter Probe 木星轨道探测器

**JOR** Joint Operational Requirement 联合运行要求

**JORP** Jet Operations Requirements Panel 喷气机运营要求专家组

**JOS** Java Operating System Java 操作系统

**JOSE** Joint Optics Structure Experiment 联合光学结构实验

**JOTR** Joint Operational Technical Review 联合运行技术审查,联合操作技术审查

**JOTS** Job Oriented Training Standards 职业培训标准

**JOVIAL** Jules Own Version of International Algorithmic Language 国际算法语言朱尔斯文本

**JP** Java Platform Java 平台
Jet Pilot 喷气航空器驾驶员
Jet Propelled 喷气推进的
Jet Propulsion 喷气发动机

**JPA** Joint Part Approval 联合零部件批准
Joint Powers Authority 联合权力机构

**JPADS** Joint Precision Airdrop System 联合精确空投系统

**JPALS** Joint Precision Approach and Landing System 联合精密进近着陆系统

**JPATS** Joint Primary Aircraft Training System 初级飞机联合训练系统

**JPB** Joint Planning Board 联合计划委员会

**JPC** Jet Propulsion Center 喷气推进研究中心

**JPDA** Joint Probabilistic Data Association 联合概率数据关联法

**JPDO** Joint Planning and Development Office 联合计划与发展办公室
Joint Program and Development Office 联合程序和发展办公室

**JPE** Java Platform for the Enterprise 面向企业的 Java 平台

**JPEG** Joint Photographic Experts Group 图像专家联合小组

**JPG** Job Proficiency Group 工作技能组
Job Proficiency Guide 工作技能手册

**JPL** Jet Propulsion Laboratory 喷气推进实验室

**JPMS** Job Performance Measurement System 职务绩效测量系统

**JPNZ** Japanese 日本的,日本人,日语

**JPO** Joint Planning Office 联合计划办公室
Joint Program Office 联合项目办公室

**JPP** Joint Program Plan 联合项目计划

**JPR** Job Performance Requirements 工作表现要求
Joint Program Review 联合项目评审
Jumper 跨接线

**JPS** Jet Plume Simulation 喷射烟流模拟
Jet Propulsion System 喷气推进系统
Joint Precision Strikes 联合精确打击
Joint Processing System 联合处理系统
Joint Procurement Specification 联合采购规范

**JPT** Jet Pipe Temperature 喷管温度

**JPTL** Jet Pipe Temperature Limiter 尾喷管温度限制器

**JPTO** Jet Propelled Take off 喷气助推起飞

**JRC** Joint Rescue Center 联合救援中心
Joint Research Centre 联合研究中心
Joint Resources Council 联合资源委员会
Joint Review Council 联合评审委员会

**JRCC** Joint Reception Coordination Center 联合接待协调中心
Joint Rescue Coordination Center 联合救援协作中心

**JRCI** Jamming Radar Coverage Indicator 干扰雷达覆盖显示器

**JRES** Journal Reporting Extract System 日志报告录取系统

**JRMET** Joint Reliability and Maintainability Evaluation Team 可靠性与维修性联合评估组

**JRNY** Journey 旅行

**JRO** Kilimanjaro, Tanzania-Kilimanjaro ( Airport Code）坦桑尼亚乞力马扎罗机场

**JRS** Jerusalem, Israel（Airport Code）以色列耶路撒冷机场

Joint Reporting System 联合报告系统

**JS** Jamming to Signal 对信号的干扰

Job Sheet 加工单，施工单

Joint Support（ICAO）联营（国际民航组织）

**JSA** Japanese Standards Association 日本标准协会

Jet Standard Atmosphere 喷射标准大气

Job Safety Analysis 工作安全分析

Joint Sales Agreement 联合销售协议

Joint Security Area 联合保密区

Joint Service Agreement 联合服务协议

**JSAE** Japan Society of Aeronautical Engineering 日本航空工程学会

**JSAT** Joint Safety Analysis Team 联合安全分析小组

Joint Security Assistance Training 联合安全协助训练

Joint System Acceptance Testing 联合系统验收测试

**JSC** Japan Science Council 日本科学委员会

Jet Support Center 喷气支援中心

Jiuquan Space Center 酒泉宇航中心

Job Scheduler Control 工作计划员控制

Johnson Space Center 约翰逊宇航中心

Johnson Spaceflight Center 约翰逊航天飞行中心

Joint Safety Committee 联合安全委员会

Joint Scientific Committee 联合科学委员会

Joint Support Committee 联合支援委员会

**JSCMM** Java Supply Chain Management Module Java 供应链管理模块

**JSEC** Joint Support Evaluation Committee 联合支援评估委员会

**JSF** Joint Strike Fighter 联合攻击战斗机

**JSG** Jump Strut（Landing）Gear 活动支柱起落架

**JSGC** Joint Service Guidance and Control 联合服务引导和管理

**JSIT** Joint Safety Implementation Team 联合安全实施小组

**JSLC** Jiuquan Satellite Launch Center 酒泉卫星发射中心

**JSME** Japanese Society of Mechanical Engineers 日本机械工程师学会

**JSP** Java Server Pages Java 服务器端页面

**JSPO** Joint System Program Office 联合系统计划室

**JSR** Jammer Saturation Range 干扰器饱和范围

**JSS** Jet Stream System 喷气带状系统

Joint Services Standard 联合服务标准

Joint Surveillance System 联合监视系统

**JSSE** Joint Service Software Engineering 联合服务软件工程

Joint Service Software Environment 联合服务软件环境

**JSSG** Jamming Signal Source Generator 干扰信号源发生器

Joint Service Specification Guides 联合服务规范指南

**JST** Jamming Station 干扰电台

Japan Science and Technology Agency 日本科技局

Japan Standard Time 日本标准时间

Jet STOL Transport 喷气式短距起落运输机

Job Site Training 工作地点训练

Johnstown, PA, USA（Airport Code）美国宾西法尼亚州约翰斯顿机场

Joint Systems Test 联合系统测试

**JSTARS** Joint Surveillance Target Acquisition and Reconnaissance System 联合侦察目标定位与监视系统

Joint Surveillance Target Attack Radar System 联合监视目标攻击雷达系统

Joint Surveillance Tracking Attack Radar System 联合监视跟踪攻击雷达系统

**JSTC** Joint Scientific and Technical Committee 联合科学与技术委员会

**JSVR** Joystick with Variable Rate 变速率驾

驶杆

**JT**　Jig Template 夹具模板
　　Jitter Tolerance 抖动容限
　　Joint 联合,加入
　　Junctor Terminal 连接线终端

**JTA**　Job Task Analysis 工作任务分析
　　Joint Test Assembly 联合测试组件

**JTAGG**　Joint Turbine Advanced Gas Generator 联合涡轮机高级气体发生器

**JTB**　Joint Bar 联结杆

**JTC**　Joint Technical Committee 联合技术委员会

**JTCG**　Joint Technical Coordinating Group 联合技术协作组

**JTCS**　Joint Tactical Control System 联合战术控制系统

**JTD**　Joint Test Document 联合测试文件

**JTDE**　Joint Technology Demonstrator Engine 联合技术演示发动机

**JTE**　Joint Technical Evaluation 联合技术鉴定
　　Joint Test and Evaluation 联合测试与评估
　　Joint Test Element 联合试验分队

**JTF**　Jitter Transfer Function 抖动传递函数
　　Joint Task Force 联合特遣人员
　　Joint Test Facility 联合测试设备

**JTFA**　Joint Time-Frequency Analysis 联合时间—频率分析

**JTG**　Joint Task Group 联合工作组
　　Joint Test Group 联合试验组

**JTIDS**　Joint Tactical Information Distribution System 联合战术情报分发系统
　　Joint Tactical Interoperable Data System 联合战术可调数据系统

**JTIS**　Joint Tactical Information System 联合战术信息系统

**JTLY**　Jointly 共同地

**JTRU**　Joint Tropical Research Unit 联合热带研究组

**JTSB**　Japan Transport Safety Board 日本运输安全委员会
　　Joint Target Selection Board 联合目标选择委员会

**JTSN**　Jettison 抛油,放水

**JTSTR**　Jet Stream 喷射气流,喷气流

**JTU**　Jet Training Unit 喷气飞机教练队
　　Joint Test Unit 连接测试组件

**JTV**　Jettison Test Vehicle 放油(水)测试车

**JTWG**　Joint Test Working Group 联合测试工作组

**JUDG**　Judgment 判断,判断力,评价

**JUDMT**　Judgement 判定,判断

**JUG**　Joint Users Group 联合用户组

**JUI**　Java Unit Interface Java 部件接口

**JUL**　July 七月

**JUN**　June 六月

**JUNCTN**　Junction 连接,接合

**JURA**　Joint-Use Restricted Area 联合使用限制区

**JURIS**　Jurisdiction 权限,管辖区域

**JUST**　Justification 证明,合理性
　　Justify 辩护

**JUT**　Jet Utility Transport 通用喷气运输机

**JV**　Joint Venture 合资企业
　　Journal Voucher 转账传票,日记账凭据

**JVC**　Jet Vane Control 燃气舵控制
　　Joint Venture Company 合资企业

**JVM**　Java Virtual Machine Java 虚拟机

**JVO**　Joint Venture Office 合资办公室

**JVT**　Joint Venture Team 合资工作组

**JVX**　Joint Vertical Lift Aircraft 联合垂直升力飞机
　　Joint Vertical Lift Experiment 联合垂直升力试验

**JW**　Jacket Water 水套
　　Jamming War 电子干扰战

**JWA**　Joint Working Agreement 联合工作协议

**JWADF**　Joint Western Air Defense Force 西部联合防空航空队

**JWL**　Jewel 宝石

**JWLRY**　Jewelry 珠宝

**JWP**　Joint Working Plan 联合工作计划

**JZ**　Jump Zone 跳伞区,空降区

# K

**K** Karat 开(黄金成色),克拉
Kelvin 开尔文,开(绝对温标)
Kerosene 煤油
Kilo 公斤,千克
Kilo-Ohm 千欧姆
Knots 节,海里/小时
Knowledge 了解,理解,知识

**KA** Kenya Airways 肯尼亚航空公司
Key Account 重要客户
Key Area 关键地区
Keyed Address 键入地址
Knowledge Acquisition 知识获取
Knowledge Architecture 知识构建

**KAC** Kaman Aircraft Corporation 卡曼飞机公司
Kawasaki Aircraft Corporation 川崎飞机公司
Kuwait Airways Corporation 科威特航空公司

**KAD** Kidde Aerospace Defense 基德航空航天防御

**KADS** Knowledge Acquisition Data System 知识获取数据系统

**KAI** Kazan Aviation Institute 喀山航空研究所
Korea Aerospace Industries 韩国航空工业

**KAL** Kaltag, AK, USA(Airport Code)美国阿拉斯加州卡尔塔格机场
Korean Air Lines, Co. Ltd. 大韩航空公司

**KAM** Knowledge Acquisition Module 知识获取模块

**KAN** Kano, Nigeria(Airport Code)尼日利亚卡诺机场

**KARI** Korea Aerospace Research Institude 韩国航空航天研究所

**KAS** Knowledge Acquisition System 知识获取系统

**KAT** Katale Aero Transport(Zaire)卡托空运公司(扎伊尔)
Key to Address Transformation 键—地址变换

**KB** Keyboard Button 键盘按钮
Keyboard 键盘
Kilobyte 千字节
Knee Brace 角撑
Knowledge Base 知识库

**KB/S** Kilo Bits per Second 每秒千比特

**KBANN** Knowledge Base Artificial Neural Networks 知识库人工神经网络

**KBCT** Kent BAC Computing Technology 波音飞机公司肯特计算技术

**KBD** Keyboard 键盘
Knowledge Base Development 知识库开发

**KB-DSS** Knowledge Based Decision Support System 基于知识的决策支持系统

**KBE** Keyboard Encoder 键盘编码器
Knowledge Based Economy 基于知识经济
Knowledge Based Engineering 基于知识工程
Knowledge Based Environment 基于知识环境
Knowledge-Based Error 基于知识的误差

**KBES** Knowledge-Based Expert System 知识库专家系统

**KBL** Kabul, Afghanistan(Airport Code)阿富汗喀布尔机场

**KBM** Knowledge Base Management 知识库管理
Knowledge Base Module 知识库模块
Knowledge-Based Machine 知识库机

**KB-MMS** Knowledge-Based Model Management System 基于知识的模型管理系统

**KBMS** Knowledge Base Management System

知识库管理系统

**KBP** Kiev, Ukraine-Borispol（Airport Code）乌克兰基辅鲍里斯皮尔机场

**KBPS** Kilobytes per Second 每秒千字节数

**Kbps** Kilo bits per second 千位每秒

**KBR** Kota Bharu, Malaysia（Airport Code）马来西亚哥达巴鲁机场

**KBS** Knowledge Base System 知识库系统
Knowledge Based System 基于知识的系统

**KBSA** Knowledge-Based Software Assistant 人工智能软件辅助

**KC** Key Conversation 关键对话,键控对话
Keyer Control 键控器控制
Kilocycle 千周
Kilocycles per Second 千周/秒

**KCAB** Korean Civil Aviation Bureau 韩国民航局

**KCAS** Knots Calibration Air Speed 校准空速（海里/小时）

**KCCU** Keyboard and Cursor Control Unit 键盘和光标控制组件

**KCH** Kuching, Malaysia（Airport Code）马来西亚古晋机场

**KCS** Key Configuration Studies 键配置研究
Keyboard Controlled Sequencer 键盘控制程序器
Knock Control System 爆震控制系统

**KD** Keep in Dark 保密
Kick Down 强迫降挡
Knowledge Discovery 知识发现

**KDAS** Keep-up-to-Date Document Accountability System 保持更新的文件责任系统
Keep-up-to-Date Drawing Accountability System 保持更新的图纸责任系统

**KDC** Key Distribution Center 密钥分配中心
Keyboard Display Console 键控显示控制台
Knock-Down Component 分解部件
Korea Design Center 韩国设计中心

**KDD** Knowledge Discovery in Database 数据库中的知识发现

**KDF** Kalamein Door and Frame 防火门与门框

Knock Down Flat 分装平板车

**KDH** Kandahar, Afghanistan（Airport Code）阿富汗坎大哈机场

**KDLY** Kindly 亲切的

**KDM** Kingdom 王国

**KDOS** Key Display Operating System 键显示操作系统

**KDP** Key Decision Point 关键性决定点
Key, Display, Printer 键,显示器,打印机

**KDR** Keyboard Data Recorder 键盘数据记录器

**KDT** Keyboard Definition Table 键盘定义表
Keyboard Display Terminal 键盘显示终端
Knowledge Development Tools 知识开发工具

**KDZ** Kilo Dozen 千打

**KE** Key Element 关键要素
Kinetic Energy 动能
Knowledge Engineering 知识工程

**KEAS** Knots Equivalent Air Speed 节当量空速,海里/小时等值空速

**KECP** Kit Engineering Change Proposal 器材包工程更改建议

**KEE** Knowledge Engineering Environment 知识工程环境

**KEF** Keflavik, Reykjavik, Iceland（Airport Code）冰岛雷克雅莫克(首都)凯夫拉维克机场
Knowledge Economy Forum 知识经济论坛

**KEG** Knowledge Engineering Group 知识工程组

**KEJ** Kemerovo, Russia（Airport Code）俄罗斯克麦罗沃机场

**KEN** Kentucky 肯塔基州(美国)
Kenya 肯尼亚

**KEOPS** Key Element for Optical Packet Switching 光分组交换的关键部分

**KEP** Key Emitter Parameters 主要发射机参数

**KER** Kerman, Iran（Airport Code）伊朗克尔曼机场

Kerosene 煤油

**KEV** Thousand Electron Volts 千电子伏

**KEY** Key Airlines（USA）关键航空公司（美国）

**KEYPWR** Key Power 主要电源

**KF** Kalman Filter 卡尔曼滤波器

Key Field 关键字段

Kruger Flap 克鲁格襟翼

**K-FACTOR** Comparison of Actual vs. Theoretical Drag K 因子(实际与理论阻力之比)

**KFI** Kruger Flaps Indicator 克鲁格襟翼指示器

**KFRP** Kevlar Fiber Reinforced Plastics 芳纶纤维复合材料

**KFS** Kalman Filtering System 卡尔曼滤波系统

Key Factor for Success 成功的关键因素

Keyed File System 键控文件系统

**kg** Kilogram 公斤，千克

**KG(S)** Keg 小桶

Kilograms 公斤，千克

**kg/hr** Kilograms per Hour 千克/小时

**kg/m³** Kilograms/Cubic Meter 千克/立方米

**KGAT** Kent Gateway 肯特网关

**kgf** Kilogram-Force 千克力

**KGH** Kilograms per Hour 公斤/小时

**KGL** Kigali, Rwanda（Airport Code）卢旺达基加利/格雷瓜尔卡伊班达机场

**kgm³** Kilograms/Cubic Meter 千克/立方米

**KGPE** Kent General Purpose Extended 肯特一般用途延伸

**KGPS** Kent General Purpose System 肯特通用系统

Kilograms per Second 千克/秒

**KGS** Kilograms 千克

**KHI** Karachi, Pakistan（Airport Code）巴基斯坦卡拉奇/金纳机场

Kawasaki Heavy Industries 川崎重工业公司

**KHV** Khabarovsk, Russia Novy（Airport Code）俄罗斯哈巴罗夫斯克诺维机场

**KHz** Kilohertz(Kilocycles) 千赫

**KIAS** Knots Indicated Air Speed 指示空速（海里/小时）

**KIF** Knowledge Interchange Format 知识互换格式

**KIFIS** Kollsman Integrated Flight Instrument System 科尔斯曼综合飞行仪表系统

**KIJ** Niigata, Japan（Airport Code）日本新潟机场

**KILD** Kilderkin 小桶(容量单位)

**kilom.** kilometer 千米

**KIN** Kingston, Jamaica Norman Manley（Airport Code）牙买加金斯敦诺曼曼利机场

**KIP** Key Intelligence Position 关键信息位置

**KIS** Kick-in Step 开始步骤，第一步

Knowledge Information Service 知识信息服务

**KISS** Korean Intelligence Support System 韩国情报保障系统

**KIT** Keep in Touch 保持接触

**KITE** Kuiper Infrared Technology Experiment 奎波红外线技术实验

**KIU** Keyboard Interface Unit 键盘接口装置

**KIV** Kishinev, Moldova（Airport Code）摩尔多瓦基希纳乌机场

**KIX** Kansai, Osaka, Japan（Airport Code）日本大阪关西机场

**KJA** Krasnojarsk, Russia（Airport Code）俄罗斯克拉斯诺达尔机场

**KJFK** Kennedy International Airport 肯尼迪国际机场(美国)

**KJRDC** Kelly Johnson Research and Development Center 凯利约翰逊研制开发中心

**KKJ** Kita Kyushu, Japan（Airport Code）日本北九州机场

**KKV** Kinetic Kill Vehicle 动能杀伤飞行器

**KL** Kilolitre 千升(容量单位)

Kit Letter 器材函件

**KLA** Klystron Amplifier 速调管放大器

**KLAAS** Kinematics Local Area Augmentation System 动态地面局域增强系统

**KLADGNSS** Kinematics Local Area Differential GNSS 动态地局域差分 GNSS

**KLCR** Knock Limited Comperssion Ratio 敲缸极限压缩比

**KLIPS** Thousands of Logical Inferences per Second 千次逻辑推理/秒

**KLO** Klystron Oscillator 速调管振荡器

**KLR** Kalmar, Sweden ( Airport Code ) 瑞典卡尔玛机场

**KLSA** Knock Limit Spark Angle 爆震极限点火( 提前 )角

**KLU** Klagenfurt, Austria( Airport Code ) 奥地利克拉根福机场

**KLW** Klawock, AK, USA ( Airport Code ) 美国阿拉斯加州克拉沃克机场

**KLY** Klystron 速调管

**KM** Kilometer 公里, 千米
Knowledge Management 知识管理

**km.** kilometer 千米

**KM/H** Kilometers per Hour 公里/小时

**KM/L** Kilometers/Liter 公里/升

**KMA** Key Management Algorithm 密钥管理算法

**KMH** Kilometer( s ) per Hour 千米/小时
Korea Multipurpose Helicopter 韩国多用途直升机

**KMI** Key Management Infrastructure 密匙管理基础设施

**KMPS** Kilometres per Second 千米/秒, 公里/秒

**KMQ** Komatsu, Japan ( Airport Code ) 日本小松机场

**KMS** Key Management System 主要管理系统
Kilometerton 公里吨, 千米吨
Knowledge Management System 知识管理系统

**KN** Kilo Newton 千牛顿
Kont 节, 海里/小时

**KNA** Kenya 肯尼亚

**KNBL** Knife Blade 刀形桨

**KND** Kind 善意的, 种类

**KNED** Knife Edge 刀口, 刀锋

**KNF** Key Notarization Facility 重要公证设备

**KNLG** Knowledge 知识

**KNPC** Kuwait National Petroleum Company 科威特国家石油公司

**KNRL** Knurl 滚花

**KNSW** Knife Switch 刀形开关

**KNU** Kanpur, India( Airport Code ) 印度坎普尔机场

**KNW** Know 知道

**KO** Kickoff 分离, 断开
Knockout 打出, 分离, 拆卸( 器 )

**KOA** Kona, HI, USA ( Airport Code ) 美国夏威夷州科纳机场

**KOD** Karaoke on Demand 卡拉 OK 点播
Kick-off Drift 消除偏流修正角

**KOEO** KEY-on Engine-off 点火开关 ON 发动机不启动

**KOER** KEY-on Engine-Running 点火开关 ON 发动机运转

**KOH** Potassium Hydroxide 氢氧化钾

**KOHM** Kilohm 千欧姆

**KOJ** Kagoshima, Japan ( Airport Code ) 日本鹿尔岛机场

**KOMA** Kansas, Oklahoma, Missouri and Arkansas 美国堪萨斯州、俄克拉荷马州、密苏里州和阿肯色州

**KOOL** Knowledge-Oriented Object Language 面向知识的目标语言

**KOPS** Thousands of Operations per Second 千次运算/秒

**KOR** South Korea 韩国

**KOST** Kent Operating System Technology 肯特操作系统技术

**KOT** Kotlik, AK, USA ( Airport Code ) 美国阿拉斯加州科特利克机场

**KP** Keep 保持, 遵守
Key Point 关键点
Key Pulse 键控脉冲
Kick Pipe 反冲管
Kick Plate 金属踏板, 反冲板

Kilopond 千克力

Kilopound 千磅

**KP&D** Kick Plate and Drip 金属踏板和滴水槽

**KPA** Key Performance Action 关键绩效行动

Key Performance Areas 关键绩效区

Key Personnel Area 关键人员领域

Klystron Power Amplifier 速调管功率放大器

**kPa** Kilo Pascal 千帕

**KPCA** Kernel Principal Component Analysis 核主元素分析

**KPG** Keeping 保管,遵守

**KPH** Kilometers per Hour 公里/小时

Knots per Hour 海里/小时

**KPI** Key Performance Index 关键绩效指数

Key Performance Indicators 关键绩效指标

**KPL** Kick Plate 反冲板

**KPLS** Key Pulsing 键控脉冲

**KPP** Key Performance Parameter 关键性能参数

Keyboard Processing Program 键盘处理程序

**KPPS** Kilo Pulses per Second 千脉冲/秒

**KPS** Kilometers per Second 公里/秒

Knowledge Processing System 知识处理系统

**KPSM** Klystron Power Supply Modulator 速调管电源调制器

**KQA** Akutan, AK, USA (Airport Code) 美国阿拉斯加州阿库坦机场

**KR** Krone 克朗

**KRA** Key Result Areas 关键结果领域

**KRAL** Koroseal 氯乙烯树脂(塑料)

**KRIA** Korea Research Institute of Aerospace 韩国航空航天研究所

**KRK** Krakow, Poland (Airport Code) 波兰克拉科夫机场

**KRM** Knowledge Resource Management 知识资源管理

**KRN** Kiruna, Sweden (Airport Code) 瑞典基律纳机场

**KRSN** Kerosene 煤油

**KRT** Khartoum, Sudan (Airport Code) 苏丹喀土穆机场

**KS** Knowledge Set 知识集合

Knowledge System 知识系统

**KSA** Knowledge, Skills and Abilities 知识、技术和能力

Knowledge, Skills and Attitudes 知识、技术和态度

Korean Standards Association 韩国标准协会

**KSANG** Kansas Air National Guard 美国堪萨斯州国民防空警卫队

**KSAO** Knowledge, Skills, Abilities and Other Characteristics 知识、技巧、能力与其他特性

**KSC** Kagoshima Space Center 鹿儿岛航天中心(日)

Kennedy Space Center 肯尼迪航天中心

Korean Service Corps 韩国服务公司

Kosice, Slovakia (Airport Code) 斯洛伐克克思雀机场

**KSEA** Korean Structural Engineers Association 韩国结构工程师协会

Korean-American Scientists and Engineers Association 韩国—美国科学家和工程师协会

**KSI** Key Success Indicators 关键成功指标

Kilo Pounds per Square Inch 千磅/平方英寸

**KSM** Key Service Management 密钥服务管理

St. Mary's, AK, USA (Airport Code) 美国阿拉斯加州圣马丽斯机场

**KSML** Kosher Meal 犹太教餐食

**KSR** Keyboard Send and Receive 键盘发送和接收

**KST** Key Seat 键槽

**KT** Kiloton 千吨

Knot(s) 节(速度单位,相当于1.852km/h)

Knowledge Tracing 知识跟踪

**Kt, K"** Knot(s) 节(海里/小时)

**KT/SEC** Knots per Second 海里/秒

**KTAS** Knots True Airspeed 海里/小时真空速,真空速节

**KTD** Killed Target Dectector 击毁目标探测

器

**KTGS** Knot Ground Speed 节(地面速度)

**KTM** Kathmandu, Nepal ( Airport Code ) 尼泊尔加德满都机场

**KTN** Kernel Transport Network 核心传送网络

Ketchikan, AK, USA ( Airport Code ) 美国阿拉斯加州凯奇坎机场

**KTS** Key Telephone System 按键电话系统

Nautical Miles per Hour 海里/小时（节）

**KTW** Katowice, Poland ( Airport Code ) 波兰卡托维兹机场

**KU** Keyboard Unit 键盘组件

**KUA** Kuantan, Malaysia ( Airport Code ) 马来西亚关丹机场

**KUL** Kuala Lumpur, Malaysia ( Airport Code ) 马来西亚吉隆坡机场

**KUS** Kulusuk, Greenland ( Airport Code ) 格陵兰库卢苏克机场

**KUV** Kunsan, South Korea ( Airport Code ) 韩国群山机场

**KUW** Kuwait 科威特

**KV** Kilo Volts 千伏

Kinematic Viscosity 运动粘度

**KVA** Kilovolt-Ampere 千伏安

**KVAH** Kilovolt-Ampere Hour 千伏安小时

**KVAHM** Kilovolt-Ampere Hour Meter 千伏安小时表

**KVAM** Kilovolt-Ampere Meter 千伏安表

**KVAR** Kilovar, Kilovolt-Amp Reactive 无功千伏安

Kilovolt Amperes Reactive 无功千伏安

**KVG** Key Video Generator 键控视频发生器

**KVIS** Kinematic Viscosity 运动粘度

**KVP** Kilovolt Peak 千伏峰值

**KW** Keyword 关键词

**KW(S)** Kilowatt(s) 千瓦

**KWD** Kuwait Dinar 科威特第纳尔(货币单位)

**kWh** Kilowatt Hour 千瓦小时

**KWHM** Kilowatt Hour Meter 千瓦小时表

**KWI** Kuwait, Kuwait ( Airport Code ) 科威特科威特机场

**KWIC** Key Word in Context 上下文内关键字

**KWOC** Key Word out of Context 上下文外关键字

**KWR** Cryptographic Receive Unit 密码的接收装置

Kilowatts Reactive 无功千瓦

**KWT** Cryptographic Transmit Unit 密码的发送装置

Key Word in Title 题头关键词

Kilowatts Thermal 千瓦热量

Kuwait 科威特

**KWY** Keyway 键槽

**KY** Kentucky 肯塔基州

Key 键, 钥匙

Keying Device 键控装置

**KYBD** Keyboard 键盘

**KYD** Kiloyard 千码

**KYODO** Kyodo News Agency 共同社

**KYPCH** Keypunch 键盘穿孔机

**KYWD** Keyword 关键字(词)

# L

**L** Lady 女士

Landing 着陆

Late 晚,迟的

Latin 拉丁语

Latitude 纬度

Left 左

Length 长度

Light 灯

Line 线,行

Litre or Liter 升

Low Altitude 低空

Low Intensity 低强度

Pound 磅

**L&U** Loading and Unloading 装卸

**L/A** Letter of Authorization 委托书

**L/C** Letter of Credit 信用证

**L/D** Length-to-Diameter(Radio) 长径比

Lift/Drag (Ratio) 升阻比

**L/G** Landing Gear 起落架

**L/H** Left Hand 左侧,左手

Local Horizontal 局部地平线

Low to High 由低到高

**L/HIRF** Lightning/High Intensity Radiated Fields 闪电/高强度辐射场

**L/L** Latitude/Longitude 纬度/经度

Lower Lobe 下波瓣

**L/M** List of Materials 材料单

Load Monitor 负载监控器

**L/O** Layout 布局

Lift-off 起飞,发射

Light off 光照终止,(发动机)熄火

**L/R** Locus of Radius 径向轨迹

**L/S** Low Speed 低速

**L/V** Local Vertical 当地垂线,局部垂线

**L/W** Lock Wire 固定索

**L/W/H** Length/Width/Height 长/宽/高

**LA** Latin America 拉丁美洲

Launch Azimuth 发射方位角

Layer Adaptation 层间适配

Lead Angle 导前角

Letter of Advice 送货通知书

Letter of Authorization 委托书

Lightning Arrester 避雷器

Line Amplifier 线路放大器

Linear Accelerometer 线性加速器

Load Adjuster 配载计算器

Loading Area 装载场,装载区

Los Angeles 洛杉矶

Louisiana 路易斯安那州

Low Altitude 低空

Low Approach 低空进近

**LAA** Lifetime Achievement Award 终身成就奖

Light Aircraft Association 轻型飞机协会

Light Attack Aircraft 轻型攻击机

Local Airport Advisory 当地机场通告,当地机场咨询

Local Airport Authority 本地机场当局

Los Angeles Airways 洛杉矶航空公司

Low Angle of Attack 小攻角

Lowest Acceptable Altitude 最低可接受高度

**LAAA** Los Angeles Art Association 洛杉矶艺术协会

**LAAB** Louisiana Adoption Advisory Board 路易斯安那州推举咨询委员会

**LAAD** Littoral Area Air Defense 沿海地区空防

Low Altitude Air Defense 低空防空

**LAADS** Low-Altitude Aircraft Detection System 低空飞机探测系统(美国)

**LAAE** Lithuanian Association of Adult Education 立陶宛成人教育协会

**LAANC** Local Authorities Aircraft Noise Council（英国）飞机噪声委员会区域机构

**LAAS** Large Advance Antenna Station 大型超前天线站

Local Area Augmentation System 局域增强系统,本地增强系统

London Amateur Aviation Society 伦敦航空业余爱好者学会

Low Altitude Alerting System 低空警告系统

**LAAU** London Accident Analysis Unit 伦敦事故分析组件

**LAB** Label 标签

Laboratory 实验室

**LABC** Large Advancing Blade Concept 大型先进桨叶概念

**LABIL** Light Aircraft Binary Information Link 轻型飞机二进制信息链

**LAC** Lineas Aereas Costarricenses 哥斯达黎加航空公司

Lineas Aereas Del Caribe 加勒比航空公司

Link Access Control 链路接入控制

Local Area Controller 局部区域控制器

Lockheed Aircraft Corporation 洛克希德飞机公司〔美〕

Low Altitude Calibration 低高度校正

**LACA** Low Altitude Control Area 低空管制区

**LACAC** Latin America Civil Aviation Commission 拉丁美洲民用航空委员会

**LACE** Laser Atmospheric Compensation Experiment 激光大气补偿试验

Liquid Air Cycle Engine 液态空气循环发动机

Low-Power Atmospheric Compensation Experiment 低功率大气补偿试验

**LACFT** Large Aircraft 大型航空器

**LACN** Local Area Cell Network 局域信元网络

Local Area Communication Network 局域通信网

Local Area Computer Network 局域计算机网络

**LACO** Laser Communications 激光通信

**LACR** Low-Altitude Coverage Radar 低空覆盖雷达

**LAD** Ladder 梯子,阶梯

Landing Distance Available 可用着陆距离

Laser Acquisition Device 激光接收器

Luanda, Angola（Airport Code）安哥拉罗安达机场

**LADAR** Laser Detection and Ranging 激光探测与测距

Laser Doppler Radar 激光多普勒雷达

Laser Radar 激光雷达

**LADC** Laser Advanced Development Center 激光高级发展中心

Left Air Data Computer 左大气数据计算机

**LADD** Low Altitude Drogue Delivery 低空制动伞空投

Low Angle Drogue Delivery 低角度制动伞空投

**LADE** State Airlines of Argentina 阿根廷国家航空公司

**LADG** Loading 装货

**LADGNSS** Local Area Differential GNSS 局域差分全球导航卫星系统

**LADGPS** Local Area Differential Global Positioning System 局域差分全球定位系统

**LADS** Local Area Differential System 本地差分系统

Local Area Distributed System 局部分布式系统

Low Altitude Defense System 低空防御系统

Low Altitude Detection System 低空探测系统

**LADT** Local Area Data Transport 局域数据传送

**LADTAC** Light Area Defense Technology Assistance Contract 小型区域防御技术援助合同

**LAE** Left Arithmetic Element 左算术单元

Licensed Aircraft Engineer 持有执照的飞机工程师

Low-Altitude Extraction 低空施投（伞投货物）

**LAES** Landing Aids Experiment Station 着陆辅助设备实验站

Latin American Economic System 拉丁美洲经济体系

**LAF** Lafayette, IN, USA（Airport Code）美国印第安那州拉斐特机场

Load Alleviation Function 载荷衰减作用

**LAFL** Landing Area Flood Lights 着陆区照明灯

**LAG** Lagging 滞后

**LAGEOS** Laser Geodynamic Satellite 激光地球动力学卫星（又名测震卫星）

**LAHS** Low Altitude and High Speed 低空和高速

**LAHSO** Land and Hold Short Operation 着陆与短暂等待操作

**LAI** Lannion, France（Airport Code）法国拉尼永机场

Linee Aeree Italiane 意大利航空公司

Low Airspeed Indicator 低空速指示器

Luxembourg Aviation Investments（South Africa）卢森堡航空投资公司（南非）

**LAIA** Latin American Integration Association 拉丁美洲一体化协会

**LAIG** London Aviation Insurance Group 伦敦航空保险集团

**LAILA** Lavatory Interface and Light Adapter 盥洗室接口及灯光适配器

**LAINS** Low Altitude Inertial Navigation System 低空惯性导航系统

**LAL** Langley Aeronautical Laboratory 兰利航空实验室

Latitude and Longitude 经纬度

Lithuanian Airlines 立陶宛航空公司

Local Address Latched 锁存的本机地址

**LALD** Low Angle Low Drag 小角度低阻力

**LAM** Laminate 分层

Landing Altitude Modification 着陆高度修改

Landing Attitude Modification 着陆姿态修正

Large Amplitude Multimode 大幅度多模式

Lateral Autopilot Module 横向自动驾驶组件

Logical Acknowledgement Message 逻辑确认报

**LAMA** Local Automatic Message Accounting 本机自动信息统计（电话计费）

**LAMCIS** Los Angeles Multiple Corridor Identification System 洛杉矶空中多路航道识别系统

**LAME** Licensed Aircraft Maintenance Engineer 有执照的航空器维修工程师

**LAMM** Lights, Audio and Miscellaneous Module 灯光音频混合组件

**LAMP** Logic Analyzer for Maintenance Planner 维修计划员逻辑分析器

Logic Analyzer for Maintenance Planning 维修计划逻辑分析器

**LAMPS** Light Airborne Multipurpose System 轻型机载多用途系统

Low-Altitude Multipurpose System 低空多用途系统

**LAMS** Light-Aircraft Maintenance Schedule 轻型飞机维修计划

Load-Alleviation Mode Stabilization 负载减轻模态稳定

**LAN** Inland 内地

Lansing, MI, USA（Airport Code）美国密执安州兰辛机场

Local Area Network 局域网，区域网

**LANA** Local Area Network Adapter 局域网络适配器

**LANC** Local Area Network Chip 局域网芯片

Local Area Network Controller 局域网络控制器

**LAND** Landing 着陆

**LANDS** Local Area Network Distribution System 局域网分配系统

Low Approach Navigation Director System 低

空进近导航指挥系统

**LANDSS** Lightweight Advanced Night-Day Surveillance System 先进的轻型昼夜监视系统

**LANE** Local Area Network Emulation 局域网络仿真

Low Altitude Navigation Equipment 低空导航设备

**LANFOX** Local Area Network Fiber-Optic Transceiver 局域网光纤收发器

**LANNET** Large Artificial Nerve Network 大型人工神经网络

**LANPS** Low Altitude Night Photo Subsystem 低空夜间摄影分系统

**LANS** Land Navigation System 陆地导航系统

**LANTIRN** Low-Altitude Navigation and Targeting Infrared for Night 低空导航和瞄准夜视红外装置

**LAO** Laboratory Assessment Office 实验室评估办公室

**LAP** LAN Access Point 局域网接入点

Large-Scale Advanced Propfan 大型先进螺桨风扇

Launcher Adaptable Platform 发射器适用平台

Line Access Point 链路接入点

Lineas Aereas Paraguayas 巴拉圭航空公司

Link Access Procedure 链路接入规程

Link Access Protocol 链路接入协议

List of Applicable Publications 适用出版物清单

**LAPB** Link Access Procedure-Balanced 链路接入规程—平衡式

Link Access Protocol-Balanced 链路接入协议—平衡式

**LAPD** Large Aircraft Product Department 大型飞机生产部门

Link Access Procedure on the D-Channel D信道链路接入规程

**LAPDC** Link Access Procedure for Digital

Cordless 数字无绳链路接入规程

**LAPM** Link Access Procedure for Modem 调制解调器链路接入规程

Link Access Protocol Model 链路接入协议方式

**LAPT** Local Apparent Time 本地视(在)时(间)

**LAQ** Local Air Quality 当地空气质量

**LAR** Laramie, WY, USA(Airport Code) 美国怀俄明州拉勒米机场

Local Acquisition Radar 本地探索雷达

**LARA** Low-Altitude Radar Altimeter 低空雷达高度表

**LARC** Low-Altitude Ride Control 低空乘坐品质控制

**LARDTS** Low Accuracy Radar Data Transmission System 低精度雷达数据传输系统

**LARP** Line Automatic Reperforator 直线自动纸带穿孔机

**LARS** Low Altitude Radar Service 低空雷达服务

Low Altitude Radar System 低空雷达系统

Lower Atmosphere Research Satellite 低空大气研究卫星

**LARs** Latin American Rules 拉丁美洲航空条例

**LARS** Local Area Radio System 局域无线电系统

**LAS** Landing Aid System 降落辅助系统

Landing Approach Simulator 着陆进场模拟器

Landing Area Security 着陆区安全措施

Las Vegas, NV, USA-McCarran ( Airport Code) 美国内华达州拉斯维加斯麦卡伦机场

League of Arab States 阿拉伯国家联盟

Light Activated Switch 光敏开关

Low-Alloy Steel 低合金钢

Lower Airspace 低空

Lower Airspeed 低速

**LASC** Legal Aid Services Council 法律援助

服务委员会

Lockheed Aeronautical Systems Company 洛克希德航空系统公司

**LASCR** Light-Activated Semiconductor Controlled Rectifier 光敏晶体管控制整流器

Light-Activated Silicon Controlled Rectifier 光起动硅控控制整流器

**LASDR** Laser Detection and Ranging 激光探测和测距

**LASH** Lighter Airborne Ship 轻型飞艇

**LASO** Low Altitude Search Option 低空搜索选择

**LASR** Low-Altitude Surveillance Radar 低空搜索雷达,低空监视雷达(美国)

**LASS** Laser Applications System Studies 激光应用系统研制

Launch Area Support Ship 发射区支援舰

Logistic Automated Support System 自动化后勤保障系统

Low Altitude Supersonic Speed 低空超音速

**LASSM** Line Amplifier and Super Sync Mixer 行放大与超同步混合器

**LASSV** Land Approach System for Space Vehicle 航天器进近着陆系统

**LAST** Low-Altitude Supersonic Target 低空超音速目标

**LASU** Large Aircraft Sector Understanding 大型飞机谅解备忘录

**LASV** Low Altitude Supersonic Vehicle 低空超音速飞行器

**LAT** Landing Approach Trainer 着陆进近练习器

Lateral 横向

Latitude 纬度

Link-Attached Terminal 与链路连接的终端

Local Area Transport 地区运输

**LATCC** London Area and Terminal Control Centre 伦敦区域与终端管制中心

**LAT DEV** Lateral Deviation 侧向偏离,横向偏离

**LAT REV** Lateral Revision 横向修正,侧向修正

**LAT/LONG** Latitude/Longitude 经度/纬度

**LATA** Local Accesss and Transport Area 本地接入和转送区

**LATAR** Laser-Augmented Target Acquisition and Recognition 激光加强目标获取与辨识

**LATARS** Laser Augmented Target Acquisition Recognition System 激光放大目标搜索识别系统

**LATCC** London Air Traffic Control Center 伦敦空中交通管制中心

**LATS** Large Aperture Technology Satellite 大孔径技术卫星

**LATT** Line Attenuation 线路衰减

**LAU** Linear Accelerometer Unit 线性加速度计组件

**LAUA** Lloyd's Aviation Underwriters Association 劳埃德航空保险商协会

**LAV** Lavatory 洗手间

Least Absolute Value 最小绝对值

Light Armored Vehicle 轻型装甲飞行器

**LAW** Lawton, OK, USA (Airport Code) 美国俄克拉荷马州劳顿机场

**LAWAS** Low Altitude Windshear Alarm System 低空风切变告警系统

**LAWN** Local Area Wireless Network 局域无线网络

**LAWRS** Limited Airport Weather Report/Reporting System 机场有限天气报告系统

Limited Aviation Weather Reporting System 航空有限天气报告系统

**LAWS** Light Aircraft Warning System 轻型航空器警报系统(英国气象局)

Lightweight Aerial Warning System 轻型空中警戒系统(美国)

**LAX** Los Angeles, CA, USA (Airport Code) 美国加利福尼亚州洛杉矶机场

**lb** Pound 磅

**LB** Lebanon 黎巴嫩

Letter Box 信箱

Libra 镑(货币单位)

Lifeboat 救生艇

Line Busy(电话)占线

Linoleum Base 油毡(漆布)基底

Load Buffer 装入(取数)缓冲器

Local Battery 本机电瓶,本机电池

Low Blower 低压增压器,电瓶电压低,电池
电压低

**LB(S)**　Pound(s) 磅(重量单位)

**LB/hr**　Pounds/Pound per Hour 磅/小时

**LB/MIN**　Poundsper Square Inch 磅/平方英
寸

**LBA**　Civil Aviation Authority of Germany 德
国联邦航空局

　　Leeds /Bradford, England, United Kingdom
　　(Airport Code) 英国利兹/布拉德福机场

　　Limited of Basic Aircraft 基本飞机极限

　　Local Boarding Application 本地登机应用

　　Logical Block Addressing 逻辑块寻址

　　Logical Bus Application 逻辑汇流条应用

**LBAC**　Local-Bus Asynchronous Controller 本
地总线异步控制板

**LBB**　Lubbock, TX, USA (Airport Code) 美国
得克萨斯州拉伯克机场

**LBBL**　Left Body Buttock Line 机身左纵剖线

**LBC**　Laser Beam Cutting 激光切割

　　Linear Block (Digital) Code 线形信息(数
字)码

　　Local Baggage Committee 本地行李委员会

**LBCM**　Locator Back Course Marker 带示位
台的后航道指点标

**LBCN**　Locator Beacon 定位信标

**LBE**　Latrobe, PA, USA (Airport Code) 美国
宾西法尼亚州拉特罗布机场

　　Low Bit Rate Encoding 低比特率编码

**LBF**　Low Birefringent Fiber 低双折射光纤

　　North Platte, NE, USA (Airport Code) 美国
内布拉斯加州北普拉特机场

　　Pounds Force 磅力

**lbf/in²**　Pounds Force per Square Inch 磅力/
平方英寸

**LB-FT**　Pound-Force-Foot (Torque) 磅—

力—英尺(力矩)

**LBI**　Low-Band Interrogator 低波段询问机

**LB-IN**　Pound-Inch 磅—英寸

**LBK**　Left Bank 左倾斜

**LBL**　Label 标签

　　Laminar Boundary Layer 层流边界层

　　Left Buttock Line 左纵剖线

**LBM**　Liquid Boost Module 液体增压模件

　　Load Buffer Memory 负载缓冲存储器

　　Locator Back Marker 带示位台的后指点标

**LBN**　Land Mark Beacon 着陆地标台

　　Line Balancing Network 线路平衡网络

**LBNP**　Lower Body Negative Pressure 下体负
压

**LBO**　Landing Burn off 着陆耗油

**LBP**　Laser Beam Printer 激光打印机

　　Length Between Perpendiculars 垂线间距离,
垂线间长度

　　Link Bus Protocol 链路总线协议

　　Low Bypass 低涵道(比),低流量(比)

**LBPR**　Low Bypass Radio 低涵道比(小于
1.5)

**LBR**　Local-Base Rescue 本场救援

　　Low Bit Rate (计算机的)低比特率

**LBRG**　Laser Beam Rider Guidance 激光束
制导

　　Laser Beam-Riding Guidance 激光乘波导引

**LBRV**　Low Bit Rate Voice 低比特率语音

**LBS**　Lateral Beam Sensor 横向波束传感器

　　Location Based Service 基于位置的服务

　　Pounds (Weight) 磅(重量)

**LBS/HR**　Pounds per Hour 磅/小时

**LBT**　Listening Before Transmission 先听后发

　　Low Band Transmitter 低波段发射机

**LBTY**　Liberty 自由

**LBUS**　Local Bus 本机总线

**LBV**　Libreville, Gabon (Airport Code) 加蓬
利伯维尔机场

**LBW**　Laser Beam Welding 激光束焊接

　　Low Band Width 下边带宽度

**LBYRPK**　Labyrinth Pack 迷宫式密封体填

充物

LC Labour Cost 人工费用

Landing Chart 着陆图

Launch Complex 发射设备

Laundry Chute 污水槽管

Lead Covered 包铅的

Leaning Curve 趋向曲线

Leased Channel 租用信道

Least-Cost 最低费用/成本

Legal Committee 法律委员会

Length of Chord 翼展,弦长

Letter of Contract 合同函件

Letter of Credit 信用证

Life Cycle 寿命周期

Light Case 轻型壳体

Limit Sales,Closed 限制售(客)票,等待名单取消

Limited Capacity 有限能力

Limited Clearance 限定容差

Limited Coordination 有限协调

Line Carrying 载线的

Line Communications 有线通信

Line Concentrator 线路集中器

Line Contactor 线连接器

Line Control 线路控制

Link Connection 链路连接

Link Control 链路控制

Link Controller 链路控制器

Liquid Crystal 液晶

Load Carrier 运货车

Load Center 载荷中心

Load Compensating 负载补偿,加载补偿

Load Control 载荷控制

Load Controller 载荷控制器

Loadline Certificate 载重线证书

Local Control 本地控制

Local Coordinator 地区协调员

Lockecl Closed 在关闭状态下锁定

Logic Circuit 逻辑电路

Logical Channel 逻辑信道

Loop Controller 环路控制器

Low Carbon 低碳的

Lower Case 下部机壳

Lower Center 中下

Lower Compartment 下货舱

LCA Larnaca,Cyprus (Airport Code) 塞浦路斯拉纳卡机场

Lateral Control Actuator 横向控制作动器

Light Combat Aircraft 轻型战斗机

Logic Cell Array 逻辑信元阵列

Logistics Control Area 后勤控制区

Low Cost Aircraft 低成本飞机

LCAC Landing Craft Air Cushion 着陆飞机气垫

LCACS Land Cargo Advance Clearance System 陆路货物预先报关安排

LCB Line Control Block 线路控制块

Line of Constant Bearing 等方位线

Liquid Cooled Brakes 液冷式刹车装置

Longitudinal Position of Center of Buoyancy 浮力中心纵向位置

LCC Landing Control Center 着陆控制中心

Launch Control Center 发射控制中心

Leadless Chip Carrier 无铅芯片载体

Leased Circuit Connection 租用电路连接

Liaison Change Commitment 联络更改委托

Liaison Commitment Center 联络委托中心

Life Cycle Center 寿命周期中心

Life Cycle Cost 寿命周期成本

Line Concentration Controller 集线控制器

Link Controller Connector 链路控制连接器

Load-Carrying Composite 载重复合板

Logic Circuit Control 逻辑电路控制

Logistic Control Center 后勤管制中心

Low Cost Carrier 低成本运营商,低成本航空公司

Low Pressure Case Cooling 低压机匣冷却

LCCA Lateral Central Control Actuator 横向中央控制作动器,侧向中央控制作动器

Life-Cycle Cost Analysis 寿命周期费用分析

LCCC Launch-Control-Center Computer 发射控制中心电脑

Leadless Ceramic Chip Carrier 无铅陶瓷芯片载体

**LCC-CTR** Liaison Change Commitment Center 联络更改委托中心

**LCCE** Life Cycle Cost Estimate 寿命周期成本估算

**LCCM** Life Cycle Cost Management 寿命周期成本管理

Life Cycle Cost Model 寿命周期成本模型

**LCCP** Life Cycle Computer Program 寿命周期计算机程序

**LCCPMP** Life Cycle Computer Program Management Plan 寿命周期计算机程序管理计划

**LCCPT** Low Cost Cockpit Procedures Trainer 低成本驾驶舱程序训练器

**LCCR** Liaison Change Commitment Record 联络更改委托记录

**LCCS** Large Capacity Core Storage 大容量磁芯存储器

List of Channels for Calls 呼叫信道表

**LCD** Large Cargo Door 大型货舱门

Launch Countdown 发射倒计时

Least Common Denominator 最小公分母

Limiting Current Density 极限电流密度

Linear Current Density 线性电流密度

Liquid Chromatograph Detector 液相色谱检测器

Liquid Crystal Detector 液晶检测器

Liquid Crystal Digital 液晶数字的

Liquid Crystal Diode 液晶二极管

Liquid Crystal Display 液晶显示器

Local Climatological Data 本地气象数据

Logic Control Devices 逻辑控制器件

**LCDP** Lateral Control Departure Parameter 横向控制离港参数

**LCDT** Liquid Crystal Display Thermometer 液晶显示温度计

Load Compressor Discharge Temperature 负载压气机排气温度

**LCE** Laser Communication Equipment 激光通信设备

Laser Communication Experiment 激光通信实验

Launch Complex Equipment 全套发射设备

Launch Control Equipment 发射控制设备

Line and Customer Efficiency 航线与客户效率

Link Control Entity 链路控制实体

Load Circuit Efficiency 负载电路效率

**LCF** Large Cargo Freighter 大型货机

Launch Control Facility 发射控制设施

Local Control Facility 局域控制设施

Local Currency Fare 当地货币票价

Longitudinal of Center of Flotation 纵向浮力中心

Low Cost Fuselage 低成本机身

Low Cycle Fatigue 低循环疲劳,低周期性疲劳

**LCFC** Low Cycle Fatigue Counter 低循环疲劳计数器

**LCFD** Low Cycle Fatigue Damage 低循环疲劳损坏

**LCFS** Last Come First Served 后来先服务

**LCG** La Coruna, Spain (Airport Code) 西班牙拉科鲁尼亚机场

Load Classification Group 载荷分类组

Longitudinal Position of Center of Gravity 重心纵向位置

**LCGN** Logical Channel Group Number 逻辑信道群号

**LCGS** Low Cost Ground Surveillance 低成本地面监视

**LCH** Lake Charles, LA, USA (Airport Code) 美国路易斯安那州莱克查尔斯机场

Latch 锁

Logical Channel 逻辑信道

**LCHG** Latching 锁住

**LCIA** Local Check-in Assistant 当地值机助理

**LCIGS** Low Cost Inertial Guidance System 低成本惯性制导系统

**LCIP** Load Compressor Inlet Pressure 负载压缩机进气压力

**LCIT** Line Circuit Interface Trunk 用户电路接口中继线

Load Compressor Inlet Temperature 负载压缩机进气温度

**LCK** Lock 锁,锁定

**LCL** Line Check List（Maintenance）航线检修清单(维修)

Local 当地的

Lower Control Limit 控制下限

**LCLOCK** Local Clock 本机时钟

**LCLU** Landing Control and Logic Unit 着陆控制和逻辑组件

**LCM** Large Capacity Memory 大容量存储器

Lead Coated Metal 包铅金属

Least Common Multiple 最小公倍数

Line Control Memory 线路控制存储器

Liquid Composites Molding 复合材料液体成型

Load Control Module 负载控制模件

Logic Control Module 逻辑控制组件

Lowest Common Multiple 最小公倍数

**LCML** Low Calorie Meal 低热量餐,低脂高纤食物

**LCN** Load Classification Number 载荷分级号

Local Communications Network 局域通信网络

Local Computer Network 计算机局域网

Logical Channel Number 逻辑信道号

Logistics Control Number 物流控制号

**LCOLNT** Low Coolant 低(温)冷却剂

**LCOS** Lead Computing Optical Sight 前视光学计算机瞄准系统

Line Class of Service 线路业务等级

Liquid Crystal on Silicon 硅片液晶

**LCP** Lateral Control Package 侧(横)向控制组件

Left Circular Polarization 左旋极化

Lighting Control Panel 照明控制板

Link Control Protocol 链路控制协议

Liquid Crystal Polymer 液晶聚合物

Local Control Panel 区域控制面板

**LCPOF** Large Core Plastic Optical Fiber 大芯径塑料光纤

**LCPT** Light Cargo Passenger Turboprop 轻型客货两用涡轮螺桨飞机

**LCR** LAC Control Register 侧向控制作动器控制寄存器

Least Cost Routing 最低成本路由

Left-Center-Right 左—中—右

Liaison Change Request 联络更改申请

Line Condition Report 航线情况报告

Liquid Crystal Reticule 液晶光栅

Locus Control Region 局部管制区

Loop Control Relay 环路控制继电器

Low Compression Ratio 低压缩比

**LCRV** Length of Curve 曲线长度

**LCS** Landing Control System 着陆控制系统,着陆管制系统

Large Capacity Storage 大容量存储器

Large Core Storage 大容量磁芯存储器

Laser Communication System 激光通信系统

Laser Correlation Spectrometer 激光相关光谱仪

Laser Crosswind System 激光侧风系统

Lateral Control System 侧操纵系统

Launch Control Sequence 发射控制程序

Launch Control System 发射控制系统

Leased Circuit Service 租用电路业务

Life-Cycle Survivability 幸存能力寿命周期

Lighting Control System 光控系统

Link Connection Subsystem 链路连接子系统

Liquid Crystals 液晶

Logic Control System 逻辑控制系统

Logical Channel Set 逻辑通道设置

Longitudinal Control System 纵向控制系统

Low Cost Strut 低成本支撑杆

**LCSD** Line Check Support Data 航线检查支援数据

**LCSO** Low Cost Systems Office 低成本系统处

**LCSS** Land Combat Support System 地面作战支援系统

Laser Communication Satellite System 激光通讯卫星系统

**LCSTB** Low Cost Simulation Testbed 低成本模拟测试台

**LCSW** Latch Checking Switch 锁定装置检查开关

**LCT** Large Cargo Transport 大型货物运输

Level Control Table 电平控制表

Line Control Table 线路控制表

Line Current Transformer 线路电流互感器

Local Civil Time 当地民用时

Locate 探测,定位,设置

Location, Command and Telemetry 定位、指挥与遥测

Log Control Table 对数控制表

Logical Control Terminal 逻辑控制终端

Longitudinal Cyclic Trim 纵向循环配平

Low Carbon Transport 低碳运输

**LCTA** Line Current Transformer Assembly 线路电流互感器组件

**LCTC** Large Civil Tandem Compound 大型民用纵列式复合

**LCTD** Located 位于

**L-CTL** Link Controller 链路控制装置

**LCTN** Location 位置

**LCTR** Large Civil Tiltrotor 大型民用倾转旋翼机

Locator 定位器

**LCU** Level Converter Unit 电平转换器

Light Control Unit 灯光控制装置

Line Control Unit 线路控制器

Liquid Cooling Unit 液体冷却装置

Logic Control Unit 逻辑控制单元

Loop Channel Unit 环路信道单元

**LCV** Light Commercial Vehicles 轻型商用飞行器

Light Contingency Vehicle 轻型应急飞行器

Load Control Valve 负载控制活门

Logic Control Variable 逻辑控制变量

Low-Cost Visual 低成本目视(系统)

**LCW** Line Control Word 行控制字

**LCY** London, England, United Kingdom-London City (Airport Code) 英国伦敦城市机场

**LD** Land 陆地,着陆

Landing Distance 着陆(滑跑)距离

Laser Demodulator 激光解调器

Laser Detector 激光探测器

Laser Diffractometer 激光衍射仪

Laser Diode 激光二极管

Laser Director 激光指向仪

Laser Disc 激光唱片

Laser Disk 激光视盘

Late Delivery 延期交付

Latent Demand 潜在需要量

Lateral Density 侧向密度

Lateral Direction 侧向

Lateral Distance 侧向距离

Lateral Drift 侧向漂移

Latitude Determination 纬度测定

Lead 铅,导线

Leading Edge Delay 前沿延迟

Left Display 左显示器

Level Detector 电平检测器

Level Discriminator 电平鉴别器

Lift-Drag Ratio 升阻比

Line Drawing 线路图,路图

Link Disconnect 链路断开

List of Drawings 线路图清单

Load 装载,载量

Logic Driver 逻辑驱动器

Long Dump 长期清除

Low Density 低密度

Low Drag 低阻力

Lower Data 较低级数据

Lower Deck 下甲板,下舱

**LDA** Land Development Aircraft 试验型陆上飞机

Landing Distance Available 可用着陆距离

Laser Doppler Anemometry 激光多普勒风速测量法

Lesson Design Approach 课程设计方法

Line Driving Amplifier 行驱动放大器

Loading Advice 装载通知

Localizer Directional Aid 定位器定向辅助装置

Locate Drum Address 磁鼓地址定位

Logical Device Address 逻辑设备地址

**LDAP** Lightweight Data Access Protocol 轻量级数据存取协议, 轻量级数据访问协议

Lightweight Directory Access Protocol 轻量级目录存取协议, 轻量级目录访问协议

**LDAR** Liaison Design Action Request 联络设计措施申请

**LDAS** Logic Design Automation System 逻辑设计自动化系统

**LDASGN** Loading Assignment 装载分配

**LDB** Light Distribution Box 光分配盒

Limited Data Blocks 有限数据程序块

Logic Data Bank 逻辑数据库

**LDBLC** Low Drag Boundary Layer Control 低阻力边界层控制

**LDBS** Large Data Base System 大型数据库系统

**LDC** Less-Developed Countries 中等发达国家

Line Directional Coupler 线路定向耦合器

Line Drop Compensation 线路电压降补偿

Line Drop Compensator 线路电压降补偿器

List of Design Changes 设计更改表

Local Domain Coordinator 本地域协调器

Local Door Controller 本地门控制器

Logistics Date Center 物流数据中心

Long Distance Call 长途电话

Long-Distance Communication 长途通信

**LDCC** Leaded Chip Carrier 有引线芯片载体

Lower Deck Cargo Compartment 下货舱

**LDCOND** Load Condition 装载情况

**LDD** Laser Development Device 激光实验装置

Laser Diode Detector 激光二极管检测器, 激光二极管检波器

Linear Delay Distortion 线性延迟失真

**LDDS** Low Density Data System 低密度数据系统

**LDEF** Long Duration Exposure Facility 长时间曝光设施

**LDET** Level Detector 电平检测器, 电平检波器

**LDF** Long Distance Flight 远程班机

Lower Deck Facilities 下甲板设施

**LDFS** Landing Direction Finding Station 着陆定向台

**LDG** Landing Gear 起落架

Landing 落地, 着陆, 降落

Leading 前缘

Loading 装载, 载量

**LDGA** Loading Area 装载区

**LDGDVC** Loading Device 装载设备

**LDGPS** Local Differential GPS 本地差分全球定位系统

**LDGS** Loading Stand 装货台

**LDGWT** Landing Weight 着陆重量

**LDI** Landing Direction Indicator 着陆方向指示器, 着陆方向标

Lean Direct Injection Combustion 贫油直接喷射燃烧

Loadable Diagnostic Information 可装载的(故障)诊断信息

**LDIL** Landing Direction Indicator Lights 着陆方向指示器灯

**LDIN** Lead-in Lighting System 导入灯光系统

Lead-in Lights 导入灯光

**LDISCR** Level Discriminator 电平鉴别器

**LDK** Lower Deck 下舱地板

**LDL** Landing Direction Light 着陆方向指示灯

Load Data Loader 负载数据输入器

Logic-Based Data-Language 基于逻辑的数据语言

Low Density Lipoprotein 低密度脂蛋白

**LDLMT** Load Limit 装载极限, 限制装载量

**LDM**　Lean Direct Mixing Combustion 贫油直接混合燃烧

　　Linear Delta Modulation 线性三角调幅

　　Load Message 载量电报,航机装载文电

**LDMCR**　Lower Deck Mobile Crew Rest 下舱可移动机组休息室

**LDMI**　Laser Distance Measuring Instrument 激光测距仪

**LDMS**　Laser Distance Mearuring System 激光测距系统

**LDMX**　Local Digital Message Exchange 本地数字报文交换

**LDN**　Local Distribution Network 本地分配网

　　Low Speed Data Network 低速数据网

**LDNA**　Long Distance Navigation Aid 远距导航辅助设施

**LDNS**　Laser Doppler Navigation System 激光多普勒导航系统

　　Lightweight Doppler Navigation System 轻型多普勒导航系统

**LDO**　Lease-Development-Operate 租赁—开发—经营

　　Laser Designator Operator 激光指示器控制器

　　Lateral-Directional Oscillation 横向振动

**LDOC**　Lifecycle Direct Operating Cost 寿命周期直接运营成本

**LDP**　Label Distribution Protocol 标记分配协议

　　Landing Decision Point 着陆决断点

　　Language Data Processing 语言数据处理

　　Laser Designator Pod 激光照射器吊舱,激光指示器吊舱

　　Load Planing 配载平衡

**LDPA**　Landing Performance Application 着陆性能应用

**LDPC**　Low Density Positive Control 低密度绝对管制

**LDPG**　Laser Diode Pulse Generator 激光二极管脉冲发生器

**LDPM**　Laser Diode Pumped Maser 激光二极管泵浦脉泽

**LDR**　Ladder 梯子

　　Landing Distance Required 所需着陆距离

　　Light Dependent Resistor 光敏电阻

　　Linear Decision Rules 线性判定规则

　　Linear Dynamic Range 线性动态范围

　　Liquid Droplet Radiator 液滴辐射器

　　Loader 装载机

　　Loader Function 装载机功能,装入程序功能

　　Loader Register 装入寄存器

　　Local Distribution Room 本地分配室

　　Low Data Rate 低数据传输速率

**LDRCL**　Low Density Radio Communications Link 低密度无线电通信线路

**LDRET**　Load Report 装载表

**LDRI**　Low Data Rate Input 低数据速率输入

**LDS**　Laser Detection System 激光探测系统

　　Load Sharing 负荷分担

　　Local Digital Switch 本地数字交换机

　　Local Distribution System 本地分配系统

**LDSL**　Low Bit-Rate Digital Subscriber Line 低比特率数字用户线

**LDST**　Load Sharing Table 负荷分担表

**LDSTB**　Local Data Strobe 本机数据选通(门)

**LDT**　Laser Data Terminal 激光数据终端

　　Laser Discharge Tube 激光放电管

　　Laser Display Technology 激光显示技术

　　Lateral Dispersion at Touchdown 着陆横向偏移

　　Lightning Data Transport 闪电数据传输

　　Linear Differential Transformer 线性差动变压器

　　Linear Displacement Transducer 线性位移传感器

　　Local Daylight Time 当地夏令时

　　Logistics Delay Time 物流延迟技术

　　Long Distance Telephone 长途电话

**LDU**　Lahad Datu, Sabah, Malaysia (Airport Code) 马来西亚沙巴州拿笃机场

　　Lamp Driver Unit 灯泡驱动器组件

Launcher Display Unit 发射台显示装置

Link Data Unit 链路数据单元

Logical Data Unit 逻辑数据单元

**LDV** Laser Doppler Velocimeter 激光多普勒测速仪

Limiting Descent Velocity 极限下降速度

**LDW** Loss Damage Waiver 损毁免责险

**LDW(T)** Landing Weight 着陆重量

**LDX** Long Distance Extender 远距离扩展器

Long Distance Xerography 远距离静电复印术

**LE** Egyptian Pound 埃及镑(货币单位)

LAN Emulation 局域网仿真

Large End 大头,大端

Launching Equipment 发射设备

Leading Edge 前缘,前沿

Lease or Leasing 租借,租约,租借物

Life Expectancy 预期寿命

Lifting Eye 吊环

Light Efficiency 光效率

Light Emission 光发射

Light Energy 光能

Light Engine 轻型发动机

Light Excitation 光激发

Limit of Error 误差极限

Local Exchanger 本地交换机

Log Entry 记录本的记录

Logic Editor 逻辑编辑部

Lumped Elements 集总元件

**LEA** Launcher Electronics Assembly 发射器电子设备

Line Equalizing Amplifier 行均衡放大器

Logistics Effects Analysis 后勤效率分析

Logistics Engineering Analysis 后勤工程分析

Long-Endurance Aircraft 续航时间长的飞机

**LEAF** Large Effective-Area Fiber 大有效面积光纤

**LEAP** Leading Edge Analysis Program 前缘分析程序

Leading Edge Aviation Program 前沿航空计划

Lightweight Extensible Authentication Protocol 轻度可扩展验证协议

**LEAPS** Laser Engineering and Application of Prototype System 原型系统的激光工程与应用

**LE-ARP** LAN Emulation Address Resolution Protocol 局域网仿真地址判定协议

**LEASEDL** LEASED Line 租借线路

**LEB** Lebanon,NH,USA(Airport Code) 美国新罕布什尔州黎巴嫩机场

Legal Bureau 法律局

Lower Equipment Bay 下部设备舱

**LEC** Local Exchange Carrier 本地交换运营商

Lumped Elements Circulator 集总元件环流器

**LECS** LAN Emulation Configuration Server 局域网仿真配置服务器

**LECT** Lecture 讲座,讲演

Lecturer 演讲者,讲师

**LED** Large Electronic Display 大电子显示器

Leading Edge Device 前缘装置

Leading Edge Devices 前缘装置

Leading Edge Down 前缘襟翼放下

Light Emitting Diode 发光二极管

Local Economic Development 地方经济发展

Pulkovo, St Petersburg, Russia (Airport Code) 俄罗斯圣彼得堡普尔科夫机场

**LEE** Latching End Effector 闩锁端操纵装置

**LEED** Leadership in Energy and Environmental Design 能源与环境设计先导

Low Energy Electron Diffraction 低能量电子衍射

**LEF** Leading Edge Flaps 前缘襟翼

**LEFAS** Leading Edge Flap Actuation System 前缘襟翼驱动系统

**LEFDU** Leading Edge Flap Drive Unit 前缘襟翼驱动组件

**LEFLP** Leading Edge Flap 前缘襟翼

**LEFT** Left Turn 左转

**LEG** Leg 航段

Legal or Legislation or Legislative 法定

**LEGPROCS** Legal Processes 法律手续,合法手续

**LEHGS** Local Electro-Hydraulic Generation System 区域电子液压发生系统

**LEI** Almeria, Spain（Airport Code）西班牙阿尔梅里亚机场

**LEID** Low Energy Ion Detector 低能离子探测器

**LEJ** Leipzig, Germany（Airport Code）德国莱比锡机场

Longitudinal Expansion Joint 纵向膨胀接头

**LEL** Lowest Effective Level 最低有效水平

**LELU** Launch Enable Logic Unit 发射启动逻辑组件

**LEM** Laser Energy Monitor 激光能量监控器

**LEMAC** Leading Edge Mean Aerodynamics Chord 平均空气动力弦前缘

**LEMF** Leading-Edge Manoeuvre Flap 前缘机动襟翼

**LEMP** Lightning Electromagnetic Pulse 雷电电磁脉冲

Limited Electromagnetic Pulse 有限的电磁脉冲

**LEMV** Long Endurance Multi-Intelligence Vehicle 长航时多智能飞行器

**LEN** Length 长度

Linear Eletrical Network 线形电网络

**LEO** Little Earth Orbit 微型环地球轨道

Logistics and Engineering Operations 物流、工程业务

Low Earth Orbit 近地轨道,低高度轨道

**LEOP** Local Emergency Operations Plans 当地的紧急行动计划

**LEOS** Local Exchange Operation System 本地交换局运行系统

Low Earth Orbiting Satellite 低轨道卫星

**LEP** Laboratory Evaluation Program 实验室评估计划

Laser Eye Protection 激光眼保护

List of Effective Pages 有效页清单

Lowest Effective Power 最小有效功率

**LEPC** Local Emergency Planning Committee 地区紧急计划委员会

**LEQ** Less than or Equal to 小于或等于

**LER** Label Edge Router 标记边界路由器

Launch Equipment Room 发射设备室

Leading Edge Radius 前缘半径

Leading Edge Rib 前缘肋部

Level Error Rate 等级错误率

**LERX** Leading Edge Root Extension 前缘翼根边条

**LES** LAN Emulation Server 局域网仿真服务器

Land Earth Station 陆地地球站

Large Eddy Simulation 大涡模拟

Launch Escape Signal 发射撤离信号

Leading Edge Slats 前缘缝翼

Leading Edge Station 前缘站位

Lincoln Experimental Satellite 林肯实验卫星

Lockheed Expert System 洛克希德专家系统

Loop Error Signal 环路误差信号

**LESO** Land Earth Station Operator 陆地地球站运营商

**LESS** Low Energy Stage Study 低能级研究

**LET** Launch and Escape Time 发射逃离时间

Leading Edge Technologies 领先的技术

Leading Edge Tracking 前沿跟踪

Leticia, Colombia（Airport Code）哥伦比亚莱蒂西亚机场

Letter 信,字母

Linear Energy Transfer 线性能量转换

Low Energy Telescope 低能望远镜

**LETFO** Letter Follows 信随后寄到

**LETS** Liaison ECR Tracking System 联络工程更改申请跟踪系统

Line Efficiency Tracking System 生产线效率跟踪系统

Low-Energy Telescope System 低能量观测系统

**LEU** Laser Electronics Unit 激光电子装置

Leading Edge Up 前缘襟翼收起

**LEVL**   Leading-Edge Vortex Lift 前缘涡升力

Licensed Electrical Work 电气工作许可

**LEX**   Leading-Edge Extension 前缘突出部分,前缘锯齿

Lexington,KY,USA（Airport Code）美国肯塔基州莱克星顿机场

**LF**   Laser Finder 激光探测器

Laser Frequency 激光频率

Last Forward 最后发出号数

Lathe Fixture 车床夹具

Launch Facility 发射设施

Ledger Folio 分类账页码

Left Front 左前

Level Flight 水平飞行

Life Float 救生筏,救生圈

Line Feed 线路馈电,换行

Line Feeder 线路馈电线

Linear Filtering 线性过滤

Load Factor 负载系数

Load/Loading Factor 载运率,负载因子

Logic Function 逻辑功能,逻辑函数

Low Frequency（30 to 300 KHz）低频

**LFA**   Landing Fuel Allowance 着陆燃油许可量

Loss of Frame Alignment 帧失步

Low Frequency Accelerometer 低频加速度计

Low Frequency Amplifier 低频放大器

Low-Flying Aircraft 低空飞行飞机

Low-Flying Area 低空飞行区

**LFAS**   Loose-Fitting Air Supplied 松配合空气补给

Low-Frequency Active Sonar 低频主动声纳

**LFB**   Linear Frame-Buffer 线性帧缓冲

Local Feed Back 局部反馈

Low Frequency Beacon 低频信标

**LFC**   Laminar-Flow-Control 层流控制

Level of Free Convection 自由对流高度

Load Frequency Control 负载频率控制

Low-Frequency Correction 低频校正

Low-Frequency Current 低频电流

**LFD**   Large Flat Display 大型平板显示器

Latest Finish Date 最晚完成日期

Longitudinal Fuselage Datum 纵向机身数据

Low Frequency Disturbance 低频扰动

**LFDF**   Low Frequency Direction Finder 低频定向仪

**LFE**   Laboratory For Electronics 电子学实验室

Low Frequency Enhanced 低频增强

**LFEC**   Low Frequency Eddy Current 低频涡流(检查)

**LFES**   Landing Field Elevation Setting 着陆机场标高设定

**LFF**   Load Factor for Flight 飞行载荷因数

Low Frequency Filter 低频滤波器

Low Pressure Fuel Filter 低压燃料过滤器

**LFFZ**   Laser-Beam Free Flight Zone 无激光光束飞行区

**LFG**   Liquid-Filled Gyroscope 注液式陀螺仪

Liquid-Floated Gyroscope 悬浮式陀螺仪

Low-Frequency Generator 低频发生器

**LFL**   Landing Field Length 跑道长度

Laser Flash Lamp 激光闪光灯

Low Frequency Loran 低频罗兰,低频远程导航系统

Lower Flight Level 低飞行高度

Lower Frequency Limit 低频极限

**LFLM**   Lowest Field Level of Maintenance 外场维修最低级别

**LFM**   Lateral Force Microscope 横向力显微镜

Linear Feet per Minute 线性英尺/分钟

Linear Frequency Modulation 线性频率调制

Low-Powered Fan Marker 低功率扇形指点标

**LFMR**   Low-Frequency Microwave Radiometer 低频微波辐射计

**LFNS**   Low Frequency Navigation System 低频导航系统

**LFO**   Light Fuel Oil 易挥发油

Low Flying Object 低空飞行物体

Lower Frequency Oscillator 低频振荡器

**LFOM** Low-Frequency Outer Marker 低频外指点标

**LFP** Late Flight Plan 迟发的飞行计划

Loaded Flank Pitch 负载侧桨距

Low Fuel Pump 低压油泵

**LFQ** Limited Flight Qualification 有限飞行资格

**LFR** Launch and Flight Reliability 发射与飞行可靠性

Local Flight Regulation 地方飞行规则

Lock-Follow Radar 自动跟踪雷达

Low Frequency Radio Range 低频无线电范围

Low Frequency Range 低频波段

**LFRD** Lot Fraction Reliability Definition 批系数可靠性定义

Lot Frequency Reliability Definition 批频率可靠性定义

**LFRE** Liquid-Fueled Ramjet Engine 液体燃料冲压喷气发动机

**LFRED** Liquid-Fueled Ramjet Engine Development 液体燃料冲压喷气发动机研制

**LFRR** Low-Frequency Radio Range 低频无线电波段

**LFRTA** Lift Cruise Fan Research and Technology Aircraft 升力巡航风扇研究与技术飞机

**LFS** Launch Facility Simulator 发射设施模拟器

Localizer Frequency Select 航向信标频率选择

Loop Feedback Signal 环路反馈信号

**LFSMS** Logistics Force-Structure Management System 后勤保障部队结构管理系统

**LFT** Lafayette, LA, USA ( Airport Code ) 美国路易斯安那州拉斐特机场

Left 左

Level Flight Time 平飞时间

**LFTOVR** Leftover 剩余的

**LFV** Low Frequency Ventilation 低频通风, 低频换气

Low Frequency Vibration 低频振动

**LFW** Lome, Togo ( Airport Code ) 多哥洛美机场

**LG** Landing Gear 起落架, 起落装置

Landing Ground 降落场

Laser Gyro 激光陀螺

Laser Gyroscope 激光陀螺( 仪 )

Lateral Guidance 侧向引导, 横向引导

Leg 航段

Length 长度

Letter of Guarantee 担保函

Level Gauge 水平规, 水准仪

Long 长

Loop Gain 环路增益

**LGA** Laguardia, NY, USA ( Airport Code ) 美国纽约拉加迪亚机场

Landing Gear Analysis 起落架分析

Landing Gear Cable 起落架钢索

Light Gun Amplifier 光电枪放大器

Low Gain Antenna 低增益天线

**LGB** Long Beach, CA, USA ( Airport Code ) 美国加利福尼亚州长滩机场

**LGC** Landing Gear Cable 起落架钢索

Landing Gear Controller 起落架控制器

**LGCIS** Landing Gear Control and Indication System 起落架控制与指示系统

**LGCIU** Landing Gear Control and Interface Unit 起落架控制和接口组件

**LGCL** Landing Gear Control Lever 起落架操纵手柄

**LGCM** Light Green Communication Manager 绿灯通信管理器

**LGCU** Landing Gear Control Unit 起落架控制组件

**LGD** Loss Given Default 违约损失率

**LGE** Language 语言

Large 大的

League 团, 同盟

**LGERS** Landing Gear Extension and Retraction System 起落架收放系统

**LGG** Light Gun Pulse Generator 光电枪脉冲

产生器

**LGH** Landing Gear Handle 起落架手柄

**LGINS** Laser Gyroscope Inertial Navigation System 激光陀螺惯性导航系统

**LGK** Langkawi, Malaysia（Airport Code）马来西亚兰卡威机场

**LGL** Legal 合法的

**LGM** Logistics Module 后勤舱

**LGMS** Landing Gear Management System 起落架管理系统

Laser Ground Mapping System 激光地形绘图系统

**LGMT** Landing Gear Maintenance Trainer 起落架维修训练器

**LGND** Landing Ground 着陆场地

Legend 图例

**LGPI** Landing Gear Position Indicator 起落架位置指示器

**LGPIU** Landing Gear Position Indicator Unit 起落架位置指示器组件

**LGR** Longer 较长的

**LGS** Landing Gear System 起落架系统

Landing Guidance System 着陆引导系统

**LGSC** Linear Glide-Slope Capture 保持直线下滑道

**LGSOWG** LANDSAT Ground Station Operators Working Group 陆地卫星地面站工作组

**LGSYS** Landing Gear System 起落架系统

**LGT** Landing Gear Tread 起落架（主）轮距

Light 灯，轻的

Lighted 有照明的

Lighting 照明

**LGTD** Lighted 有照明的

**LGTH** Length 长度

**LGTV** Landing Gear Transfer Valve 起落架转换活门

**LGV** Landing Gear Valves 起落架活门

**LGW** Landing Gross Weight 着陆总重

London, England, United Kingdom-Gatwick（Airport Code）英国伦敦盖特维克机场

**LGWB** Landing Gear Wheelbase 起落架前后轮距

**LGWSS** Landing Gear Well Surveillance System 起落架舱监视系统

**LH** Labour Hour 工时

Left Hand 左边

Left Hydraulic 左液压（系统）

Light Helicopter 轻型直升机

Light House 灯塔

Line Hunting 寻线

Link Layer Head 链路报头，链路层头

Litter Hook 杂物钩

Local Host 本地主机

Long Haul 远程（宽频带）通信

Lower Half 下半部

**LHA** Local Health Authority 地方卫生当局

Local Hour Angle 地方时角，本地时角

**LHC** Left Hand Circular 左旋

Light Helicopter Cycle 轻型直升机循环

Local Hydraulic Centre 局部液压中心

**LHCP** Left Hand Circular Polarization 左旋极化

**LHD** Large Height Deviation 大高度偏差

**LHDC** Left Hand Data Conversion 左侧数据转换

**LHDR** Left Hand Drive 左侧驱动

**LHE** Lahore, Pakistan（Airport Code）巴基斯坦拉合尔机场

Liquid Helium 液氦

**LHFEB** Left-Hand Forward Equipment Bay 左前设备舱

**LHFT** Light Helicopter Fireteam 轻型直升机消防队

**LHG** Line Hunt Group 寻线组

Liquid Heat Generator 液态热量产生装置

**LHM** Laser Hardened Materials 激光硬化材料

**LHO** Local Health Office(r) 地方健康办公室（官员）

**LHOTS** Long-Haul Optical Transmission Set 长距离光传输设备

**LHOX** Low and High Pressure Oxygen 高压

与低压氧气

**LHP** Lightning HIRF Protection 闪电高强度辐射场保护

**LHR** Landing Hour 着陆时间
Lower Hybrid Resonance 低混合共振
London, England, United Kingdom-Heathrow (Airport Code) 英国伦敦希斯罗机场

**LHS** Left Hand Side 左侧
Left Hand Stringer 左纵梁
Lightweight Hydraulic System 轻型液压系统
Load Handling System 负载处理系统

**LHSC** Left-Hand Side Console 左侧操纵台

**LHT** Left Horizontal Tail 左水平尾翼
Long-Haul Transoceanic 越洋长途通信

**LHTEC** Light Helicopter Turbine Engine Company 轻型直升机涡轮发动机公司

**LHV** Lower Heating Value 低热值

**LHX** Light Helicopter Experimental 轻型试验直升机
Load Heat Exchanger 载荷热交换器

**LI** Lane Identification 航道识别号(早期的台卡系统)
Left Inboard 左内侧
Length Identifier 长度标记符
Letter of Intent 意向书
Level Indicator 液面指示器
Limit Indicator 极限指示器
Line Interface 线路接口
Load Index 载量指数
Low Intensity 低强度

**LIAS** Liaison 联络

**LIASON** Lossless Integrated Active Splitters for Optical Network 光网络无损耗综合有源分离器

**LIB** Label Information Base 标记信息库
Least Important Bit 次要的比特
Left Inboard 左内侧
Library Acquisition Program 程序库提取程序
Library Microfilms 图书馆缩微胶卷
Library 图书馆,程序库

Loudspeaker and Indicator Box 扬声器及指示器盒

**LIBRTN** Library Renton 瑞顿图书馆(波音)

**LIBSTD** Library Standard 图书馆标准

**LIC** Level Inertial Capture 水平惯性截获
Licence 执照
Line Interface Card 线路接口插件板
Line Interface Circuit 线路接口电路
Linear Integrated Circuit 线性集成电路
Lowest Incoming Channel 最低入局信道
Low-Intensity Conflict 低强度冲突

**LICI** Link Interface Control Information 连接界面控制信息

**LID** Library Issue Document 图书馆发行文件
Lift-Improvement Device 增加升力装置
Link Interface Device 链路接口设备
Locked in Device 锁定装置

**LIDA** Load-and-Interference-Based Demand Assignment 基于负载和干扰的按需分配

**LIDAR** Laser Identification Detection and Ranging 激光识别检测与测距
Laser Imaging Detection and Ranging 激光成像检测与测距
Light Detection and Ranging 光检测与测距

**LIDS** Laser Illumination Detection System 激光照射探测系统

**LIDU** Link Interface Data Unit 链路接口数据单元

**LIE** Line Interface Equipment 线路接口设备

**LIF** Load Information 负载信息,加载信息

**LIFE** Linear Integrated Flight Equipment 线性综合飞行设备

**LIFMOP** Linearly Frequency-Modulated Pulse 线形调整脉冲频率

**LIFO** Last in, First out 后进先出

**LIFR** Limited Instrument Flight Rules 有限的仪表飞行规则
Low Instrument Flight Rules 低空仪表飞行规则

**LIFRAM** Liquid Fueled Ramjet 液体燃料冲

压喷气发动机

**LIH** Kauai Island, HI, USA（Airport Code）美国夏威夷州可爱岛机场

Light Intensity High 高强度灯光

**LIL** Light Intensity Low 低强度灯光

Lille, Franee（Airport Code）法国里尔机场

**LILO** Last in, Last out 后进后出

**LIM** Call Limiter 呼叫限制器

Laboratory Information Management 实验室信息管理

Laser Intensity Monitor 激光强度监测器

Light Intensity Medium 中亮度, 中强度灯光

Lima, Peru-Jorge CHavez（Airport Code）秘鲁利马豪尔赫查韦斯机场

Limber 可塑的, 轻快的

Limiter 限幅器, 限制器

Limiting 限制的, 极限的

Limousine Service 客车(轿车)服务

Line Interface Maintenance 线路接口维护

Line Interface Module 线路接口模块

Linear Induction Motor 线性感应电机

Locator Inner Marker 带示位台的内指点标

**LIM SW** Limit Switch 极限电门

**LIMDIS** Limited Distribution 有限分配

**LIMG** Limiting 限制的, 极限的

**LIMIFR** Limited Instrument Flight Rule 有限制的仪表飞行规则

**LIMSPD** Limited Speed 限制速度

**LIMSW** Limit Switch 极限开关, 限位开关

**LIN** Milan, Italy-Linate（Airport Code）意大利米兰莱内特机场

Linear 线形的

Link Inhibit 链路禁止

Liquid Nitrogen 液氮

**LINAC** Linear Accelerator 直线加速器

**LINBD** Left Inboard 左内侧

**LINS** Laser Inertial Navigation System 激光惯性导航系统

**LINT** Low Intensity 低强度

**LIP** Large Internet Packet 大型网际包

Line Interface Processor 线路接口处理器

**LIPBOM** Limited Initial Program Bill of Material 有限起始程序器材清单

**LIPPLE** Lowest Intelligible Phone Power Level 最低可听音量

**LIPS** Laser Image Processing Scanner 激光图像处理扫描器

Laser Image Processing Sensor 激光图像处理传感器

**LIPX** Large Inter-Network Packet Exchange 大型网间分组交换

**LIQ** Liquid 液体, 酒

**LIR** Laboratory for Insulation Research 绝缘研究实验室

Laser Intercept Receiver 激光拦截接收器

Line Integral Refractometer 线路综合折射计

**LIRL** Low Intensity Runway Lighting 低强度跑道照明

Low Intensity Runway Lights 低强度跑道灯

**LIRLY** Load Indicating Relay 负载指示继电器

**LIS** Laser Illuminator System 激光照射系统

Laser Instruction Simulator 激光指令模拟器

Laser Interferometer System 激光干涉仪系统

Laser Ion Source 激光离子源

Laser Isotope Separation 激光同位素分离

Linear Systems Analysis 线性系统分析

Lisbon, Portugal（Airport Code）葡萄牙里斯本机场

Local Information System 本机信息系统, 内部信息系统

Localizer Inertial Smoothing 定向器惯性校平, 着陆航向信标惯性平滑

Logical IP Subnet 逻辑 IP 子网

Logically-Independent IP Subnet 逻辑上独立的 IP 子网

Loop Input Signal 环路输入信号

Loran Inertial System 罗兰惯性系统

**LISA** Light Interface Standardization Adapter 灯光界面标准适配器

Limited Instruction Set Architecture 有限指令集结构

Linear System Analysis Program 线性系统分析计划

**LISC** Link Set Control 链路组控制

**LISE** Laser Integrated Space Experiment 激光器综合空间实验

**LIT** Lead into Turn 引入转动

Lead-in Training 学员培训,入门培训

Lira 里拉(货币单位)

Liter 升(容量单位)

Little Rock, AR, USA (Airport Code) 美国阿肯色州小石城机场

Little 小的

**LITAS** Low Intention Two Color Approach Slope System 低强度双色进近坡度系统

**LITE** Light 灯,轻的

**LITVC** Liquid Injection Thrust Vector Control 液体喷射推力向量控制

**LIU** Lightguide Interconnection Unit 光纤互连装置

Line Identification Unit 线路识别单元

Line Interface Unit 线路接口单元

**LIV** Left Interconnect Valve 左互连活门(阀)

**LIVE** Liquid Inertia Vibration Eliminator 液体惯性消振器

**LIW** Loss in Weight 失重

**LJ** Life Jacket 救生衣

Locating Jig 定位夹具

**LJC** Low Jet Chart 低空喷气机航图

**LJD** Laser Jamming Device 激光干扰装置

**LJR** Low Jet Route 低空喷气机航路

**LJU** Ljubljana, Slovenia (Airport Code) 斯洛文尼亚卢布尔雅那机场

**LK** Lake 湖泊

Leak 漏

Link 联络,连杆,耦合线

Lock 锁,锁定

Look 看

**LKD** Locked 锁定,上锁的

**LKF** Linear(ized) Kalman Filter 线性(化)卡尔曼滤波器

**LKG** Leakage 渗漏

Leaking 漏

Locking 锁定,同步

Looking 看

Loop Key Generator 环路主发生器

**LKHD** Lockheed 洛克希德公司

**LKO** Lucknow, India (Airport Code) 印度勒可瑙机场

**LKP** Last Known Position 最新已知位置

**LKR** Locker 锁,更衣柜

**LKROT** Locked Rotor 锁定的转子

**LKSHFT** Lockshaft 锁轴

**LKUP** Lockup 锁定

**LKWASH** Lock Washer 锁定垫圈

**LKWR** Lock Wire 保险丝,锁定线

**LL** Land Lines 陆上通信(运输)线

Landing Light 着陆灯

Latitude/Longitude 纬度/经度,横向/纵向

Lease or Loan (Aircraft) 出租或租赁(飞机)

Lebanon Pound 黎巴嫩镑(货币单位)

Left Label 左标记,左标志

Left Lateral 左侧

Left Lower 左下

Life Line 救生素,生命线

Light Lock 隔光器

Lightly Loaded 轻负荷

Limit Load 极限载荷

Line to Line 线对线(交流电压)

Link Layer 链路层

Live Load 动力负载,有效负载,人员载重

Load List 载货清单,载荷表,装载单

Local Line 本地线路

Long Lead 长引线

Loudness Level 响度级

Low Level 低空,低水平,低层

Low Limit 低限

Lower Limit 下限

**L-L** Link by Link 逐个链路

**LLA** Latitude-Longitude-Altitude 纬度—经度—高度

Launcher Loader Adapter 发射器加载适配器

Logical Layered Architecture 逻辑分层结构

Low Level Alarm 低水位报警器

Low-Level Amplifier 低电平放大器

Lulea, Sweden (Airport Code) 瑞典吕勒奥机场

**LLAC** Low Level AC Voltage 低值交流电压

**LLAD** Low-Level Air Defence 低空防空

**LLAMA** Low Level Acceleration Measurement Apparatus 低加速度测量设备

**LLAR** Lower Lobe Attendant Rest 下舱服务员休息室

**LLC** Lift-Lift/Cruise 升力—升力/巡航(发动机)

Liquid Level Control 液面控制

Logical Link Control 逻辑链路控制

Low Layer Compatibility 低层兼容

**LLCC** Lowest Life Cycle Costing 最低全寿命周期费用

**LLCCA** Low Life Cycle Cost Avionics 低寿命周期成本航空电子设备

**LLC-IE** Low Layer Compatibility Information Element 低层兼容性信息元

**LLCSC** Lower Level Computer Software Component 低级计算机软件单元

**LLD** Laser Light Detector 激光光探测器

Light Landing Device 灯光降落装置

Long Life Design 长使用寿命设计

Low-Level Detector 低空探测器

**LLDC** Low Level DC Voltage 低值直流电压

**LLDF** Low-Level Discomfort Factor 低空不适因素

**LLDR** Least-Load Deflection Routing 最小负载改向路由

**LLDS** Liquid Laser Doppler System 液态激光多普勒系统

**LLE** Laboratory for Laser Energetics 激光力学实验室

Leased Line Emulation Service 租用线仿真服务

**LLF** Lag Line Filter 延迟线滤波器

Low Layer Function 低层功能

Low Level Flight 低空飞行

**LLFMR** Least-Loaded-First Multicast Routing 最小负载优先的多播选路

**LLFP** Low Latency Floating Point 低取数时间浮点运算

**LLG** Lower Lobe Galley 下舱厨房

**LLH** Light Lift Helicopter 轻型升力直升机

Lower Left Hand 左下方

**LLI** Latitude and Longitude Indicator 经纬度指示器

Limited Life Item 有限寿命项目

Link Local Inhibit 链路本地禁止

Liquid Level Indicator 液位指示器

Logical Link Identifier 逻辑链路标识符

Long Lead Item 长期投产项目

Low Level Interdiction 低空封锁

**LLIL** Long Lead Items List 长研制周期项目清单

**LLL** Lock-Lock Loop 闭锁—闭锁(回路)

Long Lead List 长期投产清单

Long-Line Loiter 远程巡航

Loose-Leaf Ledger 活页分类账

Lossless Line 无损耗线

Low-Level Light 低照度

Low-Level Logic 低电平逻辑(电路)

Low-Low-Low 低—低—低(剖面飞行)

**LLLTV** Low Light Level Television 低亮度电视

**LLM** Launcher Loader Module 发射器装载模件

L-Band Land Mobile 频段陆地移动通信

Local Loopback Management 本地环回管理

Long Lead Material 长研制周期器材

Lower Layer Module 低层模块

**LLME** Lower Layer Management Entity 低层管理实体

**LLMM** Longer Life Motor Modification 较长寿命电机改型

**LLMS** Liquid Level Measurement System 液

位测量系统

**LLN** League for Less Noise 减少噪音联盟

Leased Line Network 租用线路网

Line Link Network 线路链接网

Linear Lightwave Network 线性光波网络

Logic Link Number 逻辑链路号码

**LLP** Laser Light Pump 激光泵

Late Lag Phase 迟缓后期

Left Lower Plug 左下插头

Life Limited Parts 有时限的零部件

Lightning Location and Protection 雷电定位和避雷

Logical Lightwave Path 逻辑光波路径

Long Line Program 远程计划

Low Level Parachute 低空伞降

Low Level Protocol 低级协议

**LLR** Leased Loaded Routing 最小负荷选路

Line of Least Resistance 最小阻力线

Load Limiting Resistor 负荷限制电阻

Location Laser Radar 定位激光雷达

Log-Likelihood Ratio 航行概率比

**LLRS** Labor Loss Reporting System 劳力损耗报告系统

Laser Lightning Radar System 激光闪电雷达系统

Laser Lightning Rod System 激光避雷装置系统

Low Level Radar System 低空雷达系统

**LLS** Left Line Select Key 左侧行选择键

Liquid Level Sensor 液面传感器

Low-Level Solids 低级别固体

**LLSAC** Laser Line Scan Aerial Camera 激光行扫描航空摄影机

**LLSK** Left Line Select Key 左侧行选择键

**LLSS** Laser Light Source Station 激光光源站

Low Level Sounding System 低空声测系统

**LLSU** Low Level Signalling Unit 低电平信号装置

**LLT** Long Lead Time 长研制周期

Low Latency Transport 低延迟传输

Low Level Terminal 低电平终端

Low Level Turbulence 低水平湍流

**LLTI** Long Lead Time Item 长引导时间项目

**LLTIL** Long Lead Time Item List 长研制周期项目清单

**LLTT** Low Level Teletypewriter Terminal 低容量电传打字机终端

**LLTV** Low-Light Television 低光度电视,低亮度级电视

**LLW** Lilongwe, Malawi Kamuzu International (Airport Code) 马拉维利隆圭卡穆组机场

**LLWAS** Low Level Wind Shear Alert System 低空风切变警告系统

**LLWS** Low Level Windshear 低高度风切变

**LLZ** Localizer 定位器,定位信标

**LM** Land Mobile 地上移动式(无线电设备)

Landmark 界际,陆标

Laser Microscope 激光显微镜

Laser Modulation 激光调制

Last-Minute (Cargo) 最后时刻(载货)

Layup Mandrel 绞合芯

Left Male 左插头,左凸模

Light Microscope 光学显微镜

Light Modulator 光调制器

Link Manager 链路管理器

List of Material 器材清单

Load Master 负载主件

Load Monitor 负载监控器

Locator, Middle 中示位台

Lockheed Martin 洛克希德—马丁公司

Logic Module 逻辑模块

Logistics Module 后勤模件

Long Module 长舱

Lumen 流明(光通量单位)

Middle Locator 中示位信标台

**LMA** Land-based Multipurpose Aircraft 陆基多用途飞机

Lean Manufacturing Assessment 精益制造评估

Lease Management Agreement 租契管理协议

Loss of Multiframe Alignment 多帧失步

**LMAC** Labor-Management Advisory Committee 劳动管理咨询委员会

Leicestershire Microlight Aircraft Club 英国莱斯特微型飞机俱乐部

**LMC** Land Mobile Channel 陆地移动信道

Last Minute Changes 最后时刻的改变

Least Material Condition 最低器材状况

Link Monitor and Control 链路监视器与控制

Local Maintenance Control 局部(本地)维护控制

Localized Multicast 局部多播

**LMCP** Load Master Control Panel 负载主件控制板

**LMCS** Linear Multivariable Control System 线性多变量控制系统

Lockheed Martin Control Systems 洛克希德马丁控制系统

**LMDS** Local Microwave Distribution System 本地微波分配系统

Local Multi-Channel Distribution System 本地多信道分配系统

Local Multi-Point Distributed Service 区域多点分散式服务

Local Multipoint Distribution System 本地多点分布式系统

Logistics Management Data System 后勤管理数据系统

**LME** Layer Management Entity 层管理实体

Line Monitoring Equipment 线路监视设备

Link Management Equipment 链路管理设备

**LMES** Land Mobile Earth Station 陆地移动地球站

Loss of Main Electrical Supply 主电源供给损失

**LMF** Large Mode Fiber 大模光纤

Liquid Methane Fuel 液甲烷燃料

Location Management Function 定位管理功能

Low Medium Frequency 中低频

**LMG** Left Main Gear 左主起落架

**LMHM** Line Maintenance Handling Manual

航线维护操作手册

**LMI** Labor Market Information 劳动力市场信息

Layer Management Interface 层管理接口

Linear Matrix Inequality 线性矩阵不等式

Load Master Instructor 负载主件引导器

Local Management Interface 本地管理接口

Local Manufactured Item 本地(制造)产品

**LMID** Logical Module Identifier 逻辑模块标识符

**LMIS** Labor Market Information System 劳动力市场信息系统

Logistic Management Information System 后勤管理信息系统

**LML** Lightweight Multiple Launcher 轻型多管发射架

Lowest Management Level 最低管理级别

**LMLF** Limit Manoeuvre Load Factor 限制机动载荷系数

**LMLG** Left Main Landing Gear 左主起落架

**LMM** Labor Match Master 劳力调配主管

Line Maintenance Manual 航线维修手册

Locator Middle Marker 带示位台的中指点标

**LMMS** Logistics Maintenance Management System 后勤维修管理系统

**LMN** Load Matching Network 负载匹配网络

Local Mach Number 局部马赫数

**LMO** Laser Master Oscillator 激光主振荡器

Lens-Modulated Oscillator 透镜调制振荡器

Light Machine Oil 轻机油

Linear Master Oscillator 线性主控振荡器

Logistics Management Office 后勤管理办公室

**LMOS** Loop Maintenance Operations System 环路维护操作系统

**LMP** Lamp 灯

Left Middle Plug 左中插头

Line Maintenance Part 航线维修零件

Logistics Management Program 物流管理程序

**LMR** Land Mobile Radio 陆地移动无线电设备

Lightweight Mobile Routing 次要移动选路

Load Monitor Relay 载荷监控继电器

Lowest Maximum Range 最低最大距离

**LMRS** Local Message Rate Service 市话按次计费业务

**LMS** Landing Monitor System 着陆监控系统

Land-Mobile Satellite 陆地移动卫星(通信)

Large Magnet Spectrometer 大型磁谱仪

Laser Mapping System 激光测绘系统

Laser Mass Spectrometer 激光质谱仪

Laser Mass Spectroscopy 激光质谱学

Laser Microscopic Spectrograph 激光显微光谱仪

Last Sent Message 最后发出电报

Layover Management System 临时滞留管理系统

Leakage Measurement System 渗漏测量系统

Learning Management System 学习管理系统

Least Mean Square 最小均方

Level Measurement System 电平测量系统,水平测量系统

Library Maintenance System 图书馆维护系统

Library Management System 图书馆管理系统

Library Media Specialist 图书馆视听资料专家

Load Measurement System 负载测量系统

Local Message Switch 本地信息交换机

**LMSI** Layer Management Service Interface 层管理业务接口

**LMSS** Land Mobile Satellite System 陆上移动卫星(通信)系统

**LMT** Klamath Falls,OR,USA(Airport Code) 美国俄勒冈州克拉马斯福尔斯机场

Limit 限制,极限

Limiter 限制器

Local Maintenance Terminal 本地维护终端

Local Mean Time 当地平均时间

**LMTD** Logarithmic Mean Temperature Difference 对数平均温差

**LMTR** Laser Marker and Target Ranger 激光标记与瞄准测距仪

Limiter 限制器

**LMU** Line Monitoring Unit 航线监测组件

Load and Measurement Unit 负载和测量单元

**LN** Left Nose 左前缘

Liaison 联络

Line 线,航线

Liquid Nitrogen 液氮

Logarithm Natural 自然对数

Lot Number 批号

**LNA** Label Not Available 不可用标签

Local Network Architecture 局部网络结构

Logical Network Address 逻辑网络地址

Low Noise Amplifier 低噪声放大器

Low Noise Antenna 低噪声天线

**LNAV** Lateral Navigation 水平导航

**LNBD** Lens Board 镜头板

**LNC** Local Node Clock 本地节点时钟

Loran Navigation Chart 罗兰导航图

Low Noise Converter 低噪声变频器

Low-Noise Cable 低噪声电缆

**LND** Land 着陆,陆地

Limiting Nose Dive 极限垂直俯冲

Logical Numerical Data 逻辑数值数据

**LNDG** Landing 着陆

Landing Weight 着陆重量

**LNE** Land Navigation Equipment 陆上导航设备

Laser Navigation Equipment 激光导航设备

**LNG** Length 长度

Lengthening 延伸,伸长

Liquid Natural Gas 液化天然气

**LNI** Local Network Interface 本地网络接口

**LNIA** Loop Network Interface Address 环形网络接口地址

**LNK** Lincoln,NE,USA (Airport Code) 美国内布拉斯加州林肯机场

**LNMS** LAN Network Management System 局域网管理系统

**LNN** Latest News Notam 最新航行通告

**LNP** Local Network Protocol 本地网络协议
Local Number Portability 本地电话号码携带

**LNR** Linear 线性的,直线的
Low Noise Receiver 低噪声接收机

**LNS** Lancaster, PA, USA（Airport Code）美国宾西法尼亚州兰卡斯特机场
Land Navigation System 地面导航系统
Laser Navigation System 激光导航系统
Laser Night Sensor 激光夜间传感器
Line Noise Simulator 线路噪声模拟器

**LNSP** Lens Speed 镜头速度

**LNSPD** Line Speed 线速度

**LNSS** Line Subsystem 线路子系统

**LNW** Local Network 本地网络

**LNY** Lanai City, HI, USA（Airport Code）美国夏威夷州拉奈机场

**LNZ** Linz, Austria（Airport Code）奥地利林茨机场

**LO** League Official 联盟官员
Left Outboard 左外侧
Level off 改平,进入平飞
Liquid Oxygen 液态氧
Local Office 当地办事处
Local Oscillator 本机振荡器
Locator Outer 远（信标）台
Lock-on 跟踪（目标）
Lockout 锁定
Long 长
Low Level 低水平,低电平,低液面
Low Observable 低可见（隐形）
Low Observables 低可见性
Low 低
Lubricating Oil 润滑油
Lubrication Order 润滑流程,润滑顺序

**LO PR** Low Pressure 低压

**LOA** Leave of Absence 休假
Length Overall 全长
Letter of Acceptance 验收单

Letter of Agency 代理书

Letter of Agreement 协议书

Letter of Apology 道歉信

Letter of Appreciation 感谢信

Letter of Approval 批准书

Letter of Authority 委托书

Letter of Authorization 授权书

Letter of Offer and Acceptance 交货验收单

Light Observation Aircraft 轻型观察飞机

Limits of Authority 授权范围

Loss of Aircraft 飞机损失

Loss of Alignment 帧失步

Lost on Assembly 组装失败

Low Observable Antenna 低探测力天线

**LOAC** Low Accuracy 低精度

**LOAD** Load Forecast 载量预报

**LOADIS** Load and Discharge 装卸

**LOAL** Lock-on after Launch 发射后锁定

**LOAMP** Logarithmic Amplifier 对数放大器

**LOB** Left Outboard 左外侧
Limited Operating Base 限定使用基地
Line of Balance 平衡线

**LOBAR** Long Baseline Radar 长基线雷达

**LOC** ILS Localizer 仪表着陆系统着陆航向信标
Limited Operational Capability 有限运行能力
Line of Communication 通信线路
Line of Source Code 源编码行
List of Components 零部件清单
List of Overhaulable Components 可大修部件清单
Local Cargo 当地货物
Local 本地（机）的,当地的
Localizer 航向信标,航向台
Located 标定位置的
Locating 定位
Loss of Consciousness 失去知觉

**LOCKG** Locking 锁

**LOCN** Location 位置

**LOCO** Locomotive 机车

**LOCS** Linear Optimal Control System 线性优化控制系统

**LOD** Learning on Demand 教学点播

Letter of Definition 定义函,定义书

**LOE** Lane of Entry 进场走廊,进场航线

Level of Effort 努力程度

Limited Objective Experiment 有限目标实验

**LOEP** Limited Objective Experiment Plan 有限目标实验计划

List of Effective Pages 有效页清单

**LOES** Low Order Equivalent System 低阶等效系统

**LOF** Lift-off 升空,离地

Line of Flight 飞行路线

Local Oscillator Frequency 本振频率

Loss of Flow 流量损耗,断流

Loss of Frame 帧丢失

Low Oil Fuel 燃油不足

Lowest Observed Frequency 最低观测频率

Lowest Operating Frequency 最低工作频率

**LOFAR** Low Frequency Acquisition and Ranging 低频采集与测距

Low Frequency Analysis and Ranging 低频分析与测距

Low Frequency Analyzer and Recorder 低频分析器和记录仪

Low Frequency Analyzing and Recording 低频分析和记录

**LOFF** Leakoff 泄漏的

Logoff 注销

**LOFI** Lubricant Oil Flow Improver 润滑油流动改进剂

**LOFT** Line-Oriented Flight Training 面向航线飞行训练

**LOG** Logarithm 对数

Logistic 后勤的

Logistics Module 后勤舱

**LOGCOM** Logistic Communication 后勤通信

**LOGFTC** Logarithmic Fast Time Constant 对数快速时间常数

**LOGO** Logogram 代表单词的字母

Logographic 航空公司图标,航徽,标志

**LOGS** Log of Pages 记录页

**LOH** Light-Observation Helicopter 轻型观察直升机

Line Overhead 线路开销

**LOI** Letter of Intent 意向书

Letter of Invitation 邀请信

Level of Involvement 牵连程度

Low Order Interface 低阶接口

**LOL** Length of Lead 引线长度

**LOLA** Low Level Wind-Shear Alert 低高度风切变警告,低高度风切变提醒

**LOM** Laser Optical Modulator 激光光学调制器

Level of Maintenance 维修水平(等级)

Light-Optic Microscope 光电显微镜

Locator with Outer Marker 示位台与外指点标

Loss of Ocular Motion 失去动眼能力

Low Order Memory 低等级存储器

**LON** League of Nations 国际联盟

London, United Kingdom (Airport Code) 英国伦敦机场

Longitude 经度,纵向

**LONG** Longitudinal 纵向的,经度的

**LONG TRIM** Longitudinal Trim 纵向配平

**LONGL** Longitudinal 纵向的,经度的

**LONGN** Longeron 纵梁

**LOO** Letter of Offer 出价书

**LOOP** Logic of Object Oriented Programming 面向对象程序设计的逻辑

Low-Cost Optimised Optical Passive Components 低成本最佳光无源元件

**LOP** Line of Position 位置线

Line Oriented Protocol 面向线路的协议

Local Operating Procedure 本机操作程序

Log of Pages 记录本

Low Oil Pressure 低滑油压力

Lubricating Oil Pump 滑油泵

**LOPA** Layers of Protection Analysis 保护层分析

Layout of Passenger Accommodation 旅客座
位布局

Layout of Passenger Area 客舱布局图

Layout per Aircraft 飞机客舱布局

Low Order Path Adaptation 低阶通道适配

**LOPAIR** Long Path Infrared 远距红外探测
（器）

**LOPAR** Low Power Acquisition Radar 低功
率搜索雷达

**LOPARS** Layout Passenger Accommodations
Retrieval System 旅客供应品配置补偿系统

**LOPD** Letter of Program Definition 程序定义
书

**LOPP** Layout of Passenger Payloads System
旅客商载配置系统

**LOPR** Low Pressure 低压

**LOPS** Layout of Payloads System 商载配置系
统

**LOR** Letter of Recommendation 推荐信

Low-Frequency Omnidirectional Range 低频
全向无线电信标

**LORA** Level of Repair Analysis 修理水平分
析

**LORAC** Long Range Accuracy 远程精度

Long Range Accuracy Navigation 远程精确
导航

Long Range Radar Accuracy 远程雷达精确
度

Long-Range Accuracy Navigation System 远
程精确导航系统

**LORAD** Long-Range Active Detection 远程
有源探测

**LORADS** Long-Range Radar and Display
System 远程雷达和显示系统

**LORAN** Long Range Aid to Air Navigation 远
距辅助空中导航（设备）

Long Range Air Navigation System 远程空中
导航系统

Long Range Navigation 远程导航

Long Range Radio Aid to Navigation 远距无
线电辅助导航

Long-Range Aid to Navigation 远距辅助导航
（设备）

**LORAN-C** Long Range Navigation System 远
距导航—C 导航系统

**LORAPH** Long-Range Passive Homing 远距
离被动自导引

**LORAS** Long-Range Airborne Surveillance 远
程空中监视

**LORASIS** Logistic Resources Acquisition
Schedule Information System 后勤资源采集
程序信息系统

**LOREC** Long Range Earth-Current Communications 远程大地电流通信

**LORRA** Land, Ocean and Rain Radar Altimeter 陆地、海洋和雨的雷达测高仪

**LOS** Lagos, Nigeria（Airport Code）尼日利亚
拉各斯机场

Lavatory Occupied Sign 盥洗室占用标记

Line of Sight 视线

Line Operational Simulation 航线操作模拟

Local Operating System 本机操作系统

Los Angeles 洛杉矶

Loss of Signal 信号丢失

Lube Oil System 润滑油系统

**LOSA** Line Operations Safety Audit 航线运
营安全审计

**LOSS** Landing Observer Signal System 着陆
观察员信号系统

**LOSU** Level of Scientific Understanding 科学
认识水平

**LOTAWS** Laser Obstacle Terrain Avoidance
and Warning System 激光地形障碍回避警
告系统

**LOTMP** Lowest Temperature 最低温度

**LOVC** Low Order Virtual Container 低阶虚
拟容器

**LOX** Liquid Oxygen 液态氧

**LOZ** Liquid Ozone 液态臭氧

**LP** Landing Performance 着陆性能

Laser Printer 激光打印机

Laser Processing 激光加工

Laser Projector 激光投影仪

Last Paid 最后支付的

Letters Patent 专利证,特许证

Licensing Plan 许可证发放计划

Life Period 寿命

Light Pen 光笔

Light Pencil 光束

Light Pulse 光脉冲

Lighting Protector 避雷器

Lightning Protector 闪电保护器

Linear Polarization 线极化

Linear Programming 线性规划,线性编程

Link Processor 链路处理器

Liquid Propellant 液体推进剂

Log Periodic 对数周期的

Long Persistence 长余辉

Long Playing 长间隙,密纹唱片

Loop 环路

Lost Part 磨损零件,损失零件

Low Pass 低通

Low Point 低点

Low Pressure 低压

Lower Order Path 低阶通道

**LPA** Linear Power Amplifier 线性功率放大器

Linear Pulse Amplifier 线性脉冲放大器

Load Planning Assistant 装载计划助理系统

Locking Phase Amplifier 锁相放大器

Log Periodic Antenna 对数周期天线

Logical Pack Area 逻辑包装区

Low Pin Actuator 锁定销作动器

Low Power Amplifier 小功率放大器

Lower Order Path Adaption 低阶通道适配

**LPB** Landing Planning and Balance 载重平衡

Language and Publications Branch 语文和出版处

Loss Prevention Bulletin 挂失通报

**LPBA** Lawyer Pilot's Bar Association (美国)律师飞行员协会

**LPC** Less Paper Cockpit (Program) 少纸化驾驶舱(程序)

Linear Polarization Component 线性极化分量

Linear Power Controller 线性功率控制器

Linear Predictive Coding 线性预测编码

Link Protocol Converter 链路协议转换器

Low Pressure Chamber 低压室

Low Pressure Compressor 低压压气机

Low Price Center 低价中心

Low-Order Path Connection 低阶通道连接

Luftfahrt Presse-Club (德国)航向报刊俱乐部

**LPCC** Low Pressure Combustion Chamber 低压燃烧室

**LPCR** Low Pressure Compressor Rotor 低压压气机转子

**LPCS** Low Pressure Compressor Stator 低压压气机静子

**LPCU** Lateral Path Control Unit 横向航路控制组件

**LPCV** Low Pressure Check Valve 低压单向活门(阀)

**LPCW** Long Pulse Continual Wave 长脉冲连续波

**LPD** List of Program Definition 程序定义清单

Lockheed Private Data 洛克希德专用数据

Low Probability of Detection 低概率检测

**LPDB** Liaison Preliminary Data Base 联络预备数据库

**LPE** Laissez-Passer Exceptionnel 特许通行证,特殊护照

Low Probability of Exploitation 低概率探测

**LPF** Liquid Pressure Filter 液压过滤器

Low Pass Filter 低通滤波器

**LPFC** Left Primary Flight Computer 左主飞行计算机

**LPFEV** Long Power Failure Event 长时间电源失效事件

**LPFM** Low-Powered Fan Marker 低功率扇形指点信标台

**LPFP** Low Pressure Fuel Pump 低压燃油泵

**LPG**    Liquid Petroleum Gas 液化石油气

**LPH**    Landing Platform for Helicopter 直升机降落平台

     Litres per Hour 升/小时

**LPI**    Lateral Path Integrator 横向通道积分器

     Low Probability of Intercept 低概率截获

**LPL**    Lightproof Louver 遮光窗

     Liverpool, England, United Kingdom-Liverpool (Airport Code) 英国利物浦机场

**LPLR**    Lock Pillar 门锁立柱

**LPM**    Landing Path Monitor 着陆轨迹监视仪

     Line Processing Module 线路处理模块

     Linearly Polarized Mode 线性偏振模

     Lines per Minute 行/分

     Liter per Minute 升/分

     Logistics Performance Management 物流绩效管理

**LPML**    Low Protein Meal 低蛋白质餐食

     Low-Pressure Mercury Lamp 低压水银灯

**LPN**    Local Packet-Switched Network 本地分组交换网

     Long Part Number 长部件号,长零件号

     Low Pass Network 低通网络

     Lumped Parameters Network 集总参数网络

**LPNVG**    Low Profile Night Vision Goggles 小剖面夜视镜

**LPO**    Landings per Overhaul 每次着陆检查

     Low Power Output 低功率输出

**LPOM**    Low Order Path Overhead Monitoring 低阶通道开销监视

**LPOX**    Low Pressure Oxygen Service 低压氧气服务

**LPP**    Lean Premixed Prevaporized 贫油预混蒸发(燃烧)

     Link Peripheral Processor 链路外围处理器

     Low Order Path Protection 低阶通道保护

**LPR**    Low Pressure Recoup 低压补偿

**LPRF**    Low Power Radio Frequency 低功率射频

     Low Pulse Repetition Frequency 低脉冲重复频率

**LPRSVR**    Life Preserver 救生器

**LPS**    Laser Power Supply 激光能源

     Lean Production Systems 低利生产系统

     Lightproof Shade 遮光罩,遮光板

     Line Program Selector 线性程序选择器

     Linear Programming System 线性程序系统

     Liters per Second 升/秒

**LPSD**    Logically Passive Self-Dual 逻辑无源自对偶

**LPT**    Last Part Tag 最新部件标签牌

     Less Part Tag 少部件标签牌

     Low Order Path Termination 低阶通道终端

     Low Pressure Turbine 低压涡轮机

**LPTACC**    Low Pressure Turbine Active Clearance Control 低压涡轮主动间隙控制

**LPTC**    Low Pressure Turbine Clearance 低压涡轮间隙

**LPTCC**    Low Pressure Turbine Case Cooling 低压涡轮机匣冷却

     Low Pressure Turbine Clearance Control 低压涡轮间隙控制

**LPTCV**    Low Pressure Turbine Cooling Valve 低压涡轮冷却活门

     LP Turbine Clearance Valve 低压涡轮间隙活门

**LPTIT**    Low Pressure Turbine Inlet Temperature 低压涡轮进口温度

**LPTN**    Low Pressure Turbine Nozzle 低压涡轮喷管

**LPTR**    Line Printer 行打印机

     Low Pressure Turbine Rotor 低压涡轮机转子

**LPTS**    Low Pressure Turbine Stator 低压涡轮机定子

**LPTV**    Low Profile Transfer Vehicle 低翼型运输机

**LPU**    Line Printer Unit 行式打印机装置

     Line Protocol Unit 线路协议部件

**LPV**    Landing Pontoon Vehicle 飞行器着陆浮桥

     Launching Point Vertical 垂直发射点

     Lightproof Vent 遮光通风口

Linear Parameter-Varying 线性变参数

Localizer Performance with Vertical Guidance 带垂直引导的航向信标性能

Lunar Probe Vehicle 月球探测器

**LPW** Lumen per Watt 流明/瓦

**LPZ** Lightning Protection Zone 雷电保护区

**LQ** Limiting Quality 极限质量

Lot Quality 批量质量

**LQA** Line Quality Analysis 线路质量分析

Link Quality Analysis 链路质量分析

**LQD** Liquid 液体

**LQGLS** Liquid in Glass 玻璃中的液滴

**LQM** Line Quality Management 生产线质量管理

Link Quality Monitoring 链路质量监控

Logistics Quality Management 物流质量管理

Logistics Queueing Network 物流排队网络

**LQMETR** Liquidometer 液体流量表,液位计

**LQR** Linear Quadratic Regulator 线性二次型调节器

Link Quality Reports 链路质量报告

Liquor 液体,酒

**LQS** Lot Quantity Schedule 大数量程序

**LQTR** Lightning Qualification Test Report 闪电鉴定试验报告

**LR** Ladder Rung 梯子横档

Last Received 最后收到的

Left Rear 左后

Link Request 链路请求

Link Restoration 链路恢复

Load Ratio 负荷比,载荷比

Load Resistor 负载电阻

Location Register 位置寄存器

Long Radius 大活动半径

Long Range 远距离,远程

Low Reverse 下反转

Lower Right 右下

**LRA** Landing Rights Airport 着陆权限航空站

Line Receiving Amplifier 行接收放大器

Line Replaceable Assembly 航线可更换组件

Load Real Address (Instruction) 装入实际地址(指令)

Locked-Rotor Ampere(s) 锁定转子电流,锁紧的转子电流

Long-Range Aviation 远程航空

Low-Range Radio Altimeter 低高度无线电高度表

**LRAACA** Long-Range Air ASW Capable Aircraft 远程反潜飞机

**LRALS** Long-Range Approach and Landing System 远距离进近和着陆系统

**LRAM** Line Replaceable Avionics Module 航线可更换电子模件

**LRAP** Long Range Acoustic Propagation 远程声传播

**LRAS** Logistics Requirements Allocation Sheets 后勤需求分配表

**LRBP** Long Range Business Plan 远程商务计划

**LRC** Langley Research Center 兰利研究中心

Load Ratio Control 载运比控制

Long Range Cruise 远程巡航,最大航程巡航

Longitudinal Redundancy Check 纵向冗余度校验

**LRCA** Location Register Coverage Area 位置寄存器有效区域

**LRCC** Long Range Climb Cruise 远程爬升巡航

**LRCF** Location Register Control Function 位置登记控制功能

**LRCO** Limited Remote Communication Outlet 输出端有限遥控通讯

**LRCP** Long Range Computing Plan 远程计算计划

**LRCS** League of Red Cross Societies 红十字会联盟

Long Range Cruise Speed 远程巡航速度

Long-Range Communication System 远距通讯系统

Low Radar Cross Section 低雷达横截面

**LRCU** Landing Rollout Control Unit 着陆滑跑控制组件

**LRD** Labelled Radar Display 示踪雷达显示器

Laredo, TX, USA (Airport Code) 美国得克萨斯州拉雷多机场

Laser Rangefinder/Designation 激光测距仪

Laser Ranger Designator 激光照射器

Lightning and Radiation Detector 闪电与辐射探测器

Lightning and Radio Detector 闪电和射电发射探测器

Lightning and Radio Emission Detector 闪电与无线电发射探测器

Lightning/Radio Emissions Detector 闪电与无线电发射探测器

Lockheed Requirements Document 洛克希德要求文件

Long Range Data 远程数据

Long Range Desinator 远程标定器

Low Rate Discharge Fire Extinguisher 低速释放灭火瓶

**LRD/T** Long Range Designator/Transceiver 远程标定器/无线电收发机

**LRDF** Location Register Data Function 位置登记数据功能

**LRE** Latest Revised Estimate 最新修订估算

List of Radioactive and Hazardous Elements 放射性和危险元素清单

Low Rate Encode 低速率编码

Low Rate Encoding 低速率编码

**L-REP** Linear Repeater 线性中继器

**LRF** Laser Range Finder 激光测距仪

Line Relative Fault 航线相关故障

Line Relevant Fault 航线相关故障

Long-Range Forecasting Research 长期天气预报研究

**LRFAX** Low Resolution Facsimile 低分辨率传真

**LRFTD** Laser Range Finder and Target Des-ignator 激光测距与目标识别

**LRG** Large 大的

Long Range 远程

**LRH** La Rochelle, France (Airport Code) 法国拉罗歇尔机场

Lower Right Hand 右下

**LRI** Left Right Indicator 左右指示器

Line Replaceable Item 航线可更换项目

Link Remote Inhibit 链路远端禁止

Liquid Resin Infusion 液体树脂注入

Long Radar Input 远程雷达输入

**LRIP** Low Rate Initial Production 初期小批量生产

**LRIRR** Low-Resolution Infrared Radiometer 低分辨率红外辐射计

**LRIRSS** Long Range Infrared Surveillance System 远程红外监视系统

**LRL** Low Intensity Runway Edge Lights 低强度跑道边缘灯

**LRM** Landslide Risk Management 滑坡风险管理

Line Replaceable Module 航线可更换模块

Liquid Radiation Monitor 液体辐射监测器

Local Resource Manager 本地资源管理器

Casa De Campo, Dominican Republic (Airport Code) 多米尼加拉罗马那机场

**LRMA** Long-Range Marine Aircraft 远程海上飞机

**LRMPF** Long Range Manpower Forecast 远程人力预报

**LRMTS** Laser Rangefinder and Marked Target Seeker 激光测距仪和标定目标探寻器

**LRN** Learn 学习,获悉

Long Range Navigation 远程导航

**LRNS** Long-Range Navigation System 远距导航系统

**LROPS** Long Range Operational Performance Standards 远程飞行性能标准

Long Range Operations 远程飞行

**LRP** Logistics Resource Planning 物流资源计划

Low Rate Production 低速率生产

**LRPA**   Long-Range Patrol Aircraft 远程巡逻机

**LRPLS**   Long Range Passive Location System 远程被动定位系统

**LRPP**   Long Range Propulsion Plan 远程推进器计划

**LRR**   Launch Readiness Review 发射准备状态检查

Long Range Radar 远程雷达

**LRRA**   Low Range Radio Altimeter 远程无线电测高计,低空无线电高度表

**LRRS**   Long Range Radar Station 远程雷达站

**LRRU**   Long Range Research Unit 远程研究单位

**LRS**   Laser-Gyro Reference System 激光陀螺基准系统

Leros, Greece(Airport Code) 希腊利罗斯机场

Load Relief System 卸荷系统

Long Range Schedule 长期计划,远景规划

Long Range Search 远程搜索

Long Range Surveillance 远程监视

Long Run Supply 长期运行补给

**LRSB**   Life Raft Storage Box 救生筏储存箱

**LRSI**   Low-Temperature Reusable Surface Insulation 重复使用低温表面绝缘

**LRT**   Laser Range Tracker 激光距离跟踪器

Light Rescue Tender 轻型急救车

Long Range Tank 远程油箱

Lorient, France (Airport Code) 法国洛里昂机场

**LRTF**   Long Range Technical Forecast 远程战术预测

**LRU**   Least Recently Used 最近最少使用

Line Replaceable Unit 航线可更换组件,外场可更换组件

Lowest Replaceable Unit 最小可替换单元

**LRUR**   Local Route Update and Recovery 本地路由更新和恢复

**LRUs**   Line Replaceable Units 航线(外场)可更换组件

**LRV**   Lunar Rover Vehicle 月球车

**LS**   Landing System 着陆系统

Landmarks Subsystem 地标子系统

Laser Scanner 激光扫描装置

Laser Sensitometer 激光感光计

Laser Sensor 激光传感器

Laser System 激光系统

Last Sending 最后发出的

Late Start Date 最晚启动日期

Least Square 最小二乘方

Left Side 左侧

Less 少于

Life Science 生命科学

Life System 救生系统

Life-Cycle Survivability 寿命期幸存能力

Light Sensitivity 光敏度

Light Sensor 光传感器

Light Source 光源

Limit Switch 终点电门,限制电门

Line of Sight 视线,瞄准线

Line Select Key 行选择键

Line Stations 航线维修站

Line System 线路系统

List of Specifications 规范目录

Local Switch 本地交换机

Logistics Support 后勤保障

Loudspeaker 扬声器

Low Speed 低速

Lower Stage 较低级,下一级

Lump Sum 总额,总值

**LSA**   Label Switched Application 标记交换应用

Light Sport Aircraft 轻型运动类航空器

Limited-Space-Charge Accumulation 限制空间电荷积累(二极管)

Link State Advertisement 链路状态广播

Loudspeaker Assembly 扬声器组件

Low Speed Aileron 低速副翼

**LSAC**   Low Speed Aerodynamic Configuration Software 低速空气动力构型软件

**LSACP**   Lower Stage Applications Computer Program 较低级应用计算机程序

**LSALT**   Lowest Safe/Safety Altitude 最低安全高度

**LSAN**   Local Service Access Node 本地业务接入节点

**LSAP**   Link Service Access Point 链路业务接入点

    Loadable Software Aircraft Parts 可装载软件的飞机部件

**LSAR**   Logistic Support Analysis Records 后勤保障分析记录本

    LSA Records 后勤保障分析记录本

**LSAS**   Longitudinal Stability Augmentation System 纵向增稳系统

**LSB**   Least Significant Bit 最低有效位

    Low Side Band 下边带

    Low Speed Buffers 低速缓冲器

    Lower Stage Battery 下一级电瓶

**LSC**   Laser Supported Combustion 激光助燃

    Least Significant Character 最低有效字符

    Line Service Center 线路服务中心

    Line Signaling Channel 线路信令信道

    Linear Slope Controlled 线性斜率控制的

    Link Set Control 链路组控制

    Link State Control 链路状态控制

    Load Signature Certificate 负载署名证书

    Logical Signaling Channel 逻辑信令信道

    Logistic Support Corporation 后勤保障公司

    Logistic Support Costs 后勤保障费用

    Loop Station Connector 环路站连接器

**LSCF**   Left Systems Card File 左系统插件卡柜

    Lower Stage Control Function 较低级控制功能

**LSCP**   Large Scale Computer Project 大型计算机工程

**LSCS**   Large Scale Computer System 大型计算机系统

**LSCU**   Logic and Speed Control Unit 逻辑与速度控制组件

**LSD**   Language for Structural Dynamics 结构动力学语言

    Large-Screen Display 大屏幕显示

    Laser Supported Detonation 激光辅助引爆

    Least Significant Digit 最低有效数字

    Library System Development 程序库系统研制

    Light-Sensitive Detector 光敏检测器

    Light-Sensitive Device 光敏器件

    Loader Serial Data 装载机串行数据

    Long Side 长边

    Low Speed Data 低速数据

**LSDA**   Low Speed Digital to Analog 低速数字—模拟

**LSDS**   Large-Screen Display System 大屏幕显示系统

**LSDU**   Link Service Data Unit 链路服务数据单元

    Load Select Display Unit 负载选择显示组件

**LSDV**   Link Segment Delay Value 链路段延迟值

**LSE**   La Crosse, WI, USA ( Airport Code ) 美国威斯康星州拉克罗斯机场

    Layer Service Element 层服务单元

    Line Signaling Equipment 线路信令设备

    Link Switch Equipment 链路交换设备

    Local System Environment 本地系统环境

**LSEC**   Life-Cycle Support Engineering Center 寿命周期保障工程中心

**LSES**   Laboratory Support Environment Segment 实验室支援环境段

**LSF**   Local Sale/Selling Fare 当地销售票价

**LSFO**   Low Sulfur Fuel Oil 低硫燃油

**LSFZ**   Laser-Beam Sensitive Flight Zone 激光光束敏感飞行区

**LSG**   Legal Service Group 法律服务组

    Logistics Solutions Group 物流解决方案

**LSH**   Least Significant Half 最低有效半部

**LSHI**   Large Scale Hybrid Integration 大规模混合集成

**LSHIC**   Large Scale Hybrid Integrated Circuit

大规模混合集成电路

**LSHT** Low Speed High Torque 低速大扭矩

**LSI** Large Scale Integrated 大规模集成的

Large Scale Integration 大规模集成

Large Scale Integrator 大规模集成器

**LSIC** Large Scale Integrated Circuits 大规模集成电路

**LSIE** Logical Support Integration Environment 逻辑支援综合环境

**LSIF** Local Subscriber Identification 本地用户识别

**LSIG** Link Signature 链路标识

**LSIT** Least Scale Integration Technology 最小规模集成技术

**LSJ** Lifesaving Jacket 救生衣

**LSK** Line Select Key 行选键,线路选择键

**LSL** Link Support Layer 链路支持层

**LSLN** Low Speed Local Network 低速局部网络

**LSM** Lateral Separation Minimum 侧向最小间隔

Linear-Synchronous Motor 线性同步马达

**LSMIS** Logistics Support Management Information System 后勤支援管理信息系统

**LSMITH** Locksmith 锁匠

**LSML** Low Salt Meal 低盐餐食

**LSMU** Lasercom Space Measurement Unit 激光通信空间测量组件

**LSN** Link Set Number 链路组号

Listen 听

**LSO** Landing Safety Officer 飞机安全降落官(管理航母探照灯)

Landing Signal Officer 降落信号员

**LSP** Label Switched Path 标签交换路径

Las Piedras, Venezuela(Airport Code) 委内瑞拉拉斯皮耶德拉斯机场

Laser Shock Peening/Processing 激光冲击强化/处理

Layer Service Primitives 层服务原语

Life Sciences Payloads 生命科学有效负载

Lifecycle Survivability Program 寿命期耐久性程序

Limit Select Panel 极限选择板

Line Select Panel 线选择板

Loadable Software Part 可负载的软件部件

Logistics Support Plan 后勤支援活动计划

Low Speed 低速

**LSPDEO** Life Sciences Payload Development Engineering and Operations 生命科学有效负载开发工程与操作

**LSPKR** Loudspeaker-Aural Warning Cockpit Speaker 扬声器—音频警告驾驶舱扬声器

**LSQ** Line Squall 飚线

**LSR** Label Switched Router 标记交换路由器

Laser Scan Reader 激光条型码阅读器

Last Significant Bit 最低(有效)位

Least Squares of Residuals 最小二乘残差

Load Shedding Relay 卸载继电器

Load Shifting Resistor 负载位移电阻

Local System Resources 本地系统资源

Logistics Support Requirement 后勤支援要求

**LSRI** Large Screen Radar Indicator 大屏幕雷达指示器

**LSRLY** Load Shed Relay 卸载继电器

**LSRT** Large-Scale Radioactive Test 大规模放射性实验

**LSS** Launch Site Support 发射场地支援

Life Support System 生命保障系统

Lightning Sensor System 闪电传感系统

Local Speed of Sound 局部音速

Local Synchronization Subsystem 本地同步子系统

Logistic Support System 后勤支援系统

Lower-Stage Simulator 较低级模拟器,较低级模拟机

**LSSAS** Longitudinal Static Stability Augmentation System 纵向稳定性增强系统

**LSSB** Lower Single Side Band 下边带

Low-Speed Serial Bus 低速串行总线

**LSSGR** LATA Switching System Generic Requirement LATA 交换系统一般要求

**LSSI**   Large Scale System Integration 大规模系统集成

Logistics Support Systems and Integration 后勤支援系统与集成

**LSST**   Large Space Systems Technology 大型空间系统技术

List of Specifications and Standards 规范与标准目录

**LSSU**   Link Status Signal Unit 链路状态信号单元

**LST**   Large Space Telescope 大型空间望远镜

Laser-Spot Tracker 激光追踪器

Last 最后的，上次的

Launceston, Australia（Airport Code）澳大利亚朗塞斯顿机场

Local Standard Time 当地标准时间

Long Wavelength Infra-Red Sensor Test 长波红外传感器试验

**LSTA**   Limited Supplemental Type Approval 限制类补充型号批准书

**LSTC**   Limited Supplemental Type Certificate 限制类补充型号合格证

**LSTE**   Large Structure Technology Experiment 大型结构技术试验

**LSTM**   Limit Switch and Position Transmitter Module 极限电门和位置传感器组件

Low Steam 低压蒸汽

**LSTP**   Low Signaling Transport Point 低级信令转接点

**LSTS**   Laser Search/Track Set 激光搜索/跟踪仪

**LSTTL**   Low-Power Schottky Transistor/Transistor Logic 小功率肖特基晶体管/晶体管逻辑

**LSTWK**   Last Week 上周

**LSU**   LAN Service Unit LAN 服务单元

Lavatory Service Unit 盥洗室服务组件

Line Signaling Unit 线路信令单元

Local Synchronization Unity 本地同步的一致性

Lone Signal Unit 孤立信号组件

**LSVCS**   Large Scale Vector Computer Study 大规模矢量计算机研究

**LSVS**   Line Service 航线服务

**LSWLT**   Landing Signal Wand Light 着陆信号指挥灯

**LT**   Lag Time 滞后时间

Language Translation 语言翻译

Large Tug 大型拖车

Laser Technique 激光技术

Laser Telemeter 激光遥测仪

Laser Tracer 激光跟踪器

Laser Transit 激光经纬仪

Lead Time 超前时间

Leading Time 送修周期

Left 左

Left Turn 左转弯

Less Than 少于

Light 灯光，照明，燃烧

Light Trap 灯盖

Line Termination 线路终端

Link Tail 链路报尾，链路层尾

Link Transfer 链路传递

Litre 升

Local Time 当地时间

Logic Theory 逻辑理论

Long Term 长时间

Lot Time 大量时间

Low Pressure Turbine 低压涡轮，低压透平

Low Temperature 低温

Low Tension 低压，低张力

Turkey Pound 土耳其镑（货币单位）

**LT（S）**   Light(s) 灯光

**Lt.**   Left Turn 左转

**LTA**   Large Transport Airplane 大型运输机

Left Trim Arm 左配平预位

Lighter Than Air 轻于空气

Line Terminal Adapter 线路终端适配器

Logic Tree Address 逻辑树地址

**LTA-HAPP**   Lighter-Than-Air High Altitude Powered Platform 轻于空气的高空动力平台

**LTBO**   Linear Time Base Oscillator 线性时基

振荡器

**LTC**　Large Test Chamber 大型试验舱

Least Total Cost 最小总费用法

Left Trim Control 左配平控制

Life Time Control 寿命时限控制

Linear Transmission Channel 线性传输波道

Load Tap Changing 负载抽头变换

Local Telephone Circuit 本地电话电路

London Teletype Circular 伦敦电传公报

Longitudinal Time Constant 纵向时间常数

Lowest Two-Way Channel 最少双向信道

**LTCC**　London Terminal Control Centre 伦敦终端管制中心

London Traffic Control Centre 伦敦交通管制中心

**LTD**　Laser Target Designator 激光目标识别器

Lighted 有照明的,发亮的

Limited 有限的,极限的,受到限制的

Long Term Disability 长期失效

**LTD PLT**　Lighted Plate 发光板

**LTE**　Late 迟的,晚的

Lightware Terminal Equipment 光端机

Line Terminating/Termination Equipment 线路终端设备

Line Test Equipment 线路测试设备

Local Thermal Equilibrium 局部热平衡

Local Thermodynamic Equilibrium 局部热动力平衡

Long Term Evolution Towards European Manned Spaceflight 欧洲载人航天长远发展研究(欧洲航天局)

Low Thrust Engine 小推力发动机

**LTEMP**　Low Temperature 低温

**LTERM**　Logical Terminal 逻辑终端

**LTF**　Laser Terrain Follower 激光地形跟踪装置

Local Training Flight 当地训练飞行

**LTG**　Lighting 照明,灯光

**LTGH**　Lightning Hole 闪电孔

**LTGN**　Lightning 闪电

**LTH**　Left Hand Thread 左旋螺纹

Length 长度

Light Training Helicopter 轻型教练直升机

Light Transport Helicopter 轻型运输直升机

Light Twin-Engined Helicopter 轻型双发动机直升机

**LTHO**　Lighthouse 灯塔,灯台

**LTID**　Light-Intensity Detector 光强检测器

**LTIT**　Low-(Pressure) Turbine Inlet Temperature 低压涡轮入口温度

**LTK**　Load Tonne Kilometer 吨公里载荷

**LTL**　Listing Time Limit 编目时限

**LTLZ**　Land on the Landing Zone 在着陆地带着陆

**LTM**　Line Type Modulation 线型调制

Livestock Transportation Manual 家畜运输手册

**LTMA**　London Terminal Control Area 伦敦终端管制区

**LTMR**　Laser Target Marker and Receiver 激光目标显示及接收器

Laser Target Marker Ranger 激光目标标记测距仪

**LTMS**　Life Time Monitoring System 寿命时限监控系统

**LT-MUX**　Line Terminating Multiplexer 线路终端复用器

**LTN**　Line Terminating Network 线路终端网络

Long Ton 长吨

Luton, London, England, United Kingdom (Airport Code) 英国伦敦卢顿机场

**LTO**　Landing and Takeoff 着陆及起飞

**LTOPT**　Low Thrust Orbital Performance Tool 低推力轨道性能工具

**LTP**　Landing/Taxi/Park 着陆/滑行/停留

Left Top Plug 左上电阻

Link Terminating Point 链路终接点

**LTPD**　Lot Tolerance Percent Defective 批量允许次品率

**LTPDR**　Long-Term Product Data Retention

长期产品数据保存

**LTPR** Lab/Laboratory Test Part Release 实验室试验零件发放

Laboratory Test Procurement Requisition 实验室试验部件请购单

Lightproof 遮光的,挡光的

Long Taper 长圆锥,长锥度

Long Term Prime Rate 长期优惠利率

**LTPUP** Long Term Power Up 长期增加动力

**LTQ** Low Torgue 低力矩

**LTR** Lab Test Request 实验室测试请求

Laboratory Technical Report 实验室技术报告

Letter 信,字母

Lettering 写字,文字

Lighter 点火器

Litre 升

Local Time Reference 本地时钟参考

**LTRI** Lightning Test Research Institute 闪电测试研究所

**LTRO** Liaison Tool Rework Order 联络工具返工单

**LTRP** Long Term Requirement Plan 长期需求计划

**LTRPRS** Letter Press 字模

**LTS** Lights 灯,灯光

Load and Trim Sheet Software 负载及配平单软件

Load Transfer Signal 负荷传递信号

Local Telephone System 本地电话系统

**LTSW** Light Switch 灯开关,光开关

**LTT** Landline Teletypewriter 有线电传打字机

Line Test Trunk 线路测试中继

**LTTK** Light Tank 灯箱

**LTU** Lateral-Thrust Unit 侧推力发动机

Line Terminating Unit 线路终接单元

Low Traffic User 低业务量用户

**LTV** Loan to Value 贷款评估

**LU** Layup 绞合,接头

Logic Unit 逻辑组件

Logical Unit 逻辑单元

**LU(B)** Lubricate,Lubrication 润滑

**LUAR** Laboratory Unit Acceptance Review 实验室组件接受评审

**LUB** Logical Unit Block 逻辑单元块

**LUBO** Lubricating Oil 润滑油

**LUBT** Lubricant 润滑剂

**LUCC** Land Use and Cover Change 土地利用/土地覆盖变化

**LUD** Lift-Up Door 上推门

**LUE** Logical Unit Equipment 逻辑单元设备

**LUF** Lowest Usable Frequency 最低可用频率

**LUG** Light Utility Glider 轻型通用滑翔机

Lugano,Switzerland（Airport Code）瑞士卢加诺机场

**LUGG** Luggage 行李

**LUH** Light Utility Helicopter 轻型通用直升机

**LUHF** Lowest Usable High Frequency 最低可用高频

**LUI** Load Update Interval 负载信息刷新间隔

**LUL** Lowest Useable Level 最低可用高度层

**LULUCF** Land Use, Land Use Change and Forestry 土地使用,土地使用变更和林业项目

**LUM** Luminous 发光的,集光的,明晰的

**LUN** Logic/Logical Unit Number 逻辑单元号

Lusaka,Zambia（Airport Code）赞比亚卢萨卡机场

**LUP** Kalaupapa, Molokai, Hi, USA（Airport Code）美国夏威夷州莫洛凯岛卡劳帕帕机场

**LUS** Load Update Subset 负载更新分集,负载更新子组件

**LUT** Local User Terminal 本地用户终端

Look-Up Table 检查表

**LUX** Luxembourg, Luxembourg（Airport Code）卢森堡卢森堡机场

**LV** Laser Velocimeter 激光测速仪

Laser Videodisc 激光视盘

Laser Vision 激光影像

Launch Vehicle 发射载具

Leave 离开,起飞

Legal Volt 法定〔国际〕伏特

Level 水平,电平,层

Light and Variable 轻而多变的(风)

Limit Value 极限值

Linear Velocity 线速度

Low Viscosity 低粘度

Low Voltage 低电压

Lower Sideband Voice 下边带话音

**LVA**　Large Vertical Aperture 大垂直孔径

Log Video Amplifier 对数视频放大器

Logic Virtual Address 逻辑虚拟地址

**LVAADR**　Log Video Amplifier Antenna and Directional Receiver 对数视频放大器天线与方向性接收机

**LVC**　Label Virtual Circuit 标记虚拟电路

**LVCD**　Least Voltage Coincidence Detector 最小电压重合检波器

**LVD**　Laser Video Disc 激光视盘

Louvered Door 百叶门

Low Voltage Drop 低压降

**LVDE**　Large Volume Data Exchange 大容量数据交换

**LVDS**　Low Voltage Data Signal 低压数据信号

Low Voltage Differential Signal 低压微分信号

**LVDT**　Linear Variable Differential Transducer 线性可调差动传感器

Linear Variable Differential Transformer 线性可调差动变压器

Linear Variable Displacement Transformer 线性可调位移变压器

Linear Voltage Differential Transformer 线性电压差动变压器

Low Voltage Differential Transducer 低压差动传感器

Low Voltage Differential Transformer 低压差

动变压器

**LVE**　Leave or Leaving 离开

Leave 离开

**LVFE**　Long Variable-Flap Ejector 长折板喷嘴

**LVG**　Leaving 离开,剩余物

**LVIS**　Low-Voltage Ignition System 低电压点火系统

**LVL**　Level 水平,层,面

Low-Velocity Layer 低速层

**LVLOF**　Level off 改平,进入平飞

**LVLS**　Low Visibility Landing System 低能见度着陆系统

**LVO**　Louver Opening 发动机盖开口,鱼鳞板开口

**LVOR**　Low-Powered Very High Frequency Omnirange 低功率甚频全向信标

**LVP**　Low Visibility Procedures 低能见度程序

Low Voltage Protection 低压保护

**LVPS**　Low Voltage Power Supply 低压电源,低压供电

**LVR**　Lever 杆,手柄

Louver 放热孔,放气孔

**LVRE**　Low Voltage Release Effect 低压释放效应

**LVRJ**　Low Volume Ramjet 小容量冲压喷气发动机

**LVRLSE**　Low Voltage Release 低压释放

**LVS**　Low Velocity Scanning 低速扫描

**LVT**　Linear Velocity Transducer 线性速度传感器

Linear Voltage Transducer 线性电压传感器

Low Volume Terminal 低容量终端

**LVTR**　Low Very High Frequency Transmitter Receiver 低甚高频收发机

**LW**　Landing Weight 着陆重量

Last Word 最后一字

Left Wing 左机翼

Light Weight 轻型的

Loaded Weight 载重

Long Wave 长波

Low 低的

**LWABTJ** Light Weight after Burning Turbo Jet 轻型后燃涡喷发动机

**LWARN** Landing Warning 着陆警告

Landing Warning Module 着陆警告模块

**LWATS** Lightweight Air-to-Surface 轻型空对地面

**LWB** Greenbrier, WV, USA（Airport Code）美国西弗吉尼亚州刘易斯堡机场

Long Wheelbase 宽轮轴距

**LWBT** Total Lift of Wing/Body/Tail 机翼、机身、尾翼的总升力

**LWC** Light Wave Communication 光波通信

Lightweight Concrete 轻型混凝土

**LWD** Left Wing Down 左大翼向下

Lowered（起落架，襟翼等）已放下，已放出

**LWF** Light Weight Fighter 轻型战斗机

**LWH** Landing Gear Warning Horn 起落架警告喇叭

**LWHC** Landing Gear Warning Horn Control 起落架报警器开关

**LWIR** Long Wave Infrared 长波红外

**LWIS** Limited Weather Information System 特定天气情报系统

**LWIX** Long Wear Index 长期磨损指数

**LWL** Load Water Line 载荷水线

Low Water Line 低水线

**LWLD** Lightweight Laser Designator 轻型激光识别器

**LWM** Lower Weather Minimum 最低安全气象

**LWMO** Low Weather Minima Operation 天气状况低于最低标准（CAT I）的操作

**LWO** Lvov, Ukraine（Airport Code）乌克兰利沃夫机场

**LWR** Laser Warning Receiver 激光报警接收器, 激光警戒接收机

Long Wave Radiation 长波辐射

Lower 下部的, 较低的

**LWR CTR** Lower Center 下部中心

**LWRU** Light Weight Radar Unit 轻型雷达组件

**LWS** Laser Warning System 激光警告系统

Left Wing Station 左翼站位

Lewiston, ID, USA（Airport Code）美国爱达荷州刘易斯顿机场

Low-Frequency Wave System 低频波系统

**LWSA** Lavatory Water Supply Assembly 盥洗室水供应组件

**LWSD** Large Wall Screen Display 大型壁挂式显示器

**LWST** Lowest 最低的

**LWT** Landing Weight 着陆重量

Large Water Tender 大型水车

Listening While Transmission 边听边发

Local Winter Time 当地冬令时

**LWTR** Licence without Type Rating（UK CAA）无机型等级执照, 基础执照（英国民航局）

**LWY** Lawas, Sarawak, Malaysia（Airport Code）马来西亚沙捞越拉瓦斯机场

**LX** Luxury 豪华的

**LXR** Luxor, Egypt（Airport Code）埃及卢克索国际机场

**LYH** Lynchburg, VA, USA（Airport Code）美国弗吉尼亚州林奇堡机场

**LYR** Layer 层, 云层

**LYS** Lyon, France（Airport Code）法国里昂机场

**LZ** Landing Zone 着陆地带

**LZCC** Landing Zone Control Center 着陆区控制中心

**LZT** Local Zone Time 本区时间

# M

**M** Mach Number 马赫数

Machine 机械,机器

Magenta 品红色,洋红色

Magnaflux 磁粉探伤法

Magnet 磁铁

Magnetic 磁的,磁性的

Magnetron 磁控管

Mail 邮件

Main Alarm 主报警信号

Maintainability 维修能力,可维修性

Maintenance Ratio 维修率

Male 男性

Maneuver Parameter 机动(飞行)参数

Maneuvering Speed (EFIS) 机动速度(电子飞行仪表系统)

Manual 手册,手工的

Marker 记号,标记,标志

Markka 马克(货币单位)

Maximum 最大

Mega 兆,百万

Memory Items 记忆项目

Metacenter 定倾中心,稳定中心

Meter 米

Metric Dimension 公制尺寸

Mid 中间

Milli 毫

Minimum 最低(的),最小(的)

Minute 分钟

Mode 方式,模式

Model 模型

Module 组件,模块

Monaural 单音,非立体声的

Month 月

**M&A** Maintenance and Administration 维修与管理

Maintenance and Assembly 维修与装配

Management and Administration 管理和行政

Manufacturing and Assembly 制造与装配

**M&C** Monitor and Control 监测和控制

**M&E** Maintenance and Engineering 维修与工程

**M&HIRS** Multi-Media and Hypermedia Information Retrieval System 多媒体和超媒体信息检索系统

**M&I** Movement and Identification 移动与识别,运动与识别

**M&IR** Manufacturing and Inspection Record 制造与检验记录

**M&LU** Monitor and Logic Unit 监控和逻辑组件

**M&O** Maintenance and Operation 维修与使用

Maintenance and Overhaul 维护与翻修

**M&P** Material and Processes 器材与工艺

**M&R** Maintenance and Repair 维护与修理

**MandS** Maintenance and Service 维修与服务

Maintenance and Supply 维护与供给

**M&T** Maintenance and Test 维修与测试

**M&W** Material and Workmanship 材料与工艺

**M/A** Mach/Airspeed 马赫数/空速

Maintenance Analysis 维修分析

Modular Assembly 模块化组件

**M/AS** Mach/Airspeed 马赫/空速

**M/ASI** Mach Air Speed Indicator 马赫空速指示器

Mach/Airspeed Indicator 马赫空速表

**M/C** Manual Control 手工控制

**M/COM** Mission Completed 完成的飞行任务

**M/D**　Miscellaneous Data 其他数据(显示)

**M/F**　Maintenance-to-Flight 维修/飞行(比例)

　　Master Folder 主纸夹

**M/H**　Man Hour 工时

**M/L**　Mood Lighting 背景照明

**M/M**　Maximum/Minimum 最大/最小

**M/P**　Machine Planning 机械加工计划

　　Machined Parts 加工零件

　　Magnetic Particle 磁性粒子

**M/REF**　Map Reference 地图坐标,地图基准,地图基线

**M/S**　Magnetostriction 磁致伸缩,磁力控制

　　Master/Slave 主/从

　　Meter per Second 米/秒

　　Mid Spar 中梁

**M/T**　Metric Ton 公吨

　　Multitrack 多通道,多信道

**M/W**　Master Warning 主警告

**m³**　Cubic Meter 立方米

**MA**　Machine Equipment 机械设备

　　Magnetic Amplifier 磁放大器

　　Main Alarm 主警报

　　Maintainability 可维修性,维修能力

　　Maintainability Analysis 可维修性分析,维修能力分析

　　Maintenance Analysis 维修分析

　　Maintenance Annunciator 维修信号器

　　Maintenance Ratio 维修率

　　Malayan Airways 马来西亚航空公司

　　Mandatory Altitude 强制高度

　　Manpower Authorization 人力管理(局)

　　Manual Answer 人工应答

　　Market Analysis 市场分析,商情分析

　　Marketing Analysis 销售分析

　　Master Altimeter 主高度表

　　Master 主人,主导装置,主控机

　　Material 材料

　　Matsushita Avionics 松下航空电子

　　Maximum Authorized Altitude 最大高度批准

　　Maximum Resolution 最大分辨率

Measure Analysis 测量分析

Measure Arrangement 测量装置

Mechanical Advantage 机械效益

Memory Address 存储地址

Mercury Arc 汞弧

Message Assembler 信息汇编程序,信息汇编器,信息收集器

Meteorological Alarm 气象警报

Microchemical Analysis 微量化学分析

Microphone Amplifier 话筒放大器

Microspectral Analyzer 显微光谱分析仪

Milliammeter 毫安表

Milliampere 毫安

Minimum Altitude 最低高度

Ministry of Aviation 航空部(英国)

Minor Airfields 次要机场

Missed Approach Point 复飞点

Mission Abort 任务中断

Mobilization Augmentee 动员扩编人员

Modulated Amplifier 调制放大器

Monitor Address 监视地址

Multichannel Analyzer 多路分析器

Multiple Access 多路存取

Multipoint Access 多点接入

My Account 我的账目

**Ma**　Mach Number 马赫数

**MAA**　Madras, India (Airport Code) 印度马德拉斯机场

Management and Administration 管理和行政

Manchester Astronautical Association 曼彻斯特宇航协会(英国)

Manufacturers of Aircraft Association 飞机制造商协会(美国)

Maximum Approach Altitude 最大进近高度

Maximum Authorized Altitude 最大批准高度

Maximum Authorized IFR Altitude 最大允许仪表飞行高度

Mean Avoiding Angle 平均规避角

Mid-America Airlines 中美洲航空公司

Minimum Approach Altitude 最低进场高度

Missed Approach Area 复飞区

Missed-Approach Action 进场失败的动作

Mission-Area Analysis 任务区分析

Monitoring Angle of Attack 监控迎角

**MAAA** Maintainability Allocation Assessment and Analysis 维修能力配置评估与分析

**MAACS** Multi-Address Asynchronous Communication System 多址异步通信系统

**MAADS** Multibus Avionics Architecture Design System 多总线电子结构设计系统

**MAAH** Minimum Asymmetric Approach Height 不对称推力进场最低高度（如单发停车）

**MAAP** Members and Associated Partners 成员和相关的合作伙伴

**MAAR** Meeting Attendance Approval Request 参加会议批准申请

Monitoring Agency for Asia Region 亚洲地区监控组织

**MAARS** Multi-Access Airline Agent Reservation System 多路存取航空代理订座系统

**MAAS** Meet and Assist (Reservations Code) 接机协助

Module Acceptance Accomplishment Summary 模块验收完成总结

**MAB** Macroaddress Bus 宏地址总线

Man and the Biosphere Programme 人与生物圈计划

Memory Address Bus 存储（器）地址总线

Mission Avionics Bus 飞行航空电子设备总线

**MABH** Minimum Approach Breakoff Height 进近最低终止高度

**MABT** Mission Avionics Bench Test 飞行航空电子设备测试台测试

**MAC** Maintenance Action Classification 维修措施分类

Maintenance Advisory Committee 维修咨询委员会

Maintenance Allocation Chart 维修项目分配图

Maintenance Analysis Center 维修分析中心

（美国联邦航空局）

Maintenance and Administration Center 维护管理中心

Management Advisory Committee 管理咨询委员会

Management Advisory Council (FAA)管理顾问理事会

Materials and Coatings 材料和涂料

Maximum Allowable Cost 最大允许费用

Mean Aerodynamic Center 平均空气动力中心

Mean Aerodynamic Chord 平均空气动力弦

Measurement and Analysis Center 测量和分析中心

Measurement and Control 测量与控制

Media Access Control 介质访问控制，媒体接入控制

Media Access Controller 媒体访问控制器

Medium Access Control 媒体访问控制

Memory Access Control 存储器存取控制

Message Authentication Check 消息认证检验

Message Authentication Code 消息认证码

Metropolitan Airports Commission 大都会机场委员会

Miami Aviation Corporation 迈阿密航空公司

Monitoring and Assessment Centre 检测及评估中心

Monitoring, Alarm and Control 监视、告警和控制

Months after Contract 合同后月数

Multi Aircraft Control 多机控制

Multi-Address Call 多址呼叫

Multi-Application Computer 多用途计算机

Multiple Access Capability 多址接入能力

Multiple Access Channel 多址接入信道

Multiple Access Computer 多路存取计算机

Multiple Address Code 多地址码

Multiplexed Analog Component 复用模拟分量（电视体制）

**MAC DEC** Magnetic Declination 磁偏差

**MACA** Multiple Access Collision Avoidance 多址冲突避免

**MACC** Maintenance Aircraft Coordination Center 飞机维修协调中心

Management and Administration Control Center 管理与行政控制中心

Material Classification and Code 器材分类和代码

Modified Air Control Center 修正大气控制中心

Modified Airworthiness Certification Criteria 修改适航审定标准

Multiple Applications Control Center 多种应用控制中心(美国)

**MACCS** Manufacturing and Cost Collection System 制造与成本收集系统

Manufacturing and Cost Control System 制造与成本控制系统

Mechanized Attendance and Cost Collection System 机械化保养与成本收集系统

Mechanized Attendance and Cost Control System 机械化保养与成本控制系统

**MACE** Maintenance Analysis Checkout Equipment 维修分析检查设备

Master Control Executive 主控制执行任务

Media-Access Controller for Ethernet 以太网的媒体存取控制器

Multiple Application Coil Engine 多用途绕线机

**MACH** Mach Number 马赫数

Machine 机械, 机器

Machine Tool 机床

**MACL** Maldives Airports Company Ltd. 马尔代夫机场有限公司

Minimum Acceptable Compliance Level 最低可接受合格标准

**MACMIS** Maintenance and Construction Management Information System 维护和工程管理信息系统

**MACNET** Multiple Access Customer Network 多址访问用户网络

**MACON** Maintenance Analysis Control 维修分析控制

**MACP** Maintenance Control Program 维修控制程序

Mission Analysis Computer Program 任务分析计算机程序

**M-ACPA** Multimedia Audio Capture and Play Back Adapter 多媒体音频抓取回放适配器

**MACR** Missing Air Crew Report 空勤人员失踪报告

Missing Aircraft Report 失踪飞机报告

Modified Advanced Capability Radar 改装先进雷达

**MACRO** Macroassembly 宏汇编程

Macroprocessor 宏处理器

**MACS** Medium Altitude Communications Satellite 中高度通信卫星

Modular Attitude Control System 模块式姿态控制系统

Module Attitude Control System 模块姿态控制系统

Monitoring and Control Station 监控站

Monitoring and Control System 监控系统

Multi-Access Communication System 多址接入通信系统

Multiple Access Satellite 多路存取卫星

Multiple Applications Control System 多用途控制系统

Multiproject Automated Control System 多元自控系统

Multipurpose Acquisition and Control System 通用采集和控制系统

**MACT** Module Application Configuration Table 模块应用构型表

**MACU** Media Access Control Unit 媒体接入控制单元

**MACVFR** Make Altitude Changes by Visual Flight Rules 按目视飞行规则改变飞行高度

**MAD** Machine Analysis Display 机器分析显示

Madrid, Spain (Airport Code) 西班牙马德

里机场

Magnetic Airborne Detection 航空磁性探测

Magnetic Airborne Detector 机载磁探测器

Magnetic Anomaly Detection 磁性异常探测

Magnetic Anomaly Detector 磁场异常探测器

Magnetic Azimuth Detector 磁方位探测器

Main Assembly Drawing 主装配图

Maintenance Access Door 维修工作口

Maintenance Analysis Data 维修分析数据

Maintenance, Assembly and Disassembly 维修、组装与分解

Major Air Disaster 重大空难事故

Major Damage 主要损坏

Management Analysis Division 管理分析部

Manufacturing Assembly Drawing 制造装配图

Material Analysis Data 器材分析数据

Material Analysis Department 器材分析处

Mathematical Analysis of Downtime 停机时间数学分析

Mean Absolute Deviation 平均绝对偏向

Mean Access Delay 平均接入延迟

Mean Administrative Delay 平均管理延迟

Memory and Display 存储(器)与显示(器)

Minimum Absolute Deviation 最小绝对偏差

Minimum Approach Distance 最短进近距离

Motor Assembly and Disassembly 发动机组装与分解

Multidisciplinary Analysis and Design 多学科分析与设计

Multiple Access Device 多路存取设备

Mutual Assured Destruction 相互确保摧毁(战略)

**MADA**　Multiple Access Discrete Address 多址接入离散地址

**MADAG**　Madagascar 马达加斯加

**MADAM**　Maintenance Analysis Data Model 维修分析数据模型

　　Multipurpose Automatic Data Analysis Machine 多用途自动数据分析机

**MADAP**　Maastricht Automatic Data Process-ing and Display System 马斯特里赫特自动数据处理和显示系统

**MADAPS**　Management Data Processing System 管理数据处理系统

**MADAR**　Maintenance Analysis, Detection and Recorder 维修分析、检测和记录器

　　Malfunction Analysis Detection and Recording 故障分析检测与记录

**MADARS**　Maintenance Analysis Detecting and Reporting System 维修分析检测与报告系统

　　Malfunction Analysis Detection and Recording Subsystem 故障分析检测和记录子系统

　　Malfunction Analysis Detection and Recording System 故障分析检测和记录系统

**MADC**　Mathematical Analysis and Downtime Computer 数学分析与停机时间计算机

　　Mini Air Data Computer 小型大气数据计算机

**MADDAM**　Macromodule and Digital Differential Analyzer Machine 微型组件与数字微分分析机

**MADDIDA**　Magnetic Drum Digital Differential Analyzer 磁鼓数字微分分析器

**MADE**　Miniature Airborne Digital Equipment 小型机载数字式设备

　　Multichannel Analogue-to-Digital Data Encoder 多路模数数据编码器

**MADGE**　Microwave Aircraft Digital Guidance Equipment 机载微波数字制导装置

**MADI**　Multichannel Audio Digital Interface 多声道数字接口

**MADIS**　Manual Aircraft Data Input System 手动飞机数据输入系统

**MADL**　Microwave Acoustic Delay Line 微波声延迟线

　　Multi-Function Advanced Data Link 多功能先进数据链

**MADM**　Maintenance Automated Data Management 维修自动化数据管理

**MADP**　Main Air Display Plot 主空中显示图

**MADRE**　Magnetic Drum Receiving Equipment 磁鼓接收设备

**MADREC**　Malfunction Detection and Recording System 故障检测和记录系统

**MADS**　Machine-Aided Drafting System 机器辅助绘图系统

Main Aircraft Data System 飞机主数据系统

Maintainability Analysis Data System 可维修性分析数据系统

Meteorological Airborne Data System 机载气象数据系统

Multiple Access Data System 多路存取数据系统

Multiple Access Digital System 多路存取数字系统

Multiply and Adds 乘与加

**MADT**　Mean Accumulated Downtime 平均积累不工作时间,平均累计停用时间

Microalloy Diffused Transistor 微合金扩散晶体管

**MAE**　Department of Mechanical and Aerospace Engineering 航空航天机械工程部

Maintenance Engineer 维修工程师

Master of Aeronautical Engineering 航空工程学硕士

Master of Aeronautics 航空学硕士

Master of Engineering 工程学硕士

Mean Absolute Error 平均绝对误差

Mean Area of Effectiveness 平均有效面积

Mechanical and Aerospace Engineering 机械与航天工程

Medical Air Evacuation 医疗空运撤离,医疗空运后送

Men and Equipment 人与机械设备

Metropolitan Area Exchange 都市交换

**MAEO**　Medium-Altitude Electro-Optic Sensor 中高度电子光学传感器

**MAEP**　Maps and Aerial Photographs 地图与航摄照片

**MAES**　Maintenance Aircraft Engineering Squadron 飞机维修工程队

Master of Aeronautical Science 航空科学硕士

**MAF**　Maintenance Access Function 维护接近功能

Maintenance Action Form 维修措施表,维修操作形式

Major Academic Field 主要学术领域

Management Application Function 管理应用功能

Master Facility Tool 主要设施工具

Matched Filter Unit 匹配滤波器组件

Maximum Amplitude Filter 最大振幅滤波器

Microwave Active Filters 微波有源滤波器

Midland,TX,USA（Airport Code）美国得克萨斯州米德兰机场

Missionary Aviation Fellowship 航空界教友团契(澳大利亚及英国)

Modular Aircraft Firefighting 组合式机载灭火

**MAFD**　Minimum Acquisition Flux Density 最小检测通量密度

**MAFFS**　Modular Airborne Fire Fighting System 模块式机载灭火系统

**MAFIS**　Mobile Automated Field Instrumented System 机动式外场自动仪表系统

**MAFT**　Major Airframe Fatigue Test 机体主要部件疲劳试验

**MAG**　Magazine 杂志

Magnetic 磁场的,磁力的,磁的

Magnetic Heading 磁航向

Magnetized Materials 磁性物质

Magnitude 幅度,巨大,重大

**MAG DEC**　Magnetic Declination 磁偏移,磁偏角

**MAG HDG**　Magnetic Heading 磁航向

**MAG LEV**　Magnetic Levitation 磁性悬浮

**MAG VAR**　Magnetic Variation 磁差

**Mag.M**　Magnetized Material 磁化材料

**MAGAMP**　Magnetic Amplifier 磁放大器

**MAGBRE**　Magnetic Bearing 磁方位(角)

**MAGG**　Modular Advanced Graphics Generation 先进模块化图形产生

**MAGI** Multiarray Gamma Indicator 复矩阵伽马指示器

**MAGIC** Machine Aided Graphics for Illustration and Composition Graph 图例和构图用计算机辅助图

**MAGIS** Marine Air-Ground Intelligence System 海上空对地情报系统

Megawatt Air-to-Ground Illuminating System 兆瓦级空对地照明系统

**MAGMOD** Magnetic Modulator 磁调制器

**MAGN** Magnetron 磁控管

**MAGSAT** Magnetic Satellite 地磁卫星

**MAGSS** Maintenance and Ground Support System 维修与地面保障系统

Multifunction Aircraft Ground Support System 多功能飞机地面保障系统

**MAGVAR** Magnetic Variation 磁差

**MAH** Menorca, Spain (Airport Code) 西班牙梅诺卡机场

**MAHF** Missed Approach Holding Fix 复飞等待定位点

**MAHO** Mobile Aided Controlled Handover 移动台辅助控制的越区切换

**MAI** Mach Airspeed Indicator 马赫数表

Maintenance Interface 维护接口

Mandatory Acceptance Inspection 指令性接收检验

Manufacturing Advanced Drawing Change Notice Inquiry 制造图纸更改预告调查

Midair (Great Britain) 中空航空(英国)

Moscow Aviation Institute 莫斯科航空学院(俄罗斯)

Multiple Access Interference 多址接入干扰

Multiple Address Instruction 多地址指令

Multiplexer Action Item 多路调制器操作项目

**MAID** Manufacturing Assembly and Installation Data 制造装配和安装数据

**MAIDS** Manufacturing Assembly and Installation Data System 制造组装与安装数据系统

Multipurpose Automatic Inspection and Diagnostic System 多功能自动检查和诊断系统

**MAILS** Multiple Antenna Instrument Landing System 多天线仪表着陆系统

**MAIN** Maintenance 维护

**MAIN MUX** Main Multiplexer 主多路调制器

**MAINF** Manifold 总管

**MAINT** Maintenance 维修

**MAINT MON** Maintenance Monitor 维护监控

**MAINTOQ** Maintenance of Equipment 设备维修

**MAIS** Maintenance Information System 维修信息系统

Major Automated Information System (US DoD) 主要的自动信息系统(美国国防部)

Military Aviation Information Service 军事航空信息服务

Multiresource Analysis and Information System 多资源分析与信息系统

**MAJ** Major 主要的

Majuro, Marshall Islands (Airport Code) 马绍尔群岛马朱罗机场

**MAJAC** Maintenance Anti-Jam Console 抗干扰维修操纵台

**MAK** Maksutov Telescope 马克苏托夫望远镜

Meal Assembly Kitchen 餐厨装配

**MAL** Maintain at Least... 至少要保持在…

Manufacturing Analysis List 制造分析清单

Maximum Allowable 最大允许的

MCDU Address Label 多功能控制显示器地址标号

**MALCT** Malicious Call Tracing 恶意呼叫追踪

**MALE** Medium Altitude Long Endurance 中空长航时

Multiaperture Logic Element 多孔逻辑元件

**MALF** Malfunction 失效, 故障

**MALIAT** Multilateral Agreement on the Liberalization of International Air Transportation 国际航空运输自由化多边协定

**MALM** Master Air Loadmaster 空勤装卸长主任

**MALS** Medium-Intensity Approach Light System 中强度进近灯光系统

Miniature Air Launcher System 小型空中发射系统

**MALSF** Medium Intensity Approach Lighting System with Sequenced Flashers 装有顺序闪光灯的中强度进近灯光系统

**MALSR** Medium Intensity Approach Light System with Runway Alignment Indicator Lights 装有跑道对准指示灯的中强度进近灯光系统

Medium Intensity Approach Light System with Runway Alignment 具有跑道中线的中强度进近灯光系统

Medium Intensity Approach Lighting System with Runway Alignment Indicator 带跑道对齐指示器的中强度进近灯光系统

**MAM** Media Access Mode 媒体存取模式

Memory Access Method 信息存取方法

Memory Allocation Manager 存储器分配管理程序

Microwave Absorption Method 微波吸收法

Miscellaneous Authorization Memorandum 综合批准备忘录

**MAMB** Mission Avionics Multiplex Bus 航空电子设备任务多路总线

**MAMDT** Mean Active Maintenance Down Time 平均有效维修停用时间

**MAMOS** Marine Automatic Meteorological Observing Station 海洋自动气象观测站

**MAMS** Manufacturing Applications Management System 制造应用管理系统

Military Airspace Management System 军事空域管理系统

**MAMSK** Multi-Amplitude Minimum Shift Keying 多幅值最小移动频键控

**MAN** Maintenance Alert Network 维护警报网

Manager 经理

Manchester, England, United Kingdom (Airport Code) 英国曼彻斯特机场

Manual 手册,手工的

Material Advisory Notice 器材咨询通知

Material as Needed 按需提供材料

Metro Area Network 城域网

Metropolitan Area Network 城域网

Microwave Aerospace Navigation 微波宇航导航

Multiple-Access Network 多址接入网络

**MANAM** Manual Amendments 人工修正

**MANF** Manifold 总管,歧管

Manufacture 制造

**MANIF** Manifest 清单,舱单

**MANL** Manual 手册,手工的

**MAN-MACH** Man-Machine 人—机

**MANOP** Manual of Operation 操作手册

Manually Operated 人工操作的,手控的

**MANOVLD** Manual Overload 人为超载

**MANS** Map Analysis System 地图分析系统

**MANT** Maintenance 维修

**MANTEC** Manufacturing Technology 制造技术

**MANTRAC** Manual Angle Tracking Capability 人工角度跟踪能力

**MANUV** Maneuver 机动,运动

**MAO** Manaus, Amazonas, Brazil (Airport Code) 巴西亚马孙州玛瑙斯机场

**MAOS** Metal-Aluminum-Oxide Silicon 金属—铝—氧化硅半导体

**MAP** Aeronautical Charts 航图

Aeronautical Maps and Charts (代号) 航空地图与航线图

Ground Mapping 地图标记,大地测绘

Machine Analyzer Package 机器分析器程序包

Machine Assembly Program 飞机装配程序

Maintenance Aid Processor 维修辅助处理器

Maintenance Analysis Procedure 维修分析程

序

Maintenance Analysis Program 维修分析程序

Maintenance and Administration Position 维修管理位置

Malfunction Analysis Procedure 故障分析程序

Management Analysis Procedure 管理分析程序

Management Assessment Program 管理评定程序

Manifold Absolute (Air) Pressure 歧管绝对(气体)压力

Manifold Air Pressure 总管气压

Manufacturing Assembly Plan 制造组装计划

Manufacturing Automation Protocol 生产自动化协议,制造自动化协议

Mapping 绘图,测绘

Mars Atmosphere Probe 火星大气探测器

Master Annunciator Panel 主信号板

Maximum Average Price 最高平均价格

Measurement Analysis Program 测量分析程序

Medium Access Protocol 媒体存取协议

Memory Access Protection 存储器存取保护

Message Acceptance Pulse 信息接收脉冲

Middle Atmosphere Program 中层大气(探测)计划

Ministry of Aircraft Production 飞机生产部

Missed Approach Point 复飞点

Missed Approach Procedure 复飞程序

Missed Approach Program 复飞程序

Missed Approach Segment 复飞阶段

Mission Audio Panel 飞行任务音频板

Mission Avionics Processor 飞行任务电子处理器

Mobile Application Protocol 移动应用协议

Mode Annunciator Panel 方式信号板

Modular Aviation Packaging 模块式航空组件

Modular Avionics Packaging 模块式航空电子组件

Multimedia Access Protocol 多媒体存取协议

Multiple Aim Point 多重瞄准点

Multiple Antenna Prose 多天线单调

Multiservice Access Platform 多服务接入平台

Municipal Airport 都市机场

Plug Map 插头图

**MAPA** Malaysia Airlines Pilots' Association 马来西亚航空公司飞行员协会

**MAPD** Material Automated Projected Demand 器材自动规划要求

Maximum Allowable Percent Defective 缺陷允许最大百分比

**MAPDU** Management Application Protocol Data Unit 管理应用协议数据单元

**MAPG** Maximum Available Power Gain 最大有效功率增益

**MAPI** Messaging Application Programming Interface 信报传递应用程序编程接口

Multimedia Application Programming Interface 多媒体应用编程接口

**MAPL** Manufacture Assembler Parts List 制造装配零件表

**MAPLE** Marketing and Product Line Evaluation 销售和生产线评估

Multipurpose Long-Endurance Aircraft 多用途远程飞机

**MAPO** Moscow Aircraft Production Organization 莫斯科飞机制造组织

**MAPPU** Multimedia Authoring and Playing Platform for UNIX UNIX 多媒体制作演播平台

**MAPRES** Mini Air Passenger Reservation System 小型航空订座系统

**MAPROS** Maintain Production Schedule 维持生产计划

**MAPS** Maintenance Analysis Procedures Simulator 维修分析程序模拟器

Major Assembly Performance System 主要部件性能系统

Management Analysis and Planning System 管理分析和计划系统

Manpower Analysis and Performance Standard 人力分析与工作标准

Measurement of Air Pollution from Satellite 卫星大气污染测量

Miniature Air Pilot System 小型飞机驾驶系统

Multiple Address Processing System 多地址处理系统

**MAPSCO** Materials and Processes Scheduling Committee 材料与加工计划委员会

**MAPSP** Master Attendant Passenger Service Panel 主乘务员旅客服务面板

**MAPT** Maintenance Activation Planning Team 维修活动计划组

Missed Approach Point 复飞点

**MAPU** Memory Allocation and Protection Unit 存储器配置与保护组件

Multiple Address Processing Unit 多址处理设备

**MAR** Main Avionics Rack 主要航空电子货架

Maintenance Action Report 维修工作报告,维修措施报告

Major Assembly Release 主要组件发放

Management Approval Routing 管理批准程序

March 三月

Marine Central Airways 海洋中央航空公司

Maximal Aggregation Ratio 最大聚集率

Mean Arrival Rate 平均到达率

Memory Address Register 存储器寻址寄存器

Midair Retrieval 空中修正,空中恢复

Minimally Attended Radar 低维护雷达,需低度护理的雷达

Mission Abort Rate (飞行的)任务中断

**MARC** Machine Readable Catalogue 机读目录

Maintenance and Repair Contract 维护与修理合同

Monitoring and Assessment Research Center 监控与评价研究中心

Multi-Micro Arrangement for Communications 多微波配置通信

Multiple Access Remote Computing 多路存取远程计算

**MARCOM** Microwave Airborne Radio Communication 机载微波无线电通信

**MARISAT** Maritime Communications Satellite 海事通信卫星

Maritime Satellite System 海事卫星系统

**MARK** Marker 信标机,标志,标记

Material Accountability and Robotic Kitting 器材责任与机器人装配

**MARPA** Modification and Replacement Parts Association 改装与更换部件协会

**MARR** Minimum Acceptable Rate of Return 最低可接受的收益率

**MARS** Machine Retrieval System 机器检索系统

Maintenance Analysis and Recording System 维护分析与记录系统

Maintenance Analysis and Repair Service 维护分析和修理服务

Maintenance Analysis and Repair Set 维护分析和修理设备

Maintenance Analysis and Reports Section 维护分析与报告科

Management Analysis Reporting System 管理分析报告系统

Manned Astronomical Research Station 载人天文研究站

Manufacturing Assembly Records System 制造组装记录系统

Manufacturing Record System 制造记录系统

Mechanical Accessory Repair Shop 机械附件修理车间

Meteorological Automatic Reporting System 气象自动报告系统

Milestone Accomplishment Reporting and Status System 重大事件完成报告与状态系统

Mission Avionics Recording System 飞行任务电子记录系统

Multiaperture Reluctance Switch 多孔磁阻开关

Multicast Address Resolution Server 多播地址解释服务器

Multimedia Audiovisual Retrieval Service 多媒体声视检索服务

Multiple Access Retrieval System 多路存取检索系统

Multipoint Aerial Refueling System 多点空中加油系统

**MART** Maintenance Analysis Review Technique 维修分析评价技术

Maximum Average Reframe Time 最大平均重组时间

Mean Active Repair Time 平均有效修理时间

**MARV** Maneuverable Reentry Vehicle 可操纵重返飞行器,机动再入飞行器

**MAS** Malta Aviation Society 马耳他航空学会

Management Advisory Service 管理咨询服务

Mass Calling 大众呼叫业务

Master 主人,主导装置

Matsushita Avionics System Corp. 松下航空电子系统公司

Metal Anchor Slots 金属固定槽

Milliampere/Second 毫安/秒

Miniature Anti-Satellite System 小型反卫星系统

Minor Autonomous Satellite 小型自动卫星

Missed Approach Segment 复飞阶段

Mission Avionics System 飞行任务电子系统

Model Analysis Station 型号分析站

Modification Approval Sheet 改装批准单

Multi-Agent System 多代理系统,多智能体系统

Multiple Access System 多路存取系统

**MASB** Mission Avionics System Bus 飞行任务航空电子系统总线

**MASC** Major Assembly Sequence Chart 主要组装顺序图

Mobitex Asynchronous Communication 移动图文信息异步通信

**MASE** Message Administration Service Element 消息管理服务单元

**MASER** Microwave Amplification by Simulated Emission of Radiation 微波激射放大器

**MASI** Mach Air Speed Indication 马赫数空速指示

Mach Airspeed Indicator 马赫空速表

**MASK** Masking 面罩

**MASOS** Mission Avionics System Operating System 飞行任务电子系统操作系统

**MASP** Material Analysis Specification Plan 器材分析规格计划

Modular Adaptive Signal Processor 模块化自适应信号处理机

**MASPS** Minimum Aircraft System Performance Specifications 飞机系统最低性能规范

Minimum Aviation System Performance Standards 航空系统最低性能标准

**MASR** Multiple Antenna Surveillance Radar 复式天线监视雷达

Multiple Antenna Moving-Target Surveillance Radar 复式天线移动目标监视雷达

**MASS** Maintenance Automation System Software 维修自动化系统软件

Maintenance Subsystem 维护子系统

Manul Analysis Scan System 人工分析扫描系统

Master Armament Selector Switch 主装备选择开关

Material Access Security System 器材存取保险系统

Minimum Aeronautical System Standards 最低航空系统标准

Mission Avionics Sensor Synergism 飞行任务电子传感器协调机构

Modular Aircraft Support System 模块化飞机

保障系统

Multimedia Application Shared Service 多媒体应用共享业务

**M-ASSY** Major Assembly 主要组合件

**MAST** Maintenance Standardization Team 维修标准化组

Management and Supervisory Training（program）管理和监督培训（程序）

Married Airmen Sharing Together 已婚空勤人员共用

Material Accounting Shortage Tracking 器材统计短缺跟踪

Mission Avionics Systems Trainer（飞行）任务航空电子系统训练器

**MASTER** Multiple Access Shared Time Executive Routine 多路存取分时执行程序

**MASU** Machined Surface 机加工表面

**MAT** Maintenance Access Terminal 维护存储终端，维护访问终端

Material 材料

Matter 事情

Maturity 到期

Message Acknowledge Time 收信时间

Metropolitan Area Trunk 大城市中继线

Microalloy Transistor 微合金扩散型晶体管

Module Acceptance Test 模块验收测试

Monitoring, Analysis and Testing 监控、分析与测试

Multidrop Access Trunk 多点分出接入中继线

Multi-Purpose Access Terminal 多用途存储终端

**MATC** Manual of Air Traffic Control 空中交通管制手册，空中交通管制细则

Maximum Aerospace Technology Controls 最高航天技术控制

**MATCEN** Military Air Traffic Control Center 军方空中交通管制中心

**MATCON** Microwave Aerospace Terminal Control 航天微波终端控制

**MATD** Maximum Acceptable Transit Delay 最大可接受转接时延

**MATE** Modular Automatic Test Equipment 模块式自动测试设备

Modular Avionics Test Equipment 模块化航空电子设备测试设备

**MATER** Material 材料

**MATH** Mathematics 数学

**MATI** Moscow Aviation Technology Institute 莫斯科航空技术学院

**MATL** Material 材料

**MATLTS** Materials Test System 材料测试系统

**MATR** Matter 事情

**MATRIX-M** Map Data Matrix Tape 地图数据矩阵磁带

**MATS** Maintenance Analysis Task Sheet 维修分析任务表

Maintenance Analysis Test Set 维修分析测试设备

Military Air Transport Service 军事空运服务

Multi-Altitude Transponder System 多高度应答系统

Multipurpose Automatic Test System 多用途自动测试系统

Mutiple Aircraft Training System 多机训练系统

**MATT** Message Analysis Test Tool 信息分析测试工具

Mobile Assembly Transport Tool 移动式装配运输工具

**MATW** Metal Awning Type Window 金属遮篷形窗

**MATWP** Missed Approach Turning Waypoint 中断进近转弯航路点

**MATZ** Military Aerodrome Traffic Zone 军用机场交通地带

**MAU** Maintenance Unit 维护单元

Media Access Unit 媒体存取设备

Media Attachment Unit 媒体连接器

Millions of Accounting Units 百万计算单位（欧洲货币）

Modular Avionics Unit 模块化航空电子单元

Multiple Access Unit 多重接入单元

Multistation Access Unit 多站点存取单元

**MAUI** Multimedia Application User Interface 多媒体应用用户接口

**MAUM** Maximum All Up Mass 最大起飞总质量

**MAUT** Multiple Attribute Utility Theory 多属性效用理论

**MAUW** Maximum All Up Weight 最大起飞全重

**MAV** Maintenance Assistance Vehicle 保养支援车

Maximum Assessed Value 最大的评估价值

Microaerial Vehicle 微型飞行器

Minimum Acceptable Value（航段价格）最低可接受值

Modern Airship Vehicle 现代空中飞船

**MAVAR** Modulating Amplifier by Variable Reactance 可变电抗调制放大器

**MAVC** Multimedia Audio Video Connection 多媒体音像连接

**MAVIX** Multimedia Audio Video Information Exchange 多媒体音频视频信息交换

**MAW** Maximum Approach Weight 最大进近重量

Mission-Adapting Wing 飞行任务自适应机翼

Modulated Arm Width 调制臂宽度,调制支路宽度

**MAWB** Master Air Waybill 航空主运单

**MAWEA** Modular Avionics Warning Electronics Assembly 模块化航空电子警告电子组件

**MAWP** Maximum Actual Working Pressure 最大实际工作压力

Maximum Allowable Working Pressure 最大容许工作压力

Missed Approach Waypoint 中断进近航路点

**MAX** Maximum 最大的,最大

**MAX CLB** Maximum Climb 最大爬升率

Maximum Engine Thrust for Two-Engine Climb 双发爬升最大发动机推力

**MAX CRZ** Maximum Engine Thrust for Two-Engine Cruise 双发巡航最大发动机推力

**MAX DES** Maximum Descent 最大下降率

**MAX END** Maximum Endurance 最大续航时间

**MAX FLT** Maximum Flight 最长飞行（距离）

**MAX PWR** Maximum Power 最大功率

**MAX SPD** Maximum Speed 最大速度

**MAXCAP** Maximum Capacity 最大能力,最大容量

**MAXHP** Maximum Horse Power 最大马力

**MAXI** Modular Architecture for the Exchange of Intelligence 交换信息的模块化结构

**MAXMIL** Maximum Mileage 最大里程

**MAY** May 五月

**MB** Magnetic Bearing 磁方位

Mail Box 邮箱

Main Battery 主电瓶

Maintenance Base 维修基地

Marker Beacon 指点信标

Mega Bytes 兆字节

Memorandum Book 备忘录

Memory Buffer 存储器缓冲器

Memory Bus 存储器总线

Metal Bond 金属焊接,金属连接

Model Block 模块

Modulation Band 调制频带

Modulation Bandwidth 调制带宽

**Mb** Megabit 兆位

**mb** Millibar 毫巴

**MBA** Marker Beacon Antenna 信标台天线

Master Builders Association 建筑大师协会

Master of Business Administration 工商管理硕士

Minimum Basic Altitude 最低基本高度

Multipoint Broadband Access 多点宽带接入

**MBAA** Multiple-Beam Adaptive Array 多波束自适应阵列

**MBAN** Multimedia Broadband Access Network 多媒体宽带接入网

**MBAR** Multiple Beam Acquisition Radar 多电波搜索雷达

**mbar** Millibar 毫巴

**MBB** Make before Break 中断前完成

**MBC** Master Brightness Control 主亮度控制
Master Bus Controller 主总线控制器
Meteor Burst Communication 流星脉冲群通信
Multicasting Balancing Circuit 多播平衡电路

**MBD** Motor Belt Drive 发动机皮带驱动

**MBDOE** Million Barrels per Day Oil Equivalent 每日百万(当量)桶石油

**MBE** Management by Excuse 理由管理
Material Bin Evaluation 器材库评估
Maximum Brake Energy 最大刹车能量
Multiple Bit Error 多位误差

**MBEU** Multi-Bundle Effectivity Update 多线束有效性更新

**MBF** Multi-Body Freighter 多机身货机

**MBFN** Multiple Beam Forming Network 多波束形成网络

**MBFS** Maximum Boom-Free Speed 最大无音爆速率

**MBHCA** Million Busy Hour Call Attempt 百万次忙时试呼

**MBI** Message Block Identifier 信息块指示器

**MBJ** Montego Bay, Jamaica ( Airport Code ) 牙买加蒙特哥贝机场

**MBK** Multiple Beam Klystron 多波束速调管

**MBL** Manistee, MI, USA( Airport Code ) 美国密歇根州马尼斯蒂机场
Mobile 可移动的

**MBLD** Manual Blind Letdown 人工盲降

**MBM** Magnetic Bubble Memory 磁泡记忆
Management by Methods 方法管理
Market Based Management 基于市场的管理
Market Based Measure 基于市场的措施

**MBMMR** Multi-Band Multi-Mission Radio 多频道多功能电台

Multi-Band Multi-Mode Radio 多频道多模式电台

**MBN** Mesh-Bonding Network 网状网络

**MBO** Mamburao, Philippines ( Airport Code ) 菲律宾曼布劳机场
Management and Budget Office 管理与预算办公室
Management by Objectives 目标管理

**MBOH** Minimum Break-off Height 最低转换高度, 最低出云高度

**MBOM** Manufacturing Bill of Material 器材加工账单

**MBP** Management by Policy 政策管理

**MBPC** Multiple Burst per Carrier 每载波多突发

**MBPS** Mega Bits per Second 兆位/秒

**MBPs** Megabytes per Second 兆字节/秒

**MBPT** Multiple Burst per Transponder 每转发器多突发

**MBR** Main Battery Relay 主电瓶继电器
Management by Rules 规则管理
Marker Beacon Receiver 指点标接收机
Member 成员

**MBRGW** Maximum Brake Release Gross Weight 最大刹车全重

**MBRV** Maneuverable-Ball Reentry Vehicle 可操纵球形重返载具

**MBRW** Maximum Brake Release Weight 最大松刹车重量

**MBS** Magnetron Beam Switching 磁控管波束转换
Main Bang Suppressor 主脉冲抑制器
Marker Beacon System 信标机系统
Material Business Systems 器材商务系统
Millibars 毫巴
Mobile Broadband System 移动宽带系统
Mobile Station 移动站
Saginaw, MI, USA ( Airport Code ) 美国密歇根州萨吉诺机场

**MBSD** Material Business Systems Development 器材商务系统开发

**MBSU** Multi Block Synchronization Signal Unit 多信息组同步信号单元

**MBSV** Main Burner Staging Valve 主燃烧室分级活门

**MBT** Main Battle Tank 主力战车

**MBTP** Multiple Buffer Transfer Protocol 多缓冲传递协议

**MBU** Manufacturing Business Unit 制造业务单位

**M-BUS** Maintenance Bus 维修总线

**MBV** Management by Values 价值管理

**MBWA** Management by Walking Around 走动管理

**MBX** Management by Exception 例外管理,按特殊情况管理

**MBYTES** Mega Bytes 兆字节

**MBZ** Mandatory Broadcast Zone 强制性广播区,强制广播地带

**MC** Magnetic Core 磁芯

Magnetic Course 磁航向

Main Computer 主计算机

Main Coolant 主冷却剂

Maintenance Center 维护中心

Maintenance Convenient 维护方便

Major Component 主要部件

Manhole Cover 检查器盖板

Manufacturing Cost 制造成本

Mass Change 质量变化

Master Caution 主提醒

Master Change 主要更改

Master Clock 主时钟

Master Computer 主计算机

Master Controller 主控制器

Master Cost 主成本

Material Control 器材控制

Material Cost 材料费用

Maximum Certificated 批准最大的(高度)

Media Control 媒体控制

Medicine Cabinet 药品柜

Megacycle 兆周

Megacycles per Second 兆周/秒

Memory Controller 存储控制器,内存控制器

Message Categories 消息种类

Message Center 消息中心

Micro Cell 微小区

Military Characteristics 军用特性

Military Computer 军用计算机

Mission Control 任务控制

Mission-Capable 能执行任务的(飞机状态标准)

Mobile Computer 移动式计算机

Mode Change 方式改变

Modular Computer 模块化计算机

Module Control 模块控制

Momentary Contact 瞬时接触

Motor Chain 发动机系列

Moving Coil 活动线圈

Multichip 多片

Multimedia Communication 多媒体通信

Multimedia Computer 多媒体计算机

Multiple Contact 复式触点,复式接触

Multipoint Connection 多点连接

Multipoint Controller 多点控制器

**MC&G** Maping, Charting, and Geodesy 测绘、制图与大地测量

**MC&W** Master Caution and Warning 主提醒和警告

**MC(M/C)** Manual Control 人工操作,手控

**MC/S** Megacycles 兆周

**MCA** Medium Combat Aircraft 中型战斗机

Micro Channel Architecture 微通道体系结构

Microwave Component Analyzer 微波组件分析仪

Minimum Crossing Altitude 最低飞越高度,最低穿越高度

Ministry of Civil Aviation 民航部

Motor Circuit Analysis 发动机电路分析

Motor Control Assembly 发动机控制组件

Multi-Channel Access 多路接入

**MCAD** Mechanical Computer-Aided Design 计算机辅助机械设计

**MCAIR** McDonnell Aircraft Company 麦道飞

机公司(美国)

**MCAL** Malicious Call 恶意呼叫

**MCAP** Manufacturing Engineering Computer-Aided Production 制造工程计算机辅助生产

**MCAS** Machinery Control and Surveillance 机械控制与监控

Midland Countries Aviation Society 内陆国家航空学会

Multi-Channel Access System 多路接入系统

**MCAT** Maintenance Cost Attribute Tool 维护成本特性工具

Multiplex Controller Aptitude Test 多重管制员能力倾向测试

**MCB** Metal Corner Bead 金属角焊缝

Microwave Circuit Board 微波电路板

Miniature Circuit-Breaker 小型断路器

**MCBF** Mean Cycles between Failures 故障间隔平均循环数

**MCC** Main Combustion Chamber 主燃烧室

Main Communication Center 主通信中心

Maintenance Control Center 维护控制中心

Maintenance Control Computer 维护控制计算机

Maintenance of Close Contact 闭合触点维修

Manchester Code Converter 曼彻斯特代码转换器

Manual Control Center 人工控制中心,手控中心

Master Control Console 主操纵台

Material Class and Code 器材等级与代码

Meteorological Communications Center 气象通信中心

Micro Computer Consortium 微型计算机协议

Microfilm Card Catalog 缩微胶片卡目录

Mission Control Center 任务控制中心

Mission Crew Commander 任务空勤组组长

Mobile Call Control 移动呼叫控制

Mobile Command Center 机动指挥中心

Mobile Country Code 移动台国家码

Mode and Configuration Control 方式与构型

(组合)控制

Modification Customer Chart 更改客户表

Modified Close Control 修改的闭合控制

Motor Case Cutter 马达壳体切割器

Motor Control Center (电动机)马达控制中心

Multicast Coordination Center 多播协调中心

Multi-Channel Communication Control 多波道通信控制

Multichannel Communication Center 多信道通信中心

Multi-Crew Cooperation 多个机组成员配合

Multimedia Communication Channel 多媒体通信频道

**MCCA** Mortgage Capital Company of America 美国抵押资本公司

Multi-Sensor Command and Control Aircraft 多传感器指挥和控制飞机

**MCCC** Major Communication Control Center 主要通信控制中心

Mission Control and Computing Center 任务控制与计算中心

Mobile Consolidated Command Center 机动综合指挥中心

Multi-Connection Call Control 多连接呼叫控制

Multi-Sensor Command and Control Constellation 多传感器指挥和控制星座

**MCCP** Manufacturing Cost Control Program 制造成本控制计划

Multimedia Computing and Communication Platform 多媒体计算及通信平台

**MCCR** Mission Critical Computer Resources 任务关键计算机资源

**MCCS** Machinery Centralized Control System 机械中央控制系统

Multifunction Command and Control System 多功能指挥控制系统

**MCCSA** Multiple Channel Communications System Analyzer 多信道通信系统分析器

**MCCU** Multiple Communication Control Unit

多路通信控制单元

**MCCV** Multi-Control Configured Vehicle 多路操纵系统飞行器

**MCD** Machine Control Data 机械控制数据

Magnetic Chip Detector 磁性金属屑检测器

Main Cabin Door 主舱门

Main Control Display 主控制显示

Main Control Drawing 主控图纸

Maintenance Case Database 维修案例库

Maintenance Control Data 维修控制数据

Master Chip Detector 主芯片检测器

Maximum Cell Delay 最大信元时延

McDonnell Douglas (USA) 麦道公司(美国)

Metal Chip Detector 金属屑探测器

Minimum Cost Design 最低成本设计

Mission and Communication Display 任务与通信显示器

Movable Class Divider 可移动等级分配器

Multifunction Cockpit Display 多功能座舱显示器

**MCDC** Maintenance Control and Display Computer 维护控制与显示计算机

Modified Condition Decision Coverage 修改条件决策覆盖

Multiple Condition Decision Coverage 多条件决策覆盖

**MCDF** Maintenance Control Display Function 维修控制显示功能

**MCDM** Multiple Criteria Decision Making 多准则决策机制

**MCDN** Micro-Cellular Data Network 微蜂窝数据通信网

**MCDP** Maintenance Control Display Panel 维护控制显示面板

Microprogrammed Communication Data Processor 微程序控制通信数据处理器

Mission Control Display Panel 飞行任务控制显示板

**MCDS** Management Communications and Data System 管理通信和数据系统

**MCDT** Multimedia CD-ROM Title 多媒体光盘标题

**MCDU** Maintenance Control Display Unit 维修控制显示组件

Management Control Display Unit 管理控制显示装置

Master Control Display Unit 主控制显示组件

Multi-Function Control Display Unit 多功能控制显示组件,多功能控制显示单元

Multipurpose Control Display Unit 多用途控制显示组件

**MCE** Manufacturing Capability Estimate 制造能力估算

Master Cascade Enable 主级联允许

Master Communication Equipment 主通信设备

Merced, CA, USA (Airport Code) 美国加利福尼亚州默塞德机场

Modular Control Equipment 模块式控制设备,积木式控制设备

Motor Control Electronic 马达电子控制

Multi Cycle Engine 多循环发动机

**MCF** Management Communication Function 管理通信功能

Master Control Facility 主控站

Media Control File 工具控制文档,中介控制文档

Message Code and Format 电报代码和格式

Message Communication Function 消息通信功能

Mobile Control Function 移动控制功能

Mutual Coherence Function 互相干函数

**MCFC** Molten Carbonate Fuel Cell 熔融碳酸盐燃料电池

**MCG** Master Control Gauge 主校准规

**MCGA** Multi-Color Graphics Adapter 多彩色图形适配卡

Multi-Color Graphics Array 多色彩图形阵列

**MCGS** Microwave Command Guidance System 微波指令制导系统

**MCH** Mail Chute 信槽

**MCHCL** Mechanically Cooled 机械冷却的

**MCHN** Machine 机器,机械

**MCHO** Mobile Controlled Handover 移动控制切换

**MCHRF** Mechanically Refrigerated 机械制冷的

**MCHRY** Machinery 机器

**MCHST** Machinist 机械师

**MCHT** Merchant 商人

**MCHY** Machinery 机械设备

**MCI** Kansas City, MO, USA (Airport Code) 美国密苏里州堪萨斯城机场

Malicious Call Identification 恶意呼叫识别

Material Condition Inspection 器材状态检测

Media Control Interface 媒体控制接口

Mobile Communication Interface 移动通信接口

Mode C Intruder 装有 C 模式应答器的入侵飞机

Module Configuration Index 模块配置索引

**MCIC** Multichip Integrated Circuit 多芯片集成电路

**MCID** Manufacturing Change in Design 设计中制造变更

**MCIN** Manufacturing Change Incorporation Notice 制造更改联合通知

**MCIS** Maintenance Control Information System 维修控制信息系统

Management Control and Information System 管理控制与信息系统

Map and Chart Information System 地图与图表信息系统

Material Certification and Information System 器材取证与信息系统

**MCK** Master Change Kit 主更改器材包

Mission and Communications Keyboard 任务与通信键盘

**MCL** Master Call(Caution) Light 主呼叫(提醒)灯

Master Change Log 主更改记录本

Master Configuration List 主配置清单

Materials Control Laboratory 材料控制试验室

Maximum Climb 最大爬升

Maximum Climb Thrust 最大爬升推力

Message Control Language 报文控制语言

Microcomputer Compiler Language 微机计算机编译语言

Mid-Canada Line 中加拿大预警线

Middle Cloud 中云

Minimum Crossing Level 最低穿越高度层

Miscrocomputer Compiler Language 微计算机编译语言

Multi-Compare List 多种对照清单

**MCLR** Mean Cell Loss Rate 平均信元丢失率

Mid-Capacity Long Range 中等载量远程

**MCLT** Maximum Climb Thrust 最大爬升推力

Maximum Cruise Level Thrust 最大巡航高度推力

**MCM** Magnetic Core Memory 磁芯存储器

Maintenance Control Module 维护控制模块

Mass Customization Manufacturing 大规模定制生产

Master Change Memorandum 主更改备忘录

Mine Countermeasures 地矿干扰,反水雷措施

Modification Coordination Memorandum 改装协调备忘录

Monte Carlo, Monaco (Airport Code) 摩纳哥蒙特卡洛机场

Multicarrier Modulation 多载波调制

Multi-Chip Module 多芯片模块

**MCMCP** Master Computer Monitor and Control Panel 主计算机监控器与控制板

**MCMF** Multicommodity Maximum Flow 多商品最大流量

**MCMM** Manufacturing Control and Material Management System 制造控制与器材管理系统

**MCMS** Multichannel Memory System 多通道

存储系统

**MCN**    Macon, GA, USA ( Airport Code ) 美国佐治亚州梅肯机场

Maintenance Control Number 维修控制编号

Management Control Number 管理控制编号

Manual Control Number 手控编号

Manufacturing Change Notice 制造更改通知

Master Change Notice 主要更改通知

Master Control Notice 主控通知

Master Control Number 主控编号

Mid Course Navigation 中程航线导航

Mobile Control Node 移动控制节点

Museum Computer Network 博物馆计算机网

**MCNS**    Multimedia Cable Network System 多媒体有线网系统

**MCO**    Maintenance Checkout 维修检查

Maintenance Control Officer 维修管理人员

Manual Change Order 人工更改指令

Manufacturer's Certificate of Origin 制造商的原产地证

Mars Climate Orbiter 火星气象卫星

Miscellaneous Charges Order 旅费证, 各种费用清单

Mission Control Operation 飞行任务控制操作

Orlando, FL, USA ( Airport Code ) 美国佛罗里达州奥兰多机场

**MCOT**    Maximum Continuous Operating Temperature 最大连续运行温度

Minimum Cost of Transport 最小运输成本

**MC-O**    Manual Control Only 仅人工操纵

**MCP**    Maintenance Control Panel 维修控制板

Management Control Plan 管理控制计划

Manual Control Panel 人工控制板

Manufacturing Change Point 制造更改点

Mass Calling Platform 大众呼叫平台

Master Change Proposal 主更改建议书

Master Change Protocol 主更改议定书

Master Control Panel 主操纵板

Master Control Program 主控程序

Maximum Climb Power 最大爬升功率

Maximum Continuous Power 最大连续功率

Message Control Program 信息控制程序

Microprogrammed Control Processor 微程控处理器

Microprogramming Central Processor 微程序化中央处理器

Mission Control Programmer 飞行任务控制编程器

Mode Control Panel 方式控制面板

Modular Component Predictor for Aircraft Noise 飞机噪音模件化部件预测器

Multi-Channel Processor 多信道处理器

Multi-Chip Package 多芯片封装

Multiflow Conversation Protocol 多流对话协议

**MCPA**    Multicarrier Power Amplifier 多载波功率放大器

**MCPC**    Multiple Channel per Carrier 多信道单载波

**MCPFD**    Multi-Crew Pictorial Format Display 多机组图像格式显示器

**MCPH**    Maintenance Cost per Hour 每小时维护费用

**MCPN**    Mobile Customer Premises Network 移动用户驻地网

**MCPS**    Maintenance Circuit Protection Status 维修电路保护状态

Million Compares per Second 百万次比较/秒

**MCPSS**    Microprocessor Controlled Page Storage System 微机控制的记录存储系统

**MCPU**    Master Central Processing Unit 主中央处理器板

Motor Control and Protection Unit 马达控制及保护组件

Multiple Call Processing Unit 多呼叫处理单元

Multiple Channel Processing Unit 多信道处理单元

**MCQ**    Multiple Choice Question 多项选择问题

**MCR** Main Control Room 主控制室

Maintenance Control Report 维修管理报告

Manual Change Request 人工更改请求

Manufacturer Change Request 制造厂更改申请

Manufacturing Change Request 制造更改申请

Master Change Record 主更改记录

Master Change Request 总更改要求

Material Credit Requisition 器材信贷申请

Material Improvement Project Commitment Record 器材改进项目提交记录

Maximum Continuous Rating 最大连续额定功率

Maximum Cruise 最大巡航

Maximum Cruise Rate 最大巡航率

Mean Cell Rate 平均信元率

Minimum Cell Rate 最低信元速率

Mission Capable Rate 能执行任务率

Multi-Channel Receiver 多通道接收机

**MCRB** Master Change Review Board 主更改审查委员会

**MCRC** Management Change Review Committee 管理更改审查委员会

Master Control and Reporting Center 主控报告中心

**MCRL** Master Cross Reference List 主交叉参考清单

**MCRO** Mission Control Readiness Operation 飞行任务控制准备工作

**MCRS** Micrographic Computer Retrieval System 缩微目录计算机检索系统

**MCRT** Maximum Cruise Thrust 最大巡航推力

**MCRWV** Microwave 微波

**MCS** Magnetic Card Storage 磁卡存储器

Magnetic Control System 磁控制系统

Magnetic Core Storage 磁心存储器

Main Control Station 主控站

Main Control System 主控系统

Maintenance and Control System 维护机控制系统

Maintenance Control Section 维修控制科

Maintenance Control Subsystem 维护控制子系统

Maintenance-Data Collection System 维护数据收集系统

Managenment Control System 管理控制系统

Maneuver Control System 机动控制系统

Manufacturing Control System 制造控制系统

Map Combination System 地图汇编系统

Maritime Communications Subsystem 海事通信子系统

Mass Calling Service 大众呼叫业务

Master Change Schedule 主更改计划

Master Circuit System 主电路系统

Master Computer System 主计算机系统

Master Control Station 主控站

Master Control Switch 主控电门

Master of Computer Science 计算机科学硕士

Material Commitment Schedule 器材提交计划

Maximum Cruise Speed 最大巡航速度

Message Control Supervisor 信息控制管理程序

Message Control System 信息控制系统

Micro-Computer System 微型计算机系统

Microgram Certification System 微程序审定系统

Microwave Carrier Supply 微波载波源

Microwave Communication System 微波通信系统

Microwave Components and System 微波组件与系统

Miniature Control System 微型控制系统

Minimal Cut Sets 最小割集

Minimum Control Speed 最小控制速度

Mission Computer Subsystem 任务计算机子系统

Mission Control Segment 飞行任务控制段

Mixed-Class Seating 混合舱级排座

Modulating Control System 模拟量控制系统

Monitoring/Control Software 监测/控制软件

Multicast Server 多播服务器

Multichannel Communication System 多通道通信系统

Multichannel Spectrometer 多通道光谱仪

Multimedia Chatting System 多媒体交谈系统

Multimedia Conferencing System 多媒体会议系统

Multipoint Communication Service 多点通信业务

Multipoint Conferencing Server 多点会议服务器

Multipurpose Communications and Signalling 多用途通信与信号

**MCSE** Magnetic Course 磁航向

**MCSG** Mathematical and Computational Sciences Group 数学与计算科学组

Mission Computer and Symbol and Map Generator 任务计算机和信号发生器

**MCSO** Multimedia Communication Service Object 多媒体通信业务对象

**MCSP** Maintenance Control and Statistics Process 维护控制和统计过程

**MCSS** Military Communication Satellite System 军事通信卫星系统

**MCSW** Motor Circuit Switch 发动机电路开关

**MCSWGR** Metalclad Switch Gear 铠装开关装置

**MCT** Main Control Tank 主控制油箱

Maritime Crew Trainer 海上机组教练机

Master Contour Template 主外形模板

Maximum Continuous Thrust 最大连续推力, 额定推力

Mean Corrective Time 平均校正时间

Mechanical Comprehension Test 机器综合试验

Memory Controller 内存控制器

Minimum Connecting Time 最短衔接时间, 最短转机时间

Minimum Connection Time 最小衔接时间

Modern Communication Theory 现代通信理论

Multicast Channel Translator 多播信道转换器

Muscat, Oman Seeb (Airport Code) 阿曼马斯喀特锡卜机场

**MCTA** Military Controlled Airspace 军事管制空域

**MCTD** Mean Cell Transfer Delay 平均信元传递时延

**MC-TDMA** Multi-Carrier Time Division Multiplexing Access 多载频时分多址接入

**MCTOM** Maxium Certificated Take-off Mass 最大允许起飞质量

**MCTOW** Maximum Certificated Take off Weight 批准最大允许起飞重量

**MCTR** Multicycle Controllable Twin Rotor 多循环可控双转子

**MCTS** Mini-Computer Test Set 小型计算机测试装置

**MCTV** Material Cycle Time Value 器材循环时间值

**MCU** Machine Control Unit 机械控制组件

Magnetic Card Unit 磁卡装置

Main Control Unit 主控制组件

Management and Communication Unit 管理和通信单元

Management Control Unit 管理控制组件, 管理控制设备(飞行数据记录仪)

Manual Control Unit 人工控制组件

Master Caution Unit 主警告组件

Master Control Unit 主控制单元

Master Controller Unit 主控制器单元

Matrix Character Unit 矩阵符号装置(显示符号)

Message Construction Unit 信息结构组件

Message Controller Unit 信息控制器组件

Micro Controller Unit 微控制器

Microcomputer Control Unit 微机控制组件

Microgram Control Unit 微程序控制组件

Microprocessor Control Unit 微处理器控制组件

Miniature Configuration Unit 微型结构组件

Minimum Coding Unit 最小编码单元

Mobile Control Unit 移动控制单元

Modular Concept Unit 模块式概念组件

Module Control Unit 模块控制单元

Monitoring Channel Unit 监控信道单元

Multifunction Concept Unit 多功能概念组件

Multipoint Control Unit 多点控制单元

Multiplexer and Conversion Unit 多路控制器与转换组件

Multipoint Conference Unit 多点会议单元

**MCUR** Mean Cycles between Unscheduled Removals 非计划拆换平均间隔循环

**MCV** Main Control Valve 主控活门

**MCVL** Manual Configuration Variation List 手册构型改变清单

**MCW** Mason City, IA, USA ( Airport Code ) 美国依阿华州梅森城机场

Modulated Continuous Wave 调制连续波

Modulated Carrier Wave 已调载波

Monitoring and Control Workstation 监视与管制工作站

Multiplex Command Word 多路指令字

**MCWG** Mission Control Working Group 任务控制工作小组

**MCWR** Mockup Coordination and Work Request 样机协调与工作申请,模型协调与工作申请

**MD** Main Drum 主鼓

Maintenance Data 维修数据

Management Directive 管理指令

Managing Director 管理主任

Manual Data 人工数据

Manufacturing Data 制造数据

Maryland 马里兰州

Master Dimension 主尺寸

Matching Device 匹配器

Material Data 器材数据

Material Development 器材研制

Matrix-Decoder 矩阵译码器

Maximum Dive Mach Speed 最大俯冲马赫速度

May Day 呼救信号

McDonnell Douglas 麦克唐纳·道格拉斯公司

Mean Deviation 平均偏差

Mean Difference 平均差

Mediation Device 协调设备,中介设备

Medium Density 中密度,中浓度

Memory Data 存储数据

Memory Data Register 存储数据寄存器

Meridian Distance 子午距

Message Data 信息数据

Message Discrimination 消息鉴别

Message per Day 每天信息(量)

Meteorological Department 气象局

Microwave Discriminator 微波鉴频器

Military Document 军事文件

Modification Drawing 改装图纸

Modulation Distortion 调制失真

Modulator 调制器

Modulator Demodulator 调制解调器

Momentum Detector 动量检测器

Multidimension Distribution 多维分布

Multimode Distortion 多模失真

Multiobjective Decision 多目标决策

**MD&T** Master Dim and Test 主调光和测试

Master Dimension and Test 主尺寸与测试

**MDA** Maintenance Depot Assistance 维修库辅助(设计)

Maintenance Design Approach 维修设计方法

Malfunction Detection Analyzer 故障检测分析器

Malfunction Detector Analyzer 故障探测器分析器

Master Diversion Airfield 主备降场

Materials Dispersion Apparatus 材料弥散装置

Maximum Domain of Attraction 最大吸引场

Mechanically Despun Antenna 机械反旋转

天线

Memory Data 存储数据

Minimum Absolute Deviation 最小绝对偏差

Minimum Decision Altitude 最低决断高度

Minimum Descending Altitude 最低下降高度

Minimum Descent Altitude 最低下降高度

Minimum Detectable Amounts 最低可探测量

Mission Data Assurance 任务数据保证

Mobile Data Association 移动数据关联

Mobile Depot Activity 机动站作业

Modification Design Approval 改装设计批准书

Monochrome Display Adapter 单色显示适配器

Multichannel Digital Autocorelator 多通道数字式自动相关器

Multi-Dimensional Analysis 多维分析

Multi-Dimensional Array 多维矩阵

Multilayer Dielectric Absorber 多层电介质吸收涂层(隐性技术)

**MDAC** McDonnell Douglas Astronautics Company 麦道航天公司

**MDAIS** McDonnell Douglas Aerospace Information Services 麦道宇航信息服务公司

**MDAP** Major Defense Acquisition Program 主要的国防采购计划

Mutual Defense Assistance Program 共同防御援助计划

**MDAR** Material Discrepancy and Action Report 器材鉴审报告

**MDAS** Miniature Data Acquisition System 小型数据采集系统

Mission Data Acquisition System 任务数据采集系统

**MDAU** Maintenance Data Acquisition Unit 维修数据采集组件

Miniature Data Acquisition Unit 小型数据采集单元

Modular Data Acquisition Unit 模块式数据获取组件

Modular Data Analysis Unit 模拟数据分析装置

**M-DAY** Manufacturing Day 制造日

**MDB** Multimedia Database 多媒体数据库

**MDBMLS** Multimedia Database Machine Learning System 多媒体数据库机器学习系统

**MDC** Emergency DC 紧急直流电

Main Display Console 主显示操纵台

Maintenance Data Collection 维护数据收集

Maintenance Dependence Chart 维修相关图表

Maintenance Diagnostic Computer 维护诊断计算机

Maintenance Direct Cost 直接维修成本

Manado, Indonesia ( Airport Code ) 印度尼西亚万鸦老机场

Master Digital Computer 主数字计算机

McDonnell Douglas Aircraft Co. 麦克唐纳道格拉斯飞机公司

McDonnell Douglas Company/Corporation 麦道飞机公司

Minimum Dispatch Configuration 最低放行标准

Mission Data Control 飞行任务数据控制

Motor Direct Connected 发动机直接连接的

Motor-Driven Compressor 电机驱动压缩机

**MDCC** Main Deck Cargo Compartment 主货舱

Master Data Control Console 主要数据控制台

**MDCD** Main Deck Cargo Door 主货舱门

**MDCHECK** Mission Design Check 飞行任务设计检查(程序)

**MDCP** Mission Data Control Panel 飞行任务数据控制板

**MDCR** Mission Data Collection Record 飞行任务数据收集记录

**MDCRS** Meteorological Data Collection and Reporting System 气象数据收集报告系统

**MDCS** Maintainability Data Collection System

可维护性数据收集系统

Maintenance Data Collection System 维护数据收集系统

Malfunction Display and Control System 故障显示与控制系统

Manufacturing Document Control System 制造文件控制系统

Master Digital Command System 总数字指令系统

Material Data Collection System 器材数据收集系统

Mission and Display Control System 任务与显示控制系统

**MDD** Master Dimensions Definition 主尺寸定义

Mission Data Display 飞行任务数据显示

Multi-Dimensional Database 多维数据库

Multimedia Database Design 多媒体数据库设计

**MDDD** Material Divisions Data Definitions 器材部门的数据定义

**MDDP** Mission Data Display Panel 飞行任务数据显示板

**MDDS** Maintenance Data Development System 维修数据开发系统

**MDDU** Multipurpose Disk Drive Unit 多功能磁盘驱动组件

**MDE** Made 制造

Mechanical Design Engineer 机械设计工程师

Minimum Detectable Error 最小可检测误差

Mockup Development Release 模型开发发布

**MDEC** McDonnell Douglas Electronics Corporation 麦道电子公司

**MDELTA** Matrix Differential Equation Linear Transform Analysis 矩阵微分方程线性变换分析

**MDF** Main Distributing/Distribution Frame 主分配帧, 主配线架, 主分配框架

Manual Direction Finder 人工定向仪

Master Data File 主数据文档

Mean Excess Function 平均超额函数

Medium Frequency Direction Finding Station 中频定向站

**MDFC** McDonnell Douglas Finance Corp. 麦道金融公司

**MDFY** Modify 修改

**MDG** Millennium Development Goals 千年发展目标

Mission Data Generation 飞行任务数据产生

**MDGT** Mission Data Ground Trail 飞行任务数据地面跟踪

**MDH** Minimum Decision Height 最低决断高度

Minimum Descent Height 最低下降高度

**MDHS** Meteorological Data Handling System 气象数据处理系统

**MDI** Magnetic Direction Indicator 磁方位指示器

Maintenance Definition Index 维修定义索引

Manual Data Input 人工数据输入

Master Dimensioning Identifier 主尺寸标识(符)

Master Dimensioning Index 主尺寸索引

Master Drawing Index 主图纸索引

Mathematical Definition Identifier 数学定义标识(符)

Mathematical Dimension Identifier 数学量纲标识(符)

Medium Dependent Interface 媒体相关接口

Miss Distance Indicator 脱靶量指示器, 误差距离指示器

Multifunction Display Indicator 多功能显示器

**MDIF** Manual Data Input Function 人工数据输入功能

**MDIS** Maintenance Diagnosis Information System 维修诊断信息系统

Manual Data Input System 人工数据输入系统

Mobile Data Intermediate System 移动数据中间系统

**MDIU** Manual Data Input Unit 人工数据输入组件

**MDIVE** Maximum Design Dive Mach Number 最大设计俯冲马赫数

**MDK** Multimedia Development Kit 多媒体开发工具

**MDL** Magnetic Delay Line 磁延迟线

Magnetic Double Layer 双磁层

Maintenance Diagnostic Logic 维护诊断逻辑

Management Data List 管理数据表

Master Data Library 主数据库

Middle 中间

Minimum Descent Level 最低下降高度层

Mirrored Data Links 镜像数据链接

Mission Data Load 飞行任务数据装载

Model 模型,样式

Modulator 调制器

Modulator Load 调制器负载

Module 组件,模件

Multipurpose Data Link 多用途数据链

**MDLP** Mobile Data Link Protocol 移动数据链路协议

**MDLR** Malaysian Dollar 马来西亚元(货币单位)

**MDM** Maintenance Diagnostic Manual 维修诊断手册

Material Data Meeting 航材数据会议

Maximum Design Meter 最大设计测量仪

Maxwell Data Management 马克斯维尔数据管理

Media Device Manager 媒体设备管理程序

Medium 中间的,媒介

Metal Dielectric Metal Filter 金属绝缘金属过滤器

Mobile Depot Maintenance 机动站维修

Multimedia Data Management 多媒体数据管理

Multiplexer/Demultiplexer 复用器/解复用器

**MDMC** Maintenance Depot Material Control 维修基地器材管理

**MDMS** Maintenance Data Management System 维修数据管理系统

**MDNEXT** Multiple-Disturber Next 多干扰近端串话

**MDO** Multidisciplinary Design Optimization 多学科设计优化

Multidisciplinary Optimization 多学科优化

**MDOB** Multidisciplinary Optimization Branch 多学科设计优化分部

**MDOF** Multi-Degree of Freedom 多自由度

**MDOP** Multimedia Data Operation Platform 多媒体数据操作平台

**MDP** Main Data Package 主数据包

Main Data Path 主数据路径

Main Display Panel 主显示板

Maintenance Data Panel 维修数据板

Maintenance Diagnostic Program 维修诊断程序

Maintenance Display Panel 维修显示板

Malfunction Detection Package 故障检测程序包

Master Data Processor 主数据处理器

Matrix Display Panel 矩阵显示板

Metal Particle Detector 金属微粒检测仪

Miniature Development Program 小型开发程序

Minimum Detectable Power 最小可检测功率

Mission Data Processor 飞行任务数据处理器

Motor-Driven Pump 电动机驱动泵

**MDPS** Maintenance Data Processing System 维修数据处理系统

Mechanized Data Processing System 机械数据处理系统

Meteoroid and Debris Protection System 流星和碎片防护系统

Metric Data-Processing System 公制数据处理系统

Mission Data Preparation System 飞行任务数据准备系统

Mission Data Program Segment 飞行任务数

据程序段

**MDR** Maintenance Data Report 维修数据报告

Maintenance Demand Rates 维修需求定额

Maintenance Design Requirement 维修设计要求

Mandatory Defect Reporting 规定的故障缺陷报告

Manufacturing Design Request 制造设计申请

Master Data Record( er) 主数据记录( 器)

Master Dimensions Request 主尺寸要求

Material Deficiency Report 器材不足报告

Median Detection Range 中值检测范围

Memory Data Register 存储器数据寄存器

Message Detail Recorder 详细消息记录器

Mission Data Reduction 飞行任务数据压缩

Mockup Development Release 样机开发发布

Multi Diagram Report 多图纸报告

Multichannel Data Record( er) 多通道数据寄存器

Multichannel Data Recorder 多路数据记录器,多通道数据记录器

Multiregional DBS ( Digital Broadcast System) Receiver 多区域数字广播系统接收器

**MDRL** Mandrel 心轴

**MDRS** Manufacturing Dates Requirements Sheet 制造日期要求页

**MDS** Main Device Scheduler 主设备调度程序

Maintenance Data System 维修数据系统

Maintenance Document System 维修文件系统

Malfunction Detection System 故障检测系统

Management Decision System 管理决断系统

Management Display System 管理显示系统

Manufacturing Data Series 生产数据序列

Manufacturing Decision Symbol 制造决断符号

Manufacturing Decision System 制造决断系统

Marine Distress Signal 海上遇难信号

Market Data System 市场数据系统

Mass Digital Storage 大容量数字存储器

Master Data Structure 主数据结构

Master Delivery Schedule 主交付时间表

Master Development Schedule 主研制计划

Master Dimensions Specification 主量纲规范

Master Dimensions Surface 主尺寸面

Mathematics Diagnostic System 数学诊断系统

Message Decoder Set 信息解码装置

Message Dropping Station 空投通信袋接收站

Meteoroid Detection Satellite 流星体探测卫星

Microcomputer Development System 微机开发系统

Microprocessor Development System 微处理器开发系统

Microprogram Development System 微程序设计系统

Minimum Detectable Signal 最低可检测信号,最小探测信号

Minimum Discernible Signal 最低可识别信号,最低可辨别信号

Minimum Discernible System 最小可辨信号系统

Minimum Discriminable Signal 最小分辨信号

Model Designator Series 型号标志系列

Modulation/Demodulation Subsystem 调制和解调子系统

Monitoring and Discovery Service 监视与发现服务

Multilevel Database System 多级数据库系统

Multiple Data Stream 多数据流

Multiple Design Series 多种设计系列

Multipoint Distribution System 多点分配系统

**MDSC** Microprocessor Design Support Center 微处理器设计支援中心

**MDSE** Message Delivery Service Element 消

息传递服务单元

**MDSL** Medium Bit-Rate Digital Subscriber Line 中比特率数字用户线

Minimum Detectable Signal Level 最低可检测信号电平

Minimum Discernible Signal Level 最低可识别信号电平

Multi-Channel Digital Synchronous Log 多通道数字同步记录仪

**MDSO** Multimedia Data Storage and Organization 多媒体数据存储和组织

**MDSR** Manufacturing Development Service Request 制造开发服务申请

**MDSS** Maintenance Decision Support System 维修决策支援系统

Marketing Decision Support System 市场营销决策支援系统

Mass Digital Storage System 大容量数据存储系统

Medical Data Surveillance System 医学数据监视系统

Meteorological Data Sounding System 气象数据探测系统

**MDT** Harrisburg, PA, USA (Airport Code) 美国宾西法尼亚州哈里斯堡机场

Maintenance Data Terminal 维护数据终端

Maintenance Display Terminal 维修显示终端

Maintenance Downtime 维修停歇期

Manchester Data Terminal 曼彻斯特数据终端

Manufacturing Data 制造日期

Mean Down Time 平均空闲时间, 平均停机时间

Mission Data Tape 飞行任务数据磁带

Moderate 中度的, 中等的

Mountain Daylight Saving Time 山区夏令时间 (美国)

Mountain Daylight Time 山区白昼时间

**MDTD** Minimum Detectable Temperature Difference 最小可探测温差

**MDTI** Material Data Transmittal Information

器材数据发射信息

**MDTOW** Maximum Design Take-off Weight 最大设计起飞重量

**MDU** Maintenance Diagnostic Unit 维护诊断设备

Management Data Unit 管理数据单元

Manual Deployment Unit 人工放出组件

Manual Drive Unit 人工驱动组件

Message Decoder Unit 信息解码组件

Multiplexer and Distributor Unit 多路复用器与分配器组件

Multiplier Divider Unit 乘法器, 除法器组件, 乘除数组件

**MDUL** Module 组件

**MDW** Chicago, IL, USA-Midway (Airport Code) 美国伊利诺伊州芝加哥米德韦机场

Maintenance Data Warehouse 维修数据仓库

Marketing Data Warehouse 市场数据仓库

**MDWL** Manufacturing Double Entry Wire List 制造双输入导线清单

**MDX** Multiplexer-Demultiplexer 复用器—解复用器

**MDY** Monday 星期一

**MDZ** Mendoza, Argentina (Airport Code) 阿根廷门多萨机场

**ME** Main Entry 主进入口

Maintenance Engineer 维修工程师

Maintenance Entity 维护实体

Maintenance Equipment 维修设备

Maintenance Expense 维修费用

Maintenance of Equipment 设备维修

Manchester Error 曼彻斯特误差

Manufacturing Engineering 制造工程 (部门)

Measurement Entity 测量实体

Mechanical Equipment 机械设备

Miter End 斜面端

Mobility Equipment 可移动设备

Multi-Engine 多发动机

**MEA** Maintenance Engineering Analysis 维护工程分析

Materials Experiment Assembly 器材实验组

装,器材实验组件

Minimum En Route IFR Altitude 最低航路仪表飞行高度

Minimum Enroute Altitude 最低航路高度

Minimum Safe En Route Altitude 航线最低安全高度

More Electric Aircraft 多电飞机

**MEAB** Maintenance Engineering Analysis Board 维修工程分析处

**MEAC** Management Estimate at Completion 完成时经营估算

**MEACN** Maintenance Engineering Analysis Code Number 维修工程分析代码

Maintenance Engineering Analysis Control Number 维修工程分析控制码

**MEAD** Maintenance Engineering Analysis Data 维修工程分析数据

**MEADS** Maintenance Engineering Analysis Data System 维修工程分析数据系统

Maintenance Event Analysis Data System 维修事件分析数据系统

**MEAPS** Maintenance Engineering Analysis Procedures 维修工程分析程序

**MEAR** Maintenance Engineering Analysis Record 维修工程分析记录

Maintenance Engineering Analysis Report 维修工程分析报告

Manufacturing Engineering Action Request 制造工程措施申请

**MEASRG** Measuring 测量

**MEB** Management Evaluation Board 管理评审委员会

**MEBAA** Middle East Business Aviation Association 中东公务航空协会

**MEC** Main Engine Console 主发动机控制台

Main Engine Control(GE Engine) 主发动机控制

Main Engine Controller 主发动机控制器

Main Engine Cutoff 主发动机停车

Main Equipment Center 主设备中心

Manual Emergency Control 人工应急操纵

Manufacturing Engineering Centre 制造工程中心

Mechanic 机械员

Mechanical 机械的

Mechanical Equipment Committee 机械设备委员会

Meteorology Engineering Center 气象工程中心

Modular Electronic Component 模块化电子部件

Most Economic Control 最经济控制

**MECA** Maintainable Electronic Component Assembly 可维修电子部件组装

Matsushita Electric Corporation of America 美国松下电气公司

Model Engine Collectors Association 模型发动机收藏协会

**MECAL** Measurement Calibration File 测量校准文档

**MECAP** Manufacturing Engineering Computer Aided Production 制造工程计算机辅助生产

**MECH** Mechanic 机械员

Mechanical 机械的

Mechanism 机构

**MECIR** Multi-Engine Command Instrument Rating 多引擎指令仪表飞行等级

**MECMA** Middle East Central Monitoring Agency 中东中央监测机构

**MECO** Main Engine Cutoff 主发动机关断

**MECR** Manufacturing Engineering Coordination Record 制造工程协调记录

**MECU** Main Engine (Electronic) Control Unit 主发动机(电子)控制组件

**MED** Aviation Medicine (ICAO) 航空医学(国际民航组织)

Main Entry Door 主进入门

Maintenance Engineering Data 维修工程数据

Manual Entry Device 人工输入器

Manufacturing Enhanced Drawing 制造放大图纸

Mechanical Equipment Description 机械设备说明

Medical 医务的,医药的

Medicine 医药

Medium 中间,媒质,中度的,中等的,中空

Micro Electronic Device 微电子器件

Minimum Engineering Development 最低工程开发

**MEDA** Maintenance Error Decision Aid 维修差错辅助决策

Maintenance Error Decision Assistance 维修差错决策辅助

**MEDB** Model/Engine Database 模型/发动机数据库,型号/发动机数据库

**MEDEVAC** Medical Evacuation 医疗后送,伤病员空运后送

**MEDI** Manufacturing Engineering Data Input 制造工程数据输入

**MEDIF** Medical Information Sheet 病患旅客调查表

**MEDL** Master Engineering Document List 主工程文件清单

**MEDMS** Machinery Electronic Documents Management System 机务维修电子文档管理系统

**MEDSVC** Medical Service 医务

**MEECN** Minimum Essential Emergency Communications Network 最低限度基本应急通信网

**MEEL** Mission Essential Equipment List 执行任务基本设备单

**MEEP** Manufacturing Engineering Exposure Program 制造工程展示程序

**MEF** Maintenance Entity Function 维护实体功能

Minimum Essential Facilities 最低限度基本设施

**MEFC** Manual Emergency Fuel Control 应急人工控制燃油

**MEFL** Minimum Essential Functional List 最低必要功能清单

**MEFT** Minimum Essential Functional Task 最低必要功能任务

**MEG** Manufacturing Engineering Graphics 制造工程图纸

**MEHP** Mean Effective Horse Power 平均有效马力

**MEHT** Minimum Eye Height over Threshold 跑道入口之上的最低眼高

**MEI** Maintenance and Equipment Inspection 维修与设备检查

Maintenance Engineering Inspection 维修工程检查

Maintenance Event Information 维护事件信息

Management Effectiveness Inspection 管理效能检查

Manual of Engineering Instruction 工程说明手册

Meridian,MS,USA(Airport Code) 美国密西西比州默里迪恩机场

**MEII** Multi-Engine Instrument Instructor 多发动机仪表教员

**MEIS** Modular Engine Instrument System 模块式发动机仪表系统

**MEL** Manufacturing Engineering Laboratory 制造工程实验室

Master Equipment List 主设备清单

Material Engineering Laboratory 材料工程实验室

Material Equipment List 器材设备清单

Maximum Emergency Load 最大应急载荷

Melbourne,Australia(Airport Code) 澳大利亚墨尔本机场

Middle East Leasing Corp.(USA) 中东租赁公司(美国)

Minimum Equipment List 最低设备清单,最少设备清单

Multi Engine Lond (Aircraft) 多发陆地飞机

**MELO** Main Engine Lubricating Oil 主发动机润滑油

**MEM** Maintenance Engineering Manual 维护

工程手册

Market Equipment Management 市场设备管理

Master Equipment Matrix 主设备矩阵

Master of Engineering Management 工程管理硕士

Materials Engineering Manual 材料工程手册

Measurement Error Model 测量误差模型

Mechanical Electrical Maintenance 机械电气维修

Memorandum 备忘录

Memory 记忆,存储

Memphis,TN,USA（Airport Code）美国田纳西州孟菲斯机场

**MEMB** Member 成员

**ME-MET** Middle East Meteorological Broadcast 中东气象广播

**MEML** Master Equipment Management List 主设备管理清单

Memorial 纪念的,记忆的,备忘录

**MEMM** Maintenance and Engineering Management Manual 维修工程管理手册

**MEMO** Memorandum 备忘录

Memory 存储,记忆,存储器

**MEMPT** Memory Point 存储点

**MEMS** Maintenance Error Management System 维修差错管理系统

Manufacturing Engineering Management System 制造工程管理系统

Micro Electronic Mechanical Sensor 微电子机械传感器

Micro-Electro-Mechanical System 微电子机械系统

**MEN** Maintenance Engineering Manual 维修工程手册

**MENS** Mission Element Need Statement 飞行任务机组需求陈述

**MENT** Mention 提及,叙述

**MENTD** Mentioned 上述的,提及的

**MEO** Major Engine Overhaul 发动机大修

Manned Earth Observatory 载人地球观测站

Manned Earth Orbit 载人地球轨道

Mass in Earth Orbit 地球轨道质量

Medium Earth Orbit 中距离地球轨道

Most Economic Observing 最经济观测

**MEOL** Manned Earth-Orbiting Laboratory 载人地球轨道实验室

**MEOM** Manned Earth-Orbiting Mission 载人地球轨道飞行任务

**MEOP** Maximum Expected Operating Pressure 预计最大运转压力

Mean Effective Operating Pressure 平均有效工作压力

**MEOS** Medium Earth Orbiting Satellite 中轨道卫星

Monitoring Engine Oil System 发动机润滑监视系统

**MEOTBF** Mean Engine Operation/Operating Time Between Failures 发动机运行故障平均间隔时间

**MEP** Main Executive Program 主要执行计划

Management Engineering Program 管理工程计划

Management Evaluation Program 管理评估计划

Manufacturing Engineering Plan(ning) 制造工程计划

Mean Effective Pressure 平均有效压力

Mean Error Probable 平均误差概率

Mission Equipment Package 飞行任务设备包

Multi-Engine Pilot 多发飞机驾驶员

**MEPL** Master Equipment Parts List 主设备零件清单

**MEPM** Manufacturing Engineering Planning Manual 制造工程计划手册

**MEPSI** Manufacturing Engineering Process and System Integration 制造工程工序与系统集成

**MEPT** Multi-Engine Pilot Trainer 多发动机教练机

**MEPU** Monopropellant Emergency Power Unit

单元推进剂应急动力装置

**MEPW** Minimum Electrical Pulse Width 最小电脉冲宽度

**MEQ/L** Milli-Equivalents per Litre 毫当量/公升

**MER** Maintenance Evaluation Report 维护评估报告

Managing Electronic Records 电子文件管理

Manpower Estimate Report 人力资源评估报告

Market Exchange Rate 市场汇率

Maximum Error Rate 最大误差率

**MER-ATS** Meteorological Applications Technology Satellite 气象应用技术卫星

**MERCH** Merchantable 有销路的

**MERDC** Mobility Equipment Research and Development Center 可移动设备研制与开发中心

**MERF** Mobile Enroute Radar Facility 移动式航线雷达设备

**MERM** Manufacturing Engineering Requirement Model 生产工程需求模型

Mechanical Engineering Reference Manual 机械工程参考手册

**MES** Main Engine Start 主发动机启动

Main Engine Strut 主发动机吊架

Manufacturing Execution System 生产制造执行系统

Master Earth Station 主地球站

Medan, Indonesia ( Airport Code ) 印度尼西亚棉兰机场

Mobile Earth Station 移动地球站

Multi Engine Sea ( Aircraft ) 多发水上飞机

**MESA** Maintenance Engineering Society of Australia 澳大利亚维修工程学会

Manufacturing Execution Systems Association 制造执行系统协会

Modular Experimental Scientific Applications 模块化试验科学应用

**MESAD** Master Equipment Schedules and Allocations Document 主设备计划与配置文件

**MESAR** Microwave Electronically Scanned Adaptive Radar 微波电扫描自适应雷达

Multifunction Electronic Scanned Adaptive Radar 多功能电子扫描自适应雷达

Multifunction Electronically Scanned Adaptive Radar 多功能电扫描自适应雷达

**MESC** Mid Electrical Service Center 中央电器服务中心

**MESH** Management of Engineering Safety Health 工程安全状态管理

**MESL** Minimum Essential Subsystems List 最低必要分系统清单

**MESSRS** Messieurs 诸位先生

**MET** Aeronautical Meteorology 航空气象学

Maintenance Engineering Technique 维修工程技术

Management Engineering Team 管理工程组

Mean European Time 欧洲平均时

Mechanical Engineering Team 机械工程组

Memory Enhancement Technology 内存增强技术

Metal 金属

Meteorological Broadcast Service 气象广播服务

Meteorology, Meteorological 气象,气象的

Meteorological Services 气象服务

Meteorological Services for Air Navigation 航行气象服务

Mission Elapsed Time 任务耗用时间

Mission Event Time 飞行任务时间

Mission Event Timer 飞行过程记时器

Multiemitter Transistor 多发射极晶体管

**META** Maintenance Engineering Training Agency 维修工程培训处

**METAG** Meteorological Advisory Group 气象咨询组

**METAR** Aviation Routine Weather Report 航空例行气象报告

Meteorological Aeronautical Ratio Code 航空气象无线电密码

Meteorological Aviation Report 航空气象报

告

Meteorological Report of Aerodrome Conditions 机场条件气象报告

Meteorological Aviation Routine Weather Report 航空例行天气报告

**METB** Metal Base 金属底座

**METC** Metal Curb 金属井栏

**METCAL** Meteorological Calibration 气象校准

**METCM** Computer Meteorological Message 计算机气象信息

**METD** Metal Door 金属门

**METEOCH** Meteorological Channel 气象信道

**METEOS** Meteorological Station 气象台,气象站

**METF** Metal Flashing 金属盖片

**METG** Metal Grill 金属格栅

**METH** Method 方法

**METINF** Meteorological Information 气象资料

**METJ** Metal Jalousie 金属百叶窗

**METO** Maximum Engine Takeoff ( Power ) 发动机最大起飞(功率)

Maximum Except Takeoff ( Power ) 除起飞外额定最大(功率)

**METON** Metropolitan Optical Network 都市光网络

**METOP** Maximum Expected Takeoff Power 预计最大起飞功率

**METP** Metal Partition 金属隔板

Metal Portion 金属部分

**METR** Metal Roof 金属顶

**METRO** Metropolitan 首都的

**METRP** Meteorological Report 气象报告

**METS** Maintenance Engineering Training Sheet 维修工程培训单

Metal Strip 金属带,金属条

Mobile Engine Test Stand 移动式发动机测试台

Modular Engine Test System 模块式发动机

试验系统

**METWNG** Meteorologist Warning 气象警报

**MEU** Message Encoder Unit 信息编码器

**MEV** Main Electro Valve 主电动活门

Million Electron-Volts 百万电子伏

**MEW** Manufacturer Empty Weight 制造厂空重

Microwave Early Warning 微波预警

**MEWS** Maintenance Engineering Work Sheet 维修工程工作单

Manufacturing Engineering Work Statement 制造工程工作陈述

Microwave Early Warning System 微波预警系统

**MEX** Mexico City , Mexico ( Airport Code ) 墨西哥墨西哥城机场

**MEXE** Mobile Station Application Execution Environment 移动基站应用执行环境

**MEXPRO** Model Extraction Program 模型录取程序

**MEY** Maximum Economic Yield 最大经济产量

**MEZs** Marine Exclusion Zones 海上限制区

**MF** Main Feed 主供给

Main Force 主力

Main Frame 主框架

Maintenance Facility 维修设施

Maintenance Function 维护功能

Mandatory Frequency 强制频率

Manual Forecasting 人工预报

Manufacture 制造

Manufacturing Fabrication 制造,装配,生产加工

Mastic Floor 胶脂地板

Mediation Function 中介功能

Medium Frequency ( 300 to 3000 KHz ) 中频

Meter Fix 表固定,表支点

Milling Fixture 研磨夹具

Minor Frame , Miniframe 小框

Modifying Factor 修正因子

Modulation Frequency 调制频率

Multi-Frame Structure 复帧结构

Multi-Frequency 多频

Multi-Frequency Code Signaling 多频编码信令

Multiplication Factor 放大系数

**MFA**　Master File Audit 主文档审查

Memorized Fault Annunciation 存储故障通告

Memorized Fault Annunciator 存储故障信号器

Memorized Fault Indicator 存储故障指示器

Minimum Flight Altitude 最低飞行高度

Minimum Fuel Advisory 最小流量咨询

Multi-Frame Alignment 多帧定位

**MFAR**　Multifunction Array Radar 多功能相控阵雷达

**MFBT**　Multicast Forward/Backward Tree 多播前向/反向树形网络

**MFC**　Magnetic Friction Clutch 磁摩擦离合器

Main Frequency Clock 主频时钟

Main Fuel Control 主燃油控制

Main Fuel Controller 主燃油控制器

Manifold Failure Controller 总管故障控制器

Manual Frequency Control 人工频率控制

Master Function Control 主功能控制

Maximum Fuel Capacity 最大燃油容量

Mechanical Fuel Control 机械燃油调节器（机械燃调）

Microfunctional Circuit 微功能电路

Motor Firing Circuit 发动机启动电路

Multi Functional Circuit 多功能电路

Multicore-Fiber Cable 多芯光缆

Multi-Frame Code 多帧编码

Multi-Frequency Code 多频码

Multi-Frequency Code Receiver 多频码接收器

Multi-Frequency Code Signaling 多频码信令

Multi-Frequency Code System 多频编码系统

Multi-Frequency Coding 多频代码

Multi-Frequency Compelled Signal 多频互控信号

Multi-Frequency Compelled Signalling 多频互控信令

Multi-Frequency Control 多频控制

Multi-Frequency Controller 多频控制器

Multiple Flight Computer 复合式飞行计算机

**MFCC**　Mel Frequency Cepstral Coefficients 梅尔频率对数倒频系数

Mission Flight Control Center 任务飞行控制中心

Multi-Function Common Console 多功能共用控制台

**MFCCS**　Motor Firing Circuit Checkout Set 发动机启动电路检查装置

**MFCD**　Multi-Function Colour Display 多功能彩色显示器

**MFCK**　Multi-Frame Clock 复帧时钟

**MFCP**　Multifunction Control Display Panel 多功能控制显示面板

Multifunction Control Panel 多功能控制板

**MFCR**　Minimum Free Capacity Routing 最小自由容量选路

**MFCS**　Manual Flight Control System 人工飞行控制系统

**MFCU**　Manual Fuel Control Unit 燃油手动控制装置

**MFD**　Magnetic Frequency Detector 磁性频率检测器

Magnetofluid Dynamics 磁性流体动力学

Manufactured 制造

Manufacturing 制造的

Microfarad 微法

Motor Firing Direction 发动机启动方向

Multifunction Data 多功能数据

Multi-Function Device 多功能设备

Multi-Function Display 多功能显示器

**MFDF**　Medium Frequency Direction Finder 中频定向机

Medium Frequency Direction Finding 中频定向

**MFDS**　Manifold Failure Detection System 总

管故障检测系统

Multi-Function Display System 多功能显示系统

**MFDT** Memory for Designs Test 设计实验用存储器

**MFDU** Multifunction Display Unit 多功能显示单元

**MFE** Manufacturing Feasibility Estimate 制造可行性预算

McAllen,TX,USA（Airport Code）美国得克萨斯州迈克艾伦机场

**MFED** Maximum Flat Envelop Delay 最大平滑包络延迟

**MFEPD** Maintenance Facility and Equipment Planning Document 维修设施和设备计划文件

**MFEPM** Maintenance Facilities and Equipment Planning Manual 维护设施及设备计划手册

**MFER** Memory Fault Extension Register 存储器故障扩充寄存器

**MFF** Master Freight File 主要运货档案
Mixed Fleet Flying 混合机队飞行

**MFG** Manufacturer 制造厂
Manufacturing 制造
Master Frequency Generator 主频发生器

**MFH** Master Folder Holder 主文件夹支架

**MFHBF** Mean Flight Hours Between Failures 平均故障间隔飞行小时数

**MFHBMA** Mean Flight Hours Between Maintenance Action 维修作业间隔平均飞行时数

**MFI** Magnetic Field Intensity 磁场强度
Master Function Interrupt 主功能中断

**MFIE** Magnetic Field Integral Equation 磁场积分方程

**MFIR** Mission Fault Isolation Record 飞行任务故障隔离记录

**MFK** Multi-Function Keyboard 多功能键盘

**MFL** Magnetic Flux Leakage 磁漏
Maintain Flight Level 保持飞行高度
Maintenance Fault List 维修故障清单

Median Fatigue Life 中值疲劳寿命

Minimum Field Length 机场最短长度

Multiple-Frequency Laser 多频激光器

**MFLI** Magnetic Fluid Level Indicator 流体液位磁指示器

**MFLOPS** Million Floating Point Operations per Second 百万浮点运算/秒

**MFLRD** Male Flared 插入式配件扩口

**MFM** Macau International Airport 澳门国际机场
Magnetic Field Modulation 磁场调制
Magnetic Force Microscopy 磁力显微镜
Maintenance Fault Memory 维修故障存储器
Mission and Flight Management 任务与飞行管理
Modified Frequency Modulation 变频调制
Motor Fire Module 发动机启动模件

**MFN** Most Favored Nation 最惠国
Most Favoured Nation 最惠国
Multi Frequency Network 多频网络

**MFNT** Most Favoured Nation Treatment 最惠国待遇

**MFO** Master Frequency Oscillator 主频振荡器

**MFOB** Minimum Fuel on Board 机上最低限度油量

**MFOP** Maintenance Free Operating Period 免维修使用期
Major Facilities Operations Plan 主设备运行计划

**MFP** Main Frame Pulse 主帧脉冲
Main Fuel Pump 主燃油泵
Maintenance Facility Planning 维修设施计划
Manufacturing Plan 制造计划
Matched Filter Performance 匹配滤波器性能
Mean Free Path 平均自由程
Minimal Flight Path 最短航迹
Mixed Fission Products 混合裂变产物
Multi Frequency Pulsing 多频率脉冲
Multi Function Probe 多功能探针

**MFPOP** Modified Full Power Operational Profile 修正全功率飞行剖面

**MFPS** Manual Flight Path Selection 人工飞行航路选择

**MFR** Malfunction Rate 故障率

Manufacturer 制造厂,制造商

Mean Failure Rate 平均故障率

Medford, OR, USA (Airport Code) 美国俄勒冈州梅德福机场

Multi-Frequency Receiver 多频接收机

Multi-Frequency Responder 多频应答机

Multifunction Radar 多功能雷达

Multifunction Receiver 多功能接收机

Multiple File Report 多文档报告

**MFR IDENT** Manufacturer Identification 制造厂家识别(码)

**MFRACAS** Maintainability Failure Reporting Analysis and Corrective Actions 维修能力失效报告分析与改正措施

**MFRE** Manufacture 制造

**MFREC** Multi-Frequency Receiver 多频接收器

**MFRN** Manufacturer's Number 制造厂号码

**MFS** Manned Flight Simulator 手控飞行模拟器

Median Fatigue Strength 中值疲劳强度

Message Formating Service 信息格式化服务

Metropolitan Fiber System 都市光纤系统

Multi Function Sensor 多功能传感器

Multi Function Spoiler 多功能扰流板

Multi-Frame Structure 复帧结构

Multi-Frame Synchronizer 复帧同步器

Multiple Frequency Shift 多频位移

**MFSK** Multi-Frequency Shift Keying 多频移键控

**MFST** Manifest 舱单

**MFSND** Multi Frequency Sender 多频发送器

**MFT** Mean Flight Time 平均飞行时间

Medium Fire Tender 中型救火车

Minimum Frame Time 最小帧时间

Mission Flight Trainer 任务飞行训练器

Multi-Function Team 多功能小组

Multi-Function Terminal 多功能终端

**MFTBF** Mean Flight Time Between Failures 飞行故障平均间隔时间

**MFTBMA** Mean Flight Time Between Maintenance Action 维修作业间隔平均飞行时间

**MFTD** Maintenance and Flight Training Device 维护与飞行训练装置,维护与飞行训练器

**MF-TDMA** Multi Frequency Time Division Multiple Access 多频时分多址接入

**MFTW** Maximum Design Fuel Transfer Weight 最大设计燃油传输重量

**MFV** Main Fuel Value 主燃油活门

**MFW** Maximum Flight Weight 最大飞行重量

Maximum Fuel Weight 最大燃油重量

**MG** Immigration 移民

Magnesium 镁

Magnetic Armature 磁电枢

Main Landing Gear 主起落架

Maintenance Group 维修组

Master Gauge 主表,主量规

Master Group 主群

Media Gateway 媒体网关

Motor Generator 电动发电机

Motor-Generator(Set) 电动—发电机(组)

Multicast Grouping 多播分组

Multigauge 多用量测仪表,多用检测仪

**mg** Milligram 毫克

**MG/L** Milligrams per Litre 毫克/升

**MGA** Managua, Nicaragua (Airport Code) 尼加拉瓜马拉瓜机场

Minimum Safe Grid Altitude 最低安全屏障高度

**MGB** Main Gearbox 主齿轮箱

**MGC** Manual Gain Control 人工增益控制

Maximum Guaranteed Capability 保证最大能力(功率)

Mean Geometric Chord 平均几何弦

**MGCO** Mars Geoscience/Climatology Orbiter 火星地球科学/气象学轨道卫星

**MGCP** Media Gateway Control Protocol 媒体网关控制协议

Mission Guidance Computer Program 飞行任务引导计算机程序

**MGCU** Main Gear Control Unit 主起落架控制组件

Main Generator Control Unit 主发电机控制装置

**MGDP** Millimeter Guidance Development Program 毫米波引导研制大纲

**MGE** Maintenance Ground Equipment 地面维修设备

Message 电报,信息

Mission Ground Equipment 飞行任务地面设备

**MGES** Maintenance Ground Equipment Section 地面维修设备科

**MGF** Magnify 放大

Mobile Gateway Function 移动网关功能

**MGG** Memory Gate Generator 存储器选通门产生器

**MGH** Minimum Guide Height 最低引导高度

**MGL** Marginal 边缘的

**MGM** Montgomery, AL, USA (Airport Code) 美国阿拉巴马州蒙哥马利机场

**MGMT** Management 管理

**MGN** Magneto 磁电机

**MGOS** Maintenance and Ground Operation System 维护和地面操纵系统

**MGPF** Mobile Geographic Position Function 移动地理定位功能

**MGQ** Mogadishu, Somalia (Airport Code) 索马里摩加迪沙机场

**MGR** Manager 经理

Micro-Graphic Recording 显微照相记录

Micro-Graphic Reporting 显微摄影报告

Miniature GPS Receiver 微型全球定位系统接收器

**MGS** Mobile Gateway Switch 移动网关交换

Mobile Ground Station 移动地面站

Multimedia Gateway Server 多媒体网关服务器

**MGSCU** Main Gear Steering Control Unit 主起落架转弯控制组件

**MGSE** Maintenance Ground Support Equipment 地面维修保障设备

Mechanical Ground Support Equipment 地面机械保障设备

**MGSS** Main Gear Steering System 主起落架转弯系统

**MGT** Major Ground Test 主要地面测试

Management 管理

Measured Gas Temperature 测量的燃气温度

Minimum Ground Time 地面最少停留时间

Mobile Ground Terminal 移动式地面终端

**MGTOW** Maximum Gross Takeoff Weight 最大起飞全重

**MGTP** Main-Gear Touchdown Point 主起落架着陆接地点

**MGTW** Maximum Gross Taxi Weight 最大滑行全重

**MGW** Maximum Gross Weight 最大全重

Media Gateway 媒体网关

Morgantown, WV, USA (Airport Code) 美国西弗吉尼亚州摩根敦机场

**MH** Magnetic Head 磁头

Magnetic Heading 磁航向

Maintenance Handbook 维修手册

Manhole 检查孔,人孔

Manhours, Man Hours 人工时,人工一小时

Message Handler 报文处理程序

Message Handling 消息处理

Microwave Hologram 微波全息图

Miles per Hour 英里/小时

Millihenry 毫亨

Minimum Height 最低高度

Mobile Host 移动式主机

**MH(D)** Magnetic Heading 磁航向

Message Handler 信息处理机

**MHA** Maintenance Hazard Analysis 维修危险性分析

Minimum Holding Altitude 最低保持高度,

最低等待高度

**MHD** Magnetic Hard Drive 硬盘驱动

Magnetohydrodynamics 磁性流体动力学

Mashhad, Iran ( Airport Code ) 伊朗马什哈德机场

Masthead 杆顶

Medium Hard Drawn 中强度拉伸

Metering Head Differential 差动测量头

**MHDA** Minimum Height for Discontinued Approach 中止进近最低高度

**MHDD** Multi-Functional Head-Down Display 多功能下视显示器

**MHDF** Medium and High Frequency Direction Finding Station 中高频定向台

**MHDG** Magnetic Heading 磁航向

**MHDI** Master Heading Pictorial Deviation Indicator 主航向图示偏离指示器

**MHDLT** Masthead Light 杆顶灯

**MHE** Material Handling Equipment 材料处理设备

Mechanical Handling Equipment 机械处理设备

Message Handling Environment 信息处理环境

**MHEG** Multimedia Hypermedia Expert Group 多媒体超媒体专家组标准

**MHF** Medium High Frequency 中高频

Microwave Height Finder 微波测高仪

**MHFDF** Medium High Frequency Direction Finder 中高频定向机

**MHFS** Maximum Horizontal Flight Speed 最大平飞速度

**MHI** Magnetic Heading Indicator 磁航向指示器

Master Heading Indicator 主航向指示器

Mitsubishi Heavy Industries, Ltd. 三菱重工业株式会社 ( 日本 )

Mitsubishi Heavy Industries 三菱重工

**MHIR** Manufacturing Hookup and Installation Report 制造线路图及安装报告

**MHL** Mast Hull Loop 机身支架环

**MHN** Multimedia Handling Node 多媒体处理节点

**MHP** Main Hydraulic Pump 主液压泵

Mechanical ( Machinery ) Horsepower 机器马力, 机械马力

Medium High Pressure 中高压

Mental Health Plan 心理健康计划

Mental Health Professional 心理健康专家

Message Handling Procedure 信息处理程序, 报文处理程序

**MHPD** Masonite Hydropress Die 绝缘纤维板水压机模

**MHPDI** Magnetic Heading Pictorial Deviation Indicator 磁向图示偏离指示器

**MHQ** Mariehamn, Aland Island, Finland ( Airport Code ) 芬兰奥兰群岛玛丽港机场

**MHR** Master Heading Recorder 主航向记录器

**MHRS** Magnetic Heading Reference System 磁航向基准系统, 磁航向参考系统

**MHS** Message Handling Systems 报文处理系统

**MHSSE** Message Handling System Service Element 报文处理系统业务单元

**MHSU** Magnetic Heading Sensor Unit 磁航向传感器装置

**MHSV** Multipurpose High Speed Vehicle 多用途高速飞行器

**MHT** Manchester, NH, USA ( Airport Code ) 美国新罕布什尔州曼彻斯特机场

Multiple Hypothesis Tracking 多重假设跟踪

**MHTV** Manned Hypersonic Test Vehicle 载人高超音速实验飞行器

**MHUC** Manufacturing Hook Up Chart 制造线路图表

**MHV** Manned Hypersonic Vehicle 载人高超音速飞行器

Mean Horizontal Velocity 平均水平速度

Miniature Homing Vehicle 小型寻的飞行器

**MHW** Magnetic Heading Warning 磁航向告警

Mean High Water 平均高潮

**MHz** Megahertz, Mega Hertz 兆周,兆赫

**MI** Magnetic Indicator 磁指示器

Magnetic Inspection 磁性检查

Maintenance Information 维护信息

Maintenance Instruction 维修指南

Malleable Iron 可锻铸铁

Management Information 管理信息

Manual Input 人工输入

Manufacturing Information 制造信息

Manufacturing Inspector 制造检查员

Manufacturing Instruction 制造说明

Master Item 主控项目

Material Identifier 器材标识牌,器材标识符

Medium Intensity (Light) 中强度(灯光)

Meridian Instrument 子午仪

Message Identification 信息识别

Meteorological Instrument 气象仪器

Metrical Information 测量信息

Metrical Instrument 测量仪器

Mica 云母

Microscope Interferometer 显微干涉仪

Microscopic Image 显微图像

Microscopic Inspection 显微检查

Microwave Inspection 微波检查

Microwave Interferometer 微波干涉仪

Mile 英里(1.609 公里)

Minor 小的

Minus 减,负

Minute 分

Modification Instruction 改装说明

Module Integration 模块集成

Mutiplex Interface 复用接口

Mutual Inductance 互感

**MI&RR** Material Inspection and Reception Report 器材检验和验收报告

**MI/S** Miles per Second 英里/秒

**MIA** Charleston, FL, USA (Airport Code) 美国佛罗里达州迈阿密机场

Mactan International Airport 麦克坦国际机场

Made in America 美国制造

Malta International Airport 马耳他国际机场

Manila International Airport 马尼拉国际机场

Melbourne International Airport 墨尔本国际机场

Metal Interface Amplifier 金属接口放大器

Miami International Airport 迈阿密国际机场(美国)

Minimum Authorized Altitude 最低批准高度

Minimum IFR Altitudes 最低仪表飞行规则高度

Minimum Instrument Altitude 最低仪表高度

Myanmar International Airways 缅甸国际航空公司

**MIAC** Maintenance Information and Control 维修信息和控制

Material Identification Accounting Code 材料识别和计算码

**MIACS** Manufacturing Information and Control System 制造信息与控制系统

**MIADS** Manufacturing Information and Display System 制造信息与显示系统

**MIAG** Management Information Analysis Group 管理信息分析小组

**MIALS** Medium Intensity Approach Lighting/Light System 中强度进近灯光系统

**MIAS** Multipoint Interactive Audiovisual System 多点交互式声视系统

**MIB** Management Information Base 管理信息库

**MIBC** Management Information Base Chip 管理信息库芯片

**MIBS** Manufacturing Interim Budgeting System 临时制造预算系统

**MIC** Management Information Center 管理信息中心

Manufacturing Inspection Controller 制造检验控制器

Maximum Input Current 最大输入电流

Media Interface Connector 媒体接口连接器

Message Identification Code 消息识别码

Micrometer 千分尺,测微表

Microphone 麦克风,话筒

Microwave Integrated Circuit 微波集成电路

Module Interface Circuit 模块接口电路

Monitoring Identification and Correction 监控识别与修正

Monolithic Integrated Circuit 单片集成电路

Multichip Integrated Circuit 多芯片集成电路

Multilayer Integrated Circuit 多层集成电路

Multimode Image Coding 多模图像编码

**MICA** Media Information Communication Application 媒体信息通信应用(程序)

**MICAPS** Meteorological Information Comprehensive Analysis and Process System 气象信息综合分析处理系统

**MICBAC** Micro-System Bus Access Channel 微系统总线存取通道

**MICD** Mechanical Interface Control Document 机械接口控制文件

**MICE** Management Information Control Exchange 管理信息控制交换

**MICELEM** Microphone Element 话筒元件

**MICH** Michigan 密歇根州

**MICOM** Micro Computer 微型计算机

**MICOS** Multifunctional Infrared Coherent Optical Sensor 多功能红外相干光传感器

**MICP** Management Incentive Compensation Plan 管理激励补偿计划

**MICR** Magnetic Ink Character Recognition 磁墨水字符识别

Manufacturing Interchange Change Request 制造互换更改申请

Manufacturing Interlink Change Request 制造结合更改申请

**MICRAD** Microwave Radiation System 微波辐射系统

Microwave Radiometry 微波辐射计,微波辐射测量

**MICRO** Microcomputer 微机

Microprocessor 微处理器

Microprogram 微程序

Microscope 显微镜

Microwave 微波

**MICROA** Microamperes 微安

**MICROP** Microprocessor 微型处理机

**MICROS** Microscope 显微镜

**MICROSW** Microswitch 微动电门

**MICROW** Microwatt 微瓦

**MICS** Management Information and Control System 管理信息和控制系统

Material Information and Control System 器材信息与控制系统

Material Inventory Control System 材料库存控制系统

Microfiche Interface Control System 缩微胶片接口控制系统

Microprocessor System 微处理机系统

Multiple Internal Communication System 多路内部通信系统

**MID** Merida, Yucatan, Mexico ( Airport Code)墨西哥尤卡坦梅里达机场

Message Identification 报文标志

Middle East 中东

Middle East Region 中东地区

Middle 中间,中间的

Midnight 午夜

Mobile Internet Device 移动互联网设备

Modification Information Document 改装信息文件

Multiplexing Identifier 复用标志

**MIDA** Middle America 中美洲

**MIDANPIRG** Middle East Air Navigation Planning and Implementation Regional Group 中东地区航行规划实施小组

**MIDAR** Microwave Detection and Ranging 微波探测与测距

**MIDAS** Maintenance Integrated Data Access System 维修综合数据存取系统

Malfunction Identification Data Acquisition System 故障识别数据获取系统

Management Information and Development

Aids System 管理信息和辅助开发系统

Management Information Data Base Automation System 管理信息数据库自动化系统

Management Information Data Flow System 管理信息数据流程系统

Management Information Data System 管理信息数据系统

Management Interactive Data Accounting System 管理交互式数据会计系统

Man-machine Integration Design and Analysis System 人机集成设计与分析系统

Manufacturing Information Distribution and Acquisition System 制造信息分配与采集系统

Measurement Information and Data Analysis System 测量信息与数据分析系统

Meteorological Integrating Data Acquisition System 综合气象资料获取系统

Meteorological Intelligence Data Analysis Study 气象情报数据分析研究

Microprogramming Design Aid System 微程序辅助设计系统

Miniature Inertial Digital Attitude System 小型惯性数字姿态系统

Modified Integration Digital Analog Simulator 改进型综合数模模拟器

Multiple Index Data Access System 多重变址数据存取系统

Multiple Index Direct Access System 多重变址直接存取系统

Multiple Input Data Access System 多路输入数据存取系统

**MIDDS** Meteorological Integrated Distribution and Display System 气象综合分配与显示系统

Meteorological Interactive Data Display System 交互式气象数据显示系统

**MIDF** Main Item Data File 主要项目数据文件

**MIDI** Musical Instrument Digital Interface 音乐指令数字化接口

**MIDL** Midland 中部地区

**MIDMS** Machine Independent Data Management System 机器独立数据管理系统

**MIDO** Manufacturing Inspection District Office 制造检查地区办公室

**MIDR** Manufacturing Inspection District Representative 制造检查地区代表

**MIDS** Management Information and Data System 管理信息和数据系统

Management Information and Decision Support 管理信息与决断保障

Management Information Dataflow System 管理信息数据流系统

Meteorological Information Display System 气象信息显示系统

Multi-Function Information Distribution System 多功能信息分配系统

Multimedia Intelligent Database Systems 多媒体智能数据库系统

Multiple Input Data Acquisition System 多路输入数据采集系统

**MIDT** Marketing Information Data Tapes 市场信息数据带

**MIDU** Multi-Input Interactive Display Unit 多输入交互式显示组件

Multi-Purpose Interactive Display Unit 多功能交互式显示组件

**MIE** Maneuver-Induced Error 操纵引致误差

Micromedia Information Exchange 微媒体信息交换

Multipurpose Internet Extensions 多用途因特网扩充

**MIF** Management Information Format 管理信息格式

Meteorological Information 气象信息,气象资料

**MIFG** Shallow Fog 浅雾

**MIFIL** Microwave Filter 微波滤波器

**MIFL** Master International Frequency List 国际频率总表

**MIFR** Master International Frequency Register 国际频率总登记册

Master International Frequency Registration 国际频率注册管理站(员)

**MIFS** Material Information Flow System 器材信息流向系统

Material Inventory Forecasting System 器材库存预报系统

**MIG** Metal Inert Gas 金属惰性气体(焊)

Mikoyan and Gurevich 米格(飞机)

**MII** Management Information Infrastructure 管理信息基础设施

Media Independent Interface 媒体独立接口

Ministry of Information Industries 信息产业部

Mobile Information Infrastructure 移动通信信息基础设施

**MIIR** Maintenance Inspection Intervals Report 维护检查间隔报告

**MIIS** Medifech Interactive Information System 美德费奇交互式信息系统

**MIKE** Microphone 麦克风

**MIL** Magnetic Indicator Loop 磁指示器回路

Malaya Indonesia Line 马来亚印尼航线

Malfunction Indicator Lamp 故障指示灯

Malfunction Investigation Laboratory 故障调研实验室

Master Index List 主要指数表

Mileage 里程

Military Specification 军用规范

Military Standards 军用标准

**MILA** Mini Light Adapter 迷你灯适配器

**MILAD** Military Advisor 军事顾问

**MILAP** Maintenance Information Logically Analyzed and Produced 逻辑分析与产生的维修信息

**MILAV** Military Aviation 军事航空

**MILE/H** Miles per Hour 英里/小时

**MILES** Multiple Integrated Laser Engagement System 多路综合激光作战系统

**MILEX** Military Exercise 军事演习

**MILFLT** Military Flight 军事飞行

**MILIRAD** Millimeter Radar 毫米波雷达

**MILN** Million 百万

**MILNET** Military Network 美国国防部网络

**MILS** Manufacturing Inventory and Location System 制造库存与定位系统

Microwave Instrument Landing System 微波仪表着陆系统

**MIM** Management Information Model 管理信息模型

**MIMD** Multiple Instruction Multiple-Data 多指令多数据(流)

**MIME** Mechanical, Industrial and Manufacturing Engineering 机械、工业和制造工程

Multipurpose Internet Mail Extension 多用途因特网邮件扩展

Multipurpose Internet Mail Extensions Protocol 多用途因特网邮件扩展协议

**MIMIC** Microwave/Millimeter-Wave Monolithic Integrated Circuit 微波/毫米波单片集成电路

**MIMO** Multi-Input Multi-Output 多输入多输出

**MIMP** Manufacturing Interface Management Plan 制造界面管理计划

**MIMU** Micro Inertial Measurement Unit 微型惯性测量装置

**MIN** Minimum 最低,最低限度

Minor 小的

Minute 分钟,分

Mobile Intelligent Network 移动智能网

Multistage Interconnected Network 多级互联

**MIN FUEL** Minimum Fuel 最低燃油量

**MIN SPD** Minimum Speed 最低速度

**MIN TIME** Minimum Time 最短时间

**MINALT** Minimum Altitude 最低高度

**MINAT** Miniature 微型

**MINEAC** Miniature Electronic Autocollimator 小型电子自动准直仪

**MINGT** Midnight 午夜

**MINIAPS** Miniature Accessory Power Supply 小型辅助电源

**MINICANE** Miniature Hurricane 小型飓风

**MINICOM** Minimum Communications 最低通信

**MININAN** Minimal Doppler Navigator 短程多普勒导航仪

**MINI-SPEC** Mini-Specification 微型规范

**MINMUX** Miniature Multiplexer 小型多路传输器

**MINN** Minnesota 明尼苏达州

**MIN-OP** Minimum Operational 最低运行的

**MINR** Minor 小的

**MINS** Minimums 最小,最低,最少
Minutes 分

**MINT** Maintenance 维护

**MINY** Ministry 部

**MIO** Manufacturing Inspection Office 制造检查办公室
Multi-Purpose Input/Output 多用途输入/输出

**MIP** Maintenance Information Printer 维修信息打印机
Manual Input Processing 人工输入处理
Master Interphone Panel 主内话板
Material Improvement Program 器材改进程序,器材改进大纲
Material Improvement Project 器材改进计划
Maximum Indicating Pointer 最大指示指针
Mean Indicated Pressure 平均指示压力
Minimum Impulse Pulse 最小动量脉冲
Multimedia Information Provider 多媒体信息提供者
Multiple Information Passageway 多信息通道方式

**MIPE** Modular Information Processing Equipment 模块化信息处理设备

**MIPO** Manually Issued Purchase Order 人工填写订货指令

**MIPS** Maintenance Information and Planning System 维修信息和计划系统
Million Instruction(s) per Second 每秒百万指令,百万指令/秒

**MIR** Maintenance Irregularity Report 非常规维修报告
Malfunction Investigation Report 故障调查报告
Manufacturing and Inspection Record 制造与检查记录
Master Inventory Record 库存总账
Material Inspection Record 器材检验报告
Memory Input Register 存储器输入寄存器
Mishap Investigation Report 事故调查报告
Monastir, Tunisia ( Airport Code ) 突尼斯莫纳斯提尔机场

**MIRA** Mission Integration and Requirements Analysis 飞行任务综合与要求分析

**MIREL** Medium Intensity Runway Edge Lights 中强度跑道边界灯

**MIRFS** Multi-Function Integrated Radio Frequency System 多功能综合射频系统

**MIRL** Middle Intensity Runway Light 中强度跑道灯

**MIRM** Manufacturing Installation Requirement Model 生产安装需求模型

**MIRR** Material Inspection and Receiving Report 材料验收报告

**MIRS** Management Information and Retrieval System 信息管理与检索系统
Miscellaneous Inspection Requirements 综合检查要求
Modular Integrated Radar System 模块化集成雷达系统
Multimedia Information Recall System 多媒体信息检索系统
Multi-Purpose Infra-Red Sight 多用途红外视域

**MIRV** Multiple Independently Reentry Vehicle 多重自主再入飞行器

**MIS** Made in Shop 在车间制造
Maintenance Information System 维修信息系统
Management Information Science 管理信息科学

Management Information Services 管理信息服务

Management Information System 管理信息系统

Manufacturing Information System 制造信息系统

Marketing Information System 销售信息系统

Mechanical Interruption Summary 机械中断摘要

Metal Insulated Semiconductor 金属绝缘半导体

Meteorological Impact Statement 气象影响通报

Microfilm Information System 缩微胶卷信息系统

Microwave Interferometric System 微波干涉测量系统

Missing 丢失的,遗漏的

Monthly International Statistics 国际运输月统计

Multiple Interactive Screen 多交互式屏幕

**MISACT** Multiple Interactive Sensors and Actuators 多相互作用传感器和致动装置

**MISC** Miscellaneous 杂项,综合

**MISDN** Mobile Integrated Service Digital Network 移动综合业务数字网

**MISEDS** Machine Independent Systems Effectiveness Data System 机器自主系统有效性数据系统

**MISG** Missing 丢失的,遗漏的

**MISHDL** Mishandle 处理错误

**MISIP** Management Information System Improvement Plan 管理信息系统改进计划

**MISO** Management Information Systems Office 管理信息系统办公室

Multiple Input, Single Output 多输入,单输出

**MISOFLD** Misoffload 误卸

**MISS** Mission 任务,使命

Mississippi 密西西比州

**MIST** Maintenance International Standardiza-tion Teams 维修国际标准化组

Study Group on Maximum Interruption Service Time 最大服务中断时间研究组

**MISTR** Management of Items Subsequent to Repair 应修项目管理

Management of Items Subject to Repair 在修项目管理

**MISWG** Management Information System Working Group 管理信息系统工作组

**MIT** Maintenance Interface Test 维护测试接口

Management Information Tree 管理信息树

Massachusetts Institute of Technology 麻省理工学院(美国)

Master Instruction Tape 主说明磁带

Mating Interface Test 配接接口测试

Miscellaneous Indexing Tool 综合转位工具

Modular Intelligent Terminal 模块化智能终端

**MITA** Multilateral Interline Traffic Agreement (IATA)航空多边协定,多边联运协议

**MITE** Master Instrumentation Timing Equip-ment 主测量仪表定时设备

Multiple Input Terminal Equipment 多输入终端设备

**MITI** Ministry of International Trade and In-dustry 国际贸易和工业部

**MITO** Minimum Interval Takeoff 最小起飞时间间隔

Minimum-Interval Take-off 最短间隔(时间)起飞

**MITRE** Miniature Individual Transmitter-Re-ceiver Equipment 小型收发报机设备

**MITS** Mechanised Interactive Tool Surveil-lance 机械化交互式工具监视

Multimedia Interactive Telelearning System 多媒体交互远程学习系统

**MIU** Maintenance Interface Unit 维修接口组件

Malfunction Insertion Unit 故障输入组件

Message Injection Unit 信息引入组件

Multi-Interface Unit 多接口装置

Multimedia Information User 多媒体信息用户

Multistation Interface Unit 多站接口装置

**MIUS** Modular Integrated Utility Systems 模件化集成共用系统

**MIW** Maintenance Information Workstation （基地级）维修信息工作站

**MIX** Fleet Mix Program 机群混合程序

Mixer 混频器，混合器

Mixture 混合物

**MJB** Main Junction Box 主接线盒

**MJC** Maintenance Job Card 维修工作卡

**MJL** Mouila，Gabon（Airport Code）加蓬穆伊拉机场

**MJO** Maintenance Job Order 维修工作指令，维修工作规程

**MJP** Micro Java Processor 微型 Java 处理器

**MJRI** Master Job Reference Index 主职业参考索引

**MK** Maintenance Kit 维修工具包

Major Key 主键，主要关键

Mark 标记

Marker 信标台，标记，标志

Master Key 主密钥

**MKE** Milwaukee，WI，USA（Airport Code）美国威斯康星州密尔沃基机场

**MKSS** Microwave Keying Switch Station 微波键控开关站

**MKT** Market 市场

**MKTG** Marketing 市场学，推销

**MKZ** Malacca，Malaysia（Airport Code）马来西亚马六甲机场

**ML** Maintenance Laboratory 维修实验室

Maintenance Levels 维修等级

Maintenance Library 维修图书馆，维修程序库

Major Lobe 主波瓣

Manual Left 人工左

Manufacturing List 制造清单

Material Laboratory 材料实验室

Material List 材料清单

Maximum Load 最大负荷

Mean Level 平均电平，平均能级，平均水平

Mean Life 平均寿命

Measuring Loop 测量回路

Microwave Laboratory 微波实验室

Migration Level 进位等级，进位电平

Mold Line 标准型线

Monolithic Laser 单片激光器

Monomode Laser 单模激光器

Monopulse Lidar 单脉冲激光雷达

Multilasered Lidar 多激光器激光雷达

Multipurpose Loader 多用装载机

**ML（ml）** Milliliter 毫升

**MLA** Malta，Malta-Luga（Airport Code）马耳他，卢卡机场

Maneuver Limited Altitude 机动限制高度

Maneuver Load Alleviation 机动减载

Manufacturing License Agreement 生产授权协议

Microwave Link Analyzer 微波交连分析仪

Multi-Line Adapter 多路转接器

Multi-Spectral Linear Array 多光谱线性阵列

**MLAN** Multichannel Local Area Network 多通道局域网

**MLAP** MAC Level Access Protocol 媒体接入控制层接入协议

**MLAT** Multilateration 多点定位技术

**MLAV** Military Aviation 军事航空

**MLB** Melbourne，FL，USA（Airport Code）美国佛罗里达州墨尔本机场

Multi-Layer Board 多层（电路）板

**MLC** Maneuver Load Control 机动载荷控制

Mobile Local Circuit 移动通信本地电路

Mood Lighting Controller 环境照明控制器

Multi-Line Controller 多路控制器

**MLCB** Multilayer Circuit Board 多层（印刷）电路板

**MLCCM** Modular Life Cycle Cost Model 模块化寿命周期成本模型

**MLCCR** Modification Level Change Commitment Record 改装等级更改委托记录

**MLD** Master Layout Duplicate 主布局复本

Mean Level Detector 平均水平探测器

Mean Logistics Delay 平均后勤延迟

Middle Landing 中间着陆

Minimum Line of Detection 最低检测线

Mixed Layer Depth 混合层深度

Molded 铸造的

Multipurpose Logic District 多功能逻辑区

**MLDG** Molding 制模,铸造物

**MLDI** Meter List Display Interval 调度表显示间隔

**MLDT** Mean Logistics Delay Time 平均后勤延误时间

Mean Logistics Down Time 平均后勤不能工作时间

**MLE** Male, Maldives (Airport Code) 马尔代夫马累机场

Manned Lunar Exploration 载人登月探索

Maximum Likelihood Estimate 最大似然估计

**MLF** Mechanical Load Fixture 机械负载夹具

Media Language and Format 媒体语言和格式

Multilateral Force 多边力量,多国部队

**ML-FRL** Mode-Locked Fiber Ring Laser 锁模光纤环型激光器

**MLFT** Magnetic Leakage Flux Test 磁泄漏通量测试

**MLG** Main Landing Gear 主起落架

Malang, Indonesia (Airport Code) 印尼玛琅机场

**MLGD** Main Landing Gear Door 主起落架舱门

**MLGS** Microwave Landing Guidance System 微波着陆引导系统

**MLGWW** Main Landing Gear Wheel Well 主起落架轮舱

**MLI** Magnetic Level Indicator 磁油面指示器

Minimum Line of Interception 最小切入线,最近截获线

Multiprocessor Link Interface 多处理器链路接口

**MLIH** Magnetic Level Indicator Housing 磁性深度指示器座

**MLLR** Multi-rate Least-Loaded Routing 多速率最低负载选路

**MLLRP** Multi-rate Least-Loaded Routing with Packing 多速率最低负载分组选路

**MLN** Middle Level Network 中级网络

**MLO** Master Layout Original 原始主布局

Material Layout 器材布局

Metal Layout 金属配置,金属布局

Milos, Greece (Airport Code) 希腊米洛斯机场

**MLP** Machine Language Program 机器语言程序

Message Link Protocol 报文链路协议

Metal Lath and Plaster 金属板条与熟石膏

Mobile Launch Platform 移动式发射平台

Multi-Layer Protocol 多层协议

Multilink Procedure 多链路规程

**MLR** Mach of Long Range 远程巡航马赫数

Mobile Laser Radar 移动式激光雷达

Most-Loaded Routing 最多负载选路

**MLS** Maintenance Landing Sites 着陆现场维修

Metal Slitting 金属切入

Microwave Landing System 微波着陆系统

Miles City, MT, USA (Airport Code) 美国蒙大拿州迈尔斯城机场

Multimedia Learning Station 多媒体学习站

Multipoint Location System 多点定位系统

**MLST** Milestone 里程碑,重大事件

**MLT** Mean Low Tide 平均低潮

Mobile Land Target 地面活动目标

Multilink Trunking 多链路骨干技术

**MLTPL** Multiplane 多翼飞机

**MLTSP** Multispeed 多(级变)速的

**MLU** Miscellaneous Live Unit 综合居住单元

Monroe, LA, USA (Airport Code) 美国路易斯安那州门罗机场

**MLW** Main Leg Weight 主起落架负载重量

Maximum Certificated Landing Weight 最大批准着陆重量

Maximum Design Landing Weight 最大设计着陆重量

Maximum Landing Weight 最大着陆重量

Mean Low Water 平均低水位

Monrovia,Liberia-Roberts（Airport Code）利比里亚蒙罗维亚罗伯茨机场

**MLWA** Maximum Landing Weight Authorized 最大允许着陆重量

**MLX** Turkey-Malatya（Airport Code）土耳其—马拉提亚机场

**MM** Main Memory 主存储器

Main Multiplexer 主多路选择器

Maintenance Management 维护管理

Maintenance Manual 维修手册

Maintenance Module 维护模块

Man-Months 人力—月

Manual Dial Manual Answer 手工拨号/手工应答

Manufacturing Manual 制造手册

Marketing Management 市场管理

Mass Memory 大容量存储器

Master Mechanic 主机械师

Master Model 主模型（号）

Master Monitor 主监控器,主监控程序

Materials Management 物资管理

Memory Module 存储器模块

Meteorological Minima 最低天气（标准）

Middle Marker 中指点标,中台

Millimeter 毫米

Mission Management 飞行任务管理

Mixed Mode 混合方式

Mobility Management 移动性管理

Modelling Multimedia 建模多媒体

Multi Mode 多模

Multimedia 多媒体

**MM（S）** Master Menuing System 主菜单化系统

**mm²** Square Millimeter 平方毫米

**mm³** Cubic Millimeter 立方毫米

**MMA** Malmo,Sweden（Airport Code）瑞典马尔默机场

Manufacuring Material Area 制造器材区

Marketing Management Association 市场管理协会

Maximum Mean Accuracy 最大平均精度

Mobile Marketing Association 移动营销协会

Money Market Account 货币市场会计

Multi Mission Aircraft 多任务飞机

Multi-mission Maritime Aircraft 多任务海上飞机

Multiple Module Access 多模块存取

**MMAC** Multimedia Access Center 多媒体访问中心

**MMALS** Multi-Mode Approach and Landing System 多模式进近和着陆系统

**MMAP** Mission Maintenance Audio Panel 任务维修音频板

Mobility Management Application Protocol 移动性管理应用协议

**MMB** Main Maintenance Base 主维修基地

Maintenance Manual Bulletin 维护手册通告

**MMBF** Mean Mile between Failure 故障平均间隔英里

**MMBMA** Mean Mile between Maintenance Abortion 维修中断平均间隔英里

**MMBOMA** Mean Mile between Operational Maintenance Abortion 运行维修中断平均间隔英里

**MMC** Maintanance Management Consultant 维修管理顾问

Maintanance,Monitoring and Control 维护和监控

Maintenance Monitoring Console 维修监测台

Man-Machine Communication 人机通信

Man-Machine Controller 人机控制器

Mass Memory Card 大容量存储器卡

Material Management Code 器材管理代码

Maximum Material Condition 最大航材条件

Metal Matrix Composite 金属基体复合材料

Mission Management Computer 任务管理计

算机

Multi Media Card 多媒体卡

Multi Media Collaboration 多媒体协作

Multi Media Communication 多媒体通信

Multi Media Communicator 多媒体通信装置

Multi Media Controller 多媒体控制器

Multimedia Marketing Council 多媒体市场委员会

Multimedia Multiparty Conferencing 多媒体多方会议

**MMCA** Message-Mode Communication Adapter 报文方式通信适配器

**MMCC** Multi Media Conference Control 多媒体会议控制

**MMCD** Multi Media Compact Disc 多媒体光盘

**MMCF** Multi Media Communications Forum 多媒体通信论坛

**MMCIAC** Metal Matrix Composites Information Analysis Center 金属基体复合材料信息分析中心

**MMCM** Multi Media Control Manager 多媒体控制管理器

**MMCP** Multimedia Mail Content Protocol 多媒体邮件内容协议

**MMCRS** Manufacturing Material Cost Reporting System 制造材料成本报告系统

**MMCS** Multimedia Mobile Communication System 多媒体移动通信系统

**MMCX** Multimedia Communicaton Exchange 多媒体通信交换

**MMD** Manufacturing Management Department 制造管理部

Master Makeup and Display 主装配与显示

Master Monitor Display 主监控器显示

Material Management Division 器材管理部

Microwave Mixer Diode 微波混频二极管

Multimode Display 多模式显示

**MMDB** Mechanical Main Data Bank 机械主数据库

**MMDD** Management Multimedia Dynamic Data 多媒体动态数据管理

**MMDS** Multichannel Microwave Distribution System 多频道微波分配系统

Multichannel Multipoint Distribution Service 多信道多点分配服务

Multipoint Multichannel Distribution System 多点多信道分配系统

**MMDT** Mean Mission Duration Time 平均任务持续时间

**MME** Maintenance Manual Extracts 维护手册摘要

Multimedia E-mail System 多媒体电子邮件系统

Teesside, England, United Kingdom (Airport Code) 英国蒂赛德机场

**MMEL** Maintenance Master Equipment List 维护主设备清单

Master Minimum Equipment List 主最低设备清单

**MMELG** Master Minimum Equipment List Group 主最低设备清单组

**MMEM** Multimedia Electronic Mail 多媒体电子函件

**MMESE** Man-Machine-Environment System Engineering 人—机—环境系统工程

**MMF** Magnetomotive Force 磁动势

Manufacturer's Maintenance Factory 制造厂的维修厂

Manufacturer's Modification Form 制造厂改装表格

Multimode Fiber 多模光纤

**MMH** Maintenance Man-Hour 维修工时

Multi Media Hub 多媒体集线器

**MMH/FH** Maintenance Man Hours per Flight Hour 每飞行小时的维修工时

**MMH/MA** Mean Manhours per Maintenance Action 每维修作业的平均工时

**MMH/OH** Mean Man-Hours per Operating Hour 工作一小时所需要的平均维修工时

**MMH/UFH** Maintenance Manhours per Unit Flight Hour 维修工时/单位飞行小时

**mmHg** millimeters of mercury 毫米汞柱

**MMHR** Maintenance Manhour(s) 维修工时

**MMI** Major Model Improvement 主模型改进
Man-Machine Interaction 人机交互
Man-Machine Interface 人机界面, 人机接口
Manual Magnetic Indicator 人工磁指示器
Multi Model Interface 多方式接口
Multimedia Input 多媒体输入
Multi-Mode Interference 多模干扰

**MMIC** Monolithic Microwave Integrated Circuit 单片微波集成电路

**MMICS** Maintenance Management Information and Control System 维修管理信息与控制系统

**MMIO** Multimedia I/O 多媒体输入/输出

**MMIP** Man/Machine Interface Processor 人/机接口处理器

**MMIPS** Multiple Motor Integrated Propulsion System 多发整体式推进系统

**MMIRA** Multi-Mission Intermeshing-Rotor Aircraft 多种任务交叉式双旋翼航空器

**MMIS** Maintenance and Material Management Information System 维修与器材管理信息系统
Maintenance Management Information System 维修管理信息系统
Material Management Information System 器材管理信息系统
Multimedia Management Information System 多媒体管理信息系统

**MMK** Murmansk, Russia (Airport Code) 俄罗斯摩尔曼斯克机场

**MML** Man-Machine Language 人机语言
Master Measurement List 主测量清单
Multimedia Mechanism Layer 多媒体结构层
Multimedia Module Library 多媒体模块库

**MMLS** Material and Maintenance Logistics System 器材和维修物流系统
Military Microwave Landing System 军用微波着陆系统
Mobile Microwave Landing System 机动微波着陆系统

**MMM** Main Memory Module 主存储器模块
Maintenance and Material Management 维修和器材管理
Maintenance Management Manual 维修管理手册
Maintenance Manpower Module 维修人力模块
Materials Management Manual 材料管理手册
Millimicron(s) 毫微米
Multi Modem Manager 多调制解调器管理器
Multimedia Mail 多媒体邮件
Multimedia Multiplexer 多媒体复用器
Multiunit Network Management and Maintenance Message 多单元网络管理和维护消息

**MMMA** Multi-Mission Maritime Aircraft 多任务海上飞机

**MMMC** Minimum Monthly Maintenance Charge 最低月度维修费用

**MMMD** Multimedia Multi-Database 多媒体多数据库

**MMMS** Maintenance and Material Management System 维修和器材管理系统
Maintenance Message Monitoring System 维修信息监控系统
Multimedia Mail Messaging Service 多媒体邮件报文服务
Multimedia Mail Service 多媒体邮政业务
Multi-Mission Management System 多用途管理系统

**MMN** Microstrip Matching Network 微波传输带线匹配网络
Module Maintenance Node 模块维修节点
Multimedia Network 多媒体网

**MMNI** Message Memory Network Interface 消息存储器网络接口

**MMO** Mach Maximum Operating 最大飞行马赫数
Mach Maximum Operating Speed 最大使用速度马赫数
Main Meteorological Office 主气象站

Materials Management Organization 器材管理组织

Maximum Operating Limit Mach 最大飞行极限马赫数

Maximum Operating Mach 最大使用马赫数

Maximum Operating Mach Number 最大使用马赫数

Maximum Operating Speed 最大飞行速度

**MMOD** Micromodule 微模块

**MMOU** Multilateral Memorandum of Understanding 多边谅解备忘录

**MMP** Main Microprocessor 主微处理器

Maintenance Management Plan 维修管理计划

Maritime Mobile Phone 海上移动电话

Module Message Processor 模块消息处理器

Money Management Program 货币管理规划

Multi-Mission Platform 多任务平台

Multiplexed Message Processor 复用信息处理器

Multipurpose Midline Processor 多功能中线处理器

**MMPA** Magnetic Materials Producers Association 美国磁性材料制造商协会

**MMPC** Multimedia PC 多媒体个人计算机

**MMPF** Microgravity and Material Processing Facility 微重力与器材处理设施

**MMPM** Multimedia Picture Management 多媒体图像管理

**MMPS** MEECN Message Processing System 最低必要紧急通信网络信息处理系统

**MMR** Mach Meter Reading 马赫数表读数

Mach of Maximum Range 最大航程马赫数

Maintenance Management Review 维修管理评审

Minimum Monitoring Requirement 最低监视要求

Multi-Mode Radar 多模式雷达

Multi-Mode Receiver 多模式接收机

**MMRH/FH** Mean Maintenance and Repair Hours per Flight Hour 每飞行小时的平均维护和修理时间

**MMRPV** Multi Mission Remotely Piloted Vehicle 多任务遥控无人驾驶飞行器

**MMRS** Maintenance Message Reporting System 维护信息报告系统

**MMS** Magnetic Mine Sweeping 磁矿扫探

Main Machine System 主机系统

Main Memory System 主存储器系统

Main Meterological Station 气象总站

Maintenance Management System 维护管理系统

Man-Machine System 人机系统

Manpower Management System 人力管理系统

Manufacturing Management System 制造管理系统

Manufacturing Message Service 加工制造报文服务

Manufacturing Monitoring System 制造监控系统

Marine Meterological Society 海洋气象协会

Market and Marketing System 市场和销售系统

Marketing Management System 销售管理系统

Mass Memory System 大容量存储系统

Master of Management Studies 管理学硕士

Mast-Mounted Sight 桅杆式(旋翼主轴式)瞄准具

Material Management System 航材管理系统

Memory Management System 存储器管理系统

Meterological Measuring System 气象测量系统

Microfilm Management System 缩微胶片管理系统

Middleware Message Service 中间件消息业务

Minimum Material Strength 材料最低强度

Mission Management System 任务管理系统

Mobile Maintenance Squadron 机动维修中队

Model Management System 模型管理系统
Module Management System 模件管理系统
Multimedia Messaging Service 彩信服务
Multimedia Service 多媒体业务
Multi-Mission Simulator 多任务模拟机
Multi-Modular Storage 多模块存储器
Multiple Microprocessor System 多微处理器系统
Multiport Memory System 多端口存储器系统

**MMSA** Multi-Mission Surveillance Aircraft 多任务监视飞机

**MMSAF** Multimedia Services Affiliate Forum 多媒体业务会员论坛

**MMSE** Man-Machine System Engineering 人机系统工程(学)
Minimum Mean Square Error 最小均方误差
Multiuse Mission Support Equipment 多用途任务保障设备

**MMSER** Minimum Mean Square Error Restoration 最小均方误差复原

**MMSE-STSA** Minimum Mean-Square Error Short Time Spectral Amplitude 最小均方误差短时谱估计

**MMSR** Master Material Support Record 主器材保障记录

**MMSS** Marine Meterological Service System 海洋气象服务系统
Maritime Mobile Satellite Service 海事移动卫星服务

**MMT** Mean Maintenance Time 平均维修时间
Mobile Maintenance Team 流动维修分队
Mobile Measurement Technology 移动测量技术
Multimedia Terminal 多媒体终端

**MMTS** Mixing Manifold Temperature Sensor 混合气管温度传感器
Multimedia Transport System 多媒体传送系统

**MMU** Main Memory Unit 主存储主件,主存储器

Manned Maneuvering Unit 载人机动装置
Mass Memory Unit 大容量存储器
Memory Management Unit 存储器管理组件,内存管理单元
Multimedia Unit 多媒体单元

**MMUI** Multi Media User Interface 多媒体用户接口

**MMUX** Main Multiplexer 主多路调制器
Minimultiplexer 微型多路选择器

**MMV** Main Metering Valve 主计量活门
Monostable Multivibrator 单稳态多谐振荡器
Multi Media Viewer 多媒体观赏器

**MMVF** Multi-Media Video File 多媒体视频文件

**MMW** Mean Maximum Weight 平均最大重量
Millimeter Wave 毫米波
Multimedia World 多媒体世界
Multi-Mega Watt 百万瓦

**MMWR** Millimeter Wave Radar 毫米波雷达
Millimeter Wave Radio 毫米波无线电

**MMX** Malmo, Sweden (Airport Code) 瑞典马尔默机场
Multi-Media Extension 多媒体扩展

**MN** Magnetic North 磁北
Main 主要的
Meganewton 兆牛顿
Memory Normal 存储器正常
Menu 菜单
Microwave Network 微波网络
Midnight Flight 深夜航班
Minute 分
Mobile Node 移动节点
Model Numbers 型号数字
Multisensor Navigation 多传感器导航

**MNA** Microwave Noise Analyzer 微波噪声分析器
Multishare Network Architecture 多共享网络结构

**MNAC** Miniature Navigation Airborne Computer 小型机载导航计算机

**MNC** Mobile Network Code 移动网络代码

Multi National Company 多国公司

**MNCS** Master Network Control Station 主网络控制站

Multipoint Network Control System 多点网络控制系统

**MNE** Mach Number Never to Be Exceeded 最大马赫数

**MNET** Measuring Network 测量网络

**MNF** Multisystem Networking Facility 多系统连网设施

**MNFCR** Manufacture 制造(商)

**MNFD** Manifold 总管

**MNFRM** Main Frame 主帧,主机架

**MNF(S)T** Manifest 舱单

**MNG** Manage 管理,处理

Mongolia 蒙古

**MNGMT** Management 管理

**MNGR** Manager 经理

**MNGT** Management 管理

**MNH** Minimum Number of Hops 最小跳频数

**MNL** Manila, Philippines (Airport Code) 菲律宾马尼拉机场

Manual 手册,手工的

**MNLOPR** Manually Operated 人工操作的

**MNM** Minimum 最小的,最低的,最低,最小

**MNMCH** Minimum Charge 最低收费

**MNO** Mobile Network Operator 移动网络运营商

**MNOS** Metal-Nitride-Oxide Semiconductor 金属氮化物—氧化物—硅(半导体)

**M-NOTE** Manufacturing Note 制造说明

**MNP** Microcom Network Protocol 微通网协议

Mobile Telephone Number Portability 移动电话号码可携

**MNPS** Minimum Navigation Performance Specification 最低导航性能规范

**MNPSA** Minimum Navigation Performance Specification Airspace 最低导航性能规范空域

**MNR** Minimum Noise Route/Routing 最低噪声航路

**MNRU** Modulated Noise Reference Unit 调制噪声参考单位

**MNS** Market Navigation System 市场导航系统

Master Navigation System 主导航系统

**MNSC** Main-Network Switching Centre 主网络交换中心

**MNT** Mach Number Technique 马赫数技术

Monitor 监视,监控,监控器

**MNTD** Mounted 安装的

**MNTG** Mounting 安装

**MNTN** Maintain 维护,保持

Maintenance 维护

**MNTR** Monitor 监视,监控,监控器

**MNU** Maintenance Unit 维护单元

**MNVR** Maneuver 机动

**MNY** Many 许多

Money 货币,钱

**MO** Macao 澳门

Machine Operation 机械运行

Magneto-Optical 磁—光的

Mail Order 邮购,邮政汇票

Maintenance and Operation 维修与使用

Maintenance Memo 维修备忘录

Maintenance Operation 维修工作

Managed Object 被管对象

Manpower and Organization 人力与组织

Manual Operation 手动操作

Manual Output 手动输出

Manufacturing Order 生产指令,制造单,生产规程

Market Order 市场订单

Marketing Objectives 销售目标

Marketing Organization 销售组织

Master Oscillator 主振荡器

Material Order 器材订单

Memory Object 存储对象

Method of Operation 操作方法

Mode 方式

Moment 片刻,此刻

Money Order 汇票

Month 月

Morse 莫尔斯(码)

Motor Operated 发动机驱动的

Multitrace Oscilloscope 多线示波器

**MO&DSD** Mission Operations and Data System Directorate 飞行操作和数据系统管理处(美国)

**MOA** Maintenance Organization Approval 维修机构批准书

Maintenance, Operation, Administration 维护、运行和管理

Matrix Output Amplifier 矩阵输出放大器

Memorandum of Agreement 协议备忘录

Military Operating Area 军事活动区

**MOAA** Ministry of Aeronautics and Astronautics (中国)航空航天部

**MOAR** Measurable One-way Attenuation Range 可测单向衰减范围

**MOAS** Multimedia Office Automatic System 多媒体办公自动化系统

**MOB** Main Operating Base 主操作基地

Management Oversight Board 管理监督委员会

Meal on Board 机上供餐

Mobile, AL, USA (Airport Code) 美国阿拉巴马州莫比尔机场

**MOBIDIC** Mobile Digital Computer 移动式数据计算机

**MOBP** Managing Organizations by Projects 按项目管理组织

**MOC** Magnetic-Optic Converter 磁光转换器

Maintenance Operational Check 维修作业检查

Maintenance Operations Center 维护运行中心

Master Operational Controller 主操作控制器

Means of Compliance 符合性方法

Memorandum of Cooperation 合作备忘录

Meridional Overturning Circulation 经向翻转环流

Minimum Obstacle Clearance 最低越障余度

Minimum Operational Characteristics 最低运行性能

Mission Operation Center 飞行运行中心

**MOCA** Minimum Obstruction Clearance Altitude 最低净空高度,最低越障高度

**MOCAN** Motor Can 电动机外罩

**MOCAT** Ministry of Civil Aviation and Tourism 民用航空和旅游部

**MOCC** Meteosat Operations Control Center 气象卫星运行控制中心(西欧)

Mission Operations Command and Control 任务运行指令与控制

Mission Operations Control Center 任务运行控制中心

Multisatellite Operations Control Center 多卫星操作控制中心

**MOCHA** Method of Characteristics 特性法

**MOCP** Mission Operation Computer Program 飞行任务运行计算机程序

**MOCR** Mission Operation Control Room 飞行任务运行控制室

**MOCS** Master Operation Control System 主操作控制系统,主运行控制系统

**MOD** Magnetic Order Document 磁指令文件

Magneto-Optic Disc 磁光碟片

Ministry of Defense 国防部

Mission Operations Directorate 飞行任务运行管理处

Mode 方式

Model 模型,样式

Moderate 中度的,适度的

Modern 现代化的

Modesto, CA, USA (Airport Code) 美国加利福尼亚州莫德斯托机场

Modify, Modification 更改,改装

Modulator 调制器

Module 组件,单元

Movies on Demand 电影点播

Music on Demand 音乐点播

**MOD/V** Modulating Valve 调节活门，随动活门

**MODACR** Modified Advanced Capability Radar 改进型先进能力雷达

**MODAL** Microwave Optical Duplex Antenna Link 微波光双工天线链路

**MODAS** Maintenance and Operational Data Access System 维修和使用数据存取系统

**MODD** Modification Drawing 改装图纸

**Mode A** ATC Mode A ( Range and Bearing ) ATC A 模式 ( 距离和方向 )

**Mode C** ATC Mode C ( Range, Bearing and Altitude ) ATC C 模式 ( 距离、方向和高度 )

**Mode S** ATC Mode S ( Range, Bearing, Altitude and Unique Identifier ) ATC S 模式 ( 距离、航向、高度和唯一识别码 )

**MODEM** Modulator-Demodulator 调制解调器

**MODES** Mode Select Beacon System 方式选择信标系统

Model Estimating 型号估算

**MODI** Module Interface 模块接口

**MODID** Module Identity 模块识别

**MODIF** Modify 更改，改装

Modular Optical Digital Interface 模块化光数字接口

**MODILS** Modular Instrument Landing System 模块化仪表着陆系统

Modular Microwave Instrument Landing System 模块化微波着陆系统

**MODIS** Moderate-resolution Imaging Spectrometer 中等分辨率成像光谱仪

**MODLTR** Modulator 调制器

**MODN** Modern 现代化的

**MODR** Monodetail Drawing 单一详图

**MODS** Modification Sheet 更改单

**MODU** Maintenance Operator Display Unit 维修操作员显示组件

Mobile Offshore Drilling Unit 可动式离岸钻探平台

**MOE** Maintenance of Equipment 设备维修

Measure of Effectiveness 效率测量

Maintenance Organization Exposition 维修机构手册

**MOEDM** Maintenance and Overhaul Equipment Data Management 维修与翻修设备数据管理

**MOEMS** Micro-Optical-Electro-Mechanical System 微光机电系统

**MOG** Maximum on Ground 大型场地

**MOGAS** Motor Gasoline 汽车汽油

**MOH** Major Overhaul 大翻修

**MOI** Maintenance Operating Instructions 维修操作说明

Moment of Inertia 惯性力矩

**MOIE** Mission Oriented Investigation and Experimentation 面向任务的调查和实验

**MOL** Manufacturing Order Location 制造规范位置，制造指令位置

More or Less 增减

**MOLELEX** Molecular Electronics 分子电子学

**MOLS** Multiple Object Location System 多目标定位系统

**MOLWT** Molecular Weight 分子量

**MOM** Magneto-Optical Modulation 磁光调制

Maintenance Operation Modules 维护操作模式

Maintenance Organization Manual 维护机构手册

Manager of Operations Models 运营模式经理

Mass Optical Memory 大容量光存储器

Maximum Operating Mach 最大使用马赫数

Message Oriented Media 面向消息的媒体

Message Oriented Middleware 面向消息中间件

Methods of Moments 力矩方法

Middle of the Month 月中

Mission Optimization Management 飞行任务最佳管理

Moment 片刻，此刻

Momentary 瞬时的

Wait a Moment 稍等

**MOML**   Media Object Markup Language 媒体目标标记语言

**MOMS**   Map, Operation and Maintenance System 地图,运行和维护系统

Meteorological Optical Measuring System 气象光学测量系统

Modular Order Management System 模块化的订单管理系统

Multiple Orbit-Multiple Satellite 多轨道,多卫星

**MON**   Above Mountains 在山区上方

Management Overlay Networks 管理覆盖网络

Monday 星期一

Monitor 监视器,监控器

Monitored 监控的

Multiservice Optical Networking 多业务光纤网络

**MONET**   Mobile Network 移动网

Multi-wavelength Optical Network 多波长光网络

**MONG**   Monitoring 监控

**MONO**   Monoaural 单声道的

Monophonic 单声道的,非立体声的

**MONS**   Monmouthshire 蒙茅斯郡

Museum of Natural Science 自然科学博物馆

**MONT**   Montana 蒙大拿州

**MOO**   Major Operating Organization 主要运营组织

**MOOS**   Maintenance out of Service 停业维护

**MOP**   Macao Pataca 澳门元

Main Oil Pressure 主滑油压力

Main Oil Pump 主滑油泵

Maintenance and Operation Processing 维护和运行处理

Maintenance Operations Program 维修运行程序

Major Outside Production 主要外部生产

Maximum Oil Pressure 最大滑油压力

**MOPA**   Master Oscillator Power Amplifier 主控振荡器功率放大器

**MOPAR**   Master Oscillator Power Amplifier Radar 主振功率放大器雷达

**MOPP**   Mission Operational Protective Posture 运行任务操作保护性姿态

**MOPR**   Minimum Operational Performance Requirements 最低运行性能要求

Mop Rack 拖把架

**MOPS**   Merchandise Order(ing) Processing System 商品订货加工系统

Millions of Operations per Second 百万次运算/秒

Minimum Operational Performance Standards 最低运行性能标准

**MOPTAR**   Multiple Object Phase Tracking and Ranging 多目标相位跟踪与定位

**MOQ**   Minimum Order Quantity 最低订货数量

**MOR**   Maintenance Operation Report 维修工作报告

Maintenance Overhaul and Repair 维修大修与修理

Mandatory Occurrence Report 强制性报告

Manufacturing and Operation Record 制造与使用记录

Memory Output Register 存储器输出寄存器

Meterological Optical Range 气象直视视距,气象光学视距

Meterology Optical Range 气象光学视程

Modulus of Rupture 断裂模数

Moment of Rupture 断裂力矩

Morocco 摩洛哥

**MORA**   Minimum off Route Altitude 最低偏航高度

**MORASS**   Modern Ramjet System Synthesis 现代冲压发动机系统综合

**MORE**   Multioptical Reconnaissance Equipment 多光学侦察设备

**MORN**   Morning 早上,上午

**MORT**   Management Oversight Review Technique 管理监督审查技术

Management Oversight Risk Tree 管理疏忽风险树

Master Operational Recording Tape 主操作记录磁带

Morse Taper 莫氏锥度

**MORTG** Mortgage 抵押

**MORV** Mixer Overpressure Relief Valve 搅拌器超压安全活门

**MOS** Maintenance Operation Subsystem 维护运行子系统

Management Operating System 管理运行系统

Manned Orbital Station 载人轨道站

Manual of Standards 标准手册

Manufacturing Operation Sheet 制造操作表

Maritime Observation Satellite 海洋观测卫星

Master Operating System 主操作系统

Mean Opinion Score 平均意见评分

Metal Oxide Semiconductor 金属—氧化物—半导体

Metal Oxide-Silicon（Semiconductor）金属—氧化物—硅（半导体）

Mission Operations System 任务操作系统

Modular Optoelectronic Scanner 模块式光电扫描仪

Monthly Operation Summary 每月工作摘要，每月作业摘要

Months 月

Multimedia Operation Software 多媒体操作软件

**MOSE** Mobile Operational Support Equipment 活动式操作保障设备

**MOSFET** Metal-Oxide Semiconductor Field-Effect Transistor 金属氧化物半导体场效应晶体管

**MOSIC** Metal-Oxide-Semiconductor Integrated Circuit 金属—氧化物半导体集成电路

**MOSOGS** Molecular-Sieve Oxygen Generating System 分子过滤氧气发生器

**MOSP** Multi-Optical Stabilized Payload 多光学传感装置稳定有效载荷

**MOSPF** Multicast Open Shortest Path First 多播开放式最短路径优先

**MOSS** Margin of Safety Summary Analysis 安全系数总结分析

Material off Site Support 现场外器材支援

Minimum Operating Security Standards 最低运行安全标准

Mission Operations Support Services 飞行任务保障服务

**MOST** Metal Oxide Semiconductor Transistor 金属氧化物半导体晶体管

**MOT** Maximum Operating Time 最大运转时间，最大操作时间

Ministry of Transport（Canada）运输部（加拿大）

Minot, ND, USA（Airport Code）美国北达科他州迈诺特机场

Motion 移动，（机械的）装置，运转

Motor 马达，电动机

Motorized 马达驱动的

**MOTA** Modernization of Tactical Airlift 战术飞机现代化

**MOTBF** Mean Operating Time between Failure 失效间隔平均工作时间

**MOTD** Message of the Day 当日消息

**MOTIS** Message-Oriented Text Interexchange System 面向报文的文本交换系统

**MOTNE** Meteorogical Operational Telecommunication Network Europe 欧洲运行气象电信网

**MOTNEG** MOTNE Regional Planning Group 欧洲运行气象电信网地区规划组

**MOTR** Multiple Object Tracking Radar 多目标跟踪雷达

**MOTV** Manned Orbital Transfer Vehicle 载人轨道转换载具

**MOU** Memorandum of Understanding 谅解备忘录

Mountain Village, Alaska（Airport Code）美国芒廷瓦利机场

**MOV** Manned Orbiting Vehicle 载人轨道飞

行器

Mechanically-Operated Valve 机械操纵活门

Motor Operated Valve 马达操纵活门

**MOVG** Moving 移动的,运动的

**MOVLAS** Manually Operated Visual Landing Aid System 人工操作目视着陆辅助系统

**MOVMT** Movement 运动,动态

**MOW** Moscow, Russia（Airport Code）俄罗斯莫斯科机场

**MOWP** Method of Working Plan 工作计划方案

**MOX** Mixed Oxide Fuel 混合氧化物燃料

**MOY** Money 货币,钱

**MP** Machine Planning 机械加工计划

Machined Parts 加工零件

Magnetic Particle 磁性粒子

Main Processor 主处理器

Maintainability Plan 可维修性计划

Maintenance Panel 维护专家,维护控制板

Maintenance Period 维护周期

Maintenance Plan 维修计划

Maintenance Platform 维修平台

Maintenance Point 维修点,维修站

Maintenance Practices 维护措施

Maintenance Prevention 预防性维修

Maintenance Procedure 维修程序,维修计划

Maintenance Processor 维护处理器

Maintenance Program 维修方案

Man Power 人力

Management Point 管理点

Management Process 管理过程

Maneuver Point 机动（飞行）点

Manufacturing Plan 制造计划

Market Price 市场价格

Marketing Policy 销售政策

Mass Property 质量特性

Material Processing 材料处理

Medium Pressure 中等压力

Melting Point 熔点

Memory Point 存储点

Memory Protection 存储器保护

Message Passing 信息传递

Message Priority 消息优先级

Meteorological Plotter 气象图描绘仪

Meter Panel 仪表板

Microprocessor 微处理机

Middle Plug 中部插头

Midpoint 中间点,航路检查点

Misrouting Probability 错接概率

Mission Planning 飞行任务计划

Mission Processor 飞行任务处理器

Modification Proposal 改装提案

Module Processor 模块处理器

Multibeam Propagation 多波束传播

Multipath Propagation 多径传播

Multiphase 多相

Multiplier Phototube 多路调制器光电管

Multipoint Processor 多点处理器

Multiprocessor 多重处理器

Multiprogram Control 多程序控制

Multipulse 多脉冲

Multipurpose 通用的

**MP&R** Manufacturing Planning and Release 制造计划与发放

Material Planning and Release 器材计划与发放

**MP（M&P）** Materials and Processing 材料与加工

**MPA** Maintenance Program Authorization 维护大纲授权

Management Problems Analysis 管理问题分析

Maneuver Propulsion Assembly 机动飞行推进组件

Man-Powered Aircraft 人力飞机

Maritime Patrol Aircraft 海上巡逻机

Mass Properties Analysis 质量特性分析

Master of Public Administration 公共管理硕士

Master Power Amplifier 主功率放大器

Maximum Permissible Amount 最大容许量

Maximum Power Assurance 最大功率保证

Maximum Pulse Amplitude 最大脉冲幅度

Megapascal 兆帕斯卡

Microwave Parametric Amplifier 微波参量放大器

Microwave Power Amplifier 微波功率放大器

Million Passengers（per）Annum 每年百万旅客

Modular Performance Analysis 单元体性能分析

Multichannel Protocol Analyser 多通道协议分析仪

Multi-Pilot Aeroplane 多机组飞行

Multiple Payload Adapter 多个有效载荷连接器

**MPAA** Minority Pilot Association Academy（美国）少数民族飞行员联合学院

**MPAC** Multipurpose Applications Console 多用途控制台

**MPAcc** Master of Professional Accounting 专业会计硕士

**MPAG** Man-Powerd Aircraft Group 有人驾驶的机群

**MPAR** Modified Precision Approach Radar 改进型精密进近雷达

Multifunction Phased Array Radar 多功能相控阵雷达

Multi-Purpose Airport Radar 多用途机场雷达

**MPAYL** Maximum Payload 最大有效载荷

**MPB** Maintenance Parts Breakdown 维修部件损坏

Material Price Book 器材价格手册

Micro Processor Bipolar 双向微处理器

Micro Processor Board 微处理机插板

Miniature Precision Bearings 微型精密轴承

**MPC** Manufacturing Planning Change 制造计划更改

Master Power Controller 主电源控制器，动力控制器

Master Program Chart 主程序图

Material Program Code 器材程序代码，器材计划代码

Maximum Productive Capability 最大生产能力

Message Passing Coprocessor 信息传送协处理器

Meteorological Prediction Center 气象预测中心

Minimal Planning Chart 最短时间计划图

Minimum Performance Criterion 最低性能指标

Motion Picture Control Panel 动画控制板

Multiple Payload Carrier 多个有效载荷运输器（美国）

Multimedia Personal Computer 多媒体个人计算机

Multimedia Product Council 多媒体产品协会

Multiple Purpose Communications 多用途通信

Multi-Processor Controller 多处理器控制器

Multi-Purpose Computer 多用途计算机

Multi-Purpose Console 多用途操纵台

**MPCD** Multi-Purpose Colour Display 多用途彩色显示器，多功能彩色显示器

**MP-CDL** Multi-Platform-Common Data Link 多平台通用数据链

**MPCI** Mobile Protocol Capability Indicator 移动协议能力指示器

**MPCMP** Mass Properties Control and Management Plan 质量特性控制与管理计划

**MPCOS** Multimedia Personal Computer Operating System 多媒体个人计算机操作系统

**MPCP** Mission Planning Computer Program 飞行任务计划计算机程序

**MPCS** Maintenance/Power Control System 维修/电源控制系统

Major Program Cost Summary 主要程序成本总结

**MPD** Magnetoplasma Dynamics 磁等离子体动力学

Maintenance Planning Data 维护计划数据

Maintenance Planning Document 维修计划文件

Maintenance Program Document 维修大纲文件

Maximum Phase Deviation 最大相位偏移

Metal Particle Detector 金属微粒检测仪

Microwave Power Device 微波功率器件

Mode Power Distribution 模功率分布

Multiple Detection 多项检测

Multiple Purpose Document 多用途运输凭证

Multi-Purpose Display 多用途显示器

**MPDB** Main Power Distribution Board 主电能分配板

Main Power Distribution Box 主电能分配盒

Manpower Distribution Board 人力分配委员会

Master Parameter Data Base 主参量数据库

**MPDM** Maintenance Planning Data Manual 维修计划数据手册

**MPDR** Mobile Pulse Doppler Radars 机动脉冲多普勒雷达

Mono-Pulse Doppler Radar 单脉冲多普勒雷达

**MPDS** Maintenance Planning Data Support 维护计划数据支援

Message Processing and Distribution System 信息处理与分发系统

Message Processing Data System 信息处理数据系统

**MPDU** Media Protocol Data Unit 媒体协议数据单元

Message Protocol Data Unit 消息协议数据单元

**MPE** Manufacturing Process Engineer 制造工艺工程师

Maximum Permissible Error 最大容许误差

Maximum Permissible Exposure 最大允许曝光量

Maximum Probability Estimator 最大概率估值器

Mean Probable Error 平均概率误差

Memory Parity Error 存储器奇偶误差

Memory Point Enable 存储点启动

Message Passing Environment 信息处理环境

Mission Planning Element 任务计划要素

Mission Processing Equipment 任务处理设备

Multimedia Processing Equipment 多媒体处理设备

Multiprocessing Executive 多重处理执行(程序)

**MPEG** Motion Picture Experts Group 运动图像专家组(编码标准)

**MPF** Maintenance Priority Factor 维修优先因素

Message Processing Facility 信息处理设施

Mobile Packet Processing Function 移动分组处理功能

**MPFD** Magnetic Particle Flaw Detection 磁粉探伤

Magnetic Particle Flaw Detector 磁粉探伤仪

**MPFI** Multi Port Fuel Injection 多点燃油喷射

**MPFS** Microwave Position Fixing System 微波定位系统

**MPG** Matched Power Gain 匹配功率增益

Microwave Pulse Generator 微波脉冲产生器

Miles per Gallon 英里/加仑

**MPGA** Mask Programmable Gate Array 掩模可编程门阵列

**MPH** Maintenance Parts Handbook 维修零件手册

Miles per Hour 英里/小时(1 英里 = 1.609千米 = 5280 英尺)

Multi-Pilot Helicopter 多机组直升机

**MPI** Magnetic Particle Inspection 磁粉探伤检查

Manufacturing Planning Input 制造计划输入

Master Pin Index 主销钉索引

Material Process Instruction 材料处理说明

Message Passing Interface 消息传递接口

Multi Purpose Interface 多用途接口

Multi-Path Interference 多径干扰

**MPIC**   Message Processing Interrupt Count 信息处理中断计数

**MPICC**   Multimedia Personal Information Communication Center 多媒体个人信息通信中心

**MPIRI**   Master Pin Index for Release Information 信息发布主销钉索引

**MPIS**   Multipurpose Information System 多用途信息系统

**MPJ**   Mono-Pulse Jammer 单脉冲干扰机

**MPK**   Maintenance Parts Kit 维修零件器材包

**MPL**   Mail Airplane 邮政飞机

Maintenance Parts List 维护零件清单

Manufacturing Parts List 制造零件清单

Master Parts List 主要零件表

Master Provisioning List 主要装备表

Materials Processing Laboratory 材料加工实验室

Maximum Payload 最大商载

Maximum Permitted Life 最大允许寿命

Message Processing Language 信息处理语言

Mid-Pacific Landing 中太平洋降落

Modular Part Library 模块化部件库

Montpellier, France（Airport Code）法国蒙彼利埃机场

Multi-crew Pilot License 多机组飞行员执照

Multiple Parallel Loop 多平行环路

Multiple-Pulsed Laser 多脉冲激光器

Multiprogramming Level 多程序设计级

**MPLE**   Multipurpose Long-Endurance（Aircraft）多用途长航时（飞机）

**MPLM**   Multi-Purpose Logistics Module 多用途后勤舱

Multiple Polarization Modulation 多偏振（极化）调制

**MPLR**   Manufacturing Problem Liaison Request 制造问题协调申请

**MPLS**   Multi-Protocol Label Switching 多协议标记交换，多协议标签交换

**MPM**   Main Propulsion Motor 主推进发动机

Maintenance and Peripherals Module 维护和外围设备模块

Maintenance Planning Manual 维修计划手册

Maintenance Program Management 维修计划管理

Major Program Memorandum 主要计划备忘录

Management of Products and Materials 产品和材料管理

Maneuver Propulsion Motor 机动推进发动机

Manufacture Procedure Manual 制造程序手册

Manufacturing Process Management 制造流程管理

Maputo, Mozambique（Airport Code）莫桑比克马普托机场

Marketing Performance Management 营销绩效管理

Maximum Permitted Mileage 最大允许里程

Message Passing Model 报文传送模型

Message Processing Module 消息处理模块

Microwave Power Meter 微波功率表

Modern Project Management 现代项目管理

Multi-Project Management 多项目管理

Multi-Processor Mode 多处理机方式

**MPMFC**   Maneuver Propulsion Motor Firing Circuit 机动推进发动机点火电路

**MPMFCA**   Maneuver Propulsion Motor Firing Circuit Assembly 机动推进发动机点火电路组件

**MPMP**   Microwave Point to Multipoint 微波一点至多点

**MPMT**   Mean Preventive Maintenance Time 平均预防性维修时间

**MPN**   Manufacturer's Part Number 制造商的零件号码

Manufacturing Part Number 制造件号码

Maximum Possible Number 最大可能数

Mean Probable Number 平均可能数

**MPNA**   Multiport Network Adapter 多端口网络适配器

**M-PNVS** Modernized-Pilot Night Vision Sensor 现代化飞行员夜视传感器

**MPO** Material Procurement Only 仅器材订购
Maximum Power Output 输出最大功率

**MPOL** Multi Problem Oriented Language 面向多种问题的语言

**MPOS** Maintenance Parts Ordering System 维修零件订购系统

**MPP** Maintainability Program Plan 可维修方案计划
Maintenance Program Proposal 维修大纲建议（书）
Massively Parallel Processor 大规模并行处理器
Master Power Panel 主电源板
Master Production Planning 主生产计划
Master Program Plan 主程序计划
Material Processing Platform 材料加工平台
Maximum Power Point 最大功率点
Message Passing Processing 报文传送处理
Message Processing Program 报文处理程序，信息处理程序
Most Probable Position 最可能位置
Motion Picture Projector 电影放映机

**MPPE** Management Performance and Planning Evaluation 管理绩效与规划评估

**MPPF** Multipoint Protocol Polling Function 多点协议查询功能

**MPPL** Multi-Purpose Processing Language 多用途程序设计语言

**MPPP** Multilink Point-to-Point Protocol 多链路点对点协议

**MPPU** Motion Picture Phonographic Unit 动画放音组件
Multi-Purpose Power Unit 多用途动力组件

**MPR** Maintainability Program Requirements 维修方案要求
Manufacturer's Problem Report 制造商的问题报告
Material Pickup Request 器材提取申请
Material Planning Requirements 器材计划要求

Medium-Power Radar 中功率雷达
Message Processing Region 信息处理区
Microwave Pulse Radar 微波脉冲雷达
Monthly Program Review 月度程序审查
Multiple Protocol Router 多协议路由器
Multipurpose Register 多功能记录器/计数器

**MPRB** Maintenance Program Review Board 维修大纲审查委员会

**MPS** Maintenance Page Snapshot 维护页快照
Maintenance Processor Subsystem 维护处理机分系统
Maintenance Programs Subcommittee 维修大纲分委员会
Manufacturing Priority Schedule 制造优先计划
Master Production Schedule 主生产计划
Maximum Performance Specifications 最大性能规范
Message Processing System 信息处理系统
Meters per Second 米/秒
Micro Processor System 微处理器系统
Microwave Phase Shifter 微波移相器
Miles per Second 英里/秒（1 英里 = 1. 609 千米）
Minimum Performance（Property）Standard 最低能标准
Minimum Performance Specifications 最低性能规范
Mission Planning Systems 任务计划系统
Mobile Phone Service 移动电话业务
Model Processing System 模型处理系统
Mono Pulse System 单脉冲系统
Multiplex Power Supply 多路电源
Multipriority System 多优先级系统

**MPSC** Multiprotocol Serial Controller 多协议串行控制器

**MPSD** Master Program Schedule Document 主程序计划文件

**MPSDF** Micro Processor Software Development Facility 微处理器软件开发设备

**MPSK** Minimum Phase Shift Keying 最小相移键控

Multiple Phase Shift Keying 多相移键控

**MPSOV** Minimum Pressure and Shut-off Valve 最小压力和关断活门

**MPSP** Mission Performance Success Probability 任务性能成功概率

**MPSPU** Multipurpose Small Power Unit 多功能小功率组件

**MPSR** Multipath Self Routing 多通路自选路由

**MPSSES** Multiple Protective Structural Support Equipment Segment 多重保护结构支援设备部分

**MPT** Magnetic Particle Testing 磁粉检测

Main Propulsion Test 主推力测试

Male Pipe Thread 管件阳螺纹

Memory Point Terminate 终端存储点

Microprocessing Programmable Terminal 微处理机可编程序终端

Microwave Power Transmission 微波功率传输

Multiple Personality Table 多重个性表

**MPTN** Multi-Protocol Transport Network 多协议传输网络

**MPTO** Maximum Performance Takeoff 最大性能起飞(特指直升机)

**MPTR** Mobile Position Tracking Radar 移动位置跟踪雷达

**MPTS** Microwave Power Transmission System 微波功率传输系统

**MPTT** Maintenance Part Task Trainer 维修部分作业训练员

Master Protocol Test Tool 主协议测试工具

Mechanical Part Task Trainers 机械部分作业训练员

**MPU** Magnetic Pickup Unit 磁传感器组件

Main Processor Unit 主处理器单元

Main Propulsion Unit 主推进装置

Message Processing Unit 信息处理组件

Microprocessor Unit 微处理器组件

Miniature Portable Unit 小型便携式装置

Miniature Power Unit 小型电源,小型动力装置

Modem Processor Unit 调制解调器处理器组件

Multimedia Processing Unit 多媒体处理单元

**MPV** Multi-Position Valve 多位置活门

**MPVCS** Multipoint Video Conferencing System 多点视频会议系统

**MPW** Maximum Permitted Weight 最大允许重量

Memory Point Wind 存储器点绕阻

**MPX** Multiplex Operation 多重运行

**MQ** Message Queue 消息队列

**MQA** Manufacture Quality Assurance 制造质量保证

**MQL** Matrix Query Language 矩阵查询语言

Maximum Queue Length 最大排队长度

Minimal Quantity Lubrication 最低数量的润滑

**MQM** Measurement Quality Monitoring 测量质量监控

**MQPB** Minimum Quantity Price Break 最低批量可接受价格

**MQT** Marquette, MI, USA (Airport Code) 美国密执安州马凯特机场

Model Qualification Test 模型鉴定试验

**MR** Main Rotor 主转子

Maintainability Review 维修能力审查

Maintenance Ration 维修工时率

Maintenance Recorder 维修记录器

Maintenance Report 维护报告

Maintenance Requirement 维修要求

Maintenance Review 维修检查

Major Repair 大修

Maneuver Radius 机动(飞行)半径

Manufacturing Record 制造记录

Manufacturing Research 制造研究

Master Record 主记录

Master Revision 主更改

Material Recycling 器材回收

Material Request 器材申请

Material Requirement 器材要求

Material Requisite 器材需求

Material Research 器材研究

Material Reserve 器材储备

Material Review 器材审查

Medium Range 中距离

Memorandum for Record 记录备忘录

Memorandum Report 备忘录报告

Memory Register 内存缓存器

Message Repeat 信息重复

Microminiature Relay 微型继电器

Microwave Radar 微波雷达

Microwave Radiometer 微波辐射计

Microwave Reflectometer 微波反射计

Milliradian 毫弧度

Mishap Report 事故报告

Mission Reliability 飞行任务可靠性

Mixture Ratio 混合比

Modem Ready 调制解调器就绪

Modification Request 改装申请

Modification Requirement 改装要求

Modification Revision 改装修正

Modulation Rate 调制速率

Monopulse Radar 单脉冲雷达

Multicast Repeater 多播复用器

**MR&D** Manufacturing Research and Development 制造研究与开发

**MR&O** Marketing Requirement and Objective 市场需求和目标

**MR/WH** Magnetic Read/Write Head 读写磁头

**MRA** Major Replaceable Assembly 主要可换件

Master Retailers Association 大零售商协会

Material Records Applications 器材记录应用

Material Review Action 器材审查措施

Material Review Authorization 器材审查授权

Materials Review Area 材料检验区

Maximum Relight Altitude 重新点火最大高度

Maximum Repair Allowance 修理最大容差

Minimum Reception Altitude 最低接收高度

Mission Risk Assessment 任务风险评估

**MRAC** Model-Referencing Adaptive Control 模型参考自适应控制

**MRAPCON** Mobile Radar Approach Control 移动雷达进近控制

**MRAS** Material Records Application System 器材记录应用系统

Model Reference Adaptive System 模型基准自适应系统

Multiframe Remote Alarm Signal 多帧远程告警信号

**MRASS** Modular Radar Analysis Software System 模块化雷达分析软件系统

**MRAT** Midrange Applied Technology 中距离应用技术

**MRB** Main Rotor Blade 主旋翼桨叶

Maintenance Report Bulletin （飞机）维修报告通报

Maintenance Review Board 维修评审委员会

Management Review Board 管理评审委员会

Manufacturing Record Book 制造记录本

Materials Review Board 材料评审委员会

Mission Requirements Board 任务需求委员会

Modification Review Board 改装评审委员会

**MRBD** Maintenance Review Board Document 维修评审委员会文件

**MRBF** Mean Rounds between Failures 平均失效周期

**MRBPB** Maintenance Review Board Policy Board 维修评审委员会政策委员会

**MRBR** Maintenance Review Board Report 维修审查委员会报告

**MRBRP** Maintenance Review Board Report Proposal 维修评审委员会报告建议书

**MRC** Machine Readable Code 机器可读代码

Maintenance and Repair Cycle 维修周期

Maintenance Requirement Card 维修要求卡

Management Research Center 管理研究中心

Manufacturing Records Control 生产记录控制

Manufacturing Reliability Center 生产可靠性中心

Manufacturing Reliability Control 生产可靠性控制

Marine Resource Conservation 海洋资源保护

Material Redistribution Center 器材再分配中心

Maximum Range Cruise 最大航程巡航

Maximum Rated Current 最大额定电流

Maximum Ratio Combining 最大比联合

Maximum Reverse Current 最大反向电流

Medicare Rights Center 医疗保险权利中心

Memory Read Command 存储器读出指令

Memory Register Clean 存储器寄存器清除

Mobile Radio Communication 移动无线电通信

Moisture Recording Controller 湿度记录控制器

Multirole Recoverable Capsule 多用途返回舱

**MRCC** Maintenance Reliability Control Committee 维修可靠性控制委员

Material Review Central Control 器材审查中心控制

Maximum Range Climbing Cruise 最大航程爬升巡航

Mobile Radio Communication Channel 机动无线电通信信道

Movement Report Control Center 移动报告控制中心

Multiple Reference Consistency Check 多基准一致性检测

**MRCL** Mercurial 水银的,汞的

**MRCP** Mobile Radio Control Post 移动无线电控制站

**MRCS** Multi-Rate Circuit Switching 多速率电路交换

**MRD** Main Rotor Diameter 主旋翼直径

Manufacturing Research Directive 研制指令

Marketing Requirement Document 市场需求文档

Memory Raster Display 存储光栅显示器

Memory Read 存储器判读

Mission Rehearsal Device 任务演练设备

Mission Requirements Document 任务要求文件

**MRDA** Mission Requirements Definition and Analysis 任务要求定义与分析

**MRDB** Mission Requirements Data Base 任务要求数据库

**MRDC** Modular Radar Display Console 模块化雷达显示控制器

**MRDS** Maintenance and Recording Data System 维修与记录数据系统

Malfunction Rate Detection System 故障率探测系统

Message Reproduction and Distribution System 电报复制与分发系统

Modular Responsive Defense System 模块化响应防御系统

**MRDY** Message Ready 信息准备就绪

**MRE** Mara Lodges, Kenya ( Airport Code ) 肯尼亚马拉洛哥斯机场

Mean Radial Error 平均径向误差

Modern Ramjet Engine 现代冲压式喷气发动机

Multivariate Regression Estimation 多元回归估计

**MREG** Multiregister 多路寄存器

**mrem/hr** Millirem per Hour 毫雷姆/小时

**MRF** Management Reporting Facility 管理报告设施

Meteorological Research Flight 气象研究飞行

Microfilm Reference File 缩微胶卷参考文档

Microwave Range Finder 微波测距仪

Minimum Radio Frequency 最低无线电频率

Music Research Foundation 音乐研究基金会

**MRFCS** Multi-Rate Fast Circuit Switching 多

速率快速电路交换

**MRFL** Master Radio Frequency List 主要无线电频率表

**MRFR** Marker Request for Revision 标记修改申请

**MRG** Maintenance Requirements Group 维修需求组

Management Research Groups 管理研究组

Master Reference Gyro（scope）主基准陀螺（仪）

Material Review Group 材料审查组

Mooring 系留

**MRH** Mechanical Recording Head 机械记录头

**MRI** Anchorage, AK, USA-Merrill Field 美国阿拉斯加州安克雷奇梅里尔菲尔德机场

Magnetic Resonance Imaging 磁共振图像

Miscellaneous Radar Input 综合雷达输入

Monopulse Resolution Improved 改进的单脉冲分辨率

**MRIL** Master Repairable Item List 主要可修复项目清单

**MRIU** Maintenance and Recording Interface Unit 维修和记录接口组件

**MRJ** Mitsubishi Regional Jet 三菱公司支线喷气机（日本）

**MRJE** Multileaving Remote Job Entry 多输出远程作业录入

**MRL** Manufacturing Readiness Level 制造成熟度等级

Manufacturing Reference List 加工基准清单

Material Requirements List 物料需求清单

Materials Research Laboratory 材料研发实验室

Medium Intensity Runway Edge Lighting 中亮度跑道边界灯光

Minimum Requirement List 最低要求清单

Minimum Risk Level 最低风险水平

**MRLS** Manufacturing Reference List System 制造基准清单系统

**MRM** Maintenance Report and Management 维修报告与管理

Maintenance Resource Management 维修资源管理

Management Review Meeting 管理评审会议

Manufacturing Resource Management 制造资源管理

Market Risk Management 市场风险管理

Monolithic Radar Module 单片雷达模件

Multi Resolution Modulation 多分辨率调制

**MRN** Material Recording Notice 器材记录通知

Materials Received Note 收料通知（单）

Materials Requisition Note 领料通知（单）

Minimal Routing Number 最小路由选择数

Multiple Reflection Noise 多点反射噪声

**MRND** Maintenance Required Not Developed 要求但未开发的维修

**MRNG** Minimum Range 最小距离，最小航程

Morning 早上，上午

**MRO** Machine Repair Operation 机器维修操作

Maintenance and Repair Organization 维修和修理组织

Maintenance, Repair and Operation 维护、修理与操作

Maintenance, Repair and Overhaul 维护、修理和大修

Maintenance Repair Order 维护修理指令

Maintenance Report Order 维修报告指令

Material Release Order 物料发放订单

Message Relay Office 报文转发处

Midrange Objective 中期目标

Movement Report Office 航行报告收集处

Multichannel Recording Oscillograph 多通道记录示波器

**MROAR** Modification and Repair Order and Acceptance Record 改型修理订单与验收记录

**MROF** Maintenance, Repair and Operation of Facility 设施维修、修理与操作

**MROM** Macro Read Only Memory 宏只读存储器

Masked ROM 带掩膜只读存储器

Microprogram ROM 微程序只读存储器

**MROPS** Maintenance, Repair, Operating and Production Supplies 维修、修理、运行和生产供应

**MROSE** Multiple-tasking Real-time Operating System Executive 多任务实时操作系统执行程序

**MRP** Machine Readable Passports 机读护照

Maintenance Real Property 维修用固定资产

Maintenance Reliability Plan 维修可靠性计划

Malfunction Reporting Program 故障报告程序

Manned Reusable Payload 载人可重复使用有效载荷

Manufacturer's Recommended Price 厂商推荐价格

Manufacturing Record Processor 加工记录处理器

Manufacturing Requirements Planning 制造需求计划

Manufacturing Resource Planning 生产资源计划

Material Request Plan 原材料需求计划

Material Requirement Planning 航材需求计划

Material Resource Planning 器材资源计划

Maximum Rated Power 最大额定功率

Message Routing Process 信息路由选择过程

Meteorological Report Point 气象报告点

Mid-Range Plan 中期计划

Mobile Repair Party 机动修理队

Monthly Report of Progress 月进度报告

**MRPF** Maintenance of Real Property Facilities 不动产设施保养

**MRPL** Material Requirements Planning List 器材需求计划单

**MRPM** Material Research and Production Methods 器材研究与生产方法

**MRP-OM** Manufacturing Resource Planning-Order Management 加工资源计划指令管理

**MRPV** Miniature Remotely Piloted Vehicle 小型遥控（无人驾驶）飞行器

**MRR** Maintenance, Repair and Replace 维护、修理和替换

Maintenance, Repair and Replacement 维护、修理和置换

Management Research Report 管理调查报告

Manufacturing Research Report 制造研制报告

Manufacturing Revision Record 制造更改记录

Manufacturing Revision Request 制造更改申请

Market Research Report 市场研究报告

Material Rejection Report 器材退回报告,器材报废报告

Material Reliability Report 器材可行性报告

Material Removal Report 器材拆除报告,器材调配报告

Material Review Record 器材审查记录

Mechanical Reliability Report 机械可靠性报告

Mechanical Research Report 机械研究报告

Medium Range Radar 中程雷达

Microelectronic Radio Receiver 微电子无线电接收机

Microfilm Read Record 缩微胶卷阅读记录

Microwave Radar Receiver 微波雷达接收机

Minimum Risk Route 风险最低航路

Miscellaneous Retrofit Requirement 其他更新要求

Mission Readiness Review 飞行任务准备审查

Monthly Review Report 审查月报

Multi-Role Radar 多用途雷达

**MRRC** Material Requirements Review Committee 器材要求审查委员会

Mobile Radio Resource Control 移动无线资

源控制

**MRS** Maintenance, Reliability, Supportability 维修、可靠性、耐久性

Maintenance, Repair and Service 维护、修理与服务

Maintenance Reporting System 维护报告系统

Maintenance Requirement Substantiated 具体维修要求

Malfunction Reporting System 故障报告系统

Management Report System 管理报告系统

Manned Reconnaissance Satellite 人控侦察卫星

Manned Repeater Station 人控中继站

Marseille, France (Airport Code) 法国马赛机场

Master Radar Station 主雷达站

Master Ranging Station 主测距站

Master Reference System 主基准系统

Master Repair Schedule 主修理日程表，主修理计划

Material Repair System 器材修理系统

Material Requirement Summary 器材需求概要

Material Retrieval System 器材检索系统

Material Returned to Stores 返回库房器材

Material Review Standards 器材审查标准

Materials Research Society 材料研究学会

Mechanical Resonant Scanner 机械谐振扫描器

Medical Records System 医学记录系统

Medium Range Surveillance 中程监视

Meeting Room System 会议室系统

Message Relay Service 消息转发业务

Minor Relay Station 小型中继站

Modification Record Sheet 改型记录单

Monitored Retrievable Storage 监控的可回收库存

Monitoring and Range Station 监视和测距站

Multiple Radar Track 多雷达跟踪

Multispectral Resource Sampler 多光谱资源抽样器

**MRSA** Mandatory Radar Service Area 指定的雷达服务区

Material Readiness Support Activity 器材准备保障活动

Material Review Segregation Area 材料检验分段区，器材审查隔离区

Microwave Radiometer, Scatterometer and Altimeter 微波辐射计、散射计和高度计

**MRSE** Message Retrieval Service Element 消息检索服务单元

Microwave Remote Sensing Equipment 微波遥感设备

Microwave Remote Sensing Experiment 微波遥感试验

**MRSL** Marconi Radar System Limited 马可尼雷达系统公司

Maintenance and Reliability Simulation Model 维修与可靠性模拟模型

**MRSP** Multifunction Radar Signal Processor 多功能雷达信号处理器

**MRSR** Multi-Role Survivable Radar 多用途可存活雷达

**MRST** Master Reset 主复位

**MRT** Maintainability Review Team 可维修性评估组

Maintenance Repair Team 维修小组

Maximum Repair Time 最长修理时间

Mean Repair Time 平均修复时间

Mean Response Time 平均响应时间

Mechanical Rubbing Test 机械摩擦试验

Medium Rescue Tender 中型援救车

Message Register Terminal 信息寄存器终端

Message Routing Table 消息路由选择表

Mildew Resistant Thread 防霉线

Miniature Receive Terminal 微型接收终端

Minimum Resolvable Temperature 最低溶解温度

Mobile Radar Target 移动雷达目标

Modification Removal Tag 改装拆卸标签

Modulator, Receiver, Transmitter 调制器、接

收机、发射机

Multi-Radar Tracking 多雷达跟踪,多雷达航迹

**MRTA** Medium Range Tactical Aircraft 中程战术飞机

Multi Role Transport Aircraft 多用途运输机

**MRTD** Machine Readable Travel Documents 机读旅行证件

Minimum Resolvable Temperature Difference 最小可分辨温差

**MRTI** Multi-Role Thermal Imagery 多用途热像

**MRTN** Motor Return 电机回线

**MRTP** Multi Radar Tracking Processing 多雷达航迹处理

**MRTR** Mobile Radio Transmit and Receive 移动无线电发射与接收

**MRTS** Master Radar Tracking Station 主雷达跟踪站

**MRTT** Modular Record Traffic Terminal 模块式记录交通终端

Multi Role Tanker Transport 多用途油轮运输

**MRU** Material Records Unit 器材记录组件

Mauritius, Mauritius-Plaisance ( Airport Code)毛里求斯普莱桑斯机场

Maximum Receive Uint 最大接收单位

Message Retransmission Unit 信息重发组件

Microwave Relay Unit 微波中继组件

Mobile Radio Unit 移动无线电组件

Mountain Rescue Unit 山区援救队

**MRUASTAS** Medium Range Unmanned Aerial Surveillance and Target Acquisition System 中程无人驾驶空中监视与目标搜索系统

**MRUR** Material Requirements and Usage Reports 器材要求与使用报告

**MRV** Machine-Readable Visa 机器可读签证

Maneuverable Reentry Vehicle 可机动再入飞行器

Measurable, Reportable, Verifiable 可测量,可报告,可验证

Multiple Reentry Vehicle 多重再入飞行器

Multi-Range Volmeter 多量程电压表

**MRW** Maximum Ramp Weight 最大机坪重量

Multiplex Response Word 多路响应字

**MRWG** Mission Requirements Working Group 任务要求工作组

**MRY** Monterey, CA, USA( Airport Code)美国加利福尼亚州蒙特雷机场

**MRZ** Machine Readable Zone 机读区

**MS** Magnetic Sensor 磁传感器

Magnetic Storage 磁存储器

Mail Stop 邮筒

Main Storage 主存储器

Main Switch 主开关

Maintenance and Supply 维修与供应

Maintenance Schedule 维修方案,维修计划

Maintenance Superintendent 维修督察官员

Maintenance Supervisor 维修督察员

Manual Setter 手动调节器

Manual Setting 人工设定(值)

Manufacture Standards 制造标准

Manufacturing Specification 制造规范

Manuscript 手抄本

Margin of Safety 安全系数(裕度)

Margin on Sales 销售净利率

Mark Sensing 标记检测,符号读出

Market Segmentation 市场分割

Market Service 销售服务

Market Share 市场份额

Market Survey 市场调查

Marketing System 销售系统

Mass Simulator 大容量模拟器

Mass Spectrometer 质谱仪

Master Scanner 主扫描器

Master Schedule 主调度程序

Master Station 主控台,总机

Master Switch 主开关,总开关

Master Synchronizer 主同步器

Material and Service 器材与勤务

Material Specification 器材规范

Material Standard 材料标准

Material Supplies 器材供应

Material Support 器材保障

Material System 器材系统

Mathematical Software 数学软件

Mathematical Subroutine 数学子程序

Maximum Stress 最大应力

Mean Square 均方(值)

Measurement Signal 测量信号

Measuring Sequence 测量顺序

Measuring Set 测量设备

Measuring System 测量系统

Mechanical Specialties 机械特性

Mechanical Stability 机械稳定性

Mechanical Strain 机械应力

Mechanical Strength 机械强度

Mechanics School 机械学校

Media Synchronization 媒体同步

Meeting of Signatories 签字者会议

Meeting Schedule 会晤时间表

Memory Synchroscope 存储同步示波器

Memory System 存储系统

Message Storage 消息存储

Message Store 消息存储器

Message Switching 信息交换

Metal Semiconductor 金属半导体

Meteorological Survey 气象观察

Meteorological System 气象系统

Meter Sensitivity 仪表灵敏度

Metric System 公制,米制

Micro Soft 微软(公司)

Micro Switch 微动开关

Microcomputer System 微型计算机系统

Microprogram Storage 微程序存储器

Microscopic Spectrometer 显微光谱仪

Microscopic Structure 显微结构

Microwave Scatterometer 微波散射计

Microwave Spectrograph 微波摄谱仪

Microwave Spectroscope 微波分光镜

Microwave Spectrum 微波频谱

Mild Steel 低碳钢

Military Specification 军用规范

Military Standard 军标,军用标准

Military Station 军用台

Miniature System 小型系统,微型系统

Minimum Scattering 最小散射

Minimum Stress 最小应力

Ministry of Supply 供应部

Minus 负,减

Minus Degrees 零下温度

Missend 误发

Mission Simulator 任务模拟机

Mission Summary 任务提要

Mixed Strategy 混合策略

Mixture Solvent 混合溶剂

Mobile Service 移动业务

Mobile Station 移动台

Mobile Subscriber 移动用户

Mode Scrambler 扰模器

Modulation Sensitivity 调制灵敏度

Modulation Sideband 调制边带

Modulation Switch 调制开关

Molecular Sieve 分子筛

Motor Start 发动机起动

Multicomputer System 多计算机系统

Multimedia System 多媒体系统

Multimicroprocessor Simulation 多微处理器模拟

Multiple Section 多路部分,复联部分

Multiplexer Storage 多路转换器存储器

Multi-Stage 多级

Mutual Synchronization 互同步

**MSA** Management Service Agreements 管理维修协议

Material Storage Area 器材存储区

Measurement System Analysis 测量系统分析

Micro Signal Architecture 微信号结构

Microsystem Analyzer 微系统分析器

Microwave Switching Assembly 微波转换组件

Minimum Safety Altitude 最低安全高度

Minimum Sector Altitude 最低扇区高度

Mobile Station Administration 移动台注册管理

Multimedia Stream Adaptive 多媒体流量自调节器

Multiplex Section Adaptation 复用段适配

**MSAFQ**   Minimum Speed for Acceptable Flying Qualities 容许飞行质量最低速度

**MSAIS**   Multiplex Section Alarm Indication Signal 复用段告警指示信号

**MSAM**   Master of Science in Applied Mechanics 应用力学硕士

**MSAPL**   Mechanical Systems and Accessories Product Line 机械系统与配件产品线

**MSAR**   Macao Special Administrative Region 澳门特别行政区

**MSAS**   Mobile Satellite Augmentation System 移动卫星增强系统

Multi-functional Satellite Augmentation System 多功能卫星增强系统

**MSAs**   Management Services Agreements 管理服务协议

**MSAT**   Microsoft Security Accessment Tool 微软安全评估工具

Mobile Satellite 移动业务卫星

**MSAT-X**   Mobile Satellite Experiment 移动卫星试验

**MSAW**   Minimum Safe Altitude Warning 最低安全高度警告，最低安全高度警报

Missing Air Waybill 未收到的航空货运单

**MSB**   Main Switch Board 主配电盘，总配电盘

Maintenance Standard Book 维修标准手册

Maximum Spare Bandwidth 最大备用带宽

Most Signficant Bit 最高有效位

Most Significant Byte 最高有效字节

**MSBLS**   Microwave Scanning Beam Landing System 微波扫描着陆系统

**MSBN**   Multi-Service Broadband Network 多业务宽带网络

**MSBP**   Multiservice Billing Protocol 多业务计费协议

**MSBS**   Multiservice Billing System 多业务计费系统

**MSC**   （Airbus）Materiel Support Centre（空客）器材支援中心

Main Switching Center 主交换中心

Maintenance Significant Component 重要维修部件

Maintenance Steering Committee 维修指导委员会

Master Supervision Center 主监控中心

Merge-Split Component 可组合分立式部件

Message Sequence Chart 信息序列图

Message Service Center 信息服务中心

Message Switching Center 信息交换中心

Miscellaneous System Controller 综合系统控制器

Mobile Service Center 移动业务中心

Mobile Switching Center 移动交换中心

Most Signficant Character 最高有效字符

Multifunction Switching Control 多功能转换控制

Multimedia Super Corridor（马来西亚）多媒体超级走廊

**MSCA**   Missing Cargo 丢失的货物，少收的货物

**MSCB**   Manufacturing Specification Control Board 制造规范控制委员会

**MSCCRD**   Miscellaneous Card 其他卡

Miscellaneous Cooling Card 综合冷却插件板

**MSCD**   Manufacturing Spares Configuration Device 制造备件构型装置

Missing Cargo Document 未收到的货物文件

**MSCH**   Multiplex Subchannel 复用子信道

**MSCM**   Multichannel Subcarrier Multiplexing 多信道副载波复用

**MSCN**   Manufacturer Specification Change Note 生产厂规范更改说明

Manufacturer Specification Change Notice 生产厂规范更改通知

Mathematics and Science Centers Network 数学和科学中心网络

Misconnect 航班衔接失误,误接

Misconnection 衔接失误

Mobile Satellite Communication Network 移动卫星通信网络

**MSCP** Mean Spherical Candlepower 平均球面烛光

Mobile Satellite Service Provider 移动卫星业务提供者

Mobility and Service Control Point 移动性和服务控制点

**MSCR** Machine Screw 机器螺丝

Manufacturing Specification Change Record 制造规范更改记录

Manufacturing Specification Control Record 制造规范控制记录

Manufacturing Specification Coordination Record 制造规范协调记录

**MSCS** Manufacturing Spares Configuration System 制造备件配置系统

Mass Storage Control System 大容量存储器控制系统

**MSCT** Message Switching Concentration Technique 消息交换集中技术

**MSCU** Microwave Sounding and Control Unit 微波探测与控制装置

**MSD** Mass Storage Device 大容量存储器

Mean Solar Day 平均太阳日

Mean Square Displacement 均方位移

Mean Standard Deviation 平均标准偏差

Medium Specific Decoder 中速专用译码器

Minimum Safe Distance 最小安全距离

Mission Scoring Data 飞行任务评分数据

Most Signification Digit 最高有效数字

Multiple Site Damage 多点损伤

Multi-Sensor Display 多传感显示器

**MSDN** Multi-Service Digital Network 多业务数字网络

**MSDS** Manpower Standard Development System 人力标准开发系统

Material Safety Data Sheet 器材安全数据单

Multi-Spectral Data System 多频谱数据系统

Multispectral Scanner and Data System 多频谱扫描器和数据系统

**MSDSE** Mobile Satellite Data Switching Exchange 移动卫星数据交换机

**MSDT** Miss Distance 偏差距离

**MSE** Maintenance Scheduling Effectiveness 维修规划的有效性

Maintenance Support Equipment 维修保障设备

Manufacturing Support Equipment 制造保障设备

Manufacturing Systems Engineering 制造系统工程

Master of Software Engineering 软件工程硕士

Materials Science and Engineering 材料科学与工程

Mean Square/Squared Error 均方误差

Mechanical Support Equipment 机械保障设备

Minimum Single-Engine Speed 单发飞行最小速度

**MSEC** Message Security 信息安全

Millisecond 毫秒

**MSEP** Maintenance Standardization and Evaluation Program 维修标准化与鉴定大纲

**MSF** Manufacturing Specification Form 制造规范表格

Mass Storage Facility 大容量存储设备

Minimum Sector Fuel 航段最低燃油量

Multichannel Selective Filter 多信道选择滤波器

**MSFC** Marshall Space Flight Center 马歇尔航天中心

**MSFN** Manned Space Flight Network 载人宇宙飞行网

**MSG** Maintenance Steering Group 维修指导小组

Maintenance System Guide 维护系统指南

Management Steering Group 管理指导组

Mapper Sweep Generator 绘图仪扫描产生器

Maximum Speed Governor 最大速度调节器

Message 电报,信息

Miscellaneous Simulation Generation 综合模拟产生

Modeling and Simulation Group 建模和仿真组

**MSGG**   Message Generator 信息产生器

**MSGS**   Message Switching 消息交换

**MSH**   Minimum Safe/Safety Height 最低安全高度

**MSHATF**   Medium Support Helicopter Aircrew Training Facility 中型支援直升机机组训练设备

**MSHDL**   Mishandled 弄错的

**MSHES**   Multiplex State of Human Errors System 人误系统复合状态

**MSHG**   Meshing 啮合

**MSI**   Maintenance Service Items 维护服务项目

Maintenance Significant Items 重要维修项目

Management System Indicator 管理系统指示器

Message Signaled Interrupt 信息信号中断

**MSIC**   Medium Scale Integrated Circuit 中规模集成电路

**MSID**   Mobile Station Identifier 移动台识别码

**MSIL**   Maintenance Significant Item List 重要维修项目清单

**MSIN**   Mobile Station Identification Number 移动站识别号码

Mobile Subscriber Indentification Number 移动用户识别号码

Multistage Interconnection Network 多级互连网络

**MSIP**   Multi-Staged Improvement Program 多阶段改进计划

**MSIS**   Maintenance Store Information System 维修存储信息系统

**MSISDN**   Mobile Station Integrated Services Digital Network (Number) 移动台综合业务数字网(号码)

**MSJ**   Meteorological Society of Japan 日本气象学会

Misawa, Japan (Airport Code) 日本三泽机场

**MSK**   Minimum Shift Keying 最小移频键控

Mission Support Kit 任务支援装备

Moscow Time Zone 莫斯科时间

**MSL**   Main Supply Line 主要电源线

Management System Laboratory 管理系统实验室

Material Support List 器材支援清单

Maximum Service Life 最大使用寿命

Maximum Service Limit 最大使用限度

Mean Sea Level 平均海平面

Message Length 消息长度

Micro Strip Laser 微带激光器

Minimum Safe Level 最低安全高度

Mirrored Server Link 镜像服务器链接

Module Support Layer 模块支持层

Multimedia Software Layers 多媒体软件层

Multiplex Section Layer 复用段层

Multi-Satellite Link 多卫星链路

Muscle Shoals, AL, USA (Airport Code) 美国阿拉巴马州马斯尔肖尔斯机场

**MSLAN**   Middle Speed LAN 中速局域网

**MSLM**   Microchannel Spatial Light Modulator 微信道空间光调制器

**MSLS**   Machine Shop Load Simulator 机械车间装载模拟器

Material Storage Location System 器材储存位置系统

Modular Site Location System 模件现场位置系统

**MSM**   Manufacturing Standards Manual 制造标准手册

Matrix Stackable Module 矩阵式堆叠模块

Media Support Module 媒体支持模块

Message Switching Multiplexing 信息交换复用

Microwave Switching Matrix 微波交换矩阵

Mobile Station Modem 移动站调制解调器

Monitoring System Module 监控系统组件

Multiwavelength Simultaneous Monitoring 多波长同步监视

**MSMB** Missing Mail Bag 未收到的邮包

**MSMC** Multiwavelength Simultaneous Monitoring Circuit 多波长同步监视电路

**MSMP** Modeling and Simulation Master Plan 建模与仿真主计划

Multi-Spectral Measurements Program 多光谱测量计划

**MSMR** Multiple Service Multiple Resource 多业务多资源

**MSMS** Maximum Safety Mach System 最大安全马赫系统

**MSN** Madison, WI, USA (Airport Code) 美国威斯康星州麦迪逊机场

Manufacturer Serial Number 厂家序号

Master Serial Number 主序号

Message Switching Network 信息转换网,消息交换网

Multi-Satellite Network 多卫星网络

Multi-Server Network 多服务器网络

Multi-System Networking 多系统网络

Mutual Synchronization Network 互同步网络

**MSNET** Microsoft Network 微软互联网络

**MSNF** Multi-System Networking Facility 多系统网络设备

**MSNS** Multimedia Service Navigation System 多媒体业务导航系统

**MSO** Missoula, MT, USA (Airport Code) 美国蒙大拿州米苏拉机场

Multiple Service Operator 多业务运营商

**MSOCC** Multisatellite Operations Control Center 多卫星操作控制中心

**MSOH** Multiplexing Section Overhead 复用段开销

**MSOV** Modulating and Shutoff Valve 调节和关断活门

**MSP** Magnetic Speed Probe 磁速率探头

Maintenance Service Plan 维修服务计划

Maintenance Support Package 维修支援组件

Management Service Provider 管理服务提供商

Manufacturing Specification Plan 制造规范计划

Master Synchronization Pulse 主同步脉冲

Media Stream Protocol 媒体流协议

Message Security Protocol 报文安全协议

Metal Splash Pan 金属防溅盘

Miniature System Program 小型系统程序

Miniature System Project 小型系统项目

Minneapolis/St. Paul, MN, USA (Airport Code) 美国明尼苏达州明尼阿波利斯/圣保罗机场

Mission Simulation Processor 飞行任务模拟处理器

Mixed Signal Processing 混合信号处理

Mock-up Support (Boeing) 波音模型支援

Mode S Specific Protocol S 模式特别协议

Mode Select Panel 方式选择板

Mosaic Sensor Program 镶嵌传感器程序

Multiplex Section Protection 复用段保护

Multi-Service Platform 多业务平台

**MSPA** Mechanically Scanned Planar Array 机械扫描平面矩阵

**M-SPD** Manual Speed 人工速度

**M-SPEC** Manufacturing Specification 制造规范

**MSPO** Management Systems Program Office 管理系统规划办公室

**MSPT** Marine Silent Power Transmission 海底寂静功率传输

Mean Shop Pocessing Time 平均车间处理时间

**MSQ** Minsk, Belarus (Airport Code) 白俄罗斯明斯克机场

Minimum Stock Quantity 最小储存量

**MSR** Magnetic Stripe Reader 磁条阅读器

Manufacturing Service Request 制造服务申请

Manufacturing Specification Record 制造规范

记录

Marginal Seat Revenue 边际座位收入

Marketing Service Representative 销售服务代表

Material Status Report 物资现状报告

Material Stores Requisition 物资存储要求

Microfiches Stores and Retrieval 微型胶卷存储与检索

Mobile Support Router 移动通信支持路由器

**MSRN** Mobile Station Roaming Number 移动通信站漫游号码

**MSRP** Minimum Skills Retention Plan 最低技能保持计划

**MSRS** Master Schedule Revision Sheet 主计划更改页

**MSS** Maintenance Service System 维修服务系统

Maintenance Support Schedule 维修保障程序

Management System Study 管理系统研究

Manned Space Station 载人空间站

Manned Space System 载人航天系统

Manual Switching System 人工开关(转换)系统

Maritime Satellite Service 海事卫星业务

Market Strategy Simulation 市场战略模拟

Mass Storage Subsystem 大容量存储器子系统

Mass Storage System 大容量存储系统

Master Schedule System 主计划系统

Master-Slave Synchronization 主—从同步

Material Support System 航材供应系统

Maximum Segment Size 最大段宽

Mechanical Systems Staff 机械系统工作人员

Message Switching System 信息交换系统

Meteorological Satellite Section 气象卫星部门

Meteorological Sounding Rocket 气象探测火箭

Metropolitan Area Network Switching System 都市网交换系统

Metropolitan Switching System 城域交换系统

Microwave Signal Source 微波信号源

Microwave Survey System 微波探测系统

Microwave Switching Station 微波中继站

Military Supply Standard 军用供应标准

Mission Simulation System 飞行任务模拟系统

Mission Software Services 飞行任务软件服务

Mobile Satellite Service 移动卫星服务

Mobile Servicing System 移动服务系统

Mobile Subscriber Station 移动用户站

Mobile Suporting Station 移动通信支持站

Mobile Surveillance Shield 移动监视屏蔽

Mode Select Switch 方式选择开关

Moored Surveillance System 停泊监视系统，系泊监视系统

Multimedia System Service 多媒体系统业务

Multispectral Scanner 多光谱扫描器

**MSSA** Maintenance Supply Services Agency 维修供应服务部

Major Supplier Stress Analysis 重要供应商应力分析

Multi Sensor Surveillance Aircraft 多传感器监视飞机

Multi-Service Storage Architecture 多服务存储结构

**MSSC** Maritime Satellite Switching Center 海事卫星交换中心

Mass Storage System Communicator 大容量存储系统通信装置

Mobile Satellite Switching Center 移动通信卫星交换中心

Mobile Service Switching Center 移动业务交换中心

**MSSCSG** Modular Spread Spectrum Code-Sequence Generator 积木式扩频码序发生器

**MSSE** Mass Storage System Extensions 大容量存储系统扩充

**MSSFU** Mobile Satellite Store-and-Forward Unit 移动通信的卫星存储转发单元

**MSSL** Modular Solid State Logic 模块化固态逻辑

**MSSN** Mobile Services Satellite Communication Network 移动业务卫星通信网

**MSSR** Monopulse Secondary Surveillance Radar 单脉冲二次监视雷达

Multiple Service Single Resource 多业务单信源

Multi-State System Reliability 多态系统的可靠性

**MSSS** Maintenance Supply Service System 维修供应服务系统

Man-Seat Separation System 人椅分离系统

Mass Spectral Search System 质谱搜索系统

Multi-Satellite Support System 多卫星支援系统

**MST** Machine Steel 机械钢

Marconi Telecommunications Systems 马可尼电信系统

Master 主控机, 主控人, 主要的, 硕士

Master Tape 主带

Mean Service Time 平均服务时间

Mean Solar Time 平均太阳时

Memory System Test 存储器系统测试

Memory System Tester 存储系统测试器

Minimum Spanning Tree 最小生成树

Minimum Stock Tag 最少储存标签

Mobile Service Tower 移动式维修塔

Monolithic System Technology 单片系统技术

Mountain Standard Time 山区标准时间

Multi Service Terminal 多业务终端

Multi-System Test 多系统测验

**MSTA** Maximum Short-Term Average 最大短期平均数

**MSTC** Mastic 胶, 膏

Microwave Sensitivity Time Control 微波时间灵敏度控制

**MSTK** Mistake 过失, 误差, 失策

**MSTR** Master 控制者, 主要的, 总的

Mathematical Sciences Technical Report 数学科学技术报告

Moisture 潮湿

**MSTS** Midcourse Surveillance Tracking System 中途监视跟踪系统

Minimal Standard for Testing Software 最小的标准检测软件

**MSU** Magnetic Sensor Unit 磁传感器组件

Main Storage Unit 主存储器

Main Switching Unit 主交换器

Maintenance Service Unit 维修勤务单元

Maintenance Signal Unit 维护信号单元

Maintenance Station Unit 维修站分队

Manual Switching Unit 人工转换组件

Mass Storage Unit 大容量存储器

Master Station Unit 主站分队, 主站组件

Material Sales Unit 器材销售单位

Media Server Unit 媒体服务器组件

Memory Storage Unit 存储器组件

Message Signal/Signaling Unit 消息信号单元

Message Switching Unit 信息交换组件

Mode Select Unit 模式选择单元, 模式选择组件

Mode Selector Unit (IRS) 方式选择器组件 (惯性基准系统)

Multiple Signal Unit 多重信号单元

Multiple Subscriber Unit 多用户单元

Multisite Update 多位置更新

**MSUP** Master Schedule Update Program 主计划更新大纲

**MSUTB** Mass Storage Unit Test Bed 大容量存储主件试验台

**MSV** Main Staging Valve 主阶变活门, 主分离活门

Manned Space Vehicle 载人宇宙飞行器

Mean Square Velocity 均方速度

Mode Select Valve 方式选择活门

**MSVC** Meta Signaling Virtual Channel 元信令虚通道

**MSVP** Miniature System Validation Phase 小型系统验证阶段

**MSW** Machine Status Word 机器状态字

Magnetostatic Wave 静磁波

Master Switch 主开关

Maximum Short Takeoff and Landing Weight 短距起落最大重量

Message Switch/Switching 消息交换器

Microswitch 微动开关

**MSY**   Maximum Sustainable Yield 最大持续捕获量

New Orleans, LA, USA (Airport Code) 美国路易斯安那州新奥尔良机场

**MSYNC**   Master Synchronizer 主同步器

**MSZ**   Major Sub Zone 主分区

**MT**   Mach Trim 马赫配平

Machine Translation 机器翻译

Magnetic Tape 磁带

Magnetic Theodolite 磁经纬仪

Magnetic Track 磁航迹

Mail Transfer 信汇

Mail Tray 邮盘

Main Tank 主油箱

Maintenance Technician 维修技术员

Maintenance Tip 维护提示

Maintenance Trailer 维修拖车

Maintenance Tree 维护树

Manufacturing Tolerance 制造公差

Master Terminal 主机终端

Matching Transformer 匹配变压器

Material Transfer 器材调拨

Material Turnover 器材周转率

Materials 器材, 材料, 物资

Maximum Torque 最大扭矩

Maximum Turnover 最高周转率

Mean Time 平均时间

Measured Time 测量时间

Mechanical Transport 机械运输

Meet 遇见, 会面

Memorandum Trade 备忘录贸易

Memory Total 存储总计

Memory Tube 存储管

Mercurial Thermometer 水银温度计

Message Transfer 消息传送

Message Type 消息类型

Metal Threshold 金属门限(检测门)

Meteorological 气象的

Metric Ton 公吨

Microwave Transmission 微波传输

Minimum Time 最少时间

Mission Trailer 维修拖车

Mission Trainer 任务训练器

Mixed Transport 混合运输

Mixed Turbine 混流式涡轮机

Mobile Terminal 移动终端

Mobile Transporter 移动运输装置

Monetary Telephone 投币式电话

Monostable Trigger 单稳触发器

Motor 发动机, 马达

Motor Transport 马达传输, 机动车运输

Mountain Time 山区时间(美国)

Moving Target 活动(移动)目标

Multi Terminal 多终端(输出)

Multilateral Trade 多边贸易

Multimedia Terminal 多媒体终端

Multimedia Toolkits 多媒体工具包

Multiple Telegram 多路电报, 复式电报

Multiple Transfer 多路传输

Multiplex Telemetering 多路遥测

My Telegram 我方电报

My Telex 我方电传

**MTA**   Mach Trim Actuator 马赫配平作动器

Mail Transfer Agent 邮件传输代理

Maintenance Task Analysis 维修任务分析

Master Timer Assembly 主定时器组装

Message Transfer Agent 消息传送代理

Message Transfer Architecture 消息传输结构

Military Transport Aircraft 军用运输机

Minimum True Altitude 最低真高

Multimedia Transport API 多媒体运送应用编程接口

Multiple Terminal Access 多终端访问

Multirole Transport Aircraft 多用途运输机

**MTA BRK**   Mach Trim Actuator Brake 马赫配平作动筒制动器

**MTAE** Message Transfer Agent Entity 报文传送代理实体

**MTAF** Mandatory Traffic Advisory Frequency 强制交通咨询频率

**MTAM** Maintenance Turnaround Analysis Model 维修周期分析模型

Maritime Tropical Air Mass 热带海洋气团

**MTB** Maintenance of True Bearing 保持真航向

Materials Transportation Bureau 物资运输局

**MTBA** Mean Time between Accidents 故障平均间隔时间

Mean Time between Alarm 警报平均间隔时间

Mean Time between Assists 援助平均间隔时间

**MTBC** Mean Time between Calls 呼叫平均间隔时间

Mean Time between Collisions 碰撞平均间隔时间

Mean Time between Complaints 申告平均间隔时间

**MTBCD** Mean Time between Confirmed Defects 确认故障平均间隔时间

**MTBCF** Mean Mission Time between Critical Failure 严重故障间隔平均任务时间

Mean Time between Component Failure 部件故障平均间隔时间

Mean Time between Critical Failure 严重故障平均间隔时间, 危险性故障平均间隔时间

**MTBCM** Mean Time between Corrective Maintenance 恢复性维修平均间隔时间

**MTBD** Mean Time between Defects 故障平均间隔时间

Mean Time between Degradation 衰变平均间隔时间

Mean Time between Demands 需求平均间隔时间

Mean Time between Detection （故障）检测平均间隔时间

**MTBDD** Mean Time between Defects Downing 发生故障平均间隔时间

**MTBDE** Mean Time between Downing Event 停用事件平均间隔时间

**MTBE** Mean Time between Errors 差错平均间隔时间

**MTBF** Mean Time before Failure 故障前平均间隔时间

Mean Time between Failure 故障平均间隔时间

Minimum Time between Flights 飞行最小间隔时间

**MTBFF** Mean Time between Functional Failures 功能失效平均间隔时间

**MTBFRO** Mean Time between Failure Requiring Overhaul 需要翻修的故障平均间隔时间

**MTBI** Mean Time between Incidents 事故征候平均间隔时间

Mean Time between Inspection 检查平均间隔时间

Mean Time between Interruption 中断平均间隔时间

**MTBIFS** Mean Time between in Flight Shutdown 空中停车平均间隔时间

**MTBM** Mean Time between Maintenance 维修平均间隔时间

Mean Time between Malfunction 误动作平均间隔时间

**MTBMA** Mean Time between Maintenance Actions 维修措施平均间隔时间

Mean Time between Mission Abort 任务中止平均间隔时间

**MTBME** Mean Time between Maintenance Events 维修事件平均间隔时间

**MTBMI** Mean Time between Maintenance Inherent 常规维修平均间隔时间

**MTBMT** Mean Time between Maintenance Total 全面维修平均间隔时间

**MTBO** Mean Time between Outage 故障停工平均间隔时间, 失效平均间隔时间

Mean Time between Overhauls 翻修平均间隔时间,翻修平均寿命

**MTBPM** Mean Time between Preventive Maintenance 预防性维修平均间隔时间

**MTBPR** Mean Time between Premature Removals 提前拆换平均间隔时间

**MTBR** Mean Time between Removal(s) 拆换平均间隔时间

Mean Time between Repairs 修理平均间隔时间

Mean Time between Replacement 更换平均间隔时间

Mean Time between Report 报告平均间隔时间

**MTBSE** Mean Time between Software Errors 软件差错平均间隔时间

**MTBSF** Mean Time between Service Failure 服务故障平均间隔时间

Mean Time between Significant Failure 严重故障平均间隔时间

Mean Time between System Failures 系统故障平均间隔时间

**MTBSM** Mean Time between Scheduled Maintenance 预定维修平均间隔时间

**MTBSO** Mean Time between Service Outage 业务中断平均间隔时间

**MTBT** Mean Time between Trouble 故障平均间隔时间

**MTBUM** Mean Time between Unscheduled Maintenance 非预定维修平均间隔时间

**MTBUR** Mean Time between Unit Removals 组件拆卸平均间隔时间

Mean Time between Unit Replacement 组件更换平均间隔时间

Mean Time between Unsatisfactory Reports 不满意报告平均间隔时间

Mean Time between Unscheduled Removals 非预定拆卸平均间隔时间

**MTBW** Mean Time between Warning 告警平均间隔时间

**MTC** Mach Trim Compensator 马赫数配平补偿器

Mach Trim Computer 马赫配平计算机

Main Test Component 主测部件

Main Trunk Circuit 主干线电路

Maintenance Task Card 维护工卡

Maintenance Task Code 维修任务代码

Maintenance Terminal Cabinet 维修终端舱,维修终端室

Maintenance Test Center 维修测试中心

Master Tape Control 主磁带控制

Mean Transinformation Content 平均传送信息量

Memory Test Computer 存储器测试计算机

Message Transmission Control 信息传输控制

Mission and Traffic Control 飞行任务与交通控制

Multimedia Telephone Communication 多媒体电话通信

**MTCA** Military Terminal Control Area 军事终端管制区

Minimum Terrain Clearance Altitude 最低地高度,最低地形超障高度

Ministry of Transport and Civil Aviation 运输及民用航空部

Multiple Terminal Communication Adapter 多终端通信适配器

**MTCD** Medium Term Conflict Detection 中期冲突探测

**MTCE** Maintenance Shop 维护车间

**MTCF** Mean Time to Catastrophic Failure 出现严重故障的平均时间

**MTCH** Match 匹配,比赛

**MTCHD** Matched 匹配的

**MTCHG** Matching 匹配

**MTCM** Message Transmission Control Module 消息传输控制模块

Multiple Trellis-Coded Modulation 多格状编码调制

**MTCS** Minimal Terminal Communications System 最小终端通信系统

Multimedia Telecommunication Conference

System 多媒体电信会议系统

**MTD** Maintenance Terminal Display 维护终端显示器

Maintenance Training Device 维护训练器

Manufacturing Technical Directive 加工技术指令

Manufacturing Technology Department 制造技术部门

Maximum Transfer Delay 最大传送延迟

Mean Temperature Difference 平均温差

Mobile Training Detachment 流动训练队

Month to Date 月到日

Mounted 安装的

Moving Target Detector 移动目标检测器

**MTDA** Mean Time between Data Access 数据存取平均间隔时间

Mean Time to Dispatch Alert 放行提醒前平均时间

Modification Table of Distribution and Allowances 修正编配定额表

**MTDE** Modern Technology Demonstration Engine 现代技术显示发动机

**MTDF** Master Tracking Data File 主跟踪数据文件

Mean Time to Degradation Failure 出现衰变的平均时间

**MTDM** Multimedia Time Division Modulator 多媒体时分调制器

**MTDTE** Mobile Telephone Data Transfer Equipment 移动电话数据传送设备

**MTE** Maintenance Targeting Equipment 维修瞄准设备

Maintenance Training Equipment 维修训练设备

Manual Test Equipment 人工测试设备

Manufacturing Test Equipment 制造测试设备

Maximum Temperature Engine 发动机最高温度

Message Transfer Event 消息传送事件

Microwave Test Equipment 微波测试设备

Mission Task Element 任务科目基元

Multi-system Test Engineering 多系统测试工程

Multi-system Test Equipment 多系统测试设备

**MTEC** Measurement and Test Equipment Certification 测量与测试设备证书

**MTEM** Maintenance Threat and Error Management 维修威胁与差错管理

**MTER** Magnetic Tape Event Recorder 磁带事件记录器

Manufacturing Test Equipment Request 制造测试设备申请

**MTES** Multiple Threat Emitter System 多威胁辐射源系统

**MTEX** Multi-Token Exchange 多令牌交换

**MTF** Maintenance Terminal Function 维护终端功能

Maintenance Test Facility 维修测试设备

Manual Terrain Following 人工地形跟随

Mean Time to Failure 故障平均间隔时间

Medical Treatment Facility 医疗处理设备

Metal Trades Federation 金属贸易联合会

Microwave Test Facility 微波测试设备

Modulation Transfer Function 调制传递函数

Multiple Technical Force 综合技术力量

**MTFF** Mean Time to First Failure 首次故障平均时间

Modulation Transfer Function Frequency 调制传递函数频率

**MTFG** Magnetic Tape Fuel Gauge 磁带式油量表

**MTG** Main Traffic Group 主话(业)务群

Main Turbine Generator 主涡轮发电机

Maintenance Test Group 维护测试小组

Manufacturing Technical Group 制造技术组

Meeting 会见,会议

Methanol to Gasoline 甲醇制汽油

Micro Turbine Generator 微型涡轮发电机

Miles to Go 待飞距离(英里)

Mortgage 抵押借款

Mounting 安装,座架

Multiple Trigger Generator 多触发产生器

Multipurpose Target Generator 多用途目标发生器

Muting 噪声抑制

**MTGC** Mounting Center 安装中心

**MTGW** Maximum Taxi Gross Weight 最大滑行全重

**MTH** Magnetic Tape Handler 磁带信息处理机

Marathon, FL, USA (Airport Code) 美国佛罗里达州马拉松机场

Mathematics 数学

Medium Transport Helicopter 中型运输直升机

Month 月

**MTHBD** Motherboard 母板

**MTHD** Method 方法,措施

**MTHLY** Monthly 每月

**MTHRD** Male Threaded 阳螺纹的

**MTI** Message Type Indicator 电报类型标志

Minimum Time Interval 最小时间间隔

Mission Time Improvement 飞行任务时间改进

Module Test Integration 模块测试集成

Moving Target Identification 活动目标识别

Moving Target Indicating 运动目标指示

Moving Target Indicator 移动目标指示器,活动目标指示器

Multi Tank Indications 多油箱指示

Multi Tank Indicator 多油箱指示器

Multimodal Transport Institute 多式联运研究所

**MTIAC** Manufacturing Technology Information Analysis Center 制造技术信息分析中心

**MTIE** Maximum Time Interval Error 最大时间间隔误差

**MTIP** Maintenance Training Improvement Program 维修训练改进大纲

**MTIR** Moving Target Indication Radar 移动目标显示雷达

**MTIS** Maintenance Task Information System 维修任务信息系统

**MTJ** Montrose, CO, USA (Airport Code) 美国科罗拉多州蒙特罗斯机场

**MTK** Magnetic Track 磁航迹

Main Tank 主油箱

**MTL** Main Telephone Line 电话主线

Master Tape Loading 主磁带录入

Material 材料

Materials Technology Laboratory 材料技术实验室

Metal 金属

Microsystems Technology Laboratories 微系统技术实验室

Minimum Triggering Level 最小触发电平

Moving Target Locator 移动目标定位器

**MTLA** Minimum Takeoff and Landing Area 最小起飞和着陆区

**MTLC** Metallic 金属的

**MTLZ** Metallize 金属化

**MTM** Maintenance Test Module 维修测试模块

Manufacturing Training Manual 制造训练手册

Master Terminal Monitor 主终端监控器

Maximum Takeoff Mass 最大起飞质量

Methods Time Measurement 时间测量方法

Million Ton-Miles 百万吨英里

Mixed Transmit Mode 混合传输方式

Modification Transmittal Memorandum 改装传送备忘录

**MTM/D** Million Ton-Miles per Day 百万吨英里/日

**MTMA** Military Terminal Manoeuvring Area 军事终端演习区

Military Terminal Movement Area 军事终端运输区

Military Traffic Management Agency 军事交通管理局

**MTMIU** Module Test and Maintenance Bus

Interface Unit 模块测试和维护总线接口单元

**MTMS** Machine Tool Management System 机械工具管理系统

Multi Sensor Multi Target 多传感器多目标

**MTMTS** Military Traffic Management and Terminal Service 军事交通管理与终点终端局

**MTN** Mega Transport Network 兆位传送网络

Mobile Telephone Network 移动电话网

Motion 运动,移动

Multilateral Trade Negotiations 多边贸易谈判

**MTO** Make-to-Order 订货生产

Manufacture Technical Order 制造技术命令

Maximum Take-off 最大起飞(推力)

Meteo 气象

Microsystems Technology Office 微系统技术办公室

Mission, Task, Objective 飞行、任务与目标

Modification Task Outline 改型任务大纲

Multichannel Trace Oscillograph 多线示波器

**MTOA** Maintenance Training Organisation Approval 维修培训机构批准

**MTOE** Maintenance Table of Equipment 设备维修表

Maintenance Training Organization Exposition 维修培训机构博览会

Mission Table of Organization and Equipment 组织和设备任务表

Modified Table of Organization and Equipment 组织和设备修改表

**MTOGW** Maximum Take-off Gross Weight 最大起飞全重

**MTOM** Maximum Take-off Mass 最大起飞质量

Mission Tradeoff Methodology 任务协调方法,任务折衷方法

**MTONS** Metal-Thick-Oxide-Nitride Silicon 金属—厚氧化氮硅

**MTOP** Maintenance Task Operating Plan 维修任务工作计划

Maximum Take-off Power 最大起飞功率

**MTOPS** Million Theoretical Operations per Second 每秒百万次理论运算

**MTOR** Man, Technology, Organisation and Risk Management 人、技术、组织和风险管理

**MTOS** Metal-Thick Oxide Silicon 金属—厚氧化硅

**MTOW** Maximum Take-off Weight 最大起飞重量

**MTP** Mail Transfer Protocol 邮件传递协议

Maintenance and Test Panel 维修和测试面板

Maintenance Training Syllabus 维修培训大纲

Mandatory Technical Publication 指令性技术出版物

Master Test Plan 主测试计划

Maximum Tyre Pressure 最大轮胎压力

Message Transfer Part 信息传递部分

Minimum Time Path 最短时间航线

Minimum Tour Price 最低团体旅游票价

Minimum Trip Path 最短旅行航线

Mission Tape Preparation 飞行任务磁带准备

Modification Task Proposal 改型任务建议

**MTPI** Multiplex Timing Physical Interface 复用定时物理接口

**MTPR** Miniature Temperature Pressure Recorder 小型温度压力记录器

Multi-Tone Power Ratio 多重音调功率比值

**MTPT** Minimal Total Processing Time 最短总处理时间

**MTR** Magnetic Tape Recorder 磁带录音机

Main and Tail Rotor (直升机)旋翼和尾桨

Master Tool Record 主工具记录

Material Return to Stores 器材返回库房

Material Test Report 材料测试报告

Material Transfer Recorder 器材调拨记录器

Matter 事件

Maximum Tracking Range 最大跟踪距离

Maximum Transmission Ratio 最大传动比

Mean Time to Repair 平均检修时间

Mean Time to Restore 平均恢复时间

Meter 米,仪表,公尺

Metering 计量

Military Training Route 军事训练航线

Miniature Temperature Recorder 微型温度记录仪

Minimum Time Rate 最小时间率

Minimum Turning Radius 最小转弯半径

Modular Tree Representation 模块化树形表示法

Mono Tiltrotor 单倾转旋翼飞行器

Motor 马达,发动机,电动机

Multimedia Task Rate 多媒体占用率

Multiple Track Radar 多目标跟踪雷达

**MTRCL**　Motorcycle 摩托车

**MTRD**　Master Test Requirement Document 主测试需求文件

Metered 计量

**MTRDN**　Motor Driven 发动机驱动的,马达驱动的

**MTRE**　Multiple Tape Recorder End 多磁带记录器终端

**MTRF**　Mean Time to Repair Fault 修复故障平均时间

Module Test and Repair Facility 模块测试与修理设备

**MTRG**　Metering 计量

**MTRK**　Magnetic Track 磁航迹

**MTRL**　Material 材料

**MTRM**　Mach Trim 马赫配平

**MTRS**　Mean Time to Restore Service 恢复服务的平均时间

**MTRSM**　Multiple Tape Recorder Start Mattress 多磁带记录器启动垫

**MTS**　Mach Trim System 马赫配平系统

Machining Technology System 加工技术系统

Magnetic Tape Storage 磁带存储器

Main Traffic Station 主交通站

Maintenance Training Simulator 维修训练模拟机

Marked-Target Seeker 标示目标寻标器

Material Test Specification 器材测试规范

Material Test System 器材测试系统

Message Telecommunication Service 信息电信业务

Message Telephone Service 信息电话业务

Message Transfer Service 信息传递服务

Message Transfer System 信息传递系统

Message Transmission Subsystem 信息传输子系统

Metering Truss Structure 测量构架结构

Microwave Test Station 微波试验站

Mobile Telephone Service 移动电话业务

Mobile Telephone Switch 移动电话交换

Mobile Telephone System 移动电话系统

Mobile Telephony Subsystem 移动电话子系统

Mobile Trainer Set 活动训练装置

Modem Termination System 调制解调终端系统

Modular Test System 模块化调制系统

Motor Operated Transfer Switch 电动机操纵转换电门

Multi-channel Television Sound 多通道电视伴音

Multiple Transaction System 多重事务处理系统

Multiplex Timing Source 复用定时源

Multiterminal Time-sharing System 多终端分时系统

**MTSAT**　Multi-function Transport Satellite 多功能运输卫星

**MTSC**　Maintenance Technical Support Center 维修技术保障中心

Mobile Telephone Switching Center 移动电话交换中心

**MTSE**　Message Transfer Service Element 消息传送业务单元

Message Transfer Service Environment 消息传送业务环境

**MTSF**    Mean Time to System Failure 系统失灵前的平均时间

**MTSO**    Mobile Telephone Switching Office 移动电话交换局

**MTSQ**    Mechanical Time and Super-Quick 机械定时与瞬发(引信)

**MTSR**    Mean Time to Service Restoral 业务恢复前的平均时间

Mean Time to System Restoration 系统恢复前的平均时间

**MTT**    Machine Tool Technology 机床技术

Magnetic Tape Terminal 磁带终端机

Magnetic Tape Transmissions 磁带变速,磁带传送

Maintenance Task Time 维修任务时间

Maritime Test Terminal 海事测试终端

Master Tooling Template 主工装模板

Mean Test Time 平均测试时间

Meteorology 气象学

Microwave Theory and Techniques 微波理论与技术

Minimum(Least)Time Track 最短时间跟踪

Moving Target Tracker 运动目标跟踪

Multi Target Trade 多目标贸易

Multi-Target Tracking 多目标追踪

**MTTA**    Machine Tool Technologies Association 机床技术协会

Machine Tool Trades Association 机床贸易行业协会

**MTTCO**    Mean Time to Check out 检查平均时间

**MTTD**    Mean Time to Detection 故障检测平均时间

Mean Time to Diagnose 故障诊断平均时间

**MTTDA**    Mean Time to Dispatch Alert 签派告警平均时间

**MTTE**    Magnetic Tape Terminal Equipment 磁带终端设备

Mean Time to Error 平均差错时间

**MTTF**    Mean Time to Failure 平均故障时间

**MTTFF**    Mean Time to First Failure 首次故障前的平均时间

**MTTFSF**    Mean Time to First System Failure 系统首次发生故障的平均时间

**MTTI**    Mean Time to Install 平均安装时间

Multi Tank Indication 多油箱指示

**MTTM**    Mean Time through Maintenance 维修平均时间

Mean Time to Maintenance 维护平均时间

**MTTMA**    Mean Time to Maintenance Alert 维修提醒前的平均时间

**MTTO**    Mean Time to Overhaul 大修平均时间

**MTTR**    Maximum Time to Repair 最长修复时间

Maximum Time to Replace 最长更换时间

Mean Time to Recovery 平均恢复时间

Mean Time to Removal 平均拆卸时间

Mean Time to Repair 平均修理时间

Mean Time to Replace 平均更换时间

Mean Time to Report 平均报告时间

Mean Time to Respond 平均响应时间

Multitarget Tracking Radar 多目标跟踪雷达

**MTTRCF**    Manual Time to Repair Critical Failures 重大故障修理工时

**MTTRF**    Mean Time to Restore Function 平均恢复功能时间

**MTTRS**    Manual Time to Restore System 修复系统的工时

**MTTS**    Mean Time to Scrap 平均报废时间

Microwave Theory and Techniques Society 微波理论和技术协会(美)

Multitask Terminal System 多任务终端系统

Multi-Task Training System 多功能训练系统

Multi-Tone Test Signal 多重音调测试信号

**MTTSF**    Mean Time to System Failure 系统故障平均发生时间

**MTTUR**    Mean Time to Unscheduled Removal 非计划拆换前的平均时间

**MTU**    Magnetic Tape Unit 磁带组件

Material Training Unit 器材训练组件(单

位)

Maximum Transfer Unit 最大传递单元

Maximum Transport Unit 最大传送单元

Metric Units 公制单位,米制单位

Mobile Test Unit 移动式测试组件

Mobile Training Unit 移动式训练组件

**MTUP** Mobile Telephone User Part 移动电话用户部分

**MTUR** Mean Time between Unscheduled Removals 非预定拆卸平均间隔时间

**MTVAL** Master Tape Validation 主磁带验证

**MTVC** Manual Thrust-Vector Control 人工推力向量控制

Motor Thrust-Vector Control 发动机推力向量控制

**MTVDB** Milestone Tracking Visible Data Base 路标目视跟踪数据库

**MTW** Maximum Design Taxi Weight 最大设计滑行重量

Maximum Takeoff Weight 最大起飞重量

Maximum Taxi Weight 最大滑行重量

Mean Time to Wait 平均等待时间

Mountain Wave 山波,地形波

**MTWA** Maximum Take-off Weight Authorized 最大批准起飞重量

Maximum Total Weight Authorized 最大批准全重

**MTWR** Mean Time Waiting for Repair 等待修复平均时间

**MTWS** Mean Time Waiting for Spares 等待备件平均时间

**MTX** Mobile Telephone Exchange 移动电话交换机

**MTY** Empty 空的,空载的

Monterrey, Mexico ( Airport Code ) 墨西哥蒙特雷机场

**MTZ** Massada, Israel ( Airport Code ) 以色列马萨达机场

Motorized 机械化的

**MTZT** Multiple Time-Zone Travel 多时区飞行

**MU** Machine Unit 机组

Machine Units 机械组件

Machine Utilization 机械效用

Main Unit 主单元

Maintenance Unit 维修单位,维修部队

Management Unit 管理单元,管理组件

Mark Up 标记出

Mass Unit 质量单位

Master Update 主更新,主升级

Media Unit 媒体单元

Memory Unit 存储器组件

Microwave Unit 微波组件

Model Usage 模拟用途,典型用途

Multiple Unit 多元

Multiuser 多用户

**MUMT** Multi-User Multi-Task 多用户多任务

**MUA** Maximum Usable Altitude 可利用的最高高度

Miniature Unmanned Aircraft 小型无人驾驶飞机

**MUAL** Manual 手册,手工的

**MUC** Multiplexer Unit 多路复用器组件

Munich, Germany-Franz Josef Strauss ( Airport Code ) 德国慕尼黑机场

**MUD** Multi-User Detection 多用户检测

Multi-User Dialogue 多用户对话

Multi-User Dimension 多用户维数

Multi-User Domain 多用户域

**MUDR** Multidetail Drawing 多详图图纸

**MUF** Maximum Usable Frequency 最大可用频率

Muffler 消音器

**MUG** Munich, Germany ( Airport Code ) 德国慕尼黑机场

**MUI** Multimedia User Interface 多媒体用户接口

Multi-User Interference 多用户干扰

**MUIV** Mobile Unit Identity Vector 移动单元特性矢量

**MUK** Multimedia Upgrade Kit 多媒体升级套

件

**MULDEM** Multiplexer-Demultiplexer 复用器—解复用器

**MULE** Modular Universal Laser Equipment 万用模件化激光设备

**MULR** Muller 研磨机

**MULT** Multiple 多重的,复合的

Multiplicand 被乘数

Multiplication 倍数,放大

Multiplier 乘法器,倍频器

**MULT-ENG** Multi-Engine 多发动机的

**MULTH** Multilith 简易平版印刷品

**MULTICS** Multiplexed Information and Computing System 信息多路传输与计算系统

**MULTR** Multimeter 万用表

**MUM** Multi-Unit Message 多单元消息

**MUN** Municipal 城市的,市政的

**MUNDI** Multiplex Network Distribution and Interactive Service 复用网络分布与交互业务

**MUP** Mockup Problem 样机问题,模型问题

**MUR** Maneuver 机动

**MURS** Multi-Use Radio Service 多用途无线电服务

Multi-Use Radio Systems 多用途无线电系统

**MUS** Machine Utilization System 机械利用系统

Minimum Use Specification 最低使用技术标准

Multi User System 多用户系统

Museum 博物馆

Music 音乐

**MUSB** Bachelor of Music 音乐学士

Mobile Unit Support Base 活动组件支援基地

**MUSDAB** Multi Source Data Base 多源数据库

**MUSE** Multi User System Environment 多用户系统环境

**MUSS** Mobile Unit Support System 活动组件支援系统

**MUT** Mean Up-Time 平均正常使用时间

Mockup Template 模型模板

Multilateral 多边的,多侧面的

Mutual 共同的

**MUTE** Mute Passenger 丧失说话能力的旅客

**MUTO** Multi-User Telecommunications Outlet 多用户电信系统插座

**MUX** Multan, Pakistan (Airport Code) 巴基斯坦木尔坦机场

Multi-Channel 多通道

Multiplex 多路的

Multiplex System 多路复用系统,多路传输系统

Multiplexer 多路转换器,多路调制器

**MV** Mean Variation 平均偏差

Measured Value 测量值

Medium Voltage 中(等电)压

Meteological Visibility 气象能见度

Metering Valve 计量活门

Millivolt 毫伏

Miniature Vehicle 微型车,微型运载工具

Move 移动

Multivibrator 多谐振荡器

Muzzle Velocity 喷口速度

**MVA** Minimum Vectoring Altitude 最低引导飞行高度,最低雷达引导飞行高度

**MVAR** Magnetic Variation 磁差

**MVB** Motor V-Belt 马达 V 型皮带

**MVBL** Movable 可移动的,可活动的

**MVC** Manual Volume Control 人工音量控制

Mobile Virtual Circuit 移动虚拟线路

Multimedia Virtual Circuit 多媒体虚拟线路

**MVCP** Miniature Vehicle Computer Program 微型飞行器计算机程序

**MVD** Microwave Video Distribution 微波视频分配

Miniature Vehicle Development 微型飞行器开发

Montevideo, Uruguay-Carrasco (Airport Code) 乌拉圭蒙得维的亚卡拉斯科机场

Motion Vector Data 移动矢量数据

Multipurpose Video Display 多用途视频显示

**MVDF** Medium and Very High Frequency Direction Finder 中频和甚高频定向台

**MVDS** Microwave Video Distribution System 微波视频分配系统

Modular Video Data System 模块化视频数据系统

Multipoint Video Distribution System 多点视频分布系统

**MVE** Miniature Vehicle Emulator 微型飞行器仿真程序器

**MVEFF** Miniature Vehicle Effectiveness 微型飞行器效率

**MVFR** Maintain Visual Flight Rules 保持目视飞行规则

Marginal Visual Flight Rules 临界目视飞行规则

**MVG** Master Vertical Gyro 主垂直陀螺

Moving 移动

**MVGA** Monochrome VGA 单色视频图形显示卡

**MVGVT** Mated Vehicle Ground Vibration Test 配套航空器地面振动试验

**MVIP** Multi-Vendor Integration Protocol 多厂商综合协议

**MVIS** Miniature Vehicle Installation Specification 小型航天器安装规范

**MVL** Motion Video Library 移动视频库

**MVM** Minimum Virtual Memory 最小虚拟存储器

Multimedia Voice Modem 多媒体语音调制解调器

**MVMT** Movement 运动,移动

**MVNO** Mobile Virtual Network Operator 移动虚拟网运营商

**MVP** Magnetic Vector Potential 磁矢量位能

Maximum Velocity Performance 最大速度性能

Media Vision Pocket 媒体视觉器

Multimedia Video Processor 多媒体视频处理器

Multi-Variable Programming 多变量程序设计

**MVPN** Mobile Virtual Private Network 移动虚拟个人网络

**MVR** Maneuver 机动,动态

Multi-channel Voice Recorder 多频道话音记录器

**MVS** Magnetic Voltage Stabilizer 磁电压稳压器

Miniature Vehicle Simulation 小型运输工具模拟

Miniature Vehicle System 小型运输工具系统

Minimum Visual Signal 最低目视信号

Multiple Virtual Storage 多虚拟存储器

Multiple Virtual System 多重虚拟系统

Multivirtual Storage 多重虚拟存储器

**MVS/SP** Multiple Virtual Storage/System Product 多虚拟存储器/系统产品

**MVS/XA** Multiple Virtual Storage/Extended Architecture 多虚拟存储器/扩展结构

**MVSF** Miniature Vehicle Support Function 小型运载工具支援功能,小型运输工具保障功能

**MVSIM** Miniature Vehicle Simulator 小型运载工具模拟器

**MVSP** Maintain Visual Separation 保持目视间隔

**MVT** Aircraft Movement Message 班机动态文电

Manual of Vehicle Training 飞行器训练手册

Manufacturing Verification Test 制造验证测试

Module Verification Test 模组件取证测试

Movement 运动,动态

Multiprogramming Variable Tasks 可变任务数量多程序设计

**MVW** Maximum VTOP Weight 最大垂直起飞和着陆重量

**MVX** Miniature Vehicle X-Model （Flight

Path Simulator) X 型小型活动装置(飞行航迹模拟器)

**MVY** Martha's Vineyard, MA, USA (Airport Code) 美国麻萨诸塞州马萨葡萄园机场

**MW** Management Workstation 管理工作站

Manual of Work 工作手册

Master Warning 主警告

Medium Wave 中波

Megawatt 兆瓦,百万瓦

Micro Wave 微波

Milliwatt 毫瓦

**MWA** Management Work Authorization 管理工作授权

Marion, IL, USA (Airport Code) 美国伊利诺伊州马里恩机场

Mobiled Wireless Access 移动式无线接入

**MWAE** Minimum Weighted-Absolute-Error 最小加权绝对误差

**MWAN** Mobile Wireless Access Network 移动无线接入网

**MWAR** Major World Air Route 世界主要航线

**MWARA** Major World Air Route Area 世界主要航线区

**MWAVE** Microwave 微波

**MWC** Main Wheel Circuit Card 主轮电路板

Master Warning Card 主警告卡

Master Warning Computer 主警告计算机

Multi-Way Calling 多方呼叫

**MWCC** Master Warning and Caution Controller 警告与警示主控制器

**MWDP** Master Warning Display Panel 主警告显示板

**MWDS** Megawords 百万字

**MWE** Manufacturer's Weight Empty 制造厂空载重量

Maximum Weight Empty 最大空载重量

**MWFL** Multiwavelength Fiber Laser 多波长光纤激光器

**MWG** Maintenance Working Group 维修工作组

Multiwavelength Grating 多波长光栅

**MWHGL** Multiple-Wheel Heavy Gear Load 多轮重型起落架负载

**MWI** Message Waiting Indicator 信息等待指示器

**MW-IC** Microwave Integrated Circuit 微波集成电路

**MWIR** Medium Wave Infrared 中红外波

**MWKS** Mechanical Workshop 机务车间

**MWL** Master Warning Light 主警告灯

Maximum Wind Level 最大风高度层

Mean Water Level 平均水平面

**MWO** Maintenance Work Order 维修工作指令

Meteorological Watch Office 气象观测室

Modification Work Order 改装工作指令

**MWP** Maneuvering Work Platform 机动工作平台

Master Warning Panel 主警告面板

Maximum Working Pressure 最大工作压力

Membrane Waterproofing 防水薄膜

Meteorologist Weather Processor 气象学家的天气处理器

**MWR** Maintenance Work Request 维修操作要求

Material Withdrawal Requisition 器材提取要求,器材回收要求

Mean Width Ratio 平均宽度比

Method of Weight Residual 加权剩余法

Microwave Radiometry 微波辐射测量术

Millimeter-Wave Radar 毫米波雷达

Millimeter-Wave Radiometer 毫米波辐射计

**MWS** Maintenance Work Station 维护工作站

Manufacturing Work Statement 制造工作说明

Master Warning System 主警告系统

Microwave Station 微波站

Multiwavelength Switch 多波长交换(机)

Multiwork Station 多工作站

**MWT** Mate with Tape 与磁带匹配

Mean Waiting Time 平均等待时间

Medium Water Tender 中型水车

Microwave Tube 微波管

**MWTC** Memory Write Command 存储器写入指令

**MWTN** Multiwavelength Transport Network 多波长传送网络

**MWV** Maximum Working Voltage 最大工作电压

**MWW** Main Wheel Well 主轮舱

**MX** Maximum 最大

Mixed Type of Ice Formation 混合型冰体

Multiplex 多路通信

**MXC** Multiplexer Channel 复用器信道

**MXCH** Multiplex Channel 复用信道

**MXD** Mixed 混合的

**MXDCR** Mode Transducer 模式转换器

**MXFL** Mixed Flow 混合气流

**MXG** Mixing 混合

**MXP** Malpensa, Italy(Airport Code) 意大利米兰机场

Mexico Peso 墨西哥比索

Midway Express (USA) 中途快递公司(美国)

Milan, Italy-Malpensa (Airport Code) 意大利米兰尔彭萨机场

**MXR** Mixer 混频器, 混合器

**MXSH** Maximum Wind Shear 最大风切变

**MXT** Mixture 混合物

**MXU** Multiplexer Unit 复用器单元

**MXWND** Maximum Wind 最大风

**MYLTR** My Letter 我方函件

**MYMSG** My Message 我方函电

**MYP** Multi-Year Procurement 多年采购

**MYR** Malaysian Ringgit 马来西亚林吉特（货币单位）

Myrtle Beach, SC, USA（Airport Code）美国南卡来罗纳州默特尔比奇机场

**MYS** Malaysia 马来西亚

**MYTEL** My Telegram 我方电报

**MYTLX** My Telex 我方电传

**MZFCG** Maximum Zero Fuel Center of Gravity 最大无燃油重心

**MZFW** Maximum Design Zero Fuel Weight 最大设计无燃油重量

Maximum Zero Fuel Weight 最大无燃油重量

**MZL** Manizales, Colombia （Airport Code） 哥伦比亚马尼萨莱斯机场

Muzzle 喷口, 喷管, 喷嘴

**MZPI** Microwave Zone Position Indicator 微波区域位置指示器, 微波目标指示器

**MZT** Mazatlan, Sinaloa, Mexico （Airport Code）墨西哥马萨特兰机场

# N

N　Navigation 导航

　　Neon 氖

　　Newton 牛顿

　　Normal 正常

　　North 北方

　　Novelty 新产品

　　Number 数字,号码

　　Nylon 尼龙

N&G　Navigation and Guidance 导航和制导

N. O. R　Normal Operating Range 正常运行范围

N/A　No Advice 未通知

　　Not Applicable 不适用的

N/B　North Bound 向北,北飞

N/C　Numerical Control 数字控制

N/C/W　Not Complied with 不符合

N/D　Non-Destruction 无损伤

N/E　Not Enough 不够

N/H　Neighbourhood 邻近

N/K　Not Known 不详

N/L　No Ledger 无总账

N/M　No Marks 无标记

　　Not Marked 未注明

N/O　Navigation Officer 导航员,领航员

N/P　New Release 新版本,新颁发

　　Nut Plate 托板螺帽

N/W　Nose Wheel 前轮

N/WPT　Next Waypoint 下一航路点

N/WS　Nose Wheel Steering 前轮转弯

NA　Name 名字

　　Namibia 纳米比亚

　　National Association 国家协会

　　Navigation Aid 助航设备

　　Network Adaptor 网络适配器

　　Network Address 网络地址

Network Analyzer 网络分析器

Network Architecture 网络体系结构

Neutral Axis 中轴

Next Assembly 下一个组合件

No Answer 未回答

No Assignment 非规定

Noise Abatement 噪声抑制

North America 北美洲

North Atlantic ( Route ) 北大西洋(航路)

Not Above 不高于,不超过

Not Applicable 不适用的

Not Authorized 未批准的,未批准,不允许

Not Available 未提供的,空缺,无现货的

Numerical Aperture 数值孔径,数字窗

NAA　Narita Airport Authority ( Japan ) 成田机场管理局( 日本 )

　　National Aeronautic Association ( USA ) 国家航空协会( 美国 )

　　National Airports Authority ( India )国家机场管理局( 印度 )

　　National Airworthiness Authority 国家适航当局

　　National Assessment Agency ( UK ) 国家评估机构( 英国 )

　　National Association of Accountants 国家注册会计师协会

　　National Aviation Authority 国家航空当局

　　North American Aviation 北美航空公司( 美国 )

NAAA　National Agricultural Aviation Association 全国农业航空协会( 美国 )

NAACS　National Association of Aircraft and Communication Suppliers 国家飞机和通信供应厂商协会

NAAQO　National Ambient Air Quality Objective（Canada）国家环境空气质量目标(加拿大)

**NAAQS** National Ambient Air Quality Standards 国家环境空气质量标准

**NAAS** Naval Auxiliary Air Station 海军辅助航空站

**NAATS** National Association of Air Traffic Specialists 全国空中交通专业人员协会

**NAB** Network Adapter Board 网络适配板
Network Address Block 网络地址块

**NAC** Nacelle 短舱,吊舱,进气道
Nacelle Air Cooling 吊舱空气冷却
National Advisory Committee 国家咨询委员会
National Aero Club 国家航空俱乐部
National Aeronautics Council 国家航空委员会
National Air Council 全国航空委员会
National Aviation Club 全国航空俱乐部(美国)
National Aviation Corporation 全国航空公司(美国)
Network Access Controller 网络接入控制器
Network Administration Center 网络管理中心
Non-Airline Carrier 非定期航线承运人
North Atlantic Council 北大西洋理事会
Nozzle Area Control 喷嘴区域控制

**NACA** National Advisory Committee for Aeronautics（美国）国家航空咨询委员会
National Air Carrier Association 国家航空承运人协会(美国)

**NACATS** North American Clear Air Turbulence Tracking System 北美晴空湍流跟踪系统

**NACBL** Nacelle Buttock Line（发动机）吊舱纵剖线

**NACC** National Aeronautic Coordinating Committee 全国航空协调委员会
National Automatic Controls Conference 全国自动控制会议

Network Administration Computer Center 网络管理计算机中心
North American, Central American and Caribbean 北美、中美和加勒比
NTC Airspace Control Center NTC 空域管制中心

**NACK** Negative Acknowledgment 出错通知

**NACLIN** Nacelle Linings 吊舱衬里,短舱衬里

**NACO** Netherlands Airport Consultants 荷兰机场咨询

**NACOA** National Advisory Committee on Oceans and Atmosphere 国家海洋与大气咨询委员会

**NACS** Nonlinear Automatic Control System 非线性自动控制系统

**NACSEM** National Communication Security/Emanation Memorandum 国家通信保密/辐射备忘录

**NACSTA** Nacelle Station（发动机）吊舱站位,短舱站位

**NACWL** Nacelle Water Line 吊舱水线

**NAD** Noise Amplitude Distribution 噪声幅度分布
North American Datum 北美数据(基准)
Not on Active Duty 非现役的
Number Average Diameter 数量平均直径

**NADAC** Navigation Data Assimilation Computer 导航数据类比计算机

**NADAR** North American Data Recorder 北美数据记录器

**NADB** Netherlands Aircraft Development Board 荷兰飞机发展委员会

**NADC** NATO Air Defense Committee 北约防空委员会
Naval Air Development Center 海军航空发展中心

**NADGE** NATO Air Defense Ground Environment 北大西洋公约组织防空地面环境

**NADIN** National Airspace/Aeronautical Data Interchange Network 国家空域数据交换网

**NADP** Noise Abatement Departure Procedures 减少噪音起飞程序

**NAE** National Academy of Engineering 国家工程科学院

National Aeronautical Establishment 国家航空研究中心

Network Addressing Extension 网络寻址扩展

Noise Augmentation Equipment 噪声增强设备

**NAE/E** Non-Avionics Electronic/Electrical 非航空电子/电气的

**NAEC** National Aerospace Educational Council 国家航空航天教育委员会(美国)

Naval Air Engineering Center 海军航空工程中心

**NAECON** National Aerospace and Electronics Conference 全国航空航天电子学会议(美国)

National Electronics Convention 国家电子公约

**NAEW** NATO Airborne Early Warning 北大西洋公约组织空中预警系统

**NAF** Naval Air Facility 海军航空设施

Naval Aviation Foundation 海军航空基金

Naval Avionics Facility 海军航空电子设施

Network Access Facility 网络存取设备

**NAFAX** National Facsimile Network 国家传真网

**NAFEC** National Aviation Facilities Experimental Centre 国家航空设备试验中心(美国)

**NAFI** National Association of Flight Instructors 飞行教员全国协会(美国)

**NAFIS** Navigational Air Flight Inspection System 导航飞行检查系统

**NAFP** National Aeronautical Facilities Program(me) 国家航空工程设施规划

**NAFS** Naval Aircraft Factory Standards 海军飞机工厂标准

**NAFTA** North American Free Trade Agreement 北美自由贸易协定

**NAG** Nagpur, India (Airport Code) 印度那格浦尔机场

Navigation and Guidance 导航和制导

Node Address Generator 节点地址发生器

**NAGC** Noise Automatic Gain Control 噪声自动增益控制

National Art Gallery of China 中国国家艺术馆

**NAH** Naha,Japan (Airport Code) 日本那霸机场

**NAHO** Network-Aided Controlled Hand-Over 网络辅助控制的越区切换

**NAI** Nacelle Anti-Icing 短舱防冰,进气道防冰

Nanjing Aeronautical Institute 南京航空学院

Negro Airmen International (美国)黑人飞行员国际组织

Netherlands Aerospace Industries 荷兰航空航天工业公司

Network Access Identifier 网络接入标识符

**NAIA** National Association of Intercollegiate Athletics 全国大学生校际体育运动协会

**NAIF** Navigation Auxiliary Information Facility 导航辅助信息设备

**NAILS** National Airspace Integrated Logistics Support 国家空域综合后勤保障

**NAIOP** Navigational Aid Inoperative for Parts 导航设备因缺零件未工作

**NAIP** Network Administration Implementation Program 网络管理应用程序

Nozzle/Afterbody Installed Performance 尾喷管/后机身安装性能

**NAIS** Navigation Accuracy Instrumentation System 导航精密仪表系统

**NAK** Negative (Technical) Acknowledgement 否认,否定应答

Not Acknowledged 否认的,未确认的,未应答的

**NAL** National Accelerator Laboratory 国家加速器实验室(美国)

National Aeronautical Laboratory 国家航空实

验室(印度)

National Aerospace Laboratory 国家航空与航天实验室(日本)

National Air Lines(USA) 国家航空公司(美国)

Network Application Layer 网络应用层

**NALF**   Naval Auxiliary Landing Facility 海军辅助着陆设施

Naval Auxiliary Landing Field 海军辅助着陆机场

**NALO**   Naval Air Logistics Office 海军后勤办公室

**NALT**   Next Available Landing Time 下一次可用的着陆时间

**NAM**   National Air Museum 国家航空博物馆

National Association of Manufacturers 国家制造厂协会(美国)

National Aviation Museum (Canada) 国家航空博物馆(加拿大)

Navigational Accuracy Module 导航精密模块

Network Access Machine 网络存取机

Network Access Method 网络访问方法

Network Access Module 网络接入模块

Network Admission Management 网络允许管理

Network Analysis Model 网络分析模型

Non-Aligned Movement 不结盟运动

Numeric Assignment Module 数字赋值模块

**NAMA**   Nigerian Airspace Management Agency 尼日利亚空域管理机构

**NAMC**   Nanchang Aircraft Manufacturing Co. 南昌飞机制造公司

**NAMFAX**   National and Aviation Meteorological Facsimile Network 国家与航空气象传真网

**NAMLBL**   Name Label 名字标签

**NAMLBLD**   Name Labeled 已贴标签

**NAMO**   Naval Aircraft Maintenance Orders 海军飞机保养指令

**NAMP**   Naval Aircraft Maintenance Program 海军飞机保养计划

Non-Avionics Monitoring Processor 非航空电子监控处理机

**NAMS**   Name and Address Management System 姓名与地址管理系统

National Air Monitoring Station 国家航空监测站

National Air Monitoring System 国家航空监测系统

Nautical Air Miles 海里航程

Network Account Management System 网络账户管理系统

Network Administration and Management System 网络经营及管理系统

Network Analysis and Management System 网络分析及管理系统

**NAMSA**   NATO Maintenance and Supply Agency 北约维修与供应处

**NAMSO**   Navy Maintenance Support Office 海军维修保障处

**NAN**   Nadi, Fiji (Airport Code) 斐济纳迪机场

Navigation 导航,领航

Network Access Node 网络访问节点

Network Application Node 网络应用节点

**NANAC**   National Aircraft Noise Abatement Council 国家航空器消减噪音委员会

**NANRC**   National Aviation Noise Reduction Committee 国家降低航空噪声委员会

**NAOCC**   Non Aircraft Operating Common Carrier 无航空经营权的公共承运人

**NAOS**   North Atlantic Ocean Station 北大西洋海洋(导航)站

**NAP**   Naples, Italy (Airport Code) 意大利那不勒斯机场

National Action Plan 国家行动计划

National Allocation Plan 国家分配计划

Naval Aviation Pilot 海军航空兵驾驶员

Network Access Point 网络接入点

Network Access Pricing 入网费

Network Access Process 网络访问过程

Network Access Processor 网络存取处理器

Network Access Protocol 网络访问协议

New Airplane Program 新飞机程序

No Agency Purchase 非部门采购

No Appropriated Purchase 非拨款购买

Noise Abatement Procedure 消音程序,噪音抑制程序

Normal Acceleration Point(超音速运输机的)正常加速点

Not at Present 非目前

**NAPA** National Academy of Public Administration 国家公共管理科学院(美国)

National Association of Purchasing Agents 国家采购代理商协会

**NAPCA** National Air Pollution Control Administration 国家空气污染控制署

**NAPD** New Airplane Product Definition 新飞机产品定义

**NAPP** National Association of Priest Pilots 全国牧师飞行员协会(美国)

**NAPPR** Not Approved 未批准的

**NAPS** Night Aerial Photographic System 夜视航空摄影系统

**NAPT** Network Address Port Translation 网络地址端口转换

**NAPTIC** National Air Pollution Technical Information Center 国家空气污染技术信息中心

**NAR** Narrow 狭窄的

National Airspace Redesign 国家空域重组

National Airspace Review 国家空域复审

No Action Required 无需采取措施

No Answer Required 不需要回答

Non-Altitude Reporting 无高度报告

North American Routes 北美航路,北美航线

North Atlantic Region 北大西洋地区

North Atlantic Route 北大西洋航路

**NARACS** National Radio Communications System 国家无线电通信系统

**NARAST** North Asia Regional Aviation Safety Team 北亚地区航空安全小组

**NARC** National Aeronautical Research Committee 国家航空研究委员会

North American Rockwell Corporation 罗克韦尔北美公司

**NARF** Nuclear Aircraft Research Facility 核动力飞机研究设施

**NARI** National Aviation Research Institute 国家航空研究学会

**NARL** National Aero Research Laboratory 国家航空研究实验室

**NARO** Naval Aircraft Repair Organization 海军飞机维修组织

**NARTE** National Association of Radio and Telecommunications Engineering 国家无线电与电信工程协会(美国)

**NAS** Narrow-band Access Server 窄带接入服务器

Nassau, Bahamas(Airport Code)巴哈马拿骚机场

National Academy of Science 国家科学院(美国)

National Aerospace Standards 国家宇航标准(美国)

National Aircraft Standards 国家飞机标准

National Airspace System 国家空域系统

Naval Air Station 海军航空站

Navy and Army Standard 海军和陆军标准

Network Access Server 网络接入服务器

Network Administration System 网络管理系统

Network Application Support 网络应用支持

Network Application System 网络应用系统

Network Attached Storage 网络附加存储

**NASA** National Aeronautics and Space Agency 国家航空航天局(美国)

**NASAO** National Association of State Aviation Officials 国家州级航空官员协会

**NASB** Netherlands Aviation Safety Board 荷兰航空安全委员会

**NASC** National Aeronautics and Space Council 国家航空航天委员会(美国)

National Aerospace Standards Committee 国

家宇航标准委员会

National Aircraft Standards Committee 国家航空器标准委员会

National Association of Spotters Clubs 全国观测员俱乐部协会

Navy Aviation Safety Center 海军航空安全中心

**NASDA** National Space Development Agency 国家太空发展局(日本)

**NASEA** National Aviation Space Education Alliance 国家航空航天教育联盟

**NASI** Non-Avionics Simulator and Integrator 非航空电子系统模拟综合器

**NASL** Naval Applied Science Laboratory 海军应用科学实验室

**NASM** National Academy of Sports Medicine 国家运动医学会(美国)

National Air and Space Museum 国家航空航天博物馆(美国)

**NASN** National Aeronautic Surveillance Network 全国航空监视网

**NASNET** National Airspace System Network 国家空域系统网络

**NASP** National Aero Space Plane 国家航空航天飞机

National Aerospace Plan 国家航空航天计划

National Airport System Plan 国家机场体系计划

National Airspace System Plan 全国空域系统计划(美国)

**NASPALS** National Airspace System Precision Approach and Landing System 国家空域系统精密进近和着陆系统

**NASS** Navigation Satellite System 导航卫星系统

Numerical Aerodynamics Simulation System 数值空气动力学仿真系统

**NASTR** National Aeronautics and Space Administration Structural Analysis Program 国家航空与航天局结构分析大纲

**NASTRAN** National Aeronautics and Space Administration Structural Analyzer 国家航空与航天局结构分析员

**NAT** Natal, Rio Grande Do Norte, Brazil (Airport Code) 巴西北里奥格兰德州纳塔尔机场

National Air Transport 国家航空运输

National or Native or Natural 国家的,本土的,自然的

Nature or Natural 自然(现象)或自然的

Nearly Air-borne Truck 勉强可空运的卡车

Network Access Table 网络存取表

Network Address Transform 网络地址转换

Network Address Translation 网络地址转换,网址解析,网址翻译

Network Attached Table 网络连接表

No Action Taken 未采取行动

North Atlantic 北大西洋,北大西洋地区

North Atlantic Tracks 北大西洋航迹

North Atlantic Traffic 北大西洋交通

**NATA** National Air Transport Association 全国航空运输协会

National Aviation Trade Association 全国航空商业协会

National Aviation Training Association 全国航空培训协会

**NATC** Naval Air Test Center 海军航空测试中心

**NATCA** National Air Traffic Control Administration 国家空中交通管制局

National Air Traffic Controllers Association 国家空中交通管制员协会

**NATCC** National Air Transport Coordinating Committee 全国航空运输协调委员会

**NatCo** National Company 国营公司

**NATCS** National Air Traffic Control System 全国空中交通管制系统

**NATF** Naval Air Test Facility 海军航空试验设施

Navigation Aids Test Facility 导航辅助测试装置

**NATL** National 国际的,国家的

**NATO** North Atlantic Treaty Organization 北
大西洋公约组织

**NAT-PT** Network Address Translation-Proto-
col Translation 网络地址转换—协议转换

**NATS** National Air Traffic Service 国家空中
交通服务

North American Telephone System 北美电话
系统

**NATSF** Naval Air Technical Service Facility
海军航空兵技术服务设备

**NAU** Network Access Unit 网络接入单元,网
络访问单元

Network Accessible Unit 网络可存取部件

Network Address Unit 网络地址单元

Noise Augmentation Unit 噪声增大装置

**NAUA** National Aircraft Underwriters' Association 全国飞机保险商协会

**NAV** Navigation 导航,领航,航行

Net Asset Value 资产净值

**NAV RAD** Navigation Radio 无线电导航

**NAVAID** Air Navigation Facility 航空导航设
施

Navigational Aid 助航设施

**NAVAR** Navigational Radar 导航雷达

**NAVARHO** Navigation and Radio Homing 导
航与无线电归航

**NAVAUX** Navigation Auxiliary 辅助导航

**NAVCM** Navigation Countermeasure and Deception 导航干扰与伪装

**NAVD** North American Vertical Datum 北美
垂直向数据

**NAVDAC** Navigation Data Assimilation Center 导航数据同化中心

**NAVLT** Navigation Light 航行灯

**NAVPUP** Navigation Performance Update
Program 导航性能更新大纲

**NAVSAT** Navigation Communication Satellite
导航通信卫星

Navigation Satellite 导航卫星

Navigation Satellite System 导航卫星系统

**NAVSIMDWR** Navigation Simulator Drawer
导航模拟机绘图员

**NAVSTAR** Navigation Satellite Timing and
Ranging 导航卫星测时测距

**NAW** Night/Adverse Weather 夜间/恶劣天
气

**NAWAU** National Aviation Weather Advisory
Unit 国家航空天气咨询处(美国)

**NAWB** Neutral Air Waybill 航空货运代理运
单

**NAWP** National Aviation Weather Processor
国家航空气象处理机

**NAWS** National Aviation Weather System 国
家航空气象(预报)系统

**NB** Narrow Band 窄带

Network Board 网络插件板

Network Bridge 网桥

Nimbus 雨云

No Bias Relay 无偏置继电器

North Bound 向北,北飞

Northbound 向北飞的

**N-B** Near-end Block 近端块

**NBAA** National Board of Accountants and Au-
ditors 会计师和审计师全国委员会

National Business Aircraft Association 国家商
用飞机协会

National Business Aviation Association 国家
商务航空协会

**N-BBE** Near-end Background Block Error 近
端背景误块

**NBC** National Broadcasting Company 国家广
播公司

National Business Communication 国内商业
通信

North Bridge Chip 北桥芯片

**NBCM** Normal Brake Control Manifold 正常
闸控制总管

**NBE** North by East 北偏东方向

**NBF** Network Bite Function 网络机内测试功
能

**NBFM** Narrow Band Frequency Modulation
窄带调频

**NBH** Network Busy Hour 网络忙时

**NBL** Nacelle Buttock Line 吊舱(短舱)纵剖线

National Business League 国家商业联盟

**NBMA** Non-Broadcast Multi-Access 非广播型多址接入

**NBN** Narrow-Band Network 窄带网络

**NBO** Nairobi, Kenya-Jomo Kenyatta (Airport Code) 肯尼亚内罗毕乔莫·肯雅塔机场

**NBPM** Network-Based Project Management 基于网络的规划管理

**NBPT** No Break Power Transfer 无间断电源转换

**NBR** Number 号码,数,人数

**NBS** National Bureau of Standards 国家标准局(美国)

Network Bus System 网络总线系统

New British Standard 英国新标准

No Buffering Shaping 无缓冲整形

**NBSV** Narrow-Band Secure Voice 窄带保密电话

Normal Brake Selector Valve 正常闸选择活门

**NBTS** National Broadcast Television Subcommittee 国家广播电视委员会分会(美国)

**NBV** Net Book Value 纯账面价值

**NBW** Noise Bandwidth 噪声带宽

North by West 北偏西方向

**NC** American National Course 美国国家航路

National Coarse Thread 国家粗螺纹

Network Card 网络适配卡,网络卡

Network Code 网络代码

Network Computer 网络计算机

Network Computing 网络计算

Network Congestion 网络拥塞

Network Connection 网络连接

Network Control 网络控制

Network Controller 网络控制器

Network Coordination 网络协调

Neuro Computing 神经元计算

New Caledonia 新喀里多尼亚

No Change 无变化

No Charge 不加费用

No Coil 无线圈

No Connection 无连接

No Cost 免费,无代价

Node Computer 节点计算机

Noise Correlation 噪声相关

Noise Criterion 噪声标准

Non-Connected 非连接的

Non-Conventional Tube Ends 非常规管端头

Nordic Council 北欧理事会

Normally Closed 正常闭合位

Not Connected 未连接

Numerical Control 数字控制

**NC&M** Network Control and Management 网络控制与管理

**NCA** National Command Authorities 国家指挥当局

National Committee for Aeronautics 国家航空委员会

Network Computer Architecture 网络计算机体系结构

Nickel Copper Alloy 镍铜合金

Node Communication Area 节点通信区

Northern Control Area 北部管制区

**NCAA** National Civil Aviation Authority 国家民用航空局

National Council of Aircraft Appraisers 全国飞机鉴定委员会(美国)

**NCAM** National Center for Advanced Materials 国家先进材料中心,国家尖端材料中心

North Carolina Aviation Museum 北卡罗莱纳州航空博物馆(美国)

**NCAMP** National Center for Advanced Materials Performance 国家先进材料性能中心

**NCAN** Network Control Access Network 网控接入网络

**NCAR** National Center for Atmospheric Research 国家大气研究中心(美国)

**NCARC** National Civil Aviation Review Commission 国家民航评审委员会

**NCAS** Non-Call Associated Signaling 非呼叫相关信令

**NCASP** National Civil Aviation Security Programme 国家民用航空保安方案

**NCAT** National Center for Advanced Technologies 全国先进技术中心

National Center for Clear Air Turbulence 国家晴空湍流中心(美国)

New Concepts and Advanced Technology 新概念和高级技术

**NCATMC** New Challenges in Aerospace Technology and Maintenance Conference 航空技术与维护新挑战大会

**NCATO** National Civil Aviation Training Organization 国家民用航空培训组织

**NCATT** National Center for Aircraft Technician Training 国家飞机技术人员培训中心(美国)

**NCB** National Certification Body 国家认证机构

Network Connect Block 网络连接块

**NCC** Nacelle Compartment Cooling 吊舱冷却

National Climate Centre 国家气候中心

National Computer Conference 国家计算机会议(美国)

Navigation Computer Control 导航计算机控制

Netscape Communications Corporation 网景通信公司

Network Communications Controller 网络通信控制器

Network Computer Center 网络计算机中心

Network Control Center 网络控制中心

Network Control Computer 网络控制计算机

Network Coordination Center 网络协调中心

Node Control Centre 主节点控制中心

**NCCAM** National Coordinating Committee for Aviation Meteorology 国家航空气象协调委员会

**NCCD** Network-dependent Call Connection Delay 网络相关呼叫连接延迟

**NCCF** Network Communications Control Facility 网络通信控制设备

**NCCS** Navigation Command and Control System 导航指挥和控制系统

Numerical Control Coordination Sheet 数字控制协调单

**NCD** Network Cryptographic Device 网络密码装置

No Computed Data 无计算数据

Nose Cargo Door 机头货舱门

Not Computed Data 未计算数据

**NCDR** Numerical Control Deck Reporting 数字控制舱报告

**NCDS** Numerical Control Distribution System 数字控制分配系统

**NCDU** Navigation Control Display Unit 导航控制显示组件

**NCE** Navigation and Command Equipment 导航与指挥设备

Net Control Element 网络控制单元

Network Connection Element 网络连接元件

Network Control Equipment 网络控制设备

Nice, France (Airport Code) 法国尼斯机场

**NCEP** National Centers for Environmental Prediction 国家环境预测中心(美国)

National Clean Energy Project 国家清洁能源项目

**NCF** Network Configuration Facility 网络配置设备

Network Connection Failure 网络连接故障

Network Control Facility 网络控制设备

Non Crimp Fabric 无皱纤维

**NCFR** No Cause for Removal 无拆卸理由

**NCH** Network Connection Handler 网络连接处理程序

Noise Compensated Headset 噪音补偿耳机

Notched 带切口的

**NCHO** Network Controlled Hand-off 网络控制的越区转接

Network Controlled Hand-over 网络控制的越区切换

**NCI** Navigation Control Indicator 导航控制指示器
Network Circuit Interface 网络电路接口
Next Change Item 下一个更改项目
Non Code Information 非编码信息

**NCIA** Native Customer Interface Architecture 本地客户接口结构

**NCID** Network Clear Indication Delay 网络拆线指示延迟

**NCIU** NEXRAD Communications Interface Unit 下一代气象雷达通信接口装置

**NCL** Network Command Language 互连网络指令语言
Network Connection Link 网络连接链路
Network Control Language 网络控制语言
Newcastle, England, United Kingdom (Airport Code) 英国纽卡斯尔机场

**NCLK** Network Clock 网络时钟

**NCLRA** National Control Line Racing Association 国家线模型(竞赛)协会

**NCLT** Night Carrier Landing Trainer 运输机夜间着陆练习器

**NCM** Network Configuration Manager 网络配置管理人
Network Control Message 网络控制消息
Network Control Module 网络控制模块
Node Computing Memory 节点运算存储空间
Noncorrosive Metal 不锈金属
Numerical Control Machine 数字控制机器
Numerical Control Media 数字控制媒介

**NCMC** National Continuous Monitoring Coordinator 国家持续监测协调员
National Crisis Management Centre 国家危机管理中心
National Critical Materials Council 国家重要物资委员会

**NCMIT** Numerical Control Miscellaneous Tool 数字控制综合工具

**NCMS** Network Channel Management System 网络信道管理系统
Numerical Control Management System 数字控制管理系统

**NCN** Network Control Node 网络控制节点

**NCO** Numerical Controlled Oscillator 数控振荡器

**NCOMBL** Noncombustible 不燃物,不燃的

**NCOMPAT** Incompatible 不相容的

**NCONF** Nonconformance 不一致

**NCP** Navigation Control Program 导航控制程序
Navigation Correction Panel 导航修正板
Netware Core Protocol 网件核心协议
Network Call Processor 网络呼叫处理器
Network Control Point 网络控制点
Network Control Processor 网络控制处理器
Network Control Program 网络控制程序
Network Control Protocol 网络控制协议
Normal Circular Pitch 正常圆齿距
Numerical Control Planning 数字控制计划

**NCPD** Numerical Control Program Description 数控程序说明

**NCPLD** Noncoupled 未耦合的

**NCPM** Network Control Processing Module 网络控制处理模块

**NC-PRMA** Non-Collision Packet Reservation Multiple Access 无冲突分组信息预留多址接入

**NCPS** Network Call Processing Subsystem 网络呼叫处理子系统

**NCPSR** Numerical Control Part Specification Record 数字控制零件规范记录
Numerical Control Programming Service Request 数字控制编程服务申请

**NCPT** Numerical Control Punch Tape 数字控制穿孔带

**NCPU** Network Call Processor Unit 网络呼叫处理器单元

**NCR** Non-Compliance Record 非符合性记录
Non-Compliance Report 非符合性报告
Nonlinear Correlation Receiver 非线性相关接收机

**NCRGET** Numerical Control Routing Get 数

字控制路线接通

**NCRHOL** Numerical Control Routing Hole 数字控制线路孔

**NCRLT** Numerical Control Reference Layout Template 数字式控制基准布局模板

**NCRN** Numerical Control Router Nesting 数字式控制嵌套路由器

**NCROUT** Numerical Control Routing 数字式控制路线

**NCRP** Non-Compulsory Reporting Point 非强制报告点

**NCRTAB** Numerical Control Routing Tabulation 数字式控制线路表

**NCRTAG** Numerical Control Routing Tagging System 数字式控制线路挂签系统

**NCRTXT** Numerical Control Routing Text 数字式控制线路报文

**NCS** National Communications System 国家通信系统(美国)

Negotiation Coordination Sheet 谈判协调页

Net Coordinating Station 通信网协调台

Netscape Communications Server 网景通信服务器

Network Communication System 网络通信系统

Network Computing System 网络计算系统

Network Control Station 网络控制站, 网络协调站

Network Control System 网络控制系统

Numerical Control System 数字式控制系统

**NCSCI** National Center for Standards and Certification Information 国家标准和认证信息中心

**NCSP** Network Compiler Simulation Program 网络编译程序的模拟程序

**NCSPL** Network Configuration Services Parameter List 网络配置服务参数表

**NCSS** Network Control Supervisor Station 网络控制监理站

**NCT** Network Control and Timing 网络控制和定时

Neutral Current Transformer 中线电流互感器

Nicoya, Costa Rica (Airport Code) 哥斯达黎加尼科亚机场

**NCTE** Network Channel Terminating Equipment 网络信道终端设备

**NCTI** National Center for Technology Innovation 国家技术创新中心

Non-Cooperative Target Identification 非协同目标识别

**NCTR** Non-Cooperative Target Recognition 非协同目标识别

Numerical Control Trouble Report 数字控制故障报告

**NCU** Navigation Computer Unit 导航计算机组件

Network Clock Unit 网络时钟单元

Network Communication Unit 网络通信单元

Network Configuration Unit 网络配置单元

Network Control Unit 网络控制单元

Nozzle Control Unit 喷管控制组件

NSS Control Unit 网络监视系统控制组件

**NCUC** Network Control Unit Controller 网络控制单元控制器

**NCUR** Navigational Computer Unit Readout 导航计算机单元读数

**NCV** Nacelle Cooling Valve 吊舱冷却活门

No Commercial Value 无商业价值

**NCWF** National Convective Weather Forecast 国家对流气象预报

**NCWX** No Change in Weather 天气无变化

**NCY** Annecy, France (Airport Code) 法国安纳西机场

**ND** Navigation Database 导航数据库

Navigation Display 导航显示

No Data 无数据

No Date 日期未定

No Drawing 无图

Nose Down 机头向下

Notching Die 开口模具, 开槽模具

**NDA** Non Disclosure Agreement 不公布协议

Non-Destructive Analysis 非破坏性分析

Non-Development Airlift 未开发空运工具

**NDAA** Non-Developmental Alternative Aircraft 未开发备用飞机

**NDAC** National Defense Advisory Committee 国防顾问委员会(美国)

**NDAR** Newly Delivered Aircraft Report 新交付飞机报告

**NDB** National Data Base 国家数据库

Navigational Data Base 导航数据库

Network Database 网络数据库

Non-Directional Beacon 无方向性信标台

**NDBC** Non-Discrete Beacon Code 连续信标码

**NDBMS** Network Database Management System 网络数据库管理系统

**NDBS** Network Database Subsystem 网络数据库子系统

Networked Database Service 网络数据库业务

**NDC** National Data Corporation 国际数据公司

National Destination Code 国家地址代码

National Development Company 国家发展公司

National Development Council 国家发展理事会

Navigational Digital Computer 数字式导航计算机

Network Diagnostic Control 网络诊断控制

New Discrete Command 新离散指令

Noninterruptible Direct Current 不间断直流电

**NDCL** New Diagram Check List 新图检查清单

**NDD** Network Data Dictionary 网络数据词典

Network Delivery Device 网络发送装置

**NDE** Network Design Engineer 网络设计工程师

Non-Destructive Evaluation 无损评估,非破坏式鉴定

**NDF** Navigation Display Function 导航显示功能

Network Design Facility 网络设计设施

Neutral Density Filter 中性滤光片

Normal Dispersion Fiber 正规色散光纤

**NDFA** Nd-Doped Fiber Amplifier 掺钕光纤放大器

**NDFFA** Neodemium-Doped Fluoride Fiber Amplifier 掺钕含氟光纤放大器

**NDI** Non Developmental Item 非开发项目

Non-Destructive Inspection 无损检测

**NDIA** National Defense Industrial Association 国防工业协会(美国)

New Doha International Airport 新多哈国际机场

**NDIS** Network Device Interface Specification 网络设备接口规范

**NDIW** Network Device Installation Wizard 网络设备安装向导

**NDJ** N' Djamena, Chad-N'djamena (Airport Code) 乍得恩贾梅纳机场

**NDL** Needle 针,针塞

Network Database Language 网络数据库语言

Network Definition Language 网络定义语言

Network Description Language 网络描述语言

**NDM** Network Design and Management 网络设计与管理

Noise Definition Manual 噪声定义手册

Normal Disconnect Mode 正常断开模式

**NDN** National Data Network 国家数据网络

New Data Network 新数据网络

**NDOS** Network Disc Operating System 网络磁盘工作系统

**NDP** Net Domestic Product 国内净产值

Network Definition Procedures 网络定义程序

Network Design Problem 网络设计问题

Normal Diametral Pitch 正常径节

**NDR** Network Data Reduction 网络数据简化

Network Data Reporting 网络数据报告

Network Data Representation 网络数据表示

**NDRB** Non Directional Radio Beacon 无方向性无线电信标台

**NDRC** National Development and Reform Commission 国家发展与改革委员会

**NDRO** Nondestructive Read-Out 无损读出

**NDS** Navigation Display System 导航显示系统

Network Directory Service 网络目录服务

Non-Dispersion Shifted 非色散位移

**NDSE** National Data Switching Exchange 全国数据交换局

**NDSF** Non-Dispersion Shifted Fiber 非色散位移光纤

**NDSS** Network Design Support System 网络设计支持系统

**NDT** Network Diagnostic Tool 网络诊断工具

Non-Destruction Test 无损探伤

Non-Destructive Testing 无损探伤,无损检测

**NDTC** Non-Destructive Testing Center 无损检测中心

**NDTM** Non-Destructive Testing Manual 无损检测手册

**NDTO** Non-Design Tool Order 非设计工具指令

**NDU** Navigation Display Unit 导航显示组件

**NE** Net Earnings 纯收益

Network Element 网元

Northeast 东北

**NEA** National Electronics Association 全国电子学会

National Engineers Association 全国工程师学会

Network Equivalent Analysis 网络等效分析

Northeast Airlines（USA）东北航空公司（美国）

Nuclear Energy Agency 核能机构

**NEAC** Noise and Emission Advisory Committee 噪声及污染排放咨询委员会

**NEACP** National Emergency Airborne Command Post 国家应急空中指挥所

**NEADS** Network Engineering Administrative Data System 网络工程管理数据系统

**NEAP** National Environmental Action Plan 国家环境行动计划

**NEB** National Energy Board 全国能源委员会

Noise Equivalent Bandwidth 噪声等效带宽

North-East Bound 东北向,飞往东北方向

**N-EBC** Near-end Errored Block Count 近端误块计数

**NEBS** Network Equipment Building System 网络设备构建系统

**NEC** National Electrical Code（US）国家电气规程（美国）

Necessary 需要

No Error Check 无差错校验

**NECS** National Electrical Code Standards 国家电气代码标准（美国）

**NEDA** Incapacitated Passenger（Medical Clearance is Required）行动不便的旅客（需提供医疗放行证明）

**NEDS** National Emission Data System 全国排放数据系统

**NEE** Near-End Error 近端错误

Noise Equivalent Exposure 噪声等效曝光量

**NEF** National Extra Fine Thread 国家标准极细牙螺纹（美国）

Network Element Function 网元功能

Noise Exposure Forecast 噪声暴露预报

Normal Economy Fare 正常经济舱票价

Not Expected Frame 非预期帧

**NEFD** Noise Equivalent Flux Density 噪声等效通量密度

**NEG** Negative 拒绝的,负的

Negotiation 谈判

**NEGO** Negotiate or Negotiation 交涉

**NEGPR** Negative Print 负电印刷

**NEGTN** Negotiation 交涉

**NEI** Noise Equivalent Intensity 噪声等效强度

Noise Equivalent Irradiance 噪声等效辐照度

**NEIC** Normal to Emergency Interlock Contactor 正常到紧急互锁接触器

**NEIL** Neon Indicating Light 氖指示灯

**NEL** Network Element Layer 网元层

**NELEC** Nonelectric 不用电的,非电的

**NEM** Noise Exposure Measure 暴噪测量

**NEMA** National Electrical Manufacturers Association 国家电气制造商协会

National Emergency Management Agency 国家应急管理机构

National Emergency Management Association 国家应急管理协会

National Equipment Manufacturers Association 国家设备制造商协会

**NEMF** Network Element Management Function 网元管理功能

Network Error Management Facility 网络错误管理设施

**NEMIS** Network Management Information System 网络管理信息系统

**NEMP** Nuclear Electro-Magnetic Pulse 核电子—磁脉冲

**NEMS** NASA Equipment Management System 国家航空航天局设备管理系统(美国)

National Emergency Management System 国家应急管理系统

Network Element Management System 网络单元管理系统

Noise Exposure Management System 噪音影响管理系统

**NEN** Network Equipment Number 网络设备编号

**NEO** Network Expansion Option 网络扩充选项

**NEOF** National Emergency Operations Facilities 国家应急指挥措施

No Evidence of Failure 无失效迹象

**NEP** Noise Equivalent Power 噪声等效功率

**NEPA** National Environment Protection Agency 国家环境保护局

National Environmental Policy Act 国家环境政策法案

Nuclear Energy for the Propulsion of Aircraft 推进飞机的原子核能

**NEPD** Noise Equivalent Power Density 噪音等效功率速度

**NERC** National Electronics Research Council 国家电子研究委员会

National Engineering Research Center 国家工程研究中心

National Environmental Research Center 国家环境研究中心

National Environmental Research Council 国家环境研究委员会

Natural Environmental Research Council 自然环境研究委员会

New En-Route Center 新航路中心

**NERD** Noise Equivalent Reflectance Difference 噪声等级反射比差

**NES** Not Elsewhere Specified 无处注明

**NESC** National Electrical Safety Code 美国电气安全代码

**NESDIS** National Environmental Satellite Data and Information Service (USA)国家环境卫星数据与信息服务(美国)

**NESHAP** National Emission Standard for Hazardous Air Pollutants 国家空中有害污染辐射标准

**NESP** Near End Signaling Point 近端信令点

**NESS** National Environmental Satellite Service 国家环境卫星局

**NET** Network 网络,网

Network Entity Title 网络实体名称

Network Equipment Technologies 网络设备技术

Noise Equivalent Temperature 噪音等效温度

Not Earlier Than 不早于

**NETBTP** Network Block Transfer Protocol 网络信息组传送协议

**NETD** Noise Equivalent Temperature Difference 噪声等效温差

**NETDDE** Network Dynamic Data Exchange 网络动态数据交换

**NETS** Negotiated Events Tracking System 谈判事件跟踪系统

Network Entity Title Sign 网络实体标号

**NETT** Network Technique 网络技术

**NETV** Network Television 网络电视

**NEUT** Neutral 中线,中性,中立的

Neutralizing 中性化

**NEVD** Networked Voice Data 网络上的语音数据

**NEWAC** New Aero Engine Core Concepts 新发动机核心机概念

**NEX(Nex)** Next Generation 下一代

**NEXCOM** National Executive Committee 全国执行委员会(英国)

Next Generation Air-to-Ground Communication 下一代空对地通信

**NEXRAD** Next Generation Radar 下一代雷达

Next Generation Weather Radar 下一代气象雷达,改进型气象雷达

**NEXST** National Experimental Supersonic Transport (Japan)国家试验超音速运输机(日本)

**NEXT** Near End Cross Talk 近端串话

New Exploratory Technologies 新的探索技术

**NF** National Fine Thread 国家标准细螺纹

National Formulary 国家公式汇编

Natural Frequency 固有频率

Navigation Function 导航功能

Negative Feedback 负反馈

Noise Factor 噪声系数

Noise Figure 噪声指数,噪声图

Noise Frequency 噪声频率

Norfolk Island 诺福克岛(澳大利亚)

**NFAC** National Full-scale Aerodynamics Complex 国家全尺寸空气动力综合试验设施

**NFAP** Network File Access Protocol 网络文件存取协议

**NFC** National Fire Codes 国家防火条例(美国)

National Freight Corporation 国家货运公司

Nose Fairing Container 头部整流罩外壳

Not Fully Closed 未全关

**NFCF** Notice for Changing Forecast 更改预估量的通知

**NFCS** Navigation and Flight Control System 导航和飞行控制系统

**NFD** No Fixed Date 无固定日期

Nodal Fault Diagnostics 节点故障诊断

**NFDC** National Flight Data Center 国家飞行数据中心

**NFDD** National Flight Data Digest 国家飞行数据摘要

**NFDPS** National Flight Data Processing System 国家飞行数据处理系统

**NFE** Network Front End 网络前端

**NFER** National Foundation for Education Research 全国教育研究基金会

**NFF** No Fault Found 未发现故障

**NFFS** National Free Flight Society 国家自由飞行学会

**NFM** Network Fault Management 网络故障管理

Network File Manager 网络文件管理器

Node Fault Model 节点故障模型

**NFMS** Navigation and Flight Management System 导航与飞行管理系统

Network Facility Management System 网络设施管理系统

**NFO** National Freight Organization 全国货运组织

Normal Frequency Offset 正常的频率偏移

Not Fully Open 未全开

**NFOT** Normal Fuel Oil Tank 额定燃油箱

**NFOV** Narrow Field of View 窄视场

**NFP** Normal Failure Period 正常损坏周期

**NFPA** National Fire Protection Association 国家消防协会

National Fluid Power Association 全国流体动力协会

**NFPMS** Noise and Flight Path Monitoring

System 噪声和飞行航路监视系统

**NFR** No Further Requirement 无进一步要求

**NFS** Network File Server 网络文件服务器

Network File System 网络文件系统

**NFSD** Nonfused 未保险的

**NFT** Network File Transfer 网络文件传递

No Fixed Time 无固定时间

Non-Functional Test 非性能测试

Normal Fuel-oil Tank 正常燃油箱

**NFU** Network Field Unit 网络字段单元

**NFY** Notify 通知

**NFZ** No-Fly Zone 禁飞区

Nuclear Free Zone 无核区

**NG** Nigeria 尼日利亚

No Good 不好的

Nose Gear 前起落架

Not Good, No Good 不好,不好的

**NG&C** Navigation, Guidance and Control 导航、制导与控制

**NGAP** Next Generation of Aviation Professionals 下一代航空专业人员

**NGATM** New Generation Air Traffic Manager 新一代空中交通管理人员

**NGATS** Next Generation Air Transportation System 下一代航空运输系统

**NGC** Next Generation Carrier 下一代运营商

**NGCU** Nose Gear Control Unit 前起落架控制组件

**NGE** Navigation Guidance Element 导航制导单元,导航制导元件

Navigation Guidance Equipment 导航制导设备

Non Ground Effect 无地效

**NGEN** Noise Generator 噪声产生器

**NGF** Navigation Flare 航行灯

**NGFV** Navigation Guidance Flight Vehicle 导航制导飞行器

**NGI** Next Generation Internet 下一代因特网

**NGL** Natural-Gas Liquid 液态天然气

**NGN** Next Generation Network 下一代网络

**NGO** Nagoya, Japan-Komaki (Airport Code) 日本名古屋小牧机场

Non-Governmental Organization 非政府机构

**NGP** Network Graphics Protocol 网络图形协议

No Glide Path 无下滑道

**NGPS** NAVSTAR Global Positioning System 导航星全球定位系统

**NGPT** Nose Gear Position Transducer 前起落架位置传感器

**NGR** Night-Goggle Readable 夜视镜可读的

**NGRS** National Geodetic Reference System 国家测地参考系统

**NGS** Nagasaki, Japan (Airport Code) 日本长崎机场

National Geodetic Survey 国家测地勘察

Navigation Guidance Segment 导航制导部分

Next Generation Switch 下一代交换机

Nose Gear Steering 前起落架转弯

**NGSA** Next-Generation Single-Aisle 下一代,单通道

**NGSO** Non-Geo Stationary Orbit 非对地静止轨道

**NGSS** Next Generation Satellite Systems 下一代卫星系统

**NGT** Negotiation 谈判,议付,转让

New Generation Trainer 新一代训练器

Next Generation Technology 下一代技术

Next Generation Transmitter 下一代发射机

Night 夜

Not Greater than 不大于

**NGTA** Next-Generation Twin-Aisle 下一代,双通道

**NGTCS** Next-Generation Target Control System 下一代目标控制系统

**NGTE** National Gas Turbine Establishment 国家燃气轮机机构

Next Generation Test Equipment 下一代测试设备

Next-Generation Telecommunications Equipment 下一代电信设备

**NGTN** Negotiation 谈判

Next Generation Transport Networks 下一代传送网络

**NGTOW** Normal Gross Takeoff Weight 正常起飞总重

**NGV** Nozzle Guide Vane 喷嘴导向叶片

**NH** Network Handler 网络处理程序
Nonhygroscopic 防潮的,不吸湿的

**NHA** National Health Association 全国卫生协会(美国)
Naval Helicopter Association 海军直升机协会
Next Higher Assembly 更高一级组合件
Nha Trang, Vietnam ( Airport Code ) 越南芽庄机场

**NHB** NASA Handbook 国家航空航天局手册

**NHC** National Hurricane Center 国家飓风中心

**NHE** Notes and Helps Editor 注解与提示编辑(器)

**NHIS** National Health Insurance Scheme 国家健康保险计划
Network Human-machine Interface Subsystem 网络人机接口子系统

**NHP** Network-Host Protocol 网络主机协议
Nominal Horse Power 额定马力

**NHRP** Next Hop Resolution Protocol 下一跳解析协议

**NHTI** Network Handler Terminal Interface 网络处理器终端接口

**NHV** Net Heating Value 净热值
Nuku Hiva, Marquesas Islands ( Airport Code ) 马克萨斯群岛努库希瓦机场

**NI** Network Identifier 网络标识符
Network Indicator 网路表示语
Network Interface 网络接口
Nicaragua 尼加拉瓜
Noise Index 噪声指数

**NIAC** National Industry Advisory Committee 国家工业咨询委员会
National Intelligence and Analysis Center 国家情报分析中心

Nuclear Information Analysis Center 核信息分析中心

**NIAM** Nijessen Information Analysis Method 尼森信息分析方法
Nijessen Information Analysis Model 尼森信息分析模型

**NIAST** National Institute for Aeronautics and System Technology 全国航空及系统技术学会(南非)

**NIB** Network Interface Board 网络接口板
Noninterference Basis 非干扰基础

**NIBS** Neutral Industry Booking System 中立空运业订座系统

**NIC** Navigation Information Center 导航情报中心
Navigation Integrity Category 导航完好性类别
Negative Impedance Converter 负阻抗转换器
Network Identification Code 网络标识代码
Network Independent Clock 与网络无关的时钟
Network Information Card 网络信息卡
Network Information Center 网络信息中心
Network Interface Card 网卡,网络接口卡
New Installation Concepts 新安装概念
Newly Industrialized Country 新工业化国家
Not in Contract 未包括在合同中

**NICAP** National Investigation Committee on Aerial Phenomena 全国空中现象调查委员会

**NICC** National Information Control Center 国家信息控制中心

**NICE** Network Information Control Exchange 网络信息控制交换
Network Integration Consultation Environment 网络综合协议环境

**NICS** National Interfacility Communications System 国家空域系统设施间的通信系统
NATO Integrated Communications System 北约一体化通信系统

**NID** Network Identification 网络识别码

Network Interface Device 网络接口装置
Node Identifier 节点标识符

**NIDL** Network Interface Definition Language 网络接口定义语言

**NIF** Network Information Files 网络信息文件
Note Issuance Facility 票据发行工具

**NIFA** National Intercollegiate Flying Association 全国大学飞行协会(美国)

**NIH** Not Invented Here 此处未虚构

**NII** National Information Infrastructure 国家信息基础设施

**NIJ** National Institute of Justice 全国司法学会

**NIL** Network Interface Layer 网络接口层
Network Interface Logic 网络接口逻辑
No Items Loaded or Manifested 无装载

**NIM** Network Interface Machine 网络接口机
Network Interface Module 网络接口模块

**NIMA** National Imagery and Mapping Agency 国家图像及测绘局
National Insulation Manufacturers Association 全国绝缘材料制造商协会

**NIMS** National Incident Management System 国家事故管理系统,国家突发事件管理系统
Network Information Management System 网络信息管理系统

**NIO** Network Interface Object 网络接口对象

**NIP** Network Interface Processor 网络接口处理器

**NIPC** National Infrastructure Protection Center 国家基础设施防护中心

**NIPCC** Nongovernmental International Panel on Climate Change 非政间国际气候变化专门委员会

**NIPO** Negative Input Positive Output 负输入正输出

**NIPP** National Infrastnucture Protection Plan 国家基本设施保护计划

**NIPS** Not in the Position Selected 不在选择的位置

**NIR** Near Infrared 近红外
Net Income Ratio 净收入比率
No Individual Requirement 无单独要求

**NIRS** Neutral Information Representation Scheme 中立位信息展示简图

**NIRT** Numerical Index and Requirements Table 数字索引与要求表

**NIRTS** New Integrated Rang Timing System 新型集成测距定时系统

**NIS** Network Information Server 网络信息服务器
Network Information Service 网络信息服务
Network Information System 网络信息系统
Network Interface Switch 网络接口交换(系统)
Not in Stock 无储备
Not-in-Service 非服务中,停业
Number Information Service 号码信息服务

**NISDN** Narrow-band Integration Services Digital Network 窄带综合业务数字网
National Integrated Services Digital Network 国家综合业务数字网

**NISL** Not in Stock Listed 未包括在库存清单中

**NISP** National Industrial Security Program 国家工业安全计划

**NISS** National Information Services and Systems 全国信息服务系统

**NIST** National Institute of Science and Technology 国家科学和技术研究院
National Institute of Standards and Technology 国家标准和技术研究院

**NIT** Nearly Intelligent Terminal 准智能终端
Network Interface Task 网络接口任务

**NITL** National Industrial Transportation League 国家工业运输联合会

**NIU** Network Interface Unit 网络接口单元
Not in Use 未使用

**NJ** Network Junction 网络节点

**NJCL** Network Job Control Language 网络作业控制语言

**NJE** Network Job Entry 网络工作入口

**NJI** Network Job Interface 网络作业接口

**NJP** Network Job Processor 网络作业处理程序

**NJR** National Job Recording 国内作业记录

**NK** Neck 颈

Not Known 不知道,不详

**NKC** Nouakchott, Mauritania-Nouakchott (Airport Code) 毛里塔尼亚努瓦克肖特机场

**NKL** Nkolo, Zaire (Airport Code) 扎伊尔恩科洛机场

**NKUTD** Not Kept Up-to-Date 到期不保存

**NL** Netherlands 荷兰

Network Layer 网络层

New Line 新航线

No Lead 无铅

No Limit 无限制

Noise Laboratories 噪声实验室

North Latitude 北纬

**NLA** Network Logical Address 网络逻辑地址

New Large Airplane 新的大型飞机

Next Level Aggregator 下层集合体

Non Listed Assembly 未标出的组件

Non-Linear Amplifier 非线性放大器

**NLAB** Non Labor 无人工,非劳务

**NLAM** Network Layer Address Management 网络层地址管理

**NLAs** New Larger Aeroplanes 新大型飞机

**NLC** Network Link Controller 网络链路控制器

Noise Level Computation (program) 噪声水平计算(程序)

Normalized Link Capacity 标称链路容量

**NLCG** Non-Linear Loop Clock Generator 非线性环路时钟发生器

**NLCHG** Normal Change 正常更改

**NLCP** Noise Level Calculation Program 噪音等级计算程序

**NLD** No Load 无负载

Non-Linear Distortion 非线性失真

**NLDM** Network Logical Data Management 网络逻辑数据管理

**NLES** Navigation Land Earth Station 导航陆地地球站

**NLF** Natural Laminar Flow 自然层流控制

Nearest Landing Field 最近降落机场

New Link Flag 新链路标志

**NLFM** Noise Level Frequency Monitor 噪声电平频率监控器

Non-Linear Frequency Modulation 非线性调频

**NLFT** No Load Frame Time 无负载帧时间,无负结构帧时间

**NLG** Non-Linear Gain 非线性增益

Nose Landing Gear 前起落架,鼻轮起落架

**NLLT** Net Laying Light 网配置灯

**NLM** Network Link Module 网络链路模块

Network Load Module 网络装入模块

Network Loadable Module 网络可加载模块

**NLN** Neural Logic Network 神经逻辑网络

**NLNR** Nonlinear 非线性的

**NLO** Non-Linear Optics 非线性光学

**NLOS** Nominal Line of Sight 标称视线

**NLP** Network Layer Packet 网络层信息分组

Non-Linear Processor 非线性处理器

Normal Link Pulse 链路正常脉冲

**NLPI** Network Layer Protocol Identifier 网络层协议标识符

**NLR** National Research Laboratory (the Netherlands) (荷兰)国家研究实验室

Netherlands National Aerospace Laboratory 荷兰国家航天实验室

Network Layer Relay 网络层转接

Noise Load Ratio 噪声负载比

**NLRB** National Labor Relations Board 国家劳动关系委员会

**NLRI** Network Layer Reachability Information 网络层可通达性信息

**NLS** Narrowband Local Switch 窄带本地交换机

National Language Support 本国语言支持

Network Layer Signaling 网络层信令(模块)

**NLSIM** Non-Linear Simulation 非线性仿真

**NLSP** Network Layer Security Protocol 网络层安全性协议

**NLT** Normal Lube Oil Tank 正常滑油箱

Not Later than 不迟于

Not Less than 不少于

**NLTS** Network Load Test System 网络负载测试系统

**NM** Name 名称,姓名

Nanometer 纳米,毫微米

Nautical Mile 海里( =1853.2 米),节

Navigation Module 导航模块

Net Manager 网络经理(管理员)

Netmeeting 网络会议

Network Management 网络管理

Network Manager 网络管理程序(器)

Network Module 网络模块

No Message 无信息

Noise Meter 噪声表

Noise Monitoring 噪声监控

Nonmetallic 非金属

Nuclear Magnetron 核磁控管

**NMA** National Management Association 全国管理协会(美国)

Network Management Application 网络管理应用

**NMAC** Near Mid-Air Collision 空中危险接近

**NMAG** Nonmagnetic 非磁性的

**NMC** National Meteorological Center 国家气象中心

Network Management Center 网络管理中心

Network Management Computer 网络管理计算机

Network Measurement Center 网络测量中心

Network Message Controller 网络消息控制器

No-Message-Crossing 无消息交叉

Non-Mission Capable 非任务能力的

**NMCC** National Military Command Center 国家军事指挥中心

Network Management Command and Control 网络管理指挥与控制

**NMCE** Network Monitoring and Control Equipment 网络监控和管制设备

**NMCS** Non Mission Capable-Supply 需要补给而不能执行任务

**NMD** Number Median Diameter 数量中值直径

**NMDPS** Network Management Data Process System 网络管理数据处理系统

**NME** Network Management Entity 网络管理实体

**NMF** Network Management Facility 网络管理设施

Network Management Framework 网络管理框架

Network Management Function 网络管理功能

**NMG** Network Management Gateway 网络管理网关

**NMHS** National Meteorological and Hydrological Service 国家气象和水文部门

**NMI** Network Management Interface 网络管理接口

Non Maskable Interrupt 非屏蔽中断,非屏蔽性中断

**NMI/HR** Nautical Mile per Hour 海里/小时

**NMIB** Network Management Information Base 网络管理信息库

New Material Introductory Briefing 新材料推广简介

**NMIC** National Meteorological Information Centre 国家气象信息中心

**NML** Network Management Layer 网络管理层

Normal 正常

**NMM** Network Management and Maintenance 网络管理和维护

Network Management Model 网络管理模型

Network Management Module 网络管理模块

Node Message Memory 节点信息存储器

**NMNL** Network Module Network Link 网络模块网络链路

**NMOD** Near Movie on Demand 准电影点播
New Model 新型号

**NMODAM** New Model Area Management 新型号区域管理

**NMP** Name Management Protocol 名字管理协议
Network Management Plan 网络管理计划
Network Management Processor 网络管理处理器
Network Management Protocol 网络管理协议

**NMPBC** Nonlinear Multipulse Block Coding 非线性多脉冲块编码

**NMR** No Maintenance Required 无维修要求
Nuclear Magnetic Resonance 核磁共振

**NMS** Navigation Management System 导航管理系统
Network Management Signal 网络管理信号
Network Management Software 网络管理软件
Network Management Subsystem 网络管理子系统

**NMSI** National Mobile Station Identify 国内移动台识别号码

**NMT** Network Management Terminal 网络管理终端
Noise Monitoring Terminal 噪声监控终端
Nordic Mobile Telephone 北欧移动电话
Not More than 不大于

**NMTBA** National Machine Tool Builders Association 全国机床制造商协会

**NMU** Navigation Management Unit 导航管理组件
Network Management Unit 网络管理单元

**NMVT** Network Management Vector Transport 网络管理向量传输

**NMWR** National Maintenance Work Requirement 国家维修工作要求

**NMWS** Network Management Web Server 网络管理 Web 服务器

**NN** National Network 国内网
National Number 国内号码
Need 需要
Network Node 网络节点
Network Number 网络号码
Neural Network 神经网络
No Name 无名称

**NNC** National Network Congestion 国内网拥塞
Normal Network Cause 正常网络条件

**NNCP** Network Node Control Point 网络节点控制点

**NND** Nation Network Dialing 国内网拨号

**NNE** North North East 东北北

**NNES** Neural Network Expert System 神经网络专家系统

**NNI** Network Node Interface 网络节点接口
Network to Network Interface 网间接口

**NNM** Network Node Management 网络节点管理
Network to Network Management Interface 网络间管理接口
Node to Node Message 节点间消息

**NNMC** National Network Management Center 全国网络管理中心

**NNP** National Navigation Plan 国家导航计划
Network Node Processor 网络节点处理器

**NNTP** Network News Transfer Protocol 网络新闻传输协议

**NNU** Normal Number of Users 规定的用户号码

**NNW** North-Northwest 西北北

**NO** Normal Operation 正常操作
Norway 挪威
Number 数,号码

**NOA** New Obligation Authority 新的支付权限
Non-linear Optical Amplifier 非线性光放大器
Non-linear Optical Antiwaveguides 非线性光反波导

Non-Operating Active 无有效操作

**NOAA** National Oceanic and Atmospheric Administration 国家海洋和大气局(美国)

**NOAO** National Optical Astronomy Observatories 国家光学天文台

**NOB** Not on Board 不在飞机上

**NOC** Network Operation Center 网络运行中心

Notice of Change 更改通知

**NOCC** National Operations Control Center 国家运行控制中心

Navigation Operational Checkout Computer 导航作业检查计算机

Network Operations Command Center 网络运行指挥中心

Network Operations Control Center 网络运行控制中心

**NOCCC** No Control Circuit Contacts 非控制电路接点,非控制电路触点

**NOCNG** No Change 不变

**NOCP** Network Operator Control Program 网络操作员控制程序

**NOD** Network Out-Dialing 网外拨号

News on Demand 新闻点播

**NOE** Nap of the Earth 贴地飞行,掠地飞行

**NODAL** Network-Oriented Data Acquisition Language 面向网络数据采集语言

**NOF** International NOTAM Office 国际航行通告室

Normal Operation Frame 正常运行帧

**NoFOCH** No Food Schedule Change（机上）供餐计划未变

**NOG** Numbering 编号,编码

**NOHOL** Not Holding 不等待

**NOI** Not Otherwise Identified 不另外标明

Notice of Inquiry 调查通知,询问通知

**NOL** Net Operating Loss 净经营损失,经营纯损额

Normal Operating Loss 正常运行损失

Normal Overload 正常超载,正常过载

**NOLD** Nonlinear Optical Laser Device 非线性激光器器件

**NOLM** Nonlinear Optical Loop Mirror 非线性光纤环形镜

**NOLO** No Live Operator 无人操纵机

**NOM** National Operations Manager 国家运行主席席位

Nomenclature 术语,名称

Nominal 标称的,额定的

Normal Operation Mode 正常运行状态,正常工作方式

**NOMAD** Nozzle Motion Attenuation Device 喷管运动衰减装置

**NOML** Normal 正常的,通常的

**NOMS** Network Operation and Management Symposium 网络营运与管理专题讨论会

**NOMSG** Number of Messages 文电数目

**NONFLAG** Non-Flammable Compressor Gas 非易燃性压气机气体

**NONFLMB** Nonflammable 不可燃的,非易燃的

**NONSKED** Non Scheduled 不定期的

**NONSTD** Non Standard 非标准

**NONSYN** Nonsynchronous 非同步的,异步的

**NOOP** No Operation 不工作

**NOP** Navigation Operating Procedures 航行操作程序

Network and Operations Plan 网络和运行计划

Network Operation Procedure 网络操作过程

No-Operational（device status）不可使用的

Non-Operative 不工作的

Number of Passes 通过次数

**NOPAC** North Pacific 北太平洋

**NOPAT** Net Operating Profit after Tax 税后净营业利润

**NOPT** No Procedure Turn 无程序转弯

No Procedure Turn Required 不需程序转弯

**NOR** No Order Required 无订货要求

Notice of Revision 改型通知,改版通知

**NORAD** North American Air Defense Command 北美防空司令部

**NORDO** No Radio 无线电讯号消失

Non-Radio Aircraft 没有无线电讯号（设备）的飞机

**NOREC** No Recognition 未认可，未承认

No Records 无（订座）记录

Not Received 未收到

**NOREP** No Reply 没回答

Not Reply 未答复

**NORM** Normal 正常的，正规的，标准的

Not Operationally Ready, Maintenance 因维修而未做好使用准备

**NORR** No Reply Received 未收到答复

**NORS** Not Operationally Ready, Spare part 因缺件而未做好使用准备

Not Operationally Ready, Supply 因供应而未做好使用准备

**NOS** National Ocean Service 国家海洋服务

Network Operating System 网络操作系统

Node Operating System 节点操作系统

Not Otherwise Specified 未另行说明

Numbers 数量

**NOSAR** No Search and Rescue Action Required 不需要进行搜寻和救援

**NOSC** Nonoscillating 非振荡

**NOSH** No Show 误机（旅客在起飞前），未到，未见面

**NOSIG** No Significant Change 无重大改变

**NOSL** Nighttime/Daylight Optical Survey of Lighting 闪电昼夜光学研究

**NOSS** National Oceanic Satellite System 国家海洋卫星系统

Normal Operations Safety Survey 正常运行安全调查

**NOSSF** No Schedule, Stops and Frequencies 非计划、停止与频率

**NOSUB** Not Subject to Load 可以预约座位的机票

**NOTAL** Not to All 无需收发

**NOTAM** Notice to Airmen 航行通告

**NOTAMC** NOTAM Cancelling a Previous NOTAM 取消前面某一航行通告

**NOTAMN** NOTAM Containing New Information 新航行通告

**NOTAMR** NOTAM Replacing a Previous NOTAM 代替前面某一航行通告

**NOTAPPL** Not Applicable 不适用

**NOTAR** No Tail Rotor 没有尾桨的（直升机）

**NOTAS** NOTAM Summary 航行通告摘要

**NOTIS** Network Operations Trouble Information System 网络操作故障信息系统

**NOTOC** Notification to Captain 告知机长之特殊物品装载讯息文件

**NOTR** No Traffic Rights (Reservations Code) 无航权

**NOTTRF** Not Transferable 不得转让

**NOU** Noumea, New Caledonia (Airport Code) 新喀里多尼亚努美阿/拉堂图塔机场

**NOV** November 十一月

**NO-VOD** NO Video on Demand 非视频点播

**NOVOLRAM** Non-Volatile Random Access Memory 非易失动态随机存储器

**NOW** Network of Workstation 工作站网

**NOWAR** Now Arriving 航班正在到达

**NOX** Nitrogen Oxides 氧化氮

**NOZ** Normal Operating Zone 正常运行区

Nozzle 喷嘴

**NP** Nepal 尼泊尔

Network Performance 网络性能

Network Processor 网络处理器

Network Provider 网络供应商

Next Page 下一页

Nickel Plated 镀镍

No Payment 拒付

Non Procedural 无程序可循的

Nonprocurable 不能获取的

Nonpropelled 非推进的

Nonprovisional 非临时的

Normal Pitch 正常桨距

North Potentiometer 北电位计

Number Portability 号码携带

**NPA** National Pilots Association 全国飞行员协会

Network Performance Analyzer 网络性能分析器

Network Planning Area 网络规划区域

Non-Precision Approach 非精密进近

Normal Pressure Angle 正常压力角

Notice for Proposed Amendment 建议增补通知

Numbering Plan Area 编号规划区

**NPAI** Network Protocol Address Information 网络协议地址信息

**NPAR** New Part Approval Request 新零件批准申请

**NPC** Network Parameter Control 网络参数控制

Notice of Proposed Change 建议更改通知单

**NPCI** Network Protocol Control Information 网络协议控制信息

**NPCID** Network Protocol Clear Indication Delay 网络协议拆线指示延迟

**NPCS** Narrow-band PCS 窄带个人通信系统

**NPD** Network Protective Device 网络保护装置

Neural Processing Demultiplexer 神经处理解复用器

New Product Definition 新产品界定

Normal Processing Delay 正常处理延误

**NPDA** Network Problem Determination Application 网络故障检测应用程序

**NPDES** National Pollutant Discharge Elimination System 国家污染物质排放清除系统

**NPDS** Network Performance Design Standard 网络性能设计标准

**NPDTE** Non-Packet Digital Terminal Equipment 非分组数字终端设备

**NPDU** Network Protocol Data Unit 网络协议数据单元

**NPE** Network Protection Equipment 网络保护设备

Non-Polluting Engine 无污染发动机

**NPET** Nonpetroleum 非石油的

**NPF** No Problem Found 未发现问题

**NPG** Network Participation Group 网络参与小组

**NPI** Network Problem Identity 网络问题标识

Network Provider Interface 网络供应商接口

Numbering Plan Identifier 编号计划标识符

Numbering Plan Indicator 编号计划指示器

**NPIAS** National Plan of Integrated Airport System 国家机场体系综合计划

**NPIC** National Passport Information Center 国家护照信息中心

National Photographic Intelligence Center 国家照片情报中心

National Photographic Interpretation Center 国家照片解读中心(美国)

**NPL** Nameplate 铭牌

National Physical Laboratory 国家物理实验室

New Plymouth, New Zealand(Airport Code) 新西兰新普利茅斯机场

Noise Pollution Level 噪声污染等级

Non-Procurement Letter 非取得信函

**NPM** National Postal Museum 国家邮政博物馆(美国)

Network Path Manager 网络路径管理程序

Network Processor Module 网络处理器模块

New Public Management 新公共管理

**NPN** National Plan for Navigation 国家航行计划

Negative-Positive-Negative NPN 三极管,负—正—负

New Public Network 新的公众网

**NPNP** Negative-Positive-Negative-Positive 负—正—负—正

**NPO** Negative Positive Zero 负正零

Network Performance Objective 网络性能目标

**NPOESS** National Polar-orbiting Operational Environmental Satellite System 国家极轨运

行环境卫星系统

**N-Port** Node Port 节点通信口

**NPP** Net Primary Productivity 净初级生产力
Network Protocol Processor 网络协议处理器

**NPPF** Non-Polarization Preserving Fiber 非偏振保持光纤

**NPPL** National Private Pilot's Licence 全国私人飞行员执照
Network Picture Processing Language 网络图像处理语言

**NPR** Narrow Pulse Rejection 窄脉冲抑制
National Performance Review 国家绩效评审
National Program Review 国家项目评审
No Power Recovery 无动力回收
Noise Power Ratio 噪声功率比
Nozzle Pressure Ratio 喷管压力比

**NPRM** Notice for Proposed Rule Making 建议规章制订通告

**NPRN** Neoprene 氯丁橡胶

**NPROC** Not Processed 未处理的

**NPS** Network Phone Server 网络电话服务器
Network Processing Supervisor 网络处理监控器
Network Product Support 网络产品支持
Numerical Plotting System 数字绘图系统

**NPSA** Nonproductive Stores Area 非生产贮存区

**NPSC** National Public Safety Commission 全国公共安全委员会

**NPSH** Net Positive Suction Head 净正抽吸压头
Net Pump Suction Head 泵的净吸压头

**NPSI** Network Packet-Switching Interface 网络包(分组)交换接口

**NPSIS** Non-Production Supplies Inventory System 非生产供应库存系统

**NPSM** Nonparametric Statistical Method 非参数统计方法

**NPT** National Pipe Thread 国家标准管螺纹
Network Planning Technique 网络规划技术
Network Product Test 网络产品测试

Nonprogrammable Terminal 非编程终端
Normal Pressure and Temperature 常压常温

**NPTN** National Public Telecommunication Network 国家公共电信网

**NPU** Network Processing Unit 网络处理单元
Node Processing Unit 节点处理单元

**NPUT** Non Position Unit Time 非位置组件时间

**NPV** Net Present Value 净现值

**NQCC** Network Quality Control Center 网络质量控制中心

**NR** Nauru 瑙鲁
Navigational Radar 导航雷达
Near 接近,靠近,近
Negative Resistance 负(电)阻
New Release 最新颁发
No Record 无记录
Noise Ratio 噪声比
Nonreactive Relay 非电抗性继电器
Not Record 无记录
Not Reported 未报告
Not Required 不需要的,未要求的
Novice Roundup 新手综述

**NRA** Network Resolution Area 网络分辨区
No Repair Action 无修理措施
Non-Repair Assembly 不可修理的配件

**NRC** National Research Committee 国家科学研究委员会(美国)
National Research Council 国家研究委员会(美国)
Network Reliability Coordinator 网络可靠性协调程序
Network Route Control 网络路由控制
Networking Routing Center 连网路径选择中心
No Record 无(订座)记录
No Record Passenger (Reservations Code) 无订座记录的乘客(订座代号)
Noise Reduction Coefficient 减噪系数
Non-Recurring Cost 无复发损耗
Non-Routine Card 非例行工卡

Nuclear Reporting Cell 核报知小组

**NRCC** National Research Council of Canada 加拿大国家研究委员会

**NRD** Negative Resistance Diode 负阻二极管

Network Raw Data 网络原始数据

Network Reference Data 网络参考数据

Nonreplenishable Demand 非补充性需求

**NRDC** National Research and Development Corporation 国家研究和发展公司

National Research and Development Council 全国研究与发展委员会

National Resource Defense Council 国家资源防御委员会

**NRE** No Record 无记录

**NREG** Non Regular 非规则

**NREN** National Research and Education Network 国家研究和教育网(美国)

**NRETN** Nonreturn 无回程,非返回

**NRF** No Reflight 不重飞

**NRIM** Network Resource Information Model 网络资源信息模型

**NRIP** Number of Rejected Initial Pickups 不作初始检查项目的数量

**NRIS** Natural Resource Information System 自然资源信息系统

**NRL** Network Restructuring Language 网络重构语言

Normal Rated Load 标准载荷,正常标定负载

**NRLA** Network Repair Level Analysis 网络式修理级别分析

**NRLSI** National Reference Library of Science and Invention 国家科学与发明参考书图书馆

**NRM** Network Resource Management 网络资源管理

Normal 正常的,正规的,标准的

Normal Response Mode 普通响应模式,正常响应方式

**NRMS** Network Resource Management System 网络资源管理系统

**NRO** Number Resource Organization 数字资源组织

**NRP** Navigation Reference Point 导航基准点

Normal Rated Power 标准额定功率

**NRPB** National Radiological Protection Board 国家放射防护局

National Resources Planning Board 国家资源规划委员会

**NRPK** Non-Revenue Passenger Kilometer 非营收客公里

**NRPM** Non-Revenue Passenger Mile 非营收客英里

**NRR** Network Restoration Ratio 网络恢复比

No Release Required 无发放要求

**NRS** National Resource Specialist 国家资源专家

Noise Reduction System 降低噪声系统

**NRSP** Non Radar Signal Processor 非雷达信号处理器

Non Radiating Signal Processor 非辐射信号处理器

**NRSS** National Renewable Security System 国家可更新的安全系统

**NRST-AVAP** Nearest Available Airport 最近可用机场

**NRT** Narita, Tokyo, Japan (Airport Code) 日本东京成田机场

Near Real Time 准实时,近实时

Network Restoration Time 网络恢复时间

Network Routine Test 网络路由测试

Nonradiating Target 无辐射目标

Nonreal Time 非实时

Normal Rated Thrust 额定推力,标称推力

**NRTDAS** Nonreal-Time Data Automation System 非实时数据自动化系统

**NRTK** Non-Revenue Tonne Kilometer 非营收吨公里

**NRTS** Not Repairable This Station 本站不能修理

**NRTT** Non-Real-Time Traffic 非实时业务

**NRT-VBR** Non-Real-Time Variable Bit Rate

非实时可变比特率

**NRV** Network Routing Vector 网络选路矢量

Non-Return Valve 单向活门

**NRVSBL** Nonreversible 不可反转的

**NRZ** Non Return to Zero 不归零

**NRZC** Non-Return-to-Zero Change 不归零变化

**NRZI** Non-Return-to-Zero Inverse 不归零倒置

**NRZL** Non-Return-to-Zero Level 不归零电平

**NRZM** Non-Return-to-Zero Mark 非归零标记

**NRZS** Non-Return-to-Zero Space 不归零空号

**NS** Nacelle Station 吊舱站位,短舱站位

Nanosecond 纳秒,毫微秒

National Special Thread 国家特殊螺纹

National Standard 国家标准

Near Side 近边

Nervous System 神经系统

Network Service 网络服务

Network Subsystem 网络子系统

Network Synchronization 网络同步

Nickel Steel 镍钢

Nimbostratus 雨层云

No Smoking 请勿吸烟

North and South 南北

Not Specified 未规定,未明确

One Billionth of a Second 纳秒,毫微秒

Starter Rotor Speed 起动机转子转速

**NSA** National Security Agency 国家安保局（美国）

National Shipping Authority 全国航运局

New Shipborne (Shipboard) Aircraft 新型舰载飞机

New Surveillance Aircraft 新型监视飞机

Node Switching Assembly 节点转接装置

North South Africa 南北非洲

**NSAC** New Strategic Airlift Concepts 新战略空运概念

**NSAF** Unsafe 不安全

**NSAP** Network Service Access Point 网络服务访问点

**NSAT** National Satellite 国内卫星

**NSB** National Secure Bureau 国家保安局（美国的）

National Standard Bureau 国际标准局

Navy Standard Board 海军标准局

**NSBF** National Scientific Balloon Facility 国家科学探测气球设备（美国）

**NSC** National Security Council 国家安保委员会

National Space Council 国家空间委员会（美国）

Navigational Star Catalogue 导航星辰表

Network Security Center 网络安全中心

Network Service Center 网络服务中心

Network Switching Center 网络交换中心

Network Systems Corporation 网络系统公司

Nil Significant Cloud 无明显的云

No Show Control 未出票控制

Noise Suppression Circuit 噪声抑制电路

Norwegian Space Center 挪威航天中心

Notice of Status Change 情况更改通知

Number Sequence Code 数字序列码

**NSCA** National Safety Council of America 美国国家安全委员会

**NSCCR** Non-Structure Configuration Control Record 非结构布局控制记录

**NSCM** NATO Supply Code for Manufacturers 北大西洋公约组织为制造厂制定的供货代码

**NSD** Network Structured Database 网络结构数据库

**NSDU** Network Service Data Unit 网络服务数据单元

**NSDW** Non Specific Design Work 非指定设计工作

**NSE** Navigation Speed Error 导航速度误差

Navigation System Error 导航系统误差

Network Switching Engineering 网络交换工程

Network Systems Engineer 网络系统工程师

North Steaming Error 向北航行误差

**NSEC** Nanosecond 纳秒,毫微秒

**NSEP** National Security and Emergency Preparedness 国家安全和应急准备

**NSF** National Science Foundation 国家科学基金会(美国)

Non-Standard Facility 非标准设施

**NSFNET** National Science Foundation Network 国家科学基金会网络(美国)

**NSFSB** No Smoking Fasten Seat Belt 请勿吸烟系紧安全带

**NSG** National System Group 国家系统组

Network Services Group 网络服务组

**NSGN** Noise Generator 噪声发生器

**NSH** Network Service Host 网络服务主机

**NSHS** Network Service Host System 网络服务主机系统

**NSI** Name Service Interface 名字服务接口

NASA Standard Indicator 国家航空航天局标准指示器

Nonsatellite Identification 非卫星识别

Nonsequenced Information frame 非定序信息帧

Nonstandard Item 非标准件

**NSIA** National Security Industrial Association 国家安全工业协会

**NSIU** Navigation Switching Interface Unit 导航电门接口组件

Navigation System Interface Unit 导航系统接口组件

Nose Static Interface Unit 机头静电接合组件

**NSL** Engine Speed Limit 发动机速度限制

Net Switching Loss 净交换损耗

Network Support Layer 网络支持层

Noise Silencer 噪声抑制器

Non-Stock Listed 非库存清单

**NSLF** Nonself 非自主的

**NSLR** Noise to Signal Loudness Ratio 噪声信号响度比

**NSM** Network Security Module 网络安保模块

Network Space Monitor 网络空间监视器

**NSME** Non Standard Measuring Equipment 非标准测量设备

**NSMS** National Safety Management Society 国家安保管理协会

National Security Management System 国家安保管理体系

Network Status and Monitoring System 网络状况及监控系统

Non-Intrusive Stress Measurement System 非接触式应力测量系统

**NSN** National Signaling Network 国内信令网

National Significant Number 国内有效号码

National Stock Number 国家库存号

Nonsynchronized Network 非同步网络

Null Steering Network 零位操纵网络,中立位操纵网络

**NSNF** Non-Strategic Nuclear Forces 非战略核部队

**NSO** Name Serve Object 名字服务对象

No Spares Ordered 无订购备件

**NSP** Name Service Protocol 名字服务协议

NASA Support Plan 航空航天局保障计划(美国)

National Signaling Point 国内信令点

Native Signal Processing 本地信号处理

Navigational Satellite Program 导航卫星计划

Network Service Part 网络服务部分,网络业务部分

Network Service Process 网络服务进程

Network Service Protocol 网络服务协议

Network Service Provider 网络服务提供者

Network Signal Processor 网络信号处理器

Network Support Processor 网络支持处理器

Non-Standard Parts 非标准件

**NSPA** National Scholastic Press Association 国家学者出版协会

National Society of Public Accountants 国家公共会计师协会

National Standard Parts Association 国家标准
零件协会

**NSPAR** Nonstandard Part Approval Request
非标准件批准申请

**NSPE** National Society of Professional Engineers 国家专业工程师协会(美国)

Network Service Procedure Error 网络服务过
程出错

**NSR** Narrow Sector Recorder 窄带记录

National Scientific Register 国家科学登记

National Security Regulation(s) 国家安保规
则

New Short Range Aircraft 新的短程飞机

No Schedule Required 无计划要求

No Scheduled Removal 不定期拆卸

**NSRCA** National Society of Radio-Controlled
Aerobatics 无线电控制航空特技运动国家
协会

**NSRP** Non-technical Support Real Property 非
技术支援不动产

**NSS** Name Space Support 名字空间支持

Network Security System 网络安全系统

Network Server System 网络服务器系统

Network Service Sharing 网络服务分享

Network Subsystem 网络子系统

Network Surveillance System 网络监视系统

Network Synchronization Subsystem 网络同
步子系统

Neutral Shift Sensor 中立轴位移传感器

Neutral Speed Stability 中性速度稳定性

Nonspread Spectrum 非扩散谱

**NSSCC** National Space Surveillance and Control Center 国家空间监视与控制中心

**NSSDC** National Space Science Data Center
国家空间科学资料中心

**NSSDU** Normal Data Session Service Data Unit 普通数据会晤业务数据单元

**NSSF** NAS Simulation Support Facility 国家
空域系统的模拟支援设施,国家空域系统
的仿真保障设施

**NSSFC** National Severe Storms Forecast Center 国家强风暴预报中心

**NSSI** Non Structural Significant Item 非重要
结构项目

**NSSL** National Severe Storms Laboratory 国家
强风暴实验室,美国国家强风暴实验室

Normal Steady State Limit 正常稳定状态极
限

**NSSP** National Security Studies Programme 国
家安保研究计划

Network Service Support Point 网络业务支持
点

**NSSR** No Stress Signature Required 无强调
签字要求

**NSSS** National Space Surveillance System 国
家空间监视系统

National Strategic Satellite Communications
System 国家战略卫星通讯系统

**NSST** National Security Science and Technology 国家安保科学与技术

National System Strategic Team 国家系统战
略组

**NST** Net Starting Torque 净起动力矩

Network Signaling Termination 网络信令终
端

Noise, Spikes and Transients 噪声、峰值和
瞬值

Nonslip Thread 防滑螺纹

**NSTA** National Safe Transit Association 国家
安保运输协会

National School Transportation Association 国
家学校运输协会

National Science Teachers Association 国家
科学教师协会

**NSTC** National Science and Technology Council 国家科学技术委员会

**NSTD** Nonstandard 非标准的

Non-Systems Training Devices 非系统训练装
置

**NSTL** National Security Threat List 国家安全
威胁清单

National Software Testing Labs 国家软件测

试实验室

National Space Technology Laboratories 国家空间技术实验室

**NSTN** Non-Standard Telephone Number 非标准电话号码

**NSTP** National Software Testing Procedure 国家软件测试程序

National Space Technology Program 国家航天技术计划

National Strategic Targeting Plan 国家战略目标计划

Non-Stop, Non Stop 直飞, 不经停, 中途不着陆

**NSTS** National Space Transportation System 国家空间运输系统

**NSU** Network Server Unit 网络服务器组件

Network Service Unit 网络服务单元

Network Statistical Utility 网络统计应用软件

**NSV** Negative Sequence Voltage 逆序电压

Noise, Shock and Vibration 噪声、震动与振动

Non-automatic Self-Verification 非自动自检验

**NSW** Nil Significant Weather 无重要天气

Nominal Specification Weight 额定要求质量

**NSWL** Nose Wheel 前轮

**NT** Net 网, 净

Network Termination 网络终端

Network Terminator 网络终接器

Network Topology 网络拓扑

Neutral Zone 中立区(沙特—伊拉克间)

No Tools 无工具

Nontight 未拧紧

Northern Telecom 北方电信(加拿大)

Note 注释

**NT/BFE** Non-Traditional Buyer Furnished Equipment 非传统的买方自购设备

**NTA** National Tax Authority 国家税务局

Network Terminal Adapter 网络终端适配器

Noise-Thrust-Altitude 噪声—推力—高度

Notice to Aircrew 航行通告

**NTAG** Network Technical Architecture Group 网络技术体系结构组

**NTAOCH** Notice to Air Operation Certificate 航行许可公告

**NTAS** Network Traffic Accounting System 网络业务量计费系统

**NTB** National Test Bed 国家试验台

Non-Tariff Barriers 非关税壁垒

**NTC** Nano-Technology Council 纳米技术协会

National Telecommunications Commission 国家电信委员会

National Television Center 国内电视中心

National Training Center 国家训练中心

National Transportation Center 国家运输中心

Navigation and Traffic Control 航行和交通管制

Negative Temperature Coefficient 负温度系数

Negative Torque Control 负扭矩控制

New Technologies Committee 新技术委员会

Non-Technical Check 非技术检查

Notice, Notification 通知

**NTCD** Nested Threshold Cell Discarding 嵌套临界点信元删除

**NTCM** Noncoherent Trellis Coded Modulation 非相干格形编码调制

**NTCME** Network Test Case Management Environment 网络测试状况管理环境

**NTD** Noted 注明的

**NTE** Nantes, France(Airport Code) 法国南特机场

Network Termination Equipment 网络终端设备

Network Testing Environment 网络测试环境

Not to Exceed 不得超出

**NTEC** Network Terminal Equipment Center 网络终端设备中心

**NTF** National Test Facility 国家试验设施

New Tactical Fighter 新战术战斗机

No Trouble Found 无异状

Non-Textile Floor Covering 无织物地板面料

Notified 已通知

Notify 通知

**NTH** New Training Helicopter 新型训练直升机

Nice to Have 欣然接受

**NTI** Network Terminating Interface 网络端接口

Non-Tactical Instrumentation 非战术仪表

**NTIA** National Telecommunications and Information Administration 国家电信与信息管理局

**NTIK** Non-Tactical Instrumentation Kit 非战术仪表器材包

**NTIP** Network Terminal Interface Program 网络终端接口程序

**NTIS** National Technical Information Service 国家技术情报局(美国)

**NTITK** Non-Tactical Instrumentation Test Kit 非战术仪表测试器材包

**NTL** Newcastle, New South Wales, Australia (Airport Code) 澳大利亚新南威尔士州纽卡斯尔机场

No Time Lost 不失时机

**NTM** Network Terminal Manager 网络终端管理程序

Night Message 夜间信息

Non-destruction Test Manual 无损探伤手册

**NTMO** National Traffic Management Officer 国家交通管理官员

**NTMS** Network Traffic Management System 网络业务管理系统

**NTN** Network Terminal Number 网络终端编号

Network Termination Number 网络终端号

Neural Tree Network 神经树网络

Neutron 中子

**NTO** Network Terminal Option 网络终端选项

No Technical Objection 非技术拒绝

**NTOL** Normal Take-off and Landing 正常起落

**NTOS** New Technology Operating System 新技术操作系统

**NTP** National Taper Pipe (Thread) 国家标准锥管(螺纹)

Network Terminal Protocol 网络终端协议

Network Terminating Point 网络终端点

Network Termination Processor 网络端处理器

Network Test Panel 网络测试板

Network Time Protocol 网络时间协议

Network Transaction Processing 网络事务处理

Normal Temperature and Pressure 正常温度与压力

Normal Transmitted Power 正常传输功率

**NTPS** National Test Pilot School 国家试飞员学院

**NTPSC** National Transportation Policy Study Commission 国家运输政策研究委员会

**NTR** Network Time Reference 网络时钟参考

No Traffic Reported 空中交通情况无报告

Noise Temperature Ratio 噪声温度比

Nothing to Report 无事奉告

**NTS** Navigation Technology Satellite 导航技术卫星

Negative Torque Sensing 负转矩检测

Negative Torque Signal 负扭矩信号

Negative Torque System 负转矩系统,顺浆系统

Network Terminal Selection 网络终端选择

Network Test System 网络测试系统

Network Transport Service 网络传送服务

New Threats Simulator 新威胁模拟器

New Transit System 新中转系统

Night-vision Targeting System 夜视瞄准系统,夜视目标抓获系统

Not to Scale 未按比例

**NTSAC** New Technical and Scientific Activities Committee 新科学技术活动委员会

**NTSB** National Transportation Safety Board 国家运输安全委员会(美国)

**NTSC** National Television Standard Committee 国家电视制式委员会(美国)

National Television System Committee 国家电视系统委员会

**NTSDB** Network Test System Database 网络测试系统数据库

**NTSTP** Night Stop 夜间经停站

**NTT** Network Transfer Table 网络转移表

Niuatoputapu, Tonga (Airport Code) 汤加纽阿托普塔普机场

**NTTAA** National Technology Transfer and Advancement Act 国家技术转让与进步条例

**NTU** Nacelle Temperature Unit 吊舱温度组件,短舱温度组件

Network Terminating Unit 网络端接设备

Network Test Unit 网络测试设备

Number of Transfer Units 传输组件数

**NTV** Network Television 网络电视

**NTW** Network 网络

**NTWG** New Technologies Working Group 新技术工作组

**NTWOA** Near-Traveling Wave Optical Amplifier 近行波光放大器

**NTWS** New Threat Warning System 新威胁警告系统

Non Track While Scan 扫描期间不跟踪

**NTWT** Net Weight 净重

**NTZ** No-Transgression Zone 不可逾越地带

**NU** National Use 国内使用

Navigation Unit 导航组件

Network Unit 网络单元

Network Utility 网络公用业务

Nose up 机头向上,抬机头

Not Usable 不能使用

**NUA** Network User Address 网络用户地址

Normal Unlock Actuator 正常解锁作动器

**NUAA** Nanjing University of Aeronautics and Astronautics 南京航空航天大学

**NUAC** Nordic Upper Area Control 北欧高空区域管制

**NUC** Navigational Uncertainty Category 导航不确定度类别

Neutral Unit of Construction 中间组合单位

**NUE** Nuremberg, Germany (Airport Code) 德国纽伦堡机场

**NUF** Network User Field 网络应用字段,网络用户字段

**NUI** Netbook/Network User Interface 笔记本/网络用户界面

Network User Identification 网络用户识别

Network User Identifier 网络用户标识

**NUICAF** Aeronautics of Physical Activity Science Institute 航空人体活动科学研究所

**NUK** Nukutavake, Tuamoto Island (Airport Code) 土阿莫土群岛努库塔瓦克机场

**NUL** Blank or Null 空白或不存在的

Null 空白,无,零,无效的

**NUM** Navigation Unit Mount 导航组件的安装

Numerical 用数字表示的

**NUSBL** Unusable 无用的,不能使用的

**NV** Network Video 网络视频

Night VFR 夜间目视飞行规则

No Value 无价值

Non-Volatile 非易失性的

**NVAL** Not Available 不可用,不提供,不具备

**NVBMM** Non-Volatile Bulk Memory Module 非易失性整体存储模件

**NVD** No Value Declared 无声明价值,不要求声明价值

**NVE** Network-Visible Entities 网络可视实体

Night Vision Equipment 夜视设备

**NVFR** Night Visual Flight Rules 夜间目视飞行规则

**NVG** Night Vision Goggles 夜视镜

**NVGM** Non-Volatile Global Memory 非易失性全存储器

**NVGS** Night Viewing Goggles 夜视镜

**NVIS** Night Vision Imaging Spectrometer 夜视成像光谱仪

Night Vision Imaging System 夜视成像系统

**NVL** Night Vision Laboratory 夜视实验室

**NVM** Node Virtual Memory 结点虚拟存储空间

Non-Volatile Memory 永久性存储器

**NVOC** Non-Volatile Organic Compound 不挥发有机化合物

**NVP** Network Voice Protocol 网络语音协议

**NVR** Networked Virtual Reality 连网的虚拟现实

Never 未曾

No Voltage Release 无电压释放

**NVRAM** Non-Volatile Random Access Memory 非易失性随机存取存储器

**NVS** Navigational Video Select 导航视频选择

Network Video System 网络视频系统

Network Virtual Schema 网络虚拟图式

Night Vision System 夜视系统

Noise and Vibration Suppression 噪声和振动抑制

**NVT** Navegantes, Santa Catarina, Brazil (Airport Code) 巴西圣卡塔琳娜州纳维根特斯机场

Network Video Technologies 网络视频技术

Network Virtual Terminal 网络虚拟终端

Night Vision Tester 夜间视觉测验器,夜间视力测验器

**NW** Navigation Warning 航行警报

Net Weight 净重

North West,Northwest 西北

North-West bound 西北向,向西北飞行

Nose Wheel 前轮

**NWA** Network Analyzer 网络分析仪

**NWAC** National Weather Analysis Center 全国气象分析中心

**NWC** Net Working Capital 净流动资本

Network Controller 网络控制器

**NWD** Number of Words 字数

**NWDEN** Number of Words per Entry 每次输入字数

**NWDS** Nation Wide Dialling System 全国拨号系统

**NWG** National Wire Gauge 国家标准线规

Network Working Group 网络工作组

**NWH** Normal Working Hours 正常工作小时数

**NWL** Nacelle Water Line 吊舱水线,短舱水线

Network Layer 网络层

**NWOW** Not Weight on Wheels 重量不在机轮上,飞机顶起

**NWP** National Work Program 国家工作计划

Newspapers, Magazines 报纸,杂志

Numerical Weather Prediction 数值天气预报

**NWPP** Numerical Weather Prediction Progress 数值天气预报进展报告

**NWR** Non-Weight Restriction 无重量限制

**NWS** National Weather Service 国家气象服务(美国)

Network Windows System 网络窗口系统

Nose Wheel Steering 前轮转弯

**NWSC** National Weather Satellite Center 国家气象卫星中心(美国)

**NWSS** National Weather Satellite System 国家气象卫星系统

**NWT** Net Weight 净重

Network Test 网络测试

Nonwatertight 非防水

**NWTTP** Network Trail Termination Point 网络跟踪终端点

**NWW** Nose Wheel Well 前轮舱

**NXT** Next 下一个

**NXWK** Next Week 下一周

**NXYR** Next Year 明年

**NYA** New York Airways (USA) 纽约航空公司(美国)

**NYARC** New York Aviation Rulemaking Committee 纽约航空法规制定委员会

**NYARTCC** New York Air Route Traffic Control Center 纽约航路空中交通管制中心

**NYC** New York, USA(Airport Code) 美国纽约机场

USA New York 纽约(简称)

**NYL** Nylon 尼龙

**NYMEX** New York Mercantile Exchange 纽约商业交易所

**NYO** Not Yet Operating 尚未工作

**NZ** New Zealand 新西兰

**NZAPA** New Zealand Airline Pilots Association 新西兰航空公司驾驶员协会

**NZCAA** New Zealand Civil Aviation Authority 新西兰民用航空局

**NZDF** Non-Zero Dispersion Fiber 非零色散光纤

**NZDSF** No-Zero Dispersion Shifted Fiber 非零色散位移光纤

**NZG** Near Zero Gravity 近零重力

**NZL** New Zealand Line 新西兰航线

Nozzle 喷嘴,尾喷管

**NZMS** New Zealand Meteorological Service 新西兰气象局

**NZS** New Zealand Standard 新西兰标准

**NZSS** New Zealand Standard Specification 新西兰标准规范

**NZST** New Zealand Standard Time 新西兰标准时间

**NZVV** Nacelle Zone Ventilation Valve 吊舱区域通风活门

# O

O  Observe, Observation 观察, 观测
Octal 八进制(的)
Off 断开, 离开, 偏离
Office 办公室, 办事处
Official 官方的, 正式的
Oil 油, 滑油
Open 打开, 开启, 开路
Operating 操作, 运行
Orange 橘子, 橙色
Ordinate 纵坐标, 有规则的
Organizational 组织的
Output 输出量, 排出量
Outstanding 突出的, 显著的
Oxygen 氧气

**O to O**  Out to out 输出至输出

**O&A**  Over and Above 在…上方并高于

**O&C**  Operation and Checkout 操作与检查

**O&I**  Ordering and Issuing 订购与发出
Organization and Intermediate 机构与中间物

**O&IRCN**  Operation and Inspection Record Change Notice 操作与检验记录更改通知

**O&M**  Operation and Maintenance 运行和维护, 运行和维修, 运营与维护保养
Overhaul and Maintenance 翻/大修与维护

**O&S**  Operation and Supply 操作与供应
Operation and Support 操作与支援

**O&SHA**  Operating and Supply Hazard Analysis 运行与供应风险分析
Operating and Support Hazard Analysis 运行与保障风险分析

**O. N.**  Octane Number 辛烷值

**O. R.**  Operational Requirement 运营要求

**O/A**  On or About 在…或大约

**O/B**  Order Book 订货簿

**O/C**  Open Circuit 开路, 断路

**O/D**  On Demand 见票即付, 按要求
On-Deck Delivery 到岸交付
Out of Detent 脱出止挡位

**O/E**  Optical/Electrical 光电变换

**O/FULL**  Overfull 过度的, 溢出的

**O/H**  Opposite Hand 异侧
Overhaul 大修, 超运, 翻修, 全面检修

**O/I**  Ordering and Issuing 订购与发出
Organization and Internediate 机构与中间物

**O/J**  Override/Jettison 超控/抛油

**O/L**  On-Line 联机, 在线, 在航线上

**O/P**  Output 输出

**O/R**  On Request 按申请, 按要求, 按请求

**O/S**  Out of Sequence 无顺序, 失序
Out of Stock 无库存
Over Size 尺寸过大, 超尺寸

**O/T**  Other Time 其他时间
Over Temperature 超温
Overtime 加班, 超时

**OA**  Obstacle Avoidance 障碍回避
Office Automation 办公自动化
Office of Application 应用办公室
Olympic Airways S. A. 奥林匹克航空公司(希腊)
On Account 暂付, 记账
Operational Analysis 操作分析, 运行分析
Operational Assurance 操作保险
Optical Amplifier 光放大器
Orbital Availability 在轨可用度
Output Amplifier 输出放大器
Output Axis 输出轴
Overall 总的
Overhead Approach 飞越机场进近

**OA BAG**  One Baggage (OB = two, OC = three, OD = four) 一件行李

**OA&M**   Operation, Administration and Maintenance 运行、管理和维护

**OAA**   Office of Aviation Affairs 航空事务办事处

Open Application Architecture 开放式应用体系结构

Open Aviation Area 开放航空空域

Orient Airlines Association 东方航空公司协会

**OAB**   Offline Address Book 离线通讯录

Operations Approval Board 运行批准委员会

**OAC**   Oceanic Area Control 海洋区域管制

Optical Area Corelation 光学区域相关

Optimal Automatic Control 最佳自动控制

Optimum Approach Course 最佳进近航线

**OACC**   Oceanic Area Control Center 海洋区域管制中心

**OAD**   Operational Availability Data 操作可用数据

Orbital Aerodynamic Drag 轨道气动阻力

Output Analog Discrete 输出模拟离散（信号）

**OADD**   Optically Amplified Direct Detection 光放大直接检测

**OADG**   Open Architecture Development Group 开放体系结构开发组

**OADM**   Optical Add-Drop Multiplexer 光分插复用器

**OADMT**   Overall Aircraft Design Management Team 航空器总设计管理小组

**OADS**   Omnidirectional Air Data System 全向大气数据系统

Optical Air Data System 光学大气数据系统

**OAE**   Office Automation Equipment 办公室自动化设备

**OAF**   Orbiting Antenna Form 轨道天线外形

Original Address Field 原始地址字段

**OAG**   Official Airline Guide 正式航班指南，官方航班指南

Operational Advisory Group 运行咨询组

Orange, New South Wales, Australia（Airport Code）澳大利亚新南威尔士奥兰吉机场

**OAI**   Omnidirectional Airspeed Indicator 全向空速指示器

Office of Aviation Information 航空信息办公室

**OAIDE**   Operational Assistance and Instructive Data Equipment 操作协助与指导性数据设备

**OAJ**   Jacksonville, NC, USA（Airport Code）美国北卡莱罗纳州杰克逊维尔机场

**OAK**   Oakland, CA, USA（Airport Code）美国加利福尼亚州奥克兰机场

**OAL**   Office of Administrative Law 行政法律办公室

Outboard Aileron Lockout 外侧副翼锁定

Overall Length 全长，总长

**OALT**   Operationally Acceptable Level of Traffic 操作可接受的交通等级

**OAM**   Operations, Administration and Maintenance 运行、管理和维护

Orthogonal Amplitude Modulation 正交振幅调制

**OAM&P**   Operations, Administration, Maintenance and Provisioning 运行、管理、维护和供给

**OAMC**   Operation, Administration and Maintenance Center 运行、管理和维护中心

**OAMC-MF**   OAMC Management Function 运行、管理和维护中心—管理功能

**OAMC-OS**   OAMC Operation System 运行、管理和维护中心—运行系统

**OAME**   Orbital Altitude and Maneuvering Electronics 轨道高度与机动电子设备

**OAMM**   Operation and Maintenance Module 运行和维护模块

**OAMP**   On-Airplane Maintenance Program 飞机维修大纲

Optical Airborne Measurement Platform 光学机载测试平台

**OAMS**   Onboard Asynchronous Messaging

Service 机载异步信息服务

Optical Attenuation Measuring Set 光衰耗测量仪

Orbital Attitude and Maneuver System 轨道姿态与操纵系统

**OAMTAS** On-Aircraft Maintenance Testability Analysis Summary 机上维护可测试性分析摘要

**OAMTRA** On-Aircraft Maintenance Testability Requirements Analysis 机上维修可测试性要求分析

**OAN** Optical Access Network 光纤接入网

**OANC** Onboard Airport Navigation Computer 机载机场导航计算机

**OANS** Onboard Airport Navigation System 机载机场导航系统

**OAO** Orbiting Astronomical Observatory 轨道天文观测台

**OAOI** On and off Instrument 仪表和非仪表飞行

**OAP** Office of Aircraft Production 飞机生产处

Oil Analysis Program 滑油分析大纲

Outlet Absolute Pressure 出口绝对压力

Output Audio Processor 输出音频处理器

Oversale Auction Plan（机票）超售竞卖计划

**OAPEC** Organization of Arab Petroleum Exporting Countries 阿拉伯石油输出国组织

**OAPWL** Overall Sound Power Level 总声音强度级,总声强级

**OAR** Office of Aerospace Research 宇航研究署

Operational Availability and Reliability 使用可用性与可靠性

Optical Automatic Ranging 光学自动测距

Optically Amplified Regenerator 光放大再生器

Overhaul and Repair 大修和修理

**OARAC** Office of Air Research Automatic Computer 航空研究局自动计算机室

**OARM** Off-Axis Rejection Model 转偏离抑制模型

**OARS** Ocean Area Reconnaissance Satellite 海洋区域侦察卫星

Open Architecture Retrieval System 开放结构检索系统

**OART** Office of Advanced Research and Technology 高级研究与技术处

**OAS** Obstacle Assessment Surface 障碍物评价面,越障判定高度面

Oceanic Automation System 海洋自动化系统

Offensive Air Support 进攻性空中支援

Offensive Avionics System 攻击性航空电子系统

Office Automation System 办公自动化系统

Office of Aircraft Services 飞机服务办公室

Omnidirectional Airspeed Sensor 全向空速传感器

Operational Announcing System 使用通告系统

Optical Access System 光接入系统

Orbit Adjust System 轨道调节系统

Organization of American States 美洲国家组织

Originating Access Situation 始端接入情况

Overhead Absorption System 顶板缓冲系统,顶板吸收系统

**OASI** Operation and Service Instruction Manual 使用和服务说明手册

**OASIS** Open Avionics System Integration Study 开放式航空电子系统综合研究

Operational Airport Safety Information System 运营机场安全信息系统

Operational and Supportability Implementation System 运行与可支持性实施系统

Operational Applications of Special Intelligence System 专用情报系统的操作使用

Organization for the Advancement of Structural Information Standards 结构化信息标准促进组织

**OASO** Overhead Absorption 顶板缓冲,顶板

吸收

**OASPL** Overall Sound Pressure Level 总声压电平

**OAST** Office of Aeronautics and Space Technology 航空与空间技术署

**OAT** Office of Aerospace Technology 航空技术办公室

Operating Ambient Temperature 工作环境温度

Operation Acceptance Test 使用验收试验

Operational Air Traffic 可供使用的空中交通

Operator Action Trees 操作员动作树

Optional Auxiliary Terminal 可选辅助终端, 选装的辅助终端

Outside Air Temperature 外界大气温度

Overall Test 总测试

Oxford Aviation Training 牛津航空训练

**OATC** Oceanic Air Traffic Center 海洋空中交通中心

Overseas Air Traffic Control 海外航线空中交通管制

**OATI** Office of Air Transport Information 航空运输情报室

**OATP** On Aircraft Test Procedure 飞机测试程序, 机上测试程序

**OATS** Optical Amplifier Transmission System 光放大器传输系统

Optical Attitude Transfer System 光学姿态变换系统

Optimum Aerial Targeting Sensor 最佳空中目标瞄准传感器

Orbit and Attitude Tracking Subsystem 轨道与姿态跟踪子系统

Oxford Air Training School 牛津飞行训练学校

**OAU** Organization of African Unity 非洲统一组织

**OAV** Operational Aerospace Vehicle 作战航天器

**OAX** Oaxaca, Mexico (Airport Code) 墨西哥瓦哈卡机场

**OB** Obscure 阴暗的, 不清楚的

Observatory 气象台, 天文台

Octave Band 倍频带

Omnibearing 全方位

Optical Bandpass 光带通(滤波)

Orbital Booster 轨道辅助推进器

Out Board 机外的, 外侧(的)

Output Buffer 输出缓冲器

Output Bus 输出总线

**OBA** Open Box Audit 开箱检查

Outbound Boom Avoidance 外传音爆防止

Oxygen Breathing Apparatus 供氧设备

**OBAP** Organization of Black Airline Pilot 航空公司黑人驾驶员组织

**OBBS** On-Board Balance System 机载平衡系统

**OBC** On-Board Checkout 机上检查

On-Board Computer 机载计算机

On-Board Controller 机载控制器

Operational Baseline Configuration 操作基线配置, 使用基线配置

**OBCE** On-Board Checkout Equipment 机载检查设备

On-Board Control Equipment 机载控制设备

**OBCR** Optical Bar Code Reader 光学条形码阅读机

**OBCS** Object Based Control Structure 对象控制结构

**OBD** Omnibearing-Distance 全方位距离

Omnibearing-Distance System 全方位距离导航系统

On-Board Diagnostics 车载诊断, 机载诊断

Optical Beam Deflection 光束偏转

Optical Branching Device 光分路器

**OBE** Off-Board Expendables 非机载消耗品

Outerback End 外后端

Overtaken by Events 偶然出现的

**OBECO** Outboard Engine Cutoff 外侧发动机停火

**OBEWS** On-Board Electronics Warfare Simulation 机上电子战模拟

On-Board Electronics Warfare System 机载电子对抗(电子战)系统

**OBF** Optical Branching Filter 光分路滤波器

**OBG** Optical Beam Rider Guidance 光束制导

**OBI** Omnibearing Indicator 全向方位指示器,全向无线电导航指示器

Optical Beat Interference 光差拍干扰

**OBIFCO** On-Board in Flight Checkout 机上飞行中检查

**OBIGGS** On-Board Inert Gas Generating System 机载惰性气体生成系统

**OBJ** Object 物件,目的

On-Board Jammer 自载式干扰机

**OBJV** Objective 物镜,目标,对象,客观的

**OBL** Oblique 倾斜

Obstruction Light 障碍灯

Octave Band Level 倍频带电平

Optimal Booking Limit 优化订座极限

**OBLGD** Obliged 感激

**OBLN** Obligation 义务,职责

**OBM** Object Behavior Model 对象行为模型

Open Business Management 开放商务管理

Organizational Behavior Management 组织行为管理

Original Brand Manufacturer 原品牌制造商

**OBND** Outbound 输出的,离岗的

**OBOGS** On-Board Oxygen Generating System 机载氧气发生系统

**OBP** On-Board Processing 机载处理

**OBPF** Optical Band Pass Filter 光带通滤波器

**OBR** Office of Budgets and Reports 预算与报告处

Operational Baseline Requirement 使用基线要求

Optical Beam Riding 光波束引导

**OBRM** On-Board Replaceable Module 机载可更换件

**OBRNR** Oil Burner 滑油燃烧器

**OBRS** On-line Bibliographic Retrieval System 联机书目检索系统

**OBS** Observer 观察员

Obsolete 废弃的,作废的

Obstruction 妨碍,阻塞,障碍物

Official Bulletin Station 官方公告台

Omnibearing Selection 全向方位选择

Omnibearing Selector 全向方位选择器

On-Board Simulation 机载模拟器

Optical Bypass Switch 光学旁通开关

Organizational Breakdown Structure 组织分解结构

Outdoor Base Station 室外基站

**OBSHT** Obstacle Height 障碍物高度

**OBSN** Obstruction 妨碍,阻塞,障碍物

**OBSNFL** Observation Flight 观察飞行

**OBSR** Observer 观察员

**OBST** Obstacle 障碍物

**OBSV** Observation 观察

**OBT** Off-Block Time 启动轮挡时间,飞机开始滑动时间

Optical Beam Transmission 光束传输

**OBTM&M** On-Board Test Monitor and Maintenance 机上测试检测器与维修

**OBV** Obviously 显然,明白地

On Board Video 机载视频

Operability Bleed Valve 可用引气活门

**OBW** Observation Window 观察窗

On-Board Wheelchair (供旅客使用的)机上轮椅

Optical Beam Waveguide 光束波导

**OBWBS** On-Board Weight and Balance System 机上重量与平衡系统

**OBWS** On-Board Weighing System 机上称重系统

**OC** Object Code 目标代码

Obstacle Clearance 越障高度

Offset Center 偏离中心

Oil Cooler 滑油冷却器

On Condition 视情况而定

Open Circuit 开路

Operating Certificate 运行合格证

Operating Characteristics 操作特性,运行特性

Operational Capability 操作能力,运行能力

Operations Channel 操作信道

Optical Circulator 光环行器

Order Cancelled 订单取消

Order Card 指令卡

Outside Circumference 外圆周

Over Current, Overcurrent 过流,过量电流

Overhead Channel 开销信道

Overload Control 过载控制

Oversale Cost (机票)过售损失

**OCA** Obstacle Clearance Altitude 越障高度

Oceanic Control Area 海域(空中交通)管制区,海洋管制区

Optical Channel Analyzer 光信道分析仪

**OCALC** Oklahoma City Air Logistics Center 俄克拉荷马城航空后勤中心

**OCAMA** Oklahoma City Air Material Area 俄克拉荷马城航空器材区

**OCAMS** On-board Check-out and Monitoring System 机上检查和监控系统

**OCAS** Operating Cost Accounting System 运营成本核算系统

Organization of Central American States 中美洲国家组织

**OCB** Oil Circuit Breaker 滑油电路跳开关

Operations Coordinating Board 作战协调委员会

Outgoing Calls Barred 呼出禁止

Override Control Bits 超控控制位

**OCC** Occupy, Occupation 占有

Occur, Occurrence 发生,发现,出现,存在

Operations/Operational Control Center 运行管制中心,操作控制中心

Optical Cable Connector 光缆连接器

Optical Cross Connect 光交叉连接

Orthogonal Convolutional Code 正交卷积码

Other Common Carrier 一般电信公司,普通运营商

Outer Communication Channel 外部通信信道

Outer Communication Control 外部通信管理

Ozone Catalytic Converter 臭氧转换器

**OCCB** Output CES Cells Buffer 输出通信工程标准信元缓冲器

**OCCC** One Connection Call Control 单个连接呼叫控制

**OCCM** Office of Commercial Communications Management 商业通信管理处

Optical Counter Countermeasures 光学反对抗措施

**OCCPD** Occupied 占用的,占有的

**OCCPNT** Occupant 占用者,乘客

**OCD** Operational Characteristics Document 操作特性文件,运行特性文件

Operational Concept Document 操作概念文件,运行概念文件

Operations Control and Display 运行控制和显示装置

**O-CDMA** Optical Code Division Multiple Access 光纤码分多址

**OCDMS** On-board Checkout and Data Management System 机上检查和数据管理系统

**OCDP** Optical Coherence Domain Polarimetry 光相干域偏振测定法

**OCDR** Optical Coherence Domain Reflectometer 光相干域反射仪

**OCDU** Optical Coupling Display Unit 光学耦合显示器

**OCE** Office of the Chief Engineer 总工程师办公室

Operational Control Element 操作控制单元

Operational Control Equipment 运行控制设备

**OCF** Operational Check Flight 操作检查飞行

Operations Control Facility 操作控制设备

Optical Cable Facility 光缆设备

Out-of-Control Flight (Aviation) 飞行(航空)失控

**OCG** Operational Coordination Group 运行协调小组

Optical Comb Generator 光梳状波发生器

Oxygen Consumption Gauge 氧气消耗量表

**OCGS** Object Code Generation System 目标码生成系统

**OCH** Obstacle Clearance Height 越障高度

**OCHIS** Optical Cell Header Interface Subsystem 光信头接口子系统

**OCHR** Oil Catcher 滑油收集器

**OCI** Object Code Insertion 目标代码插入

Office of Complaint Investigation 控告调查办公室

Office of Computer Information 计算机情报处

**OCL** Obstacle Clearance Limit 越障许可极限

Operation Control Language 运行控制语言

Operational Cable Load 光缆操作负荷

Operational Check List 操作检查单

Operational Control Level 操作管理水平

Operator Command Language 操作员命令语言

Optical Channel Layer 光纤信道层

Optimum Cruising Level 最佳巡航高度

Overall Connection Loss 总连接损耗

**OCLD** Oil Cooled 冷却的滑油

**OCLNR** Oil Cleaner 滤油器,净油器

**OCLR** Oil Cooler 滑油冷却器

**OCM** Oil Content Monitor 油量监测器

On-Chip Monitor 芯片上监测器

On-Condition Maintenance 视情维修

Ongoing Call Management 去话呼叫管理

Operating Capability Maintenance 运行能力维护

Operating Committee Meeting 运行委员会会议

Operations Cost Model 作业成本模型

Optical Countermeasures 光学干扰,光学对抗

Optimal Control Model 最优控制模型

Organizational Change Management 组织变更管理

Organizational Configuration Management 组织配置管理

Out of Control Months 控制月外

Outflow Valve Control Module 流出阀控制组件

**OCN** Open Computer Network 开放式计算机网络

**OCNI** Optimal Communications, Navigation and Identification System 最佳通信、导航和识别系统

**OCNL** Occasional or Occasionally 偶然的或不定时地

**OCO** Object Code Only 仅用于项目代码

Open-Close-Open 开—关—开

**OCONT** Oil Control 液压油调节

**OCP** Obstacle Clearance Panel 超障安全专家组

Office of Commercial Programs 商业计划处(美国航空航天局)

Open Communication Protocol 开放通信协议

Open Control Platform 开放式控制平台

Operating Control Procedure 操作控制程序

Operation Control Panel 操作控制板

Operational Computer Programs 计算机操作程序

Optional Calling Plan 任选呼叫方案

**OCR** Oceanic Control Region 海上管制区

Oil Circuit Recloser 滑油路自动开关

On-Condition Replacement 视情更换(零件)

Operation Control Reports 运行管理报告

Optical Character Reader 光学字符阅读机

Optical Character Reading 光学字符读取,光学特性读取

Optical Character Recognition 光学字符识别

Optical Code Reader 光代码读取机

Order Control Record 指令控制记录

Overcurrent Relay 过载电流继电器

Overhaul Component Requirement 部件大修要求

**OCRIT** Optical Character Recognition Intelligent Terminal 光字符识别智能终端

**OCRS** Operations Control Reporting System
运行管理报告系统

**OCS** Obstacle Clearance Surface 越障安全高
度面

Off Center Sweeps 偏心扫描

On-board Checkout System 机上检查系统

Operating Control System 运行控制系统

Operational Control Segment 操作控制段

Operations Computing System 操作计算系统

Optical Character Scanner 光字符扫描器

Optical Coherent System 相干光系统

Optical Contrast Seeker 光学对比导引头

Optical Control System 光学控制系统

Optimum-Cost Speed 最佳经济速度

Orbit Control System 轨道控制系统

Originating Call Screening 发端去话筛选

Overall Customer Satisfaction 总客户满意度

Overseas Communications Service 海外通信
业务处

**OCSM** Outflow Valve Control and Sensor
Module 外流活门控制及传感器模块

Optical Cell Selection Module 光信元选择模
块

**OCT** Octagon 八角形

Octal 八进制(的)

Octane 辛烷

October 十月

Octuple 八维

Optical Current Transducer 光流换能器

**OCTA** Outsized Cargo Tanker Aircraft 特大
型货运加油机,超大型载油加油机

**OCTAVE** Operationally Critical Threat, Asset
and Vulnerability Evaluation 可操作性关键
威胁、评估和脆弱性评估

**OCTS** Optical Cable Transmission System 光
缆传输系统

**OCU** Office Channel Unit 局内信道单元

Operational Control Unit 操作控制单元

Operational Conversion Unit 操作转换装置

**OCV** Oil Check Valve 滑油止回活门,滑油
检查活门

Open Circuit Voltage 开路电压

**OCW** Output Control Word 输出控制字

**OCWR** Optical Continuous Wave Reflectometer 光连续波反射计

**OCZ** Operational Control Zone 运行控制区

**OD** On Dock 在机库(内)

On Duty 值勤

Operation Dispatcher 签派员

Operations Directive 操作指示,作战指令

Optical Demultiplexer 光解复用器

Optical Detector 光检测器

Origin and Destination 始发站和终点站

Original Design 原始设计

Outside Diameter 外径

Outside Dimension 外部尺寸,外廓尺寸

Overdue 误点的,迟到的

**ODA** Open Document Architecture 开放式文
档结构

Organizational Designation Authorization ( 适
航)机构委任授权

Other Design Activity 其他设计活动

Output Device Address 输出装置地址

**ODADS** Outplant Drawing and Distribution
System 出厂图纸与分配系统

**ODALS** Omnidirectional Approach Lighting
System 全向进近灯光系统

**ODAP** Operation ( al ) Data Analysis Program
运转数据分析程序

**ODAPS** Oceanic Display and Planning System
海洋显示和规划系统

**ODAR** Organization Designated Airworthiness
Representatives 机构委任适航代表

**ODARS** On-line Diagnostic and Reporting
System 在线诊断和报告系统

**ODB** Object Database 目标数据库

**ODBC** Open Data Base Connection 开放数据
库互联

**ODC** Operational Discrete Command 操作离
散指令

Operational Dispatch Center 运行签派中心

Operations Data Control 操作数据控制

Optical Data Corrector 光学数据校正器

Other Direct Costs 其他直接成本

**ODCU** Optical Data Collecting Unit 光数据收集单元

**ODE** Object Database and Environment 目标数据库与环境

Open Development Environment 开放开发环境

**ODES** Office Data Exchange Standard 办公室数据交换标准

**ODF** Optical Distribution Frame 光纤配线架

Origin-Destination Fares 全程票价

**ODI** Open Design Issues 开放设计发布

**ODIAC** Operational Development of Initial Air-ground Data Communications 早期空—地数据通信运行开发

**ODID** Operational Display and Input Development 运行显示(器)与输入开发

**ODIF** Open Document Interchange Format 开放文件互换格式

**ODIFF** Oil Differential 滑油(压力)差

**ODL** Object Definition Language 目标确定语言

Oceanic Data Link 海洋数据链

Optical Data-Link 光学数据传输

Output Data Line 输出数据线

**ODLI** Open Data Link Interface 开放式数据链路接口

**ODM** Object Data Manager 对象数据管理程序

Object Data Model 对象数据模型

Office of Data Management 数据管理办公室

One-Day Mission 一日(飞行)任务

Operating Data Manual 使用数据手册,运行数据手册

Operational Data Message 操作数据通报

Operational Development Model 使用发展模型

Optimal Decision Model 最佳决策模型

Optimal Design Model 最优设计模型

Original Design Manufacturer 原始设计制造商

Original Design Manufacturing 原始设计制造

**ODMA** Open Distributed Management Architecture 开放式分布管理结构

**ODMS** Object Database Management System 目标数据库管理系统

Operational Data Management System 运营数据管理系统

Operational Databank Management System 运营数据库管理系统

Optimization Decision Management System 优化决策管理系统

**ODN** Open Data Network 开放数据网络

Optical Data Network 光学数据网络

Optical Distribution Network 光配线网

Optics Distribute Network 光纤分配网络

**ODOM** Odometer 里程表

**ODOP** Offset Doppler 多普勒偏移

**ODP** Open Distributed Processing 开放分布式处理

Open Distributed Processor 开放型分布式处理器

Operational Development Phase 运行开发阶段

Optical Data Processor 光数据处理机

Outline Development Plan 开发计划概述

**ODPSK** Oil Dipstick 测油尺

**ODR** Optical Data Recorder 光学数据记录器

Order 订货,订票,指令

**ODREP** OPMET Data Regional Exchange Point 运行气象数据区域交换点

**ODRN** Orbiting Data Relay Network 轨道数据中继网络

**ODRS** Orbiting Data Relay System 轨道数据中继系统

**ODRSS** Orbiting Data Relay Satellite System 轨道数据中继卫星系统

**ODS** Obstacle Detection System 障碍探测系统

Odessa, Ukraine (Airport Code) 乌克兰奥

德萨机场

Open Data Service 开放式数据服务

Open Data Stream 开放数据流

Operational Discrete Status 操作离散状态

Optical Data System 光学数据系统

Optical Display Sight 光学显示瞄准具

Oxide-Dispersion Strengthened 氧化扩散强度的

**ODSI**   Optical Domain Service Interconnect 光域服务互联

**ODT**   On-line Debugging Technique 在线调试技术

Operational Demonstration Test 使用演示试验

Optical Data Transmission 光数据传输

Optical Distance Terminal 光纤远程终端

Optimal Data Technologies 最佳数据技术

Organizational Design Team 组织设计组

Outside Diameter Tube 管路外径

**ODTRS**   Operational Design to Requirements 按要求的操作设计

**ODU**   Optical Disk Unit 光盘组件

Outdoor Unit 室外设备

Output Display Unit 输出显示组件

Overhead Decoder Unit 头顶解码组件

**ODV**   Open Digital Video 开放式数字视频

**ODVA**   Open Device Net Vendors Association 开放设备网供货方协会

**ODVAR**   Orbit Determination and Vehicle Altitude Reference 定轨道与飞行器高度基准

**ODVP**   Optimal Digital Voice Processor 最佳数字语音处理机

**ODXC**   Optical Digital Cross Connect 光数字交叉连接

**OE**   Open End 开口端,开路端

Operating Environment 运行环境

Operating Experience 操作经验

Operational Error 操作误差

Operator Error 操纵员误差

Opt-Electronic 光学电子的

Outlook Express 邮件收发软件

Output Enable 允许输出

Overrun Error 超转误差

**OEB**   Operating Engine Bulletin 使用(中)发动机通报

Operating Engineering Bulletin 工程运行通告

Overhead Electronics Box 顶板电子盒

**OEBS**   On-board Engine Balancing System 机载发动机平衡系统

**OEC**   Optical to Electrical Connection 光电连接

**OECD**   Organization for Economic Co-operation and Development 经济合作与发展组织

**OECF**   Overseas Economic Cooperation Fund 海外经济协作基金

**OECS**   Organization of Eastern Caribbean States 东加勒比国家组织

**OED**   Operational Environment Documentation 运行环境描述

Operational Evaluation Demonstration 运行评价示范

Operations Evaluation Department 业务评估部门

Oxford English Dictionary 牛津英语词典

**OEE**   Operating Equipment Effectivity 操作设备功效

Overall Equipment Effectiveness 设备整体效能

**OEEC**   Organization for European Economic Cooperation 欧洲经济合作组织

**OEFB**   Opto-Electronic Feed Back 光电反馈

**OEH**   Optical Exposure Head 曝光镜头

**OEI**   Office of Environmental Information 环境信息办公室

One Engine Inoperative 单发停车

**OEIC**   Opto-Electronic Integrated Circuit 光电集成电路

**OEID**   Opto-Electronic Integrated Device 光电集成器件

**OEM**   On Board Equipment Manufacture 机载设备制造

Original Equipment Manufacturer 原始设备制造商

**OEO** Optical-Electrical-Optical 光—电—光

**OEP** Office of Emergency Planning 应急计划办公室

Office of Emergency Preparedness 紧急战备处

Office of Environmental Policy 环境政策办公室

Operational Evolution Plan 运行演进计划

**OEQC** Office of Environmental Quality Control 环境质量控制处

**OES** Official Emergency Station 官方应急电台

**OET** Objective End Time 目标结束时间

Office of Education and Training 教育与培训中心

Office of Emergency Transportation 应急运输办公室

Operational Environmental Testing 使用环境试验

Optical Electronic Transducer 光学电子转换器,光学电子传感器

**OEU** Optional Equipment Unit 任选的设备组件

Overhead Electronics Unit 顶板电子组件

**OEV** Overboard Exhaust Valve 外流放气活门

**OEW** Operational/Operating Empty Weight 使用空重,运营空重

**OF** Object Finance 物品融资

Off Time 起飞时间,离地时间

Oil-Filled 已加(滑)油的

Outside Face 外表面

Over Frequency 过频

Overflow 溢流,溢出

Overfull 过满

**OFA** Off-net Access 网外接入

Optical Fiber Amplifier 光纤放大器

**OFACS** Overseas Foreign Aeronautical Communications Station 海外航空通信站

**OFBD** Optical Fiber Branching Device 光纤分路器

**OFBG** Optical Fiber Bragg Grating 光纤布拉格光栅

**OFC** Office 办公室

Office Code 局号

Operational Flight Control 运行飞行控制

Optical Fiber Communication 光纤通信

**OFCC** Optical Fiber Cable Component 光缆元件

**OFCE** Office 办公室

**OFCL** Official 官方的,正式的

Officially 正式地

**OFD** Optical Frequency Discriminator 光鉴频器

**OFDL** Optical Fiber Delay Line 光纤延时线

**OFDM** On-board Flight Data Monitoring 机载飞行数据监控

Optical Frequency Division Multiplexing 光频分复用

Orthogonal Frequency Division Multiplex 正交频分复用

**OFDMA** Orthogonal Frequency-Division Multiple Access 正交光频分多址

**OFDPS** Offshore Flight Data Processing System 近海飞行数据处理系统

**OFDR** Optical Frequency Domain Reflectometer 光频域反射计

**OFE** Optical Fiber Equalizer 光纤均衡器

**OFEP** Operation Function Element Program 运行功能元程序

**OFF** Offer, Offering 提供

Optical Fiber Facing 光纤端面

**OFF/R** Off Reset 断开/复位

**OFFLD** Offload 卸货,卸载

**OFFR** Off/Reset 关/复位

**OFG** Optical Frequency Generator 光频产生器

**OFGEN** Offset File Generation 偏置文档生成

**OFI** Operational Flight Instrumentation 飞行

操作仪表

**OFIR** Open Flight Information Region 开放飞行情报区

**OFIS** Operational Flight Information Services 飞行运行信息服务

**OFISP** Operational Flight Information Service Panel 飞行运行信息服务专家组

**OFL** Off-Line 脱机

Overflow 流出

**OFLAN** Optical Fiber LAN 光纤局域网

**OFLD** Offload or Offloaded 卸下,卸载或卸载的

**OFLTR** Oil Filter 滑油滤

**OFM** Optical Frequency Modulation 光频调制

**OFMV** Oil Flow Management Valve 滑油流量管理活门

**OFOD** On-Flight Origin and Destination 航班始发地和目的地

**OFP** Operational Flight Path 飞行运行航迹

Operational Flight Plan 飞行运行计划

Operational Flight Program 飞行运行程序, 飞行操作程序

Optical Fiber Path 光纤通道

Optimize Flight Program 最佳飞行程序

Original Flight Plan 原始飞行计划

**OFPP** Office of Federal Procurement Policy 联邦采购政策处

**OFR** Operational Failure Report 运行故障报告

Overfrequency Relay 超频继电器

**OFS** Offset 偏离

One-Function Sketch 单功能草图

Operational Flight Software 飞行运行软件

Optical Fiber Sensor 光纤传感器

Optical Frequency Shifter 光移频器

**OFSH** Operator's Flight Safety Handbook 营运人飞行安全手册

**OFST** Offset 偏置

Route Offset 航路偏移

**OFT** Operational Flight Trainer 作战飞行教练机

Orbital Flight Test 轨道飞行试验

Outfit 成套设备

**OFTF** Optical Fiber Transmission Function 光纤传输函数

**OFTS** Optical Fiber Transmission System 光纤传输系统

**OFV** Outflow Value 溢流值

Overflow Valve 溢流活门

**OFZ** Obstacle Free Zone 无障碍物地带

**OG** Object Glass 物镜

Oil Gage 油量表

Olympic Games 奥运会

On Ground 在地面

Operational Group 运行小组

**OGA** Outer Gimbal Angle 外环架角

**OGC** Organization Code 组织(机构)代码

**OGE** On-Gimbals Electronics 陀螺电子设备

Operating Ground Equipment 地面运行设备

Out of Ground Effect (飞机)离地效应

**OGES** Operating Ground Equipment Section 地面操作设备部分

**OGG** Kahului, HI, USA (Airport Code) 美国夏威夷州卡胡卢伊机场

**OGHFA** Operational Guide to Human Factors in Aviation 航空中的人为因素运行指导

**OGL** Obscure Glass 毛玻璃

**OGLC** Optical Ground Landing Control 光学地面着陆控制

**OGM** Optical Gateway Manager 光网关管理器

**OGO** Orbiting Geophysical Observatory 轨道地球物理观测台

**OGR** Outgoing Call Restriction 去话呼叫限制

Outgoing Repeater 出局中继器

**OGS** Ogdensburg, NY, USA (Airport Code) 美国纽约州奥格登斯堡机场

**OGSIG** Out Going Signaling 去话信令

**OGT** Out Going Truck 出局运货车

Out Going Trunk 去话中继电路

Outlet Gas Temperature 出口燃气温度

**OGV** Outlet Guide Vane 出口导流叶片

**OGW** Overload Gross Weight 超载总重量

**OGX** Optical Gateway Cross Connect 光纤网关交叉连接

**OH** Observation Helicopter 观察直升机

Operation Handbook 运行手册,操作手册

Operational Hardware 操作硬件

Opposite Hand 另外一侧,异侧

Over Head(Overhead) 头顶的,高架的,顶部的

Overhaul 大修

Overheat 过热,超温

**OHA** Operating Hazard Analysis 运行危害分析,操作风险分析

Outside Helix Angle 外螺旋角

Over Head Access 开销接入

**OHC** Overhead Cam 过热凸轮

**OHCO** Overhaul Change Order 翻修更改指令

**OHD** Ohrid, Macedonia (Airport Code) 马其顿王国奥赫里德机场

On Hand 现有的,现场无人领取的(行李)

On Hand Baggage Trace 手提行李跟踪,手提行李查询

Overhead Door 头顶门,高架门

**OHDS** Overheat Detection System 过热探测系统

**OHDU** Overheat Detection Unit 过热探测组件

**OHKA** Oasis Hong Kong Airlines Limited 甘泉香港航空有限公司

**OHM** Ohmmeter 欧姆表

Overhaul Manual 翻修手册

**OHME** Overhead Mechanical Equipment 高架机械设备

**OHMTADS** Oil and Hazardous Materials Technical Assistance Data System 油类及危险物品技术援助数据系统

**OHP** Overhead Panel 头顶板

**OHR** Overhaul Report 翻修报告

Office of Human Resources 人力资源办公室

Operational Hazard Report 操作风险报告

**OHS** Occupational Health and Safety 职业健康与安全

**OHSA** Occupational Health and Safety Act 职业健康与安全法案

Occupational Health Safety Agency 职业健康安全机构

**OHSAS** Occupational Health and Safety Assessment Series 职业健康与安全评估体系

Occupational Health and Safety Assessment Scheme 职业健康与安全评估计划

**OHSC** Overhead Stowage Compartment 头顶行李箱

**OHSMS** Occupational Health and Safety Management System 职业健康与安全管理系统

**OHSS** Occupational Health and Safety Standard 职业健康与安全标准

**OHU** Optical Head Unit 光学头顶装置

Overhead Unit 头顶组件

**OHV** Overhead Valve 高架活门

**OI** Office Instruction 办公室手册

Oil Immersed 浸油的

Oil Insulated 油绝缘的

On Instrument 仪表飞行

Operating Instructions 使用说明

Operational Interruption 运行中断

Optical Interconnection 光互连

Opto-Isolator 光隔离器

Output Impedance 输出阻抗

**OIA** Optical Interface Adaptor 光接口适配器

**OIAA** Office of International Aviation Affairs 国际航空事务处

**OIC** Optical Integrated Circuit 光学集成电路

**OICOOL** Oil Cooler 滑油散热器

**OID** Object Identifier 对象标识符

Optical Incremental Digitizer 光学微量化数字转换器

Outline Installation Drawing 轮廓安装图

Overseas International Distributors 国际海外分发员

**OIDA** Original Image Data Array 原始图像数据阵列

**OIDS** Operational Information Display System 运行信息显示系统

**OIF** Optical Internet Forum 光联网论坛
Optical Internetworking Forum 光互联网论坛

**OIFC** Oil Insulated Fan Cooled 滑油绝缘风扇冷却的

**OIFM** Objectives Implementation Funding Mechanism 目标实施筹资机制

**OIG** Office of Inspector General ( DOT ) 总监办公室(运输部)

**OIL** Orange Indicating Light 橘黄色指示灯

**OIM** Object Information Model 对象信息模型
Operations Interface Module 操作接口模块
Organizational and Intermediate Maintenance 机构与中间物维修
Oshima, Japan ( Airport Code ) 日本大岛机场

**OIM CF** OIM Common Function OIM 通用功能

**OIM DF** OIM Dedicated Function OIM 专用功能

**OIML** International Organization of Legal Metrology 国际法定度量衡组织

**OIP** Operation in Process 工作在进行中
Operational Improvement Plan 操作改进计划
Optical Image Processing 光学图像处理
Optical Image Processor 光学图像处理机

**OIPS** Optical Image Processing System 光学图像处理系统

**OIR** Office of Industrial Relations 工业事务联络处
Office of International Resources 国际资源处
Operation and Information Requirements 操作与信息要求
Operation and Inspection Record 操作与检验记录
Operations Integration Review 使用综合评审

Organization for Industrial Research 工业研究组织

**OIRSS** Operation and Inspection Record Supplement Sheet 操作与检验记录补充页

**OIRT** International Radio and Television Organization 国际广播电视组织

**OIS** Office Information System 办公室信息系统
On-board Information System 机载信息系统
On-line Information Service 在线的信息服务
Operational Information Service 运行信息服务
Operational Information System 运行信息系统
Operator Interface Station 操作员接口站
Order Item Split 订单项目分解

**OISA** Office of International Scientific Affairs 国际科学事务处
Office of International Student Affairs 国际学生事务处

**OISC** Oil Insulated Self Cooled 滑油绝缘自冷却的

**OIT** Oil Inlet Temperature 滑油进口温度
Oita, Japan ( Airport Code ) 日本大分机场
Onboard Information Terminal 机载信息终端
Operator Information Telex 营运人信息电报,用户信息电传

**OIU** Office Interface Unit 局内接口单元
Optical Image Unit 光学图像单元
Optical Interface Unit 光学接口单元

**OIWC** Oil Immersed Water Cooled 浸油水冷却的

**OJT** On the Job Training 在职培训

**OJTI** On-the-Job Training Instructor 在岗培训教员

**°K** Degrees Kelvin 开氏度

**OK** All Correct 无误
Correct 正确的
Okay 对,很好的
Space Confirmed 座位订妥

**OKA** Naha, Okinawa, Japan ( Airport Code )

日本冲绳那霸机场

**OKC** Oklahoma City, OK, USA ( Airport Code ) 美国俄克拉何马州俄克拉何马城机场

**OKJ** Okayama, Japan ( Airport Code ) 日本冈山机场

**OKS** Orbital Kick Stage 轨道反冲级

**OKT** Occupational Knowledge Test 职业知识测试

**OL** Off Load 卸载

Open Loop 开环

Operating Location 作业地点

Orange Line 橘色线,橘色管路

Our Letter 我方函件

Outer Loop 外环路

Outline 外形,剖面,大纲

Over Lay 外罩,涂层,镀层

Overall Length 全长,总长

Overload 超载

**OLA** Operational Level Agreement 运行级别协议

Optical Fiber Limiting Amplifier 光纤限幅放大器

Optical Laboratories Association 光学实验室协会

**OLAN** Onboard Local Area Network 机载局域网

**OLAP** On-Line Analytical Processing 在线分析处理

**OLAPFL** Left Overlap Load 左重叠负载

**OLAPFR** Right Overlap Load 右重叠负载

**OLB** Off Line Browser 离线浏览器

**OLBR** Operational Laser Beam Recorder 运行激光束记录器

**OLC** On Line Coupon 联机票联

On-Line Code 联机码,在线码

Orient Leasing Co. , Ltd. ( Japan ) 东方租赁公司 ( 日本 )

Overload Channel 过载信道

Overload Control 过载控制

**OLCA** On-Line Communication Adapter 联机通信适配器

**OLCP** On-Line Complex Processing 在线复合处理

**OLCS** On-Line Computer System 在线计算机系统

**OLCSS** On-Line Computer Shopping Services 在线电脑购物服务

**OLCTP** On-Line Complex Transaction Processing 在线复杂交易处理

**OLD** On-Line Debugging 在线调试

On-Line Diagnostics 在线诊断

**OLDAP** On-Line Data Processor 联机数据处理器

**OLDI** On-Line Data Interchange 在线数据交换

**OLDP** Office of Legislative Drafting and Publishing 立法起草和发布办公室

On-Line Data Processing 联机数据处理

**OLDS** On-Line Dynamic Server 在线动态服务器

**OLE** Object Linking and Embedding 对象链接嵌入

Optical Line Equipment 光线路设备

**OLED** Organic Light-Emitting Device 有机发光装置

Organic Light-Emitting Diode 有机发光二极管

**OLEF** Outboard Leading Edge Flap 外侧前缘襟翼

**OLEMS** Object Linking Embedding Management Service 对象链接嵌入管理服务器

**OLES** Outboard Leading Edge Station 外侧前缘站位

**O-LEVEL** Organizational Level 组织水平

**OLFS** On-Line File System 联机文件系统

**OLH** On-Line Help 在线帮助

**OLI** Optical Line Input 光线路入口

Optical Line Interface 光线路接口

**OLIVER** Output Level Intended for Very Excellent Reproduction 追求最佳复制的输出水平

**OLL**　Open-Loop Loss 开放环路损耗

**OLM**　Operational Level Maintenance 运行等级的维护

**OLMM**　Optical Loop Mirror Multiplexer 光环路镜像复用器

**OLMS**　Operational Loads Monitoring System 运行负载监控系统

**OLO**　On-Line Operation 在线操作
　　Optical Line Output 光线路出口

**OLP**　On-Line Planning 联机计划,在线计划
　　On-Line Processor 联机处理器

**OLPS**　On-Line Planning System 联机计划系统,在线计划系统

**OLR**　On-Line Replaceable 在线更换
　　Optical Line Rate 光线路速率

**OLRC**　Organization Level Repair Cycle 结构级修理循环

**OLRI**　On-Line Receivable Invoice 可接受联机发票

**OLRS**　Optical Laser Ranging System 光学激光测距系统

**OLRTS**　On-Line Real-Time System 在线实时系统

**OLS**　Oil Level Sensor 滑油油量传感器
　　On-Line Service 在线服务
　　Optical Landing System 光学着陆系统
　　Order Location and Status 指令地址与状态
　　Order Location System 指令(订单)定位系统
　　Outgoing Line Signaling 去话线路信令

**OLSS**　Order Location and Status System 指令(订单)定位与状态系统

**OLSV**　Overspeed Limiter Spill Valve 超转限制器溢流活门

**OLT**　On-Line Test 在线测试
　　On-Line Training 在线培训
　　Optical Line Terminal 光缆线路终端

**OLTM**　Optical Line Terminal Multiplexer 光线路终端复用器

**OLTP**　On-Line Transaction Processing 线上交易处理

On-Line Transition Processing 联机转换处理

**OLTT**　On-Line Terminal Testing 在线终端测试

**OLVL**　Oil Level 滑油液面,滑油量

**OLW**　Operational Landing Weight 操作着陆重量

**OLWS**　On-Line Work Station 联机工作站
　　Optical Light Wave Synthesizer 光波合成器

**OM**　ILS Outer Marker 仪表着陆系统远台
　　Mongolian Airlines 蒙古航空公司
　　Object Management 对象管理
　　Operation Mode 操作模式
　　Operational Mockup 操作样机
　　Operational Module 操作模件
　　Operations and Maintenance 使用与维护
　　Operations Manual 使用手册,飞机操作手册
　　Optical Multiplexer 光复用器
　　Order Management 订货管理
　　Ordering Material 订货器材
　　Outer Marker 外指点标
　　Outside of Metal 金属外部
　　Overbooking Management (机票)超售管理
　　Overhaul Manual 大修手册

**OM DEL**　Omnibus Delete 单向总线删除

**OM(O&M)**　Operation and Maintenance 操作与维修

**OMA**　Object Management Architecture 对象管理体系结构
　　Omaha, NE, USA (Airport Code) 美国内布拉斯加州奥马哈机场
　　Operations Management Application 操作管理应用
　　Operations Management Association 运营管理协会
　　Orthogonal Multiple Access 正交多址接入

**OMAP**　Operation, Maintenance and Administration Part 运行、维护和管理部分
　　Operation Maintenance Application Part 操作维护应用部分

**OMAT**　Operational Monitors and Tests 运行监测器与测试

**OMAV**  Omega Navigation 欧米加导航

**OMB**  Office of Management and Budget 国家管理和预算局,(联邦航空局)管理与预算办公室

Operation and Maintenance Block 运行和维护功能块

Operations Manual Bulletin 使用手册通告

Outer Marker Beacon 外指点标信标

**OMC**  On-board Maintenance Computer 机载维修计算机

Operation and Maintenance Center 运行维护中心

Operation and Maintenance Costs 运行与维护成本

Operation and Management Center 运行管理中心

Operations Monitoring and Control 运行监控和管制

Operations Monitoring Computer 操作监控计算机

Optimisation and Measurement Center 优化与测量中心

**OMC-G**  Operation and Maintenance Center-GPRS GPRS 运行维护中心

**OMD**  On-board Maintenance Data 机载维修数据

On-board Maintenance Document 机载维修文件

Operations and Maintenance Document 使用与维修文件

Operator Modifiable Data 运营人改装数据

**OME**  Nome, AK, USA (Airport Code) 美国阿拉斯加州诺姆机场

**OMEGA**  Omega Navigation System 欧米加导航系统

**OMF**  Object Management Function 目标管理功能

Open Media Framework 开放式媒体框架

**OMFI**  Open Media Framework Interchange 开放式媒体架构互换

**OMG**  Object Management Group 对象管理组

**OMI**  Omnibearing Magnetic Indicator 全向磁指示器

On-board Maintenance Information 机载维修信息

Open Messaging Interface 开放式信息界面

Operations and Maintenance Instruction 使用与维修说明

Optical Modulation Index 光调制指数

**OMIL**  On-line Maintenance Information Library 联机(在线)维护信息资料室

**OMIN**  Optical Multistage Interconnected Network 光多级互联网络

**OMJ**  Orthomode Junction 正交模耦合器

**OML**  Operational Multi-crew Limitation 多机组成员运行限制

Outside Mold Line 外模压管

**OMM**  Operation and Maintenance Manual 操作与维修手册

Operation, Maintenance and Monitoring 运营、维修和监控

Operations and Materials Management 操作和材料管理

Operations Management Meeting 运营管理会议

Optimizer Manufacturing Management 最优控制制造管理

Organizational Maintenance Manual 组织维修手册

Overhaul Manual Manufacturer 生产厂家大修手册

**OMN**  Omnibus 单向总线

Optically Multiplexed Network 光复用网络

**OMNB**  Omnidirectional Beacon 全向信标

**OMNI**  Omnidirectional 全向的

**OMNIRANGE**  Omnidirectional Radio Range 全向无线电定位

**O-MOD**  Optical Modulator 光调制器

**OMP**  Open Management Protocol 开放的管理协议

Operation and Maintenance Processor 运行维护处理器

**OMPR** Operational Maintainability Problem Reporting 运行维修能力问题报告

**OMR** Operational Management Review 运营管理评审

Optical Mark Reader 光学标记阅读器

Oradea, Romania (Airport Code) 罗马尼亚奥拉迪亚机场

Overhaul, Maintenance, and Repair 大修、维护和修理

**OMS** Object Management System 目标管理系统

On-board Maintenance System 机载维护系统,机载维修系统

Open Management System 开放管理系统

Operation and Maintenance Subsystem 运行与维护子系统

Operational Miniature System 小型操作系统

Optical Multiplexer Section Layer 光复用段层

Opto-electronic Multiplex Switch 光电子复用转换

Orbital Maneuvering System 轨道机动系统

Order Management System 订单管理系统,指令管理系统

Organization Maintenance Squadron 机构维修中队

Outage Management System 停电管理系统

Output Multiplex Synchronizer 输出多路同步器

**OMSDR** Operational Miniature System Design Review 小型操作系统设计审查

**OMSE** Open Message Switching Engine 开放平台信息交换引擎

**OMSF** Office of Manned Space Flight 载人空间飞行办公室,载人宇航办公室

**OMSP** Operational Maintenance Support Plan 使用维修保障计划

Operations Management System for Production 生产运行管理系统

**OMSR** Operational Miniature System Requirements 小型操作系统要求

**OMSRM** Operational Miniature System Requirements Mockup 小型操作系统要求样机

**OMSS** Operation Maintenance Support System 运行维护支持系统

**OMT** Object Modeling Technique 对象建模技术

Onboard Maintenance Terminal 机载维护终端

Online Master Table System 联机主表格系统

Online Master Tables 联机主表格

Orthomode Transition 正交模变换

**OMTBF** Observed Mean Time between Failure 观察到的故障平均间隔时间

**OMTS** On-board Mobile Telephony System 机载移动电话系统

Online Master Table System 联机主表格系统

Organizational Maintenance Test Sets 机构维修测试装置

**OMU** Operational Mock-up 操作模型

Operator Maintenance Unit 操作员维护组件

**OMUP** Operations Maintenance User Part 运行维护用户部分

**OMV** Orbiting Maneuvering Vehicle 轨道机动飞行器

Overhaul Manual Vendor 协作厂大修手册

**ON** Optical Network 光网络

**ONA** Office of Noise Abatement 降低噪声办公室

Off-Net Access 网外接入

Open Network Architecture 开放式网络体系结构

Optical Navigation Attachment 光学领航辅助装置

Optical Network Analyzer 光网络分析仪

Overseas National Airways 国家海外航空公司

**ONAL** Off-Net Access Line 网外接入线路

**ONC** Oceanic Navigational Chart 海洋导航图

Off-Net Calling 网外呼叫

Open Network Computing 开放式网络计算
Operational Navigation Chart 操作导航图

**ONE** Omega Navigation Equipment 欧米加领航设备
Open Net Environment 开放网络环境

**ONERA** French Aeronautics and Space Research Center 法国宇航研究中心

**ONI** Optical Network Interface 光网络接口

**ONL** On-Line 本家航线
Optical Network Layer 光网络层

**ONM** Open Network Management 开放式网络管理

**ONN** Open Network Node 开放型网络节点

**ONNC** Optical Neural Network Computer 光学神经网络计算机

**ONOZ** Oil Nozzle 喷油嘴

**ONP** Open Network Provisioning 开放式网络供给

**ONS** Omega Navigation System 欧米加导航系统
Online Notifying Server 在线通知服务器
Open Networking Supportware 开放式组网支撑件

**ONSO** Omega Navigation System Operation 欧米加导航系统运行

**ONT** Ontario, CA, USA（Airport Code）美国加利福尼亚州安大略机场
Optical Network Terminal 光网络终端

**ONTM** Open Nested Transaction Model 开放的嵌套事务模型

**ONU** Optical Network Unit 光网络单元

**OO** Object-Oriented 面向对象
Official Observer 政府观察员
Operations Order 操作指令

**OOA** Object Oriented Analysis 面向对象的分析
Open Order Audit 公开订货审计

**OOAD** Object Oriented Analysis and Design 面向对象的分析与设计

**OOAL** Operational Outline and Checklist 操作大纲与检查单

**OOAM** Object-Oriented Analysis Model 面向对象的分析模型

**OOB** Out of Band 带外

**OOBA** Out of Box Audit 开箱检查

**OOC** Optical Orthogonal Code 光正交码
Over Ocean Communication 越洋通信

**OOD** Object-Oriented Design 面向对象的设计

**OODB** Object-Oriented Database 面向对象的数据库

**OODBMS** Object Oriented Database Management System 面向对象数据库管理系统

**OODM** Object Oriented Development Method 面向对象开发方法

**OODS** Object Oriented Database System 面向对象数据库系统

**OOEU** Outboard Overhead Electronics Unit 外侧顶板电子组件

**OOF** Out of Frame 帧失步

**OOGMS** Object-Oriented Graphical Modeling System 面向对象的图形建模系统

**OOJ** Origin Open Jaw 始发地缺口

**OOK** On-off Keying 开关键控

**OOL** Gold Coast, Queensland, Australia（Airport Code）澳大利亚昆士兰州黄金海岸机场
Object-Oriented Language 面向对象的语言
Object-Oriented Layer 面向对象的层
Out of Limits 超限

**OOM** Object-Oriented Memory 面向对象的存储器
Object-Oriented Method 面向对象的方法

**OONP** Object-Oriented Network Protocol 面向对象的网络协议

**OOOI** On-Out-Off-In 登机—滑出—起飞—滑入
Out-Off-On-In 滑出—起飞—接地—停靠门位

**OOOS** Object-Oriented Operation System 面向对象操作系统

**OOP** Object-Oriented Programming 面向对象

程序设计

**OOPL** Object-Oriented Programming Language 面向对象的编程语言

**OOPS** Object-Oriented Programming System 面向对象的程序设计系统

**OORAM** Object-Oriented Role Analysis Method 面向对象的任务分析方法

**OOS** Object-Oriented Software 面向对象的软件

Orbit-to-Orbit Shuttle 轨道间穿梭机

Orbit-to-Orbit Stage 轨道间级

Out of Service 失修的，业务中止

**OOSD** Object-Oriented System Design 面向对象的系统设计

**OOT** Object Oriented Technology 面向对象技术

Oil Outlet Temperature 滑油出口温度

**OOUI** Object-Oriented User Interface 面向对象用户界面

**OOWS** Object-Oriented Window Software 面向对象的窗口软件

**OP** Observation Post 观察站

Oil Pump 滑油泵

Oilproof 防油的

Opalescent 乳色的

Open 开启，开路

Operating Procedure 操作程序，运行程序

Operation 操作，运行

Operational Program 操作程序，运行计划

Option 选装，选项

Outside Production 外部生产

Overhead Panel 舱顶仪表板

Ozalid Print 晒图

**OPA** Opaque 遮光涂料，不透明的

Operating Procedure Agreement 运行程序协议

Operational Performance Assessment 运行性能评估

Optical Parametric Amplification 光量放大

Optoelectronic Pulse Amplifier 光电脉冲放大器

**OPAG** Open Programme Area Group 开放计划领域组

**OPAL** Optical Parametric Amplification Laser 光参量放大激光器

Order Processing Automated Line 自动生产线指令处理

**OPAS** Operational Assistance 运行援助

Operational Assistant 业务助理

**OPB** Optical Power Budget 光功率分配

**OPBC** Overhead Panel Bus Controller 顶板总线控制器

**OPC** Operational Control 运行管理

Operational Program Configuration 操作程序组合

Operational Program Configuration File 操作程序组合文件

Optical Phase Conjugation 光相位共轭

Organic Photo Conductor 有机光电导体，光敏电阻

Originating Point Code 始端(信令)点代码

**OPCA** Operational Control Area 运行管制区

**OPCF** Overhead Panel Card File 顶板插件卡

**OPCODE** Operational Code 操作码

Operations Code 操作码

**OPCOM** Operations Communications 运行通信

**OPCON** Operating Condition 运行条件

Operational Control 运行控制

**OPCR** Operating Plan Change Request 运行计划更改申请

**OPD** Optical Path Difference 光程差

**OPDAR** Optical Directional and Ranging 光学定向和测距

**OPDR** Optical Polarization Domain Reflectometry 光偏振域反射测量法

**OPDU** Operation Protocol Data Unit 操作协议数据单元

Overhead Panel Display Unit 顶板显示组件

**OPEC** Organization of Petroleum Exporting Countries 石油输出国组织

**OPEN** Optical Pan-European Network 泛欧光网络

**OPER** Operation, Operating, Operative 操作,运转,有效

**OPEVAH** Operation Evaluation of Armed Helicopter 武装直升机作战评价

**OPF** Off Peak Fare 非高峰票价

Orbiter Processing Facility 轨道飞行器处理设施

Orifice 限流器

**OPGV** Optimum Practical Gas Velocity 最佳实际气体流速

**OPGW** Optical Power Ground Wire 光纤架空地线复合缆

**OPI** Oil Pressure Indicator 滑油压力指示器

Operations Process Improvement 操作工艺改进

Overall Performance Index 全性能指数

**OPIC** Overhead Panel Interface Card 顶板接口卡

**OPIL** Opalescent Indicating Light 乳白色指示灯

**OPL** Optional 供选择的,选装的

Outside Production Ledger 外部生产总账

Outside Production Line 外部生产线

**OPLE** Omega Positioning and Locating Equipment 欧米加定位设备

**OPLG** Oil Plug 滑油栓,滑油塞

**OPLL** Optical Phase Lock-Loop 光锁相环路

**OPLOGSIM** Operations and Logistics Simulation 运行与后勤模拟

**OPM** Office of Personnel Management 人员管理处

Operations and Procedures Manual 运行与程序手册

Operations per Minute 每分钟操作次数,每分钟动作次数

**OPMA** On Board Performance Monitoring and Alerting 机载性能监视与告警

**OPMET** Operational Meteorological (Information) 飞行气象(信息)

**OPMH** Operation Procedures for Message Handling 电报处理操作程序

**OPMS** Oil Pressure Monitoring System 油压监控系统

**OPMT** Operations Planning Management Team 运行规划管理小组

**OPN** Oil Pan 滑油槽

Open or Opening or Opened 开启

Operation 操作

Optimized-Profile Navigation 最佳轨迹导航

**OPNG** Opening 打开

**OPNL** Operational 可操作的,可使用的,可运行的

**OPNTKT** Open Ticket 不定期客票

**OPO** Outside Production Order 外部生产指令

Porto, Portugal (Airport Code) 葡萄牙波尔图机场

**OPOL** Optimization-Oriented Language 面向优化的语言

**OPOS** Outside Production Ordering System 外部生产指令系统

**OPP** Octal Print Punch 八进制打印穿孔机

Oppose 反对,反抗

Opposite 相反的,相对的

Optical Power Penalty 光功率损耗

**OPP CE** Opposite Commutator End 反向整流子端

**OPP HND** Opposite Hand 对手,反手,另一方面

**OPP PE** Opposite Pulley End 反转轮端

**OPPM** Overlapping Pulse Position Modulation 重叠脉位调制

**OPPOR** Opportunity 机会,时机

**OPPR** Operating Program 操作程序

**OPR** Occurrence Probability Ranking 发生概率等级

Office of Primary Responsibility 主要责任办公室

Officer of Primary Responsibility 主要责任官员

Once per Revolution 每转一次

Operate, Operated 运转,操作,操作的

Operator 操作者,报务员,经营人

Optical Preamplifier Receiver 光预放大接收

机

Optional Parts Request 选装零件申请

Overall Pressure Ratio 总压力比

**OPREP**　Operational Report 运行报告

**OPRI**　Operational Primitives 操作原语

**OPRN**　Operation 操作

**OPRND**　Operand 运算数,运算量,基数,操作数

**OPRS**　Oil Pressure 滑油压力

Overhead Performance Reporting System 顶板性能报告系统

**OPRT**　Operator Table 操作员表,算子表

**OPRTNTY**　Opportunity 机会

**OPS**　Operational Program Software 操作程序软件

Operational Program Specification 操作程序规范

Operations Manual 操作手册,运转手册

Operations or Operates 运行,运转,航行

Operations per Second 运算次数/秒

Optical Smoothing 光平整

Outside Production Service 外部生产服务

Outside Production System 外部生产系统

Overpressure Protection System 过压保护系统

**Ops Specs**　Operations Specifications 运营规范

**OPSD**　Openside 开启边

**OPSEC**　Operational Security 操作安全,操作保密

**OPSOV**　Over Pressure Shutoff Valve 超压关断活门

**OPSP**　Office of Product Standards Policy 产品标准政策办公室

Outside Production Specification Plan 外部生产规范计划

**OPSR**　Operating Plan Staffing Requirement 操作计划职员要求

**OPSUM**　Operational Summary 航务摘要,运行摘要,操作摘要

**OPSUP**　Operational Supervisor 运行监督

**OPT**　Oil Pressure Transmitter 滑油压力传感器

Onboard Performance Tool 机载性能工具

Open Packet Telephony 开放式分组电话

Operation 操作,运行,使用

Optical 光的

Optimized Production Technology 优化的生产技术

Optimum 最佳的

Optional 可选的,选装的

**OPTA**　Optimal Performance Theoretically Attainable 理论上可达到的最佳性能

**OPTAT**　Off Premise Transitional Automated Ticket 自动客票

**OPTBYP**　Optical Bypass 光旁通(器)

**OPTE**　Operator Proficiency Training Equipment 操作员熟练训练设备

**OPTI**　Office of Productivity, Technology and Innovation 生产率、技术与革新办公室

**OPTINT**　Optical Intelligence 光学信息

**OPTP**　One Page Test Plan 单页测试计划

**OPTPN**　Optional Part Number 任选部件号

**OPTS**　Oil Pressure and Temperature Sensor 油压及温度传感器

On-line Program Testing System 联机程序测试系统

**OPU**　Overhead Processing Unit 开销处理单元

Overspeed Protection Unit 超速保护组件

**OPV**　Optionally Piloted Vehicle 驾驶方式可选飞行器

Overpressure Valve 超压活门

**OPWT**　Operating Weight 使用重量,操作重量

**OPXC**　Optical Path Cross Connect 光通路交叉连接

**OQAR**　Optical Quick Access Recorder 光学快速存取记录器

**OQASK**　Offset Quadrature Amplitude Shift Keying 偏移正交幅移键控

**OQC**　Outside Quality Control 外部质量控制

**OQI**　Oil Quantity Indicator 滑油量指示器

**OQL**　Object Query Language 对象查询语言

On-line Query Language 在线查询语言

**OQMM** Operation Quality Management Manual 运行质量管理手册

**OQPSK** Offset Quadrature Phase Shift Keying 偏移正交相移键控

**OR** O-Ring O 形圈

Object Rejection 项目排除,项目取消

Oil Ring 滑油环

Old Release 旧版本

Operation Requirements 操作(运行)要求

Operational Ready 运作就绪

Operational Reliability 运行可靠性

Operational Requirements 运行要求,操作要求,飞行要求

Optical Reflectance 光反射比

OR Gate 或门

Order Register 指令寄存器

Output Register 输出寄存器

Outside Radius 外径

Overhaul and Repair 翻修与修理

Overload Relay 过载继电器

Over-Run 备用跑道,安全道,超限,超速

Oversale Risk 机票超售风险

Owner's Risk 业主风险,风险由货主负责

**°R** Degrees Rankine 兰氏度

**OR** Outgoing Route 输出路由

**ORA** Operational Requirements Analysis 运行需求分析

OR Arithmetic 或运算

Other Restricted Articles 其他限制物品

**ORACLE** Optimum Reliability and Component Life Estimator 优化的可靠性和部件的寿命估计

**ORAT** Operational Readiness and Airport Transfer 运行准备和机场移交

**ORB** Object Request Broker 对象请求代理

Omnidirectional Radio Beacon 全向无线电信标

**ORC** Optimal Retransmission Control 最佳转发控制

Originating Region Code 始发地区代码

Outbound Radar Control 离场雷达控制

**ORD** Chicago, IL, USA-O'Hare (Airport Code) 美国伊利诺州芝加哥奥黑尔机场

Operational Readiness Date 运行就绪日期

Operational Readiness Demonstration 运行准备就绪示范

Operational Requirements Document 运行要求文件

Optical Rotary Dispersion 光旋转散射

Orbital Requirements Document 轨道要求文件

Order 订单,命令,阶

Ordered 有序的,订购的

Ordering 调整,排序

Ordinance 规格,条例,布告

**ORDIR** Omnidirectional Digital Radar 全向数字式雷达

**ORDPROC** Order Processing 指令处理,订单处理

**ORDS** Online Rejection Disposition System 联机拒绝处理系统

**ORE** Overall Rreference Equivalent 全程参考当量

**O-REP** Optical Repeater 光中继器

**ORF** Norfolk, VA, USA (Airport Code) 美国弗吉尼亚州诺福克机场

Orifice 喷管,喷嘴,限流器

**ORG** Organic 有机的,组织的

Organization or Organized 机构,组织化的

**ORGTR** Originator 创作者,发明人,发起人

**ORGZ** Organize 组织

**ORH** Worcester, MA, USA (Airport Code) 美国麻萨诸塞州伍斯特机场

**ORI** Operational Readiness Inspection 运作就绪检查

Orient 向东,定向,东方的

Orientation 面向

**ORIENT** Orientation 定(取)向,方位

**ORIF** Orifice Area 小孔区,喷嘴区

**ORIG** Original 起源,原始的,创始

Originator 发明人,发起人

**ORIGN** Originated 其始的,首创的

**ORIS** Office of Scientific Information Service 科学情报服务局(美国)

On-line Rejection Tag Information System 联机拒绝标签信息系统

Open Source Information System 公开来源信息系统

**ORK** Cork, Ireland (Airport Code) 爱尔兰科克机场

**ORL** Optical Return Loss 光回波损耗

Orbiting Research Laboratory 轨道研究实验室

Orlando, FL, USA (Airport Code) 美国佛罗里达州奥兰多机场

Overrun Area Edge Lights 安全道区域边灯

**ORLA** Optimum Repair Level Analysis 最佳修理水平分析

**ORLY** Overload Relay 过载继电器

**ORM** Onboard Replaceable Module 机载可更换件

Operational Risk Management 运行风险管理

Optical Receiver Module 光接收机模块

Optical Reference Manual 光学参考手册

Organizational Record Managers 机构记录管理器

Output Reconfiguration Network 输出重构网络

Overhaul and Repair Manual 修理手册

**ORMA** Optical Reservation Multiple Access 光预留多址接入

**ORMG** Operational Risk Management Group 运行风险管理组

**ORN** Optic Remote Node 光远端节点

Oran, Algeria (Airport Code) 阿尔及利亚奥兰机场

Orange 橘,橘黄色

Ornament 装饰(品)

**OROCA** Off Route Obstacle Clearance Altitudes 偏航超障高度

**OROM** Optical Read Only Memory 光学只读存储器

**ORP** Optical Reference Point 光参考点

Output Routing Pool 输出布线区

**ORR** Omnidirectional Radar Range 全向雷达信标

Operation Readiness Review 操作就绪审查

**ORS** Offensive Radar System 攻击雷达系统

**ORSA** Operations Research and System Analysis 运筹学研究与系统分析

Operations Research Society of America 美国运筹学学会

**ORT** On-going Reliability Test 持续可靠性测试

Operational Reliability Test 运行可靠性测试

Owner Requirements Table 业主要求表

**ORTS** Operational Readiness Test System 战备测试系统

**ORU** On-line Replacement Unit 在线替换装置

Optical Reference Unit 光学基准装置

Optical Repeater Unit 光中继单元

**ORV** Object Rejection Vector 目标排斥向量

**ORY** Orly, Paris, France (Airport Code) 法国巴黎奥利机场

**ORZ** Omnirange Zero 全向无线电信标零位

**OS** Office System 办公室系统

Off-Safe 不安全,不可靠

Oil Switch 油开关

On Schedule 按时,按预定计划

One Shot 单稳(触发器)

Operating System 操作系统,运行系统

Operational Spare 运行备件

Operations Specifications 运行规范

Optical Section 光纤段

Optical Sender 光发射机

Optical Soliton 光孤子

Optical Switch 光交换

Order Sheet 订货单

Otherwise Specified 另有说明

Out of Service 不能使用,停止使用

Overshoot 测场过高,重飞(英国)

Overspeed 超速

**OS&D**  Over, Short and Damage 超额、短缺与损坏

**OS&Y**  Outside Screw and Yoke 外螺丝与线圈

**OSA**  Office of the State Auditor 国家审计办公室
Office System Automation 办公系统自动化
Open System Architecture 开放系统体系结构
Operation Suspend A-1 操作中止 A-1
Operational Safety Analysis/Assessment 运行安全分析/评估
Optical Spectrum Analyzer 光谱分析仪
Osaka,Japan（Airport Code）日本大阪机场
Out of Sequence Action 非顺序操作

**OSAN**  Optical Subscriber Access Node 光用户接入节点

**OSB**  Output Signal Balance 输出信号平衡

**OSC**  Operating System Control 操作系统控制
Optical Supervisory Channel 光监控信道
Optical Switch Core 光交换核心
Optimum Start Control 最佳启动控制
Optional Supply Code 任选供应码
Oscillate 振荡
Oscillator 振荡器
Outbound Signaling Channel 输出信令信道
Outgoing Sender Connector 出局发送器连接器

**OSCA**  Open Systems Cabling Architecture 开放系统布线结构

**OSCAR**  Orbiting Satellite Carry Amateur Radio 轨道卫星携带业余无线电设备

**OSCE**  On-line Spares and Configuration Environment 在线备件与配置环境
Organization for Security and Co-operation in Europe 欧洲安全与合作组织

**OSCG**  Oscillating 振荡

**OSCGRM**  Oscillogram 波形图

**OSCP**  On-Site Computer Programmer 现场计算机程序员

Operating System Computer Program 操作系统计算机程序
Oscilloscope 示波器

**OSCRN**  Oil Screen 滑油滤

**OSCS**  Out of Sequence Control System 无顺序控制系统

**OSCU**  Oxygen System Control Unit 氧气系统控制组件

**OSD**  Office of Secretary of Defense（美国）国防部长办公室
Operational Sequence Diagram 操作顺序图
Optical Scanning Device 光扫描装置

**OSDF**  Operational Software Development Facility 操作软件开发设备

**OSDM**  Optical Space Division Multiplexing 光空分复用

**OSDP**  On-Site Data Processor 现场数据处理机
Operating System Development Processor 操作系统开发处理器

**OSDS**  Operating System for Distributed Switching 分布式交换操作系统
Optical Space Division Switching 光空分交换

**OSE**  Observing System Experiments 观测系统试验
Open Systems Environment 开放系统环境
Operational Support Equipment 操作支援设备

**OSED**  Operational Services and Environment Description 运行勤务与环境描述

**OSF**  Office of Space Flight 宇宙飞行局
Open Software Foundation 开放式软件基金会
Open Systems Foundation 开放系统基金
Operating System Function 操作系统功能
Operational Support Facility 运行保障设施
Operations Support Flight 运行保障飞行

**OSFB**  Operation System Function Block 运行系统功能块

**OSFC**  Output Subframe Count 输出子帧计数

**OSG**  Oscillograph 示波器

Over Speed Governor 超转调节器

Overspeed Safety Governor 超速安全调速器

**OSGP**    Operational Software Generation Program 操作软件产生程序

**OSH**    Occupational Safety and Health 职业安全与卫生

**OSHA**    Occupational Safety and Health Act 职业安全与卫生法

Occupational Safety and Health Administration（美国）职业安全和卫生管理局

**OSHC**    Occupational Safety and Health Council 职业安全和健康委员会

**OSHMS**    Occupational Safety and Health Management System 职业安全健康管理体系

**OSHN**    Optical Self Healing Network 自愈光纤网络

**OSHS**    Occupational Safety and Health Standards 职业安全与健康标准

**OSI**    Open System Interconnect 开放式系统互联

Open System Interface 开放式系统接口

Operating System Interface 操作系统接口

Other Service Information 其他服务信息，其他服务资料

Other Systems Interface 其他系统接口

**OSIE**    Open System Interconnection Environment 开放式系统互连环境

**OSIR**    Organization for Scientific Investigation and Research 科学调查与研究组织

**OSIRM**    Open Systems Interconnection Reference Model 开放式系统互连参考模型

**OSIS**    Ocean Surveillance Information System 海洋监视信息系统

**OSL**    Oil Seal 油封

Operating System Layer 操作系统层

Operational Safety Pilot Limit 驾驶员安全运行限制

Optical Signal Level 光信号级

Oslo，Norway-Gardermoen（Airport Code）挪威奥斯陆加勒穆恩机场

**OSM**    Orbital Service Module 轨道服务模件

Oscillator Strength Modulation 振荡器强度调制

Outgoing Switch Module 出局交换模块

Output Signal Management 输出信号管理

**OSMF**    Open System Management Framework 开放式系统管理框架

**OSMM**    Outside Station Management Manual 外站管理手册

**OSMT**    Optical Surface Mount Technology 光表面贴装技术

**OSMV**    One Shot Multivibrator 单稳态多谐振荡器

**OSN**    Optical Shuttle Node 光信息往返节点

Optical Subscriber Network 光纤用户网

**OSNC**    Optical Section Network Conncetion 光纤段网络连接

**OSNL**    Operating System Nucleus Language 操作系统核心语言

**OSNR**    Optical Signal Noise Ratio 光信噪比

**OSO**    Offensive System Operator 攻击系统操作员

Office of Space Operations 空间活动处，航天操作处

Operations Scheduling Office 操作进度处

Orbiting Solar Observatory 轨道太阳观测卫星

**OSP**    Oil Suction Pump 滑油抽吸泵

Open Settlement Protocol 开放结算协议

Operating Steam Pressure 工作蒸汽压力

Optical Saturation Parameter 光饱和参量

Optical Signal Processing 光学信号处理

Optical-switched Service Provider 光交换业务供应商

**OSPF**    Open Shortest Path First 开放式最短路径优先（协议）

**OSPL**    Overall Sound Pressure Level 总声压电平

**OSQAM**    Operations Standards and Quality Assurance Manual 操作标准与质量保证手册

**OSQL**    Object Structured Query Language 目

标结构化查询语言

**OSR** Office of Scientific Research 科学研究处

Operating Systems Review 操作系统评审

Operation, Service and Repair 使用、维护与修理

Operation Status Report 运行状态报告

Operational Safety Review 运行安全审查

Operational Scanning Recognition 操作扫描识别

Operational Support Requirement 运行保障要求

Optical Solar Reflector 光学阳光反射器

Optimized Speed Rotor 最佳速度旋翼

Output Shift Register 输出移位寄存器

**OSRD** Office of Scientific Research and Development 科学研究与发展办公室

Office of Standard Reference Data 标准参考数据办公室

**OSS** Open Simulation System 开放式模拟系统

Open Source Software 开放源码软件

Operating System Software 操作系统软件

Operating System Storage 操作系统存储器

Operation and Supporting System 操作和支持系统

Operational Support System 运行保障系统

Operational Test Launch Signal Simulator 操作试验发射信号模拟器

Operator Service System 话务员业务系统

Optical Sight System 光学瞄准具系统

Option Selectable Software 特权选择软件

Out-Slot Signaling 隙外信令

Over Station Sensor 越台传感器

Sahara and Sahel Observatory 撒哈拉与萨赫勒观察站

**OSSA** Office of Space Science Applications 空间应用科学处

**OSSE** Observing Systems Simulation Experiment 观测系统模拟试验

**OSSG** Open Systems Software Group 开放系统软件组

**OSSL** Operating System Simulation Language 操作系统模拟语言

**OST** Office of Security (联邦航空局) 保安办公室

Optical Section Termination 光纤段终端

Optical Soliton Transmission 光孤子传输

**OSTA** Occupational Skill Testing Authority 职业技能鉴定中心，职业技能鉴定部门

**OSTD** Office of Supersonic Transport Development 超音速运输机发展处

**OSTF** Off-Site Test Facility 非现场试验设施

**OSTP** Office of Science and Technology Policy 科学与技术政策处

**OSTR** Optimum Speed Tilt Rotor 最优速度倾转旋翼机

**OSTS** Office of Space Transportation Systems 空间运输系统处

**OSU** Optical Subscriber Unit 光用户单元

**OSV** Optional Supply Vender 任选的供应商

Orbital Serving Vehicle 轨道服务飞行器

**OSW** Operational Switching 操作开关

Optical Switch 光交换

**OSWS** Operating System Work Station 操作系统工作站

**OT** Object Technology 对象技术

Oil Temperature 滑油温度

Oiltight 不漏油的

Operating Temperature 工作温度

Operational Test 工作试验

Optical Tapoff 光分接

Optical Terminal 光终端

Optical Tool 光学工具

Our Telex 我方电传

Out of Turn 脱离转弯，不按顺序

Overtone 谐波，泛音

**OT&E** Operational Test and Evaluation 运行试验和评估

**OTA** Office of Technology Assessment 技术评估办公室（美国）

Operational Transconductance Amplifier 运算

跨导放大器

Optical Telescope Assembly 光学望远镜组件

**OTBD** Outboard 外侧

**OTC** Objective, Time, and Cost 目的、时间和成本

Operating Telephone Companies 运营电话公司

Originating Toll Center 长途始发中心

Originating Toll Circuit 长途始发电路

Over the Counter 场外交易

Overseas Telecommunications Company/Corp. 海外通信公司

Oxygen Transfer Compressor 氧气输送压缩机

**OTCCC** Open Type Control Circuit Contacts 开启型控制电路接触器

**OTCP** Operational Test Computer Program 运行测试计算机程序

**OTD** On Time Delivery 按时交付

Optical Time Domain 光时域

Other Training Devices 其他训练装置

Over Temperature Detector 超温检测器

**OTDA** Office of Tracking and Data-Acquisition (美国)跟踪与数据采集处

**OTDC** Optical Target Designation Computer 光学目标标识计算机

**OTDL** Object Type Definition Language 对象类型定义语言

**OTDM** Optical Time Division Multiplexing 光时分复用

Optical Time Domain Multiplexing 光时域复用

**OTDR** Optical Time Domain Reflect 光时域反射法

Optical Time Domain Reflectometer 光时域反射计

**OTDS** Optical Time Division Switching 光时分交换

**OTE** Operational Test Equipment 运行测试设备

**OTEA** Operational Test and Evaluation Agency 运行测试与评估机构

Operational Test and Evaluation Assessment 运行测试与评估鉴定

**OTED** One-Touch Exchange of Dies 触发型模具交换

**OTF** Optical Transfer Function 光传递函数

**OTFP** Operational Traffic Flow Planning 运行的交通流量计划

**OTG** Oil Temperature Gauge 滑油温度表

**OTH** North Bend, OR, USA (Airport Code) 美国俄勒冈州北本德机场

On Time Handover 按时移交

Over-the-Horizon 超视距

**OTHB** Over the Horizon Backscatter 超视距背景噪声

**OTHR** Over the Horizon Radar 超视距雷达

**OTI** Oil Temperature Indicator 滑油温度指示器

**OTK** Oil Tank 滑油箱

**OTLM** Operational Test Launch Multiplexer 操作试验发射转换开关

**OTLSCP** Operational Test Launch Status and Control Panel 操作试验发射状态与控制板

**OTM** Operations Technology Management 运行技术管理

**OTN** Optical Transit Node 光过渡节点

Optical Transport Network 光传送网

Orthogonal Tree Network 正交树网络

**OTOW** Operational Take-off Weight 使用起飞重量

**OTP** Bucharest, Romania-Otopeni (Airport Code) 罗马尼亚布加勒斯特奥托佩尼机场

Of True Position 真实位置的

On Top 在云上

One Time Password 一次性通行口令

One Time Programmable 一次性可编程

Open Trading Protocol 开放网络贸易协议

Operational Test Profiles 运行测试剖面

Operational Test Program 运行试验程序

**OTPPP** Observer Training Program Phasing

Plan 观察员训练程序分阶段计划

**OTR** Oceanic Transit/Transition Route 海洋
过渡航线
Outer 外部的,外侧的,客观的
Oxygen Transmission Rate 氧气传输速率

**OTS** Offline Test Software 脱机测试软件
On-line Terminal System 在线终端系统
Operational Technology Satellite 实用技术卫
星
Optical Target Simulator 光学目标模拟器
Optical Technology Satellite 光学技术卫星
Orbiting Test Satellite 轨道试验卫星
Organized Track Structure 编组航迹结构
Organized Track System 编组航线/迹系统
Out-of-Service 停止工作,不工作
Over Temperature Signal 超温信号

**OTSD** Outside 外部(的)

**OTSI** Operating Time since Inspection 检修
后运行时间

**OTT** Office of Traffic and Transportation 交通
运输办公室
Office of Transportation Technologies 运输技
术办公室
Operational Training Test 操作培训测试
Optical Transmission Technology 光传输技术

**OTTN** Optical Trunk Transmission Network
光干线传输网络

**OTU** Office of Technology Utilization 技术利
用处
One Time Use 一次使用
Operational Training Unit 操作培训装置
Operational Test Unit 操作试验设备
Optical Translator Unit 光转换器单元

**OTV** Orbital Transfer Vehicle 轨道转移飞行
器

**OTW** Over the Wing 在机翼表面,机翼涵盖
的客舱座位

**OTWG** Orbiting Test Working Group 轨道试
验工作组

**OTZ** Kotzebue, AK, USA (Airport Code) 美
国阿拉斯加州科策布机场

**OU** Organization Unit 组织单元
Osaka University (Japan) 大阪大学(日本)
Outlet Unit 输出口组件
Overhead Unit 头顶组件
Oxford University 牛津大学(英国)

**OUA** Ouagadougou, Burkina Faso-Ouagadou-
gou (Airport Code) 布基纳法索瓦加杜古机
场

**OUBD** Outbound 出境

**OUC** Outbound Cargo 外运货物

**OUE** Operational Utility Evaluation 运行效用
评估

**OUESR** Questionary 调查表,询问的

**OUG** On-line User Group 在线用户群

**OUIES** Quiescent 静止,沉寂

**OUP** Originating User Prompter 主叫用户提
示器

**OUT** Outbound 外出的,离港的
Outgoing 向外
Outlet 出口
Output 输出
Outside 外部

**OUTB** Outboard 机外的,外侧的
Outbound 外出的,离港的

**OUTBD** Outboard 机外的,外侧的

**OUTDRQ** Output Data Request 输出数据请
求

**OUTDT** Outdate 过时

**OUTFLW** Outflow 流出

**OUTPL** Outplant 出厂

**OUTR** Outer 外部的

**OUTSD** Outstanding 显著的

**OUTSLTR** Output Translator 输出转换器

**OV** Open Ventilated 通风的
Orbiting Vehicle 轨道飞行器
Over 超过,在上方
Overvoltage 超压

**OVB** Novosibirsk, Russia (Airport Code) 俄
罗斯新西伯利亚机场

**OVBD** Overboard 向外,机外

**OVBK** Overbooking 超量预订

**OVC**　Overcast 阴天,密云

**OVCD**　Overcarried 卸载的

**OVCO**　Overvoltage Cutout 过压断开

**OVD**　Asturias, Spain (Airport Code) 西班牙
阿斯图里亚斯机场

　　Optical Video Disc 视频光盘

**OVDR**　Overdoor 门上的

　　Overdrive 超控

**OVE**　On Vehicle Equipment 飞行器上设备

**OVERTEMP**　Over Temperature 过热

**OVF**　Over Frequency 超频

**OVFL**　Overfill 过满,溢出

　　Overflow 溢流

**OVFLY**　Overfly 飞越

**OVH**　Oval Head 椭圆形头

　　Overhead 顶板

**OVHD**　Overhead 头顶的

**OVHG**　Overcharging 超载,过量充电

**OVHL**　Overhaul 翻修,大修

**OVHT**　Overheat 过热

**OVID**　Object-oriented Video Database 面向
对象的视频数据库

　　Obviously Visible Impact Damage 明显可见
碰撞损坏

**OVLD**　Overload 过载

**OVLPN**　Overlength Partnumber 超长部件号

**OVLY**　Overlay 外罩,覆盖(板),涂覆

**OVO**　Outside Vendor Operation 外部供货方
操作

**OVP**　Oval Point 椭圆点

　　Over Voltage Protection 过电压保护

**OVPRESS**　Overpressure 过压,超压

　　Over-pressurized 过充压的,过加压的

**OVPWR**　Overpower 过功率,过负荷,克服,
压倒

**OVR**　Over Voltage Relay 过压继电器

**OVRD**　Override 超控,优先于,越过

**OVRHT**　Overheat 过热,超温

**OVRM**　Overarm 优势,举手过肩(的)

**OVRN**　Overrun 超过,超速,溢流,备用跑道

**OVROTN**　Over Rotation 超转(速)

**OVRSPD**　Overspeed 超速

**OVRWGT**　Overweight 超重,过量

**OVS**　Overhead Video System 头顶录像系统

　　Overseas 海外,侨民

　　Oversize 超尺寸

**OVSF**　Orthogonal Variable Spreading Factor
正交可变扩频因子

**OVSL**　Oversale 机位超卖

**OVSP**　Overspeed 超速

**OVSTEER**　Oversteer 过度操纵

**OVSTK**　Overstock 过量储备

**OVT**　Over Temperature 超温

**OVTR**　Optical Video Tape Recorder 光学录
影机

　　Overtravel 过调,超层,过行程

**OVV**　Overvoltage 超压

**OVWG**　Over Wing 翼上

**OW**　Oblique Wing 斜翼

　　One Way 单程

　　Operating Weight 运行重量

　　Order Wire 指令信号线,联络线

　　Outer Wing 机翼外部

**OW-ADM**　Optical Wavelength ADM 光波长
ADM

**OWAVE**　Ordinary Wave 常规波

**OWB**　Owensboro, KY, USA (Airport Code)
美国肯塔基州欧文斯伯勒机场

**OWC**　One-Way Channel 单向信道

　　One-Way Communication 单向通信

　　Optical Wavelength Convertor 光波长转换器

**OWD**　Norwood, MA, USA-Memorial Code
(Airport Code) 美国马萨诸塞州诺伍德纪
念馆机场

　　One Way Doppler 单向多普勒

　　Open World Diode 开放世界二极管

　　Order Wire and Data unit 联络线与数据单
元

　　Overhaul Work Description 大修工作说明

**OWDM**　Optical Wavelength Division Multiplexing 光波分复用

**OWDS**　Optical Wavelength Division Switching

光波分交换

**OWE** Operating Weight Empty 使用空重

Optimum Working Efficiency 最佳工作效率

**OWF** Optimum Working Frequency 最佳工作频率

**OWN** Owner 业主

Owning 拥有

**OWPT** Order Wire Phone Trunk 联络电话中继线

**OWS** Obstacle Warning System（直升机上的）障碍告警系统

Ocean Weather Station 海洋气象站

Office Work Station 办公室工作站

Operational Weather Support 航空气象保障，运行气象保障

**OWSF** Oblique-Wing Single Fuselage 斜置翼单机身

**OWT** Outward Trunk 出局中继线

**OWTF** Oblique-Wing Twin Fuselage 斜置翼双机身

**OWU** Open Window Unit 开窗组件

**Ox** Pitch Axis 横轴（俯仰）

**OXB** Bissau，Guinea-Bissau-Osvaldo Vieira（Airport Code）几内亚比绍奥斯瓦尔多维埃拉机场

**OXC** Optical Cross Connect 光交叉连接

Optical Cross Connector 光交叉互连器

**OXCN** Optical Cross Connect Node 光交叉连接节点

**OXD** Oxide 氧化物

Oxidized 氧化的

**OXY** Oxygen 氧气

**Oxy M** Oxidizing Material 氧化材料

**OY** Optimum Yield 最佳产量

**Oy** Roll Axis 纵轴（滚转）

**OZ** Ounce 盎司（重量单位）

Ounce-inch 盎司—英寸

**Oz** Yaw Axis 立轴（偏航）

# P

**P**   Absolute Humidity 绝对湿度
   Absolute Pressure 绝对压力
   Air Pressure 大气压力
   Hydraulic Pressure Return 液压回油管
   Microprocessor 微处理器
   Page 页
   Pallet 集装板
   Panel 仪表板,面板
   Paper 文件,纸,报纸
   Parity 奇偶性,奇偶校验,同等
   Patrol Aircraft 巡逻机
   Pending Selection 在选择中
   Period 周期
   Peso(currency) 比索(货币单位)
   Physical Layer 物理层
   Pilaster 壁柱
   Pink 粉红色(的)
   Pitch 俯仰
   Plug 插头
   Polarity 极性
   Pole 极,电极,杆
   Port 港口,航空站,舱门
   Position 位置
   Positive 阳极,正的
   Pounds 英镑
   Power 功率,动力
   Presentation Layer 表示层
   Pressure 压力
   Primary Frequency 主用频率
   Private Aerodrome 私人机场
   Private Pilot 私人飞机飞行员
   Probe 探头,传感器
   Program 程序,计划,大纲,方案
   Prohibited Area 禁(航)区
   Protocol 协议

   Provisional 临时的
   Purple 紫色
   Push 推
   Roll Rate 滚转率
**P to P**   Point to Point 点到点
**P&A**   Phase and Amplitude 相位和幅度
   Planning and Analysis 计划和分析
   Price and Availability 价格和可用(供)性
   Professional and Administrative 专业的和行政的
**P&C**   Planning and Control 计划和控制
**P&D**   Planning and Design 计划与设计
   Pressurizing and Drain 增压和排放
   Production and Design 生产与设计
   Production and Development 生产与发展
**P&DD**   Plumbing and Deck Drain 管道和甲板排水
**P&ID**   Pass and Identification 通过与鉴定
**P&L**   Powers and Lighting 电源和照明
   Profit and Loss 收益与损失
   Profit and Loss Statement 损益表
**P&L DISTR**   Power and Lighting Distribution System 电源和照明分配系统
**P&M**   Plant and Machinery 工厂与机器
   Preventive Maintenance 预防性维修
   Producibility and Manufacturing 可生产和制造
   Productive Maintenance 生产维修
**P&O**   Pains and Oils 漆和油
   Pickled and Oiled 酸洗和油化的
**P&P**   Plug and Play 即插即用
**P&PP**   Pull and Push Plate 推挽板
**P&R**   Payment and Reimbursement 付款与退款
   Planning and Review 计划与评估

**P&S** Port and Starboard 左舷和右舷
Primary and Secondary 主要和次要
Principles and Standards 原则和标准
Products and Services 产品和服务
Purchase and Sale 购买和销售

**P&T** Packaging and Transportation 包装和运输
Personnel and Training 人员和培训
Posts and Timbers 支柱与木料
Products and Technology 产品和技术

**P&W** Particles and Waves 粒子和波
Pratt and Whitney（Engine Manufacturers）普拉特·惠特尼公司（发动机制造商）

**P&WA** Pratt and Whitney Aircraft 普惠飞机

**P/A** Polar to Analog 极性到模拟（转换）

**P/B** Pad and Boom（Refueling）延伸管与吊杆（加油）
Pushbutton 按钮

**P/BSW** Pushbutton Switch 按钮开关

**P/C** Paper Change 文件更改，文件变动，页号更改
Part Card 零件卡
Parts Control 零件控制或管理
Passenger Compartment 客舱
Percent 百分率
Periodic Check 周期检查
Pitch Control 变距操纵，俯仰操纵
Price Catalogue 价目表，运价表
Printed Circuit 集成电路，印刷电路
Procurement Code 采购码

**P/D** Paper Development 文件拟订
Pickup and Delivery 取货和送货
Post-Dated 迟填日期的，后填日期的
Proof of Delivery 交货凭证

**P/FLTCK** Preflight Check 飞行前检查

**P/H** Pilot Hole 定位孔

**P/L** Payload 商载，有效载荷

**P/M** Physical Measured 实际测量的
Production/Manufacturing 生产/制造

**P/N** Part Number 件号

**P/O** Pattern Orientation 规定等待航线取向，波瓣图取向

**P/P ROM** Preprocessor ROM 预处理器只读存储器

**P/PA** Preemblem and Postemblem 前标与后标
Program Procurement Authority 程序采购管理局

**P/RST** Press to Reset 按复位钮
Push to Reset 按下复位

**P/S** Passenger Information Sign 旅客信息标识
Pitot/Static 全/静压

**P/T** Power Transfer 电源转换，动力转换
Precipitation/Turbulence 降水/湍流

**PA** Parametric Amplifier 参量放大器
Part Availability 零件可用性，零件可供性
Pascal 帕斯卡
Passenger Address 旅客广播，旅客地址
Passenger Agent 旅客代理商
Peace Agreement 和平协议
Pending Availability 可用性未定，可供性未定，未到器材
Performance Analysis 性能分析
Permanent Address 永久地址
Pilotless Aircraft 无人驾驶飞机
Pilot's Association 飞行员协会（英国）
Pitch Autopilot 俯仰自动驾驶仪
Pneumatic Autopilot 气动自动驾驶仪
Power Amplifier 功率放大器
Power Approach 带油门进近着陆
Powered Ascent 动力推进上升
Pre-Amplifier 前置放大器
Prearm 前置预位
Preassignment 预分配
Preassignment Arrangement 预先分配
Precision Approach 精密进近
Preliminary Amplifier 前置放大器
Pre-programmed Autopilot 预编程自动驾驶仪
Present Altitude 现时高度，当前高度
Pressure Abnormally 压力异常

Pressure Altitude 气压高度,压力高度

Pressure Angle 压力角

Principle Axis 主要中枢,基本轴

Probabilistic Analysis 概率分析

Probability of Acceptance 合格率

Procurement Authorization 采购批准

Production Assist 生产协助

Profile Angle 剖面角

Program Address 程序地址

Program Amplifier 程序放大器

Prohibited Area 禁飞区

Property Administration 资产管理(机构)

Proportional Amplifier 比例放大器

Proposal Authorization 建议审定,提案审定

Public Address 公共广播

Public Announcement 机上广播

Pulse Amplifier 脉冲放大器

Purchase Agreement 购机协议

**PA&C** Program Assessment and Control 项目评估和控制

**PA&E** Product Acceptance and Evaluation 产品验收与评估

Program Analysis and Evaluation 项目分析与评估

Program Assessment and Evaluation 项目评估与评价

**PAA** Palletized Avionics Assembly 托架式航空电子组件

Passenger Address Amplifier 旅客广播放大器

Primary Aircraft Authorization 初级飞机审定

**PAAC** Pacific and Asian Affairs Council 太平洋与亚洲事务委员会

Program Analysis Adaptable Control 程序分析适配控制

**PAADAR** Passive Airborne Detection and Ranging 无源机载检测和测距

**PAAG** Portable Airfield Arrestor Gear 便携式飞机场制动轮

**PAAR** Precision Approach Airfield Radar 机场精密进近雷达

**PAAST** Pan American Aviation Safety Team 泛美航空安全小组

**PAB** Policy Advisory Bureau (FAO) 政策咨询局

**PABX** Private Automated/Automatic Branch Exchange 专用自动支线交换机

**PAC** Pacific 太平洋

Panama City, Panama (Airport Code) 巴拿马巴拿马城机场

Panoramic Aerial Camera 航空全景摄影机

Parker Aircraft Corporation 派克飞机公司

Passenger Address Controller 旅客广播控制器

Path Attenuation Compensation 路径穿透补偿

Performance Analysis and Control 性能分析和控制

Pitch Amplifier Channel 俯仰放大器通道

Pitch Augmentation Computer 俯仰增量计算机

Primary Address Code 基本地址码

Private Aviation Committee 私人航空委员会

Professional Advisory Committee 专业咨询委员会

Provider Advisory Committee 供应商咨询委员会

**PACC** Pacific Airlift Control Center 太平洋空运控制中心

Product Administration Contract Control 生产管理合同控制

Product Assessment and Certification Center 生产评估与认证中心

**PACE** Performance and Competency Enhancement 业绩和胜任能力强化

Performance and Cost Evaluation 性能与成本评估

Performance and Cost Evaluation Program 性能及成本评价项目

Physics and Chemistry Experiment 物理与化学试验

Pilot-Assisted Channel Estimation 导频辅助

的信道估计技术

Precision Analog Computing Equipment 精密模拟计算设备

Program Assessment and Control Environment 项目评估和控制环境

**PACIS** Passenger Address/Cabin Interphone System 旅客广播/座舱内部通话系统

**PACK** P Channel Acknowledgement P 信道确认

Pack（Air Conditioning）组件（空调）

**PACM** Passive Countermeasures 消极对抗

Pulse Amplitude Code Modulation 脉冲幅度编码调制

**PACO** Principal Administrative Contracting Office 资本管理缔约办公室

Principal Administrative Contracting Officer 资本管理缔约官员

**PACOR** Passive Corelation and Ranging 无源相关和测距

**PACOTS** Pacific Organized Track System 太平洋组织的航迹系统

**PACP** Pitch Augmentation Control Panel 俯仰增量控制板

Propulsion Auxiliary Control Panel 发动机辅助操纵板

**PACR** Performance and Compatability Requirements 性能与相容性要求

**PACS** Pacific Area Communications System 太平洋地区通信系统

Passive Attitude Control System 被动姿态控制系统

Pitch Active Control System 俯仰主动控制系统

Pitch Augmentation Computer System 俯仰增稳计算机系统

Pitch Augmentation Control System 俯仰增稳控制系统

Project Administration and Costing System 项目管理与成本系统

**PACSBB** Phased Array Concept Study and Brassboard 相控阵方案研究与试验

**PACT** Parts Accountability Control Technique 零件责任控制技术

Precision Aircraft Control Technology 飞机精确操纵技术

Production Analysis Control Technique 生产分析管理技术

Purchased Article Configuration Traveler 采购物品组合运输器

**PACU** Passengers Audio Control Unit 旅客音频控制组件

Position Actuation Control Unit 位置作动控制组件

Preprocessor Planning and Control Unit 预处理器计划与控制组件

Pressure Actuation Control Unit 压力作动控制组件

Primary Avionics Control Unit 主航空电子控制装置

**PAD** Packet Assembler Disassembler 分组拆装

Packet Attach/Detach 分组装/拆

Padder（Capacitor）微调电容器

Partner Agreement Document 合作伙伴协议文件

Parts Availability Date 零件可用日期，零件可供日期

Passive Air Defense 消极防空，被动防空

Photo Amplifier Detector 光敏放大检测器

Picture Assembly Device 图像组合仪

Pilotless Aircraft Division 无人驾驶飞机分部

Point Air Defense 要地防空，点防空

Port of Aerial Debarkation 空运卸载机场

Power Amplifier Driver 功率放大器驱动器

Program Approval Document 项目批准文件

Program Associated Data 与程序相关的数据

Programmable Automation Data 可编程自动化数据

Propellant Actuated Device 推进剂启动装置

**PADAR** Passive Detection and Ranging 无源探测和测距

**PADC** Prague Airport Development Corp. 布拉格机场发展公司

**PADDS** Portable Airborne Digital Data System 便携式机载数字式数据系统

**PADIS** Passenger and Airport Data Interchange Standards 旅客和机场数据交换标准

**PADL** Pager Application Development Language 寻呼机应用开发语言

**PADLOC** Passive Detection and Location of Countermeasures 干扰无源探测和定位

**PADRE** Portable Automatic Data Recording Equipment 便携式自动数据记录设备

**PADS** Passive Active Data Simulation 无源有效数据模拟

Performance Analysis Display System 性能分析显示系统

Planned Arrival and Departure System 计划到港和离港系统

Pneumatic Air Distribution System 气动空气分配系统

Position and Azimuth Determination System 位置和方位测定系统

Precision Aerial Deliver System 精密空投系统

Precision Aerial Display System 精密空中显示系统

Precision Attitude Determination System 精密仪态测定系统

Precision Azimuth Determination System 精确方位确定系统

Price and Availability Data Sheet 价格和可用性数据表

Product Assurance Data System 产品保证数据系统

Program Planning and Control Automated Data System 程序设计和控制自动数据系统

**PAE** Phase Angle Error 相位角误差

Port of Aerial Embarkation 启运航空港，装运机场

Preliminary Airworthiness Evaluation 初步适航性评定

Program Analysis and Evaluation 程序分析与评估

**PAFAM** Performance and Failure Assessment Monitor 性能和故障分析评估监视器

**PAFC** Phosphoric Acid Fuel Cells 磷酸型盐燃料电池

**PAFT** Product Assurance Flight Test 产品保证试飞

**PAG** Parts Acquisition Group 零件采购组

Primary Attitude Good 基本姿态良好

Pump and Governor 泵和调速器

**PAGCH** Packet Access Grant Channel 分组接入应答信道

**PAH** Holder of Production Approval 生产批准书持有人

Paducah, KY, USA (Airport Code) 美国肯塔基州帕迪尤卡机场

Pitch Attitude Hold 俯仰姿态保持

**PAI** Pacific Aerospace Index 太平洋航空航天指南

Panama Airways, Incorporated 巴拿马航空公司

Passenger Arrival Information 旅客到达通知

Principal Avionics Inspector 主任航空电子监察员

Protocol Addressing Information 协议寻址信息

**PAIR** Performance and Improved Reliability 性能与改进的可靠性

Performance and Integrated Retrofit 性能和综合更新(改进)

Phased Array Instrumentation Radar 相控阵测量雷达

Precision Approach Interferometer Radar 精密进场干涉雷达

Product Analysis Incident Report 产品分析事件报告

**PAIRS** Private Aircraft Inspection Reporting System 私人飞机检查报告系统

**PAIS** Personnel Authentication Identification System 人员身份识别系统

Project Analysis Information System 计划分析情报系统

**PAIV** Pack Anti-Ice Valve 组件防冰活门

**PAK** Pakistan 巴基斯坦(南亚国家)

Polyester Alkyd 聚酯树脂

Speed Pack 快速包装

**PAL** Pacific Aeronautical Library 太平洋航空图书馆(美国)

Pacific Air Lines 太平洋航空公司(美国)

Phase Alternation Line by Line 逐行倒相制

Pilot Activated Lighting 飞行员启动灯光

Polynesian Airline Limited 波利尼亚航空公司

Power Alarm 电源报警

Precision Approach Lighting 精密进近照明设备

Prescribed Action Link 规定的操作链

Printer Address Label 打印机地址标号

Programmable Array Logic 可编程矩阵逻辑

Pulse Alternate by Line 行交替脉冲

**PALC** Passenger Acceptance and Load Control 旅客接受与载荷控制

Precision Approach and Landing Capability 精密进近与着陆性能

**PALCS** Passenger Address Level Control Sensor 旅客广播音量控制传感器

**PALEA** Philippine Airlines Employees Association 菲律宾航空公司雇员协会

**PALL** Pallet 托架,货盘

**PALS** Passive/Active Locator System 无源/有源定位器系统

Portable Airfield Light Set 便携式机场照明装置

Precision Approach and Landing System 精密进近与着陆系统

Precision Approach Lighting System 精密进近灯光系统

**PAM** Parametric Amplifier 参量放大器

Payload Assist Module 有效载荷助推仓

Performance Assessment Monitor 性能评估监视器

Perigee Assist Module 近地点助推发动机

Picture Assembly Module 图像组合多路器

Pitch Autopilot Module 俯仰自动驾驶组件

Proactive Maintenance 主动维修

Pulse Amplitude Modulation 脉冲幅度调制

**PAMA** Professional Aviation Maintenance Association 专业航空维修协会

**PAMB** Ambient Pressure 环境压力

Ambient Pressure at the Flight Altitude 飞行高度的环境压力

**PAMC** Provisional Acceptable Means of Compliance 临时执行措施

**PAMFM** Pulse Amplitude Modulation Frequency Modulation 脉冲幅度调制频率调制

**PAMS** Patent Application Management System 专利申请管理系统

Precision Antenna Measurement System 精密天线测量系统

Predictive Aircraft Maintenance System 飞机预测维修系统

**PAN** Panoramic 全景的

Pantry 配餐室

Polled Access Network 轮询访问网络

**PANAR** Panoramic Radar 全景雷达

**PANB** Panic Bolt 应急插销,应急闩

**PANFI** Precision Automatic Noise Figure Indicator 自动精确噪声指数指示器

**PANIC** Panoramic Analysis and Homing Indicator Control 全景分析和引导指示器控制

**PANS** Procedures for Air Navigation Services 空中航行服务程序

**PANS-ABC** Procedures for Air Navigation Services-ICAO Abbreviations and Codes 航行服务程序—国际民航组织缩略语和代码

**PANS-ATC** Procedures for Air Navigation Services-Air Traffic Control 空中导航服务程序—空中交通管制

**PANS-ATM** Procedures for Air Navigation Services-Air Traffic Management 空中航行服务程序—空中交通管理

**PANS-OPS** Procedures for Air Navigation

Services-Aircraft Operations 空中航行服务程序—航空器运行

**PANS-RAC** Procedures for Air Navigation Services-Rules of the Air Traffic 空中导航服务程序—空中交通规则

**PANT** Pantograph 缩放仪

**PAO** Project Administration Office 项目管理室

**PAOAS** Parallel Approach Obstacle Assessment Surfaces 平行进近障碍物评价面

**PAP** Pierced Aluminum Plank 钻孔铝板
Port Au Prince, Haiti（Airport Code）海地太子港机场

**PAPA** Parallax Aircraft Parking Aid 视差航空器停放设备

**PAPE** Property Accounting-Program Equipment 资产会计编程设备

**PAPI** Precision Approach Path Indicator 精密进近航路指示器

**PAPL** Program Approved Parts List 计划批准的零部件清单
Project Approved Parts List 项目批准的零部件清单

**PAPR** Product Assurance Program Representative 产品保证项目代表

**PAR** Packet Arriving Rate 分组达到率
Parachute 降落伞
Parallel 纬线,平行（线）
Paris, France（Airport Code）法国巴黎机场
Parts Availability Review 零件可用性审查,零件可供性审查
Peak to Average Ratio 峰均值比
Performance and Reliability 性能和可靠性
Perimeter Acquisition Radar 远程搜索雷达
Perimeter Array Radar 环形阵列雷达
Periodic Aircraft Repair 周期性飞机修理
Planning Accountability Record 计划责任记录
Planning Action Request 计划操作要求
Planning Activity Report 计划作用报告
Precision Approach Radar 精确进近雷达

Preferential Arrival Route 优选到港航线
Preferred Arrival Routes 优选进场航路
Preventive Aircraft Repair 飞机预防性修理
Problem Analysis Report 问题分析报告
Production Action Request 生产操作要求
Program Acquisition Review 程序采集审查
Program Action Request 程序操作要求
Program Assessment Review 程序评估审查
Progressive Aircraft Rework 进展性飞机改装
Pulse Acquisition Radar 脉冲搜索雷达

**PARA** Paragraph 段落

**PARAMP** Parametric Amplifier 参量放大器数

**PARC** Pacific Asia Resources Centre 亚太资源中心
Performance-based Operations Aviation Rule-making Committee 以性能为基础的航空管理条例委员会
Periodic Aircraft Recondition/Reconditioning Cycle 飞机定期检修周期
Planning and Review Committee 计划与评审委员会

**PARD** Parts Application Reliability Data 零部件使用可靠性资料
Payload Airborne Retention Device 机上有效载荷保持装置
Periodic and Random Deviation 周期性与随机性偏差

**PARDB** Parametric Database 参量数据库

**PARDOP** Passive Ranging Doppler System 多普勒无源测距系统

**PARI** Primary Access Rights Identification 基本接入权识别

**PARIS** Passive/Active Radar Identification System 被动/主动雷达识别系统
Pulse Analysis-Recording Information System 脉冲分析与记录信息系统

**PARK** Parkerized 防锈处理的
Parking 停留
Portable Access Rights Key 可携式接入权密钥

**PARM** Parallel Approach Radar Monitor 驾驶员水平显示器
Program Analysis for Resource Management 资源管理程序分析

**PARMO** Pacific Approvals Registry and Monitoring Organization 太平洋批准注册和监视组织

**PARR** Problem Analysis, Resolution, and Ranking System 问题分析、解决与分级系统
Procurement Authorization and Receiving Report 采购授权书与验收报告
Program Analysis and Resource Review 资源评审计划分析
Program Assessment Review Report 项目评估审查报告

**PARS** Parachute Altitude Recognition System 降落伞高度识别系统
Passenger Airline Reservation System 旅客航班订票系统
Performance Analysis Reporting System 性能分析报告系统
Pilotless Aircraft Research Station 无人驾驶飞机研究站
Private Aircraft Reporting System 私人机报告系统
Programmed Airlines Reservation System 程控航班订座系统

**PART** Participant 参与者,伙伴
Participation 参与
Program Assessment Performance Tool 项目绩效评估工具
Program Assessment Rating Tool 项目等级评估工具

**PAS** Passenger Address System 旅客广播系统
Performance Advisory Station 性能咨询站
Performance Advisory System 性能咨询系统
Persion Air Service 波斯航空勤务公司(伊朗)
Phase Angle Selector 相位角选钮
Pilot Advisory Service 飞行员咨询服务

Pilot Advisory System 驾驶员咨询系统
Pilot Alert System 飞行员告警系统
Pitch Attitude Sensor 俯仰姿态传感器
Protocol Analysis System 协议分析系统

**PASAMS** Principal Axes System Angular Momentum System 主坐标系角动量系统

**PASC** Precision Adaptive Subband Coding 精度自适应子频带编码

**PASD** Property Accounting System for Depreciation 折旧率资产计算系统

**PASE** Portable Avionics Support Equipment 便携式航空电子保障设备

**PASIP** Propulsion and Airframe Structural Integration Programme 推进装置与机体结构综合计划

**PASO** Pacific Area Support Office 太平洋地区支援办公室
Pacific Aviation Safety Office 太平洋航空安全办公室

**PASS** Parked-Aircraft Security System 停机警戒系统
Passage 通道,航道
Passenger 旅客
Passive Aircraft Surveillance System 无源飞机监视系统
Performance Analysis for Surveillance System 监视系统性能分析
Private Automatic Switching System 专用自动交换系统
Procurement Automated Source Systems 自动化采集源系统
Professional Airways Systems Specialists 专业航路系统专家
Professional Aviation Safety Specialists 专业航空安全专家

**PAT** Parametric Artificial Talker 参量仿真送话器
Passenger Air Tariff 旅客航空运价
Passive Angle Track 无源角度跟踪
Patent 专利
Patna, India (Airport Code) 印度巴特那机

场

Pattern 模型,型式

Performance Acceleration Technology 性能加速技术

Physical Acceptance Test 实际验收测试

Pitch Attitude Trim 俯仰姿态配平

Primary Access Terminal 基本存取终端

Principal Agent Theory 委托代理理论

Production Acceptance Test 生产验收测试

**PATA** Pacific Asia Travel Association 太平洋亚洲旅行协会

**PATC** Philippine Aerial Taxi Company 菲律宾出租飞机公司

Precision Approach Terrain Chart 精密进近地形图

**PATCA** Panama Air Traffic Control Area 巴拿马空中交通管制区

**PATCC** Production Aircraft Test Completion Certificate 航空器生产测试完成证书

**PATCO** Professional Air Traffic Controller Organization 职业空中交通管制员组织

Professional Air Traffic Controller's Association 职业空中交通管制员协会(美国)

**PATEC** Portable Automatic Test Equipment Calibrator 便携式自动测试设备校准器

**PATM** Performance-based Air Traffic Management 基于性能的空中交通管理

Production Aircraft Test Manual 航空器生产测试手册

**PATP** Production Acceptance Test Procedure 产品验收试验规程

Production Assessment Test Procedure 产品评定试验规程

**PATPEND** Patent Pending 专利受理期

**PATRIC** Pattern Recognition Interpretation and Correlation 模型识别判断与相关性

**PATRIOT** Phased Array Tracking Radar, Intercept of Target 相控阵列跟踪雷达,目标拦截

Phased Array Tracking to Intercept of Target 用于跟踪拦截目标的相控阵列

**PATS** Passenger Air-to-ground Telephone System 旅客空地电话系统

Playback and Test Station 回放与测试台

Precision Aircraft Tracking System 精密航空器跟踪系统

Precision Attack Targeting System 精确攻击目标系统

Precision Automated Turning System 精密自动转向系统

Private Automatic Telephone System 专用自动电话系统

Product and Technology Surveillance 产品和技术监督

Proof and Transit System 检验与运输系统

**PATT** Pattern 模型,型式

Plastic Apply Trim Template 塑性应用微调板

**PATWAS** Pilots Automatic Telephone Weather Answering Service 飞行员自动电话天气问讯服务

**PAU** Passenger Address Unit 旅客广播组件

**PAV** Phase Angle Voltmeter 相位角电压表

Position and Velocity 位置和速度

Pressure Altitude Variations 压力高度差

Prototype Air Vehicle 航空器样机

**PAVE** Position and Velocity Extraction 位置和速度提取

**PAVT** Position and Velocity Tracking 位置和速度跟踪

**PAW** Plasma Arc Welding 等离子弧焊

**PAWA** Pan American World Airways, Inc 泛美世界航空公司

**PAWE** Performance Assessment and Workload Evaluation 性能评估和工作量评价

**PAWOB** Passenger Arriving without Baggage 旅客不带行李到达

**PAWRS** Private Aviation Weather Reporting System 私人航空天气报告系统

**PAWS** Phased Array Warning System 相控阵警告系统

**PAX** Passenger Announcement Entertainment

and Service Multiplex System 旅客广播娱乐和服务多路系统

Passenger(s) 旅客,乘客

Paying Passenger 付费旅客

Private Automatic Exchange 专用自动交换机

**PAX HNDST** Passenger Handset 旅客手机

**PAX TEL** Passenger Telephone 旅客电话

**PAY** Payment 支付

**PAYLD** Payload 商务载重,业载

**PAYR** Payroll 计算报告表,工资(单)

**PAYROLL** Payroll System 计算报表系统

**PB** Burner Pressure 燃烧室压力

Painted Base 上漆底座

Parts Breakdown 零件损坏,零件故障

Passenger Bridge 旅客登机桥

Passenger on Board 机上乘客

Plotting Board 标图板

Plug Board 插板,转接板

Policy Board 政策委员会

Preflight Bulletin 飞行前公告

Publication Bulletin 出版物公报

Pull Box 引线盒

Push-Button 按钮

**PB STA** Push Button Station 按钮台,按钮位置

**PB SW** Push Button Switch 按钮开关

**PBA** Performance Based Acquisition 基于绩效采集

Performance Based Agreement 基于绩效的协议

Performance Based Assessment 基于绩效评估

Price-Based Acquisition 基于价格采集

Production Base Analysis 生产基地分析

Program and Budget Analysis 计划和预算分析

Program Budget Authorization 计划预算审定

**PBAM** Performance Based Adjustment Model 基于性能的调节模型

Performance Based Assessment Model 基于性能的评估模型

**PBB** Passenger Boarding Bridge 旅客登机桥

**PBC** Printer Board Connector 印刷电路板连接器

**PBCCH** Packet Broadcast Control Channel 分组广播控制信道

**PBCCP** Patch Board Configuration Control Program 接线板组合控制方案

**PBCL** Bleed Closed Pressure 放气关闭压力

Bleed Closed Signal Pressure 放气关闭信号压力

**PBCS** Postboost Control System 后增压控制系统

**PBCT** Proposed Boundary Crossing Time 预计飞跃边界时间

**PBD** Pierce Blank Die 冲孔模

Place Bearing/Distance 方位和距离

Pressboard 压制板

Program Budget Decision 计划预算决策,计划预算决定

Program Budget Directive 计划预算指令

Program Budget Document 计划预算文件

**PBDI** Position Bearing and Distance Indicator 位置方位和距离指示器

**PBDS** Pneumatic Boot Deicing System 气动除冰系统,机械除冰系统

**PBE** Portable Breathing Equipment (Aircraft) 便携式呼吸装置

Private Branch Exchange 专用交换分机

Product Based Evaluation 基于产品的评估

Protective Breathing Equipment 保护性呼吸设备

**PBF** Pilot Briefing Facility 飞行员简报设施

**PBFD** Pierce Blank and Form Die 冲孔和成型模

**PBH** Priced by the Hour 按小时付费

**PBHD** Pound Force per Brake Horsepower 磅力/制动马达

**PBI** Parallel Block Input 并行数据块输入

Passenger Boarding Information 旅客登机通知

Pitch Boundary Indicator 俯仰极限指示器

West Palm Beach, FL, USA ( Airport Code ) 美国佛罗里达州西棕榈滩机场

**PBIT**　Periodic Built-in Test 周期性机内测试

Power-up Built-in Test 通电自检,动力装配测试

**PBL**　Passenger Boarding List 旅客登机单,舱单

Performance Based Logistics 基于性能的后勤保障

Problem-Based Learning 基于问题式学习

**PBM**　Paramaribo, Suriname ( Airport Code ) 苏里南帕拉马里博机场

Pressure Bias Modulation 偏压调节

Process-Based Management 基于过程的管理

Program Budget Memorandum 计划预算备忘录

Pulse Burst Modulation 脉冲群调制

**PBN**　Packet Based Network 分组网络

Performance-Based Navigation 基于性能的导航

Private Branch Network 专用分支网络

**PBOM**　Product Bill of Material 器材产品清单

Program Bill of Material 器材计划清单

**PBON**　Passive Branched Optical Network 无源分支光网络

**PBP**　Pulse Burst Period 脉冲群周期

Push-Button Panel 按钮板

**PBPS**　Post-Boost Propulsion System 后增压推进系统

**PBR**　Payment by Result 按业绩付酬

Policy Based Routing 基于策略路由

**PBS**　Passenger Boarding Survey 旅客登机检查

Performance Based Specification 基于性能的规范

Performance Based Standards 基于性能的标准

Personalized Basic Service 个人化基本业务

Portable Base Station 便携式基站

Product Breakdown Structure 产品分解结构

Program Breakdown Structure 程序分解结构

**PBSELV**　Park Brake Selector Valve 停留刹车选择活门

**PBSM**　Passenger Baggage Service Manual 旅客、行李运输手册

**PBSOV**　Parking Brake Shutoff Valve 停留刹车关断活门

**PBSW**　Pushbutton Switch 按钮开关

**PBU**　Protective Breathing Unit 呼吸保护装置

**PBV**　Parking Brake Valve 停留刹车活门

Postboost Vehicle 后增压运载工具

Start Bleed Close Pressure 启动放气压力

**PBW**　Parts by Weight 按重量计

Power-by-Wire 线馈动力

Pulse Burst Wave 脉冲群波

Push Button Word 按钮字

**PBWA**　Program Budget and Work Authorization 计划预算和工作审定

**PBX**　Private Branch Exchange 专用电话交换机

Public Broadcasting Exchange 公共广播交换机

**PBXL**　Private Branch Exchange Line 专用小交换机线路

**PC**　Pack Controller 组件控制器

Packing Card 包装卡

Paging Channel 寻呼信道

Pair Comparison 成对比较法

Part Card 零件卡

Parts Catalogue 零件目录,部件目录

Pay Code 支付码

Per Cent 每百中,百分率

Peripheral Control 外围控制

Personal Communication 个人通信

Personal Computer 个人电脑

Personnel Carrier 职员交通工具

Petty Cash 零用现金

Phase Coherent 相位相干

Phase Compensator 相位补偿器

Phase Conjugate 相位共轭

Phase Control 相位控制

Photoconductor 光电导体,光敏电阻

Pickup Cargo 中途卸货

Piece(s) 件(数)

Pilot Cadet 飞行学员

Pilotage Chart 领航图,地标领航图

Pitch Circle 节圆

Pitch Computer 俯仰计算机

Place Card 座位卡,场所卡,住所卡

Plug Chart 插头图

Plug Cock 旋塞

Point Code 信令点编码

Point of Curve 曲线起点

Polarization Controller 偏振控制器

Power Control 功率控制

Power Controller 动力控制器

Power Conversion 动力转换

Pressure Controller 压力控制器

Prices Catalog 价目表

Primary Contractor 原合同方

Printed Circuit 印刷电路

Private Code 专用代码

Private Communication 专用通信

Probability Computer 概率计算机

Process Condition 工艺操作条件

Process Control 工艺控制,程序控制

Processing Center (信息)处理中心

Producing Capability 生产能力

Producing Cost 生产成本

Production Certificate 生产许可证

Production Code 生产编码

Production Control 生产管理

Proficiency Check 熟练程度检查

Program Communication 程序通信

Program Computer 程序计算机

Program Console 程序控制台

Program Control 程序控制

Program Coordination 程序协调

Program Counter 程序计数器

Programmable Controller 可编程序控制器

Progress Control 进度控制

Project Control 项目管理

Projecting Camera 摄影投影机

Projection Center 投影中心

Proportionality Coefficient 比例系数

Proportioning Controller 比例控制器

Proposed Change 建议的更改

Proprietary Company 控股公司

Propulsion Coefficient 推进系数

Protocol Capability 协议性能

Protocol Conversion 协议转换

Pulsating Current 脉冲电流

Pulse Code 脉冲编码,代码

Pulse Comparator 脉冲比较器

Pulse Compression 脉冲压缩

Pulse Compressor 脉冲压缩器

Pulse Controller 脉冲控制器

Pulse Current 脉冲电流

Purchase Contract 购买合同

Purchase Control 采购控制

Purchase/Procurement Cost 采购成本

Single Paper Single Cotton 单纸单纱

**PC&H** Packing,Crating and Handling 包装、装箱和搬运

**PCA** Part Condition Assessment 零(部)件状况评估

Parts Control Area 零件控制区域

Personal Communication Assistant 个人通讯助理

Physical Configuration Audit 物理构型审查

Physics Configuration Audit 物理配置审计

Pollution Control Agency 污染控制局

Positive Control Airspace 绝对管制空域

Positive Control Area 绝对管制区域

Power Control Actuator 动力控制作动器,功率控制激励器

Power Control Address 动力控制地址

Power Control Assembly 动力控制组件

Practical Critical Area 实际重要区域

Primary Certification Authority 主要合格审定管理当局

Principal Component Analysis 主成分分析法

Printed Circuit Assembly 印刷电路组件

Production Configuration Audit 生产组合检查

Production Control Area 生产控制区域

Propulsion Controlled Aircraft 推力控制飞机

**PCAA** Philippine Civil Aeronautics Administration 菲律宾民航局

**PCAD** Parametric Computer Aided Design 参量计算机辅助设计

**PCAE** Parts Control Area Everett 爱弗瑞零件控制区(波音)

**PCAM** Punch Card Accounting Machine 卡片穿孔计数器

**PCAR** Process Control Action Request 工艺控制措施申请

Production Control Assembly Records 生产控制组件记录

**PCAS** Personnel Cost Accounting System 人员成本会计系统

Pitch Control Augmentation System 俯仰控制增稳系统

**PCAT** Procedures for the Control of Air Traffic 空中交通管制程序

**PCATD** Personal Computer-based Aviation Training Device 基于 PC 机的辅助航空训练装置

**PCB** Parts Control Board 零件控制委员会

Payload Control Board 有效载荷控制委员会

Peripheral Control Board 外围控制委员会

Planning Change Board 计划变更委员会

Play Control Block 播放控制块

Plenum Chamber Burning 增压室燃烧

Pollution Control Board 污染控制委员会

Power Circuit Breaker 电源断路器

Printed Circuit Board 印刷电路板

Process Communication Block 进程通信程序块,处理通信程序块

Process Control Block 进程控制程序块

Product Configuration Baseline 产品配置基线

Production Certification Board 生产许可审定

委员会

Program Communication Block 程序通信设备

Program Control Block 程序控制块

Program Control Board 程序控制委员会

Project Change Board 项目变更委员会

Propulsion Control Box 推进控制盒

**PCB 3.0** 3.0 Start Bleed Control Pressure 3.0 启动放气控制压力

**PCB 3.5** 3.5 Start Bleed Control Pressure 3.5 启动放气控制压力

**PCBA** Proximity Circuit Board Assembly 进近电路板组件

**PCBDP** Printed Circuit Board Design and Production 印刷电路板设计和制造

**PCBR** Printed Circuit Board Robotics 印刷电路板遥控学

**PCC** Passenger Who Changes Class (飞行中)改变座舱等级旅客

Personal Code Calling 个人代码呼叫

Pilot Controller Communication 驾驶员管制员通信

Point of Compound Curve 合成曲线点

Printed Circuit Card 印刷电路卡,印刷电路插件

Production Control Center 生产控制中心

Program Control Center 程序控制中心

Program Control Counter 程序控制计数器

Program-Controlled Computer 程序控制计算机

Punctured Convolutional Code 收缩卷积码

**PCCADS** Panoramic Cockpit Control and Display System 全景驾驶舱控制与显示系统

**PCCC** Parallel Concatenated Convolutional Code 并行链接卷积码

**PCCCH** Packet Common Control Channel 分组公共控制信道

**PCCDS** Panoramic Cockpit Control Display System 全景座舱控制显示系统

**PCCH** Physical Control Channel 物理控制信道

**PCCM** Permanent Consultative Committee of the Maghreb 马格里布常设咨询委员会

Private Circuit Control Module 专用电路控制模块

Program Control Contract Manager 计划控制合同经理

**PCCN** Part Card Change Notice 零件卡更改通知

Provisioning Contract Control Number 供应合同控制编号

**PCCS** Procurement Control Computer System 采购控制计算机系统

Program and Cost Control System 程序与成本控制系统

Program Change Control System 程序变动控制系统

**PCD** Compressor Discharge Pressure 发动机排气压力

Performance Compliance Document 性能符合文件

Personal Computing Division 个人电脑事业部

Phase Compact Disc 相变光盘

Pilot Controls and Display 驾驶员控制(器)和显示(器)

Plasma-Coupled Device 等离子体耦合器件

Point Contact PN Diode 点触型 PN 二极管

Pound per Capita per Day 磅/(人·天)

Power Control and Distribution 动力控制和分配

Precision Course Direction 精确航道方向

Precision Course Director 精确航道指引仪

Process Control Diagram 工艺控制图

Program Change Decision 项目变更决策

Program Control Document 程序控制文件

Pulse Code Demodulator 脉冲编码解调器

Pulse Counter Detector 脉冲计数检波器

Pulse-Code Dialing 脉冲编码拨号

**PCDM** Primary Contactor Driver Module 主接触器驱动器模块

Product Content and Data Management 产品目录和数据管理

**PCDS** Primary Contactor Driver System 主接触器驱动器系统

Process Control Data Systems 工艺过程控制数据系统

**PCE** Packet Concentration Equipment 分组集中设备

Passenger Cabin Entertainment Laboratory 客舱娱乐实验室

Path Core Element 通路核心单元

Personal Consumption Expenditures 人员消耗费用支出

Physical Control Element 实体控制单元

Picture Control Entity 图像控制实体

Power Control Error 电源控制错误

Power Conversion Efficiency 功率变换效率

Production Certificate Extension 生产许可证延伸

Programmed Cost Estimate 计划成本估计

**PCF** Passenger Cum Freighter 客舱混装机

Plastic Cladding Fiber 塑料包层光纤

Port Core Function 端口核心功能

Pound per Cubic Foot 磅/立方英尺

Power Cathode Follower 阴极功率跟随器

Process Control File 过程控制文件

Program Checkout Facility 程序检查设备

Program Control Facility 程序控制设备

Pulse to Cycle Fraction 脉冲周期比值

**PCFL** Pre-Cleared Flight Level 预放行高度

**PCH** Paging Channel 寻呼信道

Patch 盖板,插入码

Pitch 俯仰

Program Critical Hardware 程序关键性硬件

Punch 穿孔

**PCHBD** Patchboard 接线板,转插板

**PCHG** Punching 穿孔

**PCHT** Parachute 降落伞

Parchment 羊皮纸

**PCI** Passenger Clearance Information 旅客放行通知

Payment Card Industry 支付卡行业

Peripheral Command Indicator 外设指令显示器

Peripheral Component Interconnect 外设互连

Peripheral Component Interconnection 外围部件互连

Peripheral Component Interface 外围组件接口

Peripheral Computer Interconnect 外围计算机互连

Personal Communication Interface 个人通信接口

Physical Configuration Identification 实际构型识别

Price Consumer Index 价格消费指数

Process Control Interface 过程控制接口

Product Configuration Identification 产品配置识别

Production Cost Index 生产成本指数

Production Cost Information 生产成本信息

Program Check Interrupt 程序检查中断

Program Controlled Interruption 程序控制中断

Programmable Communication Interface 可编程通信接口

Protocol Control Indicator 协议控制指示器

Protocol Control Information 协议控制信息

**PCIA** Personal Communication Industry Association 个人通信工业协会(美国)

**PCIB** Philippine Commercial International Bank 菲律宾国际商业银行

Protocol Control Information Bus 协议控制信息总线

**PCIM** Part Cost Information Management 零件成本信息管理

**PCIN** Part Change Incorporation Notice 零件变更联合通知

Production Change Incorporation Notice 生产变更联合通知

**PCIO** Program Controlled Input/Output 程序控制输入/输出

**PCIP** Precipitation 降水

Product Cost Improvement Program 产品成本改进计划

**PCIR** Planning Change Incorporation Record 规划变更联合记录

**PCK** Pick 拾取,传感器

Pilot Check 飞行员检查

**PCL** Pallet/Container Loader 盘柜装货机

Parcel 包裹

Passive Coherent Location 被动相干定位

Pencil 铅笔,光束

Pilot Controlled Lighting 驾驶员控制的灯光,飞行员控制的灯光

Play Control List 播放控制表

Post-Conference List 协商后清单

Production Control Ledger 生产管理总账

Pulsed Chemical Laser 脉动化学激光器

**PCLS** Publication Check List System 出版物检查单系统

**PCM** Parametric Cost Model 参量成本模型

Parts Control Manual 零件管理手册

Passive Countermeasure 消极干扰

Performance Control Monitor 性能控制监视器

Phase Control Module 相位控制模件

Pilot Control Module 飞行员控制组件

Pitch Control Mechanism 俯仰控制机构,桨距控制机构

Planning Control and Control Memory 规划管理和管理记录

Power Conditioning Module 动力(电源)调节组件

Power Control Module 动力控制组件

Production Control Manual 生产管理手册

Program Control Matrix 程序控制矩阵

Project Certification Manager 项目审定主管

Pulse Code Modulation 脉冲编码调制

Pulse Control Modulated 脉冲控制调制的

**PCMA** Professional Crisis Management Association 危机管理专业协会

**PCMAS** Portable Computer-based Maintenance Aid System 基于计算机的便携式维

修辅助系统

**PCMCIA** Personal Computer Memory Card International Association 个人电脑存储卡国际协会

**PCMD** Pulse Code Modulation Device 脉冲编码调制器

Pulse Code Modulation Digital 数字式脉冲编码调制

**PCME** Packet Circuit Multiplication Equipment 分组电路倍增设备

**PCMFM** Pulse Code Modulation Frequency Modulation 脉冲编码调制频率调制

**PCMTDMA** Pulse Code Modulation Time Division Multiaccess 脉码调制时分多址存取

**PCMTS** Pulse Code Modulation Telemetry System 脉冲编码调制遥测系统

**PCN** Pacific Communications Network 太平洋通信网络

Parts Change Notice 零件更改通知

Pavement Classification Number 道面等级序号,道面等级编号

Personal Communication(s) Network 个人通信网络,个人通信网

Planning Changed Notice 计划更改通知

Plesio Chronous Network 准同步网

Position Control Number 位置控制号

Procedure Change Notice 程序更改通知

Processing Control Number 加工控制号

Procurement Control Number 采购控制号

Product Control Number 产品控制号

Production Change Number 生产变更号

Program Control Number 程序控制号

Project Control Number 项目控制号

Proposal Control Number 建议控制号

Pulse Compression Network 脉冲压缩网络

**PCNS** Polar Coordinates Navigation System 极坐标导航系统

**PCO** Parts Control Operation 零件管理操作

Phase Controlled Oscillator 相控振荡器

Planning and Global Coordination Office 规划和全球协调办公室

Point of Control and Observation 控制观察点

Principle Contracting Officer 首席合同官员

Procuring Contract Officer 鉴定合同官员

Production Control Operation 生产管理运行

Program Control Office 规划管理室

**PCOF** Plastic-Clad Optical Fiber 塑料包层光纤

**PCOS** Power Cowl Opening System 能源罩打开系统

Procurement Control On-line System 采购控制联机系统

**PCP** Peripheral Call Processing 外围呼叫处理

Pilot Call Panel 飞行员呼叫面板

Pilot/Copilot 正驾驶/副驾驶

Power Control Package 动力控制组件/电源控制组件

Primary Connection Point 一级连接点

Program Change Proposal 程序更改提议

Propulsion Control Processor 推进控制处理器

Protocol Conversion Permission 协议变换允许

Pulse Comparator 脉冲比较器

**PCPL** Production Control Print List 生产管理打印目录

**PCPS** Private Carrier Paging System 专用载波寻呼系统

**PCR** Parts Control Record 零件控制记录

Peak Cell Rate 峰值信元率

Phase Conjugate Ring 相位共轭环

Planning Change Request 规划变更申请

Preventive Cyclic Retransmission 预防性循环重发

Procedure Change Request 程序变更申请

Processor Control Register 处理器控制寄存器

Production Change Record 生产变更记录

Production Change Request 生产变更申请

Production Control Record 生产控制记录

Program Control Register 程序控制寄存器

Program Control Report 程序控制报告

Program Counter 程序计数器

Project Cost Reports 项目成本报告

Publication Configuration Record 出版组合记录

Publications Change Request 出版物更改要求

Pulsion Compression Radar 脉冲压缩雷达

**PCRAS** Passenger/Cargo Revenue Accounting System 承运客货收入结算系统

**PCS** Parts Control System 零件管理系统

Personal Communication Satellite 个人通信卫星

Personal Communication Service 个人通信业务

Personal Communication System 个人通信系统

Phase Comparison Sonar 相位比较声纳

Phase Compensation System 相位补偿系统

Phase Controlled System 相位控制系统

Physical Control Space 实际控制空间

Pitch Control System 俯仰控制系统

Plan Control Sheet 计划控制单

Planning Coordination Sheet 规划协调单

Plesiochronous Connection Supervision 准同步连接监测

Portable Communication Server 可携式通信服务器

Position, Course, Speed 位置、航向与速度

Power and Cooling Stand 动力和冷却装置

Power Conditioning System 电力调节系统

Primary Coolant System 主冷却剂系统

Prime Contract Summary 主合同摘要

Process Control System 过程控制系统，工艺控制系统

Production Control Service 生产管理服务

Production Control Source 生产控制源

Production Control Support 生产管理保障

Production Control System 生产控制系统

Production Coordination Sheet 生产协调单

Program Communication Subsystem 程序通信子系统

Program Control System 程序控制系统

Project Control System 计划管理系统

Propulsion Control System 推进控制系统

Protocol Conversion Screening 协议变换选择

**PCS&S** Production Control Systems and Services 生产管理系统和服务

**PCSAN** Permanent Commission for Safety of Air Navigation 航空安全常设委员会

**PCSB** Pulse-Coded Scanning Beam 脉冲编码扫描波束

**PCSF** Plastic-Clad Silica Fiber 塑料包层石英光纤

**PCSP** Parts Control and Standardization Program 部件管理和标准化程序

**PCSS** Parallel Combinatory Spread Spectrum 平行组合扩频

Personal Computer Software Store 个人计算机软件商店

Production Control Special Support 生产管理特殊支援

**PCT** Parachute 降落伞

Partial Compatibility Test 局部兼容性试验

Percent 百分率,百分比

Percentage 百分数

Photometric Calibration Target 光度定标测量,光度定标核准

Photon-Coupled Transistor 光子耦合晶体管

Portable Control Terminal 便携式控制终端

Power Control Test 动力控制试验

Prime Contract Termination 原合同期满

Private Communications Technology 专用通信技术

Probe Compensator Temperature 探针补偿器温度

Processing Control Table 处理控制表

Program Control Table 程序控制表,计划控制表

**PCTE** Portable Commercial Test Equipment 便携式商用测试设备

Portable Common Tool Environment 便携式

通用工具环境

**PCTM** Pulse Count Modulation 脉冲计数调制

**PCTR** Physical Constants Test Reactor 物理常数实验反应堆

Protocol Conformance Test Report 协议一致性测试报告

**PCU** Packet Control Unit 数据包控制单元

Parameter Control Unit 参数控制单元

Passenger Control Unit 旅客控制装置

Passenger Counting Unit 旅客计数装置

Piezoelectric Crystal Unit 压电晶体组件

Pilots Control Unit 驾驶员控制装置

Portable Control Unit 便携式控制组件

Power Conditioning Unit 动力调节组件

Power Control Unit 动力控制装置,电源控制装置

Power Controller Unit 电源控制器组件

Power Conversion Unit 动力转换组件

Pressure Control Unit 压力控制组件

Pressurization Control Unit 增压控制组件

Primary Control Unit 主控制组件

Priority Control Unit 优先权控制单元

Propeller Control Unit 螺旋桨控制组件

**PCV** Physical Configuration Verification 实际构型验证

Pollution Control Valve 污染控制活门

Positive Crankcase Ventilation 曲轴箱强制通风

Pressure Control Valve 压力控制活门

Primary Control Valve 主控制活门

Probability Control Voltage 概率控制电压

Propellant Control Valve 推进剂控制活门

**PCW** Pitch Control Wheel 俯仰控制轮

Program Control Word 程序控制字

Pulsed Continuous Wave 脉冲连续波

**PCWS** Pitch Control Wheel Steering 用驾驶盘作俯仰操纵

**PCZ** Physical Control Zone 实际控制区

Positive Control Zone 绝对管制地带

**PD** Packet Delay 分组延迟

Pad 法兰盘,衬垫

Paid 已付的

Panel Display 面板显示(器)

Panoramic Display 全景显示

Passive Detection 无源检测

Peak Detector 峰值检波器

Per Day 每日

Per Diem 按日给予津贴,出差津贴

Performance Data 性能数据

Peripheral Device 外围设备

Phase Defocusing 相位散焦

Phase Delay 相位延迟

Phase Demodulator 相位解调器

Phase Detector 相位检测器

Phase Discrimination 相位鉴别

Phase Discriminator 鉴相器

Phase Displacement 相位偏移

Photoconducting Device 光电导器件

Photoconductive Detector 光电导检测器

Photoelectric Detector 光电检测器

Photoelectronic Device 光电子器件

Photomultiplier Detector 光电倍增检测器

Photomultiplier Device 光电倍增器件

Photosensitive Detector 光敏检测器

Pictorial Display 图像显示

Pierce Die 冲孔膜

Pilot Direct 驾驶员操纵

Pitch Diameter 俯仰直径

Pivoted Door 枢轴门

Pix Detector 视频检测器

Planning Data 规划数据

Planning Department 规划部门

Planning Directive 计划指令

Planning Division 计划处

Planning Document 计划文件

Plasma Display 等离子显示(器)

Point Defense 保卫目标的防御

Poisson Distribution 泊松分布

Polarity Detector 极性检测器,极性检波器

Polarization Dispersion 偏振色散

Position Deltas 位置增量

Potential Device 电位器,变压装置
Potential Difference 电位差
Power Detector 功率检波器
Power Distribution 功率分配,电源分配
Power Distributor 功率分配器
Preliminary Design 初步设计
Pressure Duct 压力管路
Price Discrimination 多级票价,差别定价
Primary Display 主显示器
Prime Driver 主驱动器
Principle Diagram 原理图
Probability of Default 违约率
Probability of Detection 检测概率
Product Development 产品开发
Production Design 生产设计
Production Document 生产文件
Program Directive 程序指令,计划指令
Program Document 程序文本
Programmed Development 程序开发
Proposed Departure 已申请的离港
Protect Device 保护装置,保护设备
Protocol Discriminator 协议鉴别器
Pulse Discriminator 脉冲鉴别器
Pulse Distribution 脉冲分配
Pulse Distributor 脉冲分配器
Pulse Doppler 脉冲多普勒(雷达)
Pulse Driver 脉冲激励器
Pulse Duration 脉冲宽度,脉冲持续时间
Purchase Document 采购文件

**P-D** Punch Die 冲模

**PDA** Personal Digital Assistant 个人数字助理
Power Distribution Assembly 配电组件
Preliminary Design Approval 初步设计批准
Preliminary Design Assessment 初步设计评估
Preliminary Destination Airport 预定的目的地机场

**PDADS** Passenger Digital-Activated Display System 旅客数字启动显示系统

**PDAF** Probabilistic Data Association Filter 概率数据联合滤波器

**PDAMA** Packet Demand Assignment Multiple Access 分组按需多址访问

**PDAP** Programmable Digital Auto Pilot 可编程数字式自动驾驶仪
Public Data Access Point 公共数据接入点

**PDAR** Preferential Departure and Arrival Route 优先进出港航路
Process Data Acquisition and Reporting 工艺数据采集与报告
Producibility Design Analysis Report 生产能力设计分析报告

**PDAS** Photo Data Analysis System 图像数据信息分析系统
Programmable Data Acquisition System 可编程数字采集系统

**PDAU** Physical Delivery Access Unit 物理传递接入单元

**PDB** Performance Database 性能数据库
Physics Database 物理数据库
Power Distribution Box 配电盒
Private Database 专用数据库
Procurement Database 采购数据库
Product Database 产品数据库
Production Database 生产数据库
Public Database 公共数据库

**PDBO** Pressure Deceleration Bleed Override 减速放气超控压力

**PDC** Mueo, New Caledonia ( Airport Code ) 新喀里多尼亚穆奥埃奥机场
Panel Display Catalog 面板显示目录
Parallel Data Channel 并行数据通道
Parallel Data Communication 并行数据通信
Parallel Data Controller 并行数据控制器
Passive Dispersion Compensation 无源色散补偿
Pending Drawing Change 未定绘图更改
Performance Data Computer 性能数据计算机
Performance, Dependability, Cost 性能、可靠性、成本

Personal Digital Communication 个人数字通信

Personnel Dispatch Center 人员派遣中心

Platform Design Controller 平台设计控制器

Post Delivery Change 交付后更改

Power Distribution Control 动力分布控制

Pre-Departure Clearance 起飞前放行许可

Procurement Data Card 获得数据卡

Product Development Center 产品开发中心

Program Data Center 程序数据中心

Project Data Control 工程数据控制

Project Design Center 项目设计中心

Public Digital Cellular 公用数字蜂窝

**PDCA** Plan, Do, Check, and Act 计划、执行、检查、措施

Plan, Do, Check, and Adjust 计划、执行、检查、调整

**PDCC** Project Document Control Center 项目文件管理中心

**PDCD** Portable Data Collection Device 便携式数据采集装置

**PDCG** Pressure-Drop Control Governor 压力降控制调节器

**PDCH** Packet Data Channel 分组数据信道

**PDCN** Preliminary Design Communications Network 通信网初步设计

**PDCO** Pressure-Drop Control Orifice 压力降控制孔口

**PDCP** Pilots Display Control Panel 驾驶员显示控制板

**PDCR** Planners Drawing Change Record 规划员图纸变更记录

**PDCS** Performance Data Computer System 性能数据计算机系统

Power Distribution Control System 电源分配控制系统

Pre-Detection Combining System 探测前结合系统

Programmable Data Collection System 可编程数据收集系统

**PDCU** Power Distribution and Control Unit 电源分配和控制组件

**PDD** Performance Driven Design 性能驱动设计

Personal Digital Device 个人数字设备

Port Dialing Delay 端口拨号延迟

Preferred Delivery Date 理想交付日期

Preliminary Design and Decision 初步设计决策

Preliminary Design and Development 初步设计和研制

Priority Delivery Data 优先交付数据

Process Document Departure 工艺文件变更

Product Definition Data 产品定义数据

Program Design Data 程序设计数据

Program Design Directive 程序设计指令

Project Design Document 项目设计文件

**PDDB** Product Definition Data Base 产品定义数据库

**PDDC** Provisioning Descriptive Data Card 备件供应描述数据卡

**PDDD** Preliminary Design Data Dictionary 初步设计数据词典

**PDDI** Product Definition Data Interface 产品定义数据接口

**PDDM** Preliminary Design Decision Memorandum 初步设计决策备忘录

**PDDS** Program Design Data System 程序设计数据系统

**PDE** Path Defination Error 航迹定义误差

Product Data Exchange 产品数据交换

Product Design Engineer 产品设计工程师

Project Development Engineer 项目开发工程师

Pulse Detonation Engine 脉冲爆震发动机

**PDES** Product Data Exchange Specification 产品数据交换规范

**PDF** Pavement Depth Factor (跑道)铺筑面厚度系数

Portable Data Format 便携数据格式

Portable Document Format 便携式文件格式

Primary Display Function 主显示功能

Probability Density Function 概率密度函数

Probability Distribution Function 概率分布函数

Program Development Facility 程序开发设施

Pulse Distance Finder 脉冲测距仪

**PDFU** Pedal Damping and Friction Unit 踏板阻尼及摩擦组件

**PDG** Padang, Indonesia（Airport Code）印度尼西亚巴东机场

Parachute Drop Glider 伞投滑翔机

Polarization Dependent Gain 偏振相关增益

Procedure Design Gradient 程序设计梯度

Programmable Display Generator 程控显示发电机

**PDGN** Packet Data Gateway Node 分组数据网关节点

**PDGS** Programmable Display Generator System 可编程显示产生器系统

**PDH** Plesiochronous Digital Hierarchy 准同步数字系列

**PDI** Parallel Discrete Interface 并行离散接口

Pictorial Deviation Indicator 图示偏差指示器，偏航图表示

Picture Description Instruction 图形描述指令

Polarization Dependent Isolator 偏振相关隔离器

Product Data Index 产品数据索引

Production Data Input 生产数据输入

**PDIC** Pressure Difference Indicating Controller 压差指示控制器

**PDIO** Photo Diode 光电二极管

**PDISPL** Positive Displacement 正向位移，正向偏置

**PDIU** Propulsion Discrete Interface Unit 推进离散接口组件

**PDL** Page Description Language 页面描述语言

Parts Deletion List 零件删除清单

Picture Description Language 图像描述语言

Polarization Dependent Loss 偏振相关损耗

Portable Data Loader 便携式数据装载器

Precision Delay Line 精密延迟线

Process Design Language 过程设计语言

Program Data List 程序数据清单

Program Design Language 程序设计语言

Propulsion Design Library 推进器设计数据库

Protocol Description Language 协议描述语言

Pseudo Design Language 伪设计语言

**PDLC** Programming Development Life Cycle 项目开发生命期

**PDLCU** Privacy Door Lock Control Unit 私人门锁控制组件

**PDLS** Privacy Door Lock System 私人门锁系统

**PDM** Polarization Division Multiplexing 偏振分割复用

Possible Duplicate Message（Reservations Code）可能重复的资讯

Post Delivery Modification Material 售后改装材料

Privacy Door Module 秘密门组件

Product Data Management 产品数据管理

Product Data Manager 产品数据经理

Product Definition Management 产品定义管理

Program Decision Memorandum 计划决策备忘录

Program Design Manual 程序设计手册

Program Development Manual 程序开发手册

Program Development Memorandum 程序开发备忘录

Programmed Depot Maintenance 计划的机库维修

Project Design Memorandum 工程设计备忘录

Proposal Development Memorandum 提议开发备忘录

Pulse Density Modulation 脉冲强度调制

Pulse Duration Modulation 脉冲持续时间调制

**PDMC** Power Distribution Maintenance Computer 电源分配维护计算机

**PDME** Precision Distance Measuring Equipment 精密距离测量设备

**PDMFM** Pulse-Duration Modulation Frequency Modulation 脉冲宽度调制频率调制

**PDMI** Power Distribution Maintenance Interface 电源分配维护接口

**PDMM** Pulse Doppler Map Matching 脉冲多普勒地图匹配

**PDMPX** Primary Demultiplexer 一次群分路器

**PDMS** Plan Design Management System 计划设计管理系统

Plant Design Management System 工厂设计管理系统

Precision Distance Measuring System 精密距离测量系统

**PDN** Packet Data Network 分组数据网

Planning Discrepancy Notice 计划偏离通知

Private Data Network 专用数据网

Protocol Data Network 协议数据网

Public Data Network 公共数据网

Public Switched Data Network 公共交换数据网

**PDO** Packet Data Optimized 优化分组数据

Pendulum Dynamic Observer 振动体动力观测仪

Power Discrete Output 电源离散输出

Power Door Opening 动力开门

Procedure Development Operator 程序发展操作员

**PDOC** Primary Description of Change 更改的主要说明

**PDOP** Position Dilution of Precision 位置精度衰减因子

**PDOS** Powered Door Opening System 动力开门系统

**PDP** Packet Data Protocol 分组数据协议

Parallel Data Processing 并行数据处理

Plasma Display Panel 等离子体显示屏

Policy Decision Point 决策点

Preliminary Definition Plan 初步定义计划

Program Definition Phase 程序定义阶段

Program Development Plan 程序开发计划

Programmed Data Processor 可编程数据处理器

Programming and Display Panel 程序设计和显示专家组

**PDPC** Process Decision Program Chart 工艺决策程序图

**PDPL** Preliminary Design Program Library 初步设计程序库

**PDPS** Pack Discharge Pressure Sensor 空调组件排气口压力传感器

**PDQ** Photo-Data Quantizer 光学数字转换器

Pretty Darn Quick 马上,立刻

Price and Delivery Quotation 价格和运差报价单

Product Data Quality 产品数据质量

**PDQE** Product Development Quality Engineer 产品开发质量工程师

**PDR** Pilot's Display Recorder 飞行显示记录器

Polarization Diversity Receiver 偏振分集接收机

Power Directional Relay 电源定向继电器,定向电力继电器

Predetermined Route 预定航路

Preferential Departure Route 优选的离港航路

Preferred Departure Route 优选的离港航路

Preliminary Design Review 初步设计评审

Preliminary Design Revision 初步设计修改

Pressure Drop Regulator 压降调节器

Priority Data Reduction 优先权数据简化

Product Design Review 产品设计审查

Production Drawing Release 生产图纸发放

Program Deficiency Report 程序缺陷报告,程序故障报告

Program Design Review 程序设计审查

Program Drum Recording 程序磁鼓记录

Project Design Review 项目设计审查
Pulse Doppler Radar 脉冲多普勒雷达
Pulse Duration Ratio 脉宽比

**PDRB** Program Data Review Board 程序数据审查处

**PDRC** Pressure-Drop Ratio Control 压降比控制
Procurement Data Reference Card 获取数据参照卡

**PDRCB** Preliminary Design Review Configuration Baseline 初步设计评审构型基线

**PDRJ** Pulse-Doppler Radar Jammer 脉波多普勒雷达干扰机

**PDRS** Payload Deployment and Retrieval System 商载配置和恢复系统

**PDS** Partition Data Set 分区数据集,分段数据集,部分数据集
Passive Double Star 无源双星
Personal Digital System 个人数字系统
Physical Delivery System 物理传递系统
Predetermined Substitute 预定的替换
Premises Distribution System 宅院配线系统,综合布线系统
Primary Display System 主显示系统
Processing and Display System 处理与显示系统
Product Data Sheet 产品数据单
Product Development Schedule 产品开发计划
Production Data Sheet 生产数据单
Production Design Standard 生产设计标准
Production Development System 生产开发体系
Professional Development Schools 专业发展学校
Professional Development System 专业开发系统
Program Data Set 程序数据集
Program Debug Support 程序调试保障
Program Development Station 程序开发站
Program Development System 程序开发系统

Project Data Sheet 项目数据单
Project Design Specifications 项目设计规范
Purchasing Department Specification 采购部规范

**PDSA** Personnel Data System Airman 飞行员人事资料系统

**PDSD** Product Development Study Directive 产品开发研究指令

**PDSF** Primary Display System Function 主显示系统功能,基本显示系统功能

**PDSN** Packet Data Support Node 分组数据支持节点

**PDSR** Port Development Strategy Review 港口发展策略研究

**PDST** Pacific Daylight Saving Time 太平洋夏令时间

**PDSTT** Pulse-Doppler Single Target Track 脉冲多普勒单目标轨迹

**PDSV** Pressure Drop and Spill Valve 压力降低和溢流活门

**PDT** Pacific Daylight Time 太平洋夏令时,太平洋白昼时间
Panoramic Design Technique 全景设计技术
Pendleton, OR, USA (Airport Code) 美国俄勒冈州彭德尔顿机场
Portable Data Terminal 便携式数据终端
Preliminary Design Technique 初步设计技术
Processor Diagnostic Test 处理机诊断测试
Product Design Team 产品开发小组
Production Development Test 生产开发试验
Project Date 工程日期
Programmable Data Terminal 可编程数据终端

**PDT&E** Preliminary Design Test and Evaluation 初步设计试验和评估

**PDTCH** Packet Data Traffic Channel 分组数据业务信道

**PDTE** Packet Digital Terminal Equipment 分组数字终端设备

**PDTS** Pack Discharge Temperature Sensor 空调组件排气口温度传感器

Public Data Transmission Service 公众数据传输业务

**PDU** Packet Data Unit 分组数据单元

Pilots' Display Unit 驾驶舱显示装置, 飞行显示装置

Power Distribution Unit 电源分配组件

Power Drive Unit 动力传动装置

Pressure Drop Unit 压力下降组件, 压差组件

Processor and Distribution Unit 处理器和分配组件

Protocol Data Unit 协议数据单元

**PDV** Path Delay Value 通路延迟值

Pressurizing and Dump Valve 增压和应急放油活门

Probability of Detection and Verification 探测与证实概率

**PDW** Preliminary Design of Wing, Fuselage, Empennage 机翼、机身、尾翼的初步设计

Pulse Descriptor Words 脉冲描述字

**PDWP** Preliminary Design Word Processing 初步设计字处理

**PDX** Portland, OR, USA(Airport Code) 美国俄勒冈州波特兰机场

**PDZ** Parachute Dropping Zone 降落伞投落点

**PDZC** Pathfinder Drop-Zone Control 导航空降区控制

**PE** HP Stage Air Pressure 高压级空气压力

Ice Pellets 冰雹

Packaging Editor 装填编辑程序

Parity Error 奇偶误差

Peer Entity 对等实体

Performance Enhancement 性能增强

Performance Evaluation 性能评估

Permanent Echo 固定(目标)回波

Phase Encoding 相位编码

Phase Equalization 相位均衡

Photoconductive Effect 光电导效应

Photoelectric Diode 光电二极管

Photoelectric Effect 光电效应

Photoelectric Element 光电元件

Physical Entity 物理实体

Picture Element 像素

Piezoelectric Effect 压电效应

Piezoelectric Element 压电元件

Pilot Error 驾驶员误差, 驾驶员失误

Pilot Examiners 飞行员考试官

Pinion End 翼尖

Piston Engine 活塞式发动机

Potential Energy 位能, 势能

Power Equipment 电源设备, 动力装置

Probability of Error 差错概率

Probable Error 概率偏差, 概率误差

Processing Element 处理元件

Product Engineer 产品工程师

Professional Engineer 专业工程师

Program Element 程序单元

Protocol Entity 协议实体

Protocol Event 协议事件

Pulley End 滑轮端

Pulse Encoding 脉冲编码

Purchased Equipment 购置的设备

**PE BOOK** Purchased Equipment Book 购置的设备手册

**PE CARD** Production Estimate Card 生产预测卡

**PEAC** Photoelectric Autocollimator 光电自动准直仪

**PEAT** Procedural Event Analysis Tool 程序性事件分析工具

**PEBS** Production Engineering Business Systems 生产工程商务系统

**PEC** Personal Equipment Connector 个人设备连接器

Photoelectric Cell 光电池, 光点组件

Position Error Correction 位置误差修正

Pressure Error Correction 压力误差修正

Production Equipment Code 生产设备代号

Professional Education Center 专业教育中心

Psychology Examining Commission 心理学测试委员会

Pure Empty Cell 纯空信元

**PECLSS** Power, Environmental Control, Life Support System 能源、环境管理、生命支持系统

**PECM** Passive Electronic Countermeasures 被动电子反制

**PECO** Pitch Erection Cut-off 俯仰修正切断
Production Equipment Control Office 生产装备控制室

**PECP** Preliminary Engineering Change Proposal 初步工程更改建议

**PECR** Planning Event Compliance Report 计划事件符合报告
Program Error Correction Report 程序误差修正报告

**PECRS** PCA Expedite Control Reporting System 生产控制区域快递控制报告系统

**PED** Pedal 脚蹬,踏板
Pedestal 操纵台,基座,拖架
Peripheral Equipment Data 外围设备数据
Photoemissive Device 光电发射器件
Portable Electronic Device 便携式电子装置
Portable Electronic Document 可移植的电子文档
Production Engineering Document 生产工程文件
Program Element Directive 程序单元指令
Protocol Encoder/Decoder 协议编译码器

**PEDI** Planning, Education, Demonstration and Implementation 规划、教育、演示和实施

**PEDMU** Primary Electrical Distribution Management Unit 主电源分配管理组件

**PEDP** Production Engineering Data Processing 生产工程数据处理

**PEDR** Preliminary Engineering Design Review 初步工程设计审查

**PEDS** Purchased Equipment Development Scheduling 购置的设备开发计划

**PEDSCO** Purchased Equipment Development Schedule Committee 购置的设备开发计划委员会

**PEE** Packet Entry Event 分组进入事件

Packet Exit Event 分组离开事件
Perm, Russia (Airport Code) 俄罗斯彼尔姆机场
Photo Electron Emission 光电子放射

**PEF** Production Engineering Forecast 生产工程预测
Pylon Extension Fairing 吊架延伸整流罩

**PEG** Position Error Growth 位置误差增大
Program Evaluation Group 程序评估组

**PEI** Pereira, Colombia (Airport Code) 哥伦比亚佩雷拉机场
Preliminary Engineering Inspection 初步工程检查
Product End Item 产品终止项
Program End Item 程序终止项

**PE-IM** Phase-Encoded Intensity Modulation 相位编码强度调制

**PEJ** Premolded Expansion Joint 预铸的延长接头

**PEL** Permissible Exposure Limit 允许曝光极限
Personnel License 人员执照
Portable Emergency Lighting 便携式应急灯
Protocol Element 协议元
Purchased Equipment List 购置的设备清单

**PELEC** Photo-Electric 光电的

**PELS** Picture Elements 图像单元

**PELTP** Personnel Licensing and Training Panel 员工执照发放和培训专家组

**PEM** Performance Engineers Manual 性能工程师手册
Power Electronics Module 电力电子模块
Privacy Enhanced Mail 增强保密的邮件
Program Element Monitor 程序单元监视器
Program Evaluation Monitor 程序评估监视器
Program Exchange Meeting 程序交换会议
Project Engineering Manager 项目设计经理
Project Engineering Memorandum 项目工程备忘录

**PEMF** Plesiochronous Equipment Management Function 准同步设备管理功能

**PEMFC** Proton Exchange Membrane Fuel Cell 质子交换膜燃料电池

**PEMO** Plant Engineering Maintenance Order 工厂工程技术维修指令

**PEN** Penang, Malaysia (Airport Code) 马来西亚槟城机场

Penetrate, Penetration 穿透(率),洞察力

Photon-Exchange Network 光子交换网络

**PEN AID** Penetration Aid 穿透设备

**PENA** Penalty 罚款

**PEND** Pending 在进行中

**PENRAD** Penetration Radar 穿透雷达

**PENT** Pentode 五极管

**PEO** Program Executive Officer 计划执行官

**PEOS** Propulsion and Electrical Operating System 推进和电气操纵系统

**PEP** Packetized Ensemble Protocol 报文分组总体协议

Parametric Element Programming 参量单元程序设计

Path End Point 通路端点

Peak Envelope Power 峰值包线功率

Performance Engineers Programs 性能工程师程序

Performance Evaluation Program 绩效评估计划

Pitch Envelop Protection 俯仰包络保护

Planar Epitaxial Passivated 平面外延钝化

Policy Enforcement Point 策略执行点

Power Extension Package 功率扩大组件

Process Error Prevention 处理误差预防,工艺误差预防

Production Engineering and Planning 生产工程和计划

Production Entry Planning 产品进口计划

**PEPC** Primary External Power Contactor 外部电源主接器

**PEPDC** Primary Electrical Power Distribution Center 主电源分配中心

**PEPEC** Pilots' English Proficiency Examination of China 中国民航飞行员英语等级考试

**PER** Packed Encoding Rules 分组编码规则

Packet Error Rate 分组错误率

Part Evaluation Record 零件评估记录

Part Evaluation Review 零件评估审查

Performance 执行

Perimeter 圆周,周界

Period 期间

Permission 许可,同意

Person, Personnel 人员,全体人员

Perth, Western Australia, Australia (Airport Code) 澳大利亚佩思机场

Pitch Election Rate 俯仰选择速率

Premature Engine Removal 发动机提前拆卸

Probable Error Radial 径向概率偏差,径向概率误差

Pseudo Error Rate 伪误码率

**PERB** Performance Evaluation Review Board 绩效评估评审委员会

Program Event Review Board 程序事件审查处

**PERBD** Program Event Review Board Directive 程序事件审查处指令

**PERC** Percussion 冲击,撞击

Perishable Cargo 易腐货物

**PERD** Periodic 周期的

**PERF** Perfect 完善的,正确的

Perforate 冲孔,打眼

Perform 完成,运行

Performance 性能

Performing 完成,运行

**PERIS** Periscope 潜望镜

**PERM** Permanent 永久的,固定的

**PERMB** Permeability 渗透性,导磁率,穿透率

**PERMINVAR** Permeability Invariable 恒定导磁率,恒定穿透率

**PERP** Peak Effected Radiated Power 有效峰值辐射功率

Perpendicular 垂直,正交

**PERS** Personal 个人的

**PERSP**　Perspective 展望,观点,眼力,投影(图)

**PERT**　Performance Evaluation and Review Technique 性能评估与审查技术

Pertaining 附属物,附属的,关于

Process Engineering Review Team 工艺工程审查小组

Production Effectivity Relationship Table 生产有效性关系表

Program Evaluation and Review Technique 项目评估与审查技术

Program Evaluation Research Technology 项目评估研究技术

Progress Evaluation and Review Technique 进展评估与审查技术

**PES**　Packetized Elementary Stream 打包基本流

Passenger Entertainment System 旅客娱乐系统

Personal Earth Station 个人地球站

Photoelectric Scanner 光电扫描器

Pitch Enhancement System 俯仰增强系统

Production Engineering Services 生产工程服务

Production Engineering Specification 生产工程规范

**PESA**　Pulse Echo Spectrum Analyzer 脉冲回波频谱分析仪

**PESAR**　Passenger Entertainment System Audio Reproducer 旅客娱乐系统音频播放机

**PES-AUDIO**　Passenger Entertainment System Audio 旅客娱乐系统—音频

**PESC**　Passenger Entertainment System Controller 旅客娱乐系统控制器

**PESMMUX**　Passenger Entertainment System Main MUX 旅客娱乐系统主多路调制器

**PESS**　Passenger Entertainment Service System 旅客娱乐服务系统

**PEST**　Production Entry System Test 生产输入系统试验

**PET**　Pacific Engineering Trials 太平洋工程试验

Performance Evaluation and Tracking 性能评估和跟踪

Performance Evaluation Testing 性能评估测试

Performance Evaluation Tool 绩效评估工具

Personal Electronic Translator 专用电子翻译器

Point of Equal Time 等时点

Production Environmental Tests 生产环境试验

Propulsion Event Tracking 推进事件跟踪

**PETC**　Animal in Cabin 带进客舱的宠物

Pittsburgh Energy Test Center 匹兹堡能量试验中心

**PETRO**　Petroleum 石油

**PETS**　Production Equipment Tracking System 生产设备跟踪系统

Public English Test System 全国英语等级考试系统

**PETV**　Planar Epitaxial Tuning Varactor 平面外延调谐变容二极管

**PEV**　Peak Envelope Voltage 峰值包络电压

**PEW**　Peshawar, Pakistan ( Airport Code ) 巴基斯坦白沙瓦机场

**PEX**　PHIGS Extension for Windows 用于窗口的图形系统扩展

Private Line Exchange 专线交换机

**PF**　Package Freighter 集装箱货机

Panoramic Format 全景模式

Passenger and Freight 客货

Pattern Flight 按起落航线飞行

Physical Frame 物理帧

Pico-Farad 皮法(电容单位)

Pilot Flying 操纵飞行员,主飞驾驶员

Position Fix Error 位置固定误差

Power Factor 功率因数

Preflight 飞行前

Presentation Function 表述功能

Pressure Fuel 压力燃油

Primary Frequency 主频

Principles of Flight 飞行原理
Probability of Failure 失效概率,故障概率
Profile 剖面
Program Function 程序功能
Project Finance 项目融资
Protocol Function 协议功能
Pulse-Frequency 脉冲频率

**PFA** Pitch Feel Actuator 俯仰感力器作动筒
Popular Flying Association 飞行爱好者协会
Post Flight Analysis 飞行后分析
Private Fliers Association 私人飞行员协会（美国）
Probability of False Alarm 虚警概率

**PFAA** Professional Flight Attendants Association 职业空乘人员协会

**PFAST** Passive Final Approach Spacing Tool 被动的到达间隔工具
Powered Flight Analysis & Simulation Tool 动力飞行分析与仿真工具

**PFAT** Preflight Acceptance Test 飞行前验收试验
Preflight Assurance Test 飞行前保证试验
Preliminary Flight Approval Test 初步验证试飞

**PFAXAU** Public Fax Access Unit 公共传真接入单元

**PFB** Preformed Beam 预制横梁

**PFC** Passed Flying College 飞行学院毕业
Passenger Facility Charges 旅客附加费,旅客设施费
Passive Fiber Component 光纤无源器件
Power Factor Correction 功率因数修正
Powered Flying Control（Unit）动力飞行控制（装置）
Pre-Flight Console 飞行前问候(打招呼)
Preliminary Flight Certification 初步飞行审定
Primary Flight Computer 主飞行计算机
Primary Flight Control(s) 初级飞行控制,主飞行操纵面
Program Financial Controls 计划财政控制

**PFCA** Preliminary Functional Configuration Audit 初步功能配置审计,初步功能组合审计

**PFCC** Primary Flight-Control Computer 主要飞行控制计算机

**PFCES** Primary Flight Control Electronic System 主飞行控制电子系统

**PFCS** Primary Flight Control System 主飞行控制系统

**PFCU** Pitch Feel Control Unit 俯仰感力控制组件
Power Factor Control Unit 功率因数控制单元
Powered Flying Control Unit 动力飞行控制组件

**PFCV** Pack Flow Control Valve 组件流量控制活门

**PFD** Passive FDM Distributor 无源 FDM 分配器
Phase Flux Density 相位通量密度
Pictorial Format Display 图形格式显示
Pierce and Form Die 冲孔成形模
Pilot's Flight Display 驾驶员飞行显示器
Power Flux Density 功率通量密度
Preferred 优选的
Primary Flight Director 主飞行指引仪
Primary Flight Display 主飞行显示器
Process Flow Diagram 工艺流程图
Pulse Frequency Diversity 脉冲频率分集

**PFDA** Post Flight Data Analysis 飞行后数据分析

**PFDE** Pictorial Format Display Evaluation 图形方式显示评估

**PFDF** Primary Flight Display Function 主飞行显示功能

**PFDS** Primary Flight Display System 主飞行显示系统

**PFDU** Primary Flight Display Unit 主飞行显示组件

**PFE** Path Following Error 航路跟踪误差
Post Flight Evaluation 飞行后评估

Professional Field Engineer 专业现场工程师

Professional Flight Engineer 专业飞行工程师

Programmer's File Editor 程序员的文件编辑器

Promotion Fitness Examination 晋升合格考核

Purchaser Furnished Equipment 买方提供设备

**PFEC** Parallel FEC 并行前向纠错

**PFEV** Power Failure Event 发动机失效事故

**PFF** Pre-First Flight 预先首次飞行

Preform Fixture 预制夹具

**PFGE** Pulsed Field Gel Electrophoresis 脉冲场电泳

**PFH** Per Flight Hour 每飞行小时

**PFI** Port Fuel Injection 喷嘴燃油喷射

Post Flight Inspection 飞行后检查

Pre-Flight Inspection 飞行前检查

**PFIA** Professional Flight Instructors Association 专业飞行教员协会

**PFIB** Pre-Flight Information Bulletin 飞行前航行公告

**PFID** Preliminary Functional Interface Document 初步功能接口文件

**PFIN** Power Failure Interrupt 电源中断故障

**PFIS** Passenger Flight Information System 旅客飞行信息系统

**PF-KEY** Program Function Key 程序功能键

**PFL** Pacific Freight Lines 太平洋货运公司

Pilot Fault List 飞行员失误明细

Planned Flight Level 计划飞行高度

Practice Forced Landing 练习性迫降

Pulse Forming Line 脉冲形成线

**PFLA** Practice Forced Landing 练习性迫降场

**PFLD** Pilot Fault List Display 飞行员驾驶错误细目显示器

**PFLT** Preflight 飞行前

**PFM** Performance Figure of Merit 性能品质指数

Plan for Maintenance 维修计划

Power Factor Meter 功率因数表

Pulse Frequency Modulation 脉冲频率调制

**PFMD** Preformed 预制的,形成的

**PFN** Panama City, FL, USA (Airport Code) 美国佛罗里达州巴拿马城机场

Path Following Noise 航路跟踪噪音

Pulse Forming Network 脉冲形成网络

**PFNS** Position Fixing Navigation System 定位导航系统

**PFO** Paphos, Cyprus (Airport Code) 塞浦路斯帕福斯机场

Particle Fallout 尘粒沉降

**PFP** Pay for Performance 性能代价

Post Flight Processor 飞行后处理机

Program Familiarization Procedure 项目熟悉程序

Program Financial Plan 项目财务计划

Program Forecast Period 项目预报期

**PFPA** Potential Flight Path Angle 可能的飞行航迹角

**PFPM** Potential Flight Path Marker 潜在的飞行航迹指示器

Production Flight Procedures Manual 生产性飞行程序手册

**PFQ** Preselected Fuel Quantity 预选燃油量

**PFQT** Preliminary Flight Qualification Test 飞行资格初步测试

**PFR** Part Failure Rate 零件故障率

Passenger Flow Rate (民航的) 客流率

Performance Feedback Report 性能反馈报告

Permitted Flying Route 允许的飞行航线

Planning for Rate 速率规划

Post Flight Report 飞行后报告

Power Fail Restart 电源故障再启动

Power Failure Recovery 电源故障恢复

Preflight Review 起飞前检查

Preliminary Flight Rating 初步飞行等级

Primary Flight Reference 主飞行参考

Problem Fault Report 问题故障报告

**PFRR** Planning for Rate Review 速度计划审

查,流量计划审查

**PFRT** Performance Flight Rating Test 性能飞行等级测试

Pre-Flight Rating Test 飞行前技术等级测试

Pre-Flight Reliability Test 飞行前可靠性测试

Preliminary Field Rating Test 初步野外等级评定测试

Preliminary Flight Rating Test 初步飞行等级测试

Preliminary Flight Readiness Test 预飞准备情况测试

**PFS** Passenger Final Sales Message 最后搭机人数分析报告书,旅客最后销售报

Pitch Feel Simulator 俯仰感力模拟器

Planned Flying and Servicing 有计划的飞行和维护

Pre-Feasibility Study 预先可行性研究

Preprocessor Flight Software 预处理器飞行软件

Preprocessor Flight System 预处理器飞行系统

**PFT** Periodical Flight Training 定期飞行训练

Portable Flame Thrower 便携式喷火器

Preflight Test 飞行前试验

Profile Template 剖面模型,剖面模板

Program Flying Training 程序飞行训练

**PFTA** Payload Flight Test Article 有效载荷飞行试验件

**PFTC** Palmdale Flight Test Center 帕姆代尔试飞中心

**PFTM** Preliminary Flight Test Memo 初步飞行(预飞)试验备忘录

**PFTR** Preliminary Flight Test Report 初步飞行(预飞)试验报告

**PFTS** Primary Flying Training School 初级飞行训练学校

Production Flight Test Schedule 生产性飞行试验计划

**PFTU** Pedal Feel and Trim Unit 踏板感觉及配平组件

**PFUP** Proposed Firm Unit Price 推荐硬件价格

**PFUS** Pitch Follow-up System 俯仰随动系统

**PFV** Flight Path Vector 飞行航径引导,飞行航径矢量

Powered Flight Vehicle 有动力飞行器

Probability of Failure Vehicle 航空器失效概率

**PG** Page,Paging 页,页面,编页

Pressure Gage 压力表

Pressure Governor 压力调节器

Pulse Generator 脉冲发生器

**PG BUS** Program Bus 程序总线

**PGA** Page,AZ,USA(Airport Code) 美国亚利桑那州佩奇机场

Pin-Grid Array 引脚网格阵列

Power Gain Antenna 功率增益天线

Power Generating Assembly 发动机启动装置

Programmable Gain Amplifier 可编程增益放大器

Programmable Gate Array 可编程门阵列

**PGANE** Professional Group on Aeronautical and Navigation Electronic 航空与导航电子学专业组(美国)

**PGC** Parallel Graphics Configuration 并行图像设置

Professional Graphics Controller 专业图形控制器

Program Gain Control 程序增益控制

Program Generation Center 计划编制中心,程序编制中心

Program Group Control 程序组控制

**PGCF** Premixed Gas Combustion Facility 预混合气体燃烧设备

**PGCM** Permanent Graphic Conceptual Model 永久性图概念模型

**PGF** Perpignan,France(Airport Code) 法国佩皮尼扬机场

Pressure Gradient Force 气压梯度力

**PGG** Plan and Goal Graph 规划与目标图

**PGHM** Payload Ground Handling Mechanism

负载地面处理机构

**PGM** Program 程序

**PGMD** Programmed 编程的,计划的

**PGP** Planning Grant Program 计划批准程序
Pretty Good Privacy 可靠加密
Programmable Graphics Processor 可编程图形处理机

**PGPE** Preflight Ground Pressurization Equipment 飞行前地面增压设备

**PGPS** Packet-by-packet Generalized Processor Sharing 逐一分组通用化处理器分享

**PGRF** Pulse Group Repetition Frequency 脉冲组重复频率

**PGRV** Precision Guided Re-entry Vehicle 精密制导再入飞行器

**PGS** Power Generation System 动力系统,发电系统
Power Good Signal 电源准备好信号
Predicted Ground Speed 预测地面速度
Primary Guidance System 初级制导系统

**PGSE** Payload Ground Support Equipment 有效载荷地面保障设备
Peculiar Ground Support Equipment 特殊型号地面保障设备
Pulsed Gradient Spin Echo 脉冲梯度自旋回波

**PGT** Per Gross Ton 每毛吨

**PGV** Greenville, NC, USA (Airport Code) 美国北卡罗来纳州格林维尔机场

**PGW** Pressure Gas Welding 气压焊

**PH** Acidity or Alkalinity 酸碱性
Hydrogen Ion Concentration 氢离子浓度
Packaging and Handling 包装和处理
Packet Handling 分组处理
Packet Header 分组头
Phase 相位,阶段
Phone 电话(机)
Probability of Hit 命中率
Public Holiday 公共假日

**PHA** Preliminary Hazard Analysis 初步危险分析

Process Hazard Analysis 危险过程分析

**PHAC** Plan for Hardware Aspects of Certification 硬件合格审定计划
Public Health Advisory Committee 公共健康咨询委员会

**PHARE** Program for Harmonized ATC Research in Europe 欧洲空中交通管制研究协调计划
Program for Harmonized ATM Research in Eurocontrol 欧安局空中交通管理研究协调计划

**PHC** Probe Heat Computer 探头加温计算机
Public Health Center 公共健康中心

**PHCH** Physical Channel 物理信道

**PHCV-SD** Phase Conversion and Step/Down 相位转换和步进/步退

**PHD** Pilot Horizontal Display 驾驶员水平显示器

**PHEA** Predictive Human Error Analysis 预测性人为差错分析

**PHF** Newport News, VA, USA (Airport Code) 美国弗吉尼亚州纽波特纽斯机场

**PHH** Philips Head 菲利浦(磁)头

**PHHS** Precise Helicopter Hovering System 精确直升机悬停系统

**PHI** Packet Handling Interface 分组处理接口
Petroleum Helicopter Inc. 石油直升机公司
Position and Heading Indicator 位置及航向指示器
Position Homing Indicator 位置引导指示器
Protocol Handling Input Stream 协议处理输入码流

**PHIGS** Programmer's Hierarchical Interactive Graphics Standard 程序员的层次交互式图形标准

**PHIN** Public Health Information Network 公共卫生信息网络

**PHL** Philadelphia, PA, USA (Airport Code) 美国宾夕法尼亚州费城机场
Pilot Hole 定位孔,装配孔

Preliminary Hazard List 初步风险清单

Pressure to Horizontal Locks 水平锁定压力

**PHM** Packet Handling Module 分组处理模块

Phase Meter 相位差计

Phase Modulation 相位调制

Phase Modulator 相位调制器

Prognostic Health Management 预测健康管理

Proportional Hazards Model 比例风险模型

Public Health Management 公共卫生管理

**PHMSA** Pipeline and Hazardous Materials Safety Administration 管道与危险物品安全管理局

**PHO** Point Hope, AK, USA (Airport Code) 美国阿拉斯加州波因特霍普机场

Protocol Handling Output Stream 协议处理输出码流

**PHOC** Photocopy 照相版, 照相复制

**PHOCL** Photo-initiated Chemical Laser 光触发化学激光器

**PHOFL** Photoflash 照相闪光灯

**PHON** Phonetic 语音的

**PHONO** Phonograph 唱机

**PHOS** Phosphate 磷酸盐

Phosphorescent 磷光质

**PHOTO** Photograph 相片, 照相

**PHOTOLITH** Photolithographic 光刻的

**PHP** Physical Plane 物理平面

Physical-layer Protocol 物理层协议

Pound-force per Horsepower 磅力/马力

Propeller Horsepower 螺旋桨马力

Pure Horsepower 净马力

**PH-PC** High Pressure-Servo Pressure 高压—伺服压力

**PHPDU** Physical Protocol Data Unit 物理协议数据单元

**PHPL** Private Helicopter Pilot License 私人直升机驾驶员执照

**PHPT** Plesiochronous Higher-order Path Termination 准同步高阶通路终端

**PHR** Preheater 预热器

**PHS** Personal Handy-phone System 个人便携电话系统

Pilot Handsets 驾驶员手机

**PHSAP** Physical Service Access Point 物理业务接入点

**PHSEQ** Phase Sequence 相序

**PHSHFT** Phase Shifting 相移

**PHSP** Phase Splitter 分相器

**PHST** Packaging, Handling, Storage and Transportation 包装、处理、存储与运输

**PHT** Packaging, Handling and Transportability 包装、处理与运输能力

Phototube 光电管

Preheat 预热

**PHX** Phoenix, AZ, USA (Airport Code) 美国亚利桑那州菲尼克斯(凤凰城)机场

Primary Heat Exchanger 主热交换器

**PHY** Port Physical Layer 端口物理层

**PHY(S)** Physical 实际的, 物理的, 自然的

**PHYS** Physics 物理学

**PHYSIOL** Physiological 生理的

**PI** Parallel Interface 并行接口

Periodic Inspection 周期性检验

Peripheral Interface 外围接口

Photocell Indicator 光电管指示器

Photoelectric Indicator 光电指示器

Photoelectric Interferometer 光电干涉仪

Physical Interface 物理接口

Point Initiating 点触发

Point Insulating 点绝缘

Point of Impact 碰撞点

Point of Intersection 相交点

Polarization Independent 偏振(极化)无关

Prime Item 主要条款

Principal Inspector 主管检查员

Procedure Turn to an Intercept 程序转弯至切入点

Processing Instructions 处理指令, 加工说明

Production Illustration 生产图表

Program Information 程序信息

Program Interrupt 程序中断

Program Interrupter 程序中断器

Programmed Instruction 编程指令,编程说明

Proposal Instructions 建议说明

Protocol Identifier 协议标识符

**PIA** Pakistan International Airlines 巴基斯坦国际航空公司

Peoria, IL, USA（Airport Code）美国伊利诺州皮奥里亚机场

Peripheral Interface Adapter 辅助接口转接器

Pilot's International Associate 国际驾驶员协会

Pre-Installation Acceptance 安装前验收

Programmable Interface Adapter 程序接口适配器

**PIAC** Peak Instantaneous Aircraft Count 瞬时峰值飞机数目

**PIB** Laurel, MS, USA（Airport Code）美国密西西比州劳雷尔机场

Personal Information Bubble 个人信息泡

Plug-in Board 插件板

Polar Ionosphere Beacon 地极电离层信标

Preflight Information Bulletin 飞行前资料公告

Pulse Interference Blanker 脉冲干扰消隐器

**PI-BUS** Parallel Intermodule Communication Bus 模件间并行通信总线

**PIC** Parallel Input Channel 并行输入通道

Parallel Input Controller 并行输入控制器

Peripheral Interface Channel 外围接口信道

Personal Identification Code 个人识别码

Pilot in Command 机长

Point in Call 呼叫点

Polyethylene-Insulated Conductor 聚乙烯绝缘导体

Power Integrated Circuit 功率集成电路

Priority Interrupt Control 优先权中断控制

Priority Interrupt Controller 优先权中断控制器

Processor Interface Controller 处理机接口控制器

Program Identification Code 程序识别码

Program Information Center 程序信息中心

Program Information Code 程序信息码

Program Interrupt Control 程序中断控制

Program Interrupt Controller 程序中断控制器

**PICAO** Provisional International Civil Aviation Organization 临时国际民航组织

**PICB** Peripheral Interface Control Bus 外围接口控制总线

**PICC** People's Insurance Company of China 中国人民保险公司

**PICD** Preliminary Interface Control Drawings 初步交联控制图纸

**PICE** Programmable Integrated Control Equipment 可编程综合控制设备

**PICH** Pilot Channel 导频信道

**PICN** Proposed Interface Change Notice 建议的接口变更通知

**PICR** Pool Inventory Control Record 联合库存管理记录

**PICRS** Program Information Control and Retrieval System 程序信息控制与检索系统

**PICS** Parts Inventory and Control System 零件库存与管理系统

Parts Inventory Control System 零件存储控制系统

Production Inventory and Control System 生产库存与管理系统

Protocol Implementation Conformance Statement 协议实现一致性陈述

Pulsed Image Converter System 脉冲成像转换系统

**PICUS** Pilot-in-Command under Supervision 监视下机长

**PID** Packet Identifier 包标识符

Piping and Instrument Diagram 管道仪表流程图

Port Identification 端口识别

Process Interface Document 工序接口文件

Program Introduction Document 程序说明文件

Proportion, Integration, Differentiation/Proportional, Integral, Differential 比例、积分、微分

Protocol Identification 协议识别

**PIDA** Payload Installation and Deployment Aid 商载安装与配置辅助

**PIDB** Peripheral Interface Data Bus 外围接口数据总线

**PIDP** Pilot Information Display Panels 驾驶信息显示板

Programmable Indicator Data Processor 程序控制指示器数据处理器

**PIDS** Passenger Information Display System 乘客信息显示系统

Prime Item Development Specification 主要项目开发规范

**PIE** Pulse Interference Elimination 消除脉冲干扰

**PIEA** Production Illustration Engineering Approval 生产说明工程批准

**PIF** Packaging Information Form 包装信息表

Performance Influencing Factor 效能影响因子,效能影响因素

Pilot Information File 飞行员须知

**PIFS** Pack Inlet Flow Sensor 空调组件进气口气流传感器

**PIGT** Precision Inertial Gyro Test 精密惯性陀螺试验

**PIGS** Passive Infrared Guidance System 被动式红外制导系统,无源红外制导系统

Portable Inertial Guidance System 便携式惯性制导系统

**PIH** Pocatello, ID, USA (Airport Code) 美国爱达荷州波卡特洛机场

Pump Interstage Pressure 泵级间压力

**PII** Polarization Independent Isolator 偏振(极化)无关隔离器

**PIIC** Passenger in-Flight Information Computer 旅客飞行信息计算机

**PIIL** Path Independent Insersion Loss 通路无关插入损耗

**PIIP** Pneumatic Impulse Ice Protection 气动脉冲防冰装置

**PIL** Passenger Information List 旅客信息清单

Pilot 驾驶员,飞行员

Purple Indicating Light 紫色指示灯

**PILS** Payload Intergration Library System 有效载荷总装信息库系统

**PIM** Personal Identity Module 个人标识模块

Personal Information Management 个人信息管理

Polarization Insensitive Modulator 对偏振(极化)不敏感的调制器

Priority Interrupt Module 优先中断模块

Processor Interface Module 处理机接口模块

Programming and Indication Module 编程及指示组件

Protocol Independent Multicast 无协议多播

Pulse Interval Modulation 脉冲间隔调制

**PIMA** Production Illustration Manufacturing Approval 生产说明制造批准

**PIMS** Part Information Management System 部件信息管理系统

Product Information Management System 产品信息管理系统

Project Information Management System 工程计划资讯管理系统

**PIMU** Propulsion Interface and Monitor Unit 推力交联和监视组件

**PIN** Part Identification Number 零件识别号码

Personal Identification Number 个人识别码

Personal Information Number 个人信息代码

Pinion 游星齿轮,齿杆

Position Indicator 位置指示器

Postal Integrated Network 邮政综合网

Product Item Number 产品项目号

Program Item Number 程序项目号

Public Information Network 公共信息网络

**PIN PROG** Pin Programming 销钉程序设计

**PIN1** Boeing Portland Intergraph Number 1 波音波特兰 1 号内部图纸

**PINC** Polarization Independent Narrow Channel 偏振(极化)无关窄信道

**PIND** Partial Impact Noise Detection 粒子碰撞噪声检测

**PING** Packet Internet Groper 分组因特网探测器

**PINS** Personal Inertial Navigation System 个人惯性导航系统

Point in Space 空间内的点

Portable Inertial Navigation System 便携式惯性导航系统

Precise Integrated Navigation System 精密组合导航系统

**PINSTD** Preinserted 预插入的

**PIO** Pilot Induced Oscillations 飞行员诱发的振荡

Process Input Output 过程输入输出(通道),程序化输入输出

Programmed Input Output 可编程输入输出模式

Programmed Input/Output 编程的输入/输出

**PIOA** Polarization Insensitive Optical Amplifier 对偏振不敏感的光放大器

**PIOC** Programmed Input/Output Controller 编程的输入/输出控制器

**PION** Pioneer 先驱者

**PIOS** Pilot Instructor/Operator Station 飞行教员/操作员台

**PIP** Partner Interface Processes 伙伴接口流程

Payload Integration Plan 有效载荷综合计划

Picture in Picture 画中画

Plant in Place 安置到位

Plume Interference Prediction 羽状干扰预测

Product Improvement Package 产品改进(程序)包

Product Improvement Plan 产品改进计划

Product Improvement Program 产品改进项目

Program Improvement Plan 程序改进计划

Programmable Interface Processor 可编程接口处理机

Project Implementation Plan 项目实施计划

Proposal Instruction Package 建议说明(程序)包

**PIPA** Pulse Integrating Pendulous Accelerometer 脉冲积分摆式加速度计

**PIPC** Power Plant Illustrated Parts Catalog 动力装置图解零件目录

**PIPS** Pack Inlet Pressure Sensor 空调组件进气口压力传感器

**PIR** Packet Insert Rate 分组插入率

Pierre, SD, USA (Airport Code) 美国南达科他州皮尔机场

Pilot Incident Report 驾驶员事故征候报告

Precision Instrument Runway 有精密仪表降落设施的跑道

Pressure Interstage Return 级间回油压力

Product Improvement Record 产品改进记录

Product Improvement Request 产品改进申请

Production Inspection Record 生产检验记录

Program Incident Report 计划事件报告,程序事件报告

Program Index Register 程序索引寄存器

Property Irregularity Report 行李非正常报告,财物非正常报告

**PIREP** Pilot Report-Pilot Complaint 机长报告书

Pilot (Weather) Reports 飞行员(气象)报告

**PIREP(S)** Pilot Reports 飞行员报告

**PIRFC** Pilot Request Forecast 飞行员要求天气预报

**PIRG** Planning and Implementation Regional Group 地区规划和实施小组

Public Interest Research Group 公众利益研究组

**PIRM** Precision Instrument Runway Marking 精密仪表跑道标志

**PIRT** Precision Infrared Tracking 精密红外

跟踪

**PIS** Passenger Information Signs 旅客信息标识,旅客信息牌

Passenger Information System 旅客信息系统

Poitiers, France (Airport Code) 法国普瓦捷机场

Prime Item Specification 主要项目规范

**PISA** Passenger Interface and Supply Adapter 旅客接口及供应适配器

**PISAB** Pulse Interference Separation and Blanking 脉冲干扰分离与消隐

**PISO** Parallel in/Serial out 并行输入/串行输出

**PISTON** Piston Aircraft 活塞式航空器

**PIT** Peripheral Input Tape 外部输入磁带

Pitch 俯仰

Pittsburgh, PA, USA (Airport Code) 美国宾夕法尼亚州匹兹堡机场

**PITS** Pack Inlet Temperature Sensor 空调组件进气口温度传感器

**PIU** Passenger Information Unit 旅客信息组件

Path Information Unit 路径信息单元

Plug-in Unit 插件

Process Interface Unit 处理接口组件,工艺接口组件

Product Interface Unit 产品接口组件

**PIV** Particle Image Velocimetry 粒子图像测速技术

Peak Inverse Voltage 峰值反向电压

Personal Identity Verification 个人身份验证

Post Indicator Valve 后指示器活门

Pressure Isolating Valve 压力隔离活门

**PIWG** Pilot Implementation Working Group 飞行员执行工作组

Pilot Interface Working Group 飞行员接口工作组

Product Improvement Working Group 产品改进工作组

Propulsion Instrumentation Working Group 推进系统测量仪器工作组

**PIWM** Pulse Interval Width Modulation 脉冲间隔宽度调制

**PIX** Picture 图像

**PIXELL** Picture Cell 图像单元

**PJ** Parachute Jumper (直升机)伞降救援人员

Plasma Jet 等离子流,等离子发动机

Pressure at Jet Nozzle 喷口区压力

Pulse Jet 脉冲式喷气发动机

**PJC** Pointer Justification Count 指针调整计数

**PJCTL** Projectile 抛射体

**PJE** Parachute Jumping Exercise 跳伞训练

Parachuting Activities/Exercise 跳伞活动/练习

**PJM** Puerto Jimenez, Costa Rica (Airport Code) 哥斯达黎加吉梅内兹港机场

**PJS** Private Jet Services 私人喷气机服务公司

**PJT** Projector 投影仪

**PJTN** Projection 投射,投影

**PJTR** Projector 投影仪

**PK** Pack 包,组件

Package 成套组件,封装

Park 停放,组件

Peak 峰值

Pink 粉红色

Private Key 专用密钥

Public Key 公用密钥

Pulse Keyer 脉冲键控器

**PKC** Public Key Cryptosystems 公钥密码

**PKCS** Public Key Cryptography Standards 公钥密码标准

**PKD** Public Key Directory 公钥号码簿

Public Key Distribution 公钥分配

**PKDET** Peak Detector 峰值检波器

**PKDS** Public Key Distribution System 公钥分配系统

**PKG** Package(s) 包裹

Packing 组合,包装

Pangkor, Malaysia (Airport Code) 马来西亚

邦咯岛机场

Parking 停机,停留

**PKI** Public Key Infrastructure 公钥基础设施,公钥基础结构

**PKP** Passenger-Kilometre Performed 完成的客公里

**PKT** Packet 数据包,组群,小包

**PL** Packing List 装箱单

Padlock 挂锁

Parameter Length 参数长度

Parting Line 分型线

Parts List 部件清单

Payload 业载,商务载重,有效载重

Permanent Line 永久线路

Phase Line 相线

Physical Layer 物理层

Physical Level 物理级

Pitch Line 俯仰线,中心线

Place 放,放置

Placement 方位,位置

Plain Language 平常用语

Plan 计划,规划

Plate 板,片

Plug 插头,电嘴

Plural 复数

Power Line 动力线

Presentation Layer 表示层

Price List 价目表

Private Line 专用线

Programming Language 编程语言

Property Line 地界线

Proportional Limit 比例极限

Propulsion Laboratory 推进实验室

Protection Level 保护级

Protocol Layer 协议层

Pseudolite 伪卫星

Pulse Laser 脉冲激光器

Pulse Length 脉冲宽度

**PL&DD** Plans, Logs and Data Document 计划、记录与数据文件

**PLA** People's Liberation Army (China) 人民解放军(中国)

Post-Landing Attitude 着陆后姿态

Power Level Angle 油门杆角度

Power Light Adapter 电源灯光适配器,光能适配器

Practice Low Approach 低空进近练习

Programmable Logic Array 可编程逻辑矩阵

**PLAC** Placard 公告

**PLACE** Position, Location and Communications Experiment 定位和飞机通信试验

Programming Language for Automatic Checkout Equipment 自动检测设备编程语言

**PLACP** Pylon Applications Computer Program 悬臂(标塔)应用计算机程序

**PLACS** Production Line Automatic Checkout System 生产线自动检查系统

**PLADS** Parachute Low Altitude Delivery System 降落伞低空空投系统

**PLAS** Plaster 涂层

Plastic 塑料

**PLASI** Pulse Light Approach Slope Indicator 脉冲式光束进近坡度指示器

**PLAT** Pilot Landing Aid Television 飞行员着陆辅助电视

Platinum 铂(白金)

Programmed Lateral Assurance Test 可编程的横向保险测试

**PLATF** Platform 平台

**PLB** Passenger Loading Bridge 搭机空桥

Personal Locator Beacon 人员定位信标

Plattsburgh, NY, USA (Airport Code) 美国纽约州普拉茨堡机场

**PLBD** Plugboard 插件板

**PLBLK** Pillow Block 轴台

**PLC** Pilot Lighting Control 飞行员灯光控制

Position Line Circuit 座席线路

Power Line Communication 电力线通信方式

Power Loading Control 电源负载控制

Product Life Cycle 产品寿命周期

Product List Circular 产品目录通知

Program Level Change 程序级别变更

Program Logic Control 程序逻辑控制

Program Logic Controller 程序逻辑控制器

Programmable Logic/Logical Controller 可编程逻辑控制器

**PLCC** Plastic Leaded Chip Carriers 塑料引线芯片夹

**PLCF** Physical Layer Convergence Function 物理层会聚功能

**PLCP** Physical Layer Convergence Protocol 物理层会聚协议

**PLCRD** Placard 布告,标牌

**PLCS** Physical Layer Convergence Sublayer 物理层会聚子层

**PLCT** Product Life Cycle Theory 产品生命周期理论

**PLCY** Policy 政策,保险单

**PLD** Payload 有效载重,商务载重

Phase-Locked Demodulation 锁相解调

Phase-Locked Demodulator 锁相解调器

Phase-Locked Detection 锁相检波

Plated 电镀的,覆以金属板的

**PLDC** Preliminary List of Design Changes 设计变更的初步目录

**PLE** Path Length Efficiency 路径长度效率

Processor Latch Enable 处理器锁存允许

Production Line Engineer 生产线工程师

Prudent Limit of Endurance 合理的续航极限

Pulsed Laser Experiment 脉冲激光实验

**PLEEL** Photoluminescent Emergency Egress Lighting 电荧光紧急出口灯

**PLF** Parachute Landing Fall 跳伞降落

Passenger Load Factor 载客率

Power for Level Flight 水平飞行功率

Powered-Lift Facility 动力起重设施

**PLGL** Plate Glass 平板玻璃

**PLGR** Plunger 柱塞,波导短路插头

**PLH** Payload Handling 有效载荷处理

Plymouth, England, United Kingdom (Airport Code) 英国普利茅斯机场

Propeller Load Horsepower 螺旋桨加载马力

**PLI** Pilot Location Indicator 飞行员位置指示器

Pitch Limit Indication 俯仰限制指示

Pitch Limit Indicator 俯仰限制指示器

Preload Interrupt 预装载中断

Private Line Interface 专用线路接口

Program Language Interface 程序语言接口

**PLICF** Physical Layer Independent Convergence Function 物理层独立会聚功能

**PLINQ** Parts List Inquiry 零件目录查询

**PLIR** Photoelectric Laser Interference Rangefinder 光电激光干涉测距仪

**PLK** Plank 厚木板,基础,政策,要点

**PLL** Pallet 平板架,货盘

Phase-Locked Loop 锁相环

**PLM** Palembang, Indonesia (Airport Code) 印度尼西亚巨港机场

Passenger Load Message 旅客商载信息

Payload Level Multiplexing 净荷级复用

Payload Management 商载管理

Payload Mismatch 净荷失配

Product Lifecycle Management 产品全寿命管理

Production Line Maintenance 生产线维护

Pulse Length Modulation 脉冲宽度调制

Pulse Length Modulator 脉冲宽度调制器

Pulse Length Monitor 脉冲宽度监控

**PLMB** Plumbing 管路,管道工程

**PLMI** Parameterized Linear Matrix Inequality 参数化线性矩阵不等式

**PLMN** Public Land Mobile Network 公共陆上移动网络

**PLMR** Public Land Mobile Radio 公共陆地移动无线电

**PLN** Path Layer Network 通道层网络

Pellston, MI, USA (Airport Code) 美国密执安州佩尔斯顿机场

Plan 计划

Plane 飞机,平面

Primary Learning Net 初级学习网

Program Line Number 程序行编号

**PLNCHA** Charter Flight Plan 包机飞行计划

**PLND** Planned 计划的

**PLNG** Planning 计划

**PLNR** Planner 计划员,策划员,设计者

**PLNT** Planet 行星

**PLNTY** Planetary 轨道的

**PLO** Phase Locked Oscillator 锁相振荡器

**PLOC** Payload Operations Contractor 有效载荷操作承包商

Probability of Loss of Control 控制损失概率

**PLOT** Plotting 测绘

**PLP** Packet Layer Procedure 分组层规程

Packet Layer Protocol 分组层协议

Physical Lightwave Path 物理光波通路

**PLPT** Plesiochronous Lower Order Path Termination 准同步低阶通路终端

**PLQ** Palanga, Lithuania (Airport Code) 立陶宛帕兰加机场

Plaque 板,斑点

**PLR** Packet Loss Rate 分组丢失率

Pillar 标柱

Pliers 钳子

Production Limitation Record 生产限制记录

**PLRD** Pull Rod 拉杆

**PLRS** Pelorus 方位仪

Position Location Reporting System 位置定位报告系统

**PLRSTN** Pelorus Stand 方位台,方位站

**PLRT** Polarity 极性

**PLS** Parcels 包裹

Physical Layer Signaling 物理层信令

Point Level Sensor 点式平面传感器

Primary Landing Site 主要着陆位置

Private Line Service 专线业务

Programmable Logic Sequence 可编程逻辑顺序

Pulse 脉冲

**PLSN** Pulsation 脉动

**PLSR** Pulsator 脉动器

**PLSS** Personal Life Support System 人员生命保障系统

Portable Life Support System 便携式生命保障系统

Precision Location Strike System 精密定位攻击系统

Primary Life Support Subsystem 初级生命保障子系统

**PLST** Part List 零件目录

Plan List 计划目录

Professional Line Service Training 专业航线服务培训

**PLSTC** Plastic 塑料,塑性的

**PLT** Pilot (飞机等的)驾驶员,飞行员

Plant 设备,装置,工厂

Port Light 舱口照明

Private Line Telephone 专线电话

Procurement Lead Time 超前采购时间

**PLT LT** Pilot Light 飞行灯,指示灯

**PLTC** Port, Land and Transport Committee 港口土地及运输委员会

**PLTG** Plating 电镀

**PLTHS** Pilot House 飞行员室

**PLTR** Plotter 绘图机

**PLU** Pluggable Unit 可插入组件

Preservation of Location Uncertainty 预留位置不确定性

**PLUS** Propulsion Laboratories Unified System 推进实验室联合系统

**PLV** Presentation Level Video 显示级视频

Production Level Video 制作级视频

**PLVL** Present Level 现在高度层

**PLVRZD** Pulverized 雾化的,粉化的,喷射的

**PLW** Planned Landing Weight 计划着陆重量

**PLWS** Planned Layup Work Statement 计划接头工作说明,计划敷层工作说明

**PLYPH** Polyphase 多相的

**PLYWD** Plywood 层板

**PLZ** Polarize 极化

**PLZD** Polarized 极化的

**PLZN** Polarization 极化

**PM** Panel Maintenance 面板维修

Parallax Mapping 视差映射

Parts Model 零件模型,零件样品

Performance Management 性能管理

Performance Manual 性能手册

Performance Monitoring 性能监控

Permanent Magnet 永久磁铁

Phase Modulation 相位调制

Photo Multiplier 光倍增器

Photoelectric Microscope 光电显微镜

Physical Medium 物理介质

Pilots Manual 飞行员手册

Planned Maintenance 计划维修

Plug Map 插头图

Polarization Maintaining 偏振(极性)保持

Polarization Multiplexing 偏振(极性)复用

Position Maintenance 位置保持

Post Meridian 下午,午后

Pound-force per Minute 磅力/分

Pounds per Minute 每分钟磅

Precautionary Measures 预防措施

Pressurized Module 密封舱

Preventive Maintenance 预防性维修

Probabilistic Method 概率统计方法

Procedure(s) Manual 程序手册

Proceeding Measurement 顺序测量

Process Manual 工艺手册,处理手册

Processing Module 处理模块,处理模件

Procurement and Material 采购与器材

Production Manager 生产经理

Production Manual 生产手册

Production Means 生产方式

Production Memorandum 生产备忘录

Production Monitoring 生产监控

Productive Maintenance 生产性维修

Program Memorandum 计划备忘录,规划备忘录

Program Memory 程序存储器

Program Mode 程序方式

Programming Manual 程序设计手册

Project Management 项目管理

Project Manager 项目经理

Project Memorandum 项目备忘录

Protocol Machine 协议机

Pseudo-Multithreading 虚假多线程

Pulse Modulation 脉波调变

Purchase Memorandum 采购备忘录

**PMA** Parts Manufacturer Approval 零部件制造人批准书

Parts Manufacturing Approval 零部件制造批准证书

Permanent Magnetic Alternator 永磁交流发电机

Plane Manufacturers' Association 飞机制造商协会

Portable Maintenance Aids 便携式辅助维修

Preassigned Multiple Access 预分配多址接入

Priority Memory Access 优先级存储器存取

**PMAC** Progress Measurement and Control System 工艺测量与控制系统

**P-MAC** Packet Media Access Controller 分组媒体接入控制器

**PMAD** Portable Maintenance Aid Device 便携式维修辅助装置

Power Management and Distribution 电源管理与分配

**PMAI** Primary Mission Aircraft Inventory 主要任务飞机总量

**PMAP** Parts Manufacture Approval Procedure 零部件制造批准程序

**P-MAP** Plug Map 插头图

**PMAR** Preliminary Maintenance Analysis Report 初步维修分析报告

**PMAS** Performance Measurement Analysis System 性能测试分析系统

Performance Monitoring and Analysis System 性能监测及分析系统

**PMAT** Portable Maintenance Access Terminal 便携式维护存取终端

Portable Multipurpose Access Terminal 便携多用途存取终端

**PMB** Performance Measurement Baseline 绩效测量基准

Program Manager's Budget 计划经理的预算

**PMBA** Program Management Baseline Agreement 项目管理基线协议

**PMBOK** Project Management Body of Knowledge 项目管理知识体系

**PMBP** Program Management Best Practices 项目管理最佳实践

Project Management Business Process 项目管理业务流程

**PMBS** Packet Mode Bearer Service 分组模式承载业务

**PMC** Partial Mission Capable 可执行部分任务

Performance Computer 性能计算机

Performance Management Computer 性能管理计算机

Permanent Magnet Computer 永磁计算机

Personal Multimedia Communication 个人多媒体通信

Plane Mile Cost 飞机英里成本

Plastic/Metal Composite 金属/塑胶复合材料

Polymer Matrix Composites 聚合物基复合材料

Portable Maintenance Computer 便携式维修计算机

Power Management Computer 动力管理计算机

Power Management Control 动力管理控制

Power Management Controller 动力管理控制器,电源管理控制器

Process Material Control 材料工艺控制

Procurement Method Coding 采购方式编码,获取方式编码

Propellant Monitor and Control 推进剂监测与控制

**PMCCS** Performance Monitor Console and Control System 性能监测操纵台与控制系统

**PMCF** Post-Maintenance Check Flight 维修后的检验飞行

**PMCS** Preventive Maintenance Check and Service 预防性维修检查与服务

**PMCU** Performance Management Computer Unit 性能管理计算机组件

**PMD** Palmdale, CA, USA(Airport Code) 美国加利福尼亚州帕姆代尔机场

Physical Medium Department 物理介质门类

Physical Medium Dependent 物理媒体相关(子层)

Polarization Mode Dispersion 偏振模色散

Polarization Mode Distortion 偏振模失真

Preventive Maintenance Daily 预防性维修日报

Program Management Directive 程序管理指令

Program Memory Device 程序存储器件

Program Monitoring and Diagnosis 程序监测与诊断

**PMDB** Production Management Data Base 产品管理数据库

**PMDS** Performance Measurement Data System 性能测量数据系统

Pilot Map Display System 航路图显示系统

Power Management and Distribution System 动力管理与分配系统

Projected Map Display Set 投影地图显示装置

Projected Map Display System 投影地图显示系统

**PM-DSF** Polarization Maintaining Dispersion-Shifted Fiber 偏振保持色散位移光纤

**PME** Portable Mechanical Equipment 便携式机械设备

Precision Measuring Equipment 精密测量设备

Prime Mission Equipment 基本任务设备

Program Management Evaluation 程序管理评估

Project Management Engineer 项目管理工程师

**PMECB** Personal Message Exchange Control Block 个人报文交换控制块

**PMED** Propulsion Manufacturing Engineering Document 动力装置制造工程文件

**PMEL** Precision Measurement Equipment Laboratory 精密测量设备实验室

**PMET** Painted Metal 上漆的金属

**PMF** Performance Modification Factor 性能改善因子

Performance Module Framework 性能模块框架

Polarization Maintaining Fiber 偏振保持光纤

**PMFD** Primary Multi-Function Display 主多功能显示器

**PMFLT** Pamphlet 小册子

**PMG** Permanent Magnet Generator 永磁发电机

Prediction Marker Generator 预测标志发生器

Procurement Management Group 采购管理组

Program Maintenance Group 计划维修组

**PMGR** Permanent Magnet Generator Relay 永磁发电机继电器

**PMI** Palma Mallorca, Mallorca Island, Spain (Airport Code) 西班牙帕尔马马拉尔克机场

Payload Margin Indicator 业载极限指示器

Performance Management Indicator 性能管理指示器

Permanent Manufacturing Information 定型制造信息

Permission Management Infrastructure 授权管理基础设施

Preventive Maintenance Inspection 预防性维修检查

Principal Maintenance Inspection 主要维修检查

Principal Maintenance Inspector 主任维修检查员

Program Management Instruction 程序管理指令

Project Management Institute 项目管理协会

Purchasing Managers' Index 采购经理指数

**PMIGS** Programmers Minimal Interface to Graphics 编程器最小图形接口

**PMIS** Passive Microwave Imaging System 无源微波成像系统

Personal Mobile Information Service 个人移动信息服务

Project Management Information System 项目管理信息系统

**PMITS** Project Management Information Technology System 项目管理信息技术系统

**PMK** Pitch Mark 俯仰标志

**PML** Perfectly Matched Layer 完全匹配层

Physical Markup Language 物理标识语言

Physical Media Layer 物理媒体层

Process Modelling Language 进程建模语言

**PMM** Personalization Memory Module 个性化存储组件

Pulse Mode Multiplex 脉冲模式多路复用

**PMMC** Permanent Magnet Movable Coil 移动式永磁线圈

**PMMEL** Proposed Master Minimum Equipment List 建议的最低主设备清单

**PMMP** Permissible Mean Maximum Pressure 容许平均最大压力

Preventive Maintenance Management Plan 预防性维修管理计划

Preventive Maintenance Management Program 预防性维修管理程序

**PMMU** Paged Memory Management Unit 页面存储器管理部件

**PMN** Portable Multimedia Navigation 便携式多媒体导航

Program Management Network 程序管理网络

**PMO** Palermo, Sicily, Italy (Airport Code) 意大利巴勒莫机场

Present Mode of Operation 当前运行模式

Principal Medical Officer 负责医学的官员,主任医官

Program Management Office (专项) 大纲管理办公室

Program Management Organization 程序管理

机构

Project Management Office 项目管理办公室

Project Management Operation 项目管理运行

**PMOC** Prototype Mission Operation Center 样机任务操作中心

**PMOCCP** Prototype Mission Operation Center Computer Program 样机任务操作中心计算机程序

**PMOS** P-Type Metal Oxide Semiconductor P型金属氧化物半导体

**PMP** Performance Monitoring Program 性能监控程序

Point to Multipoint 点到多点

Power Management Panel 电源管理面板

Power Plant Monitoring Program 动力装置监控程序

Predetermined Maintenance Project 预定维修计划

Preliminary Management Plan 初级管理计划

Preventive Maintenance Period 预防性维修周期

Primary Maintenance Process 主维修工艺,基本维修工艺

Procurement Methods and Practices 采购方法与实践

Product Management Plan 生产管制计划

Program Management Plan 程序管理计划

Programmable Multimedia Processor 可编程多媒体处理器

Project Management Professional 专业项目管理

Prorate Manual-Passenger 客运分摊手册

Protected Monitoring Point 保护监视点

Pump 泵

**PMPCB** Parts, Materials and Processes Control Board 零件、器材与工序控制处

**PMPCS** Parts, Materials and Processes Control and Standardization 零件、器材与工序标准化

**PMPM** Procurement Methods and Practices

Manual 采购方法与实践手册

Project Management Procedure Manual 项目管理程序手册

**PMPP** Program Management Phaseout Plan 程序管理过渡计划

**PMPS** Performance Monitoring Point System 工作表现评分制度

Point-Multipoint System 点到多点系统

**PMPSCT** Pump Suction 泵吸管

**PMPX** Primary Multiplexer 一次群多路复用器

**PMQC** Purchase Material Quality Control 采购器材质量控制

**PMR** Peak-to-Mean Ratio 峰均比

Performance Maintenance Recorder 性能维护记录器

Performance Management Review 性能管理审查

Preventative Maintenance Recorder 预防性维护记录器

Primary Flight Reference 初级飞行参考

Private Mobile Radio 专用移动无线电

Program Management Review 程序管理审查

Project Management Review 项目管理评审

Public Mobile Radio 公用移动无线电

**PMRM** Periodic Maintenance Requirements Manual 定期维修需求手册

Power Measurement Report Message 功率测量报告消息

**PMRS** Private Mobile Radio System 专用移动无线通信系统

**PMRT** Performance Measurement and Reporting Taskforce 绩效评估和报告工作组

Program Management Responsibility Transfer 程序管理责任转移

**PMS** Path Management System 通路管理系统

Pavement Management System 铺面管理系统

Payphone Management System 付费电话管理系统

Performance Maintenance System 性能维护

系统

Performance Management System 性能管理系统

Performance Measurement/Measuring System 性能测量系统

Periodic Maintenance Squadron 定期维修中队

Planning Management System 规划管理系统

Polar Meteorological Satellite 极轨气象卫星

Portfolio Management System 组合管理系统

Power Management System 电源管理系统,动力管理系统

Precipitation Map Service 临时地图服务

Preventive Maintenance System 预防性维修系统

Probability of Mission Success (飞行)任务成功概率

Process and Material Specification 工艺和材料规范

Process Management System 工艺管理系统,工序管理系统

Production Manufacturing System 生产制造系统

Program Master Schedule 计划总进度

Project Management System 工程管理系统,项目管理系统

Projected Map System 投影地图系统

**PMSS** Program Manager Support System 计划经理支持系统

**PMT** Payment 付款,报酬,支付

Per Metric Ton 每公吨

Performance Management Team 绩效管理团队

Performance Measurement Techniques 绩效测量技术

Periodic Maintenance Team 定期维修分队

Permit 许可

Personal Mobile Telecommunication 个人移动通信

Photo Multiplier Tube 光倍增管

Program Management Team 程序管理组

Program Master Tape 程序主磁带

**PMTN** Public Mobile Telephone Network 公共移动交换网

**PMU** Portable Memory Unit 便携式存储组件

Power Management Unit 电源管理组件

**PMUX** Power Plant Multiplexer 动力装置多路调制器

Power Plant Multiplexer Specification 动力装置多路选择器规范

Programmable Multiplexer 可编程复用器

Propulsion Multiplexer 推进多路调制器

**PMV** Prototype Miniature Vehicle 微型运载器样机

**PMVR** Prime Mover 主推进器

**PMX** Packet Multiplexer 分组多路复用器

**PMZ** Palmar, Costa Rica (Airport Code) 哥斯达黎加帕尔马机场

**PN** Package No. 包装箱号码

Part Number 零件号

Permanent Phone 永久电话

Personal Number 个人号码

Positive/Negative 正/负

Pseudo Noise 伪噪声,伪噪音

Pseudorandom Noise 伪随机噪声

Pseudorandom Number 伪随机数

Public Network 公共网

Pulse Noise jamming 脉冲噪音干扰

**PNA** Pacific Northwest Airlines, Inc. (USA) 太平洋西北航空公司(美国)

Pamplona, Spain (Airport Code) 西班牙潘普洛纳机场

Part Number Application 零件号应用,部件号应用

**PNAC** Nacelle Ambient Pressure 吊舱环境压力

**PNAS** Packet Network Access Subsystem 分组网接入子系统

**PNC** Cabin Personal 随机乘务员,客舱服务员

Part Number Control 零件号管理,部件号管理

Ponca City, OK, USA (Airport Code) 美国俄克拉荷马州庞卡城机场

**PNCC** Partial Network Control Center 部分网络控制中心

Power Node Control Center 电力节点控制中心

**PNCS** Performance Navigation/Navigational Computer System 性能导航计算机系统

**PND** Primary Navigation Display 主导航显示器

**PNDB** Perceived Noise Decibels 可感噪声分贝

**PNDR** Pendular 振子的

**PNDS** Primary Navigation Display System 主导航显示系统

Prime Navigation Data Source 主导航数据源

**PNEU** Pneumatic 气动的, 气源的

**PNF** Pilot Not Flying 非操作飞行员

**PNG** Passive Night (Vision) Goggles 被动夜视镜

Portable Network Graphics 可携式网络图像

Pseudonoise Generator 伪噪声发生器

**PNH** Pan Head 皿形头

Phnom Penh, Cambodia (Airport Code) 柬埔寨金边机场

**PNHS** Part Number History System 零(部)件号记载系统

**PNI** Pictorial Navigation Indicator 图形导航指示器

Pohnpei, Caroline Islands, Micronesia (Airport Code) 密克罗尼西亚罗林群岛皮纳佩机场

**PNL** Panel 面板, 专家组, 控电板

Passenger Name List 旅客名单报

Perceived Noise Level 可感噪声级

**PNLT** Tone-corrected Perceived Noise Level 纯音修正感知噪声级

**PNMC** Provincial Network Management Center 省网络管理中心

**PNMS** Primary Navigation Mode Select 主导航模式选择

**PNN** Probabilistic Neural Network 概率神经网络

**PNNI** Private Network-to-Network Interface 专用网间接口

**PNO** Public Network Operator 公共网络运营商

**P-NOTE** Production Note 生产说明

**PNP** Positive-Negative-Positive 正—负—正

Private Numbering Plan 专用编号计划

**PnP** Plug and Play 即插即用

**PNQ** Poona, India (Airport Code) 印度浦那机场

**PNR** Part Number 零件号

Part Number Request 零件号申请, 部件号申请

Passenger Name Record 旅客订座记录, 舱单

Point of No Return 无返回点, 无返航点

Prior Notice Required 需要预先通知

**PNS** Part Number Summary 部件号汇总

Pensacola, FL, USA (Airport Code) 美国佛罗里达州彭萨克拉机场

Personal Number Service 个人号码业务

Pure Navigation System 纯导航系统

**PNSC** Packet Network Service Center 分组网业务中心

**PNT** Paint 油漆

Point 点

Pointer 指针

Polar Navigation Trainer 极地航行训练机

**PNTF** Private Network Termination Function 专用网终端功能

**PNU** Paging Network Unit 寻呼网单元

**PNUC** Part Number Usage Control 零件号使用管理, 部件号使用管理

**PNV** Potential Network Value 潜在的网络价值

**PNVG** Panoramic Night Vision Goggle 全景夜视镜

**PNVS** Pilot Night Vision Sensor 飞行员夜视传感器

Pilot Night Vision Sight 飞行员夜视视野

Pilot Night Vision System 飞行员夜视系统

**PNW** Pacific Northwest 太平洋西北部

**PNX** Private Network Exchange 专用网络交换机

**PO** Aircraft Static Air Pressure 飞机大气静压

Outside Air Pressure 外界大气压力

Parking Orbit 驻留轨道

Pay off 清算

Planetary Orbit 行星轨道

Polarity 极性

Post Office 邮局

Power Oscillator 功率振荡器

Procurement Office 采购处

Production Order 生产指令,生产规程

Program Office 计划室

Project Office 项目室

Purchase Order 订货单

**POA** Percent Open Area 部分开放区域

Power Optimized Aircraft 动力优化飞机

Production Organization Approval 生产机构批准书

Purchase Order Authorization 采购单审定

**POB** Passengers on Board 机上乘客

Persons on Board 机上人数

Post Office Box 邮政信箱

Pressure-off Brake 无压力刹车

**POC** Parallel Output Channel 并行输出通道

Parallel Output Controller 并行输出控制器

Point of Contact 接触点

Production Oversight Coordinator 生产监督协调人

Proof of Concept 概念验证,方案论证

Purchase Order Change 采购单变更

**POCC** Payload Operations Control Center 商载营运控制中心,有效载荷操作控制中心

Program Operation Control Center 计划作业控制中心

**POCN** Purchase Order Change Notice 采购单修改通知

**POCO** Point of Control and Observation 控制

与观察点

**POCSAG** Post Office Code Standardization Advisory Group 邮政总局编码标准咨询组(英国)

**POD** Pay on Delivery 交货付款

Plan of the Day 一天计划

Point of Decision 决断点

Point of Departure 起飞地点

Port of Debarkation 下机机场,卸货机场

Port of Destination 到达地

Preflight Operations Division 飞行前准备作业处

Preliminary Orbit Determination 初步轨道测定

Private Orientation Device 专用定向设备

Probability of Detection 检测概率

Project Operations Directive 项目运行指令

**PODAS** Portable Data Acquisition System 便携式数据采集系统

**PODB** Purchase Order Database 采购单数据库

**PODN** Purchase Order Discrepancy Notice 采购单差异通知

**PODS** Pneumatic Overheat Detection System 气动过热探测系统

**POE** Port of Embarkation 出境机场(港口)

Port of Entry 入境机场(港口)

Probability of Error 误码率

Production Organisation Exposition 生产组织手册

**POET** Post Operations Evaluation Tool 运行后评估工具

**POF** Plastic Optical Fiber 塑料光纤

Polymer Optical Fiber 聚合物光纤

**POFA** Polymer Optical Fiber Amplifier 聚合物光纤放大器

**POG** Port Gentil, Gabon (Airport Code) 加蓬让蒂尔港国际机场

Program Operating Guide 程序操作指南

Project Officer Group 项目官员组

**POGO** Polar Orbiting Geophysical Observatory 极轨道地球物理观测站

**POH** Path Overhead 通道开销

Pilots Operating Handbook 飞行员操作手册

**POI** Plan of Instruction 授课计划

Point of Initiation 起始点

Point of Interconnection 互相联络点

Primary Organization Identifier 基本结构识别符,基本结构识别标志

Principal Operations Inspector 主管运营检查员

Probability of Intercept 截获概率

Program of Instruction 教学大纲

Purchase Order Index 采购订单指数

**POINT** Polymer Optical Interconnect Technology 聚合物光纤互联技术

**POL** Petroleum, Oils and Lubricants 石油、油料与润滑剂

Polar 极

Polish 磨光,抛光

Probability of Loss 损耗概率

**POLCO** Petroleum Oil Company 油料公司

**POLO** Parallel Optical Link Organization 并行光链路组织

**POLTHN** Polyethylene 聚乙烯

**POLY** Polyester 聚酯

**POM** Port Moresby, Papua New Guinea (Airport Code) 巴布亚新几内亚莫尔兹比港机场

Power Plant Overhaul Manual 动力装置翻修手册

Production and Operations Management 生产与运作管理

Production Organisation Manager 生产组织经理

Production Organisation Manual 生产组织手册

Program Objective Matrix 程序目标矩阵

Program Objective Memorandum 程序目标备忘录

Project Officer's Meeting 计划官员会议

**POMAR** Position Operation Meteorological Aircraft Report 飞机位置及气象报告

Preventive Operational Maintenance and Repair 预防性操作维护与修理

**POMS** Performance Operations Management System 性能运行管理系统

Production and Operations Management Society 生产与运营管理学会

Production Order Management System 产品指令管理系统

**POMSE** Performance Operating and Maintenance Standards for Electronics 电子设备性能操作与维护标准

**PON** Passive Optical Network 无源光纤网络

**POP** Payload Optimized Program 有效载荷最佳方案

Period of Performance 性能周期

Perpendicular to the Orbit Plane 轴面垂直姿控方式

Picture out Picture 画外画

Pilot Operation Procedure 驾驶员操作程序

Plane of Position 位置平面

Point of Presence 入网点

Polar Orbiting Platform 极轨平台(日本地球遥感卫星)

Popping 爆音

Population 人口

Post Office Protocol 邮局协议

Power Operated 动力控制的

Probability of Precipitation 降水概率

Process Operation Procedure 工艺加工的操作工序

Product Optimization Program 产品优化方案

Program Operation Plan 大纲实施计划

Project Objective Plan 设计目标计划

Proof of Principle 原理论证

Purchased outside Production 采购的外部产品

**POPO** Privately Owned/Privately Operated 私有的/私人控制的

**POPSY** Protection Planning System 保护计划系统

**POR** Pacific Ocean Region 太平洋区域

Pay on Receipt 货到付款

Point of Regulation 调整点

Point of Return 返回点

Portugal 葡萄牙

Power on Reset 功率恢复

**PORC** Porcelain 瓷制品,易碎的

**PORCN** Production Order and Record Change Notice 生产指令与记录更改通知

**PORM** Plus or Minus 加或减

**PORT** Photo Optical Recorder Tracker 光学照相记录跟踪仪

**POS** Point of Sale 售出地

Point of Shipment 起运地

Point-of-Sale Terminal 销售点终端

Polar-Orbiting Satellite 极地轨道卫星

Position 位置

Positive (电池的)阳极(数)正的,正面的,确定的

Primary Operating System 主操作系统

Purchase Order Supplement 补充订货单

**POSI** Portable Operating System Interface 便携式操作系统接口

**POSN** Aircraft's Present Position 航空器当前位置

Position 位置

**POSR** Production Order Status Reporting 生产指令状态报告

**POSS** Possible 可能的

**POST** Passive Optical Seeker Technique 无源光学寻的技术

Point of Sales Terminal 销售点终端

Point of Service Test 服务测试点

Portable Operational Support Terminal 便携式运行保障终端

Portable Optical Sensor Tester 便携式光传感器测试仪

Power on Self Test 自测电源

**POT** Pick over Threshold 超阈值

Plan of Test 试验计划

Point of Terminal 终接点

Potential 电势,电位,潜在的

Potentiometer 电位计

Proof of Technology 技术验证

**POTS** Plain Old Telephone Service 普通(简单老式)电话业务

**POTW** Potable Water 饮用水

**POV** Peak Operating Voltage 峰值工作电压

**POWA** Planar Optical Waveguide Amplifier 平面光波导放大器

**POWER** Performance and Objective Workload Evaluation Research 效能与客观工作负荷评价研究

**POXIP** Passenger Oxygen Indication Panel 旅客氧气指示面板

**POZ** Poznan, Poland-Lawica (Airport Code) 波兰波兹南拉维察机场

**PP** Charges Prepaid 运费预付

Panel Point 节点

Parcel Post 包裹邮件

Patch Panel 接线板

Path Protection 通道保护

Pay Period 付款周期

Performance Parameter 性能参数

Phase Preequalization 相位预均衡

Piping 管道,配管

Point to Point 点到点

Position Pulse 定位脉冲

Post Processor 后处理器

Postage Paid 邮费付讫

Power Plant 动力装置,发电厂

Power Presentation 功率显示

Preprocessing 处理前的,预处理

Pressure Proof 压力试验

Processor Pair 处理器对

Program Pins 程序销钉

Push Pull 推挽

**P-P** Part Power 部分功率

Peak to Peak 峰值至峰值

**PP&C** Production Planning and Control 生产计划与控制

Program Planning and Control 程序规划与

管理

**PPA** Passengers per Annum 每年客运量

Plesiochronous Path Adaptation 准同步通道适配

Pollution Prevention Act 污染防治法

**PPAC** Presidential Policy and Advisory Committee 总统政策与顾问委员会

Product Performance Agreement Center 产品性能协议中心

**PPAP** Production Part Approval Process 生产件批准程序

**PPB** Parts per Billion 十亿分之几

Provisional Parts Breakdown 临时性部件故障

Provisioning Parts Breakdown 备件供应分解图

**PPBM** Power Plant Buildup Manual 动力装置装配手册

**PPBS** Program Planning and Budgeting System 程序计划与预算系统

**PPBU** Power Plant on Build up 动力装置装配

**PPC** Packet Processing Complex 分组处理复合器

Palm PC 掌上电脑

Pay per Channel 按信道计费

Phosphor Protection Circuit 荧光保护电路

Physical Port Control 物理端口控制

Pilot Proficiency Check 飞行员熟练水平检查

Policy Planning Council 政策计划委员会

Pollution Prevention and Control 污染预防与控制

Prepaid Card Service 预付费卡业务

Pre-release Part Card 预发放零件卡

Present Position Correction 现在位置修正

Production Plan Compartment 生产计划部门

Purchase Price Control 购买价格控制

**PPCE** Portable Pneumatic Checkout Equipment 便携式气动检测设备

**PPCH** Packet Paging Channel 分组寻呼信道

**PPCI** Presentation Protocol Control Information 表示层协议控制信息

**PPCR** Production Planning Change Request 生产计划变更申请

**PPCS** Pilot, Power Control, and Signalling 飞行员、功率控制与信号

Predictable Power Control Strategy 可预测的功率控制策略

Production Planning and Control System 生产计划与控制系统

**PPD** Partial Packet Discard 部分分组丢弃

Postpaid 邮费付讫

Preliminary Procurement Document 初步采购文件

Prepaid 预付的

Program Planning Directive 程序设计指令

Prototype Point Design 样机点设计

**PPDC** Provisioning Procurement Data Card 备件供应获得数据卡

**PPDES** Product Performance Data Exchange Specifications 产品性能数据交换规范

**PPDI** Pilot's Projected Display Indicator 飞行员投影显示指示器

**PPDL** Point-to-Point Data Link 点对点数据链路

**PPDN** Public Packet Data Network 公共包数据网络

**PPDP** Preliminary Project Development Plan 初步方案发展计划

Program Protection Development Plan 项目保护开发计划

**PPDU** Presentation Protocol Data Unit 表示层协议数据单元

**PPE** Personal Protective Equipment 个人防护装备

Pipet(te) 吸管，量管

Premodulation Processing Equipment 预调制处理设备

Proofreading Process Evidence 校正过程证据

Property, Plant and Equipment 财产、工厂和

设备

**PPES** Physical Performance Evaluation System 体能评估系统

Pilot Performance Evaluation System 飞行员驾驶性能评估系统

**PPF** Payload Processing Facility 有效负载处理设备

Polar Platform 极平台

Polarization Preserving Fiber 保偏光纤

**PPFRT** Prototype Preliminary Flight Rating Test 原型机初步飞行等级试验

**PPFS** Preliminary Project Feasibility Study 工程计划初步可行性研究

**PPG** Pago Pago, American Samoa ( Airport Code ) 美属萨摩亚帕果帕果机场

Pipe Plug 管道栓, 管道塞

Preplanning Group 预行计划组

Pulse Pattern Generator 脉冲图形发生器

**PPGA** Plastic Pin Grid Array 塑胶针状网阵封装

**PPH** Policy and Procedures Handbook 政策与程序手册

Pounds per Hour 磅/小时

Pulse per Hour 脉冲/小时

**PPHA** Peak Pulse Height Analysis 峰值脉冲高度分析

**PPI** Personal Planning Information 人事规划信息

Pictorial Position Indicator 图示位置指示器

Piston Position Indicator 活塞位置指示器

Pixels per Inch 每英寸像素

Plan Position Indicator 计划位置指示器

Present Position Indicator 现时位置指示器

Producer Price Index 生产者价格指数, 产品价格指数

Programmable Peripheral Interface 可编程外围接口

Pulse Position Indicator 脉冲位置指示器

**PPIAC** Plan Position Indicator and Control 平面位置显示器和控制元件

**PPIB** Programmable Protocol Interface Board

可编程协议接口板

**PPIL** Production Preflight Inspection Letter 飞行前产品检验说明书

**PPIO** Priced Provisioned Item Order 定价的储备项目订单

**PPIP** Program Protection Implementation Plan 项目保护实施计划

**PPIPC** Power Plant IPC 动力装置图解零件目录

**PPKG** Power Package 动力组件

**PPL** Power Plant Laboratory 动力装置实验室

Preferred Parts List 最佳配件表, 标准配件表

Preliminary Plan for Nonscheduled Flight 非定期航班初步飞行计划

Private Pilot License 私人飞行员执照 ( 私用飞机驾驶员执照 )

Project Products List 项目产品清单

Provisioning Parts List 临时部件清单

**PPL/H** Private Pilot's License/Helicopter 直升机私用飞机驾驶员执照

**PPL/IR** Private Pilot's License/Instrument Rating 带仪表等级的私用飞机驾驶员执照

**PPLN** Pipeline 管道, 管线

**PPLS** Pure Phase Laser System 纯相位激光系统

**PPM** Page per Minute 每分钟可打印输出的页数

Panel Position Monitor 面板位置监视器

Parts per Million 百万分之几

Performance Programs Manual 性能程序手册

Periodic Permanent Magnet 周期性永久磁铁

Pounds per Minute 磅/分钟

Product Portfolio Management 产品组合管理

Pulse Phase Modulation 脉相调制

Pulse Position Modulation 脉冲位置调制

**PPMF** Prime Part Master File 主要零件总文件

**PPMIN** Pulse per Minute 脉冲/分

**PPO** Pollution Prevention Office 污染预防办公室

President Pilot Office 总统飞机驾驶员办公室

Prior Permission Only 只限事先许可,仅限预先许可

**PPOS** Present Position 即时位置

**PPP** Peak Pulse Power 峰值脉冲功率

Point to Point Protocol 点对点协议

Precise Point Positioning 精密单点定位

Product Protection Plan 生产保护计划

Public Private Partnership 公私合作关系

Purchasing Power Parity 购买力平价

**PPPOA** Point to Point Protocol over ATM ATM 点对点协议

**PPPOE** Point to Point Protocol over Ethernet 以太网点对点协议

**PPR** Paper 证件,文件,纸

Payload Preparation Room 有效载荷准备室

Peak Power Reduction 降低峰值功率

Periodic Performance Review 定期绩效评审

Planning Pre-Release 计划预发放

Plans, Programmes, Requirement 计划、程序、要求

Preflight Parts Requisition 飞行前零件申请单,飞行前零件调拨单

Prior Permission Required 要求预先许可

Procurement Problem Report 采购问题报告

Product, Process and Resources 产品、流程和资源

Production Revision Record 生产修改记录

Program Performance Review 项目绩效评审

Prospective Price Redetermination 预期的价格重定

**PPRN** Purchased Parts Requirements Notice 购置零件要求通知

**PPS** Passenger Payloads System 旅客商载系统

Passenger Processing System 旅客处理系统

Path Protection Switching 通道保护倒换

Payload Pointing System 有效载荷定向系统

Photovoltaic Power System 光电源系统

Picture Perception System 图像感觉系统

Pilot's Performance System 飞行员考绩系统

Pixels per Second 像素/秒

Precise Positioning Service 精密定位服务

Prepaid Service 预付费业务

Production Planning System 生产计划系统

Program Project Structures 程序方案结构

Provisioning Performance Schedule 供应实施进度表

Puerto Princesa, Philippines ( Airport Code ) 菲律宾普沃托普林斯萨机场

Pulse per Second 脉冲数/秒

**PPSC** Postproduction Support Continuity 生产后连续支援

**PPSL** Preferred Parts Selection List 标准零件选择清单

Program Parts Selection List 计划零件选择清单

**PPSM** Polarization Preserving Single Mode 保偏单模光纤

**PPSN** Present Position 现时位置,当前位置

Private Packet-Switched Network 专用分组交换网

Public Packet-Switching Network 公共分组交换网

**PPSS** Production Proving Squawk System 生产检验报告系统

**PPSSP** Pre-Pre-Sync Sync Pulse 预同步前置脉冲

**PPT** Papeete, French Polynesia ( Airport Code ) 法属玻利尼西亚帕皮提机场

Passport 护照

Pay-per-Token 按令牌计费

Peak Power Tracker 峰值功率跟踪器

Plesiochronous Path Termination 准同步通道终端

Poppet 拖架,支架

Precipitate 沉积,冷凝物

Product Positioning Time 产品安置时间

Prompt 立即

Punched Paper Tape 穿孔纸带

**PPTP** Point-to-Point Tunneling Protocol 点到

点隧道协议

**PPTR** Page Printer 页码打印机

**PPU** Pay-per-Use 按每次使用计费

Peripheral Processor Unit 外设处理器组件

Pick-off Position Unit 位置传感器

Position Pickoff Unit 位置传感组件

Powerpush Unit 动力推进装置

Precision Position Update 精密位置更新

Preprocessor Unit 预处理器组件

**PPV** Pay-per-View 按次收费电视服务

**PQ** Preliminary Qualification 初步资格,初步评定

**PQA** Preliminary Qualification Assurance 初步资格保证,初步评定保证

Procurement Quality Assurance 采购质量保证

**PQAS** Program Quality Assurance System 程序质量保证系统

**PQC** Production Quality Control 产品质量控制

**PQI** Presque Isle, ME, USA (Airport Code) 美国缅因州普雷斯克岛机场

**PQMP** Product Quality Management Plan 产品质量管理计划

**PQR** Preliminary Qualification Review 初步资格审查

**PQT** Preliminary Qualification Test 初步资格测试

**PR** Panel Receptacle 板面插座

Parachute Rigger 叠(降落)伞员

Parallel Resistance 并联电阻

Part Record 零件记录

Passive Radar 无源雷达

Path Restoration 通道恢复

Pattern Recognition 模式识别

Payload Rate 净负荷速率

Payroll 工资(单),计算报告表

Pedestal Roller 旋转底座滚柱

Pen Record 笔录

Performance Rating 性能级

Performance Ratio 性能比

Performance Report 性能报告

Performance Requirement 性能要求

Periodic Report 定期报告

Photo Recorder 摄影记录器

Pilot Rating 飞行员等级

Pipe Rail 管子扶手

Pitch Ratio 螺距比

Pitch Reference 俯仰基准

Planning Reference 设计参考,设计基准

Planning Repository 规划仓库

Pneudo-Range 伪距离

Polarized Relay 极化继电器

Polysulphide Rubber 聚硫橡胶

Position Register 位置记录器

Position Report 位置报告

Posterior Ridge 后脊

Power Ready Relay 电源准备继电器

Power Return 动力返回

Precision Radar 精密雷达

Premium Rate 附加费率

Pressure 压力

Pressure Ratio 压力比

Pressure Recorder 压力记录器

Pressure Regulator 调压器

Primary 初级的,最初的,原始的,主要的,初级线圈

Primary Radar 一次雷达

Print Register 打印寄存器

Problem Report 问题报告

Procurement Request 采购要求,采购申请

Production Readiness 生产准备就绪

Production Repair 产品修理

Program Register 程序寄存器

Program Requirements 程序要求

Program Review 程序审查

Project Review 项目评审

Propulsion Range 动力推进距离,动力推进范围

Pseudo Random 伪随机

Public Relations 公共关系

Pulse Rate 脉冲速率,脉冲频率

Purchase/Procurement Request 采购申请

Radiation Pressure 辐射压力

Reduced Pressure 减压

**PRA** Particular Risks Analysis 特殊风险分析

Pitch and Roll Attitude 俯仰与横滚姿态

Popular Rotorcraft Association 大众旋翼机协会(美国)

Precision Radar Approach 精密雷达进近

Primary Rate Access 一次群速率接入

Probabilistic Risk Analysis 概率风险分析

Probabilistic Risk Assessment 概率风险评估

Product Reliability Analysis 产品可靠性分析

Production Repair Area 生产修理区域

Pulse Radar Altimeter 脉冲雷达高度表

**PRAB** Program Risk Assessment Board 项目风险评估委员会

**PRAC** Practice 实践,惯例

Public Relations Advisory Committee 公共关系咨询委员会

**PRACA** Problem Reporting and Corrective Action 问题报告与纠正措施

**PRACH** Packet Random Access Channel 分组随机接入信道

**PRAD** Public Relations and Advertising 公共关系与广告宣传

**PRAG** Performance Risk Analysis Group 性能风险分析组

Performance Risk Assessment Group 性能风险评估组

Procurement Risk Analysis Group 采购风险分析组

Program Review and Analysis Group 项目评审与分析组

**PRAIM** Predictive Receiver Autonomous Integrity Monitor 自主预测接收机完整性监视器

**PRAM** Pipeline Risk Assessment Method 管线风险评估方法

Portable Reverse Activation Monitoring 便携式反向激活监控器

Pre-Recorded Announcement and Boarding Music Reproducer 预录的通知与登机音乐放音机

Prerecorded Announcement and Music 预录的通知和音乐

Productivity, Reliability, Availability and Maintainability 生产性、可靠性、有效性和维修性

**PRAMPO** Productivity, Reliability, Availability and Maintainability Program Office 生产性、可靠性、有效性和维修性项目办公室

**PRAT** Production Reliability Acceptance Test 生产可靠性验收试验

Prorate 分摊,按比例分配

**PRAWS** Pitch/Roll Attitude Warning System 俯仰/滚转姿态告警系统

**PRB** Parabola 抛物线

Probe 探头

Procurement Review Board 采购审查委员会

Proposal Review Board 提案审查委员会

**PRBC** Pressure Ratio Bleed Control 压力比放气控制器

Pressure Regulating Bleed Controller 放气控制压力调节

**PRBD** Paraboloid 抛物面

**PRBLC** Parabolic 抛物线的,抛物面的

**PRBLM** Problem 问题

**PRBNT** Prebent 预弯的

**PRC** Part Release Card 零件发放卡

Parts Requisition Cards 零件申请卡

People's Republic of China 中华人民共和国

Pierce 穿孔,洞察

Prescott, AZ, USA (Airport Code) 美国亚利桑那州普雷斯科特机场

Price 价格

Price Redetermination Contract 重新定价合同

Primary Reference Clock 主参考时钟

Problem Review Committee 问题审查委员会

Program Review Committee 项目评审委员会,计划审查委员会

Pseudo Random Code 伪随机码

Pseudorange Correction 伪距校正

**PRCA** Pitch/Roll Control Assembly 俯仰/滚转操纵组件

**PRCG** Pricing 标价

**PRCH** Purchase or Purchasing 购置,采购

**PRCHT** Parachute 降落伞

**PRCLR** Precooler 预冷器

**PRCMT** Procurement 采购

**PRCN** Precision 精度,准确性

**PRCPM** Partial Response Continuous Phase Modulation 部分响应连续相位调制

**PRCPTN** Precipitation 降雨,降水

**PRCS** Passive and Remote Crosswind Sensor 被动式遥控侧风传感器

Personal Radio Communications System 专用无线电通信系统

Pitch Reaction Control System 俯仰反应控制系统

**PRCSG** Processing 处理,进程

**PRCSR** Processor 处理器

**PRD** Paired 配对的

Period or Periodic 周期,时期,时代或周期的

Piezoelectric Resonation Device 压电谐振装置

Planning Receipt Data 计划接收日期

Pressure Relief Device 降压设备

Primary Radar Data 原始雷达数据

Product Requirements Document 产品需求文件

Program Reference Document 程序参考文件

Program Requirements Data 程序要求数据

Program Requirements Document 程序要求文件,项目需求文档

Pseudo-Random Downstream 下行伪随机序列

Push Rod 推杆

**PRDA** Program Research and Development Announcement 程序研究与开发通告

**PRDCG** Producing 生产,制造

**PRDR** Production Reliability Design Review 生产可靠性设计评审

**PRDS** Processed Radar Display System 已处理过的雷达显示系统

**PRDT** Production Reliability Demonstration Test 生产可靠性演示试验

**PRDTY** Productivity 生产率

**PRE** Preset 预置

Protective Equipment 保护设备

Pseudo-Range Error 伪距误差

**PRE/M** Pre-pack Missing 包装前损失

**PREAMP** Preamplifier 前置放大器

**PREC** Precedence 领先

Precision 精度,准确度

Precooler 预冷器

**PRECLR** Preclearance 预先放行

**PRECOOL** Precooler 预冷器

**PRED** Predicted 预测的

Prediction 预测

Predictive 预测的

**PREF** Preference 优先,选择权,选择物

Preferring 优选的

Prefix 字首,置于前

Prefocused 预聚焦的

**PREFAB** Prefabricated 预制的

**PREFLT** Preflight 航前,飞前

**PREFMD** Preformed 预制的

**PRELIM** Preliminary 初步的,预行的

**PRELORTR** Precision Long Range Tracking Radar 精密远程跟踪雷达

**PREM** Premises 上述各点,前提

Premium 特级的,保险费

**PREMO** Presentation Environment for Multimedia Objects 多媒体对象的表示环境

**PREP** Part Reliability Enhancement Programme 部件可靠性增强计划

Pilot's Reference Eye Position 飞行员基准眼位

Product Reliability Evaluation Program 产品可靠性评估计划

Production Risk Evaluation Program 生产风

险评估计划

**PREPN** Preferred Partnumber 推荐部件号

**PRE-PRO** Pre-Processor 预处理器

**PREQ** Prerequisite 前提,必要的,先决的

**PREREC** Prerecorded 预先录制的

**PRES** Office of the President 总统办公室
Preparation for Solo Flight 单飞准备
Present or Presently 现今,当前
Presentation 介绍,展示
President 总经理,总裁

**PRESB** Prescribe 命令,规定

**PRESCR** Prescription 命令,法则,惯例

**PRESEL** Preselection 预选
Preselector 预选器

**PRESS** Prediction Error Sum of Squares 预测
误差平方和
Prediction Residual Error Sum of Squares 预
测残差平方和
Pressure 压力
Pressurization/Pressurize 增压

**PRESSRA** Presentation Equipment for Slow
Scan Radar 慢扫探雷达显示设备

**PRESSU REG** Pressure Regulator 压力调节
器,调压器

**PREV** Previous 以前的

**PREV PAGE** Previous Page 前页

**PRF** Interrogation Rate 询问率
Preferred 优选的
Primary Reference Fuel 主标准燃油
Proof 证明,验证
Pulse Recurrence Frequency 脉冲重复频率
Pulse Repetition Frequency 脉冲重复频率

**PRFCN** Purification 纯化,精炼

**PRFCS** Pattern Recognition Feedback Control
System 模式识别反馈控制系统
Prefocus 预聚焦

**PRFD** Pulse Repetition Frequency Discrimination 脉冲重复频率鉴别

**PRFJ** Pulse Repetition Frequency Jitter 脉冲
重复频率抖动

**PRFM** Performance 性能

**PRFRD** Proofread 校对

**PRFS** Pulse Repetition Frequency Stager 脉冲
重复频率摆动

**PRFT** Press Fit 压入配合
Presser Foot 压足

**PRG** Pitch Rate Gyroscope 俯仰速率陀螺仪
Pollution Research Group 污染研究组
Prague, Czech Republic (Airport Code) 捷
克布拉格机场
Program Review Group 项目评审小组,计划
审查小组
Pseudo Random Generator 伪随机码发生器
Purge/Purging 清洗,提纯,换气

**PRGM** Program 计划,方案,程序

**PRGMG** Programming 编制程序

**PRGMR** Programmer 程序设计器,程序员

**PRGN** Public Relations Global Network 公共
关系全球网

**PRI** Primary 主要的,基本的,初级的
Primary Rate Interface 基本速率接口
Prior 在前的,优先的
Prioritizing 优先序化
Priority 优先
Pulse Rate Indicator 脉冲率指示器
Pulse Repetition Interval 脉冲重复间隔

**PRICE** Programmed Review of Information for
Costing and Evaluation 程序化成本与评估
信息审查

**PRIM** Primary 首要的,主要的,基本的
Prime 主要的

**PRIMES** Product Related Information Management Enterprise System 与产品相关的信
息管理企业系统

**PRIN** Principal 主要的,主题,资本

**PRIP** Parts Reliability Improvement Program
零件可靠性改进计划

**PRIS** Project Resource Information System 项
目资源信息系统
Propeller Revolution Indicator System 螺旋桨
转数指示系统
Public Relations Information Service 公共关

系信息服务

**PRISH** Perishable 易腐烂的

**PRIV** Privilege 特许,特权

**PRK** People's Republic of Korea 朝鲜(民主主义)人民共和国

Phase-Reverse Keying 倒相键控

**PRKG** Parking 停机(车),停留,停泊

**PRL** Parallel 平行,并联

Part Record Line 零件记录线

**PRLM** Preliminary 初步的

**PRLX** Parallax 视差

**PRM** Partner Relationship Management 合作伙伴关系管理

Person with Reduced Mobility 行动不便的旅客

Portable Radiation Monitor 便携式辐射监控器

Precision Radar Monitor 精密雷达监视器

Precision Runway Monitor 精密跑道监视器

Precision Runway Monitoring 精密跑道监视

Prime 主要的

Protocol Reference Model 协议参考模型

Pulse Rate Modulation 脉冲率调制

**PRMA** Packet Reservation Multiple Access 分组预留多址

**PRMC** Pacific Regional Monitoring Center 太平洋地区监控中心

**PRML** Partial Response Maximum Likelihood 部分响应最大似然值

**PRMTR** Parameter 参数

**PRN** Packet Radio Networks 分组无线网

Parts Requirement Notice 零件需求通知

Pseudo Random Noise 伪随机噪声

Pseudo Random Number 伪随机数字

Pulse Ranging Navigation 脉冲测距导航

**PRNAV** Precision Area Navigation 精确区域导航

**P-RNAV** Precision RNAV 精密 RNAV

**PRNS** Precise Radar Navigation System 精确雷达导航系统

**PRNT** Parent 母体,起源,原因

**PRNTG** Printing 印刷

**PRO** Profession 职业,专业

Proficiency 熟练,精通

Profile 剖面

Propeller Order 推进器规程

**PROB** Probability 概率,公算

Probable 大概的

Probation 试用,检验

Problem 问题

**PROBE** Pilot Radiation Observation Experiment 飞行员辐射观测试验

Program Optimization and Budget Evaluation 程序优化和预算评估

**PROC** Procedure 程序,手续,步骤

Process or Processing 处理,过程,工艺,工序

Procurement 采购

Procurement Code 采购码,获得码

Procuring 采购,获得

Programmed Computer 程控计算机

**PROCEED** You May Proceed 你可进行,可沿航向飞行

**PROCR** Processor 处理器

**PROCT** Procedure Turn 程序转弯

**PROCU** Processing Unit 处理单元

**PROD** Production 生产

Production Decisions 生产决策

Production Value 生产值

**PROF** Professional 专业的

Profile 剖面,翼型,轮廓

**PROFI** Flight Forecast 飞行天气预报

**PROFS** Professional Office System 专业办公系统

Program for Regional Observing and Forecasting Services 地区观测和预报服务项目

**PROG** Program 程序,编程序

Programming 程序,计划

Progressive 先进的

**PROH** Prohibited or Prohibition 禁止

**PROJ** Project 方案,计划,项目

Projected 计划的

Projection 投影,投射,计划

Projector 投影仪，投射器，探照灯，发起人

**PROM** Program for European Traffic with Highest Efficiency and Unprecedented Safety 欧洲高效率和安全交通计划

Programmable Read-Only Memory 可编程只读存储器

Promote 增进，发起，推销，提升

Promotional 发起的，提升的

**PROMOTE** Program for Mobility in Transportation in Europe 欧洲运输机动性计划

**PROMPT** Project Reporting, Organization and Management Planning Technique 项目报告、组织与管理计划技术

**PRON** Parts Records On-line 联机零件记录

Pronounced 宣布

**PRONSE** Passive Ranging of Non Scanning Emitters 非扫描发射机的无源测距

**PROP** Propellant or Propelled 推进剂或推进的

Propeller 螺旋桨，推进器

Propeller Aircraft 螺旋桨航空器

Property 性质，器材，资产

Proportional 比例的

Proposal 建议，提案

Proposed 提议的，建议的

Proprietor 所有人，业主

**PROPL** Professional Pilot 职业驾驶员

**PROPN** Proportion 比率，比例，部分

**PROR** Route Forecast 航路天气预报

**PROS** Parts and Repair Ordering System 零件和维修订购系统

Parts Repair Ordering Service 零件修理订购服务

Passenger Revenue Optimization System 客运收益优化系统

Present Position 现在位置，即时位置

Prospect 展望，预期

**PROSE** Passive Ranging of Scanning Emitters 扫描发射器的无源测距

**PROT** Protect or Protected 保护

Protection 防护

Protective 保护的

**PROT DEV** Protective Device 保护器件

**PROT(R)** Protractor 量角器

**PROTO** Prototype 样机，原型

**PROV** Provide 提供，维持

Provided 提供的

Provider 提供者，供应方

Provision 预备，设备，条款 供应

Provisional 临时的，暂时的

Provisions 提供，选装

**PROX** Proximity 大约，近似，接近

**PRP** Personnel Reliability Program 人员可靠性大纲，人员可靠性计划

Power-deployed Reserve Parachute 自动打开的备份伞

Premature-Removal Period 提前拆卸周期

Production Reserve Policy 生产储备方针

Programmed Random Process 程序随机过程

Progressive Requirements Plan 逐渐增加的要求计划

Pseudo Random Process 伪随机过程

Pseudo Random Pulse 伪随机脉冲

Pulse Recurrence(Repetition) Period 脉冲重复周期

Purple 紫色(的)

Purpose 目的

**PRPA** Professional Racing Pilots' Association 职业竞赛飞行员协会

**PRPC** Periodical Route Performance Check 定期路由性能检验

**PRPD** Prepaid 预付的

**PRPHL** Peripheral 周围的

**PRPLN** Propulsion 发动机，推进(器)

**PRPLT** Propellant 推进剂

**PRPNE** Propane 丙烷

**PRPS** Pressure Rise per Stage 每级增加的压力

**PRPSD** Proposed 提议的，计划的

**PRPSL** Proposal 提议，建议

**PRR** Part Replacement Request 需换零件

Passenger Reservation Request 旅客订座要

求

Power Ready Relay 预供电继电器

Preliminary Requirements Review 初步需求评审

Premature Removal Rate 提前拆卸率

Product Revision Request 产品更改需求

Production Readiness Review 生产准备状态审查

Production Revision Record 生产修改记录

Production Revision Request 生产更改申请

Program Revision Request 程序修改申请

Pseudo Range Rate 伪距变化率

Pulse Repetition Rate 脉冲重复率

**PRRB** Program Risk Review Board 项目风险评审委员会

**PRRC** Procurement Requirements Review Committee 采购需求审查委员会

**PRR-R** Production Revision Record-Retrofit 生产修改记录—改型

**PRS** Pattern Recognition System 模式识别系统

Poverty Reduction Strategy 减贫战略

Power Relay Satellite 动力中继卫星

Press 按压

Primary Reference Source 基本参考源

Program Redefinition Study 程序重新定义研究

Provisions Requirements Statement 规定(设备)要求说明

Pseudo-Random Sequence 伪随机序列

**PRSD** Pressed 按压的,压下的

**PRSN** Prisoner 囚犯

Product Service Node 产品服务网点

**PRSODV** Pressure Raising Shutoff and Dump Valve 升压关断和安全活门

**PRSOV** Pressure Reducing Shutoff Valve 减压关断活门

Pressure Regulating Shutoff Valve 调压关断活门

**PRSOVC** Pressure Regulating and Shutoff Valve Controller 压力调节和关断活门控制

器

**PRSR** Presser 模压机

**PRSRZ** Pressurize 增压

**PRSRZG** Pressurizing 增压

**PRSTC** Prosthetic 弥补性的,替代的

**PRT** Pattern Recognition Technologies 模式(图形)识别技术

Power Recovery Turbine 功率回收透平

Print,Printed 印刷(品),打印(的)

Printer 印刷机

Product Realization Team 产品现实团队

Product Review Team 产品审核小组

Publication Requirement Tables 出版物规定表格

Pulse Repetition Time 脉冲重复时间,脉冲周期

Pulse Rise Time 脉冲上升时间

**PRTKT** Parts Kit 成套零件

**PRTL** Partial 部分的

**PRTT** Parts Transfer 零件运输

**PRTY** Priority 优先权

**PRU** Pseudo-Random Upstream 上行伪随机序列

**PRV** Peak Reverse Voltage 反向峰值电压

Pressure Reducing/Reduction Valve 减压活门

Pressure Regulating Valve 压力调节活门

Pressure Relief Valve 压力释放活门

Previous,Previously 以前(地),预先(地)

Propeller Revolution 螺旋桨转速

**PRVC** Pressure Regulating Valve Controller 调压活门控制器

**PRVDN** Private Data Network 专用数据网

**PRVT** Production Reliability Verification Test 产品可靠性验证试验

**PRVW** Preview 预审

**PRW** Darwin, Northern Territory, Australia (Airport Code) 澳大利亚北领地达尔文机场

**PRY** Pitch Roll Yaw 俯仰滚转偏航

Pretoria, South Africa( Airport Code) 南非比

勒陀利亚机场

**PS** Packet Sequencing 分组排序

Packet Switched 分组交换

Packing Sheet 装箱单

Parallel to Series 并联/串联

Part Specification 零件规范

Part Stores 零件存储

Partial Shipment 分批装运

Passing Scuttle 通气窗,进人小舱口

Patent Specification 专利说明书

Personal Station 个人站

Personnel Subsystem 人事分系统,人员子系统

Picture System 图像系统

Pilot Simulator 飞行模拟机

Pitot Static 皮托管(流速计),皮托静压管

Plate Sensitive 平面敏感的

Plus 加,正的, 加的

Point of Switch 转换点

Polystyrene 聚苯乙烯

Postscript 附启

Potentiometer Synchro 同步电位器

Power Supply 电源,供电

Pressure Sensor 压力传感器

Pressure Switch 压力开关

Print in Shop 车间内打印

Probability of Success 成功概率

Probability of Survival 生存率

Process Sheet 操作单,操作说明

Process Solution 处理决断

Process Specification 工艺操作单,工艺处理单

Procurement Specification 采购标准

Product Standard 产品标准

Production Schedule 生产计划

Production Specification 生产规范

Program Slide 程序侧滑

Proof Shot 耐冲击

Protection Switching 保护倒换

Protocol Stack 协议栈

Pull Switch 拉拔开关

Pulse Sense 脉冲检测

Pulse Stretcher 脉冲展宽电路

Purchase Standard 采购标准

Static Pressure 静压

**P-S** Parallel-to-Serial 并一串(转换)

Produce Statistics 生产统计

**PS LT** Port Side Light 左侧灯

**PS&S** Part Standards and Specifications 零件标准和规范

**PSA** Pacific Southwest Airlines, Inc. (USA) 太平洋西南航空公司(美国)

Passive Situational Awareness 被动情况了解

Phase-Sensitive Amplifier 相位敏感放大器

Pilot's Associate 副驾驶员

Pisa, Italy(Airport Code) 意大利比萨机场

Pitch Servo Amplifier 俯仰伺服放大器

Pitch Stability Augmentation 俯仰增稳

Platform Stabilization Amplifier (陀螺)平台稳定放大器

Power Supply Assembly 电源组件

Power Switching Assembly 电源转换组件

Preferred Storage Area 优选存储区

Price Surveillance Authority 价格管制局

Probability Safety Assessment 概率安全评价

Problem Statement Analyzer 问题陈述分析程序

Process Safety Analysis 工序安全分析

Professional Services Agreement 专业服务协议

Public Service Agreement 公共业务协议

**PSAC** Plan for Software Aspects of Certification 软件合格审定计划

President's Science Advisory Committee 总统科学顾问委员会

**PSACN** Process Specification Advance Change Notice 工艺规范变更预告

**PSAD** Passive Sensor for Air Defense 防空被动式传感器

Passive Sensor for Aircraft Detection 飞机探测被动传感器

Preliminary Safety Assessment Document 初

步安全评估文件

**PSAI** Public Safety Aviation Institute 公共安全航空研究所

**PSAP** Presentation Service Access Point 表示层业务接入点

**PSAR** Preliminary Safety Analysis Report 初步安全分析报告

**PSAS** Pitch Stability Augmentation System 俯仰稳定增益系统

Primary Stability Augmentation System 主增稳系统

**PSB** Power Switching Bridge 电源转换电桥

Program Specification Block 程序规范块

**PSBL** Possible 可能的

**PSBN** Public Switched Broadband Network 公共交换宽带网

**PSC** Parallel-Serial Conversion 并—串转换

Part Specification Card 零件说明卡

Pasco, Washington(Airport Code) 美国帕斯科机场

Picture Start Code 图像开始码

Pilot Status Code 飞行员状况代码

Pilot Stores Control 飞行员储备管理

Pneumatic System Controller 气动系统控制器

Power Status and Control 电源状态与控制

Preferred Systems Concept 优选系统方案

Present Course 现在航线,当前航线

Preset Course 预选航线

Program Switching Center 程序交换中心

Protection Switching Count 保护倒换计数

**PSCAF** Packet Service Control Agent Function 分组服务控制代理功能

**PSCAS** Pitch Stability Control Augmentation System 俯仰稳定控制增强系统

**PSCF** Packet Service Control Function 包服务控制功能

Pure Silica Core Fiber 纯硅芯光纤

**PSCH** Primary Synchronisation Channel 主同步信道

**PSCM** Program Schedule Change Memoran-dum 程序计划变更备忘录

**PSCMMS** Propulsion-System Control, Monitoring and Maintenance System 推力系统控制、监视与维护系统

**PSCN** Permanent System Control Number 永久性系统控制号码

Process Specification Change Notice 工艺规范变更通知

Production Schedule Completion Notice 生产计划完成通知

Program Support Communication Network 计划支援通信网络

**PSCP** Project Specific Certification Plan 专项合格审定计划

**PSCS** Personal Space Communication Service 个人空间通信业务

Photographic Sensor Control System 摄影传感器控制系统

Program Support Control System 计划支持控制系统

Public Service Communication Satellite 公共业务通信卫星

**PSCT** Packet Switched Connection Type 分组交换连接类型

**PSCU** Power Supply Control Unit 电源控制组件

Proximity Switch Control Unit 接近开关控制器

**PSD** Part Specification Departure 零件标准偏差

Phase Sensitive Demodulator 相敏解调器

Phase Sensitive Detection 相敏检波

Phase Sensitive Detector 相敏检波器

Port Select Discrete 端口选择离散量

Power Semiconductor Device 功率半导体器件

Power Spectral Density 功率谱线密度

Pressure Sensing Device 压力传感器

Process Specification Departures 工艺规范变更

Propulsion Systems Division 推进系统分部

Protection Switching Duration 保护倒换时延

Pulse Shape Discriminator 脉冲形状鉴别器

**PSDAU** Packet Switched Data Access Unit 分组交换数据接入单元

**PSDDS** Public Switched Digital Data Service 公用交换数字数据业务

**PSDN** Packet Switched Data Network 分组交换数据网

Public Switched Digital Network 公共交换数字网

**PSDP** Personnel Subsystem Development Plan 人事分系统研制计划

Preliminary System Data Package 初步系统数据包

Programmable Signal Data Processor 可编程信号数据处理器

**PSDS** Packet Switched Data Service 分组交换数据业务

Product Standards Data System 产品标准数据系统

Program Standards Distribution System 程序标准分配系统

**PSDTN** Packet Switched Data Transmission Network 分组交换数据传输网

**PSDTS** Packet Switched Data Transmission Service 分组交换数据传输业务

**PSDU** Power Supply Decoupling Unit 供电去耦组件

Presentation Service Data Unit 表示层服务数据单元

**PSE** Packet Switching Equipment 分组交换设备

Packet Switching Exchange 分组交换机

Passenger Service Equipment 旅客服务设备

Peculiar Support Equipment 特殊保障设备

Principle Structural Elements 主要结构部件

Probability of Successful Engagement 衔接成功概率

Product Support Equipment 产品保障设备

Protection Switching Event 保护倒换事件

Public Sector Entity 公共部门实体

**PSEL** Priced Support Equipment List 标价保障设备清单

**PSES** Product Support Equipment Schedules 产品保障设备计划

**PSESCO** Product Support Equipment Scheduling Committee 产品保障设备计划委员会

**PSEU** Power Supply Electronic Unit 电源电子组件

Proximity Sensor Electronic Unit 接近传感器电子组件

Proximity Switch Electronic Unit 接近电门电子组件

**PSF** Packet Switching Facility 分组交换设施

Payload-Structure-Fuel 有效载重、结构及燃料

Performance Shaping Factor 行为形成因子

Point Spread Function 点发散函数

Pounds-force per Square Foot 磅力/平方英尺

Processing and Storage Facility 处理与存储设施

Program Support Facility 程序支持设备

**PSG** Passing 经过,过往

Petersburg, AK, USA ( Airport Code ) 美国阿拉斯加州彼得斯堡机场

Phase-Shifting Grating 相移光栅

Programmable Symbol Generator 可编程符号产生器

**PSGCF** Packet Service Gateway Control Function 分组业务网关控制功能

**PSGR** Passenger 旅客

**PSGRST** Passenger Seat 旅客座位

**PSH** Phase Shift 移相

Preselected Heading 预选航向

Pre-Set Heading 预调定航向

Pressure Sensor Holder 压力传感器架

Program Support Handbook 计划支援手册

**PSI** Packet Switching Interface 分组交换接口

Pollution Standards Index 污染标准指数

Pounds per Square Inch 磅/平方英寸

Pounds-force per Square Inch 磅力/平方英

寸

Procurement Source Inspection 采购源检查

Product Suppot Instruction 产品供应说明书

Programmable Serial Interface 可编程串行接口

**PSIA**   Pounds per Square Inch Absolute 磅/平方英寸绝对压力

Pounds per Square Inch, Ambient 磅/平方英寸,环境压力

Public Security Investigation Agency 公共安全调查局

**PSIC**   Proximity Sensor Interface Card 临近传感器接口卡

**PSID**   Pounds per Square Inch Differential 磅/平方英寸压差

Preliminary Safety Information Document 初步安全信息文件

**PSIG**   Per Square Inch Gravity 每平方英寸重力

Pounds per Square Inch, Gauge 磅/平方英寸(表压)

**PSIL**   Production System Integration Laboratory 生产系统综合实验室

**PSIS**   Passenger Security Information System 旅客安全信息系统

Pool Stores Inventory System 集中存储库存系统

**PSIU**   Passenger Service Information Unit 旅客服务信息组件

**PSIV**   Passive 无源的,被动的

**PSK**   Phase Shift Keying 相移键控

**PSL**   Personnel Skill Levels 人员技术水平

Pipe Sleeve 管道套筒

Pressure Seal 压力密封

Problem Statement Language 问题说明语言

**PSM**   Packet Switching Module 分组交换模块

Passenger Service Message 旅客服务报

Passenger Statute-Mile 客英里(1英里 = 1.609千米)

Phase Shift Method 相移法

Pitch Servo Motor 俯仰伺服马达

Power Supply Module 电源模块

Power Supply Monitor 供电监视器

Power Switching Module 动力转换模件

Prism 棱镜

Process Safety Management 过程安全管理

Product Safety Management 产品安全管理

Product Supply Manual 产品供应手册

Program Security Manager 程序安全经理

Protocol Special Module 协议专用模块

Pulse Slope Modulation 脉冲斜度调制

**PSMP**   Process Safety Management Program 工艺安全管理计划

Product Support Management Planning 产品供应管理计划

Program on Short-range and Medium-range Weather Prediction Research 短期中期天气预报研究计划

Project Software Management Plan 工程软件管理计划

**PSN**   Packet Switched Network (数据)包交换网络

Packet Switching Node 分组交换节点

Personal Server Network 个人服务器网络

Phase Shift Network 相移网络

Position 位置

Pre-Solicitation Notice 征求前通知

Processor Serial Numbers 处理器序列号

Processor Sharing Node 处理器共享节点

Program Sequence Number 程序序列号

Public Switched Network 公共交换网

**PSNL**   Packet Switching Network Line 分组交换网络线路

**PSNR**   Peak Signal to Noise Ratio 峰值信噪比

Positioner 定位器

**PSO**   Pasto, Colombia( Airport Code) 哥伦比亚帕斯托机场

Pilot Systems Operator 飞行系统操纵者

Printing Service Order 打印服务指令,打印服务规程

**PSOL**　Purchased Standards on-Line 联机采购标准

Purchased Stores on-Line 联机采购存储

**PSOLA**　Pitch-Synchronous Overlap and Add 基音同步叠加技术

**PSOW**　Program Satatement of Work 工作程序说明

**PSP**　Packet Switching Processor 分组交换处理机

Packet Switching Protocol 分组交换协议

Packing Software Program 包装软件程序

Palm Springs, CA, USA（Airport Code）美国加利福尼亚州棕榈泉机场

Payload Signal Processor 有效负荷信号处理器

Performance Standards Program 性能标准程序

Performance Surveillance Plan 性能监督计划

Personal Survival Pack 个人救生包

Physical Service Port 物理业务端口

Power Steering Pump 动力转向泵

Power System Processor 电源系统处理机

Pre-Sync Pulse 前置同步脉冲

Product Support Program 生产支援计划

Production Service Provider 生产服务提供商

Professional Services Provider 专业服务提供者

Program Support Plan 程序支援计划

Programmable Signal Processor 可编程信号处理器

**PSPDN**　Packet Switched Public Data Network 分组交换公共数据网

**PSPL**　Preferred Standard Parts List 优选标准零件清单

Priced Spare Parts List 标价备件单

**PSPM**　Procurement Seminar for Project Management 项目管理采购研讨会

Product Support Procedures Manual 产品支援程序手册

Pulse Symmetrical Phase Modulation 脉冲对称相位调制

**PSPP**　Proposed System Package Plan 建议系统组装计划

**PSPT**　Passport 护照

**PSQ**　Preordering Stock Quantity 订货前库存量

**PSQM**　Propulsion System Quality Management 推进系统质量管理

**PSR**　Packed Snow on Runway 跑道上的积雪

Pad Safety Report 发射台安全报告

Path Switched Ring 通道交换环

Pescara, Italy（Airport Code）意大利佩斯卡拉机场

Phase-Sensitive Rectifier 相敏整流器

Point of Safe Return 安全返航点

Power Supply Reset 供电恢复

Precision Secondary Radar 精密二次雷达

Primary Surveillance Radar 一次监视雷达

Productive Store Requisition 生产备品调拨单

Program Status Review 程序状态审查

Program Study Request 程序研究申请

Project Status Report 工程状态报告

Purchased Shop Request 车间购置申请

**PSRCP**　Propulsion System Reliability Control Program 推力系统可靠性控制程序

**PSRE**　Propulsion System Rocket Engine 推力系统火箭发动机

**PSRM**　Power-plant Structural Repair Manual 动力装置结构修理手册

Pressurization Systems Regulator Manifold 增压系统调节器歧管

**PSRS**　Programmable Synchro/Resolver Source 可编程同步/解算器源

Public Service Resourcing System 公共服务的资源系统

**PSS**　Packet Switching Service 分组交换业务

Packet Switching System 分组交换系统

Passenger Service System 旅客服务系统

Personal Sound System 个人音响系统

Physical Security System 实际安全系统

Planning Summary Sheets 计划一览表

Pneumatic Supply Subsystem 压缩空气供应支系统

Positive Speed Stability 速度稳定性

Power System Stabilizator 电力系统稳定器

Pressure Sub-System 压力分系统

Production System Software 生产系统软件

Protection Switch Second 保护倒相秒

Proximity Sensor System 临近传感器系统

**PSSA** Pilot's Stick Sensor Assembly 飞行员驾驶杆传感器组件

Preliminary System Safety Assessment 系统安全性初步评估

**PSSD** Product Service Systems Design 产品服务系统设计

Products and Services Suppliers Directory 产品和服务供应商目录

**PSSEP** Preliminary System Safety Engineering Plan 系统安全工程初步计划

**PSSP** Pre-Sync Sync Pulse 预同步同步脉冲

**PST** Pacific Standard Time 太平洋标准时间

Paste 胶,涂胶

Pasty 糊状的,黏性的

Point of Spiral Tangent 螺线切点

Position Trunk 座席中继

Posting 邮寄的,公布的

Product Support Tools 产品支持工具

Program Sequence Table 程序序列表,计划序列表

Public Service Telephone 公共电话

**PSTC** Product Support Task Control 产品支援任务管理

Public Switched Telephone Circuit 公共交换电话电路

**PSTDD** Partially Shared Time Division Duplex 部分共享时分双工

**PSTE** Payload System Test Equipment 有效载荷系统试验设备

Personnel Subsystem Test and Evaluation 人事分系统测试与评估

**PSTL** Postal 邮政的

**PSTM** Photon Scanning Tunneling Microscope 光子扫描隧道显微镜

**PSTN** Piston 活塞

Public Switched Telecommunications Network 公共电信交换网

Public Switching Telephone Network 公共交换电话网

**PSTP** Propulsion System Test Procedure 推进系统试验程序

**PSU** Passenger Service Unit 旅客服务组件,客服组件

Power Supply Unit 供电组件,电源设备

Power Switching Unit 电源转换组件

Purchased Standards Update 采购标准更新

**PSUD** Passenger Service Unit Decoder 旅客服务组件译码器

**PSV** Preserve 保存,维持,禁区

Pressure Safety Valve 压力安全活门

**PSVC** Point to Point Signaling Virtual Channel 点到点信令虚通道

**PSVT** Passivate 钝化

**PSVTN** Preservation 保存,保护

**PSVTV** Preservative 防腐剂,保存剂

**PSW** Potential Switch 电位开关

Program Status Word 程序状态字

**PSWD** Password 口令

**PSWR** Power Standing Wave Ratio 功率驻波比

**PSYS** Pressure System(s) 压力系统

**PT** Paper Tape 纸带

Part 零件,部分

Part Transfer 部件传送

Path Terminal 通道终端

Pattern Transfer 图形(模式)转换

Payload Type 净荷类型

Performance Test 性能试验

Phasing Transformer 移相变压器

Photo Template 照相版

Photo Thyristor 光控晶闸管

Photo Transistor 光电晶体管,光敏晶体管

Photoelectric Timer 光电定时器

Photoelectric Transducer 光电传感器

Piezoelectric Transducer 压电传感器

Pint 品脱

Pipe Tap 管螺纹丝锥

Platinum 白金

Pneumatic Tube 气动管,压缩空气管

Point 要点

Point of Tangency 切点

Portable Terminals 便携式终端

Potential Transformer 测量用变压器,仪表用变压器

Power Transformer 电源变压器

Power Turbine 动力涡轮机,动力透平机

Pressure Test 加压试验

Pressure Total 全压,总压

Pressure Transducer 压力传感器

Price Type 价格类型

Primary Trainer 初级教练机

Print 印刷(品),打印

Procedure Turn 程序转向

Programmable Terminal 可编程终端

Property Transfer 资产转移

Public Transport 旅客机,公共运输机

Pulse Time 脉冲时间

Punched Tape 穿孔纸带

Switching Pressure of HP Valve 高压活门的转换压力

Total Pressure 总压

**PTA** Planned Time of Arrival 计划到达时间

Port Alsworth, USA (Airport Code) 美国阿尔斯沃特港机场

Portable Tool Accountability 便携式工具责任

Preliminary Technical Approach 初步技术途径

Prepaid Ticket Advice 预付票款通知,机票预付凭单

Programmable Test Adapter 可编程测试转接器

Public Transport Association 公共运输协会(英)

**PTAB** Project Technical Advisory Board 项目技术咨询委员会

**PTAR** Prime Time Access Rule 黄金时段准入规则

Problem Tracking and Reporting 问题追踪与报告

Production Tool Action Request 生产工具操作要求

**PTAS** Pilotless Target Aircraft Squadron 无人靶机中队

**PTB** Physical Trade Balance 物理贸易平衡

Priority Token Bank 优先令牌组

Protective Technology Branch 防护技术部门

Push the Button 按钮

**PTBL** Portable 便携式,轻便的,手提的

**PTC** Pack Temperature Controller 组件温度控制器

Part Transfer Card 零件传送卡

Passenger Type Code 旅客类型代码

Pitch Trim Compensator 俯仰配平补偿器

Polar Transit Company (England) 极地运输公司(英国)

Positive Temperature Coefficient 正温度系数

Power Transfer Coefficient 功率传递系数

Power Transmission Council 电力输送委员会

Propeller Technical Committee 螺旋桨技术委员会

Provisional Type Certificate 临时型号合格证

Push to Center 置中间位

**PTCP** Participation 参与,参加,加入

**PTCS** Pressure Transducer Calibration System 压力传感器校准系统

**PTCT** Protect 保护

**PTCV** Pilot-operated Temperature Control Valve 飞行员操纵的调温活门

Primary Temperature Control Valve 主温度控制活门

**PTD** Painted 上漆的

Parts and Tool Disposition 零件与工具配置

Passenger Transport Department 客运部

Permanent Total Disability 永久性全部失效,

永久性能力丧失

Pilot to Dispatcher 飞行员至航行调度员

Pilot Training Device 飞行员训练设备

Printed 打印的,印刷的

Proposed Time of Departure 预计起飞时间

Provisioning Technical Documentation 规定技术文件,提供技术文件

Pulse Time Demodulator 脉冲时间解调器

Push Through Die 穿通模

**PTDC** Provisioning Technical Data Card 备件供货技术数据卡

**PTDDSS** Provisioning Technical Documentation Data Selection Sheet 规定技术文件数据选择单

**PTDP** Preliminary Technical Development Plan 初步技术开发计划

**PTE** Passenger Transport Executive 客运经理

Path Terminating Equipment 通道终接设备

Periodic Table of Elements 元素周期表

Permission to Enter 允许进入

Pilot Tone Extraction 飞行员音调提取

Portable Test Equipment 便携式测试设备

Programming Tools and Environments 编程工具和环境

**PTF** Phase Transfer Function 相位传递函数

Power Transmission Fluid 动力传动液

Production Test Facility 生产测试设施

Program Temporary Fix 程序临时装配,计划临时性安排

**PTFE** Polytetrafluoroethylene 聚四氟乙烯

**PTH** Path 路径

Pitch 俯仰,节距,螺距

**PTI** Pack Temperature Indicator 组件温度指示器

Path Tracking Identifier 通道跟踪识别符

Payload Type Identifier 净负荷类型标识符

Payload Type Indicator 净负荷类型指示器

Plesiochronous Timing Interface 准同步定时接口

**PTIT** Power Turbine Inlet Temperature 动力涡轮进口温度

**PTK** Parallel Track 并行跟踪

**PTL** Part Tool List 零件工具清单

Partial 部分的

Patrol 巡查

Pintle 垂直枢轴,舵栓

Potential Term Liability 潜在项目责任

Preliminary Technical Letter 初步技术函件

**PTLST** Part Tool List 零件工具清单

**PTLY** Partly 局部地

**PTM** Packet Transfer Mode 分组传送模式

Packet Transport Mode 分组传递模式

Packet Trunk Module 分组中继器模块

Passenger Transfer Manifest 旅客转机名单

Passenger Transfer Message 旅客转港信息报

Performance Technical Memorandum 性能技术备忘录

Performance Training Manual 性能训练手册

Phase Time Modulation 相位时间调制

Point to Multipoint 点到多点

Program Timing and Maintenance 程序定时与维护

Program Trouble Memorandum 程序故障备忘录

Pulse Time Division Multiplex 脉冲时分复用

Pulse Time Modulation 脉冲时间调制

**PTML** PNPN Transistor Magnetic Logic PNPN 晶体管磁逻辑

**PTMLG** Pitometer Log 皮托压差计记录

**PTMUX** Pulse-Time Multiplex 脉冲时间多路复用

**PTN** Partition 划分,隔开,隔墙

Pattern 样式

Personal Telecommunication Number 个人通信号码

Private Telecommunication Network 专用通信网

Private Telecommunication Number 专用通信号码

Procedure Turn 程序转弯

Public Telegraph Network 公用电报网

**PTO** Part Time Operation 部分时间工作

Participating Test Organization 共享试验组织

Patent and Trademark Office 专利与商标局（美国）

Permeability Tuned Oscillator 磁导调谐振荡器

Power Take-off 动力起飞

Preliminary Technical Order 初步技术规程

Proof Test Orbiter 验证试验轨道飞船

Public Telecommunicaitons Operator 公众电信运营商

**PTOUT** Printout 印出，打印输出

**PTP** Part-Time Position 兼职

Path Termination Point 通道终接点

Picture Transfer Protocol 图片传输协议

Point-to-Point 点对点

Pointe A Pitre，Guadeloupe ( Airport Code ) 瓜德罗普岛皮特尔角城机场

Power Transfer Panel 电源转换面板

Programming and Test Panel 编程和测试面板

**PTPI** Plesiochronous Timing Physical Interface 准同步定时物理接口

**PTPN** Peer-to-Peer Network 对等网络

**PTPS** Public Transportation Priority Systems 公共运输优先系统

**PTR** Paper Tape Reader 纸带阅读器

Part Throttle Reheat 部分油门加力

Part Transfer Request 零件传送申请

Perishable Tool Replenishment 易坏工具的补充

Pilot Training Record 飞行员训练记录

Plate Type Radiator 平板型辐射器

Position Track Radar 位置跟踪雷达

Pressure Test Record 压力试验记录

Printer 打印机，印刷机，印刷工

Priority Token Ring 优先级令牌环

Processor Tape Reader ( 信息 ) 处理器纸带阅读器

Production Test Requirements 生产测试要求

Program Trouble Report 程序问题报告，计划

问题报告

Push to Reset 推复位钮，按复位钮

Push-to-Reset 按压复位

**PTRS** Program Tracking and Reporting System 计划跟踪和报告系统

**PTS** Pack Temperature Sensor 空调组件温度传感器

Packet Transit Switch 中转分组交换机

Part Task Simulation 零件任务模拟

Parts 零件

Permanant Threshold Shifts 永久性听力伤害

Petrolair Systems ( Greece ) 汽油航空系统（希腊）

Photogrammetric Target System 摄影测量目标系统

Pilots Training Syllabus 飞行员训练大纲

Plesiochronous Timing Source 准同步定时源

Pneumatic Test Set 气动试验器

Ports 排气口，端口

Practical Test Standard 实践考试标准

Precision Track Selector 准确航道选择器

Predetermined Track Structure 预定航迹结构

Programmable Terminal System 可编程终端系统

Propellant Transfer System 推进剂输送系统

Propulsion Technology Staff 推进技术工作人员

Protocol Test Specification 协议测试规范

Public Telecommunication System 公共电信系统

Public Telephone Service 公众电话业务

Purchaser Technical Specification 采购商技术规范

**PTSA** Pulse Torque Servo Amplifier 脉冲力矩伺服放大器

**PTSD** Post-Traumatic Stress Disorder 创伤后应激障碍

Production Test Specification Document 生产试验规范文件

**PTT** Part Task Trainer 局部任务训练器，零

件任务训练器

Post, Telegraph, Telephone 邮政、电报、电话

Push to Talk 按下按键通话

Push to Test 按下测试

**PTTI** Precise Time and Time Interval 精确时间与时间间隔

**PTTXAU** Public Teletex Access Unit 公共智能用户电报接入单元

**PTU** Parallel Transmission Unit 并行传输设备

Pilot Test Unit 飞行员测试装置

Planning Tracking Unit 计划跟踪装置

Power Transfer Unit 电源转换组件

Power Transformer Unit 电源变压器组件

**PTUPC** Product to Unit Product Cost 产品到组件产品成本

**PTV** Parachute Test Vehicle 跳伞试用飞行器

Particle Tracking Velocimetry 粒子迹线测速

Performance Test Vehicle 性能试验飞行器

Propulsion Technology Vehicle 推进技术运载工具

**PTVS** Passenger Transport Vehicles 旅客运送车

**PTW** Pitch Trim Wheel 俯仰配平轮

**PTWT** Photo Type Traveling Wave Tube 光电型行波管

**PTY** Panama City, Panama-Tocumen (Airport Code) 巴拿马巴拿马城托库门机场

**PU** Panel Unit 面板组件

Parts Usage 零件使用

Per Unit 每组件

Peripheral Unit 外设组件

Physical Unit 物理装置

Polyurethane 聚氨酯

Power Unit 动力装置

Propulsion Unit 推进装置

Pump Unit 泵组

**PUAR** Pulse Acquisition Radar 脉冲搜索雷达

**PUB** Published 出版的

Pueblo, CO, USA (Airport Code) 美国科罗拉多州普韦布洛机场

**PUD** Planned Unit Development 计划的组件研制

Power Unit Deicing 动力装置除冰

**PUL** Pulley 滚筒, 滑轮

**PULV** Pulverizer 粉磨机

**PUP** Performance Update Program 性能更新方案, 性能现代化大纲

Physical User Port 物理用户端口

Pick-up Pump 加速泵

**PUPS** Principal User Processor 主要用户处理器

**PUR** Purchase 采购

Purifier 清洗器

**PURP** Purpose 目的

**PURR** Pick up Repair Record 传感器修理记录

**PURS** Purser 乘务长

**PUS** Pusan, South Korea (Airport Code) 韩国金山机场

**PUT** Programmable Unijunction Transistor 可编程单结晶体管

Position Unit Time 位置单位时间

**PUW** Pullman, WA, USA (Airport Code) 美国华盛顿州普尔曼机场

**PV** Part Version 零件版本

Plan View 计划审查

Planned Value 计划值

Pressure Valve 压力值

Pressure Vessel 压力容器

Priority Valve 优先活门

Product Verification 产品检验

Protocol Version 协议版本

**P-V** Peak to Valley 峰至谷

**PVA** Patterned Vertical Alignment 图像垂直调整

**PVASI** Pulsating Visual Approach Slope Indicator 脉冲式目视进近坡度指示器

**PVB** Portametric Voltmeter Bridge 桥式分压式伏特计

Present Value of Life Cycle Benefits 寿命周

期效益现值

Pressure Vacuum Breaker 压力真空断路器

**PVC** Permanent Virtual Call 永久虚呼叫

Permanent Virtual Capacitor 永久虚电容

Permanent Virtual Channel 永久虚信道

Permanent Virtual Circuit 永久虚电路

Permanent Virtual Connection 永久虚连接

Polyvinyl Chloride 聚氯乙烯

Popular Video Coder 常用视频编码器

Private Virtual Channel 专用虚信道

**PVCL** Stator Vane Closed Pressure 静子叶片关闭压力

**PVD** Para Visual Director 旋转式着陆指引仪

Paravisual Display System 超视距显示系统

Physical Vapor Deposition 物理气相沉积

Plan Video Display 平面视频显示

Plan View Display 平面图显示

Planned Video Display 计划视频显示

Product Verification Demonstration 产品检验示范

Providence, RI, USA(Airport Code) 美国罗得岛州普罗维登斯机场

**PVE** Preset, Verify, Execute 设定,验证,执行

**PVI** Paravisual Indicator 超视距显示器

Pilot Vehicle Interface 人机界面

**PVIS** Passenger Visual Information System 旅客视觉信息系统

**PVM** Parallel Virtual Machine 并行虚拟机

**PVMTD** Preservation Method 预防方法

**PVN** Private Virtual Network 专用虚拟网

**PVNT** Prevent 防止,预防

**PVO** Stator Vane Open Pressure 静子叶片打开压力

Vanes Open Pressure 叶片打开压力

**PVOR** Precision Very High Frequency Omni-directional Range 精密甚高频全向无线电信标

Precision VHF Omnirange 精密甚高频全向信标

**PVP** Packetized Voice Protocol 分组式话音协议

Permanent Virtual Path 永久虚拟路径

Personal Video Player 个人放像机

Product Validation Plan 产品验证计划

Programmable Video Processor 可编程视频处理器

**PVPU** Programable Vertex Processing Unit 可编程顶点处理单元

**PVR** Pilot Voyage Report 驾驶员航行报告

Precision Voltage Reference 精确电压基准

Puerto Vallarta, Jalisco, Mexico ( Airport Code) 墨西哥巴亚尔塔港机场

**PVRT** Production Verification and Reliability Tests 生产验证与可靠性试验

**PVS** Personal Video System 个人视频系统

Pilot's Vision System 驾驶员观察系统

**PVT** Performance Verification Test 性能验证测试

Pivot 中枢,枢纽

Position Velocity and Time 位置速度和时间

Pressure, Volume, Temperature 压力、体积、温度

Pressure-Volume-Temperature 压力—体积—温度

Private 专用的,私人的

Production Validation Test 生产验证测试

Position, Velocity Time 位置、速度、时间

**PVU** Precision Velocity Update 精密速度更新

**PVW** Planned Value of Work 计划的工作量

**PW** Passenger Weight 旅客重量(体重)

Passing Window 通道窗

Peak Wavelength 峰值波长

Pivoted Windows 枢轴窗口

Plain Washer 平垫圈

Primary Winding 一次线圈

Printed Wire 印刷线路

Projected Windows 投影窗口

Pulse Width 脉冲宽度

**PW(P&W)** Pratt and Whitney 普拉特·惠特尼发动机公司

**PWA** Pacific Western Airlines, Ltd. ( Cana-

da）太平洋西部航空公司（加拿大）
Printed Wire Assembly 印刷线路组件
Program Work Authorization 程序工作批准
Project Work Authorization 项目工作批准

**PWAC** PWM to Analog Convertor PWM 到模拟转换器

**PWB** Printed Wiring Board 印刷线路板

**PWC** Process Water Cooler 处理水冷机
Pulse Width Coding 脉冲宽度编码

**PWC（P&WC）** Pratt and Whitney Aircraft of Canada 加拿大普惠公司

**PWCM** Pulse Width with Carrier Modulation 脉冲宽度载波调制

**PWCT** Passenger will Contact 旅客将联系

**PWCU** Potable Water Control Unit 饮用水控制单元

**PWD** Powder 火药
Predicted Wind Data 预测风数据
Pulse Width Discriminator 脉冲宽度鉴别器
Pulse Width Distortion 脉宽失真

**PWDC** PWM to Digital Convertor PWM 到数字转换器

**PWE** Pulse Width Encoder 脉冲宽度编码器

**PWI** Participation with Industry 工业参与
Pilot Warning Indicator 飞行员告警指示器
Power Warning Indicator 电源告警指示器
Proximity Warning Indicator 避撞报警显示器

**PWIP** Potable Water Indication Panel 饮用水指示面板

**PWK** Chicago, IL, USA-Pal-Waukee （Airport Code）美国伊利诺伊州芝加哥帕尔沃基机场

**PWM** Plated Wire Memory 平板线存储器
Portland, ME, USA （Airport Code）美国缅因州波特兰机场
Processor Write Monitor 处理器写入监视器
Pulse Wavelength Modulation 脉冲波长调制
Pulse Width Modulation 脉冲宽度调制
Pulse Width Modulator 脉冲宽度调制器

**PWM-AF** Pulse Width Modulation Audio Frequency 脉宽调制音频

**PWM-FM** Pulse Width Modulated Frequency Modulation 脉宽调制调频

**PWMM** Plated Wire Memory Modules 平板线存储模件

**PWO** Paper Work Only 仅为文字工作

**PWR** Power 动力,电源,电力,权力
Pressurized Water Reactor 加压水反应堆

**PWRH** Powerhouse 发电厂

**PWRP** Programme on Weather Prediction Research 天气预报研究计划

**PWS** Power Supply 电力供应
Predictive Windshear 可预测风切变
Predictive Windshear System 风切变预警系统
Private Wire Service 专用线业务
Program Work Statement 程序工作说明
Proximity Warning System 近地警告系统
Public Water Source 公共水资源
Public Weather Service 公众天气服务

**PWSP** Potable Water Service Panel 饮用水服务面板

**PWT** Passive Warning and Tracking 无源警告与跟踪
Pilot-Warning Instruments 驾驶员报警仪器
Pressure, Winds, Temperature 压力、风、温度
Propulsion Wind Tunnel 推进装置（试验）风洞

**PWTR** Pewter 锡铅合金
Pure Water 纯净水

**PX** Piston Engine 活塞式发动机
Private Exchange 专用交换机

**PXACT** Passenger Account 旅客（银行）账目

**PXO** Porto Santo, Madeira Islands, Portugal （Airport Code）葡萄牙圣港机场

**PXY** Pattaya, Thailand （Airport Code）泰国帕塔亚机场

**PYGN** Pyrogen Unit 焦化组件

**PYL** Pylon 吊架,标塔,支架

**PYLD** Payload 商载,有效负载

**PYR** Pyramid 金字塔,四面体

**PYROM** Pyrometer 高温计

# Q

**Q** Pitch Rate 俯仰率

Quadrature 正交,90°角

Quality 质量

Quantity 数量

Queue or Queuing 行列,队列或排队

**Q&A** Question and Answer 问题与答案

**Q&RA** Quality and Reliability Assurance 质量与可靠性保证

**Q/D** Quick Disconnect 快速脱开

**QA** Q-Adapter Q 适配器

Qantas Airways 澳大利亚快达航空公司

Quality Analysis 质量分析

Quality Analyst 质量分析师

Quality Assessment 质量评价

Quality Assurance 质量保证

Queue Access 队列存取

**QA/QC** Quality Assurance and Quality Control 质量保证与控制

**QAA** Quality Assessment Agency 质量评估机构

Quality Assurance Analyst 质量保证分析师

Quality Assurance Assessment 质量保证评价

Quality Assurance Assistant 辅助的质量保证

Quality Assurance Audit 质量保证审计

Questions and Answers 问题与解答

**QAAR** Quality Assurance Annual Report 质量保证年度报告

**QAC** Quality Air Check 空气质量检查

Quality Air Control 空气质量控制

Quality Assessment and Control 质量评估与控制

Quality Assurance Checklist 质量保证一览表

Quality Assurance Code 质量保证码

Quality Assurance Committee 质量保证委员会

**QACAA** Quality Assurance Configuration Accountability Auditor 质量保证组合责任审计员

**QACR** Quality Assurance Change Request 质量保证更改申请

**QAD** Quality Assurance Department 质量保证部

Quality Assurance Directive 质量保证指令

Quality Audit Division 质量审计处

Quick Attach/Detach 快速装卸,快速拆卸

Quick Attach/Detach Adaptor 快速拆装适配器

**QADI** Quality Assurance Department Instructions 质量保证部细则,质量保证部的指示

**QADS** Quality Assurance Data System 质量保证数据系统

**QAE** Quality Assurance Engineer 质量保证工程师

Quality Assurance Engineering 质量保证工程

Quality Assurance Evaluator 质量保证鉴定员

**QAET** Quality Assurance Environment Testing 质量保证环境试验

Quality Assurance Evaluation Test 质量保证鉴定试验

**QAF** Q Adapter Function Q 适配器功能

Quality Achievement Factor 质量成就因素

Quality Assessment Framework 质量评估框架

Quality Assurance Function 质量保证的功能

**QAFM** Quality Assurance Forms Manual 质量保证表手册

**QAFTS** Quality Assurance Functional Test

System 质量保证功能测试系统

**QAI** Quality Assurance Input 质量保证输入
Quality Assurance Instruction 质量保证指令
Quality Assurance International 国际质量保证
Quality Assurance Item 质量保证项目
Quality Auditing Institute 质量审计研究所

**QAIP** Quality Assurance Inspection Procedure 质量保证检查程序

**QAIR** Quality Assurance Inspection Requirements 质量保证检查要求

**QAK** Quick Action Keyboard 快速反应键盘

**QAL** Quality Assurance Laboratory 质量保证实验室
Quartz Aircraft Lamp 石英航行灯

**QALL** Quartz Aircraft Landing Lamp 飞机着陆石英灯

**QALTR** Quality Assurance Laboratory Test Request 质量保证实验测试请求

**QAM** Quadrature Amplitude Modulation 正交调幅，正交双边带调制
Quality Assurance Management 质量保证管理
Quality Assurance Manager 质量保证经理
Quality Assurance Manual 质量保证手册
Quality Assurance Monitor 质量保证监视器
Queued Access Method 排队存取法

**QAMP** Quality Assurance Management Plan 质量保证管理计划

**QANTAS** Queensland and Northern Territory Aerial Services Ltd. 昆士兰与北领地航空服务公司

**QAO** Quality Assurance Office 质量保证处
Quality Assurance Operation 质量保证操作
Quality Assurance Outline 质量保证纲要

**QAOP** Quality Assurance Operating Procedure 质量保证运行程序

**QAP** Quality Assurance Plan 质量保证计划
Quality Assurance Procedures 质量保证程序
Quality Assurance Program 质量保证计划
Quality Assurance Provisions 质量保证条款，质量保证规定

**QAPM** Quality Assurance Project Manager 质量保证项目经理

**QAPO** Quick Acquisition Purchase Order 快速导购规程

**QAPS** Quick Acquisition Purge System 快速搜索清除系统

**QAR** Quality Assurance Reliability 质量保证可靠性
Quality Assurance Report 质量保证报告
Quick Access Recorder 快速存储记录器

**QAR&D** Quality Assurance Research and Development 质量保证研究与开发

**QARF** Quick Access Recorder Function 快速存储记录器功能

**QARM** Quality Assurance and Risk Management 质量保证与风险管理

**QARR** Quality Assurance Reporting Requirements 质量保证报告要求

**QARS** Quality Assurance Record Storage 质量保证记录存储

**QAS** Quality Assurance Service 质量保证服务
Quality Assurance Specialist 质量保证专业人员
Quality Assurance Standard 质量保证标准
Quality Assurance Support 质量保证支援
Quality Assurance System 质量保证体系

**QASAR** Quality Assurance System Analysis Review 质量保证系统分析评审

**QASK** Quadrature Amplitude Shift Keying 正交幅移键控

**QASP** Quality Assurance Service Procedure 质量保证服务程序
Quality Assurance Support Plan 质量保证支援计划
Quality Assurance Surveillance Plan 质量保证监视计划

**QASS** Quality Assurance Standards for Suppliers 供应商质量保证标准

**QAST** Quality Assurance Service Test(s) 质

量保证服务测试

**QAT** Quality Assurance Technical 质量保证技术

Quatar 卡塔尔

**QATP** Quality Assurance Technical Publication 质量保证技术公布

Quality Assurance Test Procedure 质量保证测试程序

Quality Assurance Testing Plan 质量保证测试计划

**QATS** Quality Assurance and Test Services 质量保证与测试服务

Quality Assurance Test System 质量保证测试系统

Quality Audit Tracking System 质量审计跟踪系统

**QAVC** Quiet Automatic Volume Control 静噪声自动音量控制

**QB** Quick Break 快速中断,快速断路器

**QBE** Query by Example 按例查询

**QBI** Compulsory IFR Flight 强制的仪表飞行规则飞行

**QC** Quality Check 质量检查

Quality Control 质量控制

Quick Change 快速更改

Quick Cleaning 快速清除(法)

**QC&T** Quality Control and Techniques 质量控制与技术

Quality Control and Test 质量控制与试验

**QCA** Quality Control Analysis 质量控制分析

Quick Change Assembly 快速变化组件

**QCAT** Quality Control Analytical Technology 质量控制分析技术

**QCB** Quality Control Board 质量控制委员会

**QCC** Quality Control Center 质量控制中心

Quality Control Circle 质量控制循环

**QCCAA** Quality Control Configuration Accountability Auditor 质量控制组合责任审查员

**QCCR** Quality Control Change Request 质量控制更改申请

**QCD** Quality Control Data 质量控制数据

Quality Control Directive 质量控制指令

Quality Control Division 质量控制处

Quality Cost and Delivery 质量成本和交货

**QCDR** Quality Control Deficiency Report 质量管理缺陷报告

Quality Cost Data Record 质量成本数据记录

**QCDS** Quality Control and Distribution System 质量控制与分配系统

**QCDSM** Quality, Cost, Delivery, Safety and Morale 质量、成本、交货、安全与纪律

**QCE** Quality Control Engineer 质量管理工程师

Quality Control Engineering 质量控制工程

**QCERS** Quality Control Electronics Reliability System 质量控制电子设备可靠性系统

**QCFM** Quality Control Forms Manual 质量控制表手册

**QCFTS** Quality Control Functional Test System 质量控制功能试验系统

**QCG** Quality Control Group 质量管理小组

**QCGAT** Quiet, Clean General Aviation Turbofan 低噪声、低污染通用航空涡轮风扇发动机

**QCI** Quality Conformance Inspection 质量合格检查

Quality Control Information 质量控制信息

Quality Control Input 质量控制输入

Quality Control Item 质量控制项目

**QCIP** Quality Control Inspection Procedure 质量控制检查程序

**QCL** Quality Characteristics List 质量特性目录

Quality Check List 质量检查清单

Quality Control Laboratory 质量控制实验室

Quality Control Level 质量管理等级

**QCM** Quality Chain Management 质量链管理

Quality Control Manual 质量控制手册

**QCO** Quality Control Officer 质量控制(官)

员

Quality Control Organization 质量控制组织

**QCOP** Quality Control Operating Procedure 质量控制作业程序

**QCP** Quality Check Program 质量检查计划

Quality Control Procedure 质量控制程序

**QCPM** Quality Control Procedures Manual 质量控制程序手册

**QCPP** Quality Control Program Plan 质量控制程序规划

**QCR** Quality Control Reliability 质量控制可靠性

Quality Control Report 质量控制报告

Quality Control Representative 质量控制代表

Quality Control Review 质量控制评审

Quality Control Room 质量控制室

**QCR&D** Quality Control Research and Development 质量控制研究与开发

**QCRS** Quality Control Record Storage 质量控制记录存储

**QCS** Quality Control Schedule 质量控制计划

Quality Control Specification 质量控制规范

Quality Control Standard 质量控制标准

Quality Control Support 质量控制保障

Quality Control System 质量控制系统

Quality Cost System 质量成本系统

Query Control Station 询问控制台

Quick Charging System 快速充电系统

**QCSEE** Quiet, Clean Shorthaul Experimental Engine 安静、清洁的短程试验性发动机

**QCSMF** Quadruple Cladding Single Mode Fiber 四包层单模光纤

**QCT** Quality Control Team 质量控制组

Quality Control Technology 质量管理技术，质量控制技术

**QCTE** Quality Control Test Engineering 质量控制试验工程

**QCTIR** Quality Control Test and Inspection Report 质量控制试验与检查报告

**QCTR** Quality Control Test Report 质量管理

测试报告

**QD** Quadrant Depression 俯角

**QDACS** Quantized Data Attitude-Control System 量化数据姿态控制系统

**QDC** Quick Dependable Communication 快速可靠通信

Quick Disconnect Coupler 快速断开耦合器

**QDF** Quality Data Form 质量数据表

**QDH** Quick-Disconnect Handle 快速脱开手柄

**QDISC** Quick Disconnect 快速脱开

**QDM** Magnetic Bearing to Facility 向台磁方位

Magnetic Heading 磁航向

Magnetic Heading to Facility 向台磁航向

Quality Decision Management 质量决策管理

Quantitative Debris Measurement 定量磨粒管理

Quantitative Debris Monitor 定量磨粒监测器

Quantitative Debris Monitoring 定量磨粒监测

**QDP** Qualification Data Package 鉴定数据包

**QDR** Quality Data Retrieval 质量数据检索

Quality Deficiency Report 质量缺陷报告

Quality Design Review 质量设计评审

Quality Deviation Report 质量偏差报告

**QDRNT** Quadrant 象限，扇形轮

**QDRTR** Quadrature 正交

**QDS** Quality Data System 质量数据系统

**QDT** Quadrantal 象限的，扇形的

**QE** Quality Evaluation 质量评估

**QEAV** Quick Exhaust Air Valve 快速排气活门(阀)

**QEC** Quadrantal Error Correction 象限误差修正

Quick Engine Change 快速发动机更换

**QECS** Quick Engine Change Stand 快速发动机更换台

**QECU** Quick Engine Change Unit 快速更换发动机装置

**QEEL** Quality Evaluation and Engineering

Laboratory 质量鉴定与工程实验室

**QEG** Quality Evaluation Group 质量评估小组

**QEL** Quality Evaluation Laboratory 质量评估实验室

**QEO** Quality Engineering Operations 质量工程操作

**QEP** Quality Enhancement Program 质量提高计划

Quality Evaluation Program 质量鉴定计划

Quality Examination Program 质量检查计划

**QESCP** Quality Engineering Significant Control Points 质量工程重要控制点

**QEST** Quality Evaluation System Test 质量评估体系测试

**QEV** Quick Exhaust Valve 快速排气阀

**QF** Quality Factor 质量因子

Quick Firing 快速点火

**QFD** Quality Function Deployment 质量功能展开

**QFE** Altimeter Setting above Station 定位在机场上的高度表

Atmospheric Pressure at Field Elevation 场面气压

Field Elevation Atmospheric Pressure 机场标高气压

Pressure at the Airport Reference Point 机场基准点的气压,场压

Quiet Fuel Efficient 低噪声并节油的

**QFF** Atmospheric Pressure Converted to MSL 换算到平均海平面的大气压力

**QFIRC** Quick-Fix Interference-Reduction Capability 固定降干扰的性能

**QFRP** Quartz Fiber Reinforced Plastic 石英纤维增强塑料

**QFT** Quantitative Feedback Theory 定量反馈理论

**QFU** Runway Heading 跑道航向

**QHSE** Quality, Health, Safety and Environment 质量、健康、安全与环境

**QI** Quality Improvement 质量改进

Quality Increase 质量提高

Quality Index 质量指标

Quantity Indicator 数量指示器,数量显示器

**QIAW** Qualitative Impact Assessment Worksheet 定性效果鉴定工作单

**QIC** Quality Information Center 质量信息中心

Quality Information Checklist 质量信息清单

**QICC** Quality Instrument Calibration-Certification 质量仪表校准鉴定(书)

**QIE** Qualified International Executive 国际管理者资格

**QIEW** Qualitative Impact Estimate Worksheet 定性效果计算工作单

**QIK** Quick 快速(的)

**QIM** Quality Inspection Management 质量检验管理

**QIP** Quality Improvement Program (System) 质量改进程序(系统)

Quality Indicator Program 质量指示器程序

Quality Inspection Point 质量检查点

**QIS** Quality Information System 质量信息系统

**QIT** Qualification Information and Test 合格信息与测试

Quality Improvement Team 质量改进小组

Quality Information and Test 质量信息与测试

**QITS** Quality Information and Test System 质量信息和测试系统

**QJC** Yanbu, Bhutan (Airport Code) 不丹廷布机场

**QL** Quick Look 快速检查,粗查

**QLAP** Quick Look Analysis Program 快速查看分析程序

**QLC** Quantum Leasing, Corp. (USA) 量子租机公司(美国)

**QLDA** Quick Look Data Area 快速检查数据区

**QLFD** Qualified 合格的,能胜任的

**QLFR** Qualifier 鉴定员,合格者

**QLFY** Qualify 适合,合格

**QLIT** Quick Look Intermediate Tape 粗查中间带

**QLL** Quartz Landing Lamp 石英着陆灯

**QLLC** Qualified Logical Link Control 证实逻辑链路控制

**QLM** Quasi-Laser Machine 准激光器

**QLML** Quick Look Measurement List 快速查阅测量数据表

**QLT** Quality Leadership Team 质量领导小组

Quantitative Leak Test 定量的检漏测试

Queue Length Threshold 队列长度阈值

**QLTY** Quality 质量

**QM** Quadrature Modulation 正交调制

Quality Management 质量管理

Quality Manual 质量手册

Quality Memorandum 质量备忘录

**QMB** Quality Management Board 质量管理委员会

**QMC** Quality Management Center 质量管理中心

Quality Monitoring and Control 质量监测和控制

**QMCS** Quality Management and Control System 质量管理与控制系统

Quality Monitoring and Control Subsystem/System 质量监控子系统/系统

**QMF** Quadrature Mirror Filter 正交镜像滤波器

Query Management Facility 查询管理设备

**QMH** Queue Message Handling 队列信息处理

**QMI** Qualification Maintainability Inspection 合格维修性检验

Quality and Management Innovation 质量与管理改革

**QML** Qualified Manufacturer's List 合格制造商清单

Qualified Materials List 合格材料清单

**QMP** Quality Management Plan 质量管理计划

Quality Monitoring Program 质量监控程序

**QMQB** Quick Make, Quick Break 快速闭合,快速断开

**QMR** Quality Management Review 质量管理评审

Quality Monitoring and Recording 质量监控与记录

**QMS** Quality Management System 质量管理系统

**QNCATC** Queen Noor Civil Aviation Technical College 努尔皇后民用航空技术学院(约旦)

**QNCH** Quenched 淬火的,硬化的,抑制的

**QNH** Atmospheric Pressure at Nautical Height 修正海平面气压

Destination Altimeter Setting 终点高度表设定

**QNM** Queueing Network Model 排队网络模型

**QNN** Quantum Neural Network 量子神经网络

**QNS** Queue Network System 行列网络系统

**QNTY** Quantity 数量

**QOD** Quick-Opening Device 快速开启装置

**QOE** Quality of Experience 体验质量

**QOP** Quality Operating Procedures 质量操作程序

**QOR** Qualitative Operational Requirement 定性操作要求

Quality Operational Requirement 质量操作要求

Quarterly Operating Report 季度工作报告

**QOS** Quality of Service 服务质量

**QOSF** Quality of Service Forum 服务质量论坛

**QP** Qualification Plan 审定计划

Quality Planning 质量策划

Quality Policy 质量方针

Quality Procedure 质量程序

**QPA** Quantity per Application 每次施用数量

Quantity per Assembly 每次装配数量

**QPC** Quality and Process Control 质量和过程控制

Quasi Propulsive Coefficient 准推进系数

**QPI** Quality Program Instruction 合格计划指令

**QPIS** Quality Performance Instruction Sheet 质量性能指示表

Quality Planning Instruction Sheet 质量计划说明书

**QPL** Qualified Parts List 合格零件清单

Qualified Processor List 合格处理器清单

Qualified Products List 合格产品清单

Quality Parts List 优质零件清单

Quality Products List 优质产品清单

**QPM** Quality and Performance Management 质量与性能管理

Quality Project Management 质量项目管理

**QPP** Quality Program Provision 质量计划措施

Quality Protection Program 质量检定保护规划

Quantized Pulse Position 量化脉冲位置

**QPR** Qualitative Performance Requirements 质量性能要求

Qualitative Personnel Requirements 合格人员要求

Quality Performance Report 质量性能报告

Quality Progress Report 质量改进报告

Quantity Progress Report 数量进度报告

Quarterly Progress Report 季度进度报告

Quarterly Progress Review 季度进度审核

**QPS** Quick Plot System 快速绘图系统

Quick Program Search 快速程序搜索

**QPSK** Quadrature Phase Shift Keying 正交相移键控(四相移相键控)

**QPT** Quality per Task 每项任务质量

**QQP** Quick Query Program 快速询问程序

**QQPRI** Qualitative and Quantitative Personnel Requirements Information 定性与定量人员要求信息

**QR** Qualification Report 审定报告

Qualitative Research 定性研究

Quantitative Research 定量研究

Quick Response 快速反应

Quote Required 要求引用,要求报价

**QRA** Quality and Reliability Assurance 质量与可靠性保证

Quality Reliability Assurance 质量可靠性保证

Quantitative Risk Assessment 风险量化评估

Quick Reaction Aircraft 快速反应飞机

Quick Reaction Alert 快速反应警戒

Quick Reaction Antenna 快速反应天线

Quick Reaction Area 快速反应区

Quick Reaction Assessment 快速反应评估

**QRAC** Quality and Reliability Assessment Council 质量和可靠性评定委员会

**QRC** Quick Reaction Capability 快速反应能力

Quick Reaction Communications 快速反应通信

Quick Reaction Contract 快速反应合同

**QRCC** Quadripartite Research Coordination Committee 四方研究协调委员会

Query Response Communications Console 咨询通信控制台

**QRCD** Qualitative Reliability Consumption Data 质量可靠性消耗数据

**QRCG** Quasi-Random Code Generator 伪随机码产生器

**QRDN** Quality Requirement Discrepancy Notice 质量不符合要求的通知

**QRE** Quality Responsible Engineer 质量责任工程师

**QRF** Quick Reaction File 快速反应档案

Quick Review File 快速检查档案

**QRG** Quick Response Graphic 快速反应图表

**QRH** Quick Reference Handbook 快速参考手册

**QRI** Qualitative Requirements Information 质量要求资料

Quick-Reaction Interceptor 发动机快速更换组，快速反应截击机

**QRL** Quick Reference List 快速参考清单

**QRLY** Quarterly 季度的

**QRM** Quality Resource Management 优质资源管理

Quality Review Meeting 质量评审会议

Quantitative Research Methods 定量研究方法

Quantitative Risk Management 定量风险管理

Quick Release Mechanism 快速脱开机构

Quick Response Manufacturing 快速响应制造

**QRP** Low Power Transmitter 低功率发射机

QOS Reference Point 服务质量参考点

Qualified Recycling Program 合格再生计划

Qualify Review Procedure 质量评审程序

Query and Reporting Processor 询问与报告处理机

Quick Reaction Program 快速反应计划

Quick Response Package 快速响应组件

**QRR** Quality Readiness Review 启用前质量检查

Quarterly Requisition Report 季度征用报告

**QRS** Qualification Review Sheet 资格审核表

Quick Reaction Support 快速反应支援

**QRT** Quick Reaction Task 快速反应任务

Quick Reaction Team 快速反应小组

Quick Response Team 快速反应分队

Quiet Radio Transmission 静噪无线电传输

**QRV** Quick-Release Valve 快速泄放活门

**QS** Quality Standard 质量标准

Quasi-Synchronous 准同步

Query System 询问系统

**QSA** Qualitative Safety Analysis 定性安全性分析

Quality Safety Audit 质量安全审计

Quality System Assessment 质量体系评估，质量体系评定

Quantitative Safety Analysis 定量安全性分析

**QSAM** Queued Sequential Access Method 序列顺序存取方法

**QSD** Quality Safety Distance 质量安全距离

Quality Surveillance Division 质量监督部门

Quality System Database 质量系统数据库

Quality System Design 质量系统设计

**QSI** Quality Service Index 服务质量指数

**QSIC** Quality Standard Inspection Criteria 质量标准检查准则

**QSIP** Quality System Implementation Procedure 质量系统执行程序

**QSL** Qualification Status List 合格状态清单

**QSM** Quality Systems Management 系统质量管理

**QSN** Question 问题

**QSP** Quality Search Procedure 质量调查程序

Quality Surveillance Plan 质量监视计划

Quality System Procedure 质量体系流程

Quiet Supersonic Platform 低噪声超声速平台

**QSR** Qualified Status Report 审定的状态报告

Quality Survey Report 质量调查报告

Quality System Review 系统质量检查

Quarterly Statistical Report 季度统计报告

Quarterly Status Report 季度状况报告

Quarterly Summary Report 季度综合报告

**QSRA** Quiet Short-haul Research Aircraft 安静短程研究飞机

**QSSR** Quarterly Specialized Safety Report 季度专用安全报告

Quarterly Stock Status Report 季度库存状况报告

**QSST** Quiet Small Supersonic Transport 安静小型超音速运输机

Quiet Supersonic Transport 安静超音速运输机

**QSTN** Question 问题

**QSTOL** Quiet Short Take-off and Landing 低噪声短距离起降

**QT** Qualification Test 合格试验，鉴定试验

Quality Team 质量团队

Quart 夸脱(液体计量单位)

**QTA** Qualification Test Approval 鉴定试验批准

Quality Test Approval 质量试验批准

Quota 限额，份额

**QTAI** Qualification Test Article Inspection 鉴定试验物品检查

**QTE** Qualified Technological Equipment 高技术设备租赁，合格的技术设备

**QTG** Qualification Test Guide 鉴定测试指南

**QTL** Quasi-Transmission Line 准传输线

**QTM** Quality Technical Memorandum 质量技术备忘录

**QTP** Qualification Test Plan 合格试验计划，适用性试验计划

Qualification Test Procedure 合格试验程序，鉴定测试程序

Qualification Training Package 合格训练一揽子计划

Quality Test Plan 质量试验计划，质量检验计划

**QTPS** Quality Test Preparation Sheet 质量测试准备单

**QTR** Quad Tilt Rotor 四倾转旋翼机

Qualification Test Report 合格测试报告

Quality Technical Requirement 质量技术要求

Quality Test Report 质量测试报告

Quarterly Technical Report 季度技术报告

Quiet Tail Rotor 低噪声尾桨

**QTS** Qualification Test Specification 鉴定试验规范

Qualified Technology Support 合格的技术支持

Quality Technology Services 优质的技术服务

Quality Tracking System 质量跟踪系统

**QTSS** Qualification Test Summary Sheet 鉴定试验汇总单

**QTY** Quantity 量，数量

**QTZ** Quartz 石英，水晶

**QUAD** Quadrangle 四边形，四角形

Quadrant 象限，扇形板

Quadrature 正交，90°相移

**QUADR** Quadruple 四联的

**QUAL** Qualification，Qualifying 合格性，合格证明，有资格的

Qualified 合格的，适当的，限定的

Qualitative 定性的，合格的，质量的

Quality 质量

**QUANT** Quantitative 定量的

**QUAR** Quarantine 检疫

**QUAT** Quaternary 四元的，四进制的

**QUE** Call Queueing 呼叫排队

**QUEPASA** Quick User Entry-Performance Analysis and Status Assessment 用户快速进入—性能分析状态评定

**QUES(T)** Question 问题

**QUIN** Quintuple 五倍的，五重的，五联的，五元组

**QUMR** Quality Unsatisfactory Management Report 质量不合格管理报告

**QUOT** Quotation 报价，引证，引文

Quotient 商，率，份额

**QUT** Quit 退出

Utsunomiya，Japan(Airport Code) 日本宇都宫机场

**QVI** Quality Verification Inspection 质量确认检查

Quasi-Vertical Incidence 准垂直迎角

**QVL** Qualified Vendor List 合格供应商清单

**QVT** Quality Verification Testing 质量验证测试

**QVU** Vaduzvaduz，Liechtenstein ( Airport Code)列支敦士登瓦杜兹机场

**QZSS** Quasi-Zenith Satellite System 准天顶卫星系统

# R

**R**   Radar 雷达

Radio 无线电

Radius 半径,径向

Radius of Turn 转弯半径

Rail 轨道,钢轨,横杆

Range 范围,测距

Rankine 兰氏

Rapid 快速的,急剧的

Ratio 比率,系数

Recline 向后靠,靠背

Red 红色

Relay 继电器,中继

Reliability 可靠性

Remarks 备注,要点,说明

Repeat 重复

Replaceable 可更换的

Resistance 电阻

Resistor 电阻器

Revision 修改,改型,改装

Right (Runway) 向右,正确的,右(跑道)

Ring 环

Riser 垂直管,升降机,竖管

Roentgen 伦琴

Roll 滚转

Rubber(Sheath) 橡胶(封皮)

Universal Gas Constant 通用气体常数

Yaw Rate 偏航率

**R&A**   Records and Accountability 记录与责任

Reliability and Availability 可靠性与可用性

**R&D**   Research and Development 研究与发展

**R&DDD**   Research and Development Degree of Difficulty 研究和发展困难度

**R&E**   Research and Engineering 研究与工程

**R&ED**   Research and Engineering Division 研究与工程部

**R&I**   Removal and Installation 拆卸与安装

**R&M**   Reliability and Maintainability 可靠性与可维修性

Repair and Maintenance 修理与维修

**R&P**   Rules and Procedures 规则与程序

**R&R**   Remove and Replace 拆除与更换

**R,E&D**   Research, Engineering and Development 研究、工程和开发

**R.M.**   Refresh Memory 刷新存储器

**R.R.**   Rolls Royce 罗·罗公司

**R/B**   Red/Black 红/黑

**R/C**   Radio Control 无线电控制

Rate of Climb 爬升率

Ratio Changes 比率变换器

Receiving Card 接收电路卡

Record Change 改变记录

Round Convergence 圆形收敛

Round Convergent 圆形收敛

**R/D**   Radial/Distance 半径/距离

Rate of Descent 下降率

Receiver/Decoder 接收机/译码器

Required Data 要求数据

Required Date 要求日期

**R/E**   Receiver Exciter 接收机(器)激励器

**R/F/W**   Ready for Work 准备就绪

**R/G**   Rate Gyro 速率陀螺

**R/H**   Radar Height 雷达高度

**R/HR**   Roentgens per Hour 伦琴/小时

**R/I**   Radio/Inertial 无线电/惯性

Removal/Installation 拆/装

**R/L**   Reading Light 阅读灯

Red Label 红色标签

**R/M**   Reliability/Maintainability 可靠性/可维修性(维修能力)

**r/min**  Revolutions per Minute 每分钟转数

**R/O**  Rollout 滑跑

**R/P**  Record/Playback 录制/回放

**R/R**  Readout and Relay 读出与中继传送

**R/RFS**  Repeater and Radio Frequency Source 转发器与无线电频率源

**R/S**  Rack Service 支架维修,导轨维修

　　Revolutions per Second 转/秒

**R/T**  Radio Telephone 无线电话

　　Radio Transmit 无线电发射

　　Rate of Turn 转弯速度,回转速率

　　Receive/Transmit 接收/发送

　　Receiver/Transmitter 收发机

　　Research and Technology 研究与技术

　　Reverse Thrust 反推

　　Room Temperature 室温

**R/W**  Read/Write 读/写

**R/WI**  Read/Write Interface 读/写接口

**RA**  Radar Altimeter 雷达高度表

　　Radial Array 径向矩阵

　　Radio 无线电

　　Radio Altimeter 无线电高度表

　　Radio Altitude 无线电高度

　　Rain Area 雨区

　　Random Access 随机存取

　　Random Assignment 随机分配

　　Rapid Assessment 快速评价

　　Read Amplifier 读取放大器

　　Real Address 实地址

　　Reconnaissance Aircraft 侦察机

　　Reduction of Area 减小面积

　　Redundancy Assurance 余度保障

　　Registration Authority 登记中心

　　Registration Authorization 登记授权

　　Reliability Analysis 可靠性分析

　　Remote Access 远地存取

　　Resolution Advisory 决断咨询,解脱咨询

　　Resolution Authority 决断权限,决断根据

　　Return Alarm 返回告警

　　Risk Analysis 风险分析

　　Risk Assessment 风险评估

Robot Aircraft 无人驾驶飞机

Roll Axis 横滚轴

Route Analysis 路由分析

Routing Area 路由区

Runway/Final Approach 跑道进近,最后进近

**RAA**  Radar Aircraft Altitude 雷达测定的飞机高度

　　Radar Altimeter Antenna 雷达高度表天线

　　Regional Airline Association 支线航空公司协会,地区航空公司协会

　　Relocation Authorization and Agreement 重新配置批准与协议

**RAACT**  Radioactive 放射性的

**RAAG**  British Aerospace Aircraft Group 英国宇航飞机集团

**RAAP**  Risk Analysis and Abatement Plan 风险分析与降低计划

**RAAS**  Regional Area Augmentation System 区域面积增强系统

　　Remote Access Aviation System 远程访问航空系统

　　Risk Assessment and Application System 风险评估与应用系统

　　Runway Awareness and Advisory System 跑道感知与咨询系统

**RAB**  Rabaul, Papua New Guinea (Airport Code) 巴布亚新几内亚拉包尔机场

　　Rabbet 插孔,插座,槽口,嵌接

**RABS**  Reverse Actuated Bleed System 反推制动放气系统

**RAC**  Race Aviation Corp. (USA) 瑞斯航空公司(美国)

　　Radar Approach Control 雷达进近管制

　　Radiometric Area Correlator 辐射测定区域相关器

　　Random Access Control 随机存取控制器

　　Rapid Action Change 快速措施更改

　　Regional Airspace Coordinator 地区空域协调员

　　Reliability Analysis Center 可靠性分析中心

Responsibility Area Code 响应性区域码
Risk Assessment Council 风险评估委员会
Rotor Active Clearance 转子作动间隙

**RACAF** Radio Access Control Agent Function 无线接入控制代理功能

**RACC** Radar Approach Control Center 雷达进近管制中心
Radiation and Contamination Control 放射与污染控制
Reporting Activity Control Card 报告活动控制卡
Research Aviation Coordinating Committee 航空研究协调委员会
Rotor Active Clearance Control 转子主动间隙控制

**RACCA** Regional Air Cargo Carriers Association 支线航空货物运输协会

**RACCS** Rotor Active Clearance Control System 转子实际间隙控制系统

**RACD** Radar Altimeter Control Display 雷达高度表控制显示器

**RACE** Rapid Automatic Checkout Equipment 快速自动检测设备
Remote Access to Customer Equipment 用户设备的远程接入
Request Altitude Change Enroute 途中请求改变高度

**RACES** Radio Altimeter Civil Emergency Service 无线电高度表民用应急服务

**RACF** Resource Access Control Facility 资源接入控制设备
Resource Allocation Control Facility 资源分配与控制设备

**RACGAT** Russian American Coordinating Group for Air Traffic 空中交通管制俄/美协调组

**RACH** Random Access Channel 随机存取信道

**R-ACH** Reverse Access Channel 反向接入信道

**RACON** Radar Beacon 雷达信标

**RACP** Radar Altimeter Control Panel 雷达高度表控制板

**RACR** Requirements Allocation Change Request 需求配置更改请求

**RACS** Random Access Communications System 随机存取通信系统
Release Activity Collection System 释放效用收集系统
Remote Access Computing System 远程存取计算系统
Remote Automatic Calibration System 远程自动校准系统
Remote Automatic Control System 远距自动控制系统

**RACSB** Rotor Active Clearance Start Bleed 转子作动间隙启动放气

**RACT** Reactivate or Reactivation 复原,复活
Reactivity 反应度,活动性
Reverse Acting 反作用

**RACU** Remote Acquisition and Control Unit 遥控搜索与控制组件

**RAD** Radar 雷达
Radar Approach Aid 雷达进近导航设备
Radial 径向的
Radian 弧度
Radiate, Radiation 辐射
Radiation Absorbed Dose 辐射吸收剂量(拉德)
Radiation Accumulated Dose 累计辐射剂量
Radio 无线电
Radome 雷达天线罩
Ram Air Door 冲压空气门
Random Access Device 随机存储装置
Remote Antenna Driver 远端天线激励器
Request Advise 请通知
Required Availability Data 所需可用性数据
Requirement Action Directive 技术要求实施指令
Requirements and Allocation Document 需求与分配文件
Research and Development 研究与发展

Roll Anticipation Distance 转弯前置距离

Route Availability Document 航路可用文档

Running Average Detection 运行平均探测

**RADA** Random Access Discrete Address 随机存取离散地址

**RADAL** Radio Detection and Location 无线电检测与定位

**RADALT** Radar Altimeter 雷达高度表

Radio Altimeter 无线电高度表

**RADAR** Radio Detection and Ranging 无线电探测与测距,无线电探测与定位(雷达)

**RADAS** Random Access Discrete Address System 随机存取离散地址系统

**RADAT** Radar Data Transmission 雷达数据传输

**RADATA** Radar Automatic Data Transmission Assembly 雷达自动数据传输装置

**RADAY** Radio Day 无线电(工作)日

**RADBAR** Radio Barometer 无线电气压表

**RADC** Right Air Data Computer 右大气数据计算机

**RADCN** Radar Data Converter 雷达数据转换器

**RADDEF** Radiological Defense 辐射防护

**RADEM** Random Access Delta Modulation 随机存取增量调制

Rapid Airplane Drag Estimation Method 飞机阻力快速估算方法

**RADES** Realistic Air Defense Engagement System 仿真防空衔接系统

**RADFAG** Radar Fidelity and Geometry 雷达保真度与几何条件

**RADIAC** Radiation Detection Indicating and Computation 辐射检测指示与计算

**RADINT** Radar Intelligence 雷达情报

**RADISTI** Radar Distance Indicator 雷达距离显示器

**RADL** Radiological 放射性的,辐射性的,放射学的

**RADM** Radioactive Material 放射性材料

**RADMON** Radiological Monitoring 辐射监测

**RADN** Radiation 辐射

**RADNET** Radar Data Network 雷达数据网

**RADOME** Radar Dome 雷达罩

**RADOP** Radaroptical 雷达光学(的)

**RADOT** Real-time Automatic Digital Optical Tracker 实时自动数字光学跟踪仪

Recording Automatic Digital Optical Tracker 数字式自动记录光学跟踪仪

**RADPROP** Radiative Properties 辐射特性

**RADS** Radar Acquisition Data System 雷达采集数据系统

Radar Alphanumeric Display System 雷达字母和数字显示系统

Radar Simulator 雷达模拟器

Rapid Area Distribution Support 地区快速分发支援

**RADSCAT** Radar Scatter Target Facility 雷达散射器目标设备

**RADU** Radar Analysis and Detection Unit 雷达分析与探测装置

Ram Air-Driven Unit 冲压空气驱动装置

**RADVR** Random Access Digital Video Reproducer 随机读取数字录像机

**RADVS** Radar Altimeter and Doppler Velocity Sensor 雷达高度表与多普勒速度传感器

**RAE** Radar Altimeter Element 雷达高度表单元

Ram Air Exit 冲压空气出口

Range, Azimuth, Elevation 距离、方位、仰角

Regional Airworthiness Engineer (加拿大)地区适航工程师

Research Assessment Exercise 研究评估考核

Royal Aeronautical Establishment 皇家航空机构

Royal Aircraft Establishment 皇家飞机研究所(英国)

**RAEN** Radio Amateur Emergency Network 无线电业余爱好者紧急通信网

**RAES** Royal Aeronautical Society (UK) 皇家航空学会(英国)

**RAF** Royal Air Force 皇家空军

**RAFA** Royal Air Force Association 皇家空军协会

**RAFAR** Radio Automatic Facsimile and Reproduction 无线电自动传真与复制

**RAFC** Regional Area Forecast Center 区域（天气）预报中心

**RAFR** Regular Air Freight 普通航空货运

**RAFT** Remote Access Field Terminal 远端接入的现场终端

**RAG** Reliability Analysis Group 可靠性分析组
Runway Arresting Gear 跑道着陆拦阻装置

**RAGU** Redundant Rate Gyro Unit 余度速率陀螺装置

**RAHE** Ram Air Heat Exchanger 冲压空气热交换器

**RAI** Radar Altimeter Indicator 雷达高度表指示器
Radio Altimeter Indicator 无线电高度指示器
Ram Air Inlet 冲压空气进口
Random Access and Inquiry 随机存取与查询
Registro Aeronautico Italiano（Italy） 意大利航空注册局
Remote Alarm Indication 远端告警指示
Runway Alignment Indication 跑道对准指示
Runway Alignment Indicator 跑道对准指示灯

**RAIAM** Random-Access Indestructive Advanced Memory 随机存取不可毁先进存储器

**RAID** Redundant Array of Inexpensive Disks 廉价磁盘冗余阵列

**RAIL** Runway Alignment Indicator Lights 跑道对准指示灯

**RAILS** Runway Alignment Indicator Light System 跑道对正指示器灯光系统

**RAIM** Radio Autonomous Integrity Monitor 无线电自主式完好性监控器

Receiver Autonomous Integrity Monitor 接收机自主完好性监视器
Receiver Autonomous Integrity Monitoring 接收机自主完好性监视

**RAIMS** Radio and Audio Integrating Management System 无线电和音频整合管理系统

**RAIN** Regional Access Information Network 地区接入信息网

**RAIS** Rail Air International Service 国际铁路与航空服务处
Range Automated Information System 自动测距信息系统
Risk Assessment Information System 风险评估信息系统

**RAJ** Rajkot, India（Airport Code） 印度拉杰果德机场

**RAK** Marrakech, Morocco（Airport Code） 摩洛哥马拉喀什机场

**RALACS** Radar Altimeter Low Altitude Control System 雷达高度表低空控制系统

**RALS** Remote Area Landing System 遥控区着陆系统
Remote Augmented Lift System 遥控增升系统

**RALT** Radar Altitude 雷达高度
Radio Altitude 无线电高度

**RALTU** Reprogrammable Annunciator Light Test Unit 程序再编信号灯光测试组件

**RAM** Radar Absorbing Material 雷达吸收材料
Radiation Attenuation Measurement 辐射衰减测量
Random Access Memory 随机访问存储器
Rapid Area Maintenance 快速区域维修
Rapid Assessment Mode 快速评估方式
Reliability and Applications Module 可靠性和应用模块
Reliability and Maintainability 可靠性和可维护性
Reliability, Availability and Maintainability 可靠性、可用性与可维修性

Research and Applications Module 研究和应用模块

Research Aviation Medicine 航空医学研究

Responsibility Assignment Matrix 响应性测定矩阵，响应性分配矩阵

Revenue Accounting Manual 收入会计手册

Risk Assessment Methodology 风险评估方法

**RAMAC** Random Access Memory Accountability and Control 随机存取存储器责任与控制

Random Access Memory Accounting Computer 随机存取存储器会计计算机

Random Access Method of Accounting and Control 计算与控制的随机存取法

**RAMD** Random Access Memory Device 随机存取存储器

Receiving Agency Material Division 接收机构器材处

Reliability, Availability, Maintainability and Durability 可靠性、可用性、维修性与耐久性

**RAMI** Reliability, Availability, Maintainability and Inspectability 可靠性、可用性、可维护性和可检验性

**RAMIS** Rapid Access Management Information System 快速访问管理信息系统

Reliability and Maintainability Information System 可靠性与维修性信息系统

Remote Access Management Information System 远程访问管理信息系统

**RAMM** Recording Ammeter 记录安培表

**RAMMIT** Reliability and Maintainability Management Improvement Technique 可靠性与可维修性管理改进技术

Reliability, Availability, and Maintenance Management Improvements Techniques 可靠性、可用性与维修管理改进技术

**RAMP** Radar Modernization Project 雷达现代化计划

Reacting and Multi Phase 反射与多相位

Risk Abatement Management Plan 减小风险管理计划

**RAMS** Reliability, Availability, Maintainability and Safety 可靠性、可用性、可维修性与安全性

Remote Area Monitoring System 远程监控系统

Risk Analysis Management System 风险分析管理系统

**RAN** Radio Access Network 无线接入网

Radio Area Navigation 无线电区域导航

Random Access Number 随机存取号码

Regional Air Navigation 区域航空导航

Return Authorization Number 返回授权码

**RANCID** Real and Not Corrected Input Data 真实的、未校正的输入数据

**RANKG** Ranking 顺序，高级(的)

**RAO** Ram Air Outlet 冲压空气出口

Regional Affairs Office 地区事务办公室

Rudder Angle Order 方向舵角度指令

**RAOB** Rawinsonde Observation 无线电探空观测

**RAOC** Regional Air Operations Center 区域空中运行中心

**RAOD** Ram Air Overboard Dump 冲压空气机外释放

**RAP** Rapid City, SD, USA (Airport Code) 美国南达科他州拉皮德城机场

Regional Airspace Plan 地区空域规划

Requirements Analysis Processor 需求分析处理器

Ring Access Protocol 环访问协议

**RAPC** Resolution Advisory Pitch Command 决断咨询俯仰指令

**RAPCOE** Random Access Programming and Checkout Equipment 随机存取编程与检查设备

**RAPCON** Radar Approach Control 雷达进近管制

**RAPID** Rapid Airplane Program Input Design 快速飞机程序输入设计

Receive, Analyze, Process Interactive Display 接收、分析、处理与交互式显示

**RAPP** Recognised Air Picture Production 生产识别出的航拍照片

Registered Air Parcel Post 航空挂号邮包

**RAPPS** Remote Area Precision Positioning System 边远地区精密定位系统

**RAPR** Radar Processor 雷达信息处理机

Resource and Performance Report 资源与性能报告

**RAPS** Radar Automatic Plotting System 雷达自动测绘系统

Ram Air Performance System 冲压空气性能系统

Recovery, Analysis and Presentation System 回收、分析和演示系统

Risk Appraisal of Programs System 项目风险评估系统

**RAR** Radio Acoustic Ranging 无线电声测距

Rapid-Access Recording 高速读写磁带

Rarotonga, Cook Islands(Airport Code) 库克群岛拉罗汤加机场

Record and Report 记录与报告

Regulated Agent Regime 管制代理人制度

Reliability Assessment Report 可靠性评估报告

Risk Assessment Report 风险评估报告

Rules and Regulations 规则及规定

**RARC** Regional Administrative Radio Conference 地区性管理无线通信会议

**RAREP** Radar Report 雷达报告

**RARP** Reverse Address Resolution Protocol 逆向地址解析协议

**RARR** Range and Range Rate 航程与航程变化率, 距离和距离变化率

**RARS** Radio Altimeter Range Simulator 雷达高度表距离模拟器

**RAS** Radar Advisory Service 雷达咨询服务

Radar-Absorbing Structure 雷达吸收结构

Radar-Absorption Structure 雷达吸收结构

Radio Astronomy Satellite 射电天文卫星

Rectified Air Speed 修正表速, 修正空速

Reliability, Availability and Serviceability 可靠性、可用性及可维修性

Remote Access Server 远程访问服务器

Remote Access Service 远程接入服务

Repair Approval Sheet 修理批准书

Replenishment at Sea 海上补给

Replication at Sending 发端重复

Requirements Allocation Sheet 需求分配单

Revenue Accounting System 财务结算系统

Rough Air Speed 颠簸气流空速(通过阵风时)

Route Accounting Subsystem 路由记账分系统

Row Address Strobe 行地址选通

Royal Astronomical Society 皇家航空协会

**RASA** Radar Advisory Service Area 雷达咨询服务区

Russian Aviation and Space Agency 俄罗斯航空与航天局

**RASC** Reservoir Air Supply Cooler 液压油箱供气冷却器

**RASCAL** Rotorcraft Air Crew Systems Concepts Airborne Laboratory 旋翼机的机组人员系统概念机载实验室, 旋翼机的机组人员系统概念空降实验室

**RASE** Rapid Automatic Sweep Equipment 快速自动扫描器

**RASER** Radio Amplification by Stimulated Emission of Radiation 受激辐射射频放大

Radio Frequency Amplification by Stimulated Emission of Radiation 受激辐射射频放大

Revolutionary Aero-Space Engine Research 革命性的航空航天发动机研究

Research and Seeker Emulation Radar 研究与探索仿真雷达

**RASGN** Reassignment 重新指定, 重新任命

**RASGR** Row Address Strobe Graphic 行地址选通波门图

**RASGs** Regional Aviation Safety Groups 地区航空安全组

**RASH** Rain Shower 阵雨

**RASK** Revenue per Available Seat-Kilomenter 可用座公里收入

**RASMAG** Regional Airspace Safety Monitoring Advisory Group 地区空域安全监测咨询小组

**RASMSY** Row Address Strobe Memory System 行地址选通存储器系统

**RASN** Rain and Snow/Showers of Rain and Snow 雨夹雪或阵雨夹雪

**RASONDE** Radiosonde Observation 无线电探空仪观测
Radio Sonde 无线电探空仪

**RASP** Radar Absorbing Structural Panels 雷达吸收结构板
Radar Airborne Signal Processor 雷达机载信号处理器
Reliability and Aging Surveillance Program 可靠性与老化监视计划
Remote Antenna Signal Processor 远端天线信号处理器

**RASS** Radar Acoustic Sounding System 雷达声响探测系统
Radar Altitude Sensing System 雷达高度传感系统
Radio Acoustic Sounding System 无线电音响探测系统
Random Access Storage System 随机存取存储系统
Range Airspace Surveillance System 距离空域监控系统
Reliability, Availability, Serviceability, Security 可靠性、可用性、适用性、安全性

**RAST** Receive and Send Terminal 收发终端
Recover, Assist, Secure and Traverse 回收、援助、保障与运送
Recovery, Assist, Securing and Traversing 回收、辅助、固定与移动

**RASTAC** Random Access Storage and Control 随机存取存储与控制

**RASTAS** Radiating Site Target Acquisition System 辐射场所目标搜索系统

**RASV** Reusable Aerodynamic Space Vehicle 可复用空气动力航天器

**RAT** Ram Air Temperature 冲压空气温度
Ram Air Turbine 冲压空气涡轮
Redundant Axis Torquing 超静定轴向扭矩，静不定轴向扭矩
Reliability Assessment Test 可靠性保证试验，可靠性评估试验
Required Time of Arrival 到达所需时间

**RATAN** Radar and Television Aid to Navigation 雷达与电视辅助导航设备

**RATC** Radar Aided Tracking Computer 雷达辅助跟踪计算机
Remote Air Traffic Control 远程空中交通管制

**RATCC** Radar Air Traffic Control Center 雷达空中交通管制中心
Regional Air Traffic Control Center 区域空中交通管制中心

**RATCF** Radar Air Traffic Control Facilities 雷达空中交通管制设施
Regional Air Traffic Control Facility 地区性空中交通管理设备

**RATDP** Ram Air Turbine Driven Pump 冲压空气涡轮驱动泵

**RATE** Remote Automatic Telemetry Equipment 远程自动遥测设备

**RATG** Radiotelegram 无线电报
Ram-Air Turbine Generator 冲压空气涡轮发生器
Repairs Assessment Task Group 修理评估任务小组

**RATR** Radio Altimeter Transmitter/Receiver 无线电高度表/收发机

**RATS** Radar Acquisition and Tracking System 雷达采集与跟踪系统
Radar Altimeter Target Simulator 雷达高度表目标模拟器
Ram Air Turbine System 冲压空气涡轮系统
Rapid Area Transportation Support 快速地区运输支援

Rejection Advance Tracking System 报废预先跟踪系统

Relay Assembly Test Set 继电器组件测试仪

Remote Area Terminal System 远程区终端系统

**RATSCAT** Radar Target Scatter Facility 雷达目标散射设备

Radar Target Scatter Site 雷达目标散射场，雷达目标分散点

**RAU** Remote Antenna Unit 远端天线单元

**RAV** Remotely Augmented Vehicle 遥控增益飞行器

Risk Assessment Value 风险评估值

**RAVAR** Navigation Radar 导航雷达

**RAVE** Radar Acquisition Visual Tracking Equipment 雷达捕获目视跟踪设备

Radar Alignment Verification Equipment 雷达对准验证设备

Real-time Audio Visual Environment 实时音频目视环境

Research Aircraft for Visual Environment 飞机目视环境研究

**RAVEN** Ranging and Velocity Navigation 测距与测速导航

**RAWIN** Radar Wind Observation 雷达测风

**RAWINDS** Radar Wind Sounding 雷达测风仪

**RAWS** Radar Altitude Warning System 雷达高度警示系统

**RAX** Random Access Transfer Memory 随机存取转换存储器

**RAZCON** Range and Azimuth Control 距离和方位控制

**RB** Range Bin 距离单元

Readback 覆诵

Real Beam 实波束

Recovery Base 回收基地

Reference Burst 参考脉冲串

Relative Bearing 相对方位

Rendezvous Beacon 会合信标

Rescue Boat 救生艇

Return to Bias 回归偏置

Revision Block 修订区

Route Bit 路由比特

Router Board 航线板

Rubber Base 橡胶基座

**RB&S** Reserve Brakes and Steering 储备刹车与转向控制

**RBA** Rabat, Morocco (Airport Code) 摩洛哥拉巴特机场

Radar Beacon Acquisition 雷达信标捕获

Reactive Bandwidth Arbitration 反应型带宽仲裁

Relative Byte Address 相关字节地址

Rescue Breathing Apparatus 急救供氧设备

Royal Brunei Airlines 文莱皇家航空公司

**RBAC** Rule-Based Access Control 基于规则的访问控制

**RBB** Results Based Budgeting 基于成果的预算

**RBBL** Right Body Buttock Line 右机身纵剖线

**RBC** Radio Bearer Control 无线承载控制

Reservation Bandwidth Controller 预留带宽控制器

Rotating Beam Ceilometer 旋转波束云高计

**RBD** Reliability Block Diagram 可靠性数据方框图

Reservations Booking Designator 订座舱位代码

Rigid Body Dynamics 刚体动力学

**RBDE** Radar Bright Display Equipment 雷达明亮显示设备

**RBDS** Radio Broadcast Data System 无线电广播数据系统

**RBER** Residual BER 残余误码率

**RBF** Radial-Basis Function 径向基函数

Remove before Flight 飞行前排除

Rule-Based Forecasting 基于规则的预测

**RBGM** Real Beam Ground Map 实波束地图测绘

**RBI** Radar-Blip Identification 雷达回波识别

Relative Bearing Indicator 相对航向指示器

Retail Business Information 零售商业信息

**RBK** Right Bank 右坡度

**RBL** Right Buttock Line 右纵剖线

**RBM** Range and Bearing Marker 距离和方位标志

Reliability-Based Maintenance 基于可靠性的维护

Resisting Bend Moment 抗弯曲力矩

Results-Based Management 经营结果导向管理

Risk-Based Maintenance 基于风险的维护

**RBN** Radio Beacon 无线电信标

Ribbon 带状电缆,色带

**RBP** Reliable Broadcast Protocol 安全广播协议

Right Bottom Plug 右下电嘴

**RBR** Rubber 橡胶,橡皮

**RBS** Radar Beacon System 雷达信标系统,雷达信标台

Radio Base Station 无线基站

Radio Beacon Station 无线电信标台

Robbed Bit Signaling 强取比特信令

**RBSN** Regional Basic Synoptic Network 区域基本天气网

**RBT** Ringback Tone 回铃音

**RBTB AUX** Right Bus Tie Breaker Auxiliary 右辅助连接汇流条断电器

**RBTS** Radar Beacon Test Set 雷达信标测试仪

**RBV** Return Beam Vidicon 返束管

**RBW** Red-Blue-White（Signal）红—蓝—白（信号）

**RC** Radar Control 雷达管制

Radar Controller 雷达管制员

Range Connection 距离连接

Range Correction 距离修正量

Rate of Change 变化率

Rate of Climb 爬升率

Rate of Climb Indicator 爬升率指示器

Reactor Compartment 反应堆室

Read Clutch 阅读离合器

Rear Connection 后端连接

Reception Control 接收控制

Record Card 记录卡

Reference Chord 基准弦

Reference Clock 基准时钟

Reinstate Card 复原卡

Remote Concentrator 远程集中器

Remote Control 遥控

Repetitive Chime 重复谐音

Replacement Category 更换等级

Reporting Card 报告卡

Resistance-Capacitance 电阻电容

Resolver Control 解算器控制

Return Code 回归码

Reverse Course 反航向

**R-C** Resistor-Capacitor Network 电阻—电容网络

**RCA** Radio Corporations of America 美国无线电公司

Radio Council of America 美国无线电理事会

Reach Cruising Altitude 到达巡航高度

Reduced Co-ordination Airspace 减少协调空域

Root Cause Analysis 根本原因分析

**RCAG** Radio Communication Air-to-Ground 无线电空地通信

Remote Center Air/Ground Communications Facility 空对地通信设施的遥控中心

Remote Communications Air/Ground 地空遥控通信,空/地远程通信

Remote Communications Air-to-Ground 远程遥控空地通信

**RCAGL** Remote Communication Air/Ground Facility Long Range 远程遥控空/地通信设备

**RCAH** Rate Command/Attitude Hold 速度指令/姿态保持

**RCALA** Receiver Conflict Avoidance Learning Algorithm 收信碰撞预防学习算法

**RCAT** Radio Controlled Airplane Target 无线电控制的靶机

**RCB** Radar Control Box 雷达控制盒
Radio Control Bus 无线电控制总线
Reliability Control Board 可靠性控制委员会
Requirements Control Board 需求控制委员会,规格控制委员会

**RCC** Radar Control Center 雷达控制中心
Reader Common Contact 阅读器公共接点
Receiver Command Control 接收机指令控制
Regional Climate Center 区域气候中心
Regional Computer Center 区域计算机中心
Remote Charge Converter 遥控充电转换器
Remote Communication Center 远程通信中心
Remote Control Center 遥控中心
Rescue Control Center 救援控制中心
Rescue Coordination Center 救援协调中心
Rubber Converted Cable 橡胶电缆

**RCCB** Remote Control Circuit Breaker 遥控电路断路器,远程控制电路跳开关
Residual Current (Automatic) Circuit Breaker 漏电流(自动)电路断路器
Reverse Current Circuit Breaker 反向电流断路器

**RCCC** Regional Communications Control Center 地区通信控制中心

**R-CCCH** Reverse Common Control Channel 反向公用控制信道

**RCCP** Radar Control Computer Program 雷达控制计算机程序

**RCCS** Range Computation and Control System 距离计算与控制系统

**RCCU** Remote Calibration Control Unit 遥控校验控制组件

**RCD** Record 记录
Regional Cooperation for Development 区域发展合作
Reverse Current Device 反电流器件
Route Control Digit 路由控制数字

**RCDC** Ramp Cargo Door Controller 停机坪货舱门控制器

**RCDI** Rate of Climb and Descent Indicator 升降速率表

**RCDR** Recorder 记录器
Reference Conceptual Design Report 参考概念设计报告

**RCDU** Remote Control/Display Unit 遥控/显示组件

**RCE** Radio Control Equipment 无线电控制设备
Rotating Combustion Engine 旋转燃烧发动机

**RCEP** Royal Commission on Environmental Pollution (United Kingdom) (英国)环境污染皇家委员会

**RCERT** Recertification/Recertify 重新审定

**RCF** Radio Communication Failure 无线电通信故障
Remote Communication Facility 遥控通信设施

**RCG** Radar Control Group 雷达管制组
Reverberation Control of Gain 增益混响控制

**RCH** Reach or Reaching 一个航向的航程;达到
Reverse Channel 备用信道,反向信道
Riohacha, Colombia (Airport Code) 哥伦比亚里奥阿查机场

**RCHT** Ratchet 棘轮,棘齿

**RCI** Radar Coverage Indicator 雷达探测范围指示器
Range Coordinate Integrator 距离坐标积分器
Resident Cost Inspection 居民消费调查
Resident Cost Inspector 居民消费调查员

**RCID** Radio Call Identifier 无线呼叫标识符

**RCIL** Reliability Critical Item List 可靠性关键项目表

**RCIR** Request for Contractor's Initial Response 承包商初始响应请求

**RCJ** Reaction Control Jet 反应控制喷流

**RCL** Radio Communications Link 无线电通

信链路

Reaching Cruising Level 到达巡航高度层

Runway Centre Line/Runway Centerline 跑道中心线,跑道中线

**R-CLB** Reduced Thrust Climb 减推力爬升

**RCLL** Runway Centre Line Lights 跑道中心线灯光

**RCLM** Runway Center Line Marking(s) 跑道中线标志,跑道中心线标志

**RCLMG** Reclaiming 再生,精制,恢复,改造

**R-CLR** Re-Clear 再清除

**RCLS** Reclassify 再分类

Runway Centerline Light System 跑道中心线灯光系统

**RCM** Corrosive 腐蚀性物质(货运用语)

Radar Countermeasure 雷达对抗,雷达干扰,雷达反制

Radio Countermeasure 无线电对抗,无线电干扰,无线电反制

Ratio Change Module 比率更改组件,速率更改模件

Ratio Changer Module 比率变换器组件

Recommend Cruise Mach 建议巡航马赫数

Reliability Centered Maintenance 以可靠性为中心的维修

**RCMA** Reliability Centered Maintenance Analysis 以可靠性为中心的维修分析

**RCMD** Recommend 建议,推荐

**RCMPRS** Recompression 再压缩

**RCN** Recreation 改造,保养

Release Control Number 发布控制号

Remote Computer Network 远程计算机网

**RCNDT** Recondition 整形,修整,更新,复原

**RCNTR** Ring Counter 环形计数器

**RCO** Radio Control Office 无线电管理办公室

Reactor Core 反应器芯

Remote Communications Outlet 遥控通信出口,遥控通信引出线

Remote Control Oscillator 遥控振荡器

**RCONT** Rod Control 杆控制

**RCP** Radar Control Panel 雷达控制板

Radio Communication Panel 无线电通信面板

Radio Control Panel 无线电控制板

Radio Control Point 无线控制点

Radio Control Protocol 无线控制协议

Rear Cockpit 后驾驶舱

Recipient 接收器,收货人

Regional Control Point 地区控制点

Reliability Control Program 可靠性控制方案

Remote Communication Processor 远程通信处理机

Remote Control Panel 遥控板

Required Communication Performance 所需通信性能,必备通信性能

Restoration Control Point 恢复控制点

Reverse Current Protection 反流保护

Right Circular Polarization 右旋极化

**R-CPHCH** Reverse Common Physical Channel 反向公用物理信道

**RCPM** Reproduction Computer Program Master 复制计算机程序主体

**RCPT** Receipt 收据,收到

Receptacle 插座,容器

**RCPTN** Reception 接收,接见

**RCR** Radio Control Relay 无线电控制继电器

Reader Control Relay 阅读器控制继电器

Relay Controller Register 中继控制器寄存器

Reverse Current Relay 反流继电器

Runway Condition Reading 跑道情况读数

Runway Condition Report 跑道情况报告

**RCRA** Resource Conservation and Recovery Act 资源保护及恢复法案

**RCRL** Reliability Critical Ranking List 可靠性临界序列表

**RCS** Radar Cross Section 雷达扇区

Radio Carrier Station 无线电载波台

Radio Command System 无线电指令系统

Reaction Control System 喷气操纵系统

Reentry Control System 重入控制系统

Remote Computer Service 远程计算机服务

Remote Control System 遥控系统

Roll Control System 滚转控制系统

**RCSC** Red Cross Society of China 中国红十字会

**RCSM** Resident Customer Support Manager 常驻客户支援经理

**RCSS** Random Communication Satellite System 随机通信卫星系统

Remote Control Switching System 遥控交换系统

**RCT** Rear Cargo Tank 后部油箱

Recruit 补充,补给品

Remote Communications Terminal 远程通信终端

Remote Computer Terminal 远程计算机终端

Remote Control Terminal 遥控终端

Repair Correctional Time 修理排故时间

Repair Cycle Time 修理周期时间

Resolver Control Transformer 解算器控制变压器

**RCTG** Recruiting 补给员,招聘

**RCTL** Resistor Capacitor Transistor Logic 电阻电容器晶体管逻辑

**RCTN** Reaction 反应,反作用

**RCTT** Regional Center for Technology Transfer 技术转让区域中心

**RCU** Radio Control Unit 无线控制单元

Rate Construction Unit 运费构成单位,运费计算单位

Receiver Computer Unit 接收机计算机组件

Recorder Control Unit 记录器控制器

Remote Control Unit 遥控组件

**RCUI** Remote Control Unit Interface 远端控制单元接口

**RCUP** Remote Control User Part 远端控制用户部分

**RCUR** Recurrence/Recurrent 再现(的),再生(的),循环的

**RCV** Receive 接收,接受,容纳

**RCVD** Received 接收到的

**RCVG** Receiving 接收

**RCVR** Receiver 接收机

**RCVS** Receivers 接收机,接收人

**RCVY** Recovery 恢复,更新

**RCWS** Roll Control Wheel Steering 驾驶盘横滚操纵

**RCX** Explosives 1.3c 1.3c 级的易爆物(货运品用语)

**RCYC** Recycle 循环

**RD** Radar Data 雷达数据

Radiation Detection 放射检测

Random Drift 随机漂移

Rate of Descent 下降率,下降速度

R-Channel Used for Data 用于数据的 R 通道

Read 读,读作

Reference Database 基准数据库

Refuel/Defuel 加油/抽油

Register Drive 寄存器激励

Restricted Data (有)限制数据

Rework Drawing 返工图纸

Right Display 右显示器

Root Diameter 齿根直径,内径

Rotodome 旋转雷达天线罩

Round 圆形(的),环形路

**RD&E** Research, Development and Engineering 研究、开发与工程

**RD(R&D)** Research and Development 研究和发展

**RDA** Radar Data Assignment 雷达数据分配,雷达数据测定

Recommended Daily Amount 推荐日摄取量

Reliability Design Analysis 可靠性设计分析

Remote Data Access 远程数据访问

Remote Database Access 远程数据库访问

Research, Development, and Acquisition 研究、开发与采集

Runway De-icing Agent 跑道除冰剂

**RDAR** Radio Directing and Ranging 无线电定向与测距

Reliability Design Analysis Report 可靠性设

计分析报告

**RDARA** Regional and Domestic Air Route Area 地区和国内航路区

**RDAS** Radar Dome Antenna Structure 雷达天线罩天线结构

Relational Data Analysis System 关系数据分析系统

**RDB** Radio Data Broadcasting 无线电数据广播

Reference Data Base 参考数据库

Relational Data Base 关系数据库

Remote Data Base 远程数据库

**RDBC** Radar Data Bus Control 雷达数据总线控制

**RDBMS** Relational Data Base Management System 关系型数据库管理系统

**RDBS** Relational Data Base Server 关系数据库服务器

**RDC** Radar Data Correlator 雷达数据相关器

Radiac 辐射计

Reduce/Reduced 减少(的),降低(的)

Regional Data Center 区域数据中心

Remote Digital Concentrator 远程数字集中器

Repair Design Certificate 修理设计合格证

Request for Discrepancy Check 误差检查申请,亏损检查申请

Research and Development Center 研究与开发中心

**R-DCCH** Reverse Dedicated Control Channel 反向专用控制信道

**RDCN** Reduction 减少,降低,简化

**RDCP** Refuel/Defuel Control Panel 加油/放油控制面板

**RDCR** Reducer 减速器,减压器

**RDD** Radar Digital Display 雷达数据显示器

Randomize Disc Directory 随机化磁盘手册

Redding, CA, USA (Airport Code) 美国加利福尼亚州雷丁机场

**RDDMI** Radio Digital Distance Magnetic Indicator 无线电数字距离磁指示器

Radio Direction Distance Magnetic Indicator 无线电方位距离磁指示器

Radio Dual Distance Magnetic Indicator 无线电双距磁指示器

Radio Duplex DME Magnetic Indicator 无线电多路测距磁指示器

**RDDR** Rod Drive 杆驱动

**RDDU** Remote Dedicated Display Unit 远距专用显示组件

**RDEP** Redeployment 重新部署

**RDEU** Radar Display Electronics Unit 雷达显示电子设备组件

**RDF** Radar Data Function 雷达数据功能

Radar Direction Finding 雷达定向

Radio Direction Finder 无线电定向仪,无线电测向仪

Radio Direction Finding 无线电定向

Rate Decrease Factor 速率下降因素

Redefinition 重新规定,重新定义

Repeater Distribution Frame 中继配线架

**RDG** Reference Drawing Group 参考绘图组

Resolver Differential Generator 差分解算产生器

Ridge 螺纹,山脉

Rounding 圆形的,循环的

**RDH** Reference Datum Height 基准数据高度

Round Head 圆头

**RDHFL** Research Development and Human Factors Laboratory 研究开发与人为因素实验室

**RDI** Radar Data Interface 雷达数据接口

Radio Direction Indicator 无线电方位指示器

Radio-Doppler-Inertial 无线电—多普勒—惯性组合导航系统

Remote Defect Indication 远程故障指示

Routing Domain Identifier 选路域标识符

**RDJTF** Rapid Deployment Joint Task Force 快速部署合成特遣部队

**RDL** Radial 射线(的),径向(的)

Relational Database Language 关系数据库语言

**RDLU** Remote Digital Line Unit 远端数字线路单元

**RDM** Recording Demand Meter 记录指令表，记录需求量表
Reference Data Model 参考数据模型
Resource Data Management 资源数据管理
Roll Drive Motor 滚转驱动马达

**RDMI** Radio Distance Magnetic Indicator 无线电距离磁指示器
Radio Distance Measuring Indicator 无线电测距指示器

**RDMU** Range-Drift Measuring Unit 距离偏移测量设备

**RDN** Rejection Disposition Notice 报废处理通知

**RDNG** Reading 阅读，读数

**RDNS** Readiness 就绪，备用，敏捷，愿意

**RDO** Redistribution Order 再分配次序
Research and Development Objectives 研究与开发目标

**RDOS** Real-time Disc Operating System 磁盘实时操作系统

**RDOUT** Readout 读出

**RDP** Radar Data Processing 雷达数据处理
Radar Detector Processor 雷达探测仪处理机
Range Data Processor 距离数据处理器
Remaining Development Program 继续开发方案
Remote Data Processing 远程数据处理
Remote Data Processor 远程数据处理器
Remote Display Processor 远程显示处理器
Roller Drive Power 滚轴驱动动力

**RDPC** Radar Data Processing Center 雷达数据处理中心

**RDPS** Radar Data Presentation Set 雷达数据显示设备
Radar Data Processing System 雷达数据处理系统

**RDR** Radar 雷达
Radar Display Room 雷达显示室
Radar Departure Route 雷达离港航路

Rejection Disposition Report 报废处理报告
Request for Document Revision 修改文件申请

**RDRN** Rapidly Deployable Radio Network 可快速构建的无线网

**RDS** Radar Data Simulator 雷达数据模拟器
Radar Selection 雷达选择
Radial Drive Shaft 径向驱动轴
Radio Data System 无线电数据系统
Radio Digital System 无线电数字系统
Records and Design System 记录与设计系统
Regional Differential System 区域差分系统
RF Data System 射频数据系统
Rotorcraft Drive System 旋翼机驱动系统

**RDS/M** Rounds per Minute 转数/分

**RDSP** Radar Digital Signal Processor 雷达数字信号处理器

**RDSS** Radio Determination Satellite System 无线电定位卫星系统
Rapid Development Space Systems 快捷开发空间系统
Rapidly Deployable Surveillance System 快速部署监视系统

**RDT** Radar Data Terminal 雷达数据终端
Reliability Demonstration Test 可靠性论证试验
Remote Data Transmitter 远程数据发射机
Remote Digital Terminal 远程数字终端

**RDT&D** Research, Development, Trials and Demonstrations 研究、开发、试验和演示

**RDT&E** Research, Development, Test and Engineering 研究、开发、试验与工程
Research, Development, Test and Evaluation 研究、开发、试验与评估
Research, Development, Training and Evaluation 研究、开发、培训与评估

**RDTR** Radiator 辐射器

**RDU** Raleigh, NC, USA (Airport Code) 美国北卡莱罗纳州罗利机场
Remote Display Unit 遥控显示组件

**RDVP** Radar Doppler Video Processor 多普

勒雷达视频处理机

**RDY** Ready 准备,已准备的

**RDZ** Radiation Danger Zone 放射危险区

**RE** Indian Rupee 印度卢比

Radiant Efficiency 辐射效率

Radiated Emission 放射辐射,扩散的排放物

Recent 当前

Regarding 关于

Return Error 返回差错

**REA** Reason 理由

Request for Engineering Authorization 工程审定申请

**REAC** Rain Echo Attenuation Compensation 降雨回波衰减补偿

Reactance 电抗

Reactive 电抗的,反应的

Reactor 电抗器,扼流圈,反应器,反应堆

**REACH** Robotically Enabled Assembly of Cables and Harnesses 机器人启动电缆与导线组装

**REACT** Radio Emergency Associated Citizens Teams 无线电紧急援助协会市民组

Rain Echo Attenuation Compensation Technique 降雨回波衰减补偿技术

Rapid Evaluation and Correction Team 快速评估与排故组

Resource Estimating and Capacity Tracking System 资源估价与容量跟踪系统

**REACTVT** Reactivate 复原,重激活

**READ** Radar Echo Augmentation Device 雷达回波增强器,雷达回波增强装置

**REALLOC** Reallocation 重新定位,重新配置

**REASM** Reassemble 重新组装,重编

**REB** Relativistic Electron Beam 相对性电子束

**REBE** Recovery Beacon Evaluation 回收信标评估

**REC** Receive or Receiver 接收或接收机

Received 已接收

Receiving 正在接收

Recife, Pernambuco, Brazil (Airport Code) 巴西累西腓机场

Record or Recording 记录或记录的

Recordability 记录能力

Recorder 记录器

Regional Economic Communities 区域经济共同体

Request for Engineering Change 工程更改申请

Request for Extension of Contract 延长合同申请

**REC QTY** Recommended Quantity 建议的数量

**RECALC** Recalculated 重新计算的

**RECAP** Regional Economic Community Action Program 区域经济共同体行动方案

Reliability and Experience Correlation and Analysis Program 可靠性、相关经验与分析方案

Reliability Evaluation and Corrective Action Program 可靠性评估和修改工作项目

Risk Evaluation/Corrective Action Program 风险评估/修正行动方案

**RECAT** Reduced Energy Consumption of the Air Transportation System 减少航空运输系统能量消耗

**RECC** Reverse Control Channel 反向控制信道

**RECD** Received 接收的

Recorded 记录的

**RECF** Rectangular Cross-section Eccentric Core Fiber 矩形横截面偏心光纤

**RECHRG** Recharger 再充电器,再装填器

**RECIP** Reciprocal/Reciprocity 相互的,互惠

Reciprocating 往复式(的),摆动的

**RECIRC** Recirculation/Recirculate 再循环

**RECL** Reclosing 重新闭合

**RECLSS** Regenerative Environmental Control and Life Support System 再生性环境控制与生命支持系统

**RECM** Recommend 建议, 介绍, 推荐

**RECOG** Recognition 识别

**RECON** Reconnaissance 勘测, 侦察, 搜索

**RECONC** Reconciliation 调和, 和解, 和谐

**RECOND** Recondensation 再冷凝

**RECONF** Reconfiguration 改变构型, 重新配置

**RECOV** Recovery 恢复

**RECP** Receptacle 插座, 容器

**RECT** Rectangle or Rectangular 矩形 (的), 直角 (的), 正交 (的)
Rectifier 整流器

**RECY** Recoverability 可修复性

**RED** Random Early Detection 随机早期检测
Record 记录
Reduce 减少
Reduction 减少, 降低

**REDA** Ramp Error Decision Aid 停机坪差错决策辅助工具

**REDAP** Reentrant Data Processing (可) 重入数据处理

**REDARS** Reference Engineering Drawing Automated Retrieval System 参考工程图纸自动修正系统

**REDCR** Reducer 减速器, 减压器, 减振器

**REDD** Reducing Emissions from Deforestation and Degradation 减少毁林及森林退化排放

**REDL** Runway Edge Lights 跑道边灯

**REDR** Recorder 记录器

**REDSG** Redesignate 重新任命, 重新指定

**REDUND** Redundancy/Redundant 余度 (的), 冗余 (的)

**REDWN** Redrawn 重绘的, 回火, 再拉伸

**REDZ** Recent Drizzle 最近有毛毛雨

**REEP** Range Estimating and Evaluation Procedure 距离估计及评定程序
Regression Estimation of Event Probabilities 事件概率的回归估计

**REF** Refer/Reference 参考, 基准
Refer to/Reference to/Referred to 参考
Referenced 基准的, 引证的, 推荐的

Referencing 定位, 加参考符号

Referral 职业分派, 安排对象

Refinery 精炼厂

Refund 退款, 偿还

**REFL** Reference Line 基准线

Reflectance 反射率, 反射系统

Reflection 反射

Reflector 反射器

**REFLD** Reflected 反射的

**REFLING** Refueling 加油

**REFR** Refresh 刷新, 更新

Refrigerate 制冷, 冷冻

Refrigerator 冰箱, 冷柜

**REFRIG** Refrigerate or Refrigeration 冷冻, 制冷

**REFURB** Refurbishment 重新磨光, 刷新, 整修

**REG** Reggio Calabria, Italy (Airport Code) 意大利雷焦卡拉布里亚机场

Region 地区, 区域

Register 寄存器, 注册, 记录员

Regular Aerodrome 定期航班机场

Regulate 调节, 调整, 管理

Regulation 规章, 规程, 条例

Regulator 调节器

**REGAL** Range and Elevation Guidance for Approach and Landing 进场与着陆的距离、高度 (坡度) 引导

**REGEN** Regenerate 再生, 回热

**REGLTN** Regulation 规则

**REGR** Recent Hail 最近有冰雹
Regression 退化, 回归

**REI** Rang Elevation Indicator 距离、仰角指示器
Remote Error Indicator 远端误码指示
Request for Engineering Information 工程信息申请

**REID** Reidentify 再识别

**REIL** Runway End Identification Lights 跑道终端识别灯, 跑道终端标志灯

**REILS** Runway End Identification Lighting

System 跑道终端灯光识别系统

**REIMB** Reimburse 赔偿, 偿还

**REIN** Reinstatement 复原, 恢复

**REINF** Reinforce 强化, 增强

**REJ** Reject 拒绝, 拒收, 报废

**REJD** Rejected 拒收的, 报废的

**REK** Reykjavik, Iceland( Airport Code) 冰岛雷克雅未克机场

**REL** Related/Relating/Relative 相关的

Relay 继电器, 中继, 转播

Release/Releasing 发行, 发射, 释放

Released 脱开的, 松开的, 开通的, 发布的

Reliability 可靠性

Relief 释放

Reluctance 阻抗, 磁阻

Rescue Equipment Locker 救生设备柜

**RELO** Reload 重新加载

Relocatable 可重新定位的

Relocate 重新定位

**RELP** Residual Excited Linear Predictive Coding 残余激励线性预测编码

**RELQ** Relinquish 放松, 撤回

**RELSAT** Relay Satellite 中继卫星

Reliability for Satellite 卫星可靠性

**RELSIG** Release Signal 解除信号, 投射信号

**REM** Rapid Eye Movements 眼球快速移动

Reaction Engine Module 反作用发动机组件

Recovery Exercise Module 恢复行使模块

Reminder 备忘录

Remote Exchange Multiplex 远端交换局复用

Removable 可拆卸的

Remove 移动, 拆卸

**REMAD** Remote Magnetic Anomaly Detection 遥控地磁异常探测

**REMC** Resin Encapsulated Mica Capacitor 树脂封装云母电容

**REMF** Reference My Phone 请参照我方电话

**REMIS** Regional Environmental Management Information System 区域环境管理信息系统

Reliability and Maintainability Information System 可靠性与维修性信息系统

Reliability Engineering Management Information System 可靠性工程管理信息系统

Resources Management Information Systems 资源管理信息系统

Risk Evaluation and Management Information System 风险评估与管理信息系统

**REML** Reference My Letter 参阅我方信件

**REMT** Reference My Telegram or Telex 参阅我方电报, 参阅我方电传

**REN** Ringer Equivalence Number 振铃当量数

**RENL** Runway End Light( s) 跑道终端灯

**REO** Regenerated Electrical Output 再生电力输出

Repair Estimate Order 修理概算指令, 修理概算订单

**REORG** Reorganize 重组

**REP** Range Error Probability 距离误差概率

Reference Eye Position 基准眼位

Repair 修理

Repeat 重复

Reply 回答, 答复

Reporting Point 报告点

Represent 代表, 意味着

Representative 代表, 代理, 代表性的, 典型的

Republic 共和国

Roentgen Equivalent Physical 物理伦琴当量

Rotable Exchange Program 周转件交换计划

**REPBY** Replaced by Partnumber 为部件号…更换

**REPC** Regional Economic Planning Council 地区经济计划理事会

**REPL** Replace/Replacement/Replacing 更换

Replaced by 为…更换

**REPOS** Repair and Overhaul Scheme 修理和翻修方案

**REPPAC** Repetitively Pulsed Plasma Accelerator 重复脉冲等离子体加速器

**REPR** Representative 代表,代理人

Reproducer 放音机

**REPRO** Reproduce/Reproduction 再生,复制,还原

Reproducibility 可重复性,可复现性

**REPRO(L)** Reproducible 可重复的,可复现的

**REPS** Representative 代表,有代表性的

**REPT** Reference Engineering Photo Template 基准工程光学模板

Reproducible Engineering Photo Template 可复印工程光学模板

Request Engineering Photo Template 申请工程光学模板

**REQ** Dangerous Goods in Excepted Quantities 例外数量的危险品

Request 要求,请求

Require or Required 需要,必须的

Requirement 需要

Requisition 请求,需要

**REQFIN** Request Finished 要求终止

**REQID** Requirement Identity Number 需求识别号

**REQN** Requisition 要求,申请(书),调拨单,征用

**REQR** Requestor 申请人

**REQT** Requirement 必需品,必要性,要求,规格

**RERA** Recent Rain 最近有雨

**RERASN** Recent Rain and Snow 最近有雨和雪

**RERS** Radiation Event Recovery Sequence 辐射事件恢复顺序

**RES** Reference Earth Station 参考地球站

Remote Earth Station 远端站,边远地球站

Research 研究

Reservation or Reserve 保留,储备,预订

Residence 居住

Resident 居民

Resistance or Resistor 电阻或电阻器

Resolution 决议

Resolver 解算器

Restart 再启动

**RES CN** Resource Control Numbers 资源控制号

**RESA** Runway End Safety Area 跑道端安全区

**RESC** Rescind 废除,取消

**RESCH** Reschedule 重新计划

**RESEAT** Reseating 修整

**RESEP** Reentry System Environmental Protection 再入系统环境保护

**RESH** Recent Rain Showers 最近有骤雨

**RESID** Residual 清除的,残余的

**RESN** Recent Snow 最近有雪

Resonate or Resonant 谐振(的),反响(的)

**RESOLN** Resolution 分辨力,分解,溶解

**RESP** Respective or Respectively 分别的,分别地

Respiration 呼吸

Response 响应,应答

Responsibility 响应度,偿付能力,职责,责任

Responsible 可信赖的,有责任的

**REST** Radar Electronic Scan Technique 雷达电子扫描技术

**RESTR** Restorer 复位器,恢复设备

Restraint 抑制,阻尼器

Restricted 限制,制约

Restriction 限制,制约

**RESUP** Resupply 再供给

**RESV** Reserve 保留,储备,预订

**RESYNC** Resynchronize 再同步

**RESYNCING** Resynchronizing 再同步

**RET** Rapid Exit Taxiway 快速退出滑行道

Reliability Evaluation Test 可靠性评估试验

Retain 保持,记忆

Retention 保存,保持,固位,记忆(力)

Retirement 退休,收回,修复

Retracted 收回的,撤回的

Retraction 收起,缩入

Return 返回

**RET/RETD**    Retard 延时,制动

**RETO**    Revised Time Overhead 更正后的飞越时间

**RETOK**    Retest Okay 重测合格

**RETR**    Retrace 返回,回归

Retract 收起,缩入,收回

Retractable 可收起,缩入的,可收回的

Retracted 收回的,收上的

**RETRG**    Retracting 收回,收上,收缩

**RETRN**    Retraction 收回,收上,收缩

**RETRO**    Retroactive 反作用的,逆行的

Retrograde 后退(的),退化(的)

**RETROG**    Retrogressive 后退的,退化的

**RETS**    Recent Thunderstorms 最近有雷雨

**RETU**    Returned 返回的

**REU**    Radio Electronics Unit 无线电电子装置

Remote Electronic Unit 遥控电子组件

Reus, Spain ( Airport Code ) 西班牙雷乌斯机场

**REV**    Previous or Previously 先前的,先前地

Revenue 收入

Reverse 倒退,使倒转,相反,倒转

Reverse Charging 反向计费

Reverser 反向,反流,反推器,换向器

Review or Reviewed 再检查,回顾,复习,审查

Revision 改版

Revolution 旋转,运行,循环

Thrust Reverser Sleeve Position 反推套筒位置

**REV CUR**    Reverse Current 反向电流

**REVC**    Reverse Charging 反向计费

**REVD**    Revised 修订的,校正的

**REVEL**    Reverberation Elimination 混响消除

**REVOCON**    Remote Volume Control 遥控音量控制

**REVY**    Recovery 恢复

**REW**    Reward 报酬,奖金

Rework 返工,再处理

**REX**    Normally Forbidden Explosives 一般禁运物品,通常被禁带易爆物

**REXP**    Explosives 爆炸物

**REY**    Reentry 重入

**RF**    Radio Frequency 射频,无线电频率

Radius to a Fix 恒定半径到定位点

Raised Face 突起面

Ramp Fuel ( 飞机 ) 在停机坪时的油量

Range Finder 测距仪

Refill 回填,加装

Remote Fault 远端故障

Right Front 右前方

Route Forecast 航线天气预报

**RFA**    Radio Frequency Amplifier 射频放大器

Radio Frequency Authorization 无线电频率分配表册,无线电频率管理,无线电频率登记

Raman Fiber Amplifier 拉曼光纤放大器

Recurrent Fault Analysis 重复障碍分析

Remote File Access 远程文件存取

Request for Action 操作申请

Request for Analysis 分析申请

**RFA/RFI**    Radio Frequency Amplifier and Radio Frequency Interference 射频放大与射频干扰

**RFAA**    Russian Federal Aviation Authority 俄罗斯联邦航空局

**RFACA**    Royal Federation of Aero Clubs of Australia 澳大利亚皇家飞机俱乐部联合会

**RFAS**    Request for Assistance System 辅助系统申请

**RFB**    Reliability Function Blocks 可靠性功能方框图

Request for Bid 投标申请,请求报价

**RFC**    Radio Facility Chart 无线电设备图

Radio Frequency Choke 射频扼流圈

Radio Frequency Combiner 射频合成器

Reinforced Fiber Composite 纤维加强复合材料

Request for Change 更改申请

Request for Commitment 要求保证,委托申

请
Resident Fabrication Control 固有生产控制

**RFCA** Radio Frequency Cable Assembly 射频电缆组件,射频电缆组装

**RFCAP** Panel on Route Facility Cost Accounting and Cost Allocation 航路设备成本会计及成本分摊专家小组

**RFCH** Reverse Fundamental Channel 反向基本信道

**RFCP** Request for Change Proposal 更改投标申请,改变协议申请

**RFCR** Refacer 光面器

**RFCS** Regenerative Fuel Cell System 可再生燃料电池系统

**RFD** Raised Face Diameter 突面直径
Refund 退款
Request for Disposition 处理物申请
Rockford, IL, USA (Airport Code) 美国伊利诺州罗克福德机场

**RFDS** Royal Flying Doctor Service 皇家航空医疗队

**RFE** Request for Equipment 设备申请
Request for Estimate 评估申请

**RFEC** Request for Engineering Change 工程变更申请

**RFEI** Request for Engineering Information 工程信息申请

**RFF** Ready for Ferry 准备空运
Rescue and Fire Fighting 救援和消防

**RFFP** Rescue and Fire Fighting Panel 救援和消防专家小组

**RFFS** Rescue and Fire Fighting Service 救援和消防服务

**RFFTC** Rescue and Fire Fighting Training Center 救援和消防培训中心

**RFG** Flammable Gas 易燃气体
Receive Format Generator 接收格式产生器

**RFGT** Refrigerant 冷冻剂,冷却液

**RFH** Raised Face Height 突面高度

**RFI** Radio Frequency Interference 射频干扰,无线电频率干扰

Radio Frequency Investigation 无线电频率的调查

Rapid Fault Isolation 快速故障隔离

Ready for Issue 准备发放,准备发行

Reduced Flight Idle 减推力飞行慢车

Remote Failure Indication 远端失效指示

Request for Information 信息征求书

Request for Inspection 检验申请书

**RFID** Radio Frequency Identification 无线射频识别
Radio Frequency Identification Device 无线射频识别装置
Radio Frequency Identification Document 无线射频识别文件

**RFIF** Radio Frequency Interface 射频接口

**RFIT** Radio Frequency Identification Tag 无线射频识别标记
Radio Frequency Identification Technology 无线射频识别技术
Radio Frequency Information Technology 无线射频信息技术
Retrofit 改型,更新

**RFL** Flammable Liquid 易燃液体
Refuel 加油
Rough Field Landing 不平场地着陆

**RFLNG** Refueling 加油

**RFLT** Repeat Flight 重复飞行

**RFLX** Reflex 反射

**RFM** Radio Frequency Management 无线电频率管理
Reactive Factor Meter 反射系数表

**RFMC** Ram Fan Motor Controller 冲压风扇马达控制器

**RFMS** Radio Frequency Management System 射频管理系统
Risk Feedback Management Strategy 风险反馈管理策略

**RFMU** Right Fuel Management Unit 右燃油管理组件

**RFNA** Red Fuming Nitric Acid 红烟硝酸

**RFND** Refined 精炼的,精制的

**RFNI** Relative Footprint Noise Index 相关轨迹噪声指数

**RFO** Royal Flight Oman 阿曼皇家飞行公司

**RFOC** Reference Fiber Optics Connector 基准光纤连接器

**RFOG** Resonator Fiber-Optic Gyro 谐振式光纤陀螺

**RFP** Repetitive Flight Plan 重复飞行计划

Request for Planning 征求计划

Request for Proposal 需求建议书

**RFQ** Request for Quotation 征求报价单

Request for Quote 征求报价, 报价申请

**RFR** Request for Recommitment 征求重新委托

Request for Reschedule 征求重新安排

Request for Revision 请求修正

**RFRB** Refurbish 整修, 刷新

**RFRC** Refractory 耐火材料, 耐火的, 难熔的

**RFRL** Raspet Flight Research Laboratory 拉斯派特飞行研究试验室

**RFS** Flammable Solid 易燃固体

Radio Frequency Splitter 射频分配器

Radio Frequency Surveillance 射频监视

Range Filter Sample 距离滤波器取样

Ready for Sending 准备发送

Ready for Service 准备开通服务

Reason for Selection 选择的理由

Regardless of Feature Size 忽略要素尺寸

Regardless of Future Size 不考虑未来规模, 不考虑未来尺寸

Request for Service 要求维修, 要求服务

**RFSC** Renton Flight Simulator Center 瑞顿飞行模拟机中心(波音)

**RFSK** Relative Frequency Shift Keying 相对频移键控

**RFSL** Refusal 拒绝, 优先权, 购买权

**RFSMS** Radio Frequency Signal Management System 射频信号管理系统

**RFSS** Radio Frequency Simulation System 射频模拟系统

**RFT** Ready for Training 可供训练

Request for Tender 招标

Runway Friction Tester 跑道摩擦测试仪

**RFTK** Revenue Freight Tonne-Kilometers 货邮载运吨公里

**RFTO** Ready for Take-off 准备起飞

**RFTP** Request for Technical Proposal 技术建议提案

**RFTR** Radio Frequency Transmit and Receive 无线射频发射与接收

**RFTS** Radio Frequency Target Simulation 射频目标模拟

**RFU** Radio Frequency Unit 无线射频组件

**RFW** Dangerous When Wet 遭湿自燃物品

Repare Fuel Weight 备用耗油量

Request for Waiver 申请弃权

Request for Work 工作申请

Reserve Feed Water 备用给水

**RG** Radar Guidance 雷达制导

Radiation Guidance 辐射制导

Range 距离, 范围

Range (Lights) 跑到头灯, 飞行场地(灯)

Reception Good 验收合格

Reduction Gearbox 减速齿轮箱

Reset Gate 复位门

Residential Gateway 住宅网关

Reticulated Grating 网栅

Retractable Gear 收放式起落架

Reverse Gate 反转门

**RG(N)** Register (N) Stages 寄存器(N)级

**RGA** Rotation Go-around 循环复飞

**RGAT** Review Group of ATC Training 管制员培训评审组

**RGB** Red/Green/Blue 红绿蓝

**RGCSP** Review of the General Concept of Separation Panel 飞行间隔一般原理审查专家组

**RGD** Range Gate Description 距离门说明

Rigid 刚性的

**RGDS** Regards 问候, 致意

**RGH** Rough 粗糙的

**RGIC** Ranging GNSS Integrity Channel 全球

导航系统完好性信道范围

**RGL**   Regulate/Regulating 管理,调整

    Runway Guard Light 跑道警戒灯

**RGLN**   Regulation 规章,条例

**RGLR**   Regular 规则的,常规的

**RGLTD**   Regulated 调节的

**RGLTG**   Regulating 调节

**RGLTN**   Regulation 调节

**RGLTR**   Regulator 稳压器,调节器

**RGM**   Revenue Growth Management 收益增长管理

**RGN**   Region 区域,范围

    Yangon, Myanmar(Airport Code) 缅甸仰光机场

**RGNG**   Rigging 装配,调整

**RGPO**   Range Gate Pull off 距离波门拖引

**RGRT**   Regret 报歉

**RGS**   Radio Guidance System 无线电制导系统

    Range Gate Size 距离门尺寸

    Recovery Guidance System (从紧急状态)恢复的引导系统

    Remote Ground Station 远端地面站

**RGT**   Resonant Gate Transistor 谐振选通晶体管

**RGTG**   Regulating 调节

**RGTR**   Register 记录器,记录员

**RGV**   Rotating Guide Vane 可调导向叶片

**RGWO**   Range Gate Walk-off 距离波门拖引

**RGXH**   Range Crosshair 距离十字准线

**RH**   Radio Height 无线电高度

    Relative Height 相对高度

    Relative Humidity 相对湿度

    Right Hand 右手柄,右手

    Right Hydraulic 右液压系统

    Rockwell Hardness 洛克威尔硬度,洛氏硬度

**RHAW**   Radar Homing All the Way 全程雷达导引,全程雷达寻的

    Radar Homing and Warning 雷达导引与警告

**RHAWS**   Radar Homing and Warning System 雷达导引与警告系统

**RHB**   Radio Homing Beacon 无线电归航信标

**RHC**   Range Height Converter 距离高度变换器

    Right Hand Circular 右螺旋,右循环

    Robinson Helicopter Company 罗宾逊直升机公司

    Rotation Hand Control 手控旋转

**RHCL**   Renton Hydraulics and Control Laboratory 波音瑞顿液压与控制实验室

**RHCM**   Radiation Hardened Core Memory 抗辐射磁芯存储器

**RHCP**   Right Hand Circular Polarization 右旋圆极化

**RHD**   Return Head 返回头

**RHEO**   Rheostat 变阻箱

**RHF**   Radio High Frequency 无线电高频

    Red Horse Flight 红马航班

    Regular High Frequency 常用高频

    Remote Handling Facility 远程处理设施

**RHFCU**   Reheat Fuel Control Unit 加力燃油调节器

**RHI**   Radar Height Indicator 雷达高度指示器

    Rang-Height Indicator 距离—高度显示器

    Risk Hazard Index 风险危害指数

    Relative Height Indicator 相对高度指示器

    Rhinelander, WI, USA(Airport Code) 美国威斯康星州莱因兰德机场

**RHJ**   Rubber Hose Jacket 橡皮软管套

**RHK**   Reefing Hook 缩(帆)钩

**RHL**   Radiological Health Laboratory 放射性健康实验室

    Rectangular Hysteresis Loop 矩形磁滞回线

    Rudder Hinge Line 方向舵铰接线

**RHM**   Relative Humidity Monitor 相对湿度监控仪

**RHO**   Rhodes, Greece(Airport Code) 希腊罗德岛机场

**RHP**   Reduced Hard Pressure 降低的硬压

**RHPs**   Runway Holding Points 跑道等待点

**RHR** Reheater 加热器,回热器
Roughness Height Rating 粗糙度

**RHS** Remote Handset 遥控手持机,遥控电话听筒
Right Hand Side 右方,右边
Right Hand Stringer 右侧纵梁

**RHSC** Right Hand Side Console 右侧操纵台

**RHW** Receive Highway 接收公共信道

**RHWS** Radar Homing and Warning System 雷达归向与警示系统

**RI** Radar Input 雷达输入
Radio Installation 无线电装备
Radio Isotope 无线电同位素
Range Instrumentation 测距仪
Reflective Insulation 反射绝缘(信号)
Refractive Index 折射率
Remote Information 远端信息
Required Identification 需求的识别
Response Identifier 响应标识符
Revenue Integrity 收益完整性
Routing Indicator 航线指示器
Routing Information 路由信息

**RIA** Railway Industry Association (UK) 铁路行业协会(英国)
Regulatory Impact Analysis 管理效果分析
Requirements Input Analysis 需求输入分析

**RIAC** Reliability Information Analysis Center 可靠性信息分析中心

**RIAL** Runway Identifiers and Approach Lighting 跑道标志灯与进近灯光设备

**RIAS** Regulatory Impact Analysis Statement 监管影响分析报表

**RIB** Right Inboard 右内侧
Routing Information Base 选路信息库

**RIBR** Remaining Isochronous Bandwidth Ratio 保留同步带宽比

**RIBS** Residential Interactive Broadband Service 住宅用户交互式宽带业务

**RIC** Radar Input Control 雷达输入控制
Richmond, VA, USA 美国弗吉尼亚州里士满机场

Routine Inventory and Control 常规库存与控制

**RID** Radar Input Drum 雷达输入磁鼓
Relation Identity 关系标识
Request for Inspection Data 检查数据请求
Review Item Disposition 检查项目处理
Route Identification 路由识别

**RIDS** Radar Instrumentation Data System 雷达测试仪表数据系统

**RIF** Rate Increase Factor 速率增长因素
Reclearance in Flight 空中再次放行
Reliability Improvement Factor 可靠性改进系数
Routing Information Field 路由信息字段

**RIFI** Radio Interference Field Intensity 无线电干扰场强度

**RIFS** Receiving Inspection File System 接收检查文件系统

**RIG** Rate Integrating Gyroscope 速率积分陀螺仪

**RIGS** Runway Identifiers with Glide Slope 显示下滑道的跑道标志灯

**RII** Required Inspection Items 必检项目
Routing Information Indicator 路由寻径信息指示位

**RIIP** Required Inspection Item Program 必检项目方案

**RIL** Radio Interference Level 无线电干扰电平
Red Indicating Lamp 红色指示灯
Repairable Item List 可修理项目清单

**RIM** Radar Input Mapper 雷达输入绘图仪
Radiant Intensity Measurement 辐射强度测量
Reaction Injection Molding 反应式注塑,反应式压铸
Remote Interface Module 远程接口模块

**RIMS** Relational Information Management System 相关信息管理系统
Reliability Information and Monitoring System 可靠性信息与监测系统

Remote Information Management System 远程信息管理系统

**RIN** Radio Inertial 无线电惯性(导航)

**RING** Ringer 振铃器,信号器

**RINS** Radio Inertial Navigation System 无线电惯性导航系统

Redundant Inertial Navigation System 余度惯性导航系统

**RINSTL** Reinstallation 重新装配

**RINSUL** Rubber Insulation 橡皮绝缘(垫)

**RINT** Radar Intermittent 雷达间歇

Radiation Intelligence 辐射信息,辐射情报

Radio Intelligence 无线电情报

**RIO** Rio De Janeiro, Brazil-Galeao (Airport Code) 巴西里约热内卢格莱昂机场

**RIOMETER** Relative Ionospheric Opacity Meter 相对电离层不透明度仪

**RIOT** Real-time Input/Output Transducer 实时输入/输出转换器

Remote Input/Output Terminal 远程输入输出终端

**RIP** Radar Identification Point 雷达识别点

Raster Image Processor 光栅图像处理机

Receiving Inspection Plan 接收检查计划

Reference Installation Plan 参考安装计划

Refractive-Index Profile 折射率分布

Register Indicator Panel 寄存器指示板

Remote Instrument Package 遥控仪表组件

Request for Instrumentation Preflight 飞行前仪表测试要求

Routing Information Protocol 选路信息协议

Type Inspection Report 型号检查报告

**RIPC** Records Information Processing Center 记录信息处理中心

**RIPN** Reference Installation Plan Notice 参考安装计划通知

**RIPS** Radar Information Processing System 雷达信息处理系统

Recorder Independent Power Source 记录器独立电源

Runway Incursion Prevention System 机场跑道入侵预防系统

**RIQS** Remote Information Query System 远程信息访问系统

**RIR** Receiving Inspection Record 接收检查记录

Reinspection Request 重新检查申请

Reliability Investigation Request 可靠性调查申请

**RIRP** Retractable in-Flight Refueling Probe 可收回的空中加油探测器

Runway Incursion Reduction Program 减少跑道侵入计划

**RIS** Infectious Substance 传染性物质

Radar Information Service 雷达情报服务

Ramjet Inlet System 冲压喷气发动机进气系统

Range Information System 距离信息系统

Remote Instruction System 远程授课系统

Retro Reflector in Space 空间后向反射器

Revolution Indicating System 转速指示系统

Rotating Image Scanner 旋转图像扫描器

**RISA** Reduced Instruction Set Architecture 简化指令集结构

**RISC** Reduced Instruction Set Computer 简化指令集计算机

Reduced Instruction System Computer 简化指令系统计算机

Runway Incursion Severity Classification 跑道侵入严重性分类

**R-ISMA** Reserved Idle Signal Multiple Access 预留空闲信号多址访问

**RISS** Reissue 重新发布

**RIT** Receiver Incremental Tuning 接收机增量调谐

Riveting Tool 铆接工具

**RITA** Rotorcraft Industry Technology Association 旋翼机工业技术协会

**RITE** Right 右或右转

**RITS** Receiving Inspection Tracking System 接收检查跟踪系统

**RIU** Remote Interface Unit 远端接口单元

Ring Interface Unit 环接口单元

**RIV** Radio Influence Voltage 无线电干扰电压

Rapid Intervention Vehicle 快速先遣消防车

Rivet 铆接

**RIW** Reliability Improvement Warranty 可靠性改进保证

Riverton,WY,USA 美国里弗顿机场

**RIWWG** Reliability Improvement Warranty Working Group 可靠性改进保证工作组

**RIX** Riga,Latvia（Airport Code）拉脱维亚里加机场

Route Index 路由标志

**RIXT** Remote Information Exchange Terminal 远地信息交换终端

**RJ** Random Jitter 随机抖动

Regional Jet 支线喷气机

Riveting Jig 铆接夹具

**RJB** Radio Junction Box 无线电接线盒

**RJE** Remote Job Entry 远程作业输入

**RJEP** Remote Job Entry Protocol 远程作业输入协议

**RJM** Remote Jack Module 遥控插口模件

**RJP** Remote Job Processor 远程作业处理器

**RJS** Reaction Jet System 反作用喷气系统,反作用射流系统

**RJT** Technical Rejection Message 技术拒收电报

**RK** Rack 齿条,台架,支架

**RKCR** Request for Kit Commitment Record 套件提交记录申请

**RKD** Rockland, ME, USA（Airport Code）美国缅因州罗克兰机场

**RKR** Rocker 摇杆,摇臂,振动器

**RKRA** Rocker Arm 摇臂

**RKS** Record Keeping System 记录保存系统

Rock Springs, WY, USA（Airport Code）美国怀俄明州罗克斯普林斯机场

**RKVAM** Recording Kilovolt-Ampere Meter 记录式千伏特安培表

**RL** Rail 轨道,铁路

Refraction Loss 折射损耗

Return Loss 回路损耗

Revenue Leakage 收益漏洞

Rhumb Line 等角线

Roll 滑跑,滚转,横滚

Rotary Launcher 转动式发射器

Runway Light 跑道灯

**RLA** Relay 继电器,中继,转播,传递

Remote LAN Access 远程 LAN 接入

Remote Loop Adapter 远程环路适配器

Repair Level Analysis 修理水平分析,修理等级分析

Reverse Lever Angle 反推手柄角度

Reverser Lever Angle 反推杆角

**RLADD** Radar Low Angle Drogue Delivery 雷达低角度伞投(装置)

**RLASM** Reduced Lateral Separation Minima 缩减的最小横向间隔

**RLC** Radio Link Control 无线链路控制

Release Completion 释放完成

Resistance Inductance Capacitance 电阻—电感—电容

Run Length Coding 游程长度码

**RLCE** Request Level Change Enroute 请求改变航路飞行高度层,要求改变航线飞行高度层

**RLCR** Railcar 轨道车

**RLCS** Radio Launch Control System 无线电发射控制系统

**RLD** Radar Laydown Delivery 雷达瞄准空投

Reference Landing Distance 参考着陆距离

Required Landing Distance 所需着陆距离

Rolled 滚压的,轧制的

**RLDF** Red Label Design Freeze 红色标记设计冻结

**RLE** Run Length Encoding 行程编码

**RLF** Relief 释放

Reverse Line Feed 回油管

**RLG** Railing 栅形,栅栏

Ring Laser Gyro 环形激光陀螺

**RLGINS** Ring Laser Gyroscope Inertial Navigation System 环形激光陀螺惯性导航系统

**RLLS** Runway Lead-in Lighting System 跑道引进灯光系统

**RLOSM** Reduced Longitudinal Separation Minima 缩减的最小纵向间隔

**RLR** Roller 转子

**RLR BRG** Roller Bearing 滚柱轴承

**RLS** Radar Line of Sight 雷达视线

Radar Locating System 雷达定位系统

Reliable Link Service 可靠数据链服务

Reliable Link Source 可靠数据链源

Remote Light Sensor 遥控光传感器

Right Line Select Key 右行选择键

**RLSD** Received Line Signal Detect 接收线信号检测

**RLSED** Released 释放的

**RLSTN** Relay Station 中继站,转播站

**RLT** Reference Layout Table 基线布局表

Reference Layout Template 基线布局模板

Relate 相关

**RLTD** Related 有关的

**RLTK** Rhumb Line Tracker 等角线航线,等角线航迹

**RLTN** Relation 关系,方程,定律

**RLTNP** Relationship 关系,联系,关系曲线

**RLU** Red Label Unit 红色标记组件

Remote Line Unit 远端用户线单元

**RLV** Relieve 减轻,减压,救援

Reusable Launch Vehicle 重复使用运载器

**RLW** Regulated Landing Weight 额定着陆重量

**RLXN** Relaxation 放松,松弛,平衡的自动恢复,衰减

**RLY** Relay 继电器,中继,转播,传递

**RM** Radar Mile 雷达英里

Radio Magnetic 无线电电磁(的)

Range Marks 距离标志

Rapid Manufacturing 快速制造

Raw Material 原材料

Reference Mark 基准标志,参考标记

Reference Model 参考模型

Remedial Maintenance 校正维修

Rendezvous Mode 轨道会合状态,轨道会合方式

Repertory Manager 仓库经理

Resource Management 资源管理

Resource Module 资源模块

Revenue Management 收益管理

Reversion Mode 恢复方式

Room 室,房间

**RMA** Random Multiple Access 随机多址

Regional Monitoring Agency 地区监控组织,区域监测机构

Reliability, Maintainability and Availability 可靠性、可维修性与可用性

Request for Marker Artwork 标志原图申请

Resource Management Account 资源管理账户

Risk Management Association 风险管理协会

**RMAAA** Reliability, Maintainability Allocations, Assessments and Analysis 可靠性、可维修性分配、评定与分析

**RMAC** Remote Monitoring and Control 远程监视与控制

**RMAP** Risk Management and Audit Program 风险管理与审计计划

**RMAs** Regional Monitoring Agencies 地区监测机构

**RMAT** Reliability and Maintainability Assessment Tool 可靠性与维修性评估工具

**RMAX** Range Maximum 最大距离,最大航程

**RMB** Renminbi 人民币

Risk Management Board 风险管理委员会

**RMC** Regional Meteorological Center 区域气象中心

Resource Monitoring and Control 资源监视和控制

Route Marker Control 路由标志控制

**RMCP** Radio Management Control Panel 无线电管理控制板

**RMCS** Remote Maintenance Control System

远程维护控制系统

Remote Mission Control System 遥控任务控制系统

**RMCU** Reeling Machine Control Unit 矫直机控制装置

Reliable, Maintainable Computing Upgrade 可靠性、维护性计算更新升级

Remote Magnetic Compensation Unit 遥控磁性补偿组件

**RMD** Miscellaneous Dangerous Goods 杂项危险货物

Radio Management Display 无线电管理显示器

**RMDI** Radio Magnetic Direction Indicator 无线电磁航向指示器

**RMDS** Report Management Distribution System 报告管理分配系统

**RMG** Resource Management Group 资源管理组

Right Main Gear 右主起落架

Risk Management Group 风险管理组

**RMGF** Radar Mapper Gap Filler 雷达绘图仪辅助天线

**RMH** Refrigerator Mechanical Household 家用致冷器机构

**RMI** Radio Magnetic Indicator 无线电磁航向指示器

Remote Method Invocation 远程方法调用

Repair, Maintenance and Improvement 修理、维护与改进

Risk Management Information 风险管理信息

Route Monitoring Information 路由监控信息

**RMIB** Remote Management Information Base 远地管理信息库

**RMIR** Resident Manufacturing Inspection Representative 驻厂生产检验代表

**RMJ** Ramjet 冲压式喷气发动机

**RMK(S)** Remark 备注,附注,提要

**RML** Radar Microwave Link 雷达微波通信链,雷达微波链路

Requirements Modeling Language 需求建模语言

**RMLR** Radar Mapper, Long Range 远程雷达绘图仪

**RMM** Radar Map Matching 雷达地图匹配

Remote Maintenance Monitoring 远程维修监控

Remote Management Module 远程维修模块

Remote Monitoring and Management 远程监控与管理

**RMMM** Redundancy Management and Maintenance Monitor 余度管理和维护监控器

**RMMP** Reliability and Maintainability Management Plan 可靠性与可维修性管理计划

**RMMS** Remote Maintenance Monitoring System 远程维修监视系统

Routine Maintenance Management System 日常维护管理系统

**RMN** Remain 剩余,保留,保持

Remaining 保留(的),保持(的)

**RMO** Request for Modification Order 改装指令申请

Resource Management Office 资源管理办公室

Retrofit Modification Offer 翻新改装报价

Retrofit Modification Order 翻新改装指令

Risk Management Office 风险管理办公室

Risk Management Organization 风险管理组织

**RM-ODP** Reference Model for Open Distributed Processing 开放式分布处理的参考模型

**RMON** Remote Monitoring 远程监控

Remote Network Monitoring 远程网络监控

**RMP** Radar Modernization Program 雷达现代化计划

Radio Management Panel 无线电管理面板

Rated Maximum Pressure 额定最大压力

Receiver Modification Program 接收机改型计划

Reference Mode Panel 基准方式板

Regional Monitoring Program 区域监控计划

Remote Management Panel 遥控管理板

Reprogram Able Microprocessor 可重编程微处理机

Resource Management Plan 资源管理计划

Right Middle Plug 右侧中电嘴

Risk Management Plan 风险管理计划

Risk Management Program 风险管理程序

Root Mean Power 根均功率

**RMRB** Reliability Maintenance Review Board 可靠性维修评审委员会

**RMRN** Reconfigurable Multi-Ring Network 可重构多环网络

**RMRV** Radar Measurement Residual Vector 雷达测量残余向量

**RMS** Radio Management System 无线电管理系统

Range Measurement System 测距系统

Records Management Section 记录管理科

Reference Monitoring Station 基准监控站

Reliability and Maintainability Simulator 可靠性与可维修性模拟机

Reliability, Maintainability, Supportability 可靠性、维修性、保障性

Remote Manipulator System 遥控操作器系统

Remote Monitoring Subsystem 远端监视分系统

Resource Management System 资源管理系统

Revenue Management System 收入管理系统

Root Mean Square 均方根(值)

Route Management System 航线管理系统

Router Management Service 路由器管理服务

Royal Meteorological Society 皇家气象学会

**RMS&T** Reliability, Maintainability, Supportability and Testability 可靠性、可维修性、支援能力与可测试性

**RMSC** Remote Monitoring Subsystem Concentrators 遥控监视子系统集中器

**RMSE** Root Mean Square Error 均方根误差

**RMSS** Range Meteorological Sounding System 距离气象探测系统

**RMT** Reliability, Maintainability and Testability 可靠性、可维修性与可测试性

Remote Management Testing 远程管理测试

Risk Management Team 风险管理团队

Route Management Tool 航线管理工具

**RMT/RMTE** Remote 遥控的,遥远的

**RMTC** Regional Meteorological Training Centre 区域气象培训中心

**RMTN** Regional Meteorological Telecommunications Network 区域气象通信网

**RMU** Radio Management Unit 无线电管理组件

Records Management Unit 记录管理组件

Redundancy Management Unit 冗余度管理组件

Remote Maneuvering Unit 远程机动装置

Remote Measuring Unit 远程测量装置

Resource Management Update 资源管理升级

Runway Monitoring Unit 跑道监测组件

**RMUS** Rolt Master Update System 劳特主更新系统

**RMV** Remove 移动,拆卸

**RMVBL** Removable 可移动的,可拆卸的

**RMVL** Removal 拆卸

**RMX** Remote Multiplexer 远端复用器

**RN** Random Noise 随机噪声

Receiving Notice 接收通知

Reference Navigation 基准导航

Reference Noise 噪声基准

Remote Node 远端节点

Reorder Notice 再订货通知

Revision Notice 修改通知,改型通知

Reynolds Number 雷诺数

Right Nose 机头右侧

**RN** Radar Navigation 雷达导航

**RNAA** Request Network Address Assignment 请求网络地址分配

**RNAC** Royal Nepal Airlines Corporation 尼泊尔皇家航空公司

**RNAS** Relative Navigation Analytic Simulator 相关导航分析模拟器

**RNAV** Area Navigation 区域导航

Radar Navigation 雷达导航

Realisable Net Asset Value 可变现的净资产价值

Remote Navigation 远程导航

Revaluated Net Asset Value 重估的资产净值

Route Navigation 航线导航

**RNB** Ronneby, Sweden ( Airport Code ) 瑞典隆内比机场

**RNC** Radio Navigation Chart 无线电导航图

Radio Network Controller 无线网络控制器

Regional Network Center 区域网络中心

**RNCC** Regional Network Control Center 区域网络控制中心

**RND** Round 圆,环,圆形的,环绕

**RNDM** Random 随机的

**RNDV(S)** Rendezvous 交会

**RNE** Range of No Escape 无漏泄距离

Remote Network Equipment 远地网络设备

**RNET** Remote Network 远程网络

**RNF** Radio Navigation Facilities 无线电导航设备

Radio Noise Factor 无线电噪声系数

Radio Noise Figure 无线电噪声指数

Refracted Near-Field Technique 折射近场法

**RNG** Random Number Generator 随机数生成器

Range 距离,范围

**RNGG** Ringing 振铃,呼叫

**RNI** Radio Navigation Indicator 无线电导航指示器

**RNL** Rational 有理数,有理的,合理的

Renewal 更新,重复

**RNLT** Running Light 滑行灯

**RNN** Recurrent Neural Network 递归神经网络

Regional Neural Network 区域神经网络

**RNO** Reno, NV, USA ( Airport Code ) 美国内华达州雷诺机场

**RNP** Remote Network Processor 远程网络处理机

Required Navigation Performance 授权所需导航性能

**RNPC** Required Navigation Performance Capability 要求的导航性能能力

**RNR** Receive Not Ready 接收未准备就绪

Runner 转子,叶轮

**RNS** Rennes, France ( Airport Code ) 法国雷恩机场

**RNSS** Radio Navigation Satellite Services 无线电导航卫星服务

Radio Navigation Sensor System 无线电导航传感器系统

**RNST** Reinstate 复原,恢复

**RNTP** Radio Navigation Tuning Panel 无线电导航调谐板

**RNV** Radio Noise Voltage 无线电噪声电压

**RNWBL** Renewable 可更新的

**RNWL** Renewal 大修,更新

**RO** Radio Operator 无线电操作员,随机报务员

Receive Only 只接收

Relay Output 继电器输出

Remote Office 远端局

Remote Operation 远端工作

Requisition Objective 申请目标,要求对象

Roll out 滑出

Rollout 滑跑,突破

Rollover 翻转

Rough Opening 粗加工的开口

Route Order 航线指令

Runout 用完,偏转

**RO/RO** Roll-on/Roll-off 滑进/滑离

**ROA** Remotely Operated Aircraft 遥控飞机

Return on Assets 资产收益率

Roanoke, VA, USA ( Airport Code ) 美国弗吉尼亚州罗阿诺克机场

Rules of the Air 空中交通规则

**ROAM** Remote Office Access Management 远程办公访问管理

**ROB** Right Outboard 右外侧( 飞机),右机翼尖处( 飞机)

Route Operation Bulletin 航路飞行公告

**ROBEX** Regional Operational Meteorological Bulletin Exchange System 区域飞行气象通报交换系统

Regional Route Operational Meteorological Bulletin Exchange 地区航路飞行气象通报交换

Regional Route Operational Meteorological Bulletin Exchange System 地区航路飞行气象通报交换系统

**ROC** Radar Operations Center 雷达运行中心

Radar Optics Controller 雷达光学控制器

Rate of Climb 爬升率

Receiver Operating Characteristics 接收机使用特性

Reduced Operating Costs 降低运营成本

Regional Operations Centre 区域运营中心

Required Obstacle Clearance 要求越障(高度)

Required Operational Capability 要求运行能力

Residual Operational Capability 残存工作能力

Resolution of Comment 鉴定意见

Return on Capital 资本收益率

Rochester, NY, USA(Airport Code) 美国纽约州罗切斯特机场

**ROCC** Regional Operations Control Center 地区运行管制中心

**ROCE** Return on Capital Employed 使用资金的利润

**ROCP** Radar out of Commission for Parts 雷达因缺件停用

**ROD** Radar Only Display 只显示雷达(信息)

Rate of Descent 下降率

Required Operational Date 要求运行日期

**ROE** Rate of Exchange 兑换率

Remote Operating Equipment 远程操作设备

Return on Equity 资本权益报酬率

Rule of Engagement 衔接规则

**ROFOR** Route Forecast 航线天气预报

**ROH** Regular Overhaul 常规翻修

**ROI** Rovaniemi, Finland (Airport Code) 芬兰罗凡涅米机场

Return on Investment 投资收益率

**ROIC** Return on Investment Capital 投资回报率

**ROL** Repeated Optical Link 光中继链路

**ROLR** Roller 滚棒

**ROLS** Remote Oil Level Sensor 远距滑油液面传感器

**ROLT** Reorder Lead Time 重订货的提前时间

**ROM** Ramp Operation Manual 机坪保障手册

Read Only Memory 只读存储器

Reciprocal Outer Marker 反向远指标点

Romania 罗马尼亚

Rome, Italy (Airport Code) 意大利罗马机场

Rough Order of Magnitude 近似数量级

**ROMAD** ROM Address 只读存储器地址

**ROMCLK** ROM Clock 只读存储器时钟

**ROMM** Read Only Memory Module 只读存储器模块

**RON** Receiving Only 只收,仅接收

Remain Overnight (飞机)过夜

**ROOT** Relaxation Oscillator Optically Tuned 光调谐弛张振荡器

**ROP** Regional Operational Programme 区域运行方案

**ROPA** Remote Optically Pumped Amplifier 远端光泵放大器

**ROPN** Reopen 重开

**ROR** Koror, Palau (Airport Code) 帕劳科罗尔机场

Range Only Radar 仅测距雷达

Retest Okay Rate 重测合格率

**RORD** Reorder 重排,再订货

**ROS** Read-Only Storage 只读存储器

Real-time Operating System 实时操作系统

Remote Operation Service 远程操作服务

Return on Sales 销售利润

**ROSA** Rotorcraft Open Systems Avionics 旋翼机开式系统电子设备

**ROSE** Roster Optical Scanning Engine 逐行光学扫描引擎

Remote Operation Service Element 远程操作服务单元

**ROST** Regional Office Safety Team 地区办公室安全小组

**ROT** Radar on Target 雷达对准目标

Rate of Turn 转弯率

Receive-Only Terminal 单收终端

Reserve Oil Tank 储备(滑)油箱

Rotation 旋转

Runway Occupancy Time 跑道占用时间

**ROTHR** Relocatable over-the-Horizon Radar 可再定位超视距雷达

**ROTL** Remote Office Test Line 远端局测试线路

**ROTOMT** Rotometer 转速表

**ROTOS** Rotating Tool Schedule System 旋转工具程序系统

**ROTR** Rotator 转子

**ROTS** Range-on-Target Signal 距离对准目标信号

**ROUT** Routine 航线,程序安排,常规

**ROW** Right of Way 路权

**ROX** Oxidizer 氧化剂

**ROY** Royalty 特许权

**RP** Rated Power 额定功率

Reference Page 参考页

Reference Point 基准点,参考点

Reference Pulse 基准脉冲

Regional Processor 区域处理机

Reinforced Plastic 增强塑料

Remote Point 远端点

Reservation Packet 预留分组

Reverse Power 反向电源

Revision Proposal 修改建议,修改提案

Rig Pin 校装销

**RPA** Radar Performance Analyzer 雷达性能分析器

Regional Planning Association 地方规划协会

Remotely Piloted Aircraft 遥控飞机,遥控航空器

Rotocraft Pilot's Assistant 旋翼飞机驾驶员协助

**RPAG** Radiological Protection Advisory Group 放射防护咨询小组

**RPAODS** Remotely Piloted Aerial Observation Detection System 无人驾驶(遥控)空中观察探测系统

**RPB** Regional Processor Bus 区域处理机总线

**RPC** Radio Port Controller 无线端口控制器

Remote Position Control 远距位置控制

Remote Power Control 远距电源控制,远距动力控制

Remote Procedure Call 遥控程序呼叫

Repairable Processing Center 可修件处理中心

**RPCL** Repetitively Pulsed Chemical Laser 重复脉冲化学激光器

**RPCM** Remote Power Control Mechanism 远程动力控制机理

**RPCU** Residual Pressure Control Unit 剩余压力控制器

**RPD** Radar Planning Device 雷达计划装置

Radar Prediction Device 雷达预测装置

Rapid Prototyping Development 快速原型发展

Reference Point Delay 基准点延迟

Renewal Parts Data 大修部件数据

**RPDG** Rapid Product Development Group 快速产品开发组

**RPDR** Reproducer 复制器,扬声器,再现设备

**RPDS** Remote Power Distribution System 远程动力分配系统

**RPDU** Regulator, Processor and Distribution Unit 控制器、处理器和分配器

Remote Power Distribution Unit 远程组件分

配动力

**RPE** Registered Professional Engineer 注册专业工程师,注册职业工程师

Repair Parts Estimate 修理部件评估

Resource Planning and Evaluation 资源计划与鉴定

**RPF** Radiometer Performance Factor 辐射计性能因子

**RPFC** Right Primary Flight Computer 右主飞行计算机

**RPG** Regional Planning Group 区域规划组

Report Program General 报告程序概述

Report Program Generator 报告程度产生器

Toxic Gas 有毒气体

**RPH** Regional Processor Handler 区域处理机处理器

Relative Probability Hazard 灾害相关概率

Remotely Piloted Helicopter 遥控直升机

**RPI** Radar Precipitation Integrator 雷达降雨积累器

Reference Path Indicator 基准路径指示

**R-PICH** Reverse Pilot Channel 反向导频信道

**RPIE** Real Property Installed Equipment 不动产安装设备

**RPK** Revenue Passenger Kilometers 营运人公里数,收入客公里

**RPL** Radiophoto Luminescence 辐射光致发光

Rated Power Level 额定功率电平

Recreational Pilot License 休闲用飞行员执照

Renewal Parts Leaflet 大修件活页

Repetitive Flight Plan 重复性飞行计划

**RPLC** Replace or Replaced 代替

**RPLNG** Replenishing 补充,加强

**RPLNT** Repellent 防水剂,防水的,排斥的

**RPLPC** Regular Pulse Linear Predictive Coding 规则脉冲线性预测编码

**RPLR** Repeller 反射极,弹回装置

**RPLSN** Repulsion 排斥力

**RPM** Rapid Prototyping Manufacturing 快速成形制造

Reliability Planning and Management 可靠性计划与管理

Remote Performance Monitoring 遥控性能监测

Reserve Property Management 储备物资管理

Revenue Passenger Miles 收益客英里

Revolutions per Minute 每分钟转数

Runway Performance Manual 跑道性能手册

**RPMB** Remotely Piloted Mini-Blimp 遥控小型飞机

**RPMC** Remote Performance Monitoring and Control 遥控性能监测与控制

**RPNSM** Replenishment 补充

**RPOA** Recognized Private Operating Agency 认可的民营电信机构

**RPP** Receive Personality PROM 接收用可编程只读存储器

Reliability Program Plan 可靠性程序设计

**RPPI** Remote Plan Position Indicator 遥控平面位置显示器

**RPPP** Repair Parts Program Plan 修理件程序计划

**RPQ** Request for Price Quotation 报价申请,征求报价

**RPR** Repair 修理

Repairable 可修的

**RPRR** Retrofit Production Revision Request 改型生产修订请求

**RPRT** Repart 再划分

Report 报告

**RPRV** Remotely Piloted Research Vehicle 研究用遥控飞行器

**RPS** Radar Position Symbol 雷达位置符号

Rapid Prototyping System 快速原型机系统

Regional Processor Subsystem 区域处理机子系统

Remote Power Supply 远距离电源供给

Remote Processing System 远程处理系统

Requirements Processing System 要求处理系

统

Revolutions per Second 每秒钟转数

Ring Parameter Server 环参数服务器

**RPSA** Rotating Pressure Sensing Assembly 旋转压力感应组件

**RPSM** Requirements Processing System-Material 器材需求处理系统

Resources Planning and Scheduling Method 资源计划与调度法

**RPSTL** Repair Parts and Special Tool List 修理零用件和专用工具单

**RPT** Reference Point Terminate 参考点终止

Regional Processor Tester 区域处理机测试器

Regular Public Transport 定期公共运输

Repeat/Repeated 重复,重复的

Repeatability 重复性

Report/Reporting/Reported 报告(的)

Resident Provisioning Team 驻场保障组

Revenue Passenger Transport 收费客运

**RPTC** Repeating Coil 重复线圈

**RPTN** Repetition 副本,重复

**RPTR** Repeater 重复器,转发器,中继站,循环小数

**RPU** Receiver Processor Unit 接收机处理器组件

**RPV** Reactor Pressure Vessel 反应堆压力容器

Remotely Piloted Vehicle 遥控飞行器,远距离操纵车

Resource, Process and Values 资源、流程和价值观

**RPVIT** RPV Institutional Trainer 遥控飞机原理训练器

**RPVNTV** Rust Preventive 防锈剂

**RPX** Retail Package Express 零售包裹快递

**RPZ** Runway Protection Zone 跑道保护区

**RQ** Recommended Quantity 推荐数量

Reportable Quantity 应申报量

Space Requested 要求订座

Space Waitlisted 座位候补

**RQA** Request for Acknowledgment 请求确认

**RQA&SE** Reliability, Quality Assurance and Safety Engineering 可靠性、质量保证与安全性工程

**RQC** Radar Quality Control 雷达质量控制

Regional Quality Control 区域质量控制

Reliability and Quality Control 可靠性和质量控制

Ride Quality Control 乘坐品质控制

**RQCL** Request Clearance 请求放行许可

**RQL** Rich Burn-Quench-Lean Burn 富油燃烧—焠熄—贫油燃烧

**RQM** Realtime Quality Management 实时质量管理

Requirement 需求,要求

Risk and Quality Management 风险和质量管理

**RQP** Request Flight Plan Message 请求飞行计划报告

**RQRD** Required 需要的

**RQST** Request 需要

Seat Request (Reservations Code) 机位请求

**RQT** Reliability Qualification Test 可靠性鉴定试验

**RQUAL** Requalification 重新鉴定

**RR** Radar Ranging 雷达距离,雷达测距

Radio Regulations 无线电规则

Rapid Release 快速分离,快速脱开

Rapid Revision 快速修订

Receive Ready 接受准备就绪

Receiving Report 接收报告

Recurrence Rate 再现率,重复频率

Reinspection Request 重新检查要求

Rejection Report 报废报告,拒收报告

Repetition Rate 重复率

Request Response 请求应答

Requirements Review (性能)要求评审

Re-Route 重定路由

Reservations 预订

Reset Relay 复位继电器

Response Request 应答请求

Revision Record 修订记录

Right Rear 右后

Round Robin 在同一机场起落的长途飞行

Rudder Rib 方向舵翼肋

**R-R** Rolls-Royce 罗尔斯·罗伊斯(公司)

**RRA** Reservation Random Access 预留随机接入

**RRB** Requirements Review Board (性能)要求审委员会

Retrofit Review Board 改装评审委员会

**RRC** Radio Resource Control 无线资源控制

Regional Radar Center 区域雷达中心

Rudder Ratio Changer 方向舵比率改变器

**RRCA** Rudder Ratio Changer Actuator 方向舵比率改变器作动筒

**RRCM** Rolls Royce Control Module 罗·罗控制模块

Rudder Ratio Changer Module 方向舵比率变换器模块

Rudder Ratio Control Module 方向舵比率控制模块

**RRCN** Retrofit/Rework Completion Notice 改型/返修完成通知

**RRF** Rescue and Recovery Function 救援及恢复功能

Reservation Request Form 预约申请表

Risk Reduction Factor 风险减少因素

**RRG** Rejection Record Graphics 报废记录图表

Rodrigues Island, Mauritius (Airport Code) 毛里求斯罗德里格斯岛机场

**RRI** Range Rate Indicator 距离速率指示器

Removal Rate Indicator 拆卸率指示符

**RRJ** Russian Regional Jet (Program) 俄罗斯的新一代涡扇支线客机

**RRL** Remote Reload 遥控更新加载,遥控更新装入

**RRM** Radio Resource Management 无线资源管理

Receive Redundancy Management 接收冗余管理

Requirements Request Memorandum (性能)要求申请备忘录

**RRO** Rejection Rework Order 报废返修指令

Removal and Replacement Order 拆换指令

**RRP** Rapid Revision Proposal 快速修订建议

Resource Requirements Planning 资源需求计划

Resource Reservation Protocol 资源预约协议

Runway Reference Point 跑道基准点

**RRR** Rapid Runway Repair(s) 快速维修跑道

Required Rate of Return 期望收益率

Risk Reduction Report 风险降低报告

Risk-Reward Ratio 风险利润率

Routing with Resource Reservation 资源预留选路

**RRRE** Radar Range-Rate Error 雷达距离变化率误差

**RRRV** Rate of Rise of Recovery Voltage 过渡恢复电压上升率

**RRS** Radar Ranging System 雷达测距系统

Rejection Record System 报废记录系统

Release Rate for Smoke 烟雾消散率

Restraint Release System 限制器消除系统,制约解除系统

Retransmission Request Signal 重传请求信号

Roll Reference Sensor 倾斜基准传感器

**RRSA** Roll Reference Sensor Assembly 倾斜基准传感器组件

**RRSO** Roll Reference Sensor Output 倾斜基准传感器输出

**RRT** Radar Range Tracking 雷达作用距离跟踪

Radar Receiver-Transmitter 雷达收发两用机

Recovery Response Time 恢复响应时间

Refugee Review Tribunal 难民上诉审裁处

Reliability Run Test 可靠性运转试验

Requirements Review Team (性能)要求评审组

**RRTC** Roll Response Time Constant 滚转响

应时间常数

**RRTE** Reroute 绕行,改变航线

**RRTR** Risk Reduction Test Report 减小风险测试报告

**RRU** Regenerative Repeater Unit 再生中继单元

**RRW** Radioactive Material, Category Ⅰ-White Ⅰ 类放射性材料—白色

**RRWDS** Radar Remote Weather Display System 雷达远程气象显示系统

**RRY** Radioactive Material, Categories Ⅱ and Ⅲ-Yellow Ⅱ、Ⅲ 类放射性材料—黄色

**RRZ** Radar Regulation Zone 雷达管制区

**RS** Radar Set 雷达站

Radiated Susceptibility 辐射敏感度

Radio Station 无线电台

Range Search 距离搜索

Range Selector 距离选择器

Range Surveillance 距离监视

Ready Service 准备提供服务

Real Storage 实际存储器

Real System 实系统

Rear Spar 后梁

Receiver Sensitivity 接收机灵敏度

Recommended Standard 推荐标准

Reconciliation Sublayer 调和子层

Remote Sensing 遥感

Remote Switching 远程交换

Repeating Stroke 重复冲程

Repetitive Stroke 重复冲程,重复打击

Reservation System 订座系统

Restoration 恢复

Reverberation Strength 反响强度

Right Side 右侧

Rubble Stone 毛石

**R-S** Reed-Solomon R-S 码

**RS&I** Receipt, Storage and Issue 接收、存储与发送

**RS&REL** Requirement Summary and Requirement Error List 需求摘要与需求错误表

**RSA** Radar Set Antenna 雷达装置天线

Regulations, Standards and Advisory Material 规章、标准及咨询材料

Remote Station Alarm 远地站告警

Republic of South Africa 南非共和国

Request for Service Allowance 维护许可申请

Runway Safety Area 跑道安全区

**RSALT** Running Signal and Anchor Lights 航行信号及系留灯

**RSAP** Runway Safety Action Plan 跑道安全行动计划

**RSAT** Runway Safety Action Team 跑道安全行动小组

**RSB** Rudder Speed Brake 方向舵减速板(气动力减速装置)

**RSC** Radar Scan Converter 雷达扫描变换器

Radar Set Control 雷达装置控制

Receive Speech Channel 接收话音信道

Recording System Controller 记录系统控制器

Regional Support Center 区域支持中心

Remote Switching Center 远程交换中心

Rescue Sub-Center 救援辅助中心

Royal Shakespeare Company 莎士比亚皇家公司

**RSCD** Runway Surface Condition 跑道道面情况

**RSCE** Restoration Switching Control Equipment 恢复交换控制设备

**RSCF** Right System Card File 右系统插件卡目录

**R-SCH** Reverse Supplementary Channel 反向辅助信道

**RSCS** Remote Spooling Communications Subsystem 远程并行联机通信分系统

**RSCT** Reliable Scalable Cluster Technology 可靠的可伸缩集群技术

**RSCU** Representative, Supplier, Customer and User 代表、供应商、客户和用户

**RSD** Raised 上升的

Remote Sensing Data 遥感数据

Remote Storage Device 远程存储设备

Retinal Scanning Display 视网膜扫描显示

Rock Sound, Bahamas (Airport Code) 巴哈马罗克桑德机场

Rolling Steel Door 卷帘钢门

**RSE** Radar Search Equipment 雷达搜索设备

Radar Site Equipment 雷达场地设备

Reliability System Engineering 可靠性系统工程

Remote Station Equipment 远地站设备, 远端站装置

Remote Supervisory Equipment 遥控监视设备

Restoration Switching Equipment 恢复交换设备

**RSF** Routine and Standards File 常规与标准文件

**RSFS** Research Support Flight System 研究支援飞行系统

**RSG** Range Safety Generation 距离安全产生

Raster Stroke Generator 光栅笔画产生器

Raster Symbol Generator 光栅符号产生器

**RSI** Reflected Signal Indicator 反射信号指示器

Remote Server Interface 远地服务器接口

Reusable Surface Insulation 可重复使用表面隔热层

Right Scale Integration 适当规模集成(电路)

**RSIM** Radar Simulator 雷达模拟器

**RSIP** Radar System Improvement Program 雷达系统改进计划

Radar Sensitivity Improvement Program 雷达灵敏度改进计划

**RSIS** Rotorcraft System Integration Simulator 旋翼机系统综合模拟器

**RSITA** International Aeronautical Telecommunications Services Regulation 国际航空电信服务条例

**RSL** Reconnaissance and Security Line 侦察和警戒线

Runway Status Light 跑道状态灯

**RSLP** Reentry System Launch Program 重返系统发射程序

**RSLT** Result 结果

**RSLVR** Resolver 解算器

**RSM** Remote Switching Module 远地交换模块

Roll Servo Motor 横滚伺服马达

Runway Surface and Marking 跑道道面及标记

**RSMC** Regional Specialized Meteorological Centre 区域专业气象中心

**RSMM** Redundancy System Management Model 冗余系统管理模型

Resource System Management Module 资源系统管理模块

**RSN** Radio Station Node 无线电基站节点

Rearrangeable Switching Network 可再安排的交换网络

Reason 理由

**RSO** Resident Space Object 空间驻留物体

**RSOOs** Regional Safety Oversight Organizations 地区安全监督组织

**RSP** Radar Signal Processor 雷达信号处理器

Real-time Simulation Platform 实时模拟平台

Recommendation for Service Publications 服务印刷品推荐

Reduced Shortest Path 简约的最短通路

Render Safe Procedure 表演安全程序

Required Surveillance Performance 要求监视性能, 所需监视性能

Required System Performance 所需系统性能

Requirements Study Plan 需求研究计划

Requirements Summation Program 必需品总加程序

Responder 应答器

Responder Beacon 应答机信标

Reversion Select Panel 逆选板, 非常规选择板

Rotational Sampling Program 循环抽样程序

Route Selection Program 航线选择计划

**RSPA** Request for Standards Procurement Assistance 标准采购协助请求

**RSPC** Remote Speech Path Controller 远地话路控制设备

**RSPCTVLY** Respectively 分别地

**RSPD** Respond 响应

**RSPE** Radar Signal Processing Equipment 雷达信号处理设备

**RSPL** Recommended Spare Parts List 推荐的备件清单

**RSPM** Remote Speech Path Module 远地话路模块

**RSPS** Response 响应

**RSPT** Replenishment Spare Parts Team 补充备件组

**RSPTR** Respirator 呼吸面罩

**RSPV** Respective 相应的

**RSQ** Rescue 救援

**RSR** Resource Status Report 资源状况报告
Review Summary Report 评审总结报告
Route Surveillance Radar 航路监视雷达

**RSRA** Rotor Systems Research Aircraft 旋翼系统研究机

**RSRC** Resource 资源

**RSRV** Reserve 储备

**RSS** Radar Signal Simulator 雷达信号模拟器
Radar Support System 雷达支持系统
Radio Sensor System 无线电传感系统
Ready Service Spares 待命维修备件
Rear Spar Station 后梁站位
Received Signal Strength 接收信号强度
Receiver Surveillance System 接收机监视系统
Regional Surveillance System 区域监视系统
Relational Storage System 相关存储系统
Relaxed Static Stability 放宽静态稳定性
Remote Subscriber Switch 远地用户交换
Remote Surveillance System 远程监视系统，遥控监视系统
Remote Switching Stage 远地交换级
Remote Switching System 远地交换系统

Requirements Summation System 要求总加系统
Re-Set System 重置系统
Resource Security System 资源安全系统
Resource Sharing System 资源共享系统
Root Sum Square 和的平方根
Route Selection Services 选路服务
Route Switching Subsystem 路由交换子系统

**RSSI** Received Signal Strength Indication 接收信号强度指示
Received Signal Strength Indicator 接收信号强度指示器

**RSSP** Radar Software Simulation Program 雷达软件模拟程序

**RST** Radar Start 雷达启动
Readability Strength and Tone 可读性强度与音调
Regenerator Section Termination 再生段终端
Regional Signaling Terminal 区域信令终端
Reinforcing Steel 钢筋
Remote Station 远程站
Reset 复位，重置，清零
Reset-Set Trigger 复位—置位触发器
Restore 恢复，还原
Rochester, MN, USA(Airport Code) 美国明尼苏达州罗切斯特机场

**RSTG** Roasting 烘烤

**RSTM** Route Set Test Management 路由组测试管理

**RSTN** Radio Solar Telescope Network 无线电太阳望远镜网络

**RSTP** Regional Signaling Transfer Point 区域信令转换点

**RSTPF** Rustproof 防锈的

**RSTR** Raster 光栅
Restoration 修复
Restrict/Restriction 限制，约束

**RSU** Raster Shading Unit 遮光栅装置
Remote Subscriber Unit 远程用户单元
Remote Switching Unit 远端交换单元
Runway Supervisory Unit 跑道监视装置

**RSUB** Resubmitting 再建议,再提高

**RSUM** Resume 恢复,概述

**RSUS** Rack Service Update System 支架维修更新系统

**RSV** Reserve 保留,储备

**RSVD** Reserved 保留的,储备的,备用的

**RSVP** Resource Reservation Protocol 资源预留协议

**RSVR** Reservoir 储油箱

**RSW** Fort Myers, FL, USA(Airport Code) 美国佛罗里达州麦尔兹堡机场

Reset Switch 复位开关

**RT** Radar Trigger 雷达触发器

Radio Telephone 无线电话

Radio Transmission 无线电发射

Raintight 防雨的

Range Tracking 距离跟踪

Rate 速率,比例

Rate of Turn 转弯速率

Real Time 实时

Real-Time System 实时系统

Receive-Transmit 接收—发射

Receiver-Transmitter 收发机

Receiver-Transmitter(Unit) 收发机组件,收发组

Rejection Tag 报废标签

Reliable Transfer 可靠传送(转移)

Relocatable Target 可重新定位目标

Remote Terminal 远程终端

Research and Technology 研究与技术

Response Time 响应时间

Returned to 回到

Reverse Thrust 反推

Right 右

Ringing Tone 振铃音

Round Trip 来回程

Routine Test 例行测试

Routing Table 路由表

Routing Topology 选路拓扑

Run up and Taxi 试车和滑行

**RTA** Receiver-Transmitter Antenna 收发机天线

Refrigeration Trade Association 制冷贸易协会

Regional Trade Agreement 区域贸易合作协议

Reliability Test(ing) Assembly 可靠性测试组件

Required Time of Arrival 所需到达时间

Road Traffic Act 道路交通法令

Road Transport Association 道路运输协会

**RTAB** Retabulation 重新制表

**RTAC** Real-Time Adaptive Control 实时自适应控制

Regional Technical Aid Center 地区技术援助中心

Research and Technology Advisory Council 研究与技术咨询委员会

**RTAF** Radio Transmission Adaption Function 无线传输适配功能

**RTAM** Remote Terminal Access Method 远程终端接入法

**RTANG** Right Angle 直角

**RTAS** Real-Time Analysis System 实时分析系统

**RTB** Radar Test Bench 雷达测试台

Radio Technical Battalion 无线电技术营

Return to Base 返回基地

Roatan, Honduras(Airport Code) 洪都拉斯罗阿坦机场

Rotor Tracking and Balance 旋翼轨迹与平衡

**RTC** Radiotelephony Communication 无线电话通信

Rapid Transit Car 快速轨道车

Reader Tape Contact 阅读磁带触头

Real Time Clock 实时钟

Receiver Threshold Control 接收机门限控制

Reconfiguration Transfer Contactor 重构型转换接触器

Recurrent Training Center 恢复技术训练中心

Regional Training Center 区域培训中心

Reinstrumented Terrain Computer 再装备仪表的地形计算机

Return to Control 返回控制

Right Trim Control 右配平控制

Route Control 路由控制

Rudder Trim Control 方向舵配平控制器

**RTCA** Radio Technical Commission for Aeronautic 航空无线电技术委员会

Radio Technical Commission of America 美国无线电技术委员会

Requirements and Technical Concepts for Aviation 航空需求及技术概念

**RTCC** Real-Time Communications Channel 实时通信信道

Real-Time Control Center 实时控制中心

**RTCF** Real Time Conversion Facility 实时转换设备

**RTCL** Reticle 标线

**RTCM** Radio Technical Commission for Maritime 海事无线电技术委员会

Radio Technical Commission for Maritime Service 海事服务无线电技术委员会(美国)

**RTCP** Real-Time Control Protocol 实时控制协议

Real-Time Transport Control Protocol 实时传输控制协议

**RTCU** Real-Time Control Unit 实时控制组件

Receive/Transmit Control Unit 收发控制组件

**RTD** Radar Target Designator 雷达目标标示器

Range Time Decoder 距离时间解码器

Research and Technical Development 研究和技术开发

Resistance Temperature Detector 电阻式温度探测器

Resistive Temperature Device 电阻温度装置

Retard 延迟,制动

Round-Trip Delay 往返时延

Routed 指定航线的

**RTDBS** Real-Time Database System 实时数据库系统

**RTDC** Radar Target Designation Control 雷达目标标定控制

Real-Time Data Controller 实时数据控制器

Retardation Coil 延迟线圈

**RTDD** Real-Time Data Distribution 实时数据分配

Remote Timing and Data Distribution 遥控定时与数据分配

**RTDDC** Real-Time Digital Data Correction 实时数字数据校正

**RTDGPS** Real-Time Differential Global Positioning System 实时差分全球定位系统

**RTDHS** Real-Time Data Handling System 实时数据处理系统

**RTDS** Rapid Transmission Data System 快速传输数据系统

Real-Time Data System 实时数据系统

**RTE** Radiative Transfer Equation 辐射传递方程

Radiative Transport Equation 辐射输运方程

Random Timing Error 随机定时误差

Real-Time Environment 实时环境

Real-Time Executive 实时执行

Real-Time Executive System 实时执行系统

Remote Terminal Emulator 远程终端模拟器

Remote Terminal Equipment 远程终端设备

Replacement Transporter Erector 更换件输送安装架

Request to Engineering 申请工程实施

Route 航路,航线

**RTEL** Radio Telephone 无线电话

**RTEM** Radar Tracking Error Measurement 雷达跟踪误差测量

**RTF** Rubber Tile Floor 橡胶砖地板

**RTFR** Reliability, Trouble and Failure Report 可靠性、故障和失效报告

**RTG** Radio-isotope Thermoelectric Generator 放射性同位热电产生器

Radiotelegraph 无线电报

Rating 等级,规格

Routing 选定航线

**RTH** Regional Telecommunication Hub 区域电信枢纽

**RTHL** Runway Threshold Lights 跑道入口灯

**RTI** Radar Target Indicator 雷达目标指示器

Real Time Interface 实时接口

Real Time Interrupt 实时中断

Real-Time Interrogate 实时询问器

**RTIO** Real-Time Input/Output 实时输入输出

**RTIP** Radar Target Identification Point 雷达目标识别点

Real-Time Internet Protocol 实时因特网协议

**RTIR** Reliability Trend Indicator Report 可靠性倾向指示器报告

**RTIRS** Real-Time Information Retrieval System 实时情报检索系统

**RTIS** Radar Test Instrumentation System 雷达试验仪表系统

Radar Traffic Information Service 雷达交通信息服务

Remote Terminal Input Subsystem 远程终端输入子系统

Remote Traffic Information System 远程交通信息系统

**RTIU** Remote Terminal Interface Unit 远程终端接口单元

**RTK** Real-Time Kinematic Carrier Tracking 实时动态载波跟踪(技术)

Revenue Tonne Kilometer 吨公里收益

**RTL** Radar Threshold Limit 雷达门限极限

Ready to Load 准备装载

Rejection Tag Log 报废牌记录

Reverse Thrust Lever 反推手柄

Rudder Travel Limiter 方向舵偏转角限动器

Rudder Travel Limiting 方向舵行程限制

**RTLA** Rudder Travel Limiter Actuator 方向舵偏转角作动筒

**RTLG** Radio Telegraph 无线电报

**RTLP** Reference Transmission Level Point 参考传输电平点

**RTLS** Return to Launch Site 返回发射场

**RTLU** Rudder Travel Limitation Unit 方向舵行程极限组件

**RTM** Radar Target Material 雷达目标材料

Radio Transmission Management 无线传输管理

Rapid Tuning Magnetron 速调磁控管

Real Time Monitoring 实时监控

Receiver-Transmitter Modulator 收发机调制器

Recording Tachometer 记录转速表

Reference Test Method 基准测试方法

Regional Transport Model 区域运输模型

Registered Trade Mark 注册商标

Reliability Test Module 可靠性测试模块

Resin Transfer Molding 树脂传递模塑成型

Revenue Ton Mile 收费吨英里,收入吨英里

Rotterdam, Netherlands (Airport Code) 荷兰鹿特丹机场

**RTMC** Real-Time Management Coordination 实时管理协调

**RTMM** Real-Time Multimedia 实时多媒体

**RTMS** Radio Telephone Mobile System 无线电话移动系统

Real-Time Memory System 实时存储系统

**RTN** Regional Telecommunications Network 区域电信网

Remote Track Number 远程追踪号码

Remote Tracking Network 远程跟踪网

Republique Tunisienne 突尼斯共和国

Retain 保持,制动

Routing and Transit Number 行程安排与运输代码

**RTNG** Retaining 保持

**RTNMS** Real-Time Network Management System 实时网络管理系统

**RTNR** Real-Time Network Routing 实时网络选路

Retainer 保持器,止动器

**RTNV** Return Vector 返回向量

**RTO** Rejected Take-off 中断起飞,放弃起飞

Research and Technology Organization 研究与技术组织

Runway Turnoff Lights 跑道滑出口标志灯

**R-TO** Reduced Thrust Take off 减推力起飞

**RTODAH** Rejected Take-off Distance Available, Helicopter (直升机)中断起飞可用距离

**RTOK** Retest OK 再测试通过

**RTOL** Reduced Takeoff and Landing 减距起飞和着陆,减推力起飞和着陆

**RTOLW** Runway Takeoff and Landing Weight 跑道起飞及着陆重量

**RTOP** Research and Technology Operating Plan 研究与技术运行计划

**RTOS** Real-Time Operating System 实时操作系统

**RTOT** Range Track on Target 目标距离跟踪

**RTOW** Regulated Take off Weight 额定起飞重量

Regulatory Takeoff Weight Chart 标准的起飞重量图表

Runway Takeoff Weight 跑道起飞重量

**RTP** Radio Tuning Panel 无线电调谐板

Rapid Transport Protocol 快速传送协议

Real-time Transport Protocol 实时传送协议

Regional Transportation Plan 区域运输计划

Reliability Test Plan 可靠性测试计划

Request to Purchase 订货申请

Requirement and Test Procedure 需求与测试程序

Right Top Plug 右上电嘴

**RTPH** Round Trips per Hour 每小时往返程

**RTPL** Real-Time Processing Language 实时处理语言

**RTPS** Real-Time Programming System 实时编程系统

**RTQC** Real-Time Quality Control 实时质量控制

**RTR** Real-Time Reconnaissance 实时侦察

Remote Transmitter/Receiver 遥控发射机/接收机

Request to Receive 请求接收

Reserve Thrust Rating 备用推力额定值

Rotor 转子,旋翼

**RTRSW** Rotary Switch 旋转开关

**RTRY** Rotary 旋转的

**RTS** Radar Target Simulator 雷达目标模拟器

Radar Terrain Sensor 雷达地形传感器

Ready to Send 准备发送

Real-Time Simulator 实时模拟器

Real-Time System 实时系统

Remote Terminal System 远地终端系统

Remote Tracking Sensor 遥控跟踪传感器

Remote Tracking Station 遥控跟踪站

Repairable at This Station 可在本站修复

Request to Send 请求发送

Resistive Touch Screen 电阻式触摸屏

Return to Seat 返回座位

Return to Service 恢复工作,恢复服务

Return to Supplier 退回供货方

Round Trip Sample 往返样本

Run Time System 运转时间系统

**RTSE** Reliable Transfer Service Element 可靠传送服务单元

**RTSP** Real-Time Streaming Protocol 实时流协议

Required Total System Performance 必备总系统性能,所需总系统性能

**RTT** Radio Teleprinter 无线电传打字机

Radio Transmission Technology 无线传输技术

Real-Time Traffic 实时业务

Round Trip Time 往返时间

Round Trip Timing 往返定时

**RTTC** Regional Technology Transfer Center 区域技术转让中心

**RTTE** Remote Terminal Test Equipment 远地终端测试设备

**RTTL** Running Telltale Light 滑跑信号

**RTTP** Real-Time Transport Protocol 实时传送协议

**RTTS** Radio Terminal Test Set 无线电终端测试装置

Real-Time Traffic Service 实时交通服务

Real-Time Transmission System 实时传送系统

Robotic Target Training System 自动目标训练系统

**RTTV** Real-Time Television 实时电视

**RTTY** Radio Teletype 无线电电传

Radio Teletypewriter 无线电电传打字机

**RTU** Radio Tuning Unit 无线电调谐组件

Rate of Turn Unit 转弯率组件

Remote Terminal Unit 远端终端单元

Remote Testing Unit 远端测试单元

Request to Update 申请更新

Retaining Unit 保存组件

Routing Table Update 路由表更新

**RTV** Real Time Video 实时视频

Reentry Test Vehicle 再入（大气层）试验飞行器

Research Test Vehicle 科学实验飞行器

Return to Vendor 退回供货方

Room Temperature Vulcanizing 室内温度硫化

**RT-VBR** Real Time Variable Bit Rate 实时可变比特率

**RTW** Round the World 环球

**RTWS** Raw Tape Write Submodule 原始磁带写入分模块

**RTX** Relay through Telex 以电传打字机转播

**RTZ** Return to Zero 归零

**RTZL** Runway Touchdown Zone Lights 跑道接地带灯

**RU** Ready Use 备用

Recovery Unit 恢复单元

Remote Unit 远端单元

Replaceable Unit 可更换的组件

Reproduceable Unit 可复制的组件

Restricted Usage 限制使用

**RUA** Remote User Agent 远程用户代理

**RUADDS** Runway Utilization and Automated Departure Display System 跑道利用和自动离港显示系统

**RUB** Rubber 橡皮,橡胶

**RUD** Rudder 方向舵

**RUF** Revolving Underwriting Facility 循环认购工具

**RUH** Riyadh, Saudi Arabia（Airport Code）沙特阿拉伯利雅得机场

**RUM** Radar Unit Mount 雷达安装组件

**RUN** St-Denis De La Reunion, Reunion-Gillot（Airport Code）留尼汪岛圣但尼基洛机场

**RUNS** Run Stop 停止运行

**RUP** Roll up 卷起,到达,积累成

**RUSHR** Rush Reply 速复(通信简语)

**RUSS** Russet 黄褐色的

**RUT** Rutland, USA（Airport Code）美国拉特兰机场

**RV** Rated Voltage 额定电压,标称电压

Rear View 后视图

Reentry Vehicle 再入飞行器

Relief Valve 释压活门

Rendezvous 在指定地点集合

Rescue Vessel 营救船只

**RV/BCN** Rendezvous Beacon 交会信标

**RVA** Reactive Volt-Ampere Meter 无功伏安表

Relative Virtual Address 相对虚拟地址

Reverser Actuator 反作用执行机构

**RVAL** Revalidating or Revalidation 再次确认,再次批准

**RVARM** Recording Varmeter 记录无功伏安表

**RVCP** Remote Video Control Panel 远距离影像控制面板,远距离视频控制面板

**RVDT** Rotary Variable Differential Transformer 旋转式可变差动变压器

Rotary Variable Displacement Transducer(s)

旋转式可变位移传感器

**RVI** Remote Visual Inspection 远程目视检查

Residual Value Insurance 残值保险

**RVLG** Revolving 旋转的,循环式

**RVLV** Revolve 旋转,循环

**RVM** Reactive Voltmeter 反应式电压表

Recoverable Virtual Memory 可恢复的虚拟存储器

**RVNU** Revenue 收益,收入

**RVO** Runway Visibility Observer 跑道能见度观察员

**RVP** Radar Video Processor 雷达视频处理器

Reid Vapor Pressure 雷德蒸汽压

Remote Voice Protocol 远程话音协议

Requirements Verification Plan 需求验证计划

**RVPS** Rotational Vibration Protection System 旋转震动保护系统

**RVR** Runway Visual Range 跑道视程,跑道能见距离

**RVS** Real-time Vision System 实时观察系统

Refuelling Video Surveillance 加油视频监视

Remote Video Surveillance 远程视频监视

Remote Visulization System 远程可视化系统

Reverse 反向

Revise 修正

**RVSBL** Reversible 可逆的

**RVSG** Reversing 反转

**RVSL** Reversal 颠倒,反转

**RVSM** Reduced Vertical Separation Minimum 缩小的最低垂直间隔

**RVSN** Reversion 颠倒,反转,反向

**RVSP** Radar Video Signal Processor 雷达视频信号处理器

**RVSR** Reverser 换向器/反向器/反推装置

**RVT** Reliability Verification Test 可靠性验证试验

Remote Video Terminal 远程视频终端

Requirements Verification and Testing 需求验证与测试

Rivet 铆钉,铆接

Rotary Variable Transformer 可变旋转变压器

Routing Vector Table 选路矢量表

**RVTD** Riveted 铆接的

**RVV** Runway Visibility Value 跑道能见度数值

**RVW** Review/Reviewing 检查,回顾

**RW** Raw Water 生水

Rawin 无线电测风仪

Reconnaissance/Electronic Warfare 侦察/电子战

Right Wing 右机翼

Rotary Wing 旋翼

Runway 跑道

Rwanda 卢旺达

**R-W** Read-Write Head 读写磁头

**RW/ILS** Runway/ILS 跑道/仪表着陆系统

**RWA** Remote Workaround 远程工作区

Routing and Wavelength Assignment 选路及波长分配

**RWAD** Relative Wind Angle Difference 相对风向角之差

**RWC** Rainwater Conductor 雨水管

Read, Write, Compute 读、写、计算

**RWCL** Runway Center Line Lights 跑道中心线灯

**RWDS** Reactive Windshear Detection System 反应式风切变检测系统

Rearwards 后方,在后面的,向后的

**RWFM** Random Walk Frequency Modulation 随机游动频率调制

**RWG** Reliability Working Group 可靠性工作组

Requirements Working Group 需求工作组

Rigid Wave Guide 刚性波导

Roebling Wire Gage 罗布林线规

**RWH** Radar Warning and Homing 雷达警告与导航

Rapid Water Heater 快速热水器

**RWIS** Road Weather Information System 道路天气信息系统

**RWJ**　Resistance Weld Jig 电阻焊模具

**RWK(D)**　Rework(ed) 再加工,返工的

**RWM**　Read-Write Memory 读写存储器

**RWND**　Rewind 倒带

**RWO**　Routine Work Order 例行工作报告

**RWR**　Radar Warning Receiver 雷达告警接收机

　　Rear Warning Radar 后视警戒雷达

**RWREP**　Radar Weather Report 雷达天气报告

**RWS**　Radar Warning System 雷达警戒系统

　　Range While Search 边搜索边测距

　　Reactive Wind Shear 反应式风切变

　　Right Wing Station 右机翼站位

**RWSL**　Runway Status Light 跑道状态灯

**RWST**　Rotary Wing Structures Technology 旋翼结构技术

**RWT**　Ramp Weight（飞机）在机坪时的重量

**RWV**　Rotary Wing Vehicle 旋翼机

**RWW**　Right Wheel Well 右轮舱

**RWY**　Runway（机场的）跑道

**RWYCON**　Runway Condition 跑道状况

**RWYTIL**　Runway Threshold Identification Lights 跑道入口标识灯

**RX**　Receiver Input Line 接收器输入线

**RXB**　Explosives 1.4B 1.4B 级易爆物(货运用语)

**RXC**　Explosives 1.4C 1.4C 级易爆物(货运用语)

**RXD**　Explosives 1.4D 1.4D 级易爆物(货运用语)

　　Receive Data 接收数据

　　Receiver Data 接收机数据

**RXDB**　Receive Data Channel B 信道 B 接收数据

**RXE**　Explosives 1.4E 1.4E 级易爆物(货运用语)

**RXG**　Explosives 1.4G 1.4G 级易爆物(货运用语)

**RXI**　Receive Input Line 1 接收导线 1 输入

**RXIL**　Receive Input Line 接收输入行

**RXN**　Receive Input Line 2 接收导线 2 输入

**RXS**　Explosives 1.4S 1.4S 级易爆物(货运用语)

**RY**　Railway 铁路

**RYC**　Refer to Your Cable 参阅你方电报

**RYHA**　Release to You for Your Handling 由贵方处理

**RYL**　Refer to Your Letter 参阅你方信件

**RYMSG**　Reservation and Yield Management Study Group 订座与收益管理研究小组

**RYN**　Rayon 人造丝

**RYT**　Reference Your Telegram 参阅你方电报

**RZ**　Return to Zero 回零

**RZL**　Return to Zero Level 归零电平

**RZM**　Return to Zero Mark 归零标志

# S

S　Count 计算

Distance 距离

General Area 总面积

Scalar 数量(的),标量(的)

Schedule 时间表,日程安排表

Scuttle 舷窗,天窗

Seaplane 水上飞机,海上飞机

Search 搜索

Secant 正割

Second 秒,第二的

Secondary 副手,辅助的

Secret 秘诀,秘密(的)

Section 段,节,截面

Sector 扇区,扇形(轮)

Sequence 顺序,序列

Serial 系列,连续的,串联的

Series 连续,系列,串联

Siemens 西门子(公司)

Signal 信号

Silk 线

Sine 正弦

Single Silk 多股生丝

Slate 板条

Slide 滑动装置

Soft 软的

Source (数据)源

South 南方

Southern 南方的

Special 特殊的,专用的

Status 状态

Stere 立方米

Stratus (气象)层云

Sulphur 硫

Surveillance 监视

Switch 转换,开关

Swivel 旋转

**S&A**　Safety and Arming 安全与预位

Sales and Administration 销售和管理

Simulation and Analysis 仿真和分析

Studies and Analysis 研究和分析

Surveillance and Analysis 监控和分析

**S&C**　Search and Clear 搜索与消除

Sensors and Controls 传感器与控制器

Shipper and Carrier 发货人与承运人

Stabilization and Control 稳定与控制

Status and Control 状态与控制

Superseding and Cancelling 废除(取代)与注销

**S&D**　Scheduling and Distribution 调度与分配

**S&DP**　System and Data Processing 系统与数据处理

**S&E**　Salaries and Expenses 工资和开支

Science and Engineering 科学与工程

Supply and Equipment 供应与设备

**S&H LT**　Signal and Homing Light 信号与归航灯

**S&I**　Surveillance and Inspection 监视与检查

**S&P**　Stake and Platform 底架与平台

**S&QM**　Safety and Quality Management 安全和质量管理

**S&R**　Search and Rescue 搜寻与救援

Snow and Rain 雨夹雪

Storage and Retrieval 存储与检索

**S&S**　Science and Society 科学与社会

Spigot and Socket 插头与插座

System and Service 系统和服务

**S/A**　State Agent 政府代理人

Storage Area (数据)存储区

Subject to Average 平均摊算

S/B  Setback 拨回,逆转,挫折,延滞

S/C  Short Circuit 短路

Signal Conditioning 信号调节

Specified Condition 规定的条件(状态)

Step Climb 阶梯爬升

Subcarrier 副载波

Subcontractor 转包商,次承包商

System Communication 系统通信

S/D  Security Deposit 保证金

Signal-to-Distortion Ratio 信号失真比

Step Descent 阶梯下降

S/E  Standardization/Evaluation 标准化/鉴定

Stock Exchange 证券交易所

System/Equipment 系统/设备

S/F  Safety Flight 安全飞行

Scheduled Flight 定期航班

Slats/Flaps 缝翼/襟翼

S/I  Snow and Ice 冰雪

S/N  Serial Number 序号

Signal to Noise Ratio 信噪比

S/O  Shutoff 关断,断开

S/R  Send and Receive 发送和接收

S/S  Samples per Second 取样数/秒

S/T  Similar to 相似于

S/W  Side Wall 侧壁,侧板

S1S  Surfaced or Dressed One Side 磨削或打磨一面

S1S1E  Surfaced or Dressed One Side and One Edge 磨削或打磨一面及一边

S2S  Surfaced or Dressed Two Sides 磨削或打磨两侧

S4S  Surfaced or Dressed Four Sides 磨削或打磨四面

SA  Safe Arrival 安全到达

Safety Altitude 安全高度

Sail Area 迎风面积

Sale Account 销售账

Sale Agency 分销处,销售代理

Sand or Dust Storm 沙暴或尘暴

Satellite Altimeter 卫星高度计

Satellite Radio Line 卫星无线电线链路

Saudi Arabia 沙特阿拉伯

Savings Account 储蓄账户

Scanning Accuracy 扫描精度

Scanning Amplifier 扫描放大器

Secondary Airport 次级机场

Section Adaptation 段适配

Sectional Area 截面积

See Also 还可参考

Select Address 选择地址

Selection Amplifier 选择放大器

Selective Absorption 选择性吸收

Selective Availability 选择可用性

Semiannual 一年两次的

Semiautomatic 半自动的

Sense Amplifier 检测放大器

Sequential Analyzer 序列分析器

Sequential Arrangement 顺序排列

Serial Addition 串行加法

Service Adapter 业务适配器

Service Address 服务地址

Service Agent 业务代理

Service Aid 维修工具,维修设备

Service Alarm 服务告警

Service Area 维修场,勤务区,维修区

Service Availability 业务可用性

Servo Amplifier 伺服放大器

Settled Account 结账,决算

Shaft Alley 轴隧,轴通道

Shipment Advice 货运通知

Shipment Application 货运申请书

Shock Accelerometer 冲击加速度计

Signal Access 信号存取

Signal Analysis 信号分析

Signal Attenuation 信号衰减

Signal Averager 信号平均器

Silica Acid 硅酸

Simulated Annealing 模拟退火

Single Access 单存取

Single Aisle 单通道

Situation Assessment 情景评价

Situation Awareness 情景意识,势态评估,势

态察觉

Soaring Association 滑翔协会

Sonic Altimeter 声波测高计

Sonic Analyzer 声波探伤仪

Source Address 源地址

South Africa 南非

South America 南美

South Australia 南澳州

Space Application 空域应用,空间(技术)应用

Speaker Amplifier 扬声器放大器

Special Assignment 特殊任务

Special Authority 特别代理权

Spectrum Analysis 频谱分析,光谱分析

Spectrum Analyzer 频谱分析仪

Standard Accessory 标准附件

Standard Anemometer 标准风速仪

Standard Approach 标准进近,标准进场

Standard Atmosphere 标准大气

Standby Autopilot 备用自动驾驶仪

Static Analysis 静态分析

Station Address 站地址

Statistical Analyzer 统计分析员,统计分析器

Stock Account 库存账,资财账

Stock Average 库存平均数

Stockpile Availability 资源可用性,储存可用性

Store Address 存储地址

Stress Analysis 应力分析

Stress Anneal 应力退火

Subassembly 分组件

Subscriber Access 用户接入

Supplemental Agreement 补充协议

Surface to Air 地对空

Switch Access 交换接入

Symbolic Assemble (Program) 符号汇编(程序)

Synchro Amplifier 同步放大器

System Administrator 系统管理员

System Analysis 系统分析

**SAA** South African Airways 南非航空公司

Special Assignment Aircraft 专机空运

System Application Architecture 系统应用结构

System Area Address 系统区域地址

**SAAA** Sport Aircraft Association of Australia 澳大利亚体育(运动)飞机协会

**SAAAR** Special Aircrew and Aircraft Authorization Required 要求授权的特殊机组和航空器

**SAAC** Simulator for Air-to-Air Combat 空战模拟机

**SAADU** Standby Attitude Air Data Unit 备用姿态大气数据组件

**SAAF** Small Austere Airfield 小型简易机场

**SAAFI** South African Association on Flying Instructors 南非飞行教员协会

**SAAHS** Stability Augmentation Attitude Hold System 增稳及姿态保持系统

**SAAL** Signaling ATM Adaptation Layer 信令 ATM 适配层

**SAALC** San Antonio Air Logistics Center 圣安东尼奥航空后勤中心

**SAAM** Signaling ATM Adaption Module 信令 ATM 适配模块

**SAAMA** San Antonio Air Material Area 圣安东尼奥航空器材区

**SAAPA** South African Airways Pilots' Association 南非航空公司驾驶员协会

**SAAQ** State Aviation Activities Questionnaire 国家航空活动问卷调查

**SAARC** South Asian Association for Regional Cooperation 南亚区域合作联盟

**SAARU** Secondary Attitude Air Data Reference Unit 备用姿态大气数据基准组件

**SaaS** Software as a Service 软件即服务

**SAAS** Security Audit and Analysis System 安全审计和分析系统

**SAATC** Southern African Air Transport Council 南非航空运输理事会

**SAB** Scientific Advisory Board 科学咨询委

员会

Senate Advisory Board 参议院咨询委员会

Source Address Bus 源地址总线

Space Application Board 空间应用委员会

Sync Address Bus 同步地址总线

**SABM** Set Asynchronous Balanced Mode 置异步平衡方式

**SABME** Set Asynchronous Balanced Mode Extended 置扩展的异步平衡方式

**SABO** Sense Amplifier Blocking Oscillator 检测放大器阻塞振荡器

**SABR** Scalable Agile Beam Radar 可伸缩式灵巧波束雷达

**SABRE** Secure Airborne Radar Equipment 保密航空雷达设备

Self Aligning Boost and Reentry 自对准推进和返回

**SAC** Sacramento, California ( Airport Code ) 美国萨克拉门托机场

Science Advisory Committee 科学咨询委员会

Science and Astronautics Committee 科学与宇航委员会

Science Application Corporation 科学应用公司

Scientific Advisory Committee 科学咨询委员会

Scientific Advisory Committee, United Nations 联合国科学咨询委员会

Security and Access Control System 安保和门禁系统

Senate Appropriations Committee 参院拨款委员会

Service Access Controller 业务访问控制程序

Signature Approval Card 标记批准卡, 签名批准卡

Single Annular Combustor 单环形燃烧室, 常规燃烧室

Special Area Code 专用区域代码

Sprayed Acoustical Ceiling 喷涂隔音天花板

Statistical Advisory Committee 统计咨询委员会

Storage Access Channel 存储器存取通道

Storage Access Control 存储器存取控制

Storage Address Counter 存储器地址计数器

Structure and Composition 结构与组成

System Acquisition Career 系统采集历程

System Alert Controller 系统告警控制器

System Anticollision 系统防撞

**SACAA** South African Civil Aviation Authority 南非民用航空局

**SACCA** Scottish Advisory Council for Civil Aviation 苏格兰民航咨询委员会

**SACCH** Slow Associated Control Channel 缓慢相关控制信道

**SACE** Semiautomatic Checkout Equipment 半自动检查设备

System Auxiliary Control Element 系统辅助控制单元

**SACF** Service Access Control Function 业务接入控制功能

Single Association Control Function 单相关控制功能

**SACL** Strobe Anti-Collision Light 频闪防撞灯

**SACMA** Suppliers of Advanced Composite Materials Association 先进复合材料供应商协会

**SACO** Seattle Aircraft Certification Office 西雅图航空器审定处

**SACP** Satellite Aeronautical Channel Prober 卫星航空信道探测器

**SACS** Speed/Attitude Control System 速率/姿态控制系统

**SACU** Synchronization and Control Unit 同步和控制装置

**SAD** Service Action Drawing 维修操作图

Signal Azimuth Display 信号方位显示器

Spares Advance Data 备件预付数据

Stress Analysis Data 应力分析数据

System Allocation Document 系统分配文件

System Analysis and Design 系统分析和设计

**SADC** Southern African Development Community 南部非洲发展共同体

State Agriculture Development Committee 国家农业发展委员会

**SADF** Spares Advance Data Form 备件预付数据单

System Analysis and Design Facility 系统分析和设计设施

**SADI** Standby Attitude Director Indicator 备用姿态指引仪

**SADIE** Scanning Analog to Digital Input Equipment 扫描输入模/数转换设备

**SADIS** Satellite Distribution System 卫星分布系统

**SADL** Situation Awareness Data Link 态势感知数据链

**SADM** Standby Air Data Module 备用大气数据组件

**SADP** Serviceability, Availability and Debugging Procedures 可维修性、可用性和调试程序

**SADS** Solar Authority Data Set 太阳能管理局数据组

Special Air Data System 专用大气数据系统

**SAE** Society of Automotive Engineers 汽车工程师协会(美国)

Stand Alone Equipment 独立设备

Supersonic Aircraft Engine 超音速飞机发动机

**SAEC** State Administration of Exchange Control 国家外汇管理局调控

Support Analysis of Engineering Change 工程变更的保障分析

**SAEE** Service Agent Execution Environment 业务代理执行环境

**SAESC** System Analysis and Evaluation Support Contractor 系统分析和评估保障合同方

**SAF** Safety 安全(性),保安装置

Santa Fe, NM, USA(Airport Code) 美国新墨西哥州圣菲机场

Specific Access Function 特定接入功能

System Administration Facility 系统管理设备

**SAFA** Safety Assessment of Foreign Aircraft 外国航空器安全评估

**SAFCA** Safeguard Communications Agency 保卫通信处

**SAFCS** Self Adaptive Flight Control System 自适应飞行控制系统

**SAFD** Safetied 保险的

**SAFE** Selected Area for Evasion 选定脱险地区

Society of Aeronautic Flight Engineers 航空飞行工程师协会

Solar Array Flight Experiment 太阳电池阵飞行实验

South American and Far East 南美与远东空运公司

State Administration of Foreign Exchange 国家外汇管理局

Survival and Flight Equipment 救生与飞行设备

System for Automated Flight Efficiency 自动飞行效能系统

**SAFI** Semi-Automatic Flight Inspection 半自动飞行校验

**SAFSCOM** Safeguard System Command 保安系统指挥部

**SAFO** Safety Alert for Operators 运营人安全警告

**SAFT** Semi-Automatic Functional Test 半自动功能测试

**SAFTI** Secure and Facilitated International Travel Initiative 安全和简化国际旅行的倡议

**SAG** Scanner Auxiliary Group 扫描装置辅助组

Scientific Advisory Group 科学咨询组

System Analysis Group 系统分析组

**SAGAT** Situation Awareness Global Assessment Technique 情景意识全面评估技术

**SAGE** Semi-Automatic Ground Environment 半自动地面环境

**SAH** Sanaa, Yemen (Airport Code) 也门萨那机场

Semi-Automatic Homing 半自动归航

**SAHF** Semi-Automatic Height Finder 半自动测高仪

**SAHPA** South African Hang Gliding and Paragliding Association 南非悬挂式滑翔和翼伞飞行组织

**SAHRS** Standard Attitude Heading Reference System 标准姿态航向基准系统

Standby Attitude Heading Reference System 备用姿态航向基准系统

**SAHS** Swedish Aviation Historical Society 瑞典航空历史社团

**SAI** Safety Attribute Inspection 安全属性监察

Seoul Air International 首尔国际航空公司

Standby Airspeed Indicator 备用空速表

Standby Attitude Indicator 备用姿态指示仪

Static Airspeed Indicator 静压空速表

Supersonic Aerospace International 超音速航空国际

System Analysis Item 系统分析项目

System Architecture Integration 系统整体结构

Systems Analysis and Integration 系统分析和集成

Systems Architecture and Interfaces 系统结构和接口

**SAIC** Science Application International Corporation 国际科学应用公司

Shanghai Aviation Industrial Corp. 上海航空工业公司

Shijiazhuang Aircraft Industry Co., Ltd 石家庄飞机工业有限责任公司

**SAICS** Standard Avionics Integrated Control System 标准航空电子综合控制系统

**SAIG** Single-Axis Integrated Gyro 单轴积分陀螺仪

**SAIL** Shuttle-Avionics Integration Laboratories 航天飞机电子设备集成实验室

**SAILS** Simplified Aircraft Instrument Landing System 简化飞机仪表着陆系统

Software Adaptable Integrated Logic System 软件自适应综合逻辑系统

**SAIM** System Analysis and Integration Model 系统分析与综合模型

**SAIMS** Supersonic Airborne Infrared Measurement System 超音速机载红外测量系统

**SAINT** Satellite Interceptor 卫星截击器

Security Administrator's Integrated Network Tool 安全管理员集成网络工具

Systems Analysis Integrated Networks of Tasks 系统分析综合网络的任务

**SAIP** Semi-Automatic Imaging Processing 半自动成像处理

Spares Acquisition Integrated Program 备件采购综合程序

Spares Acquisition Integrated with Production 零件采购与生产一体化

**S-AIS** Section Alarm Indication Signal 段告警指示信号

**SAL** Safeguards Analytical Laboratory 安全保护分析实验室

Salinometer 盐度计

San Salvador, El Salvador (Airport Code) 萨尔瓦多圣萨尔瓦多机场

Seat Available List 可使用空位表

Semiactive Laser 半主动激光

Short Approach Light 短距离进近照明

Shortened Address Line 简化地址行

Special Test Equipment 专用测试设备

Supersonic Aerophysics Laboratory 超音速航空物理实验室(美国)

Surface and Airlift 地面与空运

Surface Mail 普通平信邮件

Symbolic Assembly Language 符号汇编语言

System Address Label 系统地址标签

**SALCOM** Sales and Compensation 销售与补

偿

**SALE** Singapore Aircraft Leasing Enterprise 新加坡飞机租赁企业

**SALGO** Simulation Algorithms 仿真算法

**SALR** Saturated Adiabatic Lapse Rate 饱和绝热递减率

Senior Logistics Aviation Representative 高级物流航空代表

**SALS** Short Approach Lighting System 短距进近灯光系统

Simple Approach Lighting System 简式进近灯光系统

**SALSA** Spot Array Light Supply Adapter 点阵灯光电源适配器

**SALSF** Short Approach Lighting System with Sequenced Flashing Lights 有顺序闪光灯的短距进近灯光系统

**SALT** Standby Altimeter 备用高度表

**SAM** Satellite Automonitor 卫星自动监控器

School of Aviation Medicine 航空医学学校

Secure Account Manager 安全账号管理

Selective Automatic Monitoring 选择性自动监测

Self-Amplitude Modulation 自调幅

Sequential Access Method 顺序存取方法

Service Access Multiplexer 业务接入复用器

Sound Absorbing Material 吸音材料

South American 南美

Special Air Mission 特殊飞行任务,专机飞行

Spoiler Actuator Module 扰流板作动器模件

Stabilizer Aileron Module 安定面副翼模件

Stabilizer/Aileron Lockout Module 安定面/副翼锁定组件

Stabilizer Trim and Aileron Lockout Module 安定面配平和副翼锁定组件

Standard Avionics Module 标准航空电子模件

Stress Analysis Manual 应力分析手册

Structural Analysis Model 结构分析模型

Sub-module Administration Module 子模块管理模块

Subsequent-Address Message 后续地址消息

Synchronous Amplitude Modulation 同步调幅

**SAM/SAT** South America/South Atlantic Regions 南美及南大西洋地区

**SAMA** Small Aircraft Manufacturers Association 小型航空器制造商协会

**SAMB** Secondary Aircraft Maintenance Base 飞机辅助维修基地

**SAMC** Subscriber Access Maintenance Center 用户接入维护中心

**SAME** Specific Area Message Encoding 特定区域信息编码

Subscriber Access Maintenance Entity 用户接入维护实体

**SAMECO** Shanghai Aviation Maintenance Engineering Company 上海航空维修工程公司

**SAMECS** Structural Analysis Method for Evaluation of Complex Structure 评估复杂结构的结构分析方法

**SAMF** Shanghai Aircraft Manufacturing Factory 上海飞机制造厂

Subscriber Access Management Function 用户接入管理功能

**SAMI** Service Access Multiplexer Interface 业务接入复用器接口

Speed of Approach Measurement Indicator 进近速度测量指示器

**SAMPE** Society for Advancement of Material and Process Engineering 材料与处理工程促进协会

Society for Aerospace Material and Process Engineer 航空航天材料与工艺工程师学会

**SAMS** Satellite Automatic Monitoring System 卫星自动监控系统

Special-use Airspace Management System 特殊用途的空域管理系统

**SAN** San Diego, CA, USA(Airport Code) 美国加利福尼亚州圣地亚哥机场

Sanitary 清洁的,卫生的

Satellite Access Node 卫星接入节点

Sensor Area Network 传感器区域网络

Storage Area Network 存储区网络

Subscriber Access Network 用户接入网

System Advisory Notice 系统咨询通告

System after Next 后一个系统

**SANC** Signaling Area/Network Code 信令区/网码

**SANDS** Standard Analog to Digital System 标准模/数转换系统

**SANET** Self-Amplified Network 自放大网络

**SANG** Standardized Aeronautical Navigation/Guidance 标准化航空导航与制导

**SAO** Sao Paulo, Brazil(Airport Code) 巴西圣宝罗机场

Single Association Object 单相关控制对象

Subassembly Order 分组件订单,分组件指令

Systems Acquisition Officer 系统采购员

**SAOCS** Submarine Aircraft Optical Communication System 潜艇飞机光学通信系统

**SAP** San Pedro Sula, Honduras ( Airport Code) 洪都拉斯圣佩德罗苏拉机场

Schedule Assessment Program 大纲鉴定程序

Secondary Audio Program 辅助音频程序

Separate Audio Program 分离音频程序

Service Access Point 业务接入点

Service Advertising Protocol 服务公告协议

Service Agent Point 业务代理点

Shop Aid Package 车间工具包

Special Audio Panel 专用音频板

Special Audio Program 专用音频程序

Speech Application Platform 语音应用平台

Stabilized Approach 稳定进近

Supportability Assessment Plan 保障能力评价计划

Symbolic Assembly Program 符号汇编程序

System Assembly Program 系统汇编程序

System Assurance Program 系统保险程序

**SAPI** Server Application Programming Interface 服务器应用编程接口

Service Access Point Identifier 业务接入点标识符

**SAPL** Service Action Parts List 维修零件一览表

**SAPS** Small Area Piloting Sheet 小区域领航作业图

Source Access Point System 信源接入点系统

Standard Airplane Parameter Services 标准飞机参数服务

**SAPWG** Strategic Action Plan Working Group 战略行动计划工作组

**SAR** Safety Action Record 安全行动记录

Safety Address Register 安全地址寄存器

Safety Analysis Report 安全分析报告

Safety Assessment Report 安全评估报告

Search and Rescue 搜寻与救援

Security Assessment Report 安全评估报告

Segmentation and Reassemble 分拆与重组

Segmentation and Reassembly 分段和重组子层

Selected Acquisition Report 选择的搜索报告

Shop Aid Recorder 车间辅助记录器

Software Acquisition Request 软件采购申请

Special Access Required 要求的特殊存取

Special Administrative Region 特别行政区

Special Aviation Requirement 特殊航空要求

Status by Airplane Report 飞机报告的状态

Storage Address Register 存储地址寄存器

Stress Analysis Report 应力分析报告

Subsequent Application Review 顺序应用检查

Surveillance Approach Radar 监视进近雷达

Synthetic Antenna Radar 合成天线雷达

Synthetic Aperture Radar 合成孔径雷达

Synthetic Array Radar 合成天线阵雷达

System Analysis Recorder 系统分析记录器

System Analysis Recording 系统分析记录

System Analysis Report 系统分析报告

**SARAH** Safety Analysis and Risk Assessment Handbook 安全分析与风险评估手册

Search and Rescue Armed Helicopter 搜寻与

救援武装直升机

Search, Rescue and Homing 搜寻、救援和归航

**SARAST** South Asia Regional Aviation Safety Team 南亚地区航空安全小组

**SARB** Search and Rescue Beacon 搜索与救援信标

Service Approval Review Board 维修批准审查委员会,勤务批准审查委员会

**SARBE** Search and Rescue Beacon Equipment 搜索与救援信标设备

**SARC** Search and Rescue Center 搜寻和救援中心

Secure Airborne Radar Control 安全机载雷达控制

Standard Aircraft Radio Case 标准航空无线电箱

Surveillance and Reconnaissance Center 监视与侦察中心

**SARCC** Search and Rescue Control Center 搜寻和救助控制中心

Search and Rescue Coordination Center 搜索与救援协调中心

**SARD** Simulated Aircraft Radar Data 模拟飞机雷达数据

Special Airlift Requirement Directive 特殊空运需求指令

**SARDA** State and Regional Disaster Airlift 国家与地区救灾空运

**SAREX** Search and Rescue Exercise 搜索及救援演习

Space Amateur Radio Experiment 宇航业余无线电试验

**SARF** Shop Aid Reference Facility 车间辅助基准设施

**SARH** Search and Rescue Homing 搜索与援救归航

Semi-Active Radar Homing 半主动雷达制导

**SARI** Small Airport Runway Indicator 小型机场跑道指示器

**SARM** Security Audit Reference Manual 保安审计参考手册

Set Asynchronous Response Mode 置异步响应方式

**SARME** Set Asynchronous Response Mode Extended 置扩充异步响应方式

**SARP** Signal Automatic Radar Processing 信号自动雷达处理

**SARPs** Standards and Recommended Practices 标准和建议措施

**SARS** Safety and Reliability Society 安全与可靠性协会

Simulated Airborne Radar System 模拟机载雷达系统

Square Area Reporting System 平方区域报告系统

Standards Automated Reschedule System 标准自动重排(时间)系统

Standby Attitude Reference System 备用姿态基准系统

Synthetic Aperture Radar System 综合孔径雷达系统,合成孔径雷达系统

Synthetic Array Radar System 合成天线阵雷达系统

**SARSAT** Search and Rescue Satellite Aided Tracking 搜索与救援卫星辅助跟踪

**SARSS** Search and Rescue Satellite System 搜索与救援卫星系统

**SART** Situation Awareness Rating Technique 情景意识等级评估技术

Structured Analysis for Real-Time Systems 实时系统结构分析

**SARTS** Switched Access Remote Testing System 交换接入远地测试系统

**SAS** Safety Assurance System 安全保险系统

Satellite Application Section 卫星应用处

Satellite Automation System 卫星自动化系统

Secure Analysis System 安全分析系统

Secure Authentication System 安全鉴定系统

Self-Adaptive System 自适应系统

Software Accomplishment Summary 软件完成总结

Specific Airworthiness Specification 特殊适航规范

Stability Augmentation System 稳定增益系统

Standard Avionic System 标准航空电子系统

Statistical Analysis Software 统计分析软件

Statistical Applications System 统计应用系统

Subscriber Authorized System 用户授权系统

Subsystem Activity Summary 分系统功效摘要

Symbolic Assembly System 符号汇编系统

Synoptic Analysis Section 天气分析处

System Analysis Study 系统分析研究

System Assembly Shop 系统组装车间

System Assessment Survey 系统评定检查

**SASCO**   ST (Singapore Technologies) Aviation Services Co. 新加坡技术航空服务公司

**SASCOM**   Southern Atlantic Satellite Communication 南大西洋卫星通信

**SASCS**   SHF Antenna Sidelobe Canceller System 超高频天线旁瓣对消器系统

**SASD**   Self-Adaptive Sequence Detection 自适应序列检测

**SASE**   Self Addressed Stamped Envelope 邮资已付并有回邮地址的信封

Specific Application Service Element 专门应用服务单元

Stand-Alone Synchronization Equipment 独立同步设备

**SASI**   Society of Air Safety Investigator 航空安全调查员协会

Standby Airspeed Indicator 备用空速指示表

Surface Air System Integration 地空系统综合使用

System on Automotive Safety Information 自动车辆安全信息系统

**SASP**   Science and Application Space Platform 科学和应用空间平台

Separation and Airspace Safety Panel 间隔和空域安全组

**SASR**   Storage Address Select Register 存储地址选择寄存器

Strategic Airport Security Rollout 战略机场安全滑行

**SASS**   Small Aerostat Surveillance System 小航空卫星监视系统

Small Airbreathing System Synthesis 小型吸气系统合成

Stability Augmentation Subsystem 稳定增益子系统

Structured Analysis and System Specification 结构分析和系统规范

Suspended Array Surveillance System 悬挂阵列监视系统

**SAT**   San Antonio, TX, USA (Airport Code) 美国得克萨斯州圣安东尼奥机场

Satellite 卫星

Satisfactory/Satisfactorily 满意的，满意地

Saturate, Saturation 饱和

Saturday 星期六

Science and Technology 科学与技术

Security Aviation Transport 安全航空运输

Shortage Analysis Team 缺件分析组, 短额分析组

Site Acceptance Test 现场验收测试

Site Activation Team 现场启动队

Software Analysis Team 软件分析组

Software Audit Team 软件审核组

Static Air Temperature 大气静温

Subscriber Access Terminal 用户接入终端

System Acceptance Test 系统验收试验

System Approach to Training 系统训练方法

**SATA**   Satellite Automatic Tracking Antenna 卫星自动跟踪天线

Student Air Travel Association 学生航空旅行协会

**SATAN**   Security Analysis Tool for Auditing Networks 审计网络用的安全分析工具

Self-contained Automatic Tactical Air Navigation 自备式自动战术空中导航

Sensor for Airborne Terrain Analysis 机载地形分析传感器

Speed and Throttle Automatic Network 速度和油门自动网络

**SATASFTDG** Structure Air Transport Association System Functional Test Data Group 结构空运协会系统功能测试数据组

**SATC** Software Assurance Technology Center 软件保证技术中心

**SATCO** Senior Air Traffic Control Officer 空中交通管制高级官员

Supervisory Air Traffic Control Organization 监督空中交通管制组织

**SATCOM** Committee on Scientific and Technical Communication 科学技术通信委员会（美国）

Satellite Command 卫星指令

Satellite Communication System 卫星通信系统

Satellite Communications 卫星通信

**SATCOMA** Satellite Communications Agency 卫星通信局

**SATCON** Satellite Condition 卫星状态

**SATCP** Systems Avionics Tester Computer Program 航空电子系统测试计算机程序

**SATDAT** Satellite Data 卫星数据

**SATELCO** Satellite Telecommunication Company 卫星电信公司

**SATF** Strike and Terrain Following Radar 打击与地形跟踪雷达

**SATFY** Satisfactory 满意的

**SATI** Standby Attitude Indicator 备用姿态指引仪

Static Air Temperature Indicator 大气静温表

**SATIN** System Architecture and Traffic Control Integration 系统体系结构和交通控制集成

**SATNAV** Satellite Navigation 卫星导航

Satellite Navigator 卫星导航仪

**SATNET** Satellite Network 卫星网络

**SATO** Scheduled Air Transportation Office 定期航空运输办公室

Software Assurance Technology Office 软件保证技术办公室

Supply and Transportation Operation 供应与运输工作

**SATOB** Satellite Observations 卫星观测

**SATORB** Satellite Orbit 卫星轨道

**SATP** Semi-Automatic Test Procedure 半自动测试程序

**SATPATT** Satellite Paper Tape Transfers 卫星纸带传送

**SATR** Schedule Air Transport Rating 空运费用表

**SATRAK** Satellite Tracking 卫星跟踪

**SATS** Singapore Airport Terminal Services Ltd 新加坡机场地勤服务公司

Small Aircraft Transportation System 小型飞机运输系统

Small Applications Technology Satellite 小型应用技术卫星

Standardized Automated Testing System 标准自动化测试系统

System Acceptance Test Specification 系统验收测试规范

**SAU** Scientific Arithmetic Unit 科学运算组件

Spectrum Analysis Unit 频谱分析器

System of Absolute Units 绝对单位制

**SAV** Savannah, GA, USA(Airport Code) 美国佐治亚州萨凡纳河机场

Starter Air Valve 启动机空气活门

**SAVE** Simulation Assessment Validation Environment 仿真评估确认环境

**SAVI** Space Active Vibration Isolation 空间有源隔振

**SAW** Society of Aviation Writers 航空作家协会

Surface Acoustic Wave 表面声波

Surveillance and Warning 监视与警告

**SAWD** Surface Acoustic Wave Device 表面声波器件

**SAWDL** Surface Acoustic Wave Delay Line 表面声波延迟线

**SAWE** Society of Aeronautical Weight Engineers 航空重力工程师协会

**SAWF** Surface Acoustic Wave Filter 表面声波滤波器

**SAWG** Schedule and Allocations Working Group 计划与分配工作组

Scientific Advisory Working Group 科学咨询工作组

**SAWR** Supplementary Airways Weather Report 辅助航线天气报告

**SAWRS** Supplementary Airway Weather Reporting Station 辅助航路气象报告站

Supplementary Aviation Weather Reporting Station 辅助航空气象报告站

Supplementary Aviation Weather Reporting System 辅助航空气象报告系统

**SAWS** Surveillance and Warning System 监视与警告系统

**SAWT** Surface Acoustic Wave Transducer 表面声波传感器

**SAWTC** Senior Airway Traffic Controller 高级航路交通管制员

**SB** Secondary Battery 备用电瓶

Secondary Buffer 备用缓冲器

Serial Binary 串行二进制

Serial Block 串行数据块,串行程序块

Service Bulletin 服务通报,服务通告

Service Bureau 维修处,服务局

Shadow Buffer 描影缓冲

Side Band 边带

Sideboard 侧板

Sleeve Bearing 套筒轴承,滑动轴承

Solid Body 固体

Solomon Islands 所罗门群岛

Sound Blaster 声霸卡

Southbound 向南行的,向南行驶的

Speed Brake 减速板

Speedbrake 减速板

Spin Block 旋转块(区)

Splash Block 防溅挡板

Standby Base 备用基地

Station Break 电台间断

Stuffing Box 填料箱

Sweeping Beam 扫描电子束

Switch Board 开关板,配电盘

Switching Box 开关盒

Synchronization Bit 同步位

System Bus 系统总线

**SBA** Santa Barbara, CA, USA (Airport Code) 美国加利福尼亚州圣巴巴拉机场

Side Band Addressing 边带寻址

Small Business Administration 小型商务管理

Standard Beam Approach 标准波束进近

**SBAC** Society of British Aerospace Companies 英国宇宙公司协会

Society of British Aircraft Constructors 英国飞机设计师协会

**SBAS** Satellite-Based Augmentation System 星基增强系统

Space Based Augmentation System 广域差分增强系统

Standard Beam Approach System 标准波束引导进近系统

**SBB** Super Broad Band 超宽带

**SBC** Single Board Computer 单片机

Sonic Boom Committee 音爆委员会

Sound Blaster Audio Card 声卡

South Bridge Chip 南桥芯片

Subband Coding 子带编码

Supplier Bar Code 供应方条(形)码

Synchronizing Binary Counter 同步二进制计数器

System Broadcast Channel 系统广播信道

**SBCCD** Schottky Barrier Charge-Coupled Device 肖特基势垒电荷耦合器件

**SBCS** Single Byte Code System 单字节编码系统

**SBCT** Schottky Barrier Collector Transistor 肖特基势垒集电极晶体管

**SBD** Schottky Barrier Diode 肖特基二极管

Simulation-Based Design 基于仿真的设计

**SBDB** Service Bulletin Data Base 服务通告

数据库

**SBDS** Santa Barbara Dual Spectrum 圣巴巴拉双谱

**SBE** Service Bulletin Engineering 服务通告工程

**SBEC** Subband Echo Canceller 子带回音消除器

**SBEW** Standard Basic Empty Weight 标准基本空重

**SBFED** Schottky Barrier Field Effect Diode 肖特基势垒场效应二极管

**SBFET** Schottky Barrier Field Effect Transistor 肖特基势垒场效应晶体管

**SBH** Safe Burst Height 安全爆炸高度
St Barthelemy, Guadeloupe (Airport Code) 瓜德罗普岛圣巴德利米机场
State Board of Health 国家卫生委员会

**SBI** Service Bulletin Index 服务通告索引

**SBIE** Service Bulletin Index and Effectiveness 服务通告索引与有效性

**SBIR** Small Business Innovation Research 小型企业创新研究

**SBJ** Supersonic Business Jet 超音速公务飞机

**SBK** Single Beam Klystron 单注速调管
Sound Blaster Kits 声霸卡工具包

**SBL** Security Block 安全块
Spaced Based Laser 天基激光器
Symbol 符号

**SBM** Subnet Bandwidth Management 子网带宽管理
Supply Base Management 供应基地管理

**SBMCS** Service Bulletin Management and Control System 服务通告管理与控制系统

**SBMPC** Software-Based Moving Picture Coding 基于软件的动态图像编码

**SBN** South Bend, IN, USA (Airport Code) 美国印第安那州南本德机场

**SBO** Specific Behavioral Objectives 具体行为目标

**SBOM** Single Bill of Material 器材单一账单

**SBP** San Luis Obispo, CA, USA (Airport Code) 美国加利福尼亚州圣路易斯欧匹斯堡机场
Sonic Boom Panel 音爆专家组
Special Business Provisions 特殊业务预留

**SBPC** Single Burst per Carrier 每载波单突发

**SBPT** Single Burst per Transponder 每转发器单突发

**SBR** Service Bit Rate 业务比特率
Space Based Radar 空基雷达
Statistical Bit Rate 统计比特率

**SBRK** Speed Brake, Speedbrake 减速板

**SBRP** Sustainable Bioenergy Research Project 可持续生物能源研究项目

**SBS** Satellite Broadcast System 卫星广播系统
Satellite Business System 卫星商业系统
Semiconductor Bilateral Switch 半导体双向开关
Sick Building Syndrome 写字楼综合症
Standard Bus System 标准总线系统

**SBSP** Single Base Solid Propellant 单基固体推进剂

**SBSS** Small Business System Sustain 小型商务系统支持
Space Based Surveillance System 空基监视系统

**SBSTA** Subsidiary Body for Scientific and Technological Advice 科学和技术咨询附属机构

**SBSTR** Substrate 基片, 基地, 衬底

**SBT** Sheet, Bar or Tubing 层、条或管形
Sound Barrier Technology 声音阻碍技术
Surface Barrier Transistor 表面势垒晶体管
System Bench Test 系统试验台测试

**SBTP** Simple Buffer Transfer Protocol 简单缓冲转移协议

**SBTS** Brazil Domestic Satellite System 巴西国内卫星通信系统

**SBTT** Small Business Technology Transfer 小型企业技术转让

**SBU** Silver Brazing Union 银青铜接头

**SBW** Sibu, Sarawak, Malaysia（Airport Code）马来西亚砂拉越诗巫机场

Spectral Bandwidth 频谱带宽

Steer-by-Wire 电传操纵

South by West 南偏西

**SBX** S Band Transponder S 频段应答机

**SBZ** Sibiu, Romania（Airport Code）罗马尼亚锡比乌机场

**SC** Sales Coupons 销售票联

Satellite Computer 通信卫星计算机

Saturable Core 饱和铁芯

Scale 比例,刻度

Schedule Change 航班更改

Score 得分,比数

Seal Cap 密封盖

Seamless Connect 无缝连接

Service Ceiling 实用升限,使用升限

Service Channel 公务信道

Service Check 勤务检查

Seychelles 塞舌尔（群岛）

Shaft Center 轴心

Shared Channel 共享信道

Shift（Control）Counter 移空（控制）计数器

Shop Complete 车间完成

Short Circuit 短路

Short Cycle 短循环

Shortage Cost 缺货成本

Signal Center 信号中心

Signal Check 信号检查

Signal Communication 信号联络

Signal Conditioner 信号调节器

Signal Conditioning 信号调节

Signaling Converter 信令变换器

Silicon Control 可控硅

Simulation Control 模拟控制

Single Chime 单谐音

Single Contact 单触点

Situation Control 情景控制

Smooth Contour 平滑轮廓

Solar Cell 太阳能电池

Source Code 源代码

Space Charge 空间电荷

Special Circuit 专用电路

Special Circular 特别通知

Special Committee 专门委员会,特别委员会

Special Condition 专用条件,特殊条件,特别状态

Specific Conductance 比电导,电导率

Specific Conductivity 电导率

Specification Control 规范（技术要求）控制

Speed Command 速度指令

Speed Control 速度控制

Standard Compass 标准罗盘

Standard Condition 标准条件

Standard Conductivity 标准电导率

Step Climb 阶梯爬升

Storage Capacity 存贮量,存贮能力

Strato-Cumulus 层积云

Subcommittee 分委员会

Sub-Component 分部件

Subscriber Connector 用户连接器

Super Charger 增压充电器

Supercharging Compressor 增压压气机

Supercomputing 超级计算

Superconducting 超导（的）

Supervision Center 监控中心

Supervisor's Console 监控台

Supervisory Channel 监测信道

Supplementary Contract 补充合同

Supply Contract 供应合同

Synchro-Cyclotron 同步回旋加速器,稳相加速器

Synchronous Clock 同步时钟

Synchronous Computer 同步计算机

System Concept 系统概念

System Control 系统控制

System Controller 系统控制器

System of Communication 通信系统

**SC&J** Signal Collection and Jamming 信号收集和干扰

**SC（UN）** Security Council of United Nations

（联合国）安理会

**SCA** Safe Clearance Altitude 安全越障高度

Satellite Communication Agency 卫星通信局

Selected Call Appearance 选定的呼叫状态

Signal Conditioning Assembly 信号整形组件,信号调节组件

Single-Crystal Alloy 单晶合金

Sneak Circuit Analysis 潜电路分析

Southern Control Area 南部管制区

Spares Control Area 备件控制区

Static Channel Allocation 静止信道配置

Static Channel Assignment 静止信道分配

Subcarrier Authorization 副载波审定,分承运人批准

Subsidiary Communications Authorization 辅助通信业务

Supersonic Cruise Aircraft 超音速巡航飞机

**SCAD** Specification Compliance Allocation Document 规范一致性定位文档

**SCADA** Supervisory Control and Data Acquisition 监督控制及数据采集

**SCADAR** Scatter Detection and Ranging 散射检测和定位

**SCADC** Standard Central Air Data Computer 标准中央大气数据计算机

**SCAG** Swedish Civil Aviation Group 瑞典民航集团

**SCAI** Switch Computer Application Interface 交换机计算机应用接口

**SCALO** Scanning Local Oscillator 扫描（本机）振荡器

**SCAN** Self-Correcting Automatic Navigation 自核式自动导航

Stereo Cabiner Area Network 立体声组合音响局域网

Switched Circuit Automatic Network 交换电路自动控制网络

**SCAO** Standing Conference of Atlantic Organization 大西洋组织常设会议

**SCAP** Silent Compact Auxiliary Power 无噪声小型辅助电源

**SCAR** Subsequent Contracts Application Review 顺序合同应用检查

Supersonic Cruise Aircraft Research 超音速巡航飞机研究

**SCARS** Serialized Control and Reporting System 序列化控制与报告系统

**SCAS** Stability and Control Augmentation System 稳定性和操纵性增强系统

**SCASS** Sino Confidential Aviation Safety Reporting System 航空安全自愿报告系统

**SCAT** Scattergram 散点图

Security Control of Air Traffic 空中交通安全管制

Selected Characteristic Acceptance Test 选定特性验收试验

Shared Clock Autonomous Terminal 共用时钟自动终端

Space Communications and Tracking 空间通信和跟踪

Special Category 特别类型

Speed Control Approach Takeoff 速度控制进近起飞

Supersonic Civil Air Transport 超音速民用航空运输

Supersonic Commercial Air Transport 超音速商用航空运输

Surface Controlled Avalanche Transistor 表面可控雪崩晶体管

**SCATANA** Security Control of Air Traffic and Air Navigation Aids 空中交通安全管制与导航设备

**SCATHA** Satellite Charging at High Altitudes 高空卫星充电

Spacecraft Charging at High Altitudes 高空充电实验航天器

**SCAT-I** Special Category-I Precision Approach 特别 I 类精密进近

**SCATS** Scheduling and Tracking System 程序安排与跟踪系统

Self-Contained Automatic Test System 自主式自动测试系统

Sequentially-Controlled Automatic Transmitter Start 自动发射机顺序控制启动

Simulation Checkout and Training System 模拟检查与训练系统

**SCAV** Scavenge 回油

**SCB** Selected Call Bandwidth 选定的呼叫带宽

Selling Cost Budget 销售成本预算

Single Cantilever Beam 单悬梁

Software Control Board 软件控制委员会

System Control Board 系统控制板

**SCBW** Standard Constant Band Width 标准恒定带宽

**SCC** Satellite Communication Center 卫星通信中心

Satellite Communication Concentrator 卫星通信集线器

Satellite Communication Controller 卫星通信控制器

Satellite Computation Center 卫星计算中心

Satellite Control Center 卫星控制中心

Security Consultative Committee 安全咨询委员会

Security Control Center 安全控制中心

Signal Collection Card 信号收集卡

Signal Consolidation Card 信号强化卡

Single Chip Computer 单片计算机

Single Conductor Cable 单芯电缆

Single Cotton Covered 单纱包的

Space Computation Center 空间计算中心

Spare Control Center 备件控制中心

Standing Consultative Commission 常设协商委员会

Status Change Code 状态改变码

Status Control Code 状态控制码

Supervisor Control Console 监督控制台

Supervisory Computer Control 监督计算机控制

Supply Control Center 补给控制中心

Synchronous Communications Controller 同步通信控制器

System Control Center 系统控制中心

**SCCA** System Configuration Control and Accountability 系统组合控制与责任

**SCCA/TD** Self-Control Cycle Access with Time Division 时分自控循环接入

**SCCB** Software Configuration Control Board 软件配置控制委员会

Supplier Configuration Control Board 供应方组合控制委员会

**SCCC** Satellite Comunications Control Center 卫星通信控制中心

**SCCCD** Supplier CAD/CAM Capability Database 供应方计算机辅助设计/计算机辅助制造能力数据库

**SCCF** Satellite Comunication Control Facility 卫星通信控制设备

**SCCH** Signaling Control Channel 信令控制信道

Synchronous Channel 同步信道

**SCCM** Standard Cubic Centimeters per Minute 标准立方厘米/分

**SCCMAS** Santa Clara County Model Aircraft Skypark 圣克拉拉国家航空模型主题公园

**SCCN** Subcontract Change Notice 分合同变更通知

**SCCO** Satellite Communications Control Office 卫星通信控制室

**SCCP** Signaling Connect Control Part 信令连接控制部分

Signaling Connection Control Part 信令连接控制部分

**SCCR** Supplier Configuration Change Request 供应方配置更改申请

**SCCS** Satellite Communications Control System 卫星通信控制系统

Standard Communications Control System 标准通信控制系统

Switching Control Center System 交换控制中心系统

**SCCW** Secure Communication Code Word 保密通信编码字

**SCD** Satellite Calling Decoder 卫星呼叫译码器

Selective Cell Discard 选择性信元丢弃

Single Channel Demultiplexer 单信道解复用器

Small Cargo Door 小型货舱门

Solar Capture Device 太阳能捕获装置

Source Control Drawing 源控制图

Specification Control Document 规范控制文件

Specification Control Drawing 规范控制图

Standard Color Display 标准彩色显示器

System Control Database 系统控制数据库

**SCDA** Supervisory Control and Data Acquisition 监督控制与数据获取

**SCDD** Super Cooled Drizzle Drops 过冷冰雨滴

**SCDL** Supplier Control Data List 供应方控制数据清单

**S-CDMA** Synchronous Code Division Multiple Access 同步码分多址接入

**SCDR** Safety Compliance Data Report 安全符合数据报告

Screwdriver 螺丝刀

Software Critical Design Review 软件关键设计检查

Subcontractor Critical Design Review 分包方关键设计检查

Supplier Critical Design Review 供应方关键设计审查

**SCDRS** Satellite Control and Data Relay System 卫星控制和数据中继系统

**SCDU** Satellite Control Data Unit 卫星控制数据单元

Satellite Control Display Unit 卫星控制显示组件

**SCE** Service Creation Environment 业务创建环境

Signal Combining Equipment 信号组合设备

Signal Conversion Equipment 信号变换装置

Source 源, 数据源, 电源

Standard Communications Environment 标准通信环境

State College, PA, USA(Airport Code) 美国宾夕法尼亚州斯塔特大学机场

**SCEA** Signal Conditioning Electronic Assembly 信号调节电子组件

Society of Cost Estimating and Analysis 成本预测和分析组织

**SCEF** Service Creation Environment Function 业务创建环境功能

**SCEO** Satellite Control Engineering Office 卫星控制工程署

**SCEP** State Committee for Environmental Protection 环境保护国家委员会

**SCF** Sampled Channel Filter 取样信道滤波器

Satellite Control Facility 卫星控制设备

Satellite Control Function 卫星控制功能

Selective Call Forwarding 有选择呼叫前转

Semiconductor Cylinder Fiber 半导体柱形光纤

Service Control Function 业务控制功能

Soliton Compression Fiber 孤子压缩光纤

Standards Control Function 标准控制功能

Synchronization and Coordination Function 同步与协调功能

**SCFH** Standard Cubic Feet per Hour 每小时标准立方英尺

**SCFLA** Semiconductor Cylinder Fiber Light Amplifier 半导体柱形光纤光放大器

**SCFM** Standard Cubic Feet Minute of Gaseous Airflow 标准立方英尺分钟气体流量

Standard Cubic Feet per Minute 标准立方英尺/分

Subcarrier Frequency Modulation 副载波调频

**SCFS** Share Control File System 共用控制文件系统

Subcarrier Frequency Shift 副载波频移

**SCG** Security Classification Guide 保密等级指南, 密级分类指南

Storage Compatibility Group 库存互换性小组

System Configuration Guide 系统构型指南，系统配置指南

**SCGD** Specification Control Group Directive 规范控制组指导

**SCH** Satellite Channel 卫星信道

Schedule（进程）表,（时间）表,程序,方案

Scheduled,Scheduling 预定的,行程安排

Socket Head 插座头

Synchronization Channel 同步信道

**SCH FLT** Scheduled Flight 定期航班

**SCHED** Schedule 计划,规划,班机时刻表

**SCHEM** Schematic 简图

**SCHG** Supercharge 对…增压

**SCHLT** Searchlight 搜索灯

**SCHM** Schematic 纲要的,概略的

**SCI** Scalable Coherent Interface 可伸缩相干接口,可变规模互连接口

Science 科学

Science Citation Index 科学引文索引

Society of Chemical Industry 化学工业协会

Software Configuration Index 软件配置索引

Switching Center Interface 交换中心接口

**SCIB** Search Compressed Index Block 路径检索简洁索引块

Selective Channel Input Bus 选择通道输入总线

**SCIC** Semiconductor Integrated Circuit 半导体集成电路

Signal Collection and Identification Card 信号收集和识别卡

**SCID** Signal Collection/Tail Identification Card 信号收集/尾部识别卡

Software Configuration Index Drawing 软件组合索引图

**SCIF** Single Console Image Facility 单控制台图像装置

**SCINT** Scintillator 闪烁仪,闪烁体

**SCIRP** Semiconductor Infra-Red Photography 半导体红外摄影术

**SCIRT** System Control in Real Time 实时系统控制

**SCIT** Standard Change Integration and Tracking 标准更改综合与跟踪

**SCK** Shift Clock 移位时钟

**SCL** Santiago, Chile（Airport Code）智利圣地亚哥机场

Section Chord Line 扇形弦线

Serial Communication Loop 连续通信环路

Simulation Control Language 模拟控制语言

Soliton Compression Laser 孤子压缩激光器

Stores Coordination Letter 库存协调函

System Control Language 系统控制语言

**SCLA** Semiconductor Laser Amplifier 半导体激光放大器

**SCLC** Space-Charge-Limited Current 空间电荷限制电流

**SCLD** Space-Charge-Limited Diode 空间电荷限制二极管

**SCLER** Scleroscope 肖氏硬度计

**SCLIB** Service Component Library 业务构件库

**SCLK** Synchronous Clock 同步时钟

**SCLT** Space-Charge-Limited Transistor 空间电荷限制晶体管

**SCM** Security Configuration Manager 安全配置管理器

Selected Coding Method 选定的编码法

Selected Communication Mode 选定的通信模式

Semiconductor Memory 半导体存储器

Service Circuit Module 服务电路模块

Signal Conditioner Module 信号调节器组件

Signal Conditioning Module 信号波形修整模块

Signature Confirmation Measurement 签字批准测量

Single-Crystal Material 单晶材料

Software Change Memorandum 软件更改备忘录

Software Configuration Management 软件配置

管理

Spoiler Control Module 扰流板控制组件

Stabilizer Control Module （水平）安定面控制组件

State Change Management 状态变换管理

Structural Condition Monitoring 结构状态监控

Subcarrier Modulation 副载波调制

Subcarrier Multiplexing 副载波复用

Subchannel Matching 子信道匹配

Supply Chain Management 供应链管理

System Configuration Monitor 系统组合监视器

**SCMA**　Subcarrier Multiple Access 副载波多址接入

**SCMP**　Single-Channel Monopulse Processor 单信道单脉冲处理器

Software Configuration Management Plan 软件配置管理计划

Specification Controlled Machined Part 规范控制机加工部件

Stream Control Message Protocol 流控消息协议

**SCMR**　Software Change Management Report 软件更改管理报告

Software Configuration Management Report 软件配置管理报告

**SCMS**　Software Configuration Management System 软件配置管理系统

Signal Command Management System 信号指令管理系统

Standard Configuration Management System 标准配置管理系统

Supply Chain Management System 供应链管理系统

**SCMT**　Supply Chain Management Tool 供应链管理工具

**SCN**　Satellite Control Network 卫星控制网络

Self-Contained Navigation 自主式导航

Service Control Node 业务控制节点

Software Change Notice 软件更改通知

Spares Cancellation Notice 备件撤销通知

Special Change Notice 特别更改通知

Specification Change Notice 规范更改通知

Specification Control Navigation 规范控制导航

Specification Control Number 规范控制号

Subscriber Connection Network 用户连接网络

Switching Center Network 交换中心网

System Change Notice 系统更改通知

System Control Number 系统控制号

**SCNA**　Sudden Cosmic Noise Absorption 突发宇宙噪声吸收

**SCNG**　Scanning 扫描，扫掠

**SCNR**　Scanner 扫描器

**SCNTN**　Self-Contained 独立的，配套齐全的

**SCO**　Sales Contracting Office 销售合同处

Single Channel Operation 单信道工作

Subcarrier Oscillator 副载波振荡器

Surface Contaminatecl Object 表面污染物

Synchronous Connect-Oriented 面向连接同步

**SCOB**　Scatter/Scattered Clouds or Better 少云或疏云

**SCODA**　Scan Coherent Doppler Attachment 相关扫描多普勒附属设备

**SCOPE**　Oscilloscope 示波器

Scheduled Cost Oriented Performance Evaluation 基于计划成本的性能评估

Scientific Committee on Problems of the Environment 环境问题科学委员会

System Analyst, Computer Operator, Program Evaluation 系统分析员、计算机操作员与程序评价

**SCOR**　Short Circuit Order Release 短路指令发布

Status Commitments and Operational Reporting System 状态支持与操作报告系统

**SCORE**　Satellite Computer Operated Readiness Equipment 卫星计算机控制备用设备

**SCOT**　Small Communication Terminal 小型通

信终端

**SCOUT** Surface Controlled Oxide Unipolar Transistor 表面控制氧化物单极晶体管

**SCP** Scratch Pad 便笺本

Sector Chief Pilot 航段总飞行师

Sensor Control Program 传感器控制程序

Service Control Point 业务控制点

Session Control Protocol 会话控制协议

Signal Control Point 信号控制点

Single Chip Processor 单片处理机

Software Change Proposal 软件更改建议

Spherical Candlepower 球面烛光

Status and Control Panel 状态与控制板

Subcontractor Change Proposal 分包方更改建议

Subscriber Call Processing 用户呼叫处理

Supplier Change Proposal 供应方更改建议

Support Computer Program 计算机支持程序

System Control Panel 系统控制板

System Control Processor 系统控制处理器

System Control Program 系统控制程序

**SCPC** Single Channel per Carrier 单路单载波

**SCPD** Suppressed Clock Pulse Duration 时钟脉冲抑制期

**SCPI** Standard Commands for Programmable Instruments 可编程仪表标准指令

**SCPL** Senior Commercial Pilot's Licence 高级商用飞机驾驶员执照

**SCPM** Service Control Point Model 业务控制点模型

**SCPU** Switching Central Processing Unit 交换中央处理器

**SCQ** Santiago De Compostela, Spain（Airport Code）西班牙圣地亚哥机场

**SCR** Schedule Commitment Record 计划约定记录

Scratch 刻痕,暂存区

Screen 屏幕,滤网

Screw 螺丝,螺钉

Selective Circuit Reservation 选择性电路预留

Semiconductor Controlled Rectifier 可控半导体整流器

Short Circuit Ratio 短路比

Signal to Crosstalk Ratio 信号串话比

Silicon Control Rectifier 硅控整流器

Software Change Request 软件更改申请

Spares Coordination Record 备件协调记录

Spares Coordination Request 备件协调申请

Special Certification Review 特别证书审查

Specific Commodity Rates 特种货物运价,指定商品运价

Specification Change Request 规范更改申请

Subcontractor Configuration Review 分包方组合检查

Supersonic Cruise Research 超音速巡航研究

Supplier Configuration Review 供应方组合审查

Support Control Room 保障控制室,支援控制室

Sustainable Cell Rate 可持续信元速率

System Change Record 系统更改记录

System Change Request 系统更改申请

System Clock Reference 系统时钟基准

**SCRAMJET** Supersonic Combustion Ramjet 超音速燃烧冲压喷气发动机

**SCRC** Sub-Committee on Radio Communications 无线电通信分委会

Supplier Change Review Committee 供应方更改审查委员会

**SCRDB** Screwed Bonnet 螺栓固定机盖

**SCRDN** Screw Down 用螺丝拧紧,螺旋压下机轮

**SCRIPT** System Contract Requirement Integration and Processing Technique 系统合同要求综合与处理技术

**SCRN** Screen 屏幕,滤网

**SCRT** Subscribers Circuit Routine Test 用户线路例行测试

Supply Chain Response Time 供应链响应时间

**SCRTD** Southern California Rapid Transit District 南加州捷运区

**SCRTERM** Screw Terminal 螺栓头,螺钉头

**SCRTHD** Screw Thread 螺纹

**SCRTY** Security 保密,安全

**SCS** Safety Control Switch 安全控制开关
Sale Contracts 销售合同
Satellite Communication System 卫星通信系统
Satellite Control Station 卫星控制站
Satellite Control System 卫星控制系统
Saving Control System 节油控制系统
Scan Converter System 扫描变换器系统
Scientific Computing System 科学计算系统
Scientific Control System 科学控制系统
Semiconductor Controlled Switch 半导体控制开关
Separate-Channel Signaling 独立信道信令
Sequence Control System 顺序(程序)控制系统
Service Creation System 业务创建系统
Signal Control Search 信号控制搜索
Silicon Controlled Switch 硅可控开关
Single Channel Control Simplex 单通道单向通讯
Single Chip System 单芯片系统
Society for Computer Simulation 计算机模拟协会
Source Control System (电)源控制系统
Space Communication Satellite 空间通信卫星
Speed Control Switch 速度控制开关
Speed Control System 速度控制系统
Stabilization and Control System 稳定与控制系统
Structured Cabling Systems 结构化布线系统
Subcontractor Support 分包方支援
Symmetric Crypto-System 对称加密系统
Synchronous Communication Satellite 同步通信卫星

**SCSFR** Stratocumulus Stratiform 层状层积云

**SCSH** Structural Carbon Steel Hard 结构用硬碳素钢

**SCSI** Small Computer Standard Interface 小型计算机标准接口
Small Computer System Interface 小型计算机系统接口

**SCSM** Structural Carbon Steel Medium 结构用中碳素钢

**SCSN** Sub-Committee on Safety Navigation 航行安全小组委员会

**SCSS** Structural Carbon Steel Soft 结构用软碳素钢

**SCT** Scattered 散射的
Scattered Clouds 散云,疏云
Sector 扇形齿轮,轴瓦
Software Compatibility Test 软件兼容性测试
Standard Clock Time 当地标准时
Surface Charge Technology 表面电荷技术
Surface Charge Transistor 表面电荷晶体管
System Communication Table 系统通信表
System Compatibility Test 系统兼容性测试
System Component Test 系统部件测试

**SCTR** Scooter 注射器,小型摩托车
Sector 扇区,扇形齿轮
Stratocumulus Translucidus 透光层积云

**SCTS** Super-C Test Station 超级 C 测试站
System Components Technical Specifications 系统零部件技术规范
System Components Test Station 系统零部件测试站

**SCU** Scanner Control Unit 扫描器控制组件
Seat Control Unit 座椅控制组件
Service Circuit Unit 业务电路单元
Servo Control Unit 伺服控制组件
Signal Conditioner Unit 信号调节器组件
Signaling and Control Unit 信令与控制单元
Signaling Channel Unit 信令信道单元
Single Conditioning Unit 单整形组件
Spares Coordination Unit 备件协调单位
Starter Converter Unit 起动机整流器组件
Station Control Unit 站控制器

Storage Control Unit 存储控制器

Subscriber Control Unit 用户控制器

Supplemental/Supplementary Control Unit 辅助控制组件

Synchronous Control Unit 同步控制器

System Capabilities Upgrade 系统性能升级

System Control Unit 系统控制器

**SCUBA** Self-Contained Underwater Breathing Apparatus 自给式水下呼吸器

**SCUC** Simulator Computer Unit Complex 模拟机计算机组件

**SCUP** Scupper 排水孔,排油口

**SCV** Source Control Valve (液压)源控制活门

Sub-Clutter Visibility 副干扰可见度

Surge Control Valve 喘振控制活门

**SCX** Specialized Communications Exchange 特殊通信交换机

**SCXI** Signal Conditioning Extensions for Instrumentation 仪表设备信号整形扩展

**SCYL** Single Cylinder 单气缸

**SD** Safety Destructor 安全自毁器

Sample Data 样本数据

Secure Digital Memory Card 安全数码存储卡

Self Destroying 自毁

Send Data 传送数据

Shipping Document 货运单

Short Dump 短清除

Shut-Down 关断,停车

Side Display 侧显示器

Signal Degrade 信号衰变

Signal Distributor 信号分配器

Source Data 源数据

Space Division 空分

Space Domain 空间域

Spark Discharger 火花放电器

Special Designation 特殊识别

Speciality Description 专用说明

Spectral Distribution 频谱分布

Standard Deviation 标准偏差

Static Detector 静电检测器

Static Discharge 静电放电

Stationary Detector 静态检测器

Statistical Decision 统计决策

Statistical Design 统计设计

Statistics Division 统计处

Step Descent 阶梯下降,步阶下降

Storm Detection 暴雨探测,暴风探测

Stress Distribution 应力分布

Student Discount 学生折扣

Subsonic Detector 亚音(速)探测器

Super Disc 超级光盘

Supply Department 供应部门

Sweep Delay 扫描延迟

Sweep Driver 扫描驱动器

Switched or Selection Diversity 交换或选择分集

Synchro to Digital 同步/数字转换

System Damage 系统损坏

System Design 系统设计

System Development 系统开发

System Diagrams 系统框图

System Display 系统显示(器)

System Dynamics 系统动态

**S-D** Synchro Digital 同步数字

**SD BL** Sand Blast 喷砂处理(设备)

**SD/PDS** Serial Digital Preprocessing Data System 串行数字式预处理数据系统

**SDA** Baghdad, Iraq (Airport Code) 伊拉克巴格达机场

Shandong Aviation Group 山东航空集团

Shared Data Administration 共享数据管理

Software Disk Array 软件磁盘阵列

Source Data Automation 源数据自动化

System Design Alternative 备用系统设计

System Design Analysis 系统设计分析

**SDAC** Satellite Data Analysis Center 卫星数据分析中心

Statistical Data Analysis Center 统计数据分析中心

System Data Acquisition Computer 系统数据

采集计算机

System Data Acquisition Concentrator 系统数据采集集中器

System Data Analog Converter 系统数—模转换器

**SDAI** Standard Data Access Interface 标准数据访问接口

Strategic Development and Integration 战略开发与综合

**SDAML** Send by Airmail 航空邮寄

**SDAP** Special Duty Assignment Pay 特别任务委派报酬

System Development and Analysis Program 系统开发与分析程序

System Development and Performance 系统开发与性能

**SDAR** Subsystem Design Analysis Report 分系统设计分析报告

**SDAU** Special Data Acquisition Unit 特殊数据采集组件

**SDB** Serial Data Bus 串行数据总线

Shared Data Base 共享数据库

Signals Data Base 信号数据库

State Database 国家数据库

Storage Data Buffer 存储数据缓冲器

Storage Data Bus 存储数据总线

Supplier Data Base 供应方数据库

**SDBD** Software Data Base Document 软件数据库文件

**SDBI** Storage Data Bus-in 存储数据输入总线

**SDBL** Sight Draft Bill of Lading 即期提货单

**SDC** Safety Design Criteria 安全性设计准则

Serial Data Channel 串行数据通道

Serial Data Controller 串行数据控制器

Serial Digital Computer 串行数字计算机

Signal Data Converter 信号数据变换器,信号资料转换器

Signal Display Converter 信号显示器转换器

Single Drift Correction 单一偏流修正

Situation Display Converter 情景显示器转换器

器

Space Date Corporation 空间数据公司

Space Development Corporation 空间开发公司

Special Devices Center 专用器件中心

Standard Departure Clearance 标准离场许可

Standard Design Criteria 标准设计准则

Standards Development Committee 标准开发委员会

Structure Design Composites Programs 结构设计组合程序

Structure Design Criteria 结构设计准则

Subsystem Design Criteria 分系统设计准则

Supplier Data Catalog 供应方数据目录

**SDCA** Stabilize-Do-Check-Act 稳定—实施—检查—行动

Standardize-Do-Check-Act 标准化—实施—检查—行动

**SDCCE** Synchronous Digital Cross Connect Equipment 同步数字交叉连接设备

**SDCCH** Separate Dedicated Control Channel 独立专用控制信道

**SDCF** Small Dispersion Controlled Fiber 小色散控制光纤

Smoke Detection Control Function 烟雾探测控制功能

**SDCPS** Safety Data Collection and Processing Systems 安全数据收集与处理系统

**SDCR** Software Development Capability Review 软件开发能力评审

System Design Compliance Report 系统设计符合性报告

**SDCU** Smoke Detection Control Unit 烟雾探测控制组件

**SDD** See Detailed Drawing 见详细图纸

Serial Digital Data 串行数字数据

Software Detailed Description 软件详细说明

Standard Delivery Date 标准交付日期

Synthetic Dynamic Display 综合动态显示器

System Definition Directive 系统定义指令

System Description Document 系统说明文件

System Development and Demonstration 系统发展和验证

System for Distributed Database 分布式数据库系统

**SDDB** Simulation Digital Data Bus 模拟数字式数据总线

**SDDBDD** Simulation Digital Data Bus Definition Document 模拟数字式数据总线定义文件

**SDDD** Software Detailed Design Document 软件详细设计文件

**SDDFRC** Sperry Dual Direction Finder Radio Compass 斯派瑞双无线电定向罗盘

**SDDL** Security Descriptor Definition Language 安全描述符定义语言

Stored Data Definition Language 存储数据定义语言

**SDDM** Secretary of Defense Decision Memorandum 国防部长决策备忘录

Software Design Description Model 软件设计描述模型

**SDDN** Secure Data Defense Network 安全数据防卫网

Sensor Data Distribution Network 传感器数据分配网

Software-Defined Digital Network 软件定义的数字网

**SDDP** Signal Data Distributor Processor 信号数据分配器处理器

**SDDR** Supplier Drawings and Data Requirement 供应方图表和数据要求

**SDDS** Software Detailed Design Specification 软件详细设计规范

**SDDSF** Specially Designed Dispersion-Shifted Fiber 专门设计的色散位移光纤

**SDDU** Simplex Data Distribution Unit 单工数据分配器

**SDDVD** Super Density Digital Video Disk 超高密度数字影碟

**SDE** Software Development Environment 软件发展环境

Standard Data Element 标准数据单元

Submission/Delivery Entity 交发/投送实体

**SDF** Louisville, KY, USA (Airport Code) 美国肯塔基州路易斯维尔机场

Self-Destruct Fuse 自毁引信

Sensor Data Fusion 传感器数据融合

Service Data Function 业务数据功能

Simplified Directional Facility 简易定向设施

Simulator Development Facility 模拟机研制设施

Software Development Facility 软件开发设施

Software Development File 软件开发文档

Software Development Folder 软件开发文件夹

Supergroup Distribution Frame 超群配线架

System Definition File 系统定义文档

System Development Facility 系统开发设施

**SDFA** System Description and Failure Analysis 系统描述和失效分析

**SDFG** Single Degree of Freedom Gyroscope 单自由度陀螺仪

**SDFIG** Single Degree of Freedom Integrating Gyroscope 单自由度积分陀螺仪

**SDFRIG** Single Degree of Freedom Rate Integrating Gyroscope 单自由度速率积分陀螺仪

**SDFTN** Soda Fountain 冷饮柜

**SDG** Secondary Drill Gauge 二次钻规

Siding 侧板,侧线

Simulated Data Generator 模拟数据发生器

Situation Display Generator 情景显示发生器

Speed-Decreasing Gearbox 减速齿轮箱

Status Display Generator 状态显示产生器

Strapdown Gyroscope 捷联式陀螺仪

System Design Group 系统设计组

**SDH** Synchronous Digital Hierarchy 同步数字系列

**SDI** Simulation Data Interface 模拟数据接口

Site Diversity Interface 位置分集接口

Source Data Identifier 源数据识别符

Source Destination Identifier 源目的地标识

符

Source Destination Indicator 源终点指示器

Speed and Drift Indicator 速度与偏流指示器

Standards Data Interface 标准数据接口

Strategic Defense Initiative 主动战略防御

Subscriber Distribution Interface 用户分配接口

Switching Device Interface 开口器件接口

System Diagram Index 系统图索引

**SDIO** Serial Digital Input/Output 串行数字输入/输出

Strategic Defense Initiative Organization 主动战略防御组织

**SDIU** Synchronous Digital Interface Unit 同步数字接口单元

**SDIV** Subdivision 分部

**SDJ** Sendai, Japan (Airport Code) 日本仙台机场

**SDK** Software Development Kit 软件开发工具包

**SDL** Saddle 滑动座架, 鞍

Software Development Laboratory 软件开发实验室

Software Development Library 软件开发库

Specification and Description Language 规格和描述语言

Stacked Disc Library 堆垛式光盘库

Structure Design Library 结构设计库

Sundsvall, Sweden (Airport Code) 瑞典松兹瓦尔机场

System Development Laboratory 系统开发实验室

**SDLC** Synchronous Data Link Control 同步数据链控制

System Data and Language Construct 系统数据和语言结构

Systems Development Life Cycle 系统开发生命周期

**SDLF** Shaft Driven Lift Fan 轴驱动升力风扇

**SDLM** Standard Depot Level Maintenance 基地级维修标准

**SDM** Satellite Data Modulation 卫星数据调制器

Security Device Manager 安全设备管理工具

Service Data Management 业务数据管理

Short Data Message 短数据消息

Software Development Manual 软件开发手册

Space Division Multiplexing 空分复用

Speaker Drive Module 扬声器驱动组件

Subscriber Demodulator Module 用户调制模块

System Definition Manual 系统定义手册

System Development Methodology 系统开发方法学

**SDMA** Security Data Management and Analysis 安全数据管理与分析

Space Division Multiple Access 空间分多址访问

**SDME** Software Development and Maintenance Environment 软件开发与维护环境

Synchronous Data Modem Equipment 同步数据调制解调设备

**SDMIS** Standardization Data Management Information System 标准化数据管理信息系统

**SDMP** Shared Data Management Procedures 共享数据管理程序

Software Development and Maintenance Plan 软件开发和维护计划

Software Development and Maintenance Procedure 软件开发和维护程序

Software Development Management Plan 软件开发管理计划

System Development Master Plan 系统开发主计划

**SDMS** Standard Database Management System 标准数据库管理系统

Scientific Data Management System 科学数据管理系统

**SDMT** Simple Discrete Mulitone 简单离散的多音调

**SDN** Secondary Distribution Network 二级分配网

Software Defined Network 软件定义网

Synchronous Digital Transmission Network 同步数字传输网

System Description Note 系统说明书

**SDNS** Software Defined Network Services 软件定义网络业务

**SDO** Standards Developing Organizations 标准制定(发展)组织

**SDOF** Single Degree of Freedom 单向自由度

**SDP** Service Data Point 业务数据点

Service Difficulty Program 维修问题大纲

Service Discovery Protocol 服务发现协议

Service Domain Processor 业务域处理器

Shut-Down Processor 关断处理器

Signal Data Processor 信号数据处理器,信号数据处理机

Software Design Plan 软件设计计划

Software Development Plan 软件开发计划

Software Development Processor 软件开发处理器

Standard Data Processor 标准数据处理机

Supplier Design Proposal 供应方设计提案

Surveillance Data Processor 监视数据处理器

System Definition Phase 系统定义阶段

System Design Proposal 系统设计建议

System Development Plan 系统研制方案

**SDPE** Special Design Protective Equipment 特殊设计保护设备

**SDPS** Surveillance Data Processing System 监视数据处理系统

**SDQ** Santo Domingo, Dominican Republic (Airport Code) 多米尼加共和国圣多明各机场

Speciality Description Questionnaire 特性说明调查表

**SDQA** Shared Data Quality Assurance 共享数据质量保证

**SDR** Santander, Spain (Airport Code) 西班牙桑坦德机场

Sender 发送端,发送器

Sensor Data Record 传感器的数据记录

Service Deviation Report 使用偏差报告

Service Difficulty Report 使用困难报告

Signal Data Recorder 信号数据记录器

Single Data Rate 单倍数据速率

Software Deficiency Report 软件缺陷报告

Software Design Review 软件设计评审

Sounder 测深器

Special Drawing Right 特殊提款权

Standard Dimension Ratio 标准尺寸比

State Dependent Routing 状态相关选路

Statistical Data Recorder 统计数据记录器

Storage Data Register 存储数据寄存器

Structural Design Review 结构设计审查

System Design Review 系统设计审查

**SDRAM** Synchronous Dynamic Random Access Memory 同步动态随机存取存储器

**SDRB** Specifications and Data Review Board 规格和数据审查委员会

Strategic Development Research Bureau 战略发展研究局

**SDRE** State Dependent Riccati Equation 状态相关黎卡提方程

**SDRL** Subcontract Data Requirements List 分包方数据要求清单

Supplier Data Requirements List 供应方数据要求清单

Supplier Data Requisition List 供应方数据征用清单

**SDRLS** Supplier Data Requirements List Status 供应方数据要求清单状态

Supplier Data Requirements List System 供应方数据要求清单系统

**SDS** Satellite Defense Systems 卫星防御系统

Schematic Drafting Systems 电路图绘制系统,简图绘制系统

Shop Distribution Standards 车间分配标准

Shop Distribution Stock 车间分配库存

Software Design Specification 软件设计规范

Space Division Switch 空分交换

Space Documentation Service 空间文献服务处

Special Detection System 专用检测系统

Supplier Data Sheet 供应方数据单

Support Data System 支持数据系统

Surveillance Distribution System 监视分配系统

Switched Data Service 交换数据业务

Switched Discrete System 交换离散系统

Synchronized Data Switch 同步数据交换

Synchronous Data Set 同步数据装置

System and Data Service 系统与数据业务

System Data Synthesizer 系统数据综合器

System Description Section 系统说明部分

System Design Specification 系统设计规范，系统设计说明书

**SDSB** Satellite Data Services Branch 卫星数据服务部

**SDSF** System Display and Search Facility 系统显示和搜索工具

**SDSL** Single Digital Subscriber Line 单数字用户线

**SDSM** Space Division Switch Module 空分交换模块

**SDSR** Simulator Data Support Request 模拟机数据支援申请

**SDT** Serial Data Transmission 串行数据传输

Serial Date 序列日期

Short Data Transfer 短数据传递

Signal Delay Time 信号延迟时间

Simulated Data Tape 仿真数据磁带

Software Development Tape 软件开发磁带

Software Development Test 软件开发测试

Software Development Tool 软件开发工具

Special Diagnostic Tape 专用诊断磁带

Standard Diagnostic Test 标准诊断测试

System Diagnostic Test 系统诊断测试

**SDTC** Servo Data Transmit Command 伺服数据发送指令

**SDTF** Standard Data Transfer Format 标准数据转换格式

**SDTM** Structure Description Training Manual 结构说明培训手册

**SDTP** Software Development Tool Plan 软件开发工具大纲

**SDTR** Supportability Design to Requirements 按要求的可支援性设计

System Design to Requirements 按要求的系统设计

**SDTV** Standard Definition Television 标准清晰度电视

**SDU** Rio De Janeiro, Brazil-Santos Dumont (Airport Code) 巴西里约热内卢桑托斯杜蒙特机场

Satellite Data Unit 卫星数据组件

Satellite Delay Unit 卫星时延组件

Seat Display Unit 座椅显示组件

SECAL Decoder Unit 选择呼叫译码器

Secure Data Unit 保密数据组件

Segment Data Unit 分段数据单元

Service Data Unit 业务数据单元

Service Delivery Unit 业务分送单元

Solenoid Driver Unit 电磁线圈驱动组件

System Display Unit 系统显示组件

**SDV** Section Delay Value 段延迟值

Slowed Down Video 视频延迟

Switched Digital Video 交换式数字视频

Tel Aviv, Israel Apt (Airport Code) 以色列特拉维夫机场

**SDVA** Switched Digital Video Access 交换式数字视频存取

**SDVB** Switched Digital Video Broadcast 交换式数字视频广播

**SDVN** Switched Digital Video Network 交换式数字视频网络

**SDVS** Software Design and Verification System 软件设计和验证系统

**SDW** Specific Design Work 特殊设计工作

**SDXC** Synchronous Digital Cross Connect 同步数字交叉连接

**SE** Safety Engineering 安全工程

Section 段, 截面

Session Exchange 会晤交换

Sight Exchange 即期汇兑

Signal Electrode 信号电极

Signal Engineer 信号工程师

Silver Electrode 银电极

Single End 单端

Single Entry 单输入,单通道

Single-Engine 单台发动机

South East 东南

Special Equipment 专用设备

Structure Element 结构单元

Support Entity 支持实体

Support Equipment 支持设备

Sweden 瑞典

Switching Element 交换单元

System Effectiveness 系统效能

System Engineer 系统工程师

System Equalizer 系统均衡器

System Error 系统误差

System Extension 系统扩充

**S-E** Sorts and Edit 分类与编辑

**SE&I** System Engineering and Integration 系统工程和综合化

**SE(S/E)** System Engineering 系统工程

**SEA** Safety Engineering Analysis 安全工程分析

Safety Equipment Association 安全设备协会

Search 搜索

Seattle, WA, USA(Airport Code) 美国华盛顿州西雅图机场

Single Engine Aircraft 单发动机飞机

Software Error Analysis 软件差错分析

South East Asia 东南亚

South East Asia Sub-area 东南亚次区

Static Error Analysis 静态误差分析

**SEA WE CEN** Southeast Asia Weather Center 东南亚气象中心

**SEAC** Safety and Environmental Advisory Council 安全与环境咨询委员会

Scientific and Engineering Advisory Committee 科学与工程咨询委员会

Signalling Engineering and Administration Center 信令工程和管理中心

**SEACP** System Exercise and Analysis Computer Program 系统使用和分析计算机程序

**SEAD** Suppression of Enemy Air Defenses 敌方空防压制

**SEADAB** Southeast Asia Data Base 东南亚数据库

**SEADS** Shuttle Entry Air Data System 航天飞机再入大气数据系统

**SEAF** Small Enterprise Assistance Funds 小企业援助基金

Systems Engineering Avionics Facility 航空电子设施系统工程

**SEAIC** Southeast Asia Information Center 东南亚信息中心

**SEAL** Sea-Air-Land 海、空、陆

Signal Evaluation Airborne Laboratory 机载信号鉴定实验室

Simple Efficient ATM Layer 简单高效的ATM层

**SEALPA** Southeast Asia Lumber Producers Association 东南亚木材生产者协会

**SEAMEO** Southeast Asian Ministers of Education Organization 东南亚教育部长组织

**SEAP** Support Equipment Acquisition Plan 保障设备采购计划

Support Equipment Action Plan 保障设备操作计划

**SEARAST** South-East Asia Regional Aviation Safety Team 东南亚地区航空安全小组

**SEAS** Satellite Early Alert System 卫星预警系统

Signaling Engineering and Administration System 信令工程和管理系统

Standard Electronic Assembly System 标准电子组件系统

Support Equipment Avionics System 支援设备航空电子系统

**SEASAT** Sea Satellite 海洋卫星

Sea Surveillance Satellite 海洋监视卫星

**SEAT** Self-Evaluation and Test 自评估和测试

Status Evaluation and Test 状态评估测试

**SEATAC** Seattle-Tacoma 西雅图—塔科马（美国华盛顿州西部港市）

**SEATAV** Seats Available 有空座位

**SEATOCC** Seats Occupied 无空座位

**SEAWBS** Southeast Asia Wideband System 东南亚宽频带系统

**SEB** Seat Electronics Box 座椅电子设备盒

Service Evaluation Bulletin 维修（服务）评估通报

Source Evaluation Board 源评估处

South-East Bound 飞往东南方向

Surface Effect Boat 表面效应船

**SEB/ST** Seat Electronic Box with Self Test 带有自测的座椅电子盒

**SEC** Secant 正割

Second 秒,第二的

Secret 秘密

Secretary 秘书

Section Emergency Coordinator 科组紧急协调员

Securities Exchange Commission 证券汇兑委员会

Security 安全

Security Screening 安全屏蔽

Software Engineering Center 系统工程中心

Space Environment Center 空间环境中心

Special Event Charter Flight 特别事件包机飞行

Spoiler Elevator Computers 扰流板升降舵计算机

Supervisory Engine Control 监控型发动机控制

Switching Equipment Congestion 交换设备阻塞

Synchronous Equipment Clock 同步设备时钟

Systems Engineering Council 系统工程委员会

**SECAL** Selected Calling System 选择呼叫系统

**SECAM** Sequential Colour and Memory System 顺序彩色传送和存储系统（欧洲电视制式）

**SECAP** System Experience Correlation and Analysis Program 系统经验相关与分析程序

**SECAR** Secondary Radar 二次雷达

Section Area 截面积

**SECB** Severely Errored Cell Block 严重信元误块

**SECBR** Severely Errored Cell Block Ratio 严重信元误块比

**SECH** Hyperbolic Secant 双曲正割

**SECI** System Evaluation/Control Instrumentation 系统评估/控制测量设备

**SECL** Symmetric Emitter-Coupled Logic (Circuit) 对称射极耦合逻辑（电路）

**SECO** Self-regulation Error-correct Coder-decoder 自调整误差修正编译码器

Sequential Coding 顺序编码

Sustainer Engine Cut off 主发动机自动停车器

**SECOM** System Engineering Communication 系统工程通信

**SECO(N)** Sequential Control 顺序控制,时序控制

**SECON** Security Condition 安全条件,安全状态

**SECP** Supplier Engineering Change Proposal 供应方工程更改建议

Support Ejector Cable Plug 支撑架电缆插头

**SECRG** Securing 紧固的,卡紧的,安全的

**SECS** Sensor Engagement Controller Subsystem 传感器衔接控制器分系统

Severe Environment Control System 恶劣环境控制系统

Solar Electric Communication Satellite 太阳能电力通信卫星

Space Environmental Control System 空间环境控制系统

Spoiler Electronic Control System 扰流板电子控制系统

**SECSY** Spin-Echo Correlated Spectroscopy 自旋回波相关谱

**SECT** Section 区段, 部分

**SECU** Spoiler Electronic Control Unit 扰流板电子控制组件

**SED** Secondary EICAS Display 发动机指示和机组警告系统辅助显示器

Slew Entry Device 旋转输入装置

Standby Engine Display 发动机备用显示器

Supplier Engineering Definition 供应商工程定义

System Engineering Documentation 系统工程文件编制

System Evaluation Display 系统评估显示器

**SEDD** System Evaluation and Development Division 系统评估与开发分部

**SEDE** Shipper Export Declaration 交运人出口申报

Software Engineering and Data Engineering 软件工程与数据工程

**SEDS** Space Electronic Detection System 空间电子探测系统

Supply Equipment Data System 供应设备数据系统

Support Equipment Delivery Schedule 保障设备交货计划

System Effectiveness Data System 系统效能数据系统

**SEE** Software Engineering Environment 软件工程环境

System and Electronics Engineering 系统和电子工程

System Effectiveness Engineering 系统效能工程(学)

System Effectiveness Evaluation 系统效能评估

System Efficiency Expert 系统效率专家

**SEEL** Space Energy and Environment Laboratory 空间能量与环境实验室

**SEER** Systems Engineering Evaluation and Research 系统工程评价与研究

**SEEX** System Evaluation Experiment 系统评估试验

**SEF** Severely Errored Frame 严重误码帧

Standard Electrical Field 标准电场

Standard Electronic Format 标准电子格式

Standard External File 标准外部文件

Support Entity Function 支持实体功能

**SEFS** Severely Errored Frame Second 严重误帧秒

**SEFT** Single Engine Flight Training 单发飞行训练

**SEG** Sealing 密封

Secondary Gauge 辅助量规

Segment 扇形体, 部分, 单元体, 段

Software Engineering Group 软件工程组

Special Effect Generator 特殊效应产生器

Standardization Evaluation Group 标准化评估小组

System Engineering Group 系统工程组

Systems Evaluation Group 系统评估组

**SEGNM** Segment Name 段名

**SEGR** Segregation 分离

**SEI** Scientific Engineering Institute 科学工程学会

Software Engineering Institute 软件工程研究所

Standby Engine Indicator 发动机备用指示器

Standby Engine Instrument 发动机备用仪表

Statistical Engineering Institute 统计工程学会

**SEIFR** Single-Engine IFR 单发动机仪表飞行规则

**SEIP** System Engineering Implementation Plan 系统工程实施计划

**SEIT** Satellite Education and Information Television 卫星教育与新闻电视

Systems Engineering, Integration and Test 系统工程集成和测试

**SEJ** Sliding Expansion Joint 滑动膨胀接头

**SEK** Select Entry Key 选择输入键

**SEL** Selcal 选叫系统(国际民航组织)

Selection 选择

Selective/Selected 选择性的

Seoul, South Korea (Airport Code) 韩国首尔国际机场

Signal Engineering Laboratory 信号工程实验室

Single Engine Land (Aircraft) 单发陆地飞机

Single Engine Licence 单发动机许可证

Sound Exposure Level 暴露噪声级

Space Environment Laboratory 空间环境实验室

Structural Engineering Laboratory 结构工程实验室

Support Equipment List 保障设备清单

**SELCAL** Selective Call/Calling 选择呼叫

Selective Calling/Call System 选择呼叫系统

Selector Panel 选择器面板

**SELCS** Self-Contained System 自主式系统

**SELF CAL** Self Calibrating 自校准

**SELFCL** Self Closing 自关闭

**SELFOC** Selffocusing 自聚焦

**SELFPROP** Self Propelled 自推进的

**SELS** Selsyn 自同步机

**SELSYN** Self-Synchronous 自同步的

**SEM** Safety and Environmental Management 安全与环境管理

Scanning Electron Microscopy 电子扫描显微镜

Search Engine Marketing 搜索引擎营销

Security Event Management 安全事件管理

Semester 半年,一学期

Sensor Emitter Manager 传感器发射管理器

Sensor Management 传感器管理

Short Edge Margin 短边间距,短边范围

Spectrographic Electron Microscope 光谱电子显微镜

Stabilizer Trim 安定面配平

Standard Electronic Module 标准电子模块

Strategic Environmental Management 战略环境管理

Subcontractor Engineering Memo 分包方工程备忘录

System Effectiveness Model 系统有效性模型

System Engineering Manual 系统工程手册

Systems Engineering Management 系统工程管理

**SEMA** Special Electronic Mission Aircraft 电子任务专用飞机

**SEMF** Synchronous Equipment Management Function 同步设备管理功能

**SEMICON** Semiconductor Conference 半导体会议

**SEMICOND** Semiconductor 半导体

**SEMP** Standard Electronic Modules Program 标准电子模块化程序

Strategic Environmental Management Program 战略环境管理程序

System Engineering Management Plan 系统工程管理计划

**SEMR** Standard Electronic Module Radar 标准电子模块雷达

**SEMS** Safety and Environmental Management System 安全与环境管理系统

Security Event Management System 安全事件管理系统

Software Engineering Management System 软件工程管理系统

Standardized Emergency Management System 标准化的应急管理体系

System Engineering Management Standard 系统工程管理标准

**SEN** Scanner Encoder 扫描编码器

Semienclosed 半封闭的

Senegal 塞内加尔

Seniority 上级,高位数

Software Error Notification 软件错误通告

**SENEL** Single Event Noise Exposure Level 单一事件噪声暴露级

**SENET** Slot Envelope Network 时隙包络网

**SENS** Sensitive 敏感的

Sensitivity 敏感性

Sensor 传感器,感应器

Software Engineering and Network Systems Laboratory 软件工程与网络系统实验室

**SEO** Satellite for Earth Observation 地球观测卫星

**SEOPSN** Select Operate Sense 选择操作传感

**SEOS** Synchronous Earth Observation Satellite 同步地球观测卫星

**SEP** Safety and Emergency Procedure 安全与应急程序

Separation 分离

Separator 分离器

September 九月

Service Exchange of Parts 维修换件

Signaling End Point 信令终结点

Single Engine Piston 单引擎活塞

Software Engineering Practice 软件工程实践

Specific Excess Power 单位剩余功率

Spherical Error Probability 球形误差概率

Support Equipment Plan 支援设备计划

Surveillance and Evaluation Program 监视与评估计划

**SEPC** Secondary External Power Contactor 备用外电源接触器

**SEPE** Sort Enhanced Processing Element 短增强处理单元

**SEPFI** Sequential Electronic Port Fuel Injection 顺序电子燃油喷射

**SEPM** Software Engineering Process Management 软件工程过程管理

System Engineering and Program Management 系统工程与项目管理

**SEPP** Secure Electronic Payment Protocol 安全电子结算协议

Stress Evaluation Prediction Program 应力评估预测程序

**SEPR** Superseding Engineering Part Release 替换工程部件发放

**SEPS** Solar Electronic Propulsion System 太阳能电力推进系统

**SEQ** Sequence/Sequential 顺序,序列

**SEQ CHK** Sequence Check 顺序检查

**SER** Safety Evaluation Report 安全评估报告

Self-Emitted Radar 自发射雷达

Series/Serial 系列的,连续的,串联的

Service Evaluation Request 维修评估,服务评估申请

Serviceable 可用的

Soft Error Rate 软件错误率

Stop End of Runway 跑道停止端

Subcontractor Engineer Review 分包方工程审查

Supplier Engineer Review 供应方工程审查

System Evaluation Report 系统评估报告

Systems Engineering Report 系统工程报告

**SER DL** Serial Data Link 串行数据链

**SER NO** Serial Number 序号

**SER/PRL** Serial to Parallel 串并联

**SERD** Supplier Equipment Requirement Documentation 供应方设备要求文件

Support Equipment Recommendation Data 保障设备建议数据

**SERDES** Serializer/Deserializer 串行器/解串器

**SERDEX** Serial Data Exchanger 串行数据交换

**SERF** Selenium Rectifier 硒整流器

Sensitivity Element Report File 敏感性元件报告文档

**SERG** Servicing 维修,服务

**SERI** Solar Energy Research Institute 太阳能研究院

**SERL** Support Equipment Requirements List 保障设备需求清单

**SERMNG** Service State Management 业务状态管理

**SERR** Serrate 齿形的

Serrated 锯齿形的,开槽

**SERS** Sensitivity Element Reporting System 敏感性元件报告系统

**SERVO** Servo Mechanism 伺服机构,伺服系

统

**SES** Satellite Earth Station 卫星地面站

Secondary Engine Start 辅助发动机启动

Shuttle Engineering Simulator 航天飞机工程模拟器

Shuttle Engineering System 航天飞机工程系统

Signal Encrypt System 信号加密系统

Signaling Equipment Subsystem 信令设备子系统

Single Engine Sea( Aircraft) 单发水上飞机

Single European Sky 单一欧洲天空

Solar Environment Simulator 太阳环境模拟器

Space Environment Simulation 空间环境模拟

Standard Engineers Society 标准工程师协会

Support Equipment Summary 保障设备综述,支援设备综述

Surface Effects Ship 气船,水面效应船

System Engineering Service 系统工程服务

System External Storage 系统外部存储器

**SESAR** Single European Sky ATM Research 单一欧洲天空空中交通管理研究

Single European Sky ATM Research Programme 单一欧洲天空空中交通管理研究项目

**SESIP** System Engineering Summary of Installation and Program 安装和程序设计系统工程摘要

**SESP** Space Experimental Support Program 空间实验支援程序

System Engineering Standard and Procedure 系统工程标准与程序

**SESR** Severely Errored Second Ratio 严重误码秒比

**SESRP** Ship Earth Station Response 舰船地球站响应

**SESS** Session 会议,会期

**SEST** Ship Earth Station Telex 舰船地球站用户电报

**SET** Science and Engineering Technology 科学与工程技术

Secure Electronic Transaction 网上安全电子交易,安全电子交易

Simulated Emergency Test 模拟紧急试验

Single-Engine Turboprop 单台螺旋桨发动机

Software Emulated Terminal 软件仿真终端

Software Engineering Technology 软件工程技术

Static Ejection Test 静电排放试验

**SETA** System Engineering and Technical Assistance 系统工程和技术协助

**SETE** Supersonic Expendable Turbine Engines 不可回收的超音速涡轮发动机

**SETG** Synchronous Equipment Timing Generator 同步设备定时发生器

**SETL** Simplified English Test Language 简化英语测试语言

**SETLG** Settling 设定值,稳定,回复,解决

**SETLT** Settlement 解决

**SETOL** Surface Effect Takeoff and Landing 起飞和着陆时的地面效应

**SETP** Service Engineering Test Program 维修工程试验程序

Society of Experimental Test Pilots 试验试飞员协会

Software Engineering Theory and Practice 软件工程理论与实践

System Engineering Test Program 系统工程测试程序

**SETPI** Synchronous Equipment Timing Physical Interface 同步设备定时物理接口

**SETR** Service Engineering Trouble Record 维修工程故障记录

Setter 调节器

Systmes Engineering Technical Review 系统工程技术评审

**SETS** Set Equation Transform System 集合方程转换系统

Synchronous Equipment Timing Source 同步设备定时源

**SEU**　Seat Electronics Unit 座椅电子组件
Servo Electronics Unit 伺服电子组件
Single Event Upset 单事件干扰
System Evaluation Unit 系统评估组件

**SEV**　Severe 严重的, 严格的
Severity 严重, 严格, 艰难, 刚度
Surface Effects Vehicle 气垫船

**SEVOCOM**　Secure Voice Communication 保密话音通信

**SEW**　Scientific Engineering Workstation 科学工程工作站
Support Evaluation Worksheet 保障支援评估工作单

**SEWA**　Study Engineering Work Authorization 研究工程工作批准

**SEWS**　Satellite Early Warning System 卫星提前报警系统
Software Engineering Work Station 软件工程工作站
Standard Emergency Warning Signal 标准应急告警信号

**SEXSAT**　Standard Experiment Satellite 标准实验卫星

**SEZ**　Mahe Island, Seychelles(Airport Code) 塞舌尔马埃岛机场
Special Economic Zone 经济特区

**SF**　Safety Factor 安全系数
Safety First 安全第一
Safety Fuel 安全燃料
Safety Fuse 保险丝
Sales Forecast 销售预测
Sampling Fraction 取样比
Sampling Frequency 取样频率
Saw Fixture 锯夹具, 锯弓
Scale Factor 比例系数
Screen Form 屏幕格式
Sealing Factor 密封系数
Select Frequency 选择频率
Self-Feeding 自动给料的
Semifinished 半成品
Semi-Finished 半精加工的

Service Factor 维修系数, 服务因素
Service Feature 业务特征
Shear(ing) Force 剪切力
Shift Forward 向上位移
Short Wave Frequency 短波频率
Side Frequency 边频频率
Signal Fail 信号失效
Signal Frequency 信号频率
Significant Figure 有效数字, 有效数位
Slip Fit 滑动配合, 紧接合
Space Factor 空间系数, 占空系数
Space Flight 空间飞行
Spare Frame 备用帧
Special Facilities 专用设施
Special Fixture 专用夹具
Special Freighter 专用货机
Special Function 特殊功能
Spot Face 局部平整面, 抽样面, 污面
Standard Form 标准型, 标准表格
Standard Frequency 标准频率
Status Field 状态字段
Storage and Forward 存储转发(系统)
Strain Gauge 应变仪
Subframe 子帧
Supersonic Frequency 超音频
Supervisory Formula 监管公式
Symbology File 字符文档
Synovial Fluid 滑液

**SFA**　Sales Force Automation 销售力量自动化
Security Failure Analysis 安全破损分析
Sfax, Tunisia (Airport Code) 突尼斯斯法克斯机场
Special Flight Approval 特许飞行批准书

**SFACT**　Aeronautical Training and Technical Standards Division (法) 航空培训与技术标准司

**SFAR**　Special Federal Aviation Regulations 特殊联邦航空条例

**SFB**　Solid Fiber Board 硬纤维板

**SFC**　Secondary Flow Control 二次流控制

Shop Floor Control 车间层控制

Space Flight Center 航天飞行中心

Special Flight Charts 特种飞行图

Special Function Card 特殊功能插件板

Specific Fuel Consumption 燃油消耗量,燃油消耗率

Surface 表面

Synchronous Forward Command 同步前向命令

**SFCA** Supplier Functional Configuration Audit 供应方功能组合审查

**SFCC** Side-Facing Crew Cockpit 有侧向机座的驾驶舱

Slat/Flap Control Computer 缝翼/襟翼控制计算机

**SFCG** Space Frequency Coordination Group 空间频率协调小组

**SFCL** Slat/Flap Control Lever 缝翼/襟翼操纵手柄

**SFCS** Safety Flight Control System 安全飞行控制系统

Secondary Flight Control System 辅助飞行控制系统

Survivable Flight Control System 抗毁飞行控制系统

**SFCT** Subsystem Functional Compatibility Tests 分系统功能兼容性测试

**SFCU** Slat Flap Control Unit 缝翼、襟翼控制装置

**SFD** Saturation Flux Density 饱和通量密度

Single Frequency Dialling 单频拨号

Software Function Description 软件功能说明

Source-to-Film Distance 源至影片距离

**SFDB** Superplastic Forming and Diffusion Bond 超塑性成型及扩散结合

**SFDF** Subsystem Fault Detection Function 子系统故障检测功能

**SFDR** Single Feeder 单级馈线

Standard Flight Data Recorder 标准飞行数据记录器

System Functional Design Review 系统功能设计评审

**SFE** San Fernando, Philippines (Airport Code) 菲律宾圣费尔南多机场

Seller Furnished Equipment 卖方装配的设备

Supplier Furnished Equipment 供应方提供的设备

Synthetic Flight Examiner 综合飞行考试员

**SFECU** Slat Flap Electronic Control Unit 缝翼襟翼电子控制装置

**SFET** Synchronous Frequency Encoding Technique 同步频率编码技术

**SFF** Swedish Aviation Historical Society 瑞典航空历史社团

**SFFL** Standard Foreign Fare Level 国外标准票价水平

**SFH** Slow Frequency Hopping 慢跳频

**SFHA** System Function Hazard Assessment 系统功能危害评估

**SFHP** Single Fluid Heat Pipe 单流体加热管

**SFI** Sequential Fuel Injection 顺序式燃油喷射

Service Feature Instance 业务特征事例

Slat/Flap Indicator 缝翼/襟翼指位器

**SFJ** Kangerlussuaq, Greenland (Airport Code) 格陵兰康克鲁斯瓦格机场

**SFIR** Specific Force Integrating Receiver 比力积分接收器

**SFIT** System Functionality Integration Team 系统功能集成组

**SFL** Sequenced Flashing Light System 顺序闪光灯系统

Sequenced Flashing Lights 顺序闪光灯

Soliton Fiber Ring Laser 孤子光纤环激光器

**SFLARE** Flare Distance 改平距离

**SFLRL** System Failure 系统故障

**SFLS** Semiflush 半齐平的

**SFLT** Seattle Flight Test 西雅图飞行试验(中心)

**SFL-V** Sequenced Flashing Lights—Variable Light Intensity 顺序闪光灯—光强可调

**SFM**    Swept Frequency Method 扫频方法
Swept Frequency Modulation 扫频调制
Switching-mode Frequency Multipliers 开关式倍频器

**SFMECA**    Software Failure Modes, Effects and Criticality Analysis 软件故障模式、影响及危害分析

**SFML**    Sea Food Meal 海鲜餐食
Standard Facility Material List 标准设施器材单

**SFN**    Single Frequency Network 单频网络

**SFO**    San Francisco, CA, USA (Airport Code) 美国加利福尼亚州旧金山国际机场
Service Fuel Oil 油料服务
Spot Face Other Side 孔口平面另一面

**SFOCS**    Single Fiber Optical Communication System 单光纤通信系统

**SFOM**    Supportability Figure of Merit 品质可提供指数,标准可提供指数

**SFP**    Selected Flight Path 预选航线
Simplified Flight Planning 简化飞行计划
Simulated Flight Plan 模拟飞行计划
Special Flight Permit 航空器特许飞行证

**SFPI**    Slats/Flaps Position Indicator 前缘缝翼/襟翼位置指示器

**SFPM**    Surface Feet per Minute 每分钟表面英尺

**SFPO**    Special Features Protection Ordering 特殊装置保护指令,特殊性能保护指令

**SFPPL**    Short Form Provisioning Parts List 短(缺)型储备件清单

**SFQR**    Supplier Formal Qualification Review 供应方正式资格审查

**SFR**    Safety of Flight Requirements 飞行安全要求
System Failure Rate 系统故障率

**SFRJ**    Solid-Fuel Ramjet 固体燃料冲压发动机

**SFROM**    Smart Flash Read Only Memory 智能快速擦除只读存储器

**SFS**    Single Frequency Smoothing 单频平滑

Start of Frame Sequence 帧序列开端

**SFSD**    Satellite Field Service Division 卫星现场服务部

**SFSG**    Sweep Frequency Signal Generator 扫频信号产生器

**SFSK**    Sinusoidal Frequency Shift Keying 正弦移频键控

**SFSS**    Satellite Field Service Station (气象)卫星机场服务站

**SFT**    Shaft 轴
Square Feet 平方英尺
Steady-Flow Turbine 稳流式涡轮机
Supplemental Flight Test 补充试飞
Surface Friction Tester 跑道表面摩擦系数测试器
System Fault Tolerance 系统容错
System Functional Test 系统功能试验

**SFTA**    Software Fault Tolerance Analysis 软件故障容差分析
Software Fault Tree Analysis 软件故障树分析

**SFTC**    Seattle Flight Test Center 西雅图试飞中心

**SFTDC**    System Flight Test Data Center 系统试飞数据中心

**SFTE**    Society of Flight Test Engineers 试飞工程师协会

**SFTI**    Special Flight Test Instrumentation 试飞专用仪表设备

**SFTP**    Supplemental Federal Test Procedure 补充的联邦测试程序
System Functional Test Procedure 系统功能测试程序

**SFTR**    System Functional Test Report 系统功能测试报告

**SFTS**    Service Flight Training School 服务飞行训练学校
Standard Frequency and Time Signal 标准频率和时间信号
Synthetic Flight Training Simulator 综合飞行训练模拟器

Synthetic Flight Training System 综合飞行训练系统

**SFU**　Store and Forward Unit 存储转发单元

**SFV**　Safety Valve 安全活门

**SFWI**　Sensor Flag Warning Indicator 传感器警告指示旗标

**SFX**　Suffix 尾缀

**SFXD**　Semifixed 半固定的

**SG**　Safety Gyroscope 安全陀螺仪

Sawtooth Generator 锯齿波产生器

Signal Generator 信号发生器

Signaling Gateway 信令网关

Sinusoidal Generator 正弦波产生器

Snow Grains 雪粒

Specific Gravity 比重

Standard Gauge 标准线规,标准量规

Stroke Generator 笔划发生器

Structural Glass 结构玻璃

Study Group 研究组

Super Group 超群

Symbol Generator 符号发生器

Sync Gap 同步间隙

System Generation 系统生成

**SGA**　Specific Gravity Adjustment 比重调整

**SGB**　Switch Gear Block 转换齿轮箱

**SGC**　Signaling Grouping Channel 信令分组信道

**SGCAS**　Study Group on Certification of Automatic Systems 自动系统审定研究组

**SGCHAS**　Second Generation Comprehensive Helicopter Analysis System 第二代直升机综合分析系统

**SGCP**　Simple Gateway Control Protocol 简单网关控制协议

**SGCT**　Self-Gauging Clock Technology 自测量时钟技术

**SGD**　Signed 标记的,签字的

**SGDE**　Steering Gear Dual Emergency 操纵齿轮双重紧急

**SGEMP**　System Generated Electromagnetic Pulse 系统产生电磁脉冲

**SGL**　Signal 信号

Single 单一的

**SGLB**　Single Room with Bath（Reservations Code）单人床附浴室

**SGLS**　Space/Ground Link Subsystem 空—地数据链子系统

Space-Ground Link System 空—地数据链系统

**SGM**　Signal Modem 信号调制解调器

Spark Gap Modulation 火花间隙调制

**SGMA**　Sporting Goods Manufacturers Association 运动物品制造商学会

Symbol Generator Memory Access 符号发生器存贮器存取

**SGML**　Standard Generalized Markup Language 标准通用标记语言

**SGMP**　Simple Gateway Management Protocol 简单网关管理协议

Simple Gateway Monitoring Protocol 简单网关监控协议

**SGN**　Ho Chi Minh City, Vietnam（Airport Code）越南胡志明市机场

Sign 符号,标记,签字

**SGP**　Singapore 新加坡

Single Ground Point 单一地面点

Stroke Generator Processor 笔划发生处理器

**SGR**　Sortie Generation Rate 出动架次率

**SGS**　Surface Geometry System 表面几何系统

**SGSUB**　Salt Glazed Structural Unit Base 上盐釉的结构盖面基础

**SGTE**　Small Gas Turbine Engine 小型燃气涡轮发动机

Supergroup Translating Equipment 超群转换设备

**SGU**　Saint George, UT, USA（Airport Code）美国尤他州圣乔治机场

Symbol Generator Unit 符号发生器组件

**SGW**　Security Guard Window 安全防卫窗

**SH**　Sample and Hold 采样和保持

Scleroscope Hardness 回跳硬度

Section Header 段信头

Session Handler 会话处理程序
Share 共用,共享
Sheeting 薄片(膜),档板
Shower 喷头,阵雨,指示器
Shunt 分流器
St. Helena 圣赫特拿岛(英国)
Switch Hubs 交换式集线器

**SHA** Safety Hazard Analysis 安全风险分析
System Hazard Analysis 系统风险分析

**SHADE** Scheduled Airbus Data Exchange 定期的空客数据交换

**SHAG** Small Hole Accelerator Grid 小孔加速栅

**SHARES** Shared Airline Reservation System 航空公司共享订票系统
Shared Resources 共享资源

**SHARP** Safety Health Awareness and Recognition Program 安全健康意识和识别程序
Standard Hardware Acquisition and Reliability Program 硬件采集和可靠性的标准程序
Standardized Hardware Reliability Program 标准化的硬件可靠性计划

**SHARPS** Ship/Helicopter Acoustic Range Prediction System 舰载直升机声波距离预测系统

**SHC** Server Headend Control 服务器前端控制
Simulation Host Computer 仿真主计算机

**SHCM** System Heating Control Module 系统加温控制模块

**SHCR** Shipping Container 集装箱

**SHD** Scheduled 计划的
Staunton, VA, USA (Airport Code) 美国弗吉尼亚州斯汤顿机场
Super High Definition 超高分辨率

**SHDBL** Sheddable 可卸载的

**SHDP** Shipper Deposit 托运人定金

**S-HDTV** Super High Definition TV 超高清晰度电视

**SHEA** Safety, Health and Environmental Affairs 安全、健康和环境事务

**SHEAR** Safety, Health and Environmental Action Request 安全性、健康与环境措施申请

**SHEC** Southern Hemisphere Exchange Center 南半球交换中心

**SHED** Shedding 卸载

**SHEL** Software, Hardware, Environment and Liveware 软件、硬件、环境和生命件

**SHEP** Solar High Energy Particles 太阳高能粒子

**SHERPA** Systematic Human Error Reduction and Prediction Approach 系统化人为差错减少与预测方法

**SHF** Shaper Fixture 磨削夹具
Side-Hole Fiber 侧孔光纤
Single-Heading Flight 单一航向飞行,固定航向飞行
Super High Frequency (3 to 30 GHz) 超高频

**SHF(T)** Shift 偏移,漂移,移位

**SHFTA** Software and Hardware Fault Tree Analysis 软件与硬件故障树分析

**SHFTGR** Shaft Gear 齿轮轴

**SHH** Shishmaref, AK, USA (Airport Code) 美国阿拉斯加州施沙木勒幅机场

**SHIP** Senior Health Insurance Program 高级健康保险项目
Shipped 已发货,已发送
State Health Insurance Advisory Program 国家健康保险咨询项目

**SHIPT** Shipment 运送,发送(货物),批货

**SHIRAN** S Band High-accuracy Ranging and Navigation S 波段高精度测距导航系统

**SHJ** Sharjah, United Arab Emirates (Airport Code) 阿拉伯联合酋长国沙迦机场

**SHK** Shank 摇把,(轮)轴,尾端
Shock 震动,振动

**SHKR** Shaker 抖动器,抖杆器

**SHL** Shell 壳,套,罩
Shellac 紫胶,虫胶

**SHLD** Shield, Shielding 屏蔽

**SHLN** Shoreline 海岸线

**SHM** Shared Memory 共享存储器

Shop Manual 车间手册

Simple Harmonic Motion 简谐运动

Structural Health Monitoring 结构健康监测

Subsystem Health Management 子系统健康管理

**SHN** Scratch Hardness Number 划痕硬度值

Self-Healing Network 自愈网

**SHNS** Self-Healing Network Services 自愈网服务

**SHODOP** Short Range Doppler 近距多普勒

**SHOP** Supplier History and Open Purchase Order 供应方历史与公开订购单

**SHORAN** Short Range Navigation 短程导航，近距离导航

**SHORT** Short Circuit 短路

**SHOT** Special Hybrid Optical Terminal 特殊混合光终端

**SHP** Shaft Horsepower 轴马力

Shape 形状

Shop 车间，商店

Standard Hardware Program 标准硬件大纲

Standard Holding Procedure 标准等待程序

Super High-band Processor 超级高频处理器

**SHPRF** Shockproof 防震

**SHR** Self-Healing Ring 自愈环

Sheridan, WY, USA(Airport Code) 美国怀俄明州谢里登机场

**SHRD** Shroud 冠，凸台

**SHRTG** Shortage 缺额，缺陷

**SHSIC** Super High-Speed Integrated Circuit 超高速集成电路

**SHSS** Short Haul System Simulation 短程系统模拟

**SHT** Sheet 篇，张，纸张，层，图表，单

Short 短的，近，短路

**SHT DN** Shut Down 关断，停车

**SHTC** Short Time Constant 短时间常数

**SHTG** Shortage 短缺

**SHTL** Shuttle 航天飞机，梭子

**SHTSD** Short Side 短边

**SHTTP** Secure Hyper Text Transfer Protocol 安全超文本转换协议

**SHUD** Should 应该

**SHV** Sheave 滑轮，凸轮盘，滑车

Shreveport, LA, USA（Airport Code）美国路易斯安那州什里夫波特机场

**SHW** Send Highway 发送公共信道

**SI** International System of Units 国际单位制

Schedule Index 进度指数

Secondary Index 辅助索引

Sensor Interface 传感器接口

Serial Interface 串行接口

Service Identifier 业务标识符

Service Indicator 业务指示符

Service Information 服务信息

Shift in 移入

Single Instruction （简）单指令

Slip Indicator 侧滑指示器

Smith Industries Limited 史密斯公司

Software Identification 软件标识

Spark Ignition 火花塞点火

Special Identification 特殊标识

Special Inspection 特殊检查

Speed Intervention 速度干预

Standby Instruments 备用仪表

Standing Instruction 长期指令，现行指令

Straight-in 直接进场

Straight-in Approach 直接进近(着陆)

Study Item 研究项目

System Integration 系统集成

System Integrator 系统集成商

**SI LT** Side Light 侧灯

**SIA** Simulator Interface Adapter 模拟机接口适配器

Singapore International Airlines 新加坡国际航空公司

Standard Instrument Approach 标准仪表进场

System Interconnect Assembly 系统互连组件

**SIAAP** Standard Interface Applicability Analysis Program 标准接口应用性分析程序

**SIAEC** Singapore International Airlines Engineering Company 新加坡国际航空工程公司

**SIAM** Signal Information and Monitoring 信号信息和监控

**SIAP** Standard Instrument Approach Procedure 标准仪表进近程序

Structural Inspection Adjustment Program 结构检查调节程序

**SIB** Security Information Base 安全信息库

Service Independent Building Block 与业务无关的构造模块

Signaling in Band 带内信令

Standard Interface Bus 标准接口总线

Super Information Base 超级信息库

**SIBF** System Information Broadcast Function 系统信息广播功能

**SIC** Secondary-in-Command 副机长

Selected Item Configuration 选定项目配置

Semiconductor Integrated Circuit 半导体集成电路

Silicon Carbide 碳化硅

Simple Interference Cancellation 简单干扰消除

Software Integration Center 软件集成中心

SONET Interface Card SONET 接口卡

Standard Industrial Classification 标准工业分类

Station Identification Code 站址识别码

System Instruction Code 系统指令码

System Isolation Contactor 系统隔离接触器

Systems Integration Center 系统集成中心

**SICAS** SSR Improvements and Collision Avoidance System 二次监视雷达增强与避撞系统

**SICD** Standard Interface Control Document 标准接口控制文件

**SICR** Selected Item Configuration Record 选项组合记录

Spares Inventory Control Report 备件库存控制报告

Study Information Commitment Request 研究信息提交申请

**SID** Seal in Device 密封装置,封口器

Security Identifier 安全标识符

Selected Item Drawing 选择项图表

Signaling Identifier 信令标识符

Simulator Interface Device 模拟器接口装置

Software Interface Drawing 软件接口图

Spares Input Document 备件输入文档

Standard Instrument Departure 标准仪表离场,标准仪表离港

Standard Interface Database 标准接口数据库

Structural Inspection Document 结构检查文件

Sudden Ionospheric Disturbance 电离层突发扰动

Supplemental Inspection Document 补充检查文件

Surveillance and Identification 监视与识别

Synchronous Identification System 同步识别系统

System Identify 系统识别

System Integration Demonstration 系统集成演示

System Interface Definition 系统接口定义

**SIDD** Software Interface Definition Document 软件接口定义文件

Supplier Interface Definition Document 供应方接口定义文件

**SIDL** Serial Input Data Line 串行输入数据(总)线

**SIDN** Selected Item Description Number 选择的信息条描述号码

**SIDP** System Installation Design Principles 系统安装设计原则

**SIDR** System Interface Design Review 系统接口设计审查

**SIDS** Standard Instrument Departure Sequence 标准仪表离港顺序

Stellar Inertial Doppler System 天文惯性多普勒系统

**SIE** Science/Scientific Information Exchange 科学信息交流处(美国)

Standard Inspection Equipment 标准检验设备

Symbols in Error 符号错误

**SIF** Selective Identification Facility 选择识别设备

Selective Identification Feature 选择性识别特性

Signaling Information Field 信令信息字段

Single Face 单面

Software Integration Facility 软件集成设施

Sound Intermediate Frequency 中音频

Spares Information File 备份信息文件

Stall Indication Failure 失速指示故障

Standard Image Format 标准图像格式

Standard Interchange Format 标准相互交换格式

Storage Information File 存储信息文档

Storage Interface Facility 存储接口设施

Stress Intensity Factor 应力强度系数

System Information File 系统信息文件

System Interface 系统接口

**SIFL** Standard Industry Fare Level 行业标准票价水平

**SIG** Signal 信号

Signaling 信令,信号设备,通信,发信号

Significant 主要的,重要的

Special Interest Group 行业集团

**SIG COND** Signal Conditioner 信号调节器

**SIG STA** Signal Station 信号台

**SIG STR** Signal Strength 信号强度

**SIG T(RA)G** Signal Training 通信训练

**SIGA** Signaling Adapter 信令适配器

**SIGC** Signaling Control 信令控制器

**SIGCEN** Signal Center 通信中心

**SIGCLD** Significant Clouds 显著云层

**SIGCOM** Signal Communication 信号通信

**SIGGRAPH** Special Interest Group for Graphics 绘图专业组

**SIGINT** Signals Intelligence 信号情报

Signals Intelligence System 信号情报系统

**SIGMET** Significant Meteorological Information 重要气象信息

Significant Meteorological Observation 重要气象观察

Significant Meteorological Advisory 重要气象咨询

**SIGN** Signature,Signing 签字,标记

Significant 显著的,有效的

Strapdown Inertial Guidance and Navigation 捷联式惯性制导与导航

**SIGND** Signed 签字的,挂牌的,标记的

**SIGP** Signal Processor 信号处理器

**SIGWX** Significant Weather 重要天气

**SII** Special Instructions Indicator 特别指令指示器

Standard Interface Item 标准接口项目

**SIIA** Software and Information Industry Association 软件与信息产业协会

Structural Item Interim Advisory 结构项目临时咨询

**SIIP** Structural Inspection Item Program 结构检查项目方案

**SIL** Sensor Integration Laboratory 传感器综合实验室

Service Information Letter 服务信函

Silence 静寂,静噪

Silver 银

Special Import License 特别进口许可证

Speech Interference Level 语音干扰电平

Surface Information Letter 面信息函件

System Integration Laboratory 系统集成实验室

**SILA** System Interconnect and Load Assembly 系统互连和负载组件

**SILG** Silencing 沉寂,静噪,抑制

**SILS** Silver Solder 银焊料

Standard for Interoperable LAN Security 互操作局域网的安全标准

**SILTF** System Integration Laboratory and Test Facility 系统集成实验室与测试设备

**SIM** Serial Interface Module 串行接口组件

Service Interaction Manager 业务交互管理器

Similar 相同,相似

Simulation 模拟

Simulator 模拟机

Subscriber Identity Module 用户识别模块

Subscriber Interface Model 用户界面模型

Subscriber Interface Module 用户接口模块

Subsystem Interface Modules 分系统接口组件

Supplemental Information Memorandum 补充信息备忘录

Synchronous Induction Motor 同步感应电机

Synchronous Interface Module 同步接口模块

System Information Management 系统信息管理

System Integration Module 系统合成模块

**SIMA** Slot Interleaved Multiple Access 时隙交错多址接入

**SIMCMP** Simple Compiler 简单编译程序

**SIMCOM** Simulator and Computer 模拟机与计算机

Simulator Compiler 模拟机编译程序

**SIMCON** Simplified Control 简化的控制

Simulation Console 模拟操纵台

Simulation Control(ler) 模拟控制(器)

**SIMD** Single Instruction Multiple Data 单指令多数据流

**SIMFAC** Simulation Facility 模拟装置

**SIMICOR** Simultaneous Multiple Image Correlation 同时多重图像相关

**SIML** Similar 相似的,类似的

**SIMMOD** Airport and Airspace Simulation Model 机场与空域仿真模型

**SIMNET** Simulation Network 仿真网络

**SIMO** Single Input, Multiple Output 单输入,多输出

**SIMP** Satellite Information Message Protocol 卫星信息报文协议

Satellite Interface Message Processor 卫星接口信息处理机

Simulation Package 模拟组件,模拟程序包

Specific Impulse 特殊冲击式(滑轮)

**SIMS** Shuttle Inventory Management System 航天飞机库存管理系统

Single-Instruction Multiple-data Streams 单指令多数据流

Stores Inventory Management System 仓库库存管理系统

**SIMSYL** Simulation Symbolic Language 模拟符号语言

**SIMSYS** Simulation System 仿真系统,模拟系统

**SIMTRACC** Simulator Trainer Command and Control 模拟训练器指挥与控制

**SIMUL** Simultaneous or Simultaneously 同时的或同时地

**SIMULA** Simulation Language 模拟语言

**SIMXI** Simulate External Interfaces Program 外部接口模拟程序

**SIN** Changi, Singapore(Airport Code) 新加坡樟宜国际机场

Satellite Inspection Network 卫星检测网

Security Information Network 安全信息网

Sine 正弦

**SINAD** Signal Plus Noise and Distortion 信号加噪声和畸变

**SINCGARS** Single Channel Ground and Airborne Radio System 单信道地面和机载无线电系统

**SINDA** Systems Improved Numerical Differencing Analyzer 系统改进数字微分分析器

**SINH** Hyperbolic Sine 双曲正弦

**SINS** Ship Inertial Navigation System 舰船惯性导航系统

Stellar Inertial Navigation System 星体惯性导航系统

Strapdown Inertial Navigation System 捷联惯性导航系统

**SINTAC** Integrated Information Systems of Civil Aviation 民航综合信息系统

**SIO** Serial I/O 串行输入/输出

Service Information Octet 服务信息八比特组

Special Input/Output (Module) 专用输入/

输出(模块)

**SIOF** Step-Index Optical Fiber 阶跃指数光纤

**SIOP** Serial Input/Output Processor 串行输入/输出处理器

**SIP** Satellite Information Processor 卫星信息处理机

Satellite Inspection Program 卫星核查计划

Satellite Interceptor Program 卫星拦截器计划

Scheduled into Production 列入生产计划

Scientific Instruction Processor 科学指令处理器(霍尼韦尔)

Scientific Instrument Package 成套科学仪器

Session Initiation Protocol 会话启动协议

Signal Identification Processor 信号识别处理器(霍尼韦尔)

Simple Internet Protocol Plus 增强的简单因特网协议

Special Identification Pulse 特殊识别脉冲

Specifications and Instruction for Production 产品规格及说明

Standard Information Package 标准信息包

Standard Inspection Procedure 标准检查程序

Step Index Plastic Optical Fiber 阶跃指数塑料光纤

Structural Inspection Procedure 结构检查程序

Structure Inspection 结构检查

Structure Inspection Program 结构检查大纲

Subscriber Interface Processor 用户接口处理器

**SIPC** Simply Interactive Personal Computer 简易交互式个人电脑

**SIPD** Structural Inspection Planning Data 结构检查计划数据

**SIPP** Standard Interline Passenger Procedure 标准联程旅客程序

Surface Impact (Impulsion) Propulsion 表面冲击推进

**SIPS** Sales Invoice Processing System 销售收据处理系统

**SIR** Scheduled Inspection and Repair 预定检查和修复

Scientific Information Retrieval 科学情报检索

Shared Interface RAM 共用接口随机存储取器

Signal Interference Ratio 信扰比

Signal-to-Intermodulation Ratio 信号互调比

Significant Item Report 重要项目报告

Snow and Ice on Runway 跑道积雪和结冰

Special Inspection Request 特殊检查申请

Special Inspection Requirement 特殊检查要求

Station Interconnection Record 站间互连记录

Structural Integrity Requirements 结构完整性要求

Structural Irregularity Report 结构缺陷报告

Submission Information Rate 信息送出率

Supplemental Inflatable Restraint 安全气囊

**SIRD** Software Implementation Requirements Document 软件执行(程序)要求文件

System Installation Requirement Document 系统安装要求文件

**SIRE** Space Infra-Red Experiment 空间(卫星)红外试验

**SIRP** Selective Information Retrieval Program 选择信息检查程序

**SIRRM** Standardized Infra-Red Radiation Model 标准化红外辐射模型

**SIRS** Shop Integrated Reporting System 车间综合报告系统

Significant Item Reporting System 重要项目报告系统

**SIRTF** Shuttle Infra-Red Telescope Facility 航天飞机红外望远镜装置

**SIS** Separation Ignition Switch 间隔点火开关

Service Interphone System 服务内话系统

Service Item Summary 维修项目纲要, 勤务项目纲要

Signaling Interworking Subsystem 信令互通子系统

Spares Investment Study 备件投资研究

Spares Item Split 备件项目分解

Stall Identification System 失速识别系统

Stall Inhibitor Systems 失速抑制器系统

Standard Instruction Set 标准指令组

Statistical Information System 统计信息系统

Supervisory Information System 监控信息系统

Supporting Implementation Strategies 辅助实施战略

System Interrupt Supervisor 系统中断管理程序

**SISC** Spy-in-Sky Camera 卫星侦察摄影机

**SISD** Single Instruction Single Data Stream 单指令单数据流

**S-ISMA** Slotted-Idle Signal Multiple Access 时隙空闲信号多址接入

**SISNOTAM** Implementation of the Aeronautical Information System 实施航行情报系统

**SISO** Single Input, Single Output 单输入,单输出

**SISP** Source Input Select Panel (信息)源输入选择板

Strategic Information Systems Plan 战略情报系统计划

**SISW** Step Index Slab Wavelength 阶跃指数平面波长

**SIT** Shuttle Interface Test 航天飞机接口测试

Sitka, AK, USA (Airport Code) 美国阿拉斯加州锡特卡机场

Situation 情况,情景

Special Investigation Test 特别调查试验

System Integration Test 系统综合测试

System Interface Test 系统接口测试

**SITA** International Society for Aeronautical Telecommunications 国际航空电信协会

Société Internationale de Télécommunications Aéronautiques 国际电信协会

Society of Information Theory and Its Applications 信息理论及其应用学会

State Information Technology Agency 国家信息技术局

**SITAN** Sandia Inertial Terrain Aiding Navigation 桑迪亚惯性地形辅助导航

**SITE** Search Information Tape Equipment 搜索信息磁带设备

**SITEA** Service Independent Test Evaluation Agency 维修独立试验评价公司

**SITN** Situation 情势

**SITP** Site Integration Test Plan 场地综合测试计划

System/Subsystem Integration Test Plan 系统/分系统综合测试计划

**SITR** Structural Integrated Test Report 结构完整性测试报告

Systems Integration Test Report 系统综合测试报告

**SITT** System Integration Test Tape 系统综合测试磁带

**SIU** Satellite Interface Unit 卫星接口单元

Secure Interface Unit 安全接口装置

Server Interface Unit 服务器接口组件

Signaling Interface Unit 信令接口单元

Simulation Interface Unit 模拟接口组件

Static Inverter Unit 静变流器装置

Subscriber Interface Unit 用户接口单元

System Interface Unit 系统接口部件

**SIV** Standard Item Variation 标准项目调整

**SIW** Say If Wrong 如有误请告

**SIWL** Single Isolated Wheel Load 单轮载荷量,当量单轮荷载

**SIYB** Start and Improve Your Own Business 开创与改善你的企业

**SJ** Slip Joint 滑动接头,伸缩接头

Soldered Joint 软焊接头

Steel Jacket 钢套

Svalbard and Jan Mayen Islands 斯瓦尔巴岛和扬马延岛(挪威)

Synaptic Junction 突触接头处

**SJA** South Jersey Airways (USA) 南泽西航空公司(美国)

**SJAC** Society of Japanese Aerospace Company 日本宇航公司协会

**SJC** San Jose, CA, USA (Airport Code) 美国加利福尼亚州圣何塞机场
Standard Job Card 标准工作卡

**SJD** San Jose Del Cabo, Baja California Sur, Mexico (Airport Code) 墨西哥圣约瑟角机场

**SJJ** Sarajevo, Bosnia and Herzegowina-Butmir (Airport Code) 波斯尼亚和黑塞哥维那萨拉热窝布特密尔机场

**SJO** San Jose, Costa Rica (Airport Code) 哥斯达黎加圣何塞机场

**SJP** Standard Jet Penetration 喷气机标准下降程序

**SJR** Signal to Jamming Ratio 信号干扰比

**SJT** San Angelo, TX, USA (Airport Code) 美国得克萨斯州圣安吉洛机场
Subsonic Jet Transport 亚音速喷气运输机

**SJU** San Juan, PR, USA (Airport Code) 波多黎各圣胡安机场
Special Job Update 特殊工种更新

**SK** Secret Key 密钥
Service Kit 维修工具箱
Sink 下水槽
Sketch 草图
Skinned 有蒙皮的
Skinner 去皮器

**SK LT** Station Keeping Lights 站定位灯

**SKA** Seat-Kilometre Available 可用座位吨公里

**SKAD** Survival Kit Air Droppable 可空投的救生包

**SKATE** Singer-Kearfott Automated Test Equipment 辛格—基尔霍夫自动测试设备

**SKC** Sky Clear 天空晴朗,天气晴朗,晴空

**SKD** Scheduled 定期的
Singer-Kearfott Division 辛格—基尔霍夫分部

Skilled 熟练的,有技能的
Skirted 有缘的

**SKED** Schedule 时刻表,计划表,程序
Scheduled 定期,班期的

**SKEL** Skeleton 构架

**SKG** Seeking 寻找
Smoke Goggles 烟雾镜
Thessaloniki, Greece (Airport Code) 希腊塞萨洛尼基机场

**SKI** Station Keeping Indicator 位置保持指示灯

**SKL** Skill 技能,技巧

**SKP** Skopje, Macedonia (Airport Code) 马其顿斯科普里机场

**SKPO** Skeleton Purchase Order 主干购货订单

**SKT** Skirt 边缘
Socket 套筒,插座

**SKY** Sandusky, OH, USA (Airport Code) 美国俄亥俄州桑达斯基机场

**SL** Safe Locker 保险柜
Safety Level 安全水平,安全等级
Safety Lighting 安全照明
Scatting Layer 散射层
Sea Level 海平面
Sealer 密封剂
Sealing 密封
Section Length 段长
Semiconductor Laser 半导体激光器
Service Letter 服务信函(为产品在使用中发生的问题提供处理办法)
Service Lifetime 使用寿命
Service Logic 业务逻辑
Sideline 侧线
Sierra Leone 塞拉利昂
Signal Lamp 信号灯
Signaling Link 信令链路
Sinc Lock 同步锁
Slide, Sliding 滑动(的),侧滑(的)
Soft Landing 软着陆
Solid Logic (Circuit) 固体逻辑(电路)

Sound Locator 声波定位器,声音测位器
South Latitude 南纬
Specialised Lending 专业贷款
Stability Loss 稳定性损耗
Standard Labels 标准标号,标准标志
Standard Length 标准长度
Stock Length 备用长度
Stock Level 库存标准
Stock List 库存清单
Storage Location 储存地点
Straight Line 直线
Submarine Lightwave 水下光波
Subordinate Ledger 附属分类账
Subscriber Line 用户线
Subsidiary Ledger 辅助分类账
Supply Lead 馈电线,电源线
Supply Line 供应线
Switch Light 电门灯
Sync Lock 同步锁定

**SL/R** Scan Left/Right 扫掠左右

**SLA** Salta, Argentina ( Airport Code ) 阿根廷萨尔塔机场
Service Level Agreement 服务水平协议,服务级别协议
Site Level Aggregator 位置层集合器
Static Load Analysis 静力分析

**SLAM** Scanning Laser Acoustic Microscopy 激光扫描声学显微术
Service Level Agreement Management 服务水平协议管理
Side Load Arresting Mechanism 侧向载荷抑制机构
Symmetrically Loaded Acoustic Module 平衡装载声学模块

**SLAMMR** Side Looking Airborne Multi-Mission Radar 机载多用途侧视雷达

**SLAP** Subscriber Line Access Protocol 用户线接入协议

**SLAR** Senior Logistics Aviation Representative 高级物流航空代表
Side Looking Array Radar 侧视天线阵雷达

Side-Looking Airborne Radar 机载侧视雷达
Sideways Looking Aperture Radar 侧视孔径雷达
Slant Range 倾斜距离

**SLAS** Super Low-cost Aluminum Structure 超低成本铝结构

**SLAT** Supersonic Low-Altitude Target 超音速低空目标

**SLAT STA** Slat Station 缝翼站位

**SLATE** Small Lightweight Altitude Transmission Equipment 小型轻量高度发送设备

**SLB** Server Load Balancing 服务器负载平衡
Side Lobe Blanking 旁瓣消隐
Signal Light Bars 信号光条

**SLBL** Soluble 可解的,可溶化的

**SLC** Salt Lake City, USA ( Airport Code ) 美国盐湖城机场
Side Lobe Cancellation 旁瓣对消
Signaling Link Code 信令链路码
Signaling Link Control 信令链路控制
Slice 薄片
Standby Letter of Credit 备用信用证
Submarine Laser Communications 水下激光通信
Subscriber Line Circuit 用户线电路
Subscriber Line Controller 用户线控制器
Subscriber Loop Carrier 用户环路载波

**SLC SAT** Submarine Laser Communication Satellite 水下激光通信卫星

**SLCC** Signaling Link Control Common 公共信令链路控制器
Software Life Cycle Cost 软件寿命期成本

**SLCL** Signaling Link Control Local 本地信令链路控制器

**SLCTD** Selected 选择的

**SLCTR** Selector 选择器, 选钮

**SLCU** Stores Logic Control Unit 存储逻辑控制器

**SLD** Sealed 封严的
Slide, Sliding 滑动
Sliding Door 滑动门

Slotting Die 切槽模

Slowdown 减速

**SLDR** Solder 钎焊

Solderability 可焊性

Soldering 焊接

**SLE** Service Life Evaluation 使用寿命评估

Small Local Exchange 小型本地交换机

Subscriber Line Equipment 用户线路设备

**SLEE** Service Logic Execution Environment 业务逻辑执行环境

**SLEM** Service Logic Execution Management 业务逻辑执行管理

**SLEP** Service Life Enhancement Program 使用寿命提高计划

Service Life Extension Program 使用寿命延长计划

Standard Life Extension Program 标准延寿大纲

**SLF** Shop Load Forecast 车间负荷预测

Spares Leader File 备件引导文档

Straight Line-Frequency 直线频率(式)

Super Low Frequency (30～300 Hz) 超低频

**SLFC** Survivable Low-Frequency Communications 残存低频通信

**SLFCLN** Self Cleaning 自清洁(的)

**SLFGEN** Self Generating 自发生(的)

**SLFIND** Self Indicating 自指示(的)

**SLFLKG** Self Locking 自锁(的)

**SLFPM** Sea Level Feet per Minute 海平面英尺/分钟

**SLFSE** Self Sealing 自密封(的)

**SLFTPG** Self Tapping 自螺纹(的),自搭接(的)

**SLG** Sludge 泥渣,污水

**SLGR** Slinger 吊环

Small, Lightweight GPS Receiver 轻小型全球定位卫星接收机,袖珍 GPS 接收机

**SLI** Shipper's Letter of Instruction 托运书

System Level Interface 系统级界面

**SLIC** Subscriber Line Interface Circuit 用户线接口电路

**SLIE** Subscriber Line Interface Equipment 用户线接口设备

**SLIM** Standards Laboratory Information Manual 标准实验室资料手册

Success Likelihood Index Methodology 成功似然指数法

**SLIP** Serial Line Interface Protocol 串行线路接口协议

Serial Line Internet Protocol 串行线路因特网协议

**SLK** Saranac Lake, NY, USA(Airport Code) 美国纽约州萨拉纳克莱克机场

**SLL** Seamless Link 无缝连接

Semi-Loop Loss 半环路损耗

**SLM** Service Level Management 服务等级管理

Signaling Link Management 信令链路管理

Single Longitudinal Mode 单纵模

Spatial Light Modulator 空间光调制器

Standard Length Message 标准长度信息

Subscriber Line Module 用户线路模块

Subscriber Loop Multiplex 用户环路复用

**SLMC** Service Life Limits and Maintenance Checks 服务寿命限制及维修检查

**SLMH** Signaling Link Message Handler 信令链路消息处理器

**SLN** Signaling Link Number 信令链路号

Standard Library Number 图书馆标准编号

**SLO** Service Level Objectives 服务等级目标

**SLOC** Sea Lines of Communication 海上通信联络线,海上交通线

**SLOP** Strategic Lateral Offset Procedure 策略横向偏置程序

**SLP** San Luis Potosi, San Luis Potosi, Mexico (Airport Code) 墨西哥圣路易斯波托西机场

Service Logic Procedure 业务逻辑规程

Service Logic Processing 业务逻辑处理

Service Logic Program 业务逻辑程序

Single Link Procedure 单链路规程

Slope 斜率,坡度,倾斜

Space Limited Payload 有限空间商载，限制体积的负载

Speed Limiting Point 速度极限点，限速点

Systematic Layout Planning 系统布置设计

**SLPC** Service Logic Program for Customization 用户化的业务逻辑程序

**SLPI** Service Logic Processing Instance 业务逻辑处理实例

**SLPN** Service Logic Program for Network Control 网络控制用业务逻辑程序

**SLPR** Speed Low Pressure Rotor 低压转子转速

**SLR** Service Level Requirement 服务等级要求

Shop Labor Register 车间劳力注册

Side Looking Radar 侧视雷达

Single Lens Reflex 单镜头反射（相机）

**SLRM** Scan Left/Right Monitor 扫掠左右监视器

**SLS** Sales 销售

Satellite-base Landing System 星基着陆系统

Sea Level Standard 标准海平面

Sea Level Static 海平面静压

Selective Laser Sintering 选择性激光烧结

Side Lobe Radiation 旁瓣辐射

Side Lobe Suppression 旁瓣抑制

Signaling Link Selection 信令链路选择

System Level Simulator 系统等级模拟器，系统电平模拟器

**SLSA** Special Light-Sport Aircraft 特殊轻型运动类航空器

**SLSI** Super Large Scale Integration 超大规模集成

**SLSIM** Service Logic Selection/Interaction Manager 业务逻辑选择/交互管理

**SLSMN** Salesman 商人，销售商

**SLSP** Slow Speed 低速

**SLSR** Side Lobe Suppression Ratio 旁瓣抑制比

**SLST** Sea Level Static Thrust 海平面静推力

Sleeper Seat 可躺座位，卧铺

**SLT** Self Loading Tape 磁带自动加载

Skylight 天窗

Slat 缝翼，板条

Slit 切口

Solid Logic Technology 固态逻辑技术

Spotlight 聚光灯

Subscriber Line Terminal 用户线路终端

Subscriber Line Trunk 用户线中继

**SLTA** Signaling Link Test Act 信令链路测试作用

**SLTC** Signaling Link Test Control 信令链路测试控制

**SLTD** Slotted 开槽的，开缝的

**SLTDP** Special Laser Technology Development Program 特殊激光技术发展计划

**SLTE** Submarine Line Terminating Equipment 海底线路终接设备

**SLTM** Signaling Link Test Message 信令链路测试报文

**SLTO** Sea Level Take-off 海平面起飞

**SLTP** Simulated Launch Training Payload 模拟发射训练业载

**SLU** Station Logic Unit 站逻辑组件

**SLV** Satellite Launching Vehicle 卫星发射运载工具

Sleeve 套，护套，轴套

Sync Lock Valve 同步锁活门

**SLVG** Sleeving 套管

**SLVT** Solvent 溶剂

**SLW** Saltillo, Coahuila, Mexico (Airport Code) 墨西哥萨尔蒂约机场

Slow 慢

Store Logical Word 存储逻辑字

**SLWL** Straight Line Wavelength 直线波长

**SM** Safety Management 安全管理

Sales Marketing 销售市场

San Marco Platform 圣马科发射平台（意大利）

San Marino 圣马力诺

Sandwich Mould 夹芯模压

Scan Monitor 扫掠监测器

Section Manager 部门经理

Security Management 安全管理,保安管理

Security Manual 安全手册

Select Matrix 选择矩阵

Selector Marker 选择指示器,选择指点标

Self-Monitoring 自监控(的)

Semiconductor Memory 半导体存储器

Semimonthly 半月刊,半月地

Service Manual 维护手册

Service Message 业务消息

Service Module 服务舱

Servomotor 伺服电动机

Sheet Metal 钣金材料

Shipping Manifest 货运舱单

Shipping Memorandum 发货备忘录

Shop Manual 车间手册

Shutter Mission 航天飞机飞行任务

Single Mode 单模

Small 小

Smart Media 智能媒体

Smooth,Smoothing 平滑(的),平稳(的)

Solid Model,Solid Modeling 实体模型,实体造型

Space Medicine 航天医学,空间医学

Special Machine 专用机器

Special Material 专用器材

Special Memorandum 特别备忘录

Special Message 特别信息,特别咨文

Special Mission 特种任务(美国)

Standard Manual 标准手册

Standard Memorandum 标准备忘录

Standard Micrometer 标准千分尺

Standard Microscope 标准显微镜

Standard of Measurement 测量标准

Standards Manual 标准手册

Static Margin 静态幅度,静余量

Station Manager 台站经理

Statistic Method 统计法

Statistical Multiplexing 统计复用

Status Manager 状态管理器

Status Matrix 状态矩阵

Status Monitoring 状态监控

Statute Mile 法定英里(5280 英尺)

Stereo/Mono 立体声/单声

Subsystem Module 分系统模件

Supervision Module 监控模块

Supply and Maintenance 供应与维修

Supply Module 供应舱

Switch Manager 交换管理器

Switch Module 交换模块

Synchronous Modulation 同步调制

Synchronous Motor 同步电机

Synchronous Multiplexer 同步复用器

System Management 系统管理

System Manager 系统经理

System Manual 系统手册

System Measurement 系统测量

System Mechanics 系统力学

**SM&P** Supply Management and Procurement 供应管理和采购

**SMA** Sales and Marketing Association 销售和营销协会

Santa Maria,Azores,Portugal(Airport Code) 葡萄牙亚速尔群岛圣玛丽亚机场

Scalable Modular Architecture 可伸缩模块化结构

Schedule Management Application 计划管理应用

Segment Minimum Altitude 航段最低高度

Shape Memory Alloy 形状记忆合金

Share Memory Architecture 共享内存结构

Software Maintenance Agreement 软件维护协议

Stress Method and Allowable 应力方法及容差

Surface Movement Advisor 地面活动顾问

Systems Management Architecture 系统管理体系结构

Systems Management Association 系统管理协会

**SMAB** Solid Motor Assembly Building 固态马达装配车间

**SMAC** Scene Matching Area Correlation 场景匹配区域相关

Scene Matching Area Correlator 场景匹配区域相关器

Spacecraft Maximum Allowable Concentration 航天器最大允许浓度

**SMACH** Sounding Machine 测高机,测深机

**SMACT** Space Material and Component Testing 空间器材与部件测试

**SMAE** System Management Application Entity 系统管理应用实体

**SMAF** Service Management Agent Function 业务管理代理功能

Special Mission Aircraft Flights 特殊任务航空器飞行

**SMALC** Sacramento Air Logistics Center 萨克拉门托航空后勤中心

**SMAMA** Sacramento Air Material Area 萨克拉门托航空器材区

**SMAP** Safety Management Action Plan 安全管理行动计划

Security Management Application Program 安全管理应用程序

Service Management Access Point 业务管理接入点

System Management Analysis Project 系统管理分析项目

System Management Application Process 系统管理应用进程

**SMART** Safety and Mission Assurance Review Team 安全和任务保障评审组

Self-Monitoring, Analysis and Reporting Technology 自动监视、分析和报告技术

Shortage Management and Record Tracking 短缺件管理与记录跟踪

Simulation, Monitoring, Analysis, Reduction and Test System 模拟、监控、分析、减少和测试系统

Smart Material Actuated Rotor Technology 智能材料旋翼技术

Standard Modular Avionics Repair and Test 标准模块航空电子修理和测试

System Monitor and Review Technique 系统监控与检查技术

**SMARTS** Sheet Metal Analysis and Retrieval System 金属板材分析和检索系统

**SMATH** Satellite Materials Hardening 卫星材料硬化

**SMATV** Satellite Master Antenna TV 卫星主控天线电视

**SMAW** Shield Metal Arc Welding 屏蔽金属电弧焊接

**SMB** Server Message Block 服务器消息块

Side Marker Board 旁侧标记牌

Small and Medium Businesses 中小型企业

**SMC** Satellite Mobile Channel 卫星移动信道

Seat Mile Cost 座英里成本

Security Monitoring Centre 安全监测中心

Stall Management Computer 失速管理计算机

Standard Mean Chord 标准平均弦长

Standards Management Committee 标准管理委员会

Stores Management Control 库存管理控制

Strategic Management Committee 战略管理委员会

Surface Mount Component (表)面安装元件

Surface Movement Control 地面活动管制

System Management and Communication 系统管理和通信

System Manager Controller 系统管理控制器

**SMC(SM&C)** System Management and Control 系统管理和控制

**SMCC** Simulation Monitor and Control Console 模拟监督和控制台

**SMCG** Safety Management Coordination Group 安全管理协调小组

**SMCM** System Management and Control Module 系统管理和控制组件

**SMCS** Spoiler Mode Control System 扰流板模式控制系统

Structural Mode Control System 结构方式控

制系统

**SMCU** Squib Monitor and Control Unit 爆炸帽监控与控制组件

Surface Movement Control Unit 地面移动控制单位

**SMD** Single Molecule Detection 单分子检测

Storage Module Device 储存模组装置

Stores Management Display 存储管理显示器

Surface Mounted Device 表面安装设备

System Management Directive 系统管理指令

**SMDS** Switched Multi-megabit Data Service 交换式多兆比特数据业务

**SME** Shipping Mechanical Equipment 装运机械设备

Signaling Message Encription 信令消息加密

Small and Medium-sized Enterprise 中小型企业

Small Message Entity 短消息实体

Society of Manufacturing Engineers 制造工程师学会

Subject Matter Expert 主题专家,行业专家

System Management Entity 系统管理单位

**SMEEP** Structural Maintenance Economic Evaluation Program 结构维修经济性评估项目

**SMEK** Summary Message Enable Keyboard 综合报文启动键盘

**SMEL** Seat-Mounted Emergency Light 装在座椅上的应急灯

Servicing Mission Equipment List 服务任务设备清单

Simulator Minimum Equipment List 模拟机最低设备清单

Single and Multiengine License 单一发动机与多发动机许可证

Super Master Equipment List 超级主要设备单

**SMF** Sacramento, CA, USA (Airport Code) 美国加利福尼亚州萨克拉门托机场

Service Management Function 业务管理功能

Single Mode Fiber 单模光纤

Software Maintenance Function 软件维护功能

Solar Magnetic Field 太阳磁场

Standard MIDI File 标准的数字化乐器接口文件

State Management Function 状态管理功能

Sub-Multiframe 子复帧

System Management Facility 系统管理设施

System Management Function 系统管理功能

**SMFCB** Subcarrier Modulated Fiber-Coax Bus 副载波调制的光纤同轴总线

**SMFD** Secondary Multi-Function Display 辅助多功能显示器

**SMG** Senior Management Group 高级管理组

Statistical Multplexing Gain 统计复用增益

System Management Group 系统管理组

**SMGC** Surface Movement Guidance and Control 地面活动引导和控制

**SMGCS** Surface Movement Guidance and Control System 场面活动引导和管制系统

**SMH** Service Message History 勤务信息记录

Signaling Message Handling 信令消息处理

Synchronous Multiplexing Hierarchy 同步复用系列

**SMHGA** Side-Mounted High Gain Antenna 侧面安装高增益天线

**SMI** Samos Island, Greece (Airport Code) 希腊萨摩斯岛机场

Specific Maintenance Instruction 专用维护说明

Standard Message Identifier 标准信息识别符

Structure of Managed Information 被管信息结构

Structure of Management Information 管理信息结构

**SMIB** Security Management Information Base 安全管理信息库

**SMIC** Special Maintenance Item Code 特殊维护的项目代码

Special Material Identification Code 特殊器材鉴别代码

**SMICG**   Safety Management International Collaboration Group 安全管理国际合作组

**SMIL**   Synchronized Multimedia Integration Language 同步多媒体集成语言

**SMIP**   Scheduled Maintenance and Inspection Program 定期维修与检验方案

**SMIS**   Safety Management Information System 安全管理信息系统
Society for Management Information Systems 管理信息系统学会
Supply Management Information System 供应管理信息系统

**SMITH**   Selected Monitoring of Integrated Trend History 综合趋势曲线的选择监控

**SMK**   Shared Management Knowledge 共享管理知识
Smoke 烟雾

**SMK GEN**   Smoke Generator 烟雾发生器

**SMKLS**   Smokeless 无烟的

**SML**   Service Management Layer 业务管理层
Small 小的
Support Material List 保障器材表

**SMLS**   Seamless 无缝的

**SMM**   Safety Management Manual 安全管理手册
Semporna, Malaysia (Airport Code) 马来西亚仙本那机场
Service Management Module 服务管理模块
Solar Maximum Mission 太阳最强活动
Standard Method of Measurement 标准测量方法
System Maintenance Management 系统维修管理
System Maintenance Manual 系统维修手册
System Management Model 系统管理模式

**SMMC**   System Maintenance Monitor Console 系统维护监视台

**SMMS**   Switched Multimegabit Data Service 交换式多兆比特数据业务

**SMN**   Switched Message Network 交换信息网
System Management Network 系统管理网

**SMNK**   Smooth Neck 平滑颈

**SMO**   Stabilized Master Oscillator 稳定主控振荡器
Stock Material Order 库存器材规程,库存器材订单
Supplementary Meteorological Office 辅助气象台
Supply Management Office 供应管理室
Synthesizer Mixer Output 混合输出
System Management Office 系统管理室

**SMOF**   Single Mode Optical Fiber 单模光纤

**SMOS**   System Management Operating System 系统管理操作系统

**SMP**   Safety Management Program 安全管理程序
Sensitized Material Print 感光材料印刷
Service Management Point 业务管理点
Service Management Process 业务管理过程
Simple Management Protocol 简单管理协议
Simulation Master Plan 仿真主规划
Software Management Plan 软件管理计划
Sound Motion Picture 声音动画
Standard Maintenance Program 标准维修程序
Standard Manufacturing Program 标准生产程序
Standard Measurement Program 标准测量程序
Stores Management Panel 仓库管理专家组
Strategic Management Plan 战略管理计划
Symmetric Multiprocessor 对称(式)多处理机,对称(式)多处理器
Symmetrical Multiprocessing 对称性多处理
System Maintenance Procedure 系统维修程序
System Management Processor 系统管理处理器
System Management Program 系统管理程序
System Modification Program 系统改装程序

**SMPHR**   Semaphore 信号杆

**SMPL**   Sample 样品

Simple 简单的

Supplier Module Parts List 供应商模块零件表

**SMPLG** Sampling 取样

**SMPTE** Society of Motion Picture and Television Engineers 影视工程师协会

**SMR** Scan Modulation Ratio 扫描调制比

Signal to Multipath Ratio 信号多径比

Single Mode Ratio 单模比

Source/Maintainability/Recoverability 资源/可维修性/可恢复性

Specialized Mobile Radio 专用移动无线电

Status Monitoring Routine 状态监控程序

Stock Management Report 库存管理报告

Substitute Material Request 代用器材要求

Surface Movement Radar 场面监视雷达

Surface Movement Radar Study Group 场面移动雷达研究组

System Malfunction Report 系统故障报告

**SMRAM** System Management RAM 系统管理内存

**SMRS** Specialized Mobile Radio System 专用移动无线电系统

**SMS** Safey Management System 安全管理系统

SDH Management Subnetwork SDH 管理子网

Service Management System 业务管理系统

Short Message Service 短信息服务

Silica-Manganese Steel 硅锰钢

Standard Modular System 标准模块系统

Stores Management Set 仓库管理设备

Stores Management System 仓库管理系统

Subcarrier Multiplexed System 副载波复用系统

Subscriber Management System 用户管理系统

Surface Mountable Semiconductors 可表面安装半导体器件

Synchronous Meteorological Satellite 同步气象卫星

System Managed Storage 系统管理仓库

System Management Server 系统管理服务器

**SMSA** Segment Minimum Safe Altitude 航段最低安全高度

**SMSC** Short Message Service Center 短消息业务中心

**SMSI** System Management Service Interface 系统管理服务接口

**SMSR** Side-Mode Suppression Ratio 边模抑制比

Surface Movement Surveillance Radar 地面活动监视雷达

**SMT** Scheduled Maintenance Task 计划维护任务

Servo Mounting Tray 伺服机构安装架

Simultaneous Multithreading 同步多线程

Square Mesh Tracking 正方网格跟踪

Station Management 站点管理

Store Material Transfer 仓库器材调动

Subject Matter Expert 主题专家

Superconductive Microtherm Technology 超导体散热材料

Surface Mount Technology 表面安装工艺,表面安装技术

Surface Mounted Technology 表面贴装技术

System Maintenance Test 系统维护测试

System Management Team 系统管理组

System Management Tool 系统管理工具

**SMTK** Sump Tank 沉淀池,废油罐

**SMTP** Simple Mail Transfer Protocol 简单邮件传送协议

Simple Message Transfer Protocol 简单消息传送协议

**SMTS** Synchronous Meteorological Test Satellite 同步气象试验卫星

**SMUT** Shrink Mockup Template 缩模模板

**SMUX** Synchronous Mutiplexer 同步复用器

**SMX** Santa Maria, CA, USA(Airport Code) 美国加利福尼亚州圣玛丽亚机场

**SMY** Summary 总结,摘要

**SMYD** Stall Management Yaw Damper 失速

管理偏航阻尼器

**SMYDC** Stall Management Yaw Damper Computer 失速管理偏航阻尼器计算机

**SN** Sequence Number 序列号

Serial Number 批号,序号

Service Network 服务网络

Service Node 业务节点

Service Note 服务通知

Service Notification 业务通知

Shuttle Network 往返网络

Sign 符号

Signal to Noise 信号噪声比

Siren 警报器

Snow 雪

Strouhal Number 斯德罗哈尔号

Subscriber Number 用户号

Switching Network 交换网络

**SNA** Los Angeles, CA, USA-Orange Country (Airport Code) 美国加利福尼亚州洛杉矶橙县机场

Software Network Analysis 软件网络分析

Standard Network Architecture 标准化网络体系结构

System Network Architecture 网络系统架构

**SNACP** Subnetwork Access Procotol 子网接入协议

**SNAP** Standard Network Access Protocol 标准网络访问协议

Subnetwork Access Point 子网络存取点

Subnetwork Access Protocol 子网接入协议

System Network Activity Program 系统网络活动程序

System Network Architecture Program 系统网络结构程序

System of Network Analysis Program 网络分析程序系统

Systems for Nuclear Auxiliary Power 核辅助动力系统

**SNB** Switching Network Block 交换网络功能块

**SNBU** Switched Network Backup 交换网络后援

**SNC** Specialist Navigation Course 导航专业教程,导航专业课程

Standard Navigation Computer 标准导航计算机

Subnetwork Connection 子网连接

Synchronous Network Clock 同步网络时钟

System Network Controller 系统网络控制器

**SNCC** Signaling Network Control Center 信令网控制中心

**SNCF** Satellite Network Control Function 卫星网络控制功能

**SNCP** Subnetwork Connection Protection 子网连接保护

**SND** Standardized Normal Distribution 标准正态分布

**SNDCF** Subnetwork Dependent Convergence Function 子网会聚功能

**SNDCP** Subnetwork Dependent Convergence Protocol 子网会聚协议

**SNDG** Sending 送出

Sounding 发声

**SNDPRF** Soundproof 隔音的

**SNDR** Signal-power to Noise-power Density Ratio 信噪功率密度比

**SNE** Satellite Navigation Equipment 卫星导航设备

**SNEC** Switch Network Command 交换网络命令

**SNEFD** Switch Network Fault Diagnosis 交换网络故障诊断

**SNES** Seven-Nation Economic Summit 七国经济最高级会议

**SNF** System Noise Factor (Figure) 系统噪声系数,系统噪声指数

**SNGL** Single 单数,单的,奇特的

**SNH** Snatch 片刻,抓住

**SNI** Sequence Number Indicator 顺序数指示器

Service Node Interface 业务节点接口

SNA (System Network Architecture) Network

Interconnection 系统网络体系结构网络互联

Subscriber Network Interface 用户网络接口

Switching Network Interface 交换网络接口

Synchronous Network Interface 同步网接口

**SNIEC** Shanghai New International Expo Centre 上海新国际博览中心

**SNIF** Sensitivity for Normal Interface 正常接口的敏感性

**SNIR** Signal to Noise Plus Interference Power Ratio 信号对噪声加干扰功率比

**SNKL** Snorkel 通气管

**SNL** Sandia National Laboratories 森地亚国家实验室

Standard Name Line 标准名称系列

Standard Nomenclature List 标准术语表,专用标准名词表

**SNLG** Signaling 信令,发信号

**SNM** Signaling Network Management 信令网管理

Simple Network Management 简单网络管理

Switching Network Module 交换网络模块

**SNMAP** Safety Net Monitoring and Aids Processing 安全网络监视辅助处理

**SNML** Subnetwork Management Layer 子网管理层

**SNMP** Simple Network Management Protocol 简单网络管理协议

**SNN** Shannon, Ireland-Shannon ( Airport Code) 爱尔兰香农机场

**SNO** Stock Number 库存号

**SNOE** Smart Noise Operation Equipment 灵敏噪声操作设备

**SNOWTAM** Snow Notice to Airmen 雪情通告

**SNP** Statistical Network Processor 统计网络处理机

Synchro Null Pulse 同步零脉冲

Synchronous Network Processor 同步网络处理机

**SNPA** Subnetwork Point of Attachment 子网连接点

**SNPDU** Subnetwork Protocol Data Unit 子网协议数据单元

**SNR** Schedule Negotiation Record 定期协商记录

Signal to Noise Ratio 信噪比

Sonar 声呐

Special Network Resource 特殊网络资源

Step Number 步骤号

**SNRME** Set Normal Response Mode Extended 置扩充正常响应方式

**SNRO** Signal to Noise Ratio for Optics 光信噪比

**SNS** Sensor 传感器

**SNSDU** Subnetwork Service Data Unit 子网业务数据单元

**SNSG** Sensing 感应,敏感

**SNSH** Snow Showers 阵雪

**SNSL** Stock Number Sequence Listing 库存号顺序表

**SNSR** Sensor 传感器

**SNSS** Stabilized Night Sight System 稳定的夜视系统

Switching Network Subsystem 交换网络子系统

**SNST** Soonest 尽速

**SNTDA** Static Noise Test Data Analysis 静态噪声测试数据分析

**SNTM** Signaling Network Traffic Manage 信令网业务管理

**SNTR** Sinter or Sintered 烧结(的)

**SNTZD** Sensitized 感光的,敏感的

**SNTZG** Sensitizing 感光,敏感

**SNUD** Stock Number User Directory 库存号用户手册

**SNW** Subnetwork 子网

**SO** Second Officer 领航员

Selected Option 选用选装件

Serial Output 串行输出

Service Order 服务顺序

Shift-out 移出

Shipping Order 运货单,发货单
Shop Order 车间工作单,车间订单
Shutoff or Shut off 关断
Signal Office 信号室
Signal Orders 信号规程
Signal Oscillator 信号振荡器
Signed Original 签字原件
Sinusoidal Oscillator 正弦振荡器
Slow Operate 慢动作
Somalia 索马里
Special Order 特殊订单,特殊指令
Standard Oil 标准油
Standard Option 选装标准
Standing Order 常备订单,标准订单
Stationary Orbit 静止轨道
Stock Order 库存订单
Stop Order 停止指令
Sweep Oscillator 扫描振荡器
Synchronous Operation 同步运行
Synthetic Oil 合成油
System Orientation 系统定向
System Output 系统输出
System Override 系统超控

**SOA** Safety Oversight Audit 安全监督审计
Scrap on Assembly 组件上碎片
Semiconductor Optical Amplifier 半导体光放大器
Service-Oriented Architecture 面向服务的体系结构
Shortage of Arrival 到达缺额
Speed of Approach 进近速度,进近着陆速度
Start of Address 地址起始

**SOAP** Spectrographic Oil Analysis Program 滑油光谱分析程序
Spectrometric Oil Analysis Program 滑油分光仪分析程序

**SOAR** Safe Operating Area 安全操作区

**SOB** Side of Body 机身侧面
Souls on Board 机上人数

**SOBS** Second Observer 第二观察员

**SOC** Satellite Orbit Control 卫星轨道控制
Seat Occupied by 放置行李的座位
Seats Occupied 无空座位
Service and Overhaul Change 维修和大修更改
Service Operating Center 服务操作中心
Social 社交
Society 学会,协会,社会
Space Operations Center 空间运行中心
Start of Climb 爬升起点,开始爬升,起始爬升点
Stop-on-Compare 停止—接通—比较
System Operation Center 航班系统运行中心
System Operational Concept 系统运行概念
System Operations Control 系统运行控制
System-on-a-Chip 集成在一块芯片上的系统,单片系统

**SOCD** Source Control Drawing 源控制图

**SOCMA** Second Order Coherent Multiple Access 二阶相干多址访问

**SOCN** Source Control Number 源控制号

**SOCO** Spares Order Control 备件订货管理

**SOCR** Sales Order Control Record 销售订单管理记录
Sales Order Control Reporting 销售订单管理报告

**SOCS** Space Operation Command System 空间运行指挥系统

**SOD** Service on Demand 按需提供服务
Service Order 业务等级
Society of Europe Aero Services (France) 欧洲航空服务协会(法国)
Statement of Direction 管理说明,指导说明
System Operation Description 系统操作说明
System Operation Design 系统操作设计,系统运行设计

**SODALS** Simplified Omni-directional Approach Lighting System 简易全向进近灯光系统

**SODAR** Sound Detection and Ranging 声波检测与定位

**SOPDAS** Satellite Operation Planning and Data Analysis System 卫星运行计划与数据分析系统

**SODL** Serial Output Data Line 串行输出数据线

**SODRS** Synchronous Orbit Data Relay Satellite 同步轨道数据中继卫星

**SOE** Sequence of Events 事件顺序
Shop Order Execution 车间工作单实施

**SOF** Safety of Flight 飞行安全
Sofia, Bulgaria (Airport Code) 保加利亚索非亚机场

**SOFA** Sub-Optimal Filter Analysis 亚最佳滤波器分析

**SOFAR** Sound Fixing and Ranging 声波定位与测距

**SOFC** Safety of Flight Certification 飞行安全许可证
Solid Oxide Fuel Cell 固体氧化物燃料电池

**SOFCS** Self-Organizing Flight Control System 自组织飞行控制系统

**SOFR** Safety of Flight Report 飞行安全报告

**SOFT** Safety of Fight Test 飞行安全试验
Suborbital Flight Test 亚轨道飞行试验

**SOH** Section Overhead 段开销
Start of Header 标题开始
Start of Heading 起始航向

**SOHO** Small Office/Home Office 小型办公室/家庭办公室

**SOI** Silicon-on-Insulator 绝缘体硅
Space Object Identification 空间物体识别
Standard Operating Instruction 标准操作指令

**SOIA** Simultaneous Offset Instrument Approach 同时偏置仪表进近

**SOIC** Small Outline Integrated Circuit 小尺寸集成电路

**SOIM** Specific Operation Instruction Manual 详细使用说明书

**SOISUM** Space Object Identification Summary 空间目标识别总结

**SOIT** Satellite Operational Implementation Team 卫星运行实施小组

**SOJ** Single Open Jaw 单缺口
Stand-off Jammer 远程干扰发射台

**SOL** Safe Operation Limit 安全运行极限,安全工作极限
Solenoid 电磁线圈
Solicitation 恳请
Solid 固体,固态,固体的,固态的
SROA on-Line Ledgers 定期需求和订货联机总账
System Output Language 系统输出语言
System-Oriented Language 面向系统的语言

**SOLB** Start of Line Block 行数据区起始

**SOLD** Simulation of Logic Design 逻辑设计模拟

**SOLE** Society of Logistics Engineers 后勤工程师学会

**SOLN** Solution 溶液,解决

**SOLO** Selective Optical Lock-on 选择性光学跟踪
Spares on-Line Ordering 运行备件联机订货

**SOLS** Synchronous Optical Line System 同步光线路系统

**SOLV** Solenoid Valve 电磁活门

**SOM** Seat Occupied Message 座位占用报
Shortest Operated Mileage 直飞里程
Software Operators Manual 软件操作者手册
Special Operations Manual 特殊运行手册
Start of Message 消息起始
System Object Model 系统对象模型
Systems Operators Manual 系统操作者手册

**SOMAD** Strategic Overview of Major Airport Developments Study 机场大型发展策略总纲研究

**SOM-H** Start of Message-High Precedence 信息起始—高优先位

**SOM-L** Start of Message-Low Precedence 信息起始—低优先位

**SOMP** Start of Message Priority 信息起始顺序

**SOMR** Supplemental Operations Manual Requirements 补充运行手册要求

**SON** Statement of Operational Need 使用（操作）要求说明

**SONAC** Sonar Nacelle（反潜飞机上）声呐吊舱

**SONAR** Sound Navigation and Ranging 声波导航与定位

Sound Operation Navigation and Range 声波导航及测距

**SONCM&D** Sonar Countermeasures and Deception 声呐干扰与诱骗

**SONCR** Sonar Control Room 声呐控制室

**SONET** Synchronous Optical Network 同步光纤网

**SONG** Satellite for Orientation, Navigation and Geodesy 定向、导航与大地测量卫星

**SONIC** Spares on-line and Non-stop Inventory Control 联机备件与不中断库存控制

Spares on-line Inventory Control System 联机备件库存控制系统

Spares Ordering Non-stop Inventory Control 备件订购与不中断库存控制

**SONOAM** System on-line Network for Information Control 信息管理系统联机网络

**SOO** Standard Operational Orders 标准作业指令

**SOON** Solar Observing Optical Network 太阳观测光学网络

**SOP** Safety Operating Procedure 安全运行程序，安全操作程序

Safety Oversight Program 安全监督计划

Set of Parts 零件组

Shop Order Planning 车间订购计划

Simulated Output Program 模拟输出程序

Simulation Operation Plan 模拟运行计划，模拟操作计划

SITA Operating Procedures 国际航空电信协会操作程序

Spares Ordering Program 备件订购大纲，备件订购程序

Standard Operating Plan 标准操作计划

Standard Operating Procedure 标准操作程序

Standard Option Pin 标准选装销钉

Synchro Output 同步输出

**SOPC** Shuttle Operations and Planning Center 航天飞机运行与计划中心

System on Programmable Chip 可编程片上系统

**SOPM** Standard Overhaul Practices Manual 标准大修施工手册

**SOPs** Standard Operating Procedures 标准作业程序

**SOR** Serial Output Register 串行输出寄存器

Source of Repair 修理资源

Special Operational Requirement 特殊操作要求

Special Order Request 特殊订单需求

Speed of Rotation 转速

Statement of Requirements 要求说明

**SORAP** Standard Omni-Range Approach 标准全向导航进近

**SORO** Scan on Receive Only 仅用于接收扫描

**SORT** Satellite Orbit Track 卫星轨道跟踪

**SORTE** Summary of Radiation Tolerant Electronics 耐辐射电子学摘要

**SORTI** Satellite Orbital Track and Intercept 卫星轨道跟踪与截获

**SOS** Distress Signal 遇险信号

Sales Order Screen 销售订单屏幕

Satellite Observation System 卫星观测系统

Save Our Shorelines 拯救海岸线

Save Our Souls 拯救我们的生命（遇险呼救信号）

Self-Organizing System 自组织系统

Sell or Scrap 卖出或报废

Service Operation System 业务运行系统

Share Operation System 共享操作系统

Shop Out of Stock 车间无库存

Silicon on Sapphire 蓝宝石上硅薄膜

Simulator Operation System 模拟机操作系统

Start of Search 搜索起始

Stores Out of Stock 仓库无库存

Symbolic Operation System 符号操作系统

Synchronous Orbit Satellite 同步轨道卫星

**SOSO** Safety Operational Supervision Office 安全运行监督办公室

**SOSTEL** Solid-State Electronic Logic 固态电子逻辑

**SOSUS** Sound and Surveillance System 声音与监视系统

**SOT** Small Outline Transistors 小尺寸晶体管

Solar Optical Telescope 太阳光学望远镜

Start of Transmission 发射起始,开始发送

Stator Outlet Temperature 定子出口温度

**SOTAS** Stand off Target Acquisition System 远距目标搜索系统

**SOTM** System Operation Training Manual 系统运行培训手册

**SOTO** Society of Transporter Owners 运输机主协会

**SOTP** System Overhaul Testing Program 系统翻修测试大纲

**SOTS** Synchronous Orbit Tracking Station 同步轨道跟踪站

**SOU** Southampton, England, United Kingdom-Eastleigh ( Airport Code ) 英国南安普顿伊斯特利机场

**SOV** Shut off Valve 断油开关

Shutoff Valve 截流阀

Solenoid Operated Valve 电磁线圈操作活门

Solenoid Valve 电磁活门

**SOW** Start of Work 工作起始

Statement of Work 工作报告,任务书

**SOWS** Standard Option Work Statement 标准选装工作说明

**SOX** Simulated Orderbase Extension 模拟指令库扩展

Solid Oxygen 固态氧

**SP** Sample Part 样品件

Sampling 取样法,抽样法

Scheduled Passenger 定期航班旅客

Seaplane Port 水上机场,水上航空港

Selected Point 选择点

Senior Pilot 高级驾驶员

Sequential Phase 顺序相位

Service Pack 服务包,补丁

Service Panel 服务面板,维护面板

Service Part 维修备件

Service Passport 公务护照

Service Position 工作位置

Service Provider 服务供应商

Set Plug 定位销,固定塞

Settlement Price 结算价格

Shadow Price 影子价格

Shear Plate 剪切板

Shipment Price 交运价格

Short Persistence 短余辉

Signal Processing 信号处理

Signal Processor 信号处理机

Signaling Point 信令点

Signaling Protocol 信令协议

Single Pilot 单一飞行员,单一驾驶

Single Pole 单极

Smart Push 智能推送

Soil Pipe 污水管

South Pole 南极

Space 空间,场地,间隔,空格键

Spare 备件,备用的

Spare Part 备件零件,备件

Special Performance 特殊性能

Special Publication 特殊出版物

Special Purpose 特殊目的

Speciality 特殊性,专长,专业

Specific 明确的,规定的

Speech Processing 话音处理

Speech Processor 话音处理器

Speed 速度

Spinal 尖晶石

Splashproof 防溅的

Splitting 分离,裂缝,蜕变

Spool 轴,短管

Standard Holding Pattern 标准等待航线

Standard Part 标准件

Standard Practice 标准常规,常规做法,标准惯例

Standard Pressure 标准压力

Standard Procedure 标准程序

Standpipe 竖管,管体式水塔

Sticker Price 标签价格

Stiffening Plate 加强板

Stop Payment 停止支付

Storage Process 存储处理

Storage Processor 存储处理器

Storage Program 存储程序

Storage Properties 保存特性

Storage Protection 存储保护

Student Pilot 飞行学员

Subprogram 子程序

Summing Point 累加点

Supercharge Pressure 增压压力

Synchronous Pulse 同步脉冲

System Product 系统产品

**S-P** Serial-to-Parallel 串—并(转换)

System and Procedure 系统与程序

**SP HT** Specific Heat 比热

**SP LT** Speed of Light 光速

**SP PH** Split Phase 分相

**SP SW** Single Pole Switch 单刀开关

**SP VOL** Specific Volume 比容,特殊值

**SPA** S Band Power Amplifier S 波段功率放大器

Sales and Purchase Agreement 销售与采购协议

Seaplane Pilots Association 水上飞机飞行员协会

Service Panel Adapter 维修面板适配器

Servo Preamplifier 前置伺服放大器

Servo-Power Assembly 伺服动力装置

Signal Processing Array 信号处理矩阵

Signal Processor Assembly 信号处理器组件

Single-Pilot Aeroplane 单个驾驶员飞机

Skill Performance Aid 技能演示设备

Software Product Assurance 软件产品保证

Solar-Powered Aircraft 太阳能飞机

Space Processing Application 空间处理应用

Special Prorate Agreement 特别拆账金额协定,特殊分摊协议

Special Purpose Aircraft 专用飞机

Substitute Part Authorization 替换零件核准

Sudden Phase Anomaly 突现相位异常

System Performance Analysis 系统性能分析

**SPAC** Space Program Advisory Council 空间计划咨询委员会

**SPACCS** Space Command and Control System 空间指挥与控制系统

**SPACECOM** Space Communications 空间通信

**SPAD** Scratchpad 暂存区,便条

Support Planning and Design 支援计划与设计

**SPADATS** Space Detection and Tracking System 空间探测与跟踪系统

**SPADATSS** Space Detection and Tracking Sensor System 空间探测与跟踪传感器系统

**SPADE** Signal Processing and Display Equipment 信号处理与显示设备

Spare Parts Analysis, Documentation and Evaluation 备件分析、文件编制与评估

**SPADS** Satellite Position and Display System 卫星定位和显示系统

Satellite Processing and Display System 卫星处理与显示系统

**SPAE** Selectable Polarization Array Element 可选择极化阵列元件

**SPAG** Sichuan Province Airport Group Company Limited 四川机场集团公司

**SPAM** Special Procurement Authorization Memorandum 特别采购授权备忘录

**SPAN** Stored Program Alphanumeric 存储的程序数字字母

**SPAR** French Light Precision Approach Radar 法式灯光精密进近雷达

Solid-state Phased Array Radar 固态相列雷达

Special Passenger Arrangement 特殊顾客安排

Statistical Product Analysis Report 统计分析报告

Super Precision Approach Radar 高精度进近雷达

Supplementary Precision Approach Radar 辅助精密进近雷达

Symbolic Program Assembly Routine 符号程序汇编程序

Synchronous Position Attitude Recorder 同步位置姿态记录仪

System Performance and Accuracy Report 系统性能与精度报告

**SPARTIS** Space Related Tactical Intelligence System 空间相关战术情报系统

**SPAS** Safety Performance Analysis Subsystem 安全性能分析子系统

Safety Performance Analysis System 安全性能分析系统

**SPASE** Single Point Audio System Earth 单点音频系统地线

**SPAT** Silicon Precision Alloy Transistor 硅精密合金晶体管

**SPATE** Special Purpose Automatic Test Equipment 专用自动测试设备

**SPATIAL** System for Acquisition and Processing of ARINC and Logic 艾瑞克和逻辑获取和处理系统

**SPBP** Split Phase Bipolar 双极分相

**SPC** Semipermanent Connection 半永久性连接

Serial-to-Parallel Converter 串并联转换

Signaling Point Code 信令点代码

Skills Process Center 技能处理中心

Software Product Control 软件产品管理

South Pacific Commission 南太平洋委员会

Spare Parts Catalog 备件种类

Spare Parts Classification 备件分类

Split Charging Service 分摊计费业务

Stall Protection Computer 失速保护计算机

Standard Precision Code 标准精度码

Standard Procedure Control 标准程序控制

Standard Products Committee 标准件委员会

Statistical Processes Control 统计过程控制

Stored Program Command 存储程序指令

Supervisory Process Control 监督进程控制

Suspended Plaster Ceiling 涂层吊顶

System Processing Computer 系统处理计算机

**SPCA** Supplier Physical Configuration Audit 供应方实际配置审计

**SPCAT** Special Category 特殊类别

**SPCB** Stream Protocol Control Block 流协议控制块

**SPCHG** Supercharge 增压

**SPCK** Spotcheck 现场检查

**SPCL** Special 特别

**SPCLN** Special Cleaning 特殊清洁

**SPCN** Satellite Personal Communication Network 卫星个人通信网

Spare Parts Change Notice 备件更改通知

**SPCONV** Speed Converter 速度转换器

**SPCP** Specification Plan 规范计划

**SPCR** Spacer 垫片,隔套

**SPCS** Spare Parts Cost System 备件成本系统

**SPCT** Special Connector Type 特型插头,特型连接器

**SPCU** Standby Power Control Unit 备用电源控制组件

**SPD** Serial Presence Detect 连续存在检测

Spare Parts Depot 备件仓库

Special Purpose Display 特殊显示器

Specified 规定的

Speed 速度

Standard Product Definition 标准产品定义

Surface Position Digitizer 舵面位置数字转换器

Synchronization Phase Distortion 同步相位失真

Synchronous Phase Demodulator 同步相位解调器

System Parameter Document 系统参数文件

System Program Directive 系统程序指令

System Program Director 系统项目主任

**SPD BK** Speed Brake 气动力减速装置，减速板

**SPD LIM** Speed Limit 速度极限

**SPD SEL** Speed Select 速度选择

**SPD/M** Speed/Mach 速度/马赫

**SPDBK** Speedbrake 减速板

**SPDL** Spindle 心轴

Supplied Part Data List 供应商零件数据表

**SPDOM** Speedometer 测速计

**SPDR** Software Preliminary Design Review 软件初步设计审查

Spider 机架

Subcontractor Preliminary Design Review 分包方初步设计审查

Supplier Preliminary Design Review 供应方初步设计审查

**SPDS** Safe Practice Data Sheet 安全技术（规章）数据单

Space-based Position Determination System 星基位置测定系统

**SPDT** Single Pole Double Throw 单刀双掷

**SPDTSW** Single Pole Double Throw Switch 单刀双掷开关

**SPDU** Secondary Power Distribution Unit 次级电源分配组件

Session Protocol Data Unit 会晤协议数据单元

**SPE** Fixed Price with Escalation 逐步上升的固定价格

Seller Purchase Equipment 卖方购买设备

Signal Processing Electronics 信号处理电子设备

Signal Processing Element 信号处理单元

Special Purpose Entity 特别目的机构

Special Purpose Equipment 专用设备

Speech Processing Equipment 语音处理设备

Spherical Probable Error 球概率误差

Static Phase Error 静相误差

Stored Program Element 存储的程序元件

Supplier Purchased Equipment 供应方购买的设备

Synchronous Payload Envelope 同步净负荷区包迹

System Performance Effectiveness 系统性能有效性

**SPEAR** Supplier Performance Evaluation and Rating Report 供应方生产能力评估与分级报告

**SPEC** Society for Pollution and Environmental Control 污染与环境控制学会

Special 特殊的，专门的

Specialist 专家

Specialized 特定的

Specification（技术）规范，说明（书）

Specified 规定的，明确的，指定的

Specimen 样品，样机，标本

Speculative 投机的

Speech Predictive Encode/Encoding Communication 语音预测编码通信

System Performance Evaluation Committee 系统性能评估委员会

**SPECI** Aviation Selected Special Weather Report 航空特选天气报告

Aviation Special Weather Report 航空特殊天气报告

**SPECLE/DA** Specification Language/Design Analysis 规范语言/设计分析

**SPECLE/DAR** Specification Language/Design Analysis Report 规范语言/设计分析报告

**SPED** Spares Pricing and Estimating Database 备件定价与评估数据库

**SPEED** Signal Processing in Evaluated Electronic Devices 评估电子装置中的信号处理

**SPES** Site Performance Evaluation System 站点性能评估系统

Supplier Performance Evaluation System 供应方生产能力评估系统

Stimulation Plan for Economic Science 经济

学鼓励计划

**SPF** Service Port Function 业务端口功能

Single Point Fueling 单口加油

Superplastic Forming 超塑性成型

**SPF/DB** Superplastic Forming/Diffusion Bonding 超塑性扩散焊

**SPG** Single Point Ground 单点接地, 单轮接地

**SPGA** Staggered Pin Grid Array 交错式针状网阵封装

**SPGR** Specific Gravity 比重

**SPH** Shortest Path Heuristic 最短路径推断

Single-Pilot Helicopter 单个驾驶员直升机

Space Heater 空间加热器

**SPHER** Spherical 球形的

**SPHINX** Space Plasma High-Voltage Interaction Experiment 空间等离子高压干涉实验

**SPHN** Siphon 虹吸

**SPI** Schedule Performance Index 进度执行效果指数, 进度性能指数

SDH Physical Interface SDH 物理接口

Serial Peripheral Interface 串行外围设备接口

Service Provider Interface 服务提供者接口

Shop Procured Item 车间采购项目

Short-Pulse Insertion 短脉冲输入

Special Position Identification 特殊位置标识, 特殊位置识别

Speech Path Interface 话路接口

Springfield, IL, USA( Airport Code) 美国伊利诺州斯普林菲尔德机场

Subsequent Protocol Identifier 后续协议标识符

Surface Position Indicator 舵面位置指示器

Symbolic Pictorial Indicator 象征图像指示器

**SPIC** Student Pilot-in-Command 学员机长

**SPICE** Simulation Program with Integrated Circuit Emphasis 集成电路强化模拟程序

**SPIF** Shuttle Payload Integration Facility 航天飞机业载集成设施

**SPINA** Subscriber PIN Access 用户 PIN 接入

**SPIP** Special Position Identification Pulse 特殊位置识别脉冲

**SPIR** Single Pilot Instrument Rating 单一飞行员仪表等级

Special Pending Interrupt Resistor 进行中专用中断寄存器

**SPIS** Software Path Instrumentation System 软件路径仪表系统

**SPK** Sapporo, Japan ( Airport Code) 日本扎幌机场

Soundspeaker 扬声器

Spark 火花

Spike 峰值

Store Packed 仓库包装的

**SPKR** Speaker 喇叭, 扬声器, 发言人

**SPL** Separate Parts List 分类零件表

Service Provider Link 服务提供商链路

Signal Processing Language 信号处理语言

Single Propellant Loading 单推进剂载荷

Sound Power Level 声音功率电平

Sound Pressure Level 声压等级

Source Program Listing 源程序表

Spare Parts List 备件单

Special 特定的, 特殊的, 专门的

Special-Purpose License 特殊用途执照

Spiral 螺旋形的

Split 分离

Split Charging 分摊计费

Sport Pilot License 运动用飞行员执照

Student Pilot Licence 学生飞行员执照

Supplementary Flight Plan 补充飞行计划

Supplementary Flight Plan Message 领航计划补充信息报

**SPLC** Splice 接头, 补块

**SPLL** Self-Propelled Launcher Loader 自推进发射器装载器

**SPLN** Spline 齿条

**SPLR** Spoiler 扰流板

**SPLTR** Splitter 分离器

**SPLX** Simplex 单纯的, 单一的

**SPLY** Supply 供应, 供给

**SPM**　Scanning Probe Microscopy 扫描探针显微镜

Seats Protected Message 座位保留文电

Self-Phase Modulation 自相位调制

Session Protocol Machine 对话(会晤)协议机

Signal Processing Module 信号处理组件

Small Power Module 小功率模块

Software Project Manager 软件项目经理

Special Practice Memorandum 特殊操作备忘录

Speech Memory 语音存储器

Stabilizer Position Module 水平安定面位置组件

Stabilizer Position Monitor 安定面位置监测器

Standard Parts Manual 标准零件手册

Standard Practice Manual 标准施工手册

Standard Procedure Manual 标准程序手册

Standard Process Manual 标准工艺规程手册

Strokes per Minute 冲程/分钟

Surface Position Monitor 舵面位置监视器

Suspend Production Memorandum 中止生产备忘录

**SPMC**　Service Process Material Control 服务进程材料控制

**SPMG**　Shorted Permanent Magnetic Generator 短接的永磁发电机

**SPML**　Special Meal 特别餐

**SPMMF**　Single Polarization Mode Maintaining Fiber 单偏振模维持光纤

**SPN**　Saipan, Northern Mariana Islands, USA (Airport Code) 美国北马里亚纳群岛塞班机场

Service Provider Network 服务供应商网络

Spin 旋转

Standard Precision Navigation 标准精密导航

Subscriber Premises Network 用户驻地网

**SPNS**　Switched Private Network Service 专用交换网络服务

**SPO**　System Program/Project Office 系统规划处

**SPOOL**　Simultaneous Peripheral Operations on-Line 假脱机(操作)

**SPOP**　Service Process Operation Procedure 服务进程工作程序

**SPOR**　Spares Pick and Order Record 备件选捡和订购记录

**SPORS**　Spares Pick and Order Record System 备件选捡和订购记录系统

**SPOT**　Spot Wind 定点风

**SPP**　Sequence Packet Protocol 顺序包协议

Sequential Packet Protocol 定序分组协议

Service Provision Point 服务提供点

Signal Processing Peripheral 信号处理外部设备

Special Purpose Processor 专用处理机

Speech Path Processing 话路处理

Standard Parallel Port 标准并行接口

Standard Practice Procedure 标准操作程序

Static Pressure Probe 静压探头

Still Picture Projector 静止图像投影仪

Supplementary Planning Procedure 补充计划程序

**SPPL**　Spark Plug 火花塞

Supplier Plan Parts List 供应商计划零件表

**SPPS**　Software Process Planning System 软件过程规划系统

**SPPT**　Spare Part 备件,部分零件

**SPR**　Safety and Performance Requirements 安全与性能需求

Sending Packet Rate 分组信息发送率

Service and Performance Report 服务和性能报告

Servo Pressure Regulator 伺服压力调整器

Short Pulse Radar 窄脉冲雷达

Short Pulse Range 窄脉冲测距,窄脉冲定位

Signal Processing Radar 信号处理雷达

Silicon Power Rectifier 硅功率整流器

Software Problem Report 软件问题报告

Spare Parts Release 备件发放

Spares 备件

Special Project Report 特别计划报告

Special Purpose Radar 专用雷达

Spring 弹簧

Sprinkler 撒布,洒水,小雨

Supplier Problem Report 供应方问题报告

Sync Phase Reversal 同步相位反转

System Parameter Record 系统参数记录

**SPRA** Special-Purpose Reconnaissance Aircraft 特殊目的侦察机

**SPRAN** Shop Part Record Adjustment Notice 车间零件记录调整通告

**SPRDR** Spreader 分配器,喷洒器

**SPRE** Service Panel Reference 维修面板参考(资料)

**SPREG** Speed Regulator 调速器

**SPRG** Sprinkling 喷洒

**SPRING** Shared Protection Ring 共享保护环

**SPRKT** Sprocket 链轮

**SPRS** Suppression 抑制,排除

**SPRT** Support 支援,支持,保障,支撑

**SPS** Samples per Second 取样数/秒

Satellite Power System 卫星动力系统,卫星电源系统

Secondary Power System 辅助动力系统

Sector Power Supply 区域电源

Service Process Solution 服务处理结果

Service Propulsion System 维修推进系统,服务推进系统

Service Provider System 服务供应商系统

Signal Processing System 信号处理系统

Software Project Specification 软件工程规范

Solar Power Satellite 太阳能卫星

Spares Pricing System 备件价格体系

Special Protection System 特殊保护系统

Speech Path Subsystem 话路子系统

Stall Protection System 失速保护系统

Standard Position/Positioning Service 标准定位服务

Standard Positioning System 标准定位系统

Standby Power System 备用动力系统

Static Power System 静电源系统

Static Pressure System 静压系统

Statistical Performance Standard 统计性能标准

Stored Program System 存储程序系统

Structured Programming System 结构程序设计系统

Support Programming System 支援程序系统

Surge Prevention System 防喘振系统

System Performance Specialist 系统性能专家

Wichita Falls, TX, USA(Airport Code)美国得克萨斯州威奇托福尔斯机场

**SPS/B** Superseded by 由…代替

**SPSA** Supplier Product Support Agreement 供应商产品支援协议,供应商产品供应协议

**SPSC** Semi-Permanently Switched Circuit 半永久交换电路

Single Point Sensor Card 单点传感器插件

Standard Performance Summary Charts 标准性能简图

System Planning and System Control 系统计划与系统控制

**SPSCR** Special Screw 特种螺钉

**SPSHP** Special Shaped 特型的,异型的

**SP-SMF** Single-Polarization Single-Mode Fiber 单偏振单模光纤

**SPSOS** Signal Processing System Operating System 信号处理系统操作系统

**SPSP** Spare Parts Support Package 备件供应包

Spread Spectrum 扩展频谱

**SPSS** Supplementary Power Supply Set 辅助供电装置

**SPST** Single-Pole Single-Throw 单刀单掷

**SPSTSW** Single-Pole Single-Throw Switch 单刀单掷开关

**SPSZ** Supersonic Zone 超音速区

**SPT** Sample Part Template 样品件模板

Septuple 七倍的

Spare Parts Transmittal 备件传送

Split 分裂

System Parameter Table 系统参数表

System Performance Test 系统性能试验

**SPTC** Specified Period-of-Time Contract 规定有效期合同

**SPTD** Supplemental Provisioning Technical Documentation 补充供给技术文件

**SPTS** Simulator Performance Test System 模拟机性能测试系统

**SPTT** Single-Pole Triple-Throw 单刀三掷

**SPU** Signal Processing Unit/ Signal Processor Unit 信号处理单元/信号处理机单元

Split, Croatia (Airport Code) 克罗地亚斯普利特机场

Start Power Unit 启动电源组件

Supplier Primary Update Data 供应方主要更新数据

**SPV** Special Purpose Vehicle 特种用途车,专用飞行器

**SPVN** Supervision 监督,指导

**SPVT** Specification Performance Verification Test 规范性能验证试验

**SPW** Signal Processing Worksystem 信号处理工作系统

**SPWG** Security Policy Working Group 安全政策工作组

Signal Processing Working Group 信号处理工作组

**SPWM** Sine Pulse Width Modulation 正弦脉冲宽度调制

**SPX** Sequenced Packet Exchange 顺序包交换协议

**SPYR** Sprayer 喷雾器

**SQ** Squadron 飞行中队

Square 平方

**SQA** Service Quality Agreement 服务质量协议

Software Quality Assurance 软件质量保证

Statistical Quality Analysis 统计质量分析

Strategy Quality Assurance 策略质量保证

Supplier Quality Assurance 供应商质量保证

System Quality Assurance 系统质量保证

**SQAAM** Software Quality Assurance Action

Memorandum 软件质量保证措施备忘录

**SQAAR** Software Quality Assurance Activity Report 软件质量保证有效性报告

**SQAP** Software Quality Assurance Program 软件质量保证程序

**SQAPP** Software Quality Assurance Program Plan 软件质量保证程序计划

**SQC** Shop Quality Control 车间质量控制

Standard Quality Control 标准质量控制

Station Quality Control 航站质量控制

Statistical Quality Control 统计质量控制

**SQCM** Square Centimeter 平方厘米

**SQCS** Shop Quality Control System 车间质量控制系统

**SQD** Service Quality Data 服务质量数据

**SQE** Supplier Quality Engineer 供应商质量工程师

Supplier Quality Evaluation 供应商质量评估

**SQEP** Software Quality Evaluation Plan 软件质量评估计划

**SQFT** Square Feet 平方英尺

**SQH** Square Head 方形头

**SQI** Safety Quality Index 安全质量指数

Service Quality Index 服务质量指数

**SQIN** Square Inch 平方英寸

**SQIP** Supplier Quality Improvement Process 供应商质量改进程序

**sqkm** Square Kilometer 平方千米

**SQL** Structured Query Language 结构询问语言

**SQL-DMF** SQL Distributed Management Framework SQL 分布式管理结构

**SQL-DMO** SQL Distributed Management Objects SQL 分布式管理对象

**SQM** Signal Quality Monitoring 信号质量监测

Software Quality Management 软件质量管理

**SQP** Software Quality Plan 软件质量计划

**SQPP** Software Quality Program Plan 软件质量项目计划

**SQPSK** Staggered Quadriphase Shift Keying

参差四相移相键控

**SQRT** Square Root 平方根

**SQS** Sales, Quality and Service 销售、质量与服务

Service Quality Statistics 服务质量统计

Stochastic Queuing System 随机排队系统

**SQUID** Superconducting Quantum Interface Device 超导量子干涉器件

**SQW** Square Wave 方波,矩形波

**SR** Safety Recommendations 安全建议

Saturable Reactor 饱和电抗器,饱和扼流圈

Schedule Review 定期审查

Scrap Rate 报废率

Selective Ringing 选择性振铃

Selenium Rectifier 硒整流器

Send Ready 准备发送

Senior 年长的,高级的

Service Request 维修申请

Shift Register 移位寄存器

Shop Responsibility 车间责任

Short Range 近距离,短程

Single Rotation 单向旋转

Slip Ring 汇流环

Slow Relay 缓动继电器,慢速中继

Slow Release 慢释放

Sound Ranging 声学测距

Special Regulations 特殊规则

Special Repair 特殊修理

Specialized Repair 专业化修理

Specific Range 比航程(燃油里程)

Specific Report 专题报告

Specific Resistance 电阻率

Specification Requirement 规范要求

Speech Recognition 语音识别

Speed Recorder 速度记录仪

Split Ring 对开环

Stack Register 堆栈寄存器

Stall Recognition 失速识别

Standard Rate 标准定额,标准速率

Standard Repair 标准修理

Starter Relay 启动继电器

Stateroom 特等舱

Static Route 静态路由

Status Register 状态寄存器

Steradian 球面度

Stochastic Retrieval 随机检索

Storage Register 存储寄存器

Stress Redistribution 应力再分配

Sunrise 日出

Supersede Release 替代发放

Supplementary Regulation 补充条例

Suppression Ratio 抑制比

Surveillance Radar 监视雷达

Surveillance Report 监视报告

Synchro Resolver 同步解算器

System Research 系统研究

**SR&O** System Requirements and Objectives 系统要求与目标

**SRA** Sierra Academy of Aeronautics（USA）希拉航空学院(美国)

Software Release Authorization 软件发行批准

Special Repair Area 特殊修理区

Special Rules Area 特殊规则区

Specialized Repair Activity 特殊修理活动

Specific Range Air 特定范围大气

Spiral Receiving Antenna 螺旋接收天线

Stimulated Raman Scattering Fiber Amplifier 受激拉曼散射光纤放大器

Stores Reclamation Area 仓库回收区

Support Requirements Analysis 支援需求分析

Surveillance Radar Antenna 监视雷达天线

Surveillance Radar Approach 监视雷达进近

System Requirements Analysis 系统需求分析

System Review and Approval 系统审查和批准

**SRADD** Software Requirements and Design Description 软件要求和设计描述

Software Requirements and Design Document 软件要求与设计文件

**SRAM** Safety, Reliability, Availability and Maintainability 安全、可靠性、可供性和可维修性

Static Random Access Memory 静态随机存取存储器

**SRAPI** Speech Recognition Application Programming Interface 语音识别应用编程接口

**SRAS** Safety Regulation and Standard 安全规则和标准

Sales Revenue Accounting System 销售收入结算系统

**SRAT** Short Range Advanced Technology 短程先进技术

**SRB** Safety Review Board 安全审查处

Software Review Board 软件审查处

Source Route Bridge 源路由网桥

Storage Resource Broker 存储资源代理

**SRC** Sampling Rate Convertor 采样率转换器

Science Research Council 科学研究委员会

Secondary Radar Code 二次雷达密码

Secondary Right Computer 右辅助计算机

Shop Resident Control 车间居留控制

Short Range Clamp 近程钳位

Signaling Route Control 信令路由控制

Source 源

Spares Release Card 备件发放卡

Special Release Card 特殊发放卡

System Resource Controller 系统资源控制器

**SRCC** Strategic Reconnaissance Communications Center 战略侦察通信中心

**SRCH** Search 寻找

**SRD** Secret Restricted Data 保密限制数据

Selected Record Data 选择记录数据

Short Range Delay 短程延迟

Shorted Rotating Diode 短接旋转二极管

Software Requirements Document 软件要求文件

Spin Reference Detector 旋转基准探测器

Spin Resistant Differential 防滑差速器

Standard Reference Data 标准参考资料

Steel Rule Die 钢制刻度模

Step Recovery Diode 阶跃恢复二极管

System Reference Document 系统参考文件

System Reliability Demonstration 系统可靠性演示

System Requirements Document 系统需求文件，系统要求文档

**SRDC** Standard Reference Data Center 标准参考数据中心

**SRDS** Standard Reference Data Service 标准参考数据服务

Standard Reference Data System 标准参考数据系统

**SRE** Surveillance Radar Element 监视雷达部件

Sucre, Bolivia( Airport Code) 玻利维亚苏克雷机场

**SRF** Security Resource Function 安全资源功能

Shop Release File 车间发放文档

Shop Relevant Fault 车间相关缺陷

Specialized Resource Function 专用资源功能

**SRFCS** Self-Repairing Flight Control System 自修复飞行控制系统

**SRFG** Surveillance Radar Functional Group 监视雷达功能组

**SRFM** Signal Reference Frequency Multiplex 信号参考频率复用

**SRG** Safety Regulation Group (英国民航局)安全规程组

Semarang, Indonesia( Airport Code) 印度尼西亚三宝垄机场

Short Range 短程，近程

Surge 喘振

**SRI** Series Resonant Inverter 串联谐振换向器

Shop Replaceable Item 车间可更换件

Special Review Insert 专项检查插入

Stanford Research Institute 斯坦福研究所

System Requirements and Implementations 系统需求和执行

**SRID** Source Routing Identifier 源选路标识符

System Requirements and Implementation Description 系统需求和执行说明

System Requirements and Implementation Document 系统需求和执行文件

**SRKL** Structural Repair Kit List 结构修理器材包清单

**SRKM** Structural Repair Kit Manual 结构修理器材包手册

**SRKN** Single Rotating Knife 单旋转刀

**SRL** Stability Return Loss 稳定回损

Storage, Retrieval and Launch 储存、恢复和发射

**SRLY** Series Relay 串联继电器

**SRM** Safety Risk Management 安全风险管理

Send Routing Message 发送选路消息

Session and Resource Manager 会晤和资源管理器

Shop Replaceable Module 车间可更换模件

Signaling Route Management 信令路由管理

Software Reliability Model 软件可靠性模型

Special Resource Management 特殊资源管理

Stabilizer Rudder Ratio Module 水平安定面方向舵比例变换组件

Stabilizer Trim/ Rudder Ratio Module 水平安定面配平/方向舵比例组件

Standard Reference Method 标准参考方法

Standard Repair Manual 标准维修手册

Structural/Structure Repair Manual 结构修理手册

Supplier Relationship Management 供应商关系管理

Switched Reluctance Motor 开关磁阻马达

System Reference Manual 系统参考手册

System Resource Manager 系统资源管理程序

**SRMC** Safety Risk Management Committee 安全风险管理委员会

Source Routing Mobile Circuit 源选路移动电路

**SRME** Specialized Resource Management Entity 专用资源管理实体

**SRMT** Signal Radio Maintenance Team 无线电信号维修队

**SRMU** Signal Radar Maintenance Unit 雷达信号维修单元

**SRN** Schedule Revision Notice 定期修改通告

Short Range Navigation 近距离导航

Spares Requiremnt Notice 备件要求通告

Strategic Road Network 重要道路网

**SRO** Singly Resonant Oscillator 单共振振荡器

**SROB** Short-Range Omnidirection Beacon 近距离全向信标

**SROD** Stove Rod 炉杆, 通条

**SRP** Seat Reference Point 座位基准点

Selected Reference Point 选择的基准点

Service Related Problem 维修相关问题

Shared Resources Processing 共享资源处理

Single Rotation Propeller 单旋转螺旋桨

Spatial Reuse Protocol 空间复用协议

Standard Repair Procedure 标准修理程序

**SRPA** Service Request and Program Authorization 维修要求和计划批准

**SRPP** Spare Repair Parts Plan 备用修理件计划

**SRPPP** Spare and Repair Parts Program Plan 备件和修理件程序计划

**SRPR** Scraper 刮刀

**SRPSU** Slide Release Power Supply Unit 滑梯释放电源组件

**SRQ** Sarasota, FL, USA(Airport Code) 美国佛罗里达州萨拉索塔机场

**SRR** Search and Rescue Region 搜寻与救援区

Shop Reallocation Request 车间再分配申请

Sound Recorder Reproducer 声音记录器放音机

Spares Requirement Reporting 备件需求报告

Specification Revision Record 规范修改记录

Specification Revision Request 规范修改申请

Supplemental Release Request 补充发放申请

Synchronous Round Robin 同步循环

System Requirements Review 系统需求评审

**SRRB** Search and Rescue Radio Beacon 搜寻和救援无线电信标

**SRRS** Spares Requirement Reporting System 备件要求报告系统

**SRS** Seat Reservation System 订座系统

Secondary Radio Station 辅助无线电台

Software Requirements Specification 软件要求说明书,软件需求规格说明

Special Revenue Sharing 特殊税收分配

Speed Reference System 速度基准系统

Statistical Reporting Service 统计报告服务

Stimulated Raman Scattering 受激拉曼散射

Supplemental Restraint System 辅助约束装置(又称安全气囊)

Synchronous Relay Satellite 同步中继卫星

**SRSG** Special Representative of the United Nations Secretary-General 联合国秘书长特别代表

**SR-SS** Sunrise-Sunset 日出—日落

**SRT** Safety Review Team 安全审查小组

Standard Rate of Turn 标准转弯率

Standard Remote Terminal 标准远程终端

Subscriber Response Time 用户响应时间

Supporting Research and Technology 支援研究和技术

**SRTA** Short Range Transport Aircraft 短程运输机

**SRTB** Source Route Transparent Bridge 源路由透明网桥

**SRTC** Search Radar Terrain Clearance 搜索雷达地形间隙

**SRTM** Shuttle Radar Topography Mission 航天飞机雷达地形测量

**SRTP** System Rig Test Plan 系统台架试验大纲

**SRTR** System Rig Test Report 系统台架试验报告

**SRTS** Synchronous Residual Time Stamp 同步剩余时间戳

System Respone Time Simulator 系统响应时间模拟器

**SRTU** Store Remote Terminal Unit 遥控存储终端组件

**SRU** Self-Recording Unit 自动记录装置

Shop Repairable Unit 车间可修件

Shop Replaceable Unit 车间可更换组件

**SRUR** System Route Update and Recover 系统路由修改和恢复

**SRV** Safety Relief Valve 安全释压活门

Service 维修,勤务,服务

**SRVL** Survival 残存,幸存

**SRWS** Stimulated Rayleigh-Wing Scattering 受激瑞利线谱侧面散射

**SRY** Secondary 次要的

Sorry 抱歉

**SRZ** Special Rules Zone 特别规则区,特别规定区

Surveillance Radar Zone 监视雷达区

**SS** Sales Support 销售支援

Same Size 同等尺寸

Sandstorm 沙暴

Satellite Switch 星上交换

Seats Sold 座位已售出

Sector Scan 扇区扫描

Security Subsystem 安全分系统

Semisteel 半钢的

Service Signal 服务信号

Session Service 会晤业务

Ship Service 装运服务

Signal Strength 信号强度

Signaling System 信令系统

Simultaneous Sampling 同步取样

Single Shot 单击,单稳触发器

Single Sideband 单边带

Single Sided 单边的

Sliding Scale 计算尺,滑动标尺

Slow Set 慢设定,慢调定

Solid State 固态

Solid-State Switch 固态电门

Space Sounding 空间探测

Space Station 空间站

Space Surveillance 空间监视

Space System 空间系统

Specification Standards 规格标准

Specification Statement 规范说明,说明语句

Speech Store 语音存储器

Spread Spectrum 扩频,发散光谱

Standard Sample 标准样品

Standard Scaling 标准刻度

Standard Scheme 标准规划

Standard Specification 标准规范

Star Sensor 星光传感器

Stationary Satellite 静止卫星

Stationary State 稳定态

Stick Shaker 抖杆器

Stock Size 库存量

Straight Shank 直柄

Strength Specification 强度规范

Structural Strength 结构强度

Subscriber Station 用户站

Subsystem 分系统,子系统

Subsystem Simulation 分系统模拟

Suggested Spares 建议的备件

Summing Selector 加法选择器

Sunset 日落

Supervision Station 监控站

Supplementary Service 辅助业务

Support Services 支援服务

Support Specification 支援规范

Switching Subsystem 交换子系统

Switching System 交换系统

Synchro Scope 同步示波器

System Schematics 系统简图(电路图)

System Sensitivity 系统灵敏度

System Software 系统软件

System Specification 系统规范,系统说明书

System Supervisor 系统管理程序

**SS&CR** Stress Standards and Certification Requirements 应力标准和审定要求

**SS&D** Synchronization Separator and Digitizer 同步分离器和数字转换器

**SSA** Safety and Security Audits Branch 安全和保安审计处

Sector Safety Altitude 扇形区安全高度

Segment Search Argument 段搜索变量

Source Selection Authority 资源选择当局

Spares Stores Area 备件存放区

Special Service Agreement 特别服务协定

Subsystem Allowed 允许的子系统

System Safety Analysis 系统安全分析

System Safety Assessment 系统安全评估

**SSAC** Shanghai Sikorsky Aircraft Company Limited 上海西科斯基飞机有限公司

Source Selection Advisory Council (资)源选择咨询委员会

**SSALF** Simplified Short Approach Light System with Sequenced Flashing Lights 有顺序闪光灯的简易短距进近灯光系统

**SSALR** Simplified Short Approach Light System with Runway Alignment Indicator Lights 带有对准跑道指示灯的简易短距进近灯光系统

**SSALS** Simplified Short Approach Light System 简易短距进近灯光系统

**SSALSRA** Simplified Short Approach Light System Runway Alignment 简化短距进近灯光系统跑道对准

**SSALT** Solid State Altimeter 固态高度表

**SSAMC** Sichuan Snecma Aero-engine Maintenance Co. , Ltd. 四川斯奈克玛航空发动机维修有限公司

**SSAP** Service Switch Access Point 业务交换接入点

Session Service Access Point 会晤业务接入点

Source Service Access Point 源服务存取点

**SSAR** Standard Safety Analysis Report 标准安全分析报告

System Safety Assessment Report 系统安全评估报告

**SSARC** Service System Acquisition Review Council 维修系统采集检查委员会

**SSB** Single Side Band 单边带

Source Selection Board 电源选择板

Split System Breaker 分离系统制动器

Super South Bridge 超级南桥芯片

Supplier Service Bulletin 供应方服务通告

Symmetric Switched Broadband 对称交换宽带

**SSBAR** Signal Stability-Based Adaptive Routing 基于信号稳定的适配选路

**SSBC** Single Sideband Circuit 单边带电路

Split System Breaker Close 分离系统跳开关闭合

**SSBJ** Supersonic Business Jet 超音速公务机

**SSBO** Single Swing Blocking Oscillator 单摆间歇振荡器

**SSB-OFS** Single-Side Band Optical Frequency Shifter 单边带光移频器

**SSBTR** Split System Breaker Trip 分离系统断路器跳闸

**SSC** Security Service Circuit 保密业务线路

Service Support Center 服务保障中心

Shipping and Storage Container 装运及存储箱

Side Stick Controller 座侧驾驶杆

Single Stroke Chime 单击谐音

Skill Specialty Code 技能专业码

Solid State Circuit 固态电路

Source Selection Criteria 源选择准则

Spread Spectrum Communication 扩频通信

Static Source Correction 静态源校正

System Simulator Complex 系统模拟器设备

System Supervisory and Control 系统监控

System Support Center 系统保障中心

**SSCF** Signal Strength Center Frequency 信号强度中心频率

**SSCH** Secondary Synchronization Channel 辅助同步信道

**SSCM** Spoilers and Stabilizer Control Module 扰流板和安定面操纵组件

**SSCMP** System Simulation Configuration Management Plan 系统模拟构型管理计划

**SSCN** Standard Specification Change Notice 标准规范更改通知

**SSCOP** Service Specific Connection Oriented Protocol 业务特定面向连接协议

**SSCP** Service Specific Convergence Protocol 业务特定会聚协议

Service Switching and Control Point 业务交换控制点

Simulation Support Computer Program 模拟支援计算机程序

System Service Control Point 系统服务控制点

**SSCS** Service Specific Convergence Sublayer 业务特定会聚子层

Strain Sensitive Cable Sensor 应力光敏电缆传感器

**SSCU** Special Signal Conditioning Unit 专用信号调整组件

Spoiler and Stabilizer Control Unit 扰流板和安定面操纵装置

**SSCVR** Solid State Cockpit Voice Recorder 固态驾驶舱话音记录器

**SSD** Seattle Services Division 西雅图维修分部

Service Support Data 业务支撑数据

Shared Security Data 共享保密数据

Solid State Detector 固态检波器,固态检测器

Solid State Disk 固态硬盘

Solid State Storage Device 固态存储器件

Status Source Document 状态源文件

Subsoil Drain 地漏

System Status Display 系统状态显示器

**SSDAR** Security Subsystem Design Analysis Report 安全分系统设计分析报告

**SSDDR** Solid State Digital Data Recorders 固态数字数据记录器

**SSDL** Space Systems Design Lab 航天系统设计实验室

Space Systems Development Laboratory 航天系统开发实验室

**SSDM** Systematic Software Development and Maintenance 系统软件研制和维护

**SSDU** Session Service Data Unit 会晤业务数据单元

**SSE** Sector Scan Enable 扇区扫描允许

Simulation Support Environment 模拟保障环境

Solid State Electronics 固体电子学

South-Southeast 南东南,南东南的

Special Support Equipment 专用保障设备

Store Safety Equipment 仓库安全设备

Surface Support Equipment 地面保障设备

System Safety Effectiveness 系统安全有效度

System Safety Engineering 系统安全工程

System Supervisory Equipment 系统监控设备

System Support Engineering 系统保障工程

System Support Equipment 系统保障设备

System Surveillance Equipment 系统监视设备

**SSEB** Source Selection Evaluation Board 资源选择评估委员会

**SSEC** Static Source Error Correction 静压源误差修正

**SSECT** Subsection 分部分,分节

**SSEM** Subsystem Evaluation Model 分系统评估模型

**SSEMP** Security System Engineering Management Plan 安全系统工程管理计划

**SSEP** System Safety Engineering Plan 系统安全工程计划

System Security Engineering Program 系统安全工程程序

**SSER** System Safety Engineering Report 系统安全工程报告

**SSES** Software Specification and Evaluation System 软件规范与评价系统

**SSET** Support Suitability Evaluation Team 保障适合性评价组

**SSF** Saybolt Seconds Furol 赛波特粘度计

Service Switching Function 业务交换功能

Single Sideband Filter 单边带滤波器

Software Support Facility 软件支援设施

Sub-Service Field 子业务字段

Super-Saver Fares 超级省钱票价

System Shop Faults 系统车间故障

System Support Facility 系统保障设施

**SSFD** Sensor Signal Fault Detection 传感器信号故障检测

Signal Selection Failure Detection 信号选择失效检测

Signal Selection Fault Detection 信号选择故障检测

Source Selection Fault Detection 信号源选择故障探测

**SSFDC** Solid State Floppy Disk Card 固态软盘卡

**SSFDR** Solid State Flight Data Recorder 固态飞行数据记录器

**SSFL** Steady State Fermi Level 稳态费米能级

**SSFM** Single Sideband Frequency Modulation 单边带频率调制

**SSG** Malabo, Equatorial Guinea-Santa Isabel (Airport Code) 赤道几内亚马拉博圣伊萨贝尔机场

Scientific Steering Group 科学指导小组

Search Signal Generator 搜索信号发生器

Standard Signal Generator 标准信号产生器

Strategic Studies Group 战略研究小组

Sweep Signal Generator 扫描信号产生器

System Safety Group 系统安全组

**SSGP** Simulation Software Generation Program 模拟软件产生程序

**SSGS** Standard Space Guidance System 标准空间引导系统

**SSH** Secure Shell 安全外壳

Sharm El Sheikh, Egypt (Airport Code) 埃

及沙姆沙伊赫机场

System Safety Handbook 系统安全手册

**SSHA** Space System Hazard Analysis 空间系统风险分析

Subsystem Hazard Analysis 子系统风险分析

System Safety Hazard Analysis 系统安全风险分析

**SSI** Sector Scan Indicator 扇区扫描显示器

Server Side Include 服务器端包含

Small Scale Integrated Circuit 小规模集成电路

Small Scale Integration 小规模集成

Special Structural Inspections 特殊结构检查

Standard Scale Integration 标准规模集成

Structural Significant Item 重要结构项目

Structure Strength Institute 结构强度研究所

Synchronous System Interface 同步系统接口

System Status Index 系统状态索引

**SSID** Service Set Identifier 服务集标志符

Standard Simulation Interface Design 标准模拟接口设计

Supplemental Structural Inspection Document 结构检查补充文件

Supplemental Structural Integrity Document 补充结构完整文件

**SSIG** Single Signal 单信号

**SSII** Solid State Image Intensifier 固态图像增强器

**SSIL** Solid State Injection Laser 固态注入式激光器

**SSIM** Standard Schedules Information Manual 标准时间信息手册

**SSIP** Supplemental Structural Inspection Program 补充结构检查大纲

**SSIU** Subsystem Interface Unit 分系统接口组件

**SSJ** Self Screening Jamming 自滤除干扰

**SSL** Secure Socket Layer 安全套接层

Segregated Storage Location 分类存储位置

Software Science Limited 软件科学有限公司

Software Specification Language 软件规范语言

Solid State Laboratory 固态实验室

Solid State Laser 固态激光器

Space Science Laboratory 空间科学实验室

Synchronization Sub-Layer 同步子层

System Service Layer 系统业务层

System Support Laboratory 系统保障实验室

**SSLL** Single Stream with Low Latency 低延迟的单独数据流传输

**SSLS** Solid State Laser System 固态激光系统

**SSLT** Starboard Side Light 右舷侧灯

**SSM** Service Session Manager 业务会晤管理程序

Servo System Monitor 伺服系统监视器

Sign Status Matrix 符号状态矩阵

Single Sideband Modulation 单边带调制

Solid State Materials 固态材料

Solid State Memory 固态存储器

Subsystem Select Matrix 分系统选择矩阵

Synchronization Status Message 同步状态信息

Synchronization Supervisor Module 同步监控模块

System Schematics Manual 系统电路图手册,系统原理图手册

System State Model 系统状态模型

System Support Manager 系统支援经理

System Support Module 系统支持模块

**SSMA** Satellite Switched Multiple Access 卫星转换多路存取

Solid-State Microwave Amplifier 固态微波放大器

Spreaded Spectrum Multiple Access 扩频多址

**SSME** Space Shuttle Main Engine 航天飞机主发动机

**SSMF** Signaling System, Multi-Frequency 多频信令系统

Standard Single Mode Fiber 标准单模光纤

**SSMM** Solid State Mass Memory 固态大容量存储器

**SSMP** System Safety Management Plan 系统安全管理计划

System Security Management Program 系统安全管理程序

**SSMTF** Space Station Mockup and Training Facility 空间站模型和训练设备

**SSN** Single Star Network 单星网络

Standard Shortage Notice 标准短缺通告

Stores Shortage Notice 库存短缺通告

Subscriber Switching Network 用户交换网络

Subsystem Number 子系统号码

Switched Service Network 交换业务网

Switched Star Network 交换式星形网络

**SSO** Steady State Oscillation 稳态振荡

**SSOF** Self-Supporting Optical Fiber Cable 自承光缆

**SSOG** Satellite System Operations Guide 卫星系统操作指南

**SSOP** Security Standard Operating Procedures 安全标准操作程序

**SSP** Satellite Service Provider 卫星服务提供商

Sensor Simulation Processor 传感器模拟处理器

Service Specific Part 业务特定部分

Service Switching Point 业务交换点

Signaling Switching Processor 信令转接处理器

Significant in-Service Problems 维修中重大问题

Source Selection Plan 供货源选择计划

Space Solar Power 空间太阳能

State Safety Programme 国家安全大纲,国家安全方案

System Safety Plan 系统安全计划

System Safety Program 系统安全程序

System Specification 系统规范

**SSPA** Solid State Phased Array 固态相控阵

Solid State Power Amplifier 固态功率放大器

**SSPB** Switching Service Processing Block 交换业务处理功能块

**SSPC** Solid State Power Contactor 固态电源接触器

Solid State Power Controller 固态电源控制器

**SSPD** Single Source of Product Data 单一产品数据源

**SSPM** Software Standards and Procedure Manual 软件标准及程序手册

**SSPP** Sustainability: Science, Practice and Policy 可持续发展:科学、务实和政策

System Safety Program Plan 系统安全项目计划

**SSPS** Satellite Solar Power Station 卫星太阳能电站

**SSQA** Sales and Service Quality Assurance 销售及服务质量保证

Standardized Supplier Quality Assessment 供应商质量评价标准

**SSR** Satellite Searching Radar 卫星搜索雷达

Secondary Surveillance Radar 二次监视雷达

Sector Scan Reverse 扇区扫描反转

Seek-Storm Radar 风暴探测雷达

Software Specification Review 软件规范审查

Software Status Review 软件状态检查

Solid State Relay 固态继电器

Source Select Relay 源选择继电器

Special Service Requirement 特殊服务要求

Standards Substitution Request 标准替代请求

Surface Sample Return 地面取样回收

Surface Search Radar 地面搜索雷达

System Safety Requirement 系统安全要求

System Safety Review 系统安全评价

**SSR RPG** SSR Regional Planning Group 二次监视雷达区域规划组

**SSRAM** System Safety, Reliability, Availability and Maintainability 系统安全性、可靠性、可供性与可维修性

**SSRD** Solid State Recording Device 固态记录装置

System Software Requirements Document 系统软件需求文件

**SSRO** Sector Scan Receive Only 仅用于扇区扫描接收

System Safety and Reliability Office 系统安全与可靠性办公室

**SSRR** Software System Requirements Review 软件系统需求审查

Supplier System Requirements Review 供应方系统需求审查

**SSRS** Start-Stop-Restart System 启动—停止—再启动系统

Supplemental Stall Recognition System 辅助失速识别系统

System Security Requirements Specification 系统安全需求规范

**SSRU** Sub-Shop Replaceable Unit 辅助车间可更换件

**SSS** Seattle Support Services 西雅图支援服务部

Silicon Symmetrical Switch 对称硅开关

Solid State Switch 固态开关

Standard Substitution 标准替代

Standard, Substitution and Supercede 标准、替代与置换

Strategic Satellite System 战略卫星系统

Subjective Sleepiness Scale 主观困倦量表

Subscriber Service System 用户业务系统

Switched Star System 交换式星形系统

Switching Subsystem 交换子系统

System Safety Society 系统安全学会(美)

System Security Standard 系统安全标准

System Segment Specification 系统分部规范

System Simulation Specification 系统模拟规范

System Simulator Software 系统模拟器软件

**SSSC** Single Sideband Suppressed Carrier 单边带抑制载波

Software Support Service Contractor 软件支援服务合同方

**SSSL** Scientific Systems Software Library 科学系统软件库

Subsonic Straight and Level 亚音速直线平飞

**SSSP** System Source Selection Procedure 系统资源选择程序

**SSSV** Solid State Stored Voice 固态存储的话音

Static Source Selector Valve 静压源选择活门

**SST** Sea Surface Temperature 海平面温度

Software System Test 软件系统测试

Soliton-Supported Transmission 孤子支持的传输

Static Structural Test 结构静态试验

Super Schottky Technology 超肖特基技术

Supersonic Transport 超音速运输机

System Self Test 系统自测试

**SS-TDMA** Satellite Switched-Time Division Multiple Access 卫星交换时分多址

**SSTO** Single Stage to Orbit 单节进轨道

**SSTOL** Super Short Take-off and Landing 超短距起飞和着陆

**SSTP** Satellite Switch Time Plan 卫星交换时间计划

System Simulation Test Plan 系统模拟试验大纲(计划)

Technical Panel on Supersonic Transport Operations 超音速运输机运行技术专家组

**SSTS** Space Surveillance and Tracking System 空间监视与跟踪系统

**SSTTP** Safeguard System Test Target Program 保卫系统测试目标程序

**SSTU** Seamless Steel Tubing 无缝钢管

Side Stick Transducer Unit 操纵杆传感器组件

**SSTV** Slow Scan Television 慢扫描电视

**SSU** Saybolt Seconds Universal 赛波特通用秒

Shared Storage Unit 共享存储器板

Single Subscriber Unit 单用户单元

Space Switching Unit 空分交换单元

Subsequent Signal Unit 辅助顺序信号装置

Subsystem Unit 分系统组件

**SSUM** System Supervision Module 系统监理

模块

**SSUS**　Spin Stabilized Upper Stage 旋转稳定上级

**SSV**　Source Select Valve 源选择活门

　　Special Shop Visit 特殊车间检查

**SSVP**　System Simulation Validation Plan 系统模拟确认大纲

**SSVT**　Software System Validation Test 软件系统批准测试

**SSW**　South-South-West 西南南

　　Synchro Switch 同步开关

**SSWG**　Software and Standards Working Group 软件和标准工作组

　　System Safety Working Group 系统安全工作组

**SSZ**　Santos, Sao Paulo, Brazil( Airport Code) 巴西桑托斯机场

　　Subsonic Zone 亚音速区

　　Subsonic Zoom 亚音速跃升

**ST**　Saw Tooth 锯齿

　　Schmitt Trigger 施密特触发器

　　Schuler Tuning 舒勒调谐

　　Search and Track 搜索与跟踪

　　Self Test 自测试

　　Self Trapping 自吸收

　　Set 一组,装置,设定,调节

　　Sidetone 侧音

　　Signaling Terminal 信令终端

　　Single Tire 单轮胎

　　Single Turn 单转

　　Single-Throw 单刀

　　Sounding Tube 探测管,测深管

　　Space Telescope 空间望远镜

　　Special Test 特种试验,专门试验

　　Special Tooling 特种工装

　　Spill Traffic 运量溢出

　　Standard Temperature 标准温度

　　Standard Test 标准试验,标准测试

　　Standard Time 标准时间

　　Standard Tolerance 标准公差

　　Standard Tool 标准工具

　　Start 开始,启动

　　Static Test 静态试验

　　Status 状态

　　Statute 法规

　　Steam 蒸汽

　　Store 仓库,存储器,商店,存储

　　Stratus 层云

　　Structural Test 结构试验

　　Synchro Transmitter 同步发送器,同步发射机

　　United Nations Secretariat 联合国秘书处

**ST LT**　Stern Light 尾灯

**ST PR**　Static Pressure 静压

**ST&SP**　Start and Stop 开始与终止,启动与停止

**STA**　Scheduled Time of Arrival 班机到港时间,预计到港时间

　　Service Trigger Agent 业务触发代理

　　Special Test Authorization 特种试验批准

　　Stationary 静止的

　　Straight-in Approach 直接进近, 直接进近着陆

　　Structural Test Article 结构测试条款

　　Supersonic Transport Aircraft 超音速运输机

　　Supplementary Travel Allowance 补助交通津贴

　　Supplementary Type Approval 补充型号批准书

　　Support Transport Aircraft 支援运输机

　　System Technology Application 系统工艺应用

**STAA**　Stabilizer Trim Actuator Assembly 安定面配平作动器组件

**STAB**　Stabilization in Roll, Pitch, Yaw 横滚,俯仰,偏航稳定

　　Stabilizer 安定装置,安定翼,水平安定面

**STABPI**　Stabilizer Position Indicator 安定面位置指示器

**STAC**　Supersonic Transport Aircraft Commission( 英国) 超音速运输机研究委员会

**STADAN**　Space Tracking and Data Acquisition Network 空间跟踪及数据采集网

**STAE**　Second Time around Echo 二次回波

**STAF**　Sales Tracking and Forecasting 销售跟踪及预测

Stealthy Terrain Avoidance Following 隐形地形回避跟随

**STAFS**　Stealthy Terrain Avoidance Following System 隐形地形回避跟随系统

**STAG**　Shuttle Turnaround Assessment Group 航天飞机检修鉴定组

Subsystem Test Adaptor Group 分系统测试接合器组

**STAHR**　Socio-Technical Approach to Assessing Human Reliability 人因可靠性社会技术评估方法

**STAL**　Station Line 站位线

**STALO**　Stable Local Oscillator 稳定本机振荡器

**STAMO**　Stabilized Master Oscillator 稳定主振荡器

**STAMP**　Small Tactical Aerial Mobility Platform 小型战术空中机动平台

**STAN**　Stanchion 支柱

**STANAG**　Standardization Agreement 标准化协议

**STANO**　Surveillance, Target Acquisition and Night Observation 监视、目标搜索和夜间观察

**STAP**　Scientific and Technical Advisory Panel 科学和技术咨询小组

Special Technical Assistance Program 特别技术援助计划

Standard Technical Avionics Package 航空电子标准技术成套装置

**STAR**　Scientific and Technical Aerospace Reports 航空航天科技报告,宇航科技报告

Security Threat Analysis and Research 安全威胁分析与研究

Shuttle Turnaround Assessment Report 航天飞机检修鉴定报告

Simultaneous Transmission and Reception 同

步发送与接收

Space Thermionic Auxiliary Reactor 空间热离子辅助反应堆

Standard Arrival Procedure 标准进场程序

Standard Instrument Arrival Route 标准仪表进场航线

Standard Terminal Arrival Route 标准终端进场航线,标准终端进场航路

Standards and Technology Annual Report 标准和技术年度报告

Stealthy Tracking and Ranging 隐形跟踪和测距

System Test Action Request 系统测试操作要求

System Threat Analysis Report 系统威胁分析报告

System Threat Assessment Report 系统威胁评估报告

**STARCO**　Shanghai Technologies Aerospace Company Limited 上海科技宇航有限公司

**STARS**　SIL Telemetry Acquisition and Reduction System 传感器集成实验室遥测搜索和衰减系统

Standard Terminal Arrival Route System 标准终端进场航路系统

Standard Terminal Automation Replacement System 标准终端区自动化替代系统

Surveillance Target Attack Radar System 目标监视攻击雷达系统

**STARs**　Standard Terminal Arrival Routes 标准终端进场航路

**STARTG**　Starting 启动,点火

**STAS**　Supplier Tool Accountability System 供应方工具责任系统

**STAT**　Photostat 照像复制,直接影印(机)

Static 静止

Station 站,位置,(导航)台

Statistic(al) 统计的

Statistical Division 统计司

Status 状态

**STB**　Set Top Box 机顶盒

Stabilization 安定性,稳定
Stabilizer 安定面
Stable 稳定的
Start Transient Bleed 启动瞬时放气
System Test Bench 系统测试台
**STBD** Starboard 朝向机头的右侧
**STBSCP** Stroboscope 闪频仪
**STBY** Standby 备用
**STC** Saint Cloud, MN, USA ( Airport Code )
美国明尼苏达州圣克劳德机场
Satellite Test Center 卫星测试中心
Science and Technology Committee 科学技术
委员会
Secure Transaction Channel 安全事务信道
Self Test Complete 自测完成
Sensitivity Time Control 灵敏度时间控制
Sound Transmission Class 声音传输等级
Space Technology Center 空间技术中心
Standard Transmission Code 标准传输码
State Technical Committee 国家技术委员会
Static 静态(的),静止(的)
Step Counter 步进计数器
Sufficient to Complete 足以完成
Supplemental Type Certificate 补充型号合格
证
Switching and Testing Center 交换和测试中
心
System Technique Center 系统技术中心
System Test Complex 系统测试设备
System Test Console 系统测试台
System Timing Controller 系统定时控制器
**STCA** Short Term Conflict Alert 短期冲突警
告
**STCM** Stabilizer Trim Control Module 安定面
配平控制组件
Systematic Trellic-Coded Modulation 系统格
形编码调制
**STCR** Stretcher Passenger 担架旅客
**STCW** System Time Code Word 系统时间编
码字
**STD** Scheduled Time of Departure 预计离开

时间,预计离港时间
Science, Technology and Development Organi-
zation 科学、技术与开发组织
Standard 标准,标准的
Standard Deviation 标准偏差
Subscriber Trunk Dialing 用户中继线拨号
Supplementary Take off Distance 辅助起飞
距离
Synchronous Time Division 同步时分
System Task Directory 系统任务目录
System Time Delay 系统时间延迟
**STD PAR** Standard Parallel 标准纬线
**STD/R** Standard Repair 标准修理
**STDD** Shared Time Division Duplex 共享时
分双工
Statistical Time Division Duplex 统计时分复
用
**STDE** Sub-Task Description Entry 子任务说
明输入
**STDF** Standoff 偏离预定航线
**STDLVY** Stores Delivery 储备运输
**STDM** Statistical Time Division Multiplexer
统计时分复用器
Statistical Time Division Multiplexing 统计时
分复用
Synchronous Time Division Multiplexer 同步
时分复用器
Synchronous Time Division Multiplexing 同步
时分复用
**STDMA** Self-organizing Time Division Multi-
ple Access ( Datalink ) 自组织时分多址
Space Time Division Multiple Access 空间时
分多址连接
**STDN** Spaceflight Tracking and Data Network
空间飞行跟踪与数据网
Standardization 标准化
**STDR** Superseding Tool Design Request 替代
工具设计要求
**STD** Standard Traffic Documents 标准运输票
证,标准运输凭证
**STDS** Standards 标准,规范

STDY Steady 稳定的,稳固的

STDZN Standardization 标准化

STE Service Testing Equipment 业务测试设备

Signaling Terminal 信令终端

Software Test Engineer 软件测试工程师

Software Test Environment 软件测试环境

Special Test Equipment 特种测试设备

Special Tool Equipment 特种工具设备

Standard Test Equipment 软件测试设备

Subscriber Terminal Equipment 用户终端设备

Subscriber Test Equipment 用户测试设备

Support Test Equipment 辅助试验设备

System Terminal Equipment 系统终端设备

System Test and Evaluation 系统测试与评估

System Test Engineer 系统测试工程师

System Test Equipment 系统测试设备

STEA System Test Equipment Assembly 系统测试设备组装

System Test, Evaluation and Assembly 系统测试、鉴定与组装

STEADES Safety Trend Evaluation, Analysis and Data Exchange System 安全趋势评价、分析与数据交换系统

STELA Simulator Test Language 模拟机测试语言

STEM Scanning Transmission Electron Microscope 扫描透射电子显微镜

Storable Tubular Extendible Member 可收藏管状可延伸部件

System Training and Exercise Module 系统培训和练习模块,系统培训和练习模件

STEN Status Enable 状态允许

Stencil 模板

STENS Standard Terrestrial Navigation System 标准地面导航系统

STEP Selector Stepping 步进式选择器

Service Test and Evaluation Program 服务试验和评估项目

Skills Training and Employment Program 技能培训和就业计划

Standard for the Exchange of Product Model Data 产品模型数据变换标准

Standard Test Equipment Procedure 标准试验设备程序

Strategic Technology Expansion Project 战略技术扩展项目

STER Seater 有…座位的交通工具

Sterilizer 消毒器

STEREO Stereophonic 立体声的

STF Satellite Tracking Facility 卫星跟踪设施

Sensor Track Function 敏感器跟踪功能

Signal Tracking Filter 信号跟踪滤波器

Signal Transfer Function 信号传递函数

Soliton Transmission Fiber 孤子传输光纤

Space Tracking Facility 空间跟踪设施

Staff 杆,标尺,职员,参谋

Stratiform 层状云

STF/TA Stealthy Terrain Following and Terrain Avoidance 隐形地貌跟随和地形回避

STFC Signaling Traffice Flow Control 信号业务流量控制

STFG Stuffing 填料

STFT Short Term Fuel Trim 短期燃油修正

Short-Time Fourier Transform 短时傅立叶变换

Stray Field Test 杂散场试验

STG Stage 阶段,级,台阶

Standard Glass 标准玻璃

Standard Taper Gauge 标准锥度规

Standard Target Generator 标准目标产生器

Standard Thread Gauge 标准螺纹规,标准线规

Standards Task Group 标准任务组

Standards Technology Group 科学与技术组

Synchronous Timing Generator 同步定时发生器

STG LT Steering Light 转弯灯

STGHED Special Tool and Ground Handling Equipment Drawings 特殊工装、地面搬运设备图纸

STGR Stringer 桁条,纵梁,加强肋

**STH** Sidewall True Horizontal 侧壁真水平

**STI** Science, Technology and Innovation 科学、技术和创新

Scientific and Technical Information 科技信息

Service and Taxes Included 包括维修费和租税

Simulation Test Interface 模拟测试接口

Space Technology Institute 空间技术研究所

Special Technical Instruction 专用技术说明书

Special Test Instruction 特殊测试指令

Special Traffic Instructions 特殊交通说明

Standard Tool Inventories 标准工具清单

Statistics Time Interval 统计时间间隔

**STIC** Scientific and Technical Information Center 科技信息(情报)中心

**STIDC** Scientific and Technical Information and Documentation Committee 科学和技术资料委员会

**STIF** Stiffener 加强板,刚性构件

**STIM** Stimulus 激励

**ST-IN** Straight-in 直线进近着陆

**STINFO** Scientific and Technical Information 科技信息

**STIR** Scientific and Technical Intelligence Review 科学与技术情报评审

Separate Track and Illumination Radar 独立跟踪和照明雷达

Short Term Interest Rate 短期利率

Sound Transimission Using Infrared Radiation 使用红外的声音传输

Specific Threat Intelligence Request 指定威胁情报要求

Stirrup 箍筋,夹头

Supplemental Type Inspection Report 补充型号检查报告

Surveillance and Target Indication Radar 监视与目标指示雷达

Surveillance Target Indicator Radar 监视目标指示雷达

**STIS** Standard Tool Index System 标准工具索引系统

**STIU** Satellite Telecommunications Interface Unit 卫星通信接口组件

Stabilizer Trim Interface Unit 安定面配平接口组件

**STK** Stack 堆栈,堆层

Standard Test Key 标准测试纲要

Stock 库存,股票,毛坯,基台

Strake 铁箍,底板

**STKF** Stock Found 库存基金,股票基金

**STL** Space Technology Laboratories, Inc. 空间技术实验室公司

St Louis, MO, USA ( Airport Code ) 美国密苏里州圣路易斯机场

Storage Life 储存寿命

**STLDD** Software Top Level Design Document 软件顶层设计文件

**STLI** Streams Transport Layer Interface 流传输层接口

**STLM** Stabilizer Trim Limit Switch and Position Transmitter Module 安定面配平限制电门和位置传感器组件

**STLP** Simulated Training Launch Payload 模拟训练发射业载

**STLR** Semitrailer 半拖车

**STLS** Standard Tool Location System 标准工具位置系统

**STLT** Satellite 卫星

**STM** Section Traffic Manager 区域交通经理

Selective Traffic Management 选择性业务管理

Signaling Traffic Management 信令业务管理

Stores Management 备件(储备)管理

Subcontract Technical Management 分合同技术管理

Switching-Type Modulator 开关式调制器

Synchronous Transfer Mode 同步转移模式

Synchronous Transport Module 同步传输模块

System Test Mode 系统试验方式

System Test Module 系统测试组件

System Training Mission 系统训练任务

**STMG** Steaming 蒸汽处理,汽化

**STML** Stimulate 刺激,激发

Synchronous Transport Module Level 同步传送模式等级

**STMT** Statement 声明,通告

**STMU** Special Test and Maintenance Unit 专用试验和维护设备

**STN** London, England, United Kingdom-Stansted (Airport Code) 英国伦敦斯坦斯德机场

Satellite Tracking Network 卫星跟踪网

Station 台,站

Stone 石

Switched Telecommunications Network 远程通信交换网络

Switched Telephone Network 电话交换网络

**STNG** Sustaining 支撑,保持

**STNLS** Stainless 不锈的

**STNR** Stationary 固定的

**STO** Short Take-off 短距起飞

Standard Tool Order 标准工具订单

Standing Tool Order 长期工具订单

Stockholm, Sweden (Airport Code) 瑞典斯德哥尔摩机场

Store 仓库,存储器,储备

Strategic Technology Office 战略技术处

**STOAL** Short Takeoff Arrested Landing 短距起飞阻拦索绊机着陆

**STODA** Supplementary Take-off Distance Available 辅助起飞可用距离

**STOGW** Short Takeoff Gross Weight 短距起飞全重

**STOL** Short Takeoff and Landing 短距起飞和着陆

**STOLA** Short Takeoff and Landing Airport 短距起飞和着陆机场

**STOLAND** STOL Navigation and Landing System 短距起飞导航与着陆系统

**STOLAS** Short Take-off and Landing Avionics

System 短距起飞和着陆电子系统

**STOLP** Short Take-off and Landing Plane 短距起飞和着陆飞机

**STOM** System Test and Operation Manual 系统测试和操作手册

**STOR** Storage 贮藏

System Test and Operation Report 系统试验和运行报告

**STORS** Stock Order Replenishment on-line System 库存订单补充联机系统

**STOVL** Short Take off and Vertical Landing 短距起飞与垂直着陆

**STOW** System for Take-off Weight 起飞重量测定系统

**STP** Satellite Test Program 卫星试验计划

Satellite Tracking Program 卫星跟踪程序

Seattle Test Program 西雅图测试程序

Selective Tape Print 选择性带式打印

Self Test Program 自测试程序

Shielded Twisted Pair 屏蔽双绞线

Signal Transfer Point 信号转送点

Signaling Transfer Point 信令转接点

Situation Target Proposal 状态目标建议

Solar-Terrestrial Physics 日地物理学

Space Test Program 空间试验程序

Spanning Tree Protocol 生成树协议

Stamp 模具,图章,邮票

Standard Telephone Part 标准电话用户

Standard Temperature and Pressure 标准温度和压力

Standard Test Procedure 标准试验程序

Standard Test Program 标准试验计划

Standardized Test Program 标准化测试程序

Standardized Training Packages 标准成套培训资料,标准训练设备

Standards Technical Panel 标准技术小组

Steer Point 控制点,导引点

Stop 停止,句号

Strip 条,带,剥去,拆卸

Subsystem Test Plan 分系统测试计划

System Test Plan 系统测试计划

System Test Procedure 系统试验程序
System Training Program 系统训练计划
Systems Technology Program 系统工艺程序
**STPA** Standard Temperature and Pressure-
Absolute 标准温度和压力—绝对
**STPDN** Step Down 辞职
**STPG** Stepping 步进
Stripping 拆卸
**STPNG** Stopping 停止
**STPR** Security Testing Procedural Review 安
全测试程序审核
Short Taper 短形锥
Software Test Procedures 软件测试程序
Stripper 脱模机
**STPS** Short Time Power Spectrogram 短时功
率谱
Shuttle Tracking Processor System 航天飞机
跟踪处理器系统
Subject to Prior Sale 需预售的
**STR** Scientific and Technological Research 科
学技术研究
Scientific Technical Report 科技报告
Search and Tracking Radar 搜索跟踪雷达
Service Trouble Report 维修故障报告
Simulation Test Report 模拟试验报告
Software Test Report 软件测试报告
Software Trouble Report 软件故障报告
State-and-Time-dependent Routing 状态与时
变路由
Store 仓库,存储器,储存
Stored or Storing 储存的
Strainer 滤网,筛网
Strength 强度
Stringer 纵梁,纵肋
Stripping 拆卸,脱模,冲洗
Structural 结构的
Stuttgart, Germany(Airport Code) 德国斯图
加特机场
Summary Technical Report 技术报告摘要
Sustained Turn Rate 持续转弯角速度
Synchronous Transfer Region 同步转移区

Synchronous Transmit and Receive 同步发送
和接收
Synchronous Transmitter Receiver 同步收发
机
**STRAAD** Special Techniques for the Repair
and Analysis of Aircraft Damage 飞机损坏修
理和分析专业技术
**STRATBULK** Strategic Planning Course for
the Bulk Sector 散装运输部门战略计划课程
**STRATO** Stratosphere 平流层
**STRATSHIP** Strategic Planning Workshops
for Senior Shipping Management 高级航运管
理战略计划讲习班
**STRAW** Simultaneous Tape Read and Write
磁带同步读写
**STRBK** Strongback 强支撑
**STRD** Stranded 绞合的,搁浅的
**STRG** Steering 转弯,操纵
Stringer 纵梁
Strong 强的
**STRIP** Statistical Tabulation and Report In-
quiry Program 统计列表和报告查询程序
**STRL** Structural 结构的
**STRLN** Streamline 流线型
**STRM** Storeroom 储藏室
**STRN** Straighten 拉直,伸直
**STRP** Strapped 搭接的,捆扎的
**STRS** Special Tooling Report System 专用工
具报告系统
System Test Rig Specification 系统试验台架
规范
**STRTR** Starter 起动机(器)
**STRUC** Structure 结构
**STRUCT** Structural 结构的
**STS** Alignment Status 校准状态
Santa Rosa, CA, USA(Airport Code) 美国
加利福尼亚州圣罗莎机场
Satellite Telecommunication System 卫星通
信系统
Self Test Source 自测试源
Severe Tropical Storm 热带强风暴

Software Test System 软件测试系统

Space Technology Satellite 空间技术卫星

Space Transportation System 空间运输系统

Spacecraft Telecommunication System 宇宙飞船通信系统

Spares Technical Specification 备件技术规范

Special Test System 专用测试系统

Special Treatment Steel 特殊处理钢

Specialty Training Standard 专业训练标准

Speed Trim System 速度配平系统

Status 状况,地位,状态

Strasto-spheric Telecommunications Service 同温层通信业务

Subscriber Transfer Service 用户转接业务

Surface Transportation System 地面运输系统

Synchronous Transfer Signal 同步转移信号

Synchronous Transport Signal 同步传送信号

System Status 系统状况

System Technical Specification 系统技术规范

System Test Set 系统测试仪

**STSA** State Technology Supervision Administration 国家技术监督局

**STSC** Software Technology Support Center 软件技术支持中心

**STSEOS** Space Transportation System Engineering Operations Support 空间运输系统工程运行支援

**STSH** Stabilized Shunt 稳定分路

**STSS** Solar Terrestrial Subsatellite 太阳地球子卫星

**STT** Set-Top Terminal 机顶终端

Single Target Track/Tracking 单目标跟踪

St Thomas Island, VI, USA(Airport Code) 美属维尔京群岛圣托马斯机场

**STTE** Special to Type Test Equipment 型号专用测试设备

**STTL** Standard Schottky Transistor Logic 标准肖特基晶体管逻辑

**STTM** Short Term Trading Market 短期贸易市场

Short-Term Traffic Management Scheme 短期交通管理计划

**STTR** Single Target Tracking Radar 单目标跟踪雷达

Stator 静子

**STU** Secure Terminal Unit 安全终端组件

Set-Top Unit 机顶装置

Short Ton Units(Tungsten) 短吨度

Signaling Terminal Unit 信令终端单元

Special Test Unit 专用测试组件

Statistical Unit 统计单元

Step up 分级爬升,分级升压

Storage Unit 存储单元

System Timing Unit 系统定时组件

**STUBR** Utility Bus Relay Status 通用汇流条继电器状态

**STU-F** Speech Transcoding Unit-Full Rate 语音变换编码单元—全速

**STU-H** Speech Transcoding Unit-Half Rate 语音变换编码单元—半速

**STV** Structure Test Vehicle 结构强度试验飞行器

**STVR** Stopover(Reservations Code) 在机票允许之约定经停地停留

**STVS** Small Tower Voice Switch 小型塔台话音交换

**STW** Solid Tap Wrench 固定丝锥扳手

Storm Water 暴雨降水

**STWD** Steward 乘务员,管理员

**STWG** Stowage 储存

**STWL** Stopway Light(s) 停止道灯

**STWY** Stairway 梯子,楼梯

**STX** Data to Follow 跟随数据

Start of Text 文本起始

Start Transmission 开始发射

**STY** Severe Typhoon 强台风

**STYPPLE** Sensitive Typical Phone Power Level 敏感的典型电话功率电平

**SU** Sampling Unit 取样单位,取样单元

Selectable Unit 可选组件,可选单元

Sensing Unit 传感器,敏感元件

Sensitivity Unit 灵敏度单位

Service Usage 业务用途

Set up 装配,建立

Sign up 签约

Signal Units 信号装置

Signaling Unit 信令单元

Signature Unit 签名单元

Sonics and Ultrasonic 声学与超声学

Standard Upkeep 标准维护

Station Unit 站单元

Subscriber Unit 用户单元

Supervision Unit 监控单元

Supply Unit 供应单位

Support Unit 支撑单元

Switching Unit 转换组件

**SUA** Special Use Airspace 特殊用途空域

**SUB** Subcontractor 分包方

Subscriber 用户,订户,签署者

Substitution 替代,取代,代理

Surabaya, Indonesia (Airport Code) 印度尼西亚泗水机场

**SUBASSY** Subassembly 分组件

**SUBCA** Subscriber Carrier 用户载波

**SUBCAL** Subcaliber 次口径的

**SUBF** Subframe 子帧

**SUBJ** Subject 主题,科目

Subject to 受制于

Subjective 主观的

**SUBMIN** Subminiature 超小型(的)

**SUBMUX** Submultiplexer 分多路转换器

**SUBOR** Subordinate 辅助的,次要的

**SUBQ** Subsequent 后来的,连续的

**SUBROUT** Subroutine 子程序

**SUBS** Subsidiary 辅助物,辅助的

Substance 物质,附属的

**SUBSTA** Substation 分站

**SUBSTR** Substructure 子结构

**SUBSY** Subsystem 分系统,子系统

**SUBTR** Subtraction 减法,减去

**SUC** Succeeding 接连的,随后的

**SUCC** Successful 成功的

**SUCH** Speed-up Channel 加速信道

**SUCT** Suction 吸入

**SUD** Session User Data 会晤用户数据

Stretched Upper Deck 延长上层

**SUERM** Signal Unit Error Rate Monitor 信号单元错误率监视器

**SUF** Lamezia Terme, Italy (Airport Code) 意大利拉默齐亚机场

Sufficient 足够的

**SUFF** Suffix 词尾,后缀

**SUG** Subuser Group 子用户群

Suggest 建议

Suggestion 建议

Surigao, Philippines (Airport Code) 菲律宾苏里高机场

**SUI** Subuser Identification 子用户标识

**SUIT** Suitability 适合性

Suitable 适合的

**SUL** Speedup Logic 加速逻辑

**SUM** Software User Manual 软件用户手册

Solar Ultraviolet Monitor 太阳紫外监测仪

Summary 摘要,总结

Summer 夏季

Summing 总和

**SUMM** Semantic Unification Meta-Model 语义单一化中间模型

**SUMT** Sequential Unconstrained Minimization Technique 顺序非约束最小化技术

**SUN** Sun Valley, ID, USA (Airport Code) 美国爱达荷州太阳谷机场

Sunday 星期日

**SUNAMAM** Superintendencia Nacional da Marinha Mercante (Brazil) 国家航运管理局（巴西）

**SUP** Supplement 附加

Supply 提供

Support 支持

**SUPCUR** Superimposed Current 叠加电流

**SUPERHET** Superheterodyne 超外差式

**SUPERSTR** Superstructure 上层建筑,上层结构

**SUPF** Superframe 超帧

**SUPHTR** Superheater 过热器

**SUPL** Supplement or Supplementary 附录,副刊,补充的,增补的

**SUP-NUM** Supernumerary 定额的,多余的

**SUPP** Supply or Supplying or Supplied 供应(的)

**SUPP or SUPPL** Supplemental or Supplementary 附加的

**SUPP(L)** Supplement or Supplemental 补角,补充(的),增补(的)

**SUPPL** Supplier 供应方,供应商

**SUPPR** Suppress or Suppression 抑制,压制 Suppressor 阻尼器,抑制器

**SUPPS** Regional Supplementary Procedures 地区补充程序
Supplementary Procedure 补充程序
Supplements 附录,增刊

**SUPSD** Supersede or Superseded 替代(的)

**SUPSD BY** Superseded by 由…代替

**SUPSENS** Supersensitive 超敏感的

**SUPT** Support or Supporting 支援,支持

**SUPVR** Supervisor 督导

**SUPVS** Supervised or Supervising 检查(的),监督(的)

**SUR** Surge 喘振
Surveillance 监视

**SURE** Stock Usage Reporting 库存使用报告

**SURF** Surface 表面

**SURV** Surveillance 监视
Survey 观察,测绘,综述

**SURVSAT** Survivable Satellite 可存活卫星

**SURVSATCOM** Survivable Satellite Communications 可生存卫星通信

**SUS** Semiconductor Unilateral Switch 半导体单向开关
Silicon Unilateral Switch 硅单向开关
Suspend or Suspension 延缓(的),悬挂(的),中止(的)

**SUST** Sustaining 支撑

**SUT** Set up Template 装配模板

**SUV** Nausori, Suva, Fiji（Airport Code）斐济苏瓦那瑠索里机场

**SUW** Superior, WI, USA（Airport Code）美国威斯康星州苏必利尔机场

**SUX** Sioux City, IA, USA（Airport Code）美国依阿华州苏城机场

**SV** Safety Valve 安全活门(阀)
Secure Voice 保密话音
Self Ventilated 自通风的
Self Verification 自验证
Servo-Valve 伺服活门
Shuttle Valve 往复活门
Side View 侧视图
Solenoid Valve 电磁活门(阀)
Space Vehicle 航天器,空间运载体(卫星),空间飞行器
Specific Volume 比容
Specified Value 给定值
Speed Variation 速度变量
Speed Variator 变速器
Spherical Valve 球形阀
Spillover Valve 溢流活门
Standard Voltmeter 标准电压表
Static Voltmeter 静电电压表
Status Valid 有效状态
Stop Valve 断流活门

**SVA** Shared Virtual Area 共享虚存区
Stator Vane Actuator 定子导向叶片作动器
Stator Vane Angle 定子叶片角
Support for Video Application 视频应用支撑
System Verification Approach 系统验证方法

**SVALT** Servo Altimeter 伺服高度表

**SVBT** Space Vehicle Booster Test 宇宙飞行器助推器试验

**SVC** Service 服务,维修
Signaling Virtual Channel 信令虚信道
Supervisory Channel 监控信道
Switch Volume Control 音量控制电门
Switched Virtual Channel 交换虚通道
Switched Virtual Circuit 交换虚电路
Switched Virtual Connection 交换式虚连接
Switched/Switching Virtual Calling 交换式虚

呼叫

Switched/Switching Virtual Circuit 交换式虚电路

**SVCBL** Serviceable 可提供服务的, 可用的

**SVCG** Servicing 维修,提供服务

**SVCI** Signaling Virtual Channel Identifier 信令虚信道标识符

**SVD** Safe Vertical Distance 安全垂直距离

Seat Video Display 座椅视频显示器

Simultaneous Voice on Data 语音数据同时传输

Software Verification Document 软件校验文件

**SVDD** Software Version Description Document 软件版本说明文件

**SVE** Service 勤务,服务,维护

Space Vehicle Electronics 航天器电子设备

**SVFR** Special Visual Flight Rules 特殊目视飞行规则

**SVG** Scalable Vector Graphics 可升级的矢量图像

**SVGA** Super Video Graphic Adapter 优质视频图形卡(适配器)

Super Video Graphics Array 超级视频图像阵列

**S-VHS** Super-high Band VHS 超级家用视频系统

**SVL** Linear Actuator 线性作动筒(器)

**SVM** Servo Valve Monitor 伺服活门监控器

Support Vector Machine 支持向量机

System Virtual Machine 系统虚拟机

**SVMF** Space Vehicle Mockup Facility 航天器模拟设备

**SVMTR** Servo Motor 伺服电机

**SVN** Space Vehicle Number 卫星号

Switched Virtual Network 交换式虚拟网络

**SVNG** Servicing 服务,维修

**SVO** Servo 伺服

Moscow, Russia-Sheremetyevo ( Airport Code) 俄罗斯莫斯科谢列梅捷沃机场

**SVP** Senior Vice President 资深副总经理

Service Processor 业务处理器

**SVQ** Safety Valve Quantity 安全活门数量

Sevilla,Spain ( Airport Code) 西班牙塞尔维亚机场

**SVR** Servo Valve Relay 伺服活门继电器

Shop Visit Rate 送修率

Slant Visual Range 斜视距离

Switching Voltage Regulator 交换式电压调节

**SVRTSTM** Severe Thunderstorm 强雷暴

**SVS** Secure Voice System 保密话音系统

Space Vision System 空间视景系统

Stage Vehicle System 多级运载系统

Stressfree Viscous System 无压力黏性系统

Synthetic Vision System 合成视景系统

**SVSCAAC** Shanghai Vocational School of CAAC 民航上海中等专业学校

**SVT** Safety Valve Tag 安全活门标签

Servo Throttle 伺服油门杆

Software Verification Test 软件校验测试

Special Verification Test 特殊鉴定测试

System Verification Test 系统校验测试

**SVTL** Semivital 半活性的

**SVU** Satellite Voice Unit 卫星话音单元

Seat Video Unit 座椅视频组件

**SVX** Ekaterinburg, Russia ( Airport Code) 俄罗斯叶卡特琳堡机场

**SW** Salt Water 盐水

Shipping Weight 装运重量

Shock Wave 冲击波,激波

Short Wave 短波

Side Wheel 侧轮

Single Weight 单一权重

Single Wheel Landing Gear 单轮起落架

Software 软件

Southwest 西南

Speeder Wrench 快速扳手

Spherical Wave 球面波

Spot Weld 点焊

Status Word 状态字

Storm Warning 风暴警报,雷雨警报

Straddle Wrench 叉形扳手

Supersonic Wave 超声波
Supersonic Welder 超声波焊机
Surface Wind 地面风
Switch 电门,开关
Switching 转换

**SW/FP** Stall Warning/Flap Position 失速警告/襟翼位置

**SWA** Safety Work Analysis 安全工作分析
Study Work Authorization 研究工作审定

**SWAMP** Severe Wind and Moisture Problem 严重风和湿度问题
Special Wind and Moisture Problem 特殊风和湿度问题

**SWAN** Satellite WAN 卫星广域网
Self-organized Wireless Adaptive Network 自组织无线适配网络

**SWANET** Self-routed Wavelength Addressable Network 自选路由波长可寻址网络

**SWAP** Severe Weather Avoidance Plan 恶劣天气回避计划

**SWAPU** South and West Asia Postal Union 西南亚邮政同盟

**SWAT** Severe Weather Alert Team 恶劣天气警戒小组
Subjective Workload Assessment Technique 主观负荷评估技术
Subjective Workload Assessment Tool 主观负荷评估工具

**SWB** Short Wheel Base or Short Wheelbase 短轴距,短纵向轮距

**SWBD** Switchboard 开关板

**SWBHD** Swash Bulkhead 缓冲舱壁

**SWC** Skywave Correction 天波校正
Stall Warning Card 失速警告卡
Switching Center 交换中心

**SWCMP** Software Configuration Management Plan 软件组合管理计划

**SWCP** Salt Water Circulating Pump 盐水循环泵

**SWCS** Strategic Warning and Control System 战略警告和控制系统

**SWD** Surface Wave Device 表面波器件
Swaging Die 型模

**SWDD** Software Design Document 软件设计文件

**SWDL** Software Development Laboratory 软件研制实验室
Surface Wave Delay Line 表面波延迟线
System/Wiring Diagram List 系统/连线图目录

**SWDR** Status by Wire Diagram Report 布线图状态报告

**SWF** Newburgh, NY, USA(Airport Code) 美国纽约州纽堡机场

**SWG** Schedule Working Group 调度工作组
Sine-Wave Generator 正弦波产生器
Sine-Wave Grating 正弦波光栅
Slotted Wave Guide 开槽波导
Standard Wire Gauge 标准线规
Steel Wire Gauge 钢丝线规
Structure Working Group 结构工作组
Stub's Wire Gauge 斯塔布线规
Sweep Waveform Generator 扫描波形产生器
Swing 旋转,摇摆,浮移

**SWGD** Swinging Door 旋转门

**SWGR** Switchgear 开关装置

**SW-HW** Software-Hardware 软件—硬件

**SWI** Short Wave Interference 短波干扰
Stall Warning Indicator 失速告警器
Standing Wave Indicator 驻波指示器

**SWID** Software Implementation Document 软件执行(程序)文件

**SWIM** System Wide Information Management 广域信息管理系统

**SWIP** Super Weight Improvement Program 超重改进计划

**SWIR** Short Wave Infra-Red 短波红外线

**SWIS** Satellite Weather Information System 卫星天气信息系统
Stall-Warning and Identification System 失速告警和识别系统

**SWL** Safe Working Load 安全工作载荷,安

全负载

Short Wave Listener 短波听众

Sidewall 侧壁

Single Wheel Loading 单轮负载

Slab Waveguide Laser 平板波导激光器

Solid Waster Litter 固体废屑

Spectrum Wave Length 频谱波长

**SWLDG** Socket Welding 承插焊接，承插焊缝

**SWM** Short-Wave Meter 短波表

Spectral Width Modulation 谱宽调制

**SWN** Switching Network 交换网络

**SWO** Shop Work Order 车间工作单

**SWOT** Strengths, Weaknesses, Opportunities and Threats 优势、劣势、机会和威胁

**SWOV** Switchover 切换，转接

**SWP** Safe Working Pressure 安全工作压力

Safety Working Party 安全工作组

Salt Water Pump 盐水泵

Simplify Working Process 简化工序

South West Pacific Sub-area 西南太平洋次区

Standard Work Procedure 标准工作程序

Strategy Working Party 战略工作小组

Sweep 扫描，扫掠

**SWPM** Standard Wiring Practices Manual 标准线路施工手册

**SWQA** Software Quality Assurance 软件质量保证

**SWR** Special Work Request 特殊工作需求

Standing Wave Ratio (Voltage) 驻波比(电压)

Steel Wire Rope 钢丝绳

Storm Warning Radar 风暴警告雷达

Synoptic Weather Report 天气预报

**SWRD** Software Requirements Document 软件要求文件

Software Requirements Drawing 软件要求图

**SWRDD** Software Requirements Design Drawing 软件要求设计图

**SWRL** Semantic Web Rule Language 语义网规则语言

**SWRS** Standardized Weights Record System 标准化重量记录系统

**SWS** Stall Warning System 失速警告系统

Structures Workstation 结构工作站

Switch Stand 开关柜

Switching Subsystem 交换子系统

**SWSG** Security Window Screen and Guard 安全窗屏蔽和保护装置

**SWSL** Supplemental Weather Service Location 补充气象服务站

**SWSP** Software Standards and Procedures 软件标准和程序

**SWSS** Share Workstation System 共用工作站系统

**SWT** Short Wave Transmitter 短波发射机

Supersonic Wind Tunnel 超音速风洞

Switch Time 转换时间

**SWTG** Switching 转换，交换

**SWTL** Surface Wave Transmission Line 表面波传输线

**SWVL** Swivel 旋转，轴承，转环

**SWY** Stopway 停止道

**SX** Simplex 单一的

**SXB** Strasbourg (France) 斯特拉斯堡(法国)

Strasbourg, France (Airport Code) 法国斯特拉斯堡机场

**SXF** Berlin, Germany-Schoenefeld (Airport Code) 德国柏林舍纳费尔德机场

**SXGA** Super Extended Graphics Array 超级扩展型图形阵列

**SXM** Symbol Extension Mode 符号扩展方式位

**SXR** Srinigar (India) 斯利那加(印度)

**SXS** Step-by-Step 一步一步的，循序渐进的

**SXT** Sextuple 六倍的

**SXTN** Sextant 六分仪

**SY** Supersonic Yawmeter 超音速偏航仪

Sustainer Yaw 主发动机偏转

Synchronize 同步化，整步

Synchroscope 同步示波器

Syrian Arab Republic 阿拉伯叙利亚共和国

System 系统,体系

**SYC** System Control 系统控制

**SYCI** System Corrected Image 系统校正图像

**SYCOM** Synchronous Communication 同步通信

Synchronous-Altitude Communication Satellite 同步高度通信卫星

System Communication Module 系统通信模块

**SYD** Sydney, New South Wales, Australia（Airport Code）澳大利亚悉尼机场

**SYDAS** Synchro Data Acquisition System 同步数据采集系统

**SYHCH** Synchronization Character 同步字符

**SYM** Symbol 符号,标志

Symmetry or Symmetrical 对称（的）

Synchromotor 同步电机

**SYM GEN** Symbol Generator 符号发生器

**SYMAN** Symbol Manipulation 符号处理

**SYMAP** Symbol Manipulation Program 符号处理程序

**SYMB** Symbol 符号,标志,象征

**SYMB AL** Symbolic Algebra 符号代数

**SYMBAL** Symbolic Algebra Language 符号代数语言

**SYMM** Symmetrical 对称的

**SYMP** Symposium 座谈会,讨论会

**SYN** Synchronizer 同步器,同步机,同步装置

Synchronous Idle 同步空闲

Synthetic 人工合成的

**SYNC** Synchronization 同步

Synchronize 使同步

Synchronous Code 同步码

Synchronous Communication 同步通信

**SYNCOM** Synchronous Communications Satellite 同步通信卫星

**SYNOP** Synoptic Chart 天气图

**SYNS** Synchronizing Signal 同步信号

Synchronous System 同步系统

Synopsis 提要,大纲

**SYNSC** Synchronous Signal Character 同步信号字

**SYNSCP** Synchroscope 同步示波器

**SYNTH** Synthetic 人工合成的

**SYNTHA-VIS** Computer System, 3 Dimensional Geometric Modeling 计算机系统,三维几何模型

**SYNTHR** Synthesizer 合成器,合成装置

**SYNTI** Synchro Tie 同步连线

**SYR** Syracuse, NY, USA（Airport Code）美国纽约州锡拉丘兹机场

Syria 叙利亚共和国（西南亚国家）

**SYS** System 系统,体制

**SYS OUT** System out 系统输出

**SYSAD** System Administration 系统管理

**SYSCON** System Control 系统控制

**SYSCONCEN** System Control Center 系统控制中心

**SYSDEP** System Development/Evolution Plan 系统开发/进展计划

**SYS-DEV** System Development 系统开发

**SYSIN** System Input 系统输入

**SYSIO** System I/O 系统输入/输出

**SYSMON** System Monitor 系统监视器

**SYSOP** System Operator 系统操作员

**SYSTO** System Staff Officer 系统参谋

**SYSTST** System Test 系统测试

**SYU** Synchronization Signal Unit 同步信号单元

**SYZ** Shiraz, Iran-Shiraz（Airport Code）伊朗设拉子机场

**SZ** Size 尺寸,大小

Swaziland 斯威士兰

**SZC** Special Zone Center 特定区域中心

**SZG** Salzburg, Austria（Airport Code）奥地利萨尔斯堡机场

**SZZ** Szczecin, Poland-Goleniow（Airport Code）波兰什切青戈列诺机场

# T

**T** Absolute Temperature 绝对温度

Airfoil Thickness 翼剖面厚度

Single Thermoplastic 单热塑料

Synthetic Thermoplastic 合成热塑料

T-Bar T 型条

Temperature 温度

Terminal 终端

Tesla 特斯拉

Test 测试

Thickness 厚度

Time 时间

Time of Flight 飞行时间

Tonne 公吨(1000 千克)

Tool 工具

Tooth or Teeth 齿

Total Thrust 总推力

Total Trim 总配平

Tourist Class 二等舱,经济座

Training Aircraft 教练机

Transfer 传递,转移,转换

Transfer Load 转港装载

Transformer 变压器

Transmitted Wave 发射波

Transmitter 传感器

Transport 运输,运输机

Transport Layer 运输层

Trim 配平,修整,调准

True 真实的

Tube 管路,管道

Turbulence 湍流

Turn 转弯,转向,圈,匝数

**T&B** Top and Bottom 顶部和底部

**T&BB** Top and Bottom Bolt 顶部和底部螺栓

**T&BI** Turn and Bank Indicator 转弯倾斜仪

**T&D** Test and Deployment 测试和展开

Transmission and Distribution 传输和分配,发送和分配

**T&DA** Tracking and Data Acquisition 跟踪和数据采集

**T&E** Test and Evaluation 试验和评估

**T&G** Tongue and Groove 舌管与槽

**T&H** Test and Handling 测试和处理

Transporting and Handling 输送和处理

**T&ID** Testability and Integrated Diagnostics 测试能力和综合诊断

**T&P** Time and Percussion 时间和撞击

**T&PPM** Tool and Production Planning Manual 工具和生产计划手册

**T&QA** Test and Quality Assurance 测试与质量保证

**T&TE** Training and Training Equipment 培训和培训设备

**T&TEP** Training and Training Equipment Plan 培训和培训设备计划

**T&WD** Time and Wavelength Division 时间和波长分割

**T/A/H** Temperature/Altitude/Humidity 温度/高度/湿度

**T/B** Test and Bias 测试和偏置

Title Block 标题区,标题组

Turn and Bank 转弯和倾斜

**T/C** Thickness-Chord Ratio 厚弦比

Top of Climb 爬升顶点

Type Certificate 型号合格证

**T/D** Target/Destination 目标/终端

**T/E** Table of Equipment 设备表

Test and Evaluation 测试和评估

Transporter Erector 传输装置升降架

**T/EMM** Thermal/Energy Management Module 热/能量管理模块

**T/F** Track of a Fixed Waypoint 固定航路点航迹

**T/FR** Top of Frame 框架顶部,帧顶部

**T/H** Tool Hole 工具孔

Track-and-Hold 跟踪与保持

**T/L** Thrust Lever 油门杆

Top-Level 顶层(设计)

Tour Leader 旅行团领队

**T/M** Telemeter 遥测器

Telemetry 遥测技术

Tool Mark 工具标记

Torque Motor 扭矩马达

Turns per Minute 转数/分

**T/O** Table of Organization 组织结构表

Takeoff 起飞

**T/O&E** Tables of Organization and Equipment 机构和设备表,组织和设备表

**T/R** Receiver-Transmitter 收发信机

Thrust Reverser 反推装置

Tooling Request 工具申请

Transfer 转换继电器

Transformer Rectifier 变压整流器

**T/R&G** Transmit Receive and Guard 发射、接收和保险

**T/S** Test Station 测试站

Transition Section 过渡部分

Troubleshooting 排故

Turn and Slip 转弯倒滑

**T/T** Tie Table 连接表

Transport/Tanker 运输加油机

**T/V** Thermal Vacuum 热真空(装置)

**T/W** Thrust-to-Weight Ratio 推力重量比

**TA** Table of Allowances 公差表

Target Acquisition 目标搜索

Target Advisory 目标咨询

Target Area 目标区

Technical Advisor 技术顾问

Technical Arrangement 技术协议

Technical Assistance 技术援助

Tentative Approval 临时批准

Terminal Acquisition 终端搜索(目标显示)

Terminal Adaptation 终端适配

Terminal Adapter 终端适配器

Terrain Avoidance 地形回避

Terrain Awareness 地形觉察

Thermal Analysis 热分析

Time to Arrival 到达时间

Traffic Advisory 交通咨询,运量咨询

Transition Altitude 过渡高度

Transmission Adapter 传输转接器

Travel Authorization 飞行批准,通过批准

Trend Analysis 趋势分析

True Altitude 真实高度,绝对高度,真高度

Twin Aisle 双过道,双通道

Type Approval 型号批准书

**TA/CE** Technical Analysis/Cost Estimate 技术分析与成本估计

**TA/CP** Technology Assessment/Control Plan 技术评估与控制计划

**TAA** Technical Airworthiness Arrangements 技术适航安排

Technical Assistance Agreement 技术援助协议

Technology Area Assessment 技术领域的评估

Terminal Advanced Automation 终端高级自动化

Terminal Arrival Area 终端进场区

Trade Agreement Act 贸易协定法案

Transport Association of America 美国运输协会

**TAAATS** The Advanced Australian Air Traffic System 澳大利亚先进的空中交通系统

**TAAD** Terrain/Obstacle Awareness Alerting and Display 地形/障碍物提醒警告与显示

**TAAF** Test, Analysis and Fix 测试、分析与修理

**TAAM** Terminal Area Altitude Measuring 终端区域高度测量

Terminal Area Altitude Monitoring 终端区域高度监视

Total Airspace and Airport Modeller 全空域

及机场模型

**TAAR** Target Area Analysis Radar 目标区域分析雷达

**TAAS** Terminal Advanced Automatic System 终端高级自动化系统

Three-Axis Attitude Sensor 三轴式传感器，三轴姿态传感器

Traffic Accident Analysis System 交通事故分析系统

**TAAT** Technical Assistance and Advisory Team 技术援助与咨询组

**TAB** Table 表，图表

Tabular 平板的，平坦的

Tabulate 列表

Tabulation and Interpolation Program 制表和插入方案

Technical Advisory Board 技术咨询委员会

Technical Assistance Board 技术援助委员会

Technical Assistance Bureau 技术援助局

Test and BITE Panel 测试和自测面板

Tobago, Trinidad and Tobago ( Airport Code ) 特立尼达和多巴哥多巴哥机场

Total Allocated Budget 总分配预算

**TABSIM** Tabulator Simulator 制表器模拟器

**TABSOL** Table Solution 列表解法，表格解释

**TAC** Taiwan Aerospace Corp. （中国）台湾宇航公司

Technical Activities Committee 技术业务委员会

Technical Advisory Committee 技术顾问委员会

Technical Assistance Center 技术援助中心

Technical Assistance Committee 技术援助委员会

Technology Application Center 技术应用中心

Telemetry and Command 遥测与指挥

Temporal Access Control 暂时接入控制

Terminal Access Controller 终端接入控制器

Terminal Area Chart 终端区域图

Test Access Control 测试存取控制，测试接入控制

Thermosetting Asbestos Composite 热固性石棉复合材料

Thrust Asymmetry Compensation 推力不对称补偿

Thrust Asymmetry Correction 推力不对称修正

Total Automatic Computer 全自动计算机

Transistorized Automatic Control 晶体管化自动控制

Translator Assembler Compiler 翻译汇编编译程序

Transport Advisory Committee 交通咨询委员会

Transport Advisory Council 交通咨询委员会

Trim Augmentation Computer 配平增稳计算机

Turbo-Alternator Compressor 涡轮交流发电机压缩机

**TAC&SD** Test Airplane Configuration and Status Document 试验飞机构型与状态文档

**TACAMO** Take Charge and Move Out 受领任务并开始行动

**TACAN** Tactical Air Navigation 战术空中导航

Tactical Air Navigation System 战术空中导航系统

Tactical Airborne Navigation 机载战术导航

**TACC** Tactical Air Command Center 战术航空指挥中心

Tactical Air Control Center 战术航空管制中心

Taipei Area Control Center( 中国)台北区域管制中心

Tracking and Control Center 跟踪和控制中心

**TACCAR** Time Average Clutter Coherent Airborne Radar 时间平均杂波相参机载雷达

**TACCOMSAT** Tactical Communication Satellite 战术通信卫星

**TACDEN** Tactical Data Entry 战术数据输入

**TACF** Terminal Access Control Function 终端接入控制功能

**TACH** Tachometer 转速表

**TACL** Terminal Assignment to Connection List 终端赋值至连接清单,终端分派至连接清单

**TACM** Threat Associate Countermeasure 威胁相关干扰

Transit Air Cargo Manifest 转运货单

**TACNAVSAT** Tactical Navigation Satellite 战术导航卫星

**TACOM** Tactical Communication 战术通信

**TACOS** Techniques and Circuits for Optical Signals 用于光信号的技术和电路

**TACS** Target and Clutter Simulator 目标和杂波模拟器

Taxiing Aid Camera System 滑行辅助摄像机系统

Technical Activities Committee 技术工作委员会

Thruster Attitude Control System 推力器姿态控制系统

Total Access Communication System 全接入通信系统

Tracking and Control Station 跟踪控制站

Travel Accounting System 旅行统计系统

**TACSAT** Tactical Satellite 战术卫星

**TACT** Table and Code Translation 表格和代码翻译

Technological Aids to Creative Thoughts 创新思维技术援助

The Air Cargo Tariff 空运货物运价表

Transonic Aircraft Technology 跨音速飞机技术

**TACTS** Tactical Aircrew Combat Training System 战术空勤人员战斗训练系统

**TAD** Target Acquisition Data 目标搜索数据

Technical Approved Demonstration 技术批准演示

Technology Availability Date 技术有效性日期,技术可用性日期

Telephone Answering Device 电话应答设备

Test Acceptance Document 试验验收文件

Test and Development 试验与研制

**TADD** Target Alert Data Display 目标警戒数据显示器

**TADIL** Tactical Data Information Link 战术数据信息链

**TADR** Tabulated Drawing 表格化图纸

**TADS** Target Acquisition and Designator and Display System 目标搜索、指示和显示系统

Tracking and Display System 跟踪与显示系统

**TAE** Taegu, South Korea (Airport Code) 韩国大邱机场

Test Access Equipment 测试接入设备

Test and Evaluation 试验和评定

Track Angle Error 航迹角误差

**TAECO** Taikoo (Xiamen) Aircraft Engineering Company Limited (厦门)太古飞机工程有限公司

**TAEM** Terminal Area Energy Management 航站区域能源管理

**TAERS** Tactical Aircrew Eye Respiratory System 战术机组人员眼部呼吸系统

**TAF** Temporary Aviation Facilities 临时航空设施

Terminal Aerodrome Forecast 终端机场气象预报

Terminal Airport Forecast 终端机场气象预报

Terminal Area Forecast 终端区气象预报

Test, Analyze, Fix 测试、分析、修理

Trim after Forming 成形后修整

**TAFCOS** Total Automatic Flight Control System 全自动飞行控制系统

**TAFCU** Timed Acceleration Fuel Control Unit 定时加速燃油管理组件

**TAFI** Turn around Fault Isolation 往返飞行中故障隔离

**TAFOR** Terminal Aerodrome Forecast in Full

Form 终点机场全式天气预报

**TAG** Technical Advisory Group 技术咨询组

Test Adapter Group 测试适配器组

**TAHAS** Terminal Avian Hazard Advisory System 终端区鸟类危险咨询系统

**TAI** International Atomic Time 国际原子时间

Technology Achievement Index 技术成就索引

Thai Airways International Public Company Ltd. 泰国国际航空有限公司

Thermal Anti-Ice 热空气防冰

Total Aircraft Inventory 总共飞机库存

Traffic Accident Investigation 交通事故调查

True Airspeed Indicator 实际空速指示器

Turkish Aerospace Industries, Inc 土耳其宇航工业公司

**TAID** Tube Assembly Inspection Drawing 管道组装检验用图纸

**TAIL** Tail Wind 顺风

**TAILRATS** Tail Radar Acquisition and Tracking System 尾部雷达搜索和跟踪系统

**TAINS** TRCOM-Aided Inertial Navigation System 有地形匹配装置的惯性导航系统

Terrain Correlation Mapping Aided Inertial Navigation System 地形相关地图辅助的导航系统

**TAJ** Thermal Arc Jet 热力弧喷气发动机

**TAK** Takamatsu, Japan ( Airport Code ) 日本高松机场

**TAL** Tailor 裁缝

Transfer Alignment 传输校准

**TALCAT** Takeoff and Landing Critical Atmosphere Turbulence 起降时临界大气湍流

**TALS** Transport Approach and Landing Simulator 运输机进近和着陆模拟器

**TALT** Tracking Altitude 跟踪高度

**TAM** Tampico, Tamaulipas, Mexico ( Airport Code ) 墨西哥坦皮科机场

Technical Acknowledgement Message 技术认可文电

Test Access Module 测试接入模块

**TAMB** Ambient Temperature 环境温度

**TAMDAR** Tropospheric Airborne Meteorological Data Reporting 对流层航空气象资料报告

**TAMPA** True Altitude Minus Pressure Altitude 绝对高度减压力高度

**TAMS** Tool Administration and Management System 工具管理机构和管理系统

Tool Administration and Manufacturing Services 工具管理和制造服务

**TAMSS** Target Area Meteorological Sensor System 目标区域气象传感器系统

**TAN** Tangent 正切, 切线

Terrain Aiding Navigation 地形辅助导航

**TANH** Hyperbolic Tangent 双曲正切

**TANS** Tactical Air Navigation System 战术空中导航系统

Terminal Area Navigation System 终端区导航系统

Terrain-Aided Navigation System 地形辅助导航系统

**TAO** Technical Applications Office 技术应用办公室

**TAP** Tapachula, Chiapas, Mexico ( Airport Code ) 墨西哥塔帕丘拉机场

Tape 磁带

Technical Area Planning 技术区规划

Test Access Path 测试接入通道

Test Access Port 测试接入端口, 测试存取端口

Testing Access Point 测试接入点

Total Air Pressure 大气全压

Trace Analysis Program 踪迹分析程序

Transient Acoustic Propagation 瞬态消音传播

Transport Access Point 传输接入点

**TAPAC** Tape Automatic Positioning and Control 磁带自动定位和控制

**TAPAK** Tape Pack 磁带盒

**TAPE** Tape Automatic Preparation Equipment

磁带自动准备设备

Technical Advisory Panel for Electronics 电子设备技术咨询小组

**TAPP** Travel Agents Protector Plus Insurance Plan 旅行社保险和保险计划

**TAPS** Target Analysis and Planning System 目标分析与计划系统

Twin Annular Premixing Swirler 双环预混旋流

**TAR** Tariffs 运价,税率

Technical Advisory Report 技术咨询报告

Technical Assessment Report 技术评估报告

Temporary Alternate Routing 暂时可选路由

Terminal Approach Radar 终端区进近雷达

Terminal Area Radar 终端区雷达

Terrain Avoidance Radar 地形回避雷达

Transfer Alignment Routine 传输对准程序

**TARAN** Test and Repair as Necessary 按需要试验和维修

**TARC** Transport Airworthiness Requirements Committee 运输机适航性要求委员会

Traffic Accident Research Centre 交通事故研究中心

**TARCS** Test and Repair Control System 测试与修理控制系统

**TARE** Telemetry Automatic Reduction Equipment 遥测自动处理设备

**TAROC** Total Airplane Related Operating Cost 与飞机相关的总运营成本

**TARP** Tarpaulin 防水布

**TARPS** Tactical Airborne Reconnaissance Pod System 机载战术侦察吊舱系统

Tactical Airborne Reconnaissance Program System 机载战术侦察程序系统

Tactical Aircraft Reconnaissance Pod System 战术飞机空中侦察吊舱系统

**TARS** Target Attack Radar System 目标攻击雷达系统

Terminal Automated Radar Services 终端自动雷达服务

Terrain and Radar Simulator 地形与雷达模拟器

Test Analysis and Retrieval System 试验分析与检索系统

Three-Axis Reference System 三轴参考系统

**TAS** Takeoff Airspeed 起飞空速

Target Acquisition System 目标搜索系统

Tashkent, Uzbekistan (Airport Code) 乌兹别克斯坦塔什干机场

Telemetry Analysis Station 遥测分析站

Terminating Access Situation 终端接入状态

Test Adapter Set 测试转接器

Tooling Advisory System 工装咨询系统

Tourism and Air Services Summit 旅游与航空峰会

Traffic Advisory System 交通咨询系统

Transceiver Administration Subsystem 收发机管理分系统

True Air Speed 真空速

**TASA** Terminal Area Support Aircraft 终端区支援飞机

Thai Aero Sport Association 泰国航空运动协会

The Aircraft Service Association 飞机勤务协会

**TASC** Telecommunications Alarm Surveillance and Control 电信警报监督与控制

Terminal Area Sequencing and Control 终端区顺序和控制

**TASCON** Television Automatic Sequence Control 电视自动顺序控制

**TASES** Tactical Airborne Signal Electronic Surveillance 机载战术电子信号监视

**TASI** Time Airspeed Inertial 时间空速惯性的

Time Assignment Speech Interpolation 时间分配语音插入,时间分配话音插空技术

True Airspeed Indicator 真空速指示器

**TASK** True Airspeed in Knots 以海里计算的真空速

**TASM** Total Available Seat-Miles 可用座英里数总数

**TASN** Time Air Speed Navigation 时间空速导航

**TASO** Television Advisory Standards Organization 电视标准咨询组织

**TASP** Technology for Advanced Space Power 先进的空间动力技术

**TASR** Terminal Air Surveillance Radar 终端空中监视雷达

Terminal Airport Surveillance Radar 终端机场监视雷达

Terminal Area Surveillance Radar 终端区域监视雷达

**TASS** Tactical Avionics System Simulator 战术航空电子系统模拟器

Technical Analytical Study Support 技术分析研究支援

Terminal Area Surveillance System 终端区域监视系统

Towed Array Surveillance System 拖曳天线阵监视系统

Transport Aircraft Servicing Specialist 运输机维修专家

**TASST** Tentative Airworthiness Standards for Supersonic Transports 超音速运输机暂行适航标准

**TAST** Terminal Assignment Sort and Test 终端指定分类和测试

**TAT** Target Aircraft Transmitter 目标飞机发射机

Technical Acceptance Team 技术验收组

Technical Assessment Team 技术评估组

Throttle Angle Transducer 油门杆角度传感器

Total Air Temperature 全温,大气总温

Total Aircraft Time 飞机总(运行)时间

Transitional Automated Ticket 过站自动机票

Truck-Air-Truck 陆空联运

True Air Temperature 真实气温

Turbine Air Temperature 涡轮大气温度

Turn around Time 周转时间

**TATC** Terminal Air Traffic Control 终端空中交通管制

**TATCA** Terminal Air Traffic Control Automation 终端空中交通管制自动化

Terminal ATC Automation 终端空中交通管制自动化

**TATCS** Terminal Air Traffic Control System 终端空中交通管制系统

**TATF** Terminal Automation Test Facility 航站自动化测试装置

**TATI** Total Air Temperature Indicator 全温指示器

**TAU** A Time Constant 一种时间常数(避撞系统)

TACAN Adapter Unit 塔康转接器组件

Terminal Access Unit 终端接入单元

Test Access Unit 测试接入单元

Token Ring Attachment Unit 令牌环连接单元

**TAV** Trans-Atmospheric Vehicle 穿越大气层航空器

**TAW** Thrust Augmented Wing 推力增强机翼

**TAWS** Target Acquisition Weather Software 目标捕获气象软件

Terrain Avoidance and Warning System 地形回避和警告系统

Terrain Awareness and Warning System 地形提示和警告系统

**TAX** Taxiing 滑行

**TB** Tabulation Block 列表数据区,列表信息区

Technical Bulletin 技术通报

Terminal Block 接线板

Terminal Board 接线板,终端板

Test Bench 测试台

Tie Bus 连接汇流条

Torsion Bar 扭杆

True Bearing 真方位

**T-B** Title Block 标题信息区

**TBA** Terminal Board Assembly 接线板组件

Time between Arrivals 到达间隔时间

To Be Added 待加上

To Be Advised or Announced 待通告

To Be Announced 待发布

To Be Assigned 待指定,待命名

Total Blade Area 叶片总面积,桨叶总面积

True Bearing Adapter 真方位测定仪

**TBAN**    Tube Axial 轴向管

**TBB**    Transfer Bus Breaker 转换汇流条断电器

**TBC**    The Boeing Company 波音公司

Thermal Barrier Coating 热障涂层

Through Bulkhead Connector 穿隔框插头

To Be Confirmed 待证实

Token Bus Controller 标记总线控制器

**TBCC**    Turbine-Based Combined Cycle 涡轮基组合循环(发动机)

**TBD**    Threshold Boundary Designator 跑道入口边界指示器

To Be Defined 待详细说明或决定

To Be Deleted 待删除

To Be Determined 待定

**TBDP**    Tie Bus Differential Protection 连接汇流条差动保护

**TBE**    Time Base Error 时基误差

**TBF**    Tie Bus Fault 汇流条故障

Time between Failures 故障间隔时间

To Be Finished 待完成

Training Base Facility 训练基地设施

**TBFB**    Tile Based Frame Buffer 碎片纹理帧缓存

**TBG**    Tubing 管道

**TBGA**    Tie Ball Grid Array 带状球形光栅阵列

**TBI**    Target Bearing Indicator 目标方位指示器

Test Bed Installation 测试台安装

Time between Inspection 检查间隔时间

To Be Initiated 待开始,待启动

**TBL**    Table 表

**TBLR**    Tumbler 拨动式开关,齿轮换向器

**TBM**    Time-Based Maintenance 定期维修方式

Time-Based Metering 基于时间的流控

**TBN**    Taxi to Be Nominated 等待滑行指示

To Be Notified 待通知

**TBO**    Time before Overhaul 翻修前时间

Time between Overhauls 大修间隔时间

**TBP**    To Be Proposed 待提出,待建议

**TBR**    To Be Resolved 待解决

To Be Revised 待修改

Transmission Bit Rate 传输比特率

**TBRL**    Terminal Balanced Return Loss 终端平衡回损

**TBS**    Task Breakdown Structure 任务分解结构

Tbilisi, Georgia (Airport Code) 格鲁吉亚第比利斯机场

Time-Based Separation 基于时间的间隔

To Be Scheduled 待计划,待安排

To Be Specified 待定,待明确

To Be Supplied 待提供

**TBSEL**    Time Base Select 时基选择

**TBSR**    Transfer Bus Sensing Relay 转换汇流条敏感继电器

**TBSS**    Test Bench Simulation Software 试验台模拟软件

**TBSV**    Time between Scheduled Visits 定期进厂间隔

**TBT**    Turbine Bearing Temperature 涡轮轴承温度

Turbine Blade Temperature 涡轮叶片温度

**TBU**    Nuku Alofa/Tongatapu, Tonga (Airport Code) 汤加努库阿洛法汤加塔布机场

Telemetry Buffer Unit 遥测缓冲器组件

Time Base Unit 时基组件

Track-Ball Unit 跟踪球组件

**TBV**    Thrust Balance Vent 推力平衡通气口

To Be Verified 待查,待验证

Transient Bleed Valve 瞬时放气活门

Turbine by-pass Valve 涡轮旁通活门

**TBVC**    Turbine Blade and Vane Cooling 涡轮叶片与导向片冷却

**TBVCA**    Turbine Blade and Vane Control Actuator 涡轮叶片与导向片控制作动器

Turbine Blade and Vane Cooling Actuator 涡轮叶片与导向片冷却作动器

**TBW** Total Bandwidth 总带宽

**TBZ** Tabriz, Iran（Airport Code）伊朗大布里士机场

**TC** Takeoff Charts 起飞图表

Tape Core 磁带盘心

Tariff Conference 运价会议

Task Card 工卡

Technical Characteristics 技术性能

Technical Circular 技术通报

Technical Commission 技术委员会

Technical Committee 技术委员会

Technical Control 技术控制

Technical Coordinator 技术协调员

Telecommunication Committee 电信委员会

Temperature Coefficient 温度系数

Temperature Compensating 温度补偿

Temperature Control 温度控制

Terminal Circuit 终端电路

Terminal Computer 终端计算机

Terminal Control 终端控制

Terminal Controller 终端控制器

Terrain Computer 地形计算机

Terrain Control 地形控制

Terrain Correction 地形校正

Terrain Correlation 地形相关

Terrestrial Channel 地面信道

Test Condition 测试条件

Test Connector 测试接头

Test Coordination 测试协调

Thermocouple 热电耦

Thread Cutting 螺纹切割

Time Closing 定时关闭

Time Coding 时间编码

Time Coherence 时间相关性

Time Compensation 时间补偿

Time Constant 时间常数

Time Coordinate 时间坐标

Time Cost 时间成本

Time Cycle 时间循环

Timing Control 定时控制

To Connector 连至插座

Toll Center 长途电话中心

Tool Classification 工具分类

Top Center 上止点

Top Chord 顶舷

Total Cost 总成本

Tourist Class 旅行舱

Traffic Channel 业务信道

Traffic Conference（航空）运输业务会议

Traffic Control 交通管制

Transaction Capability 事务处理能力

Transfer Control 转移控制

Transit Center 转接中心

Transmission Convergence 传输会聚

Transmit Command 发射指令

Transport Canada 加拿大运输航空公司

Transport Connection 传送连接

Traveler's Cheque 旅行支票

Trip Coil 跳闸线圈

Tropical Cyclone 热带气旋

True Course 真航线角,真航向

Trunk Channel 干线信道

Trunk Circuit 中继电路

Trunk Code 中继代码

Trunk Control 中继线控制

Turbo Charger 涡轮增压器

Type Certification 型号合格证

**TC&I** Task Cards and Indexes 工卡和索引

**TCA** Tactical Customer Advocate 战术客户支持

Technical Collaboration Agreement 技术合作协定

Telemetry Channel Assembly 遥测信道组件

Temperature Control Amplifier 温控放大器

Terminal Communication Adapter 终端通信适配器

Terminal Control Area 航站管制区,终端管制区

Thrust Chamber Assembly 推力燃烧室组件

Time of Closest Approach 最近进场时间

Track Crossing Angle 航迹交叉角

Traffic Conditioning Agreement 流量调节协定

Transcontinental Control Area 跨大陆管制区

Turbine Cooling Air 涡轮冷却空气

Turbine Cooling Airflow 涡轮冷却空气流

**TCAA** Technical Communication Association of Australia 澳洲技术交流协会

Trans-atlantic Common Aviation Areas 跨大西洋共同航空区

**TCAC** Tactical Collection and Analysis Center 战术收集与分析中心

Technical Control and Analysis Center 技术控制与分析中心

Tropical Cyclone Advisory Centre 热带气旋咨询中心

**TCAM** Telecommunication Access Method 远程通信存取法

Traffic Control and Monitoring 交通控制和监督

**TCAP** Transaction Capabilities Application Part 事务处理能力应用部分

**TCAS** Technical Control and Analysis System 技术控制与分析系统

Terminal Collision Avoidance System 终端防撞系统

Terminal Control Address Space 终端控制地址空间

Traffic Alert and Collision Avoidance System 交通告警与防撞系统

Traffic Collision Avoidance System 空中防撞系统

Tropical Cyclone Advisory Services 热带气旋咨询服务

**TCAS Ⅰ** Traffic（alert and）Collision Avoidance System Ⅰ 一类交通告警防撞系统（提供告警）

**TCAS Ⅱ** Traffic（alert and）Collision Avoidance System Ⅱ 二类交通告警防撞系统（提供告警并提供冲突咨询）

**TCAT** Tooling Capitalization and Taxation 工装投资与税收

**TCAV** Turbine Cooling Air Valve 涡轮冷却空气阀

**TCB** Technical Co-operation Bureau 技术合作局

Thermal Circuit Breaker 热电路跳开关

Transmit Command B 发射指令 B

Trusted Computing Base 委托计算库，赊售计算库

Type Certification Basis 型号合格审定基础

Type Certification Board 型号审定委员会

**TCBV** Temperature Coefficient of Breakdown Voltage 击穿电压的温度系数

**TCC** Technical Competency Training and Certification 技术能力培训和资格认证

Technical Co-operation Committee 技术合作委员会

Terminal Call Control 终端呼叫控制

Terminal Control Center 终端管制中心

Thrust Control Computer 推力控制计算机

Traffic Control Center 话务控制中心

Transaxle Converter Clutch 变矩器离合器，变速驱动桥

Transport Connection Clear 传送连接拆线

Transport Control Center 运输控制中心

Travel Correction Calculator 移动校正计算器

Turbine Case Cooling 涡轮机匣冷却

Turbine Clearance Control 涡轮间隙控制

Turbine Close Coupled 紧耦合涡轮

Turbine Cooling Control 涡轮冷却控制

**TCCA** Toronto City Center Airport 多伦多城市中心机场

Turbine Case Cooling Air 涡轮机匣冷却空气

**TCCB** Test Change Commitment Board 测试更改委托委员会

**TCCC** Tower Control Computer Complex 塔台管制计算机组

**TCCD** Total Call Connection Delay 总呼叫连

接延迟

**TCCF** Tactical Communications Control Facility 战术通信控制设施

**TCCS** Technical Committee on Communication Satellites 通信卫星技术委员会

Thrust Control Computer System 推力控制计算机系统

Ticket Consignment Control Sheet 证件托运控制单

Traffic Congestion Control System 业务拥塞控制系统

**TCCV** Turbine Clearance Control Valve 涡轮间隙控制活门

**TCD** Temporary Configuration Departure 临时配置离港,临时组合离港

Thyratron Core Driver 闸流管芯激励器

Time Compliance Directive 时间符合性指令

**TCDM** Type Certification Document Management 型号合格审定文件管理

**TCDS** Type Certificate Data Sheet 型号审定数据单

**TCE** Terminal Control Element 终端控制单元

Terminal Control Equipment 终端控制设备

Thermal Coefficient of Expansion 热膨胀系数

Transit Connection Element 转接连接单元

**TCEA** Training Center for Experimental Aerodynamics 实验空气动力学训练中心

**TCEP** Transport Connection Endpoint 传送连接端点

**TCF** Technical Control Facility 技术控制设施

Terrain Clearance Floor 地形净空平面

Track Characterization Function 轨迹特征函数

Transducer Calibration File 传感器校验档案

Transmission Convergence Function 传输会聚功能

Transparent Computing Facility 简明计算设施

Trillion Cubic Feet 兆立方英尺

Turbine Center Frame 涡轮中心框架

Twin-Core Fiber 双芯光纤

**TCG** Time Code Generator 时间编码产生器

Time Color Generator 时间色彩产生器

Timing and Control Generator 定时和控制产生器

Tune Controlled Gain 调谐控制的增益

**TCH** Temporary Construction Hole 临时结构孔

Threshold Crossing Height 跑道入口飞越高度,飞越跑道入口高

Tone Channel 信号音信道

Transfer in Channel 管道内传输

Type Certificate Holdr 型号合格证持有人

**TCH-EFR** Traffic Channel-Enhanced Full Rate 业务信道—增强型全速率

**TCH-FS** Traffic Channel-Full Rate Speech 业务信道—全速率话音

**TCI** Tenerife, Canary Islands, Spain(Airport Code) 西班牙特内里费机场

Terminal Control Interface 终端控制接口

Terrain Clearance Indicator 绝对高度指示器

Time/Cycle Indicator 时间/循环指示器

Turbo Charger with Inter-cooler 废气涡轮增压中冷

**TCIC** Transit Center Identification Code 转接中心识别码

Trunk Circuit Identification Code 干线电路识别码

**TCJ** Tactical Communications Jamming 战术通信干扰

**TCL** Takeoff-Cruise-Landing 起飞—巡航—着陆

Thermal Check List 热检查清单

Time-Coherent Light 时间相干光

Transistor Coupled Logic 晶体管耦合逻辑

**TCLPI** Throttle Control Level Position Indicator 油门杆位置指示器

**TCLS** Traffic Control and Landing System 交通管制和着陆系统

**TCLT** Tentative Calculated Landing Time 试算着陆时间

**TCM** Technical Change Meeting 技术更改会议

Technical Change Memorandum 技术更改备忘录

Technical Control Manager 技术管理经理

Technical Coordination Meeting 技术协调例会,技术协调会

Technician Control Mode 技术员管理方式

Telemetry Code Modulation 遥测编码调制

Temperature Compensation 温度补偿

Temperature Control Model 温度控制模型

Temperature Control Module 温度控制模块

Terrain Characteristic Matching 地形特征匹配

Test Controlling Module 测试控制模块

Thrust Control Module 推力控制模块

Time Compression Multiplexer 时间压缩复用器

Time Compression Multiplexing 时间压缩复用

Top Coat Missing 顶部涂层磨损

Total Configuration Management 总布局管理

Total Cost Management 总成本管理

Trellis Coded Modulation 网格编码调制

**TCMA** Tunable Channel Multiple Access 可调信道多址接入

**TCMS** Test Content Management System 试验内容管理系统

**TCN** Terrain Contour Navigation 地形轮廓导航,地形等高线导航

Transportation Control Number 运输控制编号

**TCO** Take Care of 照顾

Tape-Controlled Oscillator 磁带控制振荡器

Technology Cooperation Office 技术合作办公室

Temperature Compensated Oscillator 温度补偿振荡器

Temperature Controlled Oscillator 温度控制振荡器

Termination Contracting Office 终止合同办公室

Total Cost of Ownership 所有权(产权)总费用

Training Course Outline 培训课程大纲

**TCOL** Top Collector 顶部收集器

**TCP** Takeoff Charts Computation Program 起飞图表计算方案

Tape Copy Program 磁带复制程序

Task Change Proposal 任务更改建议书

Technical Change Package 技术更改包

Technical Change Proposal 技术更改建议书

Technical Cooperation Programme 技术合作方案

Termination Connection Point 终端连接点

Test Change Proposal 测试更改建议书

Traffic Control Post 交通管制台

Transfer Control Protocol 传送控制协议

Transmission Control Protocol 传输控制协议

Transmit Channel Processing 发送信道处理

Transport Control Protocol 运输控制规程(协议)

Tropical Cyclone Programme 热带气旋计划

**TCP/IP** Transmission Control Protocol/Internet Protocol 传输控制协议/互联网络协议

Transmission Control Protocol/Internetwork Protocol 传输控制协议/网络间协议

**TCPA** Time to Closest Point of Approach 到进近最近点的时间

**TCPI** To Complete Performance Index 待完成绩效指数

**TCPWR** Terrain Computer Power 地形计算机电源

**TCQ** Throttle Control Quadrant 油门操纵杆弧座

Time, Cost and Quality 时间、成本和质量

Trellis Coded Quantization 网格编码量化

**TCR** Technical Compliance Review 技术符合性审查

Temperature Coefficient of Resistance 电阻的

温度系数

Tool Change Request 工具更改要求

Transfer Control Register 传输控制寄存器

Transport Connection Request 传送连接请求

**TCRD** Time Coordination 时间协调

**TCRF** Transit Connection Related Function 转接连接相关功能

**TCS** Tanking Control System (燃料)加注控制系统

Teleconference Service 会议电话业务

Telephone Control Subsystem 电话控制分系统

Temperature Control System 温度调节系统，调温系统

Terminal Call Service 终端呼叫业务

Terminating Call Screening 终端来话筛选

Terrain Clearance Sensor 地形间距传感器

Test and Command Status 测试和指令状态

Thermal Control System 热控制系统

Tilt-Control Switch 倾转操纵电门

Tool Control Status 工具控制状态

Total Communication System 总体通信系统

Tower Communication System 塔台通信系统

Tracking and Communication Subsystem 跟踪和通信分系统

Traction Control System 牵引力控制系统

Traffic Control Subsystem 话务量控制子系统

Trajectory Control System 轨迹控制系统

Transmittal Coordination Sheet 发送协调单，传输协调单

Transportable Communication System 移动式通信系统

TV Camera Set 电视摄像机

**TCSEC** Trusted Computer System Evaluation Criteria 可信的计算机系统评估准则

**TCSG** Testing Calibration Signal Generator 测试校准信号发生器

Time Code Signal Generator 时间码信号发生器

**TCSP** Tactical Communications Satellite Program 战术通信卫星程序

Traffic Control Simulation Project 交通管制模拟计划

**TCSS** Tail Cone Structure Subsystem 尾锥结构分系统

Terminal Communication Switching System 终端通信交换系统

Tool Control Status System 工具控制状态系统

**TCST** Technical Committee for Standardization 标准化技术委员会

Turbine Case Support Temperature 涡轮机匣支撑温度

**TCT** Fan Air Valve Control Thermostat 风扇活门温度控制恒温器

Tail Control Technology 尾部控制技术

Technology Coordinating Team 技术协调小组

Temperature Control Thermostat 恒温器控制温度

Terminal Control Table 终端控制表

Time Code Translator 时间编码转换器

Time Compression Technology 时间压缩技术

Transaction Cost Theory 交易费用理论

Type Certification Team 型号合格审查组

**TCTD** Thermoplastic Composite Technology Development 热塑料复合材料技术开发

**TCTI** Time Compliance Technical Instruction 时间符合性技术说明

**TCTO** Time Compliance Technical Order 时间符合性技术规程

**TCU** Tape Control Unit 磁带控制组件

Temperature Control Unit 温度控制组件

Temperature Controller Unit 温度控制器组件

Terminal Connection Unit 终端连接单元

Terminal Control Unit 终端控制装置

Threshold Control Unit 门限控制组件

Throttle Control Unit 油门控制组件

Thrust Control Unit 推力控制组件

Timing and Control Unit 定时控制组件

Torque Control Unit 扭矩控制组件

Towering Cumulus 塔状积云，浓积云

Transfer Control Unit 转移控制单元

Transmission Control Unit 传输控制单元

**TCV**　Temperature Control Valve 温度控制活门，温度控制组件

Terminal Configured Vehicle 候机楼配置的交通车

Throttle Control Valve 节流活门

**TCW**　Terminal Controller Workstation 终端管制员工作站

Time Critical Warning 关键时间警告

**TCWP**　Threshold Crossing Waypoint 跑道入口飞越点

**TCXO**　Temperature Controlled Crystal Oscillator 温控晶体振荡器

**TCYC**　Tropical Cyclone 热带气旋

**TCZD**　Temperature Compensated Zener Diode 温度补偿的齐纳二极管

**TD**　Table of Distribution 分配表

Teacher Discount 教员折扣

Technical Data 技术数据，技术资料

Technical Department 技术部门

Technical Directive 技术指示

Technical Dossier 技术档案

Temperature Differential 温差

Terrain Display 地形显示

Test Data 试验数据

Test Directive 试验指示

Test Director 测试主任

Testing Device 测试装置

Threshold Extension Demodulator 门限扩展解调器

Time Delay 时间延迟

Time Delay Equalizer 时延均衡器

Time Difference 时差

Time Division 时分

Time Domain 时域

Time of Departure 飞离时间，离港时间

Tool Design 工具设计

Tool Document 工具文件

Touchdown 着地，接地

Touchdown Aim Point 接地瞄准点

Trace Driven 追踪驱动

Transmission Deviation 传输偏差

Transmitter Distributor 发射机分配器，发送器分配器

Transposition Docking 换位对接

Trimming Die 修整模

Tropical Depression 热带低压

Turbine Driven 涡轮驱动的

**TD&D**　Technical Data and Documents 技术数据和文档

**TD&E**　Tactics Development and Evaluation 战术研究和鉴定

**TD/TDI**　Tool Drawings/Tool Drawing Index 工具图/工具图索引

**TDA**　Technology Development Approach 技术开发途径

Trunk Distribution Amplifier 干线分布放大器

Tunnel Diode Amplifier 隧道二极管放大器

Type Design Approval 型号设计批准书

**TDAN**　Time-Division Analog Network 时分模拟网络

**TDAS**　Test-Data Acquisition System 试验数据采集系统

Tracking and Data Acquisition System 跟踪和数据采集系统

Tracking and Data Acquisition Satellite 跟踪和数据采集卫星

**TDASS**　Tracking and Data Acquisition Satellite System 跟踪和数据采集卫星系统

**TDC**　Target Designation Control 目标指示控制

Target Designator Control 目标指示器控制

Technical Data Center 技术数据中心

Technical Development Center 技术发展中心（美国联邦航空局）

Telemetry Data Center 遥测数据中心（美国）

Throttle Designator Controller 油门指示控制

器

Time Delay Closing 延时结束,延时关闭

Time Domain Coding 时域编码

Top Dead Center (活塞在气缸内的) 上死点

Type Design Change 型号设计更改

**TDCM** Transistor Driver Core Memory 晶体管激励器磁芯储存器

**TDCR** Type Design Change Request 型号设计更改申请

**TDCS** Target Detection-Conversion Sensor 目标探测转换传感器

Technical Documentation Control System 技术文件控制系统

Tone Digital Command System 单音(频)数字指令系统

Traffic Data Collection System 交通数据收集系统

**TDD** Target Detection Device 目标探测装置

Technical Design Directive 技术设计指令

Telemetry Data Digitizer 遥控数据数字转换器

Temporary Design Departure 临时设计离港

**TDDL** Time-Division Data Link 时分数据链

**TDDN** Time-Division Digital Network 时分数字网络

**TDE** Time Domain Equalizer 时域均衡器

**TDEC** Telephone Line Digital Error Checking 电话线数字误差检查

**TDFC** Technical Documentation Familiarization Course 技术文件熟知课程

**TDG** Twist Drill Gage 螺纹钻头规

**TDH** Total Dynamic Head 全动压头

**TDI** Thermal De-Ice 热电耦除冰

Time Delay Integration 延时积分,延时积累

Tool Dimensions Index 工具尺寸索引

Tooling Data Identifier 工装数据标记

Tooling Dimensions Identifier 工装尺寸标记

**TDIO** Timing Data Input Output 定时数据输入输出

**TDISTR** Tape Distributor 磁带分配器

**TDL** Time Delay Logic 延时逻辑

Tunnel Diode Logic 隧道二极管逻辑

**TDLS** Tower Data Link Service 塔台数据链服务

Tower Data Link System 塔台数据链系统

**TDM** Time Division Modulation 时分调制

Time Division Multiplex/Multiplexing 时分复用

Time Division Multiplexer 时分多路复用器

Tool Design Manual 工具设计手册

Track Definition Message 航迹定义电报,跟踪定义信息

Trouble Detection and Monitoring 故障探测和监控

**TDMA** Time Division Multiplex Address 时分多址

**TDMAS** Teletype/Data Multiplexer Address System 电传/数据多路转换器地址系统

Time Division Multiple Access System 时分多路存取系统

**TDMAT** Time Domain Multiple Access Technique 时域多路存取技术

**TDMC** Time Division Multiplexed Channel 时分复用信道

**TDMD** Time Division Multiplex Device 时分复用装置

**TDME** Tacan Distance Measuring Equipment 塔康测距仪

Test Diagnostic Measurement Equipment 测试诊断测量设备

Time Division Multiplex Equipment 时分复用设备

**TDMF** Text Data Management Facility 文本数据管理设备

**TDMM** Telecommunications Distribution Methods Manual 电信配线方法手册

Time Division Memory Module 时分存储模块

**TDMP** Test Data Management Plan 测试数据管理计划

**TDMR** Time Delay Manual Relay 人工延时继电器

**TDMS**   Technical Data Management System 技术数据管理系统

Technical Document Management System 技术文档管理系统

Terminal Data Management System 终端数据管理系统

Terminal Display Management System 终端显示管理系统

Test Data Management System 试验数据管理系统

Transmission Distortion Measuring Set 传送失真测量装置

**TDM-VDMA**   TDM-Variable Destination Multiple Access 时分复用—可变终点多址接入

**TDNN**   Time Delay Neural Network 延时神经网络

**TDNW**   Time Division Network 时分网络

**TDO**   Technical Development Objective 技术开发目标

Technical Direction Order 技术指导规程,技术指导指令

Time Delay Opening 延时打开

Tornado 龙卷风

**TDOA**   Time Delay of Arrival 到达的时间延误

Time Difference of Arrival 到达的时间差

Time or Direction of Arrival 到达的时间或方向

**TDOP**   Time Dilution of Position 时间位置衰减因子

Time Dilution of Precision 时间精度衰减因子

**TDOR**   Two-Door（Reservations Code）双门

**TDP**   Takeoff Decision Point 起飞决断点

Tag Distribution Protocol 标记分配协议

Technical Data Package 技术数据包

Technical Development Plan 技术开发计划

Thermal Design Power 热量设计功率

Touch Down Point 接地点

Touchdown Protection 接地保护

Trigger Detection Point 触发检测点

**TDPC**   Test Data Processing Center 测试数据处理中心

**TDPF**   Tail Damping Power Factor 尾翼阻尼功率因数

**TDPL**   Technical Data Package List 技术数据包清单

**TDP-N**   Trigger Detection Point-Notification 触发检测点通知

**TDP-R**   Trigger Detection Point-Request 触发检测点请求

**TDPS**   Tactical Data Processing System 战术数据处理系统

Technical Data Products and Services 技术数据产品与服务

Telemetry and Data Processing Station 遥测和数据处理站

**TDR**   Tail Damping Ratio 尾部阻尼比

Technical Data Requirement 技术数据要求

Technical Definition Report 技术定义报告

Technical Design Report 技术设计报告

Technical Development Report 技术开发报告

Technical Development Requirement 技术开发要求

Technical Directive Records 技术指令记录

Test Data Request 测试数据要求

Test Discrepancy Report 测试故障报告,测试缺陷报告

Time Delay Relay 延时继电器

Time Dependent Routing 按时间选择路由

Time Domain Reflectometer 时域反射仪

Tool Design Request 工具设计要求

Tooling Data Record 工装数据记录

Torque Differential Receiver 扭矩差接收器

Training Design Request 训练设计申请

Transfer Data Record 传输数据记录

Transfer Design Review 传输设计审查

Transmit Data Register 传输数据寄存器

Transponder 应答机

**TDRC**   Temperature Difference Recording Controller 温差记录控制器

**TDRS** Tracking and Data Relay Satellite 跟踪及数据中继卫星

**TDRSS** Tracking and Data Relay Satellite System 跟踪和数据中继卫星系统

**TDS** Tape Data Selector 磁带数据选择器

Technical Data Sheet 技术数据单

Technical Demonstration System 技术演示验证系统

Time Division Switching 时分交换

Touch Down Speed 接地速度

Track Data Simulator 航迹数据模拟器

Transit Delay Selection 转接时延选择

**TDSC** Technical Data Steering Committee 技术数据指导委员会

**TDSP** Technical Data Support Package 技术数据支援计划

Time Delay Short Pulse 延迟时间短脉冲

Time Dependent Shortest Path 依赖时间最短路径

Top Down Structured Programming 自上而下式结构性程序编制

**TDSR** Territorial Development Strategy Review 全港发展策略评审

**TDSS** Technical Data Storage System 技术数据存储系统

Technology Decision Support System 技术决策支持系统

Telemetry Data Signal Simulator 遥测数据信号模拟器

Tube Data Storage System 管道数据存储系统

**TDST** Tower Data Services Terminal 塔台数据服务终端

**TDT** Time and Date Table 时间与日期表

Transonic Dynamics Tunel 跨声速动力学风洞

**TDTL** Tunnel Diode Transistor Logic 隧道二极管晶体管逻辑

**TDU** Target Detection Unit 目标检测组件

Technical Description Unit 技术描述组件

Telephone Distribution Unit 电话分配组件

Time Display Unit 时间显示组件

**TDV** Tailored Demonstration/Validation 改装试验/批准,修整试验/批准

Touchdown Velocity 接地速度

**TDW** Tower Display Workstation 塔台显示工作站

**TDWR** Terminal Doppler Weather Radar 终端区多普勒气象雷达

**TDWT** Transonic Dynamics Wind Tunnel 跨音速动力学风洞

**TDX** Time-Division Exchange 时分交换机

**TDY** Temporary Duty 临时职责,临时任务

Today 今日

**TDZ** Touchdown Zone 接地区域,接地地带

**TDZE** Touchdown Zone Elevation 接地区标高,接地带标高

**TDZL** Touchdown Zone Lights 接地区灯光

**TE** HP Stage Air Temperature 高压级空气温度

Tangent Elevation 正切标高

Technic Engineer 技术工程师

Telecommunication Entity 电信实体

Terminal Equipment 终端设备

Test Engineer 测试工程师

Test Engineering 测试工程

Test Equipment 测试设备

Thermal Element 热元件

Thermoelectric 热电的

Time Error 时间误差

Tooling Engineer 工装工程师

Totally Enclosed 全封闭(的)

Trailing Edge 后缘(飞机的)襟翼,阻力板

Training Equipment 培训设备

Transit Exchange 转接局

Transverse Electric 横电波

Turbine Engine 涡轮发动机

**TEC** Total Electron Content 总电子容量

**TER** Trailing Edge Radius 后缘半径

**TEAC** Turbine Engine Analysis Check 涡轮发动机分析检查

**TEAMS** Test Evaluation and Monitoring System 试验评估与监控系统

Traffic Engineering and Monitoring Server 交通工程与监测服务器

**TEASE** Tracking Errors and Simulation Evaluation 跟踪误差和模拟评估

**TEC** Temporary Equipment Configuration 临时设备组合,临时设备配置

Transferred Empty Cell 传递的空信元

Turbine Exhaust Case 涡轮排气机匣

**TECEVAL** Technical Evaluation 技术评估

**TECH** Technical or Technological 技术的

Technician 技术员,技师

Technology 技术

**TECH PUB** Technical Publications 技术出版物

**TECO** Turbine Engine Checkout 涡轮发动机检查

**TECOM** Test and Evaluation Command 测试和评估指令

**TECR** Technical Reason 技术原因,技术理由

**TED** Test, Evaluation and Development 试验、鉴定和发展

Tool and Equipment Drawing 工具和设备图纸

Traffic Engineering Database 流量工程数据库

Trailing Edge Down 后缘向下

Transfer Error Detection 传送差错检测

Translation Error Detection 转换错误检测

**TEDB** Tool and Equipment Data Bank 工具和设备数据库

**TEDL** Test Equipment Description Language 测试设备描述语言

Transferred-Electron-Device Logic 电荷转移器件逻辑

**TEDM** Turbofan Engine Diagnostic Monitoring 涡轮风扇发动机诊断监控

**TEEAR** Test Equipment Error Analysis Report 试验设备误差分析报告

**TEF** Trailing Edge Flap 后缘襟翼

**TEFC** Totally Enclosed Fan Cooled 全封闭风扇冷却

**TEI** Terminal Endpoint Identifier 终端终点标识符

Terminal Equipment Identifier 终端设备标识符

Test Element Identifiers 测试单元标识

Test Equipment Interrupt 测试设备中断

Tool and Equipment Index 工具和设备索引

**TEJ** Transverse Expansion Joint 横向膨胀连接

**TEL** Taxiway Edge Lights 滑行道边灯

Telegram or Telegraph or Telephone 电报或电话

Tetraethyl Lead 四乙铅

Tool and Equipment List 工具和设备清单

Trailing Edge Left 后缘左

Training Equipment List 培训设备清单

**TELB** Telephone Booth 电话亭

**TELCO** Telephone Company 电话公司

**TELECOM** French Satellite System 法国卫星系统

Telecommunication 电信

**TELEDAC** Telemetric Data Converter 遥测数据变换器

**TELERAN** Television Radar Air Navigation 电视雷达空中导航系统

**TELEREMIT** Telegraphically Remit 电汇

**TELESAT** Canadian Domestic Satellite System 加拿大国内卫星系统

Telecommunications Satellite 实用通讯卫星

**TELINT** Telemetry Intelligence 遥测情报

**TELMTR** Telemotor 遥测马达,遥控传动装置

**TELS** Turbine Engine Load Simulator 涡轮发动机负载模拟器

**TELSCOM** Telemetry Surveillance Communications 遥测监视通信

**TEM** Technical Error Message 技术错误文电

Template 模板

Threat and Error Management 威胁与差错管

理

Thunderstorm Environmental Model 雷暴环境模型

Tool and Equipment Manual 工具和设备手册

Transmission Electron Microscopy 透射电子显微镜

Transport Environment Monitoring System 运输环境监测系统

Transverse Electromagnetic Mode 横向电磁波模式

**TEMA** Tubular Exchanger Manufacturers Association 管式热交换器制造商协会

**TEMI** Terminating Equipment Mock-up Input 终端设备模拟输入

**TEMIS** Target Engineering Management Information System 目标工程管理信息系统

**TEMP** Temperature 温度

Test and Evaluation Master Plan 测试和评估主计划

**TEMPL** Template 样板,模板

**TEMPS** Transportable Electromagnetic Pulse Simulator 移动式电磁脉冲模拟器

Transportable Empennage Simulator 可移动尾翼模拟器

**TEMS** Technical Engineering and Management Support 技术工程与管理支援

Telecommunications Management System 电信管理系统

Test and Evaluation Mission Simulator 测试和评估任务模拟器

Tracking Error Model Studies 跟踪误差模型研究

Turbine Engine Monitoring System 涡轮发动机监控系统

**TENS** Tension 张力

**TEO** Engine Oil Temperature 发动机滑油温度

**TEPI** Training Equipment Planning Information 培训设备计划信息

**TER** Tertiary 第三级的

Trailing Edge Right 右后缘

**TERC** Transportation Environmental Resource Center 交通运输环境资源中心

Turbine Engine Research Center 涡轮发动机研究中心

**TERAVD** Terrain Avoidance 地形回避

**TERCOM** Terrain Comparison 地形比较

Terrain Contour Mapping 地形轮廓测绘

Terrain Contour Matching 地形轮廓匹配

Terrain Correlation Mapping 地形相关测绘

**TERFLW** Terrain Following 地形跟踪

**TERG** Training Equipment Requirements Guide 培训设备要求指南

**TERM** Terminal 终端,接线柱

Terminate or Terminating 结束,终止

Test Requirements Model 测试要求模型

**TERMN** Termination 终止,终端

**TERN** Terminal and Enroute Navigation 终端和航线导航

**TERP** Technical Evaluation Review Panel 技术评估审查委员会

Turbine Engine Reliability Program 涡轮发动机可靠性计划

**TERPS** Terminal Instrument Procedures 终端仪表程序

**TERR** Terrain 地形

Territory 领土

**TES** Telephone Earth Station 电话地球站

Test Evaluation Squadron 测试评估组

Time Element Scrambler 时域加扰器

**TESS** Trainer External Simulation System 教员外部模拟系统

Translator Editor Software System 翻译编辑器软件系统

**TET** Technical Evaluation Team 技术评估组

Technical Evaluation Test 技术评估试验

Tetrachloride 四氯化物

Tetrode 四极管

Turbine Entry Temperature 涡轮进口温度

Turbine Exit Temperature 涡轮出口温度

**TETCP** Turbine Engine Time Control Program

涡轮发动机时间控制大纲

**TETFLEYNE** Tetrafluoroethylene 四氟乙烯

**TETRA** Terrestrial Trunked Radio 地面集群无线电

Turbine Engine Transient Response Analysis 涡轮发动机瞬态响应分析

Turbine Engine Transient Response Analyzer 涡轮发动机瞬态响应分析器

**TETS** Third Echelon Test Set 三级测试装置

Turbine Engine Technology Symposium 涡轮发动机技术研讨会

**TETWOG** Turbine Engine Testing Working Group 涡轮发动机测试工作组

**TEU** Trailing Edge up 后缘向上

Twenty-foot Equivalent Unit 二十英尺当量单位

**TEUE** Training Effectiveness User Evaluation 用户评定训练效果

**TEV** Test, Evaluation and Verification 试验、评估与验证

**TEVCS** Turbine Engine Variable Cycle Selection 涡轮发动机可变循环选择

**TEW** Transverse Electromagnetic Wave 横电磁波

**TEWS** Tactical Early Warning System 战术预警系统

Tactical Electronic Warfare System 战术电子战系统

Test and Evaluation Work Sheet 测试和评估工作单

**TEX** Telluride, CO, USA(Airport Code) 美国科罗拉多州特柳赖德机场

**TF** Tapered Fiber 锥形光纤

Target Fix 目标固定

Terminal Forecast 终端预报

Terrain Following 地形跟踪

Test Facility 测试设施

Test Flight 试飞

Test to Failure 故障试验

Thin Film 薄膜,胶片

Thread Forming 螺纹形成

Through Flight 过站飞行

Time of Flight 飞行时间

Tool Fabrication 工具制造

Training Flight 训练飞行

Transfer Function 传送功能

Transition Fiber 过渡光纤

Trust Fund 信托基金

Type of Flight 飞行方式

**T-F** Time of Fail 失效时间

**TFA** Technology Forecasting and Assessment 技术预测与评估

Trust Fund Account 信托基金账户

**TFACU** Terrain Following Avionics Control Unit 地形跟踪航空电子设备控制组件

**TFASA** Test Flying Academy of South Africa 南非试飞学院

**TFC** Terminal Flight Control 终端飞行控制

Total Fixed Costs 固定成本总额

Total Fuel Consumption 总耗油量

Traffic 交通(指避撞系统)

Traffic Flow Control 业务流量控制

**TFCP** Terrain Following Control Panel 地形跟踪控制板

**TFD** T-coded Form Die T编码成型模

Transaction Flow Diagram 业务流程图

**TFDB** Traffic Flow Database 交通流量数据库

**TFE** Threat Free Environment 无威胁环境

Turbofan Engine 涡轮风扇发动机

**TFG** Traffic Forecasting Groups 运输量预测小组

Transmit Format Generator 发射格式产生器

**TFI** Time-Frequency Interpolation 时频内插

**TFL** Teflon 特氟隆

Transfer Flight Level 转换飞行高度层

Transient Fault Locator 瞬时故障定位器

Transition Flight Level 过渡高度层,飞越高度层

**TFM** Tactical Flight Management 战术飞行管理

Tamed Frequency Modulation 平滑调频

Test Flight Manual 试飞手册

Time-quantized Frequency Modulation 时间量化频率调制

Total Facility Management 全设施管理

Traffic Flow Management 交通流量管理

Transfer Function Model 传递函数模型

Transmitter Frequency Multiplier 发射器频率倍增器

Transportation Financial Management 运输财务管理

**TFMV** Total Flow Metering Valve 总流量计量活门

**TFO** Temporary Flight Order 临时飞行命令

Test Facility Operation 测试设施作业

Tons Fuel Oil 燃油吨数

Tool Fabrication Orders 工具制造规程

**TFOM** Testability Figure of Merit 测试能力品质因素

Thin-Film Optical Modulator 薄膜光学调制器

**TFORE** Therefore 因此

**TF-OSF** Transfer Function-Operation System Function 传送功能—运行系统功能

**TFOV** Total Field of View 总视场

**TFP** Time and Frequency Processing 时间与频率处理

Total Factor Productivity 总要素生产率

**TFR** Temporary Flight Restriction 临时飞行限制

Terrain Following Radar 地形跟踪雷达

Time Frequency Reference 时间频率参考

Total Flight Restriction 总飞行限制

Transfer 转移,换车(机)

Trouble and Failure Report 故障和失效报告

**TFRC** Terrain Following Radar Control 地形跟踪雷达控制

**TFRD** Test Facility Requirements Document 测试设施要求文件

**TFRS** Terrain Following Radar Set 地形跟踪雷达装置

**TFS** Traffic by Flight Stage 按飞行段运输

Traffic Forecasting Systems 交通预测系统

**TFSDC** Terrain Following Signal Data Converter 地形跟踪信号数据转换器

**TFSF** Time to First System Failure 系统首次故障时间

Tons Force per Square Foot 吨力/平方英尺

**TFT** Temporary Facility Tools 临时设施工具

Thin Film Transistor 薄膜晶体管

Threshold Failure Temperature 故障临界温度

Trim for Take-off (飞机)起飞配平

**TFTP** Trivial File Transfer Protocol 普通文件传输协议

**TFTS** Terrestrial Flight Telecommunications System 全球飞行电信系统

Terrestrial Flight Telephone Service 全球飞行电话服务

Terrestrial Flight Telephone System 全球飞行电话系统

**TFU** Technical Follow-up 技术跟踪

Telecommunication Flying Unit 电信飞行小队

Theoretical First Unit 理论上的第一个组件

Time and Frequency Unit 时间和频率单元

Timing and Frequency Unit 定时与频率装置

Transient Fuel Unit 过渡燃油组件

**TFUEL** Fuel Temperature 燃油温度

**TFV** Tractive Force Variation 纵向强迫震动

Transient Fuel Valve 瞬时燃油阀

**TFW** Trip Fuel Weight 航段耗油量

**TFZ** Traffic Zone 交通地带

**TG** Tag 标签

Technical Group 技术组

Terminal Guidance 终端引导,候机楼引导

Time to Go 待飞时间

Transmission Gate 传输门

Turbine Generator 涡轮发电机

**TGA** Trans Global Airways,Inc.(USA) 环球航空公司(美国)

Turbine Gearbox Assembly 涡轮齿轮箱组件

**TGB** Transfer Gearbox 转换齿轮箱

**TGC**　Throttle Governor Control 油门调节器控制

Time Gain Control 时间增益控制

Transmit Gain Control 发射增益控制

Turbulence Gain Control 湍流增益控制

**TGCS**　Total Gas Control System 总燃气控制系统

Transportable Ground Communication Station 移动地面通信站

Travel Group Charters 团体旅游包机

**TGG**　Kuala Terengganu, Malaysia ( Airport Code) 马来西亚瓜拉丁加奴机场

**TGL**　Thermal Guidance Laboratory 热制导实验室

Touch-and-Go Landing 接地复飞, 连续起落

**TGLS**　Tongueless 缄默

**TGM**　Trunk Group Multiplex 中继线群复用

**TGMS**　Third Generation Mobile System 第三代移动通信系统

**TGNP**　The Greatest Noise Power 最大噪声功率

**TGOTS**　Tooling Grind Order Tracking System 工具研磨指令跟踪系统

**TGP**　Terminal Guidance Program 终端制导计划

**TGS**　Taxiing Guidance System 滑行引导系统

Telemetry Ground Station 地面遥测站

Time Generation System 时间发生系统

True Ground Speed 真地速

**TGT**　Tail Gate 尾部舱门

Target 靶标, 目标

Turbine Gas Temperature 涡轮燃气温度

**TGU**　Tegucigalpa, Honduras ( Airport Code ) 洪都拉斯特古西加尔巴机场

**TGW**　Total Gross Weight 总重

**TH**　Tail-Heavy 尾重

Test Hardness 测试硬度

Thailand 泰国

Thick 厚的

Thickness 厚度

Through-flight 过站飞行

Time Hopping 跳时

Transition Height 过渡高

Transport Helicopter 运输直升机

True Heading 真航向

True High 真高

True Horizontal 真水平( 的), 真地平( 的)

**THA**　Task Hazard Analysis 任务风险分析

Test Hazard Analysis 试验风险分析

**THC**　Temperature and Humidity Control 温度与湿度控制

Terminal Handling Charge ( 集装箱码头) 装卸作业费

Thermal Converter 热转换器

Tracking Handle Control 跟踪处理控制

**THD**　Thread 螺纹, 线索

Through-Hole Device 通孔插装元器件

Total Harmonic Distortion 总谐波失真

**THDG**　True Heading 真航向

**THDNK**　Threaded Neck 带螺纹颈, 带螺纹短管

**THDPC**　Thread Piece 螺纹件

**THEOD**　Theodolite 经纬仪

**THEOR**　Theoretical 理论上的

**THERAP**　Therapeutic 治疗的

**THERM**　Thermometer 温度计

**THERMO**　Thermostat 恒温器, 热动开关

**THERP**　Technique of Human Error Rate Prediction 人的失误率预测技术

**THF**　Berlin, Germany-Tempelhof ( Airport Code) 德国柏林滕伯尔霍夫机场

Threading Fixture 螺纹夹具

Tremendously High Frequency ( 300GHz ~ 3THz)至高频

**THK**　Thick 厚的

**THKF**　Thick Film 厚膜

**THKNR**　Thickener 浓缩剂, 浓缩器

**THKNS**　Thickness 厚度

**THL**　Threshold Light 跑道入口灯

**THMS**　Thermistor 热敏电阻

**THO**　Though 虽然

**THP**　Thrust Horsepower 推进马力

**THR** Tehran, Iran-Mehrabad (Airport Code)
伊朗德黑兰梅赫拉巴德机场
Thermometer 温度计
Threshold 跑道入口, 入口, 门限
Throttle 油门

**THR HOLD** Throttle Hold 油门入口

**THR REF** Thrust Reference 推力基准

**THR REV** Thrust Reverser 反推装置

**THRM** Thermal 热的

**THRMST** Thermostat 恒温器

**THROT PUSH** Throttle Pusher 推油门作动器

**THRSHL** Thrust Shell 推力整流罩

**THRT** Throat (风洞)试验段, 喉部

**THRU** Through 经过

**THS** Tailplane Horizontal Stabilizer 尾翼水平安定面
Trimmable Horizontal Stabilizer 可配平的水平安定面

**THSHD** Threshold 跑道入口, 门限

**THT** Through Hole Technology 插入式封装技术

**THU** Thursday 星期四

**THWR** Thrower 投掷器

**THYMOTROL** Thyratron Motor Control 闸流管马达控制

**THYR** Thyristor 可控硅整流器, 闸流管

**TI** Tape Inverter 倒带机
Target Identification 目标识别
Target Indicator 目标指示器
Technical Inspection 技术检查
Technical Instruction 技术说明, 技术指令
Technical Interchange 技术互换
Test Increment 测试增量
Texas Instruments 德克萨斯仪表
Time Index 时间索引
Track Initiate 航迹起始
Track to an Interrupt 航迹切入点
Transmission Identification 发射识别(符)
Transmit Interval 发送间隔

**TIA** Technology Investment Agreement 技术投资协议
Telecommunication Industry Association (美国)电信工业协会
Temperature Alarm Indicator 温度报警指示器
Tirana, Albania (Airport Code) 阿尔巴尼亚地拉纳机场
Training Integration Area 综合训练区
Transimpedance Amplifier 互阻抗放大器
Type Inspection Authority 型号检查授权
Type Inspection Authorization 型号检验授权书

**TIAA** New Tokyo International Airport Authority 新东京国际机场当局

**TIAC** Technical Information Advisory Committee 技术情报咨询委员会

**TIACA** The International Air Cargo Association 国际航空货运协会

**TIALD** Thermal Imaging Airborne Laser Designator 热成像机载激光照射器

**TIAS** Target Identification and Acquisition System 目标识别及获得系统
Telephone Inventory Accounting System 电话库存统计系统
True Indicated Airspeed 实际指示空速

**TIB** Tag Information Base 标记信息库
Task Information Base 任务信息库
Terminal Interface Block 终端接口功能块
Test Information Base 测试信息库

**TIBA** Traffic Information Broadcast by Aircraft 航空器交通情报广播, 飞机交通信息广播

**TIBS** Telephone Information Briefing Service 电话信息简报服务
Transcribed Information Briefing Service 手抄信息简报服务

**TIC** Technical Information Center 技术信息中心
Temperature Indicating Controller 温度指示控制器
Temperature of the Initial Combustion 起始燃烧温度

Terminal International Center 国际终端中心

Total Investment Cost 总投资成本

Traffic Information Centre 交通信息中心

Turbine Impingement Cooling 涡轮冲击冷却

**TICC** Technical Information and Communication Committee 技术信息和通信委员会

**TICS** Terminal Interface Control System 终端接口控制系统

**TICTAC** Time Compression Tactical Communications 时间压缩战术通信

**TID** Technical Impact Difference 技术影响差异

Technical Information Document 技术信息文件

Terminal Identification 终端标识

Touch Input Device 触摸式输入器

**TIDE** Transportable Interface Development Equipment 可移动接口开发设备

**TIDN** Test Identification Number 测试识别号

**TIDR** Tool Investigation and Disposition Request 工具调查与处理申请

**TIDS** Test Instrumentation Data System 测试仪表数据系统

**TIE** Technical Integration and Evaluation 技术集成和评估

Terminal Interface Equipment 终端接口设备

Terrestrial Interface Equipment 地面接口设备

Test Interface Electronics 电子测试接口设备

Time Interval Error 时间间隔误差

**TIES** Technical Information Exchange System 技术信息交换系统

Ticketing Information Exchange Standard 票务信息交换标准

Transmission and Information Exchange System 传输和信息交换系统

**TIF** Tagged Image File 标记图像文件

Technical Information File 技术信息文档

Telephone Interference Factor 电话干扰因素

Temporary Internet Files 网络临时文件

**TIFF** Tagged Image File Format 标签图像文件格式

Tagged Information File Format 标记信息文件格式

**TIFS** Total in-Flight Simulator 全飞行过程模拟机

**TIGS** Terminal Independent Graphics System 与终端无关的图形系统

**TIGT** Turbine Inlet Gas Temperature 涡轮进口燃气温度

**TIJ** Tijuana, Mexico（Airport Code）墨西哥蒂华纳机场

**TIL** Until 直到，到…为止

**TILO** Technical Industrial Liaison Office 技术工业联络处

**TILS** Tactical Instrument Landing System 战术仪表着陆系统

**TIM** Technical Interchange Meeting 技术交流会议

Terrestrial Interface Module 地面接口模块

Time Interval Meter 时间间隔计

Time Meter 计时表

Track Limitation 航迹限制

Translator Interface Module 转换器接口模块

Travel Information Manual 旅行信息手册

Trunk Interface Module 中继接口模块

**TIME** Technique for Information Management and Employment 情报管理与使用技术

Total Inventory Management for Engineering 工程全库存管理

**TIMP** Technical Information Management Plan 技术信息管理计划

**TIMR** Test Item Measurement Requirement 试验项目测量要求

**TIMS** Technical Information Management System 技术信息管理系统

Telecommunications Information Management System 通讯信息管理系统

Tool Inventory Management System 工具库存管理系统

**TINA** Telecommunication Information Network

Architecture 电信信息网络架构

**TINS** Thermal Imaging Navigation Set 热成像导航装置

**TIP** Technical Improvement Program 技术改进大纲

Terminal Interface Package 终端接口程序包

Terminal Interface Processor 终端接口处理机

Test in Progress 测试在进行中

Track Initiation and Prediction 航迹起始和预测

Training Information Points 培训信息要点

Transaction Interface Processor 事务处理接口处理机

Tripoli, Libya( Airport Code) 利比亚的黎波里机场

Turbine Inlet Pressure 涡轮入口压力

**TIPH** Taxi into Position and Hold 进入跑道等待

**TIPI** Tactical Information Processing and Interpretation 战术信息处理与说明

**TIPS** Technical Information Processing System 技术情报处理系统

Telecommunications Information Processing System 电信信息处理系统

Telemetry Integrated Processing System 遥测综合处理系统

Test Information Planning Sheet 测试信息计划页

Test Information Processing System 测试信息处理系统

Test Item Planning Sheet 测试项目计划页

Total Integrated Pneumatic System 总体综合气动系统

**TIR** Target Illuminating Radar 目标照射雷达

Technical Incident Report 技术事故报告,技术事件报告

Test Inspection Record 测试检查记录

Total Indicator Reading 全指示器读数

Type Inspection Record 型号检查记录

**TIRC** Technical Information Retrieval Center 技术情报检索中心

**TIROS** Television and Infra-Red Observation Satellite 电视及红外光观测卫星

TV Infra-Red Observation Satellite 红外电视观测卫星

**TIRS** Training Information Records System 训练信息记录系统

**TIS** Technology Integration Study 技术集成研究

Telephone Information Service 电话信息业务

Terminal Information Service Message 终端信息服务电报

Terminal Interface Subsystem 终端接口子系统

Test Increment Summary 测试增量摘要

Test Information Sheet 测试信息页

Tracking Instrumentation Subsystem 跟踪测试仪表分系统

Traffic Information Service 交通情报服务

Transport Information System 运输资讯系统

Travel Information System 旅游信息系统

**TIS-B** Traffic Information Service-Broadcasting 交通信息广播服务

**TISDL** Traffic Information Service Data Link 交通信息服务数据链

**TISEO** Target Identification System, Electro-Optical 目标识别光电系统,光电目标识别系统

**TISI** Thai Industrial Standards Institute 泰国工业标准学会

**TIT** Turbine Inlet Temperature 涡轮进口温度

**TITAS** Thermal Imaging Tracking Aid System 热成像跟踪辅助系统

**TITS** Turbine Inlet Temperature Sensor 涡轮进口温度传感器

**TIU** Tank Interface Unit 油箱接口组件

Tape Identification Unit 磁带识别组件

Tape Input Unit 磁带输入组件

Telephone Interface Unit 电话接口组件

Terrestrial Interface Unit 地面接口单元

Transmission Interface Unit 传输接口单元

**TIVV** Thermactor Idle Vacuum Valve 热控怠速真空活门(阀)

**TIWG** Technical Implementation Working Group 技术实施工作小组

Technical Integration Working Group 技术集成工作组

Test Integration Working Group 试验集成工作组

**TIZ** Traffic Information Zone 交通信息区

**TJ** Trim Jig 修饰工具,调整夹具

Turbojet 涡轮喷气发动机,涡轮喷气机

**TJB** Time-sharing Job Control Block 分时作业控制块

**TJC** Trajectory Chart 轨道图

**TJD** T-coded Joggle Die T 码榫接模具

**TJK** Tajikistan ( ISO Country Code ) 塔吉克斯坦

**TJP** Turbo Jet Propulsion 涡轮喷气推进

**TJR** Trunk and Junction Routing 干线和中继线选路

**TJRJ** Turboramjet 涡轮冲压式喷气发动机

**TK** Ground Track Angle 地面航迹角

Tank 油箱

Track 航迹

Track Angle 航迹角

**TKA** Tonne-Kilometre Available 可用吨公里

**TK CHG** Track Change 航迹改变

**TKCC** Trunk Call Control 中继线呼叫控制

**TKE** Track Angle Error 航迹角偏差

**TKG** Taking 采取

**TKIC** Trunk Interface Circuit 中继线接口电路

**TKK** Truk, Caroline Islands, Micronesia ( Airport Code ) 密克罗尼西亚罗林群岛特鲁克机场

**TKL** Tackle 滑车,滑轮

Tickler 回授线圈,打油泵

**TKM** Tonne-Kilometers 吨公里

**TKNO** Ticket Number 机票号码

**TKOF** Take-off 起飞

**TKP** Tonne-Kilometre Performed 完成的吨公里

**TKR** Tanker 空中加油机

Timekeeping 时间记录,守时

**TKRZ** Tanker Rendezvous 加油机汇合

**TKS** Thanks 多谢

Tokushima, Japan ( Airport Code ) 日本德岛机场

**TKT** Ticket 机票

**TKTL** Ticketing Time Limit 办票时间限制

**TKTNR** Ticket Number 机票号码

**TKU** Turku, Finland ( Airport Code ) 芬兰图尔库机场

**TL** Tail 尾部

Talk Listen 听说(练习)

Team Leader 组长

Test Line 测试线

Thrust Lever 推力手柄,油门杆

Thrust Line 推力作用线

Tie Line 直达通信线路

Tilt 俯仰,倾斜

Time Length 时间长度

Time Line 时间线

Time to Launch 到发射的时间

Tool 工具

Top Level 顶部电平,云顶高度,爬升高度

Total Load 总荷载,总载重量

Transition Level 过渡(高度)层

Transmission Line 传输线

Transmittal Letter 发射符号

Transport Layer 传送层

**TLA** Throttle Lever Angle 油门杆角度

Thrust Lever Angle 推力杆角度

Top Level Aggregator 最高层集合体

**TLB** Technical Log Book 技术记录本,飞行记录本

**TLC** Takeoff and Landing Computation ( Program ) 起飞和着陆计算(程序)

Tandem Link Connection 汇接链路连接

Throttle Lever Control 油门杆控制

Thrust Limit Calculation ( Computation ) 推力

极限计算

**TLCC** Thin-Line Connecting Capability 薄管连接能力

**TLCO** Telephone Line Controller 电话线路控制器

**TLCOMM** Telecommunication 电信

**TLCS** Tape Library Control System 磁带库控制系统

**TLCSC** Top Level Computer Software Component 最高等级计算机软件部件

**TLD** Time Limited Dispatch 限时放行

Troubleshooting Logic Diagram 排故逻辑图

**TLE** Temperature Limited Emission 限温发射

**TLF** Tariff Level Factor 运价系数

**TLG** Tail Landing Gear 尾起落架

Telegraph 电报

Tilting 倾斜

Tooling 工装,工具

**TLGM** Telegram 电报

**TLH** Tallahassee, FL, USA (Airport Code) 美国佛罗里达州塔拉哈西机场

Top-Level Hazards 顶级危害

**TLI** Transmission Level Interface 传输级接口

Transport Layer Interface 传输层接口

**TLIR** Thrust Lever Interlock Relay 推力手柄互锁继电器

**TLK** Test Link 测试链

**TLL** Tallinn, Estonia (Airport Code) 爱沙尼亚塔林机场

**TLLD** Total Load 总负载,全负载

**TLM** Technical Manual 技术手册

Telemeter 遥测计

Telemetry 遥测技术,遥测装置

**TLMC** Time Limits and Maintenance Checks 时间限制和维护检查

**TLN** Toulon, France (Airport Code) 法国土伦机场

Trunk Line Network 中继线网络

**TLO** Takeoff and Landing Optimization (Program) 起飞和着陆优化(程序)

**TLOF** Touchdown and Lift-off Area 接地和飞离区

**TLP** Terminal Link Processing 终端链路处理

Total Logistics Provider 全程物流商

Transmission Level Point 传输电平点

**TLPC** Tailpiece 尾部,尾段

**TLPS** Takeoff and Landing Performance System 起飞和着陆性能系统

**TLR** Thrust Lever Angle Resolver 油门杆角度解算器

Thrust Lever Resolver 推力手柄解算器

Tiller 手柄

Twin Lens Reflex 双镜头反射

**TLS** Tactical Landing System 战术着陆系统

Target Level of Safety 安全目标等级

Task List 工作清单

Telemetry Listing Submodule 遥测列表子模块

Toulouse, France (Airport Code) 法国图卢兹机场

Transponder Landing System 应答着陆系统

Transport Layer Security 传输层安全

**TLSRD** Top Level System Requirement Document 高级系统要求文件

**TLT** Temperature Limitation Thermostat 温度限制恒温器

**TLU** Terminal Logic Unit 终端逻辑单元

Time of Last Update 最近修订的时间

Transportable Laser Unit 可移动激光装置

Travel Limitation Unit 行程限制器

**TLV** Telaviv, Israel (Airport Code) 以色列特拉维夫机场

Threshold Limit Values 门限限制值

**TLW** Tailwind 尾风

**TLX** Task Load Index 任务载荷指数

Telex 电传打字电报机

**TM** Manifold Dual Temperature Sensor 多支管双温度传感器

Take-off Mass 起飞质量

Technical Manual 技术手册

Technical Memorandum 技术备忘录

Temperature Meter 温度表

Terminal Module 终端组件

Terminal Multiplexer 终端复用器

Timing Module 定时模块

Titanium Chloride 氯化钛

Token Machine 令牌机

Tone Modulation 音调调制

Tooling Memorandum 工装备忘录,工具备忘录

Trademark 注册商标

Traffic Management 交通管理

Training Manual 训练手册

Transportability Manual 运输手册

Transmission and Multiplex 传输和复用

Transmission Media 传输媒介

Transparent Mode 透明模式

Transverse Magnetic 横磁波(的)

Turbo Machinery 涡轮机构

Twisting Moment 扭矩

**TM(T/M)** Torque Motor 扭矩马达,力矩马达

**TM/PS** Technical Management/Program Status 技术管理/程序状态

**TMA** Technology Maturity Assessment 技术成熟度评估

Terminal Maneuvering Area 终端机动区,机场机动区

Terminal Movement Area 终端活动区

Thrust Mode Annunciation 推力方式信号器

Traffic Management Adviser 交通管理顾问

Transcription Mediated Amplification 转录介导扩增

Twin Motor Actuator 双马达作动筒

**TMB** Technical Management Board 技术管理委员会

Thimble 套筒,离合器,电缆接头

Torque Motor Bank 扭矩马达偏转

**TMBUS** Test and Maintenance Bus 测试和维修总线

**TMC** Technical Management Committee 技术管理委员会

Technical Measurement Corporation 技术测量公司

Technology Management Centre 技术管理中心

Telecommunication Maintenance Center 通信维护中心

Telecommunications Management Center 电信管理中心

Telemetry Controller 遥测装置控制器

Thrust Management Computer 推力管理计算机

Titanium Matrix Composite 钛基体复合材料

Torque Motor Current 力矩马达电流

Total Manufacturing Cost 总制造成本

Total Market Coverage 总市场覆盖率

Traffic Management Center 交通管理中心

Traffic Management Coordinator 交通管理协调人

Transmission Maintenance Center 传输维护中心

Transport Movement Control 运输调动管理

Transportation Management Center 运输管理中心

Travel Management Company 旅行管理公司

**TMCC** Time-Multiplexed Communication Channel 时间多路复用通信信道

Transmission and Multiplexing Configuration Control 传输和复用配置控制

**TMCF** Thrust Management Computing Function 推力管理计算功能

**TMCOMP** Telemetry Computation 遥控技术计算

**TMCS** Thrust Management Computer System 推力管理计算机系统

Thrust Management Computing System 推力管理计算系统

**TMD** Tactical Mission Data 战术任务数据

Technical Management Division 技术管理处

Theoretical Maximum Density 理论最大密度

Timed 定时的

Transportation Management Division 交通运输管理处

Tuned Mass Damper 调谐质量阻尼器

**TMDE** Test, Measurement and Diagnostic Equipment 试验、测量与诊断设备

**TMDS** Test, Measurement and Diagnostic System 试验、测量与诊断系统

Transmission Minimized Differential Signaling 传输最小差动信号

**TME** Test and Measurement Equipment 试验与测量设备

Training Management and Evaluation 培训管理与评估

Transportation Mechanical Equipment 运输机械设备

**TMF** Tele-Management Forum 电信管理论坛

Thermal Mechanical Fatigue 热机械疲劳

Thrust Management Function 推力管理功能

Traffic Management Flight 交通管理飞行

Traffic Management Function 交通管理功能

Turbine Mid Frame 涡轮中框

Two-Mode Fiber 双模光纤

**TMG** Technical Management Group 技术管理组

Technical Monitoring Group 技术监视组

Time Generator 时间产生器

Timing 定时,计时

Traffic Management Group 业务管理群

**TMGD** Timing Devices 定时器件

**TMHGA** Top Mounted High Gain Antenna 顶部安装高增益天线

**TMHS** Trusted Message Handling System 可信报文处理系统

**TMI** Technical Manual Index 技术手册索引

Telephone Multimedia Interface 电话多媒体接口

Two-Mode Interference 双模干扰

**TMIS** Technician Maintenance Information System 技术人员维修信息系统

Traffic Management Information System 交通管理信息系统

**TML** Television Microwave Link 电视微波线路

Terminal 终端区,候机楼

Total Mass Loss 总质量损失

**TMLG** Traffic Management Liaison Group 交通管理联络小组

**TMLLF** Terrain Masking Low Level Flight 地形掩护低空飞行

**TMLLFC** Terrain Masking Low Level Flight Computer 地形掩护低空飞行计算机

**TMLO** Tooling Master Layout 工装主配置

**TMM** Theory of Maintenance Manual 维修论手册

**TMMP** Technical Manual Management Program/Plan 技术手册管理大纲/计划

**TMMT** Technical Manual Management Team 技术手册管理组

**TMN** Telecommunication Management Network 电信管理网

Terminate 停止,结束,终止

True Mach Number 真马赫数

**TMO** Technology Management Office 技术管理办公室

Transportation Management Office 运输管理办公室

**TMOE** Testability Measure of Effectiveness 有效性可测试性测量

**TMOS** Telecommunication Management and Operation Supporting 通信管理与操作支持

Thermosetting 热固的

**TMP** Tampere, Finland (Airport Code) 芬兰坦佩雷机场

Technical Manual Plan 技术手册计划

Technical Monitoring Position 技术监控席位

Telecommunication Management Plan 通信管理计划

Terminal Message Processor 终端信息处理器

Terminal Monitor Procedure 终端监控程序

Test Management Plan 试验管理计划

Test Management Protocol 测试管理协议

Traffic Management Processor 交通管理处理器

Training Management Plan 培训管理计划

Transmission Maintenance Point 传输维护点

Transport Mission Pilot 运输任务飞行员

Transportation Management Plan 运输管理计划

Transportation Management Program 运输管理程序

**TMPA** Traffic Management Program Alerts 交通管理程序警告

**TMPD** Tempered 回火的

**TM-PDU** Test Management Protocol Data Unit 测试管理协议数据单元

**TMP-IL** Transmission Maintenance Point-International Line 国际线路传输维护点

**TMPRG** Tempering 回火

**TMPROC** Telemetry Processing 遥测数据处理

**TMR** Technology Management Review 技术管理评审

Technology Maturity Review 技术成熟性审核

Timer 定时器

Tooling Material Request 工具器材申请

Training Management Review 培训管理评审

Transmission Medium Requirement 传输媒介要求

Triple-Modular Redundancy 三模数余度

True Mach Reading 真马赫读数

**TMRP** Tropical Meteorology Research Programme 热带气象学研究计划

**TMS** Tape Management System 磁带管理系统

Target Management Systems 目标管理系统

Technical Maintenance Standard 技术维护标准

Teleoperator Maneuvering System 遥控机动系统

Terminal Monitoring System 终端监控系统

Test and Monitoring Station 测试与监控站

Test Management System 试验管理系统

Thermal Management System 热管理系统

Thrust Management System 推力管理系统

Thrust Modulation System 推力调节系统

Time and Motion Study 时间与运动研究

Time-Multiplexed Switching 时隙复用交换

Tooling Management System 工具管理系统，工装管理系统

Total Management System 总管理系统

Total Manufacturing System 总制造系统

Traffic Management System 交通管理系统，运量管理系统

Training Media Support 培训媒体保障

Transaction Monitoring System 业务监视系统，交易监视系统，处理监视系统

Transportation Management System 运输管理系统

**TMSI** Temporary Mobile Station Identity 临时移动台标识符

**TMSP** Thrust Mode Select Panel 推力方式选择面板

Technology Management Strategic Plan 技术管理战略计划

**TMSR** Transverse Mode Suppression Ratio 横向模式抑制比

**TMSS** Technical Manual Safety Study 技术手册安全研究

Technical Manual Specifications and Standards 技术手册规范与标准

Transportable Mission Support System 运输任务的支持系统

**TMT** Transonic Model Tunnel 跨音速模型试验风洞

**TMTR** Thermistor 热敏电阻，热元件

**TMTS** Telemetry Test Set 遥测装置测试仪

**TMV** True Mean Value 真平均值

**TMU** Traffic Management Unit 交通管理单元

**TMUX** Terrain Multiplex 地形复合

Transmultiplexer 复用转换设备

**TMX** Telemeter Transmitter 遥测仪发射机

**TMZ** Transponder Mandatory Zone 强制应答地带

**TN** Tag Number 行李牌号码

Tail Number 尾号

Technical Notes 技术说明书

Telecommunication Network 电信网络

Thermonuclear 热核的

Tone 音调

Tons 吨

Tracer Number 跟踪器号

Track Number 航迹号

Transmitting Node 发送节点

Transparent Negative 不透明(的)

Transparent Network 透明网络

Transport Network 传送网络

True North 正北

Trunk Node 中继节点

Tunisia 突尼斯

**TN&ALM** Tone and Alarm 音调和警报

**TNA** Tail Number Assignment 飞机排班

Telecommunication Network Architecture 电信网络体系结构

Time of Nearest Approach 最近进近时间

Transient Network Analyzer 瞬时网络分析器

Transition Area 过渡区

Turn Altitude 转弯高度

**TNAS** Trunk Network Administration System 干线网络管理系统

**TNAV** Time Navigation 时间导航

**TNC** Tactical Navigation Chart 战术导航图

Tactical Network Coordinator 战术网络协调员

Terminal Network Controller 终端网络控制器

Terminal Node Controller 终端节点控制器

Transport Network Controller 传送网络控制器

Transport Node Clock 传送节点时钟

**TNDU** Transport Network Data Unit 传送网络数据单元

**TNG** Tangier, Morocco（Airport Code）摩洛哥丹吉尔机场

Tongue 舌,舌簧

**TNGV** Turbine Nozzle Guide Vane 涡轮喷口导向叶片

**TNH** Turn Height 转弯高

**TNI** Terminal Network Interface 网络终端接口

Total Noise Index 总噪声指数

Traffic Noise Index 交通噪音指数

**TNIC** Transmission Network Identification Code 传输网标识码

**TNIF** Telephone Network Interface 电话网络接口

**TNIU** Trustworthy Network Interface Unit 可信网络接口单元

**TNK** Think 认为

**TNL** Terminal Net Loss 终端净损耗

Total Noise Level 总噪声级

Tunnel 隧道

**TNM** Transmission Network Management 传输网络管理

**TNN** Transfer Network Node 传送网络节点

**TNR** Antananarivo, Madagascar（Airport Code）马达加斯加安塔那利佛机场

Thinner 稀释剂

**TNRIS** Transportation Noise Research Information Service 运输噪声研究信息服务

**TNS** Tactical Navigation System 战术导航系统

Technical Service 技术服务

Transaction Network Service 交易处理网络服务

Transmission Network Surveillance System 传输网络监视系统

**TNSL** Tensile 拉力的

**TNSM** Task Numbered Shop Manual 按任务编号车间手册

**TNSN** Tension 张力

**TNTV** Tentative 试探性的

**TO** Takeoff 起飞

Technical Order 技术规范

Telecoms Operator 电信运营商

Time Out 停工时间,时间已过,暂停

Timed Opening 定时开放
Tool Order 工具指令,工具规程,工具订购
Town Office 市区办事处
Turn off 断开,关闭
Turn Out 关断,输出
**TO&E** Table of Organization and Equipment 机构和设备表,组织和设备表
**TOA** Time of Arrival 到达时间
Total Obligational Authority 总职责权限
Type of Address 编号规划指示器/地址类型
**TOAST** Team Oriented ATC Simulator Training 面向团队的空管模拟机培训
**TOB** Take-off Boost 起飞助推,起飞加速
**TOC** Table of Contents 目录
Tactical Operations Center 战术操作中心
Technical Operations Center 技术运行中心
Technical Order Change 技术指令更改
Technical Order Compliance 技术指令符合性
Terminal Operation Center 航站楼运控中心
Time of Climb 爬升时间
Top of Climb 爬升顶点
Total Operating Cost 总运营成本
Type of Call 呼叫类型
**TOCC** Technical and Operational Control Center 技术/操作控制中心
Test Operations Control Center 试验运行控制中心
**TOCN** Technical Order Change Notice 技术指令更改通知
**TOCS** Terminal Operations Control System 航站运营控制系统
**TOCU** Technical Order Control Unit 技术指令控制单元
**TOD** Take off Dry 干跑道起飞
Take-off Distance/Takeoff Distance 起飞滑跑距离,起飞距离
Task Order Directive 任务管理指令
Technical Objective Document 技术目标文件
Ticket Origin and Destination 机票起始点和终点

Time of Delivery 交付时间
Time of Departure 离港时间
Time of Descent 下降时间
Time of Dispatch 签派时间
Top of Descend 下降起点,下降顶点
**TODA** Take-off Distance Available 可用起飞距离
**TODAH** Takeoff Distance Available, Helicopter 直升机可用起飞距离
**TODFL** Top of Descent Fuel 到下降起点时油量
**TODR** Take-off Distance Required 所需起飞距离
**TODS** Test Oriented Disc System 面向测试的磁盘系统
**TOE** Thread One End 一端有螺纹
Tonne of Oil Equivalent 当量油料吨数
**TOF** Take-off Fuel 起飞时燃油存量
Time of Fall 降落时间
Time of Flight 飞行时间
**TOFCN** Technical Order Field Change Notice 技术指令范围更改通知
**TOFF** Total Fuel at Take-off 起飞总油量
**TOFL** Takeoff Field Length 起飞机场长度
**TOFP** Take-off Flight Path 起飞飞行航迹
**TOG** Take-off Gross (Weight) 起飞总重量
**TOGA** Take off, Go Around 起飞,复飞
Tropical Ocean and Global Atmosphere 热带海洋与全球大气
**TOGR** Together 共同
**TOGW** Takeoff Gross Weight 起飞总重量
**TOH** Tool Order History 工具指令历史,工具订货历史
Transport Overhead 传送开销
**TOHS** Tool Order History System 工具指令历史系统
**TOIL** Temperature Engine Oil 发动机滑油温度
Toilet 厕所
**TOJ** Track on Jamming 干扰跟踪

**TOL**   Time over Limit 时间超限

Toledo, OH, USA ( Airport Code ) 美国俄亥俄州托莱多机场

Tolerance 公差,容差,容限

**TOLA**   Take-off and Landing Analysis 起飞和着陆分析

**TOLS**   Tools on-Line System 联机工具系统

**TOM**   Take-off Mass 起飞质量

Terminal Operations Manual 航站运行手册

**TOMA**   Technical Order Management Agency 技术指令管理机构

**TONAC**   Technical Order Notification and Compliance 技术指令通知和符合性

**TONLAR**   Tone Operated Net Loss Adjuster 声控网络损耗调节器

**TOOMS**   Tooling Material Support 工具器材保障

**TOP**   Take-off Power 起飞功率

Technical and Office Protocol 技术与办公室协议

Tour Operators' Package 旅游商包机

Turbine Outlet Pressure 涡轮出口压力

**TOPAZ**   Traffic Organization and Perturbation Analyzer 交通组织与微扰分析仪

**TOPG**   Topping 上部,顶端

**TOPMS**   Take off Performance Monitoring System 起飞性能监视系统

**TOPO**   Topography 地形

**TOPSEC**   Top Secret 绝密

**TOR**   Take-off Run 起飞滑跑,起飞滑跑距离

Technical Operating Report 技术工作报告

Top of Rail 轨道顶部

Torque 扭矩

Totalizing Relay 加法继电器,求和继电器

**TORA**   Take-off Run Available 可用起飞滑跑距离

**TORB**   Technical Order Review Board 技术规程审查委员会

**TORM**   Torquemeter 扭矩表

**TORNL**   Torsional 扭力的

**TORR**   Take-off Run Required 所需起飞滑跑距离

**TORS**   Tool Order Record System 工具订购记录系统

Tool Order Reporting System 工具订购报告系统

**TORSIO**   Torsiograph 扭力计

**TOS**   Tape Operating System 磁带操作系统

Test Operating System 测试操作系统

Transfer Orbit Stage 轨道转移级

Transparent Operating System 透明的操作系统

Turbine Overspeed System 涡轮超转系统

Type of Service 业务类型

**TOSC**   Turn-on Self-Check 接通自检

**TOSS**   Take-off Safety Speed 起飞安全速度

Technical Order Status and Schedule 技术规程状态和计划

**TOT**   Time of Transmission 发播时刻

Time over Target 飞越目标时间

Total 总计

Transfer of Technology 技术转让

Transfer-Operate-Transfer 转让—经营—转让

Turbine Outlet Temperature 涡轮出口温度

Turn over Time 周转时间

**TOTLZ**   Totalize 求和,加法

**TOTS**   Tool Order Tracking System 工具订货跟踪系统

**TOV**   Technical Order Validation 技术规程批准,技术指令批准

**TOV&V**   Technical Order Validation and Verification 技术规程(指令)有效和验证

**TOW**   Takeoff Weight 起飞重量

Takeoff Wet 湿跑道起飞

Time of Week 一周时间

Towing 拖行

**TOWA**   Terrain and Obstacle Warning and Avoidance 地形障碍报警和回避

**TOWG**   Test Operations Working Group 测试运行工作组

**TOX**   Toxic 有毒的

**TOY** Toyama, Japan（Airport Code）日本富山机场

**TP** Tailplane 水平安定面,水平尾翼

Target-Practice 打靶练习

Telephone 电话

Teleprocessing 远程处理,电传处理

Teleprocessor 远程处理器

Temporary Phone 临时电话

Temporary Proposal Number 临时建议号

Terminal Point 终端点

Termination Point 终接点

Terrain Processor 地形处理器

Test Panel 测试板,测试专家组

Test Pattern 测试图形

Test Piece 测试件

Test Pilot 试飞员

Test Plan 试验计划

Test Point 测试点

Test Procedure 试验程序

Thermoplastics 热塑料

Tie Plate 连接板

Time Pulse 定时脉冲

Timing Point 定时点

Token Passing 令牌传送

Top Plug 上部电嘴

Transition Point 转接点

Translator Processor 翻译处理器

Transmission Path 传输路径

Transparent Positive 绝对透明的

Transport Pilot 运输机驾驶员

Transport Protocol 传送协议

True Position 真位置

True Profile 真剖面

Turboprop 涡轮螺旋桨,涡桨

Turn Point 转向点

Turning Point 转弯点

Twisted Pair 双绞线

**T-P** Traffic Pattern 起落航线,交通图

**TP DR** Tape Drive 磁带驱动

**TP&CD** Test Preparation and Conduct Document 预备测试和实施文件

**TP&R** Test Plans and Requirements 测试计划和要求

**TP/PS** Test Plan/Procedure Sheet 测试计划/程序页

**TPA** Tampa, FL, USA（Airport Code）美国佛罗里达州坦帕机场

Tape Pulse Amplifier 磁带脉冲放大器

Technical Publication Association 技术出版物协会

Technical Publications Agreement 技术出版物协议

Traffic Pattern Altitude 起落航线高度

Turbine Powered Aircraft( Airplane ) 涡轮动力飞机

Turbopump Assembly 涡轮泵组

**TPAIS** Transmission Path Alarm Indication Signal 传输通路告警指示信号

**TPAM** Teleprocessing Access Method 远程信息处理访问法

**TPAP** Transaction Processing Applications Program 交易处理应用程序

**TPASE** Transaction Processing Application Service Element 交易处理应用服务单元

**TPC** Technical Publication Code 技术出版物编码

Tool Program Control 工具程序控制

Tower Position Console 塔台位置控制台

Transmission Power Control 传输功率控制

**TPCI** Technical Publications Combined Index 技术出版物综合索引

**TPCV** Turbine Power Control Valve 涡轮功率控制活门

**TPD** Tapped 抽头的

Terminal Protective Device 终端保护器件

True Piston Dimension 真活塞尺寸

True Position Dimension 真位置尺寸

**TPDDI** Twisted-Pair Distributed Data Interface 双绞线分布数据接口

**TPDP** Test Package Development Plan 测试包开发计划

**TPDS** Test Plan/Demonstration Summary 测

试计划/演示摘要

**TPDT**　True Position Dimensioning and Tolerance 真位置尺寸和公差

**TPDU**　Terminal Power Distribution Unit 终端动力分配装置

Transport Protocol Data Unit 传送协议数据单元

**TPE**　Technical Personnel Examiner 技术人员考核员

**TPF**　Tactical Planning Format 战术计划格式

Theoretical Point of Fog 理论雾点

**TPG**　Tapping 攻丝,开孔

Technical Publication Guide 技术出版物指南

Test Point Ground 测试点接地

Timing Pulse Generator 定时脉冲产生器

**TPGG**　Technical Publications General Guide 技术出版物总指南

**TPHO**　Telephotograph 传真照片

**TPHP**　Typical Peak Hour Passenger 典型高峰小时旅客

**TPI**　Tail Plane Incidence 水平尾翼安装角

Target Position Indicator 目标位置指示器

Teeth per Inch 齿数/英寸

Test Program Instruction 测试程序说明

Threads per Inch 螺纹数/英寸

Tire Pressure Indicator 轮胎压力指示器

Training Planning Information 培训计划信息

Transport Programming Interface 传送编程接口

Tributary Physical Interface 支路物理接口

**TPIB**　Technical Publication Illustration Bank 技术出版物图示库

**TPIC**　Tire Pressure Indicating Computer 轮胎压力指示计算机

**TPIPT**　Technical/Technology Planning Integrated Product Team 技术规划综合产品组

**TPIS**　Tire Pressure Indicating System 轮胎压力指示系统

**TPL**　Temple, TX, USA (Airport Code) 美国得克萨斯州坦波机场

Test Problem Log 测试问题记录

Third-Party Logistics 第三方物流

Transferred Passenger List 转港旅客名单

Triple 三倍的

Tropicalized 抗湿热

Two-Party Line 双用户线

**TPLC**　Test Program and Logic Computer 测试程序和逻辑计算机

**TPM**　Technical Performance Management 技术性能管理

Technical Performance Measurement 技术性能测量

Telemetry Processor Module 遥测处理器组件

Teletype Passenger Manifest 电传旅客名单

Third-Party Maintainer 第三方维护商

Ticketed Point Mileage 客票点里程

Total Production Maintenance 全员生产维修

Total Production Management 全面生产管理

Total Productive Maintenance 全面生产性维修

**TPMM**　Technical Publications Management Meeting 技术出版物管理会议

**TPMP**　Technical Performance Measurement Plan 技术性能测定计划

**TPMS**　Technical Performance Measurement System 技术性能测量系统

Teleprocessing Management System 远程处理管理系统

Tire Pressure Monitoring System 轮胎压力监视系统

Transaction Processing Management System 事务处理管理系统

**TPMU**　Tire Pressure Monitor Unit 轮胎压力监视组件

**TPN**　Transfer Passenger Name-list 旅客转机名单

**TPO**　Technical Planning Objective 技术计划目标

Transportation Packaging Order 运输包装规程

**TPON** Telephony over Passive Optic Network 无源光纤网上传电话

**TPOP** Technical Publications Output Planning 技术出版物出版计划

**TPP** Technology Planning Projects 技术计划项目

Test Pattern Projector 测试图形投影机

Testability Program Plan 可测试性项目大纲

Tool and Production Planning 工具和生产计划

Top Program Problem 最高等级程序问题

**TPPCR** Tool and Production Planning Change Record 工具和生产计划更改记录

Tool and Production Planning Change Request 工具和生产计划更改申请

**TPPM** Technical Program Planning Manual 技术程序计划手册

**TPPP** Technical Publications Production Planning 技术出版物生产计划

**TPQ** Tepic, Nayarit, Mexico（Airport Code）墨西哥纳亚里特特克克机场

**TPQM** Total Planning Quality Maintenance 全面计划质量维修

**TPR** Taper 圆锥（的）

Teleprinter 电传打印机

Telescopic Photographic Recorder 望远照相记录器

Test Problem Report 测试问题报告

Transaction Processing Routine 事务处理例行程序

Transponder 应答机

Turbine Pressure Ratio 涡轮压力比

Turbofan Power Ratio 涡轮风扇功率比率

**TPRD** Technical Publications Reference Document 技术出版物参考文件

**TPRR** Test Plan Readiness Review 测试计划准备状态审查

**TPS** Tank Pressure Sensing 油箱压力传感装置

Task Parameter Synthesizer 任务参数合成器

Temporary Protection System 临时保护系统

Test Package Set 测试包装置

Test Procedure Set 测试程序装置

Test Procedure Sheet 测试程序页

Test Program Software 测试程序软件

Thermal Protection System 热保护系统

Transaction Processing System 事务处理系统

Transactions per Second 每秒交易数

Trapani, Sicily, Italy（Airport Code）意大利西西里岛特拉帕尼机场

**TPSMG** Technical Publications Specification Maintenance Group 技术出版物标准维护集团

**TPSN** Transposition 互换位置，易位，对换

**TPSP** Third Party Software Provider 第三方软件供应商

**TPT** Tappet 桯杆

Training Planning Team 培训计划工作组

Turn Point 转弯点

**TPTOL** True Position Tolerance 真位置公差

**TPU** Tank Processing Unit 油箱处理单元

Tape Preparation Unit 磁带准备组件

Terminal Processing Unit 终端处理单元

Transient Pressure Unit 瞬时压力组件

**TPW** Technical Publications Workstation 技术出版物工作站

**TPWG** Test Planning Working Group 测试计划工作组

**TPWS** Terrain Proximity Warning System 近地报警系统

**TPWSD** Technical Publications Work Sharing Document 技术出版物工作共享文件

**TPX** Terminal Productivity Executive 终端生产率经理

**TQA** Throttle Quadrant Assembly 油门杆扇形支架组件

Tool Quality Assurance 工具质量保证

Total Quality Assurance 全面质量保证

**TQC** Total Quality Commitment 全面质量保证

Total Quality Concept 全面质量概念

Total Quality Control 全面质量控制

**TQCS** Time, Quality, Cost, Service 周期、质量、成本、服务

**TQE** Technical Quality Evaluation 技术质量评定

**TQM** Total Quality Management 全面质量管理

**TQMS** Total Quality Management System 全面质量管理系统

**TR** Regulated Temperature 调节温度

Tactical Reconnaissance 战术侦察

Tail Rotor (直升机)尾桨

Tape Recorder 磁带记录器,录音机

Target Recognition 目标识别

Technical Regulations 技术规程

Technical Report 技术报告

Technical Review 技术审查

Temporary Revision 临时修(更)改(单)

Test Requirement 试验要求

Thrust Reverser 反推力装置

Time to Retrofire 制动发动机点火前时间

Time-delay Relay 延时继电器

Tooling Request 工装申请

Toothed Ring 齿环

Tower 指挥塔台

Trace 踪迹,跟踪

Track 航迹,跟踪

Tracking Radar 跟踪雷达

Traffic Resolution (空中)交通决断

Training Route 训练航路

Transceiver or Transmitter Receiver 收发机

Transfer 传输,转移,转换

Transformer Rectifier 变压整流器

Transistor 三极管,晶体管

Transition 过渡

Transmit/Receive 发射/接收

Transponder 转发器

Trouble Report 故障报告

Truss 桁架

Trust 信任

Trustee 受托人

Turbine Reverser 涡轮反推器

Turkey 土耳其

Turn Radius 转弯半径

Type Rating 型号等级

**T-R** Transmitter-Receiver 收发机

**TRA** Technical Requirement Analysis 技术需求分析

Technology Readiness Assessment 技术成熟度评价

Temporary Reserved Airspace 临时保留空域,暂时备用空域

Test Report Analysis 测试报告分析

Test Requirements Analysis 测试要求分析

Throttle Resolver Angle 油门杆解算角度

Thrust Reduction Altitude 减推力高度

Thrust Resolver Angle 推力解算器角度

Track Angle 航迹角

Training Requirements Analysis 培训要求分析

**TRAC** Telescoping Rotor Aircraft 可伸缩旋翼机

**TRACAB** Terminal Radar Approach Control in the Tower Cab 塔台终端雷达近近管制

**TRACALS** Traffic Control Approach and Landing System 空中交通管制进近和着陆系统

**TRACE** Task Reporting and Current Evaluation 工作报告及当前鉴定

Technical Report Analysis Condensation Evaluation 技术报告分析压缩评估

Total Risk Assessing Cost Estimate 风险评估总成本估算

**TRACOMP** Tracking Comparison 跟踪比较

**TRACON** Terminal Radar Approach Control 终端雷达进近管制

Terminal Radar Approach Control Facilities 终端雷达进近管制设施

Terminal Radar Control 终端雷达管制

**TRACS** Telemetry Receiver Acoustic Command System 遥测接收机声音指挥系统

Terminal Radar and Control System 终端雷达和控制系统

Training Record and Control System 训练记录和控制系统

Transport Advanced Control Synthesis 先进运输控制综合

Transportable Radar and Communication Simulator 可移动雷达与通信模拟器

Triangulation Ranging and Crosslix System 三角测距和交叉定位系统

**T-RAD** Turn Radius 转弯半径

**TRAD** Transmission Adapter 传输适配器

**TRAM** Test Reliability and Maintainability 试验可靠性和维修性

Tracking and Automatic Monitoring 跟踪与自动监视

**TRANS** Transaction 执行

Transformer 变压器

Transition(s) 转换,传输,过渡

Translated or Translation or Translator 转变

Transmits or Transmitter 发射或发射机

Transmittance 透射系数,透明度

Transparent 透明的,明显的

Transport 运输,转移

**TRANSA** Transaction 交易,业务,处理

**TRANSEC** Transmission Security 发射安全保护

**TRANSF** Transformer 变压器

**TRANSP** Transport or Transportation or Transportable 运输或可运输的

**TRANSV** Transverse 横(的)

**TRAP** Trapezoid 不规则四边形(的),梯形(的)

**TRAPTD** Transmit APT Data 发送自动图像数据

**TRAS** Thrust Reverser Actuation System 反推作动系统

**TRAV** Travel 旅行,冲程,迁移

Traversing 横过的,横向的

**TRB** Technical Review Board 技术评审委员会

Test Review Board 试验评审委员会

Transportation Research Board 运输研究委员会

Treble 三倍(的)

Turbulence 紊流,颠簸气流

**TRBL** Trouble 困难

**TRC** Technical Repair Center 技术修理中心

Technical Review Committee 技术评审委员会

Technical Review Criteria 技术评审规范

Technology Reports Centre 技术报告中心

Terminal Radar Control 终端雷达控制

Transmit Redundancy Controller 传输冗余控制器

**TRCR** Tracer 跟踪器,故障寻找器

**TRD** Technical Requirement Document 技术需求文件

Technical Review Document 技术评审文件

Test Requirement Document 试验需求文件

Torsional Resonance Damper 扭转谐振阻尼器

Tower Radar Display 塔台雷达显示

Transit Routing Domain 转接选路域

Tread 轮距

Trondheim, Norway (Airport Code) 挪威特隆霍姆机场

Turbine Reduction Drive 涡轮减速传动(装置)

**TRDET** Trouble Detection 故障探测

**TRDTO** Tracking Radar Data Takeoff 跟踪雷达数据起飞

**TRDV** Thrust Reverser Directional Valve 反推装置换向活门

**TRE** Telecommunication Research Establishment 电信研究设施

Thermodynamic Reliability Engineering 热动力可靠性工程

Type Rating Examiner 型号等级考试员

Type Rating Instructor 型号等级教员

**TREC** Tracking Radar Electronic Component 跟踪雷达电子部件

**TREE** Test and Repair of Electronic Equipment 电子设备的试验与修理

Transient Radiation Effects on Electronics 瞬时辐射对电子设备的影响

**TRES** Threat Radar Emitter Simulator 威胁雷达发射器模拟器

Transient Radiation Effects 瞬时辐射影响

**TRESA** Transient Radiation Effects Simulation Analysis 瞬时辐射影响模拟分析

**TRF** Sandefjord, Norway (Airport Code) 挪威桑讷菲尔德机场

Tariff 运价, 关税

Technical Reference File 技术参考文件

Tuned Radio Frequency 调谐的无线电频率

Turbine Rear Frame 涡轮后框架

**TRFC** Cargo Transfer 货物转运

Traffic 交通

**TRFM** Mail Transfer 邮件转运

**TRG** Trailing 拖尾, 后侧的

Training 训练, 培训

**TRH** Truss Head 框架头

**TRI** Bristol, TN, USA (Airport Code) 美国田纳西州布里斯托尔机场

Thrust Rating Indicator 推力比指示器

Triangle 三角

Triplex 三倍的

Type Rating Instructor 机型等级教员

**TRIB** Tributary 支流

**TRIC** Tracking Radar Input and Correlation 跟踪雷达输入和相关

**TRID** Track Identity 航迹识别

**TRIG** Trigger 扳机, 触发器, 启动器

Triggering 触发

Trigonometry 三角学

**TRIM** Test Resource Information Model 测试源信息模型

Time-Related Instruction Management 及时指令管理 (系统)

**TRIR** Tooling Request Investigation Record 工装设备申请调查记录

**TRISAT** Target Recognition through Integral Spectrum Analysis Technology 用综合频谱分析技术目标识别

**TRK** Track (Angle) 航迹 (角)

Tracking 跟踪

Truck 卡车, 转向架, 滚轴

Trunk 干线

**TRKG** Tracking 跟踪

**TRL** Technology Readiness Level 技术成熟度等级

Thrust Reverse Left 左反推

Trail 痕迹

Transistor Resistor Logic 晶体管电阻逻辑

Transport Research Laboratory 运输研究实验室

Transport Resource List 运输资源列表

Turbine Rated Load 涡轮额定载荷

**TR-LAN** Token Ring Local Area Network 令牌环形局域网络

**TRLR** Trailer 拖车, 尾部

**TRLY** Trolley 装有脚轮的小台车, 电车

**TRM** Talent Relationship Management 人才关系管理

Team Resource Management 团队资源管理

Team Risk Management 团队风险管理

Technical Reference Manual 技术参考手册

Technical Resource Management 技术资源管理

Technology Risk Management 技术风险管理

Traffic Message 业务消息

Training Resource Management 培训资源管理

Transmission Resource Management 传输资源管理

Trim 配平

**TRMA** Telecommunications Risk Management Association 电信风险管理协会

Transfer Manifest 转运单

**TRMM** Tropical Rainfall Measuring Mission 热带降雨测量卫星

**TRMP** Technical Risk Management Plan 技术风险管理大纲

**TRMR** Trimmer 调整片, 配平器

**TRMS** Test Resource Management System 试

验资源管理系统

**TRN** Terrain-Reference Navigation 地形参考导航

Test Revision Notice 测试更改通知

Token Ring Network 令牌环网

Tool Requirement Notice 工具要求通知

Trade Name 商业名称

Train 训练,火车

Transport 运输

Turin, Italy ( Airport Code ) 意大利都灵机场

**TRN CRD** Turn Coordination 转弯协调

**TRN LT** Turning Light 转弯灯

**TRNBKL** Turnbuckle 螺套

**TRND** Turned 转弯的

**TRNG** Training 训练

**TRNGR** Turning Gear 转齿轮,旋转装置

**TRNPS** Transpose 使调换位置,使互换位置

**TRNR** Trainer 教员

**TRNSN** Transition 转变

**TRNSPTN** Transportation 运输

**TRNTBL** Turntable 唱盘

**TROO** Transponder on-off 应答机接通—断开

**TROP** Troposphere 对流层

**TROPO** Tropopause 对流层顶

**TRP** Technology Reinvestment Project 技术重新投资项目

Test Requirements Package 测试要求程序包,测试要求数据包

Thrust Rating Panel 推力等级面板

Trade Pattern 贸易方式

**TRPS** Technical Reports Publication System 技术报告出版系统

**TRPU** Thrust Reverser Power Unit 反推功率单元

**TRPV** Thrust Reverser Pressurizing Valve 反推增压活门

**TRQE** Torque 扭矩

**TRQER** Torquer 扭矩器

**TRS** Technical Reference Service 技术参考服务

Technical Repair Standard 技术修理标准

Terminal Radar Service 终端雷达服务

Test Report Summary 测试报告总结

Test Requirement Specification 测试要求规范

Test Requirement Summary 测试要求摘要

Trieste, Italy ( Airport Code ) 意大利的里雅斯特机场

Tropical Revolving Storm 热带旋转风暴

Troubleshooting Record Sheet 排故记录单

**TRSA** Terminal Radar Service Area 终端雷达服务区域

Timesharing Retrieval System 分时检索系统

**TRSB** Time Reference Scanning Beam 时基扫描波束

**TRSBR** Transcriber 转录器

**TRSN** Torsion 扭力

**TRSR** Taxi and Runway Surveillance Radar 滑行和跑道监视雷达

**TRST** Throttle Reset 油门复位

**TRT** Treat 对待,处理

**TRTD** Treated 已处理的

**TRTMT** Treatment 处理,对待

**TRTO** Type Rating Training Organization 机型等级培训机构

**TRU** Test Replaceable Unit 试验可替换装置

Transformer Rectifier Unit 变压器整流组件

Transmit-Receive Unit 发射—接收装置

Transportable Radio Unit 便携式无线电装置,移动式无线电设备

True 真的

**TRUN** Trunnion 耳轴,轴颈

**TRUT** Time Remaining until Transition 发射前剩余时间

**TRV** Toxicity Reference Value 毒性参考值

Travel 旅行,行程

Trivandrum, India ( Airport Code ) 印度特里凡得琅机场

**TRVLG** Travelling 旅行的

**TRVLMT** Travel Limit 行程极限,冲程极限

**TRVLR** Traveller 旅行者

**TRW** Tarawa, Kiribati (Airport Code) 基里巴斯塔拉瓦机场

**TRWA** Trackway 轨道

**TRX** Triplex 三倍的

**TS** Tactile Sensor 触觉传感器

Tandom Switch 汇接交换机

Technical Service 技术服务

Technical Study 技术研究

Telecommunication Service 电信业务

Telegraph System 电报系统

Temperature Switch 温度开关

Tension Strength 张力强度

Tentative Specification 试行规范

Tentative Standard 试行标准

Terminal Station 终端站

Terminal Strip 终端接线条

Terminal Switch 终端开关

Test Sample 测试件

Test Set 测试装置

Test Specification 测试规范, 测试说明书

Test Summary 测试摘要

Thermal Strip 热接线条

Thesis 论点, 论题, 论文

Thunderstorm 雷暴雨

Time Sample 时间取样

Time Slot 时隙

Time Switch 时分交换

Tip Sheet 提示单

Toll Switch 长途交换

Tool Sheet 工具单

Top Secret 最高机密, 绝密

Torque Spanner 扭力扳手

Tracking Sensor 跟踪传感器

Tracking System 跟踪系统

Trade Secret 贸易秘密

Transfer Stage 传输级

Translation Server 变换服务器

Transport Service 传送业务

Transport Stream 传送流

Tree Structure 树结构

Trim Sheet 配平单

Tropical Storm 热带风暴

Trouble Shooting 故障分析

Tube Sheet 管件单

Type Specification 型号说明书, 型号规范

**TS&POP** Testing and Popping 测试与爆音

**TSA** Table of Standard Atmosphere 标准大气表

Tactical Stealth Aircraft 战术隐形飞机

Tail Strike Assembly 尾翼组件

Target Service Agent 目标服务代理

Temporary Segregated Area 临时隔离区

Time Slot Allocation 时隙分配

Time Slot Assignment 时隙安排

Transfer-Shipping Authorization 转运授权

Transportation Security Administration 运输安全局

**TSAF** Typical System Acquisition Flow 典型系统获取流程图

**TSAP** Transport Service Access Point 传送业务接入点

Transport Service Application Part 传送服务应用部分

**TSAP-ID** Transport Service Access Point Identifier 传送业务接入点标识符

**TSAU** Telephone Service Access Unit 电话服务接入单元

**TSB** Telecommunication Standardization Bureau 电信标准化局

Transportation Safety Board of Canada 加拿大运输安全委员会

**TSC** Tape Speed Compensation 磁带速度补偿

Tape Station Conversion 磁带站转换

Technology Service Corporation 技术服务公司

Transmission Control 传送控制

Transportation System Center 运输系统中心 (美国)

**TSCJ** Telecommunications Satellite Corporation of Japan 日本电信卫星公司

**TSCLT** Transportable Satellite Communications Link Terminal 可移动的卫星通信链路终端

**TSCM** Technical Surveillance Countermeasure（s）技术监视对策

**TSCN** Tooling Specification Change Notice 工装规范更改通知

**TSCR** Test Site Change Request 测试现场更改申请

**TSD** Tactical Situation Display 战术情景显示

Technical Support Document 技术支持文件

Technical System Design 技术系统设计

Test Sequence Document 测试顺序文件

Time Share Display 时分显示（器）

Total System Down 全系统故障

Traffic Situation Display 交通状态显示

Training System Definition 训练系统定义

**TSDB** Trouble Shooting Data Base 排故数据库

**TSDU** Target System Data Update 目标系统数据更新

Transport Service Data Unit 传送业务数据单元

**TSE** Kazakhstan-Astana Airport 哈萨克斯坦—阿斯塔纳机场

Technical Surveillance Equipment 技术监视设备

Terminal Station Equipment 终端站设备

Test Support Equipment 测试保障设备

Total System Error 总系统误差

Turboshaft Engine 涡轮轴发动机

**TSEC** Telecommunications Security 电信保密

**TSEL** Temperature Selector 温度选择器

**TSES** Transportable Satellite Earth Station 移动式卫星地面站

**TSF** Threat Situation Format 威胁情景格式

Tooling Support Fixture 工具支撑夹具

**TSFC** Thrust Specific Fuel Consumption 单位推力燃油消耗

**TSG** Time Signal Generator 时间信号产生器

Time Slot Generator 时隙发生器

Traffic Service Group 业务服务群

**TSGR** Thunderstorm with Hail 雷雨夹雹

**TSH** Trouble Shooting Handbook 排故指南

**TSI** Terminal Signaling Interface 终端信号接口

Thrust Setting Index 推力调定指数

Time since Installation 自安装时间

Time Slot Interchange 时隙互换

Total Solar Irradiance 太阳总辐照度

Transmitting Subscriber Identification 发送用户识别

Transportation Safety Institute 运输安全研究所（学会）

Turbine Supervisory Instrumentation 汽轮机监测仪表

**TSIMS** Telemetry Simulation Submodule 遥测模拟分模组件

**TSIR** Total System Integration Responsibility 全系统综合责任

**TSIS** Telephone Service Information System 电话业务信息系统

**TSIU** Time Slot Interchange Unit 时隙交换单元

**TSJ** Test Jig 测试夹具

**TSJF** Thermally Stable Jet Fuel 热稳定喷气燃油

**TSL** Terminal Strip List 接线柱清单，接线条清单

Time Slot 时隙

Transaction Sub-Layer 事务子层

**TSM** Tax System Modernization 税务系统现代化

Terminal Support Module 终端支持模块

Throttle Servo Mount 油门伺服机构安装架

Time Share Monitor 分时监视器

Time Sharing Multiplex 分时复用

Time-Slot Selector Module 时隙选择器模块

Transaction State Machine 事务状态机

Trouble Shooting Manual 故障诊断手册

Turbulence Simulator 湍流模拟器

**TSMD** Time Stress Measurement Device 时间

应力测量装置

**TSMO**   Tactical Satellite Management Office 战术卫星管理处

Time since Major Overhaul 大翻修后(使用) 时间

**TSMSR**   Transient Side-Mode Suppression Ratio 瞬时边模抑制比

**TSN**   Tape Serial Number 磁带、录音带等带的序列号

Time since New 服役后时间

**TSNC**   Transmission Subnetwork Controller 传输子网控制器

**TSO**   Technical Service Order 技术服务指令

Technical Standard Option 技术标准选择

Technical Standard Order 技术标准规定

Test Site Office 试验现场办公室

Time Sharing Option 分时选择

**TSO/E**   Time Sharing Option/Extension 分时选择/扩充

**TSOA**   Technical Standard Order Authorization 技术标准规定项目批准书

**TSOJ**   Turnaround Single Open Jaw 折返地单缺口程

**TSOR**   Tentative Statement of Operational Requirements 操作要求试行说明

**TSOS**   Time-Sharing Operating System 分时操作系统

**TSOW**   Technical Statement of Work 技术工作说明

**TSP**   Tank Signal Processor 油箱信号处理器

Teleprocessing Services Program 远程信息处理服务程序

Terminal Support Processor 终端支持处理机

Test Signal Processor 测试信号处理机

Thrust Setting Parameter 推力设定参数

Total Suspended Particle 总空中悬浮物

Tracking Station Position 跟踪站位置

Trade Study Plan 贸易研究计划

Transponder 应答机

Twisted Shielded Pair 屏蔽双绞线

**TSPA**   Tank Signal Processor A 油箱信号处理器 A

**TSPB**   Tank Signal Processor B 油箱信号处理器 B

**TSPD**   Total System Performance Demonstration 全系统性能演示

**TSPEC**   Test Specification 测试规范,测试说明书

**TSPI**   Telephony Service Provider Interface 电话服务器接口

Time Space Position Indication 时空位置指示

**TSR**   Technical Specification Requirement 技术规范要求

Terabit Switched Router 太比特交换路由器

Terminal Surveillance Radar 终端监视雷达,机场监视雷达

Terminate and Stay Resident 内存驻留

Test Support Request 测试支援申请

Timisoara, Romania(Airport Code)罗马尼亚蒂米什瓦拉机场

Trade Study Report 贸易研究报告

Twin Side-by-Side Rotor(直升机的)横向并列双旋翼

**TSRP**   Technical Support Real Property 技术支持实时特性

**TSRV**   Transport Systems Research Vehicle 运输系统研究飞行器

**TSS**   Tail Strike System 尾撬系统

Telecommunication Service System 电信服务系统

Telecommunication Switching System 电信交换系统

Time Standards System 时间标准系统

Time-Sharing System 分时系统

Toll Switching System 长途交换系统

Transcription Start Site 转录起始位点

Transmission Surveillance System 传输监测系统

Trunk Servicing System 干线业务系统

**TSSA**   Technical Safety Standards Association 技术安全标准协会

Thunderstorm plus Duststorm or Sandstorm 雷暴加尘暴或沙暴

**TSSF** Terminal System Support Facility 终端系统保障设施

**TSSI** Time Slot Sequence Integrity 时隙序列完整性

**T-S-S-T** Time-Space-Space-Time 时分—空分—空分—时分

**TSSU** Transmission Speed Sensor Unit 发送速度传感器单元

**TSSV** Time since Shop Visit 维修后的工作时间

**TST** T Coded Standard Tool T 编码标准工具
Test 测试
Threshold Sampling Time 界限抽样时间,临界值抽样时间
Time-Sharing Terminal 分时终端
Transonic Transport 超音速运输机

**T-S-T** Time-Space-Time 时分—空分—时分

**TSTEQ** Test Equipment 测试设备

**TSTG** Testing 测试

**TSTR** Tester 测试器
Transistor 晶体管,半导体

**TSTRZ** Transistorized 晶体管化的

**TSU** Technical Support Unit 技术支援小组
Telecommunication Switching Unit 电信交换单元
Time of Safe Unconsciousness 安全无意识时间
Traffic Simulation Unit 交通模拟装置

**TSV** Terminal Supervisory 终端监控

**TSW** Telesoftware 通信软件
Test Software 测试软件
Test Switch 测试开关

**TSWG** Test Safety Working Group 测试安全工作组

**TT** Switching Temperature of HP Valve 高压活门转换温度
Technical Term 技术术语
Technical Test 技术试验
Teletype 电传打字

Teletypewriter 电传打字机

Terminal Timing 终端定时

Terminal Transparency 终端透明性

Test Terminal 测试终端

Time and Temperature 时间与温度

Timing and Telemetry 定时和遥测

Tool Tag 工具标签

Tool Tryout 工具试验

Total Temperature 总温,全温

Total Time 总时间

Transferred to 接至,接到,转为

Transport 运输机

Trim Tank 配平油箱

Trinidad and Tobago 特立尼达和多巴哥

Tripple Thermoplastic 三层热塑料

True Track 真航迹

Your Telegram 你方来电

**TT&C** Telemetry,Tracking and Control 遥测、跟踪与控制
Tracking,Telemetry and Command 跟踪、遥测和指令

**TTA** Technical Task Agreement 技术工作协议
Technology Transfer Agreement 技术转让协议
Test Target Array 测试目标排列
Total Transducer Assembly 全压传感器组件
Travel Trade Association 旅行贸易协会

**TTB** Teletype Buffer 电传打字缓冲器

**TTC** Technical Training Center 技术培训中心
Top Temperature Control 最高温度控制
Transmission Test Center 传输测试中心

**TTCAR** Trinidad and Tobago Civil Aviation Regulations 特立尼达和多巴哥岛民用航空规章

**TTCF** Teletex/Telex Conversion Facility 智能用户电报/用户电报转换设备

**TTD** Test Tolerance Distribution 测试误差分布
Time to Delivery 到交付的时间

**TTDS**　Test Tolerance Distribution Study 测试误差分布研究

**TTE**　Telephone Terminal Equipment 电话终端设备

Terminal Transmission Equipment 终端传输设备

Termination Test Equipment 终接测试设备

**TTEL**　Tools and Test Equipment List 工具和测试设备清单

**TTERSE**　Terse Test Equipment Requirements Summary and Evaluation 简明测试设备要求摘要与评估

**TTET**　Turboprop Transport Evaluation Team 涡轮螺旋桨运输机评估组

**TTF**　Test Time Frame 测试时间帧

Test to Failure 故障试验

Transmission Test Facilities 传输测试设备

Transport Terminal Function 传送终端功能

Trend-Type Forecast 趋势型预报

**TTF&T**　Technology Transfer, Fabrication and Test 技术转让、制造和测试

**TTFF**　Time to First Fix 首次定位时间

**TTFO**　Time to First Overhaul 首次大修期限

**TTG**　Time to Go 待飞时间

**TTI**　Time to Impact 至碰撞时间

Time to Intercept 至截获时间

**TTIU**　Trustworthy Terminal Interface Unit 可信终端接口单元

**TTK**　True Track 真航迹

**TTL**　Task Table Listing 任务表清单

Temporary Traffic Lights 临时交通灯

Temporary Test Limitation 临时测试限制

Through the Lens 通过镜头

Time to Live 生存时间

Title 题目

Total Fare 运价总额

Transistor Transistor Language 晶体管—晶体管(逻辑)语言

Transistor Transistor Logic 晶体管—晶体管逻辑电路

**TTM**　Target Threat Management 目标威胁管理

Technical Training Manual 技术训练手册

Two-Tone Modulation 双音频调制

**TTMF**　Touch-Tone Multifrequency 按钮式多频率的

**TTN**　Tighten 上紧，密闭

Trenton, NJ, USA(Airport Code) 美国新泽西州特伦顿机场

**TTNG**　Tightening 使拧紧

**TTO**　Technology Transfer Office 技术转让办公室

Transmitter Turnoff 发射机关断

**TTOSS**　Totally Transparent Optical Fiber Subscriber System 全透明光纤用户系统

**TTP**　Time to Peak 至峰值时间

Technical Test Plan 技术测试计划

Technology Transfer Program 技术转让程序

Time Triggered Protocol 时间触发协议

Trusted Third Party 可信任第三方

**TTPA**　Telephone Twisted Pair Adapter 电话线扩充板

**TTR**　Target Tracking Radar 目标跟踪雷达

Technical Test Report 技术测试报告

Technical Trouble Report 技术故障报告

Teletype Translator 电传打字机译码器

Time Temperature Recorder 时间温度记录器

**TTRT**　Target Token Rotation Time 目标令牌循环时间

**TTS**　Text-to-Speech 文本言语转换

Total Technology Support 全面技术支援

Transaction Tracking System 事务跟踪系统

Transmission Test Set 传输测试设备

Trim Tank System 配平油箱系统

Turbine Technology Services 涡轮机技术服务

**TTSL**　Traditional Time-Shared Loop 传统分时环路

**TTSU**　Terminal Transmit Unit 终端发送单元

**TTT**　Tailplane Trimming Tank 水平尾翼平衡油箱

Time Temperature Transformation 时间温度

转换

Time-to-Target 到达目标时间

Total Time on Test 试验总时间

**TTTM** Total Traffic Ton-Miles 总运输吨—英里

**TTU** Tape Transport Unit 磁带传送组件

Through Transonic Ultrasonic 全跨音速超音速的

Thrust Termination Unit 推力限制组件

True Transonic Ultrasonic 真跨音速超音速的

**TTV** Termination, Test and Verification 终止、测试和验证

Text-to-Video 文像转换

**TTWS** Terminal Threat Warning System 航站威胁警戒系统, 终端威胁警报系统

**TTY** Teletype 电传模式

Teletypewriter 电传打字机

**TU** Tank Unit 油箱组件

Terminal Unit 终端单元

Transaction Units 事务处理组件

Transmission Unit 传输单位

Tributary Unit 支路单元

Tube 管

**TUA** Test Unit Adapter 测试组件转接器

**TUC** Time of Useful Consciousness 有效意识时间

**TUE** Tuesday 星期二

**TUG** Terminal Unit Group 终端单元群

Tributary Unit Group 支路单元群

**TUL** Tulsa, OK, USA (Airport Code) 美国俄克拉荷马州塔尔萨机场

**TUN** Tune or Tuning 调谐

Tunis, Tunisia (Airport Code) 突尼斯突尼斯机场

**TUNE** Tuner 频道调谐器

**TUNG** Tungsten 钨

**TUP** Telephone User Part 电话用户部分

Tupelo, MS, USA (Airport Code) 美国密西比州图珀洛机场

**TUPS** Turbine Unloading Protection System 涡轮卸载保护系统

**TUPTR** Tributary Unit Pointer 支路单元指针

**TUR** Test Unit Release 测试组件发放

Turret 回转装置

**TURB** Turbine 涡轮, 涡轮机

Turbulence 湍流

**TURN** Turning 旋转, 切削

**TUS** Tucson, AZ, USA (Airport Code) 美国亚利桑那州图森机场

**TUT** Terminal under Test 在测终端

Transistor under Test 在测晶体管

Tube under Test 在测管路

Tutorial 个别指导的

**TV** Television 电视

Terminal Velocity 极限速度, 临界速度

Throttle Valve 节流活门

Thrust Vectoring 推力转向

Transport Vehicle 运输车

Tuvalu 图瓦卢

**TVA** Target Value Analysis 目标价值分析

Tunable Vibration Absorber 可调的吸振器

**TVAC** Television Area Correlation 电视区域相关

**TVACU** Tunable Vibration Absorber Control Unit 可调的吸振控制组件

**TVAPS** Tunable Vibration Absorber Power Supply 可调的吸振器电源

**TVAS** Tunable Vibration Absorber System 可调的吸振系统

**T-VASI** T Visual Approach Slope Indicator T型目视进近坡度指示器

**TVASIS** T Visual Approach Slope Indicating System T 型目视进近下滑指示系统

**TVBC** Turbine Vane and Blade Cooling 涡轮静子和转子叶片冷却

**TVBCA** Turbine Vane and Blade Cooling Air 涡轮静子和转子叶片冷却空气

**TV-BM** TV-Based Multimedia 多媒体电视

**TVC** Thrust Vector Control 推力矢量控制

Total Variable Costs 可变成本总额

Traverse City, MI, USA（Airport Code）美国密执安州特拉佛斯城机场

Turbine Vane Cooling 涡轮叶片冷却

**TVCS** Television Communications Subsystem 电视通信子系统

Thrust Vector Control System 推力矢量控制系统

**TVD** Torsional Vibration Damper 扭振减震器

**TVE** Thermal Vacuum Environment 热真空环境

Total Vertical Error 总垂直误差

**TVEL** Track Velocity 航迹速度

**TVF** Fuji Heavy Industries 富士重工

Thief River Falls, MN, USA（Airport Code）美国明尼苏达州锡夫里弗福尔斯机场

Time Variable Filter 可随时间变化的滤波器

Transversal Filter 横向滤波器

**TVG** Time Variation of Gain 增益时间偏差

**TVI** Television Interference 电视信号干扰

Turbine Vibration Indicator 涡轮振动指示器

**TVIP** Time-Varying Image Processing 时变图像处理

**TVM** Tachometer Voltmeter 转速伏特计

Transistor Voltmeter 晶体管电压表

**TVOD** True Video on Demand 真视频点播

**TVOR** Short Range VOR 短程甚高频全向信标

Terminal Very High Frequency Omnirange 终端站甚高频全向导航信标

Terminal VOR 终端区甚高频全向信标

**TVPT** Television Point Tracking 电视点跟踪

**TVRO** Television Receive Only 电视单收天线,电视单收站

**TVS** Television Shopping 电视购物

Tornado Vertex Signature 陆龙卷涡旋特性

Transaction Visibility System 交易可见性系统,处理可见性系统

**TVSC** Television Service Center 电视业务中心

**TVSN** Television Shopping Network 电视购物网络

**TVTO** True Vertical Takeoff 真垂直起飞(飞机)

**TVV** Turbine Ventilation Valve 涡轮通风活门(阀)

**TW** Tail Warning 尾部警告

Tail Wind 顺风

Take-off Weight 起飞重量

Taxiway 滑行道

Teamwork 协作

Temperature Well 温度腔,温度舱

Tempered Water 软化水

Test Wing 试验翼

Tight Wrapped 包紧的

Tile Wainscot 装壁板

Total Weight 总重

Travelling Wave 行波

True Watt 实际瓦特

Twin Wire 双线,双心导线

Twisted 扭曲的

Typewriter 打字机

Typhoon Warning 台风警报

**TWA** Time Weighted Average 时间加权平均数

Trans World Airlines, Inc. 环球航空公司

Travelling Wave Amplifier 行波放大器

**TWC** Three Way Calling 三方通话

Truncated Whitworth Coarse Thread 锥头惠氏粗螺纹

Tunable Wavelength Converter 可调波长变换器

Two-Way Channel 双向信道

**TWCL** Taxiway Center Line Lights 滑行道中线灯

**TWDL** Two-Way Data Link 双向数据链

**TWDS** Tabulated Wiring Diagrams 列表布线图

**TWEB** Transcribed Weather Broadcast 抄录的气象广播,转录的天气广播

**TWF** Tail Warning Function 尾部警告功能

**TWFT** Truncated Whitworth Fine Thread 锥

头惠氏细螺纹

Twin Falls, ID, USA（Airport Code）美国爱达荷州特温福尔斯机场

**TWG** Test Working Group 测试工作组

Timing 定时

**TWI** Tail Warning Indicator 尾部警报指示器

Terminal Weather Information 终端区气象信息

**TWINS** Tactical Weather Intelligence System 战术气象情报系统

**TWIP** Terminal Weather Information for Pilots 飞行员用终端区气象信息

Terminal Weather Information Processor 终端气象信息处理器

**TWL** Taxiway Light 滑行道灯

**TWLM** Travelling Wave Light Modulation 行波光调制器

**TWLU** Terminal Area Wireless LAN Unit 终端区域无线局域网组件

**TWM** Travelling Wave Maser 行波微波激射器

**TWMBK** Travelling Wave Multiple Beam Klystron 多注行波速调管

**TWOV** Transit without Visa 过境无签证

**TWP** Tag with Part 带标签零件

Technological Work Plan 技术工作计划

**TWR** Aerodrome Control Tower 机场管制塔台

Tail Warning Radar 尾部警戒雷达

Tail Warning Radar System 尾部警告雷达系统

Threat-Warning Receiver 威胁警报接收器

Thrust-Weight Ratio 推力重量比

Tower（Aerodrome Control）塔台（机场管制）

**TWS** Tactical Weather System 战术气象系统

Tail Warning Set 机尾警戒雷达（装置）

Tail Warning System（飞机）尾部报警系统

Technical Work Statement 技术工作说明

Track-While-Scan 边扫描边跟踪

Tsunami Warning System 地震海啸警报系统

Two-Way Service 双向通信

**TWSB** Twin Sideband 双边带

**TWSDT** Tail Warning System Data Terminal 尾部警告系统数据终端

**TWSRO** Track While Scan on Receive Only 仅在接收时边扫描边跟踪

**TWT** Three Way Trunk 三方通话中继

Three-Wheel Turbocharger 三级叶轮增压器

Trans-sonic Wind Tunnel 跨声速风洞

Travelling Wave Tube 行波管

**TWTA** Travelling Wave Tube Amplifier 行波管放大器

**TWX** Telegraphic Message 电报信息

Teletype Transmission 电传发送

Teletypewriter Exchange 电传打字电报交换机

**TWY** Taxiway 滑行道

Taxiway-link 滑行连接道

**TWYL** Taxiway Link 滑行道联络

**TX** Time Expired 期满时间

Transmission 发射

Transmit or Transmitting 发射

Transmitter 发射机

Treatment 处理,加工,对待

**TXBL** Taxable 可征税的,应纳税的

**TXD** Transmit Data 发出数据,传送数据

Transmitter Data 发出器数据,发射机数据

**TXGS** Taxiing Guidance Sign 滑行引导信号,滑行引导标识牌

**TXK** Texarkana, AR, USA（Airport Code）美国阿肯色州特克萨卡纳机场

**TXL** Tegel, Berlin, Germany（Airport Code）德国柏林泰格尔机场

**TXMNTS** Transmit Maintenance Status 发送维护状态

**TXPDR** Transponder 应答机

**TXT** Text 电文,正文,本文

**TY** Typhoon 台风

**TYDE** Type Designator 型号代码

**TYO** Tokyo, Japan（Airport Code）日本东京机场

**TYP** Type 型号,类型
Typical 典型的
**TYPL** Type Plate 样板
**TYPPLE** Typical Phone Power Level 典型的
电话功率电平
**TYPSTG** Typesetting 排字的
**TYR** Tyler, TX, USA（Airport Code）美国

得克萨斯州泰勒机场
**TYS** Knoxville, TN, USA（Airport Code）美国田纳西州诺克斯威尔机场
**TZ** Tanzania 坦桑尼亚
Time Zone 时区
Transonic Zone 跨音速区

# U

U    Ultrasonic（Inspection Sysmbol）超声波（检查符号）

Unclassified 未分类的,不保密的

Unserviceable Container/Pallet 不能使用的集装箱板

Unspecified 未指明的,未详细说明的

Upper 上部的,较高的

**U&L**    Upper and Lower 上和下

**U. S.**    United States 美国

**U/D**    Update 更新

Upper Deck 上舱

**U/DPT**    Up/Down Potentiometer 上／下电位计

**U/L**    Up/Left 上／左

**U/S**    Unserviceable 不可靠的,不适用的

**UA**    Unattended 无人值守的

Uncontrolled Airspace 非管制空域

Unit Address 单元地址

Unit Assembly 组件组装

United Airlines Inc. 联合航空公司

Universal Asynchronous 通用异步的

Unmanned Aircraft 无人驾驶飞机

Unnumbered Acknowledgement 无编号确认（帧）

User Agent 用户代理

User Application 用户应用

User Area 用户区

Utility Aircraft 通用飞机

**UAA**    United Arab Airlines 阿拉伯联合航空公司

Update Aircraft Avionics 更新飞机电子设备

Upper Advisory Area 高空咨询区域

**UAAA**    Ultralight Aircraft Association of Australia 澳大利亚超轻型飞机协会

**UAC**    United Aircraft Corporation 美国联合飞机公司

Unsolicited Airport Control 非请求的机场控制

Upper Airspace Control 高空空域管制

Urban Area Coverage 城市覆盖区

**UACC**    Upper Airspace Control Center 高层空域管制中心

**UADF**    User Application Definition File 用户应用定义文件

**UAE**    United Arab Emirates 阿拉伯联合酋长国

User Agent Entity 用户代理实体

**UAFSS**    Universal Aircraft Flight Simulator System 通用飞机飞行模拟系统

**UAG**    Upper Atmosphere Geophysics 高层大气地球物理学

**UAI**    Universal Adaptor Interface 通用适配器接口

Universal Azimuth Indicator 通用方位指示器

**UAIDE**    Users of Automatic Information Display Equipment 自动信息显示设备用户

**UAIMS**    United Aircraft Information Management System 联合飞机公司情报管理系统（美国）

**UAK**    User Authentication Key 用户鉴定密钥

**UAL**    United Air Lines, Inc.（美国）联合航空公司

United Authorization List 联合审定清单

User Agent Layer 用户代理层

User Agent Sublayer 用户代理子层

**UAM**    User Authentication Module 用户鉴定模块

**UAN**    Unidentified Atmospheric Noise 不明的大气噪声

Unified Automatic Network 统一自动网络

Universal Access Number 通用接入号码

**UANC** Upper-Airspace Navigation Chart 高空空域导航图

**UAOR** Unit Assembly Order Request 组件组装指令请求

**UAP** Upper Atmosphere Phenomena 高层大气现象

**UAPDU** User Agent Protocol Data Unit 用户代理协议数据单元

**UAPO** Unmanned Aircraft Program Office 无人机项目办公室

**UAR** Unattended Radar 无人值守雷达

Upper Advisory Route 高空咨询航路

Upper Air Route 高空航路

Use as Required 按要求使用

**UARS** Unattended Radar Station 无人值守雷达站

Unmanned Air Reconnaissance System 无人驾驶航空侦察系统

Upper Atmosphere Research Satellite 高层大气研究卫星

**UART** Universal Asynchronous Receiver Transmitter 通用异步收发机

**UAS** Unavailable Second 不可用秒

United Aviation Services 联合航空服务(公司)

University Air Squadron 大学飞行中队(英国皇家空军一部分)

Unmanned Aerial Systems 无人驾驶航空器系统

Unmanned Aerospace Surveillance 无人宇宙空间监视

Unmanned Aircraft System 无人驾驶飞机系统

Upper Airspace Service 高空空域服务

User Aircraft System 用户飞机系统

**UASA** Upper Airspace Service Area 高空空域服务区域

**UASCC** Unit Assembly Schedule Control Card 组件组装程序控制卡

**UASE** User Application Service Element 用户应用服务单元

**UASTAS** Unmanned Airborne Surveillance and Target Acquisition System 无人机载监视与目标搜索系统

**UAT** Ultra Aperture Terminal 超小口径卫星地面接收站

Universal Access Transceiver 通用访问收发机

Universal Air Transport Sales, Ltd. (Great Britain) 通用运输机销售有限公司(英国)

Universal Avionics Tester 通用航空电子试验装置

User Acceptance Testing 用户接受程度测试

**UATC** United Aircraft and Transport Corporation 联合飞机与运输公司

**UATI** Union des Associations Techniques Internationales 国际技术协会联合会

**UATP** Universal Air Travel Plan 环球空中旅行计划

User Acceptance Test Plan 用户验收测试计划

**UAV** Uniform Annual Values 年均值

Unmanned Air Vehicle 无人驾驶飞行器,无人飞行器

**UB** Underwater Battery 水下电瓶

Unit Bond 组合焊接,组合粘接

Upper Bound 上界,上限

Upper Byte 高位字节

Usage Block 使用区

Utility Bus 公用汇流条

**UBA** Union of Burma Airways 缅甸联邦航空公司

**UBC** Universal Block Controller 通用区控制器

Unclaimed Baggage Center 无人认领行李中心

**UBCS** Unified Byte Code System 统一字符编码系统

**UBL** Unbleached 原色的,未漂白的

**UBLK** Unblock 开启,解锁,接通

**UBM** Undersea Branching Multiplexer 海底分路复用器

**UBR** Unspecified Bit Rate 未指定比特率
Utility Bus Relay 实用的汇流条继电器

**UBT** Universal Bus Transceiver 通用总线

**UC** Umbilical Cable 连接电缆,操纵电缆,临时管缆
Under Charge 充电不足
Unit Cooler 组件冷却器,设备冷却器
Upper Center 上中心
Upward Compatibility 向上兼容
User Class 用户等级
User Console 用户操纵台

**UCA** Unified Compiler Architecture 统一编译架构
Upper Control Area 高空管制区

**UCAS-D** Unmanned Combat Air System Demonstrator 无人空战系统验证机

**UCAV** Unmanned Combat Air Vehicle 无人驾驶战斗机

**UCB** Unit Control Block 单元控制块
User Control Block 用户控制块

**UCBC** US-China Business Council 美中商会

**UCC** Uniform Code Council, Inc 统一条码协会

**UCD** Universal Cockpit Display 通用驾驶舱显示器

**UCH** UNI Call Handler 用户网络接口呼叫处理器

**UCI** User Class Identifier 用户类别标识符
Utility Card Input 公用插件卡输入

**UCIF** User Console Interface 用户控制台接口

**UCL** Upper Control Limit 控制上限

**UCM** Uncommanded Motion 非指令移动
Unity Connector Mechanisms 实用连接机构
User Communication Manager 用户通信管理器

**UCNI** United Communication Navigation and Identification 联合通信导航与识别

**UCO** Utility Compiler 应用编译程序

**UCOF** Under Compensated Optical Fiber 欠补偿光纤

**UCON** Utility Control 应用控制,公用事业控制

**UCOND** Unconditional 未整形的

**UCP** User Control Path 用户控制通道
Utility Computer Program 实用计算机程序

**UCR** Unitized Consignments Rates 集装箱货物运价

**UCS** Uniform Code System 统一编码系统
Universal Card Service 通用卡业务
Universal Character Set 通用字符集
User Certifiable Software 用户认证软件
User Coordination System 用户协调系统

**UCSD** Universal Communications Switching Device 通用通信转换器件

**UCT** Universal Coordinated Time 协调世界时
Unit Cost Target 单位成本目标
Unit Compatibility Tests 装置兼容性试验

**UD** Document Unit 文件单元
Underground Distribution 地铁分布(图)
Undimensioned Drawings 无量纲图
Upper Data 上一级数据
Upper Deck 上层舱
User Data 用户数据

**UDA** Unified Driver Architecture 统一驱动程序架构

**UDAC-MB** Universal Distribution with Access Control-Media Base 通用分配存取控制媒体基准

**UDACS** Universal Display and Control System 通用显示与控制系统

**UDAM** Universal Digital Avionics Module 通用数字式航空电子模块

**UDB** Unclassified Data Bus 不保密数据总线,未分类数据总线

**UDC** Universal Decimal Classification 国际十进制分类法
Universal Disk Control 通用磁盘控制
Upper Dead Center 上死点

User Definition Character 用户定义字符

User Designation Code 用户标志码

**UDCC** Upper Deck Cargo Compartment 上层货舱

**UDCD** Upper Deck Cargo Door 上舱货舱门

**UDDS** User-Defined Data Stream 用户定义的数据流

**UDF** Ultra High Frequency Direction-Finding Station 超高频定向台

Un-Ducted Fan 无涵道风扇

Unit Development Folders 联合开发文件夹

Universal Disk Format 国际（通用）磁盘格式

User-Defined Function 用户自定义函数

**UDL** Uniform Data Link 统一数据链

Unit Detail Listing 组件详细列表

Up Data Link 上行数据链路

User Data Library 用户数据库

**UDLC** Universal Data Link Control 通用自动计算机数据链路控制

**UDMA** Ultra Direct Memory Access 极端直接内存访问

**UDOT** Ultrasharp Display Output Technology 超清晰显示输出技术

**UDP** User Datagram Protocol 用户数据报文协议，用户数据编程协议

**UDR** Udaipur, India（Airport Code）印度乌代浦尔机场

User-Defined Routing 按用户的规定选路

**UDRE** User Differential Range Error 用户差分测距误差

**UDS** Unified Development Specification System 统一开发规范说明系统

Universal Data System 通用数据系统

Universal Display System 通用显示系统

Universal Documentation System 通用文件系统

**UDSL** Ultrahigh Bit-rate DSL 超高速数字用户环路

Utility Digital Subscriber Line 通用数字用户环路

**UDT** Uniform Data Transfer 一致性数据传输

Universal Data Transcriber 通用数据转录器

Unstructured Data Transformation 非结构化数据变换

**UDTD** Updated 更新的

**UDTI** Universal Digital Transducer Indicator 通用数字式传感器指示器

**UDTS** Universal Data Transfer Service 通用数据传递业务

**UE** Under Engine 引擎底下

Under Excitation 激励不足

Unified Equipment 统一标准装置

Unit Equipment 组合装置，组件设备

User Equipment 用户设备

**UEAC** Unit Equipment Aircraft 单位装置航空器

**UEB** Underseat Electronic Box 座椅下电子盒

**UE-EI** Unplanned Event-Engineering Instruction 非计划事件工程说明

**UEET** Ultra Efficient Engine Technology 超高效发动机技术

**UEM** Universal Ethernet Module 通用以太网模块

**UEMOA** Economic and Monetary Union of West Africa 西非经济货币联盟

**UEP** Unplanned Event Pickup 非计划事件选取

**UER** Unplanned Event Record 非计划事件记录

Unsatisfactory Equipment Report 不合格设备报告

Unscheduled Engine Removal 非计划发动机拆卸

**UERE** User Equivalent Range Error 用户等效测距误差

**UERR** Unscheduled Engine Removal Rate 非计划发动机拆换率

**UES** UAV Exploitation System 无人机开发体系

User Earth Station 用户地球站

**UET** Unattended Earth Terminals 无人值守地球站

**UETS** Universal Emulating Terminal System 通用仿真终端系统

**UF** Under Frequency 频率过低
Uplink Format 上行链数据格式
User-Friendly 用户友好界面

**UFA** Ufa, Bashkortostan (Airport Code) 巴什科尔托斯坦乌法机场
Until Further Advised 待进一步通知

**UFAED** Unit Forecast Authorization Equipment Data 联合预测授权设备数据

**UFC** Unified Fuel Control 燃油统一控制(系统),统一燃油控制(系统)

**UFCP** Up Front Control Panel 前控制面板

**UFD** Unit Fault Data 单元故障数据
Up Front Display 前上方显示器
Utility Flow Diagram 公用工程流程图

**UFDR** Universal Flight Data Recorder 通用型飞行数据记录器

**UFE** Unducted Fan Engine 无涵道风扇发动机

**UFIR** Upper Flight Information Region 高空飞行信息区

**UFIS** Utility Flight Information System 通用飞行信息系统

**UFL** Upper Frequency Limit 频率上限

**UFN** Until Further Notice 在进一步通知前

**UFO** Unidentified Flying Object 不明飞行物

**UFOB** Usable Fuel on Board 机上可用燃料

**UFOC** Undersea Fiber-Optic Cable 海底光纤电缆

**UFP** Universal Flight Planning 通用飞行计划

**UFR** Under Frequency Relay 欠频继电器
Upper Flight Region 高空飞行区

**UFRD** Ultra-Fast Recovery Diode 超快恢复二极管

**UFS** Unit Fiber Structure 单元光纤结构

**UFT** Union Flights (USA) 联邦飞行(美国)

**UFTA** Universal Federation of Travel Agents 世界旅行社联合会

**UFTAA** Universal Federation of Travel Agents' Associations 世界旅行社协会联合会

**UFTR** Under Frequency Time Relay 欠频时间继电器

**UFV** Uncontrolled Fuel Volume 不可控制的燃油量

**UG** Uganda 乌干达
Users Group 用户集团

**UGA** Unity Gain Amplifier 联合增益放大器

**UGND** Underground 地下

**UGT** Underground Test 地下(爆破)试验

**UH** Unit Heater 组件加热器
Utility Helicopter 通用直升机

**UHA** Ultra High Altitude 超高空

**UHB** Ultra High Bypass 高涵道(比)

**UHC** Unburned Hydrocarbons 未燃碳氢

**UHCA** Ultra-High-Capacity Aircraft 超大容量飞机,超高性能飞机

**UHCI** USB Host Control Interface USB HOST 控制器接口

**UHCP** Uniform High-level Communication Protocol 统一的高级通信协议

**UHDT** Unable Higher Due Traffic 因空中交通原因不能再高

**UHF** Ultra High Frequency (300 ~ 3000 MHz) 特高频(300～3000MHz)

**UHFDF** Ultra High Frequency Direction Finding 超高频定向

**UHPFB** Untreated Hard Pressed Fiberboard 未处理模压纤维板

**UHPT** Undergraduate Helicopter Pilot Training 直升机飞行学员训练

**UHR** Ultra High Resistance 超高电阻,超高阻力
Ultra-High Resolution 超高分辨度,超高解晰度

**UHRRR** Ultra High Range Resolution Radar 超远程分辨雷达

**UHS** Ultra High-Speed 超高速
Unit Handling System 设备操纵系统

**UHT** Ultra Heat Treated/Treatment 超高温处理

Ultra High Temperature 超高温度

Undergraduate Helicopter Training 本科直升机培训

Underheat 低热

Universal Horizontal 通用水平(的)

**UHTV** Unmanned Hypersonic Test Vehicle 无人驾驶超高速试验飞行器

**UHV** Ultra High Vacuum 超高真空

Ultra High Voltage 超高压

**UHW** Up Highway 上行公共通道

**UI** Unique Identifier 唯一标识符

Unit Interface 组件接口

Unit Interval 单位间隔

Unit Issue 联合发布

Unlock Instruction 开锁指令

Unrecognized Information 不可识别信息

Urgent Interrupt 紧急中断

User Identification 用户识别符

User Interface 用户接口,用户界面

**UI&C** Utility Instrumentation and Control 公用仪表与控制

**UIC** U-Interface Circuit U 接口电路

Union Internationale des Chemins de Fer (International Union of Railways) 国际铁路联合会

Unit Identification Code 组件识别码

Upper Information Center 高空情报中心

User Identification Code 用户识别码

**UICN** Universal Intelligent Communication Network 通用智能通信网

**UID** User Identifier 用户标识符

**UIDS** User Interface Development System 用户接口开发系统

**UIF** Unnumbered Information Frame 非编号信息帧

**UIH** Unnumbered Information with Header 带信头的未编号信息

**UII** User-Induced Interference 用户带来的干扰

**UIM** Universal Identity Module 通用识别模块

User Identification Module 用户识别模块

**UIMF** User Identification Module Function 用户识别功能

**UIMRTD** Universal Implementation of Machine Readable Travel Documents 普遍实施机读旅行文件

**UIMS** User Interface Management System 用户接口管理系统

**UIN** Quincy, IL, USA (Airport Code) 美国伊利诺州昆西机场

User Identification Number 用户识别号码

**UIO** Quito, Ecuador (Airport Code) 厄瓜多尔基多机场

**UIP** Quimper, France (Airport Code) 法国坎佩尔机场

User Identification Program 用户标识程序

User Interface Program 用户接口程序

**UIPF** Uniform Index Profile Fiber 均匀折射率分布光纤

**UIR** Upper Flight Information Region 高空飞行情报区

User Incident Report 用户事故征候报告

User Interface Requirement 用户界面需求

**UIS** Universal Information Services 通用信息业务

Upper Information Service 高空情报服务

Urban Information System 城市信息系统

User Interface Specification 用户接口规格

User Interface System 用户接口系统

User-in-Service 用户在使用业务

**UISA** Update Involves Subsequent ADCN 包含顺序图纸变更预告的更新

**UIT** Union International Telecommunications 国际电信联盟,国际电信协会

User Interface Terminal 用户接口终端机

**UITS** Unacknowledged Information Transfer Service 未确认的信息传递业务

**UJ** Union Joint 组合接头

Universal Joint 万向接头

**UJP** Ultra Java Processor 超级 Java 处理器

**UJR** Unjustified Removal 不合理的拆卸

**UJT** Unijunction Transistor 单结晶体管

**UK** Unit Check 组件检查

United Kingdom 英国

**UKB** Kobe, Japan（Airport Code）日本神户机场

**UKCAA** United Kingdom Civil Aviation Authority 英国民航局

**UKF** Unscented Kalman Filter 无偏卡尔曼滤波器

**UKY** Kyoto, Japan（Airport Code）日本京都机场

**UL** Ultimate Load 最大负荷

Underwriter's Laboratories 保险商实验室

Unexpected Loss 非预期损失

Unlock 开锁

Up-Link 上行链路

Upper Limit 上限

Usage List 使用清单

**ULA** Ultralight Aircraft 超轻型飞机

Uncommitted Logic Array 自由逻辑阵列

Uniform Linear Array 均匀线性阵列

Universal Logic Array 通用逻辑阵列

**ULAA** Ultra Light Aircraft Association 超轻型飞机协会（澳大利亚）

**ULB** Underwater Locating/Locator Beacon 水下定位信标

**ULBM** Universal List Broadcasting Model 全局表广播模型

**ULC** Universal Logic Card 通用逻辑插件卡

**ULD** Ultra Low Drag 超低的阻力

Underwater Locator Device 水下定位装置

Unit Loading Devices 集装设备,单元装载设备

**ULEA** Ultra-Long-Endurance Aircraft 超长续航能力飞机

**ULF** Ultra Low Frequency（300～3000 Hz）特低频

**ULH** Upper Left Hand 左上方

**ULM** Ultrasonic Light Modulator 超声光调制器

Universal Logical Module 通用逻辑模块

**ULN** Ulan Bator, Mongolia（Airport Code）蒙古乌兰巴托机场

**ULP** Unleaded Petrol 无铅汽油

Upper-Layer Protocol 上层协议

**ULR** Ultra Long Range 超远程

Upper Level Radar Service 高空雷达服务

**ULS** Ultra Low Speed 超低速

**ULSI** Ultra Large Scale Integration 超大规模集成

**ULSIC** Ultra Large Scale Integrated Circuit 超大规模集成电路

**ULT** Ultimate 极限的,最终的

Ultra Low Temperature 超低温

**ULVS** Ultra Low Voltage Signal 超低电压信号

**UM** Unaccompanied Minor 无随行儿童

Unified Messaging 统一传信

Unit Manufacture 组件制造

Unit of Measure 计量单位

Unit of Measurement 测量单位

Urgent Memorandum 紧急备忘录

User's Manual 用户手册

**UMA** Unified Memory Architecture 统一内存架构

Unified Motherboard Architecture 统一主板架构

Uniform Memory Access 相同记忆存取

Unmanned Aircraft 无人机

Unscheduled Maintenance Action 不定期维修活动

Unscheduled Maintenance Analysis 不定期维修分析

Upper Memory Area 上位内存区

**UMARK** Unit Maintenance Aircraft Recovery Kit 组件维修飞机恢复工具包

**UMB** Upper Memory Block 高端存储块,上位内存块

**UMC** Unique Milestone Code 独特里程码

Universal Maintenance Center 通用维修中心

Utility Management Center 市政设施管理中心

**UMD** Universal Media Discs 通用媒体磁盘

**UME** Umea, Sweden (Airport Code) 瑞典于默真机场

UNI Management Environment UNI 管理环境

Unit Monthly Equipment 单位月度设备

User Management Entity 用户管理实体

**UMG** Universal Media Gateway 通用媒体网关

**UMIN** Unaccompanied Infant 无人陪伴婴儿

**UMIS** Urban Management Information System 城市管理信息系统

**UML** Unified Modeling Language 统一建模语言

**UMLS** Universal Microwave Landing System 通用微波着陆系统

**UMN** Unsatisfactory Material Notice 不合格航材通知

**UMNR** Unaccompanied Minors 无人陪伴儿童

**UMP** Uniformly Most Powerful 均匀最大功效的

Unscheduled Maintenance Program 非定期维修方案

**UMPDU** User Message Protocol Data Unit 用户报文协议数据单元

**UMR** Upper Maximum Range 上部最大范围,上部最大距离

**UMS** Ultrasonic Motion System 超声波运动系统

Unattended Machinery Space 无人机舱

Unified Messaging Service 统一传信服务

Universal Maintenance Standards 通用维修标准

Unmanned Multifunction Satellite 无人操纵多用途卫星

User Modifiable Software 用户可修改软件

Utility Management System 公用管理系统

**UMT** Universal Microwave Trainer 通用微波训练器

Universal Microwave Training 通用微波训练

Universal Mount 通用安装架

**UMTA** Urban Mass Transportation Administration 都市公共交通管理局

**UMTS** Universal Mobile Telecommunication System 通用移动通信系统

**UMUS** Unbleached Muslin 原色棉布

**UN** Unified 统一的,一元化的

Union 联合,联盟

Unit 组件,单元

United Nations 联合国

Unitized 统一的,成套的

Upstream Node 上行节点

Urban Network 城市网

**UNA** Unable 不能的

**UNACH** Unachievable 不可实现的,不可完成的

**UNADV** Unable, Advise 不可能,将告知

**UNAP** Unable to Approve 未批准

**UNAT** United Nations Administrative Tribunal 联合国行政法庭

**UNB** Union Bonnet 联合节,联轴节

Universal Navigation Beacon 通用导航信标

**UNBAL** Unbalance/Unbalanced 不平衡/不平衡的

**UNBLK** Unblanking 不消隐,启通

**UNC** Unified Coarse Thread 统一标准粗螺纹

Unified National Course 统一国家航线

Universal Navigation Computer 通用导航计算机

**UNCDF** United Nations Capital Development Fund 联合国资本发展基金

**UNCDPPP** United Nations Centre for Development Planning, Projections and Policies 联合国发展规划、预测和政策中心

**UNCED** United Nations' Conference on Environment and Development 联合国环境与发展大会

**UNCFMD** Unconfirmed 未确认的

**UNCHBP** United Nations Centre for Housing, Building and Planning 联合国住房、造房和

规划中心

**UNCHS** United Nations Centre for Human Settlements(Habitat) 联合国人类住区(生境)中心

**UNCID** United Nations Centre for Industrial Development 联合国工业发展中心

**UNCITRAL** United Nations Commission on International Trade Law 联合国国际贸易法委员会

**UNCLAS** Unclassified 未分类的,不保密的

**UNCLP** Unclamp 松开,放开

**UNCOND** Unconditional 无条件的,绝对的

**UNCPL** Uncouple 脱开

**UNCPLG** Uncoupling 解耦,拆离

**UNCPPP** United Nations Centre for Projection, Planning and Policy 联合国计划、设计和政策中心

**UNCRCTD** Uncorrected 未校正的

**UNCRD** United Nations Centre for Regional Development 联合国区域发展中心

**UNCSDHA** United Nations Centre for Social Development and Humanitarian Affairs 联合国社会发展和人道主义事业中心

**UNCSTD** United Nations Centre for Science and Technology for Development 联合国科学和技术促进发展中心

United Nations Commission on Science and Technology for Development 联合国科学和技术促进发展委员会

**UNCT** United Nations Country Team 联合国国家工作队

**UNCTAD** United Nations Conference on Trade and Development 联合国贸易与发展会议

**UNCTC** Unable to Contact 未能联系

United Nations Centre on Transnational Corporations 联合国跨国公司中心

**UND** Under 在…之下

**UNDA** United Nations Development Authority 联合国发展管理局

**UNDAF** United Nations Development Assis-tance Framework 联合国发展援助框架

**UNDC** Undercurrent 欠电流

United Nations Development Corporation 联合国开发公司

**UNDEF** Undefined 未定义的,不明确的

**UNDELD** Undelivered 未交付的

**UNDESA** United Nations Department of Economic and Social Affairs 联合国经济和社会事务部

**UNDET(M)** Undetermined 未确定的

**UNDF** Underfrequency 频率偏低,欠频

**UNDG** United Nations Development Group 联合国发展集团

**UNDLD** Underload 欠载

**UNDP** United Nations Development Programme 联合国开发计划署

**UNDSS** United Nations Department of Safety and Security 联合国安全和保卫部

**UNDV** Undervoltage 欠电压

**UNEDA** United Nations Economic Development Administration 联合国经济发展管理局

**UNEF** Unified Extra Fine Thread 统一超精细螺纹

United Nations Environment Fund 联合国环境基金

**UNEMG** United Nations Environment Management Group 联合国环境管理组

**UNEMP** Unemployment 失业

**UNEP** United Nations Environment Programme 联合国环境规划

**UNESCO** United Nations Educational, Scientific and Cultural Organization 联合国教科文组织,联合国科学教育及文化组织

**UNF** Unified Fine Thread 统一细螺纹

**UNFCCC** United Nations Framework Convention on Climate Change 联合国气候变化框架公约

**UNFPA** United Nations Population Fund 联合国人口基金

**UNFSTD** United Nations Fund for Science

and Technology for Development 联合国科学与技术促进发展基金

**UNGIS** United Nations Group on the Information Society 联合国信息社会小组

**UNGIWG** United Nations Geographic/Geographical Information Working Group 联合国地理信息工作组

**UNHCR** United Nations High Commissioner for Refugees 联合国难民事务高级专员署

**UNI** User Network Interface 用户网络接口

**UNIC** United Nations Information Centre 联合国新闻中心

**UNICEF** United Nations Children's Fund 联合国儿童基金会

**UNICOM** Uniform Communication 统一通信
Universal Communication Frequency 通用通信频率
Universal Components 通用部件
Universal Integrated Communications 通用集总通信

**UNID** Universal Network Interface Device 通用网络接口装置(器件)

**UNIDAP** Universal Digital Autopilot 通用数字自动驾驶仪

**UNIDO** United Nations Industrial Development Organization 联合国工业发展组织

**UNIF** Unidentified 未鉴别的,未识别的
Uniform 一致的,均匀的

**UNIF COEF** Uniformity Coefficient 均匀系数

**UNIFET** Unipolar Field Effect Transistor 单极场效应管

**UNIMG** United Nations Issue Management Group 联合国问题管理组

**UNINC** Unincorporated 未结合的,未编入的

**UNITAR** United Nations Institute for Training and Research 联合国训练和研究所

**UNIV** Universal 通用的,万能的
University 大学

**UNK** Unknown 未知数,未知的

**UNL** Unlimited 无限的,无限制的,无限

Unloading 卸载

**UNLCH** Unlatched 未锁存的

**UNLIM** Unlimited 未限制的

**UNLK** Unlock 开锁

**UNLKD** Unlocked 未锁定,开锁

**UNLKG** Unlocking 未锁定,开锁,开启,释放

**UNLS** Unless 除非

**UNMA** Unified Network Management Architecture 统一网络管理体系结构

**UNMIK** United Nations Interim Administration Mission in Kosovo 联合国科索沃特派团

**UNMKD** Unmarked 未标记的,无标牌的
Unmounted 未安装的

**UNO** United Nations Organization 联合国组织

**UNOCHA** United Nations Office for the Coordination of Humanitarian Affairs 联合国人道主义事务协调厅

**UNODC** United Nations Office on Drugs and Crime 联合国毒品和犯罪问题办事处

**UNPRESS** Unpressurized 未加压的

**UNPS** Universal Power Supply 通用电源

**UNQ** Unique 唯一的,独特的

**UNQTE** Unquote 结束引语

**UNRCVD** Un-Received 未收到的

**UNREL** Unreliable 不可靠的

**UNREP** Underway Replenishment 航途中补给

**UNRGLTD** Unregulated 未调整的,未校正的

**UNRISD** United Nations Research Institute for Social Development 联合国社会发展研究所

**UNRLZD** Unrealized 未实现的

**UNRRA** United Nations Relief and Rehabilitation Administration 联合国善后救济总署

**UNRSTR** Unrestricted 无限制的,自由的

**UNS** Unified Numbering System 统一编号方式系统
Unified Special Thread 统一特种螺纹
Universal Nesting System 通用成套系统

Universal Number Service 通用号码业务

**UNSC** United Nations Security Council 联合国安全理事会

**UNSCH(D)** Unscheduled 不定期的, 非计划的

**UNSD** Unused 未使用的

**UNSUPPR** Unsuppressed 未压制的, 未抑制的

**UNTAA** United Nations Technical Assistance Administration 联合国技术援助总署(局)

**UNTW** Untwist 解开, 捻开

**UNU** United Nations University 联合国大学

**UNXP** Unexpected 意外的

**UO(U/O)** Used on 用于

**UOC** Ultimate Operational Capability 最大使用能力

**UOD** Ultimate Oxygen Demand 极限需氧量

**UOEIC** Ultra Opto Electronic Integrated Circuit 超光电集成电路

**UOL** Underwater Object Locator 水下目标定位器

**UOM** Unit of Measurement 测量单位

**UOS** Unless Otherwise Specified 除非另有说明

User-out-of-Service 用户业务中断

**UP** Microprocessor 微处理器

User Part 用户部分

**UPA** Universal Platform Architecture 统一平台架构

**UP&P** Universal Plug and Play 通用即插即用

**UPAS** Unit Production Assembly Schedule 单价生产装配进度计划

**UPBR** Upper Berth 上铺

**UPC** Unit Production Cost 组件生产成本

Universal Product Code 通用产品编码

Usage Parameter Control 使用参数控制

**UPCH** User Packet Channel 用户分组信道

**UPD** Unpaid 未付款的

**UPDFT** Updraft 上通风, 上升气流

**UPDT** Update or Updating 更新

**UPF** User Port Function 用户口功能

**UPG** Ujung Pandang, Indonesia (Airport Code) 印度尼西亚乌戎潘当机场

**UPGR** Upgrade 升级

**UPI** User Personal Identification 用户个人识别

**UPL** Unique Parts List 专用零件目录

**UPM** Unit Price Matrix 组件价格矩阵

**UPRM** Universal Platform Resource Management 通用平台资源管理

**UPRS** Unit Production Replacement Schedule 组件生产更换计划, 设备生产更换计划

**UPS** Uninterrupted Power Supply 不中断电源

Uninterrupted Power System 不间断供电系统

Uninterruptible Power Supply 不断电电源, 不间断电源

Unit Production Schedules 组件生产计划

United Parcel Service (US) 联合包裹服务公司

User Processing Subsystem 用户处理子系统

**UPSMS** UPS Management System 不间断供电电源管理系统

**UPSR** Unidirectional Path Switching Ring 单向通道交换环

**UPSS** Unit Production Scheduling System 组件生产计划系统

**UPT** Undergraduate Pilot Training 飞行学员训练, 飞行员初级训练

Universal Personal Telecommunication 通用个人通信

**UPTS** Universal Packet Time Slot 通用分组时隙

**UPU** Universal Postal Union 万国邮政联盟

**UPWD** Upward 向上

**UQL** Unacceptable Quality Level 不合格质量水平

**UR** Unregistered 未注册的

Unsatisfactory Report 不满意报告

Utilization Rate 利用率

**URA** Ultra-Reliable Aircraft 超高可靠性飞

机

Unrestricted Article 非保密文件

User Range Accuracy 用户测距精度

User Requirement Analysis 用户需求分析

**URC** Uniform Resource Citation 统一资源引用

Unit Repair Cost 单位修理成本

Unreserved Cargo 未预定(吨位)货物

Utility Radio Communication 实用无线电通信

**URD** User Requirement Document 用户需求文件

**URE** User Range Error 用户测距误差

**URET** User Request Evaluation Tool 用户需求评价工具

**URH** Upper Right Hand 右上方

**URI** Uniform Resource Identifier 统一资源标识符

**URITS** Ultra Reliable Information Transfer System 高可靠信息传输系统

**URL** Uniform Resource Locator 统一资源定位符

User Route List 用户路由表

**URN** Uniform Resource Name 统一资源命名

**URR** Ultra Reliable Radar 高可靠雷达

**URRV** Urban Rapid Rail Vehicle 城市快速轨道车

**URSI** Union Radio-Scientifique International 国际无线电科学联盟

**URT** Universal Radar Tracker 通用雷达跟踪器

Universal Remote Terminal 通用遥控终端

Up-Right 右上

Utility Radio Transmitter 实用无线电发射机

**US** Unavailable Seconds 不可用秒

Under Speed/Underspeed 欠速

Undersize 尺寸不足,过小

Uniform System 统一系统,联合系统

Unserviceable 不适于使用的,使用不合格的

Upper Stage 前一级,上一级

User 用户

Using 使用

**USAF** United States Air Force 美国空军

**USAP** Universal Safey Audit Plan 普遍保安审计计划

Universal Security Audit Programme 普遍保安审计计划

**USARP** United States Antarctic Research Program 美国南极研究计划

**USART** Universal Synchronous Asynchronous Receiver Transmitter 通用同步非同步接收传送器

**USAS** United States of America Standard 美国标准

**USASI** United States of America Standards Institute 美国标准学会

**USAT** Ultra Small Aperture Terminal 超小口径天线终端,超小口径地球站

**USB** Unified S-Band 标准 S 波段

Universal Serial Bus 通用串行总线

Upper Side Band 上边带

Upper Surface Blowing 上表面吹风

**USBS** United States Bureau of Standards 美国标准局

User Signaling Bearer Service 用户信令承载业务

**USC** United States Code 美国法典

United States Customs 美国海关

Universal Service Circuit 通用业务电路

User Service Center 用户服务中心

**USCF** Upper Stage Control Function 上一级控制功能

**USCG** United State Coast Guard 美国海岸警卫队

**USCIS** U. S. Citizenship and Immigration Service 美国公民与移民署

**USD** Ultimate Strength Design 极限强度设计

United States Dollar 美元

**USDC** United States Digital Cellular 美国数字蜂窝

**USDM** Unified System Diagnostic Manager 统

一系统监测管理器

**USE** Usage 用途,用法,惯例

United States Steel Corp. 美国钢铁公司

User System Emulation 用户系统仿真

**USEA** User Service Environment Agent 用户业务环境代理

**USEC** United System of Electronic Computers 电子计算机联合系统

**USERID** User Identification 用户识别

**USFL** Useful 有用的

**USFS** United States Forest Service 美国林业局

**USFU** Unglazed Structural Facing Units 无光结构用表面构件

**USG** Unique Signal Generator 专用信号产生器

United States Gallon 美国加伦

United States Gauge 美国标准线规

**USGBC** United States Green Building Council 美国绿色建筑委员会

**USGPM** U. S. Gallon per Minute 美国加伦/分

**USHGA** United States Hang Gliding Association 美国悬挂式滑翔联合会

**U-SHN** Unidirectional Self-Healing Network 单向自愈网络

**USI** Ultrasonic Inspection 超声波探伤

United States Immigration Service 美国移民局

User Service Information 用户服务信息

User Service Interaction 用户业务互动

User System Interface 用户系统接口

**USIC** User Specific IC 用户专用集成电路

**USL** Upper Specific Limit 规范上限

**USM** Koh Samui, Thailand (Airport Code) 泰国苏梅岛机场

User Signaling Module 用户信令模块

**USN** Ulsan, South Korea (Airport Code) 韩国蔚山机场

**USO** Universal Service Obligation 普遍服务义务

Universal Service Order 通用服务命令

**USOAP** Universal Safety Oversight Audit Programme 普遍安全监督审计计划

**USOC** Uniform Service Order Code 通用服务指令码

Universal Service Order Code 通用服务指令码

**USP** Uncertified Service Provider 未认定的服务提供商

Unified Strategy Programme 统一战略方案

Universal Signal Processor 通用信号处理机

**USPHS** United States Public Health Service 美国公共卫生署

**USPTO** United States Patent and Trademark Office 美国专利商标局

**USR** User Service Routine 用户服务例行程序

**USS** Ultrasonic Soldering 超声波钎焊

United States Standard 美国标准

United States Steel Corp. 美国钢铁公司

Upper Stage Simulator 前级模拟机

**USSA** Underground Security Storage Association 地下安全仓储协会

Unified Systems Safety Analysis 统一系统安全分析

United States Standard Atmosphere 美国标准大气

**USSD** Unstructured Supplementary Service Data 非结构化补充业务数据

**USSFIM** United States Standard Flight Inspection Manual 美国标准飞行检查手册

**USSG** United States Standard Gauge 美国标准规格

**USSP** Universal Sensor Signal Processor 通用传感器信号处理器

**USSR** Union of Soviet Socialist Republics 苏维埃社会主义共和国联盟

**USSS** Unmanned Sensing Satellite System 无人操纵传感卫星系统

**USSST** United States Standard Screw Thread 美国标准螺纹

**UST** Ultra Sonic Testing 超声(波)测试

Unique Signal Train 专用信号序列

Universal Standard Time 世界标准时

Upper Surface Transition 上翼面转捩点

**USTB** Unstable 未稳定的,稳定解除的

**USTDA** United States Trade and Development Agency 美国贸易发展署

**USTDC** United States Travel Data Center 美国旅行数据中心

**USTND** Understand 了解

**USTOL** Ultra-Short Takeoff and Landing 超短距起飞着陆

**USTTA** U. S. Travel and Tourism Administration 美国旅游局

**USTZD** Unsensitized 未感光的,未激活的

**USUB** Unglazed Structural Unit Base 无釉结构部件基础

**USW** Ultra Short Wave 超短波

Ultrasonic Welding 超声波焊

**USWB** United States Weather Bureau 美国气象局

**USWC** Uncacheabled Speculative Write Combination 无缓冲随机联合写操作

**USWV** Uncacheable, Speculative, Write-Combining 非缓冲随机混合写入

**UT** Ultrasonic Test 超声试验

Unemployed Time 停歇时间

Unit Test 单元测试

Unit Time 单位时间,组件时间

Universal Time 世界标准时间,格林威治时间

Universal Trainer 通用练习器

User Terminal 用户终端

Utility Tool 通用工具

**UTA** Unmanned Tactical Aircraft 无人战术飞行器

Upper Control Area 高空管制区

**UTC** Coordinated Universal Time 协调世界时,世界协调时

Unit Time Coding 单一时间编码

United Technologies Corporation 联合技术公司

Universal Time Chiming 通用时间协调

Universal Time Coordinated 国际协调时,世界协调时,协调世界时

**UTCC** United Technologies Communications Company 联合技术通讯公司

Universal Traffic Control Center 世界交通管制协调中心

**UTCF** Universal Time (Coordinated) Function 世界时(协调)功能

**UTDF** Universal Tracking Data Format 通用跟踪数据格式

Universal Traffic Data Format 通用交通数据格式

**UTE** User Terminal Equipment 用户终端设备

Universal Test Equipment 通用测试设备

Unknown Traffic Environment 未知交通环境

**UTI** United Trade International, Inc. (USA) 联合国际贸易公司(美国)

**UTIL** Utility/Utilization 通用,实用

**UTM** Unified Threat Management 统一威胁管理

Universal Test Message 通用试验电报

Used to Make 用于制造

Utility Technical Manual 通用技术手册

Utmost 极端

**UTN** Upington, South Africa (Airport Code) 南非阿平顿机场

Urban Telephone Network 城市电话网

Utensil 器具

**UTP** Universal Trunk Processor 通用中继处理机

Universal Trunk Protocol 通用中继协议

Unshielded Twisted Pair 非屏蔽双绞线

**UTQGS** Uniform Tire Quality Grading System 统一轮胎分级系统

**UTR** Underwater Temperature Recorder 水下温度记录仪

Underwater Tracking Range 水下跟踪距离

**UTRA** UMTS Terrestrial Radio Access UMTS 地面无线接入

**UTRS** Universal Technical Research Services 通用技术研究服务

**UTRTD** Untreated 未处理的

**UTS** Ultimate Tensile Strength 极限抗拉强度

Units to Stock 存货单元

Universal Time Standards 标准世界时

Universal Time-sharing System 通用分时系统

**UTSI** University of Tennessee Space Institute 田纳西大学空间研究所

**UTTAS** Utility Tactical Transport Aircraft System 公用战术运输机系统

**UTTEM** University of Texas Threat and Error Management（美国）得克萨斯大学威胁和过失管理

**UU** Ultimate User 最终用户

**UUCP** Unix-to-Unix Copy Protocol Unix 到 Unix 的拷贝协议

**UUI** User to User Information 用户间信息

**UUMP** Unification of Units of Measurement Panel 统一计量单位专家小组

**UUS** User-to-User Signaling 用户间信令

**UUT** Unit under Test 被测单元

**UUTS** Unit under Test Simulator 被测件模拟器

**UUV** Unmanned Underwater Vehicle 无人潜航器

**UV** Ultraviolet 紫外线

Under Voltage 欠压

Underwater Vehicle 水下航行器

Unmanned Vehicle 无人驾驶飞行器

Upper Sideband Voice 上边带话音

**UVASER** Ultraviolet Amplification by Simulated Emission of Radiation 受激辐射紫外线放大

**UVD** Undervoltage Device 低压器件

Unified Video Decoder 通用视频解码器

**UVEPROM** Ultraviolet Erasable and Programmable Memory 紫外线的可擦可编程序存储器

Ultraviolet Erasable Programmable Read Only Memory 紫外线可抹除可编程序只读存储器

**UVEROM** Ultraviolet Erasable Read Only Memory 紫外线可擦除只读存储器

**UVFLT** Ultraviolet Floodlight 紫外线泛光灯

**UVGC** Using Video in Group Collaboration 使用视频的群协同工作

**UVI** Ultraviolet Imager 紫外线成像仪

Ultraviolet Index 紫外线指数

**UVPROM** Ultraviolet Programmable Read Only Memory 紫外线可编程只读存储器

**UVR** Ultraviolet Radiation 紫外太阳辐射

Undervoltage Relay 低电压继电器

**UVS** Ultraviolet Spectrometer 紫外线光谱仪

Universal Versaplot Software 通用绘图软件

**UVV** Ultimate Vertical Velocity 极限垂直速度

Upward Vertical Velocity 垂直上升速度

**UW** Ultrasonic Wave 超声波

Unconventional Warfare 非常规战争

Unique Word 独特码

Unit Weld 组合焊

Unit Wire 组合导线，组件电缆

Upset Welding 电阻对接焊

**UWA** User Work Area 用户工作区

**UWAL** University of Washington Aviation Laboratory 华盛顿大学航空实验室

**UWB** Ultrawideband, Ultrawide Band 超宽带

**UWBA** Ultra-Wide Bandwidth Amplifier 超宽带光纤放大器

**UWC** Under Water Cable 水下光缆

Universal Wireless Communications 通用无线通讯

**UWS** Unmanned Weather Station 无人气象站

**UWT** Unit Weight 组件重量

**UWTM** Under Water Team 水下小组

**UWTR** Under Water 水下

**UWY** Upper Airway 高空航路

**UY** Uruguay 乌拉圭

# V

V　Airspeed Vector 空速向量
Vacuum Tube 真空管,电子管
Valve 阀,活门
Varnish 清漆,釉子
Velocity 速度
Verify Control Word 校验控制字
Vertical 垂直的
Victor Aircraft 指挥机
Violet 紫色(的)
VIP Baggage 要客行李
Voice 话音,语音
Voltage 电压
Volts 伏特
VOR 甚高频全向信标

**V/A**　Video/Audio 视频/音频

**V/B**　Vertical Bearing 垂直方位

**V/CTOL**　Vertical/Conventional Takeoff and Landing 垂直或正常起落

**V/L**　VOR/Localizer 全向信标/航向台

**V/MIL**　Volt per Mile 伏/英里

**V/NAV**　Vertical Navigation System 垂直面导航系统

**V/S**　Vertical Speed 升降速度,垂直的速率

**V/STOL**　Vertical/Short Takeoff and Landing 垂直或短距起降

**V/TRK**　Vertical Track 垂直航迹,垂直跟踪

**V/V**　Verification and Validation 验证与审定
Vertical Velocity or Speed 垂直速率或速度
Volume to Volume 容积比,体积比

**VA**　Design Maneuvering Speed 设计机动速度
Vacant Auditoria 空场
Value Analysis 数值分析
Variable Area 可变区域
Variable Attenuator 可变衰减器

Video Amplifier 视频放大器
Virtual Address 虚拟地址
Visual Approach 目视进近
Volcanic Ash 火山灰
Volt Ammerter 伏安表
Volt Ampere 伏安

**V-A**　Vibro-Acoustic 振动声(学)的

**VAA**　Vaasa, Finland(Airport Code) 芬兰瓦萨机场
Virgin Atlantic Airlines (England) 维尔京大西洋航空公司(英国)
Voice Access Arrangement 话音存取设备

**VAAC**　Vectored-thrust Aircraft Advanced Flight Control 推力矢量飞机高级飞行控制
Volcanic Ash Advisory Center 火山灰咨询中心

**VAATE**　Versatile Affordable Advanced Turbine Engines 通用经济型先进涡轮发动机

**VAB**　Vehicle Assembly Building 飞行器装配大楼
Vertical Assembly Building 垂直装配大楼
Voice Answer Back 声音应答装置

**VAC**　Vacant 空闲的,空白的
Vacation 假期,空出
Vacuum 真空(的)
Value Added Carrier 增值运营商
Value Added Chain 增值链
Vaught Aircraft Company 伏夫特飞机公司
Vector Analogue Cornputer 矢量模拟计算机
Ventilation and Air Conditioning 通风与空调
Video Amplifier Channel 视频放大器通道
Visual Approach Chart(s) 可视进近图,目视进近图
Volts Alternating Current 交流电压

**VACBI**　Video and Computer Based Instruction 视频和计算机辅助教学

Visual and Computer Based Instruction 目视和计算机教学

**VACE** Verification and Checkout Equipment 验证与检查设备

**VACT** Volts Alternating Current Test 交流电压试验

**VACW** Volts Alternating Current Working 交流电压工作

**VAD** Value-Added Distributor 增值分销商

Velocity Azimuth Display 速度方位显示器

Voice Activity Detection 语音活动性检测

**VADAR** Visual Algorithm Definition for Avionics Reconfiguration 航空电子设备视觉算法定义

**VADE** Vandenberg Automatic Data Equipment 范登堡自动数据设备

Versatile Automatic Data Exchanger 多用途自动数据交换机

**VADIS** Voice and Data Integrated System 话音和数据综合系统

**VADR** Voice and Data Recorder 语音与数据记录器

**VADS** Value-Added Data Service 增值数据业务

Visual Arts Data Service 视觉艺术的数据服务

Visual Aural Digit Span Test 视觉听觉位间距检测

**VAG** Video Address Generator 视频地址发生器

**VAI** Variable-Area Intake 变截面进气口

Video Assisted Instruction 视频辅助指令

**VAL** Valid 有效的,正确的

Validation 有效,批准

Validity 有效性,确实

Valley 山谷

Valuable 贵重的

Valuable Cargo 贵重货物

Valuation 评估,鉴定,赋值

Value 价值,评价

Valve 活门,阀

Vertical Alarm Limit 垂直告警限

Visual Approach and Landing Chart 目视进近着陆图

**VALDN** Validation 有效,批准,确认

**VALT** VTOL Approach and Landing Technology 垂直起飞和着陆飞机进场与着陆技术

**VALU** Value 价值,评价

**VAM** Video Access Module 视频接入模块

Video Administration Module 视频管理模块

Virtual Access Method 虚拟存取法

Visual Approach Monitor 目视进近监控仪

Voltammeter 伏安表

**VAM/HUD** Visual Air Monitor/Head up Display 目视空中监视器/平视显示器

**VAMA** Variable Address Multiple Access 可变地址多路存取

**VAMP** Vulnerability Assessment Modeling Program 易损性评估模拟程序

**VAMS** Vector Airspeed Measuring System 空速矢量测量系统

**VAN** Runway Control Van 跑道指挥车

Value-Added Network 增值网

Video Audio Navigation 视频音频导航

Virtual Area Network 虚拟局域网

**VANS** Value-Added Network Service 增值网络服务

**VANT** Vibration and Noise Tester 振动与噪声试验器

**VAP** Video Access Point 视频接入点

Videotex Access Point 可视图文接入点

Visual Aids Panel 目视设备信号板

Visual Approach Procedure 目视进近程序

**VAPC** Vector Adaptive Predictive Coding 矢量自适应预测编码

**VAPI** Visual Approach Path Indicator 目视进近航道指示器

**VAPP** Final Approach Speed 最后进近速度(进近速度)

**VAPS** V/STOL Approach System 垂直/短距起降进近系统

Virtual Applications Prototyping System 虚拟样机应用系统

Virtual Avionics Prototyping System 航空电子虚拟样机系统

**VAR** Magnetic Variation 磁差

Value-Added Reseller 增值转卖

Variable 可变的

Variance 变化,变量,偏差,差异

Variance Analysis Report 偏差分析报告

Variant 变型,变量,不同的,差异的

Variation 变量,变化,磁差

Various 各式的

Varying 变化的,不定的

Visual-Aural Radio Range 视听式无线电信标

**VARI** Variable 可变的

**VARM** Varmeter 无功伏安计

**VARR** Visual-Aural Radio Range 视听无线电定位

**VAS** Value-Added Service 增值服务

Virtual Address Space 虚拟寻址空间

VISSR Atmospheric Sounder 可见光—红外自旋扫描辐射计大气探测器

Visual Augmentation System 目视增强系统

**VASCAR** Visual Average-Speed Computer and Recorder 可视的平均速度的计算机和录音机

**VASI** Visual Approach Slope Indicator 目视进近坡度指示器

**VASIS** Visual Approach Slope Indicating System 目视进近坡度指示系统

Visual Approach Slope Indicator System 目视进近坡度指示器系统

**VASP** Value-Added Service Provider 增值服务供应商

Virtual Analog Switching Point 虚拟模拟交换点

**VASS** Visual Analysis Subsystem 视频分析子系统

**VAST** Versatile Automatic Shop Test 通用自动车间测试

Versatile Avionics Shop Test 通用航空电子设备车间测试

Versatile Avionics System Tester 航空电子系统多样测试器

**VAT** Speed at Threshold 跑道入口速度

Value Added Tax 增值税,价值附加税

Variable Attenuator 可变衰减器

Visual Audio Tool 视觉音频工具

Voyage across Technology 航行技术

**VATE** Versatile Automatic Test Equipment 通用自动测试设备

**VATLS** Visual Airborne Target Locating System 目视空中目标定位系统

**VATS** Video-Augmented Tracking System 视频增强跟踪系统

**VATSIM** Virtual Air Traffic Simulation Network 虚拟空中交通模拟网

**VAU** Voltage Averaging Unit 电压平均组件

**VAVE** Average Velocity 平均速度

**VAX** Virtual Address Extended/Extension 虚拟地址扩充

**VB** Design Speed for Maximum Gust Intensity 最大阵风强度的设计速度

Valve Box 阀盒,活门箱

**VBA** Visual Basic for Applications 可视化BASIC应用软件

**VBAFC** Weather Bureau Area Forecast Center 气象局区域预报中心

**VBC** Video-Band-width Compression 视频宽度压缩

**VBD** Voice Band Data 话音带内数据

**VBI** Vertical Blanking Interval 垂直消隐间隔

**VBLIU** Video Balanced Line Interface Unit 视频平衡线接口组件

**VBM** Value Based Management 基于价值管理

**VBNS** Very High Bandwidth Network Service 超宽带网络服务

Very High Speed Backbone Network Service 甚高速骨干网服务

Very High-performance Backbone Network

Service 超高性能基干网服务
**VBR** Variable Bit Rate 可变比特率
**VBRG** Vertical Bearing 垂直方位
**VBR-nrt** Variable Bit Rate, non-real time 非实时可变比特率
**VBR-rt** Variable Bit Rate, real time 实时可变比特率
**VBS** Vertical Beam Sensor 垂直波束传感器
Voice Broadcast Service 语音广播业务
**VBV** Variable Bleed Valve 可变引气活门
Variable Bypass Valve 可变旁通活门
**VC** Variable Camber 可变曲度
Varnish Cambric 漆布,黄腊布
Ventilation Controller 通风控制器
Vertical Circle 地平径圈
Vicinity of Aerodrome 机场附近
Video Compressor 视频压缩器
Video Conference/Conferencing 电视会议
Video Controller 视频控制器
Virtual Call 虚呼叫
Virtual Channel 虚信道
Virtual Circuit 虚电路
Virtual Connection 虚连接
Virtual Container 虚容器
Voice 话音
Voice Coil 话音线圈
Volume of Compartment 机舱容积
**VCAIS** Virtual Channel Alarm Indication Signal 虚信道告警指示信号
**VCAP** Virtual Channel Assignment Problem 虚信道分配问题
**VCB** Visual Control Board 目视控制板
**VCC** Video Cable Communication 视频电缆通信
Video Capture Card 图像捕获卡
Video Control Center 视频控制中心
Virtual Call Capability 虚呼叫能力
Virtual Call Control 虚呼叫控制
Virtual Centralized Controller 虚拟集中控制器
Virtual Channel Connection 虚信道连接

Voltage Controlled Clock 压控时钟
**VCCE** Virtual Channel Connection Endpoint 虚信道连接端点
**VCCS** Voice Communication Center System 话音通信中心(控制)系统
**VCCT** Virtual Channel Connection Termination 虚信道连接终端
**VCD** Video Compact Disk 视频光盘
Voltage Controlled Device 电压控制器件
Vortex Control Device 涡流控制分配组件
**VCDU** Video Control Distribution Unit 视频控制分配组件
**VCE** Value Chain Engineering 价值链工程
Variable Cycle Engine 可变循环发动机
Venice, Italy(Airport Code) 意大利威尼斯机场
Vertical Centrifugal 垂直离心的
Virtual Call Evolution 虚呼叫演变
Virtual Channel Entity 虚信道实体
Virtual Collaborative Environment 虚拟合作环境
Vocational Certificate of Education 职业教育证书
**VCF** Virtual Call Facility 虚呼叫设备
Voltage Controlled Filter 电压控制滤波器
**VCG** Vertical Center of Gravity 垂直中心
Vertical Location of the Center of Gravity 重心垂直位置
Video Channel Generator 视频信道发生器
Video Command Generator 视频指令发生器
**VCH** Virtual Channel Handler 虚信道处理器
**VCHP** Variable Conductance Heat Pipes 可变导热管道
**VCHR** Voucher 票券
**VCI** Value Chain Integration 价值链集成
Virtual Call Identifier 虚呼叫识别
Virtual Call/Circuit Interface 虚呼叫/电路接口
Virtual Chain Index 虚拟链路标志
Virtual Channel Identification 虚信道识别
Virtual Channel Identifier 虚信道标识符

Virtual Circuit Identifier 虚电路标识符

Virtual Connection Identifier 虚拟连接识别符

**VCK** Variable Camber Kruger 克鲁格可变曲率(弯度)

**VCL** Vertical Centerline 垂直中心线

Virtual Channel Link 虚信道链路

Visual Check List 目视检查单

Visual Component Library 可视元件库

**VCM** Value Chain Management 价值链管理

Video Camera Module 视频照相机组件

Volatile Condensable Material 可凝性挥发物

**VCMAX** Active Maximum Control Speed 最大有效控制速度

**VCME** Virtual Channel Multiplex Entity 虚信道复用实体

**VCMIN** Active Minimum Control Speed 最小有效控制速度

**VCN** Virtual Call Network 虚呼叫网络

**VCO** Variable Crystal Oscillator 可变频晶体振荡器

Virtual Central Office 虚拟中心局

Voice Coder 话音编码器

Voltage-Controlled Oscillator 电压控制振荡器

**VCOMM** Virtual Communication 虚拟通信

**VCOP** Variable Control Oil Pressure 可控制滑油压力

Virtual Cockpit Optimization Program 虚拟座舱优化计划

**VCOS** Visible Caching Operating System 可视高速缓存操作系统

**VCOSS** Vibration Control of Space Structures 空间结构振动控制

**VCP** VHF Radio Control Panel 甚高频无线电控制板

Video Cassette Player 视频盒式放像机

Video Communication Processor 视频通信处理器

Video Control Panel 视频控制面板

Voice Channel Processor 话路处理器

Voluntary Cooperation Programme 自愿合作计划

VOR Check Point 全向信标校核点

**VCPI** Virtual Control Program Interface 虚拟控制程序接口

Virtual Control Protected Interface 虚拟控制保护接口

**VCR** Video Cassette Recorder 盒式磁带录像机

Visual Control Room 目视操纵室

**VCRDI** Virtual Channel Remote Defect Indication 虚信道远端故障指示

**VCRI** Verification Cross-Reference Index 检验交叉参考索引

**VCS** Vehicle Control System 飞行器控制系统

Vendor Coordination Sheet 协作方协调单

Vendor Coordination Supplier 协作方协调供货方

Ventilation Control System 通风控制系统

Video Communication System 视频通信系统

Video Conference Service 电视会议业务

Virtual Circuit Switch 虚电路交换

Virtual Control System 虚拟控制系统

Visual Call Sign 可视电话呼号

Visual Coupled System 视觉耦合系统

Voice Command System 语音指令系统

Voice Communication(s) System 语音通信系统,话音通信系统

Voice Control System 语音控制系统

**VCSCW** Variable Camber Supercritical Wing 可变弯度超临界翼型

**VCSR** Voltage Controlled Shift Register 压控移位寄存器

**VCSS** Voice Communication Switch System 语音通话交换系统

**VCT** Victoria, TX, USA (Airport Code) 美国得克萨斯州维多利亚机场

Video Conferencing Tool 视频会议工具

Virtual Cut Through 虚拟直通

Voltage Control transfer 压控传输

**VCTU** Variable Control Trim Unit 可变控制配平组件

**VCTV** Viewer-Controlled Cable TV 用户控制有线电视

**VCU** Valve Control Unit 活门控制组件
Video Control Unit 视频控制组件
Voice Channel Unit 话音信道单元

**VCV** Vacuum Check Valve 真空止回阀
Vacuum Control Valve 真空控制阀

**VCW** Verify Control Word 验证控制字

**VCWT** Video Call Waiting Service Trunk 视频呼叫等待业务中继线

**VCX** Virtual Container Cross Connect 虚容器交叉连接

**VCXO** Voltage Controlled Crystal Oscillator 压控晶体振荡器

**VCY** Vicinity 附近

**VD** Design Diving Speed 设计俯冲速度
Dive-Speed Maximum 最大俯冲速度
Heading to a DME Distance 到测距机距离的航向
Vandyke 锯齿边
Vault Door 拱形门
Vertical Distance 垂直距离
Video Decoder 视频译码器
Virtual Data 虚拟数据
Virtual Destination 虚拟目标
Virtual Device 虚拟设备
Voltage Drop 电压降

**VDA** Validation of Design Approval 设计认可批准书
Video Distributing Amplifier 视频分配放大器
Voice and Data Analog Adaptor 话音和数据模拟适配器

**VDAM** Virtual Data Access Method 虚拟数据存取法

**VDATS** Versatile Depot Automatic Test Station 通用场站级自动测试工作站
Virtual Deficiency Analysis and Tracking System 模拟缺陷分析与跟踪系统

**VDB** VHF Data Broadcast 甚高频数据广播

**VDC** Vendor Deficiency Change 协作方缺陷更改
Video Display Controller 视频显示控制器
Video Distribution Controller 视频传输控制器
Virtual Data Center 虚拟的数据中心
Virtual Development Center 虚拟开发中心
Virtual Device Coordinates 虚拟设备坐标
Visual Dimensional Check 目视范围检查
Volts Direct Current 直流电压

**VDD** Video Data Distribution 视频数据分配
Virtual Device Driver 虚拟设备驱动器
Visual Display Data 可视化显示数据
Voice Data Display 音频数据显示器

**VDDN** Video Data Distribution Network 视频数据分配网

**VDDP** Video Digital Data Processing 视频数字数据处理

**VDE** Voice Data Entry 话音数据输入

**VDEL** Variable Delivery 可变交付

**VDET** Voltage Detector 电压检波器

**VDF** Variable Density Filter 可变密度滤波器
Variable Digital Filter 可变数字滤波器

**VDFG** Variable Diode Function Generator 可变二极管函数产生器

**VDGS** Visual Docking Guidance System 目视停靠导引系统

**VDH** Voice and Data Hub Package 话音和数据中心组件

**VDI** Video Device Interface 视频设备接口
Video Display Interface 视频显示器接口
Virtual Device Interface 虚拟设备接口
Visual Doppler Indicator 可视多普勒指示器

**VDIG** Vertical Display Indicator Group 垂直显示指示器组

**VDL** Very High Frequency Data Link 甚高频数据链
Video Data Link 视频数据链

**VDLC** Virtual Data Link Capability 虚拟数据

链路能力

**VDM** Video Display Metafile 视频显示元文件

**VDMA** Variable Destination Multiple Access 接收站可变多址访问

**VDOP** Vertical Dilution of Precision 垂直精度衰减因子

**VDP** Vehicle Deadlined for Parts 航空器因零件损坏而停用

Video Display Processor 视频显示处理机

Visual Descent Point 目视下降点

**VDPI** Vehicle Direction and Position Indicator 飞行器方向与位置指示器

**VDPS** Voice Data Processing System 音频数据处理系统

**VDR** Vendor Data Requirement 协作方数据要求

VHF Data Radio 甚高频数据无线电

Video Dynamic Range 视频动态范围

Voice Digitization Rate 声音数字化率

**VDRAM** Video Dynamic Random Access Memory 视频动态随机存储器

**VDS** Vadso, Norway ( Airport Code ) 挪威瓦得索机场

Variable Depth Sonar 可变深度声呐

Velocity Data System 速度数据系统

Video Display System 视频显示系统

Voice Data Service 话音数据通信业务

**VDSL** Very-high-speed Digital Subscriber Line 甚高速数字用户线

**VDT** Variable Density Tunnel 可变密度风洞

Video Data Terminal 可视数据终端

Video Dial-Tone 视频拨号音

Video Display Terminal 视频显示终端

Voice Data Trunking 话音数据中继

**VDTG** Video Dial Tone Gateway 视频拨号音网关

**VDTR** Variable-Diameter Tilt Rotor 可变桨盘直径倾转旋翼机

**VDTS** Video Dial Tone Services 视频拨号音服务

**VDU** Very High Frequency Data Unit 甚高频数据组件

Video Display Unit 视频显示装置

Video Distribution Unit 视频分配组件

Visual Display Unit 目视显示装置

**VDV** Vacuum Delay Valve 真空延迟活门（阀）

Vacuum Differential Valve 真空差压活门（阀）

Vacuum Diverter Valve 真空分流阀

**VE** Value Engineering 价值工程

Velocity East 向东速度

Velocity Error 速度误差

Venezuela 委内瑞拉

Ventilating Equipment 通风设备

Video Expander 视频扩展器

Violation Error 违反误差,扰乱误差

Virtual Enterprises 虚拟企业

Virtual Environment 虚拟环境

Visual Exempted 目视除外

**VEB** Variable Elevation Beam 可变仰角波束

**VEC** Visual Enroute Chart 目视航路图

**VECO** Vernier Engine Cutoff 微调发动机关断

**VECP** Value Engineering Change Proposal 价值工程更改建议

**VECT** Vector 矢量

**VED** Voice Editor 语音编辑程序

**VEDAR** Visible Energy Detection and Ranging 可见能量检测与定位

**VEE** Visual Engineering Environment 视频工程环境

**VEF** Critical Engine Failure Speed 发动机失效临界速度

**VEG** Vegetable 植物,蔬菜

Visual Entertainment Group 视觉娱乐集团

**VEH** Vehicle 航空器,运载工具

**VEHCV** Vacuum Exhaust Heat Control Valve 真空排气热控制活门(阀)

**VEL** Vellum 精制皮纹

Velocity 速度

**VEM** Value Engineering Model 价值工程模型

**VEN** Virtual Enterprise Network 虚拟企业网
Virtual Equipment Number 虚拟设备号码

**VENT** Ventilate/Ventilation/Ventilator 通风/自由讨论/通风设备

**VEP** Video Entertainment Player 视频娱乐放像机

**VER** Veracruz, Mexico(Airport Code) 墨西哥韦拉克鲁斯机场
Version 改型, 方案, 版本
Vertical 垂直的

**VERB** Verbatim 逐字的

**VERDAN** Versatile Differential Analyzer 通用微分分析器

**VERIF** Verification 验证, 校验

**VERLORT** Very-Long-Range Tracking 超远程跟踪

**VERN** Vernier 游标

**VERS** Versed Sine 正矢

**VERST** Versatile 通用的, 万能的

**VERT** Vertical 垂直

**VERT ACC** Vertical Acceleration 垂直加速

**VERT REV** Vertical Revise 垂直修正

**VERTREP** Vertical-Replenishment 垂直补给

**VES** Video Entertainment System 视频娱乐系统
Voice Entry System 话音输入系统

**VESA** Video Electronics Standards Association 视频电子标准协会
Video Equipment Standards Association 视频设备标准协会

**VEST** Vesting 授予

**VET** Veteran 老手, 熟练工
Visual Editing Terminal 可视编辑终端

**VETAB** Vocational Education and Training Accreditation Board 职业教育与培训认可委员会

**VEV** Voice Excited Vocoder 声激励音码器

**VEWS** Very Early Warning System 极早期预警系统

**VF** Design Flap Velocity 设计襟翼速度
Validation Flight 定型飞行
Variable Frequency 可变频率
Vector Field 向量场
Velocity Factor 速度系数
Video Frequency 视频
Viscosity Factor 粘度因数(系数)
Visual Field 视野, 视界
Visual Flight 目视飞行
Voice Frequency 音频, 话频
Voltage Forward 正向电压

**VFA** Vibration Fly around 振动飞绕

**VFAT** Virtual File Allocation Table 虚拟文件分配表

**VFC** Video Frequency Carrier 视频载波

**VFCTT** Voice Frequency Carrier Teletype 音频载波电传打字机

**VFD** Vacuum Fluorescent Device 真空荧光器件
Voice Frequency Dialing 话频拨号

**VFDE** Voice Frequency Dialing Equipment 话频拨号设备

**VFDS** Vehicle Flight Dynamic Simulator 飞行器飞行动态模拟机

**VFE** Maximum Flap Extended Speed 最大襟翼放下速度
Variable Flap Ejector 可变襟翼收放器
Velocity Flaps Extended 襟翼放下速度

**VFG** Variable Frequency Generator 变频发电机
Voice Frequency Generator 话频发生器

**VFI** Visual Fault Indicator 目视故障指示器

**VFLJ** Very Low Frequency Jammer 甚低频干扰台

**VFM** Vacuum Forming Mold 真空成形模
Vertical Flight Maneuver 垂直飞行机动
Voice/Fax Mail 语音/传真信箱
Voltage Frequency Monitor 电压频率监视器

**VFN** Virtual Fax Network 虚拟传真网

**VFO** Variable Frequency Oscillator 可变频率振荡器

Variable Fuel Oil Tank 可变燃油箱

Video Frequency Oscillator 视频振荡器

Voice Frequency Oscillator 音频振荡器

**VFOP** Visual Flight Rules Operations Panel 目视飞行规则运行专家组

**VFP** Variable Factor Programming 可变因素编程

Variable Floating Point 可变浮点

**VFPR** Via Flight Planned Route 通过飞行计划(申报)的航线

**VFR** Vestibular Function Research 廊桥功能研究

Visual Flight Rating 目视飞行等级

Visual Flight Rules 目视飞行规则

**VFRAP** Visual Flight Rules Approach 目视飞行规则进近

**VFREQ CLK** Variable Frequency Clock 可变频率时钟

**VFS** Video File Server 视频文件服务器

Virtual File Storage 虚拟文件存储

Virtual File System 虚拟文件系统

Virtual Filesystem Switch 虚拟文件系统开关

Voltage Fiber Sensor 电压光纤传感器

**VFSG** Variable Frequency Starter Generator 变频启动发电机

**VFSS** Voice Frequency Signaling System 音频信令系统

**VFT** Voice-Frequency Telegraph 音频电报

**VFTO** Velocity Final Take-Off 最后起飞速度

**VFU** Velocity Flaps up 收襟翼速度

**VFXR(R)** Flap Retraction Speed 襟翼收上速度

**VFXR(X)** Flap Extension Speed 襟翼放下速度

**VFY** Verify 验证

**VG** Ground Velocity 地速

Variable Geometry 可变几何形状

Velocity Generator 测速发电机

Vertical Gyro 垂直陀螺仪

Video Graphics 视频图形

Voice Grade 音频段

Voltage Gain 电压增益

Vortex Generator 涡流发生器

**VGA** Variable Gain Amplifier 可变增益放大器

Video Graphics Accelerator 视频图形加速器

Video Graphics Adapter 视频图形适配器

Video Graphics Array 视频图形阵列卡

**VGC** Variable Gain Control 可变增益控制

Variable Geometry Combustion 可变几何燃烧室

Video Graphics Controller 视频图像控制器

Viscosity Gravity Constant 粘度比重常数

**VGCS** Voice Group Call Service 话路群呼叫业务

**VGFS** Variable Gain Feel System 可变增益人工感力系统

**VGH** Velocity, Gravity, Height 速度、重力、高度

Vertical Gust and Height 垂直阵风与高度

**VGI** Vertical Gyro Indicator 垂直陀螺指示器

**VGML** Vegetarian Meal 素食餐食

**VGND** Ground Velocity 地速

**VGO** Vigo, Spain (Airport Code) 西班牙维歌机场

**VGPI** Visual Glide-Path Indicator 目视下滑道指示器

**VGS** Video Graphics System 视频图形显示系统

Visual Glide Slope 可见滑翔道

**VGSI** Visual Glide Slope Indicator 目视下滑坡度指示器

**VGT** Variable Geometry Turbine 可变几何形状涡轮

**VGV** Variable-Geometry Vehile 变形机翼飞行器

**VGW** Variable-Geometry Wing 变形机翼

**VH** Velocity Heading 速度指向

Vent Hole 通风孔

Vertical and Horizontal 垂直与水平

Vickers Hardness 维氏硬度

**VHA** Very High Altitude 超高空,极高空
Volcanic Hazard Area 火山危险区

**VHCU** Versatile Heating Control Unit 多功能加热控制组件

**VHD** Very High Density 超高密度
Video Home Disc 家用视盘

**VHDL** VHSIC Hardware Description Language 超高速集成电路硬件描述语言
Visual Hardware Description Language 可视硬件描述语言

**VHDU** Versatile Heating Data Unit 多功能加热数据组件

**VHF** Very High Frequency (30 ~ 300MHz) 甚高频(30 ~ 300MHz)

**VHFD** Very High Frequency Direction 甚高频率定向

**VHFDF** Very High Frequency Direction Finder 甚高频率定向仪

**VHFDL** Very High Frequency Data Link 甚高频数据链

**VHFG** Very High Frequency Generator 甚高频发生器

**VHFJ** Very High Frequency Jammer 甚高频干扰台

**VHFPNL** Very High Frequency Panel 甚高频控制盒

**VHFXVR** Very High Frequency Transceiver 甚高频收发机

**VHL** Very High-Level Language 甚高级语言

**VHMS** Vehicle Health Management System 飞行器健康管理系统
Vehicle Health Monitoring System 飞行器健康监控系统

**VHOL** Very High Order Language 超高级语言

**VHP** Very High Performance 极高性能
Very High Pressure 甚高压

**VHPIC** Very High Performance Integrated Circuit 超高性能集成电路

**VHR** Voltage Holding Ratio 电压保持率

**VHRR** Very High-Resolution Radiometer 甚高分辨率辐射计

**VHS** Vertical Helical Scanning 垂直螺旋扫描
Very High Speed 甚高速
Video Helical Scan 视频螺旋扫描
Video Home System 家用视频系统

**VHSB** Very High-Speed Data Bus 甚高速数据总线

**VHSIC** Very High-Speed Integrated Circuit 甚高速集成电路

**VHSOL** Very High-Speed Optic Loop 甚高速光环路

**VHTC** Vertical Handle Toggle Clamps 侧边式肘节夹

**VHV** Very High Voltage 高压

**VI** Video Integrator 视频累加器
Virtual Image 虚拟图像
Viscosity Index 粘度指数
Volume Indicator 音量(容积)指示器

**VI&P** Visual, Intelligent and Personal 可视化、智能化和个人化

**VIA** Vancouver International Airport 温哥华国际机场
Versatile Integrated Avionics 通用综合航空电子
Video Image Analysis 视频图像分析
Videotex Interworking Architecture 可视图文互通结构
Vienna International Airport 维也纳国际机场
Virtual Interface Architecture 虚拟接口体系结构
Visual Information Access 视觉信息的访问
Voice-Interactive Avionics 人机对话式航空电子设备

**VIB** Vibrate or Vibration 振动

**VIBY** Visibility 能见度

**VIC** Very Important Cargo 极重要货物
Virtual Information Centre 虚拟信息中心
Virtual Intelligent Component 虚拟的智能组

件

**VICAR** Video Image Communication and Retrieval 视频图像通信与检索

**VICC** Visual Information Control Console 可视信息控制台

**VICI** Velocity Indicating Coherent Integrator 速度指示的相关积累器

**VICP** Virtual International Connecting Point 虚拟国际连接点

**VICS** Vehicle Information and Communication System 车辆信息和通信系统

**VID** Valid 有效的

Video 视频,录像

Video Image Display 视频图像显示

Virtual Image Display 虚拟图像显示器

Visual Identification 目视识别

**VIDAMP** Video Amplifier 视频放大器

**VIDAR** Velocity Integration Detection and Ranging 速度积累检测与定位

**VIDF** Video Frequency 视频

**VIDS** Vertical Instrument Display System 垂直仪表显示系统

Virtual Image Display System 虚拟图像显示系统

Visual Information Display System 目视情报显示系统

**VIE** Vienna, Austria (Airport Code) 奥地利维也纳机场

Virtual Information Environment 虚拟信息环境

**VIEW** Video Information Exchange Window 视频信息交换窗口

Virtual Interactive Environment Workstation 虚拟交互环境工作站

**VIFCS** VTOL Integrated Flight Control System 垂直起落飞机综合飞行控制系统

**VIG** Video Image Generator 视频图像发生器

**VIGV** Variable Inlet Guide Vanes 可变进口导向叶片

**VIH** Vancouver Island Helicopter 温哥华岛直升机公司

**VIL** Verified Item List 已验证项目清单

**VILP** Vector Impedance Locus Plotter 矢量阻抗轨迹绘图仪

**VIM** Vendor Information Management 供应商信息管理

Vendor Information Manual 供应商信息手册

Vibration Isolation Module 振动隔离模块

Virtual Item Manager 虚拟项目经理

Visual Inspection Module 目视检查模块

**VIMP** Vertical Impulse 垂直脉冲

**VINES** Virtual Networking System 虚拟网络系统

**VINT** Video Integration 视频积累

**VIO** Vertical Input Output 垂直输入输出

Violet 紫色

**VIOL** Violation 违反,扰乱

**VIP** Value in Performance 性能值

Variable Information Processing 可变信息处理

Vehicle Improvement Program 飞行器改进计划

Very Important Passenger 要客,贵宾

Very Important Person 要客,贵宾

Video Image Processor 视频图像处理机

Video Information Provider 视频信息提供者

Video Integrator Processor 视频积分处理器

Video Interface Port 视频接口

Video Interface Processor 视频接口处理机

Virtual Interface Processor 虚拟接口处理机

Virtual IP 虚拟 IP

Vision Information Processing 视觉信息处理

Visual Identification Point 目视识别点

Visual Integrated Presentation 目视综合显示

Voice Information Processor 话音信息处理机

Voice Interactive Phone 交互式音频电话

Voluntary Investment Plan 无偿投资计划

**VIPER** Variable Intensity Pulsed Effects Research 可变密度脉冲效应研究,可变强度脉冲效应研究

**VIPR** Virtual IP Route 虚拟 IP 路由

**VIPS** Voice Interruption Priority System 话音

中断优先系统

**VIR** Valves in Receiver 贮油箱阀门,储气室阀门

Virtual 虚拟的,有效的,实际的

Visible Infrared Radar 可见的红外雷达

**VIS** Lowest Selectable Airspeed 可选择的最低速度

Video Information System 视频信息系统

Video Interface System 视频接口系统

Visalia, CA, USA (Airport Code) 美国加利福尼亚州维萨利亚机场

Viscosity 粘度

Visibility 能见度

Visual 可视的,目视的

Visual Identification System 视觉识别系统

Visual Instruction Set 视算指令集

Visual Interactive Simulation 可视交互式模拟

Voice Information Service 语音信息服务

**VISC** Visualization in Scientific Computing 科学计算可视化

**VISCA** Video System Control Architecture 视频系统控制结构

**VISID** Visual Identification 视觉识别

**VISIT** Virtual Institute for Satellite Integration and Training 虚拟卫星综合培训学院

**VISMR** Viscometer 粘度计

**VISSR** Visible and Infrared Spin Scan Radiometer 可见与红外自旋扫描辐射计

**VIST** Voice Integrated Systems Technology 语音集成的系统技术

Voice Interactive System Technology 话音交互式系统技术

**VISTA** Variable-stability Inflight Simulator Test Aircraft 变稳定性空中飞行模拟试验机

**VIT** Video Image Terminal 视频图像终端

Vital 核心,生动的,重要的

Vitreous 玻璃的,透明的

**VITC** Vertical Integrated Time Code 垂直整合时间编码

Vertical Interval Time Code 垂直间隔时间编码

**VITP** Variable-Incidence Tailplane 全动式水平尾翼

**VITS** Video Image Tracking System 视频图像跟踪系统

**VIU** Vehicle Identification Unit 飞行器识别组件

Vehicle Interface Unit 飞行器接口装置

Video Interface Unit 视频接口组件

Videotex Interface Unit 可视图文接口单元

Virtual Interface Unit 虚拟接口单元

Voice Interface Unit 话音接口装置

**VIVID** Video, Voice, Image, Data 视频、语音、图像、数据

**VIVO** Video Input/Output 视频输入/输出

**VJ** Jet Velocity 喷气速度,射流速度

**VKO** Moscow, Russia-Vnukovo (Airport Code) 俄罗斯莫斯科伏努科沃机场

**VL** Vertical Ladder 垂直梯

Vertical Landing 垂直降落

Vertical Line 垂直线

Virtual Laboratory 虚拟实验室

Virtual Library 虚拟图书馆

**VLA** Very Large Aircraft 甚大型航空器

Very Light Aeroplane 甚轻型飞机

Very Low Altitude 甚低空

**VLAN** Virtual Local Area Network 虚拟局域网

**VLAR** Vertical Launch and Recovery 垂直发射与回收

**VLC** Valencia, Spain (Airport Code) 西班牙瓦伦西亚机场

Variable Length Code 可变长码

Variable Length Coder 可变长度编码器

Video Line Connector 视频线路连接器

Vital Load Center 核心加载中心

**VLCR** Variable Length Cavity Resonance 可变长度空腔共振

**VLCT** Very Large Commercial Transport 超大型商用运输机

**VLD**　Valdosta, GA, USA (Airport Code) 美国加利福尼亚州瓦尔多斯塔机场

Valid 合法的,有效的

Variable Length Decoding 可变长码译码

Video Line Driver 视频行驱动器

Visible Laser Diode 可见光二极管

**VLDB**　Very Large Database 甚大型数据库

**VLDTY**　Validity 效期

**VLE**　Maximum Landing Gear Extended Speed 最大起落架放下速度

Visual Learning Equipment 视觉学习设备

**VLEFG**　Violation Error Flag 违反误差旗

**VLF**　Variable-Length Field 变长字段

Very Low Frequency (3 ~ 30 KHz) 甚低频

**VLFS**　Variable Low Frequency Standard 可变低频标准

**VLGB**　Very Large Graphics Bases 甚大型图形库

**VLI**　Port Vila, Vanuatu-Bauerfield (Airport Code) 瓦努阿图维拉港鲍尔菲尔德机场

**VLJ**　Very Light Jet 超轻型喷气机

**VLL**　Valladolid, Spain (Airport Code) 西班牙巴利亚多利德机场

**VLLDC**　Very Low Level Direct Current 甚低电平直流电流

**VLMTRC**　Volumetric 容积的

**VLO**　Landing Gear Operating Speed 起落架操作速度

Landing Gear Operation 起落架操纵

Very Low Observable 超低能见度

**VLOF**　Lift off Speed 离地速度

**VLP**　Vertical Landing Point 垂直着陆点

**VLR**　Vertical-Looking Radar 垂直波束雷达

Very Light Rotocraft 甚轻型旋翼机

Very Long Range 超远程

Very Low Range 甚低范围

Visited Location Register 访问位置寄存器

Visitor Location Register 来访者位置登记器,外来用户位置寄存器

**VLS**　Vertical Launch Speed 垂直发射速度

**VLSI**　Very Large Scale Integration 甚大规模集成

**VLSIC**　Very Large-Scale Integrated Circuit 甚大规模集成电路

**VLSM**　Variable-Length Subnet Mask 可变长度子网掩码

**VLT**　Variable List Table 变量目录表

Very Large Telescope 甚大望远镜

Visible Light Transmittance 可见光透过率

**VLV**　Valve (机) 阀,活门

**VM**　Heading with a Manual Termination 人工终止航向

Maneuvering Speed 机动速度

Maximum Recommended Maneuvering Speed 最大推荐机动速度

Minimum Recommended Maneuvering Speed 最小推荐机动速度

Recommended Maneuvering Speed 推荐机动速度

Valid Message 有效信息,有效电报

Velocity Meter 速度计

Velocity Modulation 速度调制

Vendor Manual 供应商手册

Vertical Machine 升降机

Video Mail 视频邮件

Video Module 视频模块

Virtual Machine 虚拟机

Virtual Manufacturing 虚拟制造

Virtual Memory 虚拟存储器

Virtual Multi-access 虚拟多路存取

Voice Mailbox 语音邮箱

Voice Message 话音信息

Voice Modulation 音频调制

Volt per Meter 伏/米

Voltage Monitor 电压监视器

Voltmeter 电压表

**VM(LO)**　Minimum Maneuver Speed 最小机动速度

**VM/CMS**　Virtual Machine/Communication Facility System 虚拟机/虚拟通信设施系统

**VMA**　Valid Memory Address 有效存储地址

Virtual Memory Address 虚拟存储地址

**VMAN**　Maneuvering Speed 机动速度

**VMAS**　Visual Meteorological Avoidance System 目视气象回避系统

**VMAX**　Maximum Allowable Airspeed 最大允许飞行速度

Maximum Allowable Speed 最大允许速度

**VMB**　Maximum Break Energy Speed 最大刹车能量速度

Vehicle Management Bus 运载器管理总线

**VMC**　Minimum Control Speed 最小操纵速度

Vehicle Management Computer 飞行器管理计算机

Virtual Memory Computer 虚拟存储计算机

Visual Meteorological Condition 目视气象条件，目视气象状态

Voice Messaging Coder 语音消息编码器

**VMCA**　Minimum Air Control Speed 最小空中控制速度

**VMCB**　Virtual Machine Control Block 虚拟机控制程序块

**VMCG**　Ground Minimum Control Speed 地面最低控制速度

**VMCL**　Minimum Control Speed during Landing 着陆期间最低操纵速度

Minimum Control Speed on Landing 最小着陆操纵速度

**VMCL-1**　Minimum Control Speed during Landing Approach with One Engine 单发进近着陆最低操纵速度

**VMCL-2**　Minimum Control Speed during Landing Approach with Two Engines 双发进近着陆最低操纵速度

**VMCP**　Virtual Machine Control Program 虚拟机控制程序

**VMD**　Video Motion Detection 视频行动检测

**VME**　Virtual Machine Environment 虚拟机环境

Virtual Memory Environment 虚拟内存环境

Virtual Memory Extended 虚拟存储器扩展

Voice Message Exchange 话音信息交换

**VMG**　Voice Messaging 话音消息

**VMGE**　Voice Messaging Environment 话音消息环境

**VMGS-MS**　Voice Messaging System Message Store 话音消息系统的消息存储

**VMGS-UA**　Voice Messaging System User Agent 话音消息系统的用户代理

**VMI**　Vendor Management Inventory 供应商管理库存

**Vmin**　Minimum Operating Speed 最小使用速度

Minimum Safe Airspeed 最小安全空速

**VMIN**　Basic Clean Aircraft Minimum CAS 基本流线型飞机最小计算空速

**Vmix**　Video Mixer 视频混合器

**VML**　Virtual Memory Level 虚拟存储器级别

**VMM**　Virtual Machine Monitor 虚拟机监控程序

**VMO**　Maximum Allowance Airspeed 最大允许空速

Maximum Operating Airspeed Limit 最大使用空速极限

Maximum Operating Limit Speed 最大操作限制速度

Maximum Operating Speed/Velocity 最大飞行速度,最大使用速度

Variable Metering Orifice 可调计量孔

Velocity Maximum Operating 最大运转速度

**VMOD**　Video Modulator 视频调制器

**VMOS**　Vertical Metal-Oxide Semi-conductor 竖式金属氧化物半导体管

**VMR**　Virtual Meeting Room 虚拟会议室

**VMRR**　Vendor Material Review Report 供应商器材评审报告

**VMS**　Minimum Selectable Speed 最低可选速度

Minimum Speed in the Stall 最小失速速度

Variable-Message Sign 可变消息符号

Vehicle Management System 飞行器管理系统

Velocity Measuring System 速度测量系统

Video Modulation System 视频调制系统

Virtual Memory System 虚拟存储器系统

Voice Mail Server 语音信箱服务器

Voice Mail Service 语音信箱业务

Voice Message Server 语音消息服务器

Voice-Mail System 语音邮件系统

Volume Management System 音量管理系统,容量管理系统

Voyage Management System 航程管理系统

**VMS&F**   Voice-Mail Store and Forward 语音信箱存储转发

**VMSOS**   Vehicle Management System Operating System 运载器管理操作系统

**VMSS**   Vehicle Management System Simulator 运载器管理系统模拟机

**VMSSP**   Vehicle Management System Signal Processing 运载器管理信号处理

**VMT**   Vibration Measurement Technology 振动测量技术

Video Matrix Terminal 视频矩阵终端

Virtual Memory Technique 虚拟存储技术

**VMTP**   Virtual Machine Terminal Protocol 虚拟机终端协议

Virtual Message Transaction Protocol 虚拟信息事务协议

**VMU**   Minimum Unstick Speed/Velocity 最小离地速度

Video Modulation Unit 视频调制器

**VMX**   Voice Message Exchange 语音信息交换

**VN**   Vane 静子叶片

Variation Notice 变更通知

Virtual Network 虚拟网络

Voice Network 通话网络

Voice Notification 语音通知

**VNAP**   Vertical Noise Abatement Procedures 垂直减噪程序

Virtual Network Access Point 虚拟网接入点

**VNAV**   Variable Net Asset Value 可变资产净值

Vertical Navigation 垂直导航

**VNC**   Virtual Network Configuration 虚拟网结构

**VNE**   Velocity Never to Exceed 极限速度,不允许超过的速度

Video Network Element 视频网络元件

Virtual Natural Environment 虚拟自然环境

Virtual Network Engineering 虚拟网络工程

**VNL**   Virtual National Laboratory 虚拟国家实验室

**VNO**   Maximum Structural Cruising Speed 最大结构巡航速度

Normal Operating Speed 正常运行速度

Vilnius, Lithuania Airport (Airport Code) 立陶宛维尔纽斯机场

**VNR**   VHF Navigation Receiver 甚高频导航接收机

**VNS**   Varanasi, India (Airport Code) 印度瓦拉纳西机场

Visual Network Station 可视网络站

**VNV**   Virtual Network Visual 虚拟网络可视化

**VNZ**   Venezuela 委内瑞拉

**VO**   Visa Office 签证室

Voice 语音,话音

**VOB**   Video Outrigger Box 视频稳定箱

**VOBANC**   Voice Band Compression 音带压缩

**VOC**   Voice Order Circuit 语音指令电路

Volatile Organic Compound 挥发性有机化合物

Volts Open Circuit 电压开路

**VOCAL**   Voice-Oriented Curriculum Author Language 面向话音的课程编写语言

**VOCOM**   Voice Communication 语音通信

**VOD**   Vertical Obstruction Data 垂直障碍物数据

Video on Demand 视频点播

Voice over DDN 用 DDN 传话音

**VODACOM**   Voice Data Communication 话音数据通信

**VODAS**   Voice Operated Anti-Sing Device 声控防鸣器

**VODAT** Voice Operated Device for Automatic Transmission 自动发射声控器件

**VODIG** Voice Digitizer 话音数字转换器

**VODLIM** VOD Line Interface Module 视频点播线路接口模块

**VODS** Video on Demand Server 视频点播服务器

**VODSL** Voice over DSL 数字用户线传话音

**VOF** Variable Operating Frequency 可变工作频率

**VOFDM** Vector Orthogonal Frequency Division Multiplexing 矢量正交频分复用

**VOFR** Voice over Frame Relay 帧中继网上传语音

**VOG** Volgograd, Russia ( Airport Code ) 俄罗斯伏尔加格勒机场

**VOGAD** Voice-Operated Gain Adjusting Device 声控增益调整装置

**VOIP** Voice over Internet Protocol 互联网协议语音,网络电话

Voice over IP 基于 IP 的语音传输

**VOL** Vertical on Board Landing 舰上垂直降落

Volume 音量,容积,量,额

**VOLCAS** Voice-Operated Loss Control and Suppressor 声控损耗控制与抑制器

**VOLMET** Meteorological Information for Aircraft in Flight 供飞行中航空器用的气象情报

**VOLT** Voltage 电压,伏特数

**VOLTS** Versatile on-Line Teleprocessing System 通用联机远距处理系统

**VOM** Volt Ohm Milliammeter 万用表

**VON** Voice on the Net 语音网络

**VONC** Voice on the Net Coalition 因特网语音联盟

**V-ONU** Video Optical Network Unit 视频光网络单元

**VOR** Very High Frequency Omnirange Station 甚高频全向信标台

Voice-Operated Relay 音频控制继电器

**VORC** Vertical Rate of Climb ( 直升机的 ) 垂直上升率

**VORR** VHF Omni Range Receiver 方向无线识标

**VOS** Velocity of Sound 声速

Virtual Operating System 虚拟操作系统

Voice Operated Switch 声控开关

**VOT** Televoting 电话投票

VHF Omnidirectional Range Test 甚高频全向测距信标试验

Voice Operated Transmission 声控传输

VOR Receiver Test Facility 甚高频全向信标接收机测试设备

VOR Test Facility 甚高频全向信标测试设施

Voter 选举人

**VOX** Voice Activated 声控的

Voice Activated Carrier 语音激活载波

Voice Operated Control 声音操纵控制

Voice Operated Switching 声控开关

Voice Operated Transmitter 声控发射器

Voice Operated Transmitter Keyer Voice 声控发射机键控语音

Voice Transmission 语音传输

Voice-Operated Transmission 音控传输

**VP** Validate Process 鉴定工序

Validation Phase 鉴定阶段

Variable Pitch 变距

Variable Pitch Propeller 变距螺旋桨

Variable Pressure 可变压力

Vent Pipe 通风管

Vertical Planning 垂直计划

Vertical Polarization 垂直极化

Video Phone 可视电话

Video Processor 视频处理器

Virtual Path 虚通路

Virtual Processor 虚拟处理机

Vision Processing 视频处理

Vision Processor 可视处理机

**VPA** Vertical Path Angle 垂直进场角度

**VP-AIS** Virtual Path Alarm Indication Signal

虚通道告警指示信号

**VPATH** Vertical Path 垂直路径

**VPC** Vertical Parity Check 垂直奇偶校验

Virtual Path Connection 虚通路连接

**VPCA** Video Prelaunch Command Amplifier 发射前视频指令放大器

**VPCE** Virtual Path Connection Endpoint 虚通道连接终端

**VPD** Virtual Private Database 虚拟专用数据库

Virtual Prouduct Development 虚拟产品开发

Vital Product Data 重要产品数据

**VPDE** Video Processing and Display Engine 视频处理和显示引擎

**VPDN** Virtual Private Data Network 虚拟专用数据网

Virtual Private Dialup Network 虚拟专用拨号网

**VPE** Vector Processor Element 向量处理器单元

Virtual Path Entity 虚通道实体

**VPH** Virtual Path Handler 虚通道处理

**VPI** Vertical Path Integrator 垂直通道积分器

Virtual Path Identifier 虚通路标识符

**VPIM** Voice Profile Internet Mail 因特网函件的话音类型

**VPL** Vertical Protection Level 垂直保护级

Virtual Path Link 虚通道链路

Volume of Payload 满载容积

**VPLC** Valve Position Logic Card 活门位置逻辑卡

**VPLMN** Virtual Public Land Mobile Network 虚拟公共陆上移动网

**VPM** Vibration per Minute 振动次数/分

Virtual Product Management 虚拟产品管理

Virtual Product Modeling 虚拟产品建模

Voice Path Management 话路管理

**VPME** Virtual Path Multiplex Entity 虚通道复用实体

**VPN** Vendor Part Number 协作厂部件号

Virtual Path Network 虚通道网络

Virtual Personal Network 虚拟个人网

Virtual Private Network 虚拟专用网

**VPO** Vendor Purchase Order 供货商采购订单

**VPON** Video Phone on Network 网络视频电话

**VPP** Vendor Partnership Program 销售商合作伙伴关系程序

**VPR** Voluntary Price Reduction 自发降低价格

**VP-RDI** Virtual Path Remote Defect Indication 虚通道远端故障指示

**VPROF** Vertical Profile 垂直剖面

**VPS** Valparaiso, FL, USA(Airport Code) 美国佛罗里达州瓦尔帕莱索机场

Variable Packet Size 可变包大小

Video Playback System 视频播放系统

Video Programme System 视频节目系统

Video Projection System 视频投影系统

**VPSA** Vendor Product Support Agreement 卖方(供货方)产品保障协议

**VPT** Virtual Packet Terminal 虚分组信息终端

Virtual Path Terminator 虚通道终端设备

Virtual Private Trunk 虚拟专用中继

**VPTS** Voice Processing Training System 语音处理训练系统

**VPU** Supplier Pick up 供应方选取

Vendor Pick up 卖方(供货方)选取

Vibrator Power Unit 振动器电源组件

Video Processing Unit 视频处理单元

Video Provision Unit 视频提供单元

Virtual Processing Unit 虚拟处理机

Voice Privacy Unit 语音保密组件

Voice Processing Unit 话音处理装置

**VPX** VP Cross-Connect 虚通道交叉连接

**VQ** Vector Quantitation 矢量量化

Virtual Queuing 虚队列

**VQC** Vector Quantitation Coding 矢量量化编码

**VR** Rotation Speed 转速,抬前轮速度,旋转

速度

Takeoff Rotation Velocity 起飞抬头速度,起飞抬前轮速度

Variable Resistance 可变电阻

Velocity Rotate 抬轮速度

Video Recorder 录像机

Virtual Reality 虚拟现实

Virtual Route 虚路由

Voice Recorder 话音记录器

Voltage Regulator 电压调节器

**VRA** Velocity Rough Air 颠簸速度

**VRAC** VHF Repeater Advisory Committee 甚高频转发器咨询委员会

**VRAM** Video Random Access Memory 视频随机存取存储器

**VRAR** Voltage Regulator Annunciator Relay 电压调节报警继电器

**VRB** Variable 可变的,变化的

VHF Recovery Beacon 甚高频回收信标

Voice Rotating Beacon 音频旋转信标

**VRC** Vertical Redundancy Check 垂直冗余校验

**VRCP** Voice Recognition Call Processing 语音识别呼叫处理

**VRCTR** Varactor 变容二极管

**VRD** Vacuum Tube Relay Driver 真空管中继激励器

Variable Rate Data 变速数据

Virtual Retinal Display 虚拟视网膜显示器

**VREF** Reference Landing Speed 着陆基准速度

Reference Speed 参考速度,基准速度

**VREV** Vertical Revise 垂直修正

**VRFI** Voice Reporting Fault Indicator 语音报告故障指示器

**VRFY** Verify 证实,验证

**VRI** Visual Rule Instrument Landing 目视规则仪表着陆

Voice Response Interaction 语音响应交互

**VRL** Vertical Reference Line 垂直基准线

**VRLY** Voltage Relay 电压继电器

**VRM** Variable Range Marker 可变距离标志

Visual Raw Material 直观原始材料

Voice Recognition Module 语音识别模块

**VRMAT** Virtual Reality Maintenance Trainer 虚拟现实维修训练器

**VRML** Virtual Reality Markup Language 虚拟现实标记语言

Virtual Reality Modeling Language 虚拟现实建模语言

**VRMS** Volts Root Mean Square 电压有效值(均方根值),均方根电压

**VRN** Verona, Italy(Airport Code) 意大利维罗纳机场

**VROC** Vertical Rate of Climb 垂直爬升率

**VRP** Vehicle Routing Problem 车载路由问题

Video RISC Processor 视频精简指令集计算机处理器

**VRPS** Voltage Regulated Power Supply 稳压电源

**VRR** Vertical Refresh Rate 垂直扫描频率

**VRRA** Variable Rate Reservation Access 变速预留接入

**VRRP** Virtual Router Redundancy Protocol 虚路由器冗余协议

**VRS** Vacuum Regulator/Solenoid 真空调整器/电磁阀

Vacuum Retad Switch 真空减速开关

Velocity Response Shape 速度响应特性

Video Response System 视频响应系统

Voice Recording System 话音录制系统

Voice Response System 话音应答系统

**VRSA** Voice Reporting Signal Assembly 话音报告信号组件

**VRT** Virtual Reality Tool 虚拟现实工具

Visual Recognition Threshold 目视识别阈

**VRTG** Vertical Gyro 垂直陀螺

**VRTY** Variety 种类,变化,变形

**VRU** Velocity Reference Unit 速度参考装置

Vertical Reference Unit 垂直基准组件

Voice Response Unit 话音应答单元

Voltage Readout Unit 电压读出组件

**VRV** Vacuum Reducer Valve 真空减压活门（阀）

Vacuum Regulator Valve 真空调节器活门（阀）

**VRYG** Varying 变化,变更,修改

**VS** Reference Stalling Speed 失速基准速度

Stall/Stalling Speed 失速速度

Vapor Seal 汽密,汽封

Variable Store 可变存储器

Vehicle Sensors 航空器传感器

Vehicle Simulation 航空器模拟

Velocity Search（雷达的）速度搜索

Vent Stack 通风竖管

Versus 对抗

Vertical Speed 垂直速度

Vertical Stabilizer 垂直稳定翼

Video Server 视频服务器

Video Signal 视频信号

Virtual Source 虚拟源

Virtual Storage 虚拟存储

Virtual System 虚拟系统

Voltmeter Switch 电压表开关

**VS&T** Velocity Search and Track 速度搜索与跟踪

**V$_{S0}$** Stalling Speed 失速速度

**V$_{S1g}$** Stalling Speed at One g 一个 g 时的失速速度

**VSA** Vacuum Switch Assembly 真空开关总成

Virtual System Architecture 虚拟系统架构

Visual Software Agent 可视软件代理

**VSAM** Video Surveillance and Monitoring 视频监控

Virtual Sequential Access Method 虚拟顺序存取法

Virtual Storage Access Method 虚拟存储存取法

**VSAT** Very Small Aperture Data Terminal 极小口径数据终端

Very Small Aperture Satellite Terminal 小型卫星地面站

Very Small Aperture Terminal 甚小孔径终端

**VSB** Vendor Service Bulletin 供应商服务通告

Vestigial Sideband 残留边带

**VSBA** Vendor Service Bulletin Approval 供应商服务通告批准

**VSBL** Visible 可见的,能见的

**VSBN** Vender Service Bulletin Number 供应商服务通告号

**VSBS** Vender Service Bulletin Status 供应商服务通告状况

**VSBT** Vender Service Bulletin Title 供应商服务通告标题

**VSBY** Visibility 能见度

**VSC** Vacuum System Controller 真空系统控制器

Variable Speed Compressor 变速压缩机

Video Standards Council 视频标准委员会

Voltage Saturated Capacitor 电压饱和电容器

**VSCE** Variable Stream Control Engine 可变流量控制发动机

**VSCF** Variable Speed Constant Frequency 变速恒频

**VSCFCU** Variable Speed Constant Frequency Control Unit 变速恒频控制组件

**VSCFPS** Variable Speed Constant Frequency Power System 变速恒频电源系统

**VSCMS** Voice Switch Control and Monitoring System 话音开关控制和监视系统

**VSCS** Vertical Stabilizer Control System 垂直安定面控制系统

Video Server Control Station 视频服务器控制站

Voice Switching and Communication System 话音转换和通信系统

Voice Switching and Control System 话音转换和控制系统

**VSCU** Video System Control Unit 视频系统控制组件

**VSD** Variable Speed Drive 变速驱动

Vendor's Shipping Document 卖方航运文件

Versatile Signal Device 通用信号装置

Vertical Situation Display 垂直状况显示器

Video Smoke Detection 视频烟雾探测

Video Symbology Display 视频符号显示

**VSDL** Video Scene Description Language 视频景物描述语言

**VSDM** Variable Slope Delta Modulation 可变斜度增量调制（一种编码调制方式）

**VSEB** Video Seat Electronic Box 座椅视频电子盒

**VSEL** Autothrottle Selected Approach Speed 自动油门选择进近速度

**VSFI** Vertical Scale Flight Indicator 垂直标度飞行指示器

**VSG** Lugansk, Ukraine (Airport Code) 乌克兰卢甘斯克机场

Variable Speed Gear 变速齿轮箱

**VSI** Velocity and Steering Indicator 速度及转向指示器

Vertical Speed Indicator 垂直速度指示器

Vertical Speed Indication 垂直速度指示

Vertical System Incorporation 垂直系统公司

Video Sweep Integrator 视频扫描积累器

**VSL** Vertical Speed Limit Advisory 垂直速度极限咨询

**VSM** Variable Speed Modem 变速调制解调器

Vertical Separation Minimum 垂直间隔最低标准

Vestigial Sideband Modulation 残余边带调制

Video Service Module 视频服务模块

Virtual Service Management 虚拟服务管理

Virtual Shared Memory 虚拟共享存储器

**VSMC** Videotex Service Management Center 可视图文业务管理中心

**VSO** Very Stable Oscillator 高稳定度振荡器

**VSOUT** Vertical Synchronism Output 垂直同步输出

**VSP** Vector Scientific Processing 向量科学处理

Verification Software Program 鉴定软件程序

Vertical Separation of Aircraft Panel 航空器垂直间隔专家组

Vertical Speed 垂直速度

Video Signal Processor 视频信号处理器

Virtual Software Processor 虚拟软件处理器

Virtual Switching Point 虚拟交换点

**VSQ** Voice Switching Quality 话音交换质量

**VSR** Very Short Range 超短程

**VSRS** Variable Speed Rotor System 变速转子系统

**VSS** Stick Shaker Speed 抖杆速度

Variable Stability System 变稳定性系统

Variable Structure System 变结构系统

Vehicle Subsystem Simulation 航空器分系统模拟

Vertical Selector Service 垂直选择器维护

Vertical Stabilizer Station 垂直安定面站位

Video Signal Simulator 视频信号模拟器

Video Storage System 视频存储系统

Video Subsystem 视频子系统

**VSSA** Variable-Stability Simulator Aircraft 可变稳定模拟器飞机

**VSSE** Visual Site Structure Editor 可视化的节点结构编辑器

**VSSG** Vertical Separation Study Group 垂直间隔研究组

**VST** Variable Speed Technology 变速技术

Variable Stability Trainer 变稳教练机

Vasteras, Sweden (Airport Code) 瑞典韦斯特罗斯机场

Vertical Static Test 垂直静态试验

Video Selection 视频选择

Video System Test 视频系统测试

Virtual Storage 虚拟存储器

**VSTC** Validation of Supplemental Type Certificate 补充型号认可证

**VSTM** Valve Stem 活门杆, 活门轴

**VSTOL** Vertical or Short Takeoff and Landing 垂直或短距起降

**VSTS** Virtual Space Teleconferencing System

虚拟空间远程会议系统

**VSU** Video Signals Unit 视频信号装置
Video Switch Unit 视频开关装置
Videotex Service Unit 可视图文服务单元

**VSV** Variable Stator Valve 可调静子活门
Variable Stator Vane 可变静子叶片

**VSVA** Variable Stator Vane Actuator 可变静子叶片作动器

**VSVF** Variable Speed Variable Frequency 变速变频

**VSW** Stall Warning Speed 失速警告速度
Very Short Wave 超短波
Voltage Standing Wave 电压驻波

**VSWR** Voltage Standing Wave Radio 电压驻波比

**VSYNCHL** Vertical Synchronism Channel 垂直同步通道

**VT** Vacuum Tube 真空管
Validation Test 鉴定试验
Vaportight 气密
Vertical Tail 垂直尾翼
Video Terminal 视频终端
Video Tracker 视频跟踪器
Virtual Terminal 虚拟终端
Virtual Tributary 虚拟支路
Visual Telephony 可视电话
Voice Tube 语音管

**VTA** Variable Transfer Address 可变转移地址
Vertex Time of Arrival 到达顶点时间
Video Terminal Adapter 视频终端适配器
Voice Terrain Advisory 语音地形咨询

**VTAM** Virtual Telecommunication Access Method 虚拟通信接入法
Virtual Teleprocessing Access Method 虚拟远程处理存取方法
Virtual Terminal Access Method 虚拟终端访问方法

**VTAME** Virtual Telecommunication Access Method Entry 虚拟通信访问方法输入

**VTAS** Vertical Table of Authorization System 核准系统垂直图表
Visual Target Acquisition System 目视目标捕捉系统

**VTC** Validation of Type Certificate 型号认可证
Video Teleconference Center 电视电话会议中心
Videotex Terminal Control 可视图文终端控制
Virtual Terminal Control 虚拟终端控制
Visual Terminal Chart 目视终端图
Vocational Training Council 职业训练局

**VTCD** Videotex Terminal Control Driver 可视图文终端控制驱动器

**VTD** Vaccum Tube Detector 真空管检波器
Vendor Technical Data 协作厂技术数据
Vertial Tape Display 垂直磁带显示器

**VTE** Vertical Track Error 垂直跟踪误差
Vientiane, Laos ( Airport Code ) 老挝万象机场
Virtual Terminal Environment 虚拟终端环境
Virtual Training Environment 虚拟训练环境

**VTF** Variable Transfer 可变转移

**VTH** Target Velocity at Threshold 入口处目标速度

**VTI** Video Terminal Interface 视频终端接口

**VTIP** Visual Target Identification Point 目视目标识别点

**VTK** Vertical Track 垂直航迹
Vertical Track Distance 垂直航迹距离
Vertical Track Error 垂直航迹误差

**VTL** Vacuum-Tube Launcher 真空管发射器
Variable Threshold Logic 可变门限逻辑, 可变阈值逻辑

**VTM** Voltage Tunable Magnetron 电压调谐磁控管

**VTMAX** Maximum Threshold Speed 最大门限速度

**VTMIN** Minimum Threshold Speed 最小门限速度

**VTN** Video Teleconference Network 电视电

话会议网

**VTO** Vertical Take off 垂直起飞

Voltage Tunable Oscillator 压控振荡器

Volumetric Top off 容积关断传感器

Volumetric Topoff 容量完成

**VTOA** Voice and Telephony over ATM 语音和电话通过 ATM 传输

**VTOC** Volume Table of Contents 卷目录表

**VTOL** Vertical Take off and Landing 垂直起落(航空器)

**VTOSS** Takeoff Safety Speed for Category A Rotorcraft A 类旋翼机的起飞安全速度

**VTOVL** Vertical Takeoff and Vertical Landing 垂直起飞和垂直着陆

**VTP** Vertical Tail Plane 垂直尾翼平面

Virtual Terminal Protocol 虚拟终端协议

**VTPP** Validation Test Plan/Procedures 鉴定试验计划/程序

**VTPR** Vertical Temperature Profile Radiometer 垂直温度剖线辐射计

**VTR** Variable Takeoff Rating 可变起飞额定值

VHF Transmitter/Receiver 甚高频收发机

Video Tape Recorder 录像机,磁带录像机

Video Tape Reproducer 放像机,(录音、录影的)播放装置

**VTRP** Virtual Tree Routing Protocol 虚拟树路由协议

**VTS** Vacuum Toilet System 真空洗手间系统

Verification Test System 鉴定试验系统

Vertical Test Stand 垂直试验台

Vibration Test System 振动实验系统

Video Teleconference System 电视电话会议系统

Virtual Terminal Services 虚拟终端服务

Virtual Terminal System 虚拟终端系统

**VTTH** Video to the Home 电视至家庭

**VTVM** Vacuum Tube Voltmeter 真空管电压表

**VTX** Vertex 顶点

**VTXCF** Telex/Videotex Conversion Facility 用户电报/可视图文转换设备

**VU** Utility Speed 有效速度

Velocity Update 速度更新

Volume Unit 音量单位

**VUAC** VHF/UHF Advisory Committee 甚高频/超高频咨询委员会

**VUE** Visual User Environment 可视用户环境

**VULC** Vulcanize 硫化

**VUP** Valledupar, Colombia ( Airport Code ) 哥伦比亚巴耶杜帕尔机场

Virtual User Port 虚拟用户端口

**VUPC** Video Unblanking Protection Circuit 视频增辉(启通)保护电路

**VUTS** Verification Unit Test Set 检验组件测试仪

**VV** Verification and Validation 检验与批准,验证与批准

Vertical Velocity 垂直速度

Vertical Visibility 垂直能见度

**VVA** Variable Voltage Attenuator 可变电压衰减器

Venturi Vacuum Amplifier 文氏管真空放大器

**VV&A** Verification, Validation and Accreditation 检验、校准与鉴定

**VVC** Voltage Variable Capacitor 可变电压电容器

**VVCS** Vacuum Vent Control System 真空通风控制系统

**VVCV** Vacuum Vent Control Valve 真空通风控制阀

**VVI** Santa Cruz, Bolivia ( Airport Code ) 玻利维亚圣克鲁斯机场

Vertical Velocity Indicator 升降速度指示器,垂直速度指示器

**VVIP** Very Very Important Person 极重要人物,极重要客人

**VVO** Vladivostok, Russia ( Airport Code ) 俄罗斯符拉迪沃斯托克机场

**VVOD** Virtual Video on Demand 虚拟视频点播

**VVR** Variable Voltage Rectifier 可变电压整流器

**VVSS** Vertical Volute Spring Suspension 竖锥形弹簧悬置

**VVT** Variable Valve Timing 可变气门正时技术

Velocity Variation Tube 变速管

Venturi Vacuum Transducer 文氏管真空转换器

**VVV** Vacuum Vent Valve 真空通风活门（阀）

Vertical Velocity Vector 垂直速度矢量

**VVVac** Vent Valve Vacuum 通风阀真空

**VVVF** Variable Voltage Variable Frequency 变压变频

**VW** Virtual Window 虚拟视窗

Virtual World 虚拟世界

**VWG** Video Working Group 视频工作小组

**VWI** Virtual Workstation Interface 虚拟工作站接口

**VWL** Variable Word Length 可变字长

**VWM** Vender Warranty Manual 厂家保修手册

**VWP** Variable Width Pulse 可变宽度脉冲

Virtual Wavelength Path 虚拟波长通道

**VWS** Vertical Wind Shear 垂直风切变

Virtual Working System 虚拟工作系统

Voice Warning System 语音警告系统

**Vx** Speed for Best Angle of Climb 最佳爬升角速度

**VXB** Virtual Extended Bus 虚拟扩展总线

**VXML** Voice Extensible Markup Language 话音可扩展标记语言

**VXO** Variable Crystal Oscillator 可变晶体振荡器

Vaxjo, Sweden（Airport Code） 典韦克舍机场

**Vy** Speed for Best Rate of Climb 最佳爬升率速度

**VZD** Video Zone Distributor 录像区域分配器

# W

W　Aircraft Weight 飞机重量
Warm 温暖的
Warning 警告
Waste 浪费,耗损,废弃
Watt 瓦特
Weight 重量
Weighting Factor 载重因子,加权因子
West 西方,西方的,向西
White 白色
Wide 宽的
Width 宽度
Window 窗户(链路层的)
Wire 导线,线材
With 与,具有
Workmanship 工艺

**W&BM**　Weight and Balance Manual 载重与平衡手册

**W&BMS**　Weight and Balance Manual System 载重与平衡手册系统

**W&IR**　Work and Inspection Record 工作与检验记录

**W/A**　Walk-around 绕飞机(检查)一周
Width Average 侧边,侧列
Wrap around 卷绕,环绕

**W/B**　Wheel and Brake 机轮与刹车

**W/BO**　With Blowout 含有吹风

**W/BRAKES**　Wheel Brakes 机轮刹车

**W/C**　Wind Component 风分量

**W/D**　Waiver/Deviation 弃权/脱离
Wiring Diagram 布线图,线路图解
Withdrawn 分离的,回收的

**W/E&SP**　With Equipment and Spare Parts 含设备与备件

**W/K**　Wire Kit 导线器材包

**W/L**　Wire List 导线清单

**W/MOD**　With Modification of Vertical Profile 修改的垂直剖面

**W/O**　Water in Oil 油中含水
Without 没有

**W/P**　Warning Point 警告点

**W/STEP**　With Step Change in Altitude 有高度阶变

**W/T**　Wing Tip 翼尖
Wireless Telegraph 无线电报

**W/V**　Wind Direction and Speed 风向与风速
Wind Velocity 风速

**W/W**　Wheel Well 轮舱
Winding to Winding 绕组之间
Work with 与…合作

**W3C**　World Wide Web Consortium 万维网联盟

**WA**　Wainscot 护壁板
Waveform Analyzer 波形分析器
Weather Advisory 天气咨询
Will Advise 将咨询
Wind Angle 风向角
Wire Assembly 导线组件,导线组装
Wireless Access 无线接入
Work Authorization 工作授权
Workshop Assembly 车间组装

**WAAS**　Wide Area Active Surveillance 广域有源监视
Wide Area Application Services 广域应用服务
Wide Area Augmentation System 广域增强系统

**WAASA**　Women's Aviation Association of South Africa 南非妇女航空协会

**WABAT**　Wing and Body Aerodynamic Technique 机翼和机身气动力技术

**WABSIC** Wheel and Brake System Integrated Component 轮子和刹车系统整合部件

**WAC** Wake Analysis and Control 尾流分析与控制

Weather Analysis Center 气象分析中心

Wide Area Centrex 广域集中用户交换机

Working Alternating Current 交流工作电流

World Aerobatic Championship 世界特技飞行冠军赛

World Aeronautical Chart 世界航空图

World Aircraft Sales Corp. (Japan) 世界飞机销售公司(日本)

**WACA** World Airlines Clubs Association 世界航空公司俱乐部协会

**WACAF** Western and Central Africa 西部和中部非洲

**WACC** Weighted Average Cost of Capital 加权平均资本成本

**WACCC** Worldwide Air Cargo Commodity Classification 世界空运货物分类

**WACH** West African Cleaning House 西非清算协会

**WACO** World Air Cargo Organization 世界航空货运组织

**WACRA** Worldwide Airline Customer Relations Association 世界航空公司用户关系协会

**WACS** Wide Area Computing Service 广域计算服务

Wireless Access Communication System 无线接入通信系统

Wireless Airport Communication System 无线机场通信系统

Work Assembly Control System 组合工作管制系统

**WAD** Workload Assessment Device 工作量估算仪

**WADB** West African Development Bank 西非开发银行

**WADC** Wright Air Development Center 莱特航空开发中心

**WADGNSS** Wide Area Differential Global Navigation Satellite System 广域差分全球导航卫星系统

**WADGPS** Wide Area Differential Global Positioning System 广域差分全球定位系统

**WADS** Wide Angle Display System 大角度显示系统

Wide Area Data Service 大区域数据交换勤务,广域数据服务

Wide Area Differential System 广域差分系统

**WAE** Wireless Application Environment 无线应用环境

**WAEA** World Airline Entertainment Association 世界航空娱乐协会

**WAEC** West African Economic Community 西非经济共同体

**WAEMU** West African Economic and Monetary Union 西非经济和货币联盟

**WAEO** World Aerospace Education Organization 世界航空航天教育组织

**WAF** Rate of Fan Airflow 风扇气流的速率

Wafer (薄,圆)片,晶片

Wiring around Frame 绕线架

**WAFC** World Area Forecast Center 世界区域气象预报中心

**WAFS** World Area Forecast System 世界区域气象预报系统

**WAG** Wagon 货车,小客车

**WAGS** Windshear Alert and Guidance System 风切变警告与引导系统

**WAI** Wing Anti-Ice 机翼防冰

**WAIPS** Wing Anti-Ice Pressure Sensor 机翼防冰压力传感器

**WAIS** Wide Area Information Server 广域信息服务器

**WAIT** Waiting 等候

Wing Anti-Ice Temperature Sensor 机翼防冰温度传感器

**WAITRO** World Association of Industrial and Technological Research Organization 世界工业与技术研究组织协会

**WAITS** Wide Area Information Transfer System 广域信息传送系统

**WAIV** Wing Anti-Ice Valve 机翼防冰阀

**WAL** Western Airlines, Inc.（USA）西部航空公司（美国）

**WAM** Wide Area Multilateration 广域多点定位

Wireless Access Manager 无线接入管理器

Worth Analysis Model 价值分析模型

**WAMA** Washington Airport Management Association 华盛顿机场管理协会

**WAMOSCOPE** Wave Modulated Oscilloscope 调制波示波器

**WAMS** Wide Area Message System 广域消息系统

**WAMU** West African Monetary Union 西非货币联盟

**WAN** Wide Area Network 广域网

Wide Area Notice 广域通告

Work Assignment Notice 工作协议通知

**WAP** Wireless Access Protocol 无线入网协议

Wireless Application Protocol 无线应用协议

Work around Plan 按计划工作

Work around Procedure 按程序工作

Work Assignment Procedure 工作分配程序

**WAR** Weekly Activity Report 每周活动报告

World Aero Refueling（USA）世界航空加油公司（美国）

**WARC** World Administrative Radio Conference 世界无线电行政大会

World Administrative Radio Consortium 世界无线电协议管理组织

**WARC-MOB** World Administrative Radio Conference for the Mobile Service 世界移动服务无线电管理会议

**WARN** Warning 警告，警报

**WARP** Weather and Radar Processor 气象与雷达处理器

**WARR** Warranty 保证书，保单

**WAS** USA Washington 华盛顿（简称）

**WASA** Work Area Self Assessment 工作区域自我评价

World Air Stewardess Association 世界航空小姐协会

**WASG** World Airlines and Suppliers Guide 全球航空公司和供应商指南

**WASIS** Weather and Society Integrated Study 天气与社会综合研究

**WASTN** Wireless Auxiliary Station 辅助无线电台

**WAT** Weight-Altitude-Temperature 重量—高度—温度

Wing Activation Team 机翼启动组

Wing Activation Test 机翼启动试验

World Air Transport（USA）世界空运公司（美国）

**WATA** World Association of Travel Agents 世界旅行社协会

**WATOG** World Airlines Technical Operations Glossary 全球航空公司技术使用词汇

**WATS** Wide Area Telephone Service 广域电话业务

Wide Area Transmission System 广域传输系统

Wide-Area Telecommunications Services 广域电信业务

Wide-Area Tracking System 广域跟踪系统

World Air Transport Statistics 世界航空运输统计

**WATWS** Wide Angle Track While Scan 广角扫描跟踪

**WAU** Wireless Access Unit 无线接入单元

**WAW** Warsaw, Poland-Okecie（Airport Code）波兰华沙奥肯切机场

**WB** Water Box 水箱

Weather Bureau 气象局

Weber 韦伯

Weight and Balance 载重与平衡

West Bound 西向

Wet Bulb 湿球温度计

Wheel Base 纵向轮距

Wheelbarrow 手推车

Wide Body 宽体

Wideband 宽带,宽频带

Wire Book 导线手册

Wire Bundle 电缆,导线束

Wood Base 木基座

Work Bench 工作台

Work Book 工作手册

**WBA** Weight and Balance Application 重量和平衡应用

Wide Band Amplifier 宽带放大器

Wire Bundle Assembly 导线束组件,导线束组装

**WBADN** Wire Bundle Assembly Disposition Notice 导线束组装配置通知

**WBAI** Wire Bundle Assembly Input 导线束组装输入

**WBAMP** Wire Bundle Assembly Manufacturing Plan 导线束组装生产计划

**WBAR** Wing Bar Lights 翼排灯,侧排灯

**WBBC** Weight and Balance Backup Computation 重量和平衡备份计算

**WBC** Weight and Balance Computer 载重及平衡计算机

Wire Bundle Compare 导线束对照

**WBCD** Wire Bundle Configuration Design 导线束配置设计

**WBCLM** Weight and Balance Control Loading Manual 载重与平衡控制装载手册

**WBCS** Wide Band Communication System 宽带通信系统

Wireless Broadband Communication System 无线宽带通信系统

**WBCT** Wideband Current Transformer 宽带电流变压器

**WBD** Wide Band Data/Wideband Data 宽带数据

Wide Band Demodulator 宽带解调器

Wire Bound 导线固定

**WBDA** Wideband Data Assembly 宽带数据装置

**WBDL** Wide Band Data Link 宽带数字链路

**WBDLS** Wideband Data Links and Satellite Communications 宽带数据链与卫星通信

**WBEL** Wire Bundle Effectivity List 导线束有效性清单

**WBEM** Web-Based Enterprise Management 基于 Web 的企业管理

**WBFM** Wide Band Frequency Modulation 宽带频率调制

**WBG** Webbing 带子

**WBH** Weight Balance Handbook 载重配平手册

**WBIR** Wire Bundle Inquiry Report 导线束查询报告

**WBJ** Wing Body Join 机翼机体连接(头)

**WBL** Wide-Band Limiting 宽带限制

Wing Base Line 机翼基准线

Wing Buttock Line 机翼纵剖线

Wood Blocking 木块

World Brand Lab 世界品牌实验室

**WBM** Web Based Management 基于 Web 的网络管理模式

Weight and Balance Manual 载重和平衡手册

Wide Band Modem 宽带调制解调器

**WBN** Wide-Band Network 宽带网

**WBNL** Wideband Noise Limiting 宽带噪声极限

**WBR** Word Buffer Register 字缓冲寄存器

**WBRR** Wire Bundle Revision Record 导线束更改记录

**WBRRP** Wire Bundle Revision Record Page 导线束更改记录页

**WBS** Weight and Balance System 重量和平衡系统

West by South 西偏南

Wide Band System 宽带系统

Wireless Base Station 无线基站

Work Breakdown Structure 工作分解结构

Write Buffer Status 写入缓冲器状态

**WBSD** Work Breakdown Structure Document

工作分解结构文件

**WBSE** Work Breakdown Structure Element 工作分解结构单元

**WBSR** Wide Band Signal Recovery 宽带信号恢复

**WBST** Wideband Subscriber Terminal 宽带用户终端

**WBT** World Board of Trade 世界贸易委员会

**WBTM** Weather Bureau Technical Memoranda 气象局技术规程

**WBU** Boulder, CO, USA (Airport Code) 美国科罗拉多州博尔德机场

**WC** Water Cooler 水冷却器
Wavelength Converter 波长转换器
Wheel Card 轮插卡, 驾驶盘铭牌
Wind Component 风分量
Wording Change 改变用词
Work Center 操作中心
Write Control 写入控制

**WCA** Wind Correction Angle 风向修正角

**WCAP** World Climate Application Program 世界气候应用计划

**WCC** Weather Communications Center 气象通信中心
World Computer Congress 世界计算机大会

**WCCA** Worst-Case Circuit Analysis 电路最坏情况分析

**WCD** Web Content Distributor Web 内容分发商

**WCDMA** Wideband Code Division Multiple Access 宽带码分多址

**WCDP** World Climate Data and Monitoring Programme 世界气候资料与控制计划
World Climate Data Program 世界气候资料计划

**WCHB** Battery Operated Wheelchair 电动轮椅

**WCHC** Wheelchair to the Aircraft Seat 需要用轮椅到座位的旅客

**WCHP** Wheelchair with Manual Power 手动轮椅

**WCHR** Wheelchair to the Gate 需要用轮椅到登机口的旅客

**WCHS** Wheelchair to the Aircraft Door 需要用轮椅到登机门的旅客

**WCIP** World Climate Influence Program 世界气候影响计划

**WCMS** Web Content Management System 网络内容管理系统
Web Course Management System 网络课程管理系统

**WCO** World Customs Organization 世界海关组织

**WCP** Weather Communications Processor 气象通信处理器
Wing Chord Plane 翼弦平面
World Climate Programme 世界气候计划
WXR Control Panels 气象雷达控制板

**WCPN** Wireless Customer Premises Network 无线用户室内网络

**WCR** Weight/Capacity Ratio 重量/容量比

**WCRP** World Climate Research Programme 世界气候研究计划

**WCS** Wavelength Channel Selector 波长信号选择器
Wireless Communication System 无线通信系统
World Coordinate System 世界坐标系统

**WCT** Wireless Connect Technology 无线连接技术

**WCTV** Wired City Television 城市有线电视

**WCU** Waste Control Unit 废水控制组件

**WD** Warning Display 警告显示, 报警显示器
Wavelength Discriminator 波长鉴别器
Wavelength Dispersion 波长色散
Wavelength Division 波分
Wind Direction 风向
Wing Datum 机翼基准面
Wiring Diagram 线路图
Word 字, 词

**WDB** Wall Disconnect Box 壁板脱开盒

**WDC** World Data Center 世界数据中心

**WDCS** Wideband Digital Cross-connect System 宽带数字交叉连接系统

**WDDBS** Web Distributed Database System 万维网分布式数据库系统

**WDDM** Wavelength Division Demultiplexer 波分去复用器

**WDG** Enid, OK, USA（Airport Code）美国俄克拉荷马州伊尼德机场

**WDH** Width Depth Height 宽深高

Windhoek, Namibia（Airport Code）纳米比亚温得和克机场

**WDI** Wind Direction Indicator 风向指示器

**WDIL** Wind Direction Indicator Light 风向指示器灯

**WDL** Wireless Data Link 无线数据链路

Wiring Diagram List 布线图目录

**WDM** Wavelength-Division Multiplexing 波分复用

Wiring Diagram Manual 线路图手册

**WDMA** Wavelength Division Multiple Access 波分多址接入

**WDM-BU** WDM Branching Units 波分复用分支单元

**WDM-CCS** WDM Cross-Connect Switch 波分复用交叉连接交换

**WDM-SM** WDM Synchronously Modulated 波分复用同步调制

**WDO** Window 窗,窗口

**WDP** Wireless Datagram Protocol 无线数据报协议

**WDS** Wireless Data Server 无线数据服务器

**WDT** Watchdog Timer 监视计时器

**WDTS** World Data Transmission Service 世界数据传输业务

**WE** Weight Empty 净重,空重

Wing Leading Edges Station 机翼前缘站

Write Enable 写入允许

**WEA** Weather 气象

**WEAA** Western European Airports Association 西欧机场协会

**WEAAC** Western European Airports Authorities Conference 西欧机场管理局会议

**WEAAP** Western European Association for Aviation Psychology 西欧航空心理学协会

**WebTV** Web Television 网络电视

**WEC** World Energy Council 世界能源理事会

**WECPNL** Weighted Equivalent Continuous Perceived Noise Level 加权当量连续感觉噪声级

**WED** Wednesday 星期三

World Environment Day 世界环境日

**WEF** World Economic Forum 世界经济论坛

World Economic Fund 世界经济基金

World Environment Fund 全球环境基金

**WEFT** Wings, Engines, Fuselage, Tail 机翼、发动机、机身和尾翼

**WEG** Wireless E-mail Gateway 无线电子邮件网关

**WEL** Wall Emergency Light 壁板应急灯

**WEM** Warning Electronics Module 警告电子组件

**WEO** World Energy Outlook 世界能源展望

World Environmental Organization 世界环境组织

**WEP** Wired Equivalency Protection 有线当量保护

**WER** Weight Estimating Relationship 重量估算关系式

**WERAN** Weather Radar Data Processor and Analyzer 气象雷达资料处理分析器

**WES** Warning Electronics System 警告电子系统

**WET** Weight Effectiveness Testing 有效重量试验

**WEU** Warning Electronics Unit 警告电子组件

Wavelength Equalization Unit 波长均衡单位

Western European Union 西欧联盟

**WF** Fuel Flow 燃油流量

Weather Forecast 气象预报

Weight Flow 重量流量

Wide Frequency Band 宽频带
Wind Factor 风的因素
Wind Force 风力

**WFA** Wide Frequency Antenna 宽频天线
WXR Flat Plate Antenna 气象雷达平板天线

**WFD** Widespread Fatigue Damage 广布疲劳损伤

**WFE** Weekly Fuel Expenditure 每周燃油消耗

**WFL** Full-Load Weight 满载重量
Work Flow Language 工作流语言

**WFM** Wave Form Monitor 波形监视器
Weight of Fuel Metered 测量的燃油重量
Workflow Management 工作流管理

**WFMC** Workflow Management Coalition 工作流管理联合体

**WFO** Weather Forecast Office 天气预报办公室

**WFOV** Wide Field of View 宽视界

**WFP** Warm Front Passage 暖锋通道
World Food Program 世界粮食计划

**WFPI** Wing Flap Position Indicator 襟翼位置指示器

**WFQ** Weighted Fair Queuing 加权公平队列算法

**WFS** Web Feature Services 网络要素服务

**WFT** Fuel Flow Temperature 燃油温度
Windowed Fourier Transform 窗口傅立叶变换

**WFU** Withdrawn from Use 退出使用

**WG** Wave Guide 波导, 波导管
Wing 机翼
Working Group 工作组

**WGD** Windshield Guidance 风挡引导
Windshield Guidance Display 风挡指引显示

**WGDC** Windshield Guidance Display Computer 风挡引导显示计算机

**WGDS** Windshield Guidance Display System 风挡引导显示系统

**WGET** Working Group on Emergency Technology 应急技术工作组

Working Group on Emergency Telecommunications 应急通信工作组

**WGF** Weakly Guiding Fiber 弱导光纤

**WGLT** Winglet Station 小翼站位

**WGN** White Gaussian Noise 白高斯噪音

**WGNE** Working Group on Numerical Experimentation 数值试验工作组

**WGOCC** World Government Organization Co-ordinating Council 世界政府组织协调理事会

**WGS** World Geodetic Survey 世界大地测量勘察
World Geodetic System 世界大地测量系统

**WGT** Weight 重量

**WH** Watt-Hour 瓦时
Western Hemisphere 西半球

**WHC** Window Heat Computer 窗户加温计算机
Windshield Heat 风挡加温
Workstation Host Connection 工作站主机连接

**WHCU** Window Heat Control Unit 窗户加温控制装置

**WHL** Western Helicopters Ltd. 西部直升机有限公司

**WHLS** Wheels 车轮

**WHMIS** Workplace Hazardous Material Information System 工作场所有害物质信息系统

**WHO** World Health Organization 世界卫生组织

**WHS** Warehouse 货栈

**WHT** White 白色

**WI** Weather Information 气象信息
Within 其中

**WIBCO** Wing/Body Code 机翼/机身标码

**WIDS** Weather Information and Display System 气象信息与显示系统

**WIE** With Immediate Effect 立即生效

**WIF** Weather Instrument Flight 气象仪表飞行

**WIG** Wing-in-Ground-Effect 地面效应, 地效

Wireless Internet Gateway 无线因特网网关

**WIL** Nairobi, Kenya-Wilson（Airport Code）肯尼亚内罗毕威尔逊机场

**WIM** Weight-in-Motion 动态称重

**WIMA** Wireless Integrated Multiple Access 无线综合多址接入

**WIN** Wireless Intelligent Network 无线智能网

Worldwide Intelligent Network 全球智能网

**WINDR** Wind Direction 风向

**WINS** Windows Internet Naming Service 视窗互联网域名服务

**WINTEM** Forecast Upper Wind and Temperature for Aviation 航空高空风和温度预报

**WIP** Work in Progress 正在施工

**WIPCU** Water Ice Protection Control Unit 防水防冰控制组件

**WIPDU** Water Ice Protection Data Unit 防水防冰数据组件

**WIPO** World Intellectual Property Organization 世界知识产权组织

**WIPS** Wing Ice Protection System 大翼防冰系统

**WIRC** Warning-Inhibition-Recall-Clear 警告—抑制—重现—消除

**WIRE** Wire 导线,线

**WIRIS** Wheel-to-Rudder Interconnect System 驾驶盘到方向舵内部连接系统

**WIS** Wireless Industry Standards 无线通信行业标准

World Meteorological Organization Information System 世界气象组织信息系统

**WISE** World Information Systems Exchange 世界信息交换系统

**WISR** Weight and Inertia Status Report 重量与惯量状态报告

Worldwide Internet Service Roadmap 全球因特网服务路径图

**WISS** Weather Information Service System 气象信息服务系统

**WIT** Wire Integrity Testing 线路整合试验

**WITS** Weather Information Telemetry System 气象情报遥测系统

**WIU** Wheel Interface Unit 机轮接口组件

Wire Integration Unit 导线集成组件

**WKLY** Weekly 每周的

**WKN** Weaken or Weakening 弱

**WKND** Weekend 周末

**WL** Wave Length 波长

Warning Light 报警灯

Water Line 水线

Wideband Limiter 宽带限幅器

Work Load 工作量

**WLAN** Wave Local Area Network 波局域网

Wireless Local Area Network 无线局域网

**WLDEP** Will Depart 将离港

**WLDP** Warning Light Display Panel 警告灯显示面板

**WLE** Wing Leading Edge 机翼前缘

**WLG** Wellington, New Zealand（Airport Code）新西兰惠灵顿机场

Wing Landing Gear 机翼起落架

**WLL** Wireless Local Loop 无线本地用户环路

**WLM** Web Library Management 网络图书馆管理

Wireless LAN Manager 无线局域网管理器

**WLP** Weight Limited Payload 重量限制商载

**WLR** Wafer Level Reliability 圆片级可靠性

**WM** Wattmeter 瓦特表

Wiring Manual 线路图手册

**WMA** Web Marketing Association 网络营销协会

Windows Media Audio Windows 音频文件格式

**WMC** World Meteorological Center 世界气象中心

World Meteorological Congress 世界气象大会

**WMD** World Meteorological Day 世界气象日

**WMIC** World Meeting Information Center 世界会议情报中心

**WML** Wireless Markup Language 无线标记语言

**WMO** World Meteorological Organization 世界气象组织

**WMP** Weather Modification Programme 人工影响天气计划

**WMS** Warehouse Management Systems 仓储管理系统

Wide-area Master Station 广域主控站

Work Management System 工作管理系统

**WMSC** Weather Message Service Center 气象信息服务中心

Weather Message Switching Center 气象信息交换中心

**WMSCR** Weather Message Switching Center Replacement 气象信息交换中心更新

**WMSS** Weather Message Switching System 气象通报交换系统

**WMT** Windows Media Technology 视窗媒体技术

**WN** Week Number 星期数

**WNDR** Winder 绕线机,卷扬机

**WNG** Wideband Noise Generator 宽带噪声发生器

Wing 机翼

**WNL** Within Normal Limits 在正常限度内

**WNN** Wavelet Neural Network 小波神经网络

**WNSHR** Wind Shear 风切变

**WNW** West-Northwest 西北偏西

**WO** Wash-out 清洗

Wipe out 消除,封闭

Without 除非,没有,未经,无

Work Operate 工作操作

Work Order 动作指令

**WOC** Wireless Optical Communication 无线光通信

**WOCE** World Ocean Circulation Experiment 世界海洋环流实验

**WOD** Wind-over-Deck 甲板上风力

**WOFDM** Wavelet Orthogonal FDM 子波正交频分复用

**WOFW** Weight-off-Wheel 机轮不承重

**WOG** Water, Oil and Gas 水、滑油和汽油

**WOM** Write Only Memory 只写存储器

Write Optical Memory 光纤可写存储器

**WOR** Wavelength Optimized Routing 波长优化路由选择

Wear-out Rate 磨损率

**WORM** Write Once Read Many 单写多读,一次写入,多次读出

**WOS** Web Operating System Web 操作系统

Wireless Office System 无线办公系统

**WOTAN** Wavelength-agile Optical Transport and Access Network 波长灵活的光传送和接入网

Weather Observation through Ambient Noise 通过环境噪声的气象观测

**WOW** Weight-on-Wheel 机轮承重

**WOWS** Wire Obstacle Warning System 线路故障告警系统

**WP** Waste Pipe 废水管

Water Plane 水上飞机

Water Pump 水泵

Waterproof 防水的

Way Point 航路点

Weatherproof 全天候的

Wet Process 湿处理

White Phosphorus 白磷

Wood Pattern 木模

Word Processing 字处理

Work Package 工作包

Work Paper 工作单(文件)

Working Point 工作点,作用点

Working Pressure 工作压力

**WPA** Work Package Authorization 工作程序包批准,工作数据包批准

Works Progress(以后改为 Work Projects）Administration(美国)工程进度管理局

**WPB** Write Printer Binary 二进制写入打印机

**WPC** Watt per Candle 瓦/烛光

Wired Program Computer 插接程序计算机

**WPD** Work Package Documents 工作程序（数据）包文件

Write Printer Decimal 十进制写入打印机

**WPFC** Waterproof Fan Cooled 防水风扇冷却的

World Precision Flying Championships 世界精确飞行锦标赛

**WPG** Waterproofing 防水

Wiping 抹除

**WPI** Wholesale Price Index 批发物价指数

Wire Provisions Installation 导线预留安装

World Patent Index 世界专利索引

World Policy Institute 世界政策研究所

**WPJ** Weakened Plane Joint 弱面缝

**WPL** Wood Product Labeling 木制品标记

**WPM** Word per Minute 字数/分

**WPMS** Work Process Management System 工作流程管理系统

Work Program Management System 工作计划管理系统

**WPO** Warsaw Pact Organization 华沙条约组织

West Pacific Ocean 西太平洋

Wire Preparation Order 导线准备指令

**WPR** Waypoint Position Report 航路点位置报告

Wideband Programmable Receiver 宽带可编程接收器

Work Performance Review 工作性能审查

**WPS** Waterproof Shroud 防水罩

Welding Procedure Specifications 焊接程序规范

Word Processing System 文字处理系统

Words per Second 每秒字数

**WPSF** Way Point Selection Facility 航路点选择设备

**WPT** Waypoint 航路点

**WPX** Worldwide Package Express 环球包裹快运公司

**WR** Wall Receptacle 壁上插座

Wavelength Router 波长路由器

Weather Radar 气象雷达

Weather Resistant 不受气候影响的

Weighing Report 配重报告

Wet Runway 湿跑道

White Room 绝尘室

Wire Rope 钢缆

Work Request 工作要求

Work Revision 工作更改

Wrench 扳手

Write 写入

**WRB** Warranty Review Board 保修审查委员会

Web Request Broker Web 请求代理

**WRC** World Radiation Center 世界辐射中心

World Radiocommunication Conference 世界无线电通信会议

**WRCP** Weather Radar Control Panel 气象雷达控制面板

**WRDC** Wright Research and Development Center( U. S. ) 莱特研究与发展中心（美国）

**WRDIR** Wrong Direction 错误方向

**WRF** Weather Research and Forecasting Model 气象研究和预测模型

World Research Foundation 世界研究基金会

**WRG** Wearing 磨损

Wire Release Group 电缆交付组,钢索交付组

Wiring 线路,导线

Wrangell, AK, USA ( Airport Code) 美国阿拉斯加州兰格尔机场

**WRGR** Wringer 榨干机,绞拧器

**WRIGHT** WHO Research into Global Hazards of Travel 世界卫生组织对旅行的全球危害研究

**WRK** Work , Working 工作

**WRL** Wing Reference Line 机翼基准线

Worland, WY, USA ( Airport Code) 美国怀俄明州沃兰机场

**WRN** Wavelength Routing Node 波长路由选择节点

**WRNG** Warning 警告

**WRO** Wroclaw, Poland ( Airport Code ) 波兰弗罗茨瓦夫机场

**WRP** Wing Reference Plane 机翼基准面
Wireless Routing Protocol 无线路由选择协议

**WRR** Weighted Round Robin 加权循环法

**WRS** Weather Radar System 气象雷达系统
Wide-area Reference Station 广域基准台站
Wire Release System 电缆发放系统

**WRT** Write 写
Wrought 锻造的,精锻的

**WS** Warning System 报警系统
Water Separator 隔水器
Water Supply 供水
Water Surface 水面
Weather Service 气象服务
Weather Stripping 挡风雨条
Wetted Surface 湿润翼面,湿润表面
Wind Shear 风切变
Wind Speed 风速
Windshield, Wind Shield 风挡
Wing Span 翼展
Wing Station 机翼站位
Wire Shop 电缆车间
Wireless Set 无线组件
Work Statement 工作陈述
Work Station 工作站
Working Space 工作空间
Working Storage 工作存储
Worksheet 工作单
Wrought Steel 锻钢

**WS&D** Wind Speed and Direction 风速与风向

**WSA/I** Windshield Anti-Ice 风挡防冰

**WSAS** Wind Shear Advisory System 风切变警告系统

**WSC** Wireless Switching Center 无线交换中心

**WSCP** Warning System Control Panel 警告系统控制面板

**WSCS** Wheel Steering Control System 机轮转弯控制系统

**WSF** Weather Support Force 气象保障力量
Work Station Function 工作站功能

**WSFO** Weather Service Forecast Office 气象服务预报室

**WSHG** Washing 清洗

**WSHLD** Windshield 风挡,挡风玻璃

**WSHR** Washer 洗涤机

**WSI** Wafer Scale Integration 晶片规模集成
Weather Services International 国际气象服务

Work Station Interface 工作站接口

**WSIS** World Summit on the Information Society 信息社会世界高峰会议

**WSJ** San Juan, Columbia 哥伦比亚圣胡安机场

Wing Stub Join 机翼旋转轴接合头

**WSL** Wetted Sensing Length 湿敏感应长度

**WSM** Wide-band Subscriber Module 宽带用户模块

**WSMS** Windshear Monitor System 风切变监视系统

**WSO** Weather Service Office 气象服务办公室

World Safety Organization 世界安全组织

**WSOL** Within Safe Operating Limit 在安全操作极限内

**WSOM** Weather Service Operations Manual 气象局业务手册

**WSP** Waspam Airport, Nicaragua ( Airport Code ) 尼加拉瓜威斯帕尔机场
Weather System Processor 气象系统处理机,气象系统处理器
Wire Spares Program 电缆备件程序
Wireless Session Protocol 无线会话协议
Working Steam Pressure 工作蒸汽压

**WSPARS** Wing Skin Panel Automatic Riveting System 机翼蒙皮板组合件自动铆合系统

**WSPD** Wind Speed 风速

**WSPS** Wire Strike Protection System 电线撞击保护系统

**WSR** Weather Service Radar 气象服务雷达

Weather Surveillance Radar 气象监视雷达

Weekly Service Report 每周勤务报告

World Speed Record 世界速度记录

**WSS** Wireless Subscriber System 无线用户系统

Work Station System 工作站系统

**WSSC** Warning System Symbol Controller 警告系统符号控制器

**WSSG** Warning System Signal Generator 警告系统信号发生器

**WST** Western Standard Time 西部标准时间

Whitworth Standard Thread 惠氏标准螺纹，英制标准螺纹

Wide-band Subscriber Terminal 宽带用户终端

Wireless Subscriber Terminal 无线用户终端

World System Teletext 全球图文电视系统

**WSTL** Whistle 汽笛

**WSTPN** Wrist Pin 活塞销

**WSU** Weather Service Unit 气象服务单元

Weather Support Unit 天气保障单元

WLL Subscriber Unit 无线本地环路用户单元

**WSW** West-South-West 西南偏西

**WSW/RG** Windshear Warning, Recovery Guidance 风切变告警和恢复指导

**WSXC** Wavelength Selective Cross-Connect 波长选择性交叉连接

**WT** Water Tank 水箱

Weight 重量

Wind Tunnel 风洞

Withholding Tag 阻留卡

Wood Threshold 木门槛

**WTA** Wireless Telephony Application 无线电话应用

**WTAI** Wing and Tail Anti-Icing 机翼和尾翼防冰

Wireless Telephone Application Interface 无线电话应用接口

**WTB** Wing Tip Brake 翼尖刹车

**WTBF** Wing to Body Fairing 机翼机身整流罩

**WTC** World Trade Corporation 世界贸易公司

**WTCA** World Trade Center Association 世界贸易中心协会

**WTCC** Water Turbine Closed Coupled 水轮机闭环耦合

**WTCU** Windshield Temperature Control Unit 风挡温度控制组件

**WTDC** World Telecommunication Development Conference 世界电信发展大会

**WTDI** Wing and Tail De-Icing 机翼和尾翼除冰系统

**WTDPP** Wind Tunnel Data Postprocessor 风洞数据后处理器

**WTF** W-Type Fiber W 型光纤

**WTG** Waiting 等候

Weighting 称重

Work Type Group 工作类型组

**WTGC** Work Type Group Code 工作类型组编码

**WTHR** Weather 天气

**WTI** Wire Termination Inquiry 电缆终端审查

**WTL** Wireless Transmission Layer 无线传输层

**WTLS** Wireless Transport Layer Security 无线传输层安全机制

**WTM** Wing Tunnel Model 风洞试验模型

**WTN** Wing Tunnel Note 风洞试验记录

**WTO** Warsaw Treaty Organization 华沙条约组织(华约)

World Tourism Organization 世界旅游组织

World Trade Organization 世界贸易组织

**WTP** Wireless Transmission Protocol 无线传输协议

**WTR** Water 水

**WTRIS** Wheel-to-Rudder Interconnect System 驾驶盘到方向舵互联系统

**WTRZ** Winterize 冬季改装,使防冻

**WTS** Weights 重量

**WTSA** World Telecommunications Standardization Assembly 世界电信标准化全会

**WTSPT** Waterspout 排水管,海上龙卷风

**WTT** Wing Tunnel Test 风洞试验
Working Together Team 共同工作组

**WTURB** Water Turbine 水轮机

**WU** Warning Unit 报警装置
Wheels up 收起落架
When Used 使用后
Work Unit 作业单位,工作组(件)

**WUC** Work Unit Code 作业单位代码
Work Unit Cost 工作单元成本

**WUCM** Work Unit Code Manual 作业单位代码手册

**WUD** Would 将,可能

**WUS** Wake up Service 叫醒业务

**WUSMNG** Wake up Service Management 叫醒业务管理

**WUV** Weighted Unit Value 加权的组件值

**WV** Wall Vent 舱壁通风
Wind Vector 风速矢量
Wind Velocity 风速
Working Voltage 工作电压

**WVAC** Working Voltage Alternate Current 工作交流电压

**WVD** Waived 放弃的

**WVDC** Working Voltage Direct Current 工作直流电压

**WVG** Waveguide 波导(管)
· Wingtip Vortex Generator 翼尖涡流发生器

**WVLAN** Wireless Virtual LAN 无线虚拟局域网

**WVP** Working VP 工作的虚通道

**WVPN** Wireless Virtual Private Network 无线虚拟专用网

**WVR** Waiver 弃权(声明书)
Within Visual Range 视距内

**WVT** Wire Verification Testing 钢索验收试验,导线验收试验

**WW** Wire Way 线路
Wire Wound 线绕的
Wire Wrap 线绕(捆)

**WW(W/W)** Wheel Well 轮舱

**WWⅠ** World War Ⅰ 第一次世界大战

**WWⅡ** World War Ⅱ 第二次世界大战

**WWABNCP** Worldwide Airborne National Command Post 国家全球空中指挥所(美国)

**WWDMS** World Wide Data Management System 全球数据管理系统

**WWHL** Waterwheel 水轮

**WWIMS** Worldwide Warning Indicator Monitoring System 全球警戒指示器监视系统

**WWIS** World Weather Information Service 世界天气信息服务

**WWL** World Wide Logistics 全球物流

**WWRP** World Weather Research Program 世界天气研究计划

**WWW** World Weather Watch 世界气象观测
World Wide Web 万维网,全球信息网

**WX** Weather 天气,气象
Weather Message 气象通报
Weather Mode 气象方式(导航显示)

**WX/C** Weather/Cyclic 气象/循环

**WX/T** Weather/Turbulence 气象/湍流

**WXEQ** Weather Equipment 气象设备

**WXI** WXR Indicator 气象雷达显示器

**WXP** Weather Radar Panel 气象雷达面板

**WXR** Weather Radar 气象雷达

**WXRCP** Weather Radar Control Panel 气象雷达控制板

**WXRS** Weather Radar System 气象雷达系统

**WXRT** Weather Radar Receiver/Transmitter 气象雷达收发机

**WYPT** Waypoint Altitude 航路点高度

**WYS** West Yellowstone, MT, USA (Airport Code) 美国蒙大拿州西黄石机场

**WZ** Working Zone 工作区

# X

**X**   Cross 交叉,交输
Empty Container or Empty Pallet 空集装箱或集装板
Experimental 实验(性)的
Extension 延长,扩大,分机
Gross 总重,毛重,全部的,粗略的
On Request 按申请,按要求
Reactance 电抗
Trans 反式
Unknown 未知数
X-Axis X 轴,横轴
X-Ray X 射线

**X BLEED**   Crossbleed 交输引气

**X FEED**   Crossfeed 交输供油

**X LINE**   Crossline, Cross-Line 交输管,交输线

**XA**   Auxiliary Amplifier 辅助放大器
Extended Architecture 扩展结构

**XARM**   Cross Arm 横臂

**XBAG**   Excess Baggage（Reservations Code）超重(额)行李

**XBAR**   Cross Bar, Crossbar 指引杆,横排灯（进近灯光系统中的）

**X-BLD**   Cross Bleed 交输引气

**XBRA**   Cross Bracing 交叉撑,剪刀撑

**XBS**   Exchange Baseband Station 交换基带站

**XBT**   Crossbar Tandem 纵列式横杆

**XC**   Chief of Center 中心主任
Cross Connect 交叉连接

**X-CH**   Cross Channel 交输通道

**XCONN**   Cross Connection 横向联接,交叉联接

**XC-TB**   Cross-Connect Testbed 交叉连接测试台

**XCTR**   Exciter 激励器

**XCVR**   Transceiver 收发机
Transmitter 发射机

**XCY**   Cross Country 越野的

**XDAP**   Execute Direct-Access Program 执行直接存取程序

**XDBG**   Exchange Database Generator 交换数据库生成器

**XDCR**   Transducer 传感器

**XDSL**   Digital Subscriber Line 数字用户线

**XDSS**   Expert Decision Support System 专家决策支持系统

**XE**   Experimental Engine 实验型发动机
External Electrical 外电源

**XEC**   XBS Echo Controller 交换基带站回声控制器

**XFCN**   External Function 外部功能

**XFD**   Stratford, Ontario, Canada（Airport Code）加拿大安大略省斯特拉福德机场

**X-FEED**   Cross Feed 交叉供给(油)

**XFER**   Transfer 转换

**XFMR**   Transformer 变压器,传感器

**XFSS**   Auxiliary Flight Service Station 辅助飞行服务站

**XH**   Experimental Helicopter 实验型直升机

**XHAIR**   Cross Hair 十字标线

**XHST**   Exhaust 排气

**XHTML**   Extensible Hypertext Markup Language 可扩展超文本标记语言

**XHV**   Extremely High Vacuum 极高真空

**XHVY**   Extra Heavy 超重

**XIC**   Transmission Interface Converter 传输接口转换器

**XID**   Exchange Identification 交换机标识

**XIM**   X-Ray Inspection Module X 射线检查组件

**XIN** Extension 电话分机,延长

**XING** Crossing 交叉,越过

**XIO** Execute Input/Output 执行输入/输出
External Input/Output 外部输入/输出

**XITN** Transition 飞越,通过,转变

**XLANE** Cross Lane 交叉航线,交叉巷

**XLINK** Cross Link 交叉链

**XLOL** Expected Loss-of-Load 估计载荷损
失,预计负载损失

**XLS** Multi-mode Landing System 多模式着陆
系统

**XLTR** Translator 翻译
Transmission 传输

**XLWB** Extra Long Wheel Base 超长轮距

**XM** External Master 外部主机
Extra Marker 外加指点标
Mayotte 马约特岛(科摩罗)

**XMATN** Transformation 变换,转换

**XMFR** Transformer 变压器,变换器

**XMISSION** Transmission 传输

**XMIT** Transmit 传输,发射,传送

**XML** Extensible Markup Language 可扩展标
记语言,延伸指点标语言

**XMM** Monaco, Monaco (Airport Code) 摩纳
哥摩纳哥机场

**XMPT** Exempt, Exemption 解除,免除

**XMSN** Transmission 传输

**XMT** Transmitter 发射机

**XMTD** Transmitted 发射的,发送的

**XMTG** Transmitting 发射

**XMTL** Transmittal 发射的

**XMTR** Transmitter 发射机

**XNG** Crossing 穿越
Viet Nam-Quang Ngai Airport 越南广义机场

**XNS** Xerox Network System 静电印刷网络系
统

**XNT** Transient 瞬态,瞬时的,过渡的

**XO** Crystal Oscillator 晶体振荡器
Exchange Order 换票证

**XOR** Exclusive-OR 异或(一种逻辑运算)

**XOVR** Crossover 剖面,渡越,交叉

**XP** Cross-Polarization 横向极化
External Power 外接电源

**XPC** External Power Contactor 外电源接触器

**XPD** Cross Polarization Discrimination 交叉极
化鉴别(度/率)

**XPDR** Transponder 应答机,询答机

**XPI** Cross Polarization Isolation 交叉极化隔
离

**XPIR** Expire or Expiration 终止,期满

**XPL** External Party Line 外部用户线

**XPLN** Explain or Explanation 说明,解释

**XPNDR** Transponder 应答机

**XPORT** Transport or Transportation 运输,传
递

**XPP** Transmit Personality PROM 发送用可编
程只读存储器

**XPS** Priority Small Package 小包装优先

**XPT** Crosspoint 交叉点

**XPU** Actuator Position Variable 作动器位置
变量

**XR** Experimental Rotor 实验型旋翼机
Index Register 变址寄存器

**X-RAY** Radiographic Inspection X 射线/X
光检查

**XREF** Cross-Reference 相互参照
Extended Recovery Facility 扩展收回设施

**XRL** X-Ray Laser X 射线激光器

**XRM** Radiographic Inspection Method X 射
线/X 光检查方法

**XROE** Experimental Rotocycle (Helicopter)
实验型单座旋翼机

**XRT** X-Ray Telescope X 射线望远镜

**XRY** Jerez De La Frontera, Spain (Airport
Code) 西班牙赫雷斯机场

**XS** Cross Section 横断面
Transmission Subsystem 传输分系统,传输
子系统

**XSCC** Xi'an Satellite Control Centre 西安卫
星测控中心

**XSEC** Extra Section 外加部分
Extra Section Flight 加班飞行

**XSLC**　China Xichang Satellite Launch Center 中国西昌卫星发射中心

**XSSDU**　Expedited Session Service Data Unit 加速会晤业务数据单元

**XST**　External Start 外部启动
　　Experimental Supersonic Transport 试验超声速运输
　　Experimental Survivable Technology 试验生存技术

**XSTR**　Extra Strong 超强
　　Transistor 晶体管

**XSTRZ**　Transistorized 晶体管化的

**XT**　Communication Center 通信枢纽
　　Exchange TCAS 交通警告和避撞系统信息交流

**XTAL**　Crystal 晶体

**X-TALK**　Cross-Talk 交叉谈话,串话

**XTE**　Cross Track Error 偏航距离误差(航迹误差)

**XTK**　Cross Track Deviation 横向航迹偏离
　　Cross Track Distance 偏航距离

**XTLO**　Crystal Oscillator 晶体振荡器

**XTN**　Extension 扩展

**XTP**　Express Transfer Protocol 快速转换协议

**XUV**　Experimental Unmanned Vehicle 试验无人驾驶飞行器

**XVALVE**　Cross Valve 交输活门

**XVGA**　Extended Video Graphics Array 延伸视频图形阵列
　　X Video Graphics Adaptor X 视频图形适配器

**XVRS**　Transverse 横轴,横梁,横断的,横向的

**XVTR**　Transverter 变流机,换能器

**X-W**　X-W Compare Program X-W 比较程序

**XWAVE**　Extraordinary Wave 特别波,异常波

**XWC**　Cross Wind Component 侧风分量

**X-WIND**　Crosswind 侧风

**XX**　Cancel 取消

**XX STR**　Double Extra Strong 双倍超强

**XXNS**　Cancel Due to No-Show (旅客)未到取消

**XXRA**　Heavy Rain 大雨

**XXRASN**　Heavy Rain amd Snow 大雨夹雪

**XXSH**　Heavy Showers 大阵雨

**XXSN**　Heavy Snow 大雪

**XXSNSH**　Heavy Snow Showers 大阵雪

**XXTS**　Heavy Thunderstorm 大雷暴(雨)

**XXTSGR**　Heavy Thunderstorm with Hail 大雷暴(雨)与冰雹

**XXX**　International Urgency Signal 国际紧急信号
　　Urgency Message 紧急信号

**XYA**　X-Y Axis X-Y 轴

# Y

**Y** Admittance 准入
Crystal 晶体
Economy Class 普通舱
Full Economy Fares 完全经济运价
Non-Dimensional Acoustic Sonar 无量纲声呐
Prototype 样机
Service Test 运行(性能)试验
Transitional Aircraft Testing 过渡飞机试验
Y-Axis Y 轴,纵轴
Yaw (Axis) 航向(轴)
Yaw 偏航
Year 年
Yellow 黄色

**Y2K** Year 2000 计算机两千年问题("千年虫")

**YA** Yaw Axis 偏航轴

**YAB** Budget Fare 经济舱预算票价

**YAGI** An Antenna System 天线系统

**YAO** Yaounde, Cameroon (Airport Code) 喀麦隆雅温得机场

**YAP** Advance Purchase Excursion Fare 预购旅游票价
Yaw and Pitch 偏航与俯仰

**YAPS** Yaw and Pitch Sensing 偏航与俯仰传感

**YAW PCU** Yaw Power Control Unit 偏航动力控制组件

**YB** Year Book 年鉴
Yellowbook 黄皮书

**YBD** Y-Branching Device Y 型分路器

**YC** Tourist Class 经济舱

**YCC** Cornwall, Ontario, Canada (Airport Code) 加拿大安大略省康沃尔机场

**YCD** Nanaimo, British Columbia, Canada (Airport Code) 加拿大英属哥伦比亚省纳奈莫机场

**YCF** Economy Class-Front 前经济舱

**YCG** Castlegar, British Columbia, Canada (Airport Code) 加拿大英属哥伦比亚省卡斯尔加机场

**YCP** Yaw Coupling Parameter 偏航耦合参数

**YCR** Economy Class-Rear 后经济舱

**YCZ** Yellow Caution Zone 黄灯注意区,黄灯危险警告区

**YD** Yard 码
Yaw Damper 偏航阻尼器

**YDA** Yaw Damper Actuator 偏航阻尼器作动器
Yesterday 昨天

**YDC** Yaw Damper Computer 偏航阻尼计算机

**YDCP** Yaw Damper Control Panel 偏航阻尼器控制板

**YDM** Yaw Damper Module 偏航阻尼器组件

**YDR** Yaw Damper Confidence Test 偏航阻尼器可靠性测试

**YDS** Yaw Damper Servo 偏航阻尼器伺服
Yaw Damper System 偏航阻尼器系统

**YDV** Yaw Thrust Vector 偏航推力矢量

**YE** Excursion Fare 经济游览票价
Year 年
Yemen Republic 也门共和国

**YEA** Edmonton, Canada (Airport Code) 加拿大埃德蒙顿机场

**YEL** Elliot Lake, Ontario, Canada (Airport Code) 加拿大安大略省埃利奥特湖机场

**YEN** Japanese Yen 日元(货币单位)

**YEP** Yaw Envelope Protection 偏航包线保护

**YES** Yankee Escape System 美式撤离系统

**YF** Aeronautical Fixed Station 航行固定电台

**YFC** Fredericton, New Brunswick, Canada (Airport Code) 加拿大新不伦瑞克省弗雷德里克顿机场

**YGK** Kingston, Ontario, Canada (Airport Code) 加拿大安大略省金斯顿机场

**YGV** Group Inclusive Tour (Y Class Group Fare) 团体综合游览票价

**YHM** Hamilton, Ontario, Canada (Airport Code) 加拿大安大略省汉密尔顿机场

**YHZ** Halifax, Nova Scotia, Canada (Airport Code) 加拿大新斯科舍省哈里法克斯机场

**YIL** Yellow Indicating Lamp 黄色指示灯

**YKA** Kamloops, British Columbia, Canada (Airport Code) 加拿大英属哥伦比亚省坎卢普斯机场

**YKF** Kitchener, Ontario, Canada (Airport Code) 加拿大安大略省基奇纳机场

**YKM** Yakima, WA, USA (Airport Code) 美国华盛顿州亚基马机场

**YKZ** Buttonville, Ontario, Canada (Airport Code) 加拿大安大略省巴顿维尔机场

**YL** Your Letter 来信

**YLT** Yellow Light 黄色灯

**YLW** Kelowna, British Columbia, Canada (Airport Code) 加拿大基洛纳机场

**YM** Meteorological Office 气象室
Yawing Moment 偏航力矩

**YMS** Yield Management System 收益管理系统

**YMX** Montreal, Quebec, Canada-Mirabel (Airport Code) 加拿大魁北克省蒙特利尔米拉贝尔机场

**YNAV** Vertical Navigation 垂直导航

**YNG** Youngstown, OH, USA (Airport Code) 美国俄亥俄州扬斯敦机场

**YOK** Yokohama, Japan (Airport Code) 日本横滨机场

**YOS** Years of Service 服役年数

**YOW** Ottawa, Ontario, Canada (Airport Code) 加拿大安大略省渥太华机场

**YOX** Excursion Oneway Fare 单程经济游览票价

**YP** Yield Point 屈服点

**YPX** Advance Purchase Excursion Fare 预购经济游览票价

**YQB** Quebec, Canada (Airport Code) 加拿大魁北克机场

**YQM** Moncton, New Brunswick, Canada (Airport Code) 加拿大蒙克顿机场

**YQR** Regina, Saskatchewan, Canada (Airport Code) 加拿大萨斯喀彻温省里贾纳机场

**YQT** Thunder Bay, Ontario, Canada (Airport Code) 加拿大安大略省桑德贝机场

**YQY** Sydney, Nova Scotia, Canada (Airport Code) 加拿大新斯科舍省悉尼机场

**YR** Yard Repair 进厂修理
Year 年(份)

**YRG** Yaw-Roll Gyroscope 偏航滚转陀螺仪

**YRLY** Yearly 每年的

**YRS** Years 多年

**YS** Yield Strength 屈服强度

**YSAS** Yaw Stability Augmentation System 偏航稳定性增益系统

**YSB** Sudbury, Canada (Airport Code) 加拿大萨德伯里机场

**YSE** Yaw Steering Error 偏航驾驶误差

**YSJ** Saint John, New Brunswick, Canada-Turnbull Field (Airport Code) 加拿大新布伦瑞克省圣约翰机场

**YT** Your Telex/Telegram 来电

**YTD** Year-to-Date 最近的一年

**YTPPP** Yearly Technical Publications Processing Planning 年度技术出版物处理计划

**YU** Yugoslavia 南斯拉夫

**YUL** Montreal, Quebec, Canada-Trudeau (Airport Code) 加拿大魁北克省蒙特利尔特鲁多机场

**YUM** Yuma, AZ, USA (Airport Code) 美国亚利桑那州尤马机场

**YV** Yield Value 屈服值, 起始值

**YVR** Vancouver, Canada (Airport Code) 加拿大温哥华机场

**YWG**　Winnipeg, Manitoba, Canada（Airport Code）加拿大温尼伯机场

**YWH**　Victoria, British Columbia, Canada（Airport Code）加拿大维多利亚机场

**YWR**　White River, Ontario, Canda（Airport Code）加拿大安大略省怀特河机场

**YXC**　Cranbrook, Canada（Airport Code）加拿大克兰布鲁克机场

**YXE**　Saskatoon, Canada（Airport Code）加拿大萨斯卡通机场

**YXJ**　Fort St John, British Columbia, Canada（Airport Code）加拿大英属哥伦比亚省圣约翰堡机场

**YXS**　Prince George, British Columbia, Canada（Airport Code）加拿大英属哥伦比亚乔治王子城机场

**YXT**　Terrace, British Columbia, Canada（Airport Code）加拿大英属哥伦比亚省特勒斯机场

**YXU**　London, Ontario, Canada（Airport Code）加拿大安大略省伦敦机场

**YYC**　Calgary, Alberta, Canada（Airport Code）加拿大亚伯达省卡尔加里机场

**YYD**　Smithers, British Columbia, Canada（Airport Code）加拿大英属哥伦比亚史密瑟斯机场

**YYF**　Penticton, British Columbia, Canada（Airport Code）加拿大英属哥伦比亚彭蒂克顿机场

**YYG**　Charlottetown, Prince Edward Island, Canada（Airport Code）加拿大爱德华王子岛夏洛特敦机场

**YYJ**　Victoria, British Columbia, Canada（Airport Code）加拿大英属哥伦比亚维多利亚机场

**YYR**　Goose Bay, Newfoundland, Canada（Airport Code）加拿大纽芬兰岛古斯贝机场

**YYT**　St Johns, Newfoundland, Canada（Airport Code）加拿大纽芬兰岛圣约翰斯机场

**YYY**　Mont Joli, Quebec, Canada（Airport Code）加拿大魁北克蒙特朱利机场

**YYZ**　Toronto, Ontario, Canada-Pearson（Airport Code）加拿大安大略省多伦多皮尔森机场

**YZ**　Youth Fare 青年折扣票价

**YZP**　Sandspit, Canada（Airport Code）加拿大桑兹皮特机场

# Z

**Z** Acoustic Impedance 声抗

Ballistic Set 发射特性组件

Coordinated Universal Time（UTC）协调世界时,世界协调时(UTC)

Greenwich Mean Time 格林尼治平均时间,格林尼治标准时

Mixed Destination Load 混装

Refer to Reflectivity Factor 参见反射率因子

Z-Axis Z 轴

Zero 零

Zone 区域,地带

Zulu（Universal Coordinated Time）区域

Zulu Time 格林尼治平均时间

**ZA** Adder 加法器

Adjuster 调节器

Approach Control Office 进近管制室

Zero Adjustment 调零,零点调整

Zero and Add 零与加（指令）

**ZAC** Zenith Aircraft Company 吉尼斯飞机公司

Zero Administration Client 零管理客户机

**ZAF** Zero Alignment Feature 零校准特点

Zero Alignment Fixture 零校准定位台

**ZAG** Zagreb, Croatia（Airport Code）克罗地亚萨格勒布机场

**ZAH** Zahedan, Iran（Airport Code）伊朗扎黑丹机场

**ZAK** Zero Administration Kit 零管理工具箱

**ZAM** Zamboanga, Philippines（Airport Code）菲律宾三宝颜机场

Zero Administration for Windows 零管理视窗系统

**ZAP** Zinc Anode Plate 锌阳极板

**ZAS** Zaire Aero Service 扎伊尔航空维修公司

Zero Access Storage 立即存取存储体

Zinc Alloy for Stamping 冲压用锌合金

**ZAW** Zero Administration Workstation 零管理工作站

**ZBB** Zero-Base Budget(ing) 零基预算法

**ZBR** Zone Bit Recording 分区位存录操作

**ZBT** Zero Bus Turnaround 零转向总线

**ZBTSI** Zero Byte Time Slot Interchange 零字节时隙交换

**ZC** Zero Calibrate 零校正

Zone Center 区域中心

Zone Controller 区域控制器,区域管制员

**ZCAV** Zoned Constant Angular Velocity 区位恒定角速度

**ZCB** Zinc-Coated Bolt 镀锌螺栓

**ZCD** Zero Crossing Detector 零交叉检测仪

**ZCI** Zero Center Instrument 双向读数仪表,中心零位式仪表

**ZCL** Zacatecas, Mexico（Airport Code）墨西哥萨卡特卡斯机场

Zero Command Line 零基指令线

Zone Calibration Location 时差区域校准定位

**ZCN** Zinc-Coated Nut 镀锌螺帽

**ZCP** Zero Command Point 零基指令点

Zero Crossing Point 过零点

**ZCS** Zinc-Coated Screw 镀锌螺钉

Zone Communication Station 区域通信站

**ZD** Zener Diode 齐纳二极管

Zero Defect 零缺点

Zero Defects 无故障,无缺陷

Zero Dispersion 零色散

Zone Description 区域说明

Zone Distance 区段距离,航段距离

**ZD-DFF** Zero Dispersion-Dispersion Flatened

Fiber 零色散—色散平坦光纤

**ZDL** Zone Division Logic 分区逻辑

**ZDP** Zero Defects Program 无缺陷生产计划

Zero Delivery Pressure 零输送压力

Zero Dispersion Point 零色散点

**ZDS** Zero Defect Software 零缺陷软件

**ZDW** Zero Dispersion Wavelength 零色散波长

**ZE** Zero Effect 零效应, 无效应

Zone Evaluation 区域评价

**ZEG** Zero Economic Growth 经济零增长

**ZEV** Zero Emissions Vehicle 零排放车辆, 零污染车辆

**ZF** Zero Flag 零标志

Zero Frequency 零频率

**ZFC** Zero Failure Criteria 零故障准则

**ZFCG** Zero Fuel Center of Gravity 零燃油重心

**ZFS** Zone Field Selection 区域字段选择

**ZFT** Zero Flight Time 零飞行时间, 无飞行时间

**ZFW** Zero Fuel Weight 零燃油重量

**ZG** Zero Growth 零增长

**ZGE** Zero Gravity Effect 无重量效应

Zero Gravity Environment 无重量环境

**ZGSH** Zero Ground Speed Hovering 零地速旋停

**ZI** Zero Inventory 零库存

Zone of Interior 内部区间

**ZIA** Zone Integration Area 客舱综合区

**ZID** Indianapolis, Indiana 印第安那波利斯中心

**ZIF** Zero Incident Frequency 零事故的频率

Zero Insertion Force 零插力插座

Zero Insertion Force Package 零抗力封装插座

**ZIH** Ixtapa, Mexico (Airport Code) 墨西哥印坦巴机场

**ZIP** Zero Interest Plan 零利率计划

Zonal Inspection Program 区域检查大纲

Zone Information Protocol 区域信息协议

**ZIPS** Zero Inventory Production System 零库存产品系统

**ZIT** Zone Information Table 区信息表

**ZIU** Zone Integration Unit 区域综合组件

**ZL** Zero Lift 零升力

Zero Line 零位线, 基准线

**ZLA** Zero Lift Angle 零升力角

**ZLC** Zero Lift Chord 零升力弦

**ZLD** Zero Level Drift 零位漂移

Zero Lift Drag 零升力时阻力

**ZLIS** Zero Lift Inertial System 零升力惯性系统

**ZLO** Manzanillo, Colima, Mexico (Airport Code) 墨西哥科利马州曼萨尼约机场

**ZLU** Zone Logic Unit 区间逻辑组件

**ZM** Zero Modulation 零调制

Zone Marker 区域指点标

**ZMA** Zone Multicast Address 区(域)多播地址

**Z-Marker** Z Marker Beacon Z 指点标

**ZMDW** Zero-Material-Dispersion Wavelength 零材料色散波长

**ZMOD** Zero Module 零模组

**ZMU** Zone Management Unit 区域管理组件

**ZN** True Azimuth 真方位角

Zone 区间, 范围

**ZNA** Nanaimo, British Columbia, Canada (Airport Code) 加拿大英属哥伦比亚省纳奈莫机场

**ZNC** Zone Network Controller 局域网控制器

**ZNG** Zero Nominal Growth 零名义增长

**ZNTU** Zone Network Telephone Unit 局域网电话组件

**ZNZ** Zanzibar, Tanzania (Airport Code) 坦桑尼亚桑给巴尔机场

**ZO** Zero Output 零输出

**ZOH** Zero-Order-Hold 无指令控制

**ZOP** Zero Order Predictor 零阶预测器, 零阶指示器

**ZP** Zero Potential 零电位

Zero Power 零功率

**ZPC** Zero Point of Charge 零点电荷
Zone Power Converter 区域电源转换器

**ZPI** Zone Position Indicator 区域位置指示器

**ZPT** Zero Power Test 零功率试验

**ZQN** Queenstown, New Zealand ( Airport Code) 新西兰昆斯敦机场

**ZRE** Zero-Rate-Error 零速率误差

**ZRG** Zero Real Growth 零实际增长

**ZRH** Zurich, Switzerland(Airport Code) 瑞士苏黎世机场

**ZROC** Zero Rate of Climb 零爬升率

**ZRP** Zero Radial Play 零径向间隙

**ZRV** Zero Relative Velocity 零相对速度

**ZS** Zero Shift 零偏移

**ZSA** San Salvador, Bahamas ( Airport Code) 巴哈马圣萨尔瓦多机场
Zero Set Amplifier 归零放大器
Zonal Safety Analysis 带状安全性分析

**ZSD** Zinc Sulfide Detector 硫化锌检测器
Zebra Stripe Display 斑马条纹显示

**ZSPG** Zero-Speed Pulse Generator 零速度脉冲发生器

**ZST** Zone Standard Time 地区标准时间

**ZT** Zone Time 区时，区域时间

**ZTC** Zero Temperature Coefficient 零点温度系数
Zone Temperature Control 区域温度控制
Zero Temperature Controller 零点温度控制器

**ZTDL** Zero-Thrust Descent and Landing 零推力下降和着陆

**ZTLP** Zero Transmission Level Point 零传输电平点
Zero Test Level Point 零测试电平点

**ZTM** Zero Temperature Monitor 零点温度监控器

**ZTP** Zero-Thrust Point 零推力点

**ZTS** Zone Temperature Sensor 区域温度传感器

**ZTV** Zone Trim Valve 区域调节活门

**ZULU** Greenwich Mean Time 格林尼治平均时

**ZVP** Zoomed Video Port 视讯输出端口

**ZVS** Zero Voltage Switching 零电压切换

**ZW** Zero Wear 无磨损

**ZWG** Zonal Working Group 地区工作组

**ZWL** Zero Wave Length 零波长

**ZWV** Zero Wave Velocity 零波速

**ZXCFAR** Zero Crossing Constant False Alarm Rate 零交叉恒虚警率

**ZZD** Zig-Zag Diagram 折线图

**ZZR** Zig-Zag Riveting 交错铆接

**ZZTOP** Zero-Zero Telecoms Operating Protocol 零对零电信操作协议

**ZZV** Zero-Zero Visibility 能见度极差

# 附录一　中国民用航空运输机场（大陆地区）一览表

| 地区 | 直辖市、省、自治区 | 机场中文名称 | 机场三字代码（IATA） | 机场四字代码（ICAO） | 机场等级数 | 机场英文名称简称 |
|------|------|------|------|------|------|------|
| 华北 | 北京市 | 北京首都国际机场 | PEK | ZBAA | 4F | Capital Airport |
| | | 北京南苑机场 | NAY | ZBNY | 4C | Nanyuan Airport |
| | 天津市 | 天津滨海国际机场 | TSN | ZBTJ | 4E | Binhai Airport |
| | 河北省 | 石家庄正定国际机场 | SJW | ZBSJ | 4D | Zhengding Airport |
| | | 秦皇岛山海关机场 | SHP | ZBSH | 4C | Qinhuangdao Airport |
| | | 邯郸机场 | HDG | ZBHD | 3C | Handan Airport |
| | | 唐山三女河机场 | TVS | ZBSN | 4C | Tangshan Airport |
| | 山西省 | 太原武宿国际机场 | TYN | ZBYN | 4D | Taiyuan Airport |
| | | 运城关公机场 | YCU | ZBYC | 4D | Yuncheng Airport |
| | | 长治王村机场 | CIH | ZBCZ | 4C | Changzhi Airport |
| | | 大同倍加皂机场 | DAT | ZBDT | 3C | Datong Airport |
| | 内蒙古自治区 | 呼和浩特白塔国际机场 | HET | ZBHH | 4E | Haote Baita Airport |
| | | 包头二里半机场 | BAV | ZBOW | 4C | Baotou Airport |
| | | 呼伦贝尔海拉尔机场 | HLD | ZBLA | 4C | Hailar Airport |
| | | 通辽机场 | TGO | ZBTL | 3C | Tongliao Airport |
| | | 赤峰玉龙机场 | CIF | ZBCF | 4C | Chifeng Airport |
| | | 锡林浩特机场 | XIL | ZBXH | 3C | Xilinhaote Airport |
| | | 乌兰浩特机场 | HLH | ZBUL | 3C | Wulanhaote Airport |
| | | 乌海机场 | WUA | ZBUH | 3C | Wuhai Airport |
| | | 鄂尔多斯伊金霍洛机场 | DSN | ZBDS | 4C | Eerduosi Airport |
| | | 满洲里西郊机场 | NZH | ZBMZ | 3C | Manzhouli Airport |
| | | 二连浩特赛乌素机场 | ERL | ZBER | 4C | Erenhot Airport |
| | | 阿尔山机场 | YIE | ZBES | 4C | Aershan Airport |
| | | 巴彦淖尔天吉泰机场 | RLK | ZBYZ | 4C | Bayannaoer Airport |

| 地区 | 直辖市、省、自治区 | 机场中文名称 | 机场三字代码（IATA） | 机场四字代码（ICAO） | 机场等级数 | 机场英文名称简称 |
|---|---|---|---|---|---|---|
| 华东 | 上海市 | 上海浦东国际机场 | PVG | ZSPD | 4F | Pudong Airport |
| | | 上海虹桥国际机场 | SHA | ZSSS | 4E | Hongqiao Airport |
| | 山东省 | 青岛流亭国际机场 | TAO | ZSQD | 4E | Qingdao Airport |
| | | 济南遥墙机场 | TNA | ZSJN | 4E | Jinan Airport |
| | | 烟台莱山国际机场 | YNT | ZSYT | 4D | Yantai Airport |
| | | 威海大水泊机场 | WEH | ZSWH | 4D | Weihai Airport |
| | | 临沂沭埠岭机场 | LYI | ZSLY | 4C | Linyi Airport |
| | | 潍坊机场 | WEF | ZSWF | 4D | Weifang Airport |
| | | 东营永安机场 | DOY | ZSDY | 4C | Dongying Airport |
| | | 济宁曲阜机场 | JNG | ZSJG | 4C | Jining Airport |
| | 江西省 | 南昌昌北国际机场 | KHN | ZSCN | 4D | Nanchang Airport |
| | | 景德镇机场 | JDZ | ZSJD | 4C | Jingdezhen Airport |
| | | 赣州黄金机场 | KOW | ZSGZ | 4C | Ganzhou Airport |
| | | 九江庐山机场 | JIU | ZSJJ | 4C | Jiujiang Airport |
| | | 井岗山机场 | JGS | ZSGS | 4C | Jinggangshan Airport |
| | 安徽省 | 合肥新桥国际机场 | HFE | ZSOF | 4D | Hefei Airport |
| | | 黄山屯溪机场 | TXN | ZSTX | 4D | Huangshan Airport |
| | | 安庆机场 | AQG | ZSAQ | 4C | Anqing Airport |
| | | 阜阳机场 | FUG | ZSFY | 4C | Fuyang Airport |
| | 浙江省 | 杭州萧山国际机场 | HGH | ZSHC | 4E | Hangzhou Airport |
| | | 宁波栎社国际机场 | NGB | ZSNB | 4E | Ningbo Airport |
| | | 温州永强机场 | WNZ | ZSWZ | 4D | Wenzhou Airport |
| | | 舟山普陀山机场 | HSN | ZSZS | 4D | Zhoushan Airport |
| | | 台州路桥机场 | HYN | ZSLQ | 4C | Huangyan Airport |
| | | 义乌机场 | YIW | ZSYW | 4C | Yiwu Airport |
| | | 衢州机场 | JUZ | ZSJU | 4C | Quzhou Airport |

| 地区 | 直辖市、省、自治区 | 机场中文名称 | 机场三字代码（IATA） | 机场四字代码（ICAO） | 机场等级数 | 机场英文名称简称 |
|---|---|---|---|---|---|---|
| | 江苏省 | 南京禄口国际机场 | NKG | ZSNJ | 4E | Nanjing Airport |
| | | 常州奔牛机场 | CZX | ZSCG | 4D | Changzhou Airport |
| | | 连云港白塔埠机场 | LYG | ZSLG | 4D | Lianyungang Airport |
| | | 徐州观音机场 | XUZ | ZSXZ | 4D | Xuzhou Airport |
| | | 无锡硕放机场 | WUX | ZSWX | 4D | Wuxi Airport |
| | | 盐城南洋机场 | YNZ | ZSYN | 4C | Yancheng Airport |
| | | 南通兴东机场 | NTG | ZSNT | 4C | Nantong Nan |
| | | 淮安涟水机场 | HIA | ZSSH | 4C | Lianshui Airport |
| | | 扬州泰州机场 | YTY | ZSYA | 4C | Yangzhou Airport |
| | 福建省 | 福州长乐国际机场 | FOC | ZSFZ | 4E | Fuzhou Airport |
| | | 厦门高崎国际机场 | XMN | ZSAM | 4E | Xiamen Airport |
| | | 泉州晋江机场 | JJN | ZSQZ | 4D | Jinjiang Airport |
| | | 武夷山机场 | WUS | ZSWY | 4C | Wuyishan Airport |
| | | 连城冠豸山机场 | LCX | ZSLD | 4C | Longyan Airport |
| 新疆 | 新疆维吾尔自治区 | 乌鲁木齐地窝堡国际机场 | URC | ZWWW | 4E | Urumqi Airport |
| | | 喀什机场 | KHG | ZWSH | 4E | Kashi Airport |
| | | 和田机场 | HTN | ZWTN | 4D | Hetian Airport |
| | | 库尔勒机场 | KRL | ZWKL | 4D | Korla Airport |
| | | 克拉玛依机场 | KRY | ZWKM | 4D | Karamay Airport |
| | | 吐鲁番交河机场 | TLQ | ZWTP | 4D | Tulufan Airport |
| | | 伊宁机场 | YIN | ZWYN | 4C | Yining Airport |
| | | 阿勒泰机场 | AAT | ZWAT | 4C | Aletai Airport |
| | | 塔城机场 | TCG | ZWTC | 4C | Tacheng Airport |
| | | 阿克苏机场 | AKU | ZWAK | 4C | Aksu Airpor |
| | | 库车机场 | KCA | ZWKC | 4C | Kuqa Airport |
| | | 且末机场 | IQM | ZWCM | 3C | Qiemo Airport |
| | | 那拉提机场 | NLT | ZWNL | 4C | Nalati Airport |
| | | 布尔津喀纳斯机场 | KJI | ZWKN | 4C | Kanas Airport |
| | | 哈密机场 | HMI | ZWHM | 4C | Hami Airport |
| | | 博乐阿拉山口机场 | BPL | ZWBL | 4C | Bole Airport |

| 地区 | 直辖市、省、自治区 | 机场中文名称 | 机场三字代码（IATA） | 机场四字代码（ICAO） | 机场等级数 | 机场英文名称简称 |
|---|---|---|---|---|---|---|
| 中南 | 湖南省 | 长沙黄花国际机场 | CSX | ZGHA | 4E | Changsha Airport |
| | | 张家界荷花机场 | DYG | ZGDY | 4D | Hehua Airport |
| | | 常德桃花源机场 | CGD | ZGCD | 4C | Changde Airport |
| | | 永州零陵机场 | LLF | ZGLG | 4C | Yongzhou Airport |
| | | 怀化芷江机场 | HJJ | ZGCJ | 4C | Zhijiang Airport |
| | 湖北省 | 武汉天河国际机场 | WUH | ZHHH | 4E | Wuhan Airport |
| | | 宜昌三峡机场 | YIH | ZHYC | 4D | Yichang Airport |
| | | 恩施许家坪机场 | ENH | ZHES | 4C | Enshi Airport |
| | | 襄樊刘集机场 | XFN | ZHXF | 4C | Xiangfan Airport |
| | 广东省 | 广州白云国际机场 | CAN | ZGGG | 4F | Baiyun Airport |
| | | 深圳宝安国际机场 | SZX | ZGSZ | 4E | Baoan Airport |
| | | 珠海三灶机场 | ZUH | ZGSD | 4E | Sanzao Airport |
| | | 湛江机场 | ZHA | ZGZJ | 4C | Zhanjiang Airport |
| | | 梅县机场 | MXZ | ZGMX | 3C | Meixian Airport |
| | | 佛山沙堤机场 | FUO | ZGFS | 4C | Foshan Airport |
| | | 揭阳潮汕机场 | SWA | ZGOW | 4D | Jieyang Chaoshan Airport |
| | 海南省 | 海口美兰国际机场 | HAK | ZJHK | 4E | Haikou Airport |
| | | 三亚凤凰国际机场 | SYX | ZJSY | 4E | Sanya Airport |
| | 河南省 | 郑州新郑国际机场 | CGO | ZHCC | 4E | Xinzheng Airport |
| | | 洛阳北郊机场 | LYA | ZHLY | 4D | Luoyang Airport |
| | | 南阳姜营机场 | NNY | ZHNY | 4D | Nanyang Airport |
| | 广西壮族自治区 | 南宁吴圩国际机场 | NNG | ZGNN | 4E | Nanning Airport |
| | | 桂林两江国际机场 | KWL | ZGKL | 4E | Guilin Airport |
| | | 北海福成机场 | BHY | ZGBH | 4D | Beihai Airport |
| | | 柳州白莲机场 | LZH | ZGZH | 4D | Liuzhou Airport |
| | | 梧州长洲岛机场 | WUZ | ZGWZ | 3C | Wuzhou Airport |
| | | 百色机场 | AEB | ZGBS | 4C | Bose Airport |

| 地区 | 直辖市、省、自治区 | 机场中文名称 | 机场三字代码（IATA） | 机场四字代码（ICAO） | 机场等级数 | 机场英文名称简称 |
|---|---|---|---|---|---|---|
| 西北 | 陕西省 | 西安咸阳国际机场 | XIY | ZLXY | 4E | Xianyang Airport |
| | | 榆林榆阳机场 | UYN | ZLYL | 4C | Yulin Airport |
| | | 延安二十里堡机场 | ENY | ZLYA | 4C | Yan'an Airport |
| | | 汉中机场 | HZG | ZLHZ | 3C | Hanzhong Airport |
| | | 安康机场 | AKA | ZLAK | 3C | Ankang Airport |
| | 甘肃省 | 兰州中川机场 | LHW | ZLLL | 4D | Zhongchuan Airport |
| | | 嘉峪关机场 | JGN | ZLJQ | 4D | Jiayuguan Airport |
| | | 敦煌机场 | DNH | ZLDH | 4C | Dunhuang Airport |
| | | 庆阳机场 | IQN | ZLQY | 3C | Qingyang Airport |
| | | 天水麦积山机场 | THQ | ZLTS | 3C | Tianshui Airport |
| | | 金昌机场 | JIC | ZLJC | 4C | JinChang Airport |
| | | 张掖机场 | YZY | ZLZY | 4C | Zhangye Airport |
| | 青海省 | 西宁曹家堡机场 | XNN | ZLXN | 4D | Xining Airport |
| | | 格尔木机场 | GOQ | ZLGM | 4D | Golmud Airport |
| | | 玉树巴塘机场 | YUS | ZLYS | 4C | Yushu Airport |
| | 宁夏回族自治区 | 银川河东机场 | INC | ZLIC | 4D | Yinchuan Airport |
| | | 中卫香山机场 | ZHY | ZLZW | 4C | Zhongwei Airport |
| | | 固原六盘山机场 | GYU | ZLGY | 4C | Guyuan Airport |
| 西南 | 重庆市 | 重庆江北国际机场 | CKG | ZUCK | 4E | Jiangbei Airport |
| | | 万州五桥机场 | WXN | ZUWX | 4C | Wanzhou Airport |
| | | 黔江舟白机场 | JIQ | ZUQJ | 4C | Zhoubai Airport |
| | 四川省 | 成都双流国际机场 | CTU | ZUUU | 4E | Shuangliu Airport |
| | | 西昌青山机场 | XIC | ZUXC | 4D | Xichang Airport |
| | | 绵阳南郊机场 | MIG | ZUMY | 4D | Mianyang Airport |
| | | 攀枝花保安营机场 | PZI | ZUZH | 4C | Panzhihua Airport |
| | | 泸州蓝田机场 | LZO | ZULZ | 4C | Luzhou Airport |
| | | 宜宾菜坝机场 | YBP | ZUYB | 4C | Yibin Airport |

| 地区 | 直辖市、省、自治区 | 机场中文名称 | 机场三字代码（IATA） | 机场四字代码（ICAO） | 机场等级数 | 机场英文名称简称 |
|---|---|---|---|---|---|---|
| | | 九寨黄龙机场 | JZH | ZUJZ | 4C | Huanglong Airport |
| | | 广元盘龙机场 | GYS | ZUGU | 4C | Guangyuan Airport |
| | | 达州河市机场 | DAX | ZUDX | 4C | Dazhou Airport |
| | | 南充高坪机场 | NAO | ZUNC | 4C | Nanchong Airport |
| | | 甘孜康定机场 | KGT | ZUKD | 4C | Kangding Airport |
| | 贵州省 | 贵阳龙洞堡国际机场 | KWE | ZUGY | 4E | Guiyang Airport |
| | | 安顺黄果树机场 | AVA | ZUAS | 4C | Huang Guoshu Airport |
| | | 铜仁凤凰机场 | TEN | ZUTR | 4C | Tongren Airport |
| | | 兴义机场 | ACX | ZUYI | 4C | Xingyi Airport |
| | | 黎平机场 | HZH | ZUNP | 4C | Liping Airport |
| | | 黔南州荔波机场 | LLB | ZULB | 4C | Libo Airport |
| | 云南省 | 昆明长水国际机场 | KMG | ZPPP | 4E | Kunming Airport |
| | | 丽江三义机场 | LJG | ZPLJ | 4C | Lijiang Airport |
| | | 西双版纳嘎洒机场 | JHG | ZPJH | 4C | Jinghong Airport |
| | | 迪庆香格里拉机场 | DIG | ZPDQ | 4D | Diqing Airport |
| | | 普洱思茅机场 | SYM | ZPSM | 4C | Simao Airport |
| | | 德宏芒市机场 | LUM | ZPMS | 4C | Mangshi Airport |
| | | 大理机场 | DLU | ZPDL | 4C | Dali Airport |
| | | 保山云瑞机场 | BSD | ZPBS | 4C | Baoshan Airport |
| | | 临沧机场 | LNJ | ZPLC | 4C | Lincang Airport |
| | | 昭通机场 | ZAT | ZPZT | 4C | Zhaotong Airport |
| | | 文山普者黑机场 | WNH | ZPWS | 4C | Wenshan Airport |
| | | 腾冲驼峰机场 | TCZ | ZPTC | 4C | Tengchong Airport |
| | 西藏自治区 | 拉萨贡嘎机场 | LXA | ZULS | 4E | Lasa Airport |
| | | 林芝米林机场 | LZY | ZUNZ | 4D | Linzhi Airport |
| | | 昌都邦达机场 | BPX | ZUBD | 4D | Bangda Airport |
| | | 阿里昆莎机场 | NGQ | ZUAL | 4D | Ngari Gunsa Airport |
| | | 日喀则机场 | RKZ | ZURK | 4C | Xigaze Peace Airport |

| 地区 | 直辖市、省、自治区 | 机场中文名称 | 机场三字代码（IATA） | 机场四字代码（ICAO） | 机场等级数 | 机场英文名称简称 |
|---|---|---|---|---|---|---|
| 东北 | 黑龙江省 | 哈尔滨太平国际机场 | HRB | ZYHB | 4E | Harbin Airport |
| | | 牡丹江海浪机场 | MDG | ZYMD | 4C | Mudanjiang Airport |
| | | 佳木斯东郊机场 | JMU | ZYJM | 4C | Jiamusi Airport |
| | | 齐齐哈尔三家子机场 | NDG | ZYQQ | 4C | Qiqihaer Airport |
| | | 大庆萨尔图机场 | DQA | ZYDQ | 4C | Daqing Saertu Airport |
| | | 伊春林都机场 | LDS | ZYLD | 4C | Lindu Airport |
| | | 黑河机场 | HEK | ZYHE | 4C | Heihe Airport |
| | | 漠河古莲机场 | OHE | ZYMH | 4C | Mohe Airport |
| | | 鸡西兴凯湖机场 | JXA | ZYJX | 4C | Jixi Airport |
| | | 加格达奇机场 | JGD | ZYJD | 4C | Jiagedaqi Airport |
| | 吉林省 | 长春龙嘉国际机场 | CGQ | ZYCC | 4D | Changchun Airport |
| | | 白山长白山机场 | NBS | ZYBS | 4D | Changbaishan Airport |
| | | 延吉朝阳川机场 | YNJ | ZYYJ | 4C | Yanji Airport |
| | 辽宁省 | 沈阳桃仙国际机场 | SHE | ZYTX | 4E | Shenyang Airport |
| | | 大连周水子国际机场 | DLC | ZYTL | 4E | Dalian Airport |
| | | 锦州机场 | JNZ | ZYJZ | 4C | Jinzhou Airport |
| | | 丹东浪头机场 | DDG | ZYDD | 4C | Dandong Airport |
| | | 朝阳机场 | CHG | ZYCY | 4C | Chaoyang Airport |
| | | 鞍山腾鳌机场 | AOG | ZYAS | 4C | Anshan Airport |
| | | 长海（大长山岛）机场 | CNI | ZYCH | 1B | Changhai Airport |

注：截止到 2012 年 6 月 30 日，全国共有民用航空运输机场 182 个。

# 附录二 中国香港、澳门和台湾地区机场一览表

| 地区 | 机场中文名称 | 机场三字代码（IATA） | 机场四字代码（ICAO） | 机场英文名称简称 |
|---|---|---|---|---|
| 香港 | 香港国际机场 | HKG | VHHH | Hong Kong Airport |
| 澳门 | 澳门国际机场 | MFM | VMMC | Macau Airport |
| 台湾 | 桃园国际机场 | TPE | RCTP | Taoyuan Airport |
| | 高雄国际机场 | KHH | RCKH | Kaohsiung Airport |
| | 台中清泉岗国际机场 | RMQ | RCMQ | Ching Chung Kang Airport |
| | 台北松山机场 | TSA | RCSS | Taipei Songshan Airport |
| | 嘉义机场 | CYI | RCKU | Chiyi Airport |
| | 台东机场 | TTT | RCFN | Taitung Airport |
| | 马公机场 | MZG | RCQC | Magong or Makung Airport |
| | 台南机场 | TNN | RCNN | Tainan Airport |
| | 花莲机场 | HUN | RCYU | Hualien Airport |
| | 金门尚义机场 | KNH | RCBS | Kinmen Airport |

# 附录三　中国运输类航空公司一览表

| 地区 | 公司名称(英文) | 航徽 | 两字代码 | 公司网址 |
|---|---|---|---|---|
| 大陆地区 | 中国国际航空股份有限公司<br>(Air China) | | CA | http://www.airchina.com.cn/ |
| | 中国东方航空股份有限公司<br>(China Eastern) | | MU | http://www.ce-air.com/ |
| | 中国南方航空股份有限公司<br>(China Southern) | | CZ | http://www.csair.com/cn/ |
| | 海南航空股份有限公司<br>(Hainan Airlines) | | HU | http://www.hnair.com/ |
| | 深圳航空有限责任公司<br>(Shenzhen Airlines) | | ZH | http://www.shenzhenair.com/ |
| | 厦门航空有限公司<br>(Xiamen Airlines) | | MF | http://www.xiamenair.com.cn/ |
| | 山东航空股份有限公司<br>(Shandong Airlines) | | SC | http://www.shandongair.com.cn/ |
| | 鲲鹏航空有限公司<br>(原河南航空有限公司)<br>(Kunpeng Airlines) | | VD | http://www.kunpeng-air.com/ |
| | 昆明航空有限公司<br>(Kunming Airlines) | | KY | http://www.airkunming.com/ |
| | 重庆航空有限责任公司<br>(Chongqing Airlines) | | OQ | http://www.flycq.com/ |

| 地区 | 公司名称(英文) | 航徽 | 两字代码 | 公司网址 |
|---|---|---|---|---|
| 大陆地区 | 上海航空有限公司<br>(Shanghai Airlines) | | FM | http://www. shanghai-air. com/ |
| | 中国联合航空有限公司<br>(China United Airlines) | | KN | http://www. cu-air. com/ |
| | 幸福航空有限责任公司<br>(Joy Air) | | JR | http://www. joy-air. com/ |
| | 北京首都航空有限公司(原金鹿航空有限公司)<br>(Capital Airlines) | | JD | http://www. capitalairlines. com. cn |
| | 天津航空有限责任公司<br>(Tianjin Airlines) | | GS | http://www. tianjin-air. com/ |
| | 西部航空有限责任公司<br>(West Air) | | PN | http://www. chinawestair. com/ |
| | 云南祥鹏航空有限责任公司<br>(Lucky Air) | | 8L | http://www. luckyair. net/ |
| | 四川航空股份有限公司<br>(Sichuan Airlines) | | 3U | http://www. scal. com. cn/ |
| | 成都航空有限公司(原鹰联航空公司)<br>(Chengdu Airlines) | | EU | http://www. chengduair. cc/ |
| | 河北航空有限公司(原东北航空有限公司)<br>(Hebei Airlines) ) | | NS | http://www. hbhk. com. cn/ |
| | 春秋航空有限公司<br>(Spring Airlines) | | 9C | http://www. china-sss. com/ |

| 地区 | 公司名称(英文) | 航徽 | 两字代码 | 公司网址 |
|---|---|---|---|---|
| 大陆地区 | 上海吉祥航空有限公司<br>(Juneyao Airlines) | | HO | http://www. juneyaoair. com/ |
| | 奥凯航空有限公司<br>(Okay Airways) | | BK | http://www. okair. net/ |
| | 华夏航空有限公司<br>(China Exprss) | | G5 | http://www. chinaexpressair. com/ |
| | 西藏航空有限公司<br>(China Tibetan Airlines) | | TV | http://www. tibetairlines. com. cn/ |
| | 长城航空有限公司<br>(Great Wall Airlines) | | IJ | http://www. gwairlines. com/en/ |
| | 中国国际货运航空有限公司<br>(Air China Airlines) | | CA | http://mail. ca-cargo. com/ |
| | 中国货运航空有限公司<br>(China Cargo Airlines) | | CK | http://www. cc-air. com/ |
| | 中国邮政航空有限责任公司<br>(China Postal Airlines) | | 8Y | http://www. cnpostair. com/ |
| | 翡翠国际货运航空有限责任公司<br>(Jade Cargo International) | | JI | http://www. jadecargo. com/ |
| | 扬子江快运航空有限公司<br>(Yangtze River Express) | | Y8 | http://www. yzr. com. cn/ |
| | 深圳东海航空有限公司<br>(Shenzhen Donghai Airlines) | | J5 | http://www. donghaiair. com/ |

| 地区 | 公司名称(英文) | 航徽 | 两字代码 | 公司网址 |
|---|---|---|---|---|
| 大陆地区 | 银河国际货运航空有限公司<br>（Grandstar Cargo） | | GD | http://www.grandstarcargo.com.cn |
| | 友和道通航空有限公司<br>（Uni-Top Airlines） | | UW | http://www.uni-top.com.cn/ |
| | 顺丰航空有限公司<br>（Sf-Express） | | O3 | http://www.sf-express.com/ |
| | 上海国际货运航空有限公司<br>（Shanghai Airlines Cargo） | | F4 | http://www.shanghai-aircargo.com/ |
| 香港地区 | 国泰航空有限公司<br>（Cathay Pacific） | | CX | http://www.cathaypacific.com |
| | 港龙航空有限公司<br>（Dragonair） | | KA | http://www.dragonair.com/ |
| | 香港航空有限公司<br>（Hong Kong Airlines） | | HX | http://www.hongkongairlines.com/<br>schi/index.aspx |
| | 香港华民航空有限公司<br>（Air Hong Kong） | air Hongkong | LD | http://www.airhongkong.com.hk/<br>ahk/en/index.jsp |
| | 香港快运航空有限公司<br>（Hong Kong Express） | | UO | http://www.hongkongexpress.com |
| 澳门地区 | 澳门航空股份有限公司<br>（Air Macau） | | NX | http://www.airmacau.com.mo/ |

| 地区 | 公司名称(英文) | 航徽 | 两字代码 | 公司网址 |
|---|---|---|---|---|
| 台湾地区 | 中华航空股份有限公司 (China Airlines) | | CI | http://www. china-airlines. com/ |
| | 长荣航空股份有限公司 (EVA Air) | | BR | http://www. evaair. com/ |
| | 华信航空股份有限公司 (Mandarin Airlines) | | AE | http://www. mandarin-airlines. com/ |
| | 复兴航空股份有限公司 (Transasia Airways) | | GE | http://www. tna. com. tw/ |
| | 立荣航空股份有限公司 (UNI Air) | | B7 | http://www. uniair. com. tw/ |
| | 远东航空股份有限公司 (Far Eastern Air Transport) | | EF | http://www. fat. com. tw/ |

# 附录四　美国各州及其首府中文、英文名称一览表

| 州名简称 | 州名英文 | 州名中文 | 州首府英文 | 州首府中文 |
|---|---|---|---|---|
| AL | Alabama | 亚拉巴马州 | Montgomery | 蒙哥马利 |
| AK | Alaska | 阿拉斯加州 | Juneau | 朱诺 |
| AZ | Arizona | 亚利桑那州 | Phoenix | 菲尼克斯 |
| AR | Arkansas | 阿肯色州 | Little Rock | 小石城 |
| CA | California | 加利福尼亚州 | Sacramento | 萨克拉门托 |
| CO | Colorado | 科罗拉多州 | Denver | 丹佛 |
| CT | Connecticut | 康涅狄格州 | Hartford | 哈特福德 |
| DE | Delaware | 特拉华州 | Dover | 多佛 |
| FL | Florida | 佛罗里达州 | Tallahassee | 塔拉哈西 |
| GA | Georgia | 佐治亚州 | Atlanta | 亚特兰大 |
| HI | Hawaii | 夏威夷州 | Honolulu | 檀香山，火努鲁鲁 |
| ID | Idaho | 爱达荷州 | Boise | 博伊西 |
| IL | Illinois | 伊利诺伊州 | Springfield | 斯普林菲尔德 |
| IN | Indiana | 印第安纳州 | Indianapolis | 印第安纳波利斯 |
| IA | Iowa | 衣阿华州 | Des Moines | 得梅因 |
| KS | Kansas | 堪萨斯州 | Topeka | 托皮卡 |
| KY | Kentucky | 肯塔基州 | Frankfort | 法兰克福 |
| LA | Louisiana | 路易斯安那州 | Baton Rouge | 巴吞鲁日 |
| ME | Maine | 缅因州 | Augusta | 奥古斯塔 |
| MD | Maryland | 马里兰州 | Annapolis | 安纳波利斯 |
| MA | Massachusetts | 马萨诸塞州 | Boston | 波士顿 |
| MI | Michigan | 密歇根州 | Lansing | 兰辛 |
| MN | Minnesota | 明尼苏达州 | St. Paul | 圣保罗 |
| MS | Mississippi | 密西西比州 | Jackson | 杰克逊 |

| 州名简称 | 州名英文 | 州名中文 | 州首府英文 | 州首府中文 |
|---|---|---|---|---|
| MO | Missouri | 密苏里州 | Jefferson City | 杰斐逊城 |
| MT | Montana | 蒙大拿州 | Helena | 海伦娜 |
| NE | Nebraska | 内布拉斯加州 | Lincoln | 林肯 |
| NV | Nevada | 内华达州 | Carson City | 卡森城 |
| NH | New Hampshire | 新罕布什尔州 | Concord | 康科德 |
| NJ | New Jersey | 新泽西州 | Trenton | 特伦顿 |
| NM | New Mexico | 新墨西哥州 | Santa Fe | 圣菲 |
| NY | New York | 纽约州 | Albany | 奥尔巴尼 |
| NC | North Carolina | 北卡罗来纳州 | Raleigh | 罗利 |
| ND | North Dakota | 北达科他州 | Bismarck | 俾斯麦 |
| OH | Ohio | 俄亥俄州 | Columbus | 哥伦布 |
| OK | Oklahoma | 俄克拉何马州 | Oklahoma City | 俄克拉何马城 |
| OR | Oregon | 俄勒冈州 | Salem | 塞勒姆 |
| PA | Pennsylvania | 宾夕法尼亚州 | Harrisburg | 哈里斯堡 |
| RI | Rhode Island | 罗得岛州 | Providence | 普罗维登斯 |
| SC | South Carolina | 南卡罗来纳州 | Columbia | 哥伦比亚 |
| SD | South Dakota | 南达科他州 | Pierre | 皮尔 |
| TN | Tennessee | 田纳西州 | Nashville | 纳什维尔 |
| TX | Texas | 得克萨斯州 | Austin | 奥斯汀 |
| UT | Utah | 犹他州 | Salt Lake City | 盐湖城 |
| VT | Vermont | 佛蒙特州 | Montpelier | 蒙彼利埃 |
| VA | Virginia | 弗吉尼亚州 | Richmond | 里士满 |
| WA | Washington | 华盛顿州 | Olympia | 奥林匹亚 |
| WV | West Virginia | 西弗吉尼亚州 | Charleston | 查尔斯顿 |
| WI | Wisconsin | 威斯康星州 | Madison | 麦迪逊 |
| WY | Wyoming | 怀俄明州 | Cheyenne | 夏延 |

# 附录五　几种单位制之间的换算表

## 1. 长度单位

| | 英寸<br>(in.) | 英尺<br>(ft) | 码(yd) | 厘米<br>(cm) | 米(m) | 公里<br>(km) | 英里<br>(mile) | 国际海里<br>(n. m.) |
|---|---|---|---|---|---|---|---|---|
| 英寸<br>(in.) | 1 | 0.083333 | $2.77778 \times 10^{-2}$ | 2.54 | $2.54 \times 10^{-2}$ | $2.54 \times 10^{-5}$ | $1.57828 \times 10^{-5}$ | $1.37149 \times 10^{-5}$ |
| 英尺<br>(ft) | 12 | 1 | 0.33333 | 30.48 | 0.3048 | $3.048 \times 10^{-4}$ | $1.89394 \times 10^{-4}$ | $1.645788 \times 10^{-4}$ |
| 码<br>(yd) | 36 | 3 | 1 | 91.44 | 0.9144 | $9.144 \times 10^{-4}$ | $5.68182 \times 10^{-4}$ | $4.93737 \times 10^{-4}$ |
| 厘米<br>(cm) | $3.93701 \times 10^{-1}$ | $3.28084 \times 10^{-2}$ | $1.0936 \times 10^{-2}$ | 1 | 0.01 | $10^{-5}$ | $6.21371 \times 10^{-6}$ | $5.39957 \times 10^{-6}$ |
| 米<br>(m) | 39.3701 | 3.28084 | 1.0936 | 100 | 1 | $10^{-3}$ | $6.21371 \times 10^{-4}$ | $5.39957 \times 10^{-4}$ |
| 公里<br>(km) | 39370.08 | 3280.84 | 1093.6 | $10^5$ | 1000 | 1 | 0.621371 | 0.539957 |
| 英里<br>(mile) | 63360 | 5280 | 1760 | $1.609344 \times 10^5$ | 1609.344 | 1.609344 | 1 | 0.868976 |
| 国际<br>海里<br>(n. m.) | 72913.39 | 6076.115 | 2025.37 | $1.852 \times 10^5$ | 1852 | 1.852 | 1.150779 | 1 |

## 2. 力单位

| | 牛顿(N) | 达因(dyn) | 磅力(lbf) | 千克力(kgf) | 磅达(pdl) | 千磅(kip) |
|---|---|---|---|---|---|---|
| 牛顿(N) | 1 | $10^5$ | 0.2248 | 0.101972 | 7.23301 | $2.248 \times 10^{-4}$ |
| 达因(dyn) | $10^{-5}$ | 1 | $2.248 \times 10^{-6}$ | $1.01972 \times 10^{-6}$ | $7.23301 \times 10^{-5}$ | $2.248 \times 10^{-9}$ |
| 磅力(lbf) | 4.44822 | $4.44822 \times 10^5$ | 1 | 0.453592 | 32.174 | $10^{-3}$ |
| 千克力(kgf) | 9.80665 | $9.80665 \times 10^5$ | 2.20462 | 1 | 70.93164 | $2.20462 \times 10^{-3}$ |
| 磅达(pdl) | 0.138255 | 13825.5 | $3.1081 \times 10^{-2}$ | $1.40981 \times 10^{-2}$ | 1 | $3.1081 \times 10^{-5}$ |
| 千磅(kip) | 4448.22 | $4.44822 \times 10^8$ | 1000 | 453.592 | 32174 | 1 |

### 3. 质量单位

| | 吨(t) | 千克(kg) | 克(g) | 斯勒格(slug) | 美吨(US ton) | 磅(lbm) | 盎司(oz) |
|---|---|---|---|---|---|---|---|
| 吨(t) | 1 | 1000 | $10^6$ | 68.5 | 1.10231 | 2204.62 | 35274 |
| 千克(kg) | $10^{-3}$ | 1 | 1000 | $6.85 \times 10^{-2}$ | $1.10231 \times 10^{-3}$ | 2.20462 | 35.274 |
| 克(g) | $10^{-6}$ | $10^{-3}$ | 1 | $6.85 \times 10^{-5}$ | $1.10231 \times 10^{-6}$ | $2.20462 \times 10^{-3}$ | $3.5274 \times 10^{-2}$ |
| 斯勒格(slug) | $1.46 \times 10^{-2}$ | 14.6 | $1.46 \times 10^4$ | 1 | $1.6094 \times 10^{-2}$ | 32.3 | 514.94 |
| 美吨(US ton) | 0.907185 | 907.185 | $9.07185 \times 10^5$ | 62.136 | 1 | 2000 | 32000 |
| 磅(lbm) | $4.5392 \times 10^{-4}$ | 0.453592 | 453.592 | $3.1 \times 10^{-2}$ | $5 \times 10^{-4}$ | 1 | 16 |
| 盎司(oz) | $2.83495 \times 10^{-5}$ | $2.83495 \times 10^{-2}$ | 28.3495 | $1.94198 \times 10^{-3}$ | $3.125 \times 10^{-5}$ | $6.25 \times 10^{-2}$ | 1 |

### 4. 能量或功单位

| | 焦耳(J) | 尔格(erg) | 千克力·米(kgf·m) | 卡(cal) | 英热单位(Btu) | 英尺·磅力(ft·lbf) | 千瓦·时(kW·h) |
|---|---|---|---|---|---|---|---|
| 焦耳(J) | 1 | $1 \times 10^7$ | 0.101972 | 0.238846 | $9.47817 \times 10^{-4}$ | 0.737562 | $2.77778 \times 10^{-7}$ |
| 尔格(erg) | $1 \times 10^{-7}$ | 1 | $1.01972 \times 10^{-8}$ | $2.38846 \times 10^{-8}$ | $9.47817 \times 10^{-11}$ | $7.37562 \times 10^{-8}$ | $2.77778 \times 10^{-14}$ |
| 千克力·米(kgf·m) | 9.80665 | $9.80665 \times 10^7$ | 1 | 2.34228 | $9.29491 \times 10^{-3}$ | 7.23301 | $2.72407 \times 10^{-6}$ |
| 卡(cal) | 4.1868 | $4.1868 \times 10^7$ | 0.426936 | 1 | $3.96832 \times 10^{-3}$ | 3.08803 | $1.163 \times 10^{-6}$ |
| 英热单位(Btu) | 1055.06 | $1.05506 \times 10^{10}$ | 107.58575 | 251.996 | 1 | 778.169 | $2.93071 \times 10^{-4}$ |
| 英尺·磅力(ft·lbf) | 1.35582 | $1.35582 \times 10^7$ | 0.138255 | 0.323832 | $1.28507 \times 10^{-3}$ | 1 | $3.76616 \times 10^{-7}$ |
| 千瓦·时(kW·h) | $3.6 \times 10^6$ | $3.6 \times 10^{13}$ | $3.67098 \times 10^5$ | 859845 | 3412.14 | $2.65522 \times 10^6$ | 1 |

## 5. 功率单位

| | 瓦(W) | 千瓦(kW) | 卡/秒<br>(cal/s) | 千克力·<br>米/秒<br>(kgf·m/s) | 英制马力<br>(BHP) | 公制马力<br>(PS) | 英热单位/时<br>(Btu/h) | 英尺·<br>磅力/秒<br>(ft·lbf/s) |
|---|---|---|---|---|---|---|---|---|
| 瓦(W) | 1 | $10^{-3}$ | 0.238846 | 0.101972 | $1.34102$<br>$\times10^{-3}$ | $1.35962$<br>$\times10^{-3}$ | 3.41214 | 0.737562 |
| 千瓦(kW) | $10^3$ | 1 | 238.846 | 101.97162 | 1.34102 | 1.35962 | 3412.14 | 737.562 |
| 卡/秒<br>(cal/s) | 4.1868 | $4.1868$<br>$\times10^{-3}$ | 1 | 0.426935 | $5.61459$<br>$\times10^{-3}$ | $5.69246$<br>$\times10^{-3}$ | 14.286 | 3.08803 |
| 千克力·<br>米/秒<br>(kgf·m/s) | 9.80665 | $9.80665$<br>$\times10^{-3}$ | 2.34228 | 1 | 0.0131509 | 0.0133333 | 33.4617 | 7.23301 |
| 英制马力<br>(BHP) | 745.7 | 0.7457 | 178.107 | 76.0402 | 1 | 1.01387 | 2544.43 | 550 |
| 公制马力<br>(PS) | 735.499 | 0.735499 | 175.671 | 75 | 0.98632 | 1 | 2509.63 | 542.476 |
| 英热单位/时<br>(Btu/h) | 0.293071 | $2.93071$<br>$\times10^{-4}$ | 0.0699988 | $2.98849$<br>$\times10^{-2}$ | $3.93015$<br>$\times10^{-4}$ | $3.98466$<br>$\times10^{-4}$ | 1 | 0.216158 |
| 英尺·磅<br>力/秒<br>(ft·lbf/s) | 1.35582 | $1.35582$<br>$\times10^{-3}$ | 0.323832 | 0.138255 | $1.81818$<br>$\times10^{-3}$ | $1.8434$<br>$\times10^{-3}$ | 4.62624 | 1 |

## 6. 动力粘度单位

| | 帕·秒<br>(Pa·s) | 泊(P) | 厘泊(cP) | 磅力·秒/<br>平方英尺<br>(lbf·s/ft²) | 雷恩(reyn) |
|---|---|---|---|---|---|
| 帕·秒(Pa·s) | 1 | 10 | 1000 | $2.08854$<br>$\times10^{-2}$ | $1.45038$<br>$\times10^{-4}$ |
| 泊(P) | 0.1 | 1 | 100 | $2.08854$<br>$\times10^{-3}$ | $1.45038$<br>$\times10^{-5}$ |
| 厘泊(cP) | $10^{-3}$ | $10^{-2}$ | 1 | $2.08854$<br>$\times10^{-5}$ | $1.45038$<br>$\times10^{-7}$ |
| 磅力·秒/平方英尺<br>(lbf·s/ft²) | 47.8803 | 478.803 | $4.78803$<br>$\times10^4$ | 1 | $6.94445$<br>$\times10^{-3}$ |
| 雷恩(reyn) | $6.89476$<br>$\times10^3$ | $6.89476$<br>$\times10^4$ | $6.89476$<br>$\times10^6$ | 144 | 1 |

注：1 雷恩（reyn）=1 磅力·秒/平方英寸（lbf·s/in². )

## 7. 运动粘度单位

| | 厘斯(cSt) | 斯(St) | 平方米/秒<br>(m²/s) | 平方英寸/秒<br>(in². /s) | 平方英尺/秒<br>(ft²/s) |
|---|---|---|---|---|---|
| 厘斯(cSt) | 1 | 0.01 | $10^{-6}$ | $1.55 \times 10^{-3}$ | $1.07639 \times 10^{-5}$ |
| 斯(St) | 100 | 1 | $10^{-4}$ | 0.155 | $1.07639 \times 10^{-3}$ |
| 平方米/秒<br>(m²/s) | $10^6$ | $10^4$ | 1 | $1.55 \times 10^3$ | 10.7639 |
| 平方英寸/秒<br>(in². /s) | 645.16 | 6.4516 | $6.4516 \times 10^{-4}$ | 1 | $6.9444 \times 10^{-3}$ |
| 平方英尺/秒<br>(ft²/s) | $9.2903 \times 10^4$ | 929.03 | $9.2903 \times 10^{-2}$ | 144 | 1 |

## 8. 压力(强)单位

| | 巴(bar) | 英寸汞柱<br>(in. Hg) | 英寸水柱<br>(in. H₂O) | 千克力/<br>平方厘米<br>(kgf/cm²) | 毫米汞柱<br>(mm Hg) | 毫米水柱<br>(mm H₂O) | 磅力/<br>平方英寸<br>(PSI) | 千帕<br>(kPa) |
|---|---|---|---|---|---|---|---|---|
| 巴(bar) | 1 | 29.53 | 401.463 | 1.01972 | 750.062 | 10197.2 | 14.5038 | 100 |
| 英寸汞柱<br>(in. Hg) | 0.0338639 | 1 | 13.5951 | 0.0345316 | 25.4 | 345.316 | 0.491154 | 3.38639 |
| 英寸水柱<br>(in. H₂O) | 0.002491 | 0.073556 | 1 | 0.00254 | 1.86832 | 25.4 | 0.036127 | 0.249089 |
| 千克力/<br>平方厘米<br>(kgf/cm²) | 0.980665 | 28.959 | 393.701 | 1 | 735.559 | 10000 | 14.2233 | 98.0665 |
| 毫米汞柱<br>(mm Hg) | 0.0013332 | 0.03937 | 0.53524 | 0.0013595 | 1 | 13.5951 | 0.0193368 | 0.133322 |
| 毫米水柱<br>(mm H₂O) | 0.0000981 | 0.002896 | 0.0393701 | 0.0001 | 0.073556 | 1 | 0.0014223 | 0.009807 |
| 磅力/<br>平方英寸<br>(PSI) | 0.0689476 | 2.03602 | 27.68 | 0.070307 | 51.7149 | 703.07 | 1 | 6.89476 |
| 千帕<br>(kPa) | 0.01 | 0.2953 | 4.01463 | 0.0101972 | 7.50062 | 101.972 | 0.145038 | 1 |

## 9. 面积单位

| | 平方厘米<br>($cm^2$) | 平方英寸<br>($in^2.$) | 平方米<br>($m^2$) | 平方英尺<br>($ft^2$) | 平方码<br>($yd.^2$) | 平方英里<br>($mi.^2$) | 平方公里<br>($km^2$) |
|---|---|---|---|---|---|---|---|
| 平方厘米<br>($cm^2$) | 1 | 0.15500 | $1\times10^{-4}$ | $1.07639\times10^{-3}$ | $1.1960\times10^{-4}$ | $3.86102\times10^{-11}$ | $1\times10^{-10}$ |
| 平方英寸<br>($in^2.$) | 6.4516 | 1 | $6.4516\times10^{-4}$ | $6.94444\times10^{-3}$ | $7.71605\times10^{-4}$ | $2.49098\times10^{-10}$ | $6.4516\times10^{-10}$ |
| 平方米<br>($m^2$) | $1\times10^{4}$ | 1550 | 1 | 10.76391 | 1.19599 | $3.86102\times10^{-7}$ | $1\times10^{-6}$ |
| 平方英尺<br>($ft^2$) | 929.0304 | 144 | $9.29030\times10^{-2}$ | 1 | 0.11111 | $3.58701\times10^{-8}$ | $9.2903\times10^{-8}$ |
| 平方码<br>($yd^2$) | 8361.2736 | 1296 | 0.83613 | 9 | 1 | $3.22831\times10^{-7}$ | $8.36128\times10^{-7}$ |
| 平方英里<br>($mi.^2$) | $2.58999\times10^{10}$ | $4.01449\times10^{9}$ | $2.58999\times10^{6}$ | $2.78784\times10^{7}$ | $3.0976\times10^{6}$ | 1 | 2.58999 |
| 平方公里<br>($km^2$) | $1\times10^{10}$ | $1.55\times10^{9}$ | $1\times10^{6}$ | $1.07639\times10^{7}$ | $1.195990\times10^{6}$ | 0.38610 | 1 |

## 10. 容（体）积单位

| | 升(L) | 立方英尺 (ft³) | 立方米 (m³) | 立方英寸 (in³.) | 美加仑 (US gal) | 英加仑 (UK gal) | 桶(石油) Barrel(oil) | 美蒲式耳 (US bu.) | 英蒲式耳 (UK bu.) | 美液司 oz [US, fluid] | 英液司 oz [UK, fluid] | 立方码 (yd³) |
|---|---|---|---|---|---|---|---|---|---|---|---|---|
| 升(L) | 1 | $3.531467 \times 10^{-2}$ | $10^{-3}$ | 61.02374 | 0.26417 | 0.21997 | $6.28981 \times 10^{-3}$ | $2.837759 \times 10^{-2}$ | $2.749616 \times 10^{-2}$ | 33.81402 | 35.19508 | $1.30795 \times 10^{-3}$ |
| 立方英尺 (ft³) | 28.31685 | 1 | $2.83168 \times 10^{-2}$ | 1728 | 7.48052 | 6.22884 | 0.178107 | 0.80356 | 0.7786 | 957.50649 | 996.61367 | 0.037037 |
| 立方米 (m³) | 1000 | 35.31467 | 1 | 61023.74409 | 264.17205 | 219.96925 | 6.28981 | 28.37759 | 27.49616 | 33814.0227 | 35195.0797 | 1.30795 |
| 立方英寸 (in³.) | $1.63871 \times 10^{-2}$ | $5.78704 \times 10^{-4}$ | $1.63871 \times 10^{-5}$ | 1 | $4.329 \times 10^{-3}$ | $3.60465 \times 10^{-3}$ | $1.03071 \times 10^{-4}$ | $4.65025 \times 10^{-4}$ | $4.50581 \times 10^{-4}$ | 0.55411 | 0.57674 | $2.14335 \times 10^{-5}$ |
| 美加仑 (US gal) | 3.78541 | 0.13368 | $3.78541 \times 10^{-3}$ | 231 | 1 | 0.83267 | $2.38095 \times 10^{-2}$ | 0.10742 | 0.10408 | 128 | 133.22787 | $4.95113 \times 10^{-3}$ |
| 英加仑 (UK gal) | 4.54609 | 0.16054 | $4.54609 \times 10^{-3}$ | 277.41943 | 1.20095 | 1 | $2.8594 \times 10^{-2}$ | 0.129007 | 0.125 | 153.72159 | 160 | $5.946 \times 10^{-3}$ |
| 桶(油) Barrel(oil) | 158.98729 | 5.61458 | 0.15899 | 9702 | 42 | 34.97232 | 1 | 4.51168 | 4.37154 | 5376 | 5595.57052 | $2.07948 \times 10^{-1}$ |
| 美蒲式耳 (US bu.) | 35.23907 | 1.24446 | $3.5239 \times 10^{-2}$ | 2150.414 | 9.30918 | 7.75151 | 0.22165 | 1 | 0.96894 | 1191.57472 | 1240.24188 | $4.6091 \times 10^{-2}$ |
| 英蒲式耳 (UK bu.) | 36.36872 | 1.28435 | $3.63687 \times 10^{-2}$ | 2219.35546 | 9.6076 | 8 | 0.22875 | 1.03206 | 1 | 1229.77272 | 1280 | $4.75685 \times 10^{-2}$ |
| 美液司 oz [US, fluid] | 0.0295735 | 0.0010438 | $2.95735 \times 10^{-5}$ | 1.80469 | $7.8125 \times 10^{-3}$ | $6.50527 \times 10^{-3}$ | $1.86012 \times 10^{-4}$ | $8.39225 \times 10^{-4}$ | $8.13158 \times 10^{-4}$ | 1 | 1.04084 | $3.86807 \times 10^{-5}$ |
| 英液司 oz [UK, fluid] | 0.028413 | $1.00339 \times 10^{-3}$ | $2.84131 \times 10^{-5}$ | 1.73387 | $7.50594 \times 10^{-3}$ | $6.25 \times 10^{-3}$ | $1.78713 \times 10^{-4}$ | $8.06294 \times 10^{-4}$ | $7.8125 \times 10^{-4}$ | 0.96076 | 1 | $3.71629 \times 10^{-5}$ |
| 立方码 (yd³) | 764.55485 | 27 | 0.764555 | 46656 | 201.97403 | 168.17856 | 4.80891 | 21.69623 | 21.02232 | 25852.67532 | 26908.56918 | 1 |

## 11. 比热单位

| | 焦耳/千克·开<br>(J/kg·K) | 英热单位/<br>磅·兰氏温度<br>(Btu/lbm·°R) | 英热单位/斯勒<br>格·兰氏温度<br>(Btu/slug·°R) | 英尺·磅力/斯勒<br>格·兰氏温度<br>(ft·lbf/slug·°R) |
|---|---|---|---|---|
| 焦耳/千克·开<br>(J/kg·K) | 1 | 2.388<br>×10⁻⁴ | 7.68261<br>×10⁻³ | 5.979 |
| 英热单位/<br>磅·兰氏温度<br>(Btu/lbm·°R) | 4.1879<br>×10³ | 1 | 32.174 | 2.5038<br>×10⁴ |
| 英热单位/<br>斯勒格·兰氏温度<br>(Btu/slug·°R) | 130.164 | 3.1081<br>×10⁻² | 1 | 778.197 |
| 英尺·磅力/<br>斯勒格·兰氏温度<br>(ft·lbf/slug·°R) | 0.16725 | 0.3994<br>×10⁻⁴ | 1.28502<br>×10⁻³ | 1 |

## 12. 比焓单位

| | 焦耳/千克<br>(J/kg) | 英热单位/磅<br>(Btu/lbm) | 英尺·磅力/斯勒格<br>(ft·lbf/slug) |
|---|---|---|---|
| 焦耳/千克(J/kg) | 1 | 4.3069×10⁻⁴ | 1.0781 |
| 英热单位/磅(Btu/lbm) | 2.3218×10³ | 1 | 2.5032×10³ |
| 英尺·磅力/斯勒格(ft·lbf/slug) | 0.92756 | 3.9949×10⁻⁴ | 1 |

注:1 牛顿·米/千克(N·m/kg) = 1 焦耳/千克(J/kg) = 1 平方米/秒 (m²/s)

## 13. 热通量单位

| | 瓦/平方厘米<br>(W/cm²) | 瓦/平方英寸<br>(W/in.²) | 卡/秒·<br>平方厘米<br>(cal/s·cm²) | 英热单位/<br>秒·平方英尺<br>(Btu/s·ft²) | 英热单位/<br>时·平方英尺<br>(Btu/h·ft²) |
|---|---|---|---|---|---|
| 瓦/平方厘米<br>(W/cm²) | 1 | 6.4516 | 0.238846 | 0.880551 | 3169.98331 |
| 瓦/平方英寸<br>(W/in.²) | 0.1550 | 1 | 0.037021 | 0.136486 | 491.348395 |
| 卡/秒·平方厘米<br>(cal/s·cm²) | 4.1868 | 27.011696 | 1 | 3.686691 | 13272.086107 |
| 英热单位/<br>秒·平方英尺<br>(Btu/s·ft²) | 1.135653 | 7.326777 | 0.271246 | 1 | 3600 |
| 英热单位/<br>时·平方英尺<br>(Btu/h·ft²) | 3.1546<br>×10⁻⁴ | 2.03522<br>×10⁻³ | 7.53461<br>×10⁻⁵ | 2.77778<br>×10⁻⁴ | 1 |

## 14. 速度单位

| | 米/秒<br>(m/s) | 公里/小时<br>(km/h) | 英尺/秒<br>(ft/s) | 英里/小时<br>(mile/h) | 国际海里/小时<br>(kn) |
|---|---|---|---|---|---|
| 米/秒(m/s) | 1 | 3.6 | 3.28084 | 2.2369 | 1.94384 |
| 公里/小时(km/h) | 0.27778 | 1 | 0.9113 | 0.621371 | 0.539957 |
| 英尺/秒(ft/s) | 0.3048 | 1.0973 | 1 | 0.681818 | 0.59248 |
| 英里/小时(mile/h) | 0.44704 | 1.6093 | 1.46667 | 1 | 0.86898 |
| 国际海里/小时(kn) | 0.51444 | 1.852 | 1.68781 | 1.15078 | 1 |

## 15. 角速度单位

| | 度/秒(deg/s) | 转/分(rpm) | 转/秒(rev/s) | 弧度/秒(rad/s) |
|---|---|---|---|---|
| 度/秒(deg/s) | 1 | 0.166667 | 0.00277778 | 0.0174533 |
| 转/分(rpm) | 6 | 1 | 0.0166667 | 0.104720 |
| 转/秒(rev/s) | 360 | 60 | 1 | 6.28319 |
| 弧度/秒(rad/s) | 57.2958 | 9.54930 | 0.159155 | 1 |

## 16. 流量单位

| | 美加仑/分<br>(US gal/min) | 英加仑/分<br>(UK gal/min) | 立方米/秒<br>(m³/s) | 立方英尺/秒<br>(ft³/s) |
|---|---|---|---|---|
| 美加仑/分<br>(US gal/min) | 1 | 0.832674 | $6.30902 \times 10^{-5}$ | $2.228 \times 10^{-3}$ |
| 英加仑/分<br>(UK gal/min) | 1.20095 | 1 | $7.577 \times 10^{-5}$ | $2.676 \times 10^{-3}$ |
| 立方米/秒<br>(m³/s) | 15850.4 | 13197.8 | 1 | 35.3147 |
| 立方英尺/秒<br>(ft³/s) | 448.83 | 373.7 | $2.83168 \times 10^{-2}$ | 1 |

## 17. 温度单位

| | 摄氏温度(℃) | 开氏温度(K) | 华氏温度(°F) | 兰氏温度(°R) |
|---|---|---|---|---|
| 摄氏温度(℃) | 1 | $t℃ + 273.15$ | $1.8 \times t℃ + 32$ | $1.8 \times t℃ + 491.67$ |
| 开氏温度(K) | $tK - 273.15$ | 1 | $1.8 \times tK - 459.67$ | $1.8 \times tK$ |
| 华氏温度(°F) | $(t°F - 32)/1.8$ | $(t°F + 459.67)/1.8$ | 1 | $t°F + 459.67$ |
| 兰氏温度(°R) | $t°R/1.8 - 273.15$ | $t°R/1.8$ | $t°R - 459.67$ | 1 |

## 18. 密度单位

| | 千克/立方米<br>(kg/m³) | 斯勒格/立方英尺<br>(slug/ft³) | 磅/立方英尺<br>(lbm/ft³) |
|---|---|---|---|
| 千克/立方米<br>(kg/m³) | 1 | $1.94032 \times 10^{-3}$ | $6.24278 \times 10^{-2}$ |
| 斯勒格/立方英尺<br>(slug/ft³) | 515.379 | 1 | 32.174 |
| 磅/立方英尺<br>(lbm/ft³) | 16.0185 | $3.1056 \times 10^{-2}$ | 1 |

## 19. 扭矩单位

| | 磅力·英尺<br>(lbf·ft) | 磅力·英寸<br>(lbf·in.) | 千克力·米<br>(kgf·m) | 牛顿·米<br>(N·m) |
|---|---|---|---|---|
| 磅力·英尺<br>(lbf·ft) | 1 | 12 | 0.138255 | 1.35582 |
| 磅力·英寸<br>(lbf·in.) | 0.083333 | 1 | 0.0115212 | 0.112985 |
| 千克力·米<br>(kgf·m) | 7.23301 | 86.796514 | 1 | 9.80665 |
| 牛顿·米<br>(N·m) | 0.737562 | 8.85075 | 0.10197 | 1 |

## 20. 燃油消耗率单位

| | 英里/英加仑<br>(Mile/ UK gal) | 英里/美加仑<br>(Mile/ US gal) | 公里/升<br>(km/litre) |
|---|---|---|---|
| 英里/英加仑<br>(Mile/ UK gal) | 1 | 0.83267 | 0.354006 |
| 英里/美加仑<br>(Mile/ US gal) | 1.20096 | 1 | 0.42514 |
| 公里/升<br>(km/litre) | 2.82481 | 2.35216 | 1 |

# 附录六　世界主要城市与北京时差表

[说明：+ 表示早于北京时间；- 表示迟于北京时间]

Time Differences of World's Major Cities with Reference to Beijing Standard Time

| 城市<br>City | 时差<br>Time<br>Difference | 城市<br>City | 时差<br>Time<br>Difference |
|---|---|---|---|
| ASIA 亚洲 | | Nikolayevsk-on-Amure 尼古拉耶夫斯克（庙街）[俄罗斯] | +1 |
| Abadan 阿巴丹 [伊朗] | -4:30 | Omsk 鄂木斯克 [俄罗斯] | -3 |
| Abu Dhabi 阿布扎比 [阿联酋] | -4 | Osaka 大阪 [日本] | +1 |
| Aden 亚丁 [也门] | -5 | Phnom Penh 金边 [柬埔寨] | -1 |
| Amman 安曼 [约旦] | -6 | Pyongyang 平壤 [朝鲜] | +1 |
| Ankara 安卡拉 [土耳其] | -6 | Rangoon 仰光 [缅甸] | -1:30 |
| Baghdad 巴格达 [伊拉克] | -5 | Rawalpindi 拉瓦尔品第 [巴基斯坦] | -3 |
| Baku 巴库 [阿塞拜疆] | -6 | Seoul 首尔 [韩国] | +1 |
| Bandar Seri Begawan 斯里巴巴加湾港 [文莱] | 0 | Shanghai 上海 [中国] | 0 |
| Bangkok 曼谷 [泰国] | -1 | Singapore 新加坡 [新加坡] | 0 |
| Beirut 贝鲁特 [黎巴嫩] | -6 | Taibei 台北 [中国] | 0 |
| Bombay 孟买 [印度] | -2:30 | Tehran 德黑兰 [伊朗] | -4:30 |
| Calcutta 加尔各答 [印度] | -2:30 | Tokyo 东京 [日本] | +1 |
| Colombo 科伦坡 [斯里兰卡] | -2:30 | Ulan Bator 乌兰巴托 [蒙古] | 0 |
| Damascus 大马士革 [叙利亚] | -6 | Vladivostok 符拉迪沃斯托克（海参崴）[俄罗斯] | +2 |
| Dhaka 达卡 [孟加拉] | -2 | Yokohama 横滨 [日本] | +1 |
| Djakarta 雅加达 [印度尼西亚] | -1 | EUROPE 欧洲 | |
| Guangzhou 广州 [中国] | 0 | Aberdeen 阿伯丁 [英国] | -8 |
| Hanoi 河内 [越南] | -1 | Amsterdam 阿姆斯特丹 [荷兰] | -7 |
| Hong Kong 香港 [中国] | 0 | Antwerp 安特卫普 [比利时] | -7 |
| Irkutsk 伊尔库茨克 [俄罗斯] | 0 | Athens 雅典 [希腊] | -6 |
| Islamabad 伊斯兰堡 [巴基斯坦] | -3 | Belfast 贝尔法斯特 [英国] | -8 |
| Jerusalem 耶路撒冷 [巴勒斯坦] | -6 | Belgrade 贝尔格莱德 [南斯拉夫] | -7 |
| Karachi 卡拉奇 [巴基斯坦] | -3 | Berlin 柏林 [德国] | -7 |
| Katmandu 加德满都 [尼泊尔] | -2:30 | Birmingham 伯明翰 [英国] | -8 |
| Kuala Lumpur 吉隆坡 [马来西亚] | 0 | Bonn 波恩 [德国] | -7 |
| Kuwait 科威特 [科威特] | -5 | Brussels 布鲁塞尔 [比利时] | -7 |
| Kyoto 京都 [日本] | +1 | Bucharest 布加勒斯特 [罗马尼亚] | -6 |
| Macao 澳门 [中国] | 0 | Budapest 布达佩斯 [匈牙利] | -7 |
| Manila 马尼拉 [菲律宾] | 0 | Constantsa 康斯坦萨 [罗马尼亚] | -6 |
| Mecca 麦加 [沙特] | -5 | Copenhagen 哥本哈根 [丹麦] | -7 |
| Nagasaki 长崎 [日本] | +1 | | |
| New Delhi 新德里 [印度] | -2:30 | | |

| 城市<br>City | 时差<br>Time<br>Difference | 城市<br>City | 时差<br>Time<br>Difference |
|---|---|---|---|
| EUROPE 欧洲 | | AMERICA 美洲 | |
| Dublin 都柏林〔爱尔兰〕 | −8 | Anchorage 安克雷奇〔美国〕 | −17 |
| Edinburgh 爱丁堡〔英国〕 | −8 | Asuncion 亚松森〔巴拉圭〕 | −12 |
| Frankfurt 法兰克福〔德国〕 | −7 | Atlanta 亚特兰大〔美国〕 | −13 |
| Gdansk 格但斯克〔波兰〕 | −7 | Baltimore 巴尔的摩〔美国〕 | −13 |
| Geneva 日内瓦〔瑞士〕 | −7 | Belem 贝伦〔巴西〕 | −11 |
| Genoa 热那亚〔意大利〕 | −7 | Bogota 波哥大〔哥伦比亚〕 | −13 |
| Gibraltar 直布罗陀〔英国殖民地〕 | −7 | Boston 波士顿〔美国〕 | −13 |
| Glasgow 格拉斯哥〔英国〕 | −8 | Brasilia 巴西利亚〔巴西〕 | −11 |
| Hamburg 汉堡〔德国〕 | −7 | Buenos Aires 布宜诺斯艾利斯 | |
| Helsinki 赫尔辛基〔芬兰〕 | −6 | 〔阿根廷〕 | −11 |
| Istanbul 伊斯坦布尔〔土耳其〕 | −6 | Caracas 加拉加斯〔委内瑞拉〕 | −12 |
| Kiev 基辅〔乌克兰〕 | −5 | Chicago 芝加哥〔美国〕 | −14 |
| Leeds 利兹〔英国〕 | −8 | Churchill 丘吉尔港〔加拿大〕 | −14 |
| Leipzig 莱比锡〔德国〕 | −7 | Colon 科隆〔巴拿马〕 | −13 |
| Lisbon 里斯本〔葡萄牙〕 | −8 | Dallas 达拉斯〔美国〕 | −14 |
| Liverpool 利物浦〔英国〕 | −8 | Denver 丹佛〔美国〕 | −15 |
| London 伦敦〔英国〕 | −8 | Detroit 底特律〔美国〕 | −13 |
| Madrid 马德里〔西班牙〕 | −7 | Edmonton 埃德蒙顿〔加拿大〕 | −15 |
| Manchester 曼彻斯特〔英国〕 | −8 | Havana 哈瓦那〔古巴〕 | −13 |
| Marseilles 马塞〔法国〕 | −7 | Houston 休斯敦〔美国〕 | −14 |
| Milan 米兰〔意大利〕 | −7 | Kingston 金斯敦〔牙买加〕 | −13 |
| Moscow 莫斯科〔俄罗斯〕 | −5 | La Paz 拉巴斯〔玻利维亚〕 | −12 |
| Munich 慕尼黑〔德国〕 | −7 | Las Vegas 拉斯韦加斯〔美国〕 | −16 |
| Murmansk 摩尔曼斯克〔俄罗斯〕 | −5 | Lima 利马〔秘鲁〕 | −13 |
| Odessa 敖德萨〔乌克兰〕 | −5 | Los Angeles 洛杉矶〔美国〕 | −16 |
| Oslo 奥斯陆〔挪威〕 | −7 | Mexico City 墨西哥城〔墨西哥〕 | −14 |
| Paris 巴黎〔法国〕 | −7 | Miami 迈阿密〔美国〕 | −13 |
| Prague 布拉格〔捷克〕 | −7 | Montevideo 蒙得维的亚〔乌拉圭〕 | −11 |
| Reykjavik 雷克雅末克〔冰岛〕 | −8 | Montreal 蒙特利尔〔加拿大〕 | −13 |
| Rome 罗马〔意大利〕 | −7 | New Orleans 新奥尔良〔美国〕 | −14 |
| Rotterdam 鹿特丹〔荷兰〕 | −7 | New York 纽约〔美国〕 | −13 |
| St. Peterburg 圣彼得堡〔俄罗斯〕 | −5 | Ottawa 渥太华〔加拿大〕 | −13 |
| Sofia 索非亚〔保加利亚〕 | −6 | Panama 巴拿马城〔巴拿马〕 | −13 |
| Stockholm 斯德哥尔摩〔瑞典〕 | −7 | Paramaribo 帕拉马里博〔苏里南〕 | −11 |
| Tirana 地拉那〔阿尔巴尼亚〕 | −7 | Philadelphia 费城〔美国〕 | −13 |
| Vatican 梵蒂冈〔罗马教廷所在地〕 | −7 | Pittsburgh 匹兹堡〔美国〕 | −13 |
| Venice 威尼斯〔意大利〕 | −7 | Port-of-Spain 西班牙港〔特立尼 | |
| Vienna 维也纳〔奥地利〕 | −7 | 达和多巴哥〕 | −12 |
| Warsaw 华沙〔波兰〕 | −7 | Quebec 魁北克〔加拿大〕 | −13 |
| Zurich 苏黎世〔瑞士〕 | −7 | Rio de janeiro 里约热内卢〔巴西〕 | −11 |

续表

| 城市<br>City | 时差<br>Time<br>Difference | 城市<br>City | 时差<br>Time<br>Difference |
|---|---|---|---|
| **AMERICA 美洲** | | **AFRICA 非洲** | |
| Saint Louis 圣路易斯［美国］ | − 14 | Luanda 罗安达［安哥拉］ | − 7 |
| Salt Lake City 盐湖城［美国］ | − 15 | Lusaka 卢萨卡［赞比亚］ | − 6 |
| San Diego 圣迭戈［美国］ | − 16 | Maputo 马普托［莫桑比克］ | − 6 |
| San Francisco 旧金山（圣弗兰西 | | Mogadishu 摩加迪沙［索马里］ | − 5 |
| 斯科）［美国］ | − 16 | Mombasa 蒙巴萨［肯尼亚］ | − 5 |
| San Juan 圣胡安［波多黎各］ | − 12 | Monrovia 蒙罗维亚［利比里亚］ | − 8 |
| Santiago 圣地亚哥［智利］ | − 12 | Nairobi 内罗毕［肯尼亚］ | − 5 |
| Seattle 西雅图［美国］ | − 16 | Ndjamena 恩贾梅纳［乍得］ | − 7 |
| Toronto 多伦多［加拿大］ | − 13 | Niamey 尼亚美［尼日尔］ | − 8 |
| Vancouver 温哥华［加拿大］ | − 16 | Nouakchott 努瓦克肖特［毛里塔 | |
| Washington, D. C. 华盛顿［美国］ | − 13 | 尼亚］ | − 8 |
| Winnipeg 温尼伯［加拿大］ | − 14 | Pointe-Noire 黑角［刚果］ | − 7 |
| **AFRICA 非洲** | | Port Louis 路易港［毛里求斯］ | − 4 |
| | | Pretoria 比勒陀利亚［南非］ | − 6 |
| Abidjan 阿比让［科特迪瓦］ | − 8 | Rabat 拉巴特［摩洛哥］ | − 8 |
| Addis Ababa 亚的斯亚贝巴［埃 | | Sao Tome 圣多美［圣多美和普林 | |
| 塞俄比亚］ | − 5 | 西比］ | − 8 |
| Alexandria 亚历山大［埃及］ | − 6 | Tananarive 塔那那利佛［马达加 | |
| Algiers 阿尔及尔［阿尔及利亚］ | − 7 | 斯加］ | − 5 |
| Bamako 巴马科［马里］ | − 8 | Tripoli 的黎波里［利比亚］ | − 6 |
| Banghazi 班加西［利比亚］ | − 6 | Tunis 突尼斯［突尼斯］ | − 7 |
| Beira 贝拉［莫桑比克］ | − 6 | Windhoek 温得和克［纳米比亚］ | − 6 |
| Brazzaville 布拉柴维尔［刚果］ | − 7 | **OCEANIA AND PACIFIC LSLANDS** | |
| Cairo 开罗［埃及］ | − 6 | **大洋洲及太平洋岛屿** | |
| Cape Town 开普敦［南非］ | − 6 | Agana 阿加尼亚［关岛］ | + 2 |
| Dakar 达喀尔［塞内加尔］ | − 8 | Apia 阿皮亚［西萨摩亚］ | − 19 |
| Dar el Beida 达尔贝达（卡萨布 | | Auckland 奥克兰［新西兰］ | + 4 |
| 兰卡）［摩洛哥］ | − 8 | Brisbane 布里斯班［澳大利亚］ | + 2 |
| Dar es Salaam 达累斯萨拉姆［坦 | | Canberra 堪培拉［澳大利亚］ | + 2 |
| 桑尼亚］ | − 5 | Fremantle 弗里曼特尔［澳大利亚］ | 0 |
| Djibouti 吉布提［吉布提］ | − 5 | Hobart 霍巴特［澳大利亚］ | + 2 |
| Douala 杜阿拉［喀麦隆］ | − 7 | Honolulu 火奴鲁鲁(檀香山)［美国］ | − 18 |
| Durban 德班［南非］ | − 6 | Melburne 墨尔本［澳大利亚］ | + 2 |
| Gaborone 哈博罗内［博茨瓦纳］ | − 6 | Noumea 努美阿［法属新喀里多 | |
| Johannesburg 约翰内斯堡［南非］ | − 6 | 尼亚］ | + 3 |
| Kampala 坎帕拉［乌干达］ | − 5 | Papeete 帕皮提［法属波利尼西亚］ | − 18 |
| Kano 卡诺［尼日利亚］ | − 7 | Perth 佩思［澳大利亚］ | 0 |
| Khartoum 喀土穆［苏丹］ | − 6 | Port Moresby 莫尔兹比港［巴布 | |
| Kinshasa 金沙萨［民主刚果］ | − 7 | 亚新几内亚］ | + 2 |
| Lagos 拉各斯［尼日利亚］ | − 7 | Suva 苏瓦［斐济］ | + 4 |
| Las Palmas 拉斯帕耳马斯［加那 | | Sydney 悉尼［澳大利亚］ | + 2 |
| 利群岛］ | − 8 | Vila 维拉港［瓦努阿图］ | + 3 |
| | | Wellington 惠灵顿［新西兰］ | + 4 |

# 附录七  民用飞机注册编号前码表

| 国家/地区<br>（Country / Region） | 注册号前码<br>（Registration Prefix） |
|---|---|
| A | |
| 阿富汗（Afghanistan） | YA |
| 阿尔及利亚（Algeria） | 7T |
| 安哥拉（Angola） | D2 |
| 安提瓜和巴布达（Antigua and Barbuda） | V2 |
| 阿根廷（Argentina） | LQ，LV |
| 亚美尼亚（Armenia） | EK |
| 澳大利亚（Australia） | VH |
| 奥地利（Austria） | OE |
| 阿塞拜疆（Azerbaijan） | 4K |
| B | |
| 巴哈马群岛（Bahamas） | C6 |
| 巴林（Bahrain） | A9C |
| 孟加拉国（Bangladesh） | S2 |
| 巴巴多斯（Barbados） | 8P |
| 白俄罗斯（Belarus） | EW |
| 比利时（Belgium） | OO |
| 伯利兹（Belize） | V3 |
| 贝宁（Benin） | TY |
| 不丹（Bhutan） | A5 |
| 玻利维亚（Bolivia） | CP |
| 波斯尼亚和黑塞哥维那（Bosnia and Herzegovina） | T9 |
| 博茨瓦纳（Botswana） | A2 |
| 巴西（Brazil） | PP，PR，PT，PU |
| 文莱（Brunei） | V8 |

| 国家/地区<br>(Country / Region) | 注册号前码<br>(Registration Prefix) |
|---|---|
| 保加利亚（Bulgaria） | LZ |
| 布基纳法索（Burkina Faso） | XT |
| 布隆迪（Burundi） | 9U |
| C | |
| 柬埔寨（Cambodia） | XU |
| 喀麦隆（Cameroon） | TJ |
| 加拿大（Canada） | C, CF |
| 佛得角（Cape Verde） | D4 |
| 中非共和国（Central African Republic） | TL |
| 乍得（Chad） | TT |
| 智利（Chile） | CC |
| 中华人民共和国（including Hong Kong SAR and Macao SAR） | B |
| 哥伦比亚（Colombia） | HJ, HK |
| 刚果（Congo） | TN |
| 库克群岛（Cook Islands） | E5 |
| 哥斯达黎加（Costa Rica） | TI |
| 科特迪瓦（Ivory Coast） | TU |
| 克罗地亚（Croatia） | 9A |
| 古巴（Cuba） | CU |
| 塞浦路斯（Cyprus） | 5B |
| 捷克（Czech Republic） | OK |
| D | |
| 朝鲜（Democratic People's Republic of Korea） | P |
| 刚果民主共和国（Democratic Republic of the Congo） | 9Q |
| 丹麦（Denmark） | OY |
| 吉布提（Djibouti） | J2 |

| 国家/地区<br>（Country / Region） | 注册号前码<br>（Registration Prefix） |
|---|---|
| 多米尼克（Dominica） | J7 |
| 多米尼加共和国（Dominican Republic） | HI |
| E | |
| 厄瓜多尔（Ecuador） | HC |
| 埃及（Egypt） | SU |
| 萨尔瓦多（El Salvador） | YS |
| 赤道几内亚（Equatorial Guinea） | 3C |
| 厄立特里亚（Eritrea） | E3 |
| 爱沙尼亚（Estonia） | ES |
| 埃塞俄比亚（Ethiopia） | ET |
| F | |
| 斐济（Fiji） | DQ |
| 芬兰（Finland） | OH |
| 法国（France） | F |
| G | |
| 加蓬（Gabon） | TR |
| 冈比亚（Gambia） | C5 |
| 格鲁吉亚（Georgia） | 4L |
| 德国（Germany） | D |
| 加纳（Ghana） | 9G |
| 希腊（Greece） | SX |
| 格林纳达（Grenada） | J3 |
| 危地马拉共和国（Guatemala） | TG |
| 几内亚（Guinea） | 3X |
| 几内亚比绍（Guinea-Bissau） | J5 |
| 圭亚那（Guyana） | 8R |

续表

| 国家/地区<br>（Country / Region） | 注册号前码<br>（Registration Prefix） |
|---|---|
| **H** | |
| 海地（Haiti） | HH |
| 洪都拉斯（Honduras） | HR |
| 匈牙利（Hungary） | HA |
| **I** | |
| 冰岛（Iceland） | TF |
| 印度（India） | VT |
| 印度尼西亚（Indonesia） | PK |
| 伊朗〔Iran（Islamic Republic of）〕 | EP |
| 伊拉克（Iraq） | YI |
| 爱尔兰（Ireland） | EI |
| 以色列（Israel） | 4X |
| 意大利（Italy） | I |
| **J** | |
| 牙买加（Jamaica） | 6Y |
| 日本（Japan） | JA |
| 约旦（Jordan） | JY |
| **K** | |
| 哈萨克斯坦（Kazakhstan） | UN |
| 肯尼亚（Kenya） | 5Y |
| 科威特（Kuwait） | 9K |
| 吉尔吉斯斯坦（Kyrgyzstan） | EX |
| **L** | |
| 老挝人民民主共和国（Lao People's Democratic Republic） | RDPL |
| 拉脱维亚（Latvia） | YL |
| 黎巴嫩（Lebanon） | OD |
| 莱索托王国（Lesotho） | 7P |

| 国家/地区<br>(Country / Region) | 注册号前码<br>(Registration Prefix) |
|---|---|
| 利比里亚（Liberia） | A8 |
| 阿拉伯利比亚民众国（Libyan Arab Jamahiriya） | 5A |
| 列支敦士登（Liechtenstein） | HB plus national emblem |
| 立陶宛（Lithuania） | LY |
| 卢森堡（Luxembourg） | LX |
| M | |
| 马达加斯加岛（Madagascar） | 5R |
| 马拉维（Malawi） | 7Q |
| 马来西亚（Malaysia） | 9M |
| 马尔代夫（Maldives） | 8Q |
| 马里（Mali） | TZ |
| 马耳他（Malta） | 9H |
| 马绍尔群岛（Marshall Islands） | V7 |
| 毛里塔尼亚（Mauritania） | 5T |
| 毛里求斯（Mauritius） | 3B |
| 墨西哥（Mexico） | XA，XB，XC plus national emblem |
| 密克罗尼西亚［Micronesia（Federated States of）］ | V6 |
| 摩纳哥（Monaco） | 3A |
| 蒙古（Mongolia） | JU |
| 黑山共和国（Montenegro） | 4O |
| 摩洛哥（Morocco） | CN |
| 莫桑比克（Mozambique） | C9 |
| 缅甸（Myanmar） | XY，XZ |
| N | |
| 纳米比亚（Namibia） | V5 |
| 瑙鲁（Nauru） | C2 |

续表

| 国家/地区<br>(Country / Region) | 注册号前码<br>(Registration Prefix) |
|---|---|
| 尼泊尔（Nepal） | 9N |
| 荷兰（Netherlands） | PH |
| 荷属阿鲁巴岛（Aruba）* | P4 |
| 荷属安的列斯群岛（Netherlands Antilles）* | PJ |
| 新西兰（New Zealand） | ZK，ZL，ZM |
| 尼加拉瓜（Nicaragua） | YN |
| 尼日尔（Niger） | 5U |
| 尼日利亚（Nigeria） | 5N |
| 挪威（Norway） | LN |
| O | |
| 阿曼（Oman） | A4O |
| P | |
| 巴基斯坦（Pakistan） | AP |
| 帕劳（Palau） | T8 |
| 巴拿马（Panama） | HP |
| 巴布亚新几内亚（Papua New Guinea） | P2 |
| 巴拉圭（Paraguay） | ZP |
| 秘鲁（Peru） | OB |
| 菲律宾（Philippines） | RP |
| 波兰（Poland） | SP |
| 葡萄牙（Portugal） | CR，CS |
| Q | |
| 卡塔尔（Qatar） | A7 |
| R | |
| 韩国（Republic of Korea） | HL |
| 摩尔多瓦（Republic of Moldova） | ER |
| 罗马尼亚（Romania） | YR |

注：＊为荷兰管辖的岛屿。

续表

| 国家/地区<br>（Country / Region） | 注册号前码<br>（Registration Prefix） |
|---|---|
| 俄罗斯联邦（Russian Federation） | RA |
| 卢旺达（Rwanda） | 9XR |
| S | |
| 圣克里斯托弗和尼维斯岛（Saint Kitts and Nevis） | V4 |
| 圣卢西亚岛（Saint Lucia） | J6 |
| 圣文森特和格林纳丁斯（Saint Vincent and the Grenadines） | J8 |
| 萨摩亚群岛（Samoa） | 5W |
| 圣马力诺（San Marino） | T7 |
| 圣多美和普林西比民主共和国（São Tomé and Príncipe） | S9 |
| 沙特阿拉伯（Saudi Arabia） | HZ |
| 塞内加尔（Senegal） | 6V，6W |
| 塞尔维亚（Serbia） | YU |
| 塞舌尔（Seychelles） | S7 |
| 塞拉利昂（Sierra Leone） | 9L |
| 新加坡（Singapore） | 9V |
| 斯洛伐克（Slovakia） | OM |
| 斯洛文尼亚（Slovenia） | S5 |
| 所罗门群岛（Solomon Islands） | H4 |
| 索马里（Somalia） | 6O |
| 南非（South Africa） | ZS，ZT，ZU |
| 西班牙（Spain） | EC |
| 斯里兰卡（Sri Lanka） | 4R |
| 苏丹（Sudan） | ST |
| 苏里南（Suriname） | PZ |
| 斯威士兰（Swaziland） | 3D |
| 瑞典（Sweden） | SE |

续表

| 国家/地区<br>(Country / Region) | 注册号前码<br>(Registration Prefix) |
|---|---|
| 瑞士（Switzerland） | HB plus national emblem |
| 叙利亚（Syria Arab Republic） | YK |
| T | |
| 塔吉克斯坦（Tajikistan） | EY |
| 泰国（Thailand） | HS |
| 前南斯拉夫的马其顿共和国（The former Yugoslav republic of Macedonia） | Z3 |
| 多哥（Togo） | 5V |
| 汤加（Tonga） | A3 |
| 特立尼达和多巴哥（Trinidad and Tobago） | 9Y |
| 突尼斯（Tunisia） | TS |
| 土耳其（Turkey） | TC |
| 土库曼斯坦（Turkmenistan） | EZ |
| U | |
| 乌干达（Uganda） | 5X |
| 乌克兰（Ukraine） | UR |
| 阿拉伯联合酋长国（United Arab Emirates） | A6 |
| 英国（United Kingdom） | G |
| 安圭拉岛（Anguilla）* | VP-A |
| 百慕大群岛（Bermuda）* | VP-B |
| 开曼群岛（Cayman Islands）* | VP-C |
| 福克兰群岛［（Falkland Islands）Malvinas］* | VP-F |
| 直布罗陀（Gibraltar）* | VP-G |
| 马恩岛（Isle of Man）* | M |
| 蒙特塞拉特岛（Montserrat）* | VP-M |
| 圣赫勒拿岛/阿森松岛（St. Helena /Ascension）* | VQ-H |

注：＊为英国管辖的岛屿。

| 国家/地区<br>（Country / Region） | 注册号前码<br>（Registration Prefix） |
|---|---|
| 特克斯和凯科斯群岛（Turks and Caicos）* | VQ-T |
| 英属维尔京群岛（Virgin Islands）* | VP-L |
| 坦桑尼亚联合共和国（United Republic of Tanzania） | 5H |
| 美国（United States） | N |
| 乌拉圭（Uruguay） | CX |
| 乌兹别克斯坦（Uzbekistan） | UK |
| V | |
| 瓦努阿图（Vanuatu） | YJ |
| 委内瑞拉（Venezuela） | YV |
| 越南（Vietnam） | XV |
| V | |
| 也门（Yemen） | 7O |
| Z | |
| 赞比亚（Zambia） | 9J |
| 津巴布韦（Zimbabwe） | Z |

# 参考文献

［1］华人杰. 英汉航空航天新词典. 上海：上海科学普及出版社，1999.
［2］杨伟. 英汉航空航天缩略语词典. 北京：航空工业出版社，2009.
［3］蔡成仁，宫德深，刘连生等. 英汉民用航空缩略语词典. 北京：科学出版社，2000.
［4］周大军，周新力. 英汉航空航天缩略语词典. 北京：北京科学技术出版社，2005.
［5］贺道德，周瑞琏. 新编英汉民用航空词典. 北京：中国民航出版社，1997.
［6］马士观，范国良. 民用航空航行简缩语词典. 北京：中国民航出版社，1998.
［7］（印）库玛尔（B. Kumar），（印）德雷姆（D. DeRemer），徐元铭译. 英汉航空图解词典. 北京：航空工业出版社，2009.
［8］甘茂治. 英汉装备保障工程缩略语词典. 北京：国防工业出版社，2008.
［9］［美］C. R. 斯比策（Cary R. Spitzer），谢文涛等，译. 数字航空电子技术（上. 下册）. 北京：航空工业出版社，2010.
［10］清华大学外语系《英汉科学技术词典》编写组. 英汉科学技术词典. 北京：国防工业出版社，1991.
［11］梁炳文. 英汉航空航天工程词典. 北京：商务印书馆，2000.
［12］许学伊，马士观. 英汉对照杰普逊航图入门. 北京：中国民航出版社，2009.
［13］Bill Gunston, OBE, FRAeS. The Cambridge Aerospace Dictionary, Second Edition. Cambridge University Press.
［14］http：//acronyms. thefreedictionary. com/
［15］http：//www. icao. int/
［16］http：//www. nciku. cn/
［17］http：//www. atmb. net. cn/
［18］http：//www. faa. gov/
［19］http：//www. acronymfinder. com/
［20］http：//www. iciba. com/
［21］http：//www. caac. gov. cn/
［22］http：//www. avcodes. co. uk/
［23］http：//www. feeyo. com/
［24］http：//www. carnoc. com/
［25］http：//www. nasa. gov/
［26］http：//www. ntsb. gov/
［27］http：//www. aerospace-technology. com/
［28］http：//www. wmo. int/
［29］http：//dict. air. cc/
［30］http：//www. bts. gov/dictionary/
［31］http：//aviationweather. gov/static/info/glossary/
［32］http：//www. acronymgeek. com/
［33］http：//www. jg-jg. com/tools/szm/szmcode_ 15_ M. html